# ADVANCED ACCELERATOR CONCEPTS

# ADVANCED ACCELERATOR CONCEPTS

Eighth Workshop

Baltimore, Maryland   July 1998

EDITORS
Scientific Editor:
Wes Lawson
*University of Maryland*

Technical Editors:
Carol Bellamy
Dorothea F. Brosius
*University of Maryland*

**CD-ROM INCLUDED**

**AIP CONFERENCE PROCEEDINGS 472**

**American Institute of Physics**        **Woodbury, New York**

**Editors:**

Wes Lawson
Electrical Engineering Department
University of Maryland
College Park, MD 20742-3511
U.S.A.

E-mail: lawson@eng.umd.edu

Carol Bellamy
Electrical Engineering Department
University of Maryland
College Park, MD 20742-3511
U.S.A.

E-mail: cbellamy@glue.umd.edu

Dorothea F. Brosius
Institute for Plasma Research
University of Maryland
College Park, MD 20742-3511
U.S.A.

E-mail: dottie@glue.umd.edu

L.C. Catalog Card No. 99-61721
ISBN 1-56396-889-4–set
ISBN 1-56396-835-5–cloth
ISBN 1-56396-888-6–CD-ROM
ISSN 0094-243X
DOE CONF- 980742

Printed in the United States of America

# CONTENTS

## I. INVITED PAPERS

## II. WORKING GROUP SUMMARIES

## III. WORKING GROUP PAPERS

### Working Group 1: Beam Physics at High Energy Densities

### Working Group 2: Plasma-Based Acceleration Concepts

## Working Group 3: Novel Structure-Based Acceleration Concepts

### Working Group 4: Beam Monitoring, Conditioning, and Control at High Frequency and Ultrafast Timescales

## Working Group 5: Particle Beam Sources

## Working Group 6: Radiation Sources

## PREFACE

The Eighth Workshop on Advanced Accelerator Concepts was held at the Renaissance Harborplace Hotel in Baltimore, Maryland from July 6 to July 11, 1998. Participation was by invitation only, and the number of registered attendees was 160, including 40 graduate students; by design, the number of student participants was significantly larger than at the previous Workshops. The Workshop was sponsored by the High Energy Physics Branch of the U.S. Department of Energy and was hosted by the Institute for Plasma Research of the University of Maryland, College Park.

The purpose of the Workshop was to facilitate communication among investigators of new techniques for ultra-high gradient acceleration of charged particles. Advanced concepts for accelerators with capabilities significantly beyond the Next Linear Collider (NLC) were stressed. The Workshop addressed the critical physics and engineering issues relating to all the important features of future accelerators, including high brightness sources of electron beams; high power, efficient microwave sources at frequencies in the 17 to 94 GHz range; novel accelerator configurations; and instrumentation challenges presented by the study of advanced accelerators concepts.

The Workshop format consisted of fifteen review presentations in plenary sessions, together with smaller working group sessions in which ideas were exchanged in a more informal setting. The topics considered by the six working groups were as follows: beam physics at very high energy densities; plasma-based acceleration concepts; novel structure-based acceleration concepts; beam generation, monitoring, and control at high frequency and ultrafast time scales; particle beam sources; and radiation sources. The sessions on plasma-based acceleration concepts were especially well attended and the next Workshop will split this topic among two working groups, one on laser driven concepts and the other on electron beam driven concepts. The chairs of the various working groups reported on the highlights of their deliberations at a final plenary session of the Workshop and a summary paper for each of the working groups is included in these proceedings.

Advanced accelerator research is a rapidly advancing field with researchers from a broad spectrum of specialities interacting dynamically and synergistically. It is hoped that these proceedings give the reader the sense of breadth and excitement that characterizes the field.

Special thanks go to the members of the organizing committee for their efforts to ensure a comprehensive and stimulating program. In addition, we wish to recognize the efforts of Samar Guharay in organizing the social program and Carol Bellamy who both served as Workshop Coordinator and assisted with the preparation of these proceedings. We would also like to thank Dottie Brosius for her help with these proceedings.

Victor L. Granatstein, Workshop Chair
Wesley G. Lawson, Proceedings Editor

# I
# INVITED PAPERS

# The Equation of Motion of an Electron[*]

Kwang-Je Kim[†] and Andrew M. Sessler[‡]

[†]*Argonne National Laboratory, Argonne, IL 60439 and The University of Chicago, Chicago, IL 60637*
[‡]*Lawrence Berkeley National Laboratory, Berkeley, CA 94720*

**ABSTRACT.** We review the current status of understanding of the equation of motion of an electron. Classically, a consistent, linearized theory exists for an electron of finite extent, as long as the size of the electron is larger than the classical electron radius. Nonrelativistic quantum mechanics seems to offer a fine theory even in the point particle limit. Although there is as yet no convincing calculation, it is probable that a quantum electrodynamical result will be at least as well-behaved as is the nonrelativistic quantum mechanical results.

## INTRODUCTION

For almost 100 years there has been consideration of the proper equation of motion of an electron. Many fine physicists, beginning with Abraham (1) and Lorentz (2), have worked on this subject, and there are hundreds of papers in the literature. In this paper we review the present state of understanding, with some historical background, giving the major contributions through the years.

In contrast with what most physicists believe, it is seen that the linearized classical theory, when it is applied in the appropriate regime (nonquantum), is in fine shape: it is finite, has no contradiction with relativity, has no run-away solutions, and has no a-causal behavior.

The Abraham-Lorentz equation for a point electron involving a third derivative in time suffers from two major problems: contradiction with relativity and run-away or acausal behavior. The work of Poincaré (3) and Dirac (4) solves the problem with relativity. The run-away problem is solved by going to an extended model of an electron described by a difference-differential equation. The equation for the

[*] Work supported by the U.S. Department of Energy, Office of Basic Energy Sciences, under Contract No. W-31-109-ENG-38 and Office of Energy Research, Office of High Energy and Nuclear Physics, Division of High Energy Physics, under Contract No. DE-AC03-76SF00098.

CP472, *Advanced Accelerator Concepts: Eighth Workshop,*
edited by W. Lawson, C. Bellamy, and D. Brosius
1999 The American Institute of Physics 1-56396-889-4

nonrelativistic case was derived by Sommerfeld (5) and Page (6) and was generalized to the relativistic case by Caldirola (7). The extended model is finite and causal if the electron size a is larger than the classical electron radius $r_e = e^2/mc^2 = 2.7 \times 10^{-13}$ cm.

Classical theory is clearly not appropriate for examining behavior at a distance less than the Compton wavelength of an electron, $\lambda = \hbar/mc = 4 \times 10^{-11}$ cm. The work by Moniz and Sharp (8) indicates that in nonrelativistic quantum mechanics an electron behaves as an extended particle with a size of the Compton wavelength: The equation of motion is finite and causal even in the point particle limit as long as the Compton wavelength is larger than the classical electron radius. Furthermore, the mass correction $\delta m$ is not only finite but actually vanishes in quantum theory. We present a new quantum mechanical derivation of this interesting result (9). In quantum electrodynamic (QED) analysis, recent work by Low (10) shows that the electron motion is finite in perturbation theory in $\alpha = e^2/\hbar c$. However, a proper QED analysis has not yet been obtained (and maybe, those do *not* even exist).

There are several excellent text books (11-13) and review articles (14-16) on the classical electron theory.

## NOTATION

Notations adopted in this paper are as follows:

| | |
|---|---|
| $m_0$ | bare mass |
| m | observed mass ($m = m_0 + \delta m$) |
| c | speed of light |
| $r_c$ | classical electron radius ($e^2/mc^2$) |
| a | radius of extended electron |
| $t_e$ | $(2/3)\, r_e/c$ |
| $\lambda$ | Compton wavelength |
| **x, x′** | space coordinate (3 vectors) |
| **y**(t) | electron coordinate as a function of time (3 vectors) |
| $\dot{\mathbf{y}}$ | = dy/dt, $\ddot{\mathbf{y}} = d^2y/dt^2$, etc. |
| **u** | = d**y**/dt (three-velocity), $\dot{\mathbf{u}} = d\mathbf{u}/dt$, etc. |
| $\beta$ | = $|\mathbf{u}|/c, \gamma = 1/\sqrt{1-\beta^2}$ |
| ds | = $cdt/\gamma$ (relativistic invariant) |
| $u^\alpha$ | = $dx^\alpha/ds = \gamma\,(1, \mathbf{u}/c)$ (contra-variant four vector) |
| $u_\alpha$ | = $\gamma\,(1, -\mathbf{u}/c)$ (covariant four vector) |
| $\dot{u}^\alpha$ | = $du^\alpha/ds, \ddot{u}^\alpha = d^2u^\alpha/d^2s$, etc. |

# ABRAHAM-LORENTZ AND OTHER
# CLASSICAL EQUATIONS OF MOTION

We start by deriving various forms of equation of motion of an electron in nonrelativistic classical mechanics. Let the electron trajectory be $\mathbf{y}(t)$. The electron is really a charge distribution centered at $\mathbf{y}(t)$ represented by

$$ef(\mathbf{x} - \mathbf{y}(t)), \tag{1}$$

where e is the electron charge, and $f(\mathbf{x})$ is a spherically symmetric function, normalized so that

$$\int f(\mathbf{x}) d^3 x = 1. \tag{2}$$

The nonrelativistic equation of motion is determined by the Lorentz force law:

$$m_0 \ddot{\mathbf{y}}(t) = \mathbf{F}_{ext} + e \int f(\mathbf{x} - \mathbf{y}(t))(\mathbf{E}(\mathbf{x}) + \frac{1}{c} \dot{\mathbf{y}}(t) \times \mathbf{B}(\mathbf{x})) d^3 x, \tag{3}$$

where $m_0$ is the bare mass, the dot represents the time derivative, $\mathbf{F}_{ext}$ is the external force, and $\mathbf{E}$ and $\mathbf{B}$ are, respectively, the electric and magnet field in Gaussian units. The second term in the above equation is the electromagnetic self force.

It is convenient to work with the potentials $\mathbf{A}$ and $\phi$ in Coulomb gauge.[1] The scalar potential is given by

$$\phi(\mathbf{x}, t) = e \int \frac{f(\mathbf{x}' - \mathbf{y}(t))}{|\mathbf{x} - \mathbf{x}'|} d^3 x'. \tag{4}$$

The vector potential is obtained by solving the wave equation

$$\left( \frac{1}{c^2} \frac{\partial^2}{\partial t^2} - \nabla^2 \right) \mathbf{A} = \frac{4\pi e}{c} \dot{\mathbf{y}}_\perp(t) f(\mathbf{x} - \mathbf{y}(t)). \tag{5}$$

Here $\dot{\mathbf{y}}_\perp$ is the transverse part of $\dot{\mathbf{y}}$.

It is convenient to work in the $\mathbf{k}$ representation:

---

[1] Our discussion follows closely the derivation in Low's recent paper (10).

$$[A(x,t),f(x)] = \frac{1}{(2\pi)^3}\int e^{ik\cdot x}\left[\tilde{A}(k,t),\tilde{f}(k)\right]d^3k.$$

We solve Eq. (5) with the retarded boundary condition:

$$\tilde{A}(k,t) = 4\pi e\int_{-\infty}^{t} dt' \frac{\sin kc(t-t')}{k}e^{-ik\cdot y(t')}\tilde{f}(k), \qquad (6)$$

where $k = |k|$.

Next, we compute $E = -\nabla\phi - \partial A/(c\partial t)$ and $B = \nabla \times A$, and insert the results into Eq. (3). In doing so, we observe that $\dot{y}_\perp = \dot{y} - k\,(k\cdot\dot{y})/k^2$ can be replaced by $(2/3)$ $\dot{y}$ because of the spherical symmetry. The contribution via the scalar potential $\phi$ vanishes because of the spherical symmetry and because we are using the Coulomb gauge. The self force arising from the electric field is found to be

$$m_0\ddot{y}(t) = F_{ext}(t) - \frac{2}{3}e^2\frac{4\pi}{(2\pi)^3}\int d^3k\left|\tilde{f}(k)\right|^2\int_0^\infty d\tau Y(t,\tau)\cos kc\tau, \qquad (7)$$

where

$$Y(t,\tau) = e^{ik\cdot y(t)}e^{ik\cdot y(t-\tau)}\dot{y}(t-\tau). \qquad (8)$$

The force arising from the magnetic field is nonlinear in $y$ (and its derivative), which we neglect. In linear approximation, the exponential factors in Eq. (8) can also be replaced by 1. Therefore,

$$Y(t,\tau) = \dot{y}\,(t-\tau). \qquad (9)$$

Equations (7) and (9) are the desired expression for the classical equation of motion and can be shown to be identical to the power series expression derived by Lorentz for linearized radiation reaction in the nonrelativistic approximation (2,11).

For the case of a spherical shell of radius a,

$$f(x) = \frac{\delta(|x|-a)}{4\pi a^2}, \quad \tilde{f}(k) = \frac{\sin ka}{ka}. \qquad (10)$$

Equations (7) and (9) become

$$m_0\ddot{\mathbf{y}}(t) = \mathbf{F}_{ext}(t) + \frac{e^2}{3a^2c}\left[\dot{\mathbf{y}}(t - 2a/c) - \dot{\mathbf{y}}(t)\right]. \tag{11}$$

This differential-difference equation will be referred to as the Sommerfeld-Page equation because they derived it (5,6).

Expanding $\dot{\mathbf{y}}(t - 2a/c)$ in Eq. (11) in a Taylor series and neglecting terms that vanish as a→0,

$$m_0\ddot{\mathbf{y}}(t) = \mathbf{F}_{ext}(t) - \delta m\ddot{\mathbf{y}}(t) + \frac{2e^2}{3c^3}\dddot{\mathbf{y}}(t), \tag{12}$$

where

$$\delta m = \frac{4}{3}\frac{e^2}{2ac^2}. \tag{13}$$

Equation (12) will be referred to as the Abraham-Lorentz equation in the following. It is the simplest form of the equation of motion, taking into account the electromagnetic self force in a nonrelativistic linear approximation and in the point particle limit.

The second term on the RHS of the Abraham-Lorentz equation can be interpreted as the inertia due to the electromagnetic mass $\delta m$. However, it is in contradiction with the notion of relativity because $\delta m$ is different from the electrostatic mass $e^2/2ac^2$ by a factor of 4/3.

To see the meaning of the third derivative term in the Abraham-Lorentz equation we move the $\delta m$ term to the LHS, multiply both sides by $\dot{\mathbf{y}}$, and integrate over a finite interval of time. The LHS becomes the increase in the electron's kinetic energy. The RHS is

$$-\frac{2e^2}{3c^3}\int_{t_1}^{t_2}\ddot{\mathbf{y}}^2 dt + \frac{2e^2}{3c^3}\dot{\mathbf{y}}\cdot\ddot{\mathbf{y}}\Big|_{t_1}^{t_2}. \tag{14}$$

As long as the second term is negligible, this is the negative of the well-known dipole radiation. The last term in Eq. (12) is therefore reasonable. However, we will see later that this term gives rise to run-away or acausal behavior.

Equation (12) therefore contains two major difficulties: i) a contradiction with relativity, and ii) run-away and preacceleration behavior. We shall see in the following sections how these troubles are avoided. The bottom line is that there is no problem if the shell radius a is larger than the classical electron radius $r_e$.

7

# RELATIVITY AND POINCARÉ STRESS

Let us first consider the relativistic generalization of Eq. (12). In 1903, before special relativity was established, Abraham derived (1) the following force equation valid for relativistic speed (neglecting terms vanishing in the limit $a \to 0$):

$$m_0 \frac{d}{dt}(\gamma\mathbf{u}) = \mathbf{F}_{ext} - \frac{2}{3}\frac{e^2}{ac^2}\frac{d}{dt}(\gamma\mathbf{u}) + \mathbf{\Gamma}, \tag{15}$$

where $\mathbf{u} = d\mathbf{y}/dt$ is the three-velocity, $\gamma = 1/\sqrt{1 - \mathbf{u}^2/c^2}$, and

$$\mathbf{\Gamma} = \frac{2}{3}\frac{e^2\gamma^2}{c^3}\{\ddot{\mathbf{u}} + \frac{3\gamma^2}{c^2}(\mathbf{u}\bullet\dot{\mathbf{u}})\dot{\mathbf{u}} + \frac{\gamma^2}{c^2}[(\mathbf{u}\bullet\ddot{\mathbf{u}}) + \frac{3\gamma^2}{c^2}(\mathbf{u}\bullet\dot{\mathbf{u}})^2]\mathbf{u}\}. \tag{16}$$

As was shown by von Laue (17), the $\mathbf{\Gamma}$ in the above is $1/\gamma$ times the space part of the four vector:

$$\Gamma^\mu = \frac{2}{3}e^2 (\ddot{u}^\mu + (\dot{u}^\alpha \dot{u}_\alpha) u^\mu). \tag{17}$$

(See the explanation in the "Notation" section.)

The second term in Eq. (15) gives rise to an electromagnetic mass that is 4/3 times the electrostatic mass, in contradiction with relativity.

Abraham also found the rate of work done on an electron (neglecting terms vanishing in the limit $a \to 0$):

$$\frac{dE}{dt} = \mathbf{F}_{ext}\bullet\mathbf{u} - \frac{2}{3}\frac{e^2}{a}\frac{d}{dt}\left(\gamma - \frac{1}{4\gamma}\right) + \frac{2}{3}e^2\frac{\gamma^4}{c^3}[\mathbf{u}\bullet\ddot{\mathbf{u}} + \frac{3\gamma^2}{c^2}(\mathbf{u}\bullet\dot{\mathbf{u}})^2]. \tag{18}$$

Since $E = mc^2\gamma$ from relativity, it follows that $dE/dt = m(d\gamma\mathbf{u}/dt)\cdot\mathbf{u}$. However, the force (Eq. (15)) and the power (Eq. (18)) do not satisfy this fundamental conservation law. The bad term is $2/3\ e^2/a\ (d/dt)(1/4\gamma)$ in Eq. (18). Except for that term, Eqs. (15) and (18) would have formed a four vector equation.

Although these equations were derived before relativity was fully established, Abraham used the correct relativistic model of an electron in which the spherical charge distribution in the rest frame is contracted to a spheroid for a moving electron. The offending term in Eq. (18) arises from this relativistic change of the electron's shape. The derivations of Eq. (15) and Eq. (18) are difficult and confirmed by Schott (18) who carried out a very rigorous and complicated calculation.

We therefore have the strange situation that the relativistic equation of motion is in contradiction with relativity!

In the nonrelativistic limit, Eqs. (15) and (18) become, respectively,

$$m_0 \frac{d\mathbf{u}}{dt} = \mathbf{F}_{ext} - \frac{2}{3}\frac{e^2}{ac^2}\dot{\mathbf{u}} + \frac{2}{3}\frac{e^2}{c^3}\ddot{\mathbf{u}} \text{ , and} \tag{19}$$

$$\frac{dE}{dt} = \mathbf{F}_{ext} \cdot \mathbf{u} - \frac{5}{6}\frac{e^2}{ac^2}\mathbf{u} \cdot \dot{\mathbf{u}} + \frac{2}{3}\frac{e^2}{c^3}\mathbf{u} \cdot \ddot{\mathbf{u}} . \tag{20}$$

Once again $\mathbf{u} \cdot$ force (Eq. (19)) does not yield the power (Eq. (20)). (Note that Eq. (19) is identical to Eq. (12).) The problem was solved by Poincaré (3) with nonelectromagnetic stresses and Dirac by invoking covariance (4).

Poincaré, in his paper submitted in 1905 (3) (without knowledge of Einstein's work on special relativity)[2], observed that a purely electromagnetic model of the electron, such as the charged sphere, is not internally consistent because it will fly apart due to the electrostatic repulsion. To counteract the repulsive force, he imagined that the inside of the sphere provides a uniform negative pressure (or stress ) $p = e^2/8\pi a^4$.

To see how this Poincaré stress solves the problem mentioned in the previous section, consider the charged shell in motion. Due to the Lorentz contraction, the sphere becomes an oblate spheroid with the minor axis in the direction of the motion reduced by $\gamma$. The work done by the mechanical force is given by the pressure times the volume change. (The pressure is relativistically invariant since the force and the area element transform the same way.) Thus the mechanical system must lose energy at the rate

$$\frac{dE}{dt} = -\frac{e^2}{8\pi a^4}\frac{d}{dt}\left(\frac{1}{\gamma}\frac{4}{3}\pi a^3\right)$$

$$= -\frac{2}{3}\frac{e^2}{a}\frac{d}{dt}\frac{1}{4\gamma} . \tag{21}$$

Here the expression $4\pi a^3/3\gamma$ is the volume of the spheroid. When Eq. (21) is added to Eq. (18), the discrepancy between the force and the power equation is removed.

However, the problem of the mass—the fact that the electromagnetic mass is 4/3 times the electrostatic mass—is not solved yet. One way of solving the problem is to say that the bare mass contains a term -(1/3) times electrostatic mass. For a more

---

[2] Relative contributions of Poincaré and Einstein to special relativity is a subject of some debate and considerable historical interest. See A.A. Logunov referred to in (3).

9

formal approach (19), we can introduce a stress tensor representing the Poincaré stress by

$$\theta^{\mu\nu}_{Poincar\acute{e}} = \frac{e^2}{8\pi a^3}\left(g^{\mu\nu} - qu^\mu u^\nu\right)\Theta(a - r),\qquad(22)$$

where $g^{\mu\nu} = (1,-1,-1,-1)$, q is an arbitrary constant, $\Theta$ is the step function, and r is the radial coordinate in the rest frame of the shell. The quantity $\theta^{\mu\nu}_{Poincar\acute{e}}$ is constructed so that

$$\partial_\mu\theta^{\mu\nu} = \partial_\mu\left(\theta^{\mu\nu}_{EM} + \theta^{\mu\nu}_{Poincar\acute{e}}\right) = 0,\qquad(23)$$

where $\theta^{\mu\nu}_{EM}$ is the electromagnetic stress tensor associated with the Coulomb field of a spherical shell. Equation (23) assures that the total momentum

$$p^\mu = \int d^3x\theta^{\mu o}\qquad(24)$$

is a four vector (11). The momentum associated with the Poincaré stress alone is

$$p^\mu_{Poincar\acute{e}} = \int a^3x\theta^{\mu o}_{Poincar\acute{e}} = +\frac{e^2}{2a}\frac{1}{3}\left[-\frac{1}{\gamma} + q\gamma, q\gamma\mathbf{u}\right]\qquad(25)$$

We need to add $dp^\mu_{Poincar\acute{e}}/dt$ to the RHS of Eqs. (18) and (15). Our previous result (Eq. (21)) corresponds to the case q = 0. If, on the other hand, we choose q = 1, we obtain

$$m_0c^2\frac{d}{ds}u^u = f^\mu_{ext} + \Gamma^\mu - \frac{e^2}{2a}\frac{du^\mu}{ds}.\qquad(26)$$

We now have an equation of motion of an electron with the observed mass m = $m_0$ + $\delta$m, $\delta$m = $e^2/2ac^2$, solving the so-called 4/3 problem.

The above method of solving the 4/3 problem is clearly rather formal and arbitrary. There are also more intuitive approaches, for example by Boyer (20) who notes that the Poincaré stress may not act at the same time in all parts of the moving electron. A review of different solutions of the 4/3 problem is given by Rohrlich (16).

The saga of the 4/3 problem is in some sense a story of how special relativity proved itself as a theory of internal consistency and beauty. The investigation of the electron's equation of motion started while special relativity was still evolving. Therefore, there was doubt whether the electron theory was consistent with relativity, which lingered even after relativity was fully established. Einstein, never doubting relativity, wasted no time in checking covariance of the electron theory. He was too busy working out general relativity!

Before closing this section, we need to mention Dirac's contribution (4). He understood that the problem with the Abraham-Lorentz equation was in trying to approach too near an electron. He therefore devised an ingenious way to avoid the difficulty. Stay a finite distance away from an electron and demand relativistic covariance. He then obtained Eq. (26), which is therefore referred to as the Abraham-Lorentz-Dirac equation.

Actually, Dirac's derivation of Eq. (4) is valid only up to a term of the form $dB^\mu/ds$, where $B^\mu$ is a four vector with the restriction $u^\mu \approx dB_\mu/ds = 0$. The simplest choice, $B^\mu = ku^\mu$, gives rise to a term that can be incorporated into the inertial term ($mc\, du^\mu/ds$) in Eq. (4). The next order term involving $\dot{u}^\mu$ (the dot indicating the derivative with respect to s) can be shown to be of the form $B^\mu = k'\left(\left(\dot{u}^\alpha \dot{u}_\alpha\right)^2 u^\mu + 4\left(\dot{u}^\alpha \ddot{u}_\alpha\right)\dot{u}^\mu\right)$. Terms of such a complexity are disposed of by saying that *a simple thing like an electron cannot possibly have such a complication.*

If we are willing to make a few very plausible assumptions, the Abraham-Lorentz-Dirac equation can be derived very easily as follows (16):

$$m_0 c^2 \frac{du^\mu}{ds} = f_{ext}{}^\mu + X^\mu, \tag{27}$$

where $X^\mu$ includes radiative effects. Both sides of Eq. (27) are orthogonal to $u^\mu$. Thus we may write $X^\mu = (g^{\mu\nu} - u^\mu u^\nu)\, Y_\nu$. Try $Y^\mu$ of the form $a\, u^\mu + b\, \dot{u}^\mu + c\, \ddot{u}^\mu$. The first term does not contribute, and the second term is of the same form as the mass term, which can be incorporated into m. With the third term, it is easy to show that $X^\mu$ is proportional to $\Gamma^\mu$. (Here we are using $\dot{u}^\alpha \dot{u}_\alpha + \ddot{u}_\alpha \cdot u^\alpha = 0$.) The coefficient is determined by demanding consistency with the nonrelativistic Abraham-Lorentz equation, leading to $\Gamma^\mu = X^\mu$. End of the proof.

## RUN-AWAY SOLUTIONS, BOUNDARY CONDITIONS, AND INTEGRAL EQUATIONS

The general solution of the Abraham-Lorentz equation is

$$m\, \ddot{y}\, (t) = e^{\,t/t_e}\, [\, m\, \ddot{y}\, (0) - 1/t_e \int_0^t dt'\, e^{-t/t_e}\, F_{ext}(t')], \tag{28}$$

where

$$t_e = \frac{2}{3}\frac{r_e}{c} = \frac{2}{3}\frac{e^2}{mc^3}. \tag{29}$$

11

The solution in general exhibits exponential growth, i.e., the run-away behavior. Dirac noted that the run-away can be avoided if we choose the initial condition:

$$m \, \ddot{\mathbf{y}} \, (0) = 1/t_e \int_0^\infty dt' e^{-t'/t_e} \, \mathbf{F}_{ext}(t').$$

(30)

Then

$$m \, \ddot{\mathbf{y}} \, (t) = \int_0^\infty d\alpha e^{-\alpha} \, \mathbf{F}_{ext} \, (t + t_e\alpha).$$

(31)

This solution now exhibits preacceleration: the particle starts to move before the force is applied and the initial condition depends upon the entire future path.

Nevertheless, it satisfies the Rohrlich criteria of *"the unobservability of very small charges"* (13). That is, as the charge becomes very small we shouldn't have a solution widely different from that of an uncharged particle.

The reason why the Abraham-Lorentz equation exhibits either run-away or preacceleration is the fact that, as the electron radius a vanishes, the bare mass $m_0$ becomes negative to keep the observed mass m finite. A particle with a negative mass can clearly supply an infinite amount of energy.

The difficulty can be avoided in an extended electron model. To see this, we return to the Sommerfeld-Page Eq. (11), which can be written in the following form[3]:

$$\ddot{\mathbf{y}}(t) = \frac{\mathbf{F}_{ext}(t)}{m(1 - ct_e/a)} + \left(\frac{c}{2a}\right)\left(\frac{ct_e}{a}\right)\frac{1}{(1 - ct_e/a)} \, [\dot{\mathbf{y}}(t-2a/c) - \dot{\mathbf{y}} \, (t)].$$

(32)

In the above expression, the combination $m(1-ct_e/a)$ is simply the bare mass $m_0$. Assume that there are no external forces, and there exists a run-away solution of the form $\mathbf{y}(t) = \mathbf{y}(0) \, e^{\alpha t}$ with a positive real part of $\alpha$. From Eq. (32) it is easy to see that this is possible only when $1-ct_e/a$ is negative. Thus the run away solution is possible if and only if the electron radius a is less than $(2/3) \, r_e$. For $a > (2/3) \, r_e$ we get damped oscillatory solutions.

It is sometimes stated that the run-away behavior is due to the infinite energy associated with a point electron. This is false because the run-away occurs for even a finite a as long as it is less than $(2/3) \, r_e$. Run-away occurs if and only if the bare mass is negative (and, therefore, the Hamiltonian is no longer positive definite).

We can also show that if $a > ct_e$, the motion is causal with no preacceleration. This can be seen most easily if Eq. (32) is turned into an integral equation with a Green's function:

---

[3] This derivation here is due to Moniz and Sharp (8). See also Pearle (15).

$$\ddot{y}(t) = \int_{-\infty}^{\infty} G(t\text{-}t') \, F_{ext}(t') \, dt', \tag{33}$$

where

$$G(t\text{-}t') = \frac{1}{2\pi m} \int_{-\infty}^{\infty} \frac{e^{-i\omega(t-t')}d\omega}{B}, \text{ and} \tag{34}$$

$$B = 1 - \frac{i}{\omega}\left(\frac{c}{2a}\right)\left(\frac{ct_e}{a}\right)\left[e^{\frac{i\omega 2a}{c}} - \frac{i\omega 2a}{c} - 1\right]. \tag{35}$$

It can be shown that all poles of $1/B(\omega)$ occur in the lower half $\omega$-plane if

$$a > \frac{2}{3} \, r_e. \tag{36}$$

Then the Green's function $G(\tau)$ vanishes for $\tau < 0$. Therefore, the motion is causal. On the other hand, if $a < \frac{2}{3} r_e$, then there is in general acausal behavior.

A relativistic generalization of the Sommerfeld-Page equation was given by Caldirola (7):

$$m_0 c^2 \dot{u}^\mu = f_{ext}^{\ \mu} + \frac{2}{3}\frac{e^2}{a}\frac{1}{2a} \, [u^\mu(s\text{-}2a) - u^\mu(s)u_\alpha(s) \, u^\alpha(s\text{-}2a)]. \tag{37}$$

The Caldirola equation is fine: no run-aways and no causality problems. However, when we take a $\to$ 0, then there are problems.

Yaghjian (14) attempted to produce a causal equation with no run-aways for a point electron. He modified the Abraham-Lorentz Eq. (12) by multiplying the third derivative term by a function $\eta(t)$, which changes smoothly from 0 for $t < 0$ to 1 for $t > 2a/c$ (or by multiplying the $\Gamma^\mu$ term by a similar function $\eta(s)$), arguing that the Taylor series expansion breaks down when the force changes abruptly. However, the solution of the Yaghjian equation still exhibits acausal behavior.

## QUANTUM MECHANICAL EQUATIONS

It had been hoped that the difficulty of the classical theory in taking a point particle limit might be solved by quantum mechanics. The observation that the mass renormalization $\delta m$ is less singular in quantum electrodynamics than in classical theory seemed to reinforce this hope. However, a real quantum mechanical analysis of the electron equation has not been carried out until modern time. The reason behind this lack of activity appears to be the success of the renormalization theory in quantum electrodynamics, which established that all observable phenomena can be calculated to a finite answer order by order in perturbation theory in terms of the observed mass and charge.

In 1977, Moniz and Sharp presented a very interesting quantum mechanical calculation (8), according to which the equation of motion in nonrelativistic quantum mechanics is finite even in the point particle limit. However, the paper has not received much attention, probably because the derivation was not transparent due to the elaborate manipulations of infinite series.

One of us (KJK) was able to derive the result of Moniz and Sharp via a more understandable method (9), which will be summarized here.

The derivation of the quantum mechanical equation of motion for Heisenberg operators proceeds similar to the derivation in the classical case given in the section "Abraham-Lorentz and Other Classical Equations of Motion." Again neglecting the magnetic term, one arrives at an equation identical in form to Eq. (7), but the operator **Y** is given by the symmetrized version of the classical expression, Eq. (8), as follows:

$$\mathbf{Y}(t,\tau) = \frac{1}{4}\left\{e^{i\mathbf{k}\cdot\mathbf{y}_2}, \left\{e^{-i\mathbf{k}\cdot\mathbf{y}_1}, \dot{\mathbf{y}}_1\right\}_+\right\}_+$$

$$= \frac{1}{4}\left(e^{i\mathbf{k}\cdot\mathbf{y}_2}e^{-i\mathbf{k}\cdot\mathbf{y}_1}\right)\left(\dot{\mathbf{y}}_1 + e^{i\mathbf{k}\cdot\mathbf{y}_1}\dot{\mathbf{y}}_1 e^{-i\mathbf{k}\cdot\mathbf{y}_1}\right) + [\,\mathbf{k} \to -\mathbf{k}\,]^+. \tag{38}$$

Here $\mathbf{y}_2 = \mathbf{y}(t)$, $\mathbf{y}_1 = \mathbf{y}(t-\tau)$, $\{\ \}_+$ is the anticommutator, and $[\mathbf{k} \to -\mathbf{k}]^+$ indicates terms obtained by changing the sign of **k** and taking the Hermitian conjugate.

Noting that $\mathbf{p} = \mathbf{P} - e\mathbf{A}/c = m_0\dot{\mathbf{y}}$ is the kinetic momentum operator ($\mathbf{P}$ = canonical momentum), we obtain

$$e^{i\mathbf{k}\cdot\mathbf{y}_1}\dot{\mathbf{y}}_1 e^{-i\mathbf{k}\cdot\mathbf{y}_1} = \dot{\mathbf{y}}_1 - c\mathbf{k}\lambda, \tag{39}$$

where $\lambda = \hbar/m_0 c$ is the Compton wavelength for the bare mass $m_0$. In classical mechanics, we would have replaced the factor exp $(i\mathbf{k}\cdot\mathbf{y}_1)$ by 1 since it would at most contribute to nonlinear terms. Such a procedure is not justified in quantum mechanics as is clear from Eq. (39). Similarly, the factor exp $(i\mathbf{k}\cdot\mathbf{y}_2) \cdot$ exp $(-i\mathbf{k}\cdot\mathbf{y}_1)$ in Eq. (38) cannot simply be replaced by 1. Instead, we proceed as follows[4]:

---

[4] Similar reduction of product of exponential operators was used by Baier and Katkov (21) in their calculation of quantum synchrotron radiation. The steps used here follow closely a simplified formulation of the Baier-Katkov reduction by Cahn and Jackson (22).

$$\exp(i\mathbf{k} \cdot \mathbf{y}_2)\exp(-i\mathbf{k} \cdot \mathbf{y}_1) = \exp(i\tau H / \hbar)\exp(i\mathbf{k} \cdot \mathbf{y}_1)\exp(-i\tau H / \hbar)\exp(-i\mathbf{k} \cdot \mathbf{y}_1)$$

$$= \exp(i\tau H(\mathbf{p}) / \hbar)\exp(-i\tau H(\mathbf{p} - \hbar k) / \hbar) \quad (\tau = t - t') \quad (40)$$

Here H is the Hamiltonian. Since H is a sum of the kinetic energy $\mathbf{p}^2/2m_0$ and the electromagnetic energy, it follows that

$$H(\mathbf{p} - \hbar\mathbf{k}) = \frac{\hbar^2 k^2}{2m_0} + H(\mathbf{p}) - \frac{\hbar\mathbf{p} \cdot \mathbf{k}}{m_0}. \quad (41)$$

Now, using the Campbell-Baker-Hausdorff formula (23) to first order in $\mathbf{p}$, we can show that

$$\exp(i\mathbf{k} \cdot \mathbf{y}_2)\exp(-i\mathbf{k} \cdot \mathbf{y}_1) = \exp\left(-\frac{i\tau\hbar k^2}{2m_0}\right)[1 + i\mathbf{k} \cdot (\mathbf{y}_2 - \mathbf{y}_1)]. \quad (42)$$

Thus

$$Y = \frac{1}{2}\left\{\exp\left(-\frac{i\tau\hbar k^2}{2m_0}\right)(1 + i\mathbf{k} \cdot (\mathbf{y}_2 - \mathbf{y}_1))\left(\dot{\mathbf{y}}_1 - \frac{ck\lambda}{2}\right) + [\mathbf{k} \to \text{-k}]^+ \right\}. \quad (43)$$

Noting that odd power in $\mathbf{k}$ vanishes after $d^3k$-integration and $\langle k_x^2 \rangle = \frac{1}{3}k^2$, one obtains (keeping only terms linear in $\mathbf{y}$)

$$Y(t,\tau) = \dot{y}(t - \tau)\cos\left(\lambda c\tau k^2 / 2\right) - \frac{k^2 c\lambda}{6}(y(t) - y(t - \tau))\sin\left(\lambda c\tau k^2 / 2\right). \quad (44)$$

The linearized electron's equation of motion in nonrelativistic quantum mechanics is then given by inserting Eq. (44) into Eq. (7). Expanding the operator $y(t-\tau)$ in Eq. (44) in a Taylor series around $\tau = 0$ and inserting it in Eq. (7), an equation involving a sum of derivatives of $y(t)$ is obtained. It can be shown that the coefficients of these derivatives are exactly those derived in Moniz and Sharp[5]:

$$m_0\, y(t) = F_{\text{ext}}(t) - \frac{2}{3}\frac{e^2}{c^2}\sum_{n=0}^{\infty}\left(\frac{(-1)^n A_n}{n! c^n}\right)\frac{d^{n+2}y(t)}{dt^{n+2}}, \quad (45)$$

---

[5] The derivation in the above gives rise to a term proportional to $\dot{y}(t)$ with negative coefficient. Such a term, not discussed in Moniz and Sharp, would provide further damping of the motion.

where

$$A_n = \left(1 + \frac{\lambda_c}{3(n+2)} \frac{\partial}{\partial\lambda_c}\right)\left(1 + \frac{\lambda}{(n+1)} \frac{\partial}{\partial\lambda}\right) B_n, \text{ and}$$

$$B_n = \sum_{k=0}^{\infty} \frac{n!}{(n+2k)!} \binom{n+1-2k}{2k} \left(\frac{-\lambda_c^2}{4}\right)^k \int d\mathbf{x} d\mathbf{x}' f(\mathbf{x}) |\mathbf{x} - \mathbf{x}'|^{n-1+2k} \left(\nabla_{x'}^2\right)^{2k} f(\mathbf{x}'). \quad (46)$$

In particular, the coefficient of $\ddot{\mathbf{y}}(t)$, the quantum mechanical self-mass $\delta m$, is found to be

$$\delta m = \left(\frac{2}{3}\frac{e^2}{c^2}\right)\left(1 + \frac{\lambda}{6}\frac{\partial}{\partial\lambda}\right)\left(1 + \lambda\frac{\partial}{\partial\lambda}\right)\Omega_0. \quad (47)$$

Here

$$\Omega_0 = \frac{2}{\pi} P \int_0^{\infty} \frac{|\tilde{f}(k)|^2}{(1 - \lambda^2 k^2 / 4)} dk, \quad (48)$$

where P denotes the principal value integration. Equation (48) is remarkable in that the self-mass remains finite in the point particle limit $f(\mathbf{k}) \to 1$. *In fact, the self-mass vanishes in this limit!* $\delta m \to 0$ ($a \to 0$ with $\lambda$ fixed in quantum mechanics).

Therefore, $\lambda$ (bare mass) = $\lambda$ (observed mass). This is why we used the same notation for these two Compton wavelengths.

In another limit, $\lambda \to 0$,

$$\Omega_0 = \frac{2}{\pi} P \int_0^{\infty} |\tilde{f}(k)|^2 dk, \quad (49)$$

and the theory reproduces the classical result, as it should.

Whether the quantum mechanical equation of motion exhibits run-away or acausal behavior can be studied by writing Eq. (7) in a Green's function form, similar to that in the section on "Run-away Solutions, Boundary Conditions, and Integral Equations." Such an analysis in the point particle limit was carried out by Moniz and Sharp. It is found that the motion is causal with no run-aways if

$$\alpha = \frac{e^2}{\hbar c} = \frac{r_e}{\lambda} < 1.75. \quad (50)$$

In quantum mechanics, an electron is spread over a Compton wavelength. Thus, the above condition is reasonable in view of the classical causality condition Eq. (36), which can be written as $r_e/a < 1.5$.

Nonrelativistic quantum mechanics is not valid for rapid motion with frequency $\hbar\omega \gtrsim mc^2$ or if we approach a distance $\lambda$ near the electron, and a full QED analysis must be performed. The only calculation by QED was reported by Francis Low (13). He has not derived an equation of motion, but he has shown that the motion is finite in each order of $\alpha$. This is reasonable in view of the well-established renormalization theory that gives a finite answer for any physical process in perturbation expansion in $\alpha$. However, the perturbation theory is not suitable for taking the classical limit, and it is unclear whether QED can actually produce an equation of motion.

## CONCLUDING REMARKS

The impression one gets from reading text books is that the classical electron theory is in trouble due to run-away solutions and acausal behavior. However, we have seen that classical theory is actually fine if the electron is taken to be a spherical object of not too small a radius, greater than the classical electron radius $r_e$. The restriction is reasonable since $r_e$ is about 100 times smaller than the Compton wavelength $\lambda$, and we cannot consider distances less than $\lambda$ without considering quantum mechanics.

The nonrelativistic quantum theory also looks fine: In fact, the quantum theory is better behaved than the classical theory because it is finite and causal irrespective of the size of the electron, as long as $\alpha = r_e/\lambda < 1.75$. The inequality is certainly satisfied in the real world where $\alpha = 1/137$. The fact that the quantum theory is better behaved is also reasonable because the electron is smeared out due to the uncertainty principle. The quantum theory as reviewed here has the appropriate feature of having the correct classical limit. However, the limit of validity of the nonrelativistic quantum analysis is not really understood. The vanishing of the self-mass $\delta m$ in the point particle limit must be accordingly interpreted with care.

In fact, we know that the nonrelativistic treatment cannot strictly be valid in the point-particle limit because of vacuum polarization. It is reasonable that a full QED calculation will be at least as well-behaved as is the case in nonrelativistic quantum mechanics. Unfortunately, there exists as yet no real calculation in QED to confirm these conjectures.

Nevertheless, we are entitled to close this paper on a positive note: We started from the Abraham-Lorentz theory of point electron, which was problematic. We saw how the theory became fine by the use of an extended electron. The nonrelativistic quantum mechanics is fine even in the point-particle limit. A proper relativistic quantum electrodynamics calculation has not yet been done, and may not be possible (within QED as a perturbation theory), but should be no more singular than the nonrelativistic theory.

# ACKNOWLEDGMENTS

The authors would like to acknowledge communications with Dave Jackson, Francis Low, Robert Mills, Fritz Rohrlich, and David Sharp.

# REFERENCES

1. Abraham, M., "Theorie der Elektrizitat," Vol II: Elektromagnetische Theorie der Strahlung, Leipzig, Teubner (1905).
2. Lorentz, H.A., "The Theory of Electrons," Leipzig: Teubner (1909) (2nd edition, 1916).
3. Poincaré, H., " On the dynamics of the electron," Rendiconti del Circolo Matematico di Pulermo **21**, 129 (1906); translated by Scientific Translation Service, Ann Arbor, MI. The article is also translated into English with comments in "On the articles by Henri Poincaré," A.A. Logunov (Publishing Dept. of the Joint Institute of Nuclear Research, Dubna, 1995).
4. Dirac, P.A.M. " Classical theory of radiating electrons," Proc. Roy. Soc. Lond. **A167**, 148 (1938).
5. Sommerfeld, A., "Simplified deduction of the field and the forces of an electron moving in any given way," Akad. van Wetensch. te Amsterdam **13** (1904) (English translation **7**, 346 (1905)).
6. Page, L., Phys. Rev. **11**, 376 (1918).
7. Caldirola, P., " A new model of classical electron," Nuovo Cimento **3**, Supplemento **2**, 297 (1956).
8. Moniz, E.J. and Sharp, D.H., " Radiation reaction in nonrelativistic quantum electro-dynamics," Phys. Review **D 15**, 2850 (1977); see also Grotch, H., Kazes, E., Rohrlich, F. and Sharp, D.H., " Internal retardation," Acta Physica Austriaca **54**, 31 (1982).
9. Kim, K.-J., to be published.
10. Low, F., "Run-Away Electrons in Relativistic Spin 1/2 Quantum Electrodynamics," Preprint MIT-CTP-2522 (1997).
11. Panofsky, W.K.H., and Phillips, M., Classical Electricity and Magnetism, Reading, MA: Addison-Wesley, 1962.
12. Jackson, J.D., Classical Electrodynamics, New York: Wiley, 2$^{nd}$ Edition, 1975; 3$^{rd}$ Edition, 1998. Chapter 16 of 3$^{rd}$ Edition contains a review of the equation of motion of an electron, much more extensive than in previous editions.
13. Rohrlich, F., Classical Charged Particles, Reading, MA: Addison-Wesley (1965) (2nd edition 1990).
14. Yaghjian, A.D., "Relativistic Dynamics of a Charged Sphere," Lecture Notes in Physics **11**, Springer-Verlag, Berlin (1992)
15. Pearle, P., "Classical Electron Models," in Electromagnetism: Path to Research, D. Teplitz, Ed., New York: Plenum, 1982.
16. Rohrlich, F. " The Dynamics of a Charged Sphere and the Electron," Am. J. Phys. **65**, 105 (1997).
17. Laue, M., "Die Wellenstrahlung einer bewegten Punktladung nach dem Relativitatsprinzip," Annalen der Physik **28**, 436 (1909). Also see Podolsky, B. and Kunz, K.S., "Fundamentals of Electrodynamics," New York, Marcel Dekker, sec. 15 (1969).
18. Schott, G.A., "Electromagnetic Radiation," Cambridge University Press (1912).
19. Schwinger, J., "Electromagnetic Mass Revisited," Foundations of Physics, 13, 373 (1983).
20. Boyer, T.H., Phys. Rev. **D25**, 3246 (11982).
21. Baier, V.N., and Katkov, V.M., Sov. Phys. JETP **26**, 854 (1968).
22. Cahn, R.N., and Jackson, J.D., unpublished LBL note.
23. Dragt, A.J., and Finn, J.M., J. Math. Phys. **17**, 2215 (1976).

# Overview of Plasma Accelerators

T. Katsouleas

*University of Southern California*
*Los Angeles, CA 90089-0484*

**Abstract.** This is an overview of plasma accelerators. It is not intended to be a detailed review of the field. [See E. Esarey et al.(1) for a nice recent review, numerous recent articles in Physical Review Letters, and excellent contributions to these proceedings for a relatively complete update on the latest state of current research.] Rather, the intent of this overview is to review the concepts of plasma accelerators and provide some context for the articles that follow. We begin with a brief tutorial section on plasma wakefield accelerators based on simple physical pictures. Then we briefly assess the state of the field and its current direction. In particular, we describe the key physics and technology issues for continuing the rapid progress the field has recently enjoyed. Some other roles for plasmas in accelerators besides serving as the accelerating structure are described. We conclude with some thoughts on the future prospects for plasma accelerators.

## WAKES AND WAKEFIELDS

All current plasma accelerator experiments are in fact wakefield accelerator schemes of one form or another. Physical insight into the behavior of wakes in a plasma can be gained by considering the more familiar cases of wakes on the ocean and in other media. Consider as in Fig. 1 the comparison between the response of the ocean to a boat lowered into the water and the response of a neutral plasma to an external charge placed in it. In both cases the medium is displaced. In the plasma case, the displacement of plasma electrons shorts or shields the electric field of the external charge in a distance called the Debye length, which approaches zero in the limit of a cold plasma. If instead of a stationary disturbance, we consider as in the lower half of Fig. 1 a moving boat (or charge), we would observe that the sudden return of the displaced water (or plasma) after the passage of the boat (charge) leads to an overshoot and oscillation or wake. The phase velocity of the wake is tied to the velocity of the boat (or charge). Thus for a relativistic charge, the wake velocity is nearly c and appropriate for an accelerating structure. Similarly, the plasma disturbance can be caused by a laser pulse (laser wakefield acceleration) or a series of laser pulses (beat wave (2), self-modulated laser wakefield (3) and resonant pulses (4)). In this case the wake phase velocity is the group velocity of the laser $(v_g = c\sqrt{1 - \omega_p^2/\omega_o^2}$, where $\omega_o$ is the laser frequency, $\omega_p$ the plasma frequency = $\sqrt{4\pi n_o e^2/m}$, and $n_o$ is plasma density). The wavelength of the accelerating structure in a plasma wakefield accelerator is then $\lambda = 2\pi c/\omega_p \approx 300\mu/\sqrt{n_o/10^{16}\,\text{cm}^{-3}}$.

CP472, *Advanced Accelerator Concepts: Eighth Workshop,*
edited by W. Lawson, C. Bellamy, and D. Brosius

So how and why do ocean and plasma wakes differ? Each is an example of a Cerenkov radiated wake. Cerenkov waves can be excited in any medium by a disturbance moving faster than the phase velocity of a wave in that medium. Several familiar cases are illustrated in Fig. 2, including the shock wake of a supersonic bullet or plane. In each case, the wake properties are determined by the intersection of the Cerenkov condition ($v_b\cos\Theta=\omega/k$ where $v_b$ is the velocity of the disturbance and $\Theta$ is the angle between $v_b$ and the radiated wave vector k) and the dispersion curve for the medium. The differences between the dispersion properties of sound waves in air and space charge waves in plasma, for example, account for the markedly different form of each wake. Air is non-dispersive and the Cerenkov condition lies on the entire dispersion curve at one particular angle. Thus every frequency can be excited and each frequency travels at the same speed, conditions both necessary for supporting a sharp shock. The plasma on the other hand supports only one frequency ($\omega_p$) and this wave has zero group velocity (in a cold unmagnetized plasma), so no energy spreads outside of the path of the disturbance (i.e. there is no "vee" produced, only planar wavefronts). Ocean wakes show some behavior intermediate between the plasma and air cases.

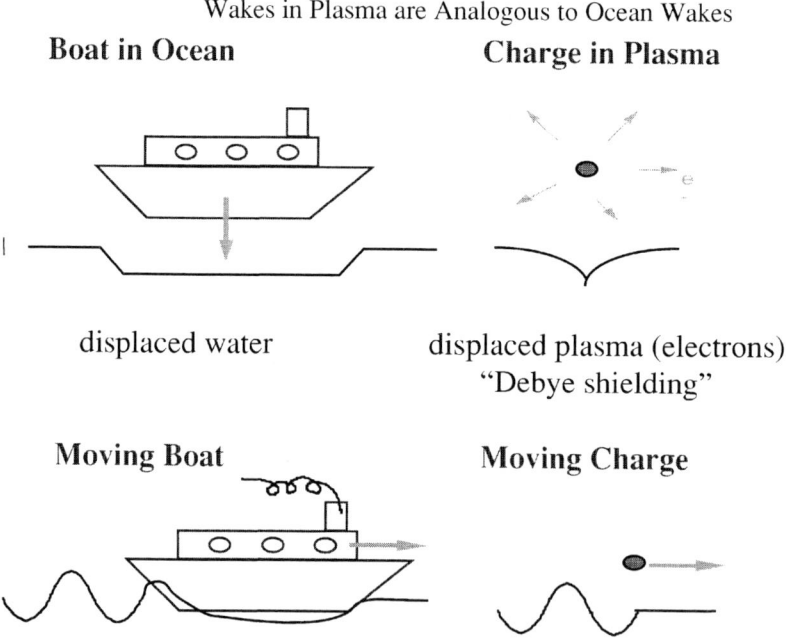

Wakes in Plasma are Analogous to Ocean Wakes

**Boat in Ocean**          **Charge in Plasma**

displaced water          displaced plasma (electrons)
"Debye shielding"

**Moving Boat**            **Moving Charge**

**FIGURE 1.** Comparison of the response of the ocean and a plasma to a disturbance

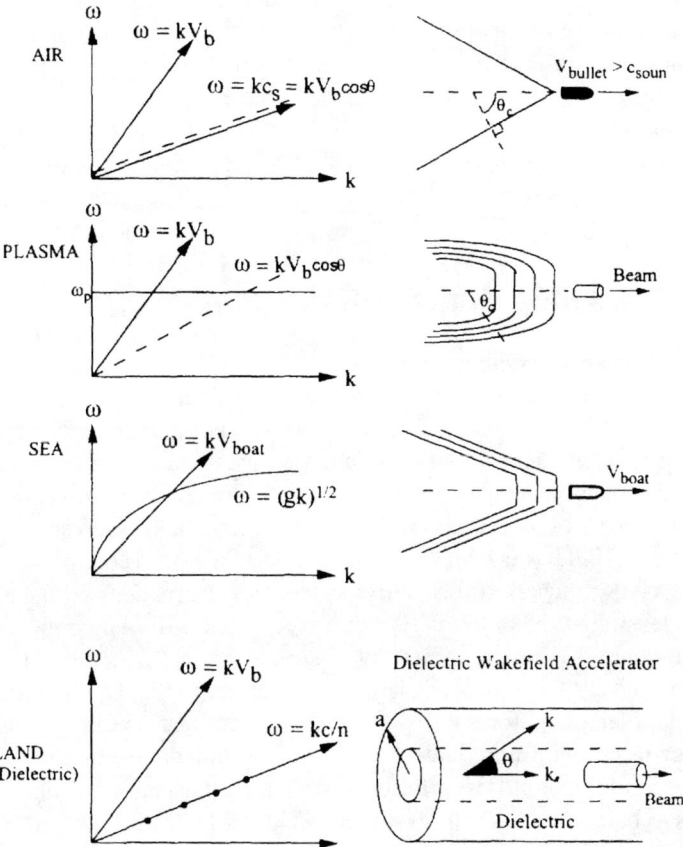

**FIGURE 2**. The differences between the forms of the Cerenkov wakes in various media (right) can be understood from the dispersion curves (left).

It is interesting to also consider electromagnetic Cerenkov wakes. The electromagnetic dispersion curve for plasma waves lies everywhere above the Cerenkov resonant curve; thus there are no electromagnetic Cerenkov waves in a cold unmagnetized homogeneous plasma. However, in a dielectric and a gas, electromagnetic wakes are the basis for two other accelerator schemes: inverse Cerenkov accelerators (5) and dielectric accelerators (6). The dielectric wakefield accelerator is illustrated in Fig. 2. It consists of a dielectric tube with a small hole through which the particle beams pass. The intersection of the dispersion curve and the Cerenkov condition defines a resonant Cerenkov angle ($\sin^{-1} (k_{\perp}/k)$). The boundary conditions at the tube edge (r=a) determine the allowed k values. Thus only discrete values of k and $\omega$ are excited as shown. For large tube radius a, the modes become more closely spaced and can support a spiky rather than sinusoidal wake as described by T. Marshall's paper in these proceedings.

## WAKE AMPLITUDE IN A PLASMA

So how large a wake is excited in a plasma by either a charged particle beam or laser beam? The answer can be easily estimated from linear fluid theory from the equations of continuity, momentum and Gauss' law. In Table 1, the derivation for the wake amplitude is shown where $n_1$ is the plasma density perturbation associated with the wake, $n_b$ is the density of a particle beam, $F_p$ is the pondermotive force (i.e., radiation pressure term) of a laser, and $V_{osc}$ is the normalized quiver velocity of electrons in the laser field ($eE_o/m\omega_o$). From the last line in Table 1 we see that wakefields approach as much as 100 GeV/m for density perturbations $n_1$ that approach $n_o$ in plasmas of density on the order of $10^{18} cm^{-3}$. We also see from the line above that the density disturbance $n_1$ is determined by a simple harmonic oscillator equation driven by a beam density term and/or a laser intensity term ($I_o$). One can easily see that $n_1$ can be of order of $n_o$ for short particle beams comparable in density to the plasma ($n_b \approx n_o$) or lasers with amplitude $V_{osc}/c = 1$ and with pulse durations of order $\pi/\omega_p$. Alternatively, multiple pulses or beats spaced at the plasma period allow similar wake amplitudes to be excited with lower driver density or laser intensity.

**TABLE 1.** Derivation of wake amplitude in a plasma

| 3-D Linear Wake Theory -- <u>Response to Laser or Particle Beam</u> |
| --- |
| $\partial/\partial t \{ \partial n/\partial t + \nabla \cdot nv \} = 0$ |
| $\partial^2 n_1/\partial t^2 + n_o \nabla \cdot \partial v/\partial t = 0$ |
| $\nabla \cdot \partial v/\partial t = \nabla \cdot (-eE/m + F_p/m) = 4\pi(n_1+n_b)e^2/m - \nabla \cdot \nabla I_o$ |
| $\partial^2 n_1/\partial t^2 + \omega_p^2 n_1 = -\omega_p^2 n_b + n_o \nabla \cdot \nabla v_{osc}^2$ |
| 1-D Gauss' law: $E = \int 4\pi e n_1 dx \sim 4\pi e n_1/k_p \sim (n_1/n_o)\sqrt{n_o}$ eV/cm |

# CURRENT LIMITATIONS AND ISSUES

The state of plasma accelerator research at the time of the last Advanced Accelerator Concepts Workshop was nicely summarized by a table from the plasma working group at Lake Tahoe (7). In many experiments around the world including those at Rutherford Appleton Laboratory (UK), Ecole Polytechnique (France), Osaka (Japan), Argonne National Laboratory, Livermore (LLNL), UCLA, U. of Michigan and U. T. Austin, the basic principles of wakefield excitation have been successfully demonstrated. Among the milestones achieved in these experiments were gradients of order 100GeV/m, energies to 100 MeV, accelerated beam charge of order nanoCoulombs and emittances on the order of a few mm-mrad. As impressive as the results have been, they are also limited. Most of the high-gradient results are over very short distances, typically a millimeter. The limits on acceleration length and hence energy gain in a single stage are illustrated in Table 2 for laser driven cases. They are diffraction, dephasing (due to the slight difference between laser group velocity and c mentioned previously) and pump depletion. Of these only the latter, pump depletion, is desirable as it also implies high energy extraction and hence potentially high efficiency. For beam driven wakefields, the limit is determined by the transformer ratio (8), the ratio of the wakefields behind and within the beam. In addition to the length limitation, there are also beam quality limitations in present experiments. Most of the experiments were in the self-modulated laser wakefield regime in which a long laser pulse becomes intensity modulated at the plasma frequency through a parametric instability (e.g., forward Raman scattering). In this regime the electrons are self-trapped at many phases and locations in the plasma, resulting in energy spreads of order 100%.

These limitations of the current experiments help to identify critical areas for further research on plasma accelerators. Some of the key issues are highlighted in Table 3: acceleration length, beam quality, efficiency. To overcome diffraction limits to stage length, a number of channel guiding schemes have been proposed and investigated (see the invited paper by J. Wurtele in these proceedings). In order to accelerate beams with a small energy spread, several ideas for injecting and phasing a short bunch of electrons on the plasma wake are being explored. (See the invited paper by E. Esarey in these proceedings). Finally, the topics of efficiency and beam dynamics in plasma wakes are just now coming to the fore.

**TABLE 2.** Approximate limits to single-stage energy gain in laser-driven plasma accelerators (linear thoery) $\sigma$ is spot size, $\lambda$ is laser wavelength, $\epsilon = n_1/n_0 < 1$

## Laser Accelerators -- Limits to Energy Gain

Diffraction:

$$E_{max} = 1\ GeV \left( \frac{n_o}{10^{18}} \right)^{1/2} \left( \frac{\sigma}{33\mu} \right)^2 \frac{1\mu}{\lambda}\ \epsilon$$

Dephasing:

$$E_{max} = 1\ GeV \frac{10^{18}}{n_o} \cdot \frac{1\mu}{\lambda^2} \cdot \epsilon$$

Pump Depletion/Instabilities:

$$E_{max} \sim 1GeV \cdot \frac{10^{18}}{n_o} \cdot \frac{1\mu}{\lambda^2}$$

**TABLE 3.** Key issues in current plasma accelerator research.

## Key Issues

| Key Issue | Experiment | Theory/Simulation |
|---|---|---|
| Acceler. Length mm ---> cm+ | *Channel Formation   *Plasma Sources | 1-to-1 models   parallel   3-D   hybrid |
| Beam Quality $\Delta\gamma$   $\epsilon$   N | *Injectors   50 fs bunch   50 $\mu$ spot   *Blowout regime | Beam Dynamics   matching $\beta$   injection phase |
| Efficiency (new) | | Drive beam evolution   Shaped driver and load   Transformer Ratio   Energy Spread |

# PROSPECTS

All of these developments make this an exciting time in plasma accelerator research. Besides their role as high-gradient electron accelerator structures, other roles are now emerging in a more serious way. For example, the plasma lens is scheduled to receive experimental testing at 30GeV at SLAC. This work will explore the possible use of a plasma as a final focus element for future colliders. There are interesting proposals for using plasmas to remove the correlated energy spread at the end of the NLC design (9), as well as proposals to accelerate protons in compact plasma accelerators for medical applications. Plasma-based photon accelerators are also proving to be fertile ground for new high-power radiation sources in the microwave to infrared range (10, 11).

As a final thought, it is interesting to take note of the rate at which new milestones in plasma accelerators are being achieved. As shown in Fig. 3, since the first high-gradient acceleration of an electron in a plasma accelerator was reported at Port Jefferson in 1992, the highest energy attained has increased by an order of magnitude every three years. With advances on key fronts such as injection, guiding, staging, and fundamental modeling, the prospects for continuing this progress are excellent.

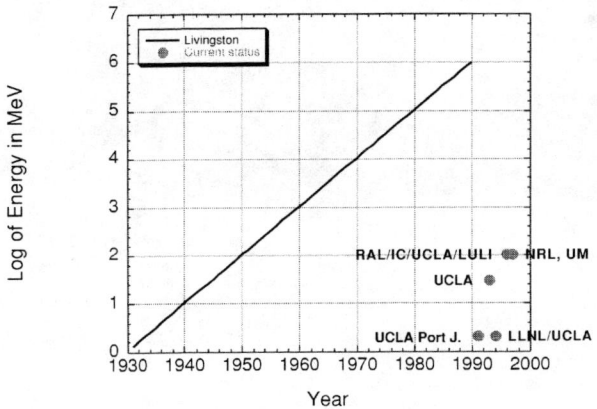

**FIGURE 3.** Traditional Livingston curve (solid) for maximum energy of accelerators vs. year and data points from some recent plasma accelerator experiments.

# ACKNOWLEDGEMENT

Work supported by the U.S. Department of Energy (Grant # DE-FG03-92ER40745).

# REFERENCES

1. Esarey, E., Sprangle, P., Krall, J. and Ting, A., *IEEE Trans. Plasma Sci.* **24**, 249 (1996).
2. Tajima, T. and Dawson, J. M., *Phys. Rev. Lett.* **43**, 267-270 (1979).
3. Esarey, E., Krall, J. and Sprangle, P., *Phys. Rev. Lett.* **72**, 2887-2890 (1990); Mori, W. B., Decker, C. D., Hinkel, D. E., and Katsouleas, T., *Phys. Rev. Lett.* **72**, 1482-1485 (1994); Andreev, N. E., Gorbunov, L. M., Kirsanov, V. I., Pogosova, A. A., and Ramazashvili, R. R., *Physica Scripta* **49**, 101-109 (1994).
4. Umstadter, D., Esarey, E. and Kim, J., *Phys. Rev. Lett.* **72**, 1224-1227 (1994).
5. Kimura, W. D. et al., Proc. Adv. Accel. Concepts Workshop, Fontana, WI, p. 131, *AIP Conf. Proc.* No. 335, P. Schoessow, Ed. (1995).
6. Gai, W. et al., *Phys. Rev. Lett.* **61**, 98 (1988).
7. Katsouleas, T., *AIP Conf. Proc.* **398**, 176, S. Chattopadhyay, Ed. (1997).
8. Bane, K. L. F., Chen, P. and Wilson, P., *IEEE Trnas. Nucl. Sci.* **32**, 3524 (1985); Katsouleas, T. *Phys. Rev. A* **33**, 2056-2064 (1986).
9. Heifets, S. and Raubenheimer, T., *AIP Conf. Proc.* **398**, 265. S. Chattopadhyay, Ed. (1997).
10. Muggli, P., Liou, R., Lai, C. H., Hoffman, J., and Katsouleas, T., *Phys. of Plasmas* **5**, 2112-2119 (1998).
11. Shvets, G., in these proceedings.

# The Stimulated Dielectric Wake-Field Accelerator: A Structure with Novel Properties

T. C. Marshall[1], T-B. Zhang[2], and J. L. Hirshfield[2,3]

[1] Department of Applied Physics, Columbia University, New York City 10027
[2] Omega-P, Inc., 202008 Yale Station, New Haven, CT 06520-2008
[3] Physics Department, Yale University, New Haven, CT 06520-8120

**Abstract.** We describe a multi-bunch wake-field accelerator in an annular, cylindrical dielectric structure in which many modes can participate in the wake-field formation, and where the wake-field period equals the period of a train of drive bunches of electrons. The structure can be designed so that the $TM_{0n}$ modes, all with phase velocities equal to the bunch velocity, have nearly equally-spaced eigenfrequencies, and thus interfere constructively. The composite wake-field is shown to cause highly peaked axial electric fields localized on each driving bunch in a periodic sequence of bunches. This permits stimulated emission of wake-field energy to occur at a rate that is larger than the coherent spontaneous emission from a single driving bunch having the same total charge. We present calcuations for an annular alumina structure which will use 2 nC bunches of electrons obtained from a 100MeV rf linac operating at 11.4Ghz. Numerical examples are given, including acceleration of a test bunch to >200 MeV in a structure 75 cm long, using three drive bunches.

## INTRODUCTION

In the conventional dielectric wakefield accelerator, a dielectric-lined waveguide supports wake fields with longitudinal electric fields induced by the passage of an electron bunch of high charge number (the "driving bunch"). Phase velocities for the modes of dielectric-lined waveguide can be less than the speed of light [1], so that Cerenkov radiation occurs[2], manifesting itself as a wake-field that fills the waveguide behind the driving bunch. If a "test bunch" of low charge number is injected at a suitable interval after the driving bunch, it can move in synchronism with the wake fields and experience net acceleration[3-5]. This approach for development of novel accelerators is appealing because no external source of rf power is required for acceleration, and because high gradient longitudinal fields are predicted for achievable high intensity driving bunches. For example, Rosing and Gai[4] consider a 100 nC, 1 mm long driving bunch passing through a dielectric lined waveguide with inner radius of 2mm and outer radius of 5mm; they took the relative dielectric constant of the outer liner to be 3.0. For this they predict a peak wake field accelerating gradient of 240 MV/m, a value about 14 times that at the Stanford Linear Collider. Experimental confirmation of wake-field generation in a dielectric-lined waveguide has been obtained[5] using 21 MeV driving bunches of 2.0 - 2.6 nC and 15 MeV test bunches of much lower charge. Acceleration gradients of 0.3-0.5 MV/m were observed in the experiments, in agreement with supporting theory. Acceleration gradients in all dielectric-lined waveguides must be below the breakdown field of the dielectric[6]. This will limit achievable gradients in any dielectric lined waveguide to a level that may not be as high as 240 MV/m.

In this, we describe a different type of dielectric wake-field accelerator, in which the superposition of wake-fields from a train of drive bunches provides the field that accelerates a trailing test bunch. Imagine a waveguide lined with a dielectric shell, along which axis are injected high energy (~100 MeV) short (~ 1 psec) electron current pulses

CP472, *Advanced Accelerator Concepts: Eighth Workshop,*
edited by W. Lawson, C. Bellamy, and D. Brosius

obtained from an rf linac. If the dielectric constant and geometrical dimensions are adjusted correctly, then the waveguide modes that travel at the axial speed of the particles can have phase velocities less than the speed of light, so that Cerenkov emission can occur; the emission manifests itself as a periodic wake-field that fills the waveguide downstream from these "driving bunches". Because of the radiative loss of energy, the drive bunches slow down. If a "test" bunch of low charge number but the same energy is injected at a suitable interval behind the driving bunches, it will move synchronously with the wake-fields and can experience net acceleration. This approach for the development of new electron or positron accelerators is appealing because no external source of rf power is required (beyond the conventional rf source to drive the injector), and because high-gradient fields may be obtained. The dielectric wake-field principle receiving greatest attention[4] uses a single high charge drive bunch which is expected to create a strong accelerating wake on its own. But, difficulties attend the creation, propagation and emittance control for such a high charge bunch; so it may be more advantageous to obtain the same accelerating gradient using a train of smaller charge bunches that should be easier to control.

The particular approach presented here enjoys two uncommon virtues. The first arises because many waveguide modes can participate in wakefield formation. These modes are designed so that, with phase velocities equal to the bunch velocity and to one another, the modes have nearly equally-spaced eigenfrequencies. This leads to a strongly-peaked spatiotemporal superposition of many co-propagating waveguide modes, so the net wake-field amplitude can be significantly larger than the amplitudes of individual modes. The second virtue arises because the near-periodic character of the wake-fields allows constructive interference of field amplitudes from successive bunches. Thus one adds to the *spontaneous* Cerenkov wake-field emission of the first drive bunch that experiences no rf fields, the *stimulated* Cerenkov emission from a train of following drive bunches that experience the wakes of preceding bunches. (The terms *spontaneous* and *stimulated* are used here to distinguish between radiation in the absence or presence of an ambient radiation field.) The bunches are assumed to be identical, and each bunch is injected to move initially with near-synchronism in the net wake-field of prior bunches. It will be shown that stimulated emission from each trailing bunch can exceed the spontaneous Cerenkov emission from a single bunch moving alone. Consequently, the drag field that decelerates a "dressed" bunch can be much larger than the drag field acting on a "bare" bunch (terms in quotes refer to the presence or absence of decelerating wake-fields from prior bunches). Thus a dressed bunch leaves behind a stronger wake than a bare bunch; and so forth for succeeding dressed bunches. Of course, successive wake maxima are not exactly periodic, and decelerating electrons can slip behind the wake-field maxima; so the cumulative superposition of wakes is less than a sum of peak values. Nevertheless, we find that a stronger wake-field might be produced by a multi-bunch train, than by a single bunch containing the total charge of all the particles. Detailed calculations of the energy loss of the drive bunches and the acceleration of the test bunch are dealt with in a paper [7] in Physical Review E. In that paper the configuration is two-dimensional rectangular dielectric slabs and a sheet beam, whereas we describe here the interaction in more practical cylindrical [8] geometry.

The wake-field concept described here should lend itself to staging, thus allowing acceleration to high energies. In the simplest staged configuration, drive bunches can be injected into a section of straight dielectric waveguide, with the spent drive bunches exiting along a collinear path. Then the waveguide modes carrying the wake fields can be diffracted around a gentle S-bend in the waveguide and coupled to a second straight section parallel to the first, wherein the test particles are synchronized to enter in the accelerating phase. This non-collinear arrangement permits the addition of energy to sustain the accelerating wake-field, prevents the spent bunches from slipping back and re-absorbing their own wake field energy, and comprises one stage of a much larger system.

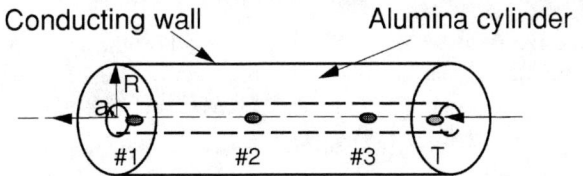

DRIVE BUNCHES: #1-3    TEST BUNCH: T

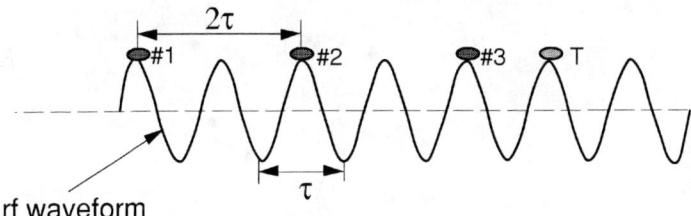

**FIGURE 1**. Schematic of the wake field element(above) and bunch timing(below).

## THEORY

The model analyzed here is a cylindrical waveguide, consisting of dielectric material with an axisymmetric hole; the outer surface of the cylinder is coated with a low-loss conductor. The dielectric constant $\kappa$ is assumed to be independent of frequency. The geometry is depicted in Fig. 1. Electrons are injected along the $z$-axis in discrete axi-symmetric bunches, with charge density given by $\rho(r,z,t) = -Ne(1/2\pi r)\delta(r)f(z - vt)$, where $e$ is the magnitude of the electron charge; $N$ is the total charge number in the bunch; $\delta(r)$ is the transverse charge distribution, assumed to be of infinitesimal width in $r$; and $f(z - vt)$ is the longitudinal charge distribution for bunch particles moving at axial speed $v$. For this geometry, orthonormal wave functions can be found for the electromagnetic fields that separate into $TE$ and $TM$ classes. Only the $TM$-modes have an axial electric field; these are the modes considered here. For the geometry shown in Fig. 1, conditions can be found where all $TM$ modes have phase velocities equal to $v$, corresponding to wake-fields that move in synchronism with the electron bunches. The field components for the complete orthonormal $TM$ mode set are given by

$$E_z(r,z,t) = \sum_{m=0}^{\infty} E_m \frac{f_m(r)}{\alpha_m} e^{-i\omega_m z_0/v} , \qquad (1)$$

where

$$f_m(r) = \frac{1}{P_0(k_{2m},R,a)} \begin{cases} P_0(k_{2m},R,a)\, I_0(k_{1m}r), & 0 \le r \le a \\ P_0(k_{2m},R,r)\, I_0(k_{1m}a), & a \le r \le R \end{cases},$$  (2)

taking $z_0 = z - vt$.

As we will see later, the field amplitude $E_m$ is expressed by the product of the Coulomb field and a structure factor that depends on the electron bunch size. In the above equations, $P_0(k,R,r) = J_0(kR)N_0(kr) - J_0(kr)N_0(kR)$ and $P_1(k,R,r) = J_0(kR)N_1(kr) - J_1(kr)N_0(kR)$; $J_m(x)$ and $N_m(x)$ are m-th order Bessel functions of the first and second kinds, and $I_m(x)$ is the modified Bessel function; $a$ and $R$ are radii of the central vacuum hole and the outer waveguide wall, respectively. The normalizing constant is found to be

$$\alpha_m = \frac{1}{2}\left\{\frac{1}{\kappa}\left(\frac{\gamma}{\gamma_\kappa}\right)^2 I_1^2(k_{1m}a)\left[\left(\frac{R}{a}\right)^2\left(\frac{P_1(k_{2m},R,R)}{P_1(k_{2m},R,a)}\right)^2 - 1\right] - (\kappa - 1)I_0^2(k_{1m}a) - I_1^2(k_{1m}a)\right\};$$

where $\beta = v/c$; $\gamma = (1 - \beta^2)^{-1/2}$; and $\gamma_\kappa = (\kappa\beta^2 - 1)^{-1/2}$. The (evanescent) transverse wave number in the vacuum is $k_{1m}$; the (real) transverse wave number in the dielectric is $k_{2m}$, and $k_{1m} = \omega_m/c\beta\gamma = k_{2m}\gamma_\kappa/\gamma$. The eigenfrequencies are $\omega_m = c\beta\gamma k_{1m} = c\beta\gamma_\kappa k_{2m}$. Since all TM modes have phase velocities equal to $v$, we also have $\omega_m = c\beta k_{z,m}$. For the fields, ortho-normalization obtains in the form

$$\int_0^R dr\, r\, E_{z,m} D_{z,n}^* = \delta_{mn} \frac{a^2}{\sqrt{\alpha_m \alpha_n}} \varepsilon_o E_m E_n \exp\left[-iz_o(\omega_m - \omega_n)/v\right],$$  (3)

where $D_{z,n}^* = \varepsilon E_{z,n}^* = \kappa\varepsilon_o E_{z,n}^*$ in the dielectric, and $D_{z,n}^* = \varepsilon_o E_{z,n}^*$ in the vacuum hole. The dispersion relation is found to be

$$\frac{I_1(k_1 a)}{I_0(k_1 a)} = \frac{\kappa k_1}{k_2} \frac{J_0(k_2 R)N_1(k_2 a) - J_1(k_2 a)N_0(k_2 R)}{J_0(k_2 R)N_0(k_2 a) - J_0(k_2 a)N_0(k_2 R)}.$$  (4)

It is noted that one can have eigenfrequencies with nearly periodic spacing, since $k_{2m}(R - a) \to (n + 1/2)\pi$ as $\kappa \to \infty$. As $m \to \infty$ the asymptotic eigenfrequency spacing approaches $\Delta\omega = \pi c\beta\left[(R - a)\sqrt{\kappa\beta^2 - 1}\right]^{-1}$. The wake-field is more strongly peaked and more closely periodic in $z_o$ as the eigenfrequencies become more nearly periodic, i.e. as a higher value of $\kappa$ is used.

To find wake-fields induced by an electron bunch, one expands in orthonormal modes the solution of the inhomogeneous wave equation,

$$\left[\frac{1}{r}\frac{\partial}{\partial r}\left(r\frac{\partial}{\partial r}\right)+\frac{\partial^2}{\partial z^2}-\frac{\kappa(r)}{c^2}\frac{\partial^2}{\partial t^2}\right]E_z(r,z,t)=S_z(r,z,t),\tag{5}$$

with the source function $S_z(r,z,t)=\mu_o\dfrac{\partial j_z}{\partial t}+\dfrac{1}{\varepsilon_o}\dfrac{\partial\rho}{\partial z}$, where the $z$-component of the current density is $j_z(r,z,t)=v\rho(r,z,t)$, and where $S_z(r,z,t)=0$ for $r\geq a$. A complete solution can be constructed from the fields, since these are solutions of Eq. 5 with $S_z(r,z,t)=0$ everywhere. We expand the solution of Eq. 5 in the interval $0\leq r\leq R$ in a Fourier integral.

$$E_z(r,z,t)=\sum_{m=0}^{\infty}\frac{1}{\alpha_m}f_m(r)\int_{-\infty}^{\infty}dk\,A_m(k)\,e^{-ikz_o}.\tag{6}$$

Inserting Eq. 6 into Eq. 5, and multiplying both sides by $w(r)D_{z,n}^*(r,z,t)$ gives

$$A_m(k)=\frac{1}{2\pi\,\alpha_m\left(k^2-\omega_m^2/v^2\right)}\int_0^R r'\,dr'\int_{-\infty}^{\infty}dz_o'\,S_z\left(r',z_o'\right)\kappa(r')f_m(r')w(r')e^{ikz_o'}\tag{7}$$

where the weighting factor is $w(r)=\left[-\gamma^2,\gamma_\kappa^2\right]$ in the intervals $\left[(0\leq r\leq a),(a\leq r\leq R)\right]$, respectively. Then, integrating over $k$, with due regard for choice of the contour of integration consistent with causality, one obtains the Green's function, with the aid of Eq. 3, and for a Gaussian bunch:

$$\rho(r,z,t)=-\frac{Ne\delta(r)}{2\pi r\Delta z}\exp\left[-(z_o/\Delta z)^2\right],\text{ one finds}$$

$$E_z(r,z,t)=-E_o\sum_{m=0}^{\infty}\frac{f_m(r)}{\alpha_m}e^{-(\omega_m\Delta z/2v)^2}e^{-i\omega_m z_o/v}.\tag{8}$$

In Eq. 8, $E_o=-Ne/4\pi\varepsilon_o a^2$ is the Coulomb field of the bunch at the edge of the hole, and causality dictates that the results are valid only behind the bunch, i.e. for $z_o\leq 0$; ahead of the bunch the fields are of course zero.

Eq. 4 has been evaluated for a waveguide with $a=0.375$ mm, $R=0.488$ cm, and $\kappa=9.43$. It is assumed that $\kappa$ is independent of frequency. The phase velocity of all the $TM$ modes corresponds to the speed of an electron bunch with $\gamma=200$, i.e. $0.999988\,c$. For this case, the first frequency interval $(\omega_2-\omega_1)/2\pi=10.797$ GHz, while the asymptotic interval $\Delta\omega/2\pi$ is 11.410 GHz. It is thus seen that conditions can be found where the eigenmodes all have equal phase velocities with nearly equal eigenfrequency spacings. This sets the fundamental period of the wake-field structure at

5.2 cm. A suitable accelerator that would power this wake-field element would use an X-band source with frequency of 11.4 GHz, or free space wavelength of 2.6 cm. A 100 MeV pulse of electrons about 1 psec in length would be produced whenever a photocathode rf gun is struck with a narrow pulse of UV light from a synchronized laser. The light pulse should be timed to arrive at the photocathode when the rf field has its maximum. The three drive bunches, to be described later (each separated by $4\pi$ in rf phase), and one test particle bunch (separated by $2\pi$), comprise the electron bunch sequence. One laser pulse will be split into 4 pulses, spaced so as to arrive at the cathode at the rf field maxima. Thus the drive pulses are to be spaced by 5.2 cm (the first subharmonic) and the test bunch follows the third drive bunch by 2.6 cm. Figure 1 shows the timing sequence referenced to the rf waveform.

Our theoretical results were evaluated numerically for comparison with wake-field experiments conducted at Argonne National Laboratory [5]. Good agreement was obtained for the wake-field structure, intensity, and mode frequencies, requiring only the first few modes that make significant contributions (also noted by the Argonne group). On the other hand, our design choice for cylindrical waveguide shows that the addition of many modes, as required here, results in a more sharply -defined wake-field structure: this is evident in Figure 2 where the wake-fields trailing a *single* bunch are shown. This is similar to the result we found for the planar slab example[7].

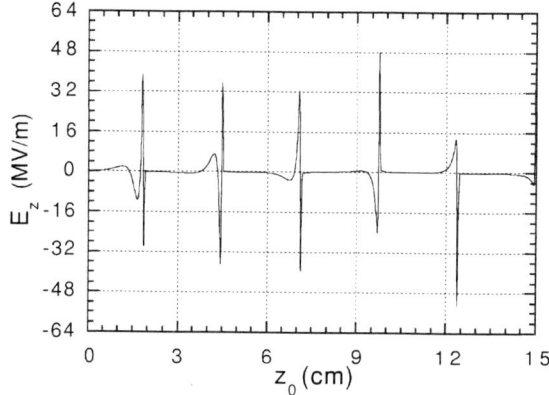

**FIGURE 2**. On-axis wake-field from a 2 nC, 100 MeV, 0.25mm long Gaussian bunch after it has traveled 15 cm left to right in the cylindrical waveguide with parameters as in Figs. 2 and 3. At $z_0 = 15$ cm, the wake-field amplitude is 57.6 MV/m.

The choice of these dielectric-lined waveguide parameters is seen to result in a spacing of 5.2 cm between the first few sharply-peaked positive-polarity wakefield features. This spacing is equal to twice the rf period of an 11.4 GHz injector gun, or for that matter, an rf linac. Therefore, if successive bunches were injected into the dielectric waveguide, the second drive bunch will find itself riding just on the crest of the first decelerating wake feature generated by the first drive bunch. Instead of generating only a spontaneous Cerenkov wake as did the first bunch, the second bunch will be decelerated in the field of the first wake, and its energy will be radiated as additional stimulated Cerenkov energy which builds up its own wake. Successive bunches will interact similarly. To make this quantitative, one calculates the incremental energy $\Delta W$ radiated into the waveguide by a bunch in advancing a distance $\Delta z$, and equates that to the energy loss rate of the bunch. This loss rate is identified with a drag field $E_{drag}$ acting on the bunch. Thus, $Q E_{drag} = \Delta W/\Delta z$. For a bare bunch, $E_{drag} = E_{spon}$, the drag field corresponding only to spontaneous emission. But for a dressed bunch that follows behind $N$ prior bunches, the drag field consists of the spontaneous drag field added to the combined wakefields of the preceding bunches. The total wakefield is next calculated by adding up the incremental wakefield energies from each bunch up to that location and then finding the resulting electric field, and so on. It is also noted that perfect synchronism is assumed in the above simplified discussion, so that wake amplitudes (and not energies) are added constructively. More highly relativistic bunches can maintain synchronism while losing a larger fraction of their initial energy, as compared with bunches of lesser energy, since velocity slip is lower, i.e. $\Delta\beta \approx \Delta\gamma/\gamma^3$.

Now we turn to considerations of energy loss by the bunch. The rate of energy accumulation with distance in wake fields behind the bunch $dW/dz$ is given by integrating the field energy density over the waveguide cross-section, thus leading to the following relation.

$$\frac{dW}{dz} = \frac{1}{4} \sum_{m=0}^{\infty} \int_0^{2\pi} d\theta \int_0^R r\, dr \left\{ \varepsilon(r)\left[E_{r,m}^2 + E_{z,m}^2\right] + \mu_o H_{\theta,m}^2 \right\}$$

$$= \frac{\varepsilon_o E_o^2 \pi R^2}{2} F \sum_{m=0}^{\infty} \frac{h^2(\xi_m)}{\alpha_m^2} \left\{ \frac{1}{R^2} \int_0^a I_0^2(k_{1m}r)r\,dr + \left(1+\beta^2\right)\gamma^2 \frac{1}{R^2}\int_0^a I_1^2(k_{1m}r)r\,dr \right.$$

$$\left. + \kappa \left( \frac{I_0(k_{1m}a)}{P_0(k_{2m},R,a)} \right)^2 \left[ \frac{1}{R^2}\int_a^R P_0^2(k_{2m},R,r)r\,dr + \left(1+\kappa\beta^2\right)\frac{\gamma_k^2}{R^2}\int_a^R P_0^2(k_{2m},R,r)r\,dr \right] \right\}$$

$$(9)$$

In Eq. 9, $F$ is a scale factor fixed by balancing the energy loss rate between drag on the bunch, and increase in the wake-field energy, as described in the prior paragraph. For a bare bunch $F = 1$. This scaling procedure assumes that the source current and charge distributions remain constant during the interaction. If not, the individual mode amplitudes must be adjusted iteratively; this amounts to introduction of a $z$-dependent structure factor $h(\xi_m)$. For the Gaussian bunch, $h(\xi_m) = \exp(-\xi_m^2)$, with $\xi_m = \omega_m \Delta z/2v$.

For the drive bunch charge of 2 nC, in Figure 2 we compute a wake-field of 57.6 MV/m and find a drag field of 11.45 MV/m (the latter is calculated by equating the spatial rate of radiation energy loss from the bunch, given by eq. 9, to the product of a "drag" electric field and the bunch charge). We note that this gives a wake-to-drag field ratio, the so-called "transformer ratio"[9] of 5.0: transformer ratios above two have been shown to be possible for multi-mode waveguides[10] (on the other hand, the ratio we quote is based on bunch average quantities, and so may differ on that account).

In the following simple model for the buildup of a *cumulative* wake-field from a multiple-drive bunch train, the bunches are taken as point charges which remain perfectly synchronized with the wake-fields. The energy radiated into the wake-field of the $n$-th bunch can be written in terms of the net drag on the charge $Q$ as

$$\frac{dW_n}{dz} = R\left[\left(\sum_{i=1}^{n} E_i\right)^2 - \left(\sum_{i=1}^{n-1} E_i\right)^2\right] = Q\left[E_{spon} + \sum_{i=1}^{n-1} E_i\right] = QE_{drag},$$

(10)

from which the combined wake-field is found to be:

$$\sum_{i=1}^{n} E_i = \left[\left(\sum_{i=1}^{n-1} E_i\right)^2 + \frac{Q}{R}\left(E_{spon} + \sum_{i=1}^{n-1} E_i\right)\right]^{1/2}.$$

(11)

The factor "$R$" in eqs. 10 and 11 is obtained from the coefficient of the electric field squared in Eq. 9(not to be confused with R, the radius). The individual wake-field from the $i$-th bunch is $E_i$. One can appreciate from Eqs.10 and 11 that participation of many co-propagating modes, and stimulated emission from a periodic sequence of driving bunches, can increase dramatically the extraction of energy from the bunches, which in turn promotes the buildup of an intense overall wake-field after passage of a relatively few moderate-charge driving bunches. These results suggest that high wake-fields can be obtained in this manner, without the use of a single, high-Q bunch and its attendant difficulties.

The conceptual model discussed above assumes perfect synchronism between driving bunches and peak wake-fields, and assumes the wake-field amplitude to be uniform over the finite spatial extent of each bunch. The problem has been examined with greater accuracy in a numerical simulation, using 100 particles per bunch, and taking slip and actual wake-field amplitude variations into account. Particles in each 0.25 mm long bunch are injected around the peak of the wake-field from prior bunches. The energy loss rate from each bunch is given by Eq. 9, with successive values of $F$ found from the drag field on that bunch, i.e. from the sum of its spontaneous drag field $E_{spon}$ and the net wake-field from prior bunches. Particles in a given bunch obey the one-dimensional equation of motion $d\gamma/dz = (e/mc^2)E_{drag}$, evaluated at each particle's instantaneous location, where $E_{drag}$ is obtained from Eq. 10. The initial energy of each bunch is chosen with $\gamma_{initial} = 200$, or about 100 MeV. In this model, the bunches-- now distributed spatially--will lose energy and thus can slip with respect to the wake-fields. Motion in the $x$-$y$ transverse plane is neglected.

# RESULTS

Figure 3a shows the buildup of the wake-fields in the device under study here, if three 2nC drive bunches moving in a wake-field structure are used. Superposition of wakes is seen to result in a wake-field amplitude at the third bunch of 188 MV/m, which is greater than the wake-field (173 MV/m) of a single bunch of three times the charge. However, a 6 nC, 1 psec bunch may be beyond current state of the art.

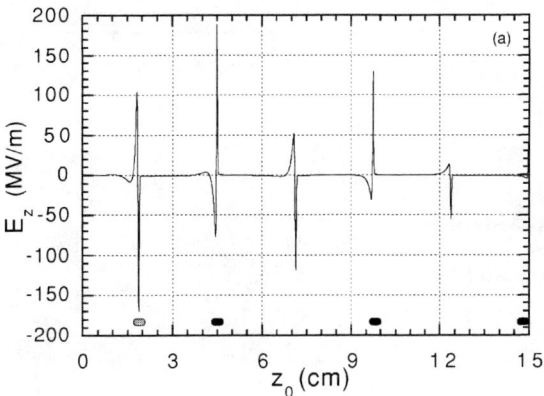

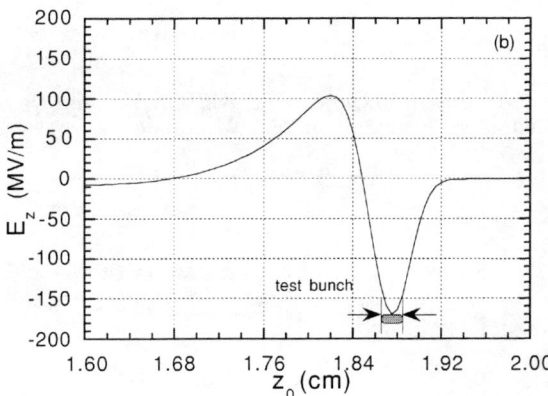

**FIGURE. 3**. (a) Cumulative wake-field set up by three identical successive drive bunches, at the time the first bunch has moved 15 cm along the waveguide. The bunch locations are indicated as black dots on the figure. (b) Location of the test bunch in the accelerating wake field behind the third drive bunch near the entrance of the accelerator, shown on an expanded scale.

Figure 4 shows the average energy increase of the test bunch (#4), about 105 MeV in 75 cm; this is less than the maximum of 170 MV/m X 0.75 m = 127 MeV because of variations in the field across the bunch, as shown in Figure 3b. Beam loading is neglected for the test bunch. As others have concluded[10], collinear transport of drive and test bunches imposes limitations on the wake-field accelerator, which suggests that the radiation and the drive particles should be separated. Nevertheless, the acceleration field of 140 Mv/m that was obtained makes this an attractive concept to explore further.

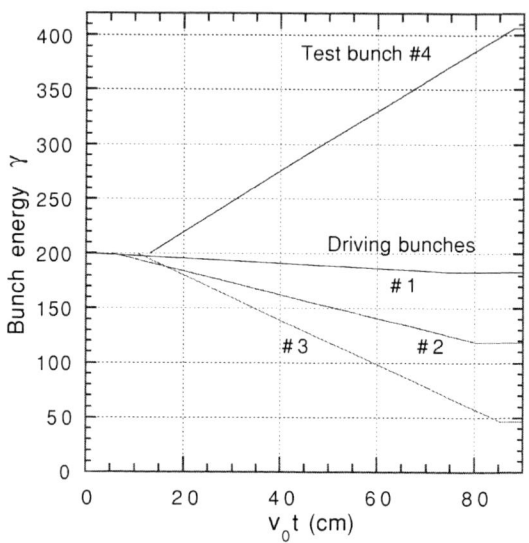

**FIGURE. 4**. History of the energy of the three drive bunches and the test bunch, up to the time the test bunch has left the 90 cm long waveguide.

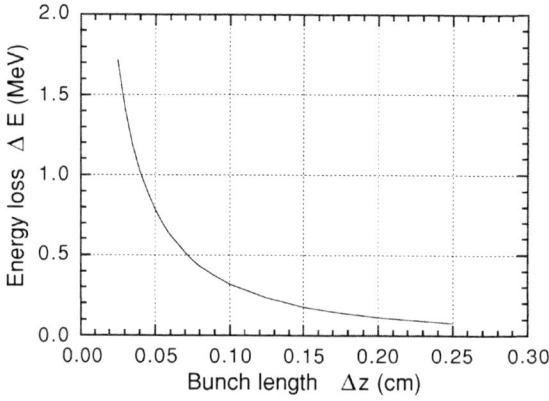

**FIGURE. 5**. Energy loss to wake-field radiation as a function of bunch length for a 2 nC, 100 MeV Gaussian bunch traversing a 15 cm long waveguide.

In Figure 5, we show calculations of the energy loss of a single 100 MeV bunch traversing a fixed length of cylindrical dielectric loaded waveguide with parameters the same as for the acceleration experiments. For a constant charge of 2 nC, the figure shows the energy loss as a function of rms axial bunch length, for a Gaussian distribution. As the energy decrease of the particles is readily monitored, and is strongly sensitive to the bunch length, it appears that there is the possibility for a simple diagnostic for bunch length. A study of this would further test our theory, and provide confidence in using it as an on-line bunch monitor (with an energy loss ~ 1%, such a diagnostic can be considered to be passive).

Our calculations have shown that a multi-bunch wake-field accelerator may have some advantages over the single bunch version (which contains a substantially larger charge). The stimulated buildup of wake-fields allows one to produce wake-fields well in excess of 100 MeV/m, using short bunches that contain modest charge (~2 nC). Nevertheless, at this time there remain some problems which can be resolved only by further analysis and experimentation: among them stability, very short-pulse dielectric breakdown, and dispersion.

# ACKNOWLEDGMENT

This research was supported by the Department of Energy, High Energy Physics Division.

# REFERENCES

[1] Chang, C.T.M., and Dawson,J.W. "Propagation of Electromagnetic Waves in a Partially Dielectric Filled Circular Waveguide", *J. Appl. Phys.* **41**, 4493 (1970)
[2] Bolotovski, B.M., "Theory of Cerenkov Radiation", *Sov. Phys. Uspekhi* **4**, 781 (1962)
[3] Ng, K-Y., "Wake Fields in a Dielectric Lined Waveguide" *Phys. Rev.* **D42**, 1819 (1990)
[4] Rosing, M., and Gai, W., "Longitudinal and Transverse Wake-field Effects in Dielectric Structures", *Phys. Rev.* **D42**, 1829 (1990)
[5] Gai, W., Schoessow, P., Cole, B., Konecny, R., Rosenzweig, J., and Simpson, J., "Experimental Demonstration of Wake-field Effects in Dielectric Structures", *Phys. Rev. Lett.* **61**, 2756 (1988)
[6] Sprangle, P., Hafezi, B., and Hubbard, R.F., "Ionization and Pulse Lethargy effects in Inverse Cerenkov Accelerators", *Phys. Rev.* **E55**, 5964 (1997)
[7] Zhang, T-B., Hirshfield, J.L., Marshall, T.C., and Hafezi, B., "Stimulated Dielectric Wake-field Accelerator", *Phys. Rev.* **E56**, 4647 (1997)
[8] Zhang, T-B., Marshall, T.C., and Hirshfield, J.L., "A Cerenkov Source of High Power Picosecond Microwaves", *IEEE Trans. Plasma Science* **26**, 787 (1998)
[9] Ruth, R.D., Chao, A.W., Morton, P.W., and Wilson, P.B., "A Plasma Wake Field Accelerator", *Particle Accelerators*, **17** 171 (1985)
[10] Bane, K.L.F., Chen, Pisin, and Wilson, P.B., "On Collinear Wake Field Acceleration", *IEEE Trans. Nuclear Sci.*, **NS-32** 3524 (1985)

# Coherent Transition Radiation-Based Diagnosis of Electron Beam Pulse Shape[1]

J.B.Rosenzweig, A. Murokh and A. Tremaine

*Department of Physics and Astronomy*
*University of California, Los Angeles*
*Los Angeles, CA 90095*

**Abstract.** The bunch shapes of an electron beams is increasingly difficult to measure, as the time-scales of interest are now sub-picosecond. We discuss here the use of coherent transition radiation CTR for such measurements. Two types of measurements are presented: the deduction of macroscopic (0.3 psec resolution) pulse profile using interferometry, and the examination of microbunch (50 fsec) structure from an FEL-bunched beam using spectral characteristics of the CTR. For the macrobunch measurement we discuss the problem of missing low frequency radiation and one solution for extracting meaningful data with this problem present. For microbunch CTR, we examing initial spectrally--resolved measurements, and some interesting deviations in the CTR spectrum from the standard theoretical predictions.

## INTRODUCTION

The standard time-domain approaches to ultra-short electron beam pulse length diagnosis, such as streak cameras and rf sweeping, are inadequate when dealing with applications such as advanced accelerators and short-wavelength free-electron lasers (FELs). In order to move paset the picosecond level of resolution, it is necessary to use other methods. Coherent transition radiation (CTR) is one of the most promising of these methods. Much effort has been recently devoted to characterization of CTR itself[1], as well as performance of the actual bunch length measurements. This measurement is often done by examination of the autocorrelation of the CTR signal with the Michelson interferometer[2], which allows one to obtain the amplitude of the

[1] This work was supported by U.S. Dept. of Energy grants DE-FG03-93ER40796 and DE-FG03-92ER40693, and the Alfred P. Sloan Foundation grant BR-3225.

CP472, *Advanced Accelerator Concepts: Eighth Workshop,*
edited by W. Lawson, C. Bellamy, and D. Brosius

beam current's Fourier Transform. Direct spectral measurements of mm-wave CTR[3], as well as coherent synchrotron radiation[4] has also been performed.

We at UCLA have implemented two versions of a CTR-based pulse length diagnostic. The first is an interferometric device which allows the autocorrelation of the pulse to be obtained. This technique has certain limitations, as for beams of a few picosecond in length, the bunch length structure information is being mostly carried in the long wavelengths which are hard to propagate and detect with full efficiency. These most informative frequencies dominating the signal spectrum are in fact strongly attenuated in the collection optics due to the finite apertures and acceptance angles. This distortion of the detected spectrum puts a great uncertainty on the beam shape in the frequency domain, and makes the analysis rather difficult. To compensate for the missing frequencies an analytical method has been developed, which allows the processing of the autocorrelation signal directly in the time domain. Even though the frequency and time domains are equally representative to the signal, the accuracy of interpretation is increased drastically, as this analysis does not require that the autocorrelation signal be processed by discrete Fourier transform.

The second CTR-based method is a spectral approach used to obtain information concerning the beam microbunching as it is manifested in an FEL or an advanced accelerator[5]. We will discuss some details of initial measurements with this method, and some unexpected physical features found in these experimental data.

## TRANSITION RADIATION DUE TO A SINGLE PARTICLE

In order to understand well the experimental measurements, we must first begin with a review of the relevant components of the theory of transition radiation and its coherent generation by short bunches of electrons. The process of a charged particle crossing the boundary of the perfect conductor can be viewed as an effective collision of the particle with its image charge (Figure 1). For the wavelengths long compare to the characteristic time of the collision process (wt<<1) the Fourier transform of the radiated magnetic field can be approximated as

$$\tilde{\mathbf{H}}(\omega) = \int_{-\infty}^{\infty} \mathbf{H}(t)e^{-i\omega t}\,dt \cong \int_{-\infty}^{\infty} \mathbf{H}(t)dt \tag{1}$$

Note that magnetic field in the integral is proportional to the full derivative of the field vector-potential $\mathbf{A}$, and therefore the spectral field produced by each particle in the collision can be written as[6]

$$\tilde{\mathbf{H}}(\omega) = \frac{1}{c}\left(\mathbf{A}_2 - \mathbf{A}_1\right) \times \mathbf{n}\ , \tag{2}$$

where $\mathbf{A_1}$ and $\mathbf{A_2}$ are the values of the vector potential before and after the collision, respectively. As the particle crosses the boundary of the conductor, its field is screened; hence, for the far field calculation we can assume that both a particle and its image counterpart come to a sudden stop, even though the real electron barely changes its velocity.

If the field of a charged moving particle far away from the source is given by the Liénard-Wiechert potential we obtain the expression for the spectral and angular distribution of the energy generated in the collision:

$$\frac{dE}{d\omega \cdot d\Omega} = \frac{e^2}{4\pi^2 c^3}\left(\frac{\mathbf{v} \times \mathbf{n}}{1-\frac{1}{c}\mathbf{n}\cdot\mathbf{v}} - \frac{\mathbf{v'} \times \mathbf{n}}{1-\frac{1}{c}\mathbf{n}\cdot\mathbf{v'}}\right)^2 , \tag{3}$$

where $\mathbf{v'}$ is a velocity of the image charge. For the case of a 45° collision, Eq. (3) can be rewritten as

$$\frac{dE}{d\omega \cdot d\Omega} = \frac{e^2 \beta^2}{4\pi^2 c}\left(\frac{\cos\theta}{1+\beta\sin\theta} - \frac{\sin\theta}{1-\beta\cos\theta}\right)^2 , \tag{4}$$

where $\theta$ is defined relative to the specular reflection angle, which is in this case 90° with respect to the initial beam axis.

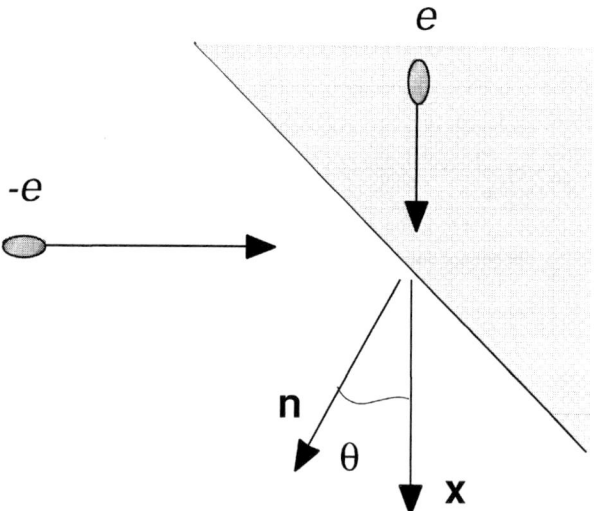

**FIGURE 1.** Transition radiation model for the far field analysis.

In the limit we have considered, where the metallic response is that of a perfect conductor, it can be seen that the angular distribution of the radiated energy does not depend on frequency. For the highly relativistic particle, the angular distribution is a narrow, slightly asymmetric double cone, centered around zero.

# MACROBUNCH CTR MEASUREMENTS

## Coherent Transition Radiation From a Bunched Beam

The total spectral energy $E_1(\omega)$ radiated by the single particle can be found by integrating Eq. (4) over all the observation angles. To find the transition radiation generated by a bunched beam of charged particles, we must add the fields due to the radiation of all the particles in the beam. For wavelengths long compare to the bunch length, the total field created by the particles is coherent, therefore for the total spectral energy, which is proportional to the square of the field, is $E(\omega) = N^2 E_1(\omega)$. However, as the wavelength approaches the size of the beam, this is no longer the case, and the total spectral energy radiated by the beam can be represented as

$$E(\omega) = |f(\omega)|^2 E_1 , \tag{5}$$

where $f(\omega)$ is a coherency function defined, for wavelengths long compared to the beam transverse width by the longitudinal beam profile $\rho(t)$. In the limit considered here, where the paraxial beam is much smaller radially than the relevant wavelengths (most beams of interest are "sausage shaped", being narrow compared to the longitudinal scale of interest), this function is

$$f(\omega) = \sum_{j=1}^{N} e^{i\omega\tau_j} \cong N \int_{-\infty}^{\infty} \rho(t)e^{i\omega t} dt , \tag{6}$$

where $\rho(t)$ is normalized to unity. A notable violation of the assumption of negligible beam width occurs during the microbunch measurements, where the beam is wide compared to the CTR wavelengths, and a more general analysis is necesssary. For low frequencies of present interest, $|f(\omega)|^2$ converges to the expected value of $N^2$. Note that the expression on the right hand side Eq. (6) is the complex conjugate of the beam profile Fourier transform, and thus $f(\omega)$ is the population-weighted Fourier transform of the bunch longitudinal profile. Thus, in terms of the Fourier transform of the bunch profile, we can write

$$E(\omega) = 2\pi\tilde{E}_1 N^2 |\tilde{\rho}(\omega)|^2 \tag{7}$$

# INTERFEROMETRIC MEASUREMENT OF CTR

The device shown in Figure 2 has been used to determine bunch lengths at UCLA[7] was built and developed at the University of Georgia by Prof. Uwe Happek. It is a polarizing Michelson interferometer based on beam splitters which use a transmission wire grid of 100 micron spacing. This spacing sets the upper limit on the spectral characteristics of the device.

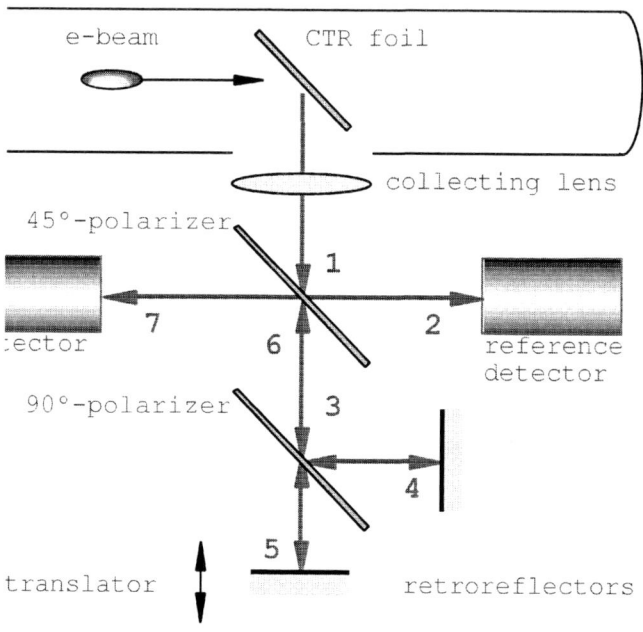

**FIGURE 2**. Interferometer device used to measure the CTR signal profile at UCLA.

The radiation is generated at a 45 °CTR foil and exits the beam vacuum through a quartz window. It is gathered by the collecting lens {1}, then split equally by the first wire grid, which is oriented at 45 ° from vertical. Half of it is sent into the reference Golay cell detector {2}, which is used to normalize the signal in the detector at the end of the interferometer arm. The other half {3} is directed through the 90 °-polarizer, where it is split back into horizontal (reflected) and vertical (transmitted) components {4-5}. Both are reflected at the end of their respective paths by retroreflectors and recombine at 45 °-polarizer. The vertical polarization path {5} is adjusted by a motorized translator which allows variation of the relative paths of the vertically and horizontally polarized waves in increments of as small as 1.4 microns. Thus, as the polarized waves recombine at {6},

$$\mathbf{E}_6(\omega) \propto e^{i\omega t}\left(\hat{\mathbf{x}} - \hat{\mathbf{y}}e^{i\omega\tau}\right) = \frac{e^{i\omega t}}{\sqrt{2}}\left[\hat{\mathbf{e}}_+\left(1 - e^{i\omega\tau}\right) + \hat{\mathbf{e}}_-\left(1 + e^{i\omega\tau}\right)\right] \qquad (8)$$

and only the $\mathbf{e}_+$ polarized fraction, whose amplitude depends on the position of the translator t, will be reflected from the polarizer and directed into the other Golay cell detector {7}. The result is an autocorrelation of the CTR signal: the total signal on the detector is effectively a combination of the signal and its t-delayed prototype. In that case, the Eq. (6) for $f(\omega)$ has to be modified:

$$f(\omega) = \frac{1}{2\sqrt{2}}\sum_{j=1}^{N}\left(e^{i\omega t_j} - e^{i\omega(t_j + \tau)}\right) \approx \frac{N}{2\sqrt{2}}\int \rho(t)e^{i\omega t}\left(1 - e^{i\omega\tau}\right)dt ,$$

$$\text{(9)}$$

$$\text{where} \quad |f(\omega)|^2 = \frac{\pi N^2}{4}\tilde{\rho}^*(\omega)\tilde{\rho}(\omega)[1 - \cos(\omega\tau)] ,$$

and the total radiation energy received by the detector is

$$E_T = E_1\int|f(\omega)|^2 d\omega \propto \int \tilde{\rho}^*(\omega)\tilde{\rho}(\omega)[1 - \cos(\omega\tau)]d\omega \qquad (10)$$

It is convenient at this point to examine the time domain picture again, by writing an explicit expression for one of the beam density Fourier transforms:

$$E_T \propto \int dt\rho(t)\int e^{i\omega t}\tilde{\rho}(\omega)\left(2 - e^{i\omega\tau} - e^{-i\omega\tau}\right)d\omega \qquad (11)$$

It follows, that the only part of the total energy that changes with the delay t is an autocorrelation function of the beam density profile, or

$$E_T(\tau) \propto \int \rho(t)\rho(t + \tau)dt + \text{constant.} \qquad (12)$$

## Signal Analysis in the Frequency Domain

Even though the problem of extracting the beam density function from the integral in Eq. (12) has no unique solution, we can make certain assumptions about the beam shape and test them with respect to the measured results. The first, simplest, and most widely used *ansatz* is a Gaussian beam profile:

$$\rho(t) = \frac{1}{\sqrt{2\pi}\sigma_t}e^{-\frac{t^2}{2\sigma_t^2}} \qquad (13)$$

43

In the case of many rf photoinjectors, it is motivated strongly by the nearly Gaussian shape of a photocathode drive laser pulse which governs the initial time profile of the emitted beam[7].

Using Eq. (14) one can see that the autocorrelation of such a beam, which can be found directly by measuring the detected energy as a function of the delay t, should give in return a Gaussian profile of width expanded by $\sqrt{2}$ :

$$E_T(\tau) \propto e^{-\frac{\tau^2}{4\sigma^2}} + \text{constant.} \tag{14}$$

Autocorrelation measurements performed have been done for the electron beam in UCLA Saturnus photoinjector beamline, an example of which is shown in Figure 4; however, one can immediately see that the signal has positive and negative maxima. That can not be interpreted as an autocorrelation of Gaussian or any other form of the unipolar function.

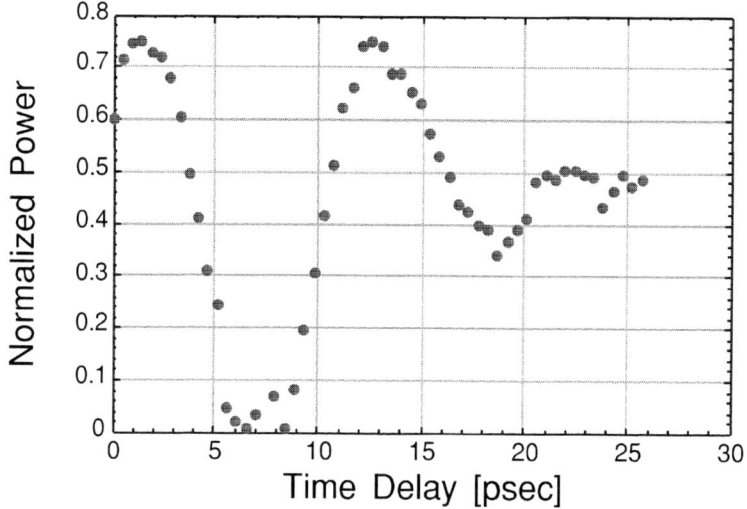

**FIGURE 3**. Normalized signal from Golay cell detector as a function of the translator position. Beam of ~200 pC at 13.25 MeV went through the 45 °aluminum foil.

Indeed, according to Eq. 14, the most general autocorrelation signal measured in the time domain must be strictly unipolar if all of the information available at the radiation generation point is preserved. We can see that this is not in fact the case in the present UCLA measurements — as well as with previously reported CTR measurements[2]. To get a more accurate understanding of the autocorrelation signal structure in the time-domain, we have to revert to looking at the beam spectrum.

As was shown above, single particle radiation contains all the frequencies in the long wavelength limit with roughly uniform efficiency of generation out to the transverse dimensions of the foil. However, in the case of the very large wavelengths (more than a few mm), where the radiation is highly coherent and carries significant part of the total energy, the interferometer acts as a high-pass filter, due to diffractive losses as well as the physical apertures associated with the optics and detectors in the device. The Fourier transform of the measured signal, which is ideally Gaussian, displays such a filtering at low frequencies. We could in principle compensate for this filtering effect by restoring the low frequency components of the spectrum, by smoothly continuing the high frequency portion into a Gaussian shape peaked at zero frequency, as is traditionally done in this technique. For our measurements, however, this process gives unsatisfactory results, however, as the missing information is too significant. The restored profile always displays notable false artifacts; it is difficult to suppress the undulations in the profile which are in the raw autocorrelation[7]. To extract more meaningful information from our data we have therefore developed a more systematic frequency filtered model which can be used directly to fit the data in the time domain.

## Time-Domain Fitting Approach

To proceed in constructing a useful model of what the frequency-filtered signal looks like in the time domain, we note first that the signal measured is in fact proportional to the autocorrelation of the filtered beam distribution function, an observation which allows an elegant and powerful formalism for model creation. We begin this analysis by introducing some analytical filter function of variable strength $g(\omega)$, which for ease of further analysis we take to be of the form

$$g(\omega) = 1 - e^{-\xi^2\omega^2},$$ \hfill (15)

which smoothly removes the low frequencies, with characteristic frequency cut-off $\xi^{-1}$. This form of the filter is physically motivated; it is obtained by the aperturing of a diffraction-limited transverse Gaussian-mode photon beam of uniform initial frequency spectrum in the far field, which is undoubtedly similar to our physical situation in this measurement..

We further assume that most of the low frequencies are attenuated before the signal arrives at the second polarizer; hence, the autocorrelation of the signal in the frequency domain is a product of a spectral beam density and the filter function $g(\omega)$

$$\tilde{\rho}_f(\omega) = \tilde{\rho}(\omega) \cdot g(\omega)$$ \hfill (16)

and the spectrum of the measured signal is

$$\tilde{s}(\omega) = \left|\tilde{\rho}_f(\omega)\right|^2 = \left|\tilde{\rho}(\omega)\right|^2 \left[1 - 2e^{-\xi^2\omega^2} + e^{-2\xi^2\omega^2}\right] \qquad (17)$$

Again assuming at this point $r(w)$ to be a Fourier transform of a Gaussian beam, we obtain, using Parseval's theorem, an analytical expression for the signal in the time domain:

$$s(\tau) \propto \left[ e^{-\frac{(\tau-\tau_0)^2}{4\sigma^2}} - \frac{2\sigma}{\sqrt{\sigma^2 + \xi^2}} e^{-\frac{(\tau-\tau_0)^2}{4(\sigma^2 + \xi^2)}} + \frac{\sigma}{\sqrt{\sigma^2 + 2\xi^2}} e^{-\frac{(\tau-\tau_0)^2}{4(\sigma^2 + 2\xi^2)}} \right]. \qquad (18)$$

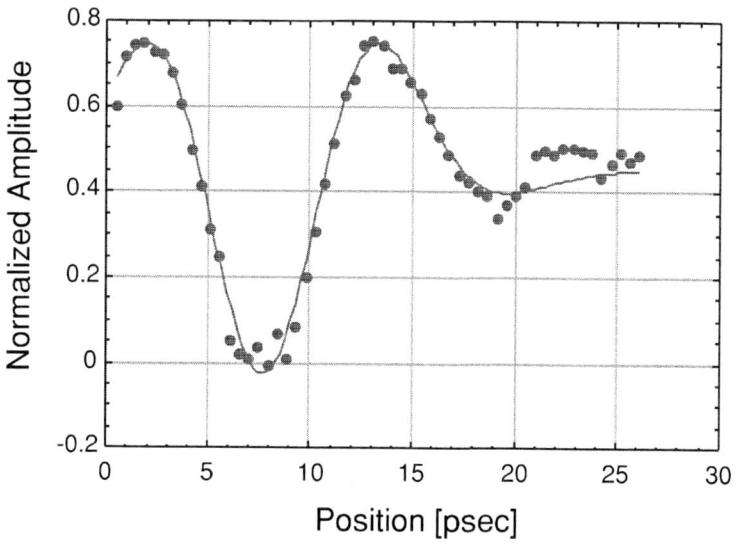

**FIGURE 4.** The fit $s(t)$ to the autocorrelated CTR signal.

This expression can be used to find the rms beam pulse length $\sigma$ and, incidentally, the cut-off frequency of the interferometer, by a simple two parameter fit $(\sigma, \xi)$ directly to our time-domain signal, without resorting to Fourier transformation of this data. On the average $z$ was found very close to 1 psec, which correspond to the filter cut-off wavelength of 3 mm. An example of this fit is shown in Figure 4, in which it was determined that $s \sim 2.8$ psec, which is quite close to the injected photocathode drive laser $s \sim 2.6$ psec (also measured by autocorrelation, in the infrared, before quadrupling to convert the pulse to ultra-violet). It can be seen that the representation of the signal

46

by the sum of three Gaussians (one is the actual unfiltered autocorrelation peak, and the other two, arising from the missing frequency components, are broader, smaller in magnitude, and of alternating sign) gives an excellent fit to the data. The characteristic signature of this form is the existence of a central peak surrounded on both sides by two alternating sign local extrema — this signature is observed in all interferograms obtained in our CTR experiments. One should not always assume this to be the case, however. For example, a flatter profile formed by two overlapping Gaussians could have four, not two secondary extrema.

## Analysis of the UCLA Measurements

Using the method described above, the number of interesting parametric studies have been performed and analyzed from the measurements made at UCLA[7]. The first consisted of measuring the dependence of the beam width on the charge $Q$, with the focusing solenoid and rf conditions held at optimum for minimized energy spread and transverse emittance. For the SASE FEL experiment at UCLA[8,9], the peak current in the beam was one of the crucial parameters which needed to be diagnosed under these conditions, and its dependence on charge was calculated using the measured$\sigma(Q)$. Another important parametric study performed was the examination of the beam pulse length as a function of the laser-induced injection phase in the rf gun. When the beam is injected so that it exits the gun ahead of the peak in the rf field, it is compressed longitudinally by the rf field gradient in both the gun, and to a lesser extent the linac. This effect was also observed using the CTR methods[7].

Up until this point we have assumed a simple Gaussian charge distribution. However, the fit function given does not cover some details in the signal. High frequency structure appears in the autocorrelation peaks in many of the measurements, indicating a fairly serious violation of the simple Gaussian model. For instance, the second (negative-going) sharp peak observed in Figure 7 at t=18.6 psec, well outside of the main signal, consistently appears in the data. This high frequency artifact in the signal cannot, of course, be generated by the low frequency filtering effect. This, along with a satellite off-momentum component of the distribution observed in momentum spectrum of the beam, led us to the conclusion that there is a satellite beam generated in the rf photocathode gun.

The effect of the satellite beam is particular well seen in the case of the compressed beam data. When the simple Gaussian model shown in Eq. (18) is used to fit the autocorrelation data, good agreement takes place only at the vicinity of the coincidence region. The mathematical formalism for the bi-Gaussian case is too extensive to be listed here; nevertheless it is fairly straightforward, following the same model as was used in the simple Gaussian case. As a result, the fit function generated provided the more accurate fit shown on Fig. 5. The beam profile deduced from the fit shows the presence of the satellite beam 10 psec away from the main peak.

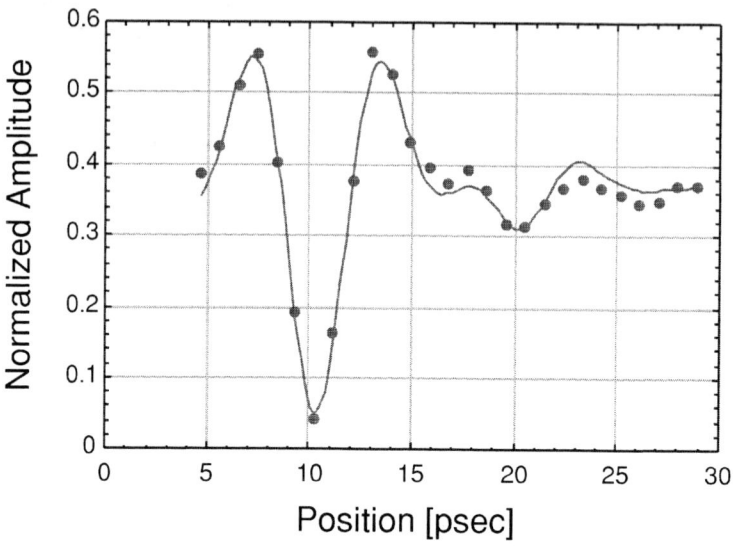

**FIGURE 5.** Two Gaussian time-domain filtered fit which takes the satellite beam into account.

# MICROBUNCH CTR MEASUREMENTS

## Microbunch-derived Coherent Transition Radiation

Charged particle beams with microbunch structure, the periodic modulation of the beam longitudinal profile, are now present in a variety of experimental scenarios, *e.g.* FELs[10], their inverse (IFELs)[11], and advanced accelerators based on laser excitation of plasmas and structures[12]. In the present experiments reported here, we focus on the microbunching that develops as a result of the self-amplified spontaneous emission FEL (SASE FEL) process[13]. This microbunching, which occurs at the wavelength of the FEL radiation, is central to the FEL gain process, as such a distribution produces radiation coherently, giving rise to exponential gain.

The traditional analysis of CTR which includes finite transverse profile and electron beam angular distribution effects, begins by writing the differential radiation spectrum due to multiparticle coherence effects as a function proportional to the single particle spectrum[14],

$$\frac{d^2U}{d\omega d\Omega} \cong N_b^2 F_L(\omega) F_T(\omega,\theta) \chi(\theta) \frac{d^2U}{d\omega d\Omega}\bigg|_{\text{single e}^-},$$

(19)

where $N_B$ is the bunch population, $F_L(\omega)$ and $F_T(\omega,\theta)$ are the Fourier transform square amplitudes of the longitudinal (time) and transverse beam profiles, respectively. The factor $\chi(\theta)$ is due to the finite divergence of the beam and is usually taken to be close to unity. For narrow band transition radiation, however, this factor is not ignorable, as is discussed below.

The case of a microbunched beam produced, *e.g.* in an FEL or IFEL, has been worked out in detail in Ref. 5. Here we need to extend the previous results to account for possible asymmetries in the beam transverse distribution. The microbunched beam distribution is therefore taken to be

$$f(r,z) = \frac{N_b}{(2\pi)^{3/2}\sigma_x\sigma_y\sigma_z} \exp\left(-\frac{x^2}{2\sigma_x^2} - \frac{y^2}{2\sigma_y^2} - \frac{z^2}{2\sigma_z^2}\right) \cdot \left[1 + \sum_{n=1}^{\infty} b_n \sin(nk_r z)\right],$$

(20)

where $k_r$ is the radiation and, therefore, beam modulation wavenumber. Because the beam has Fourier components at $k_r$ and its harmonics, an analysis following the methods of Ref. 5 predicts that the wave spectrum of CTR from the back of a 90 degree oriented foil is localized in peaks near these frequencies, with an angular spectrum of photon number at each peak $k = nk_r$ of

$$\frac{dN_\gamma}{d\theta} \cong \frac{\alpha(N_b b_n)^2}{4\sqrt{\pi}nk_r\sigma_z} \frac{\sin^3(\theta)}{(1-\beta\cos(\theta))^2} \left[\exp\left(-(nk_r\sin(\theta))^2(\sigma_x^2\sin^2(\phi)+\sigma_y^2\cos^2(\phi))\right)\right]\chi(\theta),$$

(21)

where $\theta$ and $\phi$ are the polar and azimuthal angles with respect to the beam axis, respectively. Several predictions can be deduced from Eq. 21: first, familiarly the number of photons scales as the square of the number of radiators $N_b^2$. Also, the angular spectrum is narrowed considerably (when, as in the cases of present interest, $nk\sigma_{x,y}/\gamma > 1$) by the transverse geometric factor, which expresses the diffraction-limited (as opposed to the natural transition radiation angular distribution) of the coherent radiation, which for an axisymmetric beam of size $\sigma$ gives a diffraction angle of $\theta_d = (\sqrt{2}nk_r\sigma)^{-1}$. This narrowing is a signature of coherence for the microbunched case, where the beam width is many wavelengths across. If we ignore the divergence factor $(\chi(\theta) \approx 1)$, and perform the angular integration, we obtain a predicted number of emitted photons at each harmonic (for forwad CTR, normal beam incidence),

$$N_\gamma \approx \frac{\alpha(N_b b_n)^2}{4\sqrt{\pi}\,k_r\sigma_z}\left(\frac{\gamma}{nk_r}\right)^4\left(\frac{\sigma_x^2+\sigma_y^2}{\sigma_x^3\sigma_y^3}\right),$$  (22)

which illustrates also the sensitive dependence of the CTR on beam dimensions. CTR is enhanced when the beam is dense, and there are many radiating electrons within a cubic half-wavelength.

| Beam Energy | $E$ | 17.5 MeV |
|---|---|---|
| Peak Current | $I$ | 140 A |
| Charge/bunch | $Q$ | 1.5 nC |
| Bunch length (FWHM) | $\tau$ | 11 psec |
| Energy spread | $\Delta\gamma/\gamma$ | 0.5% |
| Wiggler period | $\lambda W$ | 2 cm |
| On-axis field | $B_0$ | 7.4 kG |
| FEL Wavelength | $\lambda$ | 13 μm |
| FEL parameter | $\rho$ | 0.008 |
| Rms beam sizes | $\sigma_x, \sigma_y$ | 210, 160 μm |

**TABLE 1.** Microbunch CTR experimental parameters

A measurement of some of these effects have been carried out at BNL[15], where a 0.3 nC electron beam was strongly bunched by the IFEL interaction with a 10.6 mm CO2 laser. The electron beam was not well focused at the foil (transverse beam size 0.6 by 5.5 mm), however, and so the CTR intensity was relatively weak. In order to measure CTR in this experiment, a large signal at the IFEL fundamental had to be suppressed, by looking at the forward radiation behind the opaque $d = 63$ μm Cu foil, and use of a high-pass filter. Theprimary result of this measurement was demonstration of a quadratic dependence of $N_\gamma$ on $N_b$. Also, additional high-pass filters were used to establish CTR at or above the 4th IFEL harmonic. It is important that both effects have been previously established, as neither is easily seen in a SASE FEL experiment. The dependence $N_\gamma \sim N_b^2$ is not observable in a SASE experiment as the bunching factors $b_n$ are gain and thus $N_b$ dependent. In addition, the $b_n \propto b_1^n$ are negligibly small unless the FEL is near saturation, which is not the case despite the high gain achieved in this experiment.

## UCLA/LANL SASE FEL CTR Experiment

Because of the signal level, asymmetric beam, and calibration factors, the overall photon number was not given, nor compared to theoretical predictons for the BNL results. This exercise would have been problematic for the BNL case in any event, as scattering effects in the foil served to strongly suppress CTR production. Additionally, critical predictions of the microbunch CTR theory were not observed — the narrow-band frequency spectrum centered near the fundamental IFEL frequency and/or its harmonics, and the narrowing of the angular spectrum to the diffraction limit. Both of these attributes taken together indicate microbunching structure, meaning periodic longitudinal organization of the electrons. The BNL results indicate the presence of high frequency components, but do not strictly imply that the beam is organized into microbunches. In order to employ CTR as a method for diagnosing microbunching, all relevant aspects of the theory must be explored. The present measurements verify much of the theoretical model, and give some insight into the microbunching process in a high-gain SASE FEL.

The present experiments[16] were performed by a UCLA/LANL collaboration at the AFEL facility at Los Alamos National Laboratory, a 1300 MHz rf photoinjector which produces a 100-bunch train of low-emittance, high current electron bunches. The experimental setup is shown in Fig. 1, and the beam parameters relevant to this experiment, measured using the methods described in Ref. 17, are given in Table 1. The undulator used was the 2 m UCLA/Kurchatov[9] device employed in recent high-gain SASE FEL experiments; its parameters are also in displayed in Table 1. The 6 μm thick Al CTR foil was mounted on an insertable actuator normal to the beamline, 1 cm after the undulator exit, in a large opaque stop, to eliminate all FEL radiation when the foil is inserted. This placement of the foil allowed us to collect FEL and CTR radiation alternatively in the same optical beamline, as the object point in both cases is nearly the same. In addition, the beam defocuses transversely in 21 cm, and space charge effects are predicted to debunch the beam in roughly 50 cm from the end of the undulator[18]. These effects are avoided in our geometry. The optical beamline was set so that only diffraction limited coherent radiation passes the acceptance angle $\theta_{acc} \approx 12$ mrad, be collected and focused into the HgCdTe detector. The incoherent TR, however, with its angular peak at $\theta_{inc} \approx \gamma^{-1} = 29$ mrad, is collected with only a few percent efficiency. The detector provides an equilibrium output signal level proportional to the radiated energy per electron pulse, with the proportionality constant obtained from a calibrated laser power meter.

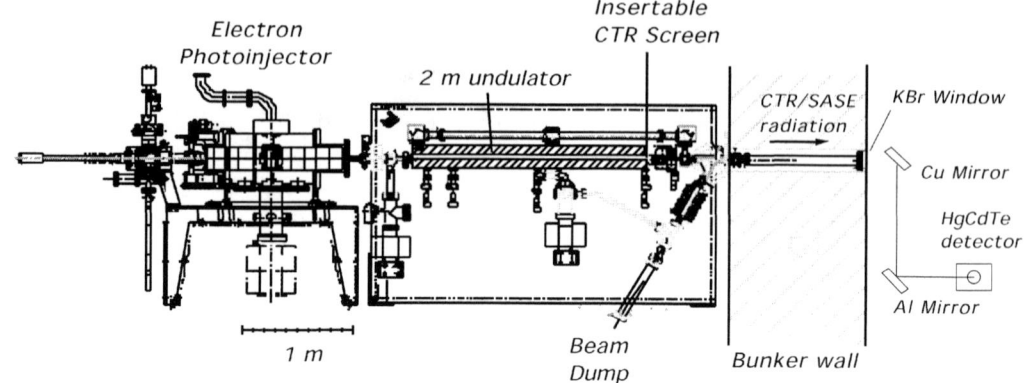

**FIGURE 6.** Electron injector, undulator, and CTR/SASE optical beamline at LANL AFEL facility.

The conditions of high SASE FEL gain with a 1.5 nC beam seen in Ref. 17 were reestablished for this experiment. The performance of the FEL was optimized by setting the beam focus at the matched condition at the undulator resonance, and fine-tuning the rf phase of the photoinjector. This procedure gave highest SASE output at relatively low injection phase, which corresponds to higher dynamical compression of the electron bunch, and thus higher peak current, FEL gain, and microbunching effect. After insertion of the foil, however, in addition to incremental changes in solenoid focusing, it was found that a small adjustment (2-3 degrees) of the rf phase was necessary to maximize in the CTR signal, as is shown in Figure 7. As the rf accelerating wave provides phase dependent focusing[19], this adjustment (which has a negligible effect on the final energy of the beam) serves to minimize the beam size attainable at the foil, thus optimizing the CTR production (*cf.* Eq. 4). The SASE signal is less sensitive to beam focusability, however, as the gain in this experiment is dominated by diffraction, which is mitigated with larger $\sigma_{x,y}$ [20]. The peak regions of the SASE and CTR signals as a function of rf phase overlap, as they must, because the CTR is dependent on the SASE-induced bunching. In our case, using the analysis of Ref. 17, the measured gain was near $10^5$. The bunching predicted for these conditions by the 3-D FEL simulation code GINGER, for a range of parameters corresponding to experimental uncertainties, was $b_1 = 0.008\text{-}0.01$, with negligible bunching at the higher harmonics.

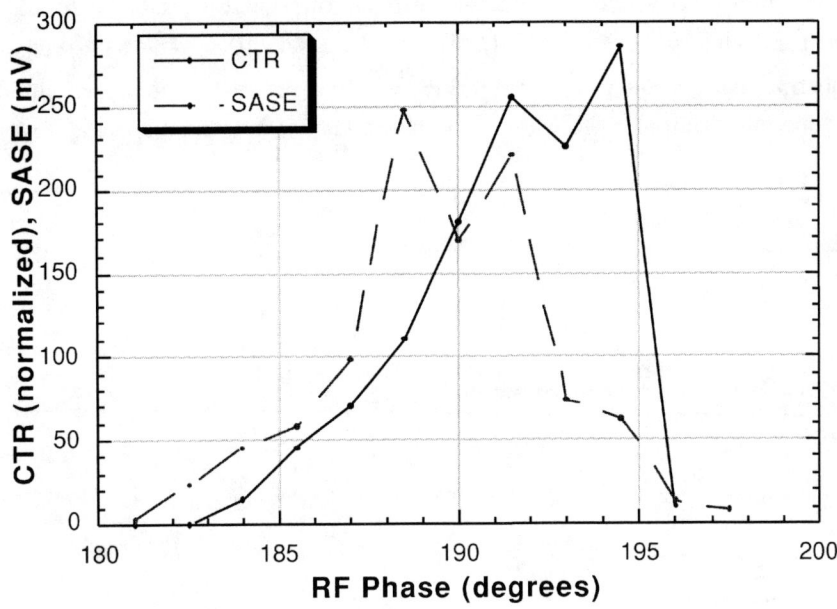

**FIGURE 7.** SASE and CTR signals as a function of rf phase, with CTR scaled to SASE amplitude.

Before discussing the data further, we remark that initially the CTR measurements were attempted with a 50 μm Al foil, with the result that the CTR signal was weaker than expected, leading us to examine the effects of foil scattering. For an uncorrelated gaussian phase space distribution typical of a scattered beam, a formalism has been developed[14] to evaluate $\chi(\theta)$. Several results of this analysis can be described. First, $\chi(\theta)$ is near unity for small angles when the angular spread of the incoherent radiation is large compared to the rms beam divergence, $\sigma' \cong \theta_{scat} \ll \gamma^{-1}$. If this condition is violated, $\chi(\theta)$ diminishes rapidly. After subsitution of the correct form of $\chi(\theta)$ into Eq. 3 and integrating, we can define a factor $\eta(\gamma\sigma')$ (keeping all other parameters constant) which indicates the degree of suppression of the CTR signal due to beam divergence. Note that since the scattering angle $\theta_{scat} \propto \gamma^{-1} d^{1/2}$, $\eta$ is independent of $\gamma$ and is a function only of foil material and thickness. For our 50 μm Al foil, where $\theta_{scat} \approx \gamma^{-1}$, $\eta \cong 0.11$, and for the highly scattered BNL case $\eta \cong 5 \times 10^{-3}$. In order to avoid this effect, we need $\theta_{scat} \ll \gamma^{-1}$, which was achieved by using the 6 μm Al foil. Integrating Eq.20, and multiplying by the factor $\eta = 0.61$ for our case yields a photon number per electron pulse, for the range of GINGER-predicted $b_1$ and other beam parameters given in Table 1, of $N_\gamma = 2.8 - 4.4 \times 10^8$. The measured photon number per pulse at the peak given in Fig. 2 is $N_\gamma = 3.5 \times 10^8$. The theory, simulation and experiment thus agree to within experimental and simulational uncertainty.

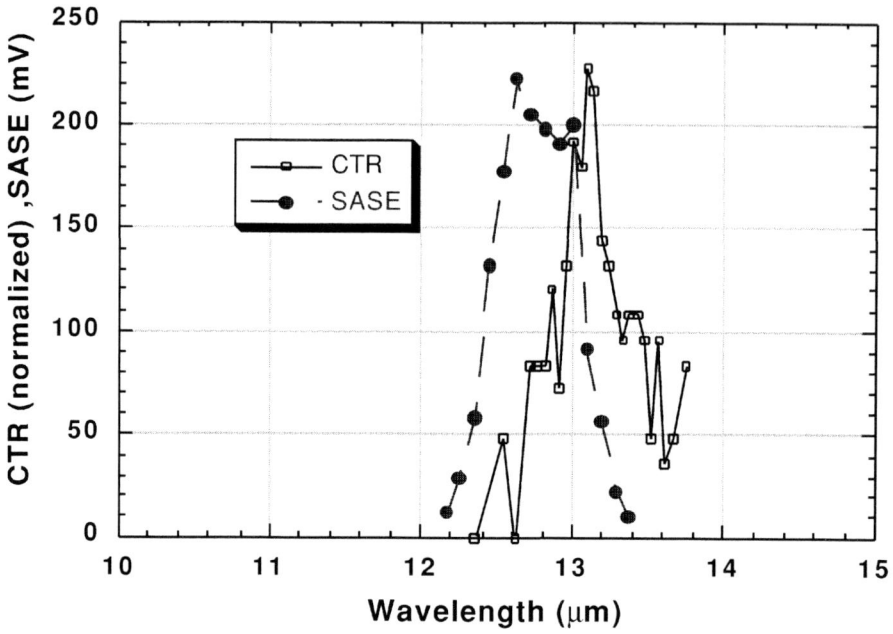

**FIGURE 8**. SASE and CTR signals as a function of wavelength as measured with monochromator, with CTR scaled to SASE amplitude.

Having established an optimization procedure for both SASE and CTR, we then undertook a spectral study of both signals by sending them through a Jerrell Ash monochromator. In order to maximize the signal through the monochromator its input collimating slits were removed, which resulted in a measured (by comparing the SASE line-width with and without the slits) intrinsic resolution of 0.177 μm . The SASE and CTR spectra thus obtained (with the SASE attenuated by a factor of 3 and the CTR multiplied by 10 to give similar scale), are shown in Fig. 3. Both the CTR and SASE signals are both localized near the same wavelength, with a small difference in the distribution centers. This discrepancy points to an inadequacy in the standard analysis of CTR[5,14]. Because the radiation components are summed by considering a temporal "snap-shot" of the beam distribution[14], possible off-axis shifting of the wave-length spectrum of the radiation, which is not created "at rest" by the foil, but over a radiation formation length[21] by the relativistic electrons, cannot be obtained. A simple calculation of energy exchange between the initially radiated wave and the microbunched beam electrons indicates that for small angles, the off-axis wavelength spectrum is shifted in analogy to the FEL shift, $\lambda \rightarrow \lambda\left(1 + (\gamma\theta)^2\right)$. While the SASE radiation is peaked at $\theta = 0$, the CTR is peaked off-axis, leading to a shift in the centroid of CTR wavelength with respect of SASE by $\Delta\lambda/\lambda \cong \left(\gamma/2k_r\sigma\right)^2 \cong 3.8\%$; the observed shift is 3.3%.

In conclusion, we have demonstrated two critical aspects of the microbunching-induced coherent transition radiation — the narrowing of the angular spectrum, and the formation of line structure in the wavelength spectrum. These observations have verified some aspects of microbunching-induced CTR theoretical analysis, but challenged others. In particular, this analysis must be redone employing a travelling beam model, and not the stationary model which is presently used[8,9]. Also, in order to have a well understood diagnostic, which produces the expected level of coherence, one must minimize the beam divergence induced by the CTR interaction (foil scattering). It should be emphasized that this diagnostic method is important not only for FEL experiments but for short wavelength advanced accelerator experiments, such as the plasma accelerators[22], cathodeless, plasma-based injectors[23], and direct laser acceleration.

It is equally useful to view the current experiments from the FEL physics view point, as these measurements were performed at a SASE FEL exit, verifying the crucial role that microbunching plays in the gain process. The narrow angular spread of the CTR signal indicates that the microbunching is fairly uniform in the transverse dimension; otherwise, the CTR signal would be have a less localized angular spectrum. Also, the agreement of measured and predicted photon number, using the microbunching given by simulations, is especially encouraging, as it provides an independent check, in addition to the FEL radiation output, on the code predictions. The CTR microbunching method will be even more useful in the next generation SASE FEL experiments, in which the FEL should saturate. In this case the signal will be larger, not only on the fundamental radiation wavelength, but on the harmonics as well. The large signal levels will allow closer investigation of off-axis Doppler shift effects (direct measurement of the CTR angular/frequency correlations). The added information from harmonics should permit a more detailed reconstruction of the beam's microbunched distribution.

# REFERENCES

1. U. Happek, A.J. Sievers and E.B. Blum, *Phys. Rev. Lett.* **67** (1991) 2962.

2 . H.-C. Lihn, P. Kung, C. Settakorn, and H.Weidemann, *Phys. Rev. E* **53** (1996) 6413.

3 . Y. Shibata, *et al.*, *Phys. Rev. E* **50** (1994) 1479.

4 . R.Lai and A.J. Sievers, *Phys. Rev. E* **52** (1995) 4576.

5. J. B. Rosenzweig, G. Travish and A. Tremaine, *Nucl. Instrum. Meth. A* **365**, 255 (1995)

6 . Landau, L.D. , *The classical theory of fields* p. 196 (Addison-Wesley, 1951).

7. A. Murokh, *et al.*, *Nucl. Instr. and Methods A.* **410**, 452 (1998)

8. G. Travish, UCLA PhD thesis (UMI Dissertation Services, #9704598, Ann Arbor, 1996)

9. M. Hogan, *et al.*, *Physical Review Letters* **80**, 289 (1998).

10. R. Bonifacio, C. Pellegrini, and L. Narducci, *Optics Comm.* **50**, 373 (1984).

11. A. Van Steenburgen, *et al.*, *Phys. Rev. Lett.* **80**, 289 (1998).

12. J. Rosenzweig, *et al.*, *Phys. Rev. Lett.***74**, 2467 (1995).

13. R. Tatchyn *et al.*, *Nucl. Instrum. Meth. A* **375**, 274 (1996), C. Pellegrini, *et al.*, *Nucl. Instr. Methods A* **331**, 223 (1993), J. Rossbach et al., *Nucl. Instr. and Meth.* **A375**, 269 (1997).

14. Yukio Shibata, *et al.*, *Phys. Rev. E* **50**, 1479 (1994)

15. Y. Liu, *et al.*, *Phys. Rev. Lett.* **80**, 4418 (1998); Y. Liu, PhD thesis, UCLA Dept. of Physics and Astronomy (1997).

16. A. Tremaine, *et al.*, submitted for publication to *Phys. Rev. Lett.*

17. M. Hogan *et al.*, accepted for publication in *Phys. Rev. Lett.* SASE FEL gain in excess of $10^5$ was first seen in this experiment.

18. J. Rosenzweig, *et al.*, *Nucl. Instr. Methods A* **393**, 376 (1997).

19. S. Reiche, *et al.*, *Phys. Rev. E* **56**, 3572 (1997) .; J. Rosenzweig and L. Serafini, *Phys. Rev. E* **49**, 1499 (1994).

20. E. Saldin, V. Schneidmuller and M. Yurkov, *Optics Comm.* **97**, 272 (1993).

21. L. Warstki, *et al.*, *J. Appl. Phys.*. **46**, 3644 (1975).

22. C.E. Clayton, *et al.*, *Phys. Rev. Lett.* **70**, 37 (1993).

23. D. Umstadter, J. K. Kim and E. Dodd, *Phys. Rev. Lett.* **76**, 2073 (1996).

# Laser-Induced Electron Trapping in Plasma-Based Accelerators

E. Esarey,[1,2] W.P. Leemans,[2] B. Hafizi,[3] R. Hubbard,[1] and A. Ting[1]

[1] *Plasma Physics Division, Naval Research Laboratory, Washington DC 20375*

[2] *Ernest Orlando Lawrence Berkeley National Laboratory*
*University of California, Berkeley CA 94720*

[3] *Icarus Research, Inc., P.O. Box 30780, Bethesda MD 20824*

**Abstract**. Trapping of plasma electrons in the self-modulated laser wakefield accelerator (LWFA) via the coupling of Raman backscatter to the wake is examined analytically and with 3D test particle simulations. The trapping threshold for linear polarization is much less than for circular and occurs for wake amplitudes of 25% which is well below wavebreaking. Self-channeling provides continuous focusing of the accelerated electrons which, along with relativistic pump laser effects, can enhance the energy gain by a factor $\geq 2$. The colliding pulse method for injecting electrons in the standard LWFA is examined. Simulations of test electrons in prescribed 1D fields indicate the production of relativistic ($\geq 25$ MeV) electrons with bunch durations as short as 3 fs, energy spreads as small as 0.3%, and densities as high as $10^{18}$ cm$^{-3}$.

## INTRODUCTION

Plasma-based accelerators [1] are capable of sustaining ultrahigh electric fields $E_z$ in excess of $E_0 = cm\omega_p/e \simeq 97 n_0^{1/2}$[cm$^{-3}$] V/m, where $n_0$ is the ambient plasma density and $\omega_p = (4\pi n_0 e^2/m)^{1/2}$ is the plasma frequency. Several recent experiments [2-6] have demonstrated the self-trapping and acceleration of plasma electrons in the self-modulated laser wakefield accelerator (SMLWFA) [1,7,8], with electron energies as high as 100 MeV [5,6] over distances $\sim 2$ mm. In the SMLWFA, $n_0$ is sufficiently high ($n_0 \sim 10^{19}$ cm$^{-3}$, $E_0 \sim 300$ GV/m) that the laser pulse extends over many plasma wavelengths $\lambda_p = 2\pi c/\omega_p$ ($\lambda_p/\lambda \sim 8$, where $\lambda = 2\pi c/\omega$ is the laser wavelength). A large amplitude plasma wave (wakefield) is generated via a self-modulation or Raman forward scattering (RFS) instability [9] with a phase velocity $v_p$ near the group velocity $v_g$ of the laser pulse, $v_p \simeq v_g$ and $\gamma_p \simeq \gamma_g = (1 - v_g^2/c^2)^{-1/2} \simeq \lambda_p/\lambda$. The wakefield rapidly grows to extreme values such that electrons are self-trapped from the background

CP472, *Advanced Accelerator Concepts: Eighth Workshop*,
edited by W. Lawson, C. Bellamy, and D. Brosius

plasma. Furthermore, the maximum electron energies observed in experiments [5,6] and simulations [6,10] are in excess of the standard dephasing limit [10,11], $W_d \simeq 2\gamma_p^2 mc^2$ ($W_d \sim 65$ MeV).

Although several recent experiments [2-6] have demonstrated self-trapping and acceleration of plasma electrons in the self-modulated LWFA, the production of electron beams with relatively low momentum spread and good pulse-to-pulse energy stability will require injection of ultrashort electron bunches into the wakefield with femtosecond timing accuracy. These requirements are beyond the current state-of-the-art performance of photo-cathode RF electron guns.

An all-optical method has been proposed for injecting electrons in a standard laser wakefield accelerator (LWFA), in which the pump laser pulse length is approximately equal to $\lambda_p$ [12]. The method of Refs. [12,13] utilizes two laser pulses which propagate perpendicular to one another. The first pulse (pump pulse) generates the wakefield, and the second pulse (injection pulse) intersects the wakefield some distance behind the pump pulse. The ponderomotive force $F_p \sim \nabla a^2$ of the injection pulse can accelerate a fraction of the plasma electrons such that they become trapped in the wakefield, where $a^2 \simeq 7 \times 10^{-19} \lambda^2 [\mu m] I [\mathrm{W/cm^2}]$, $\lambda = 2\pi c/\omega$ is the laser wavelength, and $I$ the intensity. Simulations, which were performed for ultrashort pulses at high densities ($\lambda_p/\lambda = 10$ and $E_z/E_0 = 0.7$), indicated the production of a 10 fs, 21 MeV electron bunch with a 6% energy spread. However, high intensities ($I > 10^{18}$ W/cm$^2$) are required in both the pump and injection pulses ($a \simeq 2$). An all optical electron injector would be a significant step in reducing the size and cost of a LWFA.

This paper will consider two laser-induced trapping mechanisms: (I) Self-trapping of plasma electrons in the SMLWFA due to Raman backscatter [14], and (II) a colliding pulse laser injection scheme for the standard LWFA [15,16].

## SELF-TRAPPING IN THE SMLWFA

It has been suggested that wavebreaking is the mechanism for self-trapping in the SMLWFA [2,6,10]. Wavebreaking of a cold plasma wave in 1D occurs at [17] $E_{WB} = [2(\gamma_p - 1)]^{1/2} E_0 \gg E_0$. Thermal and 3D effects can lower this value, but typically wavebreaking requires nonlinear plasma waves with $E_z > E_0$ [18]. The observed wakefield amplitude, as measured in recent experiments [19], is in the range $E_z/E_0 \sim 10 - 30\%$, well below wavebreaking. Accelerated electrons have also been observed with no evidence of wavebreaking, i.e., no broadening of the anti-stokes peaks [5]. Furthermore, wavebreaking has been suggested as the mechanism by which the electron energies exceed the dephasing limit [6,10].

Recently it was shown that self-trapping and acceleration can result from the coupling of Raman backscatter (RBS) to the wakefield [14]. Self-trapping can occur at modest wakefield amplitudes, $E_z/E_0 \simeq 25\%$, on the order of those observed experimentally, and much lower than the cold 1D wavebreaking limit, which is given [14] by $E_{WB} = [2\gamma_{\perp 0}(\gamma_p - 1)]^{1/2} E_0$, where $\gamma_{\perp 0} = (1 + a_0^2)^{1/2}$, $a_0^2 \simeq 3.6 \times 10^{-19} \lambda^2 [\mu m] I [W/cm^2]$ (circular polarization), and $I$ is the laser intensity. Furthermore, the self-trapping threshold is considerably lower for a linearly polarized laser pulse than for circular polarization. The space-charge field due to self-channeling can provide additional focusing and enhance the maximum energy gain, i.e., $W_{max} = 4\gamma_p^2 mc^2$ [14]. Nonlinear effects can further increase $W_{max}$ by a factor $F_{NL} \simeq (\gamma_{\perp 0} n_0/n)^{3/2} \sim 3$, where $n$ is the self-channeled density. This can account for the final energies observed in experiments and simulations.

As the pump laser self-modulates, it also undergoes RBS, which is the fastest growing laser-plasma instability [1,20-23]. RBS is observed in intense short pulse experiments, with reflectivities as high as 10-30% [20]. RBS generates red-shifted backward light of frequency $\omega_0 - \omega_p$ and wavenumber $-k_0$, which beats with the pump laser $(\omega_0, k_0)$ to drive a ponderomotive wave $(\omega_p, 2k_0)$. As the instability grows, the RBS beat wave, which has a slow phase velocity $v_p \simeq \omega_p/2k_0 \ll c$, can trap and heat background plasma electrons [21,22]. These electrons can gain sufficient energy and/or be displaced in phase by the RBS beat wave such that they are trapped and accelerated to high energies in the wakefield [1,5,14].

To analyze self-trapping, the motion of test particles in analytically specified fields is studied. The effects of four fields will be considered: (1) the intense pump laser field $\mathbf{a}_0$, (2) the RBS radiation field $\mathbf{a}_1$, (3) the plasma wakefield $\phi_p$, and (4) the self-channeling space charge field $\phi_s$, where $\phi = e\Phi/mc^2$ and $\mathbf{a} = e\mathbf{A}_\perp/mc^2$ are the normalized scalar and vector potentials, respectively. Once the fields are specified, the electron motion is evolved via the relativistic Lorentz equation $d\mathbf{u}/dct = \nabla\phi + \partial\mathbf{a}/\partial ct - \boldsymbol{\beta} \times (\nabla \times \mathbf{a})$, where $\mathbf{u} = \mathbf{p}/mc$ is the normalized electron momentum, $\boldsymbol{\beta} = \mathbf{u}/\gamma$, and $\gamma = (1 + u^2)^{1/2}$.

The pump laser $(i = 0)$ and RBS radiation $(i = 1)$ fields are assumed to be of the form $\mathbf{a}_i = \hat{a}_i(\zeta_i, r)(\sin\psi_i \mathbf{e}_x + \sigma\cos\psi_i \mathbf{e}_y) + a_{zi}\mathbf{e}_z$, where $\zeta_i = z - v_{gi}t$, $\psi_i = k_i z - \omega_i t$, $v_{g0} = c\beta_{g0} = c^2 k_0/\omega_0$ is the group velocity, $\sigma = 0$ ($\sigma = 1$) for linear (circular) polarization, and $\nabla \cdot \mathbf{a}_i = 0$. The wavenumbers $k_i$ and frequencies $\omega_i$ satisfy $ck_i = \delta_i\omega_i(1 - \omega_p^2/\omega_i^2 - 4c^2/r_i^2\omega_i^2)^{1/2}$ with $\delta_0 = 1$ (pump) and $\delta_1 = -1$ (RBS). The wakefield is $\phi_p = \hat{\phi}_p(\zeta_p, r)\cos\psi$, where $\zeta_p = z - v_p t$ and $\psi = (\omega_p/v_p)\zeta_p$. The field envelopes are

$$(\hat{a}_i, \hat{\phi}_p) = (a_i, \phi_0)\exp(-r^2/r_j^2)\left\{1 - \exp\left[-(\zeta_j - \zeta_{fj})^2/L_j^2\right]\right\}, \qquad (1)$$

59

for $\zeta_j \leq \zeta_{fj}$ ($j = 0, 1, p$), where $(a_i, \phi_0)$, $\zeta_{fj}$, $L_j$, and $r_j$ are constants that determine the amplitude, position of the leading edge, rise time, and spot size of the field, respectively. The average power is $P_i[\text{GW}] \simeq 21.5(1 + \sigma)(a_i r_i/\lambda_i)^2$ and equal power comparisons between polarizations require $(a_i^2)_{\sigma=1} = (a_i^2/2)_{\sigma=0}$. Furthermore, $\omega_1 = \omega_0 - \omega_p$, $v_p = v_{g0} = v_{g1}$, and $r_p^2 = r_0^2/2$ are assumed. A theory of RBS in the strong-pump regime gives a saturation amplitude [23] of $a_1 \simeq 0.046$ for linear polarization with $a_0 = 2$ and $\omega_0/\omega_p = 8.5$.

In the long pulse ($L_0 \gg \lambda_p$) SMLWFA regime, the ponderomotive force of the pump laser pulse expels electrons (self-channeling), thus creating an electron density perturbation $\delta n/n_0 = k_p^{-2}\nabla^2\phi_s$ and a space charge potential [24] via $\nabla\phi_s = \nabla\gamma_{\perp 0}$, i.e., $\phi_s = \gamma_{\perp 0} - 1$, where $\gamma_{\perp 0}^2 = 1 + (1 + \sigma)\hat{a}_0^2/2$. The radial space charge force leads to enhanced focusing of the accelerated electrons [14].

It is insightful to consider the effects of each of the waves (wake and beat wave) independently. In the absence of RBS ($a_1 = 0$), electron motion in a 1D wakefield is described by the Hamiltonian $H_w = \gamma - \beta_p(\gamma^2 - \gamma_{\perp 0}^2)^{1/2} - \phi(\psi)$, where $\gamma_{\perp 0} = (1 + a_0^2)^{1/2}$ is constant (circular polarization), $\phi = \phi_s + \phi_p$, $\phi_s = \gamma_{\perp 0} - 1$, and $\phi_p = \phi_0\cos\psi$. In phase space $(u_z, \psi)$, the boundary between trapped and untrapped orbits is given by the separatrix $H_w(\beta_z, \psi) = H(\beta_p, \pi)$. The maximum $(+)$ and minimum $(-)$ axial momenta of an electron on the separatrix are [14]

$$u_{w,m} = \gamma_p\gamma_{\perp 0}\left\{\beta_p(1 + 2\phi_0\gamma_p/\gamma_{\perp 0}) \pm \left[(1 + 2\phi_0\gamma_p/\gamma_{\perp 0})^2 - 1\right]^{1/2}\right\}. \qquad (2)$$

In the limits $2\phi_0\gamma_p/\gamma_{\perp 0} \gg 1$ and $\gamma_p \gg 1$, $u_{w,max} \simeq 4\gamma_p^2\phi_0$ and $u_{w,min} \simeq \gamma_{\perp 0}^2/4\phi_0 - \phi_0$. Notice that $u_{w,min}$ increases with $a_0$, i.e., the larger the pump laser field, the more difficult it is to trap. The background plasma electrons lie on an untrapped orbit (below the separatrix) $u_{zf}$ given by $H_w(u_{zf}, \psi) = 1$. At wavebreaking, the bottom of the separatrix $u_{w,min}$ coalesces with the plasma fluid orbit, $u_{zf} = u_{w,min}$. This occurs at a wavebreaking field of $E_{WB}/E_0 = [2\gamma_{\perp 0}(\gamma_p - 1)]^{1/2}$, e.g., $E_{WB}/E_0 = 5.1$ for $\gamma_{\perp 0} = \sqrt{3}$ and $\gamma_p = 8.5$.

In the absence of the wakefield ($\phi_p = 0$), electron motion in a 1D beat wave is described by the Hamiltonian $H_b = \gamma - \beta_{pb}\left[\gamma^2 - \gamma_\perp^2(\psi_b)\right]^{1/2} + \phi_s$. Circular polarization is assumed such that $\gamma_\perp^2 = 1 + a_0^2 + a_1^2 + 2a_0 a_1\cos\psi_b$, where $\psi_b = (k_0 - k_1)(z - v_{pb}t)$ and $v_{pb} = c\beta_{pb} = \omega_p/(k_0 - k_1) \simeq \omega_p/2k_0$ is the beat phase velocity, assuming $\omega_p^2/\omega_0^2 \ll 1$ and $1/k_0^2 r_i^2 \ll 1$. The beat separatrix is given by $H_b(\beta_z, \psi_b) = H_b(\beta_{pb}, 0)$ with a maximum and minimum axial momenta of [14]

$$u_{b,m} = \gamma_{pb}\left\{\beta_{pb}\left[1 + (a_0 + a_1)^2\right]^{1/2} \pm 2(a_0 a_1)^{1/2}\right\}. \qquad (3)$$

60

In the combined slow beat wave and the fast wakefield, self-trapping can occur as follows. Below wavebreaking (e.g., $\phi_0 < 1$), plasma electrons are oscillating on an untrapped orbit below the wake separatrix, i.e., with insufficient energy to be trapped in the fast $(v_p \simeq c)$ wakefield. They can, however, be trapped in the slow $(v_{pb} \ll c)$ beat wave. The effect of the beat wave is to displace the electrons in both momentum and phase such that a fraction of the orbits cross the wake separatrix and become trapped. Although the actual orbits in both the wakefield and beat wave are highly nonlinear, an approximate trapping threshold is given by an island overlap condition: Trapping occurs when beat and wake separatrices overlap, i.e., $u_{b,max} > u_{w,min}$. Equations (2)-(3) give a trapping threshold of [14]

$$\phi_0 > \frac{\gamma_{\perp 0}}{2} \left[ \left( 1 + \frac{u_{b,max}^2}{\gamma_{\perp 0}^2} \right)^{1/2} - \beta_p \frac{u_{b,max}}{\gamma_{\perp 0}} - \frac{1}{\gamma_p} \right]. \tag{4}$$

In the limits $\gamma_p \gg 1$, $\beta_{pb} \ll 1$, and $a_1 \ll 1$, Eq. (4) becomes $\phi_0 > (1-\beta_{pb})\gamma_{\perp 0}/2 - (a_0 a_1)^{1/2}$. The above expressions assume that, in the presence of the wakefield, well defined beat wave separatrices exist, i.e., $2k_0 a_0 a_1/\gamma_{\perp 0} > k_p|\phi_0|$. For linear polarization, $\gamma_\perp^2 = 1 + a_0^2 \sin^2 \psi_0$ oscillates on a fine scale $(\lambda_0/2)$ with a phase velocity near $c$. This in effect creates a "fuzzy" wake separatrix, the bottom of which is given by Eq. (2) with $\gamma_{\perp 0} \to 1$. Hence, it is easier to trap with linear polarization. For example, $a_0 = 1.4$, $\gamma_p = \omega_0/\omega_p = 8.5$ $(\beta_{pb} = 0.063)$, and $a_1 = 0.033$ give $u_{b,max} = 0.54$ and a threshold of $\phi_0 > 0.54$ $(\phi_0 > 0.24)$ for circular (linear) polarization.

The 1D theory neglects the effects of transverse focusing. Associated with a 3D wake $\phi_p$ is a periodic radial field which is $\pi/2$ out of phase with accelerating field, i.e., there exists a phase region of $\lambda_p/4$ that is both accelerating and focusing. If an electron is to remain in this phase region, it must lie within the "3D separatrix" defined by $H_w(\beta_z, \psi) = H_w(\beta_p, \pi/2)$ with extremum given by $u_{w,max} \simeq 2\gamma_p^2 \phi_0$ and $u_{w,min} \simeq \gamma_{\perp 0}^2/2\phi_0 - \phi_0/2$, i.e., given by Eq. (2) with the substitution $(\phi_0)_{1D} \to (\phi_0/2)_{3D}$. Similarly, the threshold for trapping on the 3D separatrix is given by Eq. (4) with $(\phi_0)_{1D} \to (\phi_0/2)_{3D}$, i.e., it takes a wake twice as large to trap on the 3D separatrix as compared to the 1D separatrix.

In the SMLWFA, additional focusing arises from the self-channeling space charge potential $\phi_s$, the radial field of which is focusing for all phases. Hence if $|\nabla_\perp \phi_s| > |\nabla_\perp \phi_p|$, the radial space-charge force dominates the radial wake and an accelerated electron is focused for all phases. This condition implies $\phi_0 < (1+\sigma)a_0^2 r_p^2/2\gamma_{\perp 0}r_0^2$, e.g, $\phi_0 < 0.58$ for $\sigma = 0$, $a_0^2 = 4$ and $r_p^2 = r_0^2/2$. When this is satisfied, the "1D separatrix" trapping results apply, i.e., Eqs. (2) and (4). The maximum energy gain is $W_{max} \simeq 4\gamma_p^2 \phi_0 mc^2$ which is twice the conventional result $W_d$. In addition, relativistic pump intensities imply $\phi_0 = \gamma_{\perp 0}^{1/2} E_z/E_0$ and $\gamma_p = \gamma_{\perp 0}^{1/2} \omega_0/\omega_p$. Likewise, self-channeling may increase $\phi_0$ and $\gamma_p$ by a factor $(n_0/n)^{1/2}$. Hence, nonlinear effects may increase $W_{max}$ by a factor as high as $F_{NL} \simeq (\gamma_{\perp 0} n_0/n)^{3/2} \simeq 3.5$, assuming $\gamma_{\perp 0} = \sqrt{3}$ and $n/n_0 = 3/4$.

Simulations of trapping were performed by pushing test (noninteracting) electrons in analytically specified 1D and 3D fields ($\mathbf{a}_0$, $\mathbf{a}_1$, $\phi_p$, and $\phi_s$ as described above). A short axial segment (typically 1 $\mu$m) of stationary electrons (typically 16,000 particles) is initiated in front of the laser pulse and is evolved in the fields via the relativistic Lorentz equation. The simulation parameters were (unless otherwise specified): flat-top field amplitudes of $a_0 = 1.4$ ($a_0 = 2$) and $a_1 = 0.033$ ($a_1 = 0.046$) for circular (linear) polarization (equal power), $\omega_0/\omega_p = 8.5$ ($n_0 = 1.5 \times 10^{19}$ cm$^{-3}$ for $\lambda_0 = 1$ $\mu$m), $\omega_1 = \omega_0 - \omega_p$ ($\beta_{pb} \simeq 0.063$), $v_p = v_{g0} = v_{g1}$, rise lengths on the field profiles of $L_0 = L_1 = L_p = 4\lambda_p$, and positions of the field fronts of $\zeta_{f0} = 0$ and $\zeta_{f1} = \zeta_{fp} = -\lambda_p$.

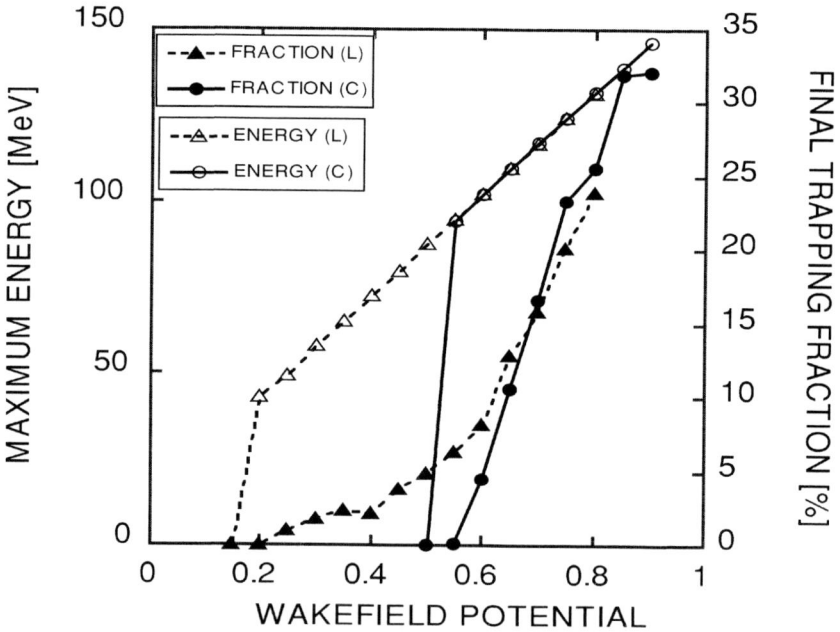

FIG. 1: Maximum electron energy (open points) and trapping fraction (solid points) versus $\phi_0$ from 1D simulations with $\omega_0/\omega_p = 8.5$ after 2 mm for circular polarization (solid curves) with $a_0 = 1.4$ and $a_1 = 0.033$ and linear polarization (dashed curves) with $a_0 = 2$ and $a_1 = 0.046$.

Figure 1 shows simulation results for 1D fields for circular (solid curves) and linear polarization (dashed curves). Plotted versus $\phi_0$ are the maximum electron

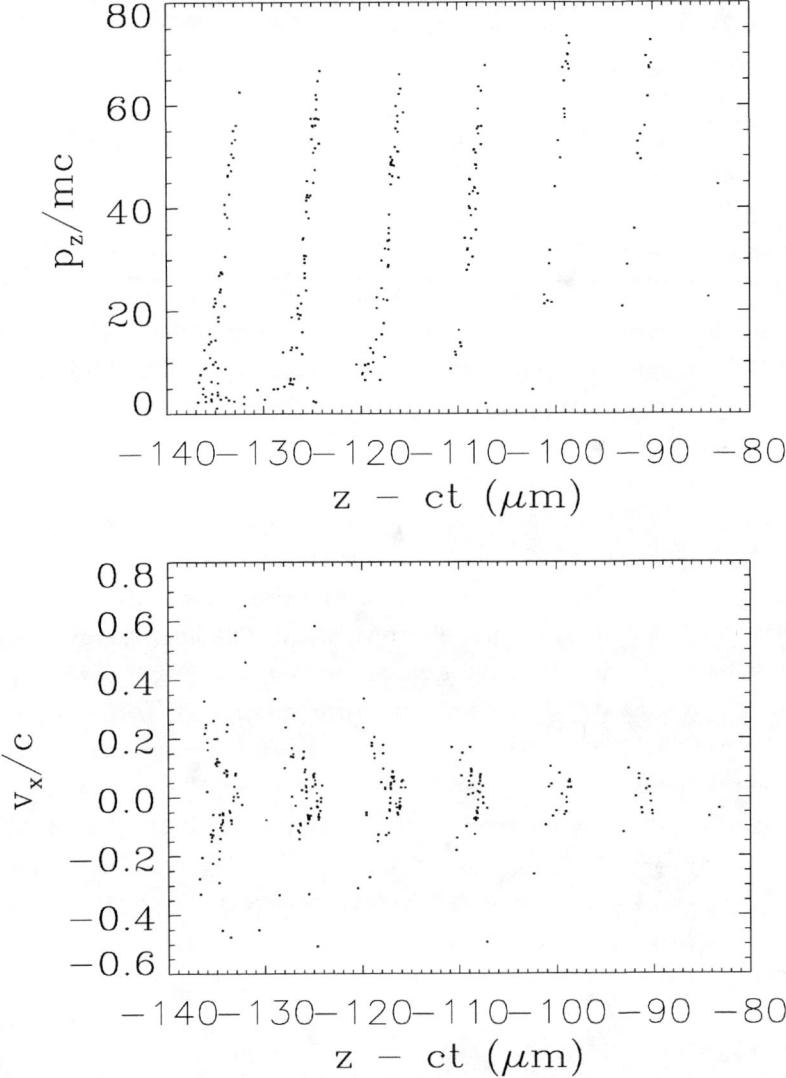

**FIG. 2**: (a) Longitudinal ($u_z$ vs $z-ct$) and (b) transverse ($\beta_x$ vs $z-ct$) phase spaces from a 3D simulation with linear polarization, $a_0 = 2$, $\omega_0/\omega_p = 8.5$, $r_0 = 4$ $\mu$m, $a_1 = 0.046$, $\phi_0 = 0.3$, after 0.5 mm.

energy $W_{max}$ (open points) and fraction of the initial electrons that become trapped $f_{tr}$ (solid points) after a distance of 2 mm. When trapping occurs, $W_{max}$

agrees with the expression from 1D theory, Eq. (2) with $\gamma_p = 8.5$. Note that the simple expression for $\gamma_p \phi_0 / \gamma_{\perp 0} \gg 1$ significantly underestimates the energy gain, i.e., $W_{max} \simeq 4\gamma_p^2 \phi_0 \simeq 59$ MeV for $\gamma_p = 8.5$ and $\phi_0 = 0.4$ as compared to 73 MeV from Eq. (2). For circular (linear) polarization, trapping occurs for $\phi_0 \geq 0.55$ ($\phi_0 \geq 0.2$), in approximate agreement with theory. A similar sudden onset in trapping was observed when $a_1$ was increase for fixed $\phi_0$.

Figure 2 shows results for linearly polarized 3D fields with $r_0 = r_1 = \sqrt{2}r_p = 4$ $\mu$m ($\gamma_p = \gamma_{g0} \simeq 7$). The initial electrons are loaded in $r \leq r_b = 2$ $\mu$m (simulations show electrons at $r > 2$ $\mu$m do not trap). Simulations with $\phi_0$ varied ($a_1 = 0.046$) show trapping with an onset at $\phi_0 \simeq 0.2$ (theory gives $\phi_0 \geq 0.23$) and a peak at $\phi_0 \simeq 0.35$ ($f_{tr} \simeq 0.66\%$). Beyond $\phi_0 \geq 0.6$, trapping is greatly reduced ($f_{tr} < 0.05\%$), in agreement with theory that predicts radial scattering (defocusing) by the wake when $\phi_0 > 0.58$. When $a_1$ is varied ($\phi_0 = 0.4$), a sudden onset for trapping was found when $a_1 \geq 0.025$. In the 3D simulations, $W_{max}$ is slightly higher than the 1D theoretical maximum, e.g., 48 MeV for $\phi_0 = 0.3$ as compared to 41 MeV from theory. For circular polarization, theory predicts trapping when $0.52 < \phi_0 < 0.58$. This behavior is confirmed in 3D simulations which show a narrow band of trapping peaked about $\phi = 0.6$ with $f_{tr} \simeq 10^{-3}$.

Longitudinal ($u_z$ vs $\zeta = z - ct$) and transverse ($\beta_x$ vs $\zeta$) electron phase spaces are shown in Figs. 2(a) and 2(b), respectively, for a 3D linearly polarized run with $\phi_0 = 0.3$ after 0.5 mm of propagation. The high energy electrons are confined to $r \leq 1$ $\mu$m with a 100% energy spread, $W_{max} \simeq 47$ MeV, $f_{tr} \simeq 0.6\%$, and RMS normalized emittance $\epsilon_n \simeq (0.5)\pi$ mm-mrad.

The test particle simulations neglect several effects such as the self-consistent evolution of the fields, space charge of the accelerated electrons, and beam loading. For example, Raman sidescatter will introduce additional waves with $v_p < c$ which will enhance trapping. Beam loading degrades the wakefield when the total number of trapped electrons $N_T$ approaches the beam loading limit [25] $N_{max} \simeq n_0(\lambda_p r_0^2/2)E_z/E_0$. In the simulations, $N_T = f_{tr}n_0\pi r_b^2 L_p$, where $L_p$ is the propagation length and $r_b = 2$ $\mu$m. For the parameters $n_0 = 1.5 \times 10^{19}$ cm$^{-3}$, $\lambda_p = 8.5$ $\mu$m, $r_0 = 4$ $\mu$m, $f_{tr} = 0.5\%$, and $E_z/E_0 = 0.5$, the beam loading limit $N_T \simeq N_{max} \simeq 5 \times 10^8$ is reached after $L_p \simeq 0.5$ mm.

## COLLIDING LASER PULSE INJECTION

The colliding pulse optical injection scheme [15,16] for a standard LWFA uses three short laser pulses: an intense pump pulse (denoted by subscript 0), a forward going injection pulse (subscript 1), and a backward going injection pulse (subscript 2), as shown in Fig. 3. The frequency, wavenumber, and normalized intensity are denoted by $\omega_i$, $k_i$, and $a_i$ ($i = 0, 1, 2$). Furthermore, $\omega_1 = \omega_0$, $\omega_2 = \omega_0 - \Delta\omega$ ($\Delta\omega \geq 0$), and $\omega_0 \gg \Delta\omega \gg \omega_p$ are assumed such that $k_1 = k_0$, and $k_2 \simeq -k_0$. The pump pulse generates a fast ($v_{p0} \simeq c$) wakefield. When the injection pulses collide (some distance behind the pump) they generate a slow ponderomotive beat wave with a phase velocity $v_{pb} \simeq \Delta\omega/2k_0$. During the time in which the two injection pulses overlap, a two-stage acceleration process can occur, i.e., the slow beat wave injects plasma electrons into the fast wakefield for acceleration to high energies. Injection and acceleration can occur at low densities ($\lambda_p/\lambda \sim 100$), thus allowing for high single-stage energy gains, with normalized injection pulse intensities of $a_1 \sim a_2 \sim 0.2$, i.e., two orders of magnitude less intensity than required in Refs. [12,13]. Furthermore, the colliding pulse concept offers detailed control of the injection process: the injection phase can be controlled via the position of the forward injection pulse, the beat phase velocity via $\Delta\omega$, the injection energy via the pulse amplitudes, and the injection time (number of trapped electrons) via the backward pulse duration.

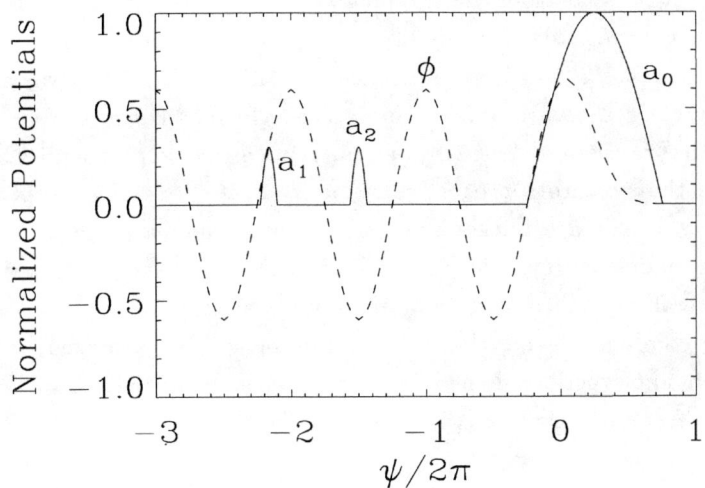

FIG. 3: Profiles of the pump laser pulse $a_0$, the wake $\phi$, and the forward $a_1$ injection pulse, all of which are stationary in the $\psi = k_p(z - v_{p0}t)$ frame, and the backward injection pulse $a_2$, which moves to the left at $\simeq 2c$.

In this paper, colliding pulse injection is analyzed in 1D. Results for 3D fields are given in Ref. [16]. The effects of three waves will be considered: a plasma wakefield $\phi = \hat{\phi}(\psi)\cos\psi$, and a forward and a backward injection laser pulse, both of the form $\mathbf{a}_i = \hat{a}_i(z - v_{gi}t)(\sin\theta_i\mathbf{e}_x + \cos\theta_i\mathbf{e}_y)$. Here, $\psi = k_p(z - v_{p0}t)$, $v_{p0} = c\beta_{p0}$ is the wake phase velocity, $\theta_i = k_iz - \omega_it$ and the amplitudes $\hat{a}_i$ and $\hat{\phi}$ are assumed to be slowly varying compared to the phases $\theta_i$ and $\psi$. Also, $k_i$ and $\omega_i$ satisfy $k_i = \sigma_i\omega_i(1 - \omega_p^2/\omega_i^2)^{1/2}$, where $\sigma_1 = 1$ and $\sigma_2 = -1$, which implies a group velocity $v_{gi} = c\beta_{gi} = c^2k_i/\omega_i$ $(v_{p0} = v_{g0} = v_{g1})$. Furthermore, $a^2 = \hat{a}_1^2 + \hat{a}_2^2 + 2\hat{a}_1\hat{a}_2\cos\psi_b$, where $\psi_b = \theta_1 - \theta_2 = \Delta k(z - v_{pb}t)$ is the beat phase, $v_{pb} = c\beta_{pb} = \Delta\omega/\Delta k$, and $\Delta k = k_1 - k_2 \simeq 2k_0$.

In the absence of the injection pulses, electron motion in the wakefield is described by the Hamiltonian [26] $H_w = \gamma - \beta_{p0}(\gamma^2 - 1)^{1/2} - \phi$, where $\phi = \phi_0\cos\psi$. The boundary between trapped and untrapped orbits is given by the separatrix $H_w(\gamma, \psi) = H_w(\gamma_{p0}, \pi)$, where $\gamma_{p0} = (1 - \beta_{p0}^2)^{-1/2}$. The minimum momentum of an electron on the separatrix is given by $u_{min} \simeq (1/\Delta\phi - \Delta\phi)/2$, where $\Delta\phi = \phi_0(1 + \cos\psi)$, assuming $\gamma_{p0}\Delta\phi \gg 1$ and $\beta_{p0} \simeq 1$. In particular at $\psi = 0$, $u_{min} = 0$ for $\phi_0 = 1/2$, which means that an electron that is at rest at the phase $\psi = 0$ will be trapped. The background plasma electrons, however, are untrapped and are undergoing a fluid oscillation with a momentum $u_f \simeq -\phi$ $(\phi^2 \ll 1)$. Hence, at $\psi = 0$, the plasma electrons are moving backward with $u_f \simeq -\phi_0$, which is far from the trapping threshold.

The beat wave leads to formation of phase space buckets (separatrices) of width $2\pi/\Delta k \simeq \lambda_0/2$, which are much shorter than those of the wakefield $(\lambda_p)$, thus allowing for a separation of time scales. In particular, it can be shown that both the transit time $2\pi/\Delta\omega$ of an untrapped electron through a beat wave bucket and the synchrotron (bounce) time $\pi/(\hat{a}_1\hat{a}_2)^{1/2}\omega_0$ of a deeply trapped electron in a beat wave bucket are much shorter than a plasma wave period $2\pi/\omega_p$. Hence, on the time scale in which an electron interacts with a single beat wave bucket, the wakefield can be approximated as static.

In the combined fields, the electron motion can be analyzed in the local vicinity of a single period of the beat wave by assuming that the wake electric field $E_z = -k_p^{-1}E_0\partial\phi/\partial z \simeq E_{z0}$ is constant. The Hamiltonian is given by [15]

$$H_b = \gamma - \beta_{pb}\left[\gamma^2 - \gamma_\perp^2(\psi_b)\right]^{1/2} + \epsilon\psi_b, \tag{5}$$

where $\epsilon = E_{z0}k_p/E_0\Delta k$ is constant and $\gamma_\perp^2 = 1 + \hat{a}_1^2 + \hat{a}_2^2 + 2\hat{a}_1\hat{a}_2\cos\psi_b$. When $\epsilon = 0$, the phase space orbits are symmetric with x-points at $\beta_z = \beta_{pb}$, $\psi_b = \pm 2\pi j$ and o-points at $\beta_z = \beta_{pb}$, $\psi_b = \pi \pm 2\pi j$ $(j = 0, 1, 2...)$. When $\epsilon \neq 0$, the separatrix distorts into fished-shape islands. When $\epsilon < 0$ $(\epsilon > 0)$, the "fish tail" of the separatrix opens to the right (left). The maximum and minimum axial momentum on the separatrix, $u_{bm}$, are given from Eq. (5) by [15]

$$u_{bm} \simeq \beta_{pb}(\gamma_0 - \pi\gamma_{pb}^2|\epsilon|) \pm 2\hat{a}_1\gamma_{pb} \left(1 - \pi\gamma_0|\epsilon|/2\hat{a}_1^2\right)^{1/2},\qquad(6)$$

where $\gamma_0 = \gamma_{bp}(1 + 4\hat{a}_1^2)^{1/2}$, $\gamma_{pb} = (1 - \beta_{pb}^2)^{-1/2}$, $\pi\gamma_0|\epsilon|/2\hat{a}_1^2 < 1$, and $\hat{a}_1 = \hat{a}_2$.

A scenario by which the beat wave leads to trapping in the wake is the following. In the phase region $-\pi/2 < \psi < 0$, the plasma electrons are flowing backward, $u_f \simeq -\phi_0 \cos\psi < 0$, and the wake is accelerating, $E_z/E_0 = \phi_0 \sin\psi < 0$. Here $\epsilon < 0$ and the beat wave buckets open to the right. An electron that is initially flowing backward resides below the beat separatrix. Since the separatrix opens to the right, there exists open orbits that can take an electron from below to above the beat separatrix. Such an electron can acquire a sufficiently large positive velocity to allow trapping and acceleration in the wake. These open phase space orbits, which provide the necessary path for electron acceleration, can exist when the beat wave resides within $-\pi/2 \le \psi \pm 2\pi j \le 0$.

An estimate for the threshold for injection into the wakefield can be obtained by considering the effects of the wakefield and the beat wave individually and by requiring (i) the maximum energy of the beat wave separatrix exceed the minimum energy of the wakefield separatrix, $u_{bmax} \ge (\Delta\phi^{-1} - \Delta\phi)/2$, and (ii) the minimum momentum of the beat wave separatrix be less than the plasma electron fluid momentum, $u_{bmin} \le -\phi$, where $u_{bmax}$, $u_{bmin}$ are given by Eq. (6) with $\epsilon = 0$. These two conditions imply that the beat wave separatrix overlaps both the wakefield separatrix and the plasma fluid oscillation, thus providing a phase-space path for plasma electrons to become trapped in the wakefield. For a given wakefield amplitude $\phi_0$, conditions (i) and (ii) imply the optimal phase location $3\phi_0 \cos\psi \simeq 3^{1/2} - 2\phi_0 - 2\beta_{pb}$ and threshold amplitude $6\hat{a}_1 > 3^{1/2} - 2\phi_0 + \beta_{pb}$ of the injection pulse, where $\phi_0^2 \cos^2\psi \ll 1$, $\hat{a}_1^2 \ll 1$, and $\beta_{pb}^2 \ll 1$ were assumed. For example, $\phi_0 = 0.6$ and $\beta_{pb} = 0.05$ imply $\psi = -1.3 - 2\pi j$ and $\hat{a}_1 > 0.11$.

To further evaluate the colliding laser injection method, the motion of test particles in the combined wake and laser fields was simulated in 1D. At $\tau = 0$, the forward (backward) pulse profile $\hat{a}_1$ ($\hat{a}_2$) is a half-period sine wave with maximum amplitude $a_{1m}$ ($a_{2m}$), centered at $\psi = \psi_1 < 0$ ($\psi_2 > 0$), of length $L_1$ ($L_2$). Test particles are loaded uniformly within $0 \le \psi \le \psi_{max}$ (initially at rest) and pushed from $\tau = 0$ to $\tau = \tau_{max}$. Also, $\hat{\phi} = \phi_0 \left[1 - \exp(-\psi^2/\pi^2)\right]$ for $\psi \le 0$.

To validate the analytical predictions for the trapping thresholds, a "near threshold" case was simulated with $a_{1m} = a_{2m} = 0.3$ ($1.2 \times 10^{17}$ W/cm$^2$), $L_1 = L_2 = \lambda_p/8$ (42 fs), $\psi_1 = -13.6$ and $\psi_2 = 21.4$ (chosen so the beat wave and test particles overlap). Also, $\omega_1/\omega_p = 100$, $\omega_2/\omega_p = 90$, and $\phi_0 = 0.6$, which for $\lambda_1 = 2\pi c/\omega_1 = 1$ $\mu$m implies $n_0 \simeq 10^{17}$ cm$^{-3}$ and $E_z = 0.6E_0 \simeq 19$ GV/m. At $\tau_{max} = 300$ (0.48 cm), the trapped bunch length is $L_b = 6.3$ $\mu$m (21 fs) and 60% of the electrons are contained within 66 MeV $\pm 8\%$. By varying the injection pulse amplitudes ($a_{1m} = a_{2m}$), trapping is observed for $a_{1m} > 0.17$, somewhat higher than the analytical prediction (0.11). Additional simulations indicate that trapping occurs when the center of the $L_1 = \lambda_p/8$, $a_{1m} = 0.3$ pulse is located within $-14.2 \leq \psi_1 \leq -13.5$. This implies that the forward pulse must be synchronized to the wake with an accuracy $< 37$ fs, which is not a serious constraint and can be relaxed somewhat by using a longer forward pulse. Simulations also show that momentum spread is relatively insensitive to variations in $L_{1,2}$. The observed momentum spread can be traced to the half-sine pulse profiles, which implies that different electrons encounter different beat amplitudes and are injected into the wake with different energies.

Important 3D beam dynamics issues — in particular, the effect of the radial electric wakefield $E_r$ — have been addressed with the 1D simulation model. The wake potential for a pump laser pulse with a Gaussian radial profile is $\phi \simeq \phi_0 \exp(-2r^2/r_0^2) \cos\psi$, where $r_0$ is the laser spot size. This implies that the radial electric field acting upon the trapped electrons will be focusing for $\cos\psi > 0$ and defocusing for $\cos\psi < 0$. For the near-threshold cases presented thus far, electrons which were initially injected within the focusing region of the wake slipped back into a defocusing region before they became highly relativistic. This is not the case in Fig. 4, which presents a simulation in which the position of the injection pulse was moved slightly forward and both the duration and amplitude of the injection pulses were increased, i.e., $\psi_1 = -12.6$, $a_{1m} = a_{2m} = 0.5$, $\phi_0 = 0.7$, $L_1 = L_2/4 = \lambda_p/4$, $\omega_1/\omega_p = 100$, and $\omega_2/\omega_p = 85$ ($\lambda_1 = 0.85$ $\mu$m, $\lambda_2 = 1$ $\mu$m, and $\lambda_p = 85$ $\mu$m). The electrons were injected at an earlier phase position with slightly higher energies and remained trapped in a focusing region of the wake. Plotted in Fig. 4 is the electron phase space (a) during the colliding pulse interaction at $\tau = 21$ and (b) after a distance of $\tau_{max} = 100$ (0.14 cm) with $\psi_{max} = 4$. Note that the front and back portions of the plasma electron segment in Fig. 4(a), which have not been significantly perturbed by the injection pulses, remain untrapped and have slipped out the back of Fig. 4(b). The results are quite dramatic: a bunch duration of 2.9 fs was obtained due to natural compression provided by the axial electric field, with a mean energy of 27 MeV and a standard deviation in energy of 0.32%. The trapping fraction is $f_{tr} \simeq 19\%$ and the bunch density is $n_b = 1.8 \times 10^{18}$ cm$^{-3}$. It should be noted that the electron slippage observed in the near threshold cases can be compensated by a slight decrease in the ambient density as a function of $z$. An appropriate reduction in $n_0(z)$ can increase the plasma wavelength such that the trapped electrons remain in the focusing region of the wake.

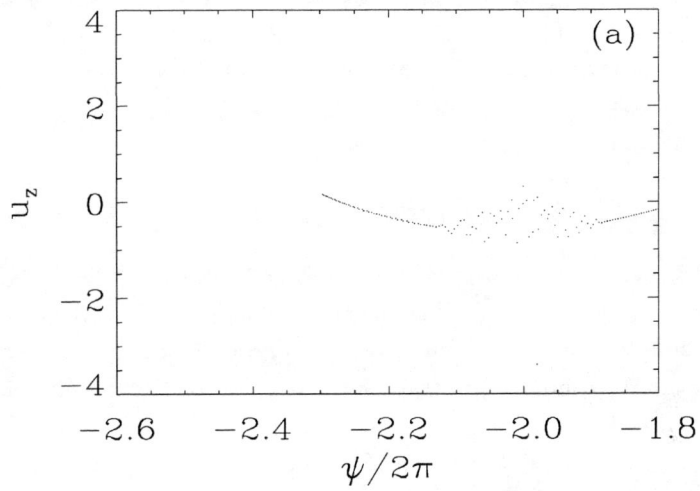

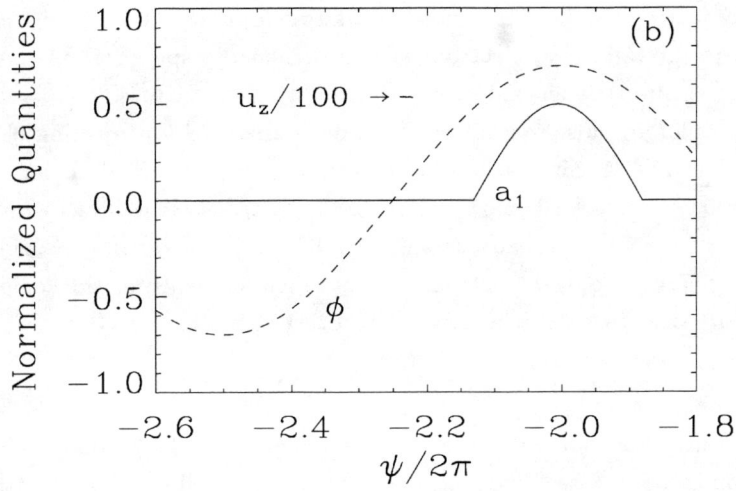

**FIG. 4**: Electron phase space $u_z$ vs $\psi$ (a) during the collision, $\tau = 21$, and (b) after the collision, $\tau = 100$, with $\omega_1/\omega_p = 100$, $\omega_2/\omega_p = 85$, $\phi_0 = 0.7$, $a_{1m} = a_{2m} = 0.5$, and $L_1 = L_2/4 = \lambda_p/4$. Shown in (b) are the trapped electron bunch (located by arrow), the injection pulse $\hat{a}_1$ (solid curve) and wake $\phi$ (dashed curve) profiles.

The bunch density is $n_b \simeq f_{tr} n_0 L_z / L_b$, where $L_z \simeq (L_1 + L_2)/2$ is the length of plasma that encounters the overlapping pulses. Assuming that the 1D results hold for a pump laser of radius $r_0$ implies a total number of trapped electrons $N_b \simeq f_{tr} n_0 L_z \pi r_0^2$, e.g., $N_b \simeq 7.7 \times 10^9$ for Fig. 4 with $r_0 = 40$ $\mu$m. Note that $N_b$ can be increased by increasing $n_0$, $r_0$, $a_{1m}$ (via $f_{tr}$) and, in particular, $L_z$ by increasing the duration of the backward pulse $L_2$. The ratio of $N_b$ to the theoretical beam loading limit $N_0$ [25] is $N_b / N_0 = f_{tr} k_p L_z E_0 / E_z$, which can easily approach unity. For $N_b$ near $N_0$, however, space-charge effects become important and a self-consistent simulation is required.

## CONCLUSION

Self-trapping and acceleration of plasma electrons in the SMLWFA via the coupling of RBS to the wake has been examined analytically and with 3D test particle simulations. A sudden onset in trapping is observed when either $a_1$ or $\phi_0$ exceeds a threshold. The trapping threshold for linear polarization is much less than that for circular polarization. This threshold occurs well below wave-breaking, e.g., $E_z / E_0 \sim 25\%$, which is consistent with experimental observations. In addition, self-channeling provides continuous focusing of the accelerated electrons which, along with relativistic pump laser effects, can enhance the energy gain by a factor of 2-7. This results in a large number ($10^9$) of short pulse ($< 1$ ps), relativistic (50 MeV) electrons with 100% energy spread and low emittance ($\epsilon_n < 1$ mm-mrad) generated over a short distance ($< 1$ mm).

The colliding pulse method has also been analyzed for injecting electrons in the standard LWFA. Simulations of test electrons in prescribed 1D fields indicate the production of relativistic ($\geq 25$ MeV) electrons with bunch durations as short as 3 fs, energy spreads as small as 0.3%, and densities as high as $10^{18}$ cm$^{-3}$. Simulations of test electrons in prescribed 3D fields for the colliding pulse injector are discussed elsewhere in this proceedings [16].

## ACKNOWLEDGMENTS

The authors acknowledge useful conversations with T. Antonsen, C. Clayton, T. Katsouleas, P.B. Lee, W.B. Mori, C. Schroeder, G. Shvets, and D. Umstadter. This work was supported by the Department of Energy.

## REFERENCES

[1] For a review, see E. Esarey et al., IEEE Trans. Plasma Sci. **24**, 252 (1996).
[2] A. Modena et al., Nature **377**, 606 (1995).

[3] K. Nakajima et al., Phys. Rev. Lett. **74**, 4428 (1995); C. Coverdale et al., Phys. Rev. Lett. **74**, 4659 (1995).

[4] D. Umstadter et al., Science **273**, 472 (1996); R. Wagner et al., Phys. Rev. Lett. **78**, 3125 (1997).

[5] A. Ting et al., Phys. Plasmas 4, 1889 (1997); C.I. Moore et al., Phys. Rev. Lett. **79**, 3909 (1997); A. Ting et al., Bull. Am. Phys. Soc. **42**, 1935 (1997).

[6] D. Gordon et al., Phys. Rev. Lett. **80**, 2133 (1998).

[7] P. Sprangle et al., Phys. Rev. Lett. **69**, 2200 (1992); E. Esarey et al., Phys. Fluids B **5**, 2690 (1993).

[8] T.M. Antonsen and P. Mora, Phys. Rev. Lett. **69**, 2204 (1992); N.E. Andreev et al., JETP Lett. **55**, 571 (1992).

[9] W.B. Mori et al., Phys. Rev. Lett. **72**, 1482 (1994); E. Esarey et al., Phys. Rev. Lett. **72**, 2887 (1994).

[10] K.C. Tzeng et al., Phys. Rev. Lett. **79**, 5258 (1997).

[11] T. Tajima and J.M. Dawson, Phys. Rev. Lett. **43**, 267 (1979).

[12] D. Umstadter et al., Phys. Rev. Lett. **76**, 2073 (1996).

[13] R.G. Hemker et al., Phys. Rev. E **57**, 5920 (1998).

[14] E. Esarey et al., Phys. Rev. Lett. **80**, 5552 (1998).

[15] E. Esarey et al., Phys. Rev. Lett. **79**, 2682 (1997); W.P. Leemans et al., Phys. Plasmas **5**, 1615 (1998).

[16] C.B. Schroeder et al., these proceedings; submitted to Phys. Rev. E.

[17] A.I. Akhiezer and R.V. Polovin, Sov. Phys. JETP **3**, 696 (1956).

[18] T. Katsouleas and W.B. Mori, Phys. Rev. Lett. **61**, 90 (1988); S.V. Bulanov et al., Phys. Rev. Lett. **78**, 4205 (1997).

[19] A. Ting et al., Phys. Rev. Lett. **77**, 5377 (1996); S.P. LeBlanc et al., Phys. Rev. Lett. **77**, 5381 (1996).

[20] C. Rousseaux et al., Phys. Rev. Lett. **74**, 4655 (1995); A. Ting et al., Opt. Lett. **21**, 1096 (1996).

[21] C. Joshi et al., Phys. Rev. Lett. **47**, 1285 (1981).

[22] P. Bertrand et al., Phys. Plasmas **2**, 3115 (1995).

[23] E. Esarey and P. Sprangle, Phys. Rev. A **45**, 5872 (1992); G. Shvets et al., Phys. Plasmas 4, 1872 (1997).

[24] For a review, see E. Esarey et al., IEEE J. Quant. Elect. **33**, 1879 (1997).

[25] T. Katsouleas et al., Particle Accelerators **22**, 81 (1987).

[26] E. Esarey and M. Pilloff, Phys. Plasmas **2**, 1432 (1995).

# Ultimate Gradient in Solid-State Accelerators

David H. Whittum

*Stanford Linear Accelerator Center, Stanford University, Stanford, California 94309*

**Abstract.** We recall the motivation for research in high-gradient acceleration and the problems posed by a compact collider. We summarize the phenomena known to appear in operation of a solid-state structure with large fields, and research relevant to the question of the ultimate gradient. We take note of new concepts, and examine one in detail, a miniature particle accelerator based on an active millimeter-wave circuit and parallel particle beams.

## INTRODUCTION

The microwave accelerator originated with Hansen's notion of a resonant accelerator cavity (1), later incorporated into a multi-cavity concept (2), implemented on the 1-GeV/100-m scale (3), and eventually the 20-GeV/3-km (4) and 50-GeV (5) scales. Today, a 1-TeV/30-km machine is ready for engineering design (6,7). Extension of colliders to higher energy and larger scale is technically feasible, but as discussion continues of the next collider, it is natural to wonder about the *last* collider. Such a machine would be limited, at least, by physical scale, and thus one is led to ask, for a particular operating wavelength, what is the ultimate accelerating gradient? And can it be employed in a collider? Efforts to conceive of a compact collider based on a high-gradient accelerator meet with numerous problems, as abstracted in Fig. 1. In this work we discuss these problems, and efforts in progress to solve them.

To appreciate the difficulties, consider the "conventional" picture of a collider (8), naively extrapolated to 5-TeV center-of-momentum, with an accelerating gradient of 1 GeV/m, and luminosity of $10^{35} \mathrm{cm}^{-2}\mathrm{s}^{-1}$. A synopsis reads something like this. There exists no tube adequate to power such a linac. If one were found, it would melt the accelerator structure surface in a single machine pulse, if operated at a conventional frequency, $10^1$ GHz or lower. At high frequencies, even if one could fabricate a small-scale structure, the correspondingly high electromagnetic wakefields would cause the intense beam to self-destruct. Ohmic heating would still be problematic due to the mechanical stress of thermal cycling. At any frequency, the average power requirements would exceed 1 GW, requiring a dedicated power plant. The optical system focusing the beams into collision would exceed 20 km in length, dwarfing the linac. In collision, the

CP472, *Advanced Accelerator Concepts: Eighth Workshop,*
edited by W. Lawson, C. Bellamy, and D. Brosius
© 1999 The American Institute of Physics 1-56396-889-4/99/$15.00

beams would destroy each other by means of their intense fields. Realizing that the problems of this collider concept are intractable, one concludes that the ultimate machine must look different.

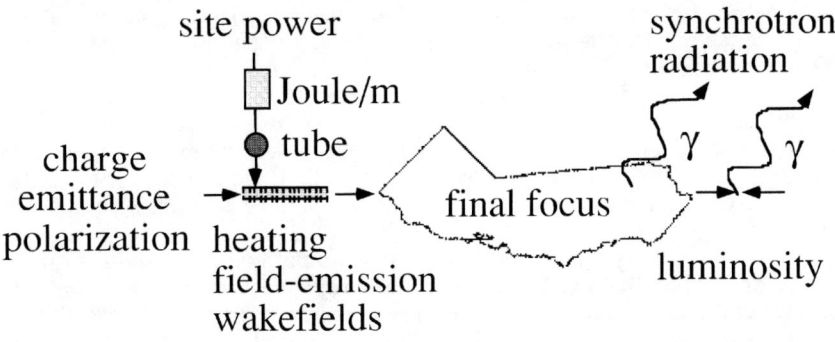

**FIGURE 1.** Problems for a compact-collider.

Work reported by F. Zimmermann in these proceedings discards the naive collider picture, and suggests that many of the problems arising after the linac may be solvable in principle (9). With his work, one can conceive, for the first time, of a compact final focus, colliding beams in an orderly, efficient fashion. To make a compact collider, one still requires a high-gradient accelerator, and it is here that we will concentrate our attention.

"High-gradient" we will take to mean 1 GeV/m or higher. We will find that high gradients favor high operating frequency, and at 1 GeV/m we will be interested in frequencies in the range of $10^2$ GHz and higher. Over the last decade, fabrication of solid-state structures on the corresponding sub-cm scale has become conceivable (10), and we will emphasize these. We pick a particular operating frequency, 91.392 GHz (W-Band), for numerical examples. This frequency is convenient in that it is a harmonic of existing beamlines ("32×SLAC"), making possible certain tests with beam. Moreover, research in W-Band benefits from years of development for applications throughout society (11), as well as recent scrutiny for accelerator applications (12-14). By our choice of frequency, we do not mean to suggest that higher frequencies are somehow less interesting. In fact application of higher-harmonics (up to $10^2$ of the linac fundamental) appears to be essential (9). We keep in mind too that for a particular frequency, there is likely an ultimate energy attainable; thus the implications of high-field research for higher frequencies are of abiding interest.

First we summarize the research problems posed by high-gradient, and then we discuss new structure concepts. We go on to explore a particular structure in detail, with parameters along the lines of (9). We emphasize at the outset that the ultimate gradient in a solid-state collider-linac is unknown; our interest is to identify the research required to determine such a figure experimentally. Eventually one hopes to answer questions along the lines of: How do we build a 1-m linac producing a beam of energy 1 GeV or higher, with a technology appropriate for a 5-TeV, $10^{35} \mathrm{cm}^{-2} \mathrm{s}^{-1}$ collider?

# PROBLEMS OF HIGH-GRADIENT

There are three problems in operating a solid-state structure with high electric field: the absence of a power source (15), field-emission (16), and pulsed-heating (17). Related empirically to field-emission are the phenomena of trapping and breakdown.

Power source development has a learning curve spanning decades, if previous experience is a guide (18,19). Thus one is delighted to find that the problem of a high-power mm-wave amplifier was solved about a decade ago. Research begun with the two-beam accelerator (TBA) concept (20) found application in plasma-heating, and eventually resulted in demonstration of GW power levels at frequencies up to 140 GHz (21). With such a source one could directly power a W-Band accelerator to gradients in excess of 1 GeV/m. Thus one may declare the problem of a power source solved "in principle". In practice, for the induction-linac TBA, one is interested in additional studies of drive beam-dynamics, phase-stability, output coupling, and overall system efficiency and reliability (15,22). Henke has proposed a beat-wave coupled TBA premised on an rf linac (23). For this or other novel TBA driver concepts, the primary challenge is drive beam stability, subject to the constraint that the drive linac should be capable of delivering a stored energy per unit length of $10^1$ J/m. For two 2.5-km linacs operated at 120 Hz, this corresponds to $10^1$ MW, and is probably the maximum consistent with a reasonable site power.

At the same time, to design a prototype linac one requires experience in the requisite manipulations of high-power mm-waves, and the complete process of structure fabrication. For this a high-power amplifier is indispensable (24). Happily, one finds that, because W-Band forms a "window" in the atmosphere, tube development there is quite active. However, the highest power commercial amplifier puts out only about 5 kW (25). The highest power W-Band research amplifier is a gyroklystron found at the Naval Research Lab (26), and is approaching the 0.1 MW level. Design of a low-voltage miniature klystron is being pursued (27,28). With power combining (29), an array of such tubes may be capable of 1 MW. With active pulse-compression (30,31), the $10^{-6}$ sec modulator pulse can be matched to the $10^{-8}$ sec W-Band structure fill-time scale, and power levels higher by a factor of $10^1 - 10^2$ are conceivable. Meanwhile, a 1-GeV/m W-Band linac will require 200-400 MW with a pulse-length of 10-20 ns; thus one prefers, if not a TBA driver, a $10^1$ MW tube, with an active pulse compression system. The short wavelength favors a fast-wave interaction, and there are quite a few to choose from (32-34); of these the gyroklystron is the most advanced and a design study for the $10^1$ MW level is complete (35). Depending on the tube-type, research problems include permanent magnet design and fabrication, three-dimensional gun design, high current density cathode materials, suppression of oscillation and spurious modes, and output coupling from an overmoded interaction region. All concepts require a microwave network implemented with quasi-optics (36) to avoid the 1-dB/ft nominal attenuation in fundamental mode guide (WR10) in transport to the linac.

With a working source, one could embark on a study of the phenomenology associated with high-gradient at the operating wavelength, and begin to observe the consequences of choices made in structure fabrication, assembly and handing. Experience at X-Band (37,38), and other frequencies (39,40) is a guide. *Field emission* refers to the tunneling of electrons from the accelerator structure, in the presence of a large electric field (41), and was first recognized in microwave linacs by P.B. Wilson (42). In practice, field-emission appears as a signal from a scintillator and photo-multiplier tube. Up-to-date discussions of field-emission are found in Padamsee,

Knobloch and Hays (43) and Wang and Loew (16). One can attempt to appreciate field-emission on elementary grounds, considering an electron tunneling through a one-dimensional potential of depth $\phi$, the work-function, perturbed by a uniform electric field, $E$. The uncertainty in the electron momentum, $\Delta p$, is given by $e\phi \approx \Delta p^2 / 2m$, with $m$ the electron mass. The uncertainty in position is $\Delta x \approx \hbar / 2\Delta p$, with $\hbar$ Planck's constant. This corresponds to an uncertainty in energy on the order of $\Delta \varepsilon \approx eE\Delta x$, with $-e$ the electron charge. When $\Delta \varepsilon \approx e\phi$, one expects to find electrons emitted. Thus on elementary grounds one expects the threshold for field-emission to lie in the range $E_0 \approx \left(8em\phi^3\right)^{1/2} / \hbar \approx 10^{11}$ V/m, for copper, with $\phi \approx 4.65$ V. A more detailed calculation shows that emitted current density emitted varies as $J \propto \left(\beta E\right)^{2.5} \exp\left(-0.6 E_0 / \beta E\right)$, where $E$ is the theoretical surface field and $\beta = 1$. This result employs a time-average valid when the rf period is longer than the tunneling time, $\Delta t \approx \hbar / 2\Delta \varepsilon$, or $10^{-14}$ s, for copper at 1 GeV/m. In practice, this result is employed with $\beta$ as a phenomenological fit parameter, referred to as the "field-enhancement factor". It is typically of order $50 - 200$.

The challenge for high-gradient operation is to understand why typical $\beta$'s are so large, and to make them smaller by an order of magnitude. The phenomenon of large $\beta$ is nowadays attributed to the presence of metallic or dielectric impurities ("flakes") on the surface; these lower the local work function and lead to enhanced "field-emission", assisted by flake geometry and transient plasma formation. The result looked for in field-emission studies is a *process* that results in a structure working reliably at high-power. This process would include a recipe for fabrication, assembly, cleaning, processing, and any other steps bearing on the problems of material behavior subject to intense electromagnetic fields (44,45). It is known that cleanliness, bakeout, high-pressure water-rinse, coatings, and sundry habits are helpful, but at present there is no generally-acclaimed recipe for normal conducting linacs (16). For a high-frequency linac, where fill-time is short, one is interested in the transient features of field emission. Recent work by Srinivasan-Rao, *et al.*, indicates that on the nanosecond time-scale lower $\beta$ is achievable (46).

Once electrons are field-emitted they are available to be accelerated, and for a large field they may be *trapped, i.e.,* accelerated to relativistic speeds, reaching synchronism with the accelerating wave. From the binding-field expression (3), one can show that for acceleration from rest in a sinusoidal wave, with phase-velocity equal to the speed-of-light, the condition for trapping is $G > \pi mc^2 / e$, with $G$ the gradient, $\lambda$ the wavelength of the sinusoid, and $c$ the speed of light. If geometry permits, this trapped current may be detected with a Faraday cup or observed on a screen (47); in a long linac, it is lost in transit through a finite bandwidth magnetic lattice, and appears as a background signal on nearby radiation-sensitive beamline instruments (48). Such *dark-current* also has implications, not thoroughly explored as yet, for anomalous loading of the rf system and higher-mode excitation. To operate a linac at high-gradient one would prefer either to avoid the presence of free-electrons in the linac, or to incorporate them into the design. For example, all plasma linacs, when operated above the trapping threshold, are dark-current linacs. (This threshold also corresponds to the regime of non-linear plasma response.) At W-Band, the onset of trapping occurs for a gradient of 0.5 GeV/m, with trapping fraction of 50% at 1 GeV/m.

We don't argue here that field-emission or trapping place limits on gradient, although such a conclusion might be warranted depending on the linac and collider concept. Instead, we point out that they appear together with *breakdown*, a phenomenon that, in practice, is disruptive to circuit behavior. Breakdown refers to three coincident symptoms in the operation of a high-power microwave circuit: intrapulse rise in reflected energy, increase in X-rays (scintillator readings), and degradation of vacuum (rise in ion-pump currents). Higher frequencies may also be observed from the circuit, although it is not common practice to look for them. Where a viewport has been provided for, breakdown events are observable by visible light. It is found in practice that breakdown is preceded by an increase in field-emission, and for this reason one is inclined to surmise that the phenomena are related. Empirically it is known that breakdown threshold depends on pulse length, and it is common practice to condition a structure up to high-power, by operation at a short pulse length, with dithering of the operating frequency. One may surmise that this pulse-length dependence also favors higher frequencies, where natural structure fill-times are shorter. It is known that breakdown thresholds are lower in klystron output cavities (45), and this invites speculation about the influence of ionizing radiation. In sum, one is inclined to view field-emission, trapping and breakdown as parts of one problem: an uncontrolled free-electron population. This transient medium disrupts and re-directs energy transfer in the otherwise orderly microwave circuit. It should be avoided, or, failing that, rendered repeatable, and incorporated into the circuit design, as is proposed in plasma accelerators.

*Pulsed-heating* refers to the heating of the structure boundary in a single pulse, due to Ohmic dissipation. Its importance to high-gradient structures was first noted by P.B. Wilson (49), and it is thought to represent the most severe limit on gradient, due to cyclic fatigue. At X-Band, and 1-GeV/m the pulsed temperature rise is enough to melt the copper surface; at W-Band, the result is 700 K. Meanwhile, it has been argued that a temperature rise of only 40 K is probably the maximum acceptable figure, due to cyclic stress and cracking of the surface (50,51). At the same time, while autopsied structures often have a finish reminiscent of the lunar surface, they still function as a circuit. To understand this, D. Pritzkau is undertaking a series of pulsed-heating studies to obtain detailed data on degradation in tune and quality factor over the requisite astronomical number of pulses, for particular materials and fabrication processes (17). Pulsed-heating confronts directly the question of the structure material and integrity and appears to be the most fundamental motivation for advanced structure concepts.

# ADVANCED STRUCTURE CONCEPTS

There are four ideas for raising pulsed-heat damage thresholds at high-field: disposable accelerators (presently, plasma accelerators), advanced materials, composite structures, and active circuits. The subject of plasma accelerators is discussed at length in these proceedings.

Advanced materials include both dielectrics and conductors. Dielectric accelerator concepts begin with the Cherenkov wakefield accelerator (52), and the dielectric RFQ of Caspers (53). The dielectric wakefield concept has since been demonstrated experimentally by Gai, *et al* (54). In the meantime, a resonantly excited dielectric accelerator has been proposed (55). For high-field applications, diamond shows promise. Diamond has a high thermal conductivity $\kappa \approx 1.5 - 2.0 \times 10^3$ W/K $-$ m (56), and can hold off fields of order 1 GV/m (57). In addition, the loss tangent is quite low, $\tan \delta < 5 \times 10^{-4}$ at W-Band (58). Not all features of diamond are convenient, however. Temperature coefficient of expansion is $\alpha \approx 1.2 \times 10^{-6} / °C$, and thus differential expansion relative to the substrate is an issue. Field emission and secondary electron emission are practical concerns. Advanced conductors include the categories of Glidcop (59), Zr-OFC (60), and other dispersion-hardened copper alloys (61). These have the strength of steel, depending on their history, and electrical and thermal properties commensurate with copper. For completeness we note too that operation of structures at low temperature offers some promise; however, one is sobered by previous experience indicating that the high-quality factors attainable can be lost in high-power operation (16).

The application of advanced materials, and the broader problems of fabrication, design for fabrication, and microwave measurements (to assess fabrication) are being pursued by D.T. Palmer (62) with emphasis on electro-discharge machining. A parallel effort is in progress by Henke (13) with emphasis on fabrication by deep X-ray lithography, "LIGA" (63), following work of Kang, et al. (64). Silicon processing technology was the first to show promise in this area (65) and may point the way to still higher frequencies.

Application of composite structures for pulsed-heat reduction has been analyzed by X. Lin, in these proceedings (66). He has shown theoretically that application of diamond to the interior of an accelerating structure can significantly reduce pulsed temperature rise. Fabrication, high-power studies at X-Band, and short-pulse field emission studies are being pursued. Application of composite structures promises other benefits at high-field (44,45).

Active circuits external to the accelerator have been studied for some years, for application at conventional wavelengths (30,31). Active elements *integrated* into the structure are a relatively new area (67). For switch operation in the presence of large fields, silicon, with its $10^7$ V/m dielectric strength will be inadequate, and thus diamond (68), and plasma (30) are of interest. For diamond, the bandgap is $\varepsilon \approx 5.5 \, eV$, so that excitation by 220-nm wavelengths or shorter is required. Carrier lifetime depends on the purity, and can range from 15 ns down to 30 ps, with a value of 1 ns noted in (69) for synthetic diamond. Application to switching benefits from the development of active elements for pulsed power (70).

In the meantime, small-scale structures have a significant disadvantage that must be addressed in structure design: large transverse wakefields. Historically there have been four approaches to the problem of higher-modes in accelerating structures: [1] ignore them, [2] use lower intensity beams, and a stronger magnetic lattice, [3] damp and detune them, and/or [4] employ them for structure alignment. The first approach was tried on the Two-Mile Accelerator, the Livermore Advanced Test Accelerator (ATA), and later again on the Stanford Linear Collider (SLC). Each time it failed, and defaulted to the second approach. As to the third approach, detuning of higher-modes encourages destructive interference in multi-bunch operation. Damping removes the microwave energy from the beam-orbit. Detuning was first employed on the Two-Mile linac, and involved detuning three cells of each 86-cell structure in a number of sectors in the linac (71). Damping was first employed operationally, and successfully on the Livermore induction linacs, ETA and ATA (72), later appearing in the choke-mode geometry of T. Shintake (73), and a travelling-wave structure ("DDS") that accomplishes both damping and detuning (74). The fourth approach was tried on ATA, the two-mile linac (75), and, most recently on the DDS structure, where M. Seidel, *et al.*, provided the first definitive measurements of "absolute beam position" in a microwave linac (76).

This all suggests that damping, detuning, and use of the structure as a self-registered beam position monitor are essential aspects of advanced structure design.

(Their scarcity in the plasma accelerator literature suggests a rich field of study.) Application of asymmetric structures requires a detailed understanding of higher multipole content (77), and, for a working collider, compensation by means of lumped multipole cavities, or incorporation into the lattice design. Ultimately reduction of a novel structure design to practice requires not just design, fabrication, and bench-measurement, but experimental study with beam. The capability to perform studies of miniature structures with beam, at MW power levels, is a prominent feature of work reported by M.E. Hill (78).

The first figure of merit for wakefields in a collider is single-bunch, wakefield-induced growth in the action variable, *i.e.*, beam orbit jitter in units of the beam size. A linac functioning as a jitter-amplifier can exhibit diverse interactions between feedback systems, ground-vibration, structure misalignments, rf kicks, and other dynamical sub-systems; where machine-protection invokes a rate-limit, diagnosis and correction of errors in linac setup can be stymied (79). Thus to avoid compounding those features of beam dynamics that are hidden from the instrumentation, the linac should not amplify incoming beam offsets. The scaling for amplification in a W-Band linac is given in (80), where end-to-end quads of equal gradient and thickness $l$ are considered. Quadrupole gradient is limited by achievable pole-tip field and pole aperture. A structure operating at free-space wavelength $\lambda$, and optimized for good $[R/Q]$, has a maximum interior transverse dimension of $\lambda/2$, and therefore can accommodate a pole-tip radius no less than $\lambda/4 \approx 0.82$ mm, at W-Band. For a 1 T pole-tip field maximum gradient is then $1.2 \times 10^3$ T/m. A gradient of 600 T/m implies, at 10 GeV, minimum $\beta_+ \approx 1.0$ m, and quad length $l \approx 0.30$ m. With acceleration, and scaling the optical functions so as to maintain constant phase-advance per period, one finds $\beta_\pm, l \propto \gamma^{1/2}$. Quad length at the linac exit (2500 GeV) is then $l \approx 4.7$ m, and $\beta_+ \approx 15.8$ m. One can show that such a linac is stable, with injection at 10 GeV and beam-port radius $R \approx 0.18\lambda$, for a 60 pC bunch with length $l_b \approx 30$ $\mu$m.

The effect of wakefields on collider parameters is to constrain efficiency and luminosity. Efficiency prefers higher structure $[R/Q]$, and this improves with *smaller* beam-port aperture. Luminosity improves with *higher* bunch-charge. Taking all considerations together, one prefers shorter bunch length and stronger focusing mechanisms. Thus manipulation and diagnosis of short bunches and design of miniature magnets are important research problems. Meanwhile, fabrication of micro-electromagnets with 0.4-T flux density, 100-$\mu$m diameter and sub-millisecond switching time has been demonstrated (81).

Closely related to these is the subject of beamline instrumentation. Monitoring of bunch lengths below 30 $\mu$m will require manipulation of THz frequencies. Where the final focus employs $\beta_* \approx 150$ $\mu$m, monitoring of beam-timing in collision will require phase-measurements accurate to a few degrees at W-Band, without averaging (82). With normalized beam emittance of $\varepsilon_{nx,y} \approx 10^{-7}$ m − rad, maximum beam-size at the exit of a 2.5 TeV machine is $\sigma_{x,y} \approx 0.6$ $\mu$m. Thus linac orbit analysis will require single-shot beam position resolution below the 100-nm level. Machine tune-up will be aided by a pulse-to-pulse emittance measurement (83), something not presently available on the SLC. Instruments for machine protection bear serious scrutiny, as given to the matter of SLC collimators (84). Ideally, to protect accelerating structures from beam-damage, one would prefer to employ *disposable* spoilers and collimators, and to this end, laser collimation and resonant collimation are being studied (9).

Structure research is critical to the injector as well. The assumed single-bunch normalized emittance $\varepsilon_{nx,y} \approx 10^{-7}$ m – rad and charge $q_b \approx 60\,\text{pC}$ are roughly consistent with expected scalings for rf photocathode guns. However, the scaled gradient approaches 1 GeV/m. Thus the problem of high-gradient appears as well in the injector, and one requires innovation comparable to that in the linac structure. A possible solution is the pulsed photocathode gun (85). In addition, phasing and laser stability requirements are severe (86). Combined with the problem of polarization, the injector poses a broad research problem.

With some glimpse of the problems of high gradient, and the research efforts past and present, let us go on to look at the details of a particular new structure concept, an example of an "active" accelerator.

## AN ACTIVE LINAC

A conventional travelling-wave accelerator consists of a linear array of resonant cavities that serve a combined function as a transmission line and an accelerator. Resonant excitation establishes a high-electric field with a minimum power level set by losses in the circuit. By design the field is largely longitudinal and suitable for acceleration. Due to the long time-scale for resonant filling this accelerator concept suffers from single-pulse temperature rise at the surface of an accelerator cell, when operated at too high a gradient. This may be expressed in terms of pulse-width $T_p$, for a rectangular accelerator cell, as (68)

$$\Delta T_{\text{max}} \approx \frac{0.84}{\sqrt{\kappa C}} \left(\frac{\delta}{\lambda}\right) \frac{G^2 T_p^{1/2}}{[R/Q]} \eta_g \eta_t$$

For room-temperature copper the thermal conductivity is $\kappa = 401$ W/K – m, and the volume specific heat capacity is $C = 3.45 \times 10^6$ J/K – m$^3$. The conductivity of copper $\sigma \approx 5.8 \times 10^7$ mho/m determines the skin-depth $\delta \approx 2.1\,\mu m\ f^{-1/2}$(GHz), with $f = c/\lambda$ the frequency, and $\lambda$ the free-space wavelength. The quantity $\eta_t$ depends on the waveform shape, with $\eta_t = 1$ for a square-wave, while $\eta_g$ depends on the cavity shape, with $\eta_g = 1$ for a symmetric rectangular pillbox. Maximum $[R/Q] \approx 221.3\,\Omega$ occurs for cell width and height $\lambda/\sqrt{2}$, and gap $0.371\lambda$. With beam ports $[R/Q]$ is easily lowered by a factor of two.

To obtain the benefits of resonant filling, with reduced pulsed heating, we must consider other than a passive circuit. For example, Fig. 2 depicts a resonant line coupled by means of fast switches to a series of parallel transmission lines composed of accelerator cavities. Operation consists of three steps: [1] resonant filling of the primary line with mm-wave power $P_1$, provided by an external power source on the natural field decrement time scale of $10^{-8}$ s [2] rapid closing of the switches on a time scale well under $10^{-9}$ s [3] propagation of a short sub-nanosecond burst of mm-waves down the secondary line, as single electron bunches arrive in parallel, to be accelerated as they pass through each secondary line, roughly orthogonal to the direction of mm-wave propagation. A detailed treatment of this concept has been given in work with Tantawi (68), with the geometrical implementation of Fig. 3.

For low loss, the primary line should be implemented as an overmoded standing-wave cavity. To simplify the problem of fabrication, we have chosen a single-depth geometry in Fig. 3, although ultimately one would prefer a greater depth for the primary. In the first approximation, both the primary cell, and the secondary cell are rectangular pillboxes, excited in the $TE_{10m}$ mode, with $m=1$ for the secondary cell, and, we suppose, $m = 15$, for the primary cell. Such an idealized geometry permits analytic calculation of the circuit parameters. Minimum stored energy density on the primary line favors maximum $[R/Q]/L$, with $L$ the primary period. However, too small a transit angle implies a lower wall quality factor, a thin and more fragile, structure, and fewer beamlines. For simplicity we optimize instead the stored energy per secondary line, maximizing the secondary-cell $[R/Q]$. For $f = 91.392$ GHz, cell depth is $2.3$ mm, secondary period $w_2 \approx 2.3$ mm and $L \approx 1.2$ mm. Secondary wall quality factor $Q_{w2} \approx 2.7 \times 10^3$, assuming a surface roughness much less than $\delta \approx 0.22 \mu$m. The pulse length discharged from the primary line is $T_p = 2m/f \approx 0.33$ ns. The primary wall quality factor is $Q_{w1} \approx 9.9 \times 10^3$, and the natural decrement time is $2Q_{w1}/\omega \approx 34.6$ ns; with critical coupling, the loaded fill-time for the primary cavity is $T_l \approx 17.3$ ns.

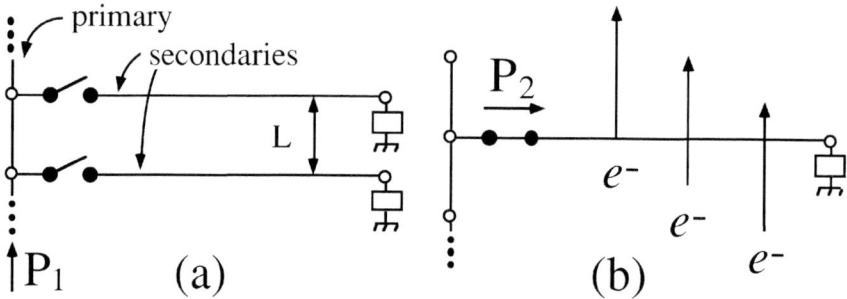

**FIGURE 2.** Concept for an active linac, consisting of a resonant storage line, coupled by means of switches to a periodic array of transmission lines. In (a) the primary resonant line is charged, and in (b) the switches have closed, discharging each primary period into the associated secondary transmission line.

An additional constraint arises due to transient operation of the secondary line. To minimize dispersion, phase-advance per cell should be near an odd multiple of $\pi/2$. Tapering is indicated to compensate wall losses. Despite these measures, the transient waveform discharged down the secondary eventually disperses and the transient peaking voltage appears on only a limited number of secondary cells (beamlines), a number set by the initial group velocity on the line, $\beta_g(0)c$, and $Q_{w2}$. For example, for $\beta_g(0) = 0.174$, maximum $N_2 \approx 35$. Detailed computation of transient waveforms is straightforward since, after switch-closure, the circuit-equivalent of Fig. 3 corresponds to a simple chain of $N_2$ coupled cavities. To illustrate the resulting scalings we consider a numerical example with $N_1 = 25$ and a design gradient of $G \approx 1.01$ GeV/m. The desired accelerating voltage in the first cell is $V_{NL} \approx 1.2 \times 10^6$ V, corresponding to stored energy $U_2 = V_{NL}^2/4k_l \approx 11.9$ mJ, with $k_l = \omega[R/Q]/4$ the loss-factor. The peak power required from the discharging primary cell is determined by the product of the initial group velocity and energy density in the secondary, $U_2/w_2 \approx 5.1 \text{J/m}$, and is $U_1/T_p \approx 2.7 \times 10^2 \text{ MW}$. The stored energy requirement in one primary period is then $U_1 \approx 88 \text{ mJ}$, and the stored energy density in the primary line prior to discharge is $U_1/L \approx 72 \text{ J/m}$. The maximum pulsed temperature rise in the circuit is $\Delta T_{max} \approx 126 \text{ K}$, for an input pulse length of 18.6 ns, corresponding to $P_1 \approx 2.9 \times 10^2 \text{ MW}$. With single-bunch charge of $q_b \approx 60 \text{ pC}$, voltage droop due to beam-loading is 5% at $n=35$. Thus in contrast to a conventional collinear structure, the beam-induced wakefield disperses and its effect on other bunches is diminished.

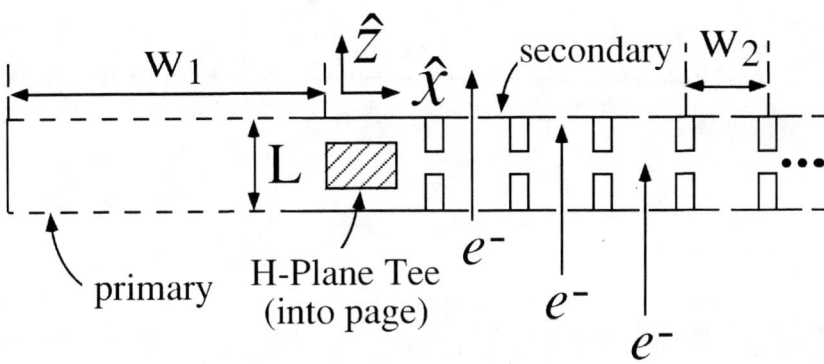

**FIGURE 3.** A single period of the geometry implementing the circuit scheme of Fig. 2. Each accelerator structure consists of $N_1$ such periods extending in $z$. The secondary line extends to the right (in $x$) for $N_2$ periods (beamlines).

An important improvement to this design would seek additional pulse compression by relaxing the single-depth constraint. In the limit of a larger primary cavity, with $Q_{w1}$ and vertical dimension larger by a factor of $O(10^1)$, one could power a 1-m, 1-GeV, $N_2 \approx 15$ beamline linac with a single power feed providing $4 \times 10^2 \text{ MW}$ in a $0.2 \mu s$ pulse. Pulsed temperature rise in the secondary would be under $100 \text{ K}$, and under $40 \text{ K}$ in the primary. The challenge of designing such a primary cavity lies in the problem of good coupling to the secondary line, so as to maintain a short discharge time-scale, equivalent to an external $Q$ after switch closure of $Q_e \approx 10^2$. To incorporate a parallel beam linac in Zimmermann's final focus picture, still more work is required, insofar as the beams considered here arrive at staggered time-intervals. For temporal beam combining, they should arrive almost simultaneously, and this implies a shorter secondary line, fewer beamlines, and higher charge per bunch. Thus additional work remains to fashion these concepts into a working parameter set for a collider.

# SUMMARY

The challenges posed by high-gradient operation are critical to the future of high-energy physics. While fundamental questions remain as to the ultimate gradient "in copper", we have indicated that there is more to life than copper, that the limits derived for passive copper structures are not in themselves fundamental, that one can do better. Advanced materials, composite structures and active elements show great promise for accelerator design. The ultimate gradient in a solid-state collider-linac is not known, but it could be several orders of magnitude beyond the 20-40 MeV/m gradients employed today.

# ACKNOWLEDGMENTS

Work with Sami G. Tantawi and Frank Zimmermann has been essential. Thanks to Angie Seymour for her support, and U.S. Department of Energy, Contract DE-AC03-76SF00515.

# REFERENCES

1. Hansen, W.W., *J. Appl. Phys.* **9** 654-663 (1938).
2. Ginzton, E.L., Hansen, W.W., and Kennedy, W.R., *Rev. Sci. Instrum.* **19** 89-108 (1948).
3. Chodorow, M. *et al.*, *Rev. Sci. Instrum.* **26** 134-204 (1955).
4. Neal, R.B., ed., *The Stanford Two-Mile Accelerator*, New York: W.A. Benjamin 1968.
5. Seeman, J.T., *Ann. Rev. Nucl. Part. Sci.* **41** 389-428 (1991).
6. Loew, G.A. and Weiland, T., eds., *International Linear Collider Technical Review Committee Report*, SLAC-R-95-471 (1995).
7. *Proceedings of the 1996 DPB/DPFWorkshop on New Directions for High Energy Physics*, New York: AIP, 1997.
8. Palmer, R.B., *Ann. Rev. Nucl. Part. Sci.* **40** 529-592 (1990).
9. Zimmermann, F., "New final focus concepts at 5 TeV and beyond", (these proceedings).
10. Chou, P.J., *et al.*, "Fabrication of millimeter-wavelength accelerating structures", *Advanced Accelerator Concepts*, AIP Conf. Proc. **398,** New York: AIP, 1997, pp. 501-517.
11. Meinel, H.H., *IEEE Trans. MTT* **43** 1639-1653 (1995).
12. Kang, Y.W., *et al.*, *Rev. Sci. Instrum.* **66** 1-5 (1995).
13. Henke, H. "Planar Millimeter-Wave RF Structures", *Advanced Accelerator Concepts*, AIP Conf. Proc. **398,** New York: AIP, 1997, pp. 485-500.
14. Willeke, T. L. and Feinerman, A. D., *J. Vac. Sci. Tech.* **B 14** 2524-2530 (1996).
15. Lidia, S.M., Whittum, D.H. and Donohue, J.T., "W-Band free electron laser for high-gradient structure research", *Proc. 1997 Particle Accelerator Conference*, New York: IEEE, (to be published).
16. Wang, J.W. and Loew, G.A., "Field Emission and RF Breakdown in High-Gradient Room-Temperature Linac Structures", *Proceedings of the Joint US-CERN-Japan School, RF Engineering for Particle Accelerators*, IOP, (to be published), SLAC-PUB-7684.
17. Pritzkau, D., *et al.*, "Experimental Study of Pulsed Heating of Electromagnetic Cavities", *Proc. 1997 Particle Accelerator Conference*, New York: IEEE, (to be published).
18. Allen, M. A., "RF power sources", *Proc. 1st European Particle Accelerator Conf.*, Rome, Italy, June 1988, SLAC-PUB-4646.
19. Caryotakis, G., *IEEE Trans. Plasma Sci.* **22** 683-691 (1994).

20. Sessler, A.M., *et al.*, *Nucl. Instrum. Methods* **A 306** 592-605 (1991).

21. Allen, S. L., *et. al.*, "Generation of High Power 140 GHz Microwaves with an FEL for the MTX Experiment", *Proceedings of the 1993 Particle Accelerator Conference*, New York: IEEE, 1993, pp. 1551-1553.

22. Li, H., *et al.*, "Design study of beam dynamics issues for a one TeV next linear collider based upon the relativistic klystron two-beam accelerator", *Advanced Accelerator Concepts*, AIP Conf. Proc. **335**, New York: AIP, 1995, pp. 817-836.

23. Whittum, D.H., Chou, P.J. and Henke, H. "High-gradient cavity beat-wave accelerator at W-Band", *Proceedings of the 1997 Particle Accelerator Conference*, New York: IEEE, (to be published).

24. Caryotakis, G., *IEEE Transactions on Plasma Science* **22** 683-691 (1994).

25. *MMW Power Klystrons for Remote Sensing and Communications*, Communications and Power Industries (CPI), 45 River Dr., Georgetown, Ontario, Canada L7G2J4.

26. Blank, M., *et. al.*, "Experimental Investigation of a W-Band Gyroklystron Amplifier", *Proc. 1997 Particle Accelerator Conference*, New York: IEEE, (to be published).

27. Caryotakis, G., *Physics of Plasmas*, **5** 1590-1598 (1998).

28. Solyga, S.A., and Bruns, W., "Cavity design for a planar mm-wave sheet beam klystron", *Proc. 5th European Particle Accelerator Conference*, Bristol: IOP, 1996, pp. 2140-2142.

29. York, R. A., "Quasi-optical power combining techniques", *SPIE Critical Review of Emerging Technologies*, New York: SPIE, 1994.

30. Petelin, M.I., Vikharev A. L., and Hirshfield, J. L., "Pulse Compressor Based on Electrically Switched Bragg Reflectors", *Advanced Accelerator Concepts*, AIP Conf. Proc. **398**, New York: AIP, 1997, pp. 822-831.

31. Tantawi, S.G., *et al.*, "Active High Power RF Pulse Compression Using Optically Switched Resonant Delay Lines", *ibid.* pp. 813-821.

32. Gold, S.H. and Nusinovich, G.S., *Rev. Sci. Instrum.* **68** 3945-3974 (1997).

33. Cheng, S., *et al.*, *IEEE Trans. Plasma Sci.* **24** 750-757 (1996).

34. Thumm, M., "State-of-the-Art of High Power gyro-Devices and Free Electron Masers Update 1996", Karlsruhe: Forschungszentrum Karlsruhe, 1996, FZKA 5877.

35. Arjona, M.R., and Lawson, W., "Design of a 7 MW, 95 GHz Three-Cavity Gyroklystron" (submitted to *IEEE Trans. Plasma Sci.*).

36. Goldsmith, P.F., *Proceedings of the IEEE* **80** 1729-1747 (1992).

37. Higo, T., *et al.*, "Precise fabrication of X-band accelerating structure", *Proc. 9th Symposium on Accelerator Science and Technology*, Tsukuba: KEK, 1993, KEK 93-57.

38. Higo, T., *et al.*, "High-gradient experiment on X-band disk-loaded structures", KEK Report 93-9.

39. Schriber, S. O., "Factors limiting the operation of structures under high-gradient", *Proc. 1986 Linear Accelerator Conference*, Stanford: SLAC, 1986, pp. 591-597.

40. Gaponov-Grekhov, A.V. and Granatstein, V.L., eds., *Applications of High-Power Microwaves*, Boston: Artech, 1996.

41. Fowler, R.H. and Nordheim, L.W., *Proc. Roy. Soc* (London) **A 119** 173 (1928) .

42. Wilson, P.B., "Power Dissipation Due to Field Emission in Electron Linacs", Stanford HEPL-TN-66-2, 1966 (unpublished).

43. Padamsee, H., Knobloch, J., and Hays T., *RF Superconductivity for Accelerators*, New York: Wiley, 1998.

44. Matsumoto, M., *et al.*, "Applications of the Hot Isostatic Pressing (HIP) for high gradient accelerator structure", *Proceedings of the 1991 Particle Accelerator Conference*, New York: IEEE, 1991, pp. 1008-1010.

45. Xu, X., *et al.*, "RF breakdown studies in X-Band klystron cavities", *Proceedings of the 1997 Particle Accelerator Conference*, New York: IEEE, (to be published).

46. Srinivasan-Rao, T., *et al.*, "Dark current measurements at field gradients above 1 GV/m" (these proceedings).

47. Wang. J.W., "RF Properties of Periodic Accelerating Structures for Linear Colliders", Ph.D. Thesis,

Stanford University (1989), SLAC-Report-339.

48. Assmann, R., *et al.*, "Observation of dark current signals in S-Band structures at the SLC linac", *Proceedings of the 1997 Particle Accelerator Conference*, New York: IEEE, (to be published).

49. Wilson, P.B., "Linear Accelerators for TeV Colliders", *Laser Acceleration of Particles*, AIP Proc. **130**, New York: AIP, 1985, pp. 560-597.

50. Wilson, I., "Surface Heating of the CLIC Main Linac Structure", Geneva: CERN, 1987, CLIC Note 52.

51. Nezhevenko, O.A., "On the Limitations of Accelerating Gradient in Linear Colliders Due to Pulse Heating", *Proceedings of the 1997 Particle Accelerator Conference*, New York: IEEE, (to be published).

52. Keinigs, R. and Jones, M., *Part. Acc.* **24** 223-229 (1989).

53. Caspers, F., "RFQ Structures with Dielectric Materials", Geneva: CERN, 1991, PS/OP 91-19.

54. Gai, W., *et al.*, *Phys. Rev. Lett.* **61** 2756-2758 (1998).

55. Rosenzweig, J., Murokh, A. and Pellegrini, C., *Phys. Rev. Lett.***74** 2467-2470 (1995).

56. Prelas, M.A., Popovici, G., and Bigelow, L.K., eds., *Handbook of Industrial Diamonds and Diamond Thin Films*, New York: Marcel Dekker, 1998, pp. 147-192.

57. Pan, L.S. and Kania, D.R., eds., *Diamond: Electronic Properties and Applications*, Boston: Kluwer, 1995, p. 254.

58. Thumm, M., "Development of Output Windows for High-Power Long-Pulse Gyrotrons and EC Wave Applications", *International Journal of Infrared and Millimeter Waves*, **19** 3-14 (1998).

59. O.M.G. Americas, 811-T Sharon Dr., Westlake, OH 44145.

60. Technical Data, TD14-975, Hitachi Cable, Ltd., 2-1-2 Marunouchi, Chiyoda-Ku, Tokyo 100, Japan.

61. Nadkarni, A.V., "Dispersion strengthened copper properties and applications", in *High Conductivity Copper and Aluminium Alloys*, Ling, E., and Taubenblat, P.W., eds., Los Angeles: AIME, 1984, pp. 77-101.

62. Palmer, D.T., *et al.*, "W-Band Structure Research at SLAC", (these proceedings).

63. Becker, E.W., *et al.*, *Microelectronic Engineering* **4** 35-56 (1986).

64. Kang, Y.W., *et al.*, *Rev. Sci. Instrum.* **66** 1-5 (1995).

65. Willeke, T.L. and Feinerman, A.D., *J. Vac. Sci. Tech.* **B 14** 2524-2530 (1996).

66. Lin, X., "Diamond Coating in Accelerating Structure", (these proceedings).

67. Bamber, C., *et al.*, *Nucl. Instrum. Meth.* **A 327** 227-252 (1993).

68. Whittum, D.H. and Tantawi, S.G., "Switched matrix accelerator" (submitted to *Rev. Sci. Instrum.*), SLAC-PUB-7848.

69. Pan, L.S., *et al.*, "Electrical Mobility and Carrier Lifetime in Single-Crystal Isotopically Pure Type IIa Synthetic Diamond", *Materials Research Society Symposium* **302** 299-304 (1993).

70. Rosen, A. and Zutavern, F., eds., *High-Power Optically Activated Solid-State Switches*, Boston: Artech, 1993.

71. Loew, G.A., *et al.*, "Linac beam interactions and instabilities", *Proc. VII Internat'l Conf. High Energy Accelerators*, Yerevan: Armenian Academy of Sciences, 1970, pp. 229-252.

72. Briggs, R.J., *et al.*, *IEEE Trans. Nucl. Sci.* **NS-28** 3360-3364 (1981); Caporaso, G.J., *et al.*, *ibid.*, **NS-30** 2507-2509 (1983).

73. Shintake, T., *Jpn J. Appl. Phys.* **31** L1567-L1570 (1992).

74. Kroll, N.M., *et al.*, "Recent Results & Planas for the Future on SLAC Damped Detuned Structures, *Advanced Accelerator Concepts*, AIP Proc. **398**, New York: AIP, 1997, pp. 455-464.

75. Seidel, M., *et al.*, "Detection of beam-induced dipole-mode signals in the SLC S-Band structures", *Proceedings of the 1997 Particle Accelerator Conference*, New York: IEEE, (to be published).

76. Seidel, M., *et al.*, *Nucl. Instrum. Methods in Phys. Res.* **A 404** 231-236 (1998).

77. Whittum, D.H. and Kolomensky, Y., "Analysis of an asymmetric resonant cavity as a beam monitor", (submitted to *Rev. Sci. Instrum.*), SLAC-PUB-7846.

78. Hill, M.E., *et al.*, "Subharmonic drive experiment at W-band" (these proceedings).

79. Assmann, R. and Zimmermann, F., "Possible sources of pulse to pulse orbit variation in the SLAC linac", *Proc. XVIII International Linac Conference*, Geneva: CERN, 1996, pp. 473-475, SLAC-PUB-

7269
80. Zimmermann, F., *et al.*, "Wakefields in a mm-wave linac" (these proceedings).

81. Rogers, J. A., Jackman, R. J., and Whitesides, G. M., *J. Microelectromechanical Systems* **6** 184-192 (1997).

82. Zimmermann, F., *et al.*, Bunch-length and beam-timing monitors in the SLC final focus, (these proceedings).

83. Kim, J.-S., *et al.*, "Pulse-to-pulse emittance measurement system", FARTECH Inc., 3146 Bunch Ave., San Diego, CA 92122 (unpublished).

84. Decker, F.-J., *et al.*, "Design and wakefield performance of the new SLC collimators", *Proc. XVIII International Linac Conference*, Geneva: CERN, 1996, pp. 137-139, SLAC-PUB-7261.

85. Srinivasan-Rao, T., *et al.*, "Optimization of gun parameters for a pulsed power electron gun" (these proceedings).

86. Piovella, N., Serafini, L., and Ferrario, M. "Multi-cell rf injectors driven by thermionic cathodes", *Proceedings of the 6th European Particle Accelerator Conference*, Stockholm, Sweden, June 1998.

# Terawatt Picosecond CO$_2$ Laser Technology for High Energy Physics Applications

I.V. Pogorelsky

*Accelerator Test Facility, BNL, Upton, NY 11973, USA*

**Abstract.** Demonstration of ultra-high acceleration gradients in the SM LWFA experiments put a next objective for the laser accelerator development: to achieve a low-emittance monochromatic acceleration over extended interaction distances. The emerging picosecond terawatt (ps-TW) CO$_2$ laser technology helps to meet this strategic goal. Among the considered examples are: the staged electron laser accelerator (STELLA) experiment, which is being conducted at the Brookhaven ATF, and the plasma-channeled LWFA. The long-wavelength and high average power capabilities of CO$_2$ lasers may be utilized also for generation of intense x-ray and gamma radiation through Compton back-scattering of the laser beams off relativistic electrons. We discuss applications of ps-TW CO$_2$ lasers for a tentative γ-γ (or γ-lepton) collider and generation of polarized positron beams.

## INTRODUCTION

The first ps-TW CO$_2$ laser is close to completion at the BNL ATF [1-3]. Benefits of using a long-wavelength laser driver for laser accelerators are two-fold. For processes based on electron quiver motion, like in the LWFA, the advantage of CO$_2$ lasers is primarily due to a gain of two orders of magnitude in the ponderomotive potential, $W_{osc}$, which is quadratically proportional to λ:

$$W_{osc} = e^2 E_L^2 / 2m\omega^2,  \qquad (1)$$

where $E_L$ is the laser electric field amplitude. The ponderomotive strength of the CO$_2$ laser is the most easy to reveal in the self-modulated LWFA. In this case, it facilitates to attain the relativistic self-focusing condition,

$$P_L \geq 17(\lambda_p/\lambda)^2 \, [GW]  \qquad (2)$$

(where $P_L$ is the laser peak power and $\lambda_p$ is the plasma wavelength), and is ultimately responsible for the net energy gain scaling per the dephasing limited accelerator stage, $\gamma_{max} \propto \lambda\sqrt{P_L}$ [4].

It is acknowledged that after the successful experimental demonstration of the ultra-high gradient (up to 100 GeV/m) laser electron acceleration [5-7], the next goal in the laser accelerator development is to design and demonstrate extended quasi-monoenergetic acceleration of appreciably bulk electron charges (~0.1 nC).

Similar to a conventional RF accelerator, a high monochromaticity can be achieved in the laser driven accelerator when the electron bunch is a small fraction of the accelerating field period. Using the expansion of a sine function, it is easy to estimate

CP472, *Advanced Accelerator Concepts: Eighth Workshop,*
edited by W. Lawson, C. Bellamy, and D. Brosius

that the electron beam energy spread, $\Delta E / E$ drops quadratically with the reduction in the bunch length. For example, $\Delta E / E$ reaches 0.1% when the longitudinal and transverse dimensions of the quasi-relativistic electron bunch are equal to ~2% of the driving field period.

In the case of RF linacs, with their ~100 MHz accelerating fields, the required electron bunches are generated using photocathode RF guns driven by picosecond lasers. Orders of magnitude higher frequencies of the drive field in laser accelerators may require proportionally short electron bunches. Producing of sufficiently short, femtosecond bunches presents a serious technical problem. In addition, bunch shortening leads to reduction of the number of electrons per bunch (see below). Under these circumstances, a prudent selection of the laser driver wavelength is required.

In the present paper, we illustrate the importance of the wavelength optimization for both structure scaling and ponderomotive wake enhancement by two practical examples: a prospective plasma channeled LWFA and the far-field STELLA experiment which is ongoing at the ATF.

In Section 2, we start with a brief overview of the ATF, its ps-TW $CO_2$ laser project, and review some prospects of the ps-TW $CO_2$ laser technology.

In Section 3, we discuss the dependence of the LWFA performance upon the laser wavelength and apply our conclusions to optimization of the plasma-channel LWFA operating in a linear regime. In this case, the electron beam energy spread, emittance, and luminosity improve with longer $\lambda_p$ which tends to be proportional to $\lambda$. This enables design of two-stage monochromatic GeV laser accelerators as well as future high-luminocity, multi-stage TeV electron linacs.

Electron bunch miniaturization is required for far-field laser accelerators based on direct electron acceleration in the laser beam. In this case, the electron bunch shall fit into a small fraction of $\lambda$. Such ultra-short periodical micro-bunches synchronized to the laser accelerating field can be produced by no means but energy modulation of the initial macro-bunch by the same laser source that drives the laser accelerator. A number of considerations point to the long-wavelength $CO_2$ laser as a practical choice for such an accelerator. In Section 4, we illustrate this approach by the example of STELLA experiment that is under way at the ATF.

The long-wavelength and large average power capabilities of $CO_2$ lasers may be utilized also for generation of intense x-ray and gamma radiation through Compton back-scattering of the laser beams off relativistic electrons. In Section 5, we discuss applications of TWps-$CO_2$ lasers for a tentative $\gamma$-$\gamma$ (or $\gamma$-lepton) collider and generation of polarized positron beams.


# OVERVIEW OF THE ATF


The approach to the monochromatic laser acceleration discussed in this paper is based on using a combination of the emerging ps-TW $CO_2$ laser technology with conventional high-brightness picosecond and subpicosecond electron injectors. Both these ingredients become available at the ATF.

The ATF's electron linac delivers up to 3 nC, 10-0.3 ps, 50 MeV electron bunches with a peak current of 100-300 A, energy spread of 0.2%, and a normalized emittance $\varepsilon_n$=0.5 mm.mrad. Note, that with the present 10-ps UV photocathode laser driver, the electron bunch compression to 370 fs is achieved by the proper electron phasing to the RF field in the gun [8]. A shorter, down to 100 fs electron bunch duration may be attained with the faster, ~1 ps, laser driver.

Upgrade of the presently operational at the ATF 100-ps 10-GW $CO_2$ laser to the terawatt level is close to completion. The power increase will be attained via both energy boost in the additional big-aperture laser amplifier and pulse shortening to 1-10 ps [1-3].

Shown in the floor plan in Fig.1, is the new $CO_2$ laser system called PITER I (I$^{st}$ PIcosecond TERawatt). Positioning of the high-pressure laser amplifier in Room C1, and its 1-MV Marx generator behind the partition, helps to isolate these potential sources of the EM noise from the computer-interfaced control and diagnostics equipment located in Room C2. By means of in-vacuum laser beam transport tube, the control Nd:YAG laser located close to the linac front end is optically connected with the $CO_2$ laser system. The optical table in room C2 accommodates the oscillator, picosecond pulse slicer, and preamplifier.

The oscillator and semiconductor switch supply a seed pulse into the regenerative preamplifier with the active discharge volume of 1.2 liters (optical aperture 2×5 cm$^2$, discharge length 1.2 m). Extracted by a Pockels cell at several millijoules, the laser pulse is redirected for additional passes through the same discharge cell to acquire ~100 mJ energy. The preamplifier output is directed to the final high-pressure amplifier stage that is the main component of the ATF $CO_2$ laser upgrade.

In the 10-atm, x-ray preionized, transverse electric discharge amplifier, pressure broadening of the $CO_2$ gain spectrum into the 1 THz wide quasi-continuum permits direct amplification of picosecond terawatt laser pulses. For a $\tau_L$=1 ps pulse, the small-signal gain is 3-4%/cm, saturation fluence 500 mJ/cm$^2$, and the extractable specific energy ~20 mJ/cm$^3$. Taking into account that the total discharge volume of the amplifier is 10 liters, the possibility of extracting of the order of 100 J of energy in a picosecond pulse looks possible. However, the limiting factor to the high energy extraction is the damage of the output window at the level of 0.5 J/cm$^2$. An optical window of the 80 cm$^2$ size permits ~30 TW peak power extraction at the 1-ps laser pulse duration.

The tests of the new laser system are close to completion. Fig.2 and 3 illustrate the last preparations of the amplifier high-pressure discharge cell to operation.

An alternative way to achieve gain smoothing is to reduce the spectrum modulation period using a $CO_2$ gas mixture with a combination of the isotopes $O^{16}$ and $O^{18}$. Due to the isotopic shifts, the combined spectrum has 4-times denser rotational line structure than with a regular $CO_2$ molecule. That permits to reach gain smoothing at a lower gas pressure. This approach will be used in the UV-preionized preamplifier that operates at the 5 atm gas pressure.

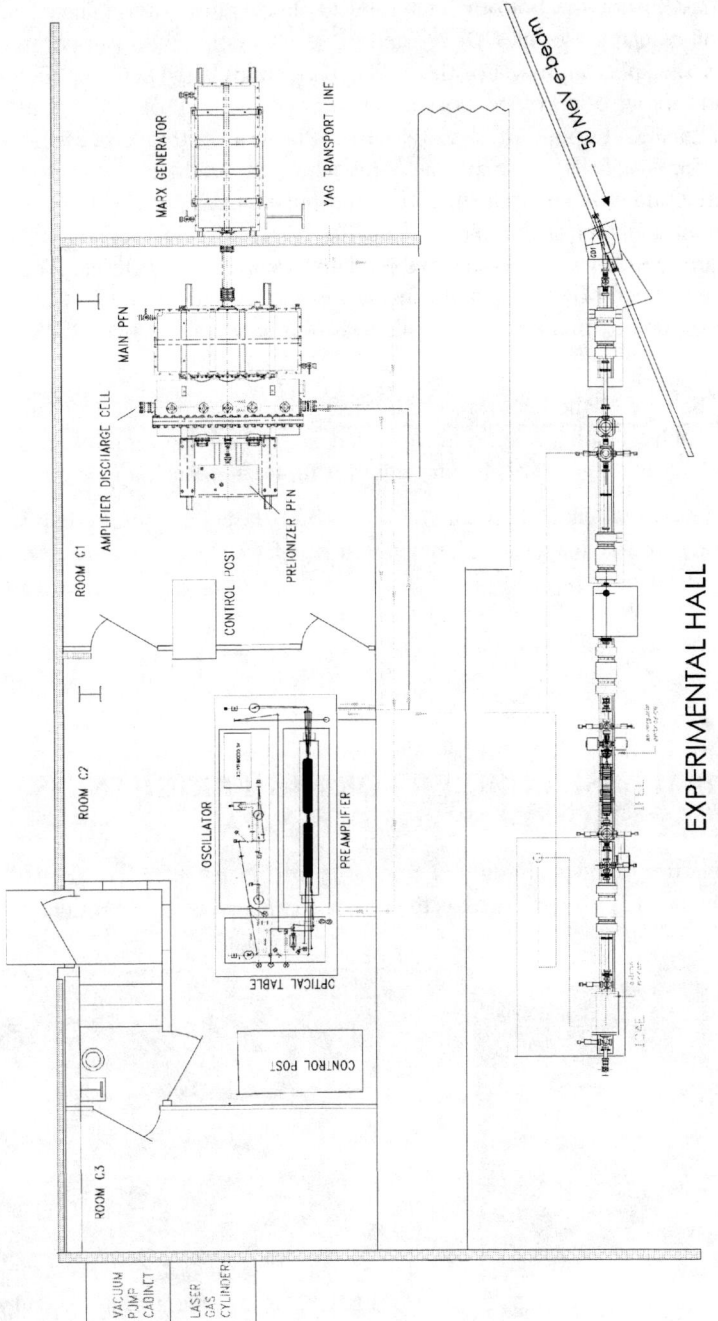

**FIGURE 1.** Fragment of the ATF floor plan showing $CO_2$ laser rooms and a portion of the experimental hall

89

However, 1 ps is still not the bottom limit for $CO_2$ laser pulse shortening. Taking a combination of all available isotopes $O^{(16,18)}$ and $C^{(12,13,14)}$ a continuum can be produced that extends over the spectral region of the 7 THz bandwidth [9]. Then amplification of as short as 100 fs pulse becomes possible. Another method of $CO_2$ laser pulse shortening uses ipulse chirping in a gas [9,10]. Physics of this process is rather straightforward. Intense laser pulse propagating in a gas produces ionization. As a result the refractive index is nonuniform across the pulse envelope, and the tail of the pulse propagates at a higher speed catching up the front. After a propagation over a certain distance in a gas, or in the dispersive medium, the pulse will shrink. This effect has been observed already by P. Corkum inside the laser amplifier [10]. The desired chirping may be organized in a more controlled fashion inside a gas-filled hollow fiber [11].

Because of the ease of the heat removal by fast gas exchange in the $CO_2$ amplifier, it is potentially capable to high repetition rates that are difficult to attain with massive glass or crystal active elements. This is important for future advanced applications.

A review of potential capabilities of the TWps-$CO_2$ laser technology implies that relatively moderate efforts may lead to development of $CO_2$ lasers with ~10 kW of average power, ~10 J per pulse, as short as 100 fs pulse duration, with the resulting peak power of ~100 TW [2,3]. This set of parameters looks even more attractive if to take into account additional advantages of the $CO_2$ lasers due to their longer wavelength.

## OPTIMUM LASER DRIVER FOR HIGH BRIGHTNESS MONO-ENERGETIC LWFA

Due to a relatively high stability and regularity of the produced wake field, the linear LWFA scheme [12] is considered as the most promising candidate for the

**FIGURE 2.** At the amplifier output window

**FIGURE 3.** Alignment of the ground mesh electrode inside the amplifier discharge cell

90

prospective laser-driven accelerator in the GeV energy range. For this scheme, the laser pulse duration and plasma wavelength satisfy the resonance condition $\tau_L \approx \lambda_p/2c$.

The amplitude of the accelerating field is

$$E_{a0}[V/cm] = 0.4\frac{a^2}{\sqrt{1+a^2/2}}\sqrt{n_e}[cm^{-3}], \tag{3}$$

where $a$ is the normalized vector-potential

$$a = eE_L/m\omega \tag{4}$$

and

$$n_e[cm^{-3}] = \frac{\pi mc^2}{e^2\lambda_p^2} = 10^{19}/\lambda_p^2[\mu m] \tag{5}$$

is the plasma density. By definition of the linear regime, $a<1$.

The net acceleration attainable over the dephasing distance

$$L_{ph} = \lambda_p\left(\frac{\lambda_p}{\lambda}\right)^2 \tag{6}$$

is

$$\Delta W[MeV] = \int_0^{L_{ph}} E_a dz \approx \frac{E_a L_{ph}}{2} = 0.6\frac{a^2}{\sqrt{1+a^2/2}}\left(\frac{\lambda_p}{\lambda}\right)^2. \tag{7}$$

In the uniform plasma, the acceleration distance within the laser focus waist is defined by the Rayleigh length, $z_0 = \pi r_L^2/\lambda$. To extend the acceleration over several Rayleight distances, some form of laser channeling needs to be used. As long as the relativistic self-focusing regime is not allowed for the standard LWFA scheme, we assume that a plasma channel is pre-formed and has a density step at the "wall" to satisfy the waveguide condition

$$\Delta n_e[cm^{-3}] = 1.13 \times 10^{20}/r_L^2[\mu m]. \tag{8}$$

Assuming an evacuated channel "drilled" in the uniform plasma with $\lambda_p$ defined by Eq.(5) we come to the conclusion that the radius of the vacuum channel is equal to $\lambda_p/3$. If we assume just a partially rarefied plasma channel with the 10% stepped-down density, $\Delta n_e = 0.1n_e$, then $r_L \approx \lambda_p$. These estimates define pragmatic limits for $r_L$.

Using Eqs. (3)-(7), we make a quick estimate for the maximum net gain. We assume $a=1$, the length of the channel $L_{ch} \approx L_{ph}$ and $\lambda_p/r_L \approx 2$. Then,

$$\Delta W^{max}[GeV] \approx 0.06P_L[TW]. \tag{9}$$

Another scaling law follows from the resonance condition and Eq. (7):

$$\tau_L^{min}[\lambda] \approx 20\sqrt{\Delta W}[GeV]. \tag{10}$$

For example, the 3 GeV LWFA requires $P_l \approx 50$ TW and $\tau_L \geq 1$ ps at $\lambda=10$ μm. The length of the accelerating channel, $L_{ch} \approx 1$ m. Such compact accelerator is attractive as a stand-alone device or as a construction block for a future staged linac of the TeV energy range.

For any application, such an accelerator shall provide a reasonable monochromaticity ($\Delta E / E \leq 1\%$) and an appreciably high charge ($N_e \geq 10^9$ electrons/bunch). The maximum number of the electrons per bunch is defined by the condition that the self-field of the bunch does not alter the plasma wakefield structure,

$$N_e \leq n_e \left( c/\omega_p \right)^3 = 4 \times 10^6 \lambda_p [\mu m]. \tag{11}$$

That means that the $N_e^{max}$ scales proportionally to $\lambda_p$ reaching $10^9$ at $\lambda_p \approx 250$ μm. A similar conclusion may be driven from the energy conservation condition. Indeed, as follows from Eqs.(9) and (10), with the increase of $\lambda_p$ the proportionally higher laser pulse energy is required to drive the LWFA in the linear regime. A higher laser energy permits to accelerate the proportionally bigger electron bunch charge.

We know that another important condition for the small energy spread requires the electron bunch to be much shorter than the plasma wake period, $\tau_b \ll \lambda_p/c$. This again favors bigger $\lambda_p$. At the other hand, there is an apparent downside to the $\lambda_p$ increase: drop in the acceleration gradient that, according to Eqs.(3) and (5), is

$$E_a^{max} [GV / m] \approx 10^3 / \lambda_p [\mu m]. \tag{12}$$

Thus, we see a certain tradeoff between two trends ($\lambda_p$ increase for better $\Delta E / E$ and $N_e$ but its reduction for higher $E_a$) that has to be resolved depending upon the accelerator design goals.

The main cumulative performance characteristic of the collider is a luminosity,

$$\Lambda = \frac{N_e^2 f}{4\pi\sigma_\perp^2}, \tag{13}$$

where $f$ is the repetition rate of electron bunches (laser pulses), and $\sigma_\perp$ is the e-beam radius at the interaction point. The luminosity-targeted optimization of the 2 GeV LWFA design for two possible types of lasers is attempted in Table 1 that shows the proportionality between $N_e$ and $\lambda$ under the optimum conditions. According to Eq.(13), $\lambda$-proportional increase in $N_e$ allows $\lambda^2$-proportional reduction in $f$. Then, the average laser power calculated from

$$P_{av} = P_L \tau_L f \tag{14}$$

scales as $P_{av} \sim 1/\lambda$. For example, to meet the extremely luminosity requirements for the future $e^- e^+$ collider, $\Lambda = 10^{35}$ s$^{-1}$cm$^{-2}$ [13], the picosecond $CO_2$ laser of the 10 kW average power may be required. The equivalent 1-μm laser driver must be of the 100 kW power! Thus, $CO_2$ laser appears to be the only possible choice to drive the high-luminosity LWFA.

This statement is illustrated by the examples in columns 1 and 2 of Table 1. These examples, together with the generic analytical approach elaborated in this paper, show

**TABLE 1.**  Prospective comparative characteristics for multi-stage LWFA driven with 1-μm and 10-μm lasers

| PARAMETERS | Column number | | | |
| --- | --- | --- | --- | --- |
| | 1 | 2 | 3 | 4 |
| Wavelength, $\lambda$ [μm] | 10 | 1 | 10 | 1 |
| Pulse Energy [J] | 50 | 5 | 1.5 | 15 |
| Pulse length, $\tau_L$ [ps] | 1 | 0.1 | 0.3 | 0.3 |
| Peak power, $P_L$ [TW] | 50 | 50 | 5 | 50 |
| Focal spot radius, $r_L$ [μm] | 360 | 36 | 120 | 120 |
| Peak intensity, $I$ [W/cm$^2$] | $10^{16}$ | $10^{18}$ | $10^{16}$ | $10^{17}$ |
| Laser field, $E_L$ [TV/m] | 0.33 | 3.3 | 0.33 | 1 |
| Dimensionless laser strength, $a$ | 1 | 1 | 1 | 0.3 |
| Plasma density, $n_e$ [cm$^{-3}$] | $3\times10^{15}$ | $3\times10^{17}$ | $3\times10^{16}$ | $3\times10^{16}$ |
| Plasma wavelength, $\lambda_p$ [μm] | 600 | 60 | 200 | 200 |
| Critical laser power, $P_{cr}$ [TW] | 60 | 60 | 6 | 600 |
| Acceleration gradient, $E_a$ [GeV/m] | 2 | 20 | 6 | 0.8 |
| Free space interaction length, $\pi z_0$ [cm] | 10 | 1 | 1 | 10 |
| Phase detuning length, $L_{ph}$ [cm] | 220 | 22 | 7 | 700 |
| Assumed channel length, $L_{ch}$ [cm] | 125 | 12.5 | 5.5 | 125 |
| Energy gain (with channeling), $\Delta W_{ch}$ [GeV] | 2 | 2 | 0.2 | 0.8 |
| Electrons/bunch, $N_e$ | $3\times10^9$ | $3\times10^8$ | $10^9$ | $10^9$ |

that the ponderomotively strong long-wavelength laser tends to improve the LWFA performance (within the selected spectral range). To abate any concerns regarding the "arbitrary" choice of the input laser parameters, we evaluate two extra cases (columns 3 and 4). These examples illustrate that deviations from the optimum laser driver parameters result in deterioration of the accelerator performance.

For example, in column 3 we estimate the impact of the laser pulse reduction below the minimum calculated by Eq.(10). We see that in order to stay in the linear regime we need to reduce the laser peak power in proportion $P_L \sim \tau_L^2$. Because of the higher plasma density, the acceleration gradient improves inversely proportional to $\tau_L$. However, due to the significant reduction in $L_{th}$, the net gain per stage drops dramatically. Together with the reduction in $N_e$, this leads to less attractive accelerator performance (consider the seven times higher number of stages and two times higher total plug-in power consumption by the laser system).

In other example, illustrated by column 4, we increase the laser pulse duration above the calculated optimum (minimum) value. However, due to the recognized solid state laser technology limitations, we can not keep up with the $\tau_L$ increase by upscaling $P_L \sim \tau_L^2$ that is required to maintain the acceleration gradient at the same level. In spite of the longer interaction length, the net acceleration per stage is also down. The number of the acceleration stages and the total accelerator length is up. The requirements to the average laser power will be relaxed, but still beyond any realistic expectations for the solid state laser technology.

Returning to discussion on the electron monochromaticity attainable with the $CO_2$ laser driven LWFA, we address recent simulations for electron bunch acceleration by plasma wake excited in the matched parabolic plasma channel [14]. The following input parameters have been used in simulations: $CO_2$ laser with $\tau_L = 1$ ps, $P_L = 50$ TW, $a = 0.71$; plasma channel parameters $k_p r_L = 3.8$, $k_p R_{ch} = 14.3$, $\lambda_p = 800$ μm, where $\lambda_p$ and $k_p$ are defined at the axis of the plasma channel that has the profile $n_e(r) = n_e^0 \left( 1 + \left( r / R_{ch} \right)^2 \right)$.

The injected electron bunch has the initial energy 50 MeV, energy spread 0.2%, geometric emittance 3 mm.mrad, the 30 μm (100 fs) length, and 80-160 μm radius at the entrance to the plasma channel. These conditions can be practically realized using a combination of the prospective ps-TW $CO_2$ laser, linac with the photocathode RF gun, and a plasma channel produced by the high-current capillary discharge in vacuum.

Simulations [14] revealed a need for further reduction in $\tau_b$ in order to maintain the electron energy spread below 10%. Accordingly, a feasible monoenergetic LWFA scheme (see Fig.4) includes: the LWFA bunch compressor to 20-15 fs and the energy gain stage. Result of 3D test particle simulations for these stages are shown correspondingly in Fig.5 and 6. In this configuration, 1,7 GeV energy gain with 2% of the energy spread is predicted.

# STELLA - FAR-FIELD STAGED ELECTRON LASER ACCELERATOR EXPERIMENT

The problem of the monoenergetic electron acceleration looks even more demanding for far-field accelerator schemes driven directly by the laser electromagnetic field with the wavelength 10-100 times shorter than the plasma wake wavelength. The important step towards a solution of this problem is the STELLA experiment at the ATF. In STELLA, the inverse free electron laser (IFEL) accelerator serves as a prebuncher and inverse Cherenkov accelerator (ICA) provides the main energy gain, similar to the two-stage LWFA discussed in Section 3.

IFEL is based on the accelerating action of a linearly polarized laser beam onto electrons having an oscillating trajectory inside the wiggler [15]. Transverse laser field has an electrical component along the local direction of the e-beam propagation. Electrical field produces an additional kick to the electrons in the direction of propagation if the laser field is in phase with wiggling. Electrons injected in the form of a long bunch will experience both acceleration and deceleration, however, in a very regular way as is shown in Fig. 7 [16].

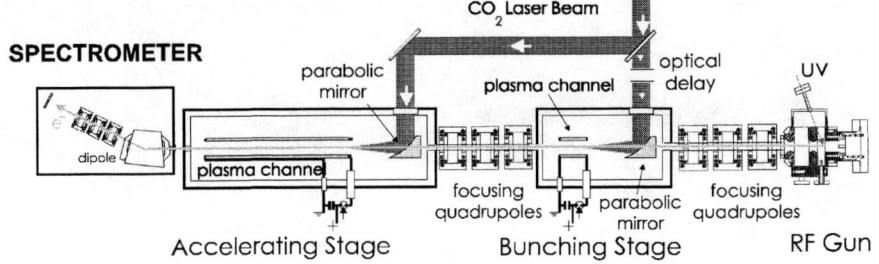

**FIGURE 4.** Principle diagram of the two-stage monochromatic LWFA

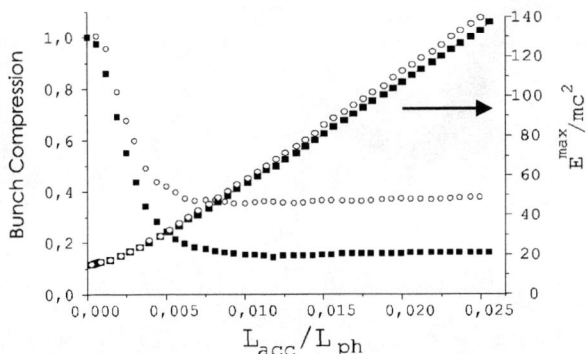

**FIGURE 5.** Energy modulation and bunch compression in the LWFA bunching stage; initial bunch length $\tau_{b0}$=100 fs; black boxes - $r_b$=50 μm, $\varepsilon_0$=0.6 mm.mrad; open circles - $r_{b0}$=100 μm, $\varepsilon_0$=0.3 mm.mrad

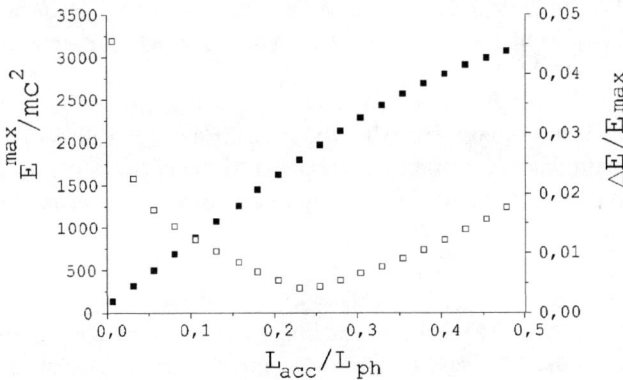

**FIGURE 6.** Net acceleration (filled boxes) and energy spread (open boxes) in the LWFA acceleration stage after prebuncher; initial electron bunch parameters: $\tau_{b0}$=100 fs; $r_{b0}$=50 μm, $\varepsilon_0$=0.6 mm.mrad [14]

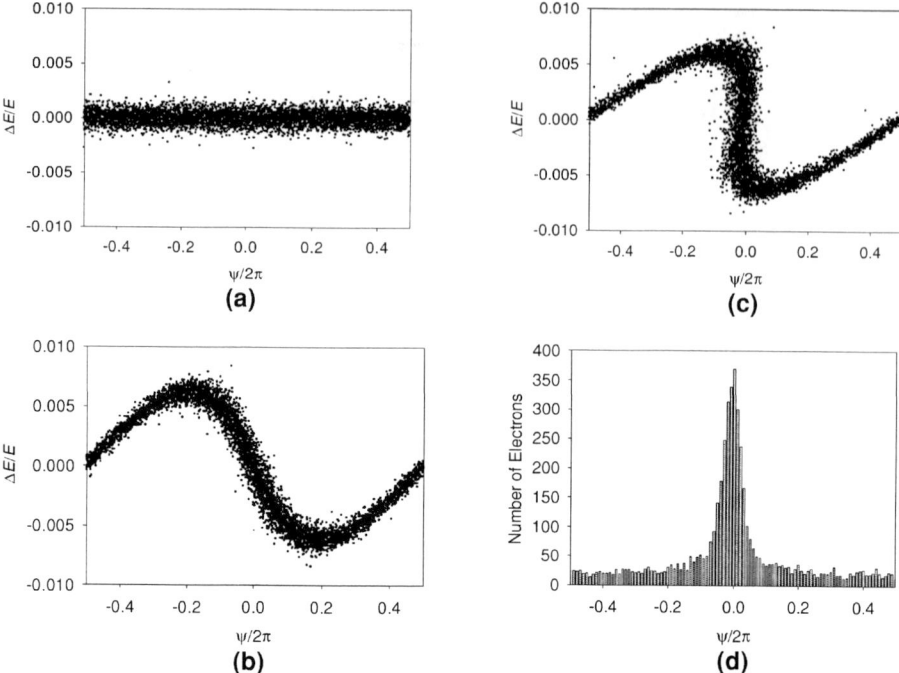

**FIGURE 7.** Simulations of electron beam bunching at $\Delta E/E=1.2\%$ energy modulation in IFEL wiggler: a) initial uniform energy distribution; b) energy modulation at the wiggler exit; c) energy distribution at the entrance to the ICA cell; d) longitudinal density distribution in which 50% of the electrons are bunched into FWHM=0.63 µm. [16]

Being allowed to propagate in a free space, faster electrons will catch up with slower electrons thus developing periodical micro-bunches exactly at the laser wavelength spacing (Fig.7 c-d). This micro-bunch train is injected into the inverse Cherenkov accelerator stage.

In the ICA scheme we start with a radially polarized laser beam [17,18]. Using an axicon, the laser beam is converged to the e-beam axis. Because of the wave front inclination, a longitudinal accelerating component of the electric field is developed. The length of the interaction region depends upon the axicon angle and the input beam diameter. The interaction cell is filled with gas to satisfy the Cherenkov phase matching condition,

$$\beta n \cos\theta = 1. \tag{15}$$

In the separate ATF ICA experiment, the used 10 ps e-bunch uniformly covered 300 $CO_2$ laser periods. As a result a diffuse electron energy modulation was observed on the spectrometer [19]. In order to make acceleration monochromatic, the electrons need to be injected precisely at the maxima of the accelerating field. That is what the IFEL prebuncher is for.

The principle diagram in Fig.8 illustrates the layout of the STELLA experiment [16]. The $CO_2$ laser beam is split into two beams. A relatively small amount of the laser power is sent into the IFEL in order to modulate the energy just enough so that the peak bunching occurs at the end of the 190 cm long drift region at the entrance to the ICA stage.

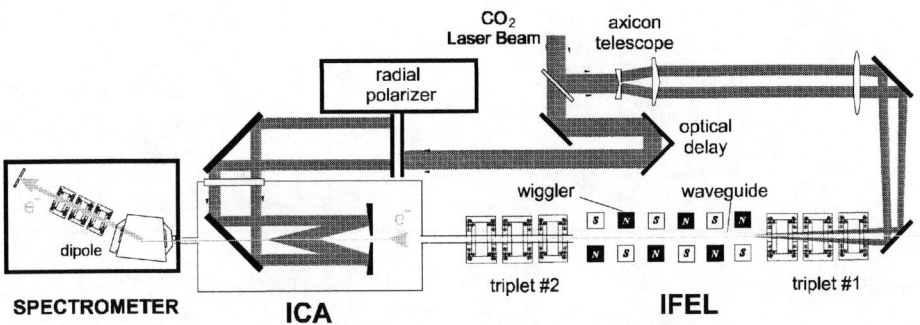

FIGURE 8. Principle diagram of STELLA experiment [17]

Another portion of the split $CO_2$ laser beam, after passing through a radial polarization converter, is introduced into the ICA gas cell and focused by an axicon mirror along the e-beam axis. To satisfy the phase matching Cherenkov condition, the ICA cell is filled with ~1.7 atm of hydrogen gas. The interaction cell is separated from the vacuum beamline by the 1-μm thick diamond windows transparent for the electron beam.

Let us reveal how the laser driver wavelength affects the STELLA performance. Self-interference of the axicon-focused radially polarized laser beam results in the $J_0$ Bessel distribution for the longitudinal (accelerating) field and $J_1$ distribution for the radial (focusing) component of the electric field. The amplitude of the accelerating field, $E_0(z)$, at the axicon axis is

$$E_0(z) = 2\pi\theta\sqrt{\frac{2zI(R)}{\varepsilon_0 c\lambda}} \qquad (16)$$

where $z = R/\theta$ and $I(R)$ is the laser intensity distribution at the axicon surface. The radial position of the first minimum of the $J_0$ Bessel distribution is observed at

$$r_{min} = 0.38\lambda/\theta. \qquad (17)$$

To ensure monochromatic acceleration, $r_{min}$ shall be set much bigger than the realistic size of the e-beam The results of Monte-Carlo computer simulations for the realistic low-emittance bunched ATF e-beam acceleration in the ICA cell induced by the 1 TW $CO_2$ laser pulse (see Fig.9) demonstrate the importance of the condition $r_{min} > r_b$.

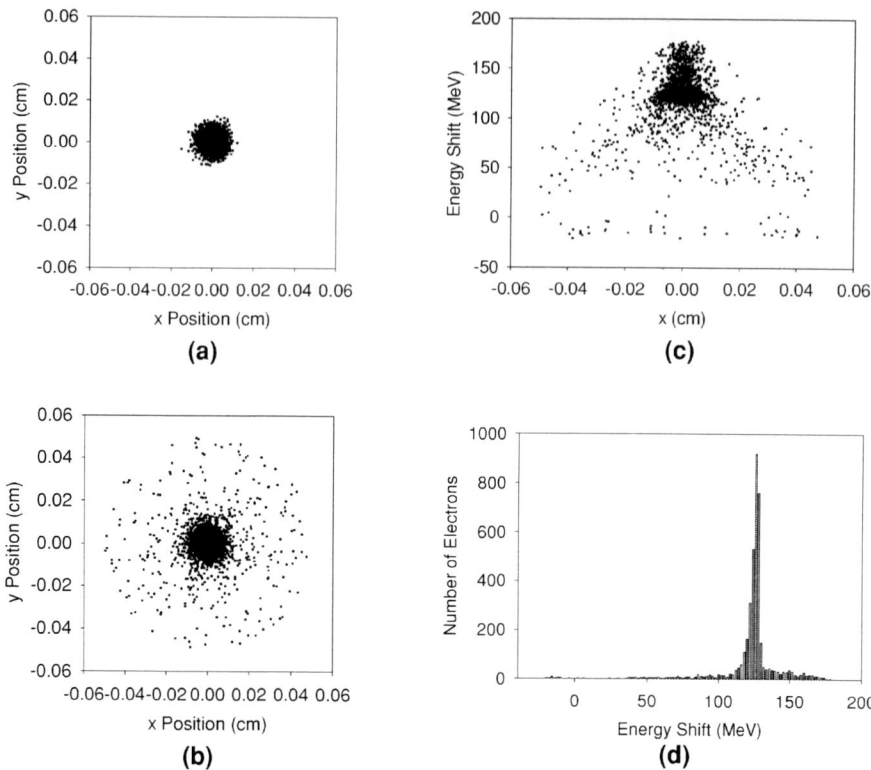

**FIGURE 9.** Simulations for electron bunch acceleration over 19.4-cm ICA interaction length at $P_L=1$ TW: a) e-beam spot size at axicon; b) e-beam spot size exiting ICA interaction cell; c) radial energy distribution; d) electron spectrum

Electrons farther away from the axis do not experience the peak laser field and, therefore, do not gain the maximum possible amount of energy. However, Fig. 9(c) shows that, due to the initially small electron beam radius and focusing in the laser field, most of the accelerated electrons are confined within the central portion of the beam. As a result, most of the electrons gain up to 125 MeV energy. The FWHM of the electron spectrum shown in Fig. 9(d) is 6 MeV and contains ~50% of all the electrons.

From Eq.(17) we see that the longer $\lambda$ permits to maintain the proper $r_{min}$ at the proportionally bigger angle $\theta$. Combining this condition with Eq.(16), we come to the conclusion that $E_z \sim \lambda^{1/2}$. Thus, the acceleration gradient achievable with the $CO_2$ laser is three times higher that with the solid state laser, due to stronger inclination of the laser wavefront to the e-beam propagation.

Another advantage of using the relatively long-wavelength $CO_2$ laser radiation for STELLA is due to the favorable proportion of the longitudinal microbanch dimension to the laser wavelength. Simulations show that the finite electron beam emittance and energy spread, together with the difference in the drift distance of the axial and off-axis electrons between the IFEL and ICA stages results in the microbunch smearing to the minimum 0.63 μm size (see Fig.7). Due to a jitter of the optical components in the laser beam transport to the IFEL and ICA, it is not conceivable to ensure phase synchronization between the microbunches and the accelerating laser field in the ICA stage to the accuracy better than 0.5 μm. All these and similar considerations add to the conclusion that the long-wavelength $CO_2$ laser is the only practical choice for the staged far-field laser accelerator, such as STELLA.

# GAMMA AND POSITRON SOURCES BY COMPTON SCATTERING OF $CO_2$ LASER BEAMS

By Compton backscattering of the laser photons from the relativistic electron beam, a high brightness x-ray and gamma photon beams can be created. By this process, $e^-e^+$ collider can be reconfigured into the lepton-gamma collider. It opens an opportunity to study a variety of interaction processes by colliding $e^-$, $e^+$ and $\gamma$ beams in any combination and at independently controlled polarization.

The expression for the maximum gamma photon energy for linear (single photon) Compton backscattering is

$$\hbar\omega_\gamma = \frac{x}{x+1}E_e,\qquad(18)$$

where $E_e$ is the electron energy, and $x = 4E_e\hbar\omega/m^2c^4$. At $x \gg 1$, the Compton photon energy approaches the electron energy, $\hbar\omega_\gamma \approx E_e$. For $CO_2$ laser, $x=1$ at $E_e=0.5$ TeV.

Another strong requirement to the laser wavelength is set by rescattering of gamma photons on the laser beam into pairs through the reaction $\gamma+\lambda \Rightarrow e^-+e^+$. This occurs when $\omega\omega_\gamma > m^2c^4/\hbar^2$. Based on this condition and using Eq.(18), the optimum laser wavelength is derived:

$$\lambda[\mu m]=4.2E_e[TeV].\qquad(19)$$

Then, for the 2.5 TeV collider the laser with $\lambda=10.5$ μm is the right choice.

For $\tau_L=1$ ps, probability of the $e^\pm \Rightarrow \gamma$ conversion,

$$\chi = \sigma_C E_L/\hbar\,\tau_L c^2,\qquad(20)$$

where $\sigma_C=1.9\times10^{-25}$ cm$^2$ is the Compton scattering cross-section, reaches unity at the laser pulse energy $E_L \approx 1$ J.

The polarization of electrons and positrons in the future linear colliders will play an important role for experimental verification of the standard model and for a search of new phenomena beyond the standard model. A prospective high-intensity polarized positron source [20] is based on production of the electron-positron pairs when Compton scattered gamma-quanta are stopped at the foil target. Polarization of the produced particles is easy controlled by the input laser beam. The recent observation of the first polarized positrons produced by this method proves the viability of this approach [21]. Capable to high average power and delivering ten times more photons than solid state lasers of the similar energy, picosecond $CO_2$ lasers become the optimum choice for this application as well. The requirements for the high peak flux and short pulse duration of the polarized gamma rays specify the high-brightness electron accelerator and the picosecond subterawatt $CO_2$ laser as essential components of the projected Compton source for the Japan Linear Collider (JLC). A preliminary proof-of-principle Compton scattering experiment using a ps-TW $CO_2$ laser is initiated at the BNL ATF.

The JLC project requires up to 1 kHz repetition rate of the delivered positron bunches. To obtain such a cumulative repetition rate, the plan is to deliver the electron and laser pulses in trains of 85 pulses with the 1.4 ns separation between the pulses and 0.1 s intervals between the trains. It will employ 40 $CO_2$ lasers of 10 ps pulse duration and ~ 1 kW of average power. This project may become the most massive apllication of $CO_2$ lasers in fundamental science, to date.

## CONCLUSIONS

In the previous sections, by examples of two laser accelerator schemes, STELLA and LWFA, we demonstrate potential advantages of the picosecond $CO_2$ lasers for high gradient monochromatic laser accelerators. This advantages are primarily due to the favorable scaling of the characteristic accelerator structures (e.g., the accelerating field period) with the laser wavelength, or the quadratic increase in the ponderomotive potential (important for LWFA).

Similarly, big amplitude of electron oscillation in the low-frequency $CO_2$ laser beam results in the $\lambda$-proportional intensity of the Compton (Thomson) scattered radiation. This feature will be utilized in the prospective polarized positron source for JLC.

The proposed schemes can be practically realized using a combination of the conventional electron beam injectors and the emerging ps-TW $CO_2$ laser technology. The first laser of this kind will be operational at the Brookhaven ATF in 1999.

## ACKNOWLEDGMENTS

The author wish to thank I. Ben-Zvi, W. Kimura, N. Andreev, T. Hirose, A. Tsunemi, and X.J. Wang for valuable input and discussions and all members of the multi-institutional team of scientists, engineers, and students who participate in design and preparation of the STELLA experiment. We hail the OPTOEL Co. (St. Petersburg, Russia) and the BNL staff who developed the first ps-TW $CO_2$ laser.

This work was supported by the U.S. Dept. Of Energy under the Contract DE-AC02-76CH00016.

# REFERENCES

1. Pogorelsky I.V., Ben-Zvi I., Skaritka J., Segalov Z., Babzien M., Kusche K., Meshkovsky I.K., Lekomtsev V.A., Dublov A.A., Boloshin Yu.A., Baranov G.A., 7th Workshop on Advanced Accelerator Concepts, October 12-18, 1996, Lake Tahoe, CA, *AIP Conference Proceedings* **398**, 937-950 (1997).

2. Pogorelsky I.V., Ben-Zvi I., Babzien M., Kusche K., Skaritka J., Meshkovsky I., Dublov A., Lekomtsev V., Pavlishin I., and Tsunemi A., "The first picosecond terawatt $CO_2$ laser at the Brookhaven Accelerator test facility", in *Proceedings of Conference LASERS '97*, New-Orleans, LA, December 15-19, 1997, STS Press, McLean, 861-867 (1998)

3. Pogorelsky I.V., Ben-Zvi I., Babzien M., Kusche K., Skaritka J., Meshkovsky I.K., Dublov A.A., Lekomtsev V.A., Pavlishin I.V., Boloshin Y.A., Deineko G.B., and Tsunemi A., " The first picosecond terawatt $CO_2$ laser", *to be published in Conference Proceedings of Laser Optics '98*, St. Petersburg, June 22-26, 1998

4. Pogorelsky I.V., van Steenbergen A., Fernow R.,. Kimura W.D, Bulanov S.V., 7th Workshop on Advanced Accelerator Concepts, October 12-18, 1996, Lake Tahoe, CA, *AIP Conference Proceedings* **398**, 923-936 (1997)

5. Madena A., Najmudin Z., Dangor A.E., Clayton C.E., Marsh K.A., Joshi C., Malka V., Darrow C.B., Danson C., Neely D., and Walsh F.N., *Nature*, **377**, 606-608 (1995)

6. Nakajima K., Fisher D., Kawakubo T., Nakanishi H., Ogata A., Kato Y., Kitagawa Y., Kodama R., Mima K., Shiraga H., Suzuki K., Yamakawa K., Zhang T., Sakawa Y., Shoji T., Nishida Y., Yugami, N. Downer M., and Tajima T., *Phys. Rev. Lett.*, **74**, 4428-4431 (1995)

7. Everett M., Lal A., Gordon D., Clayton C., Marsh K., Joshi C., *Nature*, **368**, 527-529 (1994)

8. Wang X.J., Qui X., and Ben-Zvi I., *Phys. Rev.* **E54**, R3121-R3124 (1996)

9. Biglov Z.A. and Gordienko V.M., *Current Problems in Laser Physics* **4**, Moscow, 1991

10. Corkum P.B., *IEEE J. Quant. Electron.* **QE-21**, 216-232 (1985)

11. Corkum P.B., private communication

12. Tajima T. and Dawson J.M., *Phys. Rev. Lett.* **43**, 267-270 (1979)

13. Downer M. and Siders C.W., *Proceedings of Advanced Accelerator Workshop, Lake Tahoe, CA, 1996*, AIP, **398**, 214-229 (1997); Xie M., Tajima T., Yokoya K., and Chattopadhyay S., *same Proceedings*, 233-242

14. Pogorelsky I.V., Andreev N.E., and Kuznetsov S.V., "Practical approach to monochromatic LWFA", in *these Proceedings*

15. Palmer R.B, *J. Appl. Phys.* **43**, 3014-3023 (1972)

16. Pogorelsky I.V., vanSteenbergen A., Gallardo J.C., Yakimenko V., Babzien M., Kusche K.P., Skaritka J., Kimura W.D., Quimby D.C., Robinson K.E., Gottschalk S.C., Pastwick L.J., Steinhauer L.C., Cline D.B., Liu Y., He P., Camino F., Ben-Zvi I., Fiorito R.B., Rule D. W., Pantell R.H., and Sandweiss J., *JAERI-Conf 98-004*, in *Proceedings of the 1st JAERI-Kansai International Workshop on Ulrashort-Pulse Ultrahigh-Power Lasers and Simulation for Laser-Plasma Interactions*, Kyoto, Japan, July 14-18, 1997, pp. 25-30 (1998)

17. Fontana J.R. and Pantell R.H., *J. Appl. Phys.* **54**, 4285-4288 (1983)

18. Kimura W.D., Kim G. H., Romea R. D., Steinhauer L. C., Pogorelsky I. V., Kusche K. P., Fernow R. C., Wang X., and Liu Y., *Phys. Rev. Lett.* **74**, 546-549 (1995)

19. Kimura W. D., Babzien M., Ben-Zvi I., Cline D. B., Fiorito R. B., Fontana J. R., Gallardo J. C., Gottschalk S.C., He P.., Kusche K. P., Liu Y., Pantell R. H., Pogorelsky I. V., Quimby D. C.,

Robinson K. E., Rule D. W., Sandweiss J., Skaritka J., van Steenbergen A., and Yakimenko V., "Design and Model Simulations of Inverse Cerenkov Acceleration Using Inverse Free Electron Laser Prebunching," in *Proceedings of the 1997 Particle Accelerator Conference*, May 12-16, 1997, Vancouver, BC, http://www.triumf.ca/pac97/papers/pdf/7V013.PDF

20. Okugi T., Kurihara Y., Chiba M., Endo A., Hamatsu R., Hirose T., Kumita T., Omori T., Takeuchi Y., and Yosioka M., *Jpn. J. Appl. Phys.*, **35**, 3677 (1996)

21. T. Hirose, "Generation of Polarized Positrons via Laser-Compton Scattering at the KEK Damping Ring", in *Proceedings of the First Asian Particle Accelerator Conf., APAC98*, March 23-27, 1998, KEK, Tsukuba, Japan

# New Final Focus Concepts at 5 TeV and Beyond[1]

## Frank Zimmermann

*Stanford Linear Accelerator Center*
*Stanford University*
*Stanford California 94309*

**Abstract.** At multi-TeV energies, the length of conventional beam-delivery systems becomes excessive, raising doubts about the value of a compact, high-gradient accelerator to future high-energy physics. In this paper, the reasons for the unfavorable length scaling are discussed, and alternative design concepts are described, for which final focus and collimation systems are orders of magnitude shorter and which produce higher luminosity at lower beam power than conventional approaches. These concepts include a sextupole-free final focus, linac energy-spread compensation, bunch combination and laser collimation. They are compatible with novel acceleration techniques, such as an active matrix linac. A consistent parameter set for a 5 TeV collider is presented.

## MOTIVATION

It is a striking feature of the proposed design for the 1-TeV Next Linear Collider (NLC) [1] that a third of its length (10 km) is occupied not by the linac, but by the beam delivery system, which consists of collimation section and final focus. For colliders at higher energies the beam delivery system could easily dwarf the linac. A second remarkable feature of the NLC design is that a large portion of the beam power is not converted into luminosity.

## FINAL FOCUS

### Conventional System

A conventional final-focus system consists of a final telescope, *i.e.*, a few strong focusing quadrupoles producing the small spot at the interaction point (IP), and an upstream chromatic correction section, with sextupole magnets placed at locations of large dispersion generated by bending magnets. Figure 1 shows a schematic of

---

[1] Work supported by the U.S. Department of Energy under contract DE-AC03-76SF00515.

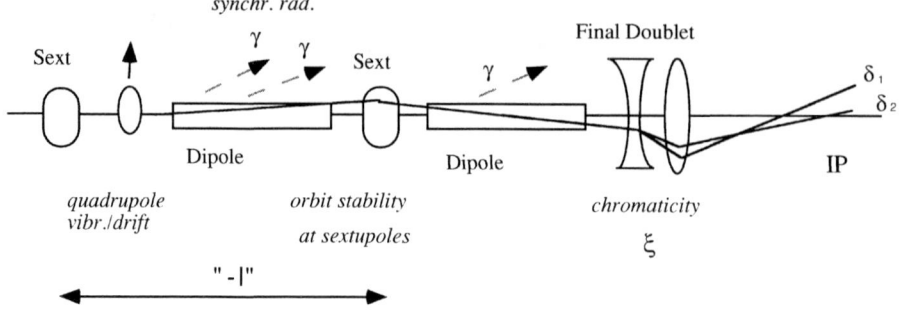

**FIGURE 1.** Schematic of a conventional final focus.

such a final focus, illustrating its main building blocks and indicating some of the physical processes which dilute the spot size at the interaction point (IP) and give rise to the unfavorable length scaling.

A characteristic property of the final focus is the chromaticity of the last quadrupoles, defined by $\xi = \int K\beta \sin^2 \phi \, ds$, where $K$ is the quadrupole strength and $\phi$ the betatron phase advance to the IP. The chromaticity $\xi$ describes the variation of the focal length with energy. In a conventional final focus, the large chromaticity of the final quadrupoles is compensated by sextupoles in the chromatic correction section. These sextupoles introduce a chromaticity of opposite sign such that particles of different initial energy are focused at the same point. The sextupoles are usually grouped in pairs, separated by an optical $-I$ transform, an arrangement which cancels geometric aberrations. However, this correction scheme is not effective for energy errors generated in the final focus itself: If an additional energy spread is introduced after the first sextupole, the chromaticity of the last quadrupoles is not fully compensated, the focal points for different energies will vary, and the interaction-point spot size will increase.

In the final focus, energy spread is primarily generated by synchrotron radiation in the bending magnets. The higher the beam energy, the weaker and longer the bending magnets of the chromatic correction section must be in order to confine this energy spread. The length $l$ of the final focus is then roughly proportional to the length of the bending magnets, which already in the NLC occupy more than half of the available space. The induced energy spread scales with beam energy $\gamma$ (energy in units of the rest mass), total bend angle $\theta_B$ and length $l$ as [2]

$$\Delta\delta_{\rm rms} \propto \frac{\gamma^{5/2}\theta_B^{3/2}}{l} \tag{1}$$

As discussed above, energy spread induced in and behind the chromatic correction section increases the IP spot size $\sigma^*$. The relative blow-up is given by $\Delta\sigma^*/\sigma^* = \xi\,\Delta\delta_{rms}$, which is added in quadrature to the unperturbed spot size. In a proper design, it is small:

$$\xi\,\Delta\delta_{rms} \ll 1. \tag{2}$$

Next, since the cross section for most reactions decreases inversely with the square of the energy, we assume that, to obtain reasonable reaction rates, the collider luminosity increases as $\gamma^2$. If the free length between the interaction point and the last quadrupole as well as the current and the normalized beam emittances are held constant, the chromaticity then increases in proportion to the energy:

$$\xi \propto \frac{1}{\beta^*} \propto \gamma. \tag{3}$$

In addition, the chromatic correction of the final focus for incoming energy errors can be expressed by

$$2\eta_S\,(k_S l_S)\beta_S \approx \xi, \tag{4}$$

where $\eta_S$ and $\beta_S$ denote the dispersion and the beta function at the sextupoles, the term $(k_S l_S)$ is the integrated sextupole strength, and the factor of 2 accounts for the two sextupoles of a pair. The $\xi$ is the quadrupole chromaticity defined above. The dispersion $\eta_S$ in Eq. (4) is proportional to bending angle and system length:

$$\eta_S \propto \theta_B l. \tag{5}$$

A further constraint arises from the orbit stability. Horizontal orbit changes at the second sextupole of a pair, caused by vibrations or position drifts of quadrupoles between the two sextupoles, shift the longitudinal location of the beam waist. This changes the beta function at the collision point, and thereby increases the IP spot size. The achievable orbit stability limits the product of integrated sextupole strength and beta function to

$$(k_S l_S)\,\beta_{S,y} \leq \frac{1}{\Delta x}, \tag{6}$$

where $\Delta x$ denotes the tolerance on the orbit motion. If we assume that the value of $\Delta x$ cannot be pushed much below the tolerances assumed in the NLC design [1], combining Eqs. (1)–(6) finally yields the scaling law [3]:

$$l \propto \gamma^2 \tag{7}$$

Thus, the length of a conventional final focus increases roughly as the 2nd power of the energy. Counting both sides of the IP, the length of the 1.5-TeV NLC final focus is 4 km. The final-focus length for a 5 TeV collider would approach 40 km.

# Compact Final Focus

The final focus can be made much more compact, if one omits chromatic correction in favor of energy-spread compensation in the linac. This eliminates the long bending magnets and strong sextupoles, as well as the associated tight alignment and stability tolerances on the sextupole orbit. The final focus then consists of quadrupoles only, and can be quite short (one or two hundred meters). Without chromatic correction the incoming beam energy spread must be small: $\delta_{rms} \leq 1/\xi \propto 1/\gamma$. Typical values of $\xi$ for 5 TeV require a relative energy spread smaller than $10^{-5}$. For comparison, the present rms energy spread at the end of the SLAC linac is about $8 \times 10^{-4}$.

One possibility to attain a smaller energy spread is by employing rf sections operated at harmonics of the fundamental [3]. The energy kick imparted by a linac with such a harmonic acceleration and single-bunch beam loading takes the form

$$V(t) = \sum_h V_h \cos(h\omega_1 t + \phi_h) - \int_0^t dt' \, I(t')W_{\parallel}(t - t'), \qquad (8)$$

with $V_h, \phi_h$ the voltage and phase for the harmonic $h$, and $\omega_1$ the angular frequency for the fundamental. The function $W_{\parallel}$ describes the longitudinal wakefield of the linac and $I > 0$ the bunch current waveform. The total rf input energy per pulse for the harmonic sections relative to that for the fundamental mode rf system is $U_h/U_1 \approx 1/(\rho h^4)$, where $\rho$ is the fractional contribution to the loss factor from the harmonic sections. For $h = 10$ and $\rho = 10\%$, $U_h/U_1 \approx 10^{-3}$.

For the sake of definiteness let us consider a model wakefield, varying with time as $W_{\parallel} \propto t^{-1/2}$, and a flat-top current profile turning on at $t = 0$ and extending to $t = T$, with $\omega_1 T = 0.2$. The strength of the beam loading is characterized by $\hat{Q} = 1.5 \, k_l Q/V_1 \approx 1.3 \times 10^{-2}$, where $k_l = \int_{-\infty}^{\infty} dt \, I(t) \int_{-\infty}^t dt' \, I(t)W_{\parallel}(t - t')$ is the loss factor. In this case, adding a 2nd ($h = 10$) and a third frequency ($h \approx 30$), and optimizing five parameters: fundamental mode phase, harmonic phases and harmonic amplitudes, the energy spread can be reduced to $9 \times 10^{-6}$ excluding the front 5% of the beam. For this scheme, the pulse-to-pulse intensity fluctuation, $\Delta Q/Q$, must be less than $1/(\hat{Q}\xi) \approx 0.1\%$.

A second method for reducing the energy spread is tailoring the longitudinal bunch distribution [4]. The energy spread along the bunch is reduced to zero, if the bunch distribution $I(t)$ is a solution of

$$I(t) = \frac{\omega_1 V_1}{W_{\parallel}(0)} \sin(\omega_1 t + \phi_1) - \int_0^t \frac{I(t')\frac{dW_{\parallel}}{dt}(t - t')}{W_{\parallel}(0)} \, dt' \qquad (9)$$

which can be found numerically. The front of the bunch distribution ($t = 0$) must be sharp edged, with an initial value that depends on the phase $\phi_1$: $I(0) = \omega_1 V_1/(W_{\parallel}(0)) \sin \phi_1$. If the beam is generated by an rf photo cathode, its longitudinal distribution can be adjusted by manipulating the laser pulse shape.

# BUNCH COMBINATION

In addition to its extreme length, a second drawback of the conventional collider layout is that a large portion of the beam power is not converted into luminosity. Due to transverse and longitudinal wakefields in the linac, the beam charge is split into $n_b$ bunches, which are collided separately at a loss in luminosity by a factor of $n_b$. One can recover some of this luminosity by combining individual bunches into superbunches prior to the collision.

If, for example, the beam consists of bunches propagating in parallel channels, each of small energy spread, but slewed across an energy full width of about 10%— a natural configuration for an active matrix linac [5]—, this combination is easily accomplished. Making use of the energy variation the individual bunches can be combined in a half-chicane, as depicted in Fig. 2. The half chicane consists of two horizontal bending magnets, each of length $l_0$ and with opposite deflection angles $\pm\theta$ and bending radii $\pm\rho$. The condition for bunch combination is

$$\Delta x = l_0 \theta \, \Delta \delta, \tag{10}$$

where $\Delta x$ denotes the inter-channel distance, and $\Delta\delta$ the bunch-to-bunch energy difference.

The length of the half chicane is determined by synchrotron radiation, inducing an rms energy spread of [2] $\delta_{rms}^2 = 55 r_e \lambda_c/(12\sqrt{3})\gamma^5\theta^3/l_0^2$, with $r_e$ the classical electron radius and $\lambda_c$ the Compton wavelength. Since for a compact final focus there is no chromatic correction, this energy spread increases the IP spot size by interacting with the uncompensated final-focus chromaticity $\xi$. Requiring $\delta_{rms} \leq 1/\xi$, and using Eq. (10), the minimum half length of the half-chicane combiner is

$$l_0 \geq \gamma \left( C_\delta \xi^2 \left( \frac{\Delta x}{\Delta \delta} \right)^3 \right)^{1/5} \tag{11}$$

where $C_\delta = 55 r_e \lambda_c/(12\sqrt{3}) \approx 2.8 \times 10^{-27}$ m$^2$. If $\xi$ increases in proportion to $\gamma$, and the interchannel spacing $\Delta x$ decreases as $1/\gamma$ (assuming that the linac rf wavelength is decreased inversely proportional to the beam energy), the length $l_0$ increases as the 4/5th power of energy.

The multi-bunch combination in a half-chicane takes advantage of the different bunch energies, but this energy difference also implies that the optics for each bunch must be matched individually to obtain the same IP beta function. This can be achieved by means of a multi-passband final-focus optics and fine-matching using quadrupole magnets in the still separated beam lines.

# COLLIMATION

In general, the beam entering the beam delivery system is not of the ideal shape, but it can have a significant halo extending to large amplitudes, both transversely

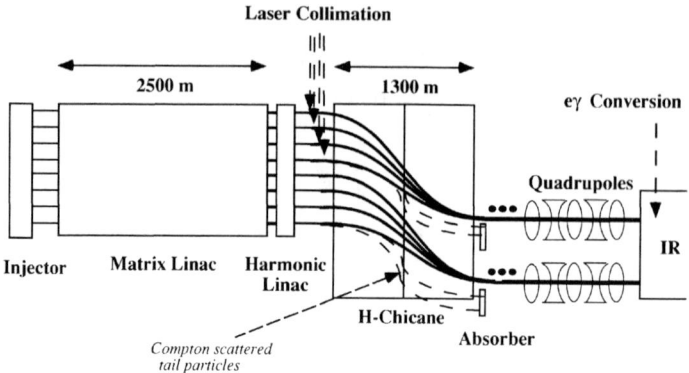

**FIGURE 2.** Schematic of a 5 TeV collider with linac energy compensation, laser collimation, bunch combination, sextupole-free final focus, and $e\gamma$ conversion.

and longitudinally. There are many sources of beam halo: (1) beam-gas Coulomb scattering, (2) beam-gas bremsstrahlung, (3) Compton scattering on thermal photons, (4) linac wakefields, (5) the source or the damping ring, respectively. The halo generation due to (1) will be reduced by a higher accelerating gradient, while the halo formation due to (2) and (3) scales with the length of the accelerator. The contributions of (4) and (5) to the halo size depend on many parameters; in a first, very rough approximation, if measured as a fraction of the bunch population, they could be considered as constant, independent of energy.

If halo particles hit the beam pipe or a magnet aperture close to the interaction point, they can cause unacceptable background. At the Stanford Linear Collider (SLC), collimation upstream of the final focus was found to be essential for smooth operation and for obtaining clean physics events in the detector. The same is expected to be true for future linear colliders.

## Conventional System

A conventional collimation system consists of a series of spoilers and absorbers, which serve two different functions: they remove particles from the beam halo to reduce the background in the detector, and they also protect downstream beamline elements against missteered or off-energy beam pulses. The spoilers increase the angular divergence of an incident beam so that the absorbers can withstand the impact of an entire bunch train [6]. A schematic is shown in Fig. 3.

An important requirement determining the system length is that the collimators have to survive the impact of a bunch train. This requires a minimum spot size, in

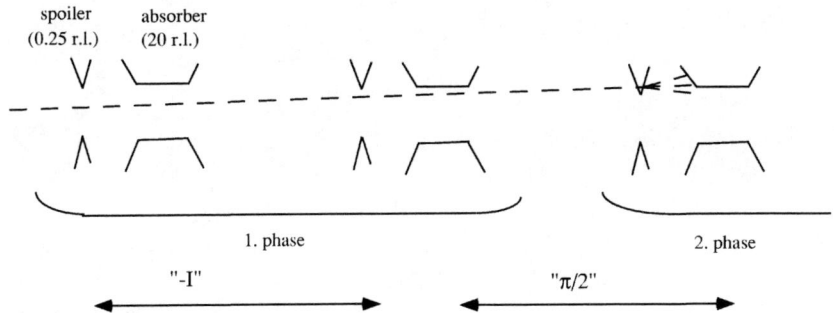

**FIGURE 3.** Schematic of a conventional collimation system, consisting of a series of spoilers and absorbers. The size of the spoilers and absorbers is approximately 1/4 and 20 radiation lengths, respectively.

order that the collimator surface does not fracture or that the collimator does not melt somewhere inside its volume. For the NLC parameters, fracture and melting conditions give rise to about the same spot-size limit (roughly $10^5/\mu m^2$ for a copper absorber at 500 GeV [1]). While the surface fracture does not depend on the beam energy, the melting limit does, since the energy of an electromagnetic shower deposited per unit length increases in proportion to the beam energy. Therefore, the beam area at the absorbers must increase linearly with energy. Since, in addition, the emittances decrease inversely proportional to the energy, the beta functions must increase not linearly but quadratically. Assuming that the system length $l$ scales in proportion to the maximum beta function at the absorbers, this results in a quadratic dependence, $l \propto \gamma^2$, *i.e.*, the same scaling as for the final focus. Counting both sides of the IP, the NLC collimation system is 5 km long. At 5 TeV the length of a conventional collimation system could be 50 km.

## Laser Collimation

The length of the collimation section can be substantially shortened, if, instead of a solid material, a laser beam is employed as a spoiler. Laser collimation would consist in Compton scattering of particles in the transverse beam tails on a high-power laser beam. At shorter wavelengths also pair production is possible, which would enhance the collimation efficiency. The Compton scattered particles lose a substantial amount of energy, and can be intercepted easily in a dispersive region downstream. Since the energy distribution of the scattered electrons extends over a wide range, the local density of the Compton-scattered part of the beam, which impinges on the energy interceptor, can be very low, without requiring large beta functions in this region. In addition, the laser beam cannot be 'destroyed', and, hence, the beta functions can be much smaller than for a conventional collimator.

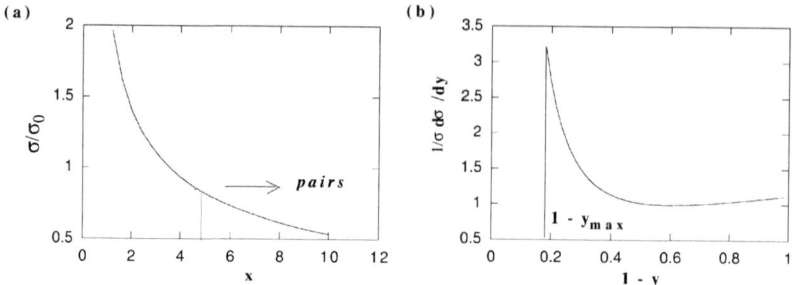

**(a)**

σ/σ₀

pairs

x

**(b)**

1/σ dσ /dy

1 - y_max

1 - y

**FIGURE 4.** (a) Total Compton cross section $\sigma/\sigma_0$ as a function of $x$; for the scattering of 10-$\mu$m photons off a 2.5-TeV electron beam $x = 4.5$. If $x \geq 4.8$, pair production is possible, a regime which should be avoided. (b) Energy spectrum of cross section, $1/\sigma \, d\sigma/dy$, as a function of the fractional scattered-electron energy $E/E_0 = 1 - y = 1 - \hbar\omega/E_0$ for $x = 4.5$, and assuming unpolarized beams. The energy spectrum of the scattered electrons extends from 0 to $1/(1 + x)$. The maximum photon energy after scattering is $\hbar\omega_{max} = y_{max}E_0 = x/(1 + x) \, E_0$. The cross sections are from Ref. [7].

Indeed, it is advantageous to have a small beta function at the laser spoiler, since this reduces the area to be covered by the laser, and, thereby, also the laser energy needed for efficient collimation. The small beta functions at the spoiler and large energy spread of the scattered particles suggest that a laser based collimation system could be very short.

The total Compton cross section [7] for an unpolarized photon beam with a few-micron wavelength, scattering off a multi-TeV electron beam, is of the order of the Thomson cross section $\sigma_0 = 2.5 \times 10^{-25}$ cm². It is illustrated in Fig. 4 (a) as a function of $x$. For head-on collisions of laser and particle beam $x$ is given by $x = 4E_0\hbar\omega_0/(m_e^2c^4)$, with $E_0$ the beam energy and $\hbar\omega_0$ the photon energy. The energy spectrum of the scattered electrons extends from almost unperturbed (100%) to a minimum value of about 20%, as shown in Fig. 4 (b).

To obtain a sharp collimation boundary and to collimate both sides of the beam at once, it is best to use not the fundamental but a higher-order laser mode. The electric field of an arbitrary TEM$_{lm}$ mode is given by [8] $|\hat{E}_{l,m}(x, y, z)| = E_0 \, w_0/w(z) \, H_l\left(\sqrt{2}x/w(z)\right) H_m\left(\sqrt{2}y/w(z)\right) \exp\left(-(x^2 + y^2)/w^2(z)\right)$ where $H_l$ is the Hermite polynomial of order $l$, and $w$ is defined by $w^2(z) = w_0^2 \, (1 + z^2/Z_R^2)$ with $w_0 = (Z_R\lambda/\pi)^{1/2}$, with $Z_R$ the Rayleigh length, $\lambda$ the laser wavelength, and $z$ the longitudinal distance from the laser waist. To be specific, let us consider the TEM$_{10}$ mode, for which

$$|\hat{E}_{1,0}| = \sqrt{8} \, E_0 \frac{w_0 x}{w^2(z)} \, \exp\left(-\frac{x^2 + y^2}{w^2(z)}\right). \tag{12}$$

The field amplitude $E_0$ is related to the energy $A_{\text{laser}}$ and the total length $\tau_l$ of

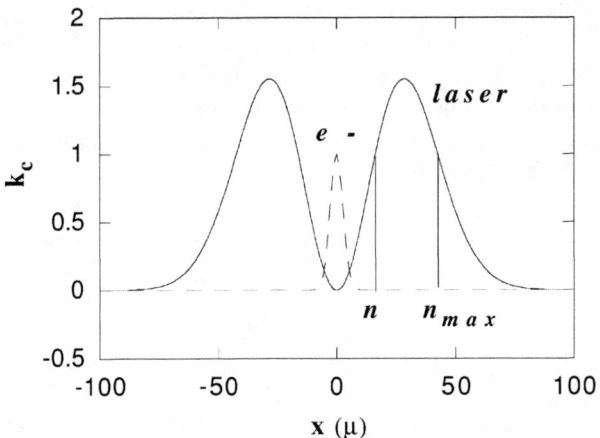

**FIGURE 5.** Conversion efficiency parameter $k_c$ for Compton scattering of a 5-J 10-$\mu$m TEM$_{10}$-mode laser beam off a 5-TeV electron beam as a function of horizontal position. The beam shape is also indicated by a dashed line. Other parameters are given in Table 2.

the laser pulse via $A_{\text{laser}} = \pi\epsilon_0|E_0|^2\tau_l w_0^2$. The conversion efficiency $K_c$, *i.e.*, the probability that an electron (or positron) in the beam will Compton scatter, is given by $K_c = 1 - \exp(-k_c)$, where

$$k_c \approx \sigma\frac{dN_\gamma^2}{dx\,dy} = \sigma\frac{\epsilon_0|\hat{E}_{1,0}|^2}{\hbar\omega}\tau_l = \sigma\frac{A_{\text{laser}}}{\hbar\omega w_0^2\pi}\left(\frac{|\hat{E}_{1,0}|^2}{E_0^2}\right). \tag{13}$$

Via $|\hat{E}_{1,0}|^2$, it is a function of the transverse position. For a TEM$_{10}$ mode, the conversion efficiency is zero at the center of the laser beam, then increases quadratically, and falls off at large amplitudes like a Gaussian with an rms width $w(z)/2$. If the laser intensity is high enough, $k_c$ is equal to 1 at two different amplitudes. The smaller of these amplitudes may be considered as the effective collimation depth, the larger one as the maximum amplitude which is still collimated. We denote these two amplitudes, in units of the rms beam size, by $n$ and $n_{max}$, respectively. The position dependence of $k_c$ is illustrated in Fig. 5.

Even if the particle beam is of purely Gaussian shape and is well-centered in the node of the laser field, there is an unavoidable fraction of particles that is lost due to Compton scattering. For a collimation depth $n$ this fraction is

$$\frac{\Delta N_b}{N_b} \approx \frac{1}{n^2} \tag{14}$$

where $N_b$ is the bunch population. For example, if we collimate at $35\sigma_x$ ($n \approx 35$) less than $10^{-3}$ of the particles in the centered Gaussian beam core are scattered, while at $6\sigma_x$ the scattered fraction would be 3% (per plane).

If the laser collimation is situated at the end of the linac, the scattered tail particles, which are off energy, can be intercepted downstream of the bunch-combining half chicane (see Fig. 2), where the dispersion is nonzero, $\eta = \theta l_0$. The density of the scattered particles must stay below the melting limit of the absorbing material: $N_b/(\sigma_x \sigma_y) \leq n_{\text{limit}}/\gamma$, where $n_{\text{limit}} \approx 10^{12}/\mu m^2$. The horizontal size of the Compton-scattered beam at the absorber is determined by the rms energy spread of the scattered particles, $\delta_C$ ($\delta_C \approx 0.1$), and by the dispersion, $\eta$, as $\sigma_x \approx \eta \delta_C$. The vertical size follows from the angular spread of the scattered particles, $y'_{\text{rms}} \approx 1/\gamma$, and the length of the combiner: $\sigma_y \approx 2l_0/\gamma$. Combining these equations with the constraint on the energy spread induced by synchrotron radiation, $\delta_{rms} \leq 1/\xi$, we derive a lower limit on the combiner half length $l_0$ which guarantees survival of the absorber, even if the entire electron beam is missteered and Compton scattered:

$$l_0 \geq \gamma^{11/8} \, N_b^{3/8} \, \xi^{1/4} \left( C_\delta \frac{1}{\delta_C^3 n_{\text{limit}}^3} \right)^{1/8} \approx 10^{-12} \, [\text{m}] \, \gamma^{11/8} \, N_b^{3/8} \, \xi^{1/4} \tag{15}$$

For the 5-TeV parameters discussed below, this limit is a factor 6 shorter than the minimum combiner length of Eq. (11). So it is automatically fulfilled. If on the other hand the laser is located behind the combiner and off-energy scattered particles are to be absorbed in a dedicated full chicane downstream, the total length of that chicane would be 4.4 $l_0$ with $l_0$ as given in Eq. (15).

## 5-TEV COLLIDER

As an illustration, we now consider the beam delivery system for a 2.5-TeV parallel-beam accelerator operating at W-band (91 GHz) [5]. The linac consists of a primary energy storage line running parallel to the beam axes, secondary lines running roughly orthogonal to the beam axes, and a switch coupling between the lines every 1/3 of a wavelength, all as depicted in Fig. 6. The temporally coincident beams propagate in parallel channels, spatially separated by $\Delta x = 1.4$ mm.

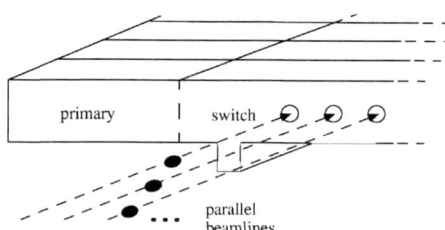

**FIGURE 6.** The linac employs a 'primary' storage line, a switch and a series of secondary transmission lines [5].

We assume that the two colliding beams consist of 50 bunches with a charge of 60 pC each, and with transverse emittances of $\gamma \epsilon_{x,y} \approx 100$ nm, as could be produced by an advanced rf gun. The rms bunch length is chosen as about 10 $\mu$m. These and other beam parameters are compiled in Table 1.

A relative energy spread across each bunch smaller than $10^{-5}$ is achieved either by harmonic acceleration, or bunch shaping, or a combination thereof. For example, harmonic acceleration could employ the 10th and 30th harmonic of the fundamental frequency, where the 10th harmonic (0.91 THz) might be provided by a 100-m matrix rf section with a gradient of 200 MeV/m. The $h = 30$ (2.7 THz) section would correspond to a 1-m plasma linac at the linac exit, with a plasma density of $n_e \approx 10^{17}$ cm$^{-3}$ and a 'modest' accelerating gradient of about 3 GV/m.

Behind the linac, groups of 10 bunches are combined into superbunches, for higher luminosity. Assuming a 1% energy difference $\Delta\delta$ between adjacent channels, the minimum length of the bunch-combining half chicane, Eq. (11), is $2l_0 \geq 1300$ m.

The laser collimation can be performed at two different locations. The first possibility is to place the laser at the end of the linac and to intercept the Compton scattered electrons after the 1.3-km long bunch combiner, where the dispersion is nonzero. The second possibility is to install the laser behind the combiner, in which case an additional bending section downstream is required, for example a 500-m long full chicane, where a collimator can intercept the scattered off-energy particles. The second option would simplify the laser system, by reducing the number of laser pulses, but it would increase the system length. Sample laser and beam parameters, applicable for either option, are listed in Table 2. A $CO_2$ laser fulfilling all the requirements is considered within the reach of the $CO_2$ laser technology [9].

The final-focus optics must have a multiple-energy passband, individually matched for each accelerating channel. Fig. 7 shows 3 matched beta functions over the last 80 m prior to the IP, spanning a total energy range of 10%, with initial optical functions (on the right) identical to the FODO lattice at the end of the linac. In Fig. 7 a series of final-focus quadrupoles were adjusted to obtain the same IP beta function for each energy. For perfect matching the optics can be fine-tuned using quadrupoles in the linac, where the bunches are still separated.

For the assumed emittances, an IP free length of $l^* = 2$ m and a maximum final quadrupole strength of $K \approx 1$ m$^{-2}$, the effect of synchrotron radiation in the final two quadrupoles (Oide effect) [10] limits the IP spot size in both planes to about 1.7 nm. Monte Carlo simulations show that, for this spot size, the ideal luminosity is reduced by about 20%. This factor increases rapidly, when the IP beta functions are lowered further.

Synchrotron radiation in the other final-focus quadrupoles is not an issue: At a beam energy of 2.5 TeV, each electron radiates about $dN_\gamma/ds \approx 5/(3\sqrt{2})\alpha\gamma K\sigma_{x,y} \approx 6\times10^{-3}[\text{m}^{1/2}] \, K\sqrt{\beta_{x,y}}$ photons per unit length, with $K$ the quadrupole gradient (in m$^{-2}$), $\sigma_{x,y}$ the rms beam size and $\beta_{x,y}$ the beta function (both in m), and assuming emittances of $\gamma\epsilon_{x,y} \approx 100$ nm. For $K \approx 0.5$ m$^{-2}$, $\beta_{x,y} \approx 10$ km, and a length of 60 m, this amounts to $N_\gamma \approx 18$ photons per electron, a sizable number, but the critical relative photon energy is only $\delta_c \approx 1.5\lambda_c K\sigma_{x,y}\gamma^2 \approx 2\times10^{-8}[\text{m}^{3/2}] \, K\sqrt{\beta_{x,y}} \approx 10^{-6}$,

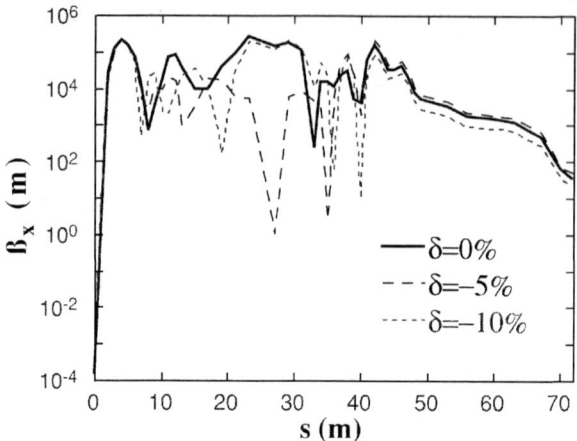

**FIGURE 7.** Final-focus beta functions for 10% different beam energies; on the right are linac-like FODO cells with $\beta \approx 20$ m, on the left the IP with $\beta = 150$ $\mu$m.

a factor of 10 smaller than the final-focus energy bandwidth.

Finally, during the collision, particles emit synchrotron radiation in the field of the opposing beam. The strength of this 'beamstrahlung' is characterized by the parameter $\Upsilon$, which is proportional to the average critical energy, $\Upsilon \approx 5\gamma r_e^2 N_b/(6\alpha\sigma_z(\sigma_x + \sigma_y))$, where $\alpha \approx 1/137$ is the fine structure constant, and $\sigma_z$ the rms bunch length. For the parameters of Table 1, $\Upsilon \approx 500$, implying an enormous energy loss per electron, a large number of photons per electron, and coherent pair production as the dominant background source. Two possible remedies are: $\gamma\gamma$ collisions and charge compensation.

## $\gamma\gamma$ Collisions

Photon-photon collisions are a very attractive option for a 5-TeV high-luminosity collider, because of two reasons: (1) there is no luminosity degradation and no background due to beamstrahlung, and (2) unpolarized low-current electron beams with the required emittances in both tranverse planes may be produced by advanced photocathode rf guns.

Photon-photon collisions are realized by converting the 2.5-TeV electrons into high-energetic photons via Compton scattering on a high-power laser beam [11]. The laser parameters are very similar to those assumed for collimation. The optimum laser wavelength with regard to conversion efficiency and photon energy spectrum depends on the beam energy as [7] $\lambda = 4.2$ $E_0$[TeV] $\mu$m. At a beam energy $E_0$ of 2.5 TeV, the required wavelength is about 10 $\mu$m; this corresponds to a $CO_2$ laser. After conversion the photon beam diverges as $1/\gamma$. This limits the maximum distance $b$ between conversion point and collision point (CP) to $b \leq \sigma_{x,y}^*/(2\gamma) \sim 5$ mm, where $\sigma_{x,y}^*$ denotes the electron-beam spot size without Compton scattering.

To obtain a reasonable conversion efficiency of 65% or higher, the laser energy flux per pulse must exceed $I\tau_l \geq \hbar\omega/\sigma \sim 10^5$ Ws/cm$^2$, where $I$ denotes the intensity (in W/cm$^2$), $\tau_l$ the laser pulse length, and $\sigma$ the Compton cross section.

If the laser density is too high, multiphoton processes occur, and, at the same time, the maximum photon energy acquired in a single photon process is reduced [7,12]. For this reason, the laser intensity $I$ for a 10 $\mu$m wavelength should be smaller than $I \leq 10^{16}$ W/cm$^2$. Combined with the above limit on $(I\tau_l)$ this requires a total laser pulse length of $\tau_l \geq 10$ ps (or $c\tau_l \geq 3$ mm).

The Rayleigh length should not be much shorter than the pulse length, e.g., $R_L \approx 1$ mm. The transverse laser extent $w_0 \approx (\lambda Z_R/\pi)^{1/2} \approx 56$ $\mu$m then determines the energy of the laser pulse: $A \approx I\tau_l w_0^2 \pi/2 \approx 5$ J. The divergence of the laser light, $\alpha_\gamma \approx \lambda/(2\pi w_0)$, and a typical damage threshold for mirror materials of 1 J/cm$^2$, yield the minimum focal distance of the last mirror: $F \geq (A_{\text{laser}} 2\pi w_0^2/\lambda^2/(1\text{J/cm}^2))^{1/2} \approx 30$ cm. Finally, the F-number '$FN$', defined as the ratio of focal length and incoming laser-beam diameter, relates the wavelength and the spot size at the focus: $w_0 \approx 2.4\lambda\, FN$. For our example, $FN$ is 2, a realistic value. Parameters relevant to $e\gamma$ conversion are summarized in Table 2.

## Charge Compensation

An alternative to photon-photon collisions is the suppression of beamstrahlung via charge compensation [13]. Here electron and positron bunches are combined into neutral bunches, prior to the collision. Since the net charge is greatly diminished (ideally there is none), the electromagetic fields and, hence, the beamstrahlung can be reduced by orders of magnitude. However, if the two oppositely charged bunches are initially offset with respect to one another, a charge-separation instability can develop [13,14]. For a single collision point, this effect was analyzed both analytically and by a computer simulation [15]. For the moderate collision strength considered here, with disruption parameters $D_{x,y} = 2N_b r_e \sigma_z/(\gamma\sigma_{x,y}^*(\sigma_x^* + \sigma_y^*)) \approx 9$, this instability is not a problem.

## CONCLUSIONS

Extrapolation of present-day final-focus and collimation systems to higher energies results in excessive site length and beam power. In this report, we have outlined an alternative design concept, which provides for a much shorter system length and for higher luminosity at lower beam power than the conventional approach. This concept was illustrated by means of a sample design for a 5-TeV $\gamma\gamma$ collider, whose parameters are summarized in Table 1. The proposed collider achieves a $\gamma\gamma$ luminosity of $1.5 \times 10^{34}$ cm$^{-2}$s$^{-1}$ for an average beam power as low as

**TABLE 1.** Parameters for a 5-TeV $\gamma\gamma$ collider.

| variable | symbol | value |
|---|---|---|
| beam energy | $E_0$ | 2.5 TeV |
| particles per superbunch at CP | $N_b^*$ | $3.8 \times 10^9$ |
| number of superbunches | $n_b^*$ | 5 |
| number of linac bunches | $n_b$ | 50 |
| charge per linac bunch | $Q$ | 60 pC |
| repetition frequency | $f_{rep}$ | 120 Hz |
| average beam power (per side) | $P$ | 0.9 MW |
| rms linac energy spread | $\delta_{rms}$ | $10^{-5}$ |
| rms bunch length | $\sigma_z$ | 10 $\mu$m |
| transverse emittance | $\gamma\epsilon_{x,y}$ | 100 nm |
| IP spot size w/o Oide effect | $\sigma_{x,y}^*$ | 1.7 nm |
| IP beta function | $\beta_{x,y}^*$ | 150 $\mu$m |
| $\gamma\gamma$ luminosity w. 65% conversion efficiency | $L_{\gamma\gamma}$ | $1.5 \times 10^{34}$ cm$^{-2}$s$^{-1}$ |

**TABLE 2.** Parameters for laser collimation and $e\gamma$ conversion.

| variable | symbol | value (collimation) | value ($e\gamma$ conversion) |
|---|---|---|---|
| laser energy / pulse | $A_{\text{laser}}$ | 5 J | 5 J |
| laser wavelength | $\lambda$ | 10 $\mu$m | 10 $\mu$m |
| Compton scattering parameter | $x$ | 4.5 | 4.5 |
| laser pulse length | $\tau_l$ | 2 ps | 10 ps |
| laser mode TEM | $(l, m)$ | (1,0) | (0,0) |
| laser parameter | $w_0$ | 40 $\mu$m | 56 $\mu$m |
| laser Rayleigh length | $Z_R$ | 500 $\mu$m | 1 mm |
| laser intensity | $I$ | $10^{17}$ W/cm$^2$ | $10^{16}$ W/cm$^2$ |
| min. focal distance of last mirror | $F$ | 16 cm | 30 cm |
| number of photons / pulse | $N_\gamma$ | $2.7 \times 10^{20}$ | $2.7 \times 10^{20}$ |
| rms beam size at laser IP | $\sigma_{x,y}$ | 2.7 $\mu$m | N/A |
| beta function at laser IP | $\beta_{x,y}$ | 365 m | N/A |
| collimation depth | $n_{x,y}$ | 6 | N/A |
| collimation limit | $n_{max,\ x,y}$ | 16 | N/A |
| fraction of 'core' scattered | $\Delta N_b/N_b$ | 6% | N/A |
| rms beam size at $e\gamma$ CP | $\sigma_{x,y}$ | N/A | 60 nm |
| distance between $e\gamma$ CP and IP | $b$ | N/A | 5 mm |
| conversion efficiency | $K_c$ | N/A | 65% |

0.9 MW. A beam power of 6 MW would yield a luminosity of $10^{35}$ cm$^{-2}$s$^{-1}$. The total collider length is less than 10 km.

## ACKNOWLEDGEMENTS

I am grateful to David Whittum for many of the ideas, and to John Irwin, Tor Raubenheimer and Dick Helm for sharing their insights into linear-collider final-focus and collimation systems. I thank Theo Kotseroglou for confirming the feasibility of the laser parameters.

## REFERENCES

1. "Zeroth Order Design Report for the Next Linear Collider," *SLAC-Report* **474** (1996).
2. M. Sands, "The Physics of Electron Storage Rings," *SLAC*-**121** (1979).
3. F. Zimmermann and D.H. Whittum, "Final-Focus System and Collision Schemes for a 5-TeV W-Band Linear Collider," in *Proc. of 2nd Int. Workshop on $e^-e^-$ Interactions at TeV Energies, Santa Cruz 1997*, and *SLAC-PUB*-**7741** (1998).
4. G.A. Loew and J.W. Wang, "Minimizing the Energy Spread with a Single Bunch by Shaping its Charge Distribution," *IEEE Trans. Nucl. Sc.*, **32**, no. 5 (1985).
5. D. Whittum and S. Tantawi, "Active Millimeter-Wave Accelerator with Parallel Beams," subm. to *Phys. Rev. Lett.* (1998).
6. H. DeStaebler, D. Walz, and J. Irwin, in Ref. [1].
7. V. Telnov, "Principles of Photon Colliders", *Nucl. Instr. Methods A* **355**, 3 (1995).
8. A. Yariv, "Quantum Electronics", Wiley (1976).
9. I. Pogorelsky, "Terawatt Picosecond $CO_2$ Laser Technology for Future TeV Colliders", submitted to *Laser and Particle Beams, BNL*-**64281** (1997).
10. K. Hirata, K. Oide, B. Zotter, "Synchrotron Radiation Limit of the Luminosity in TeV Linear Colliders," *Phys. Lett. B224*, **437** (1989).
11. I. Ginzburg, G. Kotkin, V. Serbo, V. Telnov, "Colliding $\gamma e$ and $\gamma\gamma$ Beams based on the Single-Pass $e^-e^-$ Colliders (VLEPP type), *Nucl. Instr. Methods A* **205**, 47 (1983).
12. K.J. Kim it et al., "A Second Interaction Region for Gamma-Gamma, Gamma-Electron and Electron-Electron Collisions for NLC", in Ref. [1] and *LBNL*-**38985** (1986).
13. V.E. Balakin and N.A. Solyak, *Proc. XIII Int. Conf. on High Energy Accelerators*, Novosibirsk, USSR, p. 151 (1986).
14. J.A. Rosenzweig, B. Autin and P. Chen, *Proc. 1989 Lake Arrowhead Workshop on Advanced Accelerator Concepts*, p. 324 (1989).
15. D.H. Whittum and R.H. Siemann, "Neutral Beam Collisions at 5 TeV," in *Proc. of IEEE PAC97 Vancouver* (to be published).

# The Laser Driven Particle Accelerator Project: Theory and Experiment

T. Plettner*, J.E. Spencer[†], Y.C. Huang[††], R.L Byer*, R.H. Siemann[†], and T.I. Smith[‡]

*Department of Applied Physic, and [‡]Hansen Experimental Physics Laboratory, Stanford University, Stanford, CA 94305,
[†]Stanford Linear Accelerator Center, Stanford, CA 94309,
[††]Department of Atomic Science, National Tsing-Hua University,Taiwan 30043

**Abstract**. A proof of principle experiment for laser driven electron acceleration from crossed laser beams in a dielectric loaded vacuum is being carried out at Stanford University. We seek to measure a maximum energy gain of about 250 keV for a 30-35 MeV electron beam in one accelerator cell. We use laser pulses of a few picoseconds of duration from a regenerative Ti:sapphire laser amplifier at a wavelength of 800 nm in a laser-electron interaction distance of ~ 1 mm.

## MOTIVATION

The rapid development of laser technology has made it possible to produce sub-picosecond pulses with fields in excess of $10^{10}$ V/m at wavelengths of the order of 1 μm with commercial table top lasers. At these wavelengths and pulse durations dielectric materials can tolerate very intense fields. A 1 psec pulse of 1053 nm light on a $SiO_2$ has been measured to have a damage threshold of ~ $2J/cm^2$, equivalent to about $10^{10}$ V/m (1). Fields of this magnitude could produce an average gradient of about 1 GeV/m in a suitably designed dielectric based accelerator structure.

## THE EXPERIMENT

### The site

The experiment is carried out at the SCA-FEL center at Stanford University because this facility is capable of providing a low energy spread beam (50-100 keV). It has an energy of 35 MeV and consists of 1 psec bunches that contain ~$10^8$ electrons. A high

CP472, *Advanced Accelerator Concepts: Eighth Workshop,*
edited by W. Lawson, C. Bellamy, and D. Brosius

peak power Ti:sapphire laser system is also available there. The experiment is located downstream from an FEL. In order to avoid a large background from unwanted electron bunches the experiment is placed in a separate line. A fast deflector located in the main line selects the bunch that is timed with the laser. Fig. 1 shows the basic layout of the experiment at the SCA-FEL center.

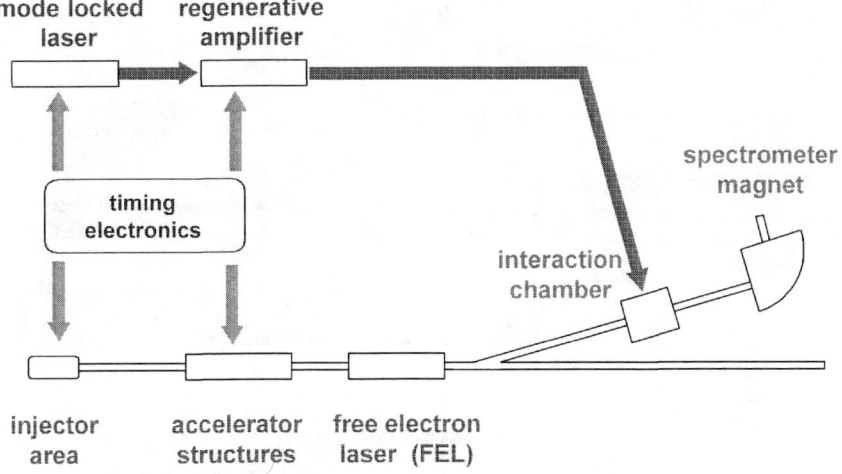

**FIGURE 1**. Diagram of the experiment

The accelerator cell is located inside the interaction chamber and any change in energy produced in the cell is measured with a high resolution spectrometer magnet located downstream.

## The energy spectrometer and some aspects of the beam

The energy spectrometer is a dipole sector magnet designed to bend the beam by 90 degrees in a radius of ~1/2 m. Its first order resolving power $R$ has been determined to be

$$R = \sqrt{\left\langle \left(\frac{P}{\Delta P}\right)^2 \right\rangle} = \frac{\eta_x}{\sigma_x} = \frac{1.12 \quad m}{30 \quad \mu m} = 4 \cdot 10^4 \tag{1}$$

where $\eta_x$ is the dispersion and $\sigma_x$ is the monochromatic spot size at the focal plane. At 30 MeV this would correspond to an energy resolution of 750 eV, but due to the much larger initial energy spread of the beam the effective energy resolution is the initial energy spread of the beam itself: about 50 keV. The FEL has a detrimental effect on the energy spread of the beam. Fig. 2 shows this effect clearly. The deflector has to be timed such that it selects an electron bunch that passed through the wiggler when no FEL action was present in order to obtain a narrow energy spread pulse for the experiment.

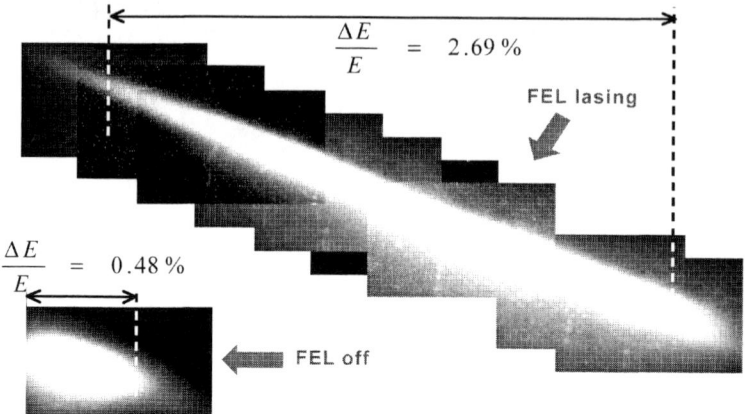

**FIGURE 2.** Effect of the FEL action on the electron beam

## The accelerator cell

The accelerator cell consists of a set of four fused silica prisms with a dielectric high reflecting coating on their surface. One pair of prisms is used to couple the pair of laser beams into the cell and the other pair of prisms takes the laser beams out. The gap between the prisms in each pair is the slit that allows the electron beam to transverse the cell. Fig. 3 shows a perspective view of the cell.

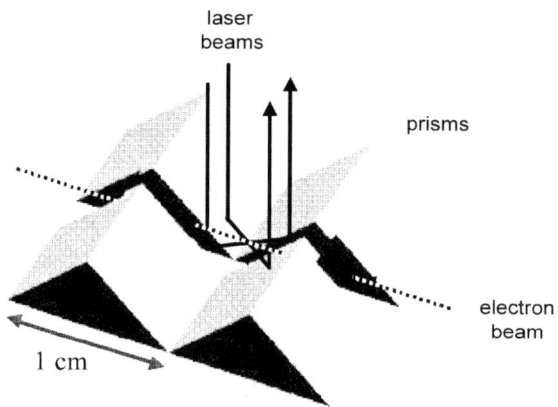

**FIGURE 3.** The accelerator cell

# ANALYSIS

A pair of properly phased crossed laser beams yields an interference pattern that at certain regions produces an electric field oriented half way between the crossing laser beams. For small angles of crossing between the laser beams this effective longitudinal field travels at a phase velocity which is close enough to the speed of light so that a beam of particles with a large $\gamma$ traveling in that direction sees a reasonably long slippage distance[1]. This acceleration scheme has been studied in the past (2,3) and possible accelerator structure designs have been proposed (4,5). The analytical expression for the longitudinal electric field $E_z$ seen by a particle travelling at the speed of light is given by

$$E_z(z,t) = -\frac{2E_0 \sin\theta}{\left(1+\hat{z}^2 \cos^2\theta\right)^{3/2}} \cdot \exp\left[-\frac{(z/\theta_d)^2 \sin^2\theta}{\left(1+\hat{z}^2 \cos^2\theta\right)}\right] \cdot \cos\psi_t$$

$$\psi_t = k \cdot z \cdot \cos\theta - \omega \cdot t + \frac{\hat{z}^3 \cos^3\theta \cdot \tan^2\theta}{\theta_d^2 \left(1+\hat{z}^2 \cos^2\theta\right)} - 2\tan^{-1}(\hat{z}\cos\theta) + \phi_0$$

(2)

where $\theta$ is the angle of crossing of the laser beam, $\theta_d$ the diffraction angle, $\omega$ the frequency of the laser, $\hat{z}$ the position (normalized to the Rayleigh range) of the particle down the axis of propagation, $k$ the wave vector and $\phi_0$ an arbitrary initial phase of the laser field. A diagram showing the electric fields and the corresponding $E_z$ field down the axis of propagation of the particles is shown in Fig. 4. Because of the phase velocity mismatch between the laser field and the particle $E_z$ shows an oscillatory behavior.

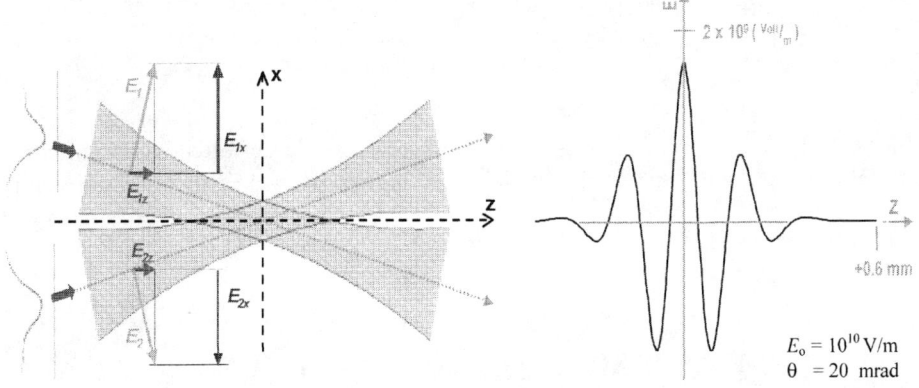

$E_0 = 10^{10}$ V/m
$\theta = 20$ mrad

**FIGURE 4.** Illustration of the $E_z$ field generated from a pair of gaussian beams.

---

[1] the distance required for a particle beam to see a phase shift of $\pi$ in the phase of the electric field

By the Lawson-Woodward theorem (6) the particle gains no energy if it interacts with the laser field linearly over an infinite distance. Therefore to obtain a nonzero energy gain the particle-laser interaction distance has to be made finite by the use of an accelerator cell with suitable optical components to couple the light in and out of the path of the electron beam in a short enough distance. The energy gain in one cell of inner length $a$ is given by

$$U(a) = \int_{-a/2}^{+a/2} E_z(z) \cdot dz \qquad (3)$$

From equation (2) the energy gain depends on the length of the cell $a$, the angle of crossing of the laser beams $\theta$, their relative phase, their polarization and their intensity. Fig. 5 shows the effect of the length of the cell and the angle of crossing between the laser beams.

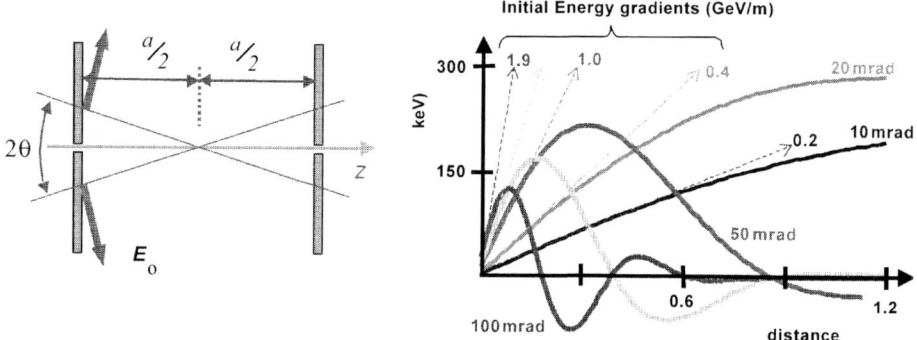

**FIGURE 5.** Schematic diagram of an accelerator cell and curves for Energy gain versus size of accelerator cell for various laser beam crossing angles

At steeper angles of crossing the laser beams create a larger effective longitudinal field larger and hence the initial gradient is higher. However the phase velocity mismatch between the laser and the particle beams is bigger and therefore slippage distance is decreased. This explains the large initial gradient and oscillatory behavior for the 100 mrad curve when compared to the 10 mrad curve in Fig. 5.

## EXPERIMENTAL CONSIDERATIONS

There are several non-ideal experimental conditions that affect the outcome of the experiment, and have to be taken into account in order to make the correct predictions for the acceleration. These factors are:

- low energy electron beam with a γ of about 70
- significant initial energy spread
- the electron beam is not optically bunched
- ultra short laser pulses comparable to the e-beam bunch length
- effects caused by the slit of the accelerator cell

### finite γ and initial energy spread

In Fig. 5 an infinite γ was assumed. For a low energy electron beam the slippage distance is reduced further. The effect is shown in Fig. 6. For an electron beam with an energy of ~35 MeV, laser beams crossing at 20 mrad and a cell 1 mm long the energy gain is reduced by about 30% when compared to that of a beam with infinite γ.

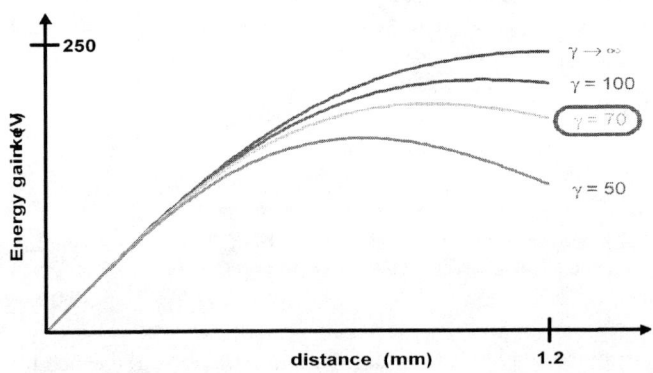

**FIGURE 6.** Effect of finite γ on the energy. The circled value corresponds to the approximate energy of the electron beam used in this experiment

### Effect of initial energy spread and of no optical bunching

The previous graphs were obtained from single particle calculations that assumed the optimum phase of the laser beams for acceleration. The electron beam is not optically bunched and therefore it will see all the phases of the optical field. Hence an energy distribution showing accelerated as well as decelerated electrons is expected. Fig. 7 shows the predicted energy profiles at the exit of the accelerator cell for different input energy spreads of the electron beam for a time independent optical field. Assuming an equal distribution of mono energetic electrons with energy E over all phases of the laser field of constant laser intensity an energy distribution $I(U-E)$ for the electron beam can be calculated. For a particular initial energy spread $S(E)$ of the beam the expected histogram $H(E)$ is the convolution of $S(E)$ and $I(E)$.

$$H(E) = \int_{0}^{\infty} S(E') \cdot I(E - E') \cdot dE' \tag{4}$$

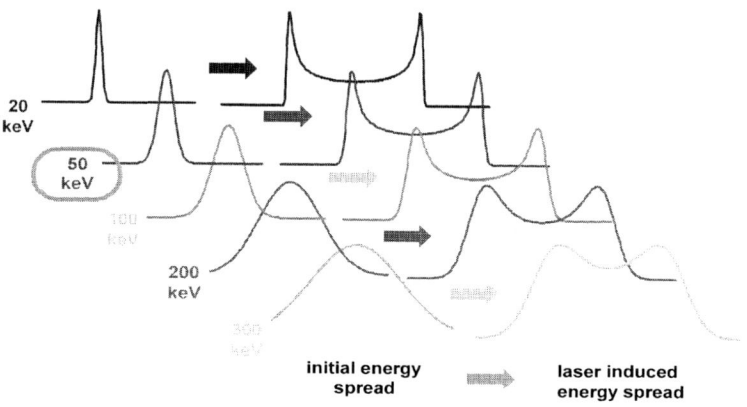

**FIGURE 7.** Energy profile of the e-beam at the exit of the accelerator cell. Energy spreads for single electron beam bunches below 50 keV have been observed.

The laser beam pulses have a time duration comparable to the e-beam bunches, and as Fig. 8 illustrates the energy profile is strongly dependent on the relative time duration between the e-beam and the laser beam.

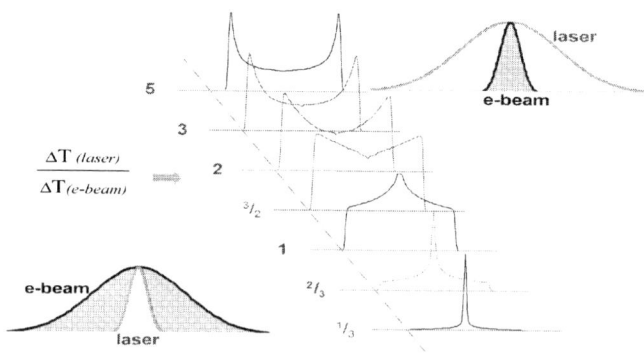

**FIGURE 8.** Effect of a finite laser pulse on the energy profile of the e-beam.

# Effect of the slit of the accelerator cell

In order to prevent the electron beam from travelling through matter a slit has to be built into the accelerator cell. A slit of a few microns is already larger than the wavelength of the laser field and hence diffraction effects due to the slit have to be considered. At the entrance and the exit of the accelerator cell some light leaks out of the cell. The leaking field at the entrance decreases the acceleration field inside the accelerator cell, while that at the exit removes some energy from the electrons. The net effect is a reduction of the electron energy gain. Slit effects have been studied in a previous paper (7). A 4-$\mu$m slit is expected to reduce the energy gain by 17% and a 6$\mu$m slit by 40%. Hence very small slit widths are preferred in order to reduce the leakage field.

When the electrons traverse the slits, the impact of the electron self-field upon the dielectric boundaries causes transition radiation at the cost of electron kinetic energy. A high impact field may also cause dielectric damage. In an attempt to obtain first-order estimation, we model the propagating electron self-field as an optical pulse and calculate the reflected radiation energy as the electron energy loss. The maximum line charge density subjected to a 10 GV/m damage field and radiation loss = 10% energy gain is ~ $10^7$ electrons/mm for our design parameters.

# TIMING BETWEEN THE LASER AND THE ELECTRON BEAM

Relative timing between the laser beam and the electron beam to the picosecond level is essential for the success of the experiment. A timing monitor that produces an error signal that controls the timing of the laser is required. We are seeking a reliable method for timing that is non-destructive to the laser or the electron beam. Compton scattering is a very precise method for confirming spatial and temporal overlap between the two beams, however it cannot provide a signed error signal and hence is no useful timing monitor when the beams are not overlapped. Furthermore only a maximum of ~$10^3$ photons produced by the Compton can be expected in this experiment.

One simple coarse timing scheme involves the use of a photo detector and a separate electron beam detector. A 6 picosecond rise time photo detector that monitors the laser has been installed. For the electron beam we will test the performance of several different monitors for lowest timing jitter: a beam position monitor (BPM), a resonant RF cavity, transition radiation and a multi channel plate (MCP). In order to obtain a signed timing error signal an RF mixer multiplies the signals from the photo diode and from the e-beam monitor. Once the beams are timed with this coarse timing technique a blind search for the precise overlap of the two beams will be performed. To keep this range reasonably small we hope to find the scheme suitable to get a timing uncertainty below 100 psec. Figure 9 is a diagram for this method of coarse timing.

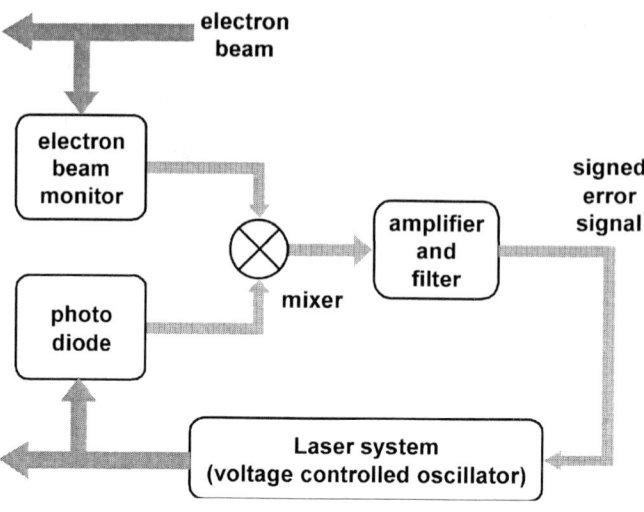

**FIGURE 9.** Block diagram of the coarse timing scheme

## SUMMARY

During the first year of the experiment the electron beam line and the laser beam transport line were installed. The energy spectrometer was tested and calibrated and furthermore we have achieved the ability to select single electron bunches and thus keep background noise levels low. Presently work is concentrated on finding a reliable scheme to synchronize the laser beam and the electron beam in order to overlap them in the accelerator cell. Once we have reliable method for achieving spatial and temporal overlap between the two beams we will attempt to confirm acceleration by measuring the change in the energy of the beam as a function of the various laser and accelerator cell parameters.

## ACKNOWLEDGEMENTS

The authors thank R. Route and M. Hennessy for their valuable help. This work is supported by the U.S. Department of Energy under the contract number DE-FG03-98ER41043.

# REFERENCES

1   Stuart, B.C. et al, Phys. Rev. Lett. **74**, 2248 (1995)
2   Haarland, C.M., *Optics Comm.* 114 (1995), pp 280-284
3   Sprangle, P., E. Esarey, J. Krall, A. Ting, *Optical Comm.* 124, 1996, pp. 69-73
4   Huang, Y.C., D. Zeng, W.M. Tulloch and R.L. Byer, Appl. Phys. Lett 68 (1996), pp 753-755
5   Huang, Y.C., R.L. Byer,, Appl. Phys. Lett. **69**, 2175 (1996).
6   Lawson, J.D., IEEE Trans. Nucl. Sci. **NS-26**, 4217 (1979). P.M. Woodward, J. IEE **93**, 1554 (1947)
7   Huang, Y.C. and R.L. Byer, to be published in *Rev. Sci. Ins.* July (1998).

# High Power Operation of the University of Maryland Coaxial Gyroklystron Experiment

W. Lawson, M. Arjona, M. Castle, B. Hogan, V. Granatstein, and M. Reiser

*Institute for Plasma Research and Electrical Engineering Department*
*University of Maryland, College Park, MD 20742*

**Abstract.** We report the experimental studies of high power amplification in a coaxial three-cavity X-band gyroklystron. A single-anode magnetron injection gun (MIG) is used to produce a 520 A beam of 470 keV electrons with an average ratio of perpendicular-to-parallel velocity of about one. The voltage flat top is nearly 2 μs. All cavities are designed to operate in the $TE_{011}$ coaxial mode near 8.6 GHz. The input cavity is driven by a 150 kW, 3 μs coaxial magnetron through a single slot in the radial wall. Peak powers of 75-85 MW are measured with a conversion efficiency of nearly 32% and a large signal gain of about 30 dB. This performance is in good agreement with simulations and represents approximately a tri-fold increase in the peak power capability of pulsed X-band gyroklystrons. We also report on the design of a three cavity second harmonic gyroklystron which is expected to produce 100 MW at 17.14 GHz. We close with a general discussion of scaling our designs to higher frequencies.

## INTRODUCTION

There has been considerable interest recently in the development of high power microwave and millimeter wave sources for a number of applications, including plasma heating and diagnostics, radars, materials processing, and particle accelerators. Gyrotron oscillators and amplifiers have been leading candidates for many of these applications. (1-5) For example, gyrotron oscillators have achieved unparalleled average powers at frequencies above 100 GHz with highly-overmoded cavities for electron-cyclotron resonance heating of plasmas. Efficient, single mode operation has been achieved by magnetic-field tuning and careful placement of the beam. In applications where narrowband amplifiers are required, such as drivers for linear colliders, gyroklystrons appear to be a leading candidate from X-band to W-band. (6-12) For these applications, only moderately overmoded cavities can be used effectively because the beam tunnels must be cut off to the operating mode in order to isolate the cavities from each other. Nonetheless, gyroklystrons have shown the potential to achieve much higher powers at high frequencies than conventional linear-beam tubes.

Sources for collider applications have received considerable attention from a number of groups in recent years. One design under consideration for a 500 GeV elec-

CP472, *Advanced Accelerator Concepts: Eighth Workshop,*
edited by W. Lawson, C. Bellamy, and D. Brosius

tron-positron linear collider requires the use of 4500 amplifiers, each with a peak power of 50 MW at an output frequency of 11.424 GHz. (13) Significant increases in energy beyond the 0.5 TeV energy level of this "Next Linear Collider" (NLC) will likely require amplifiers at higher frequencies and higher powers. In addition to gyroklystrons, possible drivers include klystrons, (14-17) magnicons, (18,19) traveling-wave tubes, (20) and free-electron lasers. (21-23) Considerable work on conventional klystrons has resulted in a tube that has met the requirements for the NLC and has made it the leading contender. At frequencies above X-Band, no tube has met all of the expected requirements.

Over the past decade, we have been investigating the suitability of a variety of high power gyrotron amplifier tubes as drivers for linear colliders. (11) Previous results include about 30 MW of peak power with efficiencies near 30% at both 9.87 and 19.7 GHz. The X-band results were achieved with a three-cavity tube operating at the fundamental cyclotron resonance and the K-band results were achieved with a two-cavity second harmonic tube. Efficiency was limited by instabilities in the beam tunnel preceding the input cavity and beam power was limited by the electron gun performance. In the past several years we have been upgrading the system with the goal of achieving peak powers above 100 MW. The latest results of this effort are described in this paper. First we discuss the experimental test facility, which includes the modulator, the electron gun, the three cavity microwave circuit, and the microwave diagnostics. Next we discuss the computer simulations. After that, we describe the experimental procedure and the amplification results. Then we discuss the design of the current second harmonic experiment. Finally, we describe our effort to scale our designs to higher frequencies.

## THE GYROKLYSTRON TEST BED

A schematic of the test bed is shown in Fig. 1. The voltage pulse is generated by a line-type modulator with 8 pulse-forming networks in parallel which feed a 22:1 transformer. The modulator can produce 2 μs flat-top pulses at up to 2 Hz with voltages and currents up to 500 kV and 800 A, respectively. A capacitive voltage divider and a current transformer are used to measure voltage and current pulses, respectively.

The electron gun is a single-anode magnetron injection gun (MIG). (11) According to EGUN simulations, (24) the single-anode MIG is capable of producing a 500 kV, 720 A beam with an average orbital-to-axial velocity ratio of $\alpha = 1.5$ and an RMS axial velocity spread of less than 10%. Below a current of 500 A, the simulations predict an axial velocity spread less than 6%. The 7.5 cm radius emitter is run temperature-limited with a current density of 9 A/cm$^2$ at 720 A. The emitter angle is large to promote laminar flow and minimize the dependence of velocity spread on current. During gun acceptance testing, the MIG produced about 670 A in a zero magnetic field, impedance-limited test, but the circuit was designed to operate at lower currents where the beam quality is better. The nominal beam parameters are given in Table 1 for the operating point where maximum amplification occurs. The beam voltage and current are measured quantities. All other quantities come from the EGUN

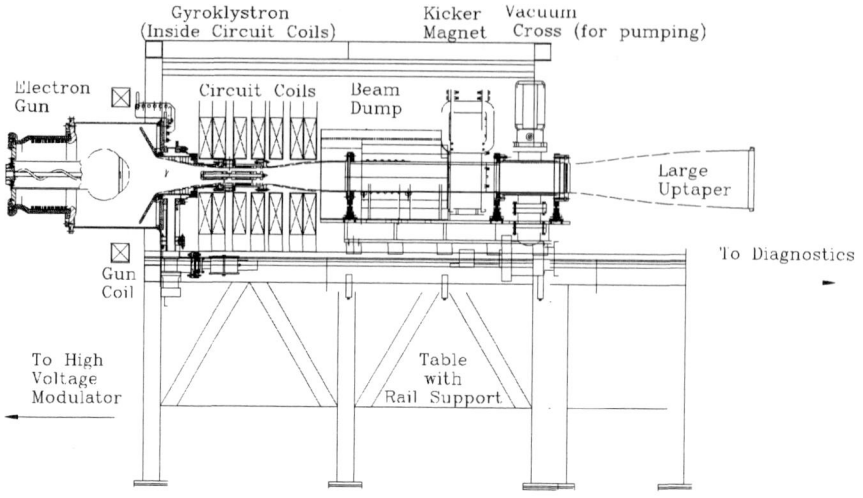

**FIGURE 1.** The University of Maryland gyroklystron facility.

simulation and are based on the MIG geometry and the magnetic field profile. The MIG is pumped from behind the anode by two large ion pumps which are conductance-limited by the vacuum connections. The electron gun has a gate valve which is used to keep the gun under vacuum during microwave tube changes; this is critical because reactivation time for the gun is several weeks due to its size and the pumping limitations. The gate valve region is covered during tube operation by the downtaper, which has a large bellows as part of the vacuum jacket to allow insertion of the downtaper into the gate valve. The downtaper has a large number of carbon-impregnated aluminosilicate (CIAS) rings on the outer wall which are designed to eliminate spurious oscillations. (25)

The magnetic field is produced by 8 water-cooled magnets which are driven by four separate power supplies to allow for flexible tapering of the axial field profile. One coil is centered over the emitter strip and the other seven are in the circuit region. About 400 amperes are required in each

**TABLE 1.** System parameters at the nominal operating point.

| Beam parameters | |
|---|---|
| Beam voltage (kV) | 470 |
| Beam current (A) | 505 |
| Average velocity ratio | 1.05 |
| Axial velocity spread (%) | 4.4 |
| Average beam radius (cm) | 2.38 |
| Beam thickness (cm) | 1.01 |
| Magnetic field parameters | |
| Cathode axial field (kG) | 0.59 |
| Input cavity field (kG) | 5.69 |
| Buncher cavity field (kG) | 5.38 |
| Output cavity field (kG) | 4.99 |
| Compression length (cm) | 48.2 |

circuit region coil to produce the nominal magnetic field of 5.3 kG. The axial fields at the centers of each cavity for the nominal operating point are given in Table 1. The field decreases by a total of 12%. throughout the circuit These values represent a resonance detuning of -3.5% in the input cavity, -8.8% in the buncher cavity, and -15.4% in the output cavity. This detuning level is similar to what was used to obtain optimum performance in our earlier work with circular cavities. The total magnetic compression from the center of the emitter to the start of the circuit is less than 10.

The input cavity power is supplied by a coaxial magnetron which produces 150 kW in 3 μs pulses. The power is injected through a double-disk window configuration into the input cavity via a single slot in the radial wall. Incident and reflected power are sampled with a dual directional coupler near the input window; in this paper we report the incident power as the drive power.

The output waveguide is also shown in Fig. 1. A non-linear taper brings the outer radius up to 6.35 cm, which is the radius of the beam dump and the output window. The inner conductor has a linear taper which transforms the coaxial system into a simple circular system in 1.3 cm. The water-cooled beam dump is about 35 cm long. A kicker magnet insures that no electrons hit the half-wavelength, alumina output window. The circuit and output waveguide are pumped through a set of 72 small holes at the end of the output waveguide. A second nonlinear uptaper, which brings the waveguide radius to 12.7 cm, must be placed between the output window and our anechoic chamber so that the peak electric field at the nominal output power level is below the threshold for breakdown in air.

Two main microwave diagnostics (not shown in Fig. 1) have been used to determine peak output power. The main diagnostic is the anechoic chamber, because of its

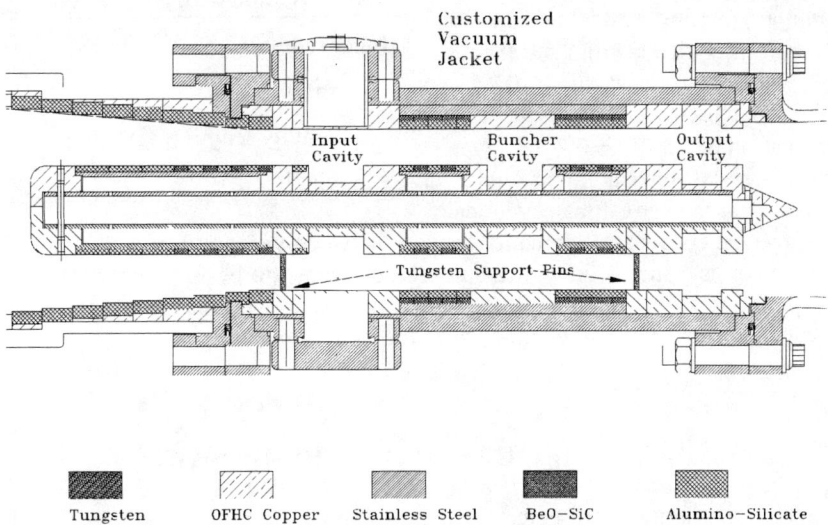

**FIGURE 2.** The three-cavity, first harmonic microwave tube schematic.

131

ability to detect and characterize spurious oscillations. The receiving antenna is simply made from an open-ended X-band waveguide which can be remotely swept about 1 m in the radial direction. This uptaper is filled with sulfur-hexaflouride. The receiving antenna is not in the far field, but *in situ* calibration of the anechoic chamber is used to determine a peak coupling of -30 ± 0.4 dB at the location of maximum radiation for a $TE_{01}$ mode. An additional attenuation with calibrated directional couplers and attenuators results in a net attenuation of 95 dB ± 0.7 dB. The power is measured with a crystal detector that is calibrated with a power meter. A small loop in the anechoic chamber is used as a secondary antenna to detect low frequency (< 6.56 GHz) spurious oscillations.

The combination of a directional coupler and a liquid calorimeter is used as a secondary diagnostic. The main arm of the directional coupler is a 6.35 cm radius circular guide and the secondary arm is a cut-down X-band waveguide. The coupler must be filled with sulfur-hexaflouride to avoid breakdown. Because the main arm is highly overmoded, the coupler has a relatively narrow bandwidth. Twenty holes are placed along a 1.05 m common wall in a pattern that maximizes the directivity of the $TE_{01}$ mode and selectivity over other modes near the operating frequency. The measured nominal coupling at 8.60 GHz of 62.3 dB ± 0.4 dB differs from the theoretical value by less than 0.2 dB. Additional hardware results in a total attenuation of 95 dB ± 0.7 dB. A calibrated crystal measurement yields a second peak power estimate. The calorimeter load consists of methanol flowing between two polyethylene cones. The average power is estimated by measuring the temperature rise with a thermopile and comparing it to *in situ* calibrations with a dc heater. The third peak power estimate is obtained by integrating the signal from the directional coupler and comparing it to the average power.

TABLE 2. First harmonic tube parameters.

| Drift tube parameters | |
|---|---|
| Inner radius  (cm) | 1.83 |
| Outer radius (cm) | 3.33 |
| Length (between I-B) (cm) | 5.18 |
| Length (between B-O) (cm) | 5.82 |
| Input cavity parameters | |
| Inner radius  (cm) | 1.10 |
| Outer radius (cm) | 3.33 |
| Length (cm) | 2.29 |
| Coupling slot size (cm) | 0.32 x 1.52 |
| Output cavity parameters | |
| Inner radius  (cm) | 1.01 |
| Outer radius (cm) | 3.59 |
| Main cavity length (cm) | 1.70 |
| Lip length (cm) | 0.90 |

## THEORETICAL TUBE DESIGN

The first-harmonic three-cavity microwave circuit is shown in Fig. 2 and the principle dimensions are given in Table 2. The primary function of the inner conductor is to force the drift tubes to be cutoff to the operating mode. The dimensions of the drift tube are given in Table 2; the 1.5 cm radial gap between the conductors yields a cutoff frequency of 10.12 GHz for the $TE_{01}$ mode. To minimize the size and weight of the inner conductor, it only extends a few centimeters into the downtaper and is rapidly terminated after the output cavity. Lossy ceramics are placed in the drift regions in order to suppress spurious modes. The rings on the inner conductor alternate between CIAS rings and nonporous rings comprised of 80% BeO - 20% SiC. Two layers of lossy ceramics are placed along the outer conductor in the drift regions. The outer layer is BeO-SiC and the inner layer is CIAS. The nominal attenuation in the drift region of the $TE_{11}$ mode in X-band, which was found in previous experiments to be the most prevalent spurious mode, is at least 4.7 dB/cm. The inner conductor is supported by two 2 mm diameter tungsten pins. The first is located just upstream from the input cavity and the other one is just before the output cavity. The pins intercept approximately 2.5% of the electron beam. To avoid overheating these support pins, the nominal repetition rate is typically 0.5 Hz. For the repetition rates that will be required by a linear collider, the microwave tube would have to be mounted vertically and the inner conductor would be supported in the beam dump.

The input cavity is formed via a step in the inner conductor and has a length that matches that of WR 90 waveguide. The cold-test values for the resonant frequency and Q are 8.566 ± 0.005 GHz and 65 ± 10, respectively. The cavity loss is divided roughly evenly between the diffractive loss of the coupling slot and the ohmic loss of the cavity. The latter loss is provided by a CIAS ring on the inner conductor which is placed adjacent to the cavity and is separated by a thin copper section that is adjusted to produce the required Q. The return loss on the coupling arm is more than -20 dB at the operating frequency.

The buncher cavity has identical dimensions as the input cavity for the metal components. However, the Q is generated entirely by the ohmic loss of the nearby CIAS ceramics. The measured cold resonant frequency and Q of this cavity are 8.563 ± 0.005 GHz and 75± 10, respectively.

The output cavity is defined by radial transitions on both conductor walls and the lip radii are equal to the drift tube radii. The quality factor is essentially equal to the diffractive Q, which is adjusted by changing the length of the coupling lip. The measured resonant frequency and Q of this cavity are 8.565 ± 0.005 GHz and 135 ± 10, respectively.

A partially self-consistent large-signal code is used to design the circuit and magnetic field configuration and to estimate the performance of the tube at the actual operating point. (26-27) A small-signal start-oscillation code is used to determine the stability properties of the cavities and set limits on the cavity Qs. (28) Theoretical and experimental results are plotted together in Fig. 3. The solid line shows the expected performance of the tube as a function of velocity spread when the velocity ratio is 1.5 (the design point). The simulation predicts a zero-spread efficiency of 43%, and an efficiency of 34% for 10% spread. For the 6% spread predicted for a 500 A beam, the simulated interaction efficiency is about 40%. All microwave cavities are expected to be stable at the design operating point for the quality factors discussed above.

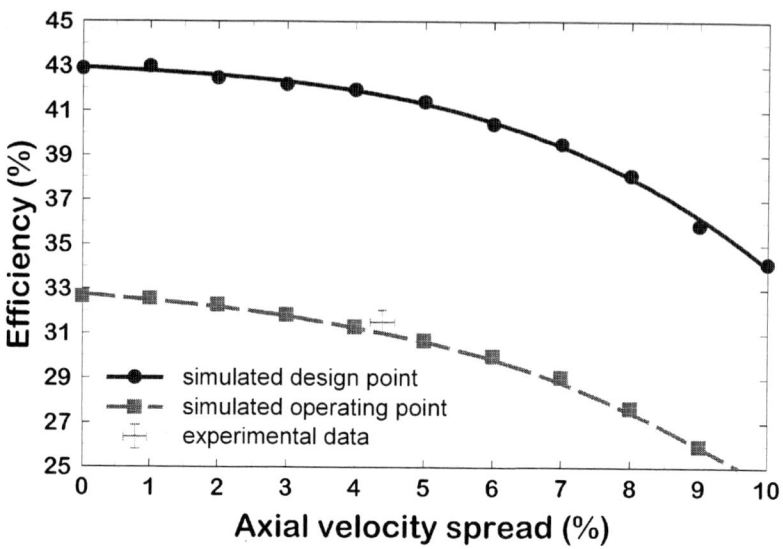

**FIGURE 3.** The simulated efficiency of the microwave circuit for the ideal beam parameters (solid line) and for the actual beam parameters (dashed line). The cross indicates the experimental measurement.

## EXPERIMENTAL RESULTS

To begin the search in parameter space for maximum amplified power, the beam voltage, beam current, and drive frequency are brought to their design values. The magnetic field profile is varied systematically and the average velocity ratio at a given point is increased by decreasing the current in the cathode magnet until the onset of an instability is reached, or beam scraping occurs and the vacuum pressure degrades, or the microwave signal otherwise degrades. Once a local maximum has been established, all other parameters are varied to further enhance the output power. The whole process is repeated until the maximum amplified power has been found.

The experimental results are listed in Table 3 for the maximum amplified power. The values given in this paper are all taken from anechoic chamber measurements; the measurements from all three diagnostics are comparable. The time dependence of the beam voltage, beam current, and the amplified signal are shown in Fig. 4. There is a slight droop on the voltage flat top due to a mismatch between the load and the

**TABLE 3.** Amplifier experimental results.

| Amplifier results | |
| --- | --- |
| Drive Frequency (GHz) | 8.60 |
| Output Power (MW) | 75 |
| Pulse length (FWHM) ($\mu$s) | 1.7 |
| Gain (dB) | 29.7 |
| Efficiency (%) | 31.5 |

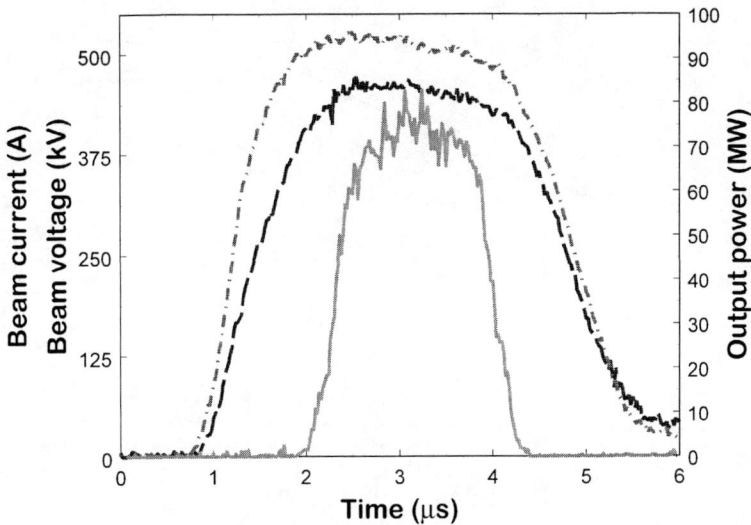

**FIGURE 4.** Time dependence of the amplified microwave signal (solid line). The beam voltage is indicated by the dashed line and the beam current is shown by the dot-dashed line.

modulator impedances. There is also some ground loop noise on the pulses which is most noticeable on the microwave signal (due to its relatively low amplitude). The peak power of 75 MW listed in Table 3, which represents an efficiency of about 32%, indicates conservatively the average value of the signal in the flat top region. The corresponding saturated gain is nearly 30 dB and the pulse width is about 1.7 μs (FWHM). Attempts to increase the peak power further by raising the beam's velocity ratio result in a sharp cut in the output signal near the maximum value which is usually indicative of an instability.

An EGUN simulation using the parameters of the operating point indicate that the beam's velocity ratio at the entrance to the circuit is near one. There is a reasonably large uncertainty in this ratio due to the incomplete treatment of the self-axial magnetic field in EGUN and the uncertainty in the applied field at the cathode. The average velocity ratio was not measured in this tube. However, in a previous experiment at the University of Maryland, the measured average velocity ratio according to a capacitive probe was consistently higher than the simulated ratio by about 15%. (29) Calculations showed that a significant part of the discrepancy was due to the diamagnetic effect. Simulations of the amplifier performance at the operating point are given by the dashed line in Fig. 3. The simulated cathode magnetic field is adjusted slightly to produce the best match between the theoretical efficiency and the measured efficiency, which is indicated by the cross. The required field is about 20 G lower than the calculated ideal field and well within the uncertainties of the experimental data. Furthermore, efficiencies of about 27% were predicted by our large-signal code when the simulated velocity ratio was used.

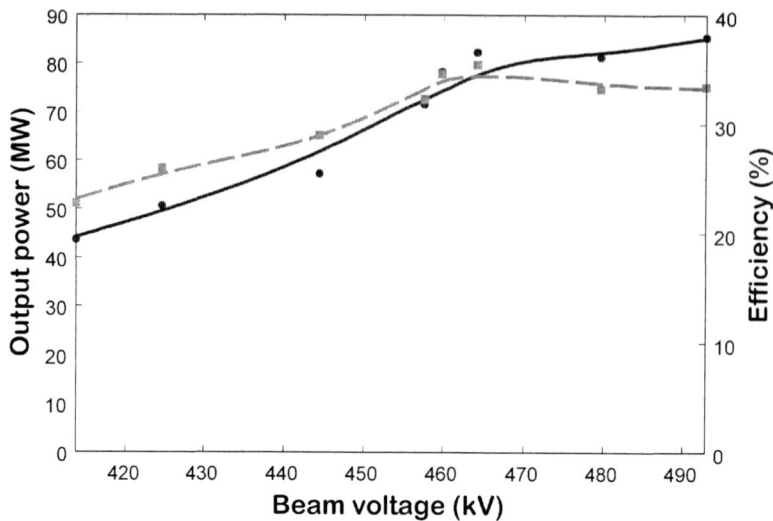

**FIGURE 5.** Peak power measurement (solid line) and interaction efficiency (dashed line) versus the beam voltage.

Figure 5 shows the dependence of output power (solid line) on beam voltage. The current increases somewhat with voltage because of the dependence of the temperature-limited emission on electric field. The cathode magnetic field is adjusted to maximize efficiency, but all other parameters are maintained at the optimal values. The efficiency is given by the dashed line. Peak efficiency is seen to roll off at the maximum beam power, no doubt due to a decrease in beam quality.

## SECOND HARMONIC SYSTEM

The second harmonic tube is realized by keeping the first harmonic tube's input cavity but replacing the buncher and output cavities with ones that resonate at twice the drive frequency in the $TE_{021}$ mode. Such cavities are normally difficult to realize, because the cavities' end walls generate other radial modes due to the beam tunnel opening. For inter-cavity isolation, the fields must not leak substantially into the drift regions, yet the operating frequency is well above the cutoff for the $TE_{01}$ mode. In circular waveguide systems, the usual way to avoid this problem is to introduce smoothly-varying wall radii, but the added length of these transitions is often unacceptable. Fortunately, in coaxial tubes, making the radial wall transitions that form the $TE_{021}$ cavity approximately equal on the inner and outer walls naturally leads to a mode with very little conversion to the $TE_{01}$ modes and subsequent leakage fields.

An outline of our three cavity second harmonic tube layout and a sketch of the optimal axial magnetic field profile are given in Fig. 6. The principal design parameters for the tube are given in Table 4 along with the simulated performance estimates.

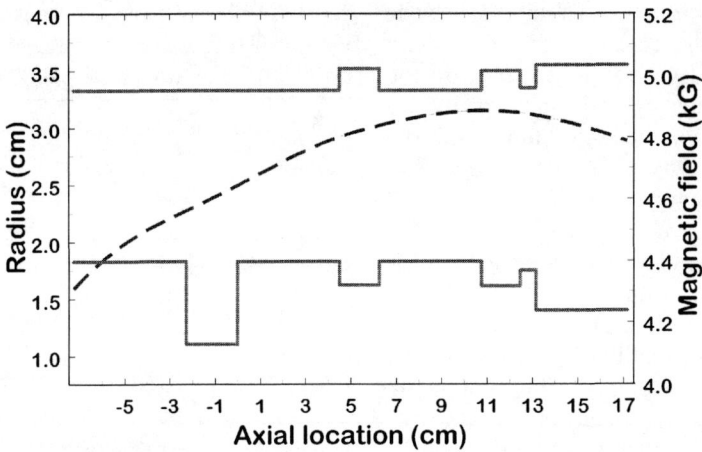

**FIGURE 6.** Schematic of the second harmonic cavity design (solid line) and theoretical optimal axial magnetic field profile (dashed line).

The drift tube radii and the input cavity parameters have been omitted from the table because they are the same as for the first harmonic tube. The drift tube lengths are slightly longer than the corresponding lengths in the first harmonic tube. The quality factor of the input cavity has been lowered to about 50 by moving the lossy ceramics in the drift tube closer to the input cavity. This was done, because the anticipated operating current is higher for this tube.

The buncher cavity has a measured resonant frequency and Q of 17.136 GHz and 390, respectively. The required quality factor is achieved by placing lossy ceramics in

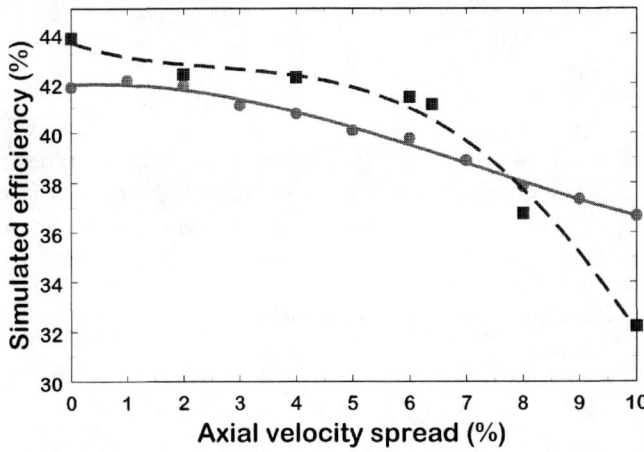

**FIGURE 7.** Comparison of the theoretical performance curves for the first (solid line) and second harmonic (dashed line) tubes.

the drift tube and adjusting the distance between the ceramics and the cavity region. (30) The dimensions of the main section of the output cavity are virtually identical to those of the buncher cavity. Abrupt transitions are again used, with the lip radii adjusted to minimize mode conversion to the $TE_{01}$ mode. The measured cold resonant frequency and Q of this cavity are of 17.136 GHz and 325, respectively. the Q is essentially entirely diffractive. The output signal is simulated to be 99% in the $TE_{02}$ mode and the percentage of total microwave power which enters the drift region (rather than the output waveguide) is less than 0.5%.

Large signal simulations estimate the best efficiency to be about 41% when the beam power is 385 MW. Thus the estimated output power is above 150 MW. The saturated gain is estimated to be nearly 50 dB. The dependence of efficiency on axial velocity spread for the second harmonic tube is indicated in Fig. 7 by the dashed line. The efficiency of a first harmonic design is shown in the figure (solid line) for comparison purposes. The efficiency of the second harmonic design is slightly higher at low spreads, but falls off more rapidly as the velocity spread increases beyond 6%, as is typical with harmonic tubes.

**TABLE 4.** Second harmonic tube parameters and design simulation.

| Drift tube parameters | |
|---|---|
| Length (between I-B) (cm) | 5.52 |
| Length (between B-O) (cm) | 6.15 |
| Buncher cavity parameters | |
| Inner radius  (cm) | 1.61 |
| Outer radius (cm) | 3.50 |
| Length (cm) | 1.69 |
| Output cavity parameters | |
| Inner radius  (cm) | 1.61 |
| Outer radius (cm) | 3.51 |
| Main cavity length (cm) | 1.71 |
| Lip inner radius (cm) | 1.76 |
| Lip outer radius (cm) | 3.36 |
| Lip length (cm) | 0.692 |
| Simulated results | |
| Beam voltage (kV) | 500 |
| Beam current (A) | 770 |
| Velocity ratio | 1.508 |
| Gain (dB) | 49 |
| Efficiency (%) | 41% |

At this point, the tube has been entirely constructed, and the hot testing has recently commenced. The electron gun has only been able to put out about 500 A. At this low beam power, the tube performance is predicted to be considerably worse. We have just begun an attempt to re-activate the cathode and expect to resume high current testing of the second harmonic tube in the near future.

## SCALING TO HIGHER FREQUENCIES

We have taken a look at scaling two of our previous designs to higher frequency. The method for scaling depends on the current limitations of the system. Potential limitations include peak RF electric fields, average power in the cavities, beam dump, or window, beam power density, electrostatic potential depression, cathode loading, magnetic compression, and DC electric fields in the electron gun. For our gyroklystron tubes, peak and average power levels in the tube are not predicted to be an issue at the expected required power levels and duty factors. On the contrary, the main limitation on power in the past has been the electron gun.

Scaling of MIGs with frequency has been considered in the past. (31) When the magnetic compression and the cathode loading are not too large, MIG designs can be scaled so that the power is proportional to the wavelength. While it is true that the circuit dimensions scale with the wavelength so that the beam cross section must scale with the square of the wavelength, the current density can be increased with the wavelength. This is true because the cathode loading can be increased in proportion to the cube root of the wavelength and the compression can be increased with the two-thirds power of the wavelength. The

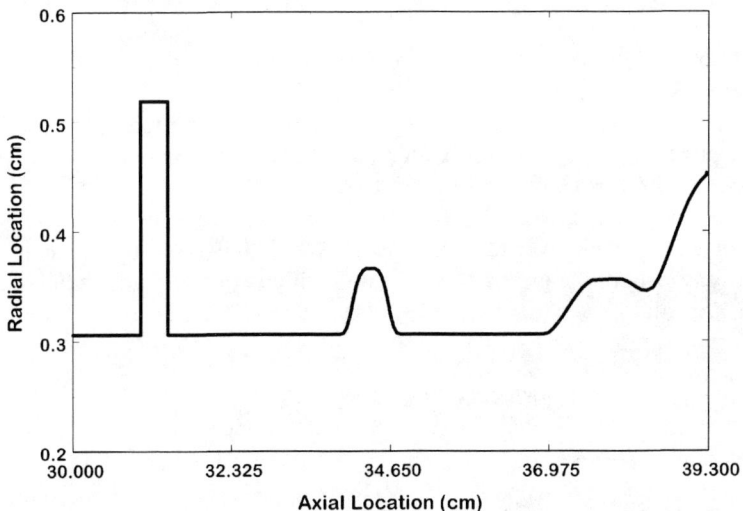

**FIGURE 8.** Circuit schematic for the 7 MW, 95 GHz, three cavity, second harmonic gyroklystron.

maximum cathode loading depends on the cathode material, required lifetime, and other issues but is the same for klystrons as it is for gyroklystrons. The maximum magnetic compression depends on many beam parameters, perhaps the most important being the beam quality as measured by the axial velocity spread. The magnetic compression is typically kept below a value of 40.

Scaling in proportion to the wavelength can't be continued indefinitely. Eventually, limits on magnetic compression, cathode loading, or some other beam or circuit constraint will be reached. However, for two interesting cases we have achieved wavelength scaling to much higher frequencies. The first example involves scaling our 32 MW, second harmonic, two cavity circular tube to 95 GHz. (32, 33) A picture of the proposed circuit is shown in Fig. 8. A $TE_{021}$ buncher cavity was added to improve gain and efficiency. A MIG was designed that produced a high quality 500 kV, 45 A beam (according to simulations). The simulated large signal efficiency and gain were 35% and 56 dB, respectively. The efficiency corresponds to an output power of 7.5 MW.

The second example involves scaling the second harmonic coaxial design to 12 times the current SLAC frequency (34.32 GHz). Again, both an electron gun and a microwave circuit were designed. The simulations predicted an output power of 60 MW via a 40% efficient interaction with a 500 kV, 300 A beam.

## CONCLUSIONS

In summary, we have developed an X-band, three-cavity, first harmonic coaxial gyroklystron which has increased the state-of-the-art in peak power for gyroklystrons to over 75 MW. Peak efficiencies were about 32% and the gain was near 30 dB. We have just begun hot testing a three-cavity second harmonic tube, with the goal of obtaining over 100 MW of peak power at 17.136 GHz. We are currently designing an "advanced concept" tube which we hope will be compatible with high repetition rate operation with a pulse compressor / accelerator load. (34) In the long term we expect to build and power a 17.136 GHz accelerator structure with an accelerating gradient near 200 MV/m. Finally, we have scaled two of our existing tubes to 35 GHz and 95 GHz via equations that result in a reduction in power which is proportional to the wavelength. Designs of both the electron guns and the microwave circuits indicate that the scaling rules are valid.

## ACKNOWLEDGEMENTS

The authors would like to thank J. Anderson, N. Ballew, J. Cheng, J. Calame, P. Chin, P. E. Latham, C. Liu, G. Nusinovich, G. P. Saraph, and X. Xu for their contributions to this effort. This work was supported by the US Department of Energy.

# REFERENCES

1.  Schneider, J., *Phys. Rev. Lett.* **2**, 504 (1959).
2.  Denisov, G. G., *et al.*, *Int. J. Electronics*, **72**, 1079 (1992).
3.  Dammertz, G., *et al.*, *IEEE Trans. Plasma Sci.* **24**, 586 (1996).
4.  Read, M. E., *et al.*, *IEEE Trans. Plasma Sci.* **24**, 586 (1996).
5.  Granatstein, V. L., *et al.*, *IEEE Trans. Plasma Sci.* **25**, 1322 (1997).
6.  Chu, K. R., *et al.*, *IEEE Trans. Plasma Sci.* **PS-13**, 424 (1985).
7.  McCurdy, A. C., *et al.*, *Phys. Rev. Lett.* **57**, 2374 (1986).
8.  Lawson, W., *et al.*, *Phys. Rev. Lett.* **67**, 520 (1991).
9.  Zasypkin, E. V., *et al.*, *IEEE Trans. Plasma Sci.* **24**, 666 (1996).
10. Lawson, W., *et al.*, *Phys. Rev. Lett.* **71**, 456 (1993).
11. Granatstein, V. L. and W. Lawson, *IEEE Trans. Plasma Sci.* **24**, 648 (1996).
12. Blank, M., *et al.*, in 1997 ICOPS Conference Record, p.240.
13. Raubenheimer, T. O., *et al.*, SLAC-PUB-474 (1996).
14. Allen, M. A., *et al.*, *Phys. Rev. Lett.* **63**, 2472 (1989).
15. Westenskow, G. A. and T. L. Houck, *IEEE Trans. Plasma Sci.* **22**, 750 (1994).
16. Caryotakis, G., *IEEE Trans. Plasma Sci.* **22**, 683 (1994).
17. Haimson, J., *et al.*, in Proc. Wkshp. Pulsed RF Sources for Linear Colliders, Montauk, N. Y., 1994, p. 146. AIP Conf. Proc. vol. 337.
18. Karliner, M. M., *et al.*, *Nucl. Inst. and Meth. Phys. Res.* **A-269**, 459 (1988).
19. Gold, S. H., *et al.*, *IEEE Trans. Plasma Sci.* **24**, 947 (1996).
20. Shiffler, D., J. A. Nation, and G. S. Kerslick, *IEEE Trans. Plasma Sci.* **18**, 546 (1990).
21. Orzechowski, T. J., *et al.*, *Phys. Rev. Lett.* **57**, 2172 (1986).
22. Conde, M. E. and G. Bekefi, *Phys. Rev. Lett.* **67**, 3082 (1991).
23. Balkcum, A. J., *et al.*, *IEEE Trans. Plasma Sci.* **24**, 802 (1996).
24. Herrmannsfeldt, W. B., SLAC-PUB-226, Nov. 1979.
25. Calame, J. P. and W. Lawson, *IEEE Trans. Electron Devices*, **38**, 1538 (1991).
26. Latham, P. E., W. Lawson, and V. Irwin, *IEEE Trans. Plasma Sci.* **22**, 804 (1994).
27. Saraph, G. P., *et al.*, *IEEE Trans. Plasma Sci.* **24**, 671 (1996).
28. Latham, P. E., S. M. Miller, and C. D. Striffler, *Phys. Rev. A*, **45**, 1197 (1992).
29. Calame, J. P., *et al.*, *IEEE Trans. Plasma Sci.* **22**, 476 (1994).
30. Castle, M., *et al.*, "An Overmoded Coaxial Buncher Cavity for a 100 MW Gyroklystron", *IEEE Microwave and Guided Wave Letters*, **8**, 302 (1998).
31. Lawson, W., "Magnetron Injection Gun Scaling," *IEEE Trans. Plasma Sci.* **16**, 290 (1988).
32. Matthews, H. W., *et al.* "Experimental Studies of Stability and Amplification in a Two-Cavity Second Harmonic Gyroklystron," *IEEE Trans. Plasma Sci.* **22**, 825 (1994).
33. Arjona, M. R. and W. Lawson, "Design of a 7 MW, 95 GHz, Three-Cavity Gyroklystron," *IEEE Trans. Plasma Sci.* (in press).
34. Xu, Xiaoxi, *et al.*, "New Concept Input and Output Systems for High Power Gyroklystron," in these proceedings.

# Polarized RF Guns*

## J. E. Clendenin

*Stanford Linear Accelerator Center,*
*Stanford University, Stanford, CA 94309*

**Abstract**. RF guns employing photocathodes are now well established as viable electron sources for accelerator applications. For high-energy accelerators, the desirable properties of such sources include the relative ease of pulse formation (using the source laser system to establish the pulse shape and number) and a low transverse beam emittance, eliminating the need for rf chopping and bunching systems and reducing the demands on emittance-reducing damping rings. However, most high-energy accelerators now require polarized electrons. Polarized electron beams can in principle be generated by substituting a III-V semiconductor, such as GaAs, for the traditional photocathode in a conventional rf gun structure. In the past, the principal criterion for selection of photocathodes for rf guns has been the ability to maintain a reasonable quantum yield from the cathode over a relatively long operating period. It is well known that GaAs is significantly more sensitive to the vacuum environment than any of the cathode types that have been used to date. However, advances in our understanding of how to operate GaAs cathodes in hostile environments combined with similar advances in our ability to prepare gun structures that will significantly reduce dark current and rf breakdown lend credibility to the prospect of successfully operating an rf gun with an activated GaAs cathode. The status of research related to polarized rf guns is reviewed and the outline of a future polarized rf source combining the best of what is known today is presented.

## INTRODUCTION

Since their initial introduction in 1985 at LANL by Fraser, Sheffield, and Gray (1), rf photoinjectors have made rapid progress. Because of the strong acceleration of the beam in the first rf cavity, the effects of space charge are quickly reduced with the result that beams of up to several nanocoulombs per bunch can be readily produced with low emittance. In addition, since the electrons emitted from the cathode generally follow closely the pulse shape of the optical source, pulse shaping is readily accomplished with the optical beam, allowing single- and multi-bunch formation over a wide range of parameters. These features have made rf photoinjectors popular for a variety of applications today including free electron lasers (2).

Given the unique properties presented by photocathode rf guns, it was realized very early that such an electron source would be ideal for linear colliders if the beam could

---

* Work supported by Department of Energy contract DE-AC03-76SF00515.

CP472, *Advanced Accelerator Concepts: Eighth Workshop,*
edited by W. Lawson, C. Bellamy, and D. Brosius

be polarized (3). Present collider designs typically require a fairly complex pulse train with a reasonably low emittance. RF guns for this purpose are already being used for the TESLA Test Facility (TTF) at DESY and for both the drive and probe beams of the CLIC Test Facility (CTF) at CERN. However, the primary electron beam for colliders must be polarized (4). Polarized electron sources for accelerators have also made rapid progress. Since the introduction of the polarized source for the SLC at SLAC in 1992, the world's first linear collider has operated exclusively with a polarized electron beam, resulting in >30,000 hours of beam time with >95% availability of the source (5). The SLC source utilizes a dc-biased gun with a solid-state (GaAs or its derivatives) photocathode. Today, all linac-based polarized electron sources are of a similar design (6). The success of the SLC source is directly correlated with the achievement of an extremely good vacuum environment for the cathode, both during the cathode activation (which is done in a separate vacuum chamber) and during the source operation.

Because of its high quantum efficiency (QE) in the visible, GaAs was actually considered as a possible cathode material for the first rf photoinjector (7) but rejected because of multiple concerns. The concerns expressed at that time included the sensitivity of GaAs to the vacuum environment and the emission time of the photoelectrons. Extensive discussion of using GaAs in an rf gun did not begin until 1993 at the Workshop on High Intensity Electron Sources in Legnaro, Italy, where two papers were presented with proposals for such a source based on the achievements of both rf guns and polarized dc guns (8). The Legnaro workshop was followed soon after by SOURCES '94, organized by DESY to examine electron and positron source possibilities for future linear colliders. At the time there was some excitement concerning the possibility that a polarized rf gun might obviate the need for an e⁻ damping ring for TESLA. However, the final recommendation of the relevant working group was that the impact of lower emittance on the cost and performance of damping rings needed further study (9).

Assuming that a polarized rf gun could be successfully produced and operated, what would be its advantages compared to a dc polarized gun plus rf compressor? The principal advantage is that an rf photoinjector produces an inherently lower beam emittance. While for a collider this emittance is not low enough to eliminate the electron damping ring, it will certainly simplify both the damping ring and the intervening beam transport system. Because of the rapid acceleration to high energy, bunch shaping is much more feasible and can be used to further reduce the beam emittance. Since the transmission will be higher, the charge required to be extracted is reduced by up to a factor of two, mitigating the restrictions of the cathode charge limit (10). The QE will also be higher (due to the Schottky effect), reducing the demands on the source laser system. Finally, beam-loading compensation is somewhat more effective and certainly more straightforward for an rf gun.

## ACTIVATED GaAs CATHODES

GaAs is a direct-gap III-V semiconductor that crystallizes in the zincblende structure. The energy levels and transition probabilities in the vicinity of the $\Gamma$ point are shown in

Fig. 1. At the band-gap minimum there is a two-fold degeneracy of the valence band (VB) maximum at which the heavy-hole (hh) and light-hole (lh) bands are mixed. The VB maximum has $P_{3/2}$ symmetry while for the conduction band (CB) minimum it is $S_{1/2}$. Therefore if circularly polarized light is used that is at least the energy of the band gap, but less than the energy of the next lower VB, $E_g+\Delta_{SO}$, the electrons promoted to the CB will be highly polarized. For bulk GaAs the polarization, $P = \left(N^+ - N^-\right)/\left(N^+ + N^-\right)$, is limited to 50% as indicated by the figure. However it is possible to break the degeneracy of the VB maximum sufficiently to promote electrons exclusively from the higher of the hh or lh states, thus offering the possibility of up to 100% polarization.

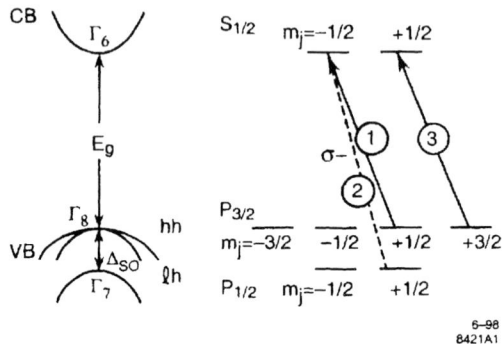

**FIGURE 1.** Energy level diagram (left) and transition probabilities (right) at $\Gamma$ point for bulk GaAs. Only the transitions for left circularly polarized light ($\sigma$-) are shown. The solid-line transitions are for $E_g < \left(E_g - \Delta_{SO}\right)$.

For use as a photoemissive material, the GaAs crystal is typically heavily p-doped, which results in band bending at the surface. The electron affinity, $\chi$, defined as the difference between the vacuum and the CB minimum at the surface, is about 4 eV for a clean surface with flat bands. Activation of the GaAs surface requires depositing approximately a monolayer of an alkali (usually Cs) and an oxide (usually $O_2$ or F) on the atomically clean surface, resulting in a reduction of $\chi$ to nearly zero. During the deposition of the (Cs,O) layer, the surface of the p-GaAs becomes positively charged, repelling the positively-charged holes near the surface but not the distributed negatively-charged ionized acceptors, resulting in a high field that bends the band edges down at the surface by as much as a third of the band gap. Since the band gap for GaAs is about 1.4 eV at room temperature, this results in a negatively electron affinity (NEA) surface that greatly enhances the probability of photoemission. The photoemission process from an NEA semiconductor is usually described using the three-step model, which is illustrated in Fig. 2.

# PROBLEMS AND ISSUES

In addition to vacuum and emission-time issues that were noted very early, outstanding problems that have been identified include the effects of dark current and rf breakdown on the QE of the cathode, the effect of the rf on the electron polarization, the effect of an activated GaAs crystal on the rf performance of the cavity in which it resides, and the question of the charge limit under conditions of very short, high-intensity optical excitation.

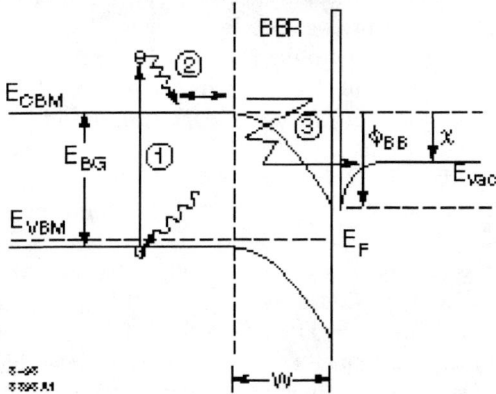

**FIGURE 2**. Schematic energy diagram near the surface for GaAs illustrating the three-step emission process where step 1 is absorption of a photon creating an electron-hole pair, step 2 is the thermalization and diffusion of the conduction band electron to the band bending region (BBR), and step 3 is the emission of the electron to vacuum. $E_{VBM}$, $E_{CBM}$ and $E_{vac}$ are valence band maximum, conduction band minimum, and vacuum level energies respectively. $E_{BG}$ is the band gap, $E_F$ the Fermi energy, $\chi$ the electron affinity, and $W$ and $\phi_{BB}$ the width and depth of the BBR respectively.

It is well known that the QE of an activated GaAs crystal can be maintained only in an extremely good vacuum. For a QE lifetime on the order of a few hours, a vacuum of at least $10^{-10}$ Torr is required. The SLC source, which has achieved lifetimes of several thousand hours under SLC operating conditions, maintains a vacuum of $10^{-11}$ Torr that is entirely dominated by $H_2$. Some gas species that are commonly found in high vacuum systems such as CO are relatively benign, while $O_2$, $CO_2$, and $H_2O$ are especially harmful. Many of the most common background gases are pumped by the excess Cs on the GaAs surface following an activation. Although the details of how the initial (Cs,O) layer is formed are important, it appears that the thickness of the layer can be increased many fold without any detriment to performance. In fact, periodically adding a small amount of Cs to the surface is a common technique used with dc guns to restore the initial QE. Following the initial activation, the QE for a GaAs cathode drops with time. For the SLC source, up to about 50 re-cesiations have been applied without re-activating and without adding any additional oxide, each re-cesiation completely restoring the initial QE. Assuming no loss of Cs from the cathode surface (11), the 50 re-cesiations represent the accumulation of 1 additional monolayer.

To prepare an rf or dc-biased gun for operation, it must be rf or high-voltage (HV) processed to eliminate rf or HV breakdowns respectively before the activated semiconductor cathode is installed. This is readily done if the cathode is activated in a separate chamber, then installed in the gun maintaining a vacuum by using a "load-lock" system. RF guns generally require additional processing to reach the highest rf fields whenever a new cathode is installed. To preserve the QE of a newly installed GaAs crystal, a cathode-plug design must be found that eliminates this additional processing.

Once a gun is operational, there is some level of dark current, defined as the average current emitted when the extraction voltage (dc or rf) is on but the laser beam is off. Surfaces other than the photoemitter surface can and do contribute to the total dark current. For a dc gun, the dark current is easily measured by noting the average current drawn from the HV power supply with the laser off. In the case of the SLC source, dark currents of >100 nA has been found to destroy the QE within a few hours, while <20 nA has almost no effect on the QE. External dark current is easily measured using downstream charge monitors. Internal dark current, meaning electrons that are both field emitted and re-absorbed inside the gun cavities, is more difficult to measure. High levels of dark current can be observed in the beam loading on the reflected rf pulse signal.

Band bending at the surface for p-GaAs is about 0.5 eV while the width of the band-bending region (BBR) for a dopant density $>2 \times 10^{19}$ cm$^{-3}$ is <5 nm, resulting in an internal dc field >100 MV/m but confined to the BBR. By contrast, rf fields will penetrate well beyond the BBR. Since the conductivity of even highly-doped GaAs is less than that of Cu, the skin depth, which is inversely proportional to the square root of the conductivity, will be on the order of tens of micrometers at 3 GHz, but still considerably less than the ~1-mm thickness of the crystal (12). Thus a polarized rf gun may constitute a field-assisted photoemitter that will result in both enhanced emission with a shorter bunch length and also possibly higher polarization (13).

The bunch length of the electrons extracted from an rf gun should be kept below about 20° of rf phase. This is important not only to maintain a low emittance beam but also to minimize both loss of electrons during acceleration and also backward acceleration into the photocathode. However, the delay between promotion and emission in a semiconductor can be on the order of nanoseconds. For an NEA semiconductor, the emission time is governed by the diffusion time for electrons in the conduction band. One can easily calculate the time, $t_D$, for an electron promoted to the conduction band to diffuse to the surface from a depth of $x$ within the epilayer:

$$t_D = \frac{x^2}{D},$$
(1)

where D is the electron diffusion constant. If a conservative value of $D \approx 50$ cm$^2$ s$^{-1}$ for GaAs at room temperature and a starting depth equal to one optical absorption length, $l_a \approx 0.7$ μm at 750 nm, are assumed, then $t_D$ should be ~1 ns. For a thin epilayer of 100 nm, ~2 ps would be expected.

The cathode charge limit measured with the SLC source is ~4 A cm$^{-2}$ for an extraction field of 1.8 MV/m, a dopant density of $5 \times 10^{18}$ cm$^{-3}$, and a QE of ~0.25% at the polarization peak (14). At low fields, the charge limit increases linearly with field. If this relationship holds for high fields, then the charge limit for a field of 30 MV/m (see final section) should be on the order of 65 A cm$^{-2}$. The charge limit imposes a severe restriction on the product of bunch length and cathode radius. Based on a limit of 65 A cm$^{-2}$, extracting 1 nC of charge in 20 ps would require the cathode radius to be 5 mm.

# REVIEW OF R&D PROGRESS

Up to the present there has not been a successful demonstration of photoemission from an activated GaAs cathode using an rf gun. However there has been experimental progress with most of the problems and issues outlined above, including major advances with vacuum and emission-time issues.

Recently-built S-band guns have pressures (at the ion pumps) after rf processing of about $10^{-10}$ ($10^{-9}$) Torr with rf off (on). Dark currents on the order of 1 nC per $\mu$s of rf (after filling cavity) are typical for conventionally cleaned S-band guns operating at ~130 MV/m (15), which would result in >100 nA average dark current at 180 Hz. At KEK it has been demonstrated that the dark currents from test cavities that are specially cleaned with high-pressure ultra-pure water (UPW) and then assembled in a Class 100 clean room can be reduced to ~130 (~13) pC/($\mu$s of rf) at 130 (90) MV/m (16). More recently, field emission dark current from Cu electrodes–fabricated from Class-1 Cu using hot isotatic pressing (HIP), then electro-polished and UPW rinsed–has been reduced to the level of 1 nA for a dc field of 28 MV/m (17).

A complete GaAs test requires an rf gun equipped with a load-lock and an associated activation chamber. A relatively high-power pulsed laser tuned to a wavelength between 750 and 850 nm is also required although initial tests can use a pulsed laser system operating at shorter wavelength such as a doubled Nd:YAG or YLF. There are at present only two guns being used. Testing of GaAs cathodes has been underway at CERN since 1995 (18). The 1.5-cell S-band gun for the CTF drive beam is based on the BNL design. The gun and following booster section, which have been manufactured and assembled under clean room conditions (19), normally exhibits relatively low dark current. Since the gun was designed to be used with a $Cs_2Te$ photocathode, it was already equipped with a separate activation chamber with a cesium source and a transporter for moving the cathode into the gun through a load-lock. For the initial tests, no attempt was made to activate the GaAs. The GaAs crystal, cut from a bulk-grown wafer with dopant density of $5\times10^{18}$ $cm^{-3}$, was simply glued to an inset in the face of a detachable Mo nose for the rf plug using In as the glue. After chemical cleaning and a mild bake, the rf plug was inserted into the gun. Given the high dopant density, it is not surprising that the resonant frequency of the cavity was found to be unaffected. After vigorous processing of the first sample, a field at the cathode of 87 MV/m was achieved, but the cathode plug was later found to be seriously damaged as a result of the rf processing. At the maximum field, no dark current was detected, however from the light visible on a screen, an upper limit of 60 pC/($\mu$s of rf) was estimated. For an S-band gun operating with a 2-$\mu$s rf pulse at 180 Hz, this limit corresponds to an average current of ~10 nA, well within the criterion mentioned earlier. A second sample, not activated but with a thin layer of Cs, was processed more gently to 60 MV/m without visible damage and again without detectable dark current.

At INP, Novosibirsk, a prototype S-band polarized rf gun has been constructed for the injector of the VEPP-5 complex (20). A cross section of this gun is shown in Fig. 3. As originally commissioned the base pressure of the gun was $<10^{-11}$ Torr. In the initial test the lifetime of a poorly activated GaAs cathode (QE~0.05%) was measured to be

147

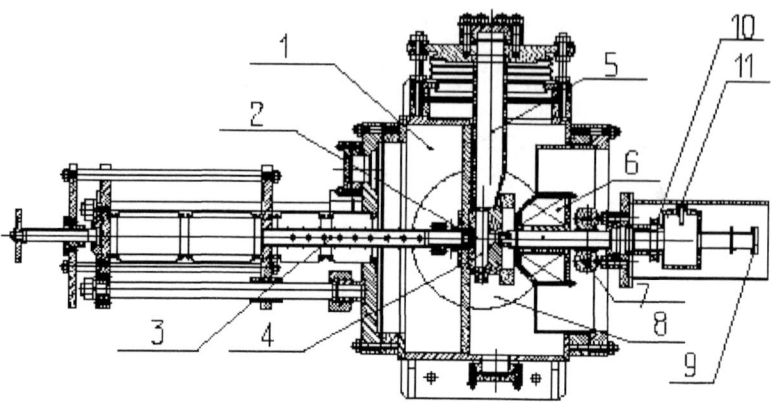

**FIGURE 3**. Cross section of the INP prototype polarized rf gun showing: 1) activation chamber, 2) photocathode assembly, 3) manipulator, 4) accelerating cavity, 5) waveguide, 6) focusing lens, 7) transverse corrector, 8) working chamber, 9) vacuum window for laser beam, 10) ceramic insulator, and 11) cavity for measuring bunch length. (From Aleksandrov et al., ref. 21.)

~0.5 h. During the test, dark current for a field of 60 MV/m was observed to be about 2 orders of magnitude larger than for a clean cathode (21). In a later test, a well activated GaAs cathode (QE~5%) was inserted. In this case the dark current was high enough to completely kill the QE within a few rf pulses (22). Additional testing is planned.

Although the INP results are disturbing, there is actually no fundamental reason that an NEA semiconductor should field emit since there are essentially no free CB electrons when optical excitation is absent unless the external field is large enough to create an inversion layer at the surface. An inversion layer begins when the bands at the surface are bent downward to at least mid-gap. Since the behavior of highly-doped p-GaAs is similar to a metal, the additional bending necessary can be estimated from the usual Schottky-effect analysis. Assuming band bending of ~0.5 eV at low field, a weak inversion layer should begin at an external field of ~50 MV/m although rapid accumulation of free electrons would require significantly higher fields (23)!

The experimental determination of the level of field emission that can be attributed to an NEA photocathode in an rf gun is a complex problem because of uncertainties in the field emitting properties of the rf plug. Developing a technique to mount the crystal in an rf cell so as not to generate either rf breakdown or dark current is crucial. In preparation for a new set of tests at the CTF, a clever crystal mounting scheme–shown in Fig. 4–has been developed at SLAC (24). A slurry cutter is used to cut a "top hat" cross section from the original ~1-mm thick GaAs wafer with a precision of a few micrometers. The active surface can be installed parallel to the Mo face within a few milliradians if care is taken. The Ta leaf spring anticipates the effects of thermal expansion when the crystal is heat-cleaned as part of the activation process.

The latest tests at the CTF using this crystal-mounting scheme were conducted between November 1997, and January 1998 (25). There being no heat-cleaning capability in the CTF system, an attempt was made to clean the GaAs crystals by ionic controlled

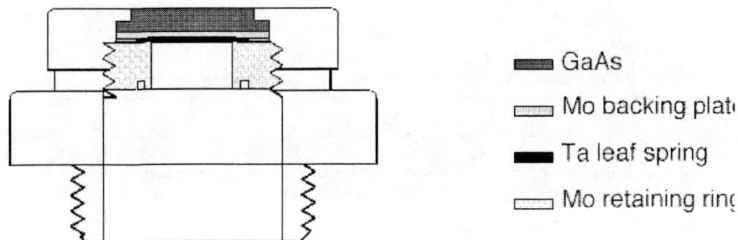

**FIGURE 4**. Cross section of Mo nose assembly for mounting GaAs crystal on end of cathode rf plug. The identified components are listed vertically in the order of their appearance in the cross section.

etching (ICE), a technique developed at CERN for cleaning other types of cathodes (26). Unfortunately it was not possible to achieve a high QE. Nonetheless the cesiated crystals were again tested in the rf gun in fields up to 73 MV/m. In at least one case low-level dark current was observed only beginning at ~60 MV/m, and in any case was lower than observed for $Cs_2Te$ crystals. Reliable performance with a non-activated cathode should be established before one can determine the added characteristics of an activated surface.

Recent measurements at Mainz of the electron photoemission time from a thin (epilayer of 150 nm), strained-layer GaAsP crystal indicate an emission time of no more than ~10 ps for the lowest charge (~0.5 fC) measured (27). The measurements were made using a dc gun biased at 100 kV. The crystal was illuminated by a 100-fs laser tuned to the polarization peak, which was at 836 nm. The laser spot diameter at the cathode was ~0.6 mm. The resolution of the rf analyzer was ~1 ps at 2.45 GHz. An observed increase of bunch duration with charge was attributed to an external space charge effect, but more recently the Novosibirsk-Legnaro group has shown that this effect is internal to the gun and is a current limitation not attributable to a gun space-charge effect (28). In fact the limit may be the same cathode charge limit described in refs. 5 and 10. The limit for the Mainz measurement, ~20 mA $cm^{-2}$, is only about a factor of ten below the limit found for the SLC source under similar conditions after scaling for differences in QE, epilayer thickness, and extraction field (29).

There is increasing evidence that the effect of the cathode charge limit can be greatly reduced by using a very high dopant density at the surface of the GaAs (30).

## FUTURE POSSIBILITIES

Clearly much work remains to be done to establish the feasibility of a polarized rf gun. The resources required to build and test a new rf gun that would optimize polarization are sizable–well beyond the resources of most laboratories. An informal international collaboration of physicists from major accelerator laboratories was formed in 1996 to promote the investigation of the known problems and to maximize the use of existing resources (31). The GaAs testing at the CTF described above continues as part of this collaboration. In addition, the KEK-Nagoya group are utilizing the manufacturing and

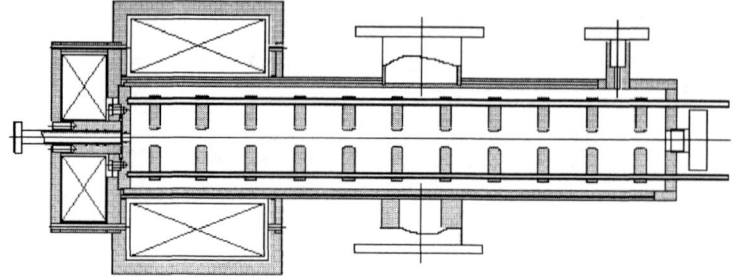

**FIGURE 5**. Cross section of the plane wave transformer S-band photoinjector being fabricated as part of an SBIR collaboration between UCLA and DULY Research. The cathode plug is on the left surrounded by the emittance compensating solenoid and the cathode field bucking coil. The rf feed and vacuum pump-out ports are in the center. An rf probe port is to the right. The laser beam enters and the electron beam exits at the far right end. (From Rosenzweig, ref. 32.)

assembly techniques developed to reduce dark current to construct a CERN-designed rf gun for use at the CTF. If the dark current proves especially low, this gun will be an excellent platform for testing an activated GaAs photocathode.

A promising candidate design for a future polarized rf gun is the plane wave transformer (PMT) S-band structure now being developed at UCLA (33). A cross section of the UCLA design is shown in Fig. 5. The features of this design that are appealing for operation with a GaAs photocathode are threefold. 1) The optimum transverse emittance is achieved with a peak accelerating field of 60 MV/m (minimum inversion layer) and a nominal launch phase of ~30° (reasonably high charge limit). For a carefully fabricated gun, this low field should essentially eliminate dark current and rf breakdown problems including those originating with the cathode plug. 2) The open structure allows more efficient vacuum pumping in the vicinity of the photocathode. 3) The volume between the disks and the outer wall increases the stored energy, which reduces the beam-loading problem for a collider multi-bunch photoinjector.

## REFERENCES

1. Fraser, J.S., Sheffield, R.L. and Gray, E.R., in *Laser Acceleration of Particles*, AIP Conf. Proc. **130**, 598 (1985), and *Nucl. Instrum. and Meth.* A **250**, 71 (1986).
2. Clendenin, J., *Proc. of the XVIII Int. Linear Accelerator Conf.*, Geneva, CH, 1996, p. 298.
3. The possibility of operating an rf gun with a GaAs cathode to produce a polarized beam was suggested by the author during one of the discussion sessions at the 8th Int. Symp. On H.E. Spin Physics, Minneapolis, MI, 12-17 Sep. 1988.
4. Clendenin, J., "Highly Polarized Electron Beams for Linear Colliders," SLAC-PUB-7726 (1998), presented at the 2nd Int. Workshop on Electron-Electron Interactions at TeV Energies, Santa Cruz, CA, 22-24 Sep. 1997.
5. The SLC source is used for fixed-target as well as colliding beam experiments. The source is described in detail in Alley, R.A. et al., *Nucl. Instrum. and Meth.* A **365**, 1 (1995).

6. Linear accelerators utilizing polarized electron beams for at least part of their physics program include those at SLAC, TJNAF and MIT/Bates in the U.S.A., and Mainz, Bonn, and NIKHEF in Europe.

7. Fraser, J.S. et al., *Proc. 1987 IEEE Particle Accelerator Conf.*, Washington, DC, p. 1705.

8. One proposal, Aleksandrov, A.V. et al., *Nucl. Instrum. and Meth.* A **340**, 118 (1994), was motivated by the desire to produce a "cold" electron beam, while the second, Clendenin, J. et al., *Nucl. Instrum and Meth.* A **340**, 133 (1994), had production of polarized electrons for colliders explicitly as a goal.

9. *Proc. of the Int. Workshop on e⁻e⁻ Sources and Pre-Accelerators for Linear Colliders*, Schwerin, Germany, 29 Sep.-4 Oct. 1994, p. 239ff.

10. Tang, H. et al., "Polarized Electron Sources for Future e⁺e⁻ Linear Colliders," presented at the 1997 Particle Accelerator Conf., Vancouver, BC, 12-16 May 1997.

11. The oxides on the surface of an activated cathode have been observed to increase slowly with time, while the Cs remains constant. See Reichert, E. in *Polarized Beams and Polarized Gas Targets*, Singapore: World Scientific, 1996, p. 285.

12. Penetration of the fields to the conducting support would presumably result in field emission.

13. Tang, B. et al., *Il Nuovo Cimento* **106 A**, 231 (1993).

14. This value of the charge limit only applies to excitation at the band gap edge as is required for high polarization. The charge limit is much higher if the excitation energy is significantly greater than the band gap as is usually the case when using bulk GaAs.

15. Schmerge, J., SLAC, private communication (1998).

16. These values scaled from data given in Yoshioka, M. et al., *Proc. of the 1994 Int. Linac Conf.*, Tsukuba, Japan, p. 302. Progress in reducing dark currents has been summarized in Matsumoto, M., *Proc. of the XVIII Int. Linear Accelerator Conf.*, Geneva, CH, 1996, p. 626.

17. Suzuki, C. et al., "Reduction of the Field Emission Dark Current from Copper Surface in a High Gradient DC Field," presented at the 1st Asian Particle Accelerator Conf., Tsukuba, Japan, 23-27 March 1998.

18. Aulenbacher, K. et al., "RF Guns and the Production of Polarized Electrons," CLIC Note 303 and NLC-Note 20 (May 1996); Braun, H.H. et al., *Fifth European Particle Accelerator Conf.*, Stiges, Spain, 1996, p. 42; and Aulenbacher, K. et al., *Spin 96 Proceedings*, Singapore: World Scientific, 1997, p. 674.

19. Bossart, R. et al., "A 3 GHz Photoelectron Gun for High Beam Intensity," CLIC Note No. 297 (1995).

20. Aleksandrov, A.V. et al., *Fifth European Particle Accelerator Conf.*, Sitges, Spain, 1996, p. 1538.

21. Aleksandrov, A.V. et al., *VII Int. Workshop on Linear Colliders*, Zvenigorod, Russia, Sept. 29-Oct. 3, 1997, http://www.vlep.serpukhov.su/LC97/proceed/html/027/13.htm.

22. Aleksandrov, A.V., INP, private communication (1998).

23. Sze, S.M., *Physics of Semiconductor Devices*, 2nd ed., New York: John Wiley, 1981, p. 366 ff.

24. Mulhollan, G.A., SLAC, private communication (1998).

25. Braun, H. et al., "Gallium Arsenide photocathodes used in high electric fields–Results of experiment from November 1997 to January 1998," CTF Note 98-13 (1998).

26. Chevallay, E. et al., *Nucl. Instrum. and Meth.* A **340**, 146 (1994).

27. Hartmann, P. et al., *Nucl. Instrum. and Meth.* A **379**, 15 (1996).

28. Aleksandrov, A.V. et al., *Fifth European Particle Accelerator Conf.*, Stiges, Spain, 1996, p. 1535.

29. The current limit found for the Novosibirsk-Legnaro source was ~25 A cm⁻². The higher value can be attributed to excitation with much higher-energy laser pulses (524 nm), which, it should be pointed out, cannot be used for a high-polarization cathode.

30. See Togawa, K. et al., "Surface charge limit in NEA superlattice photocathodes of polarized electron source," DPNU-98-11 (4 March 1998), submitted to Nucl. Instrum. and Meth.

31.The initial collaboration was formed from a meeting at CERN of CERN, KEK, and SLAC physicists interested in developing a polarized rf gun for a future collider. Since then physicists at DESY and INP have indicated a desire to cooperate.

32. Rosenzweig, J.B., UCLA, private communication (1998), and presentation at this Workshop.

33. Rosenzweig, J.B. et al., "Beam Dynamics in an Integrated Plane Wave Transformer Photoinjector at S- and X-Band," presented at the 1997 Particle Accelerator Conf., Vancouver, BC, 12-16 May 1997.

# A Nonlinear Particle Dynamics Map of Wakefield Acceleration in a Linear Collider

T. Tajima, S. Cheshkov, W. Horton and K. Yokoya*

*Department of Physics and Institute for Fusion Studies*
*The University of Texas at Austin*
*Austin, Texas 78712 USA*
* *KEK National Laboratory for High Energy Physics, Japan*

**Abstract.** The performance of a wakefield accelerator in a high energy collider application is analyzed. In order to carry out this task, it is necessary to construct a strawman design system (no matter how preliminary) and build a code of the systems approach (a typical systems code approach was used, for instance, in SSC studies [1]). A nonlinear dynamics map built on a simple theoretical model of the wakefield generated by the laser pulse (or whatever other method) is obtained and we employ this as a base for building a system with multi-stages (and components) as a high energy collider. The crucial figures of merit for such a system other than the final energy include the emittance (that determines the luminosity). The more complex the system is, the more "opportunities" the system has to degrade the emittance (or entropy of the beam). Thus our map guides us to identify where the crucial elements lie that affect the emittance. We find that a strong focusing force of the wakefield coupled with a possible jitter of the axis (or laser aiming) of each stage and a spread in the betatron frequencies arising from different phase space positions for individual particles leads to a phase space mixing. This sensitively controls the emittance degradation. We show that in the case of a uniform plasma the effect of emittance growth is large and may cause serious problems. We discuss possibilities to avoid it and control the situation.

## INTRODUCTION

The use of strong plasma waves excited by laser beams for electron acceleration has been considered [2]. Since then the subject has grown tremendously and become an area of intense research. Many variant acceleration schemes have been suggested: plasma beat wave accelerator, laser wakefield accelerator, plasma wakefield accelerator. These differ in details and how one excites the wakefield or accelerating structure, but the basic idea is common. They also share common mathematical treatment when considered as an element of a system of a large scale accelerator. There are many possible applications of plasma based accelerators and

CP472, *Advanced Accelerator Concepts: Eighth Workshop,*
edited by W. Lawson, C. Bellamy, and D. Brosius

one of them, perhaps the most challenging, is for construction of a linear collider. It is believed that conventional linear colliders can go up to 1 TeV center of mass energy of colliding particles [3]. However, beyond that probably some new technology needs to enter. A feature of plasma based accelerators is their ability to sustain extremely large acceleration gradients ($\sim 100\,\mathrm{GV/m}$ ). Correspondingly we can hope to achieve high energy gains on short distance. Such a machine, no doubt, consists of a large number of elements to reach for a desired energy and forms a complex system. In order to identify the most crucial research questions in such a complex problem, it is necessary to design a strawman's system, no matter how simplistic it may be, and to try to analyze and deduce the fundamental properties of such a system. Such a strawman's design was in fact carried out for the first time for a wakefield acceleration-based collider in the last AAC meeting [4]. For an accelerator for high energy physics the energy is one of the important parameters, but there are many others that are crucial for such an accelerator. For instance, low emittance is also necessary. This is because the cross-section of collisions decreases inversely proportional to the energy of the beams and thus we need high luminosity to detect new physics. The requirement for luminosity, in turn, demands for low emittance. The geometrical luminosity is given by

$$\mathcal{L} = \frac{f_c N^2}{4\pi\sigma_x\sigma_y}, \tag{1}$$

where $f_c$ is the collision frequency, $N$ is the particle number per bunch and $\sigma_x$ and $\sigma_y$ are the r.m.s. beam sizes at the IP (interaction point). Thus, it is important to analyze the performance of laser wakefield accelerators not only energy-wise but also with respect to all other relevant beam parameters. The first thing to notice is that the whole acceleration process takes place over a period too short for ions to move. Therefore the analysis may be limited to considering the electron motion only in a background of immobile ions. This not only simplifies analysis, but (generally speaking) stabilizes the system, as immobile ions anchor electrons and photons. In most scenarios the desired final energy of accelerating particles ($\sim$ TeV) cannot be achieved over a single acceleration stage. Thus we need to evaluate the effects associated with multistaging and to analyze the complete acceleration process. In the present investigation we limit ourselves to the linear regime of wakefield generation. A major simplification arises from the separate treatment of beam electrons and plasma electrons. The plasma electrons are supporting the wakefield but not trapped by it. This is the sequence of the linear regime. On the other hand, the beam electrons are affected (accelerated and focused) by the wakefield.

In Sec.2 of this paper we analyze the Laser Wakefield Accelerator (LWFA) in the case of an initially homogeneous plasma and obtain expressions in a similar way as in [5] for the longitudinal and radial wakefields in the case of cylindrical geometry. The motion of an accelerated particle beam is considered in Sec.3. With the help of some simplifying assumptions valid in the ultrarelativistic case, we integrate the single particle motion. Based on the single stage analytical results, in Sec.4 we derive a map for multistage LWFA, which is used as a base for a tracking code.

Considering the target center of mass energy $E_{cm} = 5$ TeV and luminosity $\mathcal{L}_g = 10^{35}$ cm$^{-2}$s$^{-1}$ in [4], it is shown that for a fixed power $P_b$ of the colliding beams there are only two independent parameters describing the beam at IP: the number of particles per bunch $N$ and the longitudinal beam size $\sigma_z$ (we assume a round beam: the aspect ratio $R = 1$). All other beam parameters at IP can be expressed as functions of the above. So we know what the requirements on the normalized emittance are.

In Sec.5 we analyze the effect of random dislocations of the accelerating stages on the beam emittance. These dislocations combined with unavoidable spread in individual particle betatron frequencies lead to a considerable emittance growth which may require alternative LWFA design. Conclusions are drawn in Sec.6.

## WAKEFIELD MODEL

A short high power laser pulse propagating in plasma can excite wakefields which may be used for acceleration of charged particles. In the linear regime the plasma responds to the ponderomotive force acting on plasma electrons:

$$\mathbf{F} = e\nabla\Phi_l(r, z, t), \tag{2}$$

where the ponderomotive potential $\Phi_l$ is connected with the laser vector potential $A_l$:

$$\Phi_l(r, z, t) = -\frac{m_e c^2}{2e} a_0^2(r, z, t),$$

$$a_0 = \frac{eA_l}{m_e c^2}, \tag{3}$$

and $a_0$ is the normalized vector potential. The plasma response to this force can be obtained from the cold fluid equations ( [5]):

$$\frac{d}{dt}\mathbf{v} = -\frac{e}{m_e\gamma}\left\{\mathbf{E} + \frac{1}{c}\mathbf{v}\times\mathbf{B} - \frac{1}{c^2}\mathbf{v}(\mathbf{v}\cdot\mathbf{E})\right\}, \tag{4}$$

$$\frac{\partial}{\partial t}n + \nabla\cdot n\mathbf{v} = 0, \tag{5}$$

where $n$ is the electron density and $\gamma$ is the Lorenz factor. These equations could be solved perturbatively, assuming that the density perturbation is relatively small.

For a laser pulse of the form (with a pulse length $l_l$, spot radius $r_s$)

$$\mathbf{A_l} = \begin{cases} \mathbf{A_{l0}}\sin\left(\frac{\pi\xi}{l_l}\right)\exp\left(-\frac{r^2}{r_s^2}\right), & 0 < \xi < l_l, \\ 0, & \text{otherwise}, \end{cases} \tag{6}$$

where $\xi = z - ct$, the maximum electric field in the $z$-direction behind the pulse ($\xi < 0$) is

$$E_z(\xi, r) = \frac{\pi^2}{l_l} \Phi_{l0} \exp\left(-\frac{2r^2}{r_s^2}\right) \cos\left(2\pi\xi/l_l\right),\tag{7}$$

where $\Phi_{l0}$ is given by

$$\Phi_{l0} = -\frac{m_e c^2}{2e} a_0^2.\tag{8}$$

The maximum field (7) is reached when the resonance condition is satisfied: $\lambda_p \approx l_l$, where $\lambda_p$ is the plasma wave wavelength $2\pi c/\omega_p$. The transverse electric field $E_r$ and magnetic field $B_\theta$ are generated according to the Panofski–Wenzel theorem [6]

$$\frac{\partial E_z}{\partial r} = \frac{\partial(E_r - B_\theta)}{\partial \xi},\tag{9}$$

leading to

$$(E_r - B_\theta) = -\frac{2\pi}{r_s^2} \Phi_{l0} r \exp\left(-\frac{2r^2}{r_s^2}\right) \sin\left(2\pi\xi/l_l\right).\tag{10}$$

We observe that for a relativistic particle ($v_z \approx c$) the transverse force will be proportional to $(E_r - B_\theta)$. There is a region in the wake (quarter period) where a relativistic electron experiences simultaneous acceleration and transverse and longitudinal focusing. This feature of the LWFA makes it different from the conventional accelerators and we do not need the alternating gradient principle here. At the same time, however, it may pose a problem, as we shall see in Sec.5. The wakefield structure of this model is common to other sisters of wakefield accelerators such as PBWA and PWA (See, for example, [7]).

## ELECTRON MOTION IN THE PLASMA WAKEFIELD

The main acceleration follows an electron injector which can be used as a charged particle source for our accelerator. Designing such an injector is a task itself (e.g. [8], [9], [10]), but we leave it to investigation elsewhere.

We consider the motion of high energy electrons of the beam in the plasma wakefield. The complicated problem may be simplified to obtain the essential dynamics as far as the beam particle dynamics is concerned with the following assumptions:

1. The region in the phase space occupied by the beam is small.

2. The wakefield is not affected by the beam (the beam loading effect, however, will be considered in a follow up paper to be published in the same proceedings).

3. The particles in the beam are highly relativistic and move predominantly in $z$-direction (which is the direction of propagation of the laser pulse).

$$\dot{z} \gg \dot{x}, \dot{y}$$

$$\dot{z} \approx c$$

4. The motions in $x$ and $y$ are decoupled.

5. No interaction among the beam particles.

6. The laser pulse does not evolve.

These assumptions do not take into account a significant part of the physics of all the processes. Work in progress on the problem will relax some of these assumptions. The wakefield generated by the beam can be included in the considerations using the results in [11]. This will be carried out in a follow-up paper to be published in the present proceedings. Assumption 5 is probably good enough for high energy particles and relatively low currents, because the space charge force diminishes by a factor of $1/\gamma^2$. Assumption 6 is related to the pump-depletion problem [12] and also should be taken into account in the future.

Starting with the single particle equation of motion

$$\frac{d\mathbf{p}}{dt} = -e\left(\mathbf{E} + \frac{\mathbf{v} \times \mathbf{B}}{\mathbf{c}}\right), \tag{11}$$

and assuming that the beam particles are close to the $z$-axis, we obtain the following basic system of differential equations for the longitudinal motion

$$\frac{d\gamma}{d\hat{z}} = \Phi_0 \cos \Psi, \tag{12}$$

$$\frac{d\Psi}{d\hat{z}} = 1 - \frac{\beta_p}{\beta}, \tag{13}$$

where $\Psi$ is the particle phase with respect to the wakefield. For the transverse motion

$$\frac{d\hat{p}_u}{d\hat{z}} = -\frac{4\Phi_0}{r_s^2}\hat{u}\sin\Psi, \tag{14}$$

$$\frac{d\hat{u}}{d\hat{z}} = \frac{\hat{p}_u}{\gamma k_p^2}, \tag{15}$$

where

$$\beta_p = \sqrt{1 - (v_p/c)^2}, \quad \Phi_0 = \frac{e\pi E_0 a_0^2}{4k_p m_e c^2} = \frac{\pi a_0^2}{4}, \tag{16}$$

$$\hat{z} = k_p z \,, \quad \hat{u} = u \,, \quad \hat{p}_u = \frac{k_p}{m_e c}\, p_u \,. \tag{17}$$

Here $E_0 = cm_0\omega_p/e$, $k_p = \omega_p/c$, $v_p$ is the phase velocity of the wake and $u$ and $p_u$ stand for transverse variables $x$ and $p_x$ or $y$ and $p_y$. After convenient normalizations, the important points are that we use $k_p z$ as our time coordinate and the energy and the phase of the particles with respect to the wake are our 'longitudinal' variables.

Equations (12), (13) decouple from (14) and (15) and we can consider these two sets independently. The first set is conveniently analyzed using the following one-dimensional Hamiltonian [13]:

$$H = \gamma(1 - \beta\beta_p) + \Phi(\Psi), \tag{18}$$

where

$$\Phi(\Psi) = -\Phi_0 \sin \Psi. \tag{19}$$

In the phase space formed by the first pair of variables $(\gamma, \Psi)$ we have stable fixed points: $\gamma = \gamma_p$ and $\Psi = \pi/2 + 2n\pi$ and unstable fixed points: $\gamma = \gamma_p$ and $\Psi = -\pi/2 + 2n\pi$.

There are two phase space regions – trapped, where the particles execute bounded motion and untrapped one, where the motion is unbounded in $\Psi$ direction [1] (see Figure 1). Because we are primarily interested in high energy physics applications of wakefield accelerator here, we consider the untrapped case, where the particle orbits are well above the separatrix. It is important to understand that even initially we have very high energy particles, so the upper untrapped region is our operating region.

We can further simplify the equations of motion for $\gamma$ and $\Psi$ by putting $\beta = 1$ to obtain:

$$\frac{d\gamma}{d\hat{z}} = \Phi_0 \cos(\Psi) \,, \tag{20}$$

$$\frac{d\Psi}{d\hat{z}} = \frac{1}{2\gamma_p^2}, \tag{21}$$

where $\gamma_p = 1/\sqrt{1 - \beta_p^2}$. These equations are integrated directly to give

$$\Delta\gamma = 2\Phi_0\gamma_p^2(\sin \Psi - \sin \Psi_0) \,, \tag{22}$$

$$\Psi = \Psi_0 + \frac{\hat{z}}{2\gamma_p^2} \,, \tag{23}$$

where $\hat{z} = k_p z$ and $\Psi_0$ is the initial phase of the particle with respect to the wakefield. First we observe that the maximum energy gain is $2\Phi_0\gamma_p^2$ and taking

---

[1] the graph is not to scale.

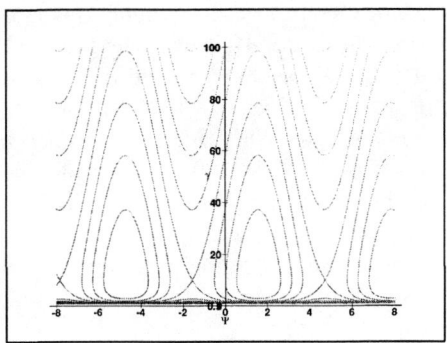

**FIGURE 1.** The phase space: electron Lorentz factor $\gamma$ vs. the phase with respect to the wakefield $\Psi$.

typical values of the parameters $\Phi_0 = 0.2$ (which corresponds to $a_0 = 0.5$ [2]) and $\gamma_p = 100$ we see that the above gain is about $4 \cdot 10^3$ in units of electron's rest energy or in other words about 2 GeV. This energy is achieved over a 'distance' $\hat{z}$ of $3 \cdot 10^4$ which is about 50 cm. The actual gain will be smaller if the pump depletion [12] is taken into account. Keeping in mind this tentative 'design' of one stage, we now look at the multistaging effects.

## MULTISTAGE ACCELERATION

When we try to accelerate particles to TeV energies, we need to investigate problems associated with the multistaging. In order to carry this out, we would like to obtain a map which describes the one to one correspondence between the entrance coordinates and the exit coordinates of the beam particles during the propagation of the beam trough many accelerating stages and use it to build a systems code. As in the standard RF linac theory, we have a reference particle moving along the ideal (design or synchronous) orbit. All other particles in the bunch will be described by their relative position with respect to the reference one.

The linearized equations of motion for the longitudinal degrees of freedom are:

$$\delta\Psi_{n+1} = \delta\Psi_n \tag{24}$$

$$\delta\gamma_{n+1} = 2\gamma_p^2 \Phi_0 (\cos(\Psi_s + \Delta) - \cos(\Psi_s))\delta\Psi_n + \delta\gamma_n, \tag{25}$$

where $n$ enumerates the stage, $\Psi_s$ is the 'synchronous' phase, $\Delta$ is the phase slippage per accelerating stage (actually it can also depend on $n$). Because of the fact that we are considering extremely high energy particles ($\gamma \sim 10^5 - 10^7$) the equation (24) decouples from (25) (There is a very small coupling term $\sim 1/\gamma^2$). Formally

---

[2] We take $a = 0.5$ to be still in the 'controlled' linear regime.

the equations look the same as in standard linac theory when the synchrotron oscillation frequency approaches zero. However, the physical regime of operation is quite different from the RF linac – we have a significant phase slippage over a stage (it is precisely this slippage which gives us the energy gain). And it also limits the maximum possible gain per stage. This difference comes from the fact that the plasma wave is relatively "slow" ($\gamma_p \approx 100$, instead of $\infty$). From equations (24) and (25) we see that in the approximation we are working in the phases of the particles do not change and the absolute energy spread increases linearly with the stage number (actually this is the beginning of a very slow synchrotron oscillation which happens on a time scale much greater than the time it takes a particle to travel the whole accelerator).

Now let us consider the transverse motion. It is described by

$$\ddot{\tilde{u}} + \left[ \omega_\beta^2 \sin\left( \omega_s \hat{z} + \Psi_s + \delta\Psi_n \right) - \frac{1}{2}\frac{\ddot{\gamma}}{\gamma} + \frac{1}{4}\frac{\dot{\gamma}^2}{\gamma^2} \right] \tilde{u} = 0 \,, \tag{26}$$

where

$$\omega_s = \frac{1}{2\gamma_p^2} \,, \tag{27}$$

$$\omega_\beta = \left( \frac{4\Phi_0}{\gamma (k_p r_s)^2} \right)^{1/2} \,, \tag{28}$$

$$\tag{29}$$

and $\tilde{u} = \sqrt{\gamma}\, u$. In the high energy regime the third term in the square brackets in (26) is negligible and the second term is usually also small because of the proportionality to $1/\gamma_p^2$. Still, in the cases of very weak focusing we have to take it into account (for instance, in plasma channel).

More generally, the equation we are considering is Hill's equation:

$$\ddot{\tilde{u}} + f(\hat{z})\tilde{u} = 0. \tag{30}$$

An analytic solution can be found when some additional approximations are used. We treat two simple cases where we can obtain analytical results easily before we discuss the general case. The simplest and the first thing to do is to approximate the focusing function in (30) by some constant value and then the equation describes just a simple harmonic oscillator. We adopt this, and also assume just a free drift (which can be replaced by magnets, if necessary) of the particles between the stages. Let us forget for a moment that the particles are being accelerated, and that the strength of the focusing force actually depends on the stage even if the stages are physically identical. To get a stable solutions we need to satisfy:

$$|\mathrm{Tr}\, M| < 2, \tag{31}$$

where $M$ is the transfer matrix:

$$M = \begin{pmatrix} \cos(\frac{\omega}{\omega_s}\Delta), & \frac{1}{\omega}\sin(\frac{\omega}{\omega_s}\Delta) \\ -\omega\sin(\frac{\omega}{\omega_s}\Delta), & \cos(\frac{\omega}{\omega_s}\Delta) \end{pmatrix} \cdot \begin{pmatrix} 1 & \hat{L} \\ 0 & 1 \end{pmatrix}, \tag{32}$$

where $\Delta$ is the phase advance of the particle per stage with respect to the wakefield and $\hat{L}$ is the length of the distance between the stages in units of $k_p^{-1}$. ($1/\omega$ is betatron length in units $k_p^{-1}$). The matrix (32) can be written as

$$M = \begin{pmatrix} \cos(\frac{\omega}{\omega_s}\Delta), & \frac{1}{\omega}\sin(\frac{\omega}{\omega_s}\Delta) + \hat{L}\cos(\frac{\omega}{\omega_s}\Delta) \\ -\omega\sin(\frac{\omega}{\omega_s}\Delta), & -\hat{L}\omega\sin(\frac{\omega}{\omega_s}\Delta) + \cos(\frac{\omega}{\omega_s}\Delta) \end{pmatrix}. \tag{33}$$

The map $\mathcal{M}$ for the total accelerator system is

$$\mathcal{M} = M^N, \tag{34}$$

where $N$ is the total number of stages. When we do not have any drift space, the solutions are always stable. If we increase $\hat{L}$ keeping the other parameters fixed at some point we reach a "blow-up" of emittance. So the maximum distance between the stages is limited. The trace of $M$ is:

$$\mathrm{Tr}M = 2\cos\left(\frac{\omega}{\omega_s}\Delta\right) - \omega\hat{L}\sin\left(\frac{\omega}{\omega_s}\Delta\right) \tag{35}$$

and for stability it should satisfy (31). This result was used to check the map code – up to a some value of $L$ there is no emittance growth (see Figure 2) and after that we indeed find the "blow up".

Now, this calculation does not take into account the fact that particles accelerate and $\omega_\beta$ is decreasing ( $\omega_\beta = \sqrt{\frac{4\Phi_0}{\gamma(k_p r_s)^2}}$ ). Also in reality particles have different (random) phases with respect to the wakefield and different energies which causes a spread in the individual particle betatron frequencies. Therefore , the above analysis should be done for each particle separately, but if the differences in their phases are small the conditions for stable motion are practically the same for all the particles.

The other case is when the argument of the sine in (26) is relatively small and we can use $\sin(x) \approx x$. In this case (26) reduces to

$$\ddot{\tilde{u}} - z'\tilde{u} = 0, \tag{36}$$

whose solutions are well known – they are expressible as linear combination of the Airy functions $Ai(z')$ and $Bi(z')$.

In the general case we can do the following analysis. Suppose we know two fundamental solutions to (30) when $\delta\Psi_n = 0$. Say $\tilde{u}^1(\hat{z})$ and $\tilde{u}^2(\hat{z})$. Then

$$\tilde{u} = C_1\tilde{u}^1(\hat{z}) + C_2\tilde{u}^2(\hat{z}). \tag{37}$$

161

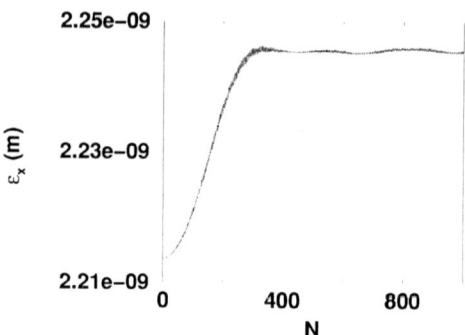

**FIGURE 2.** The normalized $x$-emittance $\epsilon_x$ vs. stage number $N$; $\gamma_p = 100$, $\hat{L} = 10000$, $\epsilon_x^0 = 2.2$ nm, $r_s = 0.5$mm, $a_0 = 0.5$, no acceleration, no dislocations, $\delta\gamma/\gamma = 0.01$, $\delta\Psi = 0.01$.

And let us assume that

$$\tilde{u}^1(0) = 1, \qquad \dot{\tilde{u}}^1(0) = 0 \tag{38}$$

$$\tilde{u}^2(0) = 0, \qquad \dot{\tilde{u}}^2(0) = 1, \tag{39}$$

so that

$$\tilde{u}_{n+1} = \tilde{u}_n \tilde{u}^1(\hat{l}) + \dot{\tilde{u}}_n \tilde{u}^2(\hat{l}) \tag{40}$$

$$\dot{\tilde{u}}_{n+1} = \tilde{u}_n \dot{\tilde{u}}^1(\hat{l}) + \dot{\tilde{u}}_n \dot{\tilde{u}}^2(\hat{l}). \tag{41}$$

For a particle which has some finite $\delta\Psi_n$, equation (30) becomes

$$\ddot{\tilde{u}} + f(\hat{z} + \delta\Psi_n/\omega_s)\tilde{u} = 0, \tag{42}$$

where $f(\hat{z})$ is the same (as a function) as in (30). In this case

$$\tilde{u} = C_1 \tilde{u}^1(\hat{z} + \delta\Psi_n/\omega_s) + C_2 \tilde{u}^2(\hat{z} + \delta\Psi_n/\omega_s). \tag{43}$$

It can be shown easily that

$$\tilde{u}_{n+1} = \tilde{u}_n \left( \dot{\tilde{u}}^2(\delta\Psi_n/\omega_s)\tilde{u}^1(\hat{l} + \delta\Psi_n/\omega_s) - \dot{\tilde{u}}^1(\delta\Psi_n/\omega_s)\tilde{u}^2(\hat{l} + \delta\Psi_n/\omega_s) \right) +$$

$$+ \dot{\tilde{u}}_n \left( \tilde{u}^1(\delta\Psi_n/\omega_s)\tilde{u}^2(\hat{l} + \delta\Psi_n/\omega_s) - \tilde{u}^2(\delta\Psi_n/\omega_s)\tilde{u}^1(\hat{l} + \delta\Psi_n/\omega_s) \right), \tag{44}$$

and

$$\dot{\tilde{u}}_{n+1} = \tilde{u}_n \left( \dot{\tilde{u}}^2(\delta\Psi_n/\omega_s)\dot{\tilde{u}}^1(\hat{l} + \delta\Psi_n/\omega_s) - \dot{\tilde{u}}^1(\delta\Psi_n/\omega_s)\dot{\tilde{u}}^2(\hat{l} + \delta\Psi_n/\omega_s) \right) +$$

$$+ \dot{\tilde{u}}_n \left( \tilde{u}^1(\delta\Psi_n/\omega_s)\dot{\tilde{u}}^2(\hat{l} + \delta\Psi_n/\omega_s) - \tilde{u}^2(\delta\Psi_n/\omega_s)\dot{\tilde{u}}^1(\hat{l} + \delta\Psi_n/\omega_s) \right). \tag{45}$$

The point here is that even though in general $\tilde{u}^1(\hat{z})$ and $\tilde{u}^2(\hat{z})$ cannot be found in analytical form and we need to find them numerically it has to be done only once, because they are the same for all the particles. This is an exact statement only if we neglect spread in $\gamma$, in general it is not possible, and then the solution is to linearize (30) assuming $\delta\gamma$ and $\delta\Psi$ small and then do perturbation analysis. Of course the above analysis would modify the map but it is not going to change the final results significantly. So, in general (of course, we again incorporate the drift space),

$$\mathcal{M} = M_N M_{N-1}...M_2 M_1. \tag{46}$$

We note that because of the common structure of the wakefield in all plasma based accelerators the obtained map, with just slight modifications, can be used to analyze their performance as well.

We coded the map in the case of a constant betatron frequency (a different constant for each stage and each particle depending on the stage number and particle distribution in the $(\gamma, \Psi)$ phase space) and tests showed that when we start with a normalized r.m.s. emittance of $\epsilon_u = 2.2$ nm (and in the code particles of course get accelerated) up to some value of the drift space the emittance is preserved (See Figure 3). This emittance corresponds to the 5 TeV design I presented in [4].

## JITTER AND NOISE

For a complex system cumulative errors can give rise to a surprising (and often unpleasant) result. We identify that one of the most important such effects stems from the transverse jitter of the aligned wakefield (by whatever mechanism) stage by stage. The problem here is that up to this point we have not considered possible dislocation of the consequent stages. This, combined with the fact that the focusing force is different for different particles can lead to a severe emittance growth. Basically, what happens is that all particles rotate at different angular velocities in the phase space and if there is a stage position shift present, we get a characteristic "banana" shaped distribution (it is "banana" shaped only if the dislocation size is larger than the beam size, but in any case the particle distribution gets diluted because of the misalignments). This process critically depends on the magnitude of the betatron frequency spread which means that the typical strength of the focusing force is of great importance. The effect of plasma noise (or other noise, such as laser or the boundary) on the particle dynamics over a stage may be incorporated in a map similar to the stage-by-stage jitter. Such dynamics results in a fuzzy or stochastic [14] map.

We consider the case of stage jitter (in reality there are longitudinal stage jitters as well, but in this paper we restrict ourselves to the transverse case only). The dislocation of the aligned position of each stage is given in our code as a stochastic variable which has a Gaussian distribution. The map is modified according to:

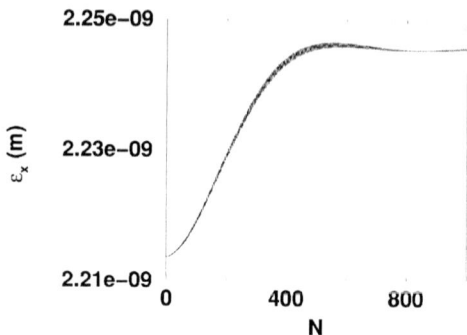

**FIGURE 3.** The normalized $x$-emittance $\epsilon_x$ vs. stage number $N$; $\gamma_p=100$, $\hat{L}=10000$, $\epsilon_x^0=2.2$ nm, $r_s=0.5$mm, $a_0 = 0.5$, no dislocations, $\delta\gamma/\gamma=0.01$, $\delta\Psi = 0.01$.

$$\begin{pmatrix} \tilde{x}_{n+1} \\ \dot{\tilde{x}}_{n+1} \end{pmatrix} = M_n \begin{pmatrix} \tilde{x}_n - \tilde{\mathcal{D}} \\ \dot{\tilde{x}}_n \end{pmatrix} + \begin{pmatrix} \tilde{\mathcal{D}} \\ 0 \end{pmatrix}, \tag{47}$$

where $\mathcal{D}$ is the misalignment size ($\tilde{\mathcal{D}} = \sqrt{\gamma_n}\mathcal{D}$). The longitudinal degrees of freedom are not affected. Run with random dislocations of magnitude $\sigma_\mathcal{D} = 1 \cdot 10^{-7}$ m is presented in Figure 4 and Figure 5. We see that in this case (corresponds to design I [4]) we have a severe emittance growth (the initial normalized emittance is 2.2 nm). We have to point out that even though there are cases corresponding to large spot sizes which preserve the normalized emittance quite well their practical realization would require a huge laser power probably well above any future experimental limits. The other possible way to cure the situation is to increase the number of stages (and decrease $\gamma_p$) which also does not seem plausible (it may be difficult practically to have more than a thousand stages). To control the laser aiming and beam position better than 0.1 $\mu$m during the acceleration process is also not promising. In general, the way to alleviate the situation is to decrease the strength of the focusing force. Here comes the idea of possible use of a preformed hollow plasma channel [15]. The conclusion is that the problem is quite serious and should be analyzed in detail. We do this in a follow up paper to appear in the same proceedings. Here we just note that in the case of a initially uniform plasma, for typical values of parameters, we encounter a severe emittance growth in the presence of small stage jitters. The difficulty is primarily due to the fact that the wakefield focusing force is too large in this case. We should also remember that the above considerations do not include any nonlinear effects which also contribute to the phase area increase, not to mention that in addition to all of the above we have to come up with a mechanism for guiding of the laser pulse (when the laser spotsize is small), otherwise the diffraction may limit significantly the acceleration length which is incompatible with our design.

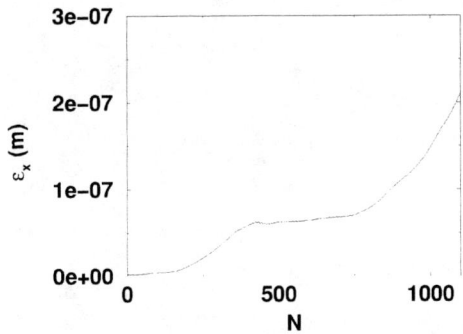

**FIGURE 4.** The normalized $x$-emittance $\epsilon_x$ vs. stage number $N$; $\gamma_p=100$, laser wavelength $\lambda = 1\mu$m, $\hat{L}=10000$, $\epsilon_x^0=2.2$ nm, $r_s=0.5$mm, $a_0 = 0.5$, dislocation size $= 0.1\mu$m, $\delta\gamma/\gamma=0.01$, $\delta\Psi = 0.01$. The final energy is 2.5 TeV.

# CONCLUSION

We studied the cumulative effects of the successive acceleration, transport, and focusing in the laser wakefield (or its sister methods) over many stages. Such cumulative dynamical behaviors are important to investigate for real world accelerators such as a high energy collider. Errors arising from the jitter of each stage or equivalently (in our map approach) the noise in the system can accumulate in such a way to degrade some of the parameters of the beam. The most crucial of these may be the emittance (or the entropy) of the beam. We showed that a set of stages with an ideal wakefield acceleration, drift, and focusing can preserve even a very small emittance over a thousand stages while the energy of the accelerated particles reaches the goal of 5 TeV center of mass energy.

When we have stochastic variables on the wakefield (we chose the stage jitter of the axis of the wakefield in particular), the emittance can significantly increase over the many stages due to the strong focusing of the wakefield. This is probably the most serious effect on the long range behavior of the beams in this kind of accelerator for high energy applications. Search for the best parameter set in the multidimensional parameter space of a large scale accelerator should be performed taking into account, to our best notion, future experimental limits and restrictions which might come from them. The last point we would like to make is that the proposed map for LWFA gives us an opportunity to study not only effects of stage jitter, but also effects which come from other random sources i.e. plasma noise. The noise enter our map equations in the form of random kicks in the r.h.s. of (24) and (25). To be able to do this, we need the statistical distributions of these kicks, which will be presented in a future publication. The work is supported by US DoE.

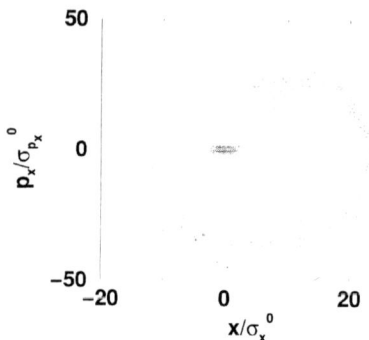

**FIGURE 5.** The phase space $p_x$ vs. x. corresponding to Figure 4. The original distribution is taken Gaussian (beam is assumed initially matched, so $\sigma_x^0 = \sqrt{\epsilon_x^0 \beta_x^0 / \gamma^0}$ and $\sigma_{p_x}^0 = \sigma_x^0 / \beta_x^0$) and the final after 1100 acceleration stages.

# REFERENCES

1. Yan, Y., Naples, J. and Syphers, M., eds., *Accelerator Physics at the SSC* (AIP, New York 1995).
2. Tajima, T. and Dawson, J., *Phys. Rev. Lett.* **43**, 267-270 (1979).
3. The NLC Design Group and NLC Physics Working Group. *Physics and technology of the Next Linear Collider.* A Report Submitted to Snowmass 96 (1996).
4. Xie, M., Tajima, T., Yokoya, K. and Chattopadhyay, S., "Studies of laser-driven 5 TeV $e^+e^-$ colliders in strong quantum beamstrahlung regime", *Advanced Accelerator Concepts 7* (AIP, New York 1997conf), p.233-242.
5. Esarey, E., Ting, A., Sprangle, P. and Joyce, G., *Comments Plasma Phys. Controlled Fusion* **12**, 191–204 (1989).
6. Chen, P., *Part. Accelerators* **20**, 171–182 (1985).
7. Breizman, B., Fisher, D., Chebotaev, P. and Tajima, T., *IFS report, University of Texas at Austin* **502**, (1991).
8. Rau, B., Tajima, T. and Hojo, H., *Phys. Rev. Lett.* **78**, No. 17, 3310–3313 (1997).
9. Umstadter, D., Kim, J. and Dodd, D., *Phys. Rev. Lett.* **76**, No. 12, 2073–2076 (1996).
10. Esarey, E., Hubbard, R., Leemans, W., Ting, A. and Sprangle, P., *Phys. Rev. Lett.* **79**, No. 14, 2682–2685 (1997).
11. Katsouleas, T., Wilks, S., Chen, P., Dawson, J. and Su, J., *Part. Accelerators* **22**, 81–98 (1987).
12. Horton, W. and Tajima, T. *Phys. Rev. A* **34**, 4110–4119 (1986).
13. Esarey, E., Sprangle, P., Krall, J. and Ting, A., *IEEE Trans. Plasma Science* **24**, No. 2, 252–288 (1996).
14. Van Kampen, N., *Phys. reports* **24**, No. 3, 171–228 (1976).
15. Chiou, T., Katsouleas, T., Decker C., Mori, W., Wurtele, J., Shvets, G. and Su., J., *Phys. Plasmas* **2**, No. 1, 310–318 (1995).

# II

# WORKING GROUP SUMMARIES

# Summary of Working Group 1
# Beam Physics at High Energy Densities

Swapan Chattopadhyay* and Pisin Chen[†]

*Lawrence Berkeley National Laboratory. Berkeley. California 94720
†Stanford Linear Accelerator Center. Stanford. California 94309

## THE PROGRAM

The topics explored in the Working Group 1 at the Advanced Accelerator Concepts Workshop in Baltimore, Maryland (July 5 - 11, 1998) were concerned with Beam Physics at High Energy Densities. Particle beams of high phase space density and the associated environment of high energy density electromagnetic radiation — both appear at various stages of existing and envisioned colliders. Accordingly, in addition to general discussions on nonlinear Compton scattering, phase-space cooling, coherent electromagnetic emissions and plasma wakefield acceleration schemes, there were focussed discussions and presentations on two major collider frontiers of the day: (a) the severe electromagnetic and hadronic background environment at the IP of high energy multiple-TeV electron-positron-gamma colliders; and (b) muon beams for high energy muon colliders.

In the context of multiple-TeV $e^+e^-$ colliders, there were general discussions on Interaction Point (IP) considerations at high energies and energy densities. These included discussions on dynamic focusing using the electron and positron bunches themselves as lenses and various other novel approaches to the final focus at high luminosities. Such approaches include four-beam space-charge and current neutralization schemes at the IP, matrixed linacs, etc. There were also discussions on nonlinear scattering involving multiple photon interactions with electrons.

In the context of phase space cooling of muons, there were also discussions on laser cooling of ion beams, pick-ups and kickers for optical stochastic cooling techniques and the promise of exciting atomic physics experiments at the Relativistic Heavy Ion Collider at BNL using supercooled ion beams.

In the domain of advanced acceleration concepts not specific to any particular collider, there were discussions on high-brightness laser synchrotrons, coherent radio Cerenkov emission, laser-plasma and beam-plasma wakefield accelerators and a most interesting exploration of ideas for plasma wakefield accelerators for protons. Finally a speculative concept of testing Unruh Radiation using ultra-intense lasers was vigorously debated.

CP472, *Advanced Accelerator Concepts: Eighth Workshop,*
edited by W. Lawson, C. Bellamy, and D. Brosius
© 1999 The American Institute of Physics 1-56396-889-4/99/$15.00

The participants of Working Group I and its four-day agenda outside of the general sessions of the Workshop, are listed in Appendix A and Appendix B respectively.

# SYNOPSIS OF WHAT WE LEARNED

In the context of multiple-TeV high energy (5 TeV center-of-mass energy, say) electron-positron colliders, in-depth studies and computations of Xie, Ohgaki and Murayama indicate that indeed at such high energies the beamstrahlung is severely "quantum-suppressed" and provides a domain of collisions compatible with higher luminosities, shorter bunch lengths and in general, with most "laser acceleration" schemes. This is in sharp contrast with the situation for "classical colliders" at lower energies and luminosities where the beamstrahlung poses one of the major limitations to the collider performance. It is also fair to say that one now has reasonable estimates of not only the electromagnetic, but also hadronic backgrounds in 5-TeV class $e^+e^-$ colliders of tomorrow. This allows good-faith phenomenological estimates of the $e^+e^-$ physics at such energies. We refer the readers to the various papers by M. Xie, T. Ohgaki, H. Murayama, K. Thompson and P. Chen in these proceedings for further details.

Conventional magnetic final focussing of electron and positron bunches is a major awkward element in linear colliders, leading to multiple-kilometers long final focus lines dominating the region between the high energy accelerator end and the interaction point. Various clever schemes of plasma focussing generated by the dynamic focussing effects of stored beams on colliding bunches, four-beam compensation, multiple-matrixed linacs, etc. promise to provide alternatives which are attractive enough for further exploration. We refer the reader to articles by J. Irwin, D. Whittum and F. Zimmerman in these proceedings for these new schemes.

In the context of Muon Colliders, while the general layout of the collider complex is still evolving, much remain to be tested in the domain of phase space cooling of muons. Ionization Cooling offers the only promise to date of generating muon beams of adequate transverse density for collision purposes. There is a significant amount of modelling work done that has already yielded a reasonable muon cooling scenario and led to the definition of a prototype R & D experiment on muon cooling. Significant challenges remain, e.g., controlling tails of multiple scattering, energy straggling, transverse-longitudinal exchange without emittance dilution, matching cooling sections, etc. We refer the reader to articles by D. Neuffer and R. Fernow in these proceedings for further details. For studies of polarization effects in muon colliders, we refer to the article by Z. Parsa.

The possibility of studying parity nonconservation in relativistic hydrogenic ions can be made into a reality if one could implement a super-cooled ion beam $\left(\Delta p/p \sim 10^{-6}\right)$ based on a proton-filled relativistic heavy-ion collider. Such possibilities were discussed by M. Zolotorev. Paul Schoessow discussed coherent Cerenkov radio emission from high energy electromagnetic showers expected in detection studies of ultrahigh energy cosmic ray neutrinos. This has led to a proposed proof-of-principle experiment for the Fermilab Main Injector.

Various institutions and laboratories around the world are planning on using Terawatt table top lasers in conjunction with relativistic electron beams for various vacuum laser acceleration, plasma wakefield acceleration, chirped pulse IFEL, Thomson- and Compton-scattered x-ray and $\gamma$-ray sources for medical and nuclear spectroscopic applications, etc. We refer to articles by K. Nakajima, T. Tajima, and F. Hartemann in these proceedings for further details.

Finally, proton acceleration by plasma waves was discussed and advantages and limitations explored. Proton acceleration by backward Raman scattering in a plasma is possible up to a few hundred MeVs using tapered plasmas with maximum densities reaching up to $4 \times 10^{20}$ cm$^{-2}$. We refer to the articles by A. Ogata and T. Katsouleas in these proceedings. Finally D. Finley of FNAL provided much of the motivation for proton acceleration studies and guided a whole afternoon session on plasmas and proton beams. Issues of energy, learning from electrons and present and future work were all touched upon and articulated in D. Finley's article in these proceedings. While acceleration of protons from $\gamma > 10$ to $\gamma \sim 50,000$ can be contemplated using plasmas, the nuclear scattering of plasma ions against protons for long channels required for high energy applications may pose a severe limitation eventually.

Overall we would summarize by stating that the tools to understand beams of high energies and energy densities are getting increasingly refined, leading to alternative and novel applications in manipulation of beams — be it in high energy colliders, radiation sources, atomic physics or gravitational physics.

# ACKNOWLEDGEMENT

This work was supported by the U.S. Dept. of Energy under Contract Nos. DE-AC03-76SF00098 and DE-AC03-76SF00515.

171

# APPENDIX A

Swapan Chattopadhyay, LBNL
Pisin Chen, SLAC
Sergey Cheshkov, UT Austin
David Cline, UCLA
Rick Fernow, BNL
David Finley, FNAL
Wei Gai, ANL
Fred Hartemann, LLNL
John Irwin, SLAC
Alexander Mikhailichenko, Cornell University
Kazuhisa Nakajima, KEK, Japan
David Neuffer, FNAL
Tomomi Ohgaki, LBNL
Soo Young Park
Zoreh Parsa, BNL
John Power, ANL
Martin Reiser, Univ. of Maryland
Paul Schoessow, ANL
Toshiki Tajima, UT Austin
Kathleen Thompson, SLAC
David Whittum, SLAC
Mark Wilson, US DOE
Ming Xie, LBNL
Frank Zimmermann, SLAC
Ping Zan, SRRC, Taiwan, ROC
Max Zolotorev, LBNL

# APPENDIX B

AGENDA
Working Group 1
Beam Physics at High Energy Densities
Leaders: S. Chattopadhyay and P. Chen

## Tuesday:  July 7, 1998
1:30 - 2:00   M. Xie. "Quantum Suppression of Beamstrahlung"
2:00 - 2:30   T. Ohgaki, "Hadronic Backgrounds in $e^+e^-$ Collisions"
2:30 - 3:00   K. Thompson. "$e^+e^-$ Pair Production in Deep Quantum Regime"
3:00 - 3:30   F. Hartemann, "Nonlinear Compton Scattering"
3:30 - 4:00   BREAK
4:00 - 4:30   H. Nakajima. "A Compact High Brightness Laser Synchrotron Source for Medical Applications"
4:30-5:00   P. Schoessow, "Coherent Radio Cerenkov EmissionNeutrino Wakefields"
5:00-5:30   M. Zolotorev. "Pick-up and Kicker for Optical Stochastic Cooling"

## Wednesday: July 8, 1998
9:30 - 10:00  T. Tajima. "Accelerator Systems Approach to Wakefield Colliders"
10:00 - 10:30 J. Irwin. "Dynamic Focusing Using $e^+e^-$ Beams as Lenses"
10:30 - 11:00 BREAK
11:00 - 11:30 D. Whittum. "IP Considerations at High Energies"
11:30 - 12:00 F. Zimmermann. "Novel Approach to Final Focus for High Luminosities"

## Thursday: July 9, 1998
9:30 - 10:00  P. Chen. "Testing Unruh Radiation Using Ultra-Intense Lasers"
10:00 - 10:30 N. Madsen, "Laser Cooling of Ion Beams"
10:30 - 11:00 BREAK
11:00 - 11:30 M. Zolotorev. "Atomic Physics at RHIC"
12:00 - 1:30  LUNCH
1:30 - 2:00   R. Fernow. "Transverse Emittance Cooling in Muon Colliders"
2:00 - 2:30   Z. Parsa. "Muon Collider Technique"
2:30 - 3:00   D. Neuffer, "Muon Cooling Studies"
3:00 - 4:00   BREAK
4:00 - 4:30   D. Finley, "Proton Plasma Accelerator"
4:30 - 5:00   General Discussion on Proton Plasma Accelerator

## Friday: July 10, 1998
9:30 - 10:30  Round Table Discussion on Applications of High Energy Density Beams
10:30 - 11:00 BREAK
11:00 - 12:00 Discussion Continues
Afternoon FREE

# Summary Report: Working Group 2 on "Plasma Based Acceleration Concepts"

W.P. Leemans[1] and E. Esarey[1,2]

[1]*Center for Beam Physics, Accelerator and Fusion Research Division, Lawrence Berkeley National Laboratory, University of California, Berkeley CA 94720*

[2]*Plasma Physics Division, Naval Research Laboratory, Washington DC 20375*

**Abstract.** A summary of the talks, papers and discussion sessions presented in the Working Group on Plasma Based Acceleration Concepts is given within the context of the progress towards a 1 GeV laser driven accelerator module. The topics covered within the Working Group were self-modulated laser wakefield acceleration, standard laser wakefield acceleration, plasma beatwave acceleration, laser guiding and wake excitation in plasma channels, plasma wakefield acceleration, plasma lenses and optical injection techniques for laser wakefield accelerators. An overview will be given of the present status of experimental and theoretical progress as well as an outlook towards the future physics and technological challenges for the development of an optimized accelerator module.

## INTRODUCTION

The Working Group on "Plasma Based Acceleration Concepts" consisted primarily of presentations on experimental /theoretical progress on the following topics:

(1.)    Laser wakefield acceleration (LWFA) including self-modulated regime (SMLWFA) and plasma beatwave acceleration (PBWA);

(2.)    Laser guiding including relativistic self-guiding and plasma channel guiding;

(3.)    Electron beam driven wake excitation including plasma wakefield acceleration (PWFA) and plasma lenses;

(4.)    Plasma based radiation sources.

CP472, *Advanced Accelerator Concepts: Eighth Workshop,*
edited by W. Lawson, C. Bellamy, and D. Brosius
© 1999 The American Institute of Physics 1-56396-889-4/99/$15.00

More than 45 papers were presented, 16 of them in poster format, during the Monday - Thursday sessions. On Friday, discussion groups were formed on the following subjects:

(1.) Laser guiding experiments: progress and challenges;
(2.) SMLWFA experiments: summary;
(3.) Particle trapping in SMLWFA: wavebreaking mechanisms, dephasing length and maximum energy gain;
(4.) LWFA experiments: summary;
(5.) Optimum wavelength choice for laser drivers.

Next, we present a brief summary of various papers presented in the Working Group as well as summaries of the discussion sessions. The emphasis is on highlighting new experimental/theoretical developments that address major issues as well as new challenges relevant to the development of a compact laser driven, plasma based accelerator structure (1).

## LASER GUIDING AND WAKEFIELD EXCITATION

Various groups presented experimental and theoretical progress on laser guiding in plasma channels (2). Table 1 summarizes the present status of experiments on relativistic self-guiding of high power laser pulses at University of Michigan (3) and NRL (4); and of experiments on laser guiding in preformed plasma channels at LBNL (5), University of Maryland (6), University of Texas at Austin (7), and at Hebrew University/NRL (8). The experiments using preformed channels rely on plasma channel formation through hydrodynamic shock expansion in a heated plasma column (5-7) or on a capillary discharge (8).

The Michigan group (Umstadter et al. (3)) presented results on guiding of high power beams in relatively dense plasmas by relying on relativistic self-focusing. Waveguide profiles were obtained through interferometry.

The LBNL group (Leemans et al. (5)) presented results on laser guiding at vacuum intensities of up to $5 \times 10^{17}$ W/cm$^2$ in plasma channels produced using a novel ignitor-heater scheme in a cylindrical gasjet, with a length of 1 mm. The ignitor pulse is a short intense pulse ($I > 10^{14}$ W/cm$^2$) that ionizes the gas and the heater pulse is an energetic (>200 mJ) relatively long (>150 ps) laser pulse that heats the plasma through inverse Brehmsstrahlung. This dual-pulse technique allows use of low Z gases which alleviates the concern of further ionization by the intense pulse that is to be guided.

The Maryland group (Milchberg et al. (6)) presented results on guiding intense pulses ($10^{17}$ W/cm$^2$) in 1-1.5 cm long plasma channels, produced in an Ar/N$_2$O slit gasjet which is ionized using a long pulse focused with an axicon lens. Results were presented of tunnel coupling of radiation into the fiber, a double heater pulse approach to improve control of the radial plasma profile, and fiber-end visualization using

TABLE 1: Summary of experiments on laser guiding presented at the Advanced Accelerator Concepts 1998 Workshop.

| | U Mich | NRL | NRL | LBNL | Maryland | UT-Austin | Hebrew/NRL |
|---|---|---|---|---|---|---|---|
| 1. Length [cm] | 0.1 | 0.3 | 0.3 | 0.1 | 1.5 | 1-2 | 1,2,3,6 |
| 2. Diameter [m] | 10 (30) | 5 | 10-20 | 16 | variable | 30 | 300 |
| 3. $n_e(r=0)$ [cm$^{-3}$] | $3 \times 10^{19}$ ($3 \times 10^{18}$) | $1-3 \times 10^{19}$ | $1.5 \times 10^{19}$ | $7 \times 10^{18}$ | $2 \times 10^{18} - 6 \times 10^{19}$ | $5 \times 10^{18}$ | $5-10 \times 10^{18}$ |
| 4. Method | Relativistic | Relativistic | Relativistic | Hydro | Hydro | Hydro | Hydro |
| 5. $I_{max}$ [W/cm$^2$] | $2 \times 10^{17}$ | $5 \times 10^{18}$ | $3 \times 10^{16}$ | $2 \times 10^{17}$ | $10^{17}$ | - | $10^{16} - 10^{17}$ |
| 6. Gas/Target | He | $H_2$/He | He | $N_2$/$H_2$ | Ar+$N_2$O | He | $CH_2$ |
| 7. Rep. Rate | 7 min/shot | 3 min/shot | 3 min/shot | 10 Hz | 10 Hz | 20 Hz | min/shot (few 100 shots/capill) |
| 8. Gasjet/Backfill | Jet | Jet | Jet | jet | jet | back-fill | - |
| 9. Transmission | 60% | ? | 70% | 25% | 52% | - | 15% (1cm) 50% (3 cm) 10% (6 cm) |
| 10. Cost | $$ | $$ | $$ | $$ | $$ | $$ | $ |

176

interferometry. The development of long gasjets is an important issue for extending the acceleration length in LWFA schemes.

An alternative approach for producing long plasma channels was presented by the University of Texas at Austin group (Downer et al. (7)). An initially low temperature plasma is generated using an electric discharge and a relatively long (100-400 ps), energetic (250-400 mJ) laser pulse is focused with an axicon to heat the plasma through inverse Brehmsstrahlung. Diagnostics based on frequency domain interferometry were also proposed to study the longitudinal wake excited in the channel.

The NRL group (Ting et al.), in collaboration with the Hebrew University (Zigler et al.), has been using a double capillary discharge scheme for guiding high intensity pulses (8). The main capillary discharge is preceded by a small initial electric discharge to seed the main discharge in the $CH_2$ capillary. Guiding of intense pulses was reported ($<10^{17}$ $W/cm^2$). The lifetime of the capillaries is at the present time is limited to a few 100 shots.

The short term (next 2 years) goals for guiding experiments are:

(A.)    guiding of laser pulses with $I>10^{18}$ $W/cm^2$
(B.)    improvement of coupling and transmission efficiency
(C.)    generation/ measurement of laser wakefields in channels

During the discussion session on guiding the following issues were raised:

(1.)    Production of long channels: To achieve a 1 GeV net acceleration, few cm-long channels are needed. By extending current work, production of channels with lengths up to 10 cm seems achievable in the next few years. Some of the technology developed for z-pinches used in x-ray laser work have produced 10 - 20 cm long plasma columns (albeit in a different density parameter regime). This might be worth considering.

(2.)    Efficient production of channels: There is a need for high efficiency in the channel production. Various schemes are currently being evaluated using laser ionization and heating (e.g., ignitor-heater) as well as electric discharge based methods. An open question remains on whether the laser beam can be re-used after producing a plasma channel, since only a small fraction of the laser energy is deposited in the channel.

(3.)    Production of low on-axis density: There is a definite need to push $n_e(r=0)$ to lower values ($1-10x10^{17}$ $W/cm^2$). As can be seen in Table 1, the on-axis density in all experiments reported at the Workshop was in the 0.2 - $6x10^{19}$ $cm^{-3}$ range. To operate in the standard LWFA regime, such high densities would require laser pulses between 5 - 30 fs and would result in a linear dephasing length on the order of 13.8 mm (30 fs pulse) and 64 μm (5 fs pulse), where a laser wavelength of 1 μm is assumed. The maximum energy in GeV after such a dephasing length is then on the order of 0.336 P[TW] for a 30 fs pulse focused to a 10 μm radius spot size and 0.003 P[TW] for a 5 fs pulse, where P is the laser power in TW. From this argument it is clearly advantageous to keep the plasma density on axis as low as possible.

(4.)    Plasma density profile control: There is a need for ideas/technology that would enable control of the radial plasma density profile (e.g., the production of hollow channels). It has been shown that a hollow channel supports an electromagnetic mode whereas a wide parabolic channel supports a predominantly electrostatic mode (9). This implies that the electron beam phase space properties will be superior (i.e., lower emittance) for a hollow channel since the transverse focusing forces will be linear and nearly cancel. The effects of channel shape and uniformity on the wake amplitude and wake temporal decay will need to be examined in detail to help in assessing the laser–to–wake coupling efficiency for the laser driven accelerator schemes.

# SELF-MODULATED LASER WAKEFIELD ACCELERATION

Experiments on self-modulated laser wakefield acceleration (SMLWFA) over the last several years have (I) shown acceleration to high energies, (II) provided a platform for the development of experimental techniques and diagnostics, and (III) allowed detailed comparison with analytic theory and simulations. In the SMLWFA regime, the initial laser pulse is many plasma periods long. As the laser pulse propagates through the plasma it gets temporally and spatial modulated at the plasma frequency via a self-modulation or Raman forward scattering instability and thereby efficiently excites a large amplitude plasma wave (10). Self-trapping of plasma electrons in the wake can occur.

Table 2 summarizes the results presented at the Workshop by the Collaboration between Rutherford Appleton Laboratories, Imperial College and UCLA (11); the NRL group (12); and the Michigan group (13). In all these experiments, electron acceleration to high energies (30 - 100 MeV) has been observed. The high power laser pulse (ranging from 2 - 20 TW) was observed to be self-guided by relativistic focusing in all experiments. Large numbers (up to $10^{10}$) of electrons have been observed, most with energies around a few MeV with an exponential drop-off towards higher energies. The laser intensity in all these experiments was on the order of 3 ($\pm$ 1) x $10^{18}$ W/cm$^2$ and the wakefield amplitude $\Delta n/n$, measured using Thomson scattering, ranged from 0.3 - 1.0, i.e., approaching but still below the cold wavebreaking limit. Raman scattering spectra of the pump or probe beams, which are a measure of the plasma wave temporal structure, show significant broadening at laser powers above the onset of self-trapping. The NRL and Michigan groups reported observation of electron

178

beams with a divergence less than the laser divergence. The Michigan group also performed measurements of spatial profiles versus laser power, obtaining electron beams with a divergence angle less than 1.5° from which a transverse geometric emittance less than $0.1\,\pi$ mm-mrad was inferred.

In all these experiments, the accelerated electrons were self-trapped from the background plasma. Self-trapping limits the amplitude of a wakefield and thus the maximum acceleration field gradient which can be sustained by the structure. Aside from its basic plasma physics interest, this self-trapping or "uncontrolled" acceleration of background electrons is equivalent to production of "dark-current" in conventional structures, and is therefore a very relevant issue affecting the development of future laser driven accelerator modules. Several potential mechanisms have been proposed to explain the observations:

(1.)  Trapping without pre-heating: Possible if the amplitude of the EPW was actually higher than inferred from Thomson scattering. The RAL/IC/UCLA group has proposed direct wavebreaking as the trapping mechanism (11).

(2.)  Trapping with pre-heating by Raman side- or backscattering (Esarey et al. (14)): Analytic modeling and simulations of the NRL experiments were presented showing reasonable agreement between theory and experiment.

(3.)  Two-dimensional wavebreaking in the density channel associated with relativistic self-guiding, due to the curvature of the wavefronts of the plasma wave (15): The Michigan group (J.K. Kim et al. (16)) reported analytic calculations of 2-D wavebreaking, identifying the transverse momentum as the key parameter and calculating the number of usable accelerating "buckets" behind the laser pulse front prior to the onset of wavebreaking. Results from fluid model calculations presented by Shadwick and Wurtele (Berkeley) indicate that phasefront curvature, which occurs also in parabolic channels, can be mitigated by the use of sufficiently steep channel walls, the limiting case being the hollow channel. The wake phasefront curvature that results from the radial dependence of the plasma wavelength (via the local density) in a parabolic channel, which leads to wavebreaking at relatively low amplitudes, can be avoided in hollow channels.

Another topic of discussion was on the maximum electron energies observed in SMLWFA experiments. In the RAL/IC/UCLA experiments, the observed energy of approximately 100 MeV exceeds the simple linear dephasing limit, $W_d = 2\gamma_p^2 \varepsilon mc^2$, where $\varepsilon = \Delta n / n$ is the wake amplitude and $\gamma_p$ is the relativistic factor associated with the phase velocity of the wake ($\gamma_p = \omega / \omega_p$ in the linear limit). Likewise, in particle simulations performed at UCLA, the resulting maximum energies also exceeded $W_d$.

Esarey et al. (14) pointed out several nonlinear effects that could directly enhance $W_d$. They argued that in the self-modulated regime, the space charge force that results from electron self-channeling provides a radial force that is focusing for all wake phases. This can double the dephasing (and phase slippage) length resulting in a maximum energy gain of $W_{max} = 2W_d$. Furthermore, relativistic effects and self-channeling can substantially decrease the effective value of $\omega_p$, which results in higher wake phase velocities and higher energy gains.

**TABLE 2:** Summary of results on self-modulated laser wakefield acceleration experiments as reported at the AAC98 meeting (B/FRS = backward/forward Raman scattering, TS = Thomson scattering, EMS = electromagnetic spectrometer). The wakefield amplitude is $\Delta n/n$, the dephasing length is $\lambda_p^3/\lambda^2$, and the accelerating field is $\Delta n/n\,(n_p)^{1/2}$.

| | RAL/UCLA | NRL | U Michigan |
|---|---|---|---|
| **Laser** | | | |
| Wavelength [μm] | 1.05 | 1.05 | 1.05 |
| Pulse length [fs] | 800 | 400 | 400 |
| Peak power [TW] | 20 | 2 | 4 |
| Intensity [W/cm$^2$] | $4\times10^{18}$ | $2.5\times10^{18}$ | $3\times10^{18}$ |
| Rayleigh length [μm] | 300 | 75 | 135 |
| Rep. Rate | single shot | One/3 min | One/7 min |
| **Plasma** | | | |
| Source | gas jet | Gas jet | gas jet |
| Plasma species | H, He | He | He |
| Plasma density [cm$^{-3}$] | $5\times10^{18}$ - $2\times10^{19}$ | $3\times10^{19}$ | $3\times10^{19}$ |
| Plasma length [mm] | 4 | 1 | 1 |
| Laser guiding | self-guided | Self-guided | Self-guided |
| P/P$_{crit}$ | 6-20 | 3 | 6 |
| **Wakefield** | | | |
| Wavelength [μm] | 7-8 | 6 | 6 |
| Wakefield amplitude | 0.5 | $\approx1$ | 0.3 |
| Dephasing length [μm] | 500 | 200 | 200 |
| Acc. Field [GV/m] | 160 | 500 | 160 |
| Duration [ps] | not measured | 5 | 2 |
| **Trapping Mechanism** | self-trapped | 2 stage acc w/BRS | Self-trapped |
| **Accelerated electrons** | | | |
| Max. gain [MeV] | 96 | 120±50 | 70±20 |
| Total # of el. Acc. | $10^{10}$ | $10^8$ (> 1 MeV) | $10^{10}$ |
| Flux at $\Delta E_{max}$ [/MeV/sr] | $10^3$ | $10^3$ | $3\times10^4$ |
| S/N at $\Delta E_{max}$ | 2 | 2 | 3 |
| Divergence of acc. el. | | less than laser divergence | 1.5 degrees |
| **Diagnostics** | | | |
| Plasma | Stokes/Anti-S TS of Probe; TS of self-gen 2ω | 0° FRS; 90° TS | Collective TS |
| Electrons | EMS | 8 ch EMS; scintil-lating fiber/PMT | 1 ch EMS; wire chamber detector |

The explanation proposed by the UCLA group (11) was that local wavebreaking near the front of the laser pulse led to the self-trapping and acceleration of a dense electron bunch. This bunch quickly reached velocities exceeding the wake phase velocity. When this occurs, a secondary wake produced by the trapped bunch itself is generated with a phase velocity greater than that of the initial wake. This secondary wake could then accelerate trailing bunches to energies exceeding $W_d$.

In summary, it is too early to say definitively what mechanisms are leading to self-trapping in SMLWFA experiments. There is a need for 2-D analytical theories (16) that can be compared with simulation and experiment. On the experimental side, wavebreaking and particle trapping might be most optimally studied by relying on wake excitation with resonantly driven LWFA, which, in principle, would allow better control and characterization of experiments. A parametric study of particle trapping versus $\lambda/\lambda_p$ and $\gamma_g$, providing spatially and temporally resolved information on the electron distribution in the plasma and the plasma temperature, would allow direct comparison with theory.

## LASER WAKEFIELD ACCELERATION

Two groups (Ecole Polytechnique, France (17) and KEK/JAERI, Japan (18)) presented experimental results on acceleration of externally injected electrons in a standard laser wakefield accelerator (LWFA). The experimental parameters are summarized in Table 3. Whereas the French group reported an absolute energy gain of 1.5 MeV, the Japanese group obtained an energy gain of approximately 300 MeV.

The French group reported results of an extensive study of their electron beam optics and spectrometer detection system. After careful measurement of contributions to the signal on the detectors from electron beam scattering in the plasma and wakefields, a detailed understanding was obtained of 3-D effects in electron trapping and acceleration. Good agreement between experiment and theoretical modeling was obtained.

The Japanese result is more than one order of magnitude larger than the expected value from linear LWFA theory. This was explained by the KEK/JAERI group by invoking channeling of the laser pulse over a distance of more than 1 cm. The low peak power and short pulse duration of the injected laser beam precludes relativistic guiding as being the guiding mechanism. Without the creation of a plasma channel, the laser beam would refract due to the presence of the plasma rather than being guided (19). This anomalous channeling result is therefore not well understood although nonlinear effects in the neutral gas could provide a contribution to self-focusing at low intensity. The effect of electron scattering, as studied in the French experiment, could also contribute to the detected signal in the spectrometer and needs further evaluation.

*Laser Injection.* An issue of relevance to the standard LWFA is that of laser injection. The self-modulated LWFA demonstrated acceleration of electrons to high

**TABLE 3**: Summary of results on laser wakefield acceleration experiments with external electron injection as reported at the AAC98 meeting. (FRS = forward Raman scattering, TS = Thomson scattering, FDI = frequency domain interferometry). The wakefield amplitude is $\Delta n/n$, the dephasing length is $\lambda_p^3/\lambda^2$, and the accelerating field is $\Delta n/n\ (n_p)^{1/2}$.

| | KEK/JAERI | Ecole Polytechnique |
|---|---|---|
| **Laser** | | |
| Wavelength [$\mu$m] | 0.79 | 1.057 |
| Pulse length [fs] | 90 | 400 |
| Peak power [TW] | 1.8 | 3.5 |
| Intensity [W/cm$^2$] | $7 \times 10^{17}$ | $4 \times 10^{17}$ |
| Rayleigh length [$\mu$m] | 670 | 2000 |
| Rep. Rate | 10 Hz | one/5 min |
| **Plasma** | | |
| Source | Backfilled | Backfilled |
| Plasma species | He | He |
| Plasma density [cm$^{-3}$] | $1.4 \times 10^{18}$ | $2.2 \times 10^{16}$ |
| Plasma length [mm] | 20 | 25 |
| Laser guiding | Self-guided | no |
| $P/P_{crit}$ | 0.14 | $\approx 0$ |
| **Wakefield** | | |
| Wavelength [$\mu$m] | 29 | 226 |
| Wakefield amplitude | 0.11 (calculated) | 0.1 (calculated) |
| Dephasing length [mm] | 40 | >> |
| Acc. Field [GV/m] | 15 (calculated) | 1.5 |
| Wakefield duration | $\approx 1$ ps | $\approx 1$ ps |
| Injection | | |
| Injector | 3 GHz RF linac | VandeGraaff (CW) |
| Energy [MeV] | 17 | 3 |
| El./bunch | 1 nC | 300 A (CW) |
| Phase occupied | 360° | 360° |
| Accelerated electrons | | |
| Max. gain [MeV] | 300 | 1.5 |
| Total # of el. acc. | $2 \times 104$ (>10 MeV) | 200 |
| Flux at $\Delta E_{max}$[/MeV/sr] | 250 | 6 |
| S/N at $\Delta E_{max}$ | 1 | 1 |
| Divergence of acc el. | not reported | Not reported |
| Diagnostics | | |
| Plasma | TS; FDI-wakefield | 0° FRS; 90° TS |
| Electrons | Desmarques screen-spot size Cerenkov light-pulse length, timing 32 ch scintillator and magnet-energy | high acceptance 2-focus spectrometer/17 ch scintillating fiber/PMT at 0.15 MeV binning |

energies (near 100 MeV), however, since the electrons are self-trapped the resulting beam has 100% energy spread. To achieve acceleration in a standard LWFA with small energy spread requires the injection of an ultrashort electron bunch (short compared to the plasma wavelength) at the optimum phase location with respect to the wakefield. This cannot be achieved with present RF photoinjectors, since the duration of the wakefield bucket is typically <300 fs.

Umstadter et al. (20) suggested using a second laser pulse (the injection pulse), propagating transversely to the pulse driving the wake, to inject background plasma electrons into the wakefield. The transverse ponderomotive force of the tightly focused injection pulse would depart sufficient axial momentum to the background electrons such that they become trapped in the wake. Particle simulations of this scheme indicate the production of a 10 fs, 21 MeV electron bunch with a 6% energy spread. High intensities, however, are required in both the drive and injection pulses ($a \approx 2$).

The following schemes have been proposed for laser injection:

(1.) The transverse LILAC scheme originally proposed by Umstader et al. (20) (described above). Hemker et al. (21) also performed PIC simulations of this process, and pointed out the importance of the wake from the high intensity injection pulse.

(2.) The longitudinal LILAC scheme (Dodd et al. (22)). In this case the injection pulse propagates in the same direction as the drive pulse. The injection pulse is tightly focused with a much shorter Rayleigh range such that the wake produced by the injection pulse adds to that of the drive pulse to produce a local region of wavebreaking and hence trapping.

(3.) The colliding pulse scheme (Esarey et al. (23)). This concept uses two injection pulses, one propagating in the same direction and the other opposite to the drive pulse. When the two injection pulses collide some distance behind the drive pulse, they create a ponderomotive beat wave with a slow phase velocity. This slow beat wave can displace the plasma electrons in both phase and momentum such that they become trapped in the fast wake. Trapping can occur at low injection intensities ($a \approx 0.2$) and the colliding pulse geometry offers detailed control over the injection process via the phasing, duration, and amplitude of the injection pulses. Test particle simulations in 3D indicate the production of ultrashort (3 fs) bunches with low energy spread (1%) and emittance (1 mm-mrad) (24).

Experiments on laser injection are being pursued at Michigan, LBNL, and NRL.

## LWFA SCALING LAWS

Working group discussions, initiated by presentations given by I. Pogorelsky (25), commenced on the topic of the scaling of various wakefield quantities as a function of wavelength. In particular, how a 1 micron laser driver compares with a 10 micron laser driver. During his presentations, Dr. Pogorelsky gave examples in which a 10 micron laser driver may have advantages over a 1 micron driver. In this section, simple scaling laws for LWFA quantities are presented under idealized assumptions.

These idealized scaling laws assume the following:

(1.) A standard LWFA that is channel guided.
(2.) The mildly relativistic regime, $a^2 \ll 1$.
(3.) The acceleration length is limited by electron dephasing.
(4.) The plasma channel is sufficiently broad such that the formula describing wakefield generation in a uniform plasma apply.
(5.) The transverse size of the laser pulse is $2c/\omega_p$ and the transverse size of the electron bunch is $c/\omega_p$.
(6.) The total electrons per bunch is the beam loading limit.

In the following, when equations are presented in practical form (with numerical coefficients), $E_z$ is in V/m, $n$ is in $\text{cm}^{-3}$, $\lambda$ is in microns, $I$ is in W/cm$^2$, $W_L$ is in J, $L_d$ is in m, $\Delta W$ is in GeV, $Lum_s$ is in cm$^{-2}$, and $a^2$ is dimensionless.

In the mildly relativistic limit within a broad channel, the axial electric field of the wake can be written as $E_z = 0.38 a^2 E_0$, where $E_0 = mc\omega_p / e = 96 n^{1/2}$, i.e.,

$$E_z = 2.7 \times 10^{-17} I \lambda^2 n^{1/2}$$

$$= 3.4 \times 10^{-25} W_L \lambda^2 n^2$$

This assumes a linearly polarized laser pulse with Gaussian profiles in the radial and axial directions. This also assumes that the laser pulse length is optimized to maximize the wakefield amplitude, i.e., $L = \lambda_p / \sqrt{2\pi} = 0.4\lambda_p$, where the pulse length L is defined such that $W_L = (1/8\pi) A_L E_L^2 L$ is the pulse energy, $\lambda_p = 2\pi c / \omega_p$ is the plasma wavelength, $E_L$ is the peak laser electric field, $A_L = \pi r_0^2 / 2$ is the cross-sectional area of the Gaussian pulse, and $r_0$ is the laser spot size. The laser spot size is assumed to be $r_0 = 2c/\omega_p$ in order to ensure high efficiency of energy transfer between the wake and the accelerated electrons (26), since electrons loaded near the axis will absorb wake energy out to a radius of approximately $c/\omega_p$. Furthermore,

$$a^2 = 9.4 \times 10^{-27} W_L \lambda^2 n^{3/2}$$

The acceleration length is assumed to be equal to the electron dephasing length $L_d = \lambda_p^3 / \lambda^2$,

$$L_d = 3.7 \times 10^{25} \lambda^{-2} n^{-3/2}$$

The ideal maximum energy gain is given by $\Delta W = e E_z L_d$,

$$\Delta W = I / n$$

$$= 1.3 \times 10^{-8} W_L n^{1/2}$$

The number of electrons accelerated per bunch is assumed to be equal to the beam loading limit (26) $N_b = E_z A_b / 4\pi e$, where $A_b$ is the effective cross-sectional area of the beam which is assumed to be $A_b = \pi c^2 / \omega_p^2$,

$$N_b = 1.7 \times 10^{-9} W_L \lambda^2 n$$

Another figure of merit is the luminousity $Lum = (k_b f_b / 4\pi) N_b^2 / \sigma_x \sigma_y$, where $k_b$ is the number of bunches per linac, $f_b$ is the linac rep rate, and $\sigma_{x,y}$ are the transverse rms bunch sizes, which are assumed to be equal to $c / \omega_p$. For scaling purposes, it is convenient to define the "single bunch" luminousity as $Lum_s = N_b^2 / \sigma_x \sigma_y$,

$$Lum_s = 9.9 \times 10^{-30} W_L^2 \lambda^4 n^3$$

Next, to determine scaling with wavelength, several examples are given. In all these examples, the laser pulse energy $W_L$ is assumed constant.

(A.)  Constant $E_z$ : The axial electric field of the wake is held fixed (in addition to the pulse energy). This implies:
$n \propto \lambda^{-1}$, $L_d \propto \lambda^{-1/2}$, $\Delta W \propto \lambda^{-1/2}$, $N_b \propto \lambda$, $Lum_s \propto \lambda$

(B.)  Constant $L_d$ : The acceleration length is held fixed (in addition to the pulse energy). This implies:
$n \propto \lambda^{-4/3}$, $E_z \propto \lambda^{-2/3}$, $\Delta W \propto \lambda^{-2/3}$, $N_b \propto \lambda^{2/3}$, $Lum_s \propto const$

(C.)  Constant $\Delta W$ : The electron energy gain is held fixed (in addition to the pulse energy). This implies:
$n \propto const$, $E_z \propto \lambda^2$, $L_d \propto \lambda^{-2}$, $N_b \propto \lambda^2$, $Lum_s \propto \lambda^4$

(D.)  Constant $N_b$ : The number of electrons per bunch is held fixed (in addition to the pulse energy). This implies:
$n \propto \lambda^{-2}$, $E_z \propto \lambda^{-2}$, $L_d \propto \lambda$, $\Delta W \propto \lambda^{-1}$, $Lum_s \propto \lambda^{-2}$

(E.)  Constant $Lum_s$ : The single bunch luminousity is held fixed (in addition to the pulse energy). This implies:
$n \propto \lambda^{-4/3}$, $E_z \propto \lambda^{-2/3}$, $L_d \propto const$, $\Delta W \propto \lambda^{-2/3}$, $N_b \propto \lambda^{2/3}$

In making comparisons between 1 and 10 micron drivers, care must be taken so as not to violate the above assumptions, in particular, $a^2 \ll 1$. Note that $a^2 \propto W_L \lambda^2 n^{2/3}$. Hence, when making comparisons at constant density and pulse energy, as in Case (C.), the assumption $a^2 \ll 1$ may be violated at long wavelengths. On the other hand, for short wavelengths, operation at high density is valid. A definitive conclusion regarding an optimum driver wavelength is problematic. For example, at sufficiently low density (such that $a^2 \ll 1$), a design for a fixed energy gain favors longer wavelengths, as implied by Case (C.). On the other hand, a design for a fixed number of electrons per bunch favors short wavelengths, as implied by Case (D.). Furthermore, a design for a fixed acceleration distance (and fixed luminosity) allows higher energies to be obtained for short wavelengths, however, a higher bunch number is obtained for long wavelengths. The above scaling laws all assume a fixed laser pulse energy. A rigorous study of a LWFA for various wavelength drivers must also include other properties of the driver, such as repetition rate, pulse stability, and average power. Since laser technology is rapidly progressing, a rigorous design study is premature. In terms of physics experiments, invaluable information can be obtained at both 1 and 10 micron.

## WORKING GROUP PRESENTATIONS

On the topic of laser guiding in plasmas, the following presentations were given (titles and authors are approximate): Evolution of plasma waves and channels in self-guided laser pulse experiments (S.Y. Chen et al., Michigan); Experiments on two pulse laser channel formation (P. Volfbeyn et al., LBNL); Guiding in preformed plasma channel experiments (S.P. Nikitin et al., Maryland); Mode control in plasma waveguide experiments (H.M. Milchberg et al., Maryland); Generation and diagnosis of a preformed plasma channel in pure helium (E.W. Gaul et al., Texas); Laser guiding experiments at NRL/Hebrew U. (A. Ting et al.); Finite pulse effects on the stability of laser pulses (P. Sprangle et al., NRL); Long-wavelength laser hosing (K.C. Tzeng et al., UCLA); Ionization induced scattering of short laser pulses (T.M. Antonsen et al., Maryland); Electromagnetically-induced guiding of counter-propagating lasers in plasmas (G. Shvets et al., PPPL); Multimode analysis of the hollow plasma channel accelerator (C.B.Schroeder et al., LBNL); Simulations of pulse propagation in capillary discharge plasma channels (R.F. Hubbard et al., NRL); Plasma channels as accelerating structures (B.A. Shadwick, LBNL); Quasi-modes and continuum damping in plasma channels (G. Shvets et al., PPPL).

On the topics of LWFA, SMLWFA, and PBWA, the following presentations were given (titles and authors are approximate): Observation of LWFA of electrons (D. Bernard et al., Ecole Polytechnique); Laser wakefield acceleration of an injected electron beam (H. Dewa et al., JAERI); LWFA experiments at Imperial College (K. Krushelnick et al.); Status of the NRL LWFA experiment (A. Ting et al.); PBWA experiments at UCLA (C. Clayton et al.); High energy electrons from PW laser-solid interactions (T. Cowan et al., LLNL); Cold wavebreaking of 2D wakefields (J.K. Kim et al., Michigan); Suppression of electron blowout and self-focusing by Raman scattering and heating (W.B. Mori et al., UCLA); Optimal laser pulse shaping for LWFA (P. Chen et al., SLAC); LWFA with CO2 drivers (I. Pogorelsky et al., BNL); Experimental characterization of laser wakefields (R. Wagner et al., Michigan); Ultrafast optical diagnostics for LWFA (S.P. Le Blanc et al., Texas); Analysis of the electron spectrum in SMLWA (A. Charman et al., Berkeley); Electron beam characteristics from wavebreaking (W.B. Mori et al., UCLA); Particle dynamics map for LWFA (S. Cheshkov et al., Texas); Generation of ultrashort electron bunches by colliding laser pulses (C.B. Schroeder et al., LBNL).

On the topics of PWFA and plasma lens, the following presentations were given (titles and authors are approximate): PWFA experiments using the Neptune photoinjector (J. Rosenzweig et al., UCLA); Design for a 1 GeV PWFA at SLAC (R.

Assmann et al.); Meter long plasma sources for advanced accelerators (P. Muggli et al., USC); Relativistic electron beam focusing by very overdense plasma lenses (R. Govil et al., LBNL); Underdense plasma lens experiment at UCLA (C.E. Clayton et al.); High energy plasma lens experiment at SLAC (P. Chen et al.); Acceleration in the blowout regime of the PWFA (N. Barov et al., ANL); Resonant excitation of plasma wakefields by multiple electron bunches (M. Conde et al., ANL); PWFA in the blowout regime with mobile ions (S. Lee et al., USC); Envelope equation for a magnetically self-focused beam in a plasma (K. Backhaus et al., Berkeley); Test results of the plasma source for underdense plasma lens experiments at UCLA (H. Suk et al.); Simulations of the SLAC E150 plasma lens experiment (S. Masudea et al.).

On the topic of plasma based radiation sources, the following presentations were given (titles and authors are approximate): Cerenkov radiation from electrostatic wakes in magnetized plasmas (P. Muggli et al., USC); Theory of laser-driven undulator radiation (G. Shvets et al.).

# CONCLUSION

There has been tremendous progress over the last two years on experiments, analytic theory, simulations (fluid and PIC) for laser driven acceleration in plasmas as evidenced by the numerous publications in Science, Nature, Phys. Rev. Lett., Phys. Rev. E, Phys. Plasmas, etc. During this Workshop, various issues were discussed related to the development of a 100 MeV - 1 GeV compact, high brightness, plasma based laser driven accelerator module. The discussions were centered on a) laser guiding, b) self-modulated laser wakefield acceleration, c) standard laser wakefield acceleration, and d) power sources for wakefields in plasmas.

On power sources (laser systems), the topic of the scaling of various wakefield quantities as a function of wavelength was raised. There have been notable developments at BNL towards the generation of a picosecond, TW $CO_2$–based laser system. From simple scaling laws presented in this summary paper, the optimum choice of laser driver clearly depends on the quantity desired to be optimized. It therefore seems essential to maintain the complementarity in the area of parameter regimes that can be studied by the long and short wavelength laser drivers, to further enhance the field.

Several groups reported progress on channel guiding of intense laser pulses. The use of gasjets allowed an improved coupling of the laser beam into the plasma channel at high intensities. The intensity of guided pulses is now exceeding $10^{17}$ $W/cm^2$. Various methods of producing the plasma channels have been implemented: channels produced through hydrodynamic expansion and channels produced in capillary discharges. Multi-pulse laser schemes (e.g., ignitor-heater) are being studied to efficiently produce channels in gases with a high ionization potential. Discharge based techniques are being examined to produce channels at low cost. Optical diagnostics have been used to diagnose the spatial density profile of the channel, and are being designed and studied to measure the laser excited wakefields in the channel.

Various groups reported new results on self-modulated laser wakefield acceleration. These experiments are serving as a platform for development of experimental diagnostics and know-how, as well as a test-bed for theory/simulation tools. They have provided insight into the basic physics of wake excitation, laser beam propagation (self-guiding) and electron production. The measurements also indicate a further need to study the physics of wavebreaking and particle dephasing. Parametric measurements of maximum energy gain versus plasma and laser parameters will enable the evaluation of the maximum sustainable wakefield amplitude prior to electron self-trapping (the equivalent of dark current emission in RF structures) and the dephasing length. This in turn determines the length of the structure that needs to be produced for guiding the laser pulses and the energy gain per stage that can be expected.

Two groups reported results on standard laser wakefield acceleration of externally injected unbunched electrons. The experiments demonstrate the need for careful characterization of the experimental apparatus. The beam dynamics seemed to be well understood and modeled when including all 3-D effects. More experiments are needed to address some of the discrepancies that exist between some of the experimental results and theory.

Novel ideas on laser triggered injection of electrons were also discussed: the so-called LILAC and Colliding Pulse schemes. These schemes show great promise for producing high brightness ultrashort electron bunches. Results of proof-of-principle experiments are expected before the next Workshop.

Beam-driven plasma accelerators/devices are being pursued by several groups. Results were shown of a study of plasma lens focusing in the very overdense or return current cancellation regime where the plasma skin depth is comparable to the electron beam size (27). Upcoming experiments on plasma lens focusing (SLAC E-150) and plasma wakefield acceleration (SLAC E-157) at SLAC with the 30 GeV electron beam were discussed. These experiments are expected to produce results in the summer of 1999.

## ACKNOWLEDGMENTS

We would like to thank all the participants in this working group for taking part in the discussions and for sharing their most recent results. In particular, thanks to Paul LeBlanc for compiling the results on laser guiding in plasma channels, Tony Ting for preparing summary tables for the SMLWFA and LWFA, Don Umstadter for summarizing the discussion on wavebreaking and particle trapping issues, Igor Pogorelsky, Tom Katsouleas, and Chan Joshi for the discussion on optimum laser driver parameters and Warren Mori for his view on dephasing lengths. We also apologize for any presentation s not explicitly discussed in this summary due to a lack of space. Please refer to the papers in these proceedings for a more complete discussion. This work was supported by the Department of Energy.

# REFERENCES

1. For a review of plasma based accelerator concepts see, Esarey, E., et al., IEEE Trans. Plasma Sci. **24**, 252 (1996).
2. For a review of laser guiding in plasmas see, Esarey, E., et al., IEEE J. Quant. Electron. **33**, 1879 (1997).
3. Chen, S.Y., et al., Phys. Rev. Lett. **80**, 2610 (1998).
4. Krushelnick , K., et al., Phys. Rev. Lett. **78**, 4047 (1997).
5. Volfbeyn, P., and Leemans, W.P., "Guiding of high intensity ultrashort laser pulses in plasma channels produced with the dual laser pulse ignitor-heater technique," these proceedings; Leemans , W.P., et al., Phys. Plasmas **5**, 1615 (1998).
6. Nikitin, S.P., et al., "High efficiency coupling and guiding of intense femtosecond laser pulses in preformed plasma channels in an elongated gas jet," these proceedings; Nikitin, S.P., et al., Opt. Lett. **22**, 1787 (1997); Clark, T.R., and Milchberg, H.M., Phys. Rev. Lett. **81**, 357 (1998).
7. Gaul, E., et al., "Efficient excitation and measurement of plasma channels," these proceedings; Le Blanc, S.P., et al., "Excitation and measurement of laser induced wakefields," these proceedings; Le Blanc, S.P., et al., Phys. Rev. Lett. **77**, 5381 (1996); Siders, C.W., et al., Phys. Rev. Lett. **76**, 3570 (1996).
8. Hubbard, R.F., et al., "Intense laser pulse propagation in capillary discharge plasma channels," these proceedings; Ehrlich, Y., et al., J. Opt. Soc. Am. B. **15**, 2416 (1998).
9. Shvets, G., and Li, X., "Collisionless damping of laser wakes in plasma channels," these proceedings; Shvets, G., et al., IEEE Trans. Plasma Sci. **24**, 351 (1996); Schroeder, C.B., et al., "Multimode analysis of the hollow plasma channel accelerator," these proceedings; Volfbeyn, P., et al., Phys. Plasmas **4**, 3403 (1997).
10. Esarey, E., et al., Phys. Rev. Lett. **72**, 2887 (1994); Mori, W.B., et al., Phys. Rev. Lett. **72**, 1482 (1994).
11. Gordon, D., et al., Phys. Rev. Lett. **80**, 2133 (1998); Tzeng, K.C., et al., Phys. Rev. Lett. **79**, 5258 (1997).
12. Moore, C.I., et al., Phys. Rev. Lett. **79**, 3909 (1997); Ting, A., et al., Phys. Plasmas **4**, 1889 (1997).
13. Chen, S.Y., et al., "Acceleration of electrons in a self-modulated laser wakefield," these proceedings; Wagner, R., et al., Phys. Rev. Lett. **78**, 3125 (1997).
14. Esarey, E., et al., Phys. Rev. Lett. **80**, 5552 (1998).
15. Bulanov, S.V., et al., Phys. Rev. Lett. **78**, 4205 (1997).
16. Kim, J.K., et al., "Two-dimensional theory of cold plasma wave breaking," these proceedings.
17. Amiranoff, F., et al., "Laser wakefield acceleration of electron at Ecole Polytechnique," these proceedings; Amiranoff, F., et al., Phys. Rev. Lett. **81**, 995 (1998).
18. Dewa, H., et al., "Recent developments for the 2nd generation LWFA experiments," these proceedings; Dewa, H., et al., Nucl. Instr. Meth. A **410**, 357 (1998).
19. Leemans, W.P., et al., Phys. Rev. Lett. **68**, 321 (1992); Leemans, W.P., et al., Phys. Rev. A **46**, 1091 (1992).
20. Umstadter, D., et al., Phys. Rev. Lett. **76**, 2073 (1996).
21. Hemker, R.G., et al., Phys. Rev. E **57**, 5920 (1998).
22. Dodd, E., et al., "Electron injection by dephasing electrons with laser fields," these proceedings.
23. Esarey, E., et al., Phys. Rev. Lett. **79**, 2682 (1997); Leemans, W.P., et al., Phys. Plasmas **5**, 1615 (1998).
24. Schroeder, C.B., et al., "Generation of ultrashort electron bunches by colliding laser pulses," these proceedings.

25. Pogorelsky, I.V., "Terawatt picosecond CO2 laser technology for high energy physics application," these proceedings; Pogorelsky, I.V., et al., "Practical approach to monochromatic LWFA," these proceedings.

26. Katsouleas, T., et al., Particle Accel. **22**, 81 (1987).

27. Govil, R., and Leemans, W.P., "Experimental observation of return current effects in passive plasma lenses," these proceedings.

# Summary Report: Working Group 3 on "Novel Structure-Based Acceleration Concepts"

J. L. Hirshfield* and W. D. Kimura[†]

*Omega-P, Inc., 202008 Yale Station, New Haven, CT 06520-2008;
and Dept. of Physics, Yale University, New Haven, CT 06520-8120.
[†] STI Optronics, Inc., 2755 Northup Way, Bellevue, WA 98004-1495

**ABSTRACT.** The Working Group 3 papers were divided into two general categories, those related to laser-based schemes and those based on microwave technology. In the laser-based area highlights include ongoing theoretical and experimental efforts to demonstrate electron acceleration in vacuum based either on ponderomotive or nonponderomotive mechanisms. Papers on advanced issues such as staging the laser acceleration process, compensating for space charge effects, improving inverse bremsstrahlung acceleration, and possible laser linear collider designs were also presented. In the microwave area, noteworthy contributions were reviewed on slow-wave accelerating structures employing dielectric loading; smooth-wall, fast-wave structures with a helical wiggler for inverse FEL acceleration; multistaging of smooth-wall, fast-wave structures with an axial magnetic field for increased energy cyclotron autoresonance acceleration; and technological approaches for fabrication and diamond coating of W-band structures to achieve high acceleration gradients. Dielectric-loaded structures to support multimode, multibunch wakefield excitations for trailing bunch acceleration were also discussed.

## INTRODUCTION

Structure-based acceleration schemes are those that rely on some external structure to modify the electromagnetic fields in appropriate ways to permit net acceleration of the particles. This distinguishes itself from plasma-based schemes where the plasma plays a key role in the acceleration process. Thus, structure-based schemes include any microwave-driven systems where the waveguide walls and nearby dielectrics control the electromagnetic wave propagation; inverse processes, such as inverse free electron lasers (IFEL), inverse Cerenkov accelerators (ICA), inverse bremsstrahlung electron accelerators (IBEA), and inverse diffraction accelerators (IDA); and so-called vacuum laser accelerators where structures are needed to properly terminate the laser beam. Good progress since the 1996 AAC Workshop at Lake Tahoe (1) was reported in

CP472, *Advanced Accelerator Concepts: Eighth Workshop*,
edited by W. Lawson, C. Bellamy, and D. Brosius
© 1999 The American Institute of Physics 1-56396-889-4/99/$15.00

several areas, indicating that the field of novel structure-based acceleration enjoys a healthy measure of activity, with continued ingenuity shown by the roughly two-dozen or so contributors.

Within the structured-based schemes, the various approaches and investigations can be divided into two general categories, those based on lasers and those based on microwaves. The large difference in wavelengths between these two categories leads to fundamentally different issues and concepts. Therefore, the papers presented in this working group summary have been divided into these two categories.

Wakefield acceleration is in a category of its own, since the drive bunch or bunches induce the accelerating fields for a trailing bunch, and no external source of radiation is employed. The structure dimensions and excited wavelength spectrum of the wakefield accelerator work presented at this Workshop are in the cm- and mm-ranges, so this work is discussed in the microwave section of this summary.

# LASER-BASED CONCEPTS

The papers presented under laser-based concepts can be further divided into three common areas: staging, vacuum acceleration, and advanced analysis and concepts. These are discussed below

## Staging of Laser Acceleration Process

With routine acceleration of electrons being performed, both by structure-based schemes and plasma-based ones, the technology has reached a new point in its evolution where the important issue of staging the acceleration process needs to be examined. As with conventional microwave linacs, staging, whereby the electrons and accelerating field interact repeatedly, allows one to accumulate high net energy gain. However, also like microwave linacs, for this process to be efficient it is necessary for the electrons to be prebunched with a bunch length a small fraction of the accelerating wavelength. For the case of laser acceleration using, say, a 10.6-μm laser, these microbunches would be of order 1-2 μm in length (FWHM).

The primary purpose of the Staged Electron Laser Acceleration (STELLA) experiment being conducted at the BNL Accelerator Test Facility (ATF) is to demonstrate staging between two laser accelerator systems. Specifically, the BNL IFEL (2) will be used as a prebuncher to create a train of microbunches that will be accelerated by an inverse Cerenkov accelerator (3) located $\approx 2$ m downstream of the IFEL. Kimura, et al., (4) presented model simulations for the STELLA experiment, which include the IFEL prebuncher, drift space and focusing triplet after the IFEL, and the ICA stage. The simulations show that for the IFEL prebuncher it is important to have small initial $e$-beam energy spread ($\Delta E/E < 0.3\%$ FWHM) and low $e$-beam energy jitter ($< \pm 0.11\%$). A large energy spread leads to bunch smearing and energy jitter causes the microbunch phase to vary with respect to the laser field in the ICA stage. For the ICA stage it is important to have very low emittance ($\varepsilon_n < 0.85 \pi$ mm-mrad), low $e$-beam energy jitter ($< \pm 0.11\%$), small phase (pathlength) jitter ($\Delta \ell < \pm 0.45$ μm), and accurate alignment of the $e$-beam through the ICA stage ($\Delta r \leq 25$ μm).

Kusche, $et\ al.$, (5) discussed the STELLA hardware system and how it is being designed to satisfy the stringent requirements for the experiment. Noteworthy features include a new energy spectrometer with a $\pm 20\%$ energy acceptance, the usage of a coherent transition radiation (CTR) diagnostic to detect the presence of the microbunches, and a laser transport system designed to minimize pathlength jitter.

## Vacuum Laser Acceleration Processes

Laser acceleration in a vacuum has already been demonstrated in devices such as an IFEL where the external magnetic field permits phase matching to occur. Another approached being investigated at NRL in a collaboration with Omega-P, Inc. is the vacuum beat wave accelerator (VBWA) (6), which utilizes the ponderomotive force produced by the beating of two co-propagating laser beams at wavelengths $\lambda_1$ and $\lambda_2$. Related to this effort, Hafizi, $et\ al.$, (7) performed an analysis to compare the relative performance of Gaussian beams versus axicon-focused Bessel beams for the VBWA. Equal energy gain is theoretically predicted for both the Gaussian and Bessel beams; however, more detailed simulations show that the Gaussian beam is ~3 times better in performance than the Bessel beam.

Preparations are underway at NRL (8) to demonstrate a VBWA using a 3-TW $T^3$ laser in which $\lambda_1 = 2\lambda_2 = 1$ μm, and the laser powers $P_1 = P_2/2 = 1.5$ TW. Factors of 3 energy gain are predicted, corresponding to gradients >15 GV/m. Other noteworthy features of this experiment include a new high sensitivity electron spectrometer and testing of an alternative injector based upon laser-ionized and ponderomotively accelerated electrons created when focusing a portion of the laser beam in Kr gas.

A fundamentally different approach to vacuum laser acceleration is to limit the laser beam interaction with the $e$-beam, which limits the phase slippage between the electrons and light wave thereby achieving net acceleration. One approach was described in a plenary talk by Plettner (9), in which the $e$-beam travels through two crossing laser beams such that the interaction length is $\approx 1.2$ mm. Huang, $et\ al.$ (10), presented an alternative scheme for creating the two crossing beams by using a single Gaussian beam where the Gaussian profile is split in half and the two halves are crossed over each other. Numerical simulations indicate this scheme yields comparable energy gain as the two-beam case, but has the advantage of inherently more stable operation.

Liu, $et\ al.$, (11) proposed a vacuum acceleration experiment at the BNL ATF that can utilize much of the existing equipment developed for the ICA experiment (3). The basic scheme is to use the ICA gas cell without gas and focus the terawatt radially-polarized $CO_2$ laser beam colinearly with the $e$-beam. To achieve net acceleration the laser beam must be terminated shortly after the waist of the focus using a mirror. Although it is highly probable the mirror will damage, this proposed experiment will nonetheless be a significant proof-of-principle demonstration that so-called far-field laser acceleration in vacuum is possible.

## Advanced Analysis and Concepts

An important issue related to prebunching the $e$-beam to optical scales is space charge spreading. In microwave linacs, where the transverse $e$-beam dimensions can be orders of magnitude smaller than the longitudinal dimensions, space charge spreading of the beam diameter is typically of concern. In laser accelerators, the microbunch longitudinal length can be orders of magnitude smaller than the transverse beam size. Thus, space charge spreading of the bunch length now becomes the major concern.

Steinhauer, $et$ $al.$, (12) has shown theoretically that it is possible to compensate for this space charge spreading of the bunch length by slightly over modulating the amount of energy imparted on the electrons within the prebuncher, e.g., the IFEL in the STELLA experiment. This extra degree of energy is enough to compensate for the space charge forces tending to keep the electrons apart as they attempt to bunch while drifting away from the prebuncher. As an added benefit, the extra work done by the electrons to counter the space charge force also reduces the energy spread induced by the laser. This should help improve subsequent trapping and acceleration of the microbunches in the ICA stage.

Turning to a completely different laser acceleration approach, an inverse Bremsstrahlung electron accelerator (IBEA) (13) uses a small electrostatic field $E_{app}$ applied transversely to a co-propagating $e$-beam and laser beam. In nonlinear amplification of IBEA (NAIBEA) (14), the direction of $E_{app}$ is periodically switched $180°$ at phases corresponding to $\varphi = (2n + 1)\pi/2$, $n = 1, 2,\ldots$ Switching the direction of $E_{app}$ permits higher energy gains to be obtained, but efficient energy transfer requires high electron densities. Pakter (15) showed that beam plasma effects become pronounced at high densities and limit the energy gain. It is possible to compensate for these effects, thereby achieving efficient energy gain at high densities, by modifying when the phase $\varphi$ for switching the sign of $E_{app}$ occurs to account for the phase shift caused by plasma effects.

Parsa (16) presented a way of enhancing the performance of both NAIBEA and IFEL by utilizing a scheme where the applied fields have a square wave distribution instead of the typical sinusoidal pattern. For an IFEL this can result in an improvement in energy gain by up to two times, which is equivalent to four times higher laser power. Similar improvements would be expected for an NAIBEA. While this Workshop primarily dealt with electron acceleration processes, it is noteworthy that the forward process, i.e., FEL emission, could also be enhanced using this square wave distribution.

On a more advanced issue, Mikhailichenko (17) presented a study of a possible design for laser linear collider, where an open structure (i.e., grating) laser accelerator is assumed as the basic acceleration method. Electro-optic elements would be used to "sweep" the laser pulse along the surface of each structure as the electron bunch propagates through the structures. This helps minimize the amount of laser power deposited on the structures. For a $2 \times 1$ km long linac, 30 TeV is predicted for 300-J (total), 100-ps laser pulses at $\lambda = 1$ μm, and 3 TeV at $\lambda = 10$ μm. For a 160 Hz repetition rate, the luminosity associated with the colliding beams could reach $10^{33}$ $cm^{-2}s^{-1}$ per bunch at $10^7$ electrons. Wall plug power requirements to drive the system would be ~2 MW.

## MICROWAVE-BASED CONCEPTS

Accelerating structures operating in the microwave regime obviously cannot achieve accelerating gradients comparable to the high values projected for laser-based accelerators. Nevertheless, the study of novel structures for microwave accelerators remains attractive since multistage operation to achieve a high ultimate beam energy is conceptually straightforward, and since proof-of-principle experiments can often be conducted using available or newly-emerging rf source technology. However, for a given novel structure to remain attractive, it should probably compete favorably in terms of acceleration gradient, complexity, and cost with advanced versions of traditional structures, such as the NLC 11.4 GHz linac structure developed at SLAC, which has an acceleration gradient of ~50 MV/m, increasing in the future to about 80 MV/m. Furthermore, higher frequency operation of any proposed structure should lead to accelerating gradients that scale roughly in direct proportion to frequency.

The microwave structure papers presented at this Workshop fell into three categories: proof-of-principle concepts being tested with high-power rf sources at 2.8, 11.4, or 17 GHz; structure development for 94 GHz, where high-power sources are so far not available; and wakefield accelerator structures, where microwave accelerating fields are excited by the passage of one or more drive bunches. An overview of recent developments was provided in a wide-ranging invited paper by T. C. Marshall (18).

Reports on four categories of novel microwave accelerating structures were presented. These are dielectric-loaded, smooth-wall slow wave structures as an alternative to cavity-loaded linac structures (19, 20); smooth-wall, fast-wave structures with an applied wiggler magnetic field for IFEL acceleration (21); smooth-wall, fast wave structures with an applied axial magnetic field for cyclotron autoresonance acceleration (22); and modified racetrack cavities for a 17 GHz high-gradient rf gun (23).

# Dielectric-Loaded Structures

Dielectric-loading to achieve a slow phase velocity axisymmetric $TM_{01}$-mode in a cylindrical waveguide has long been recognized as an alternative to the conventional cavity-loading used in all rf linacs. The simplest such structure is a ceramic tube with an axial hole for the beam and a conducting outer coating. This structure has the virtue of simplicity, low cost, and ruggedness, but is subject to limitations due to rf breakdown and poor thermal conductivity of the ceramic. Two groups are currently developing prototype accelerators using this structure. One experiment, dubbed MICA (microwave inverse Cerenkov accelerator), is being carried out in a collaboration between Yale and Columbia Universities, and Omega-P, Inc. (19). Here, up to 15 MW of rf power at 2.8 GHz will be coupled into the structure, consisting of a copper-coated alumina tube with inner and outer diameters of 8.9 mm and 31.4 mm, respectively. A train of 6-MeV electron bunches from an rf gun phase-synchronized with the rf drive wave will be injected in a narrow phase window at the maximum accelerating field. All particles are expected to enjoy the same acceleration up to about 12 MeV in a four-section 1.2-m structure designed and built by Titan Beta. This experiment is designed to test the underlying theory, to determine rf breakdown limits, and to measure emittance growth. First operation of MICA is scheduled for Fall 1998. The potential for this structure to operate at higher rf powers and higher rf frequencies will be assessed once the prototype results are analyzed.

A second closely related study is underway at Argonne National Laboratory (20), using advanced ceramics for operation at 11.4 GHz. For both structures, input and output coupling of the rf is a serious design issue, but Titan-Beta has achieved an acceptable design at 2.8 GHz, which has been implemented in MICA. The significance of transverse wakefields on beam stability in a MICA structure is being assessed using newly-emerging theory (24), and results are expected in the near future.

# Fast-Wave Structures

Two accelerating mechanisms using fast-wave structures were described. The structures both consist of smooth-wall cylindrical waveguides excited in the $TE_{11}$ mode. Phase synchronism with an $e$-beam is achieved in both by use of an external static magnetic field. For an axial magnetic field, cyclotron autoresonance acceleration (CARA) can be achieved (25), but an upper energy limit exists due to beam stalling in the axially-rising magnetic field. For energies up to a few MeV, CARA has been shown capable of rf-to-beam power efficiencies >95%. One way to overcome the upper energy stalling limit in CARA is to use a multistage structure, and analysis for a zero-emittance beam shows the possibility of acceleration to GeV energies (26), albeit with relatively low acceleration gradient. A beam accelerated in this manner would constitute a powerful synchrotron light source. However, preliminary results for a finite-emittance beam show that efficient phase trapping is clearly essential if multistage operation is to succeed, but this may be difficult to achieve in practice (22).

For an imposed wiggler magnetic field, inverse free-electron laser acceleration can be achieved (27). A status report was presented on operation of a 2.8-GHz microwave inverse FEL accelerator (MIFELA) (21). This proof-of-principle experiment utilizes an injected 6-MeV bunched e-from an rf gun, with an expected acceleration to about 7 MeV. This relatively small acceleration gradient is not intrinsic to MIFELA, but results from the limited available rf power and low rf driver frequency. Gradients comparable to that of NLC can be achieved at higher frequency and higher rf drive power. , Because of its simplicity and relative immunity to higher-order wakefields, MIFELA remains an attractive concept for acceleration to energies of perhaps up to a few GeV as an injector into a higher energy machine or storage ring light source. But, the acceleration gradient in MIFELA is predicted to fall as the beam energy increases, and synchrotron radiation energy loss will pose an ultimate upper limit.

## Cavity Structures

High acceleration gradient structures are receiving close attention in connection with rf guns, which are becoming the injectors of choice for many advanced accelerator concept experiments. S-band rf guns that are in wide use operate with acceleration gradients in the range of 80-100 MV/m, made possible using precision fabrication to produce ultra-smooth surfaces, and by applying about 3 MW of rf power per rf cell in the gun. Higher gradients at higher rf drive frequencies have been recently produced, notably in tests at MIT with a 17-GHz Haimson klystron driver, where a gradient of over 250 MV/m was demonstrated. Degradation in rf gun performance will occur if care is not taken to symmetrize the rf fields in the cavities. In a 2-1/2 cell gun being designed by Haimson (23), it has been shown that reliance upon only opposing waveguide feeds with a conventional cylindrical cavity results in a field imbalance between the first and second cavities, but that use of a "racetrack" configuration, in which an oblate cross-section is employed, can result in equality of the peak fields at 250 MV/m in the first two cells, and a field diminution by only 10% in the last cell. This work exemplifies the care needed to build high performance structures at high operating frequencies.

## W-Band Structures

Acceleration gradients approaching 1 GeV/m are predicted to be possible in principle in rf accelerators operating near 90 GHz (W-band), if only scaling in direct proportion to frequency is considered. No high power rf source is available in this frequency range, but if successful high gradient structures can be developed, then strong motivation to develop the sources should follow. Serious issues to be overcome at W-band involve the sheer complexity of reproducible fabrication of the microstructures, and development of an acceptable means of dissipating the heat arising from ohmic losses in the copper. Two presentations were made to provide status reports on these critical issues, the first by Palmer, et al., (28) on structure fabrication and testing, the second by Lin (29) on diamond coatings for improved heat dissipation.

Palmer presented test methods and results on 7- and 25-cell muffin tin accelerating structures fabricated using EDM by Ron Witherspoon, Inc. The tests were carried out using a "home-built" 90-GHz vector network analyzer assembled by R. Siemann. Details of mechanical design, rf simulations, and rf measurements were presented. A dimensional precision has been achieved of better than 1 μm for the 7-cell structure, and better than 3 μm for the 25-cell structure; and a dimensional accuracy of better than 7 μm have been achieved for both. The goal for these is 1 μm, which is believed to be possible with the techniques under refinement.

Diamond coating of copper surfaces to aid in dissipation of heat is a promising approach to allow operation of a non-cryogenic copper accelerating structure with an accelerating gradient greater than 300 MV/m. Otherwise, Nezhevenko (30) has estimated for acceleration gradient and pulse width levels following conventional scaling, that pulsed temperature rises of over 150°C will occur causing metal fatigue and leading to unacceptably short structure lifetimes. Since a gradient of 300 MV/m is projected to be possible at about 40 GHz, operation at W-band to achieve a higher gradient will perforce require enhanced heat dissipation. Otherwise the pulsed temperature rise would approach 900°C. Diamond coatings are computed to satisfy the requirements because their thermal conductivity is about 5 times that of copper. Methods for CVD application of diamond to copper have been developed, but an intermediate layer, such as Ti-W, has been shown necessary to obtain good adherence. Lin (29) presented computations for multilayer structures indicating that reductions of nearly a factor of four in pulsed temperature rise are possible. Experiments at X-band are underway, which have already shown in preliminary runs a reduction in pulsed temperature rise by nearly a factor of three. This important work is definitely a critical element in the program to develop a W-Band accelerating structure.

## Dielectric Wakefield Structures

The dielectric-lined, slow-wave coaxial structure discussed above for the microwave inverse Cerenkov accelerator MICA (19) is, of course, that employed for many years in experiments on wakefield acceleration (31). Recently, it has been discovered (32) that this structure can be configured so that all the $TM_{0n}$ modes will have nearly the same phase velocity, slightly below the speed of light. Thus, many modes can be excited by a passing electron drive bunch. Furthermore, when a periodic train of drive bunches is employed, the eigenfrequencies of these modes can be adjusted so that the superposition of excited modes results in a space-time localized periodic wave whose period can equal the bunch train period. This allows for constructive superposition of wakes from successive drive bunches, with the possibility of building up high acceleration gradients without need for exceptionally high charge in a single bunch. This approach to high-gradient acceleration using a simple smooth-wall structure was reviewed by T. C. Marshall (18) in a Workshop

Plenary session. He showed an example of acceleration of a 100-MeV test bunch up to 200 MeV in a 75-cm test structure driven by a train of three 2-nC bunches. A preliminary proof-of-principle experiment on multimode excitation by one or more drive bunches is to be performed in Fall 1998 at the Brookhaven ATF by the Yale/Columbia/Omega-P collaboration.

A further presentation based on the same idea was given by Power (33), which showed computed wakefields similar to those published previously by Zhang, *et al.* (32). However, Power raised questions on some aspects of the prior work, which need to be resolved given the exciting potential now stimulated by this new approach to wakefield acceleration. All the work described above assumes the drive bunches to be moving along the waveguide axis. But, small deviations off the axis are bound to occur, and these will lead to excitation of higher-order waveguide modes with transverse deflecting forces that could lead to beam instability. This issue could have significance for MICA as well. A report on the formulation of a theory for higher-order mode excitation by a nonaxisymmetric drive bunch was presented (24). This work is complicated by the fact that nonaxisymmetric modes of this structure are not pure TE- or TM-modes, but are hybrid modes. Prior to the present work, the orthonormalization proof for these modes had not been published, nor had the normalization constants been obtained. Preliminary estimates for transverse wakefield deflections on drive and test bunches in a multibunch, multimode configuration suggest that, since the higher-order modes have a variety of phase velocities, one may not have significant space-time superposition of the higher order fields, so that transverse forces may not be too serious. But clearly, more work is needed on this issue.

## JOINT SESSION BETWEEN WORKING GROUPS 3, 4, AND 5

This working group held an afternoon joint session meeting with Working Groups 4 "Beam Generation, Monitoring, Conditioning..." and 5 "Particle Beam Sources." Advanced diagnostic needs were identified and are listed in Table 1. The ones particularly critical for experiments, such as STELLA, are ultralow emittance measurements, diagnostics for ultrashort microbunches, temporal synchronization, and

**TABLE 1.** Advanced diagnostic needs.

- Minimally invasive and ultralow (<0.1 mm-mrad) emittance (phase-space) measurement.
- Diagnostics for ultrashort pulses (<10 fs resolution).
- Real-time measurement for control.
- Temporal synchronization of systems (1 degree of RF phase).
- Phase-space mapping.
- Ultrafine position monitoring (<10 μm).
- Energy spread monitoring.

ultrafine position monitoring. A number of ideas were discussed, which hopefully provided the seeds for further discussions after the Workshop.

## CONCLUSIONS

Laser-based acceleration schemes have matured to the point where the practical problems of scaling these devices to higher energies are now being investigated. This gives rise to solving important issues such as synchronizing fsec microbunches, compensating for space-charge longitudinal spreading, and developing advanced diagnostics for detecting these microbunches. In parallel, exciting efforts are in progress to demonstrate far-field vacuum laser acceleration, which is fundamentally different from ponderomotive-based approaches such as the recent demonstration by Malka, et al., (34). These schemes are attractive because of the possibility to eliminate all media (e.g., gas or plasma) and remove the need to operate very near the surface of an acceleration structure. Finally, novel ideas continue to emerge to further enhance the basic processes for laser-based acceleration.

Vigorous study of a variety of microwave structures continues, as evidenced by the presentations at this Workshop. Clearly, the full potential of all the schemes discussed has been far from fully realized, and additional work remains. Microwave structures with notable promise are the inverse FEL accelerator and the inverse Cerenkov accelerator, since both employ smooth-wall cylindrical structures that should exhibit fewer problems with transverse wakefields than do conventional cavity-loaded rf structures. New ideas with multibunch dielectric wakefield acceleration appear to offer the possibility of high-gradient acceleration without the need for either a high-frequency, high-power rf driver (other than the source that provides the injected electron bunches), nor of exceptionally high-charge driving bunches. Early work on W-Band structures shows promise, but it is important to perfect diamond coating techniques or alternatives, in order to dissipate heat if the desired acceleration gradients of 1 GeV/m are to be achieved. Otherwise, there would be little point in operating an rf accelerator using a conventional cavity-loaded structure at a frequency higher than about 40 GHz, where a gradient of about 300 MV/m should be achievable without damaging pulsed heating.

By the AAC'00 Workshop, it is anticipated that further experimental results will be in hand for several of the new ideas presented here, and that this subfield of accelerator research will continue to exhibit the vigor shown at AAC'98.

## ACKNOWLEDGMENTS

We wish to thank the authors for their contributions to our working group and all those who participated in our meetings, which at times made for lively and simulating discussions. We wish to also thank T. Smith and C. Clayton, co-chairs of Working

Group 4, and P. O'Shea and L. Spentzouris, co-chairs for Working Group 5, for helping to arrange a joint session during the Workshop with our group.

# REFERENCES

1. Rosenzweig, J. B. in *Advanced Accelerator Concepts*, Lake Tahoe, CA, AIP Conference Proceedings No. 398, S. Chattopadhyay, Ed., (American Institute of Physics, New York, 1997), 181-186.
2. van Steenbergen, A., Gallardo, J., Sandweiss, J., Fang, J.-M., Babzien, M., Qiu, X., Skaritka, J., and Wang, X.J., *Phys. Rev. Lett.* **77**, 2690 (1996).
3. Kimura, W. D., Kim, G. H., Romea, R. D., Steinhauer, L. C., Pogorelsky, I. V., Kusche, K. P., Fernow, R. C., Wang, X. J., and Liu, Y., *Phys. Rev. Lett.* **74**, 546-549 (1995).
4. Kimura, W. D., Ben-Zvi, I., Babizen, M., Campbell, L. P., Cline, D. B., Fiorito, R. B., Gallardo, J. C., Gottschalk, S. C., He, P., Kusche, K. P., Liu, Y., Pantell, R. H., Pogorelsky, I. V., Quimby, D. C., Robinson, K. E., Rule, D. W., Sandweiss, J., Skaritka, J., Steinhauer, L. C., van Steenbergen, A., and Yakimenko, V., "STELLA Experiment - Design and Model Predictions," in these Proceedings.
5. Kusche, K. P., Ben-Zvi, I., Babizen, M., Campbell, L. P., Cline, D. B., Fiorito, R. B., Gallardo, J. C., Gottschalk, S. C., He, P., Kimura, W. D., Liu, Y., Pantell, R. H., Pogorelsky, I. V., Quimby, D. C., Robinson, K. E., Rule, D. W., Sandweiss, J., Skaritka, J., Steinhauer, L. C., van Steenbergen, A., and Yakimenko, V., "STELLA Experiment - Hardware Issues," in these Proceedings.
6. Sprangle, P.; Esarey, E.; Krall, J.; Ting, A., *Optics Comm.*, **124**, 69-73 (1996).
7. Hafizi, B., Ting, A., Moore, C. I., Ganguly, A. K., and Sprangle, P., "Vacuum Beat Wave Acceleration with an Axicon Beam," in these Proceedings.
8. Moore, C. I., Ting, A., Hafizi, B., Sprangle, P., and Hirshfield, J., "Status of the NRL/Omega-P Vacuum Beatwave Accelerator Experiment," in these Proceedings.
9. Plettner, T., "The Laser Electron Acceleration Project: Theory and Experiment," in these Proceedings.
10. Huang, Y. C., Plettner, T., and Byer, R. L., "The Proposed Interferometric-Type Laser-Driven Particle Accelerators," in these Proceedings.
11. Liu, Y., Cline, D., and He, P., "Laser Acceleration in Vacuum Using a Donut-Shaped Laser Beam," in these Proceedings.
12. Steinhauer, L. D. and Kimura, W.D., "Space Charge Compensation in Laser Particle Accelerators," in these Proceedings.
13. Kawata, S.; Maruyama, T.; Watanabe, H.; Takahashi, I., *Phys. Rev. Lett.*, **66**, 2072-5 (1991).
14. Hussein, M.S.; Pato, M.P., *Phys. Rev. Lett.*, **68**, 1136-9 (1992).
15. Pakter, R., "Intensity Effects on the Inverse Bremsstrahlung Electron Accelerator," in these Proceedingsl.
16. Parsa, Z., "Enhanced Inverse Free Electron Laser and Nonlinear Amplification of Inverse Bremsstrahlung Electron Acceleration," in these Proceedings.
17. Mikhailichenko, A. A., "A Laser Linear Collider Design," in these Proceedings.
18. Marshall, T. C., Zhang, T-B., and Hirshfield, J. L., "Novel Structure-Based RF Accelerators," in these Proceedings.
19. Zhang, T-B., Marshall, T. C., LaPointe, M. A., Hirshfield, J. L., and Ron, A., *Phys. Rev. E* **54**, 1918 (1996).
20. Gai, W., Zou, P., and Konecny, R., "Externally RF Powered dielectric Acceleration Structure at 11.4 GHz," in these Proceedings.

21. Yoder, R. B., Marshall, T. C., Wang, M., and Hirshfield, J. L., "Current Status And Preliminary Results Of The Microwave Inverse FEL Experiment," in these Proceedings.
22. Wang, C., Hafizi, B., and Hirshfield, J. L., "Multi-stage Cyclotron Autoresonance Accelerator," in these Proceedings.
23. Haimson, J., *et al.*, "A Field Symmetrized Dual Feed 2 MeV RF Gun for a 17 GHz Electron Linear Accelerator," in these Proceedings.
24. Hirshfield, J. L., Park, S. Y., and Zhang, T-B., in these Proceedings.
25. LaPointe, M. A., Yoder, R. B., Wang, C., Ganguly, A. K., and Hirshfield, J. L., *Phys. Rev. Lett.* **76**, 2718 (1996).
26. Wang, C. and Hirshfield, J. L., *Phys. Rev. E* **57**, 7184 (1998).
27. Hirshfield, J. L., Marshall, T. C., Zhang, T-B., Ganguly, A. K., and Sprangle, P. A., *Nucl. Instr. Meth. Phys. Res. A* **358**, 129 (1995).
28. Palmer, D., Chou, P. J., Schwartzkopf, S., Witherspoon, R., and Siemann, R. H., "W-Band Structure Work at SLAC," in these Proceedings.
29. Lin, X. E., "Diamond Coating in Accelerator Structure," in these Proceedings.
30. Nezhevenko, O. A., in Proceedings of PAC'97, to be published.
31. Gai, W., Schoessow, P., Cole, B., Konecny, R., Rosenzweig, J., and Simpson, J. *Phys. Rev. Lett.* **61**, 2756 (1988).
32. Zhang, T-B., Hirshfield, J. L., Marshall, T. C., and Hafizi, B., *Phys. Rev. E* **56**, 4647 (1997).
33. Power, J., Gai, W., and Schoessow, P., "Multimoded Wakefield in Dielectric Tube," in these Proceedings.
34. Malka, G., Lefebvre, E., and Miquel, J. L., Phys. Rev. Lett. **78**, 3314-17 (1997).

# Summary Report: Working Group 4 on "Beam Monitoring, Conditioning, and Control at High Frequencies and Ultrafast Timescales"

Todd I. Smith

*Hansen Experimental Physics Laboratory*
*Stanford University*
*Stanford, CA 94305-4085*

**Abstract.** Working Group 4 at the 8th Advanced Accelerator Concepts Workshop (AAC'98), held July 5-11,1998 in Baltimore, Maryland hosted more than fifteen scheduled or impromptu talks (all punctuated with lively discussion) on the general topic of "Beam Monitoring, Conditioning, and Control at High Frequencies and Ultrafast Timescales". This report is a summary of these talks and discussions.

## INTRODUCTION

Our ability to measure and control many of the key properties of particle beams, already tasked by present accelerators, is going to be severely challenged by accelerator concepts currently under test or discussion. The charter of Working Group 4 (WG4) at the 8th Advanced Accelerator Concepts Workshop (AAC'98, Baltimore, Maryland, July 5-11, 1998) was to discuss these issues under the title "Beam Monitoring, Conditioning, and Control at High Frequencies and Ultrafast Timescales".

The emphasis of the sixteen talks presented during the meetings of the working group, and the ensuing discussions, was on techniques of measuring and monitoring beam parameters. As all of the techniques involve observation of some aspect or another of the six dimensional phase space occupied by the beam, the summaries in the following sections will be sorted by the amount of information obtained by a particular technique on details of the 6D space. Thus, the first section contains a discussion of one method for obtaining information on density in three of the six dimensions, and a discussion of two techniques for measuring the density as projected onto two of the six dimensions. The second section discusses techniques which measure the phase space projection onto a single dimension. Transverse measurements will be discussed first, followed by longitudinal measurements.

CP472, *Advanced Accelerator Concepts: Eighth Workshop,*
edited by W. Lawson, C. Bellamy, and D. Brosius
© 1999 The American Institute of Physics 1-56396-889-4/99/$15.00

# MEASUREMENTS OF PHASE SPACE DISTRUBUTIONS

The three talks summarized in this section are unique in that they deal with measurements of density distributions <u>within</u> phase space-not just projections along a single axis. The phase space of a real beam is seldom well represented by the usual model of a hyper-gaussian distribution, and knowledge of its actual shape can be valuable in helping understand causes of brightness dilution and in finding ways to combat this dilution.

The description by I. Ben-Zvi of the slice emittance measuring system [1] at BNL provides an excellent example of the importance of such information. In this system a short time slice is selected out of an energy chirped beam by a slit in a dispersive region. The area, inclination angle, and eccentricity of the ellipse that best fits the transverse phase space of the slice is then obtained with a quadrupole scan technique. The slice emittance is, of course, the area of the ellipse. After all of the slices of the beam were measured, the space charge induced rotation of the inclination angle of the ellipses of different slices through the length of the beam was evident. Within experimental error the emittance of each slice was constant. However, the rotation caused the total emittance to appear larger. When the emittance compensation technique proposed by Carlston [2] was applied the misalignment of the ellipses could be seen to decrease, and the overall emittance of the beam was reduced.

A report given by V. Yakimenko [3], also of BNL, described results obtained with a system which maps out transverse phase space using tomography. In this system the betatron phase advance of the beam through a transport line with nine adjustable quadrupoles was adjusted in steps, and the beam intensity profile was measured on a CCD camera for each step. This information is then used to reconstruct the transverse phase space distribution at the entrance to the quadrupole channel. An initial use of this system has been to study the effect of the spatial profile of a photocathode's drive laser on the emittance of the electron beam emitted. Differences in the transverse phase space distributions produced by gaussian, flat, and donut shaped laser beams were dramatic. (The intensities of each were adjusted to produce the came charge per bunch.)

T.I. Smith gave a review of the longitudinal phase space reconstruction technique developed at Stanford [4,5]. In this method the synchrotron phase advance of a beam is altered through adjustment of the phase angle between an electron bunch and the field of an accelerating section just ahead of an energy spectrometer. (An adjustable non-isochronous transport section ahead of the accelerator section is useful, but not mandatory.) Energy spectra taken as a function of the accelerator phase angle (and strength of the non-isochronous section) can then be used to recover the longitudinal phase space distribution of the beam. The Stanford group has used the technique to characterize a beam that proved to have three local density maxima in the longitudinal phase space distribution. In this case the temporal profile of the beam obtained by projecting the distribution onto the time axis gave an auto-correlation signal in good agreement with that obtained from a transition radiation interferometer.

# ONE DIMENSIONAL MEASUREMENTS

In this section the techniques are characterized by the fact that they measure only the projection of phase space along on a single axis (time or position), or even only the location of the centroid of the distribution in time or space. These techniques tackle some very tough problems, and the word "only" is used only in comparison with the previous section.

## Transverse Measurements

J. Irwin gave a report [6] on work at SLAC on techniques for Model-Independent Analysis (MIA) of a beamline using correlation matrices of physical variables and Singular Value Decomposition of a beamline beam position (BPM) monitor matrix. The beamline matrix is formed from the readings of many BPMs for a large number of pulses. The method takes advantage of the high degree of correlation between BPM readings and provides many advantages over other techniques. To name a few: the resolution of BPMs can be measured directly and can be improved by using multiple pulses and BPMs; systematic BPM errors can be identified and removed; BPM noise can be mostly cut by performing the SVD; and primary effects, such as betatron motion, can be identified and easily separated from other effects.

A talk addressing the need for absolute beam position measurements in an accelerating structure [7,8] was presented by M. Seidel, now at DESY. The problem is that the use of the high frequency structures now under serious consideration for high gradient acceleration will require exquisite control of the beam position relative to the structure to avoid the excitation and effects of transverse deflecting fields. The talk reported the measurement of such absolute (with respect to the accelerator structure) beam position by analyzing beam induced excitation of these modes in a 206 cell damped-detuned X-band structure (DDS) at SLAC. Among other achievements, they were able to demonstrate absolute beam position along the axis of the structure with a precision at the level of 40 μm.

The third topic discussed under transverse measurements was that of the use of diffraction radiation for non-invasive measurement of various beam parameters. As diffraction radiation (DR) is closely related to transition radiation (TR), the beam line portion of a DR system is not much more than a foil (a TR system) with an aperture in it. It can thus be quite compact and can be placed at virtually any place along a beam line. R.B. Fiorito [9] of the Catholic University gave a presentation describing ways in which three DR observables (the far-field angular distribution of the DR, the near field image of the DR at the screen surrounding the aperture, and the polarization measured at the aperture) can be utilized to measure beam position, size, and divergence. (Beam energy and temporal information is also available, but the emphasis here was on transverse aspects.) These systems have the potential to become quite useful, particularly for high energy beams.

# Longitudinal Measurements

The summary of talks and discussions of longitudinal measurement techniques is divided into one section dealing with direct measurements and one dealing with indirect measurements

## *Direct Measurements*

These discussions began with a timely talk by M. Uesaka of the University of Tokyo in which he presented results of a direct comparison between electron pulse lengths measured by a femtosecond streak camera and those obtained via coherent transition radiation interferometry. He began by outlining the steps [10] necessary to preserve femtosecond information as it makes its way from the source (electron beam) to the 200 fs streak camera under test (Hamamatsu FESCA-200), and then presented data obtained as the camera recorded a 83 fs Ti:Sapphire pulse. The measured width was 258 fs, implying a response function of 243 fs. This camera and a TR interferometer were then used to make simultaneous measurements of the electron bunch length produced by their sub-picosecond linac. Finally, the bunch distribution was reconstructed from the TR data under various assumptions about the bunch form factor (symmetric and anti-symmetric gaussian and exponential extrapolation), and the results were compared with the streak camera data-which of course, needed no reconstruction. The streak camera recorded a FWHM of 550 fs, while three of the four bunch form factor combinations gave values between 570 fs and 520 fs. The fourth combination, symmetric exponential extrapolation, produced a FWHM of 420 fs.

D. Oepts of FOM Rijnhuizen, The Netherlands (currently at CEBAF) presented a talk on differential optical gating [11,12] (DOG) as a technique to measure pulse shapes with a resolution approaching that of the 60 fs Ti:Sapphire laser used to drive an optical gate. Although the technique is a sampling one, and requires a repetitive signal, its unique feature is that it does not require synchronization between the sampling laser and the signal to be measured. Any of several reasonably standard methods (sum-frequency generation, optical Kerr effect, saturable absorption,...) can be used as the optical gate, but the trick is to use a differential technique to simultaneously measure the optical intensity and its time derivative. Since the two signals are not synchronized, after a sufficiently large number of samples a complete map of the signal intensity vs. time derivative is obtained. For reasonably simple pulse shapes reconstruction of the actual pulse from this map is straightforward. Modifications of the technique can allow reconstruction of complex shapes. The technique has been used to measure, with a resolution of a few hundred femtoseconds, the detailed pulse shape of a picosecond mid-infrared FEL at Stanford University. Plans exist to use the DOG technique to measure electron bunch lengths, either with TR or Cherenkov radiation as a light source, or possibly with a ZnTe crystal as a direct electro-optic probe [13,14] of the electron beam's electric field.

The last of the presentations in this section was delivered by T.I. Smith of Stanford on the possibility of achieving femtosecond resolution using a non-isochronous (referred to as dispersive for the rest of this paragraph) transport system followed by an accelerator section phased at 90° with respect to the beam, and an energy spectrometer [15]. The use of a dephased accelerator section to produce a correlation between longitudinal position and energy, and the subsequent use of an energy spectrometer to extract temporal information is not new. The primary limitation of this simple technique is the degradation of the temporal resolution caused by the inherent energy spread of the beam. It turns out that if the product of the temporal dispersion of the transport system and the temporal energy slew rate of the structure is -1 (In TRANSPORT notation $R_{56}*R_{65}= -1$), then to first order the time axis is identically transformed onto the energy axis. The result is that the intrinsic energy spread of the beam has no influence whatever on the temporal resolution. An example was presented showing that, including second order effects, the temporal details of a 100 MeV beam with an energy spread as large as 1% could be measured with a resolution of 30 fs, using a 0.01% energy spectrometer, a 10 MeV linac section at 10 GHz, and a 0.5 mm/% dispersive section. Without the dispersive section, the resolution of the same beam would only be 1.6 ps, a factor of ~50 degradation.

## *Indirect Measurements*

### Destructive Methods

Two talks were given on coherent transition radiation (CTR) diagnostics for obtaining longitudinal information. D. Rule of the Carderock Division NSWC gave an analysis of the effect of the finite bandwidth of a detector on the observed intensity of CTR produced by a long train of micro bunches [16]. Using the parameters of the STELLA [17] experiment on the ATF linac at BNL as an example, he concluded that meaningful CTR bunch length measurements were possible with the use of fairly wide band pass filters (~5%) and that micropulse widths of ~1-3 μm could be distinguished by observing multiple harmonics of the 10.6 μm spacing. His analysis also showed how to include the effects of beam divergence and bunch spacing jitter in the treatment of CTR from trains of micro bunches. P. He of BNL discussed improvements [18] in the existing CTR micro-bunching diagnostic [19] for STELLA. It is expected that a signal improvement of about four orders of magnitude will be achieved, primarily by better focussing.

A novel concept for micro-bunch diagnostics using Compton scattering from a standing laser wave was presented by I. Pogorelsky of BNL. The possibility relies on a recent analysis by Amatuni and Pogorelsky [20] which shows that the intensity of the nonlinear Compton scattering is a function of the phase of the standing wave when an electron enters the field.

## Nondestructive methods

F. Zimmermann of SLAC presented a talk on bunch-length and bunch-timing monitors [21] in the SLC final focus. Both the length and timing monitors used the same remarkably straightforward mechanism for getting beam information. Basically, they coupled the beam electromagnetic fields radiated through a ceramic gap in the beam pipe to a long (~50 m) piece of X-band waveguide and simply processed the radiation at the other end. The coupling mechanism was no more complicated than aiming the open of the guide at the gap from a distance of about 0.5 cm. Since the processing end of the guide is outside the ionizing radiation environment of the accelerator, component lifetime (including humans) is high and noise level is low. For bunch length monitoring purposes the signal power is split into four separate frequency bands between 7 and 110 GHz. They have been able to monitor pulse-to-pulse changes in the interaction point (IP) bunch length of both colliding beams at the SLC with a precision of a few percent (out of a nominal length of ~1 mm). To monitor the relative timing of the two beams they narrow-band filter the beam-induced signals, delay the first beam into coincidence with the second, and then homodyne mix the signals. Timing shifts between the beams then show up as a voltage shift at the mixer output. To date they have achieved a short term resolution better than 5° X-band (~1 ps).

Two talks were presented on what are at present conceptual ideas for nonintercepting bunch length monitors. The ideas clearly have merit, but whether or not they will prove practical is uncertain. During his invited presentation [22] at one of the plenary sessions during the workshop, T.C. Marshall of Columbia showed calculations illustrating the strong bunch length dependence of the interaction strength in a dielectric wakefield accelerator. As the structure of such an accelerator is quite simple, being only a dielectric tube with a hole in it, it was appealing to imagine it being used as a bunch length monitor. Marshall agreed to discuss the concept with the group, and it does appear to have possibilities, especially for sub-millimeter bunch lengths. Novel ideas were proposed by M.C. Lampel [23] for producing Smith-Purcell radiation outside the beam line, and then using it as a pulse length diagnostic [24]. After some discussion it appeared as though exploration of transversely slotted beampipes might be fruitful, the idea being to encourage the image wall current to interact with the outside world directly and thus to maximize the ease of coupling the radiation into the rest of the diagnostic equipment.

A series of talks and discussions took place throughout the workshop on fluctuation based beam diagnostics. This technique, proposed by Zolotorev and Stupakov [25], is based on measuring fluctuations in the visibility of interference fringes in the spectrum of incoherent electromagnetic radiation emitted by the beam. The method has many potential advantages, such as being a single shot technique, and in actually becoming somewhat easier to implement as the pulse length becomes shorter. One of the most intriguing aspects of the concept is that the ratio of the participants who clearly understood it to those who didn't appeared to be in the range of a few percent. Nonetheless, P. Catravas of LBL presented convincing results of proof of principle experiments [26,27] done at BNL, and would seem to have converted most of the skeptics. Although the experiment emphasized bunch length measurements, it appears as though the technique has promise for obtaining bunch shape information as well.

## CONCLUSION

The talks and discussions hosted by WG4 throughout the workshop were stimulating and often provocative. It's likely that everyone learned some useful things, and that many of the participants were inspired to explore new paths. On the other hand, there will soon be a need for beam diagnostics with capabilities which are not now available, and it's not at all clear at this point just how these needs will be met. Various forms of laser accelerators will need some form of bunch length with at least 10 fs resolution, and this is roughly an order of magnitude better than has been achieved to date. Transverse diagnostics are in a similar state. The emittance of beams will become more and more difficult to measure by present methods as they get smaller than the present best values of about 1 $\pi$-mm-mr, especially at high energies when beam dimensions and divergences become miniscule.

The obvious conclusion is that new ideas are needed.

## ACKNOWLEDGEMENTS

Chris Clayton was the co-chair of Working Group 4. His help and advice in identifying and contacting potential contributors to WG4 was very much appreciated, as was his help throughout the workshop. Thanks are also due to the dedicated working group participants. Those who gave presentations and/or whom I remember as having been active during the discussions are (I apologize for any omissions): Ilan Ben-Zvi, Palma Catravas, Chris Clayton, Ralph Fiorito, Michael Fitch, Ping He, John Irwin, Wayne Kimura, Mike Lampel, Greg LeSage, Tom Marshall, Dick Oepts, Igor Pogorelsky, Don Rule, Mike Seidel, Mitsuru Uesaka, Vitaly Yakimenko, Frank Zimmermann, and Max Zolotorev.

## REFERENCES

1. X. Qui, K. Batchelor, I. Ben-Zvi, and XJ Wang, "Demonstration of Emittance Compensation through the Measurement of the Slice Emittance of a 10-ps Electron Bunch", PRL **76**, No. 20, 3723, (1996).
2. B. E. Carlston, "New Photoelectroc Injector Design for the Los Alamos National Laboratory XUV FEL Accelerator", Nucl. Instr. and Meth. 285, 313-319 (1989).
3. V. Yakimenko, M. Babzien, I. Ben-Zvi, R. Malone, and X.-J. Wang, "Emittance Control of a Beam by Shaping the Transverse Charge Distribution Using a Tomography Diagnostic", Proceedings of EPAC'98, June 22-27, Stockholm, Sweden.
4. E.R. Crosson, K.W. Berryman, B.A. Richman, T.I. Smith and R.L. Swent, "A Technique for Measuring an Electron Beam's Longitudinal Phase Space with Sub-picosecond Resolution", Micro Bunches Workshop, eds. E. B. Blum, M. Dienes and J. B. Murphy, AIP Conference Proceedings **367**, 397 (1996).

5. E.R. Crosson, K.W. Berryman, B.A. Richman, T.I. Smith and R.L. Swent,, "The Determination of an Electron Beam's Longitudinal Phase Space Distribution Through the Use of Phase-Energy Measurements", Nucl. Instr. and Meth., A375, 87 (1996).

6. J. Irwin, C.X. Wang, Y.T. Yan, K.L.F. Bane, Y. Cai, F.-J. Decker, M.G. Minty, G.V. Stupakov, and F. Zimmermann, "Model-Independent Analysis with BPM Correlation Matrices", SLAC-PUB-7863 (1998) and Proceedings of EPAC'98, June 22-27, Stockholm, Sweden.

7. M. Seidel, C. Adolphsen, K.L.F. Bane, R.M. Jones, N.M. Kroll, R.H. Miller, and D.H. Whittum, "Absolute Beam Position Measurement in an Accelerator Structure", Nucl. Instr. And Meth., A404, 231-236 (1998).

8. M. Seidel and C. Adolphsen, "Precision Beam to Structure Alignment in Linear Accelerators", Proceedings of the Advanced Accelerator Concepts Workshop, Baltimore Maryland, July 5-11, 1998.

9. R.B. Fiorito, D.W. Rule, and W.D. Kimura, "Noninvasive Beam Position, Size, Divergence and Energy Diagnostics Using Diffraction Radiation", Proceedings of the Advanced Accelerator Concepts Workshop, Baltimore Maryland, July 5-11, 1998.

10. M. Uesaka, T. Ueda, T. Kozawa, and T. Kobayashi, "Precise Measurement of a Subpicosecond Electron Single Bunch by the Femtosecond Streak Camera", Nucl. Inst. and Meth. A406, 371-379 (1998).

11. C.W. Rella, G.M.H. Knippels, D. Palanker, T.I.Smith, and H.A. Schwettman,, "Pulse Shape Measurements of the Stanford Mid-Infrared Free Electron Laser Using Differential Optical Gating", Nucl. Inst. And Meth. A407 (1998).

12. C. W. Rella, G.M.H. Knippels, D. Palanker and H.A. Schwettman, "Pulse Shape Measurements Using Differential Optical Gating of a Picosecond Free Electron Laser Source with an Unsynchronized Femtosecond Ti:Sapphire Gate", Optics Communication, Optics Comm. (To be published).

13. Q. Wu, M. Litz, X.-C. Chang, "Broadband detection Capability of ZnTe Electro-Optic Field Detectors", Appl. Phys. Lett. Vol 68, 2924-2926 (1996).

14. Z. Jiang and X.-C. Zhang, "Electro-Optic Measurement of THz Field Pulses with a Chirped Optical Beam", Appl. Phys. Lett. Vol 72, 1945-1947 (1998).

15. K.N. Ricci, E.R. Crosson and T.I. Smith, "Sub-picosecond Electron Bunch Profile Measurement Using Magnetic Longitudinal Dispersion and Off-phase RF Acceleration", Proceedings of the Advanced Accelerator Concepts Workshop, Baltimore Maryland, July 5-11, 1998.

16. D.W. Rule, R.B. Fiorito, and W.D. Kimura, "The Effect of Detector Bandwidth on Micro Bunch Length Measurements made with Coherent Transition Radiation", Proceedings of the Advanced Accelerator Concepts Workshop, Baltimore Maryland, July 5-11, 1998.

17. W. D. Kimura, et al., "STELLA Experiment: Design and Model Predictions", Proceedings of the Advanced Accelerator Concepts Workshop, Baltimore Maryland, July 5-11, 1998.

18. P. He, Y. Liu, D.B. Cline, M. Babzien, K. Kusche, I. Pogorelsky, J. Skaritka, A. van Steenbergen, V. Yakimenko, and W. Kimura, "Micro-Bunching Diagnostic for STELLA Experiment", Proceedings of the Advanced Accelerator Concepts Workshop, Baltimore Maryland, July 5-11, 1998.

19. Y. Liu, D.B. Cline, X.J. Wang, M. Babzien, J.M. Fang, and V. Yakimenko. "Micro-Bunching Diagnostics for the IFEL by Coherent Transition Radiation", AIP Conference Proceedings 398, 664-672 (1996).

20. A. Ts. Amatuni and I. V. Pogorelsky, "Nonlinear Compton scattering and electron acceleration in interfering laser beams", Phys. Rev. ST Accel. Beams 3, 034001 (1998)

21. F. Zimmermann, G. Yocky, D.H. Whittum, M. Seidel, C. Ng, D. McCormick, K.L.F. Bane, "Bunch-Length and Bunch-Timing Monitors in the SLC Final Focus", Proceedings of the Advanced Accelerator Concepts Workshop, Baltimore Maryland, July 5-11, 1998.

22. T.C. Marshall, T_B. Zhang, and J.L. Hirshfield, "The Stimulated Dielectric Wakefield Accelerator: A Structure with Novel Properties", Proceedings of the Advanced Accelerator Concepts Workshop, Baltimore Maryland, July 5-11, 1998.

23. M.C. Lampel, "Coherent Smith-Purcell Radiation for use in Electron Beam Diagnostics", Proceedings of the Advanced Accelerator Concepts Workshop, Baltimore Maryland, July 5-11, 1998.

24. M.C. Lampel, "Coherent Smith-Purcell Radiation as a Pulse Length Diagnostic", Nucl. Inst. And Meth. A385, 19-25 (1997)
25. M.S.Zolotorev and G.V.Stupakov, "Fluctuational Interferometry for Measurement of Short Pulses of Incoherent Radiation", Submitted to Phys Rev Lett., SLAC-PUB-7132 (1996)
26. P. Catravas, M. Babzien, I. Ben-Zvi, W. P. Leemans, Z. Segalov, X.-J. Wang, J. S. Wurtele, V. Yakimenko, and M. S. Zolotorev, "Non-destructive, Single Shot Electron Beam Diagnosis Based on Fluctuational Characteristics of Microwiggler Emissions", Proceedings of the Advanced Accelerator Concepts Workshop, Baltimore Maryland, July 5-11, 1998.
27. P. Catravas, "Non-perturbative Electron Beam Characterization with a Microwiggler", Ph.D. thesis, Massachusetts Institute of Technology, ( Feb. 1998).

# Particle Beam Sources Working Group Report

## Patrick G. O'Shea
*Department of Physics*
*Duke University*
*Durham NC 27708*
and
## Linda Spentzouris
*Beams Division*
*Fermi National Accelerator Laboratory*
*Batavia IL 60510*

**Abstract.** We report of the proceeding of the Particle Beam Sources Working Group. The group discussed recent developments in high-performance electron sources and related beam dynamics issues.

## INTRODUCTION

The goals of the working group on Particle Beam Sources were two-fold:
- Report on the state of the art of electron beam sources and promising new concepts
- Develop an outline plan for important research topics that should be addressed in the coming two-year period.

There were twenty formal presentations in the working sessions. These presentations can be loosely divided into three categories:

- Photocathode guns (11)
- Novel sources (6)
- Beam transport and beam dynamics (4)

A complete list of the presentations is given at the end of this report.

The above distribution of papers continues the trend of the past several conferences where photocathode sources (RF photocathodes in particular) dominated our presentations and discussions. In spite of the relative maturity of RF photocathode technology, we are delighted to note that a number of technical firsts were reported. We also heard of the exciting work toward the development of new and better sources, and the experimental and theoretical studies on emittance growth and its mitigation.

CP472, *Advanced Accelerator Concepts: Eighth Workshop,*
edited by W. Lawson, C. Bellamy, and D. Brosius
© 1999 The American Institute of Physics 1-56396-889-4/99/$15.00

# ADVANCES IN PHOTOCATHODE GUNS

In many respects RF photocathode technology has achieved a certain level of maturity. In particular there have been no significant breakthroughs in measured emittance in recent years. This is evident from the emittance versus time plot in Fig. 1 for 1 nC bunches, where $\varepsilon_{n,rms}$ of 2 mm-mrad represents a barrier that has yet to be substantially breached.

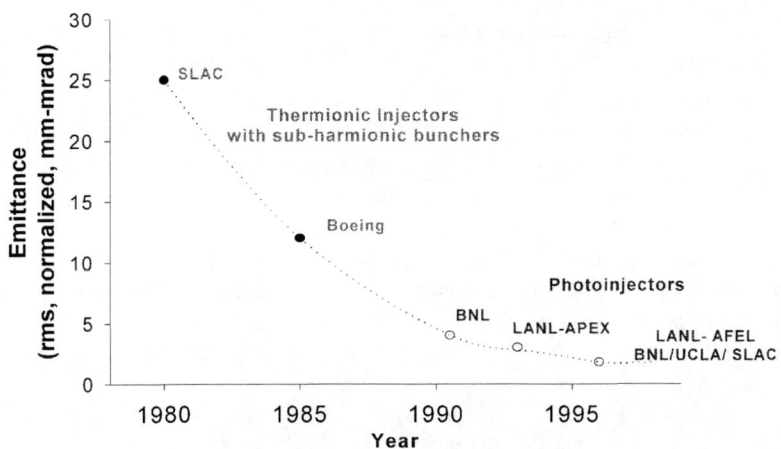

**Fig 1**.    Measured emittance (rms, normalized) from the leading thermionic and RF photocathode injectors. All data are for bunched beams approximately 1 nC bunch charge, and measured at several MeV in energy in L- and S-band systems. The data for thermionic injectors were taken after sub-harmonic bunchers and initial RF acceleration. Note that there has been little reduction in emittance in the past several years.

In previous workshops the issue of photocathode lifetime was one of grave concern. The goal of high quantum efficiency, prompt electron emission and long lifetime in practical accelerator systems has been elusive. Substantial steps have been made towards that goal with the development of $Cs_2Te$ cathodes (with a CsBr surface coating in some cases). These cathodes have been used at LANL and Fermilab for some time, and now exhibit life times of hundreds of hours with quantum efficiencies of several percent in the UV [1,2].

We note a number of other technical firsts:

A new record has been set for the highest frequency and highest accelerating gradient RF gun – 17 GHz at MIT [3]. The device is a 1.5 cell π-mode gun, and produces 0.1 nC per pulse with 1 MeV final energy. The RF driver is a 17.15 GHz klystron with a 50 ns pulse. The cathode is Cu, and the drive laser is Ti:SAF. The very high accelerating gradient of 300 MV/m at the cathode allows a peak current density of 8.8 kA/cm$^2$.

The record highest charge per bunch has been pushed to 100 nC at the ANL-AWA experiment [4]. The AWA accelerator uses a L-band gun with Cu cathode.

The Jefferson Labs IRFEL DC photocathode gun has demonstrated a new record average electron beam brightness 5 x 10$^8$ A/(m-rad)$^2$ (normalized) at an average current of 1 mA [5]. The cesiated GaAs cathode exhibits a lifetime of 200 hours in the operating gun.

The nitrogen laser-driven S-band RF gun at Duke has produced electrons that have been accelerated to 270 MeV - the highest energy for a linac with an RF photocathode gun [6]. The gun uses a LaB$_6$ cathode that can be operated in a thermionic or photoelectric mode. The 100-μJ, 337-nm laser pulse is 600ps long, and ps electron bunches are produced in an α-magnet buncher at 1 MeV just after the gun.

A novel laser pulse-shaping scheme is under construction at Fermilab [7]. It is called a "pulse stacker" and uses multiple compact delay lines around a 50/50 beamsplitter (M. Fitch). The device will take a 1-ps Gaussian input pulse and produce a 10 ps flat-top output pulse. The device looks promising as a source of pulses with various longitudinal distributions, and will be useful for studying the impact of the longitudinal distribution on emittance growth. The new gun at Fermilab will meet the TTF requirements of 10 nC, 10 ps bunches with 1250 A/cm$^2$ [8].

## NOVEL ELECTRON SOURCES

A variety of novel electron sources for use as drivers of RF generation devices and for high-brightness accelerators were discussed.

Self-bunching RF guns [9]

A new type of self-bunching RF gun has been tested at FMT. These guns use the secondary emission effect to produce electrons. In an L-band gun 40 ps pulses with 1.1 nC have been produced, with a current density of 22 A/cm$^2$. This device has shown excellent reliability and has been operating continuously for 1.5 years with 5 μs macropulses at 300 Hz. A S-band gun has also been tested and an X band gun is in design. These secondary emission guns show promise as sources for RF devices and for industrial and medical accelerators.

Ferro-electric cathode gun [10,11]

Substantial advances in the state of the art of ferro-electic sources were reported by the Cornell group. An operational device was described with an energy of

550 keV, 300 A peak current in a 300ns pulse. The peak current density is on the order of 125 A/cm$^2$. Such devices show promise for reaching 1 kA/cm$^2$ with 100ns pulses.

Pulsed HV photocathode gun [12]

A new 1 GV/m photocathode gun is nearing operation at BNL. The device achieves a very high accelerating gradient by using a 1-ns, 1-MV pulse across a 1-mm anode cathode gap. The device will produce up to 1 nC in 1-10 ps pulses from a Cu photocathode. The predicted normalized brightness is > 3000 x 10$^{14}$ A/(m-rad)$^2$ at 1 nC and 100 A. If this brightness is achieved, it will exceed the present art by about four orders of magnitude. The device has already been cold tested, and the advertised gradient has been demonstrated.

Thermionic RF gun with FEL buncher [13]

Beams from thermionic RF guns usually require some form of magnetic buncher before the beam can be efficiently injected in to an RF linac. A new type of multi-cell thermionic gun that uses an FEL buncher is under consideration at INFN. The s-band device will be capable of producing a 10 MeV beam with 150 A peak current, a 2 mm-mrad rms emittance.

All-laser injector: laser injector linear accelerator (LILAC)[14]

A new all-laser injector for use in laser acceleration schemes is under construction at the University of Michigan. The pulses produced by conventional photocathode electron sources are far too long ($\cong$ 1 ps) for laser acceleration schemes that require much shorter pulses ($\cong$ 1 fs) for efficient operation. The device uses two laser beams crossed at right angles in a plasma. One laser generates free electrons, while the other will accelerate them to several MeV. The device will produce approximately 0.01 nC with a normalized rms emittance of 1 mm-mrad and an energy spread of a few percent.

## Beam dynamics and beam transport

In recent years there have been substantial advances in the understanding of emittance growth and emittance compensation [15-17]. These advances are already leading to a more efficient design process for photocathode guns. At this workshop, there was a shift in focus to higher energy phenomena. A number of papers were presented dealing with issues of beam transport beyond the first few MeV.

Experimental results from the University of Tokyo were reported where a 2 nC bunch was compressed from 13 ps to 440 fs [18]. Studies are underway to investigate emittance growth from coherent synchrotron radiation in the magnetic chicane buncher.

A new concept for emittance cooling in electron linacs using wigglers was presented [19]. The scheme relies on a phenomenon similar to synchrotron radiation cooling in storage rings. In the case of a linac, wigglers are added between accelerator sections to produce enhanced radiative cooling.

In studies of the LCLS injector it was shown that space charge forces were still important up to beam energies of 130 MeV for low emittance, high peak-current beams [20].

Halo formation and instabilities leading to emittance growth were discussed in detail [21]. A new experiment called the E-Ring is under construction at the University of Maryland to study space-charge dominated beam dynamics in a quadrupole-focusing channel with dispersion (bends) [22]. Halo formation still remains an important and poorly understood issue for high duty factor machines.

## Unresolved issues and future directions

In recent years a general theoretical understanding of space charge induced emittance growth and its mitigation by emittance compensation has been developed. However, there are still two substantial issues that need to be resolved. The first is the issue of the intrinsic or thermal emittance of the cathode, and the second is the question of how accurately the PARMELA code models the space charge dominated beam dynamics near the cathode.

As noted previously in reference to figure 1, the measured emittance of 1 nC beams from the newest RF photocathode guns has not improved in recent years. In one well-documented case the measured emittance is approximately a factor of two higher than predicted by PARMELA [23]. There is some uncertainty in regard to the intrinsic emittance of the cathode, and also some concern about the modeling of the low energy beam transport.

Because the thermal emittance of the source ($\varepsilon_{th}$) represents the minimum emittance for a given cathode size, a detailed knowledge of its value in various cases is of great significance. In many cases, however, the value of the intrinsic emittance ($\varepsilon_{th}$) is neglected in computer simulations of the beam dynamics. In the case of the new BNL pulsed HV photocathode [12], for example, the expected emittance growth is much less than the estimates for the intrinsic emittance. Estimates for $\varepsilon_{th}$ vary widely, but in most cases are < 0.5 mm-mrad for a 1-mm radius cathode. The relationship between the intrinsic normalized rms emittance and the effective source temperature is as follows:

$$\varepsilon_{th} = \frac{r_c}{2} \sqrt{\frac{k_b T_e}{mc^2}}$$ , where $r_c$ is the cathode radius, $k_b$ is Boltzmann's Constant,

$T_e$ is the effective temperature of the electrons leaving the cathode, and $mc^2$ is the electron rest mass equivalent energy.

The data on the intrinsic emittance of RF photocathode sources under ps laser illumination is not comprehensive. Most data is for GaAs cathodes illuminated with CW laser sources. For laser photon energies between 1.5-2 eV the measured $T_e \cong$ 0.025-0.04 eV for room-temperature GaAs cathodes, i.e. the emitted electron distribution appears to be approximately in thermal equilibrium with the crystal lattice of the source [24]. The inferred $\varepsilon_{th} \cong 0.11$-$0.14$ mm-mrad for a 1-mm radius cathode. For pulsed ps laser sources, there are indications that $T_e$ may on the order of 0.2 eV, and $\varepsilon_{th} \cong 0.3$ mm-mrad per mm cathode radius [24]. The reason for the discrepancy between the CW and the ps-pulsed laser data is not known.

In the case of metallic cathodes, there is very little data. Theoretical considerations imply that $T_e \cong 0.27$ eV for 4 eV incident photons, and $\varepsilon_{th} \cong 0.35$ mm-mrad per mm cathode radius [25]. Measurements on the BNL/ UCLA/SLAC gun at 5.7 MeV give an upper bound of $\varepsilon_{th} \leq 0.8$ mm-mrad for a 1-mm cathode [26], and one infers that $T_e \leq 1.3$ eV.

In general there exists no comprehensive data set for the intrinsic emittance of commonly used cathodes (e.g. $Cs_2Te$, $LaB_6$, Cu) under the typical conditions encountered in accelerators.

There was some discussion concerning the observation that the workhorse design code PARMELA does not simulate the space-charge induced emittance growth at the 1 mm-mrad level in the low-energy end of the rf guns. The measured emittance from the BNL/UCLA/SLAC 1.6 cell gun at a charge of 1 nC gave an rms normalized emittance approximately a factor of two larger than predicted by PARMELA [23]. Related issues of virtual cathode formation and plasma oscillations near the cathode [27] were discussed, but remain unresolved.

We had a joint meeting with the Beam Diagnostics and Control Group where we discussed the need for the following types of beam diagnostics:

- Minimally invasive emittance measurement
- Laser based diagnostics for short pulses ($< 10$ fs resolution)
- Real-time measurements for control
- Temporal synchronization of systems (1 deg, of RF phase)
- Ultra-low emittance measurement ($< 0.1$ mm-mrad)
- Phase space mapping

We laid out the following outline for important source-related research topics for the next two years:

- Measure intrinsic emittance of sources
- Study low-energy space charge effects (codes and experiments)
- Techniques for synchronizing/stabilizing high-frequency RF and laser systems.
- Better diagnostics for measuring emittance and temporal structure of bunches.
- Investigate polarized electron photocathodes for use in RF guns
- Achieve $\varepsilon_n = 0.5$ mm-mrad for a 1-nC 1 MeV beam

| Authors | Title |
|---|---|
| **J. Nation**, L Schachter, D Fletcher G. Golkowski, J.D. Ivers | Ferro-Electric Cathodes |
| **L Schachter**, D Fletcher G. Golkowski, J.D. Ivers and J.A. Nation | A Diode with Dielectric-Gridded Cathode |
| **F. Mako**, L.K Len, and W. Peter | Self-bunching electron guns |
| **Bruce Dunham** | Advances in DC Photocathode Electron Guns |
| **A.R. Fry**, E. Hahn, W.H. Hartung, M. Kuchnir, (FNAL); P. Michelato, D.Sertore, (INFN, Milano) | Performance of $Cs_2Te$ Photocathodes in the FNAL Photoinjector |
| **J. Rosenzweig** | Ultra-high Brightness Integrated Photoinjectors |
| **Mitsuru Uesaka** | Low-emittance Femtosecond Electron Single Bunch by the S-band Laser Photocathode RF gun and Linac |
| **M.A.Shapiro**, W.J.Brown, S. Trotz, K.E.Kreischer, M.Pedrozzi, and R.J.Temkin | Design, analysis and cold test of a 17 GHz rf gun. |
| **W. J. Brown**, S. Trotz , K. E. Kreischer, M. Pedrozzi, M. A. Shapiro, R. J. Temkin | Results of a 17 GHz RF Gun Experiment |
| **W. Gai**, X. Li, M. Conde, J. Power and P. Schoessow, | A high charge and short pulse RF photocathode gun for wakefield acceleration |
| **P.L. Colestock** | Plasma Wakefield Accelerator Test at the Fermilab Photoinjector |
| **E. Dodd**, J.-K. Kim and D. Umstadter | Electron Injection by Dephasing electrons with laser fields |
| **A. Mikhailichenko** | Injector for a laser linear collider |
| *Michael Fitch* | Manipulating the temporal structure of a single laser pulse |
| **A.R. Fry**, M.J. Fitch, B.D. Taylor, A.C. Melissinos | Multipulse Laser System for the FNAL Photoinjector |
| **M. Ferrario**, | Relativistic and non-relativistic multi-bunch beam dynamics |
| **A. D. Yeremian**, P. Emma, R. H. Miller, D. Palmer, M. Woodley | Low Emittance 150 MeV S-band Injector using RF GUN |
| **K. Batchelor**, T. Srinivasan-Rao, J. Smedley, , J.P. Farrell, G. Dudnikova, Jack E. Boers | Optimization of a 1-GV/m electron gun |
| **Chiping Chen** and Renato Pakter | Equilibrium and Non-equilibrium Properties of Periodically Focused High-Brightness Beams |
| **M. Reiser** and S. Bernal | Beam Transport in quadrupole and bending systems |

Papers presented at the Particle Beam Sources Working Group. Name of presenting author in bold.

We would like to thank all of those who contributed to lively and stimulating working sessions.

# REFERENCES

1. R. Sheffield, workshop discussion
2. A. Fry et al., these proceedings
3. W. Brown et al., these proceedings
4. W. Gai et al. these proceedings
5. B. Dunham, these proceedings
6. P.G. O'Shea et al. Appl. Phy. Lett. **73**, 411 (1998)
7. M. Fitch et al., these proceedings
8. P. Colestock et al., these proceedings
9. F. Mako et al, these proceedings
10. J. Nation et al., these proceedings
11. L. Schachter et al these proceedings
12. K. Batchelor et al., these proceedings
13. M. Ferrario et al., these proceeding
14. E. Dodd et al., these proceedings
15. X. Qui et al Phys. Rev. Lett. **76**, 3723 (1996)
16. L. Serafini and J.B. Rosenzweig, Phys. Rev. E **55**, 7565 (1997)
17. P.G. O'Shea, Phys. Rev. E **57**, 1081 (1998)
18. M. Ueseaka et al., these proceedings
19. A.A. Michailichenko, these proceedings
20. A.D. Yeremian et al., these proceedings
21. C. Chen, these proceedings
22. M. Reiser and S. Bernal these proceedings
23. D.T. Palmer et al., proceedings of the 1997 Particle Accelerator Conference, Vancouver, Canada, page 2687 (1998)
24. S. Pastuseka wt al., Appl. Phys. Lett. **71** 2967 (1997)
25. B. Dunham, workshop discussion
26. J.E. Clendenin and G. A. Mulhollan, presented at 15th ICFA Advanced Bean Dynamics Workshop, Quantum Aspects of Beam Physics, Monterey CA, USA January 4-9, 1998, SLAC publication SLAC-PUB-7760.
27. D.H. Dowell et al. Phys. Plasmas **4**, 3369 (1997)

# Radiation Sources
# Working Group Summary Report

Michael V. Fazio

*Los Alamos National Laboratory*
*High Power Microwaves, Advanced Accelerators, and Electrodynamic Applications Group*
*MS-H851, Los Alamos, New Mexico 87545*

**Abstract.** The Radiation Sources Working Group addressed advanced concepts for the generation of RF energy to power advanced accelerators. The focus of the working group included advanced sources and technologies above 17 GHz. The topics discussed included RF sources above 17 GHz, pulse compression techniques to achieve extreme peak power levels, component technology, technology limitations and physical limits, and other advanced concepts. RF sources included gyroklystrons, magnicons, free-electron masers, two beam accelerators, and gyroharmonic and traveling wave devices. Technology components discussed included advanced cathodes and electron guns, high temperature superconductors for producing magnetic fields, RF breakdown physics and mitigation, and phenomena that impact source design such as fatigue in resonant structures due to pulsed RF heating. New approaches for RF source diagnostics located internal to the source were discussed for detecting plasma and beam phenomena existing in high energy density electrodynamic systems in order to help elucidate the reasons for performance limitations.

## INTRODUCTION

The Radiation Sources Working Group sessions consisted of eighteen short talks by working group members and discussion sessions on specific topics. Two joint sessions were held with the working groups on Novel Structure-Based Acceleration Concepts (on W-band structure work and diamond coating) and Particle Beam Sources (on ferroelectric cathodes). Approximately one-third of the 15 hours of time spent in the working group sessions was dedicated to discussing specific topics of importance to advanced radiation sources and systems. Each discussion session had a designated leader to stimulate and guide the discussions.

The four discussion sessions explored the following questions and issues:
• What are the engineering and physics limitations to the generation of high power and high frequencies? Where should our R&D efforts be focused?
• Slow wave device scaling issues.
• What in-situ diagnostics could be useful for high power devices? Should there be a marriage between the high power conventional tube and the plasma physics communities?

CP472, *Advanced Accelerator Concepts: Eighth Workshop,*
edited by W. Lawson, C. Bellamy, and D. Brosius
1999 The American Institute of Physics 1-56396-889-4

• Fast wave device scaling issues.
The topical area discussions and the individual talks are summarized below.

## TOPICAL AREA DISCUSSION SUMMARIES

### What Are The Engineering And Physics Limitations To The Generation Of High Power And High Frequencies? Where Should Our R&D Efforts Be Focused? Slow Wave Device Scaling Issues

This set of discussion topics, led by Danly, covered a number of subject areas. These areas are summarized.

### *Beam Confinement and Compression*

As frequency increases, RF sources generally shrink in size. The result is that the beam needs to be confined to a smaller cross-sectional area in the RF tube. As beam size shrinks, we would like the beam power (beam voltage x beam current) to remain high and incur no degradation in beam quality. The limitation in emission current density from thermionic cathodes to 5-20 $A/cm^2$, necessary to achieve reasonable cathode lifetimes consistent with reliable accelerator operation (greater than 50,000 hours), dictate that one use large cathode areas if high beam currents are needed. Therefore it is essential to do beam compression in the electron gun to get hundreds to thousands of amperes of beam with 10 $A/cm^2$ cathode loading. The questions arising are

• How does the maximum value of beam power decrease as operating frequency increases?

• What is the limit to area convergence in solid beam electron guns such as klystrons and magnicons? In annular beam guns such as magnetron injection guns (MIGs)?

In magnicons an area compression of 3000 appears to be do-able, with 2300 actually demonstrated in Nezhevenko's 7 GHz magnicon at Budker INP. New magnicons at 11.4 and 34 GHz are being designed and built for 2500-3000 compression ratios. These beams are focused down to roughly 1-mm-diameter, and in the case of the 11.4 GHz magnicon, represent a current density of 12 $kA/cm^2$. Gold's 11.4 GHz thermionic gun magnicon requires a 25 μm tolerance between the cathode and focus electrode. With a bias on the focus electrode of $10^{-3}$ of the cathode voltage, this tolerance goes from 25 μm to 2000 μm.

In magnetron injection guns the compression limit appears to be around 25-40 because of pole tip saturation. Can these limits be pushed and if so how far? These limits are based on experience with thermionic cathodes. New cathodes currently under development could be better or worse in terms of beam quality and ultimate compression ratios. The evidence from this workshop is that very little development is occurring on advanced cathodes for microwave tubes. One exception is the PZT ferroelectric cathode work reported by John Nation, but this work is in its infancy.

## *Mode Competition*

As tubes go up in power and as overmoded structures become more common, the problem of spurious modes competing with the desired mode for proper beam interaction becomes a serious issue. How can one avoid dipole modes in linear beam devices (e.g. BBU modes)? In practice it has been observed that tubes with gains in excess of 60 dB (such as Balakin's 15 GHz klystron) have considerable difficulty achieving stable operation. High order modes need to be carefully suppressed and loaded-out. Designers need to employ suppression and loading in the drift tubes, cavities, and also external to the device. Non-magnetic stainless steel cavities and drift tubes have proven useful for mode suppression in numerous applications. Some of the experience with damping accelerator structures (Shintake cavity, SLAC damped/detuned NLCTA structure) should be examined for applicability to RF sources. It was suggested by Danly that perhaps a secondary beam could be introduced into the tube and tuned to the appropriate voltage to act as a load for suppression of undesired modes; or multi-beam tubes could place beamlets to achieve the desired mode selectivity. This suggestion is not a simple one to implement, but certainly causes one to think about the mode suppression problem in a new and different way.

Haimson provided some guidance for avoiding BBU modes. When building a circuit one should study the R and Q of the next 3 or so modes. When round trip time in a structure is ~1/2 ns, one needs to couple HOMs (high order modes) out from the ends. Other alternatives are to change the circuit geometry in a way to alter the growth of HOMs or to introduce HOM incoherence. Damping can also be added to the body of the circuit, the cavities, or the mode can be extracted in the drift tubes.

## *Annular Beams*

Annular beams present a wealth of possibilities for the generation of extreme power levels, but need a lot more research. For one to consider 100's of megawatts to several gigawatts of RF out of a single tube, the use of an annular beam is practically a necessity. Much more beam power can be transported in an annular beam than in a solid beam of similar diameter without approaching the space charge limiting current. Approaching the space charge limiting current results in potential depression of the beam. In other words, since the total beam energy (kinetic + potential) must be constant, increasing the potential component means a reduction in kinetic energy. Since the kinetic is what is converted to microwave energy, the efficiency of the device can be strongly impacted by a large beam potential depression. Using a larger diameter annular beam (rather than a solid pencil beam) allows one to open up the structure bore resulting in a reduction in energy density in the structure.

Several interesting ideas were suggested:
• Why not marry the annular beam MIG with a linear interaction circuit?
• Are inverted MIGs a realistic approach? In this configuration the emitting surface is on the inside surface of the outer conductor. The cathode could be at ground potential and the center conductor anode could be pulsed positively.

There was a discussion of annular beams versus sheet beams. The relatively uncontroversial conclusion reached was that the annular configuration was preferable over the sheet. The principle reason cited was the higher tolerance of the annular geometry to misalignment. Mechanical misalignment of sheet beam guns tends to result in beam instabilities much more readily than annular beams.

Very high power single pulse klystrons have been demonstrated at Los Alamos (Fazio) and NRL (Friedman) with intense annular beams using explosive field emission cathodes. At Los Alamos, 500 MW in a long pulse (0.5 μs half-power pulse width) was produced. NRL produced around 3 GW in a much shorter (100 ns) pulse. These sources are not viable for accelerator applications as long as they employ explosive emission cathodes, but if these sources can be implemented with high quality, constant impedance, high current cathodes they could become viable options for accelerator drivers.

Using an annular beam effectively lowers the perveance, or the perveance per square, of the beam. CPI takes advantage of this in building the high order mode Inductive Output Tube. This is a low peak power, high average power (~1MW) tube that uses a segmented annular cathode and hopes to reach 70% efficiency.

## *Fabrication Issues and Components*

Submicron accuracy will be required for high frequency structures. For example the KEK cavity's tolerance is 0.5 μm at X-band. Tubes at 94 GHz have a tolerance of +/- 5 μm, which results in a frequency tuning variation of 118 MHz. Electric discharge machining (EDM) may be a good fabrication approach and diffusion bonding needs to be considered as an alternative to brazing.

W-band components have an attenuation of 1 dB/ft and better ways to couple in and out of devices are needed. It was pointed out by Nusinovich that an extensive amount of work on HOM devices, couplers, etc., has been done by the fusion community for high peak power applications and we should not re-invent what has already been developed.

## What In-Situ Diagnostics Could Be Useful For High Power Device Development? Should There Be A Marriage Between The High Power Conventional Tube And The Plasma Physics Communities?

The second discussion group led by Gold explored the topic of in-situ diagnostics for high power tubes that could be utilized to develop an understanding of problems like RF pulse shortening. Are there common problems shared between the high power conventional tube community and the high power microwave (HPM) community and are there any new approaches for solving them? Things that need to be measured include RF fields, plasmas, electrons, x-rays, and light (including the spectrum). Sophisticated spectroscopy techniques, such as measuring Stark shift, is being done by Carmel at the University of Maryland. It was noted that intense photon generation that accompanies high power beams could dramatically reduce RF breakdown voltage levels. Possible causes of pulse-shortening were discussed including time dependent beam parameters such as cathode plasma expansion, field emission and breakdown in the microwave circuit, plasma loading of cavities or slow-wave structures by gas desorption and multipactor, space charge buildup by trapping of low-energy electrons, growth of parasitic modes, and beam disruption caused by instabilities, ExB drift, or halo formation. Pulse shortening may be mitigated by: good vacuum; proper choices of materials, fabrication techniques, and vacuum processing such as sputter cleaning and high temperature bakeout; RF conditioning of the circuit; designing to limit RF fields at surfaces; and suppression of BBU and other HOM modes.

## Fast Wave Device Scaling Issues

The final discussion group led by Fazio covered fast wave device topics including limits to device power and efficiency at the higher frequencies of 35 and 90 GHz. Other topical questions included the phase stability of fast-wave devices, pulse heating, and the adequacy of current design tools. More questions were raised than were answered and many of the questions can provide directions for future research. The question was posed that if higher order modes such as the 28, 8 mode work well for gyrotron oscillators, why won't they work for amplifiers as well? Can a gyroklystron run at the 4th harmonic? Can an FEL compete with a gyrotron at millimeter wavelengths? It was stated that MAGY is a better code than it was 10 years ago and is pretty good now.

Goals were discussed and the consensus was that a 100 MW source at 35 GHz is a desirable goal for the community. Sources are also needed for testing structures at 90 GHz and some effort should be put on meeting this need.

# PRESENTATION SUMMARIES

## RF Sources

RF source development has experienced significant recent progress. Both gyroklystrons and magnicons in particular have demonstrated very encouraging results. The gyroklystron effort at the University of Maryland led by Lawson has resulted in both designs and experiments over the impressive range of 8 to 95 GHz. Gyroklystrons scale favorably to higher frequencies. The linear dimensions scale $\sim \lambda$, the magnetic compression as $\sim (\lambda)^{-2/3}$, and the cathode loading as $\sim (\lambda)^{-1/3}$. Because $TE_{0np}$ modes have zero surface electric fields, peak electric fields leading to breakdown are not an issue. Peak power scales $\sim \lambda$ as long as the magnetic compression limit ($\sim 40$) and the cathode loading is not exceeded.

The X-band (8.6 GHz), first harmonic, three cavity coaxial gyroklystron has produced 75 MW at X-band. Castle described a 2nd harmonic, three cavity coaxial design about to undergo testing which seeks to produce 150 MW at 17GHz with a beam voltage of 500 kV and current of 770 A at 41 % efficiency. The magnetic compression ratio in the magnetron injection gun is 9x. A 35 GHz, 60 MW, 40 % efficient design is being developed that is scaled from the 17 GHz/150 MW design. Beam voltage and current are 500 kV and 300 A.

For high rep-rate operation at 17 GHz, Xu described a new design with several dramatic innovations including a triaxial input system. In order to obtain the high mode purity of a $TE_{011}$ mode in the overmoded X-band input cavity while maintaining high coupling efficiency, a coaxial dual-cavity input structure with an outer $TE_{411}$ mode and an inner $TE_{011}$ mode coaxial cavity has been designed to get a low Q while avoiding mode distortion from the coupling aperture between the input waveguide and the input cavity. There are four coupling slots between the input cavity and the coupling cavity. The design uses a single $TE_{021}$ buncher cavity. The coaxial $TE_{021}$ output cavity employs radial extraction into the inner conductor that is designed to convert the $TE_{02}$ mode to the $TE_{01}$ in coaxial waveguide. Another approach being considered includes the use of two buncher cavities and a 4th harmonic penultimate cavity at 35 GHz. There is also a design for a 35 GHz, fourth harmonic output cavity.

Arjona presented a design for a 94 GHz, three-cavity cylindrical 8 MW gyroklystron that should produce a peak power at least 100 times the current state of the art. The design approach is to scale the MIG and circuit design of an experimentally successful 20 GHz, two-cavity, 2nd harmonic tube. The beam power and cavity dimensions are scaled proportional to wavelength to 95 GHz. This tube has a predicted gain of 52 dB and 34% efficiency with three cavities and a 9-cm-length. The voltage is 500 kV and the current is 45 A.

Progress on magnicon development has also been encouraging. Nezhevenko reported that the Budker INP 7 GHz 2nd harmonic magnicon has produced 55 MW with 72 dB of gain and an efficiency of 56% operating at a 3 Hz rep-rate with a 1.1 µs pulse length. A beam area compression of 2300 is used. A 34 GHz, 3rd harmonic, 45 MW magnicon is in the design phase with a 3000:1 beam compression, 55 dB gain, 1.5 µs pulse length, and a 10 Hz PRF. The beam diameter is 0.8 mm and the output cavity has a 2.3 T magnetic field. The cathode loading has been kept to 15 A/cm$^2$ and peak surface E < 800 kV/cm. To avoid mechanical stress, the temperature rise due to pulse heating of the conductor walls is kept to < 150 C.

Gold described the project underway at the Naval Research Laboratory that is preparing to test a 2nd harmonic 11.4 GHz magnicon with a thermionic cathode. The voltage is 500 kV and the current is 600 A. Projected peak power is 63 MW. The beam radius is 0.75 mm with a compression ratio of 2500:1 and a current density of 12 kA/cm$^2$. Litton built the high convergence electron gun. The tube is designed for a 10 Hz PRF, 62 dB gain, and 60% efficiency.

While resonant ring systems have enabled the high peak power testing of specialized waveguide components, the testing of high gradient accelerating structures and the investigation of coupler cavity RF breakdown at very high peak powers has been limited by the availability of suitable 200 to 300 MW RF test facilities. Haimson described an elegant, compact, economical, high peak power amplification system based on a dual hybrid bridge configuration that eliminates the need for power splitters at the accelerator dual feed couplers and also provides a convenient interface for installing high gradient accelerator test structures. Using a 75 MW, 1/2 µs RF source, and two hybrid bridges, one can produce a 280 MW, 1/4 µs pulse for testing dual feed traveling wave accelerator sections. An additional benefit is that the RF source always sees a matched load.

Schachter presented a theoretical investigation of a 35 GHz TWT amplifier that had the aperture diameter comparable to an X-band structure. This results in a high group velocity and low interaction impedance. The lower impedance requires a longer interaction length which leads to a higher sensitivity to beam quality, but at the same time lowers the probability of electrons hitting the structure's walls. In comparing the theoretical performance of two amplifiers, one at 35 GHz and one at X-band, the sensitivity of the former to beam quality is higher, but its operation is feasible.

LaPointe discussed experiments involving the generation of harmonic radiation from a 300 kV electron beam produced by a cyclotron autoresonant accelerator (CARA). A $TE_{411}$ cavity tuned to the fourth harmonic of the CARA drive frequency (11.4 GHz) has produced 0.5 MW. Beam quality and mode competition are major obstacles to generating the expected 3.5 MW at X-band. A third harmonic experiment is underway using a $TE_{311}$ cavity tuned to 8.5 GHz which shows only the $TE_{311}$ spectrum on a 3.3

MW beam. The theoretical efficiency estimates are 25 to 45% depending on beam quality with up to 3.6 MW possible under optimum conditions from a 9.2 MW beam.

Gyrating electrons can excite standing waves in resonators. Nusinovich described a double resonance condition that can exist in relativistic gyrodevices because the standing wave consists of the superposition of forward and backward waves that have opposite Doppler shifts of the operating frequency in the moving electron frame of reference. Therefore for certain axial wavenumbers both forward and backward waves can simultaneously be in cyclotron resonance with gyrating electrons at different cyclotron harmonics. In the case of azimuthally symmetric modes the efficiency of the double resonance interaction can be higher than in the case of a single resonance. With non-symmetric modes the efficiency is always lowered.

Fazio presented results on a 17 GHz free electron maser amplifier experiment. The experimental goal was several hundred MW at 17 GHz with phase stability. The 6.5-cm-dia, 8-mm-thick intense annular electron beam supplied by an explosive field emission cathode was 600 kV and 5 kA. A 6-port input structure was used to couple input power into the rippled wall structure to modulate the beam for the FEL interaction. At full beam current it proved impossible to produce a stable interaction. The RF input drive power coupled into the tube disappeared when the electron beam was injected. The reason for this phenomenon was undetermined. When the beam current was reduced to 1.5 kA by an aperture placed in the drift tube, a phase stable interaction was observed, although the power level was only a few megawatts. The performance of the tube would probably benefit from the availability of a constant impedance thermionic electron gun and a better understanding of intense annular beam physics. Further work is needed to evaluate the potential of the free electron maser.

Bogacz discussed a scheme for generating electromagnetic radiation in the ultraviolet and x-ray regions using a relativistic beam of hydrogen-like ions having a single bound electron in which a population inversion is achieved by the application of laser radiation tuned to the Doppler-shifted 1s to 2p transition. When the laser beam and ion beam are moving in opposite directions the required pump laser frequency is reduced by a factor of $2\gamma$ (for example in the infrared). Subsequently short wavelength radiation moving in the same direction as the inverted population ion beam will be amplified by stimulated emission. The emitted radiation in the laboratory frame is shorter than the original laser wavelength by a factor of $1/(2\gamma)$.

The relativistic klystron two-beam accelerator (RK-TBA) prototype was described by Westenskow. The challenge in this effort is in the propagation of the drive beam over long distances through multiple extraction sections. A 1 MeV, 1.2 kA, electron induction prototype injector is under construction with a 3.5-in-dia, flat surface, thermionic cathode. The pulse length flat top is 150 ns with 1% energy variation. The flat cathode is used to achieve a lower emittance because with a curved cathode the emission is lower in the center because of the lower voltage seen by the beam, causing the emittance to rise. The RK-TBA approach scales favorably to 30 GHz and the wall-plug efficiency for a 5 TeV design appears to be in the range of tens of percent. Also described was the new TBA approach recently developed by the CERN Linear Collider group. The novelty lies in storing the energy for RF production in the form of a long-pulse electron beam that is efficiently accelerated to 1.2 GeV by a fully loaded, conventional linac operating around 1 GHz. The beam pulse length is about two times the length of the high–gradient linac. Portions of the drive beam are compressed using combiner rings to generate a sequence of higher peak power drive beams with gaps between them. The train of drive beams is appropriately distributed to low impedance decelerator structures where the drive beam energy is converted to RF power which is used to accelerate the high-energy beam in the main high-gradient linac.

## Pulse Compression

The high peak power per meter in short pulses required by linear colliders and the limitations of microwave sources such as klystrons to deliver that power has led to the development of pulse compression technology. These schemes allow one to store a lower peak power from the source over a longer time (1-2 μs) and then switch it to the accelerator in a pulse that is a hundred or so nanoseconds long with a peak power level many times higher (3x-8x) than the klystron output. Important parameters of pulse compression schemes are pulse flatness and efficiency.

Vikharev described an approach at IAP which stores the energy in oversized Bragg resonators operating in the axially symmetric $TE_{0m}$ modes in circular waveguide. The use of Bragg reflectors allows one to use oversize waveguide resonators while retaining control of the spatial mode structure. Electrically controlled gas discharge tubes in the output Bragg reflector are used to switch the stored energy to the load. Demonstrated parameters are 100 MW, 100 ns output at 11.4 GHz, 60% efficiency and power gain of 17. The waveguide storage section is 1.5 m long. Phase stability needs to be measured.

An alternative approach by Tantawi for the NLC uses a delay line distribution system (DLDS), rather than compressing pulses in time. This approach reduces the length of cylindrical waveguide for the NLC from 560 km for SLED II pulse compression to 115 km. The power of several klystrons (for example, eight) can be combined into a single delay line using a multi-mode launcher. The output mode of the launcher is determined by the phase coding of the klystron output power signals. The combined power is extracted at the appropriate places along the linac using several mode extractors. Each extractor extracts only a single mode. Hence the phase coding of the klystrons determines which output mode converter is employed at any given time. The DLDS allows pulses, much shorter than the klystron output pulse length, to be time multiplexed down the length of the linac to particular accelerating sections. A 600 MW peak power system has been designed. In conjunction with the DLDS a high power phase shifter/switch has been designed using a series of six three-port networks. The active element in the switch is a silicon wafer, which can be optically switched using a short pulse laser.

## Pulse Heating in Structures

Nezhevenko presented a discussion on RF induced pulse heating and the resultant mechanical stress induced in the 3 to 20 µm surface layer of a conductor. Once pulse heating exceeds a "safe" temperature determined by the elastic limit, coefficient of linear expansion, and Young's modulus, plastic deformation occurs which leads to accumulation of defects causing destruction of the surface after some number of pulses. The destruction has been observed in microwave tube collectors. Calculations indicate that a temperature rise of several hundred degrees C, at frequencies above 45 GHz, can reduce structure lifetimes to hours or even minutes. These issues need to be carefully considered when designing high peak power devices, especially at higher frequencies.

## High Temperature Superconductor Progress

Fortgang provided a brief review of high temperature superconductor (HTSC) technology. Since most microwave power tubes require some type of magnetic field, the availability of HTSC magnet technology could greatly impact the viability of particular RF sources for advanced accelerator applications. The state of HTSC technology is such that in ten years we will probably be using HTSC magnets equivalent to copper magnets currently operating at 1 $kA/cm^2$. At Los Alamos, HTSC YBCO ribbon, with dimensions of 2.0-µm-thick by 1-cm-wide by 83-cm-long, is now operating at 210 A corresponding to a current density of 2.1 $MA/cm^2$. Researchers are working with industrial partners to develop commercial fabrication methods for producing long lengths of the tape. Danly noted that closed-cycle cryocooled magnets with no cryogenic liquids are commercially available that use about 1 kW in steady state. He also noted that PPM focused coupled cavity TWTs at 94 GHz are commercial products.

## SUMMARY

The Radiation Sources Working Group had some lively discussions on a wide variety of topics. Some of these discussions and topics may provide issues for the next Advanced Accelerator Concepts Workshop to delve into two years from now. Several sources, particularly the gyroklystron and the magnicon have shown considerable progress in the last several years and their futures look encouraging. A number of other source concepts look interesting, but need further investigation and development to establish their feasibility for accelerator applications. These sources include: the two beam accelerator and the ability to do multiple extraction and reacceleration; the free-electron maser and the ability to generate high power with phase stability and efficiency; high order mode concepts such as a high harmonic gyroklystron and very high order (such as 28,8) gyro-amplifier devices. RF breakdown in components is a problem that requires more development and testing. Annular beams may prove to be very useful for future very high power tubes, but much R&D is needed on generating and transporting intense annular beams. We also need to develop an understanding of the beam physics and it's impact on tube performance.

Experimental efforts should focus on producing 100-200 MW at 35 GHz and designs should be developed for 90 GHz sources. A 1 MW source for testing components and structures at 90 GHz is needed. The Working Group believed that the accelerator structure people and RF source people should be more closely integrated. This could be useful for helping to address some of the accelerator issues such as frequency choice, and for applying microwave solutions to problems such as energy spread compensation in the final focus.

# III

# WORKING GROUP PAPERS

## Working Group 1:

## Beam Physics at
## High Energy Densities

# Transverse Cooling in the Muon Collider

R.C. Fernow, J.C. Gallardo, H.G. Kirk &  R.B. Palmer

*Physics Department, Bldg. 901A*
*Brookhaven National Laboratory, Upton, NY 11973*

**Abstract**. Ionization cooling is the preferred method for reducing the emittance of muon beams in a muon collider. The method described here uses passive liquid hydrogen absorbers and *rf* acceleration in an alternating lattice of solenoids. We consider the basic principles of ionization cooling, indicating our reasons for selecting various parameters. Tracking simulations are used to make  detailed examinations of effects on the beam, such as transmission losses, transverse cooling,  bunch lengthening, and introduction of energy spread. The system reduces the overall 6-dimensional emittance to 44% of its initial value.

## IONIZATION COOLING

The most likely process for achieving a large reduction in the emittance of muon beams in a future muon collider (1-5) is ionization cooling (6-8).[1] In this process the muon loses transverse and longitudinal momentum by dE/dx in a material, and then has the longitudinal momentum (but not the transverse momentum) restored in a subsequent  *rf* cavity. The combined effect is to reduce the beam divergence and thus the emittance of the beam. The process is complicated by the simultaneous presence of multiple scattering in the material, which acts as a source of "heat" and increases the emittance. The cooling effect can dominate for low Z materials in the presence of strong focusing fields. A lattice of solenoids with alternating direction is the method considered here for achieving the required focusing.

A charged particle traversing matter loses energy because of electromagnetic interactions with the atomic electrons. The energy loss falls dramatically as the particle energy increases, reaching a minimum value for muons with energy around 300 MeV.

---

[1]Conventional forms of cooling, such as electron cooling or stochastic cooling, are too slow compared to the muon lifetime.

CP472, *Advanced Accelerator Concepts: Eighth Workshop,*
edited by W. Lawson, C. Bellamy, and D. Brosius
© 1999 The American Institute of Physics 1-56396-889-4/99/$15.00

Above this is the region of relativistic rise where the energy loss increases logarithmically. Consider a diffuse beam of muons focused onto a block of material. Muons traveling at an angle through the material lose both transverse and longitudinal components of momentum. Operation at any part of the dE/dx curve could be used for transverse cooling.

In order to cool the longitudinal emittance the higher energy particles in the beam must lose more energy than the lower energy particles. On the dE/dx curve this only occurs for muon energies greater than $\approx$ 500 MeV (9). This natural longitudinal cooling is very inefficient. A more practical idea is to introduce dispersion into the beam, so that the muons receive a transverse displacement proportional to their deviation from the mean momentum. Then a wedge shaped absorber can be used to cause the higher momentum muons to lose more energy, and thus reduce the momentum spread in the beam. However, this system results in a corresponding increase in the transverse direction, and thus only exchanges longitudinal for transverse emittance.

The theory of ionization cooling predicts (7,8) that the *rms* normalized transverse emittance is changed by traveling a step dz into a material at the rate

$$\frac{d\varepsilon_{xN}}{dz} = -\frac{1}{\beta^2}\frac{\varepsilon_{xN}}{E}\left|\frac{dE}{dz}\right| + \frac{\beta_\perp}{2}\frac{E_S^2}{\beta^3 E m c^2}\frac{1}{L_R} \tag{1}$$

where $\beta$ is the relativistic velocity factor, E is the energy of the muon, dE/dz is the local value of the ionization energy loss function, $\beta_\perp$ is the betatron focusing parameter, $E_S \approx 15$ MeV is a characteristic energy from multiple scattering theory, m is the mass of the muon, c is the velocity of light, and $L_R$ is the radiation length of the material.

The minimum achievable normalized emittance, reached when the cooling rate equals the heating rate, is

$$\min \epsilon_{xN} = \frac{\beta_\perp E_s^2}{2\beta m c^2 L_R \left|\frac{dE}{dz}\right|} . \tag{2}$$

Thus cooling to small emittance values requires a combination of strong focusing (small $\beta_\perp$) together with the choice of a material with a large value for the product $L_R$ dE/dz.

The cooling method described here uses solenoidal focusing, which has a betatron function given by

$$\beta_\perp = \frac{2 p_Z}{c B_Z} \tag{3}$$

We can obtain small values of $\beta_\perp$ by using a low momentum or by using a strong magnetic field.

It is important that any focusing scheme proposed for ionization cooling be able to transmit large angle (non-paraxial) particle trajectories. The required angular acceptance of the lattice is given by

$$\theta_{acc} = j \left\{ \frac{n E_S^2}{2\beta^2 \gamma m c^2 L_R \left| \frac{dE}{dx} \right|} \right\}^{1/2} \tag{4}$$

where $j$ is the required number of standard deviations for the angular acceptance, i.e. $\theta_{acc} = j \, \sigma_\theta$, and $n$ is the size of the initial transverse normalized emittance compared to the equilibrium value, i.e. $\epsilon_N = n \, \epsilon_N^{eq}$. As an example for $j=n=4$, 186 MeV/c muons in liquid hydrogen absorber require a lattice with an angular acceptance of 0.39 radians. Higher Z materials would require a larger acceptance.

The normalized longitudinal emittance also changes in a material according to

$$\frac{d}{dz} \epsilon_{zN} \approx \frac{\sigma_{P_Z}}{mc} \frac{d}{dz} \sigma_Z + \frac{\sigma_Z \sigma_E}{\beta mc^2} \frac{d}{dE}\left( \frac{dE}{dz} \right) + \frac{K_S \sigma_Z}{2 \beta mc^2 \sigma_E} \gamma^2 (1 - \tfrac{1}{2}\beta^2) \tag{5}$$

The terms on the right side of the equation come from bunch lengthening, the curvature of the dE/dx relation with energy, and straggling, respectively. The coefficient $K_S$ of the straggling term is given by

$$K_S = 4\pi (r_e mc^2)^2 \frac{N_A Z \rho}{A} \tag{6}$$

where $r_e$ is the classical radius of the electron, $N_A$ is Avogadro's number, and $\{Z, \rho, A\}$ are the {atomic number, density, atomic mass} of the material.

# CHOICE OF PARAMETERS

Many conditions must be simultaneously satisfied in order to produce a useful cooling system. We have chosen to use liquid hydrogen as the absorber since it provides the lowest possible minimum achievable emittance and makes the least demands on angular acceptance of the focusing lattice. We have chosen to use high field solenoids for the focusing since they have large angular acceptance, they focus both transverse dimensions simultaneously and the technology exists for achieving high field strengths. The space between absorbers must be filled with $rf$ cavities to replace the energy lost in the absorbers. The particular cooling section described here is designed to reduce the transverse emittance of the muon beam. Because of straggling and path length variations, the longitudinal emittance will be simultaneously growing. We use the 6-dimensional normalized emittance ($\epsilon_6$) as the figure of merit for the overall design. We seek a design that rapidly decreases the value of $\epsilon_6$. We continue the channel until the longitudinal growth becomes excessive and the $\epsilon_6$ cooling rate begins to saturate.

We have chosen 186 MeV/c as the reference momentum for the muon beam, corresponding to a velocity factor $\beta = 0.87$. There are at least three advantages in choosing a low working momentum. 1) We see from Eq. 3 that this helps achieve a smaller value of $\beta_\perp$. 2) We see from Eq. 1 that, ignoring heating, the fractional change in emittance is equal to the fractional loss of energy. Thus the absolute amount of $rf$ re-acceleration required for the same fractional cooling is higher at larger momenta. 3) The amount of straggling, which is proportional to $\gamma^2$, is less at lower momenta. However, if one chooses too small a momentum (<100 MeV/c), one has to deal with the unfavorable curvature of the dE/dx curve, which produces very large longitudinal heating.

One of the main, non-intuitive design decisions was to periodically alternate the direction of the solenoid field. Simulations using a constant solenoid channel gave worse cooling performance than the alternating system proposed here[2]. In addition the alternating field arrangement is cheaper, since the high field solenoids are only placed over the absorber regions. A third reason for alternating the field direction concerns the build up of angular momentum in the channel.

---

[2]The simulation looked at cooling inside a 20 m long solenoid channel with a constant field of 15 T. The initial beam distributions, liquid hydrogen absorber and $rf$ cavity structure were identical to that used for the alternating cooling system described in this paper. The transverse cooling factor saturated at 79% after 11 m. The losses at 15 m were ~3 times worse and the longitudinal phase space was ~50% worse than the alternating solenoid example.

The canonical angular momentum is given by

$$L_z^C = r p_\phi - \frac{e}{2} |B_z(s)| r^2 \qquad (7)$$

Consider first a particle with 0 angular momentum starting outside a solenoid that is devoid of absorbing material. As a consequence of the symmetry the canonical angular momentum is a conserved quantity with an initial value of 0. As the particle crosses the fringe field entering the solenoid, a rotational angular momentum is developed represented by the first term in Eq. 7. However, the canonical angular momentum is still 0 since the second term cancels the first. This follows because the azimuthal momentum given to the particle by the fringe field is

$$p_\phi = \frac{e}{2} B_z r \qquad . \qquad (8)$$

When the particle leaves the solenoid the exit fringe field stops the rotational angular momentum. Now consider the case when absorbing material is present inside the solenoid. Energy loss causes the particle to suddenly obtain a $p_\phi$ that does not satisfy Eq. 8. As a result the canonical angular momentum develops a non-zero value. The problem with this is that when the particle finally leaves the solenoid, the first term in Eq. 7 no longer vanishes and the particle beam has to diverge from the end of the magnet. However, we have found that breaking the solenoid up into an alternating sequence of shorter solenoids prevents the build up of canonical angular momentum, as shown in Fig. 1.

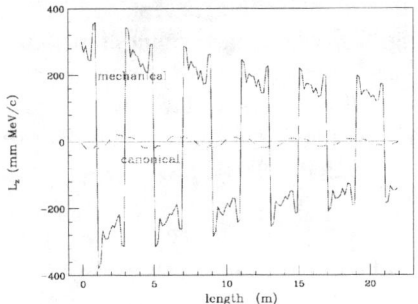

**FIGURE 1.** The mechanical (rotational) angular momentum and canonical angular momentum vs. length in the alternating solenoid system.

One cell of our proposed cooling arrangement is shown in Fig. 2a. The magnetic field peaks at a magnitude of 15 T in the center of the absorbers. The field falls to 0 and alternates direction in the center of the *rf* cavities. Matching the betatron focusing between cells is a very important issue in an alternating solenoid lattice. Without

237

careful matching unacceptable particle losses occur after the zero field crossing points. The matching was accomplished by using three independent sets of solenoidal coils in

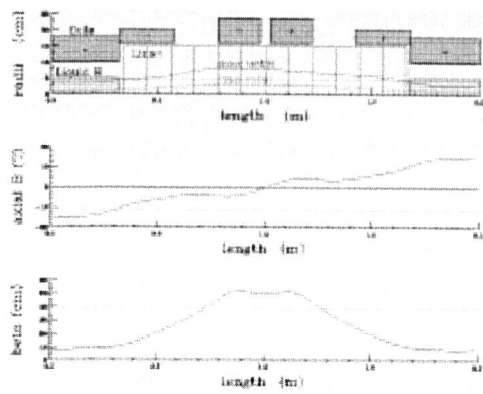

FIGURE 2. (a) Cross section of one cell of the alternating solenoid cooling system; (b) axial magnetic field vs. length; (c) betatron function vs. length.

each cell. Let us refer to these as the absorber, matching, and bucking coils. The axial field produced by these coils is shown in Fig. 2b. High field solenoids are used over the absorber region to provide the low $\beta_\perp$ required for efficient cooling. Immediately following the absorber the betatron function begins growing rapidly, as shown in Fig. 2c. We adjust the matching coils so that $\beta_\perp$ rises linearly in this region, since this gives the best matching and the largest momentum acceptance. We want to do the field reversal in a region with high $\beta_\perp$, where the amount of modulation in the betatron function from the rapidly varying magnetic field is minimized. The bucking coil was adjusted to symmetrize the betatron function across the 0-crossing point. The matching coils on opposite sides must have different currents in order to account for the energy gain of the particles going through the *rf* cavities.

The channel was constructed from a series of identical 2 m long cells. The length of each cell was chosen to avoid resonances that were known to seriously affect cooling in an earlier scheme, known as the FOFO lattice (1,4). These resonances can lead to large particle losses when the betatron wavelength equals the period of the magnetic field or the synchrotron oscillation wavelength (~14 m).

The total length of liquid hydrogen absorber was chosen to be 64 cm per cell. The liquid hydrogen was assumed to be contained in a vessel with thin end windows. Studies of the effects of window material and thickness show that 2 mil stainless steel or 5 mil Al windows have negligible effects on the cooling performance. The absorber is centered under the high field portion of the solenoidal field, so that the low $\beta_\perp$

region is in the absorber. The length was chosen so that the increase in $\beta_\perp$ near the ends of the vessel was not significant. The mean energy loss (including small contributions from the vessel and *rf* windows) is 20.6 MeV. For modeling purposes the cell was chosen to begin at the center of the absorber region, as shown in Fig. 2a.

Reacceleration is done by a series of pillbox *rf* cavities. In order to get simultaneous acceleration and phase stability the *rf* cavity wavelength must be sufficiently long to totally include the bunch within a quarter *rf* wavelength, i.e. $\sim 6\sigma_z < \lambda_{rf}$. For the initial bunch length given in Table 1, this matches an *rf* frequency of 805 MHz. The cavity phase was selected so that the center of the bunch lies near the center of the good *rf* quarter wavelength. The exact phase of 39° up from the 0-crossing point of the electric field was chosen to optimize performance. The peak gradient of 36 MV/m was chosen to be as large as possible, consistent with expectations for breakdown. Finally, the cavity length was chosen to provide sufficient acceleration to match the energy loss in the liquid hydrogen. Once the reference momentum and *rf* frequency were specified, the cavity cell length was determined to be 8.1 cm and the total cavity length was broken into 16 cells. The transit time factor is 0.90. The *rf* was modeled using exact fields for a $TM_{010}$ cylindrical pillbox cavity. This turns out to an excellent approximation in this case, since the cavity design for the muon collider includes thin beryllium end windows. (10) Studies of the effects of window material and thickness show that 5 mil Be windows have negligible effects on the cooling performance. After the initial 32 cm of liquid hydrogen, the *rf* drives the reference particle momentum up beyond what is required by the loss in the initial liquid hydrogen, and then it gets restored to the nominal value by the subsequent 32 cm of liquid hydrogen.

## COOLING PERFORMANCE

The Monte Carlo simulation was done using the program ICOOL. Energy loss was modeled using the Vavilov distribution function, while multiple scattering was modeled using the Moliere distribution.

The properties of the initial beam used in the simulation are shown in Table 1. All the initial phase space dimensions are assumed to be gaussian except $P_z$. Several correlations must be imposed on the variables of the initial particles. 1) Since we begin the simulation inside a strong solenoidal field, each particle must start with the angular momentum the particle would have received from crossing the entrance solenoid fringe field. 2) We have seen that the lattice must transmit large angle tracks for efficient cooling. However, large angle tracks have a significantly longer path length in a solenoidal channel than paraxial tracks. This would lead to significant bunch lengthening unless a correlation between the particle's longitudinal velocity and its transverse amplitude is imposed. We believe this correlation could be introduced from energy loss in a radially varying absorber, placed in the cooling channel immediately preceding the alternating cooling section. We then introduce this correlation in the initial $P_Z$ distribution and this leads to a non-gaussian, high momentum tail. 3) Other possible correlations, such as a longitudinal emittance distribution matched to the asymmetric shaped (alpha) *rf* bucket, have not yet been introduced into the simulation.

**TABLE 1.** Performance summary

|  |  | z = 0 m | z = 22 m |
|---|---|---|---|
| p | McV/c | 186 | 186 |
| Transmission |  | 1.000 | 0.990 |
| $\sigma_X = \sigma_Y$ | mm | 7.92 | 5.96 |
| $\sigma_{PX} = \sigma_{PY}$ | MeV/c | 25.4 | 17.3 |
| $\sigma_t$ | ps | 43.4 | 68.1 |
| $\sigma_{PZ}$ | MeV/c | 5.74 | 9.87 |
| $\epsilon_{XN}$ | mm mr | 1400 | 650 |
| $\epsilon_{ZN}$ | mm mr | 1000 | 2040 |
| $\epsilon_6$ | $x10^{-12}$ (m rad)$^3$ | 1960 | 865 |

The *rms* beam size in the channel oscillates in response to the variations in the alternating magnetic field and has a maximum value of 2 cm. The maximum absolute radius seen by any particle is 7 cm. Fig. 3 shows a comparison of the transverse phase space at the beginning and end of the channel.

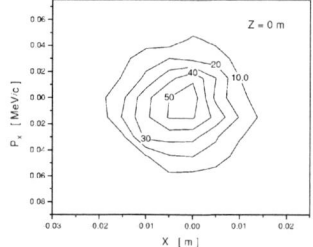 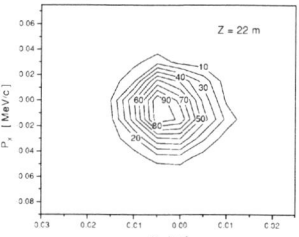

**FIGURE 3.** Contour plots of transverse normalized emittance at the beginning and at the end of the 22 m long channel.

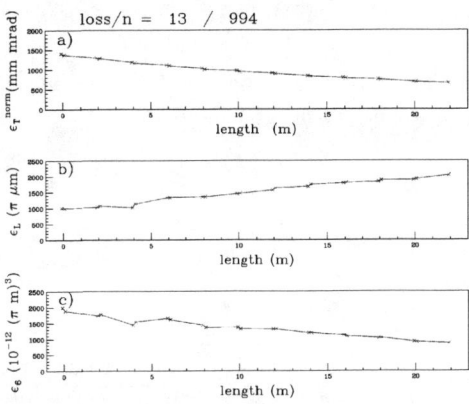

**FIGURE 4.** Normalized emittances vs. length in 11 alternating solenoid cells; (a) transverse emittance; (b) longitudinal emittance; (c) 6-dimensional emittance.

For our chosen parameters the equilibrium normalized transverse emittance from Eq. 3 is 420 mm mr. The results of the simulation are shown in Table 1 and in Fig 4. The transverse emittance in both the x and y planes is reduced to 46% of its initial value, while the longitudinal emittance grows by a factor of ~2. The overall 6-dimensional emittance is reduced to 44% of its initial value. Fig. 4c shows that the cooling rate for 6-dimensional emittance is noticeably smaller after 22 m. This happens because the longitudinal emittance has grown to fill the acceptance of the system. At this point significant numbers of particles begin to fall out of the *rf* bucket and get lost. In a real muon collider one would terminate the alternating solenoid cooling here and begin an emittance exchange system to bring the longitudinal emittance back down to a useable level. In addition to the transmission losses given in Table 1, the muon beam suffers decay losses of 1.9 % while traversing the 22 m long section. The major features of this simulation have been independently confirmed using DP-GEANT (11) and Parmela.

A similar cooling system, suitable for the final stage of a Higgs particle factory, has been designed using 31 T solenoids (5). The minimum $\beta_\perp$ for this case is 4 cm and the minimum equilibrium emittance is 195 mm mr. The simulation achieved a transverse emittance of 240 mm mr and a 6-dimensional emittance of 95 x $10^{-12}$ (m rad)$^3$, both within specifications for the Higgs factory design.

We anticipate making further refinements in this design, for example in the longitudinal matching of the initial beam to the channel. Another topic that we have only begun to study is the effects of space charge and wakefields on the cooling performance.

# ACKNOWLEDGMENTS

We like to thank Paul LeBrun and Yasuo Fukui for useful discussions. This research was supported by the U.S. Department of Energy under Contract No. DE-ACO2-98CH10886.

# REFERENCES

1. Muon Collider Collaboration, μμ Collider: a feasibility study, in *Proc. 1996 DFF/DPB Summer Study on High Energy Physics*, Snowmass'96, New Directions for High Energy Physics, SLAC, 1997.
2. Cline, D. (ed), *Physics Potential and Development of μμ Colliders*, AIP Conf. Proc. 352, 1996.
3. Gallardo, J. C. (ed), *Beam Dynamics and Technology Issues for μμ Colliders*, AIP Conf. Proc. 372, 1996.
4. Cline, D. (ed), μμ Colliders, *Nuc. Phys. B (Proc. Suppl.)* 51A, 1996.
5. Muon Collider Collaboration, Status of muon collider research and development and future plans, (submitted for publication), 1998.
6. Skrinsky, A. and Parkhomchuk, V. , Methods of cooling beams of charged particles, *Sov. J. Part. Nuc.* 12:223-247, 1981.
7. Neuffer, D. , Principles and applications of muon cooling, *Part. Acc.* 14:75-90, 1983.
8. Fernow, R. C. and Gallardo, J. C. , Validity of the differential equations for ionization cooling, in Ref. 2, op. cit., p. 170-177.
9. Particle Data Group, Review of particle properties, *Phys. Rev. D* 54:132, 1996.
10. Moretti, A. et al., RF system concepts for a muon cooling experiment, submitted to proceedings of EPAC 98.
11. LeBrun, P. , private communications.

# Dynamic Focusing for Linear Colliders Using a Stored Beam as the Lens *

## J. Irwin

*Stanford Linear Accelerator Center, SLAC, 2575 Sand Hill Rd. Menlo Park, CA 94025, USA*

**Abstract.** By using the intense fields of a demagnified bunch for a final lens, one can completely eliminate the conventional final focus and collimation systems of linear colliders. In the dynamic focusing scheme described here, the bunches of the lens beam are stored in a ring. Lens beam re-use, plus arranging that the lens beam be divergent as it collides with the main beam, allows much longer lengths for the main beam bunches and larger wall-plug power ratios between the main and lens beam. Under these circumstances dynamic focusing can be used for next generation linear colliders. Design details of such systems are discussed, and IP parameters for energies from .5 to 10 TeV c.m. are presented. All parameters for next generation systems appear achievable, and we argue that the tolerances on beam profile and stability can also be achieved.

## INTRODUCTION

Dynamic focusing refers to the use of bunches in a secondary beam as the final focus lenses for the primary beam. See Fig 1. The principle constraint equations have been presented in a previous report (1). Successfully implemented, such a scheme would completely eliminate the need for the 10 km long, sensitive conventional final focus and collimation system of the NLC (2) and would provide a path to higher energies, where conventional schemes get longer yet and may be impossible because of the sharp decrease in the normalized emittances.

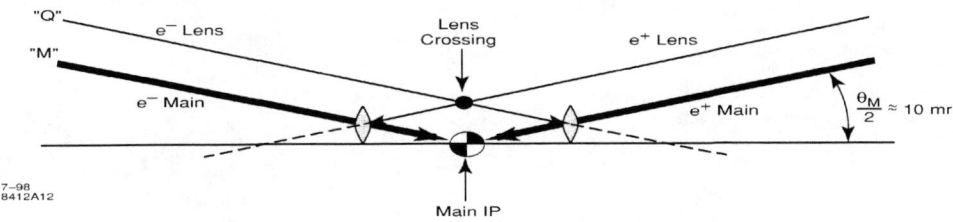

**Figure 1.** Two beam enter the IP from each direction. One beam from each side serves as a final focus lens for the main beam on the opposite side.

The issues that must be addressed to implement a dynamic focusing scheme are

- main-beam bunch-length limits,
- lens beam currents and power,

CP472, *Advanced Accelerator Concepts: Eighth Workshop*,
edited by W. Lawson, C. Bellamy, and D. Brosius
© 1999 The American Institute of Physics 1-56396-889-4/99/$15.00

- lens-beam jitter (bunch-to-bunch and train-to-train),
- uniformity of lens-beam bunches at collision,
- timing and alignment,
- multi-bunch instabilities, and
- lens system operations procedures.

The most favorable geometry is to have all collision geometries be cylindrical.

## MAIN-BEAM BUNCH LENGTH

The main-beam bunch-length constraint arises from a pinch which concentrates the lens-beam distribution during passage through the main beam and shortens the focal length for later particles in the main beam. See Fig. 2. As a guide to the luminosity loss we use (3)

$$\frac{\Delta L}{L} \approx C_P \left( \frac{f_M}{f_Q} \frac{\sigma_z}{\beta_M^*} \right)^2 \tag{1}$$

where $C_P = 2/15$ when the lens beam has a beta minimum at the lens-main collision, or $C_P = 1/120$ if the lens beam is optimally divergent. Here $f_M \approx l^*$, is the focal length of the lens-main collision as experienced by the main beam ($l^*$ is the distance from this collision to the main beam IP), and $f_Q$ is the focal length of the this collision as experienced by the lens beam.

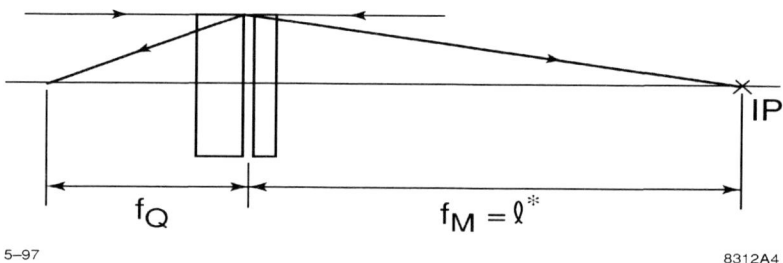

Figure 2. At the collision of the main beam with the lens beam, the lens beam is focused to $f_Q$. The lens beam distortion must be constrained to maintain focusing for the whole main beam.

The ratio $f_M / f_Q = P_M / P_{Q0} = \gamma_M N_M / (\gamma_Q N_{Q0})$, where $N_{Q0}$ would be the number of particles in the lens if it had a Gaussian shape matching the main beam. We assume that efforts are made to achieve a uniform disk distribution for the lens beam, so that the final total charge in the lens is $N_Q = (N_Q / N_{QD})(N_{QD} / N_{Q0}) N_{Q0}$. The first ratio represent the excess arising from charge outside the disk (taken to be about 1.4), and the second ratio is the charge excess in the disk (chosen so that only a few percent of main beam is not properly focused). Typically $f_M / f_Q > 4 \, P_M / P_Q$, where the latter is the ratio of main-beam power to lens-beam power. The bottom line is the ratio $PW_M / PW_Q$ of the wall-plug power being used for each beam. The NLC linac has a wall-plug-to-beam efficiency of 9%, whereas at LEP and CESR the storage ring wall-plug-to-beam efficiency is 5.5 times this value. Or a low gradient linac with a higher rf-to-beam beam efficiency could be used for the lens beam. By using a divergent lens beam (see Fig. 3) at the main-lens interaction and using a more efficient wall-plug-to-beam system for the lens beam acceleration, we can relax the original constraint on the bunch length by a factor of 20. With this improvement, dynamic focusing appears feasible for next generation linear colliders.

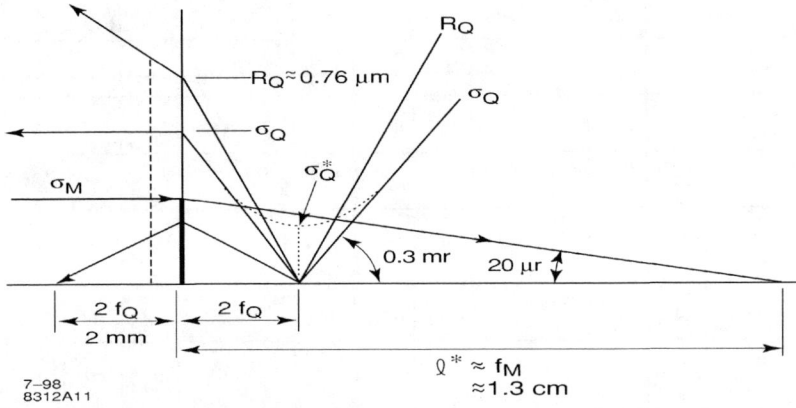

**Figure 3.** At the collision of the main beam with the lens beam, the lens beam diverging. Typical parameters for this geometry are shown here.

## IP BEAM PARAMETERS

The parameters listed in Table 1 have been derived by

1) assuming a luminosity that scales as the square of energy and a main beam power that scales as the square root of energy, with the nominal values at 1.0 TeV c.m. of $10^{34}/(\text{cm}^2 \text{ sec})$ and 8 MW respectively,

2) assuming the horizontal and vertical emittances and IP beam sizes are equal,

3) choosing a disruption parameter to attain the desired main-beam bunch length,

4) choosing $n_\gamma \leq 1$ so that $\delta_B \leq 0.2$,

5) choosing a normalized emittance somewhat below the Oide limit, iterating the "lens-beam to main-beam" wall-plug power ratio until it equals the fractional luminosity loss from the pinch effect, or until the total loss

245

**TABLE 1.** Parameters of the primary and lens beams for cm energies from 0.5 to 10 TeV

| E(cm) | TeV | 0.5 | 0.5 | 1.0 | 1.0 | 1.5 | 1.5 | | 3.0 | 3.0 | 5.0 | 5.0 | 10 | 10 |
|---|---|---|---|---|---|---|---|---|---|---|---|---|---|---|
| $\sigma_M^z$ | μm | 100 | 50 | 100 | 50 | 100 | 50 | | 50 | 25 | 25 | 10 | 10 | 5 |
| D | | 0.7 | 0.5 | 1.2 | 0.8 | 1.5 | 1.0 | | 1.7 | 1.1 | 4.8 | 1.9 | 4.9 | 2.3 |
| H | | 2.1 | 1.4 | 3.1 | 2.5 | 4.9 | 3.4 | | 5.7 | 4.6 | 2.1 | 2.1 | 2.4 | 2.4 |
| Y | | .01 | .05 | .24 | .52 | .35 | .90 | | 1.5 | 3.6 | 8.8 | 2.6 | 59 | 59 |
| $n_\gamma$ | | 1.0 | 1.0 | 1.0 | 1.0 | 1.0 | 1.0 | | 0.8 | 0.8 | 0.9 | 0.8 | 0.7 | 0.7 |
| $\delta_B$ | | .04 | .07 | .06 | .10 | .08 | .13 | | .12 | .16 | .21 | .21 | .19 | .20 |
| $\sigma_M^*$ | nm | 47 | 33 | 14 | 12 | 8 | 6 | | 1.6 | 1.6 | .28 | .35 | .08 | .11 |
| $N_M$ | $10^9$ | 2.4 | 1.9 | .74 | .71 | .45 | .41 | $10^7$ | 9.2 | 11 | 2.7 | 4.1 | 1.0 | 1.0 |
| $n_B^{120}$ | $10^3$ | .47 | .62 | 1.1 | 1.2 | 1.5 | 1.6 | | 5.1 | 4.4 | 14 | 8.9 | 26 | 26 |
| $\gamma\varepsilon_M$ | $10^{-7}$ m | 8.0 | 8.0 | 3.0 | 2.0 | 1.0 | 1.0 | $10^{-9}$ m | 17 | 12 | 1.5 | 1.0 | .13 | .12 |
| $\beta_M^*$ | mm | 1.4 | 0.7 | .6 | .7 | 1.0 | .6 | | .47 | .62 | .26 | .61 | .45 | .99 |
| $\dfrac{PW_M}{PW_Q}$ | | 30 | 30 | 21 | 30 | 25 | 28 | | 30 | 70 | 30 | 180 | 30* | 130* |
| $\ell^*$ | cm | 2.7 | 2.5 | 1.3 | 1.5 | 2.0 | 1.3 | | 1.2 | 1.4 | 0.6 | 1.4 | .9 | 2.0 |
| $\xi$ | | 20 | 22 | 22 | 22 | 21 | 22 | | 26 | 22 | 22 | 22 | 20 | 20 |
| $f_Q$ | mm | 1.1 | .58 | .90 | .58 | 1.0 | .57 | | .52 | .26 | .25 | .10 | .06 | .03 |
| $\beta_Q$ | mm | .74 | .39 | .60 | .39 | .69 | .38 | | .35 | .17 | .16 | .06 | .04 | .02 |
| $\gamma\varepsilon_Q$ | $10^{-6}$ m | 4.0 | 3.2 | 1.0 | 1.2 | .70 | .67 | $10^{-8}$ m | 14 | 16 | 4.4 | 6.7 | 1.6 | 1.6 |
| $\sigma_Q'$ | mr | .78 | 1.2 | .28 | .42 | .15 | .23 | | .07 | .13 | .024 | .08 | .02 | .06 |
| $N_Q$ | $10^{10}$ | 2.7 | 3.2 | 1.2 | .80 | .35 | .39 | $10^8$ | 7.2 | 4.5 | 0.6 | 0.4 | .04 | .04 |
| $E_{lens}$ | GeV | 4.6 | 3.0 | 11 | 9.0 | 24 | 17 | | 39 | 32 | 230 | 88 | 400 | 200 |
| $R_{lens}$ | μm | 2.4 | 2.0 | .71 | .69 | .43 | .37 | $10^{-8}$ m | 11 | 9.2 | 1.7 | 2.1 | .39 | .56 |
| Cost | % | 9 | 8 | 15 | 9 | 12 | 10 | | 13 | 8 | 8 | 8 | 11 | 12 |

(from pinch, from unfocused particles outside the disk, and the wall-plug energy expended on the lens-beam) is about 10%, and

6) choosing an $l^*$ so that the chromaticity of the lens system is about $\xi = 20$.

There are no free parameters in this process. For the round beams assumed here, the vertical beam size is twice as large as for the flat beams. The normalized emittances of the main beam for the 0.5 and 1.0 TeV c.m. are thought to be achievable. The number of particles per bunch is reduced significantly from flat-beam parameter sets and the number of bunches is correspondingly larger. These many-bunch trains could be accelerated with only modest adjustment of present rf designs by reducing the inter-bunch spacing. The short-range wakes will be much smaller. Concerning the lens beam we note that for next generation energies:

1) The total system "cost", which includes the wall-plug power of the lens-beam plus the luminosity loss from the pinch effect and particles being outside the lens beam disk, hovers around 10%, and is similar to the gains arising from the elimination of the conventional final focus and collimation systems,

2) the lens-beam divergences are acceptable,

3) the lens-beam charges are acceptable and suitable for storage in a ring,

4) the lens-beam beta-minimum for optimal divergence is achievable,

5) the lens-beam normalized emittance is achievable, and

6) the lens-beam energies are comfortable for a storage ring implementation.

A surprising fact is that for energies near 1.0 TeV c.m. the wall-plug power ratios do not depend on bunch length. This is because the smaller disruption associated with the shorter bunches yields smaller enhancements.

Center-of-mass energies of 3.0 TeV and larger are difficult, first and foremost, because of the requirements on the normalized emittance of the main beam. For example, the 3 TeV c.m. energy requires a normalized emittance of $10^{-8}$ m-rad. Such a value is presently out of reach. However, were it available, the lens-system parameters appear workable: the storage ring energy is 30 to 40 GeV, the bunch charge is low, and the normalized emittance of the lens beam is achievable.

Using beams as lenses raises the question as to whether adiabatic focusing techniques could be used, circumventing the need for such small main-beam normalized emittances.

## BEAM AS OPTICAL BENCH

Having the beam come around a double loop after extraction from a damping ring or acceleration in a linac allows one the beam to pass it over an optical table where at least the lower frequency train variations are measured. See Fig. 4. A feed-forward system can then correct for the just-measured variations. The bunch-to-bunch and train-to-train jitter needs to be no larger than about 1 %. The bunch-to-bunch jitter is removed by looking at many cycles and presuming that the intra-beam structure is pretty much constant. The parameters of these feed-forward systems need study, but at first glance this appears feasible.

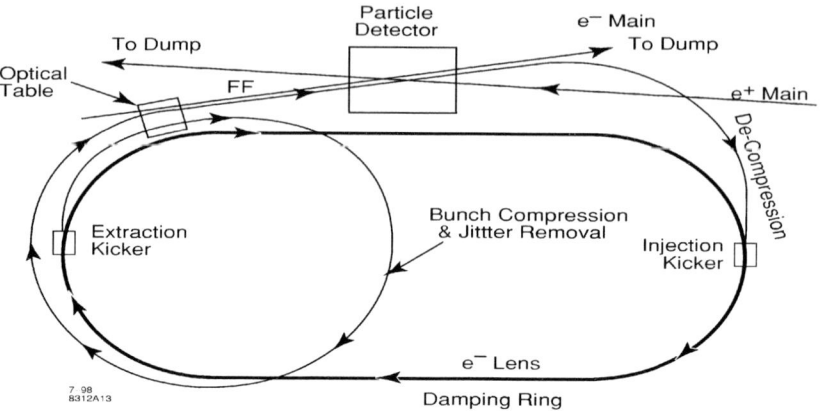

**Figure 4.** A schematic layout of one the necessary storage rings.

Once the overall beam placement is normalized to the coordinates of the optical bench, it remains to locate the final doublet of the lens-beam final focus system with an optical anchor, such as the one planned for the NLC. The tolerances for this system are identical to the system currently under development at SLAC (5).

There appears to be an alternative to the optical anchor based on increasing the length of the lens-beam and observing the positions of pre-cursor bunches. If the two lens beams are not aligned, they will give one another a rather large kick that can be observed by a BPM system located within the lens-beam final doublets. The magnification is several thousand, and normal micron BPM sensitivities are adequate. Following the measurement, the beam trajectory through the IP can be aligned with a fast feed-forward system.

The longitudinal timing is important because of the crossing angles, the divergent lens beam, and the waist location. The crossing-angle problem is handled by "crabbing" the lens beam. Two objectives are accomplished: i) the crabbed lens beam will automatically and properly crab the main beam, and ii) the longitudinal timing tolerance is relieved by the fact that, because of the crab, the bunches from opposing directions will be aligned if the two crab cavities are synchronized. The two cavities are presumed to be driven with the same klystron. The tolerances in this system are about 0.01 degrees, tighter by an order-of-magnitude than the NLC system, because of the round IP beam profile. We do not know to what precision two cavities driven by one klystron can be synchronized but we believe this is an achievable tolerance by using the rf transport as arms of an interferometer.

248

Finally one would like the transverse lens-beam distribution at the collision with the main beam to be a round uniform disk. The tolerance on the uniformity is determined by the chromaticity, which for this system, is identical to the demagnification. We have chosen this parameter to have a value of about 20, which implies that the integral of the charge distribution from the axis to any radius r < R, where R equal the radius of the disk region, should be accurate to a few percent. There is some indication in reference (4) that multipoles can be used to create a uniform distribution from a Gaussian distribution. The result of this study using two octupoles is illustrated in Fig. 5. It appears that 3 octupoles can produce a round distribution, and that higher order multipoles can further flatten the distribution.. A lattice is under development.

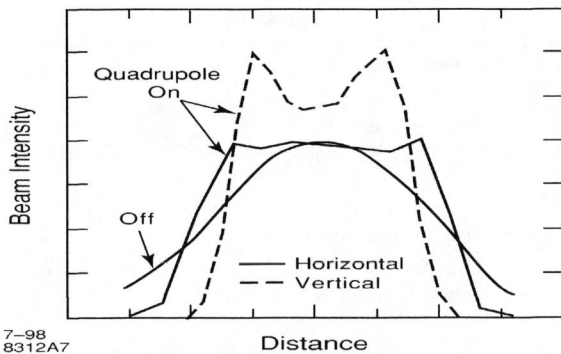

**Figure 5.** An illustration of the potential for shaping the bunch distribution using octupoles in the final focus system.

## SUMMARY

Arguments have been presented to motivate and justify the use of a secondary beam as a final focus lens for linear colliders in next generation machines. The lens beam can be stored in a ring for a damping time between main-beam collisions or accelerated in a specially designed linac, to improve rf-to-beam efficiencies and enhance the removal of jitter in the beam. Many aspects of this idea need further study, including: i) producing uniform distributions with multipoles, ii) suppressing jitter to 1% with fast feed-forward systems, iii) synchronization of crab cavities, and iv) an increase by a factor of 10 in the number of bunches while keeping the train length almost constant. Further details such as operationals issues will be addressed in a subsequent report.

# ACKNOWLEDGMENT

I wish to thank many people for helpful discussions: John Fox on feed-forward systems, Perry Wilson on train-length changes, Tor Raubenheimer on damping ring capabilities, Gordon Bowden on IR design issues, Albert Hoffman on storage ring current limits, and Eberhard Keil and Joe Rogers on wall-plug to beam rf efficiencies. And I wish to thank the Institute of Theoretical Physics at Santa Barbara for their hospitality and congenial environment where these ideas were first developed.

This work was supported by Department of Energy Contract DE-AC03-76SF00515.

# REFERENCES

[1] Irwin, J., Dynamic focusing schemes for linear collider, *Proceedings IEEE Particle Accelerator Conference (PAC97)*, Vancouver, BC (1997).

[2] *Zeroth-order design report for the Next Linear Collider*, SLAC-Report 474 (1996)

[3] Ref [1], Section 3.3, Pinch Effects. At higher energies with higher disruption, the dynamic lens advantageously produces a "traveling focus". Such effects are not included in this simple estimate.

[4] N. Tsoupas, R. Lankshear, C.L.Snead, Jr., and H.A. Enge, Uniform beam distributions using octupoles, *Proceedings IEEE Particle Accelerator Conference (PAC95)*, San Francisco, CA, p1695 (1995)

[5] M. Woods, Optical anchor development group, SLAC (98)

# Polarization and Luminosity Requirement for the First Muon Collider*

Zohreh Parsa

*Physics Department 901A,*
*Brookhaven National Laboratory,*
*Upton, New York 11973-5000*

**Abstract.** Muon Polarization and Luminosity requirement for physics studies at a muon collider are discussed. An overview of a muon collider concepts and design parameters for 0.1, 0.5, and 4 Tev muon colliders are also presented.

## I   INTRODUCTION

We discuss the effects of beam polarization and luminosity in Higgs resonance studies, for improving precision measurements and Higgs resonance "discovery" capability at a muon collider (e.g. at FMC). In the following, we provide a brief overview of a muon collider concepts and parameters (for 0.1, 0.5 and 4 TeV Center of mass energy muon collider rings), and discuss Higgs physics at the first muon collider.

## II   MUON COLLIDER

In a muon Collider (concepts) complex, a high intensity proton source is bunch compressed and focussed on a heavy metal target. The pions generated are captured by a high field solenoid and transferred to a solenoidal decay channel within a low frequency linac. The linac reduces, by phase rotation the momentum spread of the pions and of the muons into which they decay. Subsequently, the muons are cooled by a sequence of ionzation cooling stages. Each stage consists of energy loss, acceleration, and emittance exchange by energy absorbing wedges in the presence of dispersion. Once they are cooled the muons must be rapidly accelerated to avoid decay. Muon collisions occur in a separate high field collider storage ring with a single very low beta insertion. Fig. 1, shows a schematic of a Muon collider and Fig. 2, illustrates the relative sizes of Muon Collider, Large Hadron Collider (LHC), and Next Linear Collider (NLC), relative to the sizes of the existing laboratory (BNL and FNAL) sites [1].

---

*) Supported by U.S. Department of Energy .
†) E-mail: parsa@bnl.gov

CP472, *Advanced Accelerator Concepts: Eighth Workshop,*
edited by W. Lawson, C. Bellamy, and D. Brosius
© 1999 The American Institute of Physics 1-56396-889-4/99/$15.00

It is expected that the first stage, proton driver would be 20 to 30 GeV (e.g., AGS at Brookhaven); but would be much faster pulsed, keeping the number of protons per pulse the same or smaller than the AGS. Which is about $6 \times 10^{13}$ protons per pulse and would go to about $10^{14}$ protons per pulse in a year or so. Roughly one expect to get 1 muon/proton on target which would give Luminosity between $10^{34}$ to $10^{35}$ for a 4-TeV envisioned muon collider.

In recent studies a 50 GeV $\times$ 50 GeV Muon Collider is being considered as the First Muon Collider (FMC) which could serve as a test of the technology for muon colliders. Some of the parameters [1-2] of the collider rings under study are given in Table 1, for 0.1 TeV, 0.5 TeV and 4 TeV center of mass (C.M.) energy $\mu^+\mu^-$ colliders.

**TABLE 1.** Parameters of $\mu^+\mu^-$ collider Rings.

| Energy (C.M.) TeV | 4 | 0.5 | 0.1 |
|---|---|---|---|
| Beam Energy TeV | 2 | 0.25 | 0.05 |
| Beam $\gamma$ | 19,000 | 2,400 | 473 |
| Rep. rate Hz | 15 | 2.5 | 15 |
| p Energy GeV | 30 | 24 | 16 |
| p/pulse | $10^{14}$ | $10^{14}$ | $5 \times 10^{13}$ |
| $\mu$/bunch | $2 \times 10^{12}$ | $4 \times 10^{12}$ | $4 \times 10^{12}$ |
| Bunches/sign | 2 | 1 | 1 |
| Beam Power MW | 38 | 0.7 | 1.0 |
| $\epsilon_N$ $\pi$ mm-mrad | 50 | 90 | 195 |
| Bending Field T | 9 | 9 | |
| Circumference km | 8 | 1.3 | 0.3 |
| Ave. ring field B T | 6 | 5 | 3.5 |
| Effective turns | 900 | 800 | 450 |
| $\beta^*$ mm | 3 | 8 | 9 |
| IP beam size $\mu$m | 2.8 | 17 | 187 |
| Chromaticity | 2000-4000 | 40-80 | |
| $\beta_{max}$ km | 200-400 | 10-20 | 1.5 |
| Lumin. $cm^{-2}s^{-1}$ | $10^{35}$ | $10^{33}$ | $2 \times 10^{31}$ |

Although muon colliders remain a promising complement, to $e^+e^-$ and hadron colliders, much work is still needed, including demonstration of $\mu$ production and cooling, detector, and radiation.

# III   HIGGS PHYSICS AT THE FIRST MUON COLLIDER

If the Higgs boson has a mass $\lesssim 160$ GeV (i.e. below the $W^+W^-$ decay threshold), it will have a very narrow width and can be resonantly studied in the s-channel via $\mu^-\mu^+ \to H$ production at the First Muon Collider (FMC) [1,2]. A strategy for "light" Higgs physics studies would be to first find the Higgs particle at LEPII, the Tevatron, or the LHC and then thoroughly scrutinize its properties on resonance

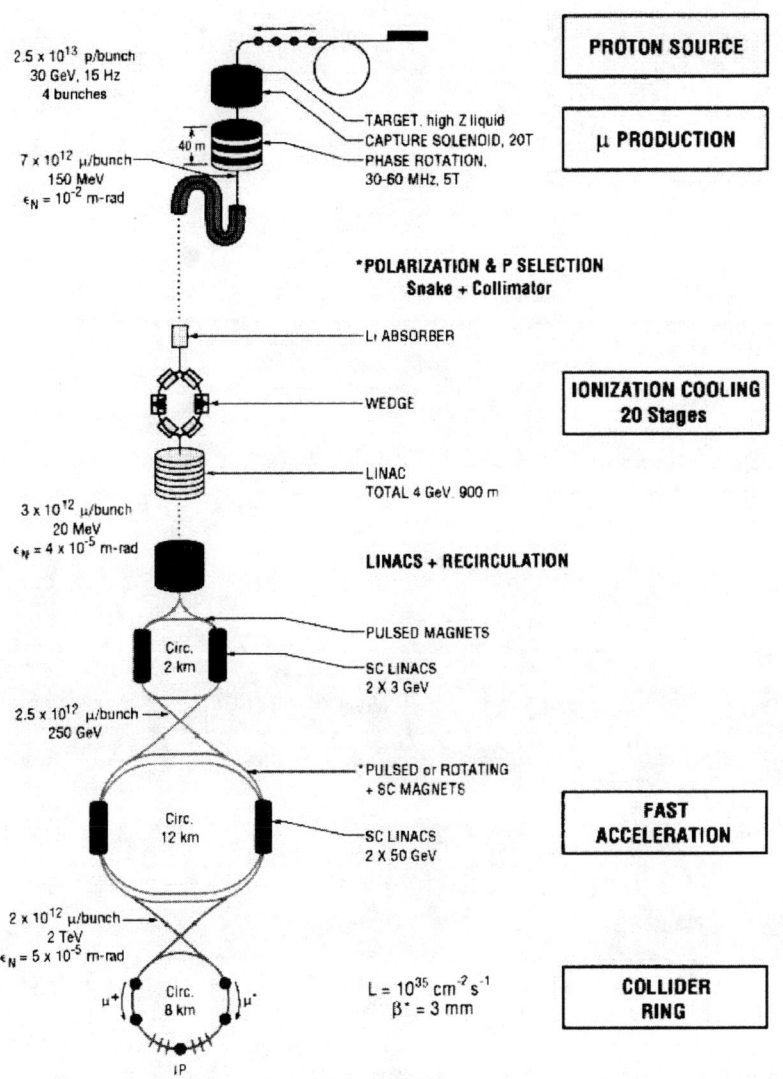

 contains the following labels:

2.5 x 10$^{13}$ p/bunch
30 GeV, 15 Hz
4 bunches

**PROTON SOURCE**

TARGET, high Z liquid
CAPTURE SOLENOID, 20T
PHASE ROTATION,
30-60 MHz, 5T

**μ PRODUCTION**

40 m

7 x 10$^{12}$ μ/bunch
150 MeV
$\epsilon_N = 10^{-2}$ m-rad

*POLARIZATION & P SELECTION
Snake + Collimator

Li ABSORBER

WEDGE

**IONIZATION COOLING
20 Stages**

LINAC
TOTAL 4 GeV, 900 m

3 x 10$^{12}$ μ/bunch
20 MeV
$\epsilon_N = 4 \times 10^{-5}$ m-rad

**LINACS + RECIRCULATION**

PULSED MAGNETS

Circ.
2 km

SC LINACS
2 X 3 GeV

2.5 x 10$^{12}$ μ/bunch
250 GeV

*PULSED or ROTATING
+ SC MAGNETS

Circ.
12 km

SC LINACS
2 X 50 GeV

**FAST
ACCELERATION**

2 x 10$^{12}$ μ/bunch
2 TeV
$\epsilon_N = 5 \times 10^{-5}$ m-rad

μ$^+$    Circ.
8 km    μ$^*$

$L = 10^{35}$ cm$^{-2}$ s$^{-1}$
$\beta^* = 3$ mm

**COLLIDER
RING**

IP

**FIGURE 1.** Schematic of a Muon Collider [1].

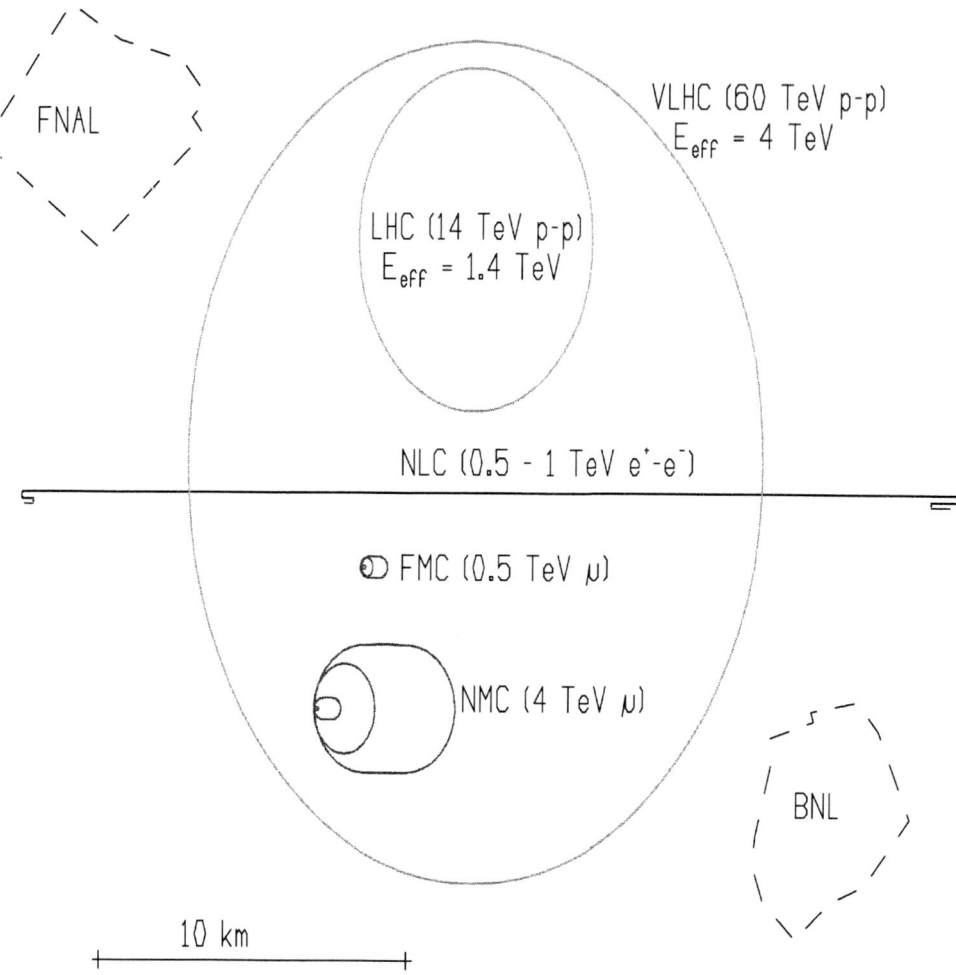

**FIGURE 2.** Schematics of the relative sizes of Muon Collider (.5 and 4 TeV), Large Hadron Collider (LHC), and Next Linear Collider (NLC) are shown relative to the BNL and FNAL sites [1] for illustration. Noting that, the present studies [1], envisions a 0.1 TeV (not the 0.5 TeV shown, center of mass) energy, for the First Muon Collider (FMC).

254

at the FMC. There, one would hope to precisely determine the Higgs mass, width, and primary decay rates [3].

The FMC Higgs resonance program would entail two stages: 1) "Discovery" via an energy scan which pinpoints the precise resonance position and (perhaps) determines its width. Since pre-FMC efforts may only determine the Higgs mass to $\sim \pm 0.2$-1 GeV and its width is expected to be narrow $\mathcal{O}(1 \sim 30 \text{ MeV})$ for $m_H \lesssim 160$ GeV, the resonance scan may be very time consuming [3]. 2) Precision measurements of the primary Higgs decay modes. Deviations from standard model expectations could point to additional Higgs structure or elucidate the framework of supersymmetry [3].

The Higgs resonance "discovery" capability and scan time will depend on $N_S/\sqrt{N_B}$ (the scan time is proportional to $N_B/N_S^2$), where $N_S$ is the Higgs signal and $N_B$ is the expected background. The precision measurement sensitivity will be determined by $N_S/\sqrt{N_B + N_S}$. For both, it will be extremely important to enhance the signal and suppress backgrounds as much as possible. To that end, one should employ highly resolved $\mu^+\mu^-$ beams with a very small energy spread. The proposed $\Delta E/E \simeq 3 \times 10^{-5}$ is well matched to the narrow Higgs width. It allows $N_S/N_B \sim \mathcal{O}(1)$ for the primary $H \to b\bar{b}$ mode. Unfortunately, high resolution is accompanied by luminosity loss.

Expectations for $m_H = 110$ GeV are illustrated in Table 2 for luminosity $\mathcal{L}_{\text{ave}} \simeq 5 \times 10^{31} \text{cm}^{-2}\text{s}^{-1}$ .

**TABLE 2.** Expected signals and backgrounds (fully integrated) for a standard model Higgs with $m_H = 110$ GeV, $\Gamma_H \simeq 3$ MeV.

| $H \to$ | $b\bar{b}$ | $c\bar{c}$ | $\tau\bar{\tau}$ |
|---|---|---|---|
| $N_S$ (events) | 24,000 | 1,200 | 2,700 |
| $N_B$ (events) | 25,200 | 24,160 | 9,450 |
| $\pm\sqrt{N_S + N_B}/N_S$ | ±0.009 | ±0.13 | ±0.04 |

In Table 2, we assumed muon collider resonance conditions with no polarization, $\Delta E/E \simeq 3 \times 10^{-5}$, and $L = 0.5$ fb$^{-1}$. The total number of Higgs scalars produced is $\sim 30,000$. Realistic efficiency and acceptance cuts are likely to dilute signal and backgrounds for $b\bar{b}$ and $c\bar{c}$ by a 0.5 factor. In this table $c\bar{c}$ branching ratios have been reduced compared to those given previously [4,7], since for values given here a smaller charm quark mass was assumed. The prediction is quite sensitive to the mass value assumed.

The selection of the energy and luminosity depends on 1) the reduced scan time to normal time needed, and 2) to improve precision to do physics. For example, to measure $c\bar{c}$, a factor of 10 increase in luminosity results in the improvement from 42%, (at $5 \times 10^{30} \text{cm}^{-2}\text{s}^{-1}$) to about 13%, (at $5 \times 10^{31} \text{cm}^{-2}\text{s}^{-1}$, in Table 2). Further, a factor of 10 increase in Luminosity ( Table 2, as compared to the example of $5 \times 10^{30} \text{cm}^{-2}\text{s}^{-1}$ Luminosity) reduces the running (scan) time by factor 10 less. Thus instead of a "3 years" running time, it will be reduced to ($\frac{3}{10}$ year) over "3 months".

Other decays such as $WW^*$ and $ZZ^*$ are very small for this mass regions and to measure them need to improve precision. The parameter $\pm\sqrt{N_S + N_B}/N_S$ in Table 2 was included for convenient. Further note, the values listed in Table 2 should be reduced by $\frac{1}{\sqrt{2}}$ to include the effect of acceptance.

# IV ENHANCING THE HIGGS SIGNAL TO BACKGROUND RATIO

In the following we discuss additional ways of potentially enhancing the Higgs signal to background ratio. The Higgs signal $\mu^-\mu^+ \to H \to f\bar{f}$ results from left-left (LL) or right-right (RR) beam polarizations and leads to an isotropic (i.e. constant) $f\bar{f}$ signal in $\cos\theta$ (the angle between the $\mu^-$ and $f$). Standard model backgrounds $\mu^-\mu^+ \to \gamma^*$ or $Z^* \to f\bar{f}$ result from LR or RL initial state polarizations and give rise to $(1 + \cos^2\theta + \frac{8}{3}A_{FB}\cos\theta)$ angular distributions. Similar statements apply to $WW^*$ and $ZZ^*$ final states, but those modes will not be discussed here.

To illustrate the difference between signal, $\mu^-\mu^+ \to H \to f\bar{f}$, and background, $\mu^-\mu^+ \to \gamma^*$ or $Z^* \to f\bar{f}$, we give the combined differential production rate with respect to $x \equiv \cos\theta = 4\mathbf{p}_{\mu^-} \cdot \mathbf{p_f}/s$ for polarized muon beams and fixed luminosity

$$\frac{dN(\mu^-\mu^+ \to f\bar{f})}{dx} = \frac{1}{2}N_S(1 + P_+P_-) \tag{1}$$
$$+ \frac{3}{8}N_B[1 - P_+P_- + (P_+ - P_-)A_{LR}](1 + x^2 + \frac{8}{3}xA_{eff}).$$

$P_+(P_-)$ is the $\mu^+(\mu^-)$ polarization with $P = -1$ pure left-handed, $P = +1$ pure right handed, and $P = 0$ unpolarized. $N_S$ is the fully integrated $(-1 < x \le 1)$ Higgs signal and $N_B$ the integrated background for the case of unpolarized beams, $P_+ = P_- = 0$. In that general expression,

$$A_{LR} \equiv \frac{\sigma_{LR \to LR} + \sigma_{LR \to RL} - \sigma_{RL \to RL} - \sigma_{RL \to LR}}{\sigma_{LR \to LR} + \sigma_{LR \to RL} + \sigma_{RL \to RL} + \sigma_{RL \to LR}}, \tag{2}$$

where, for example, $LR \to LR$ stands for $\mu_L^-\mu_R^+ \to f_L\bar{f}_R$. The effective forward-backward asymmetry is given by

$$A_{eff} = \frac{A_{FB} + P_{eff}A_{LR}^{FB}}{1 + P_{eff}A_{LR}}, with \tag{3}$$

$$P_{eff} = \frac{P_- - P_-}{1 - P_+P_-}, \tag{4}$$

$$A_{FB} = \frac{3}{4}\frac{\sigma_{LR \to LR} + \sigma_{RL \to RL} - \sigma_{LR \to RL} - \sigma_{RL \to LR}}{\sigma_{LR \to LR} + \sigma_{RL \to RL} + \sigma_{LR \to RL} + \sigma_{RL \to LR}}, \tag{5}$$

$$A_{LR}^{FB} = \frac{3}{4}\frac{\sigma_{LR \to LR} + \sigma_{RL \to LR} - \sigma_{LR \to RL} - \sigma_{RL \to RL}}{\sigma_{LR \to LR} + \sigma_{RL \to LR} + \sigma_{LR \to RL} + \sigma_{RL \to RL}}. \tag{6}$$

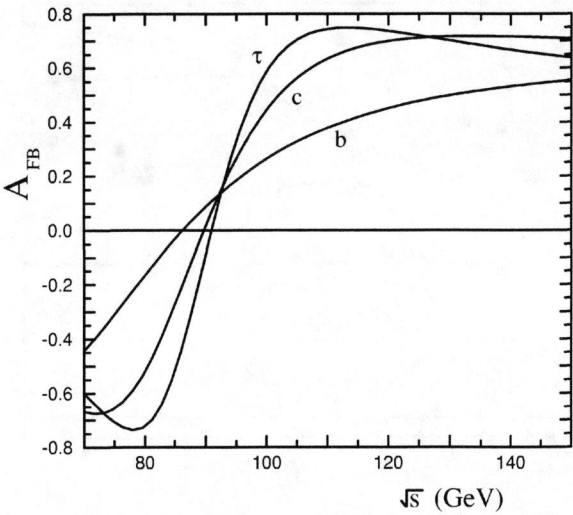

**FIGURE 3.** Forward-backward asymmetry for $\mu^-\mu^+ \to f\bar{f}$.

Realistic cuts, efficiencies, systematic errors etc, will not be considered. They are likely to dilute the $b\bar{b}$ and $c\bar{c}$ event rates by a factor of 0.5. In addition, we ignore the radiative $Z$ production tail under the assumption such events are vetoed.

The (unpolarized) forward-backward asymmetries are illustrated in Fig. 3. Note that $A_{FB}$ is large (near maximal) for $\tau\bar{\tau}$ and $c\bar{c}$ in the region of interest. As we shall see, that feature can help in discriminating signal from background.

## V  POLARIZATION - ENHANCEMENT FACTOR

In principle, large polarization in both beams can be important for enhancing "discovery" and precision measurement sensitivity for the Higgs. From Eq. (1), we find for fixed luminosity that $N_S/\sqrt{N_B}$ is enhanced (for integrated signal and background) by the factor

$$\kappa_{\text{pol}} = \frac{1 + P_+P_-}{\sqrt{1 - P_+P_- + (P_+ - P_-)A_{LR}}} \,, \tag{7}$$

where the $A_{LR}$ are shown in Fig. 4. That result generalizes the $P_+ = P_-$ case [5]. For natural beam polarization [1], $P_+ = P_- = 0.2$ (assuming spin rotation of one beam), the enhancement factor is only 1.06. For larger polarization, $P_+ = P_- = 0.5$, one obtains a 1.44 enhancement factor (statistically equivalent to about a factor of 2 luminosity increase). Similarly, $P_+ = P_- = 0.7$ leads to a factor of 2 enhancement or equivalently a factor of 4 scan time reduction. Unfortunately, obtaining even 0.5 polarization simply by muon energy cuts reduces each beam intensity [1] by a

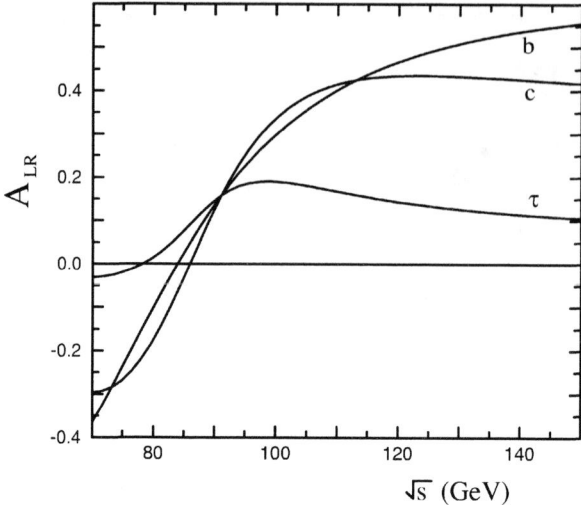

**FIGURE 4.** Left-right asymmetry for $\mu^- \mu^+ \to f\bar{f}$.

factor of $1/4$, resulting in a luminosity reduction by $1/16$. Such a tradeoff is clearly unacceptable. Polarization will be a useful tool in Higgs resonance "discovery" and studies only if high polarization is achievable with little luminosity loss. If $P_+ = P_- = 1$, you have completely spin 0 state, pure signal (and no background) in which case $\kappa_{\text{pol}}$ (formula 8) no longer apply.

Ideas for increasing the polarization are still being explored [1,6]. Tau final state polarizations can also be used to help improve the $H \to \tau\bar{\tau}$ measurement.

# VI "DISCOVERY" FROM ANGULAR DISCRIMINATION

Some "discovery" or sensitivity enhancement can also be obtained from angular discrimination. A proper study would include detector acceptance cuts and maximum likelihood fits. Here, we wish to only approximate the gain. For that purpose, we assume perfect (infinitesimal) binning and obtain a (maximal) measurement sensitivity enhancement factor

$$\frac{1}{2}(1 + P_+ P_-)\sqrt{N_S + N_B} \left[ \int \frac{dx}{dN/dx} \right]^{1/2}, \tag{8}$$

which becomes, from Equations (1) and (7),

$$\kappa_{\text{pol}}\sqrt{\frac{2}{3}}\sqrt{\frac{N_S + N_B}{N_B}} \left( \frac{\tan^{-1}\left(\frac{2}{\zeta}\sqrt{1 - \frac{16}{9}A_{eff}^2 + \zeta}\right)}{\sqrt{1 - \frac{16}{9}A_{eff}^2 + \zeta}} \right)^{1/2}, \qquad \zeta \equiv \frac{4}{3}\frac{N_S}{N_B}\frac{\kappa_{\text{pol}}^2}{1 + P_+ P_-}. \tag{9}$$

In the case of "discovery", high polarization and/or a near maximal forward-backward asymmetry can significantly reduce the scan time. Additional analysis and detail will be given in [4,7].

# VII  CONCLUSION

We investigated the effect of beam polarization on Higgs resonance signals and backgrounds ($b\bar{b}$, $\tau\bar{\tau}$, $c\bar{c}$), angular distributions (forward-backward charge asymmetries) and the resulting effective enhancement of the Higgs signal relative to the background, as well as the reduction in scan time required for Higgs "discovery" [4,7,8]. We conclude that $\mathcal{L}_{ave} \simeq 5 \times 10^{31} \text{cm}^{-2}$ is (about the minimum Luminosity), needed for the Higgs resonance studies at the First Muon Collider. And that a factor of 10 in luminosity reduces the scan time by a factor of 10 and increases the resolution by about a factor of 3. The choice of energy and luminosity depends on the scan time needed, and how precise a measurement is needed to do physics. Other decays such as $WW^*$ and $ZZ^*$ are very small for this mass regions and to measure them need to improve precision, thus the need for increase in Luminosity, etc.

We have shown that polarization is potentially useful for Higgs resonance studies, but only if the accompanying luminosity reduction is not significant. Large forward-backward asymmetries can also be used to enhance the Higgs "discovery" signal or improve precision measurements, particularly for $\tau\bar{\tau}$. However, to make the $s$-channel Higgs "factory" a compelling facility, we must attain a very good beam resolution and the highest luminosity possible. An additional "discovery" or sensitivity enhancement can be obtained from angular discrimination. For additional discussion see [7].

# REFERENCES

1. Muon Collider Feasibility Study, BNL Report BNL-52503 (1996); Muon Collider Collaboration; Mu-Cool Collaboration; Figs: Norem Priv. Comm., K. McDonald Priv. Comm.
2. Cline, D., "The Problems and Physics Prospects for a $\mu^+\mu^-$ Collider", in *Future High Energy Colliders*, edited by Z. Parsa, AIP Conference Proceedings **397**, 1997, pp. 203–218.
3. Barger, V., Berger, M.S., Gunion, J.F., and Han, T., "The Physics Capabilities of $\mu^+\mu^-$ Colliders", in *Future High Energy Colliders*, edited by Z. Parsa, AIP Conference Proceedings **397**, 1997, pp. 219–233; *Phys. Rep.* **286**, 1–51 (1997); *Phys. Rev. Lett.* **75**, 1462–1465 (1995).
4. Kamal, B., Marciano, W., Parsa, Z., BNL Report BNL-65193 (1997), hep-ph/9712270.
5. Parsa, Z., $\mu^+\mu^-$ Collider and Physics Possibilities (1993) (unpublished).
6. Skrinsky, A., *Presentation, Dec 1997.*
7. Kamal, B., Marciano, W., Parsa, Z., to be published.
8. Parsa, Z., Polarization Effects at a Muon Collider, *Presentation at EPAC98, Stokholm, Sweeden, June 1998.*

# Electron-Positron Pair Production in the Deep Quantum Regime

Kathleen A.Thompson and Pisin Chen

*Stanford Linear Accelerator Center, MS 26,*
*P.O.Box 4349, Stanford, CA 94309, USA*
*E-mail: kthom@SLAC.Stanford.edu*

**Abstract.** Electron-positron pair production via real and virtual photons is significant to the design of linear colliders, especially in the deep quantum regime (i.e., beamstrahlung parameter $\Upsilon >> 1$). In this regime, pair production via a virtual photon (the trident process) can become comparable in rate to pair production via a real beamstrahlung photon. We derive characteristics of the e+e- pairs produced via the trident process, using the quasi-classical approach of Baier, Katkov, and Strakhovenko [1]. We have also examined some of the implications of e+e- pair production for the design of very high energy (several TeV in the center of mass) linear colliders in the deep quantum regime, both in this paper and elsewhere [2].

## INTRODUCTION

In extremely high energy linear collider designs (several TeV in the center of mass), a tightly focused bunch consisting of $\sim 10^8$ electrons is to pass through a similar bunch of positrons travelling in the opposite direction. Individual high energy electrons and positrons will radiate photons due to their interaction with the collective electromagnetic field of the oncoming bunch. Some of these beamstrahlung photons convert to $e^+e^-$ pairs as they continue moving through the collective field. The strength of the interaction with the external field is characterized by the usual $\Upsilon$ parameter (see next section). Very high energy collider designs, for example those using laser acceleration [3], typically need very short bunch lengths and thus tend to be in the deep quantum beamstrahlung regime ($\Upsilon \gg 1$). Coherent pair production (i.e. that due to the interaction with the strong collective field produced by the other bunch) may occur through a real beamstrahlung photon (we shall refer to this as the *cascade process*), or the intermediate photon may be virtual, in which case the pair production is said to occur by the *trident process*. The virtual photon process becomes comparable to the cascade process in the deep quantum regime. Thus the main motivation of this paper is to investigate further the possible impact of the trident process on the design of very high energy linear

CP472, *Advanced Accelerator Concepts: Eighth Workshop,*
edited by W. Lawson, C. Bellamy, and D. Brosius
1999 The American Institute of Physics 1-56396-889-4

colliders.

Pair production via the cascade process was first treated by Klepikov [4] and by Nikishov and Ritus [5]. The first correct treatment of the trident process was given by Ritus [6]. Useful approximate formulas for the total pair production probability via the trident process were given by Baier, Katkov, and Strakhovenko (BKS) [1] including one for the high $\Upsilon$ limit that is of interest to us, but almost no details of the derivation that would enable one to easily obtain an expression for the energy spectrum of the pairs were given. Thus the main work of our paper is to reconstruct a derivation (which is presumably along the lines of one already carried out by BKS) that will allow us to obtain an explicit expression for the energy spectrum of pairs produced via the trident process at high $\Upsilon$. We shall also compare our result for the total trident pair production rate with that obtained by BKS, as well as with that for the rate via the cascade process; both of which have already been widely used in the literature on linear colliders.

## CALCULATION OF PAIR PRODUCTION RATE

Consider an electron or positron of very high energy $E$ traversing a strong electromagnetic field. Such a situation may be characterized by the Lorentz invariant parameter $\Upsilon$, defined by

$$\Upsilon \equiv \frac{e\hbar}{m^3 c^4} \sqrt{|F_{\mu\nu} p^\nu|^2} = \gamma \frac{B}{B_c} \quad . \tag{1}$$

Here $p^\nu = (E, \overrightarrow{p})$ is the 4-momentum of the incoming electron or positron, $m$ is the electron mass, $\gamma \equiv E/mc^2$ is the usual Lorentz factor, $F_{\mu\nu}$ is the energy-momentum tensor of the electromagnetic field, $B = |\overrightarrow{B}| + |\overrightarrow{E}|$, and $B_c \equiv m^2 c^3/\hbar e \approx 4.4 \times 10^{13}$ Gauss is the Schwinger critical field.

We follow the quasi-classical approach of BKS, whereby the very high energy electron can be regarded as following a classical trajectory through the magnetic field. The quantum nature of the photon emission and the corresponding recoil of the electron are, however, taken into account. Under such assumptions, BKS derive the following expression for the total pair production probability (per unit time) via a virtual intermediate photon:

$$W_{tot} = -\frac{\alpha^2 m^2}{8\pi^2 E} \frac{c^4}{\hbar} \int_0^\infty \frac{du}{(1+u)^2} \int_0^\infty \frac{d\xi}{\cosh^2 \xi} \cdot I_{\sigma\tau} \quad . \tag{2}$$

Here $\alpha \equiv e^2/\hbar c \approx 1/137$ is the fine-structure constant. The variables $u$ and $\xi$ are defined by

$$u \equiv \frac{\hbar\omega}{E - \hbar\omega} = \frac{y}{1 - y},$$

$$\cosh^2 \xi \equiv \frac{(\hbar\omega)^2}{4E_+ E_-} = \frac{y^2}{4x(y - x)}. \tag{3}$$

261

where $\hbar\omega$ is the energy of the intermediate virtual photon, $E_+$ is the energy of the positron of the produced pair, $E_- = \hbar\omega - E_+$ is the energy of the produced electron, $y \equiv \hbar\omega/E$ is the fractional energy of the intermediate virtual photon, and $x \equiv E_+/E$ is the fraction of the initial energy carried by the positron in the produced pair.

For the moment, it is most convenient to express $I_{\sigma\tau}$ in the following form:

$$I_{\sigma\tau} = \int_{-\infty}^{\infty} d\sigma \int_{-\infty}^{\infty} d\tau\, B_\sigma B_\tau \left\{ \frac{u\Upsilon}{(1+u)\cosh^2\xi} \frac{\delta(\sigma-\tau)}{\sigma\tau} \right.$$

$$+ \left[ A_{-1,-1}\frac{1}{\tau\sigma} + A_{-1,1}\frac{\sigma}{\tau} + A_{1,-1}\frac{\tau}{\sigma} + A_{1,0}\tau + A_{1,2}\tau\sigma^2 \right.$$

$$\left. + A_{2,-1}\frac{\tau^2}{\sigma} + A_{2,1}\tau^2\sigma \right] \cdot [\theta(\sigma-\tau) - \theta(\tau-\sigma)] \right\} \cdot$$

$$\exp\left[ -i\frac{u}{\Upsilon}(\sigma + \frac{\sigma^3}{3}) - i\kappa(\tau + \frac{\tau^3}{3}) \right] \quad . \tag{4}$$

where the $\tau$ and $\sigma$ dependences are shown explicitly, and the coefficients $A_{i,j}$ still depend on the remaining integration variables $u, \xi$ (or $x, y$) and are defined below. Here $\delta(z)$ is the Dirac delta function, $\theta(z)$ is the Heaviside step function, and we have defined

$$\kappa \equiv \frac{4\cosh^2\xi\,(1+u)}{u\Upsilon} = \frac{y}{\Upsilon x(y-x)} \quad . \tag{5}$$

The integrals over $\sigma$ and $\tau$ are regularized for $\sigma, \tau \to 0$ via the operator $B_\tau$, which is defined by:

$$B_\tau \tau^n e^{a\tau^3} = \begin{bmatrix} \tau^n e^{a\tau^3} & (n \geq 0) \\ \tau^n(e^{a\tau^3} - 1) & (n = -1) \end{bmatrix} . \tag{6}$$

The quantities $A_{i,j}$ depend on $\Upsilon$, as well as on the fractional energies $x$ and $y$ (through the variables $u$ and $\xi$), and are given by

$$A_{-1,-1} = \frac{-i}{\cosh^2\xi}$$

$$A_{-1,1} = \frac{-id(u)}{(1+u)\cosh^2\xi}$$

$$A_{1,-1} = \frac{ib(\xi)d(u)}{3u^2}$$

$$A_{1,0} = -A_{2,-1} = \frac{2(1+u)}{3u\Upsilon}b(\xi)$$

$$A_{1,2} = -A_{2,1} = \frac{2(1+u)}{3u\Upsilon}\left( \frac{b(\xi)d(u)}{1+u} - 3 \right) \tag{7}$$

262

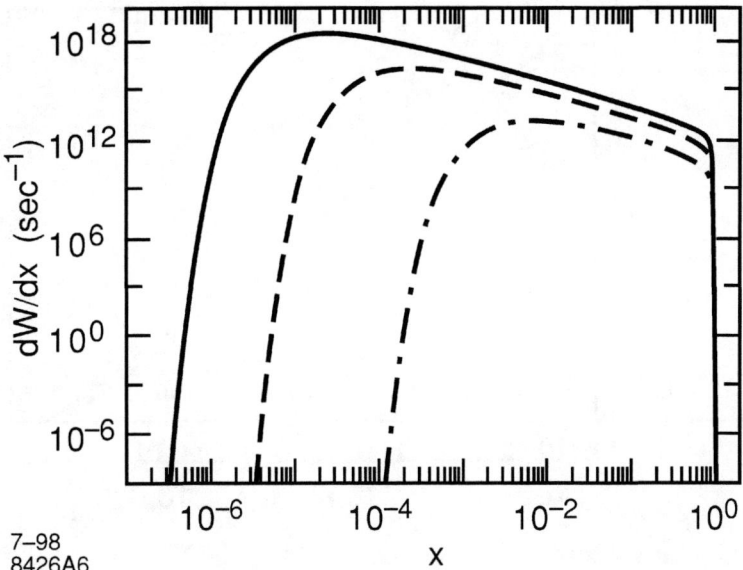

**FIGURE 1.** Spectrum of probability per unit time $[\sec^{-1}]$ for pair production via the trident process, as a function of $x \equiv E_+/E$, for $\Upsilon = 100$ (dot-dashed curve), $\Upsilon = 3000$ (dashed curve), and $\Upsilon = 30000$ (solid curve). The vertical axis, which scales as $1/E$, assumes $E = 2.5$ TeV.

where $d(u) \equiv 1 + (1+u)^2$ and $b(\xi) \equiv 8 \cosh^2 \xi + 1$.

After a lengthy calculation, in which the assumption $\Upsilon \gg 1$ is used, the integrals over $\sigma$ and $\tau$ in Eq. (4) may be carried out in terms of the Airy function $\mathrm{Ai}(z)$ and the related Airy function $\mathrm{Gi}(z)$:

$$\mathrm{Ai}(z) \equiv \frac{1}{\pi} \int_0^\infty \cos(\frac{v^3}{3} + zv)dv \quad ,$$

$$\mathrm{Gi}(z) \equiv \frac{1}{\pi} \int_0^\infty \sin(\frac{v^3}{3} + zv)dv \quad . \tag{8}$$

The full result for $I_{\sigma\tau}$ is given in the Appendix. There are three terms in $I_{\sigma\tau}$ that are significant for large $\Upsilon$:

$$I_{\sigma\tau} \approx -\frac{8\pi}{9u^2} b(\xi)d(u)\kappa^{-2/3} \mathrm{Ai}'(\kappa^{2/3}) \ln(u/\Upsilon)$$

$$+ 8\pi \frac{(1+u)}{3u} \left(\frac{b(\xi)d(u)}{1+u} - 3\right) u^{-1}\kappa^{-2/3} \mathrm{Ai}'(\kappa^{2/3})$$

$$- \frac{4\pi}{3}(\frac{2}{3}\mathcal{C} + \frac{1}{3}\ln 3)\frac{1}{u^2} b(\xi)d(u)\kappa^{-2/3} \mathrm{Ai}'(\kappa^{2/3})$$

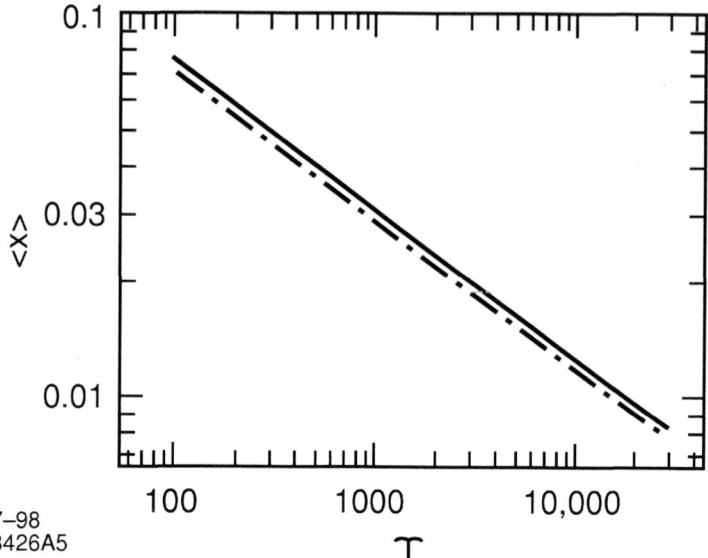

**FIGURE 2.** Mean value of $x \equiv E_+/E$ for pair production via the trident process, as a function of $\Upsilon$. The lower curve includes only the dominant term in Eq. (9), while the solid curve includes all three terms in Eq. (9).

$$
= -\frac{8\pi}{9}[(1-y)^2 + 1]\left[\frac{2y^2}{x(y-x)} + 1\right]y^{-8/3}[x(y-x)]^{2/3} \cdot
$$
$$
\Upsilon^{2/3}\mathrm{Ai}'(\kappa^{2/3}) \ln\left[\frac{y}{(1-y)\Upsilon}\right]
$$
$$
+ \frac{8\pi}{3}\left\{[(1-y)^2 + 1]\left[\frac{2y^2}{x(y-x)} + 1\right] - 3(1-y)\right\}y^{-8/3}[x(y-x)]^{2/3} \cdot
$$
$$
\Upsilon^{2/3}\mathrm{Ai}'(\kappa^{2/3})
$$
$$
- \frac{4\pi}{3}\left(\frac{2}{3}\mathcal{C} + \frac{1}{3}\ln 3\right)[(1-y)^2 + 1]\left[\frac{2y^2}{x(y-x)} + 1\right]y^{-8/3}[x(y-x)]^{2/3} \cdot
$$
$$
\Upsilon^{2/3}\mathrm{Ai}'(\kappa^{2/3}) \quad . \tag{9}
$$

Here $\mathcal{C}$ is Euler's constant ($\approx 0.577$). The first term shown dominates. The second two terms give a correction of order 10% for parameters of interest for very high energy linear colliders. Note that all three terms depend on $\Upsilon$ through $\Upsilon^{2/3}\mathrm{Ai}'(\kappa^{2/3})$. The main reason for the dominance of the first term is its additional dependence on $\ln \Upsilon$.

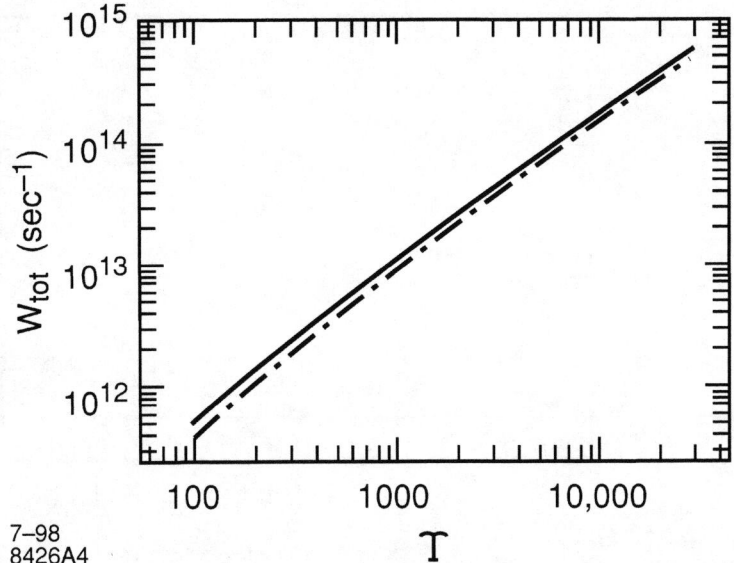

**FIGURE 3.** Total probability per unit time [sec⁻¹] for pair production via the trident process, as a function of $\Upsilon$, for $E \equiv \gamma mc^2 = 2.5$ TeV. The lower curve includes only the dominant term in Eq. (9), while the solid curve includes all three terms in Eq. (9). [This figure may of course be scaled to arbitrary energy since the vertical scale is proportional to $1/\gamma$.]

## Energy spectrum of pairs

In order to obtain the energy spectrum, we express the remaining integrations in terms of $x$ and $y$ rather than $u$ and $\xi$. In terms of $x$ and $y$, the total probability per unit time for producing pairs at any energy between 0 and $E$ is:

$$
\begin{aligned}
W_{tot} &= -\frac{\alpha^2 m\, c^2}{8\pi^2}\frac{1}{\hbar}\frac{1}{\gamma}\int_0^{1/2} dx \int_{2x}^1 dy \frac{\frac{1}{y}(y-2x)}{[y^2/4 - x(y-x)]^{1/2}} \cdot I_{\sigma\tau} \\
&= -\frac{\alpha^2 m\, c^2}{16\pi^2}\frac{1}{\hbar}\frac{1}{\gamma}\int_0^1 dx \int_x^1 dy \frac{\frac{1}{y}|y-2x|}{[y^2/4 - x(y-x)]^{1/2}} \cdot I_{\sigma\tau} \quad .
\end{aligned}
\tag{10}
$$

(The last equality follows from the symmetry properties of the integrand when $x$ and $(y-x)$ are interchanged.) Given Eq.(10), it is straightforward to do the integration over $y$ numerically to get the spectrum $dW/dx$, and to do both integrations numerically to get $W_{tot}$. The spectra for $\Upsilon = 100$, $\Upsilon = 3000$, and $\Upsilon = 30000$ are shown in Figure 1. Here we have assumed $E = 2.5$ TeV, but again the particular value of the energy only affects the vertical scale through the $1/\gamma$ factor. Using the result for $dW/dx$, the mean value of $x$ as a function of $\Upsilon$ may also be computed, as is shown in Figure 2.

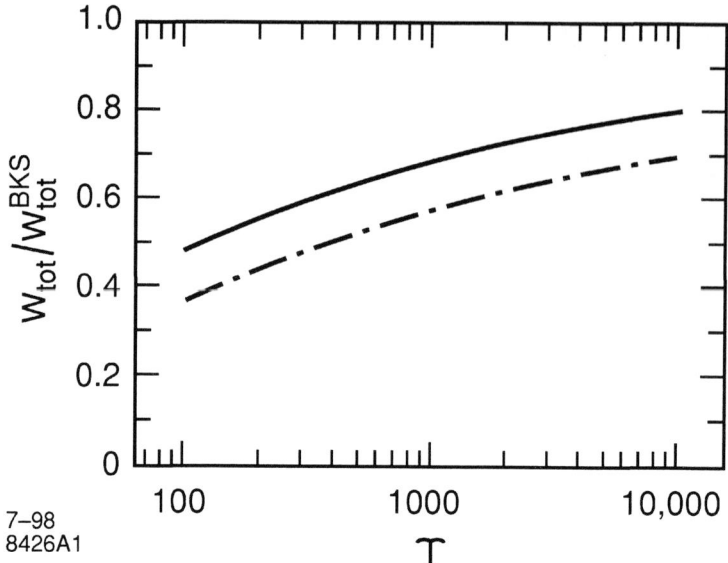

**FIGURE 4.** Ratio of $W_{tot}$, the total probability per unit time for pair production via the trident process to $W_{tot}^{BKS}$, the approximation of Baier, Katkov, and Strakhovenko, as a function of $\Upsilon$. The lower curve includes only the dominant term in Eq. (9), while the solid curve includes all three terms in Eq. (9).

## Total rates of pairs

The total probability $W_{tot}$ as a function of $\Upsilon$ is shown in Figure 3, for $E \equiv \gamma mc^2 = 2.5$ TeV. (The figure may be scaled to arbitrary $\gamma$ since the vertical scale is simply proportional to $1/\gamma$.) The upper curve includes all three terms in Eq. (9), while the lower curve includes only the first term.

Next we compare our result for $W_{tot}$ with the result given by BKS [1]. These authors appear to have made an approximation to the first term of Eq. (9) which allows them to carry out the integrations over $u$ and $\xi$, obtaining the following convenient analytic expression:

$$W_{tot}^{BKS} = \frac{13\alpha^2 m}{9\sqrt{3}\pi} \frac{c^2}{\hbar} \frac{1}{\gamma} \Upsilon \ln \Upsilon \quad . \tag{11}$$

In Figure 4 we show the ratio of $W_{tot}$ to $W_{tot}^{BKS}$, as a function of $\Upsilon$. The lower curve includes only the dominant term in Eq. (9), while the solid curve includes all three terms in Eq. (9). The overall $\Upsilon$ dependence of our result is in fact quite close to $\Upsilon \ln \Upsilon$ for very large $\Upsilon$, but we do not know what approximations BKS made to obtain precisely their result (or indeed whether or not our derivation agrees with theirs up to the point where such an approximation would be made).

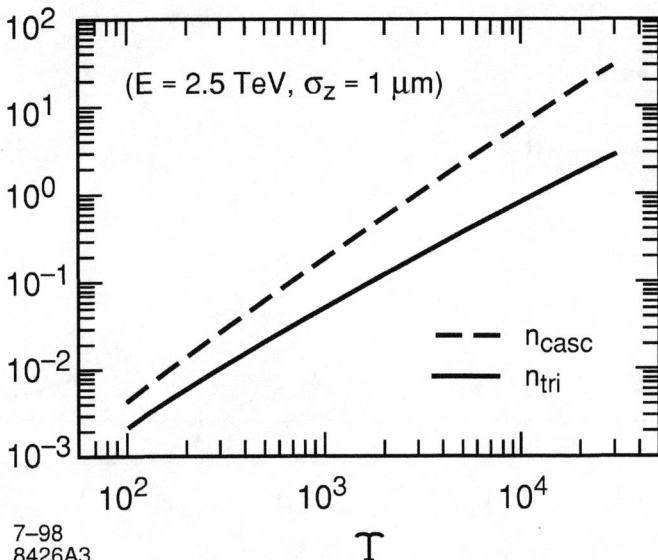

**FIGURE 5.** Number of pairs per particle produced at an energy $E = 2.5$ TeV and for a bunch length $\sigma_z = 1\mu$m, via the cascade process (dashed curve) and the trident process (solid curve).

Consider now a linear collider in which the external field is that created by the oncoming bunch, and in which the bunch lengths are $\sigma_z$. The bunches are assumed to be Gaussian in all three space dimensions, and the charge and transverse size of the bunch are then the remaining determinants of the effective $\Upsilon$. The total (integrated over $x$) number of pairs produced via the trident process, per incoming electron or positron, is $n_{tri} \approx \frac{\sqrt{3}\sigma_z}{c}W_{tot}$. For comparison, an estimate [6,7] of the number of pairs per particle produced via the cascade process is

$$n_{casc} = (0.295)\left[\frac{\alpha\sigma_z\Upsilon}{\gamma\lambda_e}\right]^2 \Upsilon^{-2/3}(\ln\Upsilon - 2.488) \qquad (\Upsilon \gg 1) \quad . \qquad (12)$$

Here $\lambda_e = \hbar/mc$ is the Compton wavelength of the electron.

In Figure 5 we show the number of pairs per particle produced via the cascade process (dashed curve) and the trident process (solid curve) as a function of $\Upsilon$, for energy $E = 2.5$ TeV and bunch length $\sigma_z = 1$ $\mu$m.

In Figure 6 we show the ratio of the number of pairs per particle produced via the trident process to the number of pairs produced via the cascade process, normalized to an energy $E = 2.5$ TeV and for a bunch length $\sigma_z = 1$ $\mu$m. To obtain the ratio $n_{tri}/n_{casc}$ for arbitrary energy and bunch length, one would multiply by energy in TeV and divide by the bunch length in microns. Keep in mind, of course, that changing the bunch length will change the effective $\Upsilon$, unless the bunch charge and transverse size are adjusted to compensate.

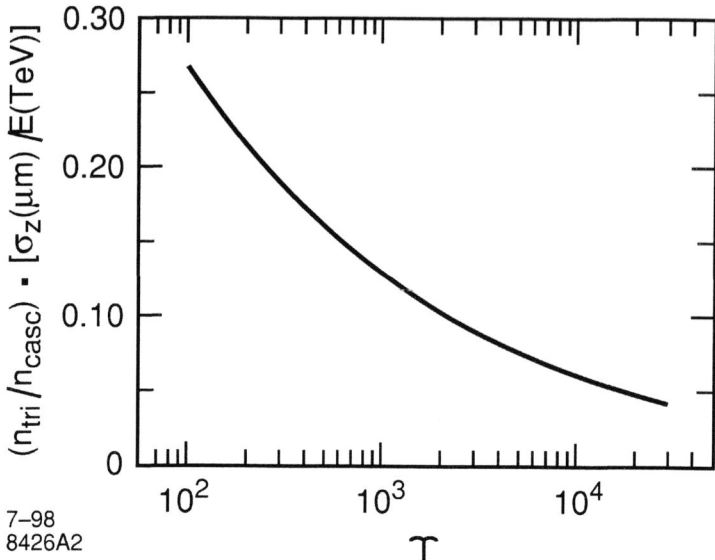

**FIGURE 6.** Ratio of number of pairs per particle produced via the trident process to the number of pairs produced via the cascade process, normalized to an energy $E = 2.5$ TeV and for a bunch length $\sigma_z = 1$ $\mu$m. (To obtain $n_{tri}/n_{casc}$ for arbitrary energy and bunch length, multiply by $E$ in TeV and divide by $\sigma_z$ in microns.)

# CONCLUSIONS AND ACKNOWLEDGMENTS

Our results for the total trident pair production rate $W_{tot}$ are in reasonable agreement with that predicted by the approximate formula of Baier, Katkov and Strakhovenko, although their approximate expression is somewhat larger than ours. The agreement is within 20% at extremely high $\Upsilon$ ($\Upsilon > 10000$ up to the limit at which the assumptions in our calculations break down), but there is a discrepancy of about a factor of two for $\Upsilon$ values of a few hundred.

For reasonable linear collider parameters, it does not appear that the total number of trident pairs would exceed the total number of cascade-process pairs, although the trident pairs can can comprise a significant fraction of the total. Another possible issue, which we have addressed elsewhere, is whether there are a significant number of trident pairs with sufficiently low energy that the pair particle with the same sign as the oncoming beam can be deflected to large angles in the beam-beam field. Our conclusion [2] is that there is much less than one trident pair per bunch crossing that would reach an outgoing angle of a radian or more, assuming very high energy (several TeV) linear collider parameters similar to those proposed in Ref. [3].

We thank John Irwin for correcting a factor two error in $n_{tri}$.

Work supported by Department of Energy Contract DE-AC03-76SF00515.

# REFERENCES

1. Baier, V.N., Katkov, V.M., and Strakhovenko, V.M., *Soviet Journal of Nuclear Physics* **14**,572,1972. See also Baier, V.N. and Katkov, V.M., *Soviet Physics JETP* **26**,854,1968.
2. Thompson, K.A. and Chen, P., in Proceedings of Advanced ICFA Beam Dynamics Workshop on Quantum Aspaects of Beam Physics, Monterey, CA, January 4-9, 1998; SLAC-PUB-7776.
3. Xie, M., Tajima, T., Yokoya, K., and Chattopadhyay, S., *Proceedings of Seventh Workshop on Advanced Accelerator Concepts*, Lake Tahoe, CA, October 12-18, 1996.
4. Klepikov, N.P., *Zh.Eksp.Teor.Fiz* **46**,28,1964.
5. Nikishov, A.I. and Ritus, V.I., *Soviet Physics JETP* **19**,529,1964.
6. Ritus, V.I., *Nuclear Physics* **B44**,236,1972.
7. Yokoya, K. and Chen, P., "Beam-beam phenomena in linear colliders", in M.Dienes, et.al. (Eds.), *Frontiers of Particle Beams: Intensity Limitations, Lecture Notes in Physics, Vol. 400* (Springer-Verlag, 1992).

# APPENDIX

We give here the full expression for $I_{\sigma\tau}$:

$$
\begin{aligned}
I_{\sigma\tau} = {} & -\frac{8\pi}{9u^2}b(\xi)d(u)\kappa^{-2/3}\mathrm{Ai}'(\kappa^{2/3})\ln\left(\frac{u}{\Upsilon}\right) \\
& +8\pi\frac{(1+u)}{3u^2}\left(\frac{b(\xi)d(u)}{1+u}-3\right)\kappa^{-2/3}\mathrm{Ai}'(\kappa^{2/3}) \\
& -\frac{4\pi}{3u^2}(\frac{2}{3}\mathcal{C}+\frac{1}{3}\ln 3)b(\xi)d(u)\kappa^{-2/3}\mathrm{Ai}'(\kappa^{2/3}) \\
& +\frac{4\pi^2}{3u^2}b(\xi)d(u)\kappa^{-2/3}\mathrm{Ai}'(\kappa^{2/3})\int_0^{(\frac{u}{\Upsilon})^{2/3}}\mathrm{Gi}(v)dv \\
& +8\pi^2\frac{(1+u)}{3u^{5/3}\Upsilon^{1/3}}\left(\frac{b(\xi)d(u)}{1+u}-3\right)\kappa^{-1/3}\mathrm{Ai}(\kappa^{2/3})\mathrm{Gi}'((\frac{u}{\Upsilon})^{2/3}) \\
& +8\pi^2\frac{(1+u)}{3u^{4/3}\Upsilon^{2/3}}b(\xi)\kappa^{-1/3}\mathrm{Ai}'(\kappa^{2/3})\mathrm{Gi}((\frac{u}{\Upsilon})^{2/3}) \\
& -8\pi^2\frac{(1+u)}{3u^{4/3}\Upsilon^{2/3}}\left(\frac{b(\xi)d(u)}{1+u}-3\right)\kappa^{-2/3}\mathrm{Ai}'(\kappa^{2/3})\mathrm{Gi}((\frac{u}{\Upsilon})^{2/3}) \\
& -8\pi^2\frac{(1+u)}{3u\Upsilon}b(\xi)\kappa^{-1/3}\mathrm{Ai}(\kappa^{2/3})\int_0^{(\frac{u}{\Upsilon})^{2/3}}\mathrm{Gi}(v)dv \\
& +8\pi(\frac{2}{3}\mathcal{C}+\frac{1}{3}\ln 3)\frac{(1+u)}{3u\Upsilon}b(\xi)\kappa^{-1/3}\mathrm{Ai}(\kappa^{2/3}) \\
& +\frac{16\pi}{3}\frac{(1+u)}{3u\Upsilon}b(\xi)\kappa^{-1/3}\mathrm{Ai}(\kappa^{2/3})\ln\left(\frac{u}{\Upsilon}\right) \quad.
\end{aligned}
\tag{13}
$$

The $\Upsilon$ dependence of the terms beyond the first three is through non-positive powers of $\Upsilon$ and through the Airy and related Airy functions.

# Wake Fields in a mm-Wave Linac[1]

## F. Zimmermann, D.H. Whittum, C.K. Ng and M.E. Hill [†] [2]

*Stanford Linear Accelerator Center*
*Stanford University, Stanford California 94309*

[†] *Harvard University, Cambridge, MA 02138*

**Abstract.** We estimate the short-range wake fields in the W-band active matrix linac of a 5-TeV collider, and demonstrate that for the assumed 60-pC bunch charge and 10-$\mu$m rms bunch length they are acceptable.

## INTRODUCTION

We consider an active matrix linac as described in Ref. [1,2], operating at at a wavelength of $\lambda = 3.28$ mm (91 GHz). A single cavity of such an accelerator is sketched in Fig. 1. Viewed in the beam direction, the transverse dimension of the cavity is square with a full width of $\lambda/\sqrt{2}$ or a half width of $b = 1.16$ mm. The full cavity gap is $g = 0.37\lambda = 1.21$ mm, and the iris radius $a = 0.1\ \lambda = 0.328$ mm. We assume that the full period $l$ is 25% larger than $g$, or $l = 1.51$ mm. The bunch charge is 60 pC ($3.75 \times 10^{10}$ electrons per bunch) and the rms bunch length is taken to be 10 $\mu$m. Cavity and beam parameters are summarized in Table 1. The calculations presented in this paper do not pay attention to the rectangular geometry of the cells. This is justified, since the irises are round and we are only concerned with the short-range wake fields.

## LONGITUDINAL WAKE FIELD

### Geometric Wake

The longitudinal geometric wake field is maximum at the position of the drive particle, where it assumes the value [3]

$$W_L(0) = \frac{Z_0 c}{\pi a^2} \tag{1}$$

[2] Work supported under a National Science Foundation Graduate Fellowship.
[1] Work supported by the U.S. Department of Energy under contract DE-AC03-76SF00515.

CP472, *Advanced Accelerator Concepts: Eighth Workshop,*
edited by W. Lawson, C. Bellamy, and D. Brosius
© 1999 The American Institute of Physics 1-56396-889-4/99/$15.00

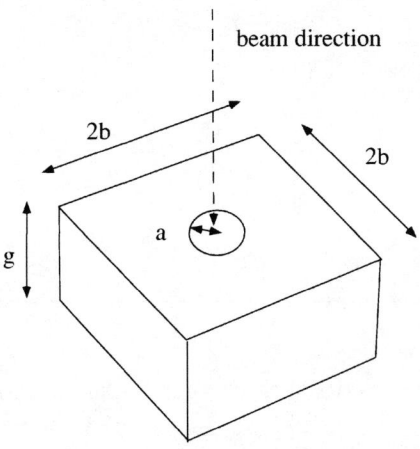

**FIGURE 1.** Schematic of a single cavity in a 2.5-GeV active matrix linac [1,2]

**TABLE 1.** Single-cell and beam parameters for the linac of a 5-TeV collider [1,2].

| variable | symbol | value |
|---|---|---|
| charge per bunch | $Q$ | 60 pC |
| rms bunch length | $\sigma_z$ | 10 $\mu$m |
| wavelength | $\lambda$ | 3.28 mm |
| full gap | $g$ | 1.21 mm |
| iris radius | $a$ | 0.328 mm |
| full period | $l$ | 1.51 mm |
| cavity half width | $b$ | 1.16 mm |

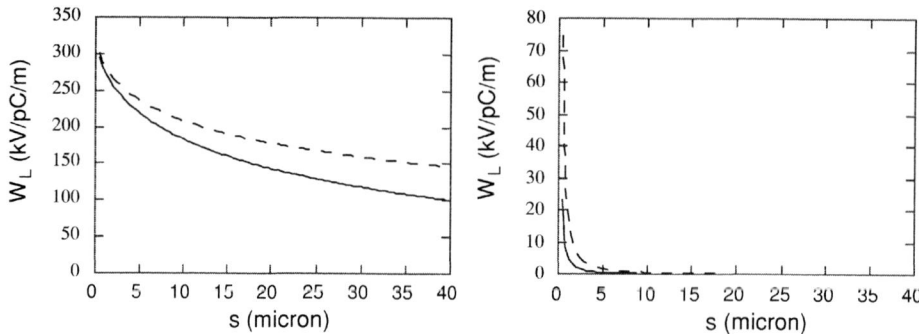

**FIGURE 2.** Left: longitudinal geometric wake field vs. distance for $a/\lambda = 0.1$; dashed: Eq. (2), solid: Eq. (3). Right: longitudinal resistive-wall wake field vs. distance; dashed: cavity walls, solid: iris.

with $Z_0 = 377\ \Omega$. Thus $W_L(0)$ depends only on the iris radius of the cavity. For our cell, it evaluates to 336 kV/pC/m. As a worst case, we could assume that the wake field is constant across the bunch, equal to $W_L(0)$. If we then consider a beam of charge $Q$ equal to 60 pC, the induced voltage is 20 MV/m, a factor 50 smaller than the accelerating gradient of 1 GV/m.

Let us now include the $s$-dependence of the short-range wake field. An inverse Fourier transform of the high-frequency impedance derived in Ref. [3] yields [4]

$$W_L(s) \approx \frac{Z_0 c}{\pi a^2} \exp\left(\frac{2\pi \alpha^2 l^2 s}{a^2 g}\right)\ \mathrm{erfc}\left(\frac{\alpha l}{a}\sqrt{\frac{2\pi s}{g}}\right) \tag{2}$$

where, for $g/l \to 1$, the coefficient $\alpha$ approaches 0.46 [4].

In Ref. [4] an alternative approximation to the short-range wake field was given: Over a wide parameter range, the wake fields from a complex-frequency domain calculation are well reproduced by a quasi-exponential decay [4],

$$W_L(s) = \frac{Z_0 c}{\pi a^2}\ \exp\left(-\sqrt{s/s_e}\right) \tag{3}$$

where

$$s_e = 0.41\ \frac{a^{1.8} g^{1.6}}{l^{2.4}} \tag{4}$$

is the decay length. In our example, $s_e \approx 27\ \mu$m.

Equations (2) and (3), evaluated for the parameters of Table 1, are plotted in Fig. 2 (left). The two curves are quite different at large $s$. Based on the results of Ref. [4], we expect that Eq. (3) provides the more accurate description.

# Resistive-Wall Wake

The character of the resistive wall wake field is determined by the ratio between bunch length and the characteristic distance [5]

$$s_0 = \left( \frac{cb_p^2}{2\pi\sigma} \right)^{1/3} \tag{5}$$

where $\sigma$ denotes the conductivity in cgs units (for copper at room temperature $\sigma = 5.8 \times 10^{17}$ s$^{-1}$) and $b_p$ is radius of the beam pipe. In this case $b_p$ is either equal to iris radius $a$ or the cavity half width $b$, for which $s_0 \approx 2$ $\mu$m and $s_0 \approx 5$ $\mu$m, respectively. The bunch length $\sigma_z \approx 10$ $\mu$m is a factor 2–5 larger than $s_0$. Hence we can, in a good approximation, use the formula for the resistive-wall wake field of a long bunch, derived by Chao [6]:

$$W_L(s) = \frac{Z_0 c}{2(2\pi)^{3/2} b_p^2} \left( \frac{s_0}{s} \right)^{3/2} \tag{6}$$

Since the iris walls occupy about 20% of the total length, and their radius $a$ is about 1/3 of $b$, we find that the contributions to the resistive-wall wake field from iris and cavity wall are comparable.

These two wake fields, for iris and walls, are depicted in Fig. 2 (right), where we have multiplied by the different filling factors of about 20% and 80%, respectively. The resistive wake field falls off rapidly over a distance much shorter than the bunch length, and its magnitude is small compared with the geometric wake. Thus, the longitudinal resistive-wall wake field can be neglected.

## Coating and Surface Roughness

The effects of a dielectric coating or surface roughness can be described by [8]:

$$W_L(s) = \frac{Z_0 c}{\pi b_p^2} \cos k_0 s \tag{7}$$

where

$$k_0 = \left( \frac{2\epsilon}{b_p \delta(\epsilon - 1)} \right)^{1/2}, \tag{8}$$

$b_p$ is the radius of the beam pipe (*e.g.*, equal to $b$ or $a$), and $\delta$ the thickness of the dieletric layer or corrugation.

It is foreseen as an option to coat the inside of the W-band cells with a 5-$\mu$m layer of diamond ($\epsilon = 5.5$) [7]. According to a recipe put forward in Ref. [8], to describe the effect of surface roughness we should use Eqs. (7) and (8) with $\epsilon \approx 2$. Using $b_p = b$, we then find, $1/k_0^{(1)} \approx 49$ $\mu$m for the dielectric, and $1/k_0^{(2)} \approx 17$ $\mu$m for a pessimistic 1-$\mu$m surface roughness. The corresponding wake functions are shown in Fig. 3. They are comparable to the geometric wake field.

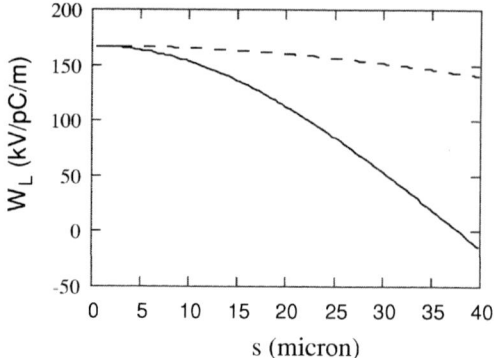

**FIGURE 3.** Longitudinal wake field for a 5 $\mu$m diamond coating (dashed) and a 1 $\mu$m surface roughness (solid) vs. distance.

## Total Longitudinal Wake

The total wake is now the sum of the geometric. the dielectric and the surface roughness wakes (we neglect the resistive-wall wake field, since it is much smaller). If we fold the Green function wake $W_L(s)$ with a (Gaussian) charge distribution, we obtain the beam-induced voltage for the entire bunch:

$$V_L(s) = \frac{Ne}{\sqrt{2\pi}\sigma_z} \int_{-\infty}^{s} W_L(s-s') e^{-\frac{s'^2}{2\sigma_z^2}} \, ds' \tag{9}$$

with

$$W_L(s) \approx \frac{Z_0 c}{\pi} \left( \frac{e^{-\sqrt{s/s_e}}}{a^2} + \frac{\cos k_0^{(1)} s}{b^2} + \frac{\cos k_0^{(2)} s}{b^2} \right) \tag{10}$$

In Fig. 4 (left) we compare the total beam induced voltage. the bunch wake field without dielectric coating and the geometric wake field alone, all obtained by numerical integration of Eq. (9). For the latter case, we also present the result of a MAFIA calculation, which is in reasonable agreement, and thus confirms the approximation of Eq. (3). As can be seen, the dielectric and roughness components contribute a little more than half the total.

Figure 4 (right) shows the beam-induced voltage $V_L$ at a distance $\sigma_z$ behind the bunch center, due to the geometric wake field only, vs. the ratio $a/\lambda$. In calculating $V_L$ we have used the fit result of Eq. (3). This figure demonstrates that opening the iris radius from $0.1\lambda$ to $0.17\lambda$ would decrease $V_L$ by about a factor of 3.

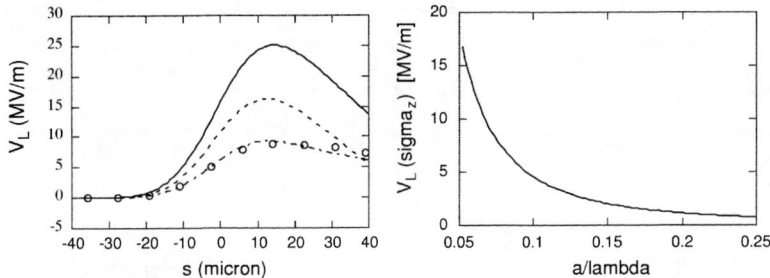

**FIGURE 4.** Left: beam-induced voltage $V_L$ vs. distance from the bunch center, for a Gaussian bunch with 60 pC charge and 10 $\mu$m rms length, and $a/\lambda = 0.1$. The total voltage arising from geometric, dielectric and roughness wake fields (solid) is compared with the induced voltage without dielectric (dashed) and that due to the geometric wake field only, using the approximation of Eq. (3) (dot-dashed) and calculated by MAFIA (the circles). Right: beam-induced voltage $V_L(\sigma_z)$ due to the geometric wake field only as a function of $a/\lambda$, according to Eq. (3), again for a Gaussian bunch with 60 pC charge and 10 $\mu$m rms length.

# TRANSVERSE WAKE FIELD

## Resistive Wall

The transverse effect of the wall resistivity is described by the wake function[3] [6]

$$W_T(s) = \frac{Z_0 c}{4\pi^2 b_p^3} \sqrt{\frac{c}{\sigma}} \frac{1}{\sqrt{s}}, \tag{11}$$

where as for the longitudinal case $\sigma$ denotes the conductivity in cgs units (for copper at room temperature $\sigma = 5.8 \times 10^{17}$ s$^{-1}$) and $b_p$ is the radius of the beam pipe. This formula applies for distances $s$ larger than $(c/(4\pi\sigma b_p))^{1/3} b_p$, which amounts to 1.6 $\mu$m for the iris, and 3.8 $\mu$m for the cavity walls. The resistive-wall transverse wake fields due to iris and wall are illustrated in Fig. 5.

## Geometric Wake

The slope of the transverse geometric wake field at the origin is determined by the iris radius [9]:

$$W_T'(0) = \frac{2Z_0 c}{\pi a^4} \tag{12}$$

and evaluates to 6 TV/m$^3$/pC, or, in Gaussian units, to $7 \times 10^6$ cm$^{-4}$. At higher frequencies, the impedance can be derived from a diffraction model and decays as

---

[3] We here adopt a sign convention opposite to that in Ref. [6].

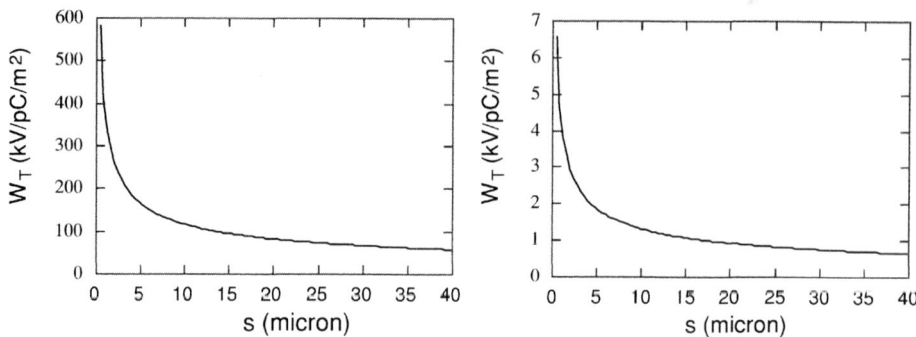

**FIGURE 5.** Transverse resistive wall wake field vs. distance; left: contribution from iris; right: contribution from cavity wall.

$\omega^{-3/2}$ [10]. However, no general formula exists for the low-frequency part of the impedance, and, to compute the wake field, we resort to a numerical calculation using MAFIA [11]. As the bunch passes through a periodic array of cells, the wake field reaches a steady state after about [12] $n_{crit} \approx a^2/(4\sqrt{3}g\sigma_z)$ number of cells. In our case, $n_{crit} \approx 1$. Using MAFIA we calculated the wake fields for an array of 4 and for 3 cells, and took the difference to obtain an estimate of the steady state wake field per cell.

The result of this calculation for a 10 $\mu$m rms bunch length and $a/\lambda = 0.1$ is shown in Fig. 6 (left), where it is also compared with the bunch wake field

$$V_T(s) = \frac{Ne}{\sqrt{2\pi}\sigma_z} \int_{-\infty}^{s} (s - s') \, W_T'(0) e^{-\frac{s'^2}{2\sigma_z^2}} \, ds', \tag{13}$$

expected for a purely linear wake with slope given by Eq. (12). Figure 6 (right) shows the bunch wake field $V_T(s)$, for three different values of $a/\lambda$, as calculated by MAFIA. From these curves, we can deduce the effective slope $W_T'$. For $a/\lambda = 0.1$ this slope is about 2.5 TV/m$^3$/pC, or, in Gaussian units, $3 \times 10^6$ cm$^{-4}$, and thus a factor 2–3 smaller than the point-bunch slope. For $a/\lambda = 0.15$ the effective slope is about 0.7 TV/m$^3$/pC, or $8 \times 10^5$ cm$^{-4}$, and for $a/\lambda = 0.2$ it is 0.25 TV/m$^3$/pC, or $3 \times 10^5$ cm$^{-4}$.

## Coating and Surface Roughness

The longitudinal impedance corresponding to the wake function of Eq. (7) is

$$Z_L(k) = \int_0^\infty ds \, W_L(s)e^{iks} = \frac{Z_0 c}{\pi 2b^2} \left[\delta(k - k_0) + \delta(k + k_0)\right] \tag{14}$$

The transverse impedance of a small perturbation is related to the longitudinal impedance via [6] $Z_T(k) = 2Z_L(k)/(kb^2)$, so that

276

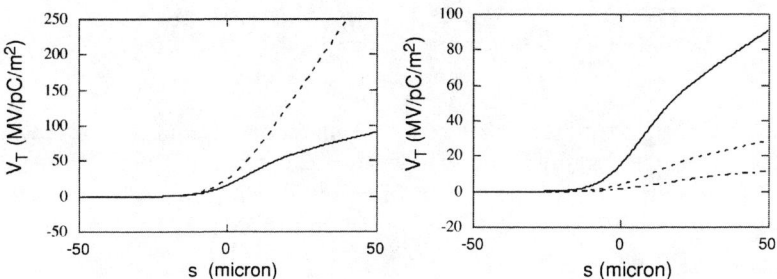

**FIGURE 6.** Transverse geometric wake field $V_T(s)$ for a 10 $\mu$m rms bunch length vs. distance $s$. Left: wake field for $a/\lambda = 0.1$, according to a MAFIA calculation (solid), and for a simplified linear wake (dashed) with slope as in Eq. (12). Right: geometric wake field calculated by MAFIA for a 10 $\mu$m rms bunch length, considering $a/\lambda = 0.1$ (solid), 0.15 (dashed) and 0.2 (dot-dashed).

$$Z_T(k) = \frac{cZ_0}{kb^4} \left[ \frac{\delta(k - k_0) + \delta(k + k_0)}{k} \right] \tag{15}$$

We obtain the transverse wake field by Fourier transform

$$W_T(s) = \frac{i}{2\pi} \int_{-\infty}^{\infty} dk \; Z_T(k)e^{-iks} = \frac{Z_0 c}{\pi b^4} \left[ \frac{\sin k_0 s}{k_0} \right] \tag{16}$$

Its slope at the origin is $W_T'(0) = \frac{2Z_0 c}{\pi b^4}$, independent of $k_0$, and this evaluates to 40 GV/m$^3$/pC or, in Gaussian units, to $4 \times 10^4$ cm$^{-4}$.

## Beam Break Up

Single bunch charge in the W-band linac is constrained by beam break up. The motion of a beam-slice centroid takes the form $x(s,z) = \text{Re}(\chi(s,z)B(s))$, where $B = (\beta/\gamma)^{1/2}e^{j\psi}$ describes the machine lattice in terms of the beta function $\beta$, Lorentz factor $\gamma$ and betatron phase $\psi$. The coordinate $s$ is the position along the linac, and $z$ is the longitudinal distance from the bunch head. The asymptotic solution for a unit initial offset in the presence of a linear (in $z$) wake field is [13,14]

$$\chi \approx \frac{3^{1/4}}{2^{3/2}\pi^{1/2}} \frac{1}{A^{1/2}} \exp\left\{ A\left(1 - \frac{j}{3^{1/2}}\right) + j\frac{\pi}{12} \right\} \tag{17}$$

where, for a lattice with $\beta \propto \sqrt{\gamma}$,

$$A = \frac{3^{3/2}}{2^{5/3}} \left(\frac{\beta_0}{GL^2}\right)^{1/3} \left(\sqrt{\frac{\gamma}{\gamma_0}} - 1\right)^{1/3}. \tag{18}$$

The characteristic length $L$ is related to the strength of the wake field,

$$L = \left( \frac{e^2 N_b W_T'(0) l_b}{mc^2} \right)^{-1/2}, \qquad (19)$$

where $l_b$ is the (flat-top) bunch length, taken to be 30 $\mu$m, and $r_e$ the classical electron radius. The largest transverse wake is the geometric one. The left picture of Fig. 7 shows the variation of $L$ as a function of $a/\lambda$, inferred from the effective slope $W_T'(0)$ provided by MAFIA. The characteristic length $L$ is 1 cm for $a/\lambda = 0.1$, and about 3.5 cm for $a/\lambda = 0.2$.

Figure 7 (right) compares the analytical solutions of the oscillation growth for $L$ equal to 1 and 3 cm with the result of a macroparticle simulation, where we have assumed an initial beta function $\beta_0 \approx 1.6$ m at 10 GeV, increasing along the linac as $\gamma^{1/2}$, an accelerating gradient of $G \approx 1$ GV/m, and 60 pC bunch charge. With $L \approx 1$ cm, an initial offset gets amplified by more than a factor of 10. For $L > 3$ cm, there is negligible growth in the linac, and this value of $L$ would be obtained for $a/\lambda \geq 0.18$.

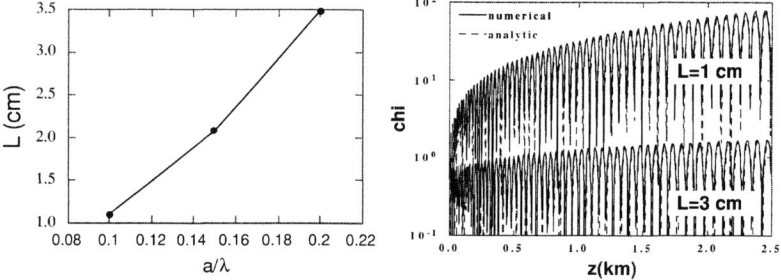

**FIGURE 7.** Left: the characteristic length $L$, inferred from MAFIA calculations, for an rms bunch length of 10 $\mu$m and a 60 pC charge as a function of $a/\lambda$. Right: simulated and analytical beam break up for two different values of $L$, without BNS damping.

# CONCLUSION

We have estimated the longitudinal and transverse wake fields in a W-band (91 GHz) accelerating structure. The transverse wake field is almost completely determined by the structure geometry (iris radius). For a 60 pC charge, and $a/\lambda \geq 0.18$, the transverse beam break up is negligible. In the longitudinal plane, the effect of a dielectric coating and of surface roughness could become as significant as the geometric wake field. The single-bunch beam loading due to the geometric wake field is much less than 1% of the accelerating gradient. The resistive-wall wake fields are insignificant in all cases.

# REFERENCES

1. D.H. Whittum and S.G. Tantawi, "Active Millimeter Wave Accelerator with Parallel Beams", *SLAC-PUB*-**7845** (1998).
2. D. Whittum, these proceedings.
3. R. Gluckstern, *Phys. Rev. D* **39**, p. 960 (1989).
4. K.L.F. Bane, A. Mosnier, A. Novokhatskii, K. Yokoya, "Calculations of the Short-Range Longitudinal Wake Fields in the NLC linac", presented at EPAC98, Stockholm, and *SLAC-PUB*-**7862** (1998).
5. K.L.F. Bane and M. Sands, "The Short-Range Resistive Wall Wake Fields", in *Micro Bunches Workshop*, Upton, *AIP Conference Proceedings* **367** (1995).
6. A.W. Chao, "Physics of Collective Beam Instabilities in High Energy Accelerators", John Wiley & Sons, Inc., New York (1993).
7. E. Lin, private communication (1998).
8. A. Novokhatskii and A. Mosnier, "Wake Fields of Short Bunches in the Canal Covered with Thin Dieletric Layer".
9. K. Bane, in chapter 8 of the "Zeroth Order Design Report for the Next Linear Collider", *SLAC Report* **474**, p. 368 (1996).
10. S.A. Heifets and S.A. Kheifets, "Coupling Impedance in Modern Accelerators", *Rev. Mod. Physics*, **63**, 631 (1991).
11. "MAFIA RELEASE 3", *DESY* **M-90-05K** (1990).
12. R. Palmer, "Radiation from a Ring Charge Passing Through a Resonator", *Particle Accelerators*, **25**, p. 107 (1990).
13. F. Zimmermann, D.H. Whittum and M.E. Hill, "New Concepts for a Compact 5 TeV Collider", presented at the *6th European Particle Accelerator Conference (EPAC 98)*, Stockholm, and *SLAC-PUB*-**7856** (1998).
14. D.H. Whittum, M.E. Hill and F. Zimmermann, "Transverse Beam Dynamics in a mm-Wave Planar Accelerator", unpublished.
15. F. Zimmermann and D.H. Whittum, "Final-Focus System and Collision Schemes for a 5-TeV W-Band Linear Collider", *Proc. of 2nd Int. Workshop on $e^-e^-$ Interactions at TeV Energies*, Santa Cruz 1997 and *SLAC-PUB*-**7741**(1998).

# A Compact High Brightness Laser Synchrotron Light Source for Medical Applications

Kazuhisa Nakajima

*High Energy Accelerator Organization, KEK.*
*1-1 Oho, Tsukuba, Ibaraki 305-0801, Japan*

**Abstract.** The present high-brightness hard X-ray sources have been developed as third generation synchrotron light sources based on large high energy electron storage rings and magnetic undulators. Recently availability of compact terawatt lasers arouses a great interest in the use of lasers as undulators. The laser undulator concept makes it possible to construct an attractive compact synchrotron radiation source which has been proposed as a laser synchrotron light source. This paper proposes a compact laser synchrotron light source for mediacal applications, such as an intravenous coronary angiography and microbeam therapy.

## INTRODUCTION

The present high-brightness hard X-ray sources have been developed as third generation synchrotron light sources based on large high energy electron storage rings and magnetic undulators. A compact, tunable, narrow bandwidth, high-brightness, hard X-ray source has basic and industrial applications in a number of fields, such as solid-state physics, material, chemical, biological and medical sciences. Particularly for medical applications, such as X-ray diagnostics and therapy, it is essential to downscale the present synchrotron light sources into the environment of a hospital, keeping properties of X-ray radiations. Recently availability of compact terawatt lasers arouses a great interest in the use of lasers as undulators of which a period is $\sim 10^4$ shorter than the conventional undulator. This feature of laser undulators allows the use of $\sim 100$ times less energetic electrons to generate X-rays of a particular wavelength. The laser undulator concept makes it possible to construct an attractive compact synchrotron radiation source which has been proposed as a laser synchrotron light source [1]

It is known that intense monochromatic X-rays delivered by the synchrotron light sources take the advantage of X-ray imaging and medical diagnoses over conventional X-ray tubes. In particular the intravenous coronary angiography using synchrotron radiation is of great interest in medical applications as a powerful and

CP472, *Advanced Accelerator Concepts: Eighth Workshop,*
edited by W. Lawson, C. Bellamy, and D. Brosius

safer diagnostic method for assesment of the coronary heart diseases. The method is known as iodine K-edge digital subtraction angiography. In this method the different absorption of quasi-monochromatic X-rays just above and below the K-edge of iodine at 33 keV photon energy is used to eliminate the contrast due to nonvascular body structures. It is necessary to provide sufficient radiation intensity at 33 keV for the short exposure times required to avoid blurring of images by periodic movement of the heart. The development of the coronary angiography has been undertaken on the synchrotron light sources based on large high energy electron storage rings [2]. Recently compact electron storage rings dedicated and optimized to the coronary angiography have been proposed to fit it into a clinical environment in a hospital within a diameter of 10 to 15 m [3] [4]. These compact storage rings are based on the synchrotron radiation from magnetic multipole wigglers made from superconducting magnets with a high magnetic field of 5 T to 8 T. The beam energy more than 1.5 GeV is needed. Because of a broad spectrum of the synchrotron radiation, the second and the third harmonic of 33 keV lead to an undesirable background on the angiography.

The features of laser synchrotron radiation can be applied to the coronary angiography with the much lower electron beam energy which leads to a smaller facility. This paper proposes a compact laser synchrotron light source capable of generating a high-brightness hard X-ray beam for medical applications. Two schemes, the single-pass and the intracavity schemes, are described to generate a sufficient photon flux required for the coronary angiography.

# X-RAY GENERATION VIA THOMSON SCATTERING

When a laser beam interacts with an electron beam at an angle $\phi$, Thomson scatterings of relativistic electrons in the laser undulator field generate frequency up-shifted radiation with the peak frequency given by

$$\omega_X = \frac{2\gamma^2(1 - \cos\phi)}{1 + a_0^2/2}\omega_0. \tag{1}$$

where $\gamma$ is the Lorentz factor of the electrons, $\omega_0$ the incident laser frequency and $a_0$ the normalized vector potential of the laser field, which corresponds to the undulator strength parameter $K$, given by

$$a_0 = 0.85 \times 10^{-9}I^{1/2}[\text{W/cm}^2]\lambda_0[\mu\text{m}] \tag{2}$$

for the peak intensity $I$ in units of W/cm$^2$, the laser wavelength $\lambda_0 = 2\pi c/\omega_0$ in units of $\mu$m. In Thomson scatterings the radiation frequency is correlated to the scattering angle $\theta \ll 1$ as $\omega = \omega_X/(1 + \gamma^2\theta^2)$. The spectral width within the cone of half angle $\theta$ is $\delta\omega/\omega_X \simeq \gamma^2\theta^2$. The maximum radiation photon energy is written as

$$E_X[\text{keV}] = \hbar\omega_X = \frac{0.0095(1 - \cos\phi)E_b^2[\text{MeV}]}{\lambda_0[\mu\text{m}](1 + a_0^2/2)}. \tag{3}$$

where $E_b$ is the electron beam energy. The radiation wavelength is $\lambda_X[\text{Å}] = 12.4/E_X[\text{keV}]$. For a conventinal synchrotron radiation from a undulator magnet, the radiation photon energy is given by $E_X[\text{keV}] = 0.95E_b^2[\text{GeV}]/\lambda_u[\text{cm}]$, where $\lambda_u$ is the undulator magnet wavelength and $K^2 \ll 1$ is assumed. As an example, in order to produce 33 keV radiation, the electron beam energy of $E_b = 12$ GeV is needed using the undulator magnet with a period of $\lambda_u = 4$ cm, while the same photon energy is generated via the backward Thomson scattering from the electron beam of $E_b = 12$ MeV in the laser field with $\lambda_0 = 1\mu\text{m}$.

The synchrotron radiation power radiated by a single electron interacting with a laser beam is $P_s = (64r_e^2\gamma^2/3r_0^2)P_0$, where $r_e = e^2/m_ec^2$ is the classical electron radius, $r_0$ the laser spot radius of the Gaussian laser profile, and $P_0$ the incident laser power. The energy loss of the electron beam due to the radiation is obtained by $U_L = P_sL/(ec)$, where $L = \min[2Z_R, L_0/2]$ is the laser-electron interaction length, $Z_R = \pi r_0^2/\lambda_0$ the Rayleigh length, and $L_0$ the laser pulse length. The beam energy loss is given by

$$U_L[\text{keV}] = 4.2 \times 10^{-5}E_b^2[\text{MeV}]P_0[\text{GW}]\lambda_0^{-1}[\mu\text{m}](L/Z_R). \tag{4}$$

The total radiation power radiated by an electron beam with current $I_b$ is given by $P_T = U_LI_b$, i.e.

$$P_T[\text{W}] = 4.2 \times 10^{-2}E_b^2[\text{MeV}]P_0[\text{GW}]I_b[\text{A}]\lambda_0^{-1}[\mu\text{m}](L/Z_R). \tag{5}$$

In the backscattering configuration, i.e. $\phi = 180°$, the intensity distribution of the fundamental $n = 1$ radiation in the forward direction $\theta = 0$ is

$$\left[\frac{d^2I(\omega)}{d\omega d\Omega}\right]_{\theta=0} = \frac{e^2\omega^2}{8\pi c^2}\lambda_0N_0a_0^2G_1(\omega). \tag{6}$$

$$G_1(\omega) = \frac{N_0}{\omega_X}\left[\frac{\sin[\pi(\omega - \omega_X)N_0/\omega_X]}{\pi(\omega - \omega_X)N_0/\omega_X}\right]^2. \tag{7}$$

where $\Omega$ is the solid angle, $N_0 = L_0/\lambda_0$ is the number of laser wavelengths within the laser pulse length $L_0$. The frequency width of the intensity distribution is given by $(\delta\omega/\omega_X)_0 = 1/N_0$, noting that $G_1(\omega) \to \delta(\omega - \omega_X)$ as $N_0 \to \infty$. The spectral flux within the spectral width $\Delta\omega/\omega_X$ is obtained by integrating over the frequency range $\Delta\omega$ and the radiation cone solid angle $\Omega \simeq 2\pi\Delta\omega/\gamma^2\omega_X$ as [1]

$$F \simeq 2\pi\alpha a_0^2(I_b/e)N_0(\Delta\omega/\omega_X). \tag{8}$$

where $\alpha = 1/137$, $I_b$ is the electron beam current, and $I_b/e$ is the electron flux interacting with the laser field assuming the laser spot size is larger than the electron beam size. The spectral flux in terms of practical units is given by

$$F[\text{photons/s}] = 8.4 \times 10^{16}(L/Z_R)I_e[\text{A}]P_0[\text{GW}](\Delta\omega/\omega_X) \qquad (9)$$

$$= 2 \times 10^3 W[\text{J}]N_e f_I/Z_R[\mu\text{m}](\Delta\omega/\omega). \qquad (10)$$

where $N_e$ is the total number of electrons per bunch, $W$ the laser pulse energy, $f_I$ the interaction frequency. In the backscattered radiation, only the odd harmonic intensities are finite along the $\theta = 0$ axis and the even harmonics vanish. An estimate of the ratio of the 3rd harmonic intensity $I_3$ to the fundamental intensity $I_1$ is $I_3/I_1 \sim (81/64)(a_0^4/(1 + a_0^2/2)^2$, assuming $a_0^2 \ll 1$.

The radiation spectrum is broadened by effects of emittance and energy spread of actual electron beams. The total natural spectral width of the radiation will be given by

$$(\delta\omega/\omega_X)_T = [(\delta\omega/\omega_X)_0^2 + (\delta\omega/\omega_X)_\epsilon^2 + (\delta\omega/\omega_X)_i^2]^{1/2}. \qquad (11)$$

where $(\delta\omega/\omega_X)_0 = 1/N_0$ is the spectral width due to the finite interaction length, and $(\delta\omega/\omega_X)_\epsilon = \epsilon_n^2/(\gamma^2 r_b^2)$ is the emittance broadened spectral width for the normalized beam emittance $\epsilon_n$, and $(\delta\omega/\omega_X)_i = 2(\delta E/E_b)$ is the intrinsic energy spread broadening effect. The radiation with the total spectral width is confined to the angle $\theta_T \simeq (\delta\omega/\omega_X)_T^{1/2}/\gamma$. If the specified bandwidth is $\Delta\omega/\omega_X \gg (\delta\omega/\omega_X)_T$, all the radiation within a cone of half-angle $\theta_s \simeq (\Delta\omega/\omega_X)^{1/2}/\gamma$ can be used.

# DESIGNS OF THE LASER SYNCHROTRON LIGHT SOURCE

The intravenous coronary angiography using iodine as a contrast agent requires the photon flux of $2 \times 10^{15}$ photons/sec in the bandwidth of 0.3% at the photon energy of 33 keV [4]. The exposure area of the radiation should be at least 150 mm × 150 mm. In order to produce such sufficient photon flux by means of the laser synchrotron radiation , two schemes are conceivable: the single-pass and the intracavity schemes

## Single-pass generation scheme

In the single-pass generation, the multi-bunch electron beams of 46.2 MeV are delivered by the linac or the microtron with the photoinjector producing the beam charge of 1.6 nC per bunch (i.e. the number of electrons $N_e \simeq 10^{10}$ with the emittance of $\epsilon_n = 1\mu$mrad and the intrinsic energy spread of 0.1%. Fig. 1 shows a schematic configuration of the sigle-pass generation scheme. For a 2 ps laser pulse with $\lambda_0 = 1\mu$m, $N_0 = 600$. As the natural spectral width is 0.2%, there is no need to filter all the radiation within the cone $\theta = 1$ mrad to obtain the bandwidth of 0.3%. The required photon flux for the angiography can be achieved by interaction of the electron beam pulse consisting of 160 bunches at 100 Hz with the laser pulse energy of 2 J within a focal spot size $r_0 = 10\mu$m. In order to generate a uniform radiation distribution at the location of the patient over a width of at least 150 mm, the deflecting magnet with the field of 3.3 T is placed in the interaction region of $2Z_R = 0.63$ mm to create a fan X-ray beam horizontally spread over 15 mrad. This deflecting magnet is used to separate electron beams from X-rays by a bending angle of 30°.

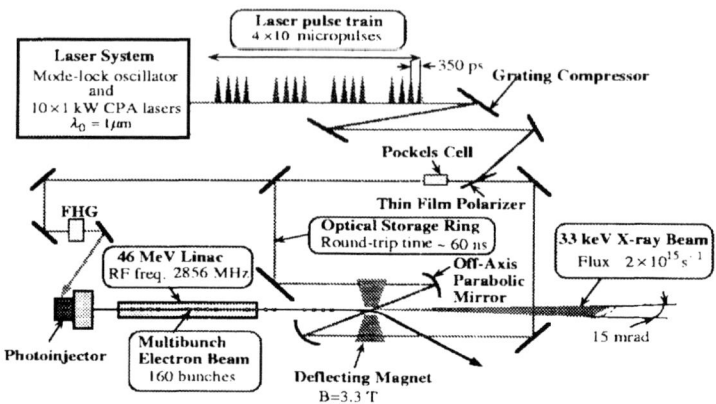

**FIGURE 1.** Schematic diagram of the single-pass laser synchrotron light source using the electron linac.

A laser system suitable for the single-pass X-ray source is a high average power solid-state laser (e.g. Nd:YAG) based on the chirped pulse amplification (CPA). The laser consists of a single mode-lock oscillator and 10 laser amplifiers each of which produces the average power of 1 kW. In each amplifier, four pulses seeded from the mode-lock oscillator through the pulse stretcher are amplified to energy of 2.5 J/pulse at the repetition rate of 100 Hz. The output pulses of 10 amplifiers are combined to form the pulse train consisting of 40 pulses through the pulse combiner. Afterwards the pulse duration is compressed through a grating pulse compressor to obtain 1 ps, 2 J pulses. Finally the pulse train is injected into the optical storage ring to collide the electron bunches during its round-trip. The optical storage ring is composed of the thin-film polarizers, the Pockels cells for polarization control, and the focusing mirror arrangement. In order to achieve synchronization between the laser pulses and the RF linac electron beam operated at 2856 MHz, each separation of the michropulse is 350 ps. 10 macropulses are separated from one another by 4.9 ns and the full laser pulse train circulates in the ring resonator configuration with optical path length of 59.5 ns. When the laser pulse train circulates in the optical storage ring, a small part of the pulse energy is diverted to illuminate the laser photocathode at the 4th harmonic of the $1\mu m$ wavelength. The parameters of the single-pass X-ray generation scheme are summarized in Table 1. In this scheme, the third harmonic contamination of 99 keV is estimated to be $\sim 18\%$ with respect to the 33 keV radiation.

**TABLE 1.** Design parameters of the single-pass laser synchrotron light source using the electron linac.

| X-ray parameters | |
|---|---|
| Photon energy | 33 keV |
| Photon flux | $2 \times 10^{15}$ s$^{-1}$ |
| Natural spectral width | 0.2% |
| Electron beam parameters | |
| Beam energy | 16.2 MeV |
| Charge per bunch | 1.6 nC/bunch |
| Bunch length | 2 ps |
| Number of bunches | 160 |
| Repetition frequency | 100 Hz |
| Normalized emittance | 1 mm mrad |
| Beam radius at I.P. | $10\mu m$ |
| Beam energy spread | 0.1% |
| Laser parameters | |
| Wavelength | $1\mu m$ |
| Pulse energy | 2 J/pulse |
| Pulse duration | 2 ps |
| Peak power | 1 TW |
| Spot radius | $10 \ \mu m$ |

## Intracavity generation scheme

Electron beams in the storage ring interact with the CW laser inside the optical cavity made of high reflectivity mirrors referred to as a supercavity. Fig. 2 shows a schematic configuration of the intracavity generation scheme. The electron storage ring with the circumference of 10m stores the average beam current 1.2 A at 132 MeV to produce a sufficient radiation of 33 KeV X-rays via Thomson scattering on the laser at $\lambda_0 = 10\mu m$ (e.g. $CO_2$ laser). The required photon flux can be obtained by the interaction of the circulating electron beam with the laser field of the average power of 10 MW in the interaction length $2Z_R = 63$ mm for the focal spot radius $r_0 = 100\mu m$. As $N_0 = 6.3 \times 10^4$, the spectral width is determined by the intrinsic beam energy spread in the storage ring. The beam energy loss due to interaction with the laser field is $U_L = 1.5$ eV. The energy loss per turn due to the normal synchrotron radiation is $U_M = 27$ eV in the arc with the bending radius of $\rho_M = 1$ m. The damping time is given by

$$\tau_d = \frac{E_b T_r}{U_M + U_L} = \frac{1}{1/\tau_{dM} + 1/\tau_{dL}}. \tag{12}$$

where $T_r$ is the revolution time of the electron beam in the ring, $\tau_{dM} = E_b T_r / U_M$ and $\tau_{dL} = E_b T_r / U_L$ is the damping time contributed by the energy loss in the laser interaction and in the bending magnets, respectively. Letting $T_r = 33$ ns, these damping times are: $\tau_{dM} = 0.16$ s, $\tau_{dL} = 3.0$ s, and $\tau_d = 0.15$ s.

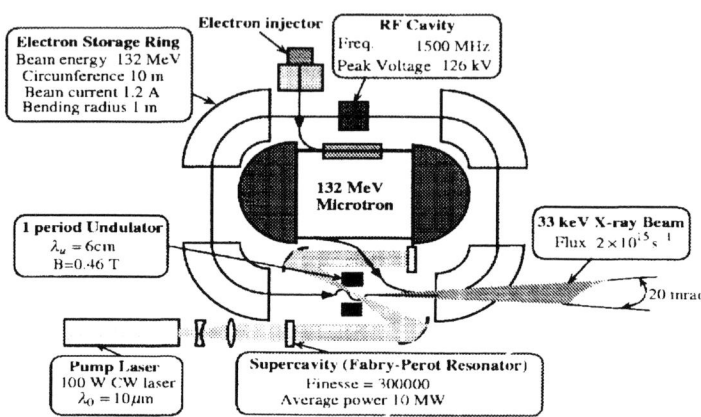

**FIGURE 2.** Schematic diagram of the intracavity laser synchrotron light source using the electron storage ring.

Consider the equilibrium emittance of the electron beam in the storage ring with the laser-electron interaction. The equilibrium normalized emittance resulting from the balance between damping and the quantum excitation via Thomson scatterings is given by [5]

$$\varepsilon_{nL} = \frac{3}{10} \frac{\lambda_c}{\lambda_L} \beta^* = 0.73 \times 10^{-6} \frac{\beta^*}{\lambda_0 [\mu m]}. \tag{13}$$

where $\lambda_c = h/mc \simeq 2.43 \times 10^{-12}$ m is the Compton wavelength of the electron, and $\beta^*$ is the beta function at the interaction point. For $\beta^* = 1$ m and $\lambda_0 = 10 \mu m$, $\varepsilon_{nL} = 7.3 \times 10^{-8}$ m. The equilibrium normalized emittance determined from the synchrotron radiation in the bending magnets is approximately given by [6]

$$\varepsilon_{nM} \approx \frac{J_E}{J_x} \frac{\alpha R}{\nu_x} \frac{\sigma_{EM}^2}{E_b^2} \gamma. \tag{14}$$

where $J_E \approx 2+\alpha R/\rho_M$ and $J_x = 1-\alpha R/\rho_M$ are the energy and horizontal damping partition numbers in the electron storage ring with the momentum compaction $\alpha$, the average ring radius $R$, and the bending radius $\rho_M$, $\nu_x$ is the horizontal betatron tune, and $\sigma_{EM}/E_b$ is the fractional energy spread due to the synchrotron radiation in the bending magnets. The appropriate design parameters become $\varepsilon_{nM} = 4.1 \times 10^{-7}$ m. The average quantum excitation rate to the normalized emittance due to both effects of Thomson scatterings and synchrotron radiation in the storage ring is written as

$$\langle \frac{d\varepsilon_n}{dt} \rangle_{QE} = \langle \frac{d\varepsilon_{nL}}{dt} \rangle_{QE} + \langle \frac{d\varepsilon_{nM}}{dt} \rangle_{QE} = -\frac{\varepsilon_{nL}}{\tau_{dL}} - \frac{\varepsilon_{nM}}{\tau_{dM}} = -\frac{\varepsilon_n}{\tau_d}. \tag{15}$$

Hence, the equilibrium normalized emittance is given by

$$\varepsilon_n = \frac{\tau_d}{\tau_{dL}}\varepsilon_{nL} + \frac{\tau_d}{\tau_{dM}}\varepsilon_{nM}. \tag{16}$$

This becomes $\varepsilon_n = 3.9 \times 10^{-7}$ m.

Consider the energy spread of the electron beam in the storage ring with the laser-electron interaction. The equilibrium energy spread caused by Thomson scatterings is given by [5]

$$\frac{\sigma_{EL}}{E_b} = \left(\frac{7\lambda_c}{5\lambda_0}\gamma\right)^{1/2} = 1.84 \times 10^{-3} \left(\frac{\gamma}{\lambda_0[\mu m]}\right)^{1/2}. \tag{17}$$

For $\lambda_0 = 10\mu m$, the energy spread due to Thomson scatterings is $\sigma_{EL}/E_b = 9.4 \times 10^{-3}$. The energy spread caused by the bending magnet is given by [6]

$$\frac{\sigma_{EM}}{E_b} = 0.44 \times 10^{-6}\frac{\gamma}{\sqrt{\rho_M[m]}}. \tag{18}$$

i.e. $\sigma_{EM}/E_b = 1.1 \times 10^{-4}$ for $\rho_M = 1$ m. The average quantum excitation rate to the energy spread due to both effects of Thomson scatterings and synchrotron radiation in the storage ring is written as

$$\langle \frac{d\sigma_E^2}{dt} \rangle_{QE} = \langle \frac{d\sigma_{EL}^2}{dt} \rangle_{QE} + \langle \frac{d\sigma_{EM}^2}{dt} \rangle_{QE} = -\frac{2}{\tau_{dL}}\sigma_{EL}^2 - \frac{2}{\tau_{dM}}\sigma_{EM}^2 = -\frac{2}{\tau_d}\sigma_E^2. \tag{19}$$

Hence, the energy spread due to both effects is given by

$$\frac{\sigma_E}{E_b} = \left[\frac{\tau_d}{\tau_{dL}}\left(\frac{\sigma_{EL}}{E_b}\right)^2 + \frac{\tau_d}{\tau_{dM}}\left(\frac{\sigma_{EM}}{E_b}\right)^2\right]^{1/2} \tag{20}$$

For the design parameters, the equilibrium energy spread becomes $\sigma_E/E_b = 2.1 \times 10^{-3}$. In order to obtain a sufficient quantum beam lifetime, say 1 hour, the energy aperture for the synchrotron oscillation must be 1.1%. The required peak RF

**TABLE 2.** Design parameters of the intra-cavity laser synchrotron light source using the electron storage ring.

| | |
|---|---:|
| X-ray parameters | |
| Photon energy | 33 keV |
| Photon flux | $2 \times 10^{15}$ s$^{-1}$ |
| Natural spectral width | 0.4% |
| Electron beam parameters | |
| Beam energy | 132 MeV |
| Beam current | 1.2 A |
| Number of bunches | 50 |
| Circumference | 10 m |
| Bending radius | 1 m |
| Energy loss per turn | 28.5 eV |
| Damping time | 155 msec |
| Equil. energy spread | 0.2% |
| Equil. norm. emittance | $3.9 \times 10^{-7}$ m |
| rms Bunch length | 4.9 mm |
| RF frequency | 1500 MHz |
| Peak RF voltage | 126 kV |
| Energy aperture | 1.1% |
| Quantum beam lifetime | 1 hour |
| Beam radius at I.P. | 30 $\mu$m |
| Laser parameters | |
| Wavelength | 10 $\mu$m |
| Average power | 10 MW |
| Spot radius at I.P. | 100 $\mu$m |

voltage becomes 126 keV. For this condition, the bunch length is 4.9 mm. Since the energy aperture is large enough to keep the electron beam in the phase space, effects of the diffusion provided by intrabeam scattering are negligible.

The high finesse Fabry-Perot resonator, a supercavity, can be made of high reflective mirrors with reflectivity of $R = 99.999\%$ [7]. The finesse of this optical resonator is Finesse $= \pi\sqrt{R}/(1 - R) = 3.1 \times 10^5$. In this supercavity the laser power builds up to $P_0 \simeq P_i/(1 - R) = 10$ MW with the incident power $P_i = 100$ W pumped by the CW laser. In order to produce a fan of the X-ray radiation horizontally spread over 20 mrad, a single period of the linear undulator magnet with the field of 0.46 T and the wavelength of $2Z_R \simeq 60$ mm is placed in the interaction region. The design parameters for the intracavity generation scheme are summarized in Table 2. In this scheme, the third harmonic contamination of 99 keV is negligible because of $a_0 = 2.2 \times 10^{-3} \ll 1$.

# CONCLUSIONS

A compact high brightness hard-X ray source can be achieved by Thomson scatterings of relatively low energy electron beams in laser fields. Two laser-electron beam interaction schemes has been proposed for application to the coronary angiography. Tunability, compactness, high enrgy photon capability, and relatively low cost may provide an important useful tool for the other medical applications and the basic sciences.

# REFERENCES

1. E. Esarey, P. Sprangle and A. Ting, Nucl. Instr. Meth. **A 331**, 545 (1993).
2. W. Thomlinson et al., Rev. Sci. Instr. **63**, 625 (1992).
3. H. Wiedemann et al., Nucl. Instr. Meth. **A 347**, 515 (1994).
4. M. Torikoshi et al., Proc. of of the 1997 IEEE PAC, to be published.
5. Z. Huang and R. D. Ruth, Phys. Rev. Lett. . **80**, 976 (1998).
6. M. Sands, SLAC-121, UC-28, 1 (1970).
7. J. Chen et al., Nucl. Instr. Meth. **A 358**, 14 (1995).

# Quantum Suppression of Beamstrahlung for Future $e^+e^-$ Linear Colliders: an Evaluation of QED Backgrounds

Ming Xie

*Lawrence Berkeley National Laboratory, Berkeley, CA94720, USA*

**Abstract.** Beamstrahlung at interaction point may present severe limitations on linear collider performance. The approach to reduce this effect adopted for all current designs at 0.5 TeV range in center-of-mass energy will become more difficult and less effective at higher energy. We discuss the feasibility of an alternative approach, based on an effect known as quantum suppression of beamstrahlung, for future linear colliders at multi-TeV energy.

## Introduction

One of the most important constraints on the performance of an $e^+e^-$ linear collider is that imposed by the QED processes [1], in particular beamstrahlung [2-8], at the Interaction Point (IP). Beamstrahlung is the synchrotron radiation produced by particles of one beam as they pass through the electric and magnetic fields of the oncoming beam. The fields can be so strong due to extremely high charge density that colliding particles may lose significant amount of their energy, causing severe luminosity degradation. The photons generated by beamstrahlung may also turn to copious $e^+e^-$ pairs, or even hadrons through QCD processes, causing troublesome background problem to the detector and hence the particle physics under study. Therefore, a crucial task to assess the potential of future linear colliders is to identify the operation regimes and approaches for beamstrahlung suppression with which the impact of these deleterious effects on collider performance can be minimized, taking into account other collider constraints, of course.

To suppress beamstrahlung, the so-called flat-beam approach has been adopted for all current designs of linear collider at 0.5 TeV c.m. energy range [9]. However, this approach will become more difficult technically and less effective at higher energy, as will be explained later. In addition, several other methods have been proposed. Charge compensation method [10,11] requires the mixing of beams of opposite charge to neutralize the beam fields before collision. But, due to a beam

CP472, *Advanced Accelerator Concepts: Eighth Workshop,*
edited by W. Lawson, C. Bellamy, and D. Brosius
1999 The American Institute of Physics 1-56396-889-4

instability, imperfection in mixing could cause luminosity degradation. The beam fields may also be reduced by the return current in a plasma [12] introduced at the IP. The problem of concern in this case is the hadronic backgrounds due to collisions of beams with dense plasma ions. Instead of colliding charged particle beams, one may also convert them into photon beams to make a $\gamma\gamma$ collider [13]. However, it seems unlikely that a $\gamma\gamma$ collider could scale more favorably to higher energy than its $e^+e^-$ counterpart due to other types of technical constraints. Apart from that, $\gamma\gamma$ collision is not meant to be a substitute for $e^+e^-$ annihilation in terms of physics discovery potential [14]. Regardless of the variety, the working philosophy behind all these methods is the same that is to reduce or eliminate if possible the strong beam fields. Nevertheless, there is an exception.

In this paper, we discuss an effect known as Quantum Suppression of Beamstrahlung (QSB). Unlike all other approaches, QSB is effective only when the beam field is sufficiently strong. In that regard, it is compatible with the ever-increasing beam density required of a linear collider at higher energy, thus deserves a careful investigation. A brief description of beamstrahlung and the rational behind the interest in QSB are given in Section II. Monte-Carlo IP simulation for a 5 TeV collider case is presented in Section III to illustrate the main characteristics of QSB and in particular to evaluate the collision induced QED backgrounds. This then leads to a discussion in Section IV of major issue and uncertainty involved in establishing QSB as a feasible IP approach for linear colliders at higher energy.

## Beamstrahlung

Beamstrahlung can be classified into three regimes [1] according to the magnitude of the beamstrahlung parameter, $\Upsilon = \gamma B/B_c$, where $\gamma = E_b/mc^2$, $E_b$ is the beam energy, $B$ the beam field and $B_c$ the Schwinger critical field. The three regimes are, respectively, the classical regime if $\Upsilon \ll 1$, the extreme or strong or deep quantum regime if $\Upsilon \gg 1$, and in between the transition regime. In the classical regime, beamstrahlung can be calculated with the usual synchrotron radiation formula derived from classical electrodynamics. Alternatively, the beamstrahlung parameter in this regime may be expressed as $\Upsilon = 2\epsilon_c/3E_b$ in terms of a classical quantity known as critical photon energy, $\epsilon_c$. The classical theory is valid only if the energy of the radiated photon, characterized by $\epsilon_c$, is much less than the kinetic energy of the radiating particle. This condition corresponds to $\Upsilon \ll 1$.

So far all the designs of linear colliders at 0.5 TeV range have managed to stay in the regime with $\Upsilon < 1$ [9], where beamstrahlung and its deleterious effects can be reduced by having smaller $\Upsilon$. Therefore reducing $\Upsilon$ by reducing the beam field, has been adopted as a guideline and that is made possible by taking the flat-beam approach. However, as we know the required luminosity for a collider has to rise as the square of its energy, thus to keep wall plug power under control the beams have to be focused to smaller size with higher charge density. This will unavoidably raise $\Upsilon$ and put a linear collider into the deep quantum regime. As a result, the flat-beam approach will become more difficult and less effective at higher energy.

More difficult for technical reason, as the flat-beam approach requires beam size in one transverse direction to be much smaller (for given beam area), thus pushing the limit for tight beam positioning control at higher energy. For current designs at 0.5 TeV, vertical beam size is already down to a few nanometers. Less effective for physical reason, as it has been shown [3] the dependence of spread in luminosity spectrum on beam shape is very week in the deep quantum regime.

Question then arises: what would be the approach to suppress beamstrahlung at higher energy if a linear collider will be unavoidably pushed into the deep quantum regime? Fortunately, the very nature itself offers help. As $\Upsilon$ increases due to either stronger fields or higher beam energy, the radiated photons become more energetic. Quantum theory has to be used to take into account radiation recoil and the fact that photon spectrum beyond the particle energy is kinetically forbidden. A full quantum treatment of synchrotron radiation was given by Sokolov et al. [15] for arbitrary value of $\Upsilon$ in a constant field. This result was later applied to and extended for the study of beamstrahlung [2–8].

According to the quantum theory, beamstrahlung scales differently in the regimes $\Upsilon \ll 1$ and $\Upsilon \gg 1$. It was shown [2] that advantage may be taken of this behavior in the deep quantum regime to extend collider energy to multi-TeV without excessive beamstrahlung. It was also made clear that the beam parameters required to take advantage of this effect, such as very short bunch or small emittance, are not readily achievable, and the flat-beam approach is a much better choice at 0.5 TeV energy range.

However, one should not forget that 0.5 TeV is only a near term goal for linear collider development, very much limited by the current technologies. Considering competitions from hadron or even muon colliders, it would be much more compelling for linear collider to go beyond that energy. Recently, high energy physics community has been emphasizing the importance of higher energy reach (up to 5 TeV) for a linear collider [16]. There is also a need to explore drastically different collider parameter regime that might potentially be reached with the advanced acceleration techniques currently under active investigation [17]. It is now becoming increasingly important to search for more feasible IP approaches at higher energy.

In particular, possibility of employing quantum suppression as an IP approach was explored over a wide range of beam parameters at 5 TeV by Xie et al [18]. It was shown in this study that when major accelerator and IP constraints are taken into account, it becomes increasingly necessary to operate linear colliders in high $\Upsilon$ regime and use to our advantage the quantum effect to suppress beamstrahlung. Monte-Carlo simulation was performed to study luminosity spectrum. The results were surprisingly encouraging. To carry this study a step further, in this paper we present a detailed evaluation of QED backgrounds for a representative case of collider parameters in high $\Upsilon$ regime. The analysis of hadronic backgrounds due to QCD processes will be presented in a companion paper [19] in these proceedings.

What is quantum suppression and how could it be realized in a linear collider? Before going to the simulation next let's address this question by giving at least one scenario where QSB could manifest itself in a linear collider. Consider a case when all beam parameters are fixed except bunch length. As bunch length decreases, beam density hence the beam field and $\Upsilon$ increase, radiative energy loss per unit time will also increase either in the deep quantum regime or classical regime. However when multiplied by bunch length, radiative energy loss per bunch crossing decreases in the deep quantum regime while still increases in the classical regime. This effect thus may be called the quantum suppression of beamstrahlung. The QSB so defined calls for short bunch length. This is again compatible with the trend of reducing the wavelength of acceleration field from current microwave accelerators to future laser-driven accelerators.

## Simulation in High $\Upsilon$ Regime

We now present full-blown IP simulations using CAIN developed by Yokoya and co-workers [20]. CAIN is capable of handling all major electromagnetic and QED processes occurred at the IP, including disruption, beamstrahlung, bremsstrahlung, coherent and incoherent pair creation. It is a Monte-Carlo code which follows beam particles, photons and pairs in six-dimensional phase space, as well as their spins and polarization. In comparison, previous studies of beamstrahlung in the deep quantum regime [2–8] were concentrated mainly on obtaining analytical and semi-analytical results to understand the physics, thus were limited to treating only simple, idealistic models. In these early studies, either disruption or multiple beamstrahlung or both were neglected, and none was able to treat simultaneously pair production and give angular-momentum distributions. However this information is essential to background analysis and overall assessment of collider performance, especially in high $\Upsilon$ regime. Beam parameters used in simulation are given in Table 1, which are taken from the CASE II by Xie et al [18], along with their definitions.

Figure 1 shows the luminosity spectrum for $e^+e^-$ and $\gamma\gamma$ collisions. For $e^+e^-$ case the spectrum is characterized by an outstanding core at full energy and a very broad tail two orders of magnitude below the peak. Seen from Table 3, the core itself within 1% of full energy accounts for 65% of the geometrical luminosity, even though on average the beam loses 26% of its energy and has a rms energy spread of 36%. The sharpness and high peak value of the core is surprisingly encouraging. Upon careful examination, it is found that nearly half of the primary particles went through beam crossing without having enough probability to suffer energy loss through any QED process, even though their trajectories are bent significantly by the beam field. Because of quantum suppression, number of beamstrahlung photons defined in terms of $n_\gamma$ is even lower than most of the designs at 0.5 TeV [9]. Although cross sections for background events are generally higher at lower energy, this effect is significantly suppressed. The products from most collisions in the low energy region are highly boosted due to the asymmetry in energy of the collision partners, thus are confined mostly within small angular cones along the beam pipe.

Angle spectrum and angle-energy distribution of the photons are given in Figure 2. In the right plot we see features of two distinct distributions. The photons

293

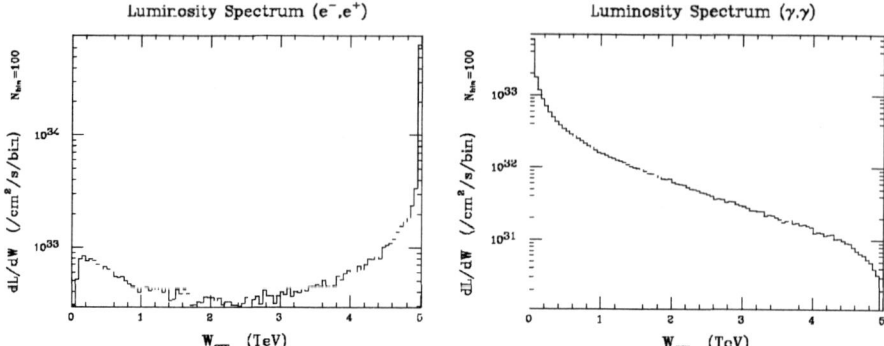

**FIGURE 1.** Luminosity spectrum for $e^+e^-$ (left) and $\gamma\gamma$ (right) with 100 bins.

generated by primary particles at full energy occupy the band below 0.2 mrad, roughly. This number corresponds to the characteristic disruption angle of primary particles given by $\theta_d = D_y\sigma_y/\sigma_z$. The photons with angle larger than 0.2 mrad are generated either through secondary beamstrahlung or by pair particles to be discussed later. The angle-energy correlation, shown more remarkably above the lower band, is due to the fact that the lower the energy of the radiating particle, the larger the angle it is deflected by the beam field, and the larger the angle of the radiated photon.

Another major source of backgrounds at high $\Upsilon$ is the copious coherent $e^+e^-$ pairs created by beamstrahlung photons traveling in the strong field of the opposing beam

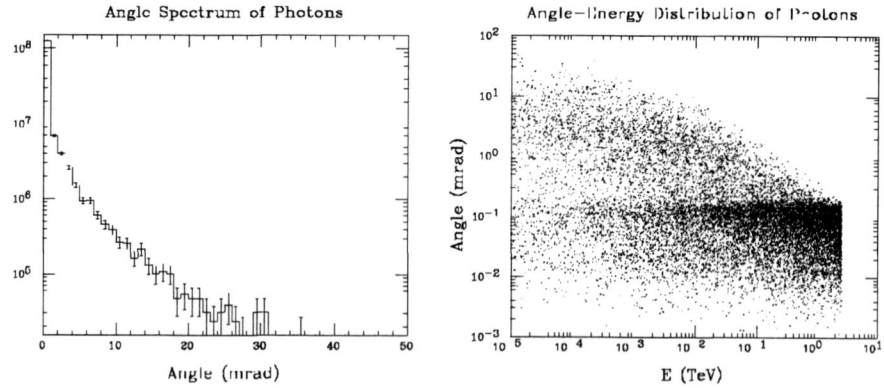

**FIGURE 2.** Angle spectrum of photons with bin size of 1 mrad (left). Scatter plot of photons in angle-energy space (right).

294

**TABLE 1.** Beam parameters used for simulation.

| $P_b$(MW) | $N(10^8)$ | $f_c$(kHz) | $\varepsilon_y$(nm) | $\beta_y(\mu m)$ | $\sigma_y$(nm) | $\sigma_z(\mu m)$ |
|:---:|:---:|:---:|:---:|:---:|:---:|:---:|
| 20 | 1.6 | 156 | 25 | 62 | 0.56 | 1 |

**TABLE 2.** Results given by the formulas.

| $\Upsilon$ | $D_y$ | $n_\gamma$ | $\delta_E$ | $n_b$ | $n_v$ | $\mathcal{L}_g(10^{35}\mathrm{cm}^{-2}\mathrm{s}^{-1})$ |
|:---:|:---:|:---:|:---:|:---:|:---:|:---:|
| 631 | 0.29 | 0.72 | 0.2 | 0.094 | 0.026 | 1 |

[1,21]. In high $\Upsilon$ regime, coherent pair partners are more likely to share the photon energy asymmetrically, giving rise to particles with significantly lower energy. These low energy pair particles, if deflected to large enough angle, may enter the detector and cause background or even damage problems. The pair partner with the same sign as the co-moving beam sees a focusing field, while the opposite sign pair partner sees a defocusing field of the opposing beam. As a result, the angle characteristics can be quite different for different pair partners. Angle-energy distributions of coherent pairs together with beam particles are shown in Figure 3 (left). The beam particles are concentrated mostly in the area near full energy. Notice the split of two bands in the lower energy region. The band with larger angle corresponds to the opposite sign pair partners. The band with smaller angle corresponds to the same sign pair partners and beam particles. Deflection angle for opposite sign pair particle is up to 100 mrad for energy as low as several hundred MeV.

Because of the angle-energy correlation, the number of detector hits by the charged particles may be further reduced with a solenoid magnetic field along the beam pass. In this situation, rather than particle energy, a more relevant variable is transverse momentum, $P_t$, which determines radius of the helical orbit in given solenoid field. Angle-$P_t$ distributions of coherent pairs and beam particles are shown in Figure 3 (right). Here again the band with larger angle corresponds to the opposite sign pair partners. On this plot, only those particles in the top right corner with large enough angle and $P_t$ will fall outside of a given forward cone and have a chance of hitting the detector directly. The detector planned for NLC

**TABLE 3.** Results given by CAIN simulations.

| $n_\gamma$ | $\delta_E$ | $\sigma_e/E_b$ | $n_b$ | $\mathcal{L}/\mathcal{L}_g(\mathrm{W}_{\mathrm{cm}} \in 1\%)$ | $\mathcal{L}/\mathcal{L}_g(\mathrm{W}_{\mathrm{cm}} \in 10\%)$ |
|:---:|:---:|:---:|:---:|:---:|:---:|
| 0.97 | 0.26 | 0.36 | 0.12 | 0.65 | 0.80 |

has a half angle of 100 mrad [22], seemingly large enough to swallow all coherent pairs and photons in our case. More definitive assessment on these backgrounds requires enhancement in simulation statistics especially for the particles distributed in the low energy, large angle tails. It is also necessary to know more details about masking scheme and design of the Interaction Region (IR).

Coherent pairs can also be produced from virtual photons (as opposed to real photons from beamstrahlung) through a process known as trident cascade. Current version of CAIN does not include this process, but its production rate can be estimated with a simple formula [1,21]. Seen from Table 2 and Table 3 for our case the number of pairs per primary particle due to virtual photon process, $n_v$, is somewhat lower than the real photon pair production, $n_b$. Recently Thompson and Chen [23] have checked this process in more detail for our parameter set and found it does not seem to cause extra problem.

In addition to coherent pairs produced in collective beam field, incoherent pairs can also be created through individual particle-particle scattering processes. The following processes are included in the simulation: Breit-Wheeler: ($\gamma + \gamma \longrightarrow$

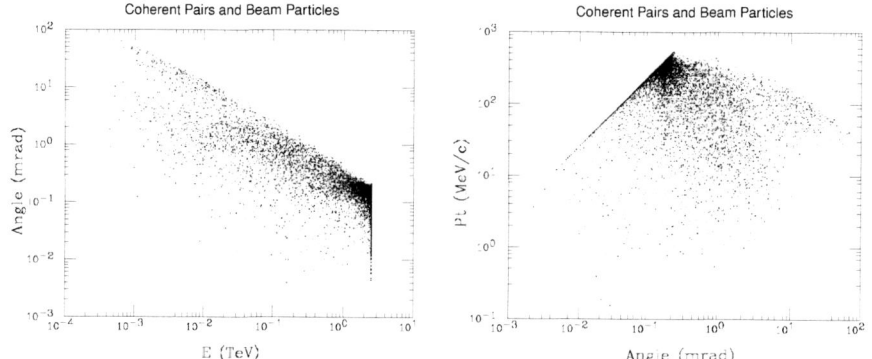

**FIGURE 3.** Coherent pairs and beam particles in angle-E (left) and angle-$P_t$ space (right).

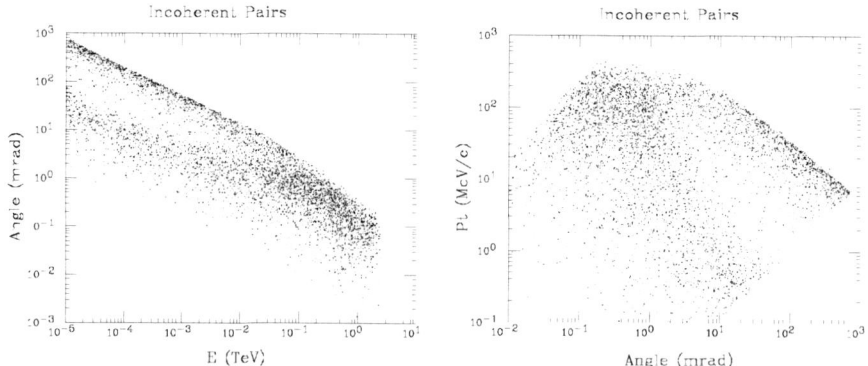

**FIGURE 4.** Incoherent pairs in angle-E (left) and angle-$P_t$ space (right).

$e^+ + e^-$); Bethe-Heitler: $(\gamma + e^{\pm} \longrightarrow e^{\pm} + e^+ + e^-)$; and Landau-Lifshitz: $(e^+ + e^- \longrightarrow e^+ + e^- + e^+ + e^-)$. Figure 4 shows the scatter plot of incoherent pairs (without beam particles) in angle-energy space (left) and in angle-$P_t$ space (right). The simulation used a 10 MeV cut on pair member energy. The two bands seen in the left plot corresponds to the opposite sign partners in the larger angle region and the same sign partners in the smaller angle region. Similarly, in the right plot the band with larger angle on the right corresponds to opposite sign partners. Comparing with the coherent pair distribution, incoherent pairs spread much more to the lower energy region and thus are deflected to larger angles. However the total number of incoherent pairs, about 5 thousands for our case, is more than 3 orders of magnitude below that of the coherent pairs. In fact each macro-particle in this Figure corresponds to a real pair particle. With angle and $P_t$ cuts similar to NLC case [22] the situation here does not seem to be much worse than the 0.5 TeV machine.

## Major Issue and Uncertainty

A major issue involved in establishing QSB as a feasible IP approach is the assessment of various sources of backgrounds. We have shown in the preceding section that it seems collision products from QED processes could all be confined within a cone of reasonable opening angle along the beam path. However, the detector may still be affected by secondary particles generated by the spent beam hitting other components such as quadrupole magnets within the forward cone. In addition, the spent beam may induce damage or even radioactivation on these components. A detailed analysis of these issues requires more specific detector design and realistic detector environment simulation, which is beyond the scope of this paper. It is hoped the situation could somehow be managed with appropriate masking scheme and IR design.

Collisions of beamstrahlung photons can also produce hadrons through QCD processes, giving rise to yet another source of backgrounds. Conceptually, the hadronic interaction between two colliding photons can be separated into two parts: the soft and hard scattering depending on the energy and transverse momentum involved. Cross section for the hard scattering can be calculated in part with perturbative QCD as in the minijet model [24], while the soft part is nonperturbative in nature and has to be treated with different model. For detector background consideration, the hard part is more serious as it could generate hadronic jets with higher transverse momentum. However, current theory on $\gamma$-$\gamma$ minijet cross section is subject to significant uncertainty.

Generally speaking, the uncertainty is due to inherent difficulty in nonperturbative QCD calculations. In particular, it comes from two major sources. First of all, as the hard scattering between two photons occurs among their respective partonic constituents, the knowledge of parton distribution of a photon or photon structure function is required. The photon structure function is available only in the form of parametrizations that are empirical and model dependent. Secondly, the minijet cross section is infrared divergent, thus a cutoff at low transverse momentum is necessary. As the transition between the hard and soft scattering is not well defined, this cutoff has to be determined also empirically for each parametrization used. So far experimental data on process $\gamma\gamma \rightarrow hadrons$ is available only up to 100 GeV. As a result, when extrapolating to multi-TeV, predictions for minijet cross section based on different parametrization are in nowhere near converging.

Nevertheless, the situation can still be improved by plausible arguments, phenomenological considerations and empirical scaling laws. Taking such an approach, recently Ohgaki et al. [19] have attempted to reach a more realistic, upper bound estimation of minijet cross section at higher energy. Furthermore, a complete case study was conducted to evaluate hadronic backgrounds for our parameter set. Monte-Carlo simulations were carried out in steps from beam-beam interaction to hadronic event generation, from minijets fragmentation to detector selection performance. The final result: both background event rate and energy deposits on detector per bunch crossing are quite small.

Last but not the least, backgrounds due to standard model processes, such as W pair production in two-photon collisions, may also have to be dealt with for exploration of physics beyond the standard model.

Constrained by the sheer size and cost of a modern collider, scaling of current technology and approach to higher energy is becoming prohibitive. Thus more than ever before, the future of high energy colliders will depend critically on innovative concepts and techniques, and more important on the successful integration of these concepts and techniques into a collider system, from acceleration to collision, to detection, and all the way to the origination of a discovery experiment. Should the approach of quantum suppression of beamstrahlung be proven acceptable for high energy physics community and viable technically, it will make a strong scientific case with potentially significant strategic value for the future developments of high energy physics and accelerator technology.

### Acknowledgments

I wish to extend many thanks to K. Yokoya for help on understanding and using the simulation code CAIN; to J. Siegrist for making available the computer facility with which the simulations for this paper were performed; to S. Chattopadhyay, J. Siegrist and B. Barletta for program support; and to K. Yokoya, T. Tajima, S. Chattopadhyay, T. Ohgaki, H. Murayama, K.-J. Kim, J. Siegrist and P. Chen for helpful comments and discussions. This work was supported by the U.S. Department of Energy under contract No.DE-AC03-76SF00098.

# REFERENCES

1. For a review on beam-beam interaction and QED processes in a linear collider, see K. Yokoya and P. Chen, *Frontiers of Particle Beams: Intensity Limitations*, Lecture notes in Physics, **400**, eds. M. Dienes et al., (Springer-Verlag, 1992).

2. T. Himel and J. Siegrist, AIP Conf. Proc., **130**, 602(1985).
3. K. Yokoya, *Nucl. Instr. Meth.* **A251**, 1(1986).
4. R. Noble, *Nucl. Instr. Meth.* **A256**, 427(1987).
5. V. Baier, V. Katkov and V. Strakhovenko, INP-88-168, Novosibirsk, (1988).
6. P. Chen and K. Yokoya, Phys. Rev. Lett., **61**, 1101(1988).
7. R. Blankenbecler and S. Drell, Phys. Rev. D,**37**, 3308(1988).
8. M. Jacob and T. Wu, *Nucl. Phys.*, **B318**, 53(1989).
9. International Linear Collider Tech. Review Report, SLAC-R-95-471, (1995).
10. N. Solyak, INP-preprint 88-44, Novosibirsk, (1988).
11. J. Rosenzweig, B. Autin and P. Chen, AIP Conf. Proc., **193**, (1989).
12. D. Whittum, A. Sessler, J. Stewart and S. Yu, LBL-25759, (1989).
13. V. Telnov, *Nucl. Instr. Meth.* **A294**, 72 (1990).
14. H. Murayama and M. Peskin, Ann. Rev. Nucl. Part. Sci. **46**, 533(1996).
15. A. Sokolov, N. Klepikov and I. Ternov, Soviet Phys. Doklady, **89**, 665(1953).
16. Proc. of 1996 DPF/DPB Summer Study on New Directions for High Energy Physics, Snowmass, Colorado, (1996).
17. Proc. of the 1996 Workshop on Advanced Accelerator Concepts, AIP conf. Proc. 398, (1997).
18. M. Xie, T. Tajima, K. Yokoya and S. Chattopadhyay, AIP conf. Proc., **398**, 233(1997).
19. T. Ohgaki, M. Xie and H. Murayama, 'Estimates of hadronic backgrounds in a 5 TeV $e^+e^-$ linear collider', these proceedings.
20. P. Chen, T. Ohgaki, A. Spitkovsky, T. Takahashi and K. Yokoya, *Nucl. Instr. Meth.* **A397**, 458 (1997).
21. P. Chen and V. Telnov, Phys. Rev. Lett., **63**, 1796(1989).
22. Zeroth-order design report for the NLC, SLAC Report-474, (1996).
23. K. Thompson and P. Chen, Proc. of ICFA Workshop on QABP, 1998.
24. For a review, see M. Drees and R. Godbole, J. Phys. G**21**, 1559(1995).

# Working Group 2:

# Plasma-Based
# Acceleration Concepts

# Laser Wakefield Acceleration of Electrons at Ecole Polytechnique

## Presented by D. Bernard

F. Amiranoff[1], S. Baton[1], D. Bernard[2], B. Cros[3], D. Descamps[1],
F. Dorchies[1], F. Jacquet[2], V. Malka[1], J. R. Marquès[1],
G. Matthieussent[3], P. Miné[2], A. Modena[1], P. Mora[4], J. Morillo[5],
Z. Najmudin[6]

[1] *Laboratoire pour l'Utilisation des Lasers Intenses, Ecole Polytechnique, CNRS, 91128 Palaiseau, France*

[2] *Laboratoire de Physique Nucléaire et des Hautes Energies, Ecole Polytechnique, IN2P3 & CNRS, 91128 Palaiseau, France*

[3] *Laboratoire de Physique des Gaz et des Plasmas, CNRS, Université Paris Sud, 91405 Orsay, France*

[4] *Centre de Physique Théorique, Ecole Polytechnique, CNRS, 91128 Palaiseau, France*

[5] *Laboratoire des Solides Irradiés, CEA/DSM/DRECAM,URA-CNRS 1380, Ecole Polytechnique, 91128 Palaiseau, France*

[6] *Imperial College, Blackett Laboratory, SW7 2AZ London, UK*

**Abstract.** The acceleration of electrons injected in a plasma wave generated by the laser wakefield mechanism has been observed. A maximum energy gain of 1.6 MeV has been measured and the maximum longitudinal electric field is estimated to 1.5 GV/m. The experimental data agree with theoretical predictions when 3D effects are taken into account. The duration of the plasma wave inferred from the number of accelerated electrons is of the order of 1 ps.

CP472, *Advanced Accelerator Concepts: Eighth Workshop,*
edited by W. Lawson, C. Bellamy, and D. Brosius
© 1999 The American Institute of Physics 1-56396-889-4/99/$15.00

# I  INTRODUCTION

The generation of large amplitude electric fields in plasmas by high-power lasers has been studied for several years in the context of high-field particle acceleration [1]. The ponderomotive force of the laser excites a longitudinal electron plasma wave (EPW) with a phase velocity close to the speed of light [2]. Two mechanisms have been considered to excite the EPW.

In the Laser Beat Wave Acceleration (LBWA) approach, the beating of a two frequency laser creates a modulation of its intensity. When the frequency difference is equal to the natural oscillation frequency of the plasma electrons $\omega_p$, an EPW is excited resonantly. This can lead to large amplitude electric fields. A precise tuning of the electron density is therefore mandatory in LBWA experiments. LBWA has been extensively studied during the 90's with 1 $\mu$m [3] and 10 $\mu$m [4–6] lasers.

In the "standard" Laser Wake Field Acceleration (LWFA) approach, a single short laser pulse excites the EPW [2,7,8]. As the ponderomotive force associated with the longitudinal gradient of the laser intensity exerts two successive pushes in opposite directions on the plasma electrons, the excitation of the EPW is maximum when the laser pulse duration is of the order of $1/\omega_p$.

At high electron density, and high laser intensity, a long — with respect to $1/\omega_p$ — laser pulse breaks into short pulselets at $1/\omega_p$ through the stimulated Raman scattering instability [9–11]. In this self-modulated mode (SM LWFA), the very high longitudinal electric field of the EPW traps plasma electrons and accelerates them to high energies [12–16]. However, SM LWFA may not be the best candidate for very high energy accelerators, in particular because the EPW grows from an instability so that its phase is unpredictable, and also because of the low Lorentz factor $\gamma_p \approx \omega/\omega_p$ of the phase velocity of the EPW at high electron density.

Standard LWFA seems particularly suited for particle acceleration. It is not affected by saturation (e.g. relativistic detuning [5] or modulational instability [3]) as is LBWA, and operates at low density, where $\gamma_p$ can be quite high. The excitation of radial EPW by laser wake field has already been observed by two-pulse frequency-domain interferometry [17,18].

We present the first observation of LWFA of injected electrons. Part of the material presented here will be published elsewhere [19]. A particular emphasis has been given to the separation of the signal from the background (BG) noise in the design of the experimental apparatus [23] and in the analysis of the data. In the case of LBWA experiments, Clayton et al. have shown that magnetic and/or transverse electric fields, due to a Weibel-like instability [20], still exist in the plasma a long (a few nanoseconds) time after the excitation of the EPW. Electrons deflected by such fields can scatter on the walls of the vacuum chamber and provide spurious signal, as is possibly the case in [21] and in the surprising result of [22].

# II  LASER WAKEFIELD ACCELERATION

The transverse and longitudinal components of a linear EPW created by laser wakefield, for a laser beam with a gaussian radial profile and a gaussian time distribution, can be expressed [1,7] as : $E_r = (4r/w^2)A\sin(\omega_p t - k_p z)$ and $E_z = k_p A\cos(\omega_p t - k_p z)$, with

$$A = \sqrt{\pi}\omega_p\tau_0 \exp\left(-\frac{\omega_p^2\tau_0^2}{4}\right)\frac{I_{max}e}{2\epsilon_0 mc\omega^2}\exp\left(-\frac{2r^2}{w^2}\right), \tag{1}$$

where the time variation of the laser intensity is described by $\exp(-(t/\tau_0)^2)$, $I_{max}$ is the maximum intensity, $w$ the $1/e^2$ radius in intensity, and $k_p = \omega_p/c$. At a given value of $\tau_0$, $E_z$ varies like $(\omega_p\tau_0)^2\exp(-(\omega_p\tau_0)^2/4)$. This gives a broad maximum close to $\omega_p\tau_0 = 2$, i.e. $\omega_p\tau = 4\sqrt{\ln 2}$, where $\tau$ is the pulse duration at FWHM. With $\tau = 400$ fs, this corresponds to an electron density $n = 2.2\ 10^{16}\text{cm}^{-3}$, an EPW wavelength $\lambda_p = 226\ \mu$m, and an EPW Lorentz factor $\gamma_p = 214$. The corresponding Helium pressure is 0.4 mbar for a fully ionized plasma. Finally, the maximum electric field at resonance is $E_z[\text{GV/m}]=1.35\ 10^{-18}I_{max}[\text{W/cm}^2](\lambda[\mu\text{m}])^2/\tau[\text{ps}]$. The relative longitudinal perturbation of the electron density is $\delta_\parallel = E_z/E_0$, where $E_0 = mc\omega_p/e$.

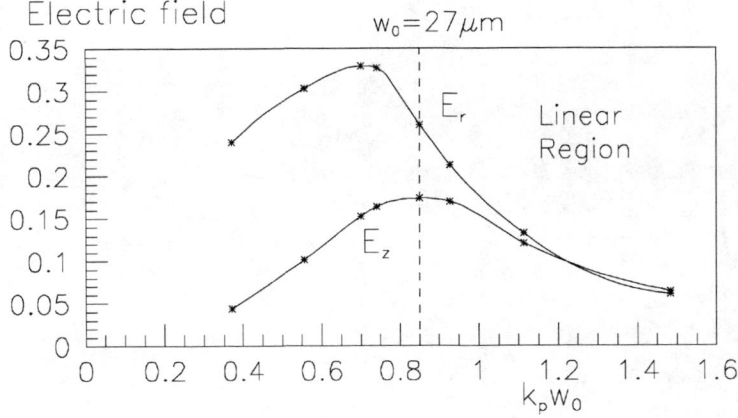

**FIGURE 1.** Variation of the longitudinal ($E_z$) and transverse ($E_r$) electric field in units of the cold wave breaking limit $E_0$, as a function of the laser spot size $w$ in units of $2\pi/\lambda_p$. The laser energy is 3 J, duration 350 fs, and $\omega_p\tau_0 = 2$ ($E_0=16$ GV/m). The limit of the linear region is obtained for $\delta_\perp \approx 2$, (dashed line), and is approximately independent of laser energy.

The ratio of the transverse to the longitudinal electric field, at $r \approx w_0/\sqrt{2}$ is $E_r/E_z = \sqrt{2}\lambda_p/\pi w_0$, here equal to 4, $w_0$ being the laser beam size $w$ at the waist. We obtain the value of the relative transverse perturbation of the electron density [17] by $\delta_\perp/\delta_\parallel = (E_r/E_z)^2$, here equal to 16. This means that, in our conditions, the EPW is mainly excited in the radial regime : the transverse electric field is stronger than the longitudinal electric field.

Particle simulations using the model described in Ref. [24] show (Fig. 1) that with our parameters, $E_z$ is actually lower than the linear value given above, when the laser energy is so high that $\delta_\perp \geq 2$. The cavitation created by the radial oscillation affects the development of the longitudinal oscillation. The corresponding limit value of $\delta_\parallel$ is here $\approx 10 - 20\%$.

## III  EXPERIMENTAL APPARATUS

The experimental apparatus is based on the existing facility already used for the study of LBWA [3], and is presented in Ref. [23]. A sketch of the present experiment can be found in Fig. 2. We use the 400 fs, 1.057 $\mu$m chirped pulse amplification laser at LULI. The 80 mm diameter beam is injected into a pulse compressor, and

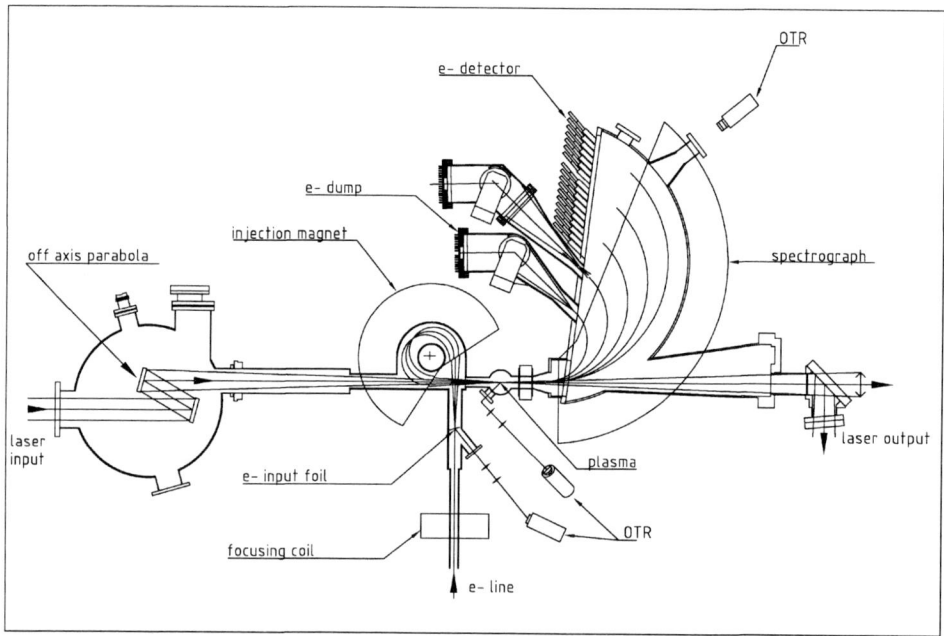

**FIGURE 2.** Layout of the experiment. See text and Ref. [23].

focused in a gas filled chamber by a 1.4 m focal length 30° off-axis parabola. A fraction of the compressed beam is collected before focusing and sent to a single-shot second-order autocorrelator for pulse duration measurement. A low intensity fraction of the beam is collected after the plasma and sent to a focal spot monitor. A 300 $\mu$A cw electron beam is injected in the plasma at a total energy of 3 MeV with an RMS spot size of $30\mu$m and an RMS divergence of 10 mrad [23]. The accelerated electrons are measured by a magnetic spectrograph and 17 detectors in the range 3.3 to 5.9 MeV. The linear gates have been withdrawn, and the voltage of the photomultipliers was tuned so that the calibration factor was equal to 2.5 ADC (analog to digital converter) count per electron. The duration of the gate was set to 20 ns.

## IV EXPERIMENTAL RESULTS

A series of 250 shots has been performed, most of them with a laser energy in the range 4–9 J. On average, after compression, 20% of this energy is focused to a spot with typical size $w_{0,H} = 30$ $\mu$m (horizontal waist) and $w_{0,V} = 19$ $\mu$m (vertical waist), with Rayleigh length of $z_{0,H} = 2.3$ mm and $z_{0,V} = 2.0$ mm. With a central spot energy of 1.5 J, the values of the maximum power, intensity, electric field, EPW amplitude, and of the expected linear energy gain are $P_{max}$=3.5 TW, $I_{max} = 4 \cdot 10^{17}$W/cm$^2$, $E_z$=1.5 GV/m, $\delta_{\parallel}$=10%, $\Delta W = \pi e z_0 E_z$=10 MeV. The main source of fluctuation is due to the laser pulse duration. For shots for which the quantities $\tau$, $E$, $w_{0,H,V}$ could be measured, the amplitude varies in the range $\delta_{\parallel}$=1–15%. Electron acceleration was observed in all of these shots.

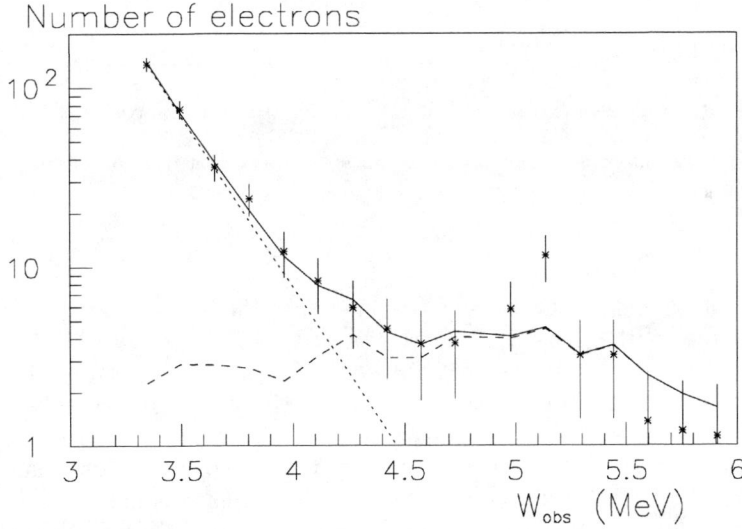

**FIGURE 3.** Spectrum of a typical shot (dots). The fit is described in the text. The continuous line shows the sum of the two contributions.

A typical spectrum is presented in Fig. 3 (dots). It shows a peak at low electron energy, that can be fitted by a decreasing exponential (dotted line) and a high

307

energy tail (dashed line) that has the same shape as the BG noise spectrum, as explained below.

To check the energy of the electrons impinging on a given channel, we have inserted stainless steel filters with various thicknesses in front of some scintillators. The signal of the corresponding channel is reduced by a factor which depends on the mean electron energy. The transmission factor for laser shots and BG noise runs is compared with the result of a simulation [25] at the electron energy corresponding to the channel (Fig. 4). From the low transmission factor in channel 12, with

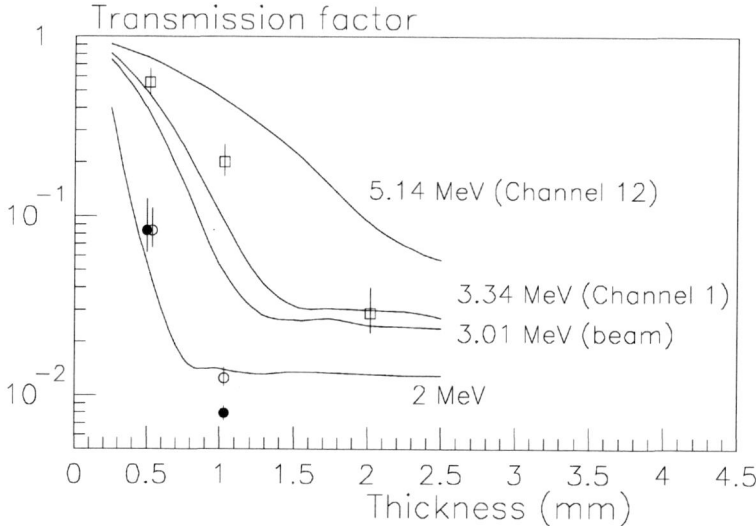

**FIGURE 4.** Transmission factors with stainless steel filters as a function of their thickness; channel 1 : □ (laser shots); channel 12 : ○ (laser shots), ● (gas BG noise runs). The error bars indicate the dispersion on several shots or runs. Continuous lines : simulation [25].

nominal electron energy of 5.14 MeV, we infer that the high energy tail is actually due to electrons with an average energy of about 2 MeV.

We now examine the various contributions to the BG noise. The BG noise due to Coulomb scattering of the beam electrons in the gas, has been substracted in Fig. 3. This noise has been studied in separate runs, without the laser. For each channel, the average value scales with pressure with a typical proportionality factor of 8 e⁻/mbar. This factor does not decrease with the channel number as for simple Coulomb scattering. This "gas" BG is due to electrons deflected at low angle in the gas, that impact on the flange of the bottle neck of the dump. Part of these are back-scattered, re-enter the magnetic field of the spectrograph, and may fly back into the detector [23].

The tail in Fig. 3 is due to an excess of BG noise. It is observed only for shots with accelerated electrons, i.e. *in correlation with the EPW*. We call it "EPW"

BG noise. It is due to electrons deflected in the plasma close to the waist, while Coulomb scattering occurs along the whole path of the electrons, with a different geometry. To simulate the former, we have introduced a 11 $\mu$m Al foil at focus, in vacuum. The obtained noise spectrum has a shape similar to the shape of the gas spectrum. The electrons scattered at large angle in the foil are blocked by the $d_1$ collimator (See Fig. 12 of Ref. [23]). Few of them are re-scattered at the edge of the collimator. As the latter is not at focus, some of them impact on the flange of the dump. This is the reason for the similar shape of the two distributions.

The signals of three channels have also been recorded on a storage oscilloscope for each shot. A peak, about 10 ns in duration at 10%, is observed in correlation with the ADC recording, for channels 1 (signal), 8 and 12 (EPW BG noise). Therefore both the EPW BG noise and the signal are shorter than 10 ns, while the gas BG noise is obviously continuous. The EPW BG noise level is too high to be due only to the electrons deflected by the transverse electric field of the EPW, because of its short (ps) life-time, and because of the high rejection power of the collimation system [23], as shown by the low noise level induced by the foil. An effect like the Weibel instability already observed in ref. [20] is a good candidate to explain a long term (ns) deviation of the electrons. It could thus explain this BG. The signal is separated from the EPW BG noise by the process of the simultaneous fit of the exponential peak and of the tail (Fig. 3). We define the end point $W_{obs}$ of the

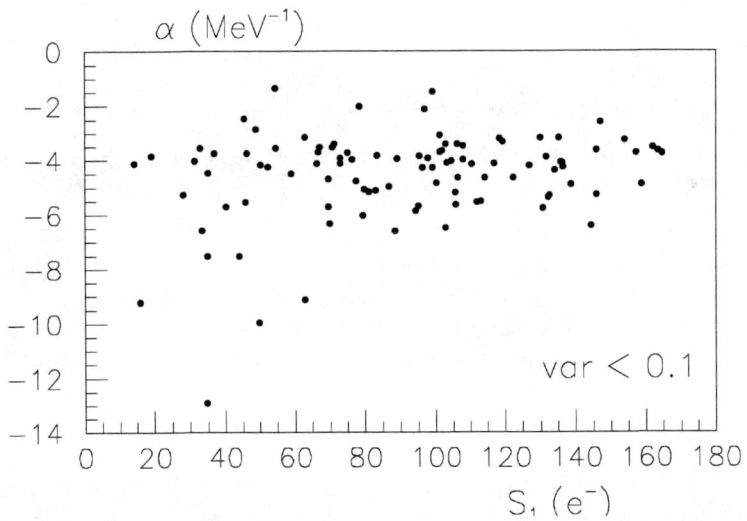

**FIGURE 5.** Variation of the slope $\alpha$ with the number of $e^-$ in channel 1.

spectrum of the signal as the energy for which the exponential peak decreases to one electron. For the shots for which enough channels have a signal to make a fit, the slope $\alpha$ is found equal to $\alpha_0 = -4.4 \pm 1.1$ MeV$^{-1}$, a number that is observed not to depend on the parameters of the laser pulse or of the plasma (as an example, the variation of $\alpha$ with the number of $e^-$ in channel 1 is given in Fig. 5). Therefore we have used the same value $\alpha = \alpha_0$ to compute $\Delta W_{obs}$ for all the shots.

The variation of the signal $S_1$ in channel 1 with $\omega_p \tau_0$ is presented in fig. 6(left), where both $\omega_p$ and $\tau_0$ have been varied. As expected, the data show a maximum close to $\omega_p \tau_0 = 2$. The spectrum of $\Delta W_{obs}$ is much broader (right), as $\Delta W_{obs}$ varies like $\log S_1$ in the exponential peak. Here, $\delta_\parallel$ is low, and the length of the high gradient region, of the order of $2z_0$, is smaller than the dephasing length of the electrons with respect to the EPW, equal to 8 mm. Therefore, $\Delta W_{obs}$ should have the same resonant dependance with $\omega_p \tau_0$ as $A$ (Eq. 1, and curve in Fig. 6 right). Note also that the maximum value of $\Delta W_{obs}$, close to 1.6 MeV, is smaller than the value obtained from the linear approximation in 1D geometry, close to 10 MeV.

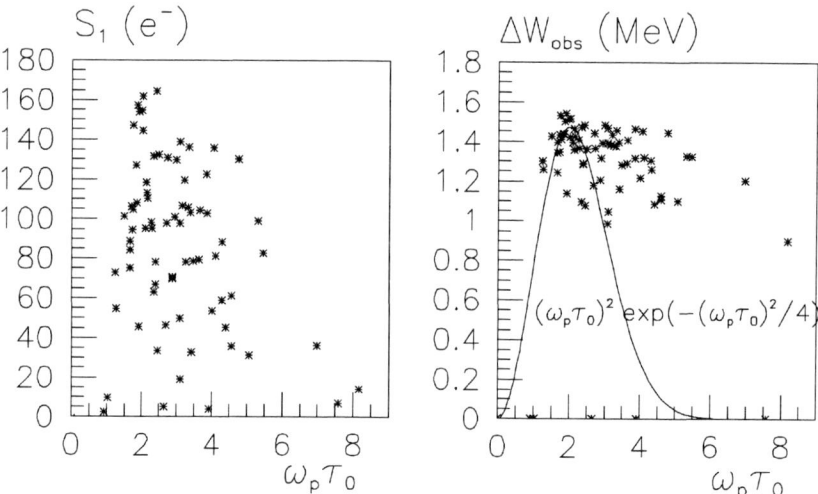

**FIGURE 6.** Variations of $S_1$(left) and of $\Delta W_{obs}$ (right) with $\omega_p \tau_0$. The fitting procedure ($S_1 > 10$ e$^-$) introduces a cut off at $\Delta W_{obs}=0.85$ MeV. The pressure is varied in the range 0–2 mbar with half of the data taken at "resonance". The main source of variation of $\omega_p \tau_0$ is the fluctuation of $\tau_0$. Only data with $\tau < 1$ ps are used. The curve describes the $\omega_p \tau_0$ dependence of $E_z$ in linear LWFA theory.

The transverse electric field of the EPW affects the trajectories of the electrons. Depending on their phase, electrons undergo a focusing or defocusing force when they enter the EPW. The defocused electrons are expelled radially before they enter the high accelerating region. On the contrary, the focused electrons are transversely trapped in the EPW, and should be accelerated in it efficiently [26].

In fact, a numerical tracking of the trajectories of electrons in the EPW, using the code described in [27], shows that most of them miss the waist transversely.

This can be understood in the simple model of ref. [26], where the trajectory of an electron is described by a three domain approximation : a drift in free space, an "adiabatic" region where the electron is trapped by the transverse field, and another drift on exit. Trapping occurs very far from the waist, at a location where $\delta_\parallel$ is equal to a critical value $\delta_c = \gamma(w_0/z_0)^2/4$ [26], here equal to $10^{-3}$, $\gamma$ being the electron Lorentz factor. Then, in the central region, the evolution of the envelope of the electron beam is determined by the evolution of the betatron function in the EPW, so that the beam size at the waist is $\sigma_w = \sigma_0 \sqrt[4]{\delta_\parallel/\delta_c} z_0/\beta^*$ where $\sigma_0$ and $\beta^*$ are the beam size and the betatron function at the waist in vacuum. For $\sigma_0 = 30\mu$m and $z_0 \approx \beta^*$, and for $\delta_\parallel = 10\%$, we have $\sigma_w = 90\mu$m, much larger than the size of the plasma wave $\sigma_{\rm EPW} = w_0/2 \approx 10\mu$m. The key point is that after trapping in the EPW, the $e^-$ beam size varies like $\sqrt{\beta} \propto \delta_\parallel^{-1/4} \propto [1 + (z/z_0)^2]^{1/4}$, while in vacuum it varies like $w \propto [1 + (z/z_0)^2]^{1/2}$.

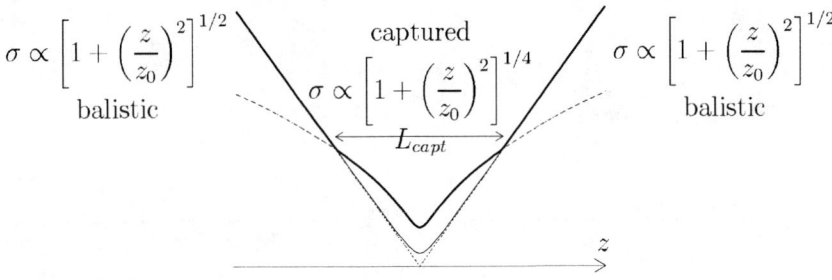

**FIGURE 7.** Scheme of the enveloppe of the electron beam (solid thick line). Solid thin line : balistic beam, dashed line : captured beam.

In the presence of the EPW, the decrease of the beam size while approaching the waist is much slower. A more precise description of this effect (Fig. 8, left) is obtained using the simulation; electrons are tracked [27] through an EPW computed in the linear regime, created by a laser beam, in the gaussian approximation with cylindrical symmetry, ie. according to eq. 1. Electrons injected on axis (curve a) undergo an acceleration or a deceleration, depending on their injection phase $\varphi$. Electrons injected with a tiny emittance (b) in the focusing part of the wave are not affected, while those that are defocused are expelled before the high accelerating gradient region is reached. Electrons injected with real emittance (c) miss the EPW waist even in the focusing part of the EPW. Note also that both the length of the completely ionized plasma, $L_{pl} \approx 25$ mm, and the length on which the electrons are captured by the EPW [26] $L_{capt} = 2z_0\sqrt{\delta/\delta_c} \approx 40$ mm, are larger than the dephasing length $L_{dephas} \approx 8$ mm : most of the electrons have the occasion to be expelled from the EPW by a defocusing field during their path through the EPW.

The corresponding fraction of the electrons accelerated throughout the plasma is low (Fig. 8, right) and the maximum energy gain observed in the simulation is much lower than the maximum possible energy gain $\Delta W = \pi e z_0 E_z$. The slope of the simulated spectrum is in agreement with the observed value. The accelerated electrons are contained in a divergence angle of $\pm 70$ mrad, well inside the acceptance of the detector. To reach the maximum possible energy gain, the increase of the radial size of the accelerated electron beam could be overcome either by an injection at a higher energy, or by a limitation of the EPW length, by using a gas jet.

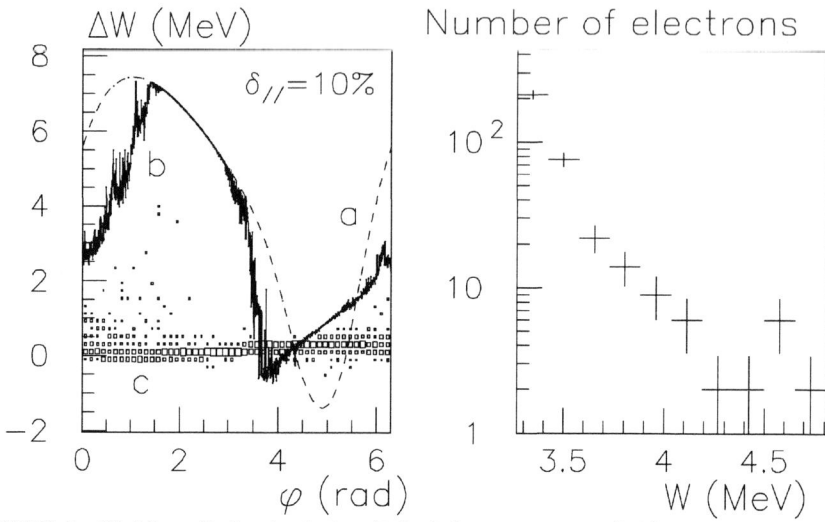

**FIGURE 8.** 3D MonteCarlo simulation [27] of the energy gain (left) of 1000 electrons as a function of their phase with respect to the EPW. a) beam on axis; b) small emittance beam (30 nm$\times$10 $\mu$rad RMS); c) real emittance beam (30 $\mu$m$\times$10 mrad RMS). The corresponding spectrum in the 10 first channels (right) shows an exponential peak with a slope of -6.1 MeV$^{-1}$.

Figure 9 shows electron spectra at three laser central energies. As the electron flow delivered by the Van de Graaf is constant during the life-time $T$ of the EPW, we infer an estimate of $T$ from a comparison of the normalisations of the observed and simulated spectra. The obtained value is of about 1 ps, in agreement with particle simulations using the model of Ref. [24].

# V  CONCLUSION

In conclusion, we have observed the acceleration of electrons injected in an EPW generated by laser wakefield, with a maximum energy gain of 1.6 MeV. We also observe a tail in the high energy channels. Our cross-check using stainless steel filters proves that this tail is actually due to low energy deflected electrons. This BG, clearly correlated with the plasma wave, can fake accelerated electrons in this kind of experiments. The experimental data agree with theoretical predictions when 3D effects are taken into account.

# VI ACKNOWLEDGEMENTS

We gratefully acknowledge the help of the technical staff of the LULI, LPNHE, LSI, and CEA/DSM/DAPNIA-SEA for this experiment. This work has been partially supported by Ecole Polytechnique, IN2P3-CNRS, SPI-CNRS, and by the EU Large Facility Program under Contract No. FMGE CT95 0044.

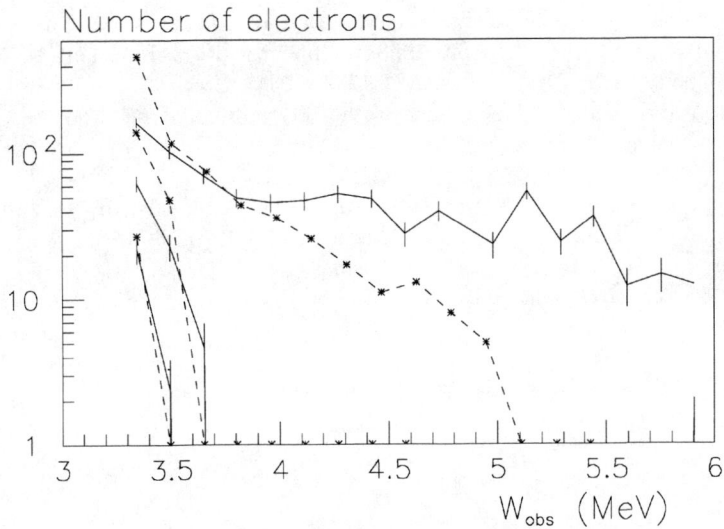

**FIGURE 9.** Electron spectra with $E = 0.25, 0.49, 2.1$ J (continuous lines) compared to simulated spectra (2000 incident electrons, dashed lines). At 2.1 J, the high energy tail is due to EPW BG noise.

# REFERENCES

1. A review can be found in : E. Esarey *et al.*, IEEE Trans. Plasma Science, **24** 252 (1996).
2. T. Tajima and J. M. Dawson, Phys. Rev. Lett. **43**, 267 (1979).
3. F. Amiranoff *et al.*, Phys. Rev. Lett. **74**, 5220 (1995);
   F. Amiranoff *et al.*, IEEE Trans. on Plasma Sci., **24** 296 (1996).
4. Y. Kitagawa *et al.*, Phys. Rev. Lett. **68**, 48 (1992).
5. C. E. Clayton *et al.*, Phys. Rev. Lett. **70**, 37 (1993).
6. N. A. Ebrahim, J. Appl. Phys. **76**, 7645 (1994).
7. L. M. Gorbunov and V. I. Kirsanov, Zh. Eksp. Teor. Fiz. **93** 509 (1987); Sov. Phys. JETP **66** 290 (1987).
8. P. Sprangle *et al.*, Appl. Phys. Lett. **53** 2146 (1988).
9. N. E. Andreev *et al.*, Sov. Phys. JETP **55** 571 (1992).
10. T. M. Antonsen Jr. and P. Mora, Phys. Rev. Lett. **69** 2204 (1992).
11. P. Sprangle *et al.*, Phys. Rev. Lett. **69** 2200 (1992).
12. A. Modena *et al.*, Nature, **377** 606 (1995).
13. C. Coverdale *et al.*, Phys. Rev. Lett. **74** 4659 (1995).
14. K. Nakajima *et al.*, Phys. Rev. Lett. **74** 4428 (1995).
15. A. Ting *et al.*, Phys. Plasmas **5** 1889 (1997).
16. R. Wagner *et al.*, Phys. Rev. Lett. **78**, 3125 (1997)
17. J. R. Marquès *et al.*, Phys. Rev. Lett. **76** 3566 (1996);
    C. W. Siders *et al.*, Phys. Rev. Lett. **76** 3570 (1996).
18. J. R. Marquès *et al.*, Phys. Rev. Lett. **78** 3463 (1997).
19. Observation of Laser Wakefield Acceleration of Electrons, F. Amiranoff *et al.*, preprint X-LPNHE 98/02, April 1998, to appear in Phys. Rev. Lett.
20. C. E. Clayton *et al.*, Phys. Plasma, **1** 1753 (1994).
21. K. Nakajima *et al.*, in Proceedings of the International Workshop on Acceleration and Radiation Generation in Space and Laboratory Plasmas, Kardamily, Greece, 1993, Published in Phys.Scripta T52:61-64,1994.
22. M. Kando *et al.*, KEK Preprint 97-10, April 1997, submitted to Phys. Rev. Lett.
23. F. Amiranoff *et al.*, Nucl. Instr. and Meth. A **363** 497 (1995).
24. P. Mora, T. M. Antonsen Jr, Phys. Plasma **4** 217 (1997).
25. W. R. Nelson *et al.*, The EGS4 code system, SLAC report 265, Dec. 1985.
26. D. Bernard, in Proceedings of the 13th Advanced ICFA Beam Dynamics Workshop and 1st ICFA Novel and Advanced Accelerator Workshop, Kyoto, Japan, 1997, Nucl. Instrum. Methods Phys. Res. A 410 (1998) 418.
27. P. Mora, J. Appl. Phys. **71** 2087 (1992).

# Progress on Plasma Lens Experiment at the Final Focus Test Beam[*]

S. Chattopadhyay[1], P. Chen[2], D. Cline[3], P. Colestock[4], W. Craddock[2],
F.-J. Decker[2], J. Holt[4], R. Iverson[2], T. Katsouleas[5], F. King[2], R. Kirby[2],
P. Kwok[3], S. Masuda[3], D. D. Meyerhofer[6], K. Nakajima[7], J. Ng[2],
R. Noble[4], A. Ogata[7], A. Sessler[1], D. Walz[2], A. Weidemann[8].

**Abstract.** The Plasma Lens Collaboration proposed and has been approved to perform the Plasma Lens Experiment at the Final Focus Test Beam (SLAC E-150*bis*). The experiment is a revised and simplified proposition which reflects the pertinent scientific and technological advances since an earlier proposal (1) was submitted. The goals of the experiment are to study plasma focusing of high energy, high density particle beams; to investigate plasma beamsstrahlung suppression; and to develop technologies for plasma lens applications in present and future linear colliders.

## INTRODUCTION

Plasma focusing devices are compact, simple, and very strong focusing elements. The focusing strengths for typical parameters are equivalent to focusing magnets with gradients of $10^9$ Gauss/cm. In principle, such strong fields are capable of focusing beams to very small spot sizes (2-7) and perhaps even capable of avoiding (8) the inherent (Oide) limitation (9) in discrete strong focusing. Several proof-of-principle experiments using low density particle beams have demonstrated plasma focusing (10-12). Our goal is to demonstrate the effectiveness of plasma lenses in the parameter regime of interest for SLC and the next generation of high energy linear colliders. The experience gained is expected to yield new final focus designs capable of generating spot sizes smaller than ever produced before.

A facility such as the FFTB offers a unique environment to test all aspects of plasma focusing for high energy, low emittance beams. In our earlier proposal (1) we intended to test plasma focusing with both round and flat beams. In the latter case, the intention

[*] Work supported by DOE contract DE-AC03-76SF00515 and grant DE-FG03-92ER40695.

[1] Lawrence Berkeley Laboratory, Berkeley, California.
[2] Stanford Linear Accelerator Center, Stanford, California.
[3] University of California, Los Angeles, California.
[4] Fermi National Accelerator Laboratory, Batavia, Illinois.
[5] University of Southern California, Los Angeles, California.
[6] University of Rochester, Rochester, New York.
[7] National Laboratory for High Energy Physics (KEK), Tsukuba, Japan.
[8] University of Tennessee, Knoxville, Tennessee.

was to demonstrate a focused beam size which would be even smaller than that achieved by the FFTB. We believe that the plasma lens can indeed deliver such an unprecedented beam size at the FFTB. With that in mind we will test the plasma focusing of round beams and concentrate on demonstrating the power of a plasma lens through its demagnification power, as a first step of experimentation.

We will perform a round beam focusing experiment at the FFTB. It would utilize the round beam aspect ratio of $\varepsilon_{nx} = \varepsilon_{ny} = 3 \times 10^{-5}$ m-rad from the FFTB (13), and a charge of $10^{10}$ particles per bunch. The plasma lens would have electron densities adjustable from $10^{16}$ to $10^{18}$ cm$^{-3}$ and a thickness of 3 mm. The plasma will be generated by laser ionization immediately prior to the arrival of the beam. With optimized conditions, the plasma lens will be capable of demonstrating the focusing of a 5 μm round beam to a final spot size of less than 2 μm in both dimensions, or a luminosity enhancement factor of about 10. The simplicity and economy of such a scheme will prove the plasma lens to be a worthy add-on device for luminosity enhancement in linear colliders (7).

## PARAMETER STUDIES

The prospect of testing plasma lens concepts at the FFTB was investigated (14). In the study, plasma lens parameters for round beam focusing, flat beam focusing, and "adiabatic focusing" (8) were provided. Here we refine our treatment and limit ourselves to round beam focusing only. We consider beam energies of 50 and 30 GeV, which correspond to the current configuration of the accelerator and its future configuration as the PEP-II (B factory) injector, respectively. In either case, the effectiveness of the plasma lens can be demonstrated.

Ignoring the effects due to the return current, the focusing strength K for underdense plasma lenses is governed by the plasma density $n_p$,

$$K = \frac{2\pi r_e}{\gamma} n_p . \tag{1}$$

where $r_e$ is the classical electron radius and $\gamma$ is the relativistic kinetic energy factor of the beam. For overdense plasma lenses the strength is determined by the beam density $n_b$,

$$K = \frac{2\pi r_e}{\gamma} n_b . \tag{2}$$

The plasma return current tends to reduce the focusing effect of the lens (15). The effect is approximately given by

$$K_{rc} = \frac{K}{1 + (k_p \sigma_r)^2} , \tag{3}$$

**TABLE 1.** Parameters for Round Beam Focusing.

| Beam Parameters | Case A | | Case B | | Case C | |
|---|---|---|---|---|---|---|
| $E_0$ [GeV] | 30 | 50 | 30 | 50 | 30 | 50 |
| N [$10^{10}$] | 1 | 1 | 1 | 1 | 1 | 1 |
| $\varepsilon_n$ [$10^{-5}$ m-rad] | 3.0 | 3.0 | 3.0 | 3.0 | 3.0 | 3.0 |
| $\beta_0^*$ [mm] | 45 | 75 | 45 | 75 | 45 | 75 |
| $\sigma_{r0}^*$ [$\mu$m] | 4.74 | 4.74 | 4.74 | 4.74 | 4.74 | 4.74 |
| $\beta_0$ [mm] | 48.1 | 80.3 | 48.1 | 80.3 | 48.1 | 80.3 |
| $\sigma_0$ [$\mu$m] | 4.91 | 4.91 | 4.91 | 4.91 | 4.91 | 4.91 |
| $\sigma_z$ [mm] | 0.47 | 0.47 | 0.47 | 0.47 | 0.47 | 0.47 |
| $n_{b0}$ [$10^{16}$ cm$^{-3}$] | 5.3 | 5.3 | 5.3 | 5.3 | 5.3 | 5.3 |

| Lens Parameters | Case A | | Case B | | Case C | |
|---|---|---|---|---|---|---|
| $n_p$ [$10^{16}$ cm$^{-3}$] | 2 | 2 | 10 | 10 | 100 | 100 |
| $k_p\sigma_z$ | 12.5 | 12.5 | 28 | 28 | 88.5 | 88.5 |
| $s_0$ [mm] | -20 | -20 | -20 | -20 | -20 | -20 |
| $\ell$ [mm] | 3 | 3 | 3 | 3 | 3 | 3 |
| f [mm] | 22.8 | 47 | 17.4 | 35.7 | 21.6 | 36 |

| Focused Beam | Case A | | Case B | | Case C | |
|---|---|---|---|---|---|---|
| $\beta_r^*$ [mm] | 22.2 | 37 | 12.6 | 21 | 20.4 | 34 |
| $\sigma_r^*$ [$\mu$m] | 2.01 | 3.35 | 1.53 | 2.55 | 1.94 | 3.23 |
| $s^*$ [mm] | 2.8 | 28.5 | -2.6 | 10.7 | 1.6 | 17.5 |

where $\sigma_r$ is the rms size of the beam and $k_p = \sqrt{4\pi r_e n_p}$ is the plasma wave number. The effect is appreciable only when the plasma is considerably denser than the beam.

The response of the plasma occurs in a time $\tau_p = 1/\omega_p$ and therefore in the overdense regime the requirement for response within the time of a beam pulse is $\tau_p \ll \tau_b \sim \sqrt{2\pi}\sigma_z$, or $k_p\sigma_z \gg 1/\sqrt{2\pi}$, where $\sigma_z$ is the rms bunch length. In the underdense regime, the wave breaking limit is quickly reached and the focusing is less subjected to the limit.

For the plasma lens experiment, the FFTB emittance is adjusted to a round beam configuration of $\varepsilon_{nx} = \varepsilon_{ny} = 3 \times 10^{-5}$ m-rad for a bunch population of $10^{10}$. With the relaxed beam properties, many of the chromatic corrections at the FFTB can be less stringent as well. A laser ionized plasma with a thickness of 3 mm and adjustable plasma density from $10^{16}$ to $10^{18}$ cm$^{-3}$ is placed at a distance of 2 mm from the vacuum focus.

Parameters for three typical cases with different plasma densities are shown in Table 1. Here, $E_0$ is the beam energy, N is the number of particles per bunch, $\varepsilon_n$ is the normalized emittance, and the bunch length is $\sigma_z$. The focused beam size in vacuum is $\sigma_{r0}^*$ with corresponding beta function $\beta_0^*$. The density of the plasma lens is $n_p$, and the focal length is $f = s^* - s_0 - \ell/2$, where $s^*$ is the distance of the new focal point from that of the vacuum, $\ell$ is the lens thickness, and $s_0$ is the location of the entrance to the plasma lens with respect to the vacuum focal point. At the entrance to the plasma lens, the beam size, beta function, and beam density are $\sigma_0$, $\beta_0$, and $n_{b0}$, respectively.

In Case A, an underdense plasma lens with $n_p = 2 \times 10^{16}$ cm$^{-3}$ focuses the beam from about 5 μm to 3.4 μm and to 2.0 μm, for 50 GeV and 30 GeV beams, respectively. Increasing the plasma density to $10^{17}$ cm$^{-3}$, as shown in Case B, allows the study of overdense plasma focusing without degradation due to return currents. In this case, the plasma focusing is strongest in this experiment, and the beam size decreases from 5 μm to 2.6 μm and to 1.5 μm, for 50 GeV and 30 GeV beams, respectively. In Case C, the plasma is much denser than the beam, and the beam-plasma interaction is in the extreme overdense regime, where the focusing strength diminishes due to return currents. It is important to note that the extreme overdense regime in Case C is the basis for the concept of "plasma beamsstrahlung suppression" (3,15).

In addition, a completely new particle-in-cell code has been developed by the plasma lens collaboration (16) to augment the numerical calculations discussed. Results from the new simulations have confirmed the parameters and provides refined guidance to the research and development of the experimental design.

## EXPERIMENTAL DESIGN

The apparatus for the plasma lens experiment is to be installed at Station 1027 on the FFTB (13). The location is within the final focus region of the FFTB and has ample space available for installation of all accessories. The design is to minimize the impact on the existing FFTB beam line, as well as to allow simultaneous presence of other scheduled experiments. At the conclusion of the plasma lens experiment, it will be possible to revert the FFTB beam line configuration to its original state.

The arrangement consists of a plasma chamber sandwiched between two identical differential pumping systems, and has ports for ionization laser, plasma diagnostics, gas supply, and beam size measuring devices. The differential pumping stages are separated by thin beryllium windows, with millimeter-sized holes for electron beam passage, to maximize pumping efficiency. The ionization laser is the same 1 μm wavelength laser system as was used for the non-linear Compton scattering experiment at the FFTB (E-144) (17). A small portion of the laser will be doubled in frequency for plasma diagnostics using Thomson scattering. Conventional wire scanners will be used for all beam size measurements, and the bremsstrahlung signal is monitored by the existing detectors downstream. Hydrogen gas will be used as the plasma source to minimize background radiation from beam-plasma interaction. The outline of the plan is shown as Figure 1.

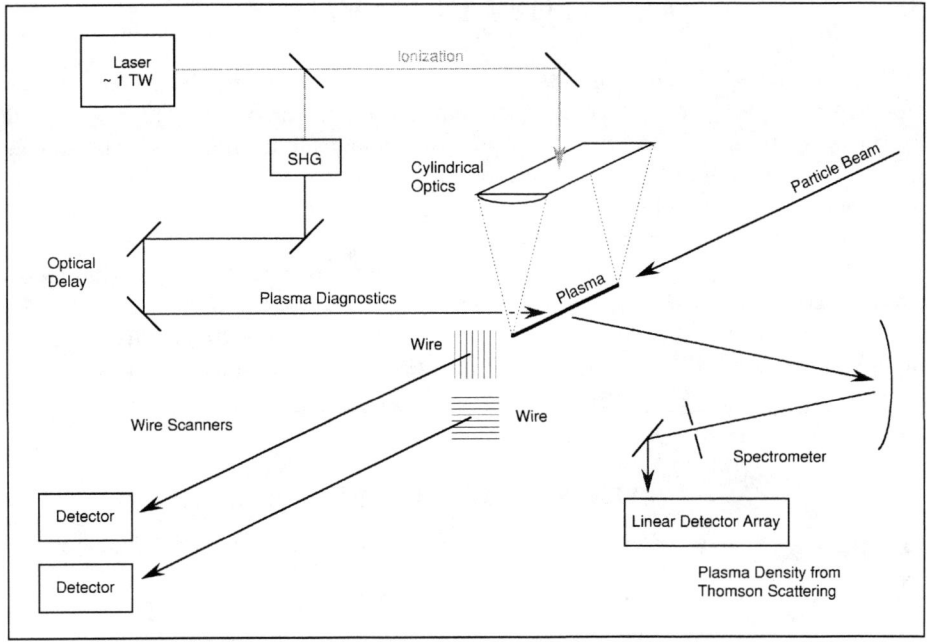

**FIGURE 1.** Outline of the Plasma Lens Experiment at the FFTB.

## RECENT DEVELOPMENTS

Detailed designs of the experiment began in October 1997, immediately after the approval from SLAC Experimental Program Advisory Committee. With guidance from the new particle-in-cell simulations (16), a preliminary design was constructed and reviewed in March 1998.

At the March 1998 review, plasma diagnostics compatibility issues were raised. It was suggested that laser interferometry should be used in place of Thomson scattering for plasma density measurements, the former which is much easier to setup and provides more reliable performance. However, the change requires pathways free of high pressure gas for the interfering laser beams. Consequently, a minor change has to be made to the gas delivery method.

With the change in plasma diagnostics, a pulsed gas nozzle is used as the plasma gas source now in place of a static gas cell as suggested in the proposal. The gas nozzle is procured from a commercial enterprise, which cuts development time significantly. A similar nozzle has been proven to perform satisfactorily at LBL (18) in the parameters of interest to the plasma lens experiment.

As synchrotron radiation induced by plasma focusing provides essential information on focusing strength, the collaboration is developing a synchrotron radiation monitor as an additional diagnostic tool.

# ACTIVITIES IN PROGRESS

As the planned operational date of April 1999 for the plasma lens experiment is approaching rapidly, the activities of the plasma lens collaboration are intensifying. Currently, the testing of the new gas nozzle is underway, along with the commissioning of the vacuum system, and will be completed by the end of August 1998. The experimental design will be completed following the gas nozzle test, which is expected to be near the close of September 1998.

Meanwhile, other systems such as plasma diagnostics, beam size measurements, and data acquisition are being developed independently. Fabrication of the plasma lens chamber, and the laser transport system is expected to take three months from October through December 1998. All systems will be integrated and installed in the first quarter of 1999.

# ACKNOWLEDGMENTS

We appreciate the contributions from Wim Leemans, our former collaborator, on plasma diagnostics and gas nozzle research and development.

# REFERENCES

1. Barletta, W., et al., "SLAC Proposal E-150," SLAC, 1993.
2. Chen, P., *Part. Accel.* **20**, 171 (1987).
3. Chen, P., et al., *IEEE Trans. Plasma Sci.* **15**, 218 (1987).
4. Rosenzweig, J. B., and Chen, P., *Phys. Rev. D* **39**, 2039 (1989).
5. Chen, P., Rajagopalan, S., and Rosenzweig, J. B., *Phys. Rev. D* **40**, 923 (1989).
6. Su, J. J., et al., *Phys. Rev. A* **41**, 3321 (1990).
7. Chen, P., *Phys. Rev. A* **45**, R3398 (1992).
8. Chen, P., Oide, K., Sessler, A., and Yu, S., *Phys. Rev. Lett.* **64**, 1231 (1990).
9. Oide, K., *Phys. Rev. Lett.* **61**, 1713 (1988).
10. Rosenzweig, J. B., et al., *Phys. Rev. Lett.* **61**, 98 (1988).
11. Nakanishi, H., et al., *Phys. Rev. Lett.* **66**, 1870 (1991).
12. Hairapetian, G., et al., *Phys. Rev. Lett.* **72**, 2403 (1994).
13. Tenenbaum, P., et al., "SLAC-PUB-6593," SLAC, 1994.
14. Betz, D., et al., *Proc. IEEE Part. Accel. Conf.* **88-647453**, 619 (1991).
15 Whittum, D. H., et al., *Part. Accel.* **34**, 89 (1990).
16. Masuda, S., et al., Wkshp. Adv. Accel. Concepts, 1998.
17. Heinrich, J. G., et al., "SLAC PROPOSAL E-144," SLAC, 1991.
18. Leemans, W., Private Communications, 1998.

# Optimal Laser Pulse Shaping in Laser Wakefield Accelerators

Pisin Chen[*] and Anatoly Spitkovsky[†]

[*] *Stanford Linear Accelerator Center, Stanford University, Stanford, CA 94305*
[†] *Department of Physics, University of California at Berkeley, Berkeley, CA 94720*

**Abstract.** We show that, in analogy to the electron beam driven Plasma Wakefield Accelerator (PWFA), an optimal transformer ratio can be attained in the laser driven Laser Wakefield Accelerator (LWFA), by properly shaping the longitudinal profile of the driving laser pulse. The concept of transformer ratio is introduced for accelerating scheme comprised of different particle species, and then applied to LWFA. The optimal laser pulse shape is derived under general conditions including the nonlinear regime of laser-plasma interaction. We show that, while the scaling for optimal transformer ratio is the same in the nonlinear regime, the corresponding laser intensity distribution is different from that for the electron beam. However, when our nonlinear result is reduced to the linear regime of plasma perturbation, the requirement for the laser pulse shape, which we previously derived, is identical to that for the electron beam.

## I  INTRODUCTION

In terms of fundamental mechanisms involved, there are basically two classes of plasma accelerator concepts – the laser driven scheme [1] and the electron beam driven scheme [2]. Within the laser driven scheme, the two major concepts have been the Laser Beat-Wave Accelerator (LBWA) and the LWFA. Historically, although both LBWA and LWFA were introduced in the original paper by Tajima and Dawson [1], due largely to the laser technology of the time, the research activities in the laser driven scheme were mostly focused on the LBWA during the 1980s [3]. Unlike LBWA, where plasma wakefield is resonantly excited by a beat-wave, the electron beam driven scheme invokes the Langmuir (shock) excitation of the plasma and, therefore, a genuine "wakefield" excitation from the outset [2] It was not until the discovery of the chirp pulse amplification (CPA) [4] in the late 80s, which revolutionized the laser technology, before the LWFA received a more serious consideration [5,6]. In this scheme one no longer relies on resonant excitation of the plasma wave. Instead, akin to electron driven scenario, the plasma wave is shock excited, and moves with a phase velocity equal to the driving beam (laser) group velocity. One curiosity therefore naturally arises: On the one hand the plasma wakefields in both cases look very similar. Yet on the other hand the

CP472, *Advanced Accelerator Concepts: Eighth Workshop,*
edited by W. Lawson, C. Bellamy, and D. Brosius
© 1999 The American Institute of Physics 1-56396-889-4/99/$15.00

detailed dynamics of laser-plasma interaction should in principle be different from that of electron beam-plasma interaction. How similar are they?

One salient feature of any collinear wakefield accelerator concept, where the driving and the accelerated beams propagate in the same direction, is that the accelerated beam is constrained to gain at most twice the driving beam energy, dictated by the Fundamental Theorem of Beam Loading [7]. It was, however, pointed out [8,9] that this theorem can be evaded provided that the driving beam current is not symmetric around the beam centroid. Furthermore, an optimal bunch (current) distribution was obtained by the same authors where *transformer ratio* much larger than two is shown to be in principle possible.

The implication of such a driving beam intensity shaping actually lies beyond what stated above. It was found that the condition that gives rise to the optimal transformer ratio corresponds to the situation where the driving beam distribution is such that the associated retarding plasma wakefield within the driving beam is essentially constant. As such all driving beam particles slow down at the same rate. This helps to suppress the two stream instability and the driving beam can thus propagate a much longer distance and be extracted with maximal possible energy into the plasma before the beam degrades too much to be usable.

Quite independently of these motivations, earlier there has been a suggestion to tailor the laser pulse *transversely* such that the laser spot size is tapered from a large value at the front to a small value at the back , so that the laser power remains constant throughout the pulse and equals to the critical laser power, $P_c$. [10] This tailored laser pulse was found to be able to propagate in a plasma for many Rayleigh lengths where large wakefields are excited and self-modulation suppressed. While encouraging, this study through computer simulations was somewhat empirical with no explicit guidance on how the pulse can be optimally shaped. Besides, its attention was mainly in the attempt to achieve a sustained relativistic self-guiding of laser pulse.

Inspired by these earlier studies and the similarities between PWFA and LWFA, we recently investigated the issue of transformer ratio in the LWFA and derived the optimal laser pulse shape which should give rise to the largest possible transformer ratio, and therefore the maximal possible efficiency of laser-plasma coupling. [11] That analysis, however, was carried out in the linear regime of laser-plasma interaction. Within the linear regime, we found that the resultant optimal transformer ratio and its corresponding pulse shape in the LWFA is identical to that in the PWFA. In this paper we extend this investigation to the nonlinear regime, and show that in the nonlinear regime (and, therefore, in general) the optimal pulse shapes for the laser and the electron beam turn out to be indeed different, although the associated optimal transformer ratios do obey the same asymptotic scaling.

## II  PLASMA WAKEFIELD GENERATION

We consider a homogeneous unmagnetized plasma which is charge neutral, at rest, and has density $n_0$ in the absence of the electromagnetic wave. Laser propagates in the $\hat{z}$ direction with initial frequency $\omega_0 \gg \omega_p \equiv \sqrt{4\pi e^2 n_0/m}$, and group and phase speed $v_g \approx v_p \approx c$. To focus on the essense of the issue, we limit ourselves to one (longitudinal) dimensional approximation, where the system is governed by the following complete set of equations:

$$\partial n_e/\partial t + \partial/\partial z(n_e v_\parallel) = 0, \tag{1}$$

$$\partial \mathbf{u}/\partial t + v_\parallel \partial \mathbf{u}/\partial z = -(e/mc)[\mathbf{E} - (\mathbf{v} \times \mathbf{B})/c], \tag{2}$$

$$\hat{z} \times (\partial \mathbf{E}/\partial z) = -c^{-1}(\partial \mathbf{B}/\partial t), \tag{3}$$

$$\hat{z} \times (\partial \mathbf{B}/\partial z) = -4\pi e c^{-1} n_e \mathbf{v} + c^{-1}(\partial \mathbf{E}/\partial t), \tag{4}$$

$$\partial E_\parallel/\partial z = -4\pi e(n_e - n_0). \tag{5}$$

Here, $n_e$ is the plasma electron number density, $\mathbf{u} = \mathbf{p}_e/mc$ is the dimensionless electron momentum, $\mathbf{v} = c\mathbf{u}/\gamma$ is the electron velocity, and $\gamma = (1 + u^2)^{1/2}$ is the Lorentz factor, $\mathbf{E}$ and $\mathbf{B}$ are the fields. We obtain a nonlinear wave equation by taking the curl of (3) and substituting from (4)

$$\partial^2 E_\perp/\partial t^2 - c^2 \partial^2 E_\perp/\partial z^2 = 4\pi e \partial(n_e v_\perp)/\partial t, \tag{6}$$

$$\partial^2 E_\parallel/\partial t^2 = 4\pi e \partial(n_e v_\parallel)/\partial t, \tag{7}$$

where $\perp$ and $\parallel$ refer to components perpendicular and parallel to direction of propagation. One fundamental feature of collinear wakefield excitations is that the phase velocity of the wakefield is equal to driving beam's group velocity $v_g \equiv \beta_g c$. Thus every quantity under consideration in LWFA is stationary in the laser co-moving frame, described by the coordinates $\zeta = z - v_g t$, $\tau = t$. Furthermore, it is convenient to introduce dimensionless, normalized scalar and vector potentials, $\phi(\zeta)$ and $a(\zeta)$, such that $E_\parallel \equiv -(mc^2/e)\partial\phi/\partial\zeta$, $E_\perp = -(mc/e)\partial a/\partial t = (mc^2\beta_g/e)\partial a/\partial\zeta$.

Since the system is invariant under perpendicular translations, the transverse canonical momentum is conserved. Therefore, $u_\perp = a$. In terms of $\phi$ and $a$, equations (6) and (7) now become:

$$\frac{2}{c}\beta_g \frac{\partial^2 a}{\partial\zeta\partial\tau} + (1 - \beta_g^2)\frac{\partial^2 a}{\partial\zeta^2} - \frac{1}{c^2}\frac{\partial^2 a}{\partial\tau^2} = k_p^2 \frac{n_e}{n_0}\frac{a}{\gamma}, \tag{8}$$

$$\frac{d^2\phi}{d\zeta^2} = k_p^2\left(\frac{n_e}{n_0} - 1\right), \tag{9}$$

where $k_p = 2\pi/\lambda_p$ is the plasma wavenumber. From continuity equation and equation of motion (Eqs.(1) and (2)), $\gamma$ and $n_e/n_0$ can be expressed in terms of $a$ and $\phi$:

$$1 + \phi = \gamma(1 - v_{\parallel}/c), \tag{10}$$

$$\gamma = \frac{1 + a^2 + (1 + \phi)^2}{2(1 + \phi)}, \tag{11}$$

$$\frac{n_e}{n_0} = \frac{\gamma}{1 + \phi} = \frac{1 + a^2 + (1 + \phi)^2}{2(1 + \phi)^2}. \tag{12}$$

Substituting in (8) and (9), retaining leading order derivatives in (8), and defining the variable $x \equiv 1 + \phi$, we finally arrive at two fully nonlinear, self-consistent coupled equations describing laser pulse evolution and plasma wakefield generation:

$$\frac{\partial^2 a}{\partial \zeta \partial \tau} = \frac{c}{2} \frac{k_p^2}{x} a, \tag{13}$$

$$\frac{d^2 x}{d\zeta^2} = \frac{k_p^2}{2} \left( \frac{1 + a^2}{x^2} - 1 \right). \tag{14}$$

Let us first concentrate on the generation of plasma waves (14), and later consider the back-reaction of this plasma wakefield on the laser pulse. Let the laser pulse have length $L_p$, extending from $\zeta = 0$ to $\zeta_f = L_p$ (pulse moving to the left). For convenience, we split Eq.(14) into two equations corresponding to regions within and outside (behind) the laser pulse:

$$\frac{d^2 x}{d\zeta^2} = k_p^2 \left( \frac{1 + a^2}{2x^2} - \frac{1}{2} \right) \quad \text{for} \quad 0 \leq \zeta \leq \zeta_f, \tag{15}$$

$$\frac{d^2 x}{d\zeta^2} = k_p^2 \left( \frac{1}{2x^2} - \frac{1}{2} \right) \quad \text{for} \quad \zeta < \zeta_f. \tag{16}$$

The initial condition for (15) is that the plasma is unperturbed prior to the arrival of the laser ($\zeta < 0$). Thus both $v_{\parallel}$ and the electric field $\mathcal{E} \equiv eE_{\parallel}/mc\omega_{p0} = -k_p^{-1}dx/d\zeta$ are zero initially. This establishes initial conditions $x(0) = 1$, and $x'(0) = 0$. For Eq.(16), initial conditions are matched to the solution within the laser: $x(\zeta_f) = x_f$, and $k_p^{-1}x'(\zeta_f) = -\mathcal{E}(\zeta_f)$.

The laser wakefields can now be determined if we choose $x$ as an independent variable for the system and treat $\zeta$ as a function of $x$ only, for $0 \leq \zeta \leq \zeta_f$ or $1 \leq x \leq x_f$. Then Eq.(15) can be integrated to give

$$[\mathcal{E}^{in}(x)]^2 = \left[ \frac{1}{k_p} \frac{dx}{d\zeta} \right]^2 = 2 - \frac{1}{x} - x + \int_1^x \frac{a^2(y)}{y^2} dy. \tag{17}$$

for $1 \leq x \leq x_f$, while outside the laser, Eq.(16) gives:

$$[\mathcal{E}^{out}(x)]^2 = 2 - \frac{1}{x} - x + \int_1^{x_f} \frac{a^2(y)}{y^2} dy. \tag{18}$$

We can now use these solutions to investigate the efficiency of energy transfer between the laser and accelerated beam.

# III  TRANSFORMER RATIO IN LWFA

In the particle (electron)-beam driven wakefield acceleration the transformer ratio is simply defined as the maximum accelerating plasma wakefield behind the driving beam (experienced by the accelerated particles), divided by the maximum retarding plasma wakefield within the driving beam (experienced by the driving beam particles). At least two hidden assumptions are involved in this notion. First, being both electron beams, the energy gains and losses, which are linearly proportional to the wakefield, are represented by the wakefield strengths. Second, being electrons, under the condition where collisional loss is negligible, the total number of driving beam particles is conserved. Therefore, the definition of the transformer ratio, which is a measure of the efficiency of energy transfer from the driving beam to the accelerated beam, is represented by the ratio of the wakefields. Let's generalize this definition to the case where the driving and accelerated beams are not necessarily of the same species of particles. We start with the following definition of transformer ratio:

$$R = \frac{\text{Max. gradient of energy gain/particle of accelerated beam}}{\text{Max. gradient of energy loss/particle of the driver beam}}, \tag{19}$$

within the same traversed distance. If the energy loss (or gain) of both species is linearly proportional to the distance traveled, then we can reformulate the transformer ratio by expressing the rate of energy loss per particle in terms of the depletion length $l_d$ of the driver beam, i.e., "Max. gradient of energy loss/particle" $= (\gamma_d - 1)m_d c^2/l_d$ where $\gamma_d m_d c^2$ is the initial energy of the driver particle. In other words, $l_d$ is the distance in which the particle of the driving beam that experiences maximum deceleration will loose all its kinetic energy. By noting that the numerator of (19) multiplied by $l_d$ is equal to the maximum energy gain of the particle of accelerated beam attainable before the driver beam comes to complete stop, we can re-write $R$ as:

$$R = \frac{\Delta\gamma_a}{(\gamma_d - 1)} \frac{m_a c^2}{m_d c^2}, \tag{20}$$

where $\Delta\gamma_a m_a c^2$ is the maximum energy gain of accelerated particle of mass $m_a$ after traveling one driver depletion length. By defining the transformer ratio in this way, we can isolate the optimizable characteristic of the accelerator - the ratio of relativistic factors given above, from the rest mass ratio which is fixed from the outset. Equations (19) and (20) then help us pave the way to a definition of transformer ratio in LWFA.

Obviously in LWFA the driving and accelerated beam consist of different species of particles. A natural choice would be to use laser photons as the driving particles and electrons as the accelerated particles. Although in principle the number of photons in a laser pulse need not be conserved, in the setting of plasma accelerators the scattering cross sections are so low that the photon number is effectively

325

constant. The transfer of energy between the laser and the plasma is therefore predominantly in the form of frequency shifts, or, the so-called *photon acceleration* [13]. This photon energy loss should be in the denominator of eq. (19). Let's determine it quantitatively. The energy density $U$ of the laser pulse is the sum of electric and magnetic energy densities:

$$U = \frac{1}{8\pi}(|E|^2 + |B|^2) = \frac{1}{4\pi}\left(\frac{mc^2}{e}\right)^2\left(\frac{\partial a}{\partial \zeta}\right)^2. \tag{21}$$

The change of energy density with distance $z = c\tau$, or $dU/dz$, can be calculated using the laser evolution equation (8):

$$\frac{dU}{dz} = \frac{1}{4\pi}\left(\frac{mc^2}{e}\right)^2\left[\frac{k_p^2}{2}\frac{\partial}{\partial\zeta}\frac{a^2}{x} - \frac{k_p^2}{2}a^2\frac{\partial}{\partial\zeta}\frac{1}{x}\right]. \tag{22}$$

The first term, which vanishes upon integration over all space, represents the redistribution of energy within the laser due to the contraction of the pulse during propagation, while the second term constitutes the net energy density loss to the plasma. In order to obtain the rate of energy loss per photon, we should divide the second term of (22) by the photon density in the laser or, equivalently, by $1/\hbar$ times the action density $\mathcal{A} = Im(A^*E/4\pi c) = a^2 m\omega/4\pi r_e$, where $A$ is laser vector potential, $E$ – electric field, $r_e$ is classical electron radius, and $\omega$ is the laser frequency. Then the energy loss per photon is given by

$$\frac{\partial\epsilon}{\partial z} \equiv \frac{\partial\hbar\omega}{\partial z} = -\frac{1}{2}\left(\frac{\omega_p}{\omega}\right)^2\hbar\omega\frac{\partial}{\partial\zeta}\frac{1}{x} = -\frac{1}{2}\left(\frac{\omega_p}{\omega}\right)^2\hbar\omega\frac{\partial}{\partial\zeta}\frac{n_e}{\gamma n_0}, \tag{23}$$

which is the same expression as derived in [15]. Physically, the laser frequency red-shift is triggered by a time- dependent change of the index of refraction of the perturbed plasma in comoving frame. Since the index of refraction of plasma is $[1 - (\omega_p(\zeta)/\omega)^2]^{1/2} = [1 - (\omega_p/\omega)^2 n_e(\zeta)/\gamma n_0]^{1/2}$, variations of plasma density and electron inertia both contribute to photon acceleration/deceleration. Due to the varying index of refraction, phase speed of the laser $\beta_p = v_p/c \approx 1 + \omega_p^2(\zeta)/2\omega^2$, is not the same over the spatial extent of the pulse. Hence, separation between phase peaks changes, which modifies the frequency of the laser.

Before we can use formula (20) for transformer ratio, we should determine the other driver parameters such as $\gamma_d$, $m_d c^2$ and the depletion length $l_d$. Since the laser in plasma obeys dispersion relation $\omega^2 = k^2 c^2 + \omega_p^2$, its group velocity is $c(1 - \omega_p^2/\omega^2)^{1/2}$, and we can directly calculate initial $\gamma_d$ to be : $\gamma_d = (1 - \beta_g^2)^{-1} = \omega_0/\omega_p$, where $\omega_0$ is the initial laser frequency. As for the mass, the laser photon in plasma gains an effective mass $m_{eff}c^2 = \hbar\omega_p$, such that $\gamma_d m_{eff}c^2 = \hbar\omega$ as expected (this can also be seen by multiplying the dispersion relation by $\hbar^2$ and comparing term by term with relativistic energy equation $E^2 = (pc)^2 + (mc^2)^2$). This analogy can be carried even further as we find that the most energy the laser can loose to the plasma is given by $(\gamma_d - 1)m_{eff}c^2 = \hbar(\omega - \omega_p)$, because laser ceases to propagate below plasma frequency. Therefore, we can determine the driver depletion length by solving equation (23) for $\omega(z)$ and setting it equal to $\omega_p$. This produces

$$l_d = \frac{(\omega_0/\omega_p)^2 - 1}{\left|\frac{\partial}{\partial\zeta}\frac{1}{x}\right|_{max}},$$

(24)

where denominator should be evaluated at a maximum inside the pulse, which would correspond to the depletion length of maximally decelerated photons in the laser.

For accelerated electrons the energy gain mechanism is the same as in PWFA case. Therefore, the numerator of equation (20) can be formulated as the product of the maximum electrostatic force in the wake and the laser depletion length: $\Delta\gamma_a mc^2 = eE_{max}^{out}l_d = \mathcal{E}_{max}^{out}k_pl_d\, mc^2$. We are now ready to put all the pieces together and define the transformer ratio for LWFA using formula (20):

$$R^{\text{LWFA}} = \frac{\Delta\gamma_a}{(\gamma_d - 1)}\frac{m_ac^2}{m_dc^2} = \frac{\mathcal{E}_{max}^{out}k_pl_d}{(\omega_0/\omega_p - 1)}\frac{mc^2}{\hbar\omega_p}$$

(25)

# IV   OPTIMAL SHAPE OF LASER DRIVER

We now proceed to maximize the wakefield-dependent part of the transformer ratio. Upon substitution for $\mathcal{E}$ and $l_d$, we get:

$$\frac{\Delta\gamma_a}{\gamma_d - 1} = \left(1 + \frac{\omega_0}{\omega_p}\right)\frac{\left|\frac{\partial}{\partial\zeta}x(\zeta)\right|_{max}^{out}}{\left|\frac{\partial}{\partial\zeta}\frac{1}{x(\zeta)}\right|_{max}^{in}}$$

(26)

The numerator of this expression is proportional to the maximum accelerating field in the wake, while denominator is proportional to the maximum photon deceleration inside the laser. From Eq.(18) it follows that the maximum value of the electric field in a free plasma wave is achieved at $x = 1$. Then $(|x'|_{max}^{out})^2 = k_p^2\int_1^{x_f}a^2(y)/y^2dy \equiv k_p^2b(x_f)$. To find the maximum value of laser deceleration inside the pulse we use conclusions of earlier studies in PWFA [9,16], and impose the following ansatz that, *in order that the transformer ratio be optimized, the photon deceleration within the laser pulse should be a non-decreasing function of $\zeta$*, i.e., the rate of energy deposition into plasma should not oscillate inside the laser. The maximum photon deceleration, therefore, should occur at the end of the pulse at $\zeta = \zeta_f$, and

$$\left(\frac{d\omega}{dz}\right)_{max} \propto \left|\frac{\partial}{\partial\zeta}\frac{1}{x(\zeta)}\right|_{\zeta=\zeta_f} = \frac{k_p}{x_f^2}\left\{2 - x_f - \frac{1}{x_f} + \int_1^{x_f}\frac{a^2(y)}{y^2}dy\right\}^{1/2}.$$

(27)

Now the transformer ratio can be expressed as:

$$\frac{\Delta\gamma_a}{\gamma_d - 1} = \left(1 + \frac{\omega_0}{\omega_p}\right)\sqrt{\frac{x_f^4 b(x_f)}{2 - x_f - x_f^{-1} + b(x_f)}} \tag{28}$$

Maximizing this expression with respect to $x_f$, we get $b(x) = (x - 1)^2/x$, or a laser pulse shape $a_c^2(x) = b'(x)x^2 = x^2 - 1$, for $1 \le x \le x_f$. This is in a sense a trivial solution, since for the laser amplitude $a(x) = a_c(x)$ the right hand side of (14) becomes identically zero, and, therefore, $x(\tau) = 1$. Such solution corresponds to an infinitely long laser pulse, and, hence is unphysical. For a finite-length laser we introduce a trial function $g(x)$ [16] such that

$$a_l^2(x) = a_c^2(x) + g(x), \tag{29}$$

with $g(x) \ge 0$ and $\int_1^x g(y)/y^2 dy > 0$ for $1 \le x \le x_f$. Following our ansatz, to have an optimal transformer ratio $g(x)$ must be chosen such that the photon deceleration is remains constant over the length of the pulse. Namely, we require that $|d\omega/dz| \propto x'/x^2 \equiv \kappa = \text{const}$. Therefore, $x' = \kappa x^2$ for $x > 1$. This, however, has to match with the boundary condition for the unperturbed plasma at the beginning of the pulse, or $x'(x = 1) = 0$, which can be satisfied by introducing the Heaviside step-function into the solution: $x' = \kappa x^2 \theta(x - 1)$. The laser intensity variation can now be found by substituting this $x'$ into (17):

$$a_l^2(x) = x^2 - 1 + 4\left(\frac{\kappa}{k_p}\right)^2 x^5 \theta^2(x - 1) + 2\left(\frac{\kappa}{k_p}\right)^2 x^6 \delta^+(x - 1) \tag{30}$$

where $\delta^+(x - 1) = \delta(x - 1)\theta(x - 1)$ is the right half of a delta-function centered at $x = 1$. Equation $x'/x^2 = \kappa$ can be solved for $x$ as a function of $\zeta$: $x(\zeta) = 1/(1 - \kappa\zeta)$. Substituting this expression into (30) we finally obtain the optimal laser shape as a function of the normalized comoving coordinate $\hat{\zeta} \equiv k_p\zeta$:

$$a_l^2(\hat{\zeta}) = \frac{2\hat{\kappa}\delta^+(\hat{\zeta})}{(1 - \hat{\kappa}\hat{\zeta})^6} + \frac{4\hat{\kappa}^2}{(1 - \hat{\kappa}\hat{\zeta})^5}\theta^2(\hat{\zeta}) + \frac{1}{(1 - \hat{\kappa}\hat{\zeta})^2} - 1, \quad \text{[LWFA]} \tag{31}$$

where $\hat{\kappa} \equiv \kappa/k_p$ is the normalized photon deceleration parameter. From (25), the corresponding optimal transformer ratio is then

$$R^{\text{LWFA}} = \left(1 + \frac{\omega_0}{\omega_p}\right)\sqrt{\frac{1 + (k_p L_p)^2(1 - \hat{\kappa}(k_p L_p))^3}{(1 - \hat{\kappa}(k_p L_p))^4}}\left(\frac{mc^2}{\hbar\omega_p}\right) \tag{32}$$

# V  COMPARISON OF LWFA AND PWFA

A schematic drawing of the optimal laser intensity variation and its associated plasma wakefield is shown in Fig. 1. We note that in order to attain the optimal condition, the laser intensity is required to increase dramatically near the end of the pulse. This is due to the fact that the plasma density perturbation, $n_e$, and the electron inertia, $\gamma$, work against each other in contributing to the change of the index of refraction, and thus the photon deceleration. Near the end of the laser pulse the plasma electrons become very relativistic, and therefore to maintain a constant deceleration it is necessary that the laser intensity increase dramatically so as to sustain further plasma perturbations. This is in sharp contrast to situation in PWFA. As shown by Yan and Chen [16], the optimal electron beam intensity in the nonlinear regime is

$$\frac{n_b(\hat\zeta)}{n_0} = \hat\kappa\delta^+(\hat\zeta) + \frac{1}{2}\Big[1 - \frac{1}{(1 + \hat\kappa\hat\zeta)^2}\Big], \qquad [\text{PWFA}] \qquad (33)$$

where $\hat\kappa$ is analogous deceleration parameter for electron driver beam. In the long pulse limit this expression reaches an asymptotic limit of $1/2$ (see Fig. 2). This means the driving electron beam reaches a constant density. As Rosenzweig [14] has shown, a constant beam density in the nonlinear regime can already provide a transformer ratio greater than 2, though not optimized. As a delta-function pulse is technically hard to attain, it was shown in [16] that the optimal transformer ratio can be well approached if one introduces a bump near the front of the beam, in place of the delta function.

On the contrary, the asymptotic form of the optimal laser pulse shape (31) should approach a single sharp spike, $a^2 \sim 4\hat\kappa^2/[1 - \hat\kappa\hat\zeta]^5$, which is the dominant term in Eq.(31). Note that unlike the case of PWFA, in LWFA the optimal pulse cannot be arbitrarily long. Indeed, for a given photon deceleration $\kappa$, the laser pulse length is limited to a critical length, $\zeta_c = 1/\kappa$. Thus the way of increasing transformer ratio in LWFA is, instead of extending the length of the beam with a fixed intensity as in PWFA, to stagger the ever-increasing laser intensity near the end of the

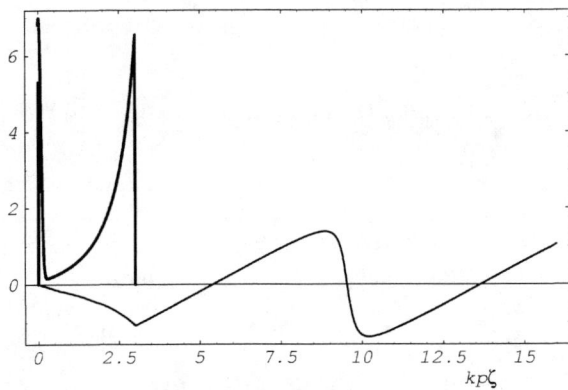

**FIGURE 1.** General shape of nonlinear laser intensity profile and corresponding wakefield (arbitrary units on y-axis)

329

pulse, but with the pulse length fixed to $\zeta_f \lesssim \zeta_c$. Remembering that $\kappa$ is inversely proportional to the depletion length of the laser pulse (Eq.(24)), the length of the pulse is uniquely determined from the total available energy of the pulse and required depletion length. Conversely, given the energy and length of the pulse that should satisfy the optimum shape, we can calculate the depletion length and corresponding transformer ratio.

Learning from Ref. [16], we can also envision replacing the leading delta function by a small intensity bump near the front of the laser. This construction, which is a practical compromise, then leads to a single sharp spike at the end of the pulse. We should emphasize, however, that an untailored short pulse is not the same as what we described here. Head-tail asymmetry of the laser intensity is an essential ingredient in the theory. To ensure an ultra-large wakefield behind the laser pulse, it is still necessary that the pulse be terminated abruptly.

It is interesting to note that, although the optimal shapes are very different between PWFA and LWFA in the nonlinear regime, in the linear regime where $n_b/n_0 \ll 1$ or $a^2 \ll 1$, the two cases do converge . This is mainly due to the fact that in the linear regime the plasma perturbation is nonrelativistic. Therefore, the Lorentz factor $\gamma$ does not effectively participate in the change of index of refraction, and driving beam deceleration relies essentially on the plasma density perturbation, same as in PWFA [11]. In this limit $\hat{\kappa} \to 0$, and two shapes demonstrate equivalent pulse shape characterized by the initial half-delta function and linear rise of intensity/beam current:

$$\frac{n_b}{n_0} = \frac{a_l^2}{2} = \hat{\kappa}(\delta^+(\hat{\zeta}) + \hat{\zeta}), \qquad (34)$$

where $\hat{\kappa}$ stands for electron or photon deceleration depending on the accelerating scheme. In this limit, the transformer ratios for both schemes reduce to

$$R^{\mathrm{PWFA}} \to \sqrt{1 + (k_p L_{beam})^2}, \qquad [\mathrm{PWFA}] \qquad (35)$$

$$R^{\mathrm{LWFA}} \to \left(1 + \frac{\omega_0}{\omega_p}\right)\sqrt{1 + (k_p L_{pulse})^2}\left(\frac{mc^2}{\hbar\omega_p}\right). \qquad [\mathrm{LWFA}] \qquad (36)$$

Both transformer ratios show equivalent dependence on the length of the driver, and $R^{\mathrm{LWFA}}$ has expected factor due to the different rest masses of driver and accelerated beam in LWFA. In terms of the energy transfer efficiency characterized by $\Delta\gamma_a/(\gamma_d - 1)$, we see that LWFA is intrinsically more efficient than PWFA by a factor of $(1 + \omega_0/\omega_p)$.

# VI  DISCUSSION

We have shown that in analogy to the PWFA, the concept of transformer ratio can also be introduced in the LWFA. In particular, we show that the optimal transformer ratio corresponds to a laser pulse shape which is composed of two parts: a half-delta function followed by a ramp of laser intensity increasing as $(1 - \kappa\zeta)^{-5}$. In such an arrangement the laser photon deceleration is constant throughout the laser, except for the very head of the pulse where the delta function locates.

When all photons red-shift, or "slow down", at the same rate, the laser self-modulation, or Raman, instability is suppressed. Indeed, the suppression of the Raman instability is indicative of high efficiency in transferring the laser energy to the plasma (instead of being reabsorbed by other laser photons). This is the physical reason why the plasma wakefield behind the laser can be much larger than what an untailored driving pulse can attain, and therefore an optimal transformer ratio much higher than what the Fundamental Theorem of Beam Loading dictates.

Although what we asserted above has been already demonstrated in the PWFA, we find it still worthy of reiterating. The significance of driving beam tailoring in wakefield accelerators lies not only in its promise to excite the largest possible accelerating wakefield for a given total driving beam current (in the case of an electron beam) or laser power. It also provides a means to maximally suppress the two-stream instability in the case of PWFA and its counterpart, the Raman instability in the case of LWFA.

There are issues related to pulse shaping which have not been discussed in this paper. In reality it is hard to construct a delta-function pulse. Earlier studies of pulse shaping in the PWFA have shown that other physically more realistic bunch shapes, such as a "triangle", a "door-step" [9], or a "half-Gaussian" [17] in the linear regime, and a "bump" in the nonlinear regime, can still retain a very decent transformer ratio. We believe that the situation must be the same for the LWFA. One very important issue which was not covered in this study is the stability of our solution. It is an open question whether a small deviation of the laser intensity distribution from the optimal solution would trigger an instability that would further drive the distribution away from the desired one. One other very important issue is the influence of the beam-plasma interaction in the transverse dimension. Studies have shown [18] that if the laser spot size is much larger than the plasma wavelength, certain beam filamentation instabilities are very severe with growth rates faster than instabilities in the longitudinal dimension. It is therefore very important that one studies these issues before one can be confident in the utility of the laser pulse shaping concept that we have introduced in this paper.

# VII   ACKNOWLEDGMENTS

We appreciate the contributions of T. Katsouleas and W. Mori in the earlier collaboration on the subject in the linear regime. A.S. would also like to thank J. Wurtele and J. Arons for very useful discussions, suggestions and support during the preparation of this work.

# REFERENCES

1. T. Tajima and J. M. Dawson, *Phys. Rev. Lett.*, (1979).
2. P. Chen, J. M. Dawson, R. Huff, and T. Katsouleas, *Phys. Rev. Lett.* **54**, 693 (1985).
3. C. Joshi, editor, *Advanced Accelerator Concepts, Lake Arrowhead*, AIP (1989)
4. P. Maine, D. Strickland, P. Bado, M. Pessot, and G. Mourou, *IEEE J. Quantum Elec.* **QE-24**, 398 (1988).
5. L. M. Gorbunov and V. I. Kirsanov, *Sov. Phys. JETP* **66**, 290 (1987).
6. P. Sprangle, E. Esarey, A. Ting, and G. Joyce, *Appl. Phys. Lett.* **53**, 2146 (1988).
7. A. Chao, *Physics of Collective Instabilities in High Energy Accelerators*, John Wiley & Sons (1993).
8. K. Bane, P. Chen, and P. Wilson, *IEEE Trans. Nucl. Sci.* **NS-32**, 3524 (1985).
9. P. Chen, J. J. Su, J. M. Dawson, K. Bane, and P. Wilson, *Phys. Rev. Lett.* **56**, 1252 (1986).
10. P. Sprangle, E. Esarey, J. Krall, and G. Joyce, *Phys. Rev. Lett.*, **69**, 2200 (1992).
11. P. Chen, A. Spitkovsky, T. Katsouleas, and W. Mori, *Nucl. Instr. Meth.*, **410**, no.3, 488, 1998.
12. E. Esarey, P. Sprangle, J. Krall, A. Ting, *IEEE Trans. Plasma Sci.*, **24**, no. 2, 252, (1996).
13. S. Wilks, J. M. Dawson, W. Morri, T. Katsouleas, and M. Jones, *Phys. Rev. Lett.* **62**, 2600 (1989).
14. J. B. Rosenzweig, *Phys. Rev. Lett.* **58**,555 (1987)
15. E. Esarey, A. Ting, and P. Sprangle, *Phys. Rev. A*, **42**, 3526 (1990).
16. Y. T. Yan, H. Chen, *Phys. Rev. A*, **38**, 1490 (1988)
17. T. Katsouleas, *Phys. Rev. A*, **33** 2056 (1986).
18. C. D. Decker et al., *IEEE Trans. Plasma. Sci.* **24**, 379 (1996).

# Acceleration of Electrons in a Self-Modulated Laser Wakefield

S.-Y. Chen, M. Krishnan, A. Maksimchuk and D. Umstadter

*Center for Ultrafast Optical Science, University of Michigan, Ann Arbor, MI 48109*

**Abstract.** Acceleration of electrons in a self-modulated laser-wakefield is investigated. The generated electron beam is oberved to have a multi-component beam profile and its energy distribution undergoes discrete transitions as the conditions are varied. These features can be explained by simple simulations of electron propagation in a 3-D plasma wave.

Understanding the dynamics of electron acceleration in an electron plasma wave is important for developing plasma-based electron accelerators [1]. Of the several methods for driving large-amplitude plasma waves, the laser wakefield accelerator (LWFA) and the self-modulated LWFA, have recently received considerable attention because of the reduction in size of terawatt class laser systems [2]. In the LWFA, an electron plasma wave is driven resonantly by a short laser pulse, and an additional injection mechanism is required [3,4]. In the self-modulated LWFA, an electron plasma wave is excited by a relatively long laser pulse undergoing stimulated Raman forward scattering instability [5], and the injection of electrons is achieved by trapping hot background electrons which are preheated by other processes such as Raman backscattering and sidescattering instabilities [6–8].

Several groups observed the generation of MeV electrons from the self-modulated LWFA [7–11]. A two-temperature distribution was reported in the electron energy spectrum [11]. R. Wagner *et al.* [10] observed that the generated electron beam has a multi-component beam profile, and that the temperature of the electrons in the low energy range undergoes abrupt change, coinciding with the onset of the extention of the laser channel by self-channeling of the laser pulse, when the laser power or plasma density is varied. Several 1-D and 2-D simulations [6,12–14] have also been done to study the electron beam characteristics of a self-modulated LWFA, in addition to the 1-D theoretical analysis [15,16]. However, no explanation were given for these experimental observations. In this Letter, we reported the observation of multi-component electron-beam profiles and discrete changes in the slope of electron energy distribution. These phenomena show more complicated behaviors compared to those reported before [10], mainly due to the ability to reach higher laser intensity and plasma density. Most importantly, by using a

CP472, *Advanced Accelerator Concepts: Eighth Workshop,*
edited by W. Lawson, C. Bellamy, and D. Brosius

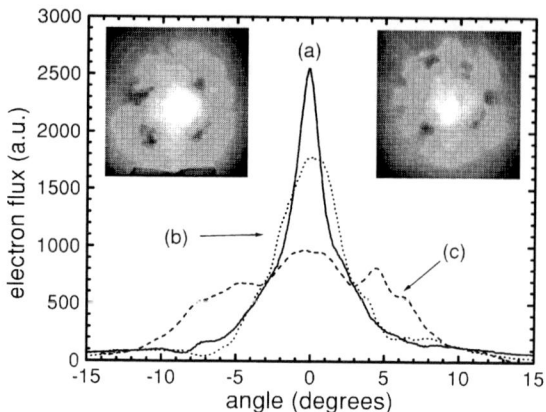

**FIGURE 1.** Lineouts of the electron beam profiles for different peak laser powers and plasma densities: (a) 2.9 TW, $3.4 \times 10^{19}$ cm$^{-3}$, (b) 3.5 TW, $6.2 \times 10^{19}$ cm$^{-3}$, and (c) 0.6 TW, $2.3 \times 10^{19}$ cm$^{-3}$. Inset on the left side shows the electron beam profile at 3.5 TW and $6.2 \times 10^{19}$ cm$^{-3}$. Inset on the right side shows the electron beam profile at 2.0 TW and $2.3 \times 10^{19}$ cm$^{-3}$.

simple 3-D simulation, we are able to explain these phenomena for the first time.

The experiment was done by using a laser system that produced 400-fs laser pulses at 1.053 μm with a maximum peak power of 4 TW. The 50-mm-diameter laser beam was focused with an f/3.3 parabolic mirror onto the front edge of a supersonic helium gas jet. The focal spot is a 7 μm FWHM Gaussian spot, which contains 60 % of the total energy, and a large dim spot (100 μm FWHM). The helium gas was fully ionized by the foot of the laser pulse. At the laser power of $\geq 2$ TW and the plasma density of $\geq 2 \times 10^{19}$ cm$^{-3}$, the laser pulse undergoes relativistic-ponderomotive self-channeling [10,17], and the laser channel extends to 750 μm long, the length of the gas jet. The length of the laser channel was monitored by side imaging of Thomson scattering of the laser pulse in the plasma.

The generated electron beam can be characterized by its energy distribution (the longitudinal emittance) and beam divergence (the transverse emittance). The electron energy spectrum in the low energy range ($< 8$ MeV) was measured using a dipole permanent magnet with a LANEX scintillating screen imaged by a CCD camera as the detector. Higher-energy electron energy spectrum was obtained by using a dipole electromagnet and a multi-wire proportional chamber (MWPC). The electron beam profile was measured using a LANEX screen imaged by a CCD camera at 16 cm away from the gas jet. Since the source size of the generated electron beam is small, $\sim 10$ μm (determined by the diameter of the laser channel), the electron beam profile on the LANEX is actually a measurement of the electron beam divergence (angular pattern).

The electron beam profile (angular pattern) is observed to contain several concentric Gaussian-like-profile beams, and the number of the beam components depends on the laser power and plasma density. At the plasma density of $2.3 - 6.2 \times 10^{19}$

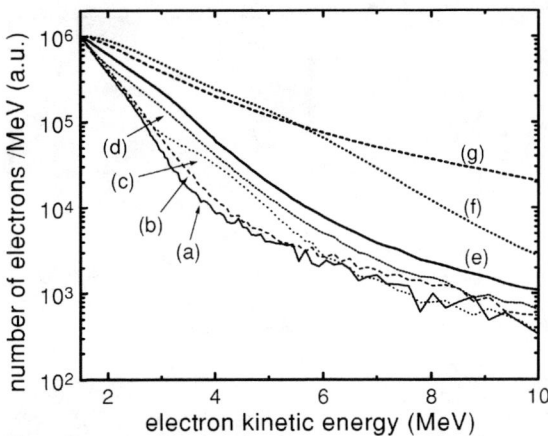

**FIGURE 2.** Electron energy spectra for different peak laser powers and plasma densities: (a) 2.6 TW, $3.4 \times 10^{19}$ cm$^{-3}$, (b) 2.9 TW, $3.5 \times 10^{19}$ cm$^{-3}$, (c) 3.3 TW, $4.8 \times 10^{19}$ cm$^{-3}$, (d) 3.9 TW, $4.8 \times 10^{19}$ cm$^{-3}$, (e) 1.7 TW, $6.2 \times 10^{19}$ cm$^{-3}$, (f) 2.7 TW, $6.2 \times 10^{19}$ cm$^{-3}$, , and (g) 3.5 TW, $6.2 \times 10^{19}$ cm$^{-3}$.

cm$^{-3}$, only one beam component ($\sim 15°$ FWHM) exists in the electron beam for 0.6-TW laser power. For the laser power larger than 1 TW, generally two beam components were observed, which have the divergence of 15° and 7.5° FWHM, respectively. Under the condition of $2.0 - 3.5$ TW laser power and $2.3 - 3.4 \times 10^{19}$ cm$^{-3}$ plasma density, a third beam component was observed and its divergence varies from 1.2° to 2.5°. The electron flux of this third component can be as high as 10 times that of the second component. Figure 1 shows the lineouts of the electron beam profiles at three different conditions, corresponding to the cases of one-, two-, and three- component beam. Furthermore, when the second beam component shows up, there are usually some holes appearing in the first (widest) beam component. These holes form regular patterns, such as TEM$_{10}$, TEM$_{11}$, and TEM$_{12}$ modes, as shown in the insets of Fig. 1.

Figure 2 shows the normalized electron energy spectra in the low energy range for different laser powers and plasma densities. The spectra are found to have Maxwellian-like distributions, i.e., $\exp(-\alpha\gamma)$, where $\gamma$ is the relativistic factor of the electron energy and $\alpha$ is a fitting parameter ($(511 \text{ keV})/\alpha$ is the temperature). The slope, $\alpha$, of the spectrum is found to change discretely with varying the laser power and plasma density. For instance, at a fixed plasma density, the slope stays the same with increasing laser power until a certain laser power is reached. Then the slope $\alpha$ changes to a lower value, and stays the same with further increase of laser power until the next jump. The same behavior occurs for varying plasma density at a fixed laser power. Three $\alpha$ values (two jumps) were observed in this experiment: 1.0, 0.6, 0.3. The abrupt jump of the energy slope and the emergence of the second or third component in the electron beam profile do not have direct link

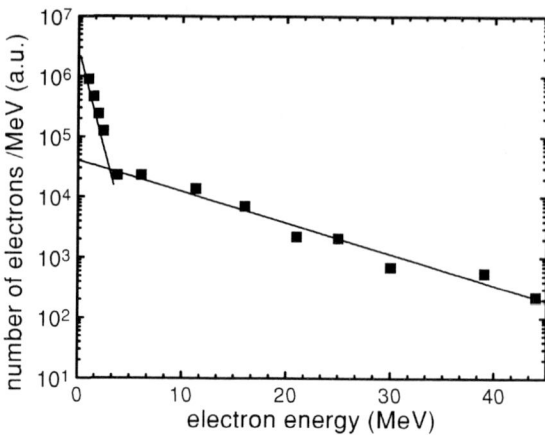

**FIGURE 3.** Electron energy spectrum for 3-TW peak laser power and $3 \times 10^{19}$-cm$^{-3}$ plasma density.

with the occurrence of laser self-channeling. Another important observation in this experiment is the two-temperature distribution in the electron energy spectrum. As shown in Fig. 3, which is obtained using high-energy electron spectrometer, the slope of electron energy distribution in the low energy range ($\leq 5$ MeV) is steep, and the slope in the high energy range is much less steep (almost flat).

To understand the physical origin of these phenomena, we run a simple 3-D particle simulation code and compare the results with the experimental observations. In this simulation, we inject monoenergetic electrons (with longitudinal kinetic energy $T_{ez}$) into presumed longitudinal and transverse electric fields of an electron plasma wave. The magnetic field is neglected in this simulation and the transverse electric field is derived from the longitudinal field by $\partial E_r / \partial z = \partial E_z / \partial r$ (results from the Maxwell's equations with B equaling to zero or a constant). The electric field assumed is

$$
\begin{aligned}
\overrightarrow{E}\,(r, \phi, z) = \ &\hat{z} E_0 \exp\left(-r^2/r_0^2\right) \cos\left(k_p z - \omega_p t\right) \\
&+ \hat{r} E_0 k_p^{-1} \left(-2r/r_0^2\right) \exp\left(-r^2/r_0^2\right) \\
&\cdot \cos\left(k_p z - \omega_p t - \pi/2\right).
\end{aligned} \tag{1}
$$

, where $r_0$ is the radius of the plasma wave, $E_0$ is the peak longitudinal electric field, $k_p$ is the wave number of the plasma wave, and $\omega_p$ is the plasma frequency. The key feature in a self-modulated LWFA with self-trapping is that the electrons are injected everywhere along the channel of the plasma wave (determined by the laser channel). 2000 electrons are injected into a region of $r_0$ (x) $\times$ $r_0$ (y) $\times$ $\lambda_p$ (z) in one plasma period (one bucket). The momentum of each electron is saved whenever $t = n \cdot t_p$, where $t$ is the time since the injection, $t_p$ is the plasma period, and $n$ is an integer. The final result is the summation of all electrons saved, which is equivalent to injecting electrons uniformly over the entire channel.

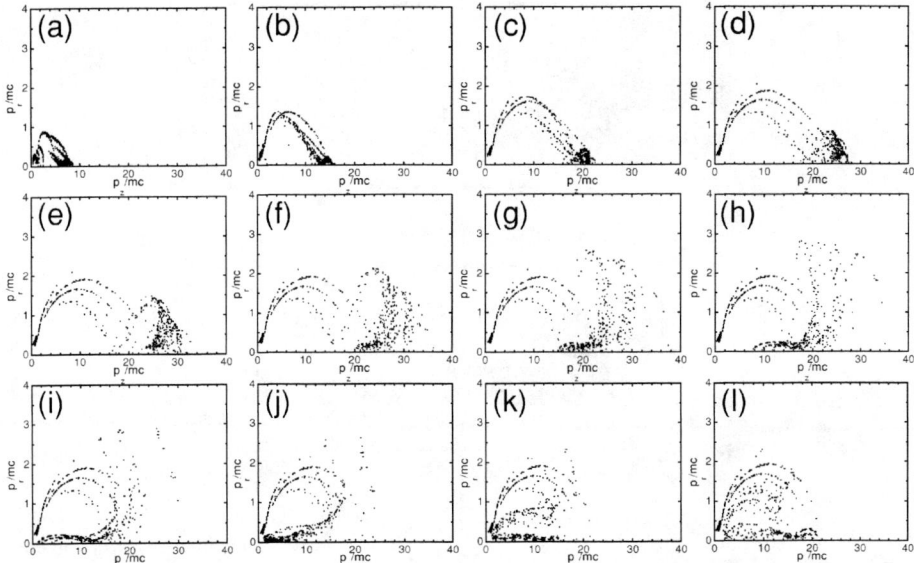

**FIGURE 4.** Simulational result of momentum distribution of electrons injected in one plasma period after propagating different distances for $\epsilon = 0.3$, $r_0 = 5$ $\mu$m, $\omega_p = 3.4 \times 10^{14}$ rad/s, and $T_{ez} = 200$ keV: (a) 22, (b) 44, (c) 65, (d) 87, (e) 109, (f) 131, (g) 152, (h) 174, (i) 196, (j) 218, (k) 240, and (l) 261 $\mu$m.

Figure 4 shows the evolution of momentum distribution of electrons injected in the first plasma wave bucket for $E_0/E_b = 0.3$, $r_0 = 5$ $\mu$m, $\omega_p = 3.4 \times 10^{14}$ rad/s, and $T_{ez} = 200$ keV, where $E_b = e\omega_p v_p/c$ is the nonrelativistic cold wavebreaking limit. After the injection, electrons that are not trapped (inside the separatrix [1,16]) are expelled by the transverse field to a contour defined by $(p_r/mc)^2/2(p_z/mc) = \epsilon$ $(= E/E_b)$, where $p_r$ is the transverse momentum. The trapped electrons are mainly confined near $p_r = 0$ and move toward higher $p_z$ (higher energy) with time. When they reach the maximum energy (the upper limit of the separatrix) after propagating one electron-detuning-length, $L_d \sim \gamma_p^2 \lambda_p$, where $\gamma_p$ is the relativistic factor of phase velocity of the plasma wave, the electrons turn back and move toward the decreasing $p_z$ direction (lower energy). After the electrons reach the lower limit of the separatrix (the trapping threshold), they turn again and move toward higher $p_z$, and so on. While the trapped electrons move in an oscillatory trajectory inside the separatrix, they also drag a tail which spreads to the region confined by the contour, as a result of the transverse defocusing field of the plasma wave. As a result, less and less electrons are confined (guided by the transverse focusing field) in the region near $p_r = 0$ (also the region near $r = 0$), as they oscillate inside the separatrix. The time it takes for all the electrons to lose their confinement increases with increasing plasma wave amplitude. For example, for

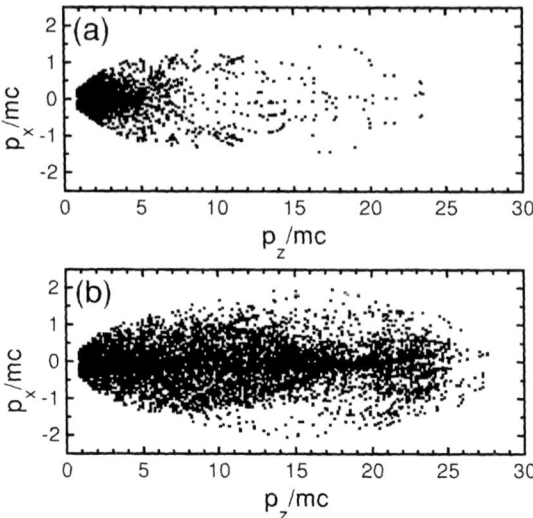

**FIGURE 5.** Simulational result of momemtum distribution of electrons injected over the entire 400-$\mu$m-long channel for different field amplitudes at $r_0 = 5$ $\mu$m, $\omega_p = 3.4 \times 10^{14}$ rad/s, and $T_{ez} = 200$ keV: (a) $\epsilon = 0.15$, and (b) $\epsilon = 0.3$.

the parameters used in Fig. 4, the confinement time of the electrons in the plasma wave with $\epsilon = 0.15$, 0.2 and 0.3 is about one-fifth, twice and four times electron detuning length, respectively. The contour observed in the simulation results from the conservation of canonical momentum for the acceleration of an electron, which is at rest initially, by a plasma wave. This contour line is identical to the $p_r - p_z$ relation of the electrons accelerated by laser ponderomotive force (direct laser acceleration) when $\epsilon = 1$ is used. Therefore, the appearance of electrons that satisfying $(p_r/mc)^2/2(p_z/mc) = 1$ in laser-plasma interaction does not guarantee that it is the result of direct laser acceleration. It can come from the wavebreaking of plasma waves excited through Raman instability or resonant absorption.

The momentum distribution of the electrons injected over the entire channel for a channel length of 400 $\mu$m are shown in Fig. 5. Generally, two groups of accelerated electrons are produced by the plasma wave, one is in the whole region confined by the contour, and another is in the region near $p_r = 0$. These two groups of electrons results in the first and second component of the electron beam profile observed in the experiment. That is, the first beam component (with larger divergence) results from electrons expelled by the transverse field before they exit the channel, and the second beam component (with smaller divergence) results from electrons that are still confined in the channel transversely when they exit the channel. A third component is also observed in certain conditions in which the length of the channel is less than one electron detuning length. Figure 6 shows the electron beam profiles for several cases. As can be seen, the ratio between the divergences of different beam

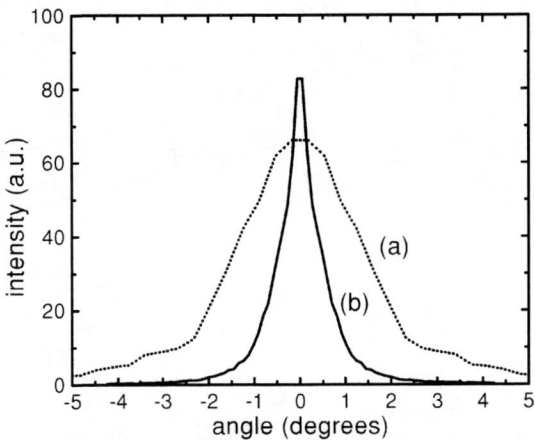

**FIGURE 6.** Simulational result of angular profile of the electron beam for different field amplitudes at $r_0 = 5$ $\mu$m, $\omega_p = 3.4 \times 10^{14}$ rad/s, $T_{ez} = 200$ keV, and $L = 400\mu$m: (a) $\epsilon = 0.3$, and (b) $\epsilon = 0.15$.

components and between the intensities of these components can be reproduced in the simulation. In fact, the ratio between the divergence of the first and second beam components are found to be roughly a constant ($\sim 2$) in both the results of the simulation and the experiment. The absolute value of the divergence is lower in the simulation (about a factor of 5), compared to the experimental results. The fact that the beam divergence is larger in the experiment can be explained by several reasons. The most importance factor is the transverse space charge effect occurring during the acceleration and after exiting the channel. This is not considered in our simulations. Other effects such as nonlinear plasma wave and possible errors in measuring the plasma wave amplitude will also affect the result.

The electron energy spectra obtained from the simulations show a Maxwellian distribution in the low energy range and a flat region in the high energy range (and a high energy cutoff), as shown in Fig. 7. This is consistent with the experimental results. Furthermore, such a two-temperature distribution also appears in the 1-D simulation, as shown in Fig. 7 by setting $r_0/\lambda_p \simeq 100$. The exponential distribution in the low energy range is found to be composed of the untrapped but accelerated electrons (those outside the separatrix), and the newly trapped electrons which are trapped at the end of the channel. The energy distribution of the trapped electrons injected in a single bucket is a narrow band with the central energy sweeping up and down inside the separatrix as the electrons propagate down the channel, as seen in Fig. 4. In the case of the self-modulated LWFA discussed here, the electrons are injected over the entire channel, and thus the spectrum of the electrons is the summation of all these narrow bands, leading to a flat-topped distribution in the high energy range.

Figure 8 shows the electron energy spectra for different channel lengths at $\epsilon = 0.3$,

339

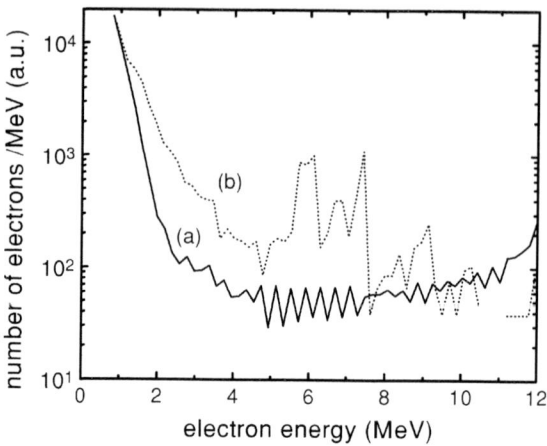

**FIGURE 7.** Simulational result of electron energy spectrum for different channel radii at $\epsilon = 0.15$, $r_0 = 5$ $\mu$m, $\omega_p = 3.4 \times 10^{14}$ rad/s, $T_{ez} = 200$ keV, and $L = 400\mu$m: (a) $r_0 = 500\mu$m, and (b) $r_0 = 5\mu$m.

$r_0 = 5$ $\mu$m, $\omega_p = 3.4 \times 10^{14}$ rad/s, and $T_{ez} = 200$ keV. When the channel length is very short, the energy spectrum is an exponential distribution in the low energy range. With increasing the channel length, while the slope of energy distribution in the low energy range maintains the same, the energy distribution in the high energy range becomes a flat-top with its maximum energy extending to higher energy. The flat-topped region reaches the upper limit of the separatrix after one electron-detuning-length, and then more electrons are added into the flat-topped region toward the lower energy direction with increasing channel length. After two electron-detuning-lengths, as the earliest-injected electrons travel back to the bottom (the low energy region) of the separatrix, the addition of these electrons to the low energy spectrum leads to a change in the slope of the exponential distribution. After the channel length is larger than two electron-detuning-lengths, the increase of channel length results in the increase of the electron number in the high energy region once again, and the slope of the energy distribution in the low energy range stays the same until the next jump which occurs at four electron-detuning-lengths.

The simulational results in Fig. 8 do not match well with the experimental results quantitatively. This is because that we consider only the electrons injected at 200 keV energy. In reality, the injected electrons should have a continuous spectrum with a lot more electrons at lower energy. These electrons with lower injection energy will have more electrons in the low-energy exponential distribution, which slope becomes steeper, and less electrons in the high-energy flat-top region, because less electrons are in the trapped (and confined) regions. Therefore, when injected with electrons with a continuous spectrum, we expect that the slope to be much steeper and the ratio between the numbers of electrons in the low energy range and the high energy range to be much larger, as in the experimental results. The discrete

340

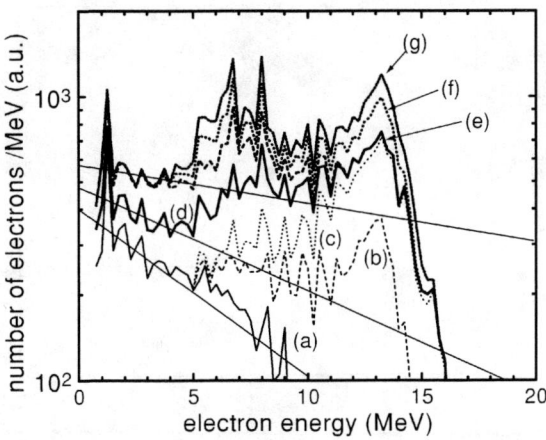

**FIGURE 8.** Simulational result of electron energy spectrum for different channel lengths at $\epsilon = 0.3$, $r_0 = 5$ $\mu$m, $\omega_p = 3.4 \times 10^{14}$ rad/s, and $T_{ez} = 200$ keV: (a) 54, (b) 109, (c) 163, (d) 218, (e) 272, (f) 327, and (g) 381$\mu$m.

jump of the slope of the low energy spectrum will still occur every twice electron-detuning-lengths, since the motion of the electrons trapped in the separatrix are basically the same regardless of the injection energy. To compare this with the experimental results, we plotted the experimental data on a $\alpha - (L/2L_d)$ diagram, as shown in Fig. 9, in which $L$ is the channel length, and $L_d$ is determined from the plasma density. The results show jumps occurring when $L/2L_d$ equals to an integer, as expected from the simulations. Qualitatively, the increase in the channel length or the plasma density (decreasing $L_d$) change $L/2L_d$ to a larger value, and abrupt changes of the slope is expected to occur at the integers. For the cases in which the laser power is increased at a fixed plasma density and a long channel length (fixed $L/2L_d$), the jump of slope still happens because the confinement time of electrons depends on the plasma wave amplitudes. In these cases, $L$ should be replaced by the confinement length, which increases with increasing the amplitude of the plasma wave (with increasing the laser power or plasma density).

In conclusion, the characteristics of the electron beam generated from a self-modulated LWFA are measured in the experiment, and the main features in the beam profile and the energy spectrum can be understood with the help of a simple 3-D particle simulation. Because of its simplicity, this simulation can not be used to explain the observed dark mode structures in the first beam component. There are at least two possibilities for such phenomena. The mode structure could be a result of the complicated transverse structure of the plasma wave induced by the nonlinearity of a large-amplitude plasma wave. It can also result from electron beam instabilities induced by magnetic fields, such as the Weibel instability [18] and the Kelvin-Helmholtz instability [19]. This work is supported by NSF PHY 972661 and NSF STC PHY 8920108. The author would like to thank E. Dodd,

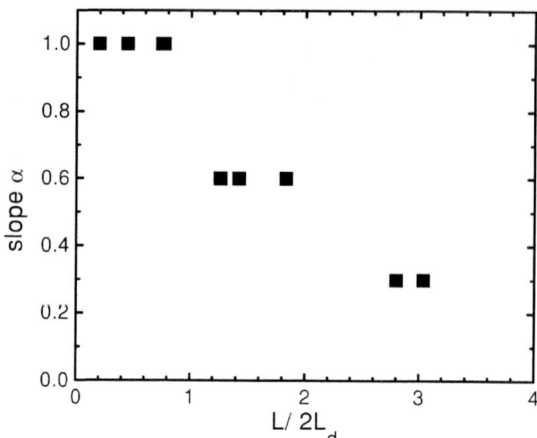

**FIGURE 9.** Slope of the electron energy spectrum in the low energy range as a function of $L/2L_d$.

J.-K. Kim, G. Mourou, and R. Wagner for their useful discussion.

# REFERENCES

1. See references in E. Esarey *et al.*, IEEE Trans. Plasma Sci. **PS-24**, 252 (1996).
2. P. Maine *et al.*, IEEE J. Quantum Electron. **24**, 398 (1988).
3. D. Umstadter, J. K. Kim, and E. Dodd, Phys. Rev. Lett. **76**, 2073 (1996).
4. E. Esarey *et al.*, Phys. Rev. Lett. **79**, 2682 (1997).
5. P. Sprangle *et al.*, Phys. Rev. Lett. **69**, 2200 (1992); T. M. Antonsen and P. Mora, Phys. Rev. Lett. **69**, 2204 (1992); N. E. Andreev *et al.*, JETP Lett., **55**, 571 (1992).
6. P. Bertrand *et al.*. Phys. Rev. E **49**, 5656 (1994).
7. D. Umstadter *et al.*, Science **273**, 472 (1996).
8. C. I. Moore *et al.*, Phys. Rev. Lett. **79**, 3909 (1997).
9. A. Modena *et al.*, Nature **377**, 606 (1995).
10. R. Wagner *et al.*, Phys. Rev. Lett. **78**, 3125 (1997).
11. G. Malka *et al.*, Phys. Rev. Lett. **79**, 2053 (1997).
12. C. D. Decker, W. B. Mori, and T. Katsouleas, Phys. Rev. E **50**, R3338 (1994).
13. N. E. Andreev, L. M. Gorbunov, and S. V. Kuznetsov, IEEE Trans. Plasma Sci. **24**, 448 (1996).
14. K.-C. Tzeng *et al.*, Phys. Rev. Lett. **79**, 5258 (1997).
15. P. Mora and F. Amiranoff, J. Appl. Phys. **66**, 3476 (1989).
16. E. Esarey and M. Pilloff Phys. Plasmas **2**, 1432 (1995).
17. S.-Y. Chen *et al.*, Phys. Rev. Lett. **80**, 2610 (1998).
18. E. S. Weibel, Phys. Rev. Lett. **2**, 83 (1959).
19. C. F. Driscoll and K. S. Fine, Phys. Fluids B **2**, 1359 (1990).

# Particle Dynamics and its Consequences in Wakefield Acceleration in a High Energy Collider

S. Cheshkov, T. Tajima, W. Horton and K. Yokoya*

*Department of Physics and Institute for Fusion Studies*
*The University of Texas at Austin*
*Austin, Texas 78712 USA*
* *KEK National Laboratory for High Energy Physics, Japan*

**Abstract.** The performance of a wakefield accelerator in a high energy collider application is analyzed by use of a nonlinear dynamics map built on a simple theoretical model of the wakefield generated by the laser pulse (or whatever other method) and a code based on this map [1]. The crucial figures of merit for such a system other than the final energy include the emittance (that determines the luminosity). The more complex the system is, the more "opportunities" the system has to degrade the emittance (or entropy of the beam). Thus our map guides us to identify where the crucial elements lie that affect the emittance. If the focusing force of the wakefield is strong when there is a jitter in the position (or laser aiming) of each stage coupled with the spread in the individual particle betatron frequencies, particles experience a phase space mixing. This effect sensitively controls the emittance degradation. We investigate these effects both in a uniform plasma and in a plasma channel. We also study the effect of beam loading. Further, we briefly consider collision point physics issues for a collider expected or characteristic of such a construction based on a scenario for the multi-staged wakefield accelerators.

## INTRODUCTION

The use of plasma waves excited by laser beams for electron acceleration was proposed by Tajima and Dawson [2]. There are many possible applications of plasma based accelerators and one of them, perhaps the most challenging, is for construction of a linear collider. It is believed that conventional linear colliders can go up to 1 TeV center of mass energy of colliding particles [3]. However, beyond that probably some new technology needs to enter. A feature of plasma based accelerators is their ability to sustain extremely large acceleration gradients ($\sim 100\,\mathrm{GV/m}$). Correspondingly we can hope to achieve high energy gains on short distance. Such a machine, no doubt, consists of a large number of elements to reach the desired energy and forms a complex system. In order to identify the most

CP472, *Advanced Accelerator Concepts: Eighth Workshop,*
edited by W. Lawson, C. Bellamy, and D. Brosius
© 1999 The American Institute of Physics 1-56396-889-4/99/$15.00

important research questions in such a complex problem, it is necessary to design a strawman's system, no matter how simplistic it may be, and to try to analyze and deduce the fundamental properties of such a system. Such a strawman's design was in fact carried out for the first time for a wakefield acceleration-based collider in the last AAC meeting [4]. In Sec.2 of this paper we analyze the performance of Laser Wakefield Accelerator (LWFA) in the case of an initially homogeneous plasma using the map developed in [1]. Considering the target center of mass energy $E_{cm} = 5$ TeV and luminosity $\mathcal{L}_g = 10^{35}$ cm$^{-2}$s$^{-1}$ in [4], it is shown that for a fixed power $P_b$ of the colliding beams there are only two independent parameters describing the beam at IP: the number of particles per bunch $N$ and the longitudinal beam size $\sigma_z$ (we assume a round beam: the aspect ratio $R = 1$). All other beam parameters at IP can be expressed as functions of the above. So we know what the requirements on the normalized emittance are.

In Sec.2 and 3, respectively, we analyze two different scenarios with an initially uniform plasma density and with a preformed 'hollow' channel. In Sec.4 we consider the effect of beam loading. In Sec.5 we consider possible impact of such a collider system and its performance on IP physics issues, discovery potential and possible limits on future machines. Conclusions are drawn in Sec.6.

# MULTISTAGE ACCELERATION. UNIFORM PLASMA

When we try to accelerate particles to TeV energies, we need to investigate problems associated with the multistaging. In order to carry this out, we obtained a map [1] which describes the one to one correspondence between the entrance coordinates and the exit coordinates of the beam particles during the propagation of the beam trough many accelerating stages and used it to build a systems code. The linearized equations [1] of motion for the longitudinal degrees of freedom are:

$$\delta\Psi_{n+1} = \delta\Psi_n \tag{1}$$

$$\delta\gamma_{n+1} = 2\gamma_p^2\Phi_0(\cos(\Psi_s + \Delta) - \cos(\Psi_s))\delta\Psi_n + \delta\gamma_n, \tag{2}$$

where $n$ enumerates the stage, $\Psi_s$ is the 'synchronous' phase, $\Delta$ is the phase slippage per accelerating stage (actually it can also depend on $n$).

The transfer matrix $M$ for the transverse coordinates is:

$$M = \begin{pmatrix} \cos(\frac{\omega}{\omega_s}\Delta), & \frac{1}{\omega}\sin(\frac{\omega}{\omega_s}\Delta) \\ -\omega\sin(\frac{\omega}{\omega_s}\Delta), & \cos(\frac{\omega}{\omega_s}\Delta) \end{pmatrix} \cdot \begin{pmatrix} 1 & \hat{L} \\ 0 & 1 \end{pmatrix}, \tag{3}$$

where $\Delta$ is the phase advance of the particle per stage with respect to the wakefield and $\hat{L}$ is the length of the distance between the stages in units of $k_p^{-1}$ ($1/\omega$ is the betatron length in units $k_p^{-1}$ and it depends on the stage number (because the energy

---
[1] Notation and normalizations in this paper are the same as in [1]

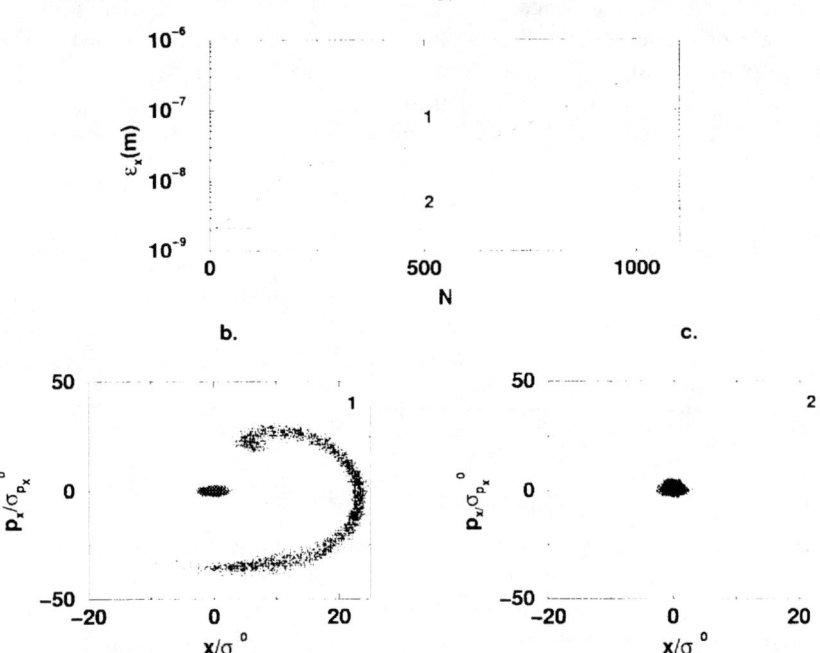

**FIGURE 1.** The normalized $x$-emittance $\epsilon_x$ vs. stage number $N$ and the transverse phase space before and after the acceleration; 1. $\gamma_p=100$, $\hat{L}=10000$, $\epsilon_x^0=2.2$ mm, laser wavelength $\lambda = 1\mu$m, laser spotsize $r_s=0.5$mm, $a_0 = 0.5$, dislocation size $= 0.1\mu$m, $\delta\gamma/\gamma=0.01$, $\delta\Psi = 0.01$. The total number of stages is N=1100 which gives the required final energy of 2.5 TeV. 2. Channeled case, for parameters see the corresponding section

is increasing) and on the individual particles positions in $(\gamma.\Psi)$ phase space). The matrix (3) can be written also as

$$
M = \begin{pmatrix} \cos(\frac{\omega}{\omega_s}\Delta), & \frac{1}{\omega}\sin(\frac{\omega}{\omega_s}\Delta) + \hat{L}\cos(\frac{\omega}{\omega_s}\Delta) \\ -\omega\sin(\frac{\omega}{\omega_s}\Delta), & -\hat{L}\omega\sin(\frac{\omega}{\omega_s}\Delta) + \cos(\frac{\omega}{\omega_s}\Delta) \end{pmatrix} . \tag{4}
$$

So, for $N$ stages

$$
\mathcal{M} = M_N M_{N-1}...M_2 M_1 . \tag{5}
$$

We note that because of the common structure of the wakefield in all plasma based accelerators the obtained map, with just slight modifications, can be used to analyze their performance as well.

For a complex system cumulative errors can give rise to a surprising (and often unpleasant) result. We identify that one of the most important such effects stems from

The problem here is that up to this point we have not considered possible dislocation of the consequent stages. This, combined with the fact that focusing force is different for different particles can lead to a severe emittance growth. Basically, what happens is that all particles rotate at different angular velocities in the phase space and if there is a stage position shift present, we get a characteristic "banana" shaped distribution (it is "banana" shaped only if the dislocation size is larger than beam size, but in any case the particle distribution gets diluted because of the misalignments). This process critically depends on the magnitude of the betatron frequency spread which means that the typical strength of the focusing force is of great importance. The effect of plasma noise (or other noise, such as laser or the boundary) on the particle dynamics over a stage may be incorporated in a map similar to the stage-by-stage jitter. Such dynamics results in a fuzzy or stochastic [7] map. We consider the case of stage jitter. The dislocation of the aligned position of each stage is given in our code as a stochastic variable which has a Gaussian distribution. The map is modified according to:

$$
\begin{pmatrix} \tilde{x}_{n+1} \\ \dot{\tilde{x}}_{n+1} \end{pmatrix} = M_n \begin{pmatrix} \tilde{x}_n - \tilde{\mathcal{D}} \\ \dot{\tilde{x}}_n \end{pmatrix} + \begin{pmatrix} \tilde{\mathcal{D}} \\ 0 \end{pmatrix}, \tag{6}
$$

where $\mathcal{D}$ is the misalignment size ($\tilde{\mathcal{D}} = \sqrt{\gamma_n}\mathcal{D}$). The longitudinal degrees of freedom are not affected. Run with random dislocations of magnitude $\sigma_D = 1 \cdot 10^{-7}$ m is presented in Figure 1a and 1b. We see that in this case (corresponds to design I [4]) we have a severe emittance growth (the initial r.m.s. normalized emittance is 2.2 nm). We have to point out that even though we can find cases corresponding to large laser spot sizes which preserve the normalized emittance quite well their practical realization would require a huge laser power probably well above any future experimental limits. The other possible way to cure the situation is to increase the number of stages (and decrease $\gamma_p$) which also does not seem plausible (it may be difficult to have more than a thousand stages). To control the laser aiming and beam position better than 0.1 $\mu$m during the acceleration process is also not promising. If the initial normalized emittance is larger than the relative growth will be smaller, thus we can hope that, for instance, the design III emittance [4] will work better. This is really the case, Figure 2a and 2b shows basically the best we can do if we go as high as $r_s = 0.5$ mm and using the design III [4] normalized emittance of 330 nm. We see the characteristic dilution of the particle distribution and the corresponding emittance growth which is rather small here. The problem is that the spot-size is large and would require an immense laser power. If we limit ourselves to something a little bit more "reasonable" as $r_s = 0.2$ mm, keeping the other parameters the same, the result is discouraging, the emittance growth is very large in this case (see Figure 2a and 2c). From the computer simulations the conclusion is that in the case of initially homogeneous plasma it is very difficult based on reasonable parameters (laser spot size, dislocation size and number of stages) to avoid a severe emittance growth of the accelerated beam in the presence of small jitters stage-by-stage. The difficulty is primarily due to the fact that the

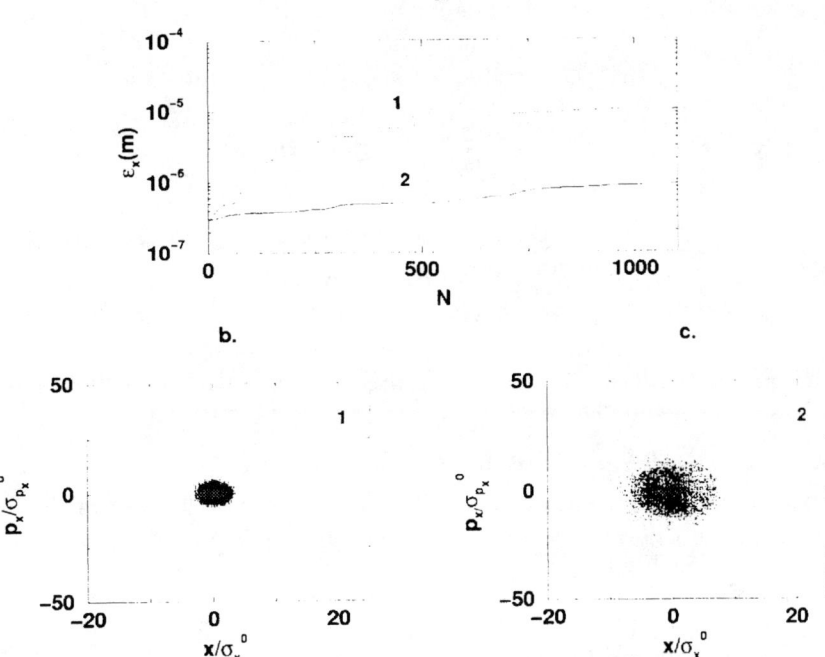

**FIGURE 2.** The normalized $x$-emittance vs. stage number and the transverse phase space before and after the acceleration for different laser spot sizes. $\gamma_p = 100$, $\hat{L} = 10000$, $\epsilon_x^0 = 330$nm, $r_s = 0.5$mm (1) and $r_s = 0.2$mm (2), $a_0 = 0.5$, dislocation size$=0.1\mu$m, $\delta\gamma/\gamma = 0.01$, $\delta\Psi = 0.1$.

wakefield focusing force is too large in this case. We should also remember that the above considerations do not include any nonlinear effects which also contribute to the phase area increase, not to mention that in addition to all of the above we have to come up with a mechanism for guiding of the laser pulse (in the cases of small laser spot size). Otherwise the diffraction limits the acceleration length.

## HOLLOW PLASMA CHANNEL

A possible way to avoid these difficulties is the 'hollow channel' design [8] in which we use a preformed vacuum channel in an underdense plasma (the overdensed case was studied in [9]). In this case we get several important features: the focusing force is exactly (because the phase velocity of the wakemode is very close to the speed of light) linear and weak in the channel (the weak focusing is a very important improvement over that of a uniform plasma case); there exists a stable solution for the laser mode; the acceleration gradient is uniform in transverse coordinates within the channel. An obvious drawback is a loss in the magnitude of the acce for the wakefield in the channel are [8]:

$$E_z(\xi, r) = -k_{ch}^2 \int_\xi^\infty \cos(k_{ch}(\xi - \xi')) \; \Phi_l(a, \xi') \; d\xi', \qquad (7)$$

$$E_r(\xi, r) - B_\theta(\xi, r) = \frac{k_{ch}T}{4\gamma_p^2} k_{ch}^2 \int_\xi^\infty \sin(k_{ch}(\xi - \xi')) \; \Phi_l(a, \xi') \; d\xi', \qquad (8)$$

where $a$ is the channel radius, $\Phi_l$ is the ponderomotive potential and $k_{ch}$ is given by: $k_{ch} = k_p / \sqrt{1 + k_p a \frac{K_0(k_p a)}{2K_1(k_p a)}}$, where $K_0$ and $K_1$ are the modified Bessel functions of zeroth and first order, respectively. For instance, if we choose $k_p a = 1$ then the electric field in $z$-direction will be reduced by a factor of 0.6 [8] compared to the initially uniform plasma. So, formally there are no major changes to our previous map scheme. There is a reduction in $\Phi_0$ and the magnitude of the focusing changes: $\omega = \frac{k_{ch}}{k_p} \left( \frac{\Phi_0}{2\gamma \gamma_p^2} \right)^{1/2}$ (in units of $k_p$). Importantly, we have a decrease in the magnitude of the focusing force. Making these minor changes to the map, we are able to investigate the accelerator performance in this case.

Run shown on Figure 1a and 1c indicates a significant improvement over the previous design. Here we are able to preserve even design I emittance of 2.2 nm. The stage considered is: $\gamma_p = 150$, the channel radius $a = 30\mu m$, the laser spot size $r_s = 50\mu m$, the plasma density (outside the channel) $n = 5 \cdot 10^{16} cm^{-3}$ and the laser wavelength $\lambda \sim 1\mu m$. The size of the stage dislocations was increased in this case to $0.5\mu m$. From the graphs we see that the emittance growth of the accelerated beam is now much smaller and the design is more promising. The total number of stages was 900 which gives a total energy of 2.5 TeV. The drift space is 20000 (in units of $1/k_p$). Unfortunately, there is an additional effect: because in reality we have a finite density gradient it leads to a resonant absorption where the local plasma frequency matches the wakefield frequency. This effect has been studied in [10], where an expression for the quality factor of the hollow channel is derived. Possible low values of this factor can limit the acceleration of multiple bunches in a single shot created wakefield. We note that the real problem we have to deal with is the strong focusing of the wakefield. Channel is one way to reduce it, but there may be other possibilities as well, for instance, "flat top" laser pulse profile, etc. Still more work should be done on the subject and on the closely related problem of beam loading to come up with the best possible design of LWFA for collider applications. The purpose of the present paper is to introduce the systems approach to LWFA so that more future work on a complex collider machine may be theoretically carried out in detail.

## BEAM LOADING EFFECT

All of our previous considerations are strictly applicable only in the case of weak beams which correspondingly produce weak wakefields. In a real case we may want to extract a significant portion of the plasma wave energy. This can be accomplished when we load a large number of particles into the wave (from the requirements for high luminosity it already follows that we need relatively intense beams). We now need to take into account the wakefield produced by the accelerated beam itself. The total accelerating field (in a simple 1-d model), including the beam induced wakefield [6], is:

$$E_{z\,tot} = E_w \cos\left(k_p \xi\right) - 4\pi \int_{\xi}^{\xi_0} d\xi' \, \rho(\xi') \, \cos k_p(\xi - \xi'),\tag{9}$$

where the first term is due to the accelerating plasma wave and the second is the wakefield produced by the accelerated particle bunch ($\xi_0$ is the position of the bunch head). If we assume a constant bunch density profile, (9) is easily integrated to give

$$E_{z\,tot} = E_w \cos\left(k_p \xi\right) - \frac{4\pi Ne}{k_p l d_{eff}^2} \sin k_p(\xi - \xi_0),\tag{10}$$

where $d_{eff}$ is the effective transverse beam size (of order $1/k_p$), $l$ is the bunch length and $N$ is the number of electrons in the bunch.

We employ (10) in our map formulation and obtain

$$\delta\Psi_{n+1} = \delta\Psi_n \,,\tag{11}$$

$$\delta\gamma_{n+1} = 2\gamma_p^2 \Phi_0(\cos(\Psi_s + \Delta) - \cos(\Psi_s))\delta\Psi_n + 2\Delta\gamma_p^2 \Phi_0 \frac{N}{N_0}\frac{1}{k_p l}\delta\Psi_n + \delta\gamma_n \,.\tag{12}$$

The second term in (12) corresponds to the differences in energy gains due to the beam self-wakefield, $N/N_0$ is the percent of beam loading and $N_0 = \frac{k_p \Phi_0 d_{eff}^2 m_e c^2}{4\pi e^2}$. In our case the second term is much greater then the first term in (12) and it appears that we have a large energy spread due to this. One way to avoid this is to use longer beam length (and then to compensate the loading by proper choice of the initial phase) and another is to decrease the number of particles per bunch (for instance to use microbunching). Yet another possibility is bunch shaping [6], which however may not be realistically manageable for such short bunches. In any case future studies on this subject are necessary. The above does not take into account the radial self-wakefield [5]. For dense beams it should be included as well and it gives a correction to the radial focusing force. The correction vanishes for the particles in the head of the bunch and is maximum for the particles in the bunch tail. We also need to compensate for this simultaneously with the longitudinal compensation. Fortunately, our operating region in the wake has the required features – acceleration gradient decreases and focusing force increases with $\xi$. Thus, in principle, such compensation is possible.

# COLLISION POINT PHYSICS

One of the main purposes of the next generation high energy accelerators - the Large Hadron Collider (LHC) and the Next Linear Collider (NLC) is to check the predictions of the weak scale supersymmetry (SUSY) which if correct should lead to the discovery of light Higgs particles and light superpartners. Preservation of beam quality during the acceleration is extremely important, however it is not the only problem we have to solve. There are fundamental difficulties associated with the interaction point (IP) physics. Colliding of beams gives rise to the well known phenomenon of beamstrahlung [11]. It is often undesirable because it effectively decreases colliding particles energies and can also lead to a possible photon contamination of the detectors. The way to avoid the former is to either work in the classical regime with a very small beamstrahlung parameter $\Upsilon \ll 1$ or to move to the "quantum suppression" regime [4] characterized by $\Upsilon \gg 1$. We carried out QED simulations based on the CAIN code [12], [4]. This shows that the photon contamination is relatively easy to avoid due to the fact that the photon angular distribution has a very sharp peak (see Figure 3, in this example all the beamstrahlung photons are in a 20 mrad angle around the beam axis) in the forward (backward) direction (with respect to the beam). Secondary particles created via coherent and incoherent pair production are also sharply peaked in the forward (backward) direction. So we just need to place our detectors at large enough angles.

However, there are additional major obstacles to do discovery physics on such a machine. We should be able to extract the new physics signatures from the experimental data. At these energies, however, there are strong contributions from Standard Model processes which create a significant background to deal with. One of the most important ones, due to the high $\gamma - \gamma$ luminosity (see Figure 4), is that of $\gamma\gamma \to W^+ W^-$. We need to impose additional cuts to eliminate background as much as possible and to improve the signal to noise ratio. The analysis presented in [13] shows that in the region of center of mass energies of a few hundreds GeV it should be possible to observe signatures of new physics. We hope to be able to do Higgs physics in this range (if the light Higgs exists) and also to find the lightest supersymmetric partners (if SUSY is correct). A major method to eliminate the background is $b$-tagging (for Higgs sector). Consider the following example (if $m_h < 2m_Z$): $e^-e^- \to Z \to Z h$. The background consists of $ZZ$ and $WW$ and the way to deal with it may be the double $b$-tagging and jet-jet mass reconstruction followed by visible energy cut. In the search of s-leptons a possible plan is to use polarized electron beams. For instance, $e^-e^- \to \tilde{e}^-\tilde{e}^-$ is practically free of the $W$-background if we use right handed polarized electron beams [14]. The produced s-electrons can decay in several different modes, all of them ending in neutralino (invisible) production and lepton pairs which leads to a clean missing energy signature of the reaction. High luminosity to be achieved in the high $\Upsilon$ regime is more than welcome in this case. The great opportunity provided by $e^+e^-$ linear machine is that it can be relatively easy changed into $e^-e^-$ or $\gamma\gamma$ linear collider. In this sense

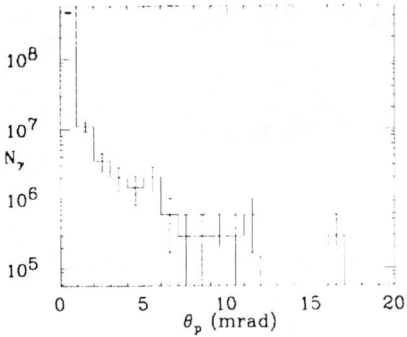

**FIGURE 3.** The number of photons $N_\gamma$ vs. $\theta_p$ (angle with respect to the beam axis) at IP of a 5 TeV $e^+e^-$ linear collider. (parameters used correspond to design III [4]; this and the next graphs are obtained using the CAIN )

linear colliders will be of great importance even after the start of LHC. In the TeV range the new physics we can expect is largely unknown. The predictions are strongly model dependent and probably modifications will occur after LHC and NLC become operational. If the light Higgs and superpartners are not discovered at 500 GeV center of mass energy we can hope that moving to TeV range will enable us to do this.

# CONCLUSION

We studied the cumulative effects of the successive acceleration, transport, and focusing in the laser wakefield (or its sister methods) over many stages. Errors arising from the jitter of each stage or equivalently (in our map approach) the noise in the system can accumulate in such a way to degrade some of the parameters of the beam. The most crucial of these may be the emittance (or the entropy) of the beam. When we have stochastic variables on the wakefield (we chose the stage jitter of the axis of the wakefield in particular), the emittance can significantly increase over the many stages due to the strong focusing of the wakefield. This is probably the most serious effect on the long range behavior of the beams in this kind of accelerator for high energy applications. In order to ameliorate this situation, we consider the weak focusing of a hollow plasma channel and find that in fact the emittance may be well preserved. However, there are many details still need to be investigated. These include the effect of resonance of the wakefield in a channel, and alternative methods to decrease the focusing force. In order to make an efficient accelerator it is desired to have heavy beam loading. In this case the effect of beam loading must be considered. It can have degrading effect on the emittance, as the head and the tail of the beam may experience different forces. Search for the best parameter set in the multidimensional parameter space of a large scale accelerator should be performed taking into account, to our best notion, future experimental limits and restrictions which might come from them. The work is supported by US DoE.

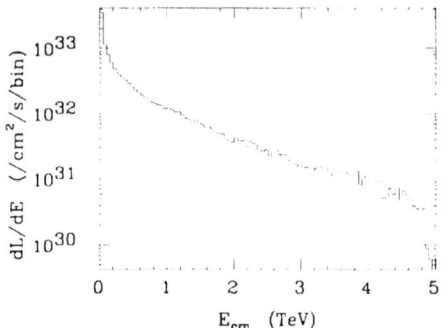

**FIGURE 4.** The differential $\gamma - \gamma$ luminosity $dL/dE$ vs. the center of mass energy $E_{cm}$ at IP (again design III [4])

# REFERENCES

1. Tajima, T., Cheshkov, S., Horton, W. and Yokoya, K., *This proceedings* (AIP, New York 1998conf).
2. Tajima, T. and Dawson, J., *Phys. Rev. Lett.* **43**, 267-270 (1979).
3. The NLC Design Group and NLC Physics Working Group. *Physics and technology of the Next Linear Collider*. A Report Submitted to Snowmass 96 (1996).
4. Xie, M., Tajima, T., Yokoya, K. and Chattopadhyay, S., "Studies of laser-driven 5 TeV $e^+e^-$ colliders in strong quantum beamstrahlung regime", *Advanced Accelerator Concepts 7* (AIP, New York 1997conf), p.233-242.
5. Chen, P., *Part. Accelerators* **20**, 171–182 (1987).
6. Katsouleas, T., Wilks, S., Chen, P., Dawson, J. and Su, J., *Part. Accelerators* **22**, 81–98 (1987).
7. Van Kampen, N., *Phys. reports* **24**, No. 3, 171–228 (1976).
8. Chiou, T., Katsouleas, T., Decker C., Mori, W., Wurtele, J., Shvets, G. and Su, J., *Phys. Plasmas* **2**, No. 1, 310–318 (1995).
9. Zaidman, E., Tajima, T., Neuffer, D., Mima. K., Ohsuga, T. and Barnes, D., *IEEE Trans. Nucl. Science* **NS-32**, No. 5, 3545–3547 (1985).
10. Shvets, G., Wurtele, J., Chiou, T. and Katsouleas, T., *IEEE Trans. Plasma Science* **24**, No. 2, 351–362 (1996).
11. Yokoya, K. and Chen, P., *Frontiers of Particle Beams* **400**, 415- (1992).
12. Yokoya, K. *CAIN21b*. http://jlcux1.kek.jp/subg/ir/Program-e.html
13. Murayama, H. and Peskin, M., *Annu. Rev. Nucl. Part. Sci.* **46**, 533-608 (1996).
14. Cuypers, F., Oldenborgh, G. and Ruckl, R., *Nucl. Phys.* **B409**, 128-143 (1993).

# Resonant Excitation of Plasma Wakefields using Multiple Electron Bunches

Manoel E. Conde and Wei Gai

*Argonne National Laboratory, Argonne, Illinois  60439*

**Abstract.**   We plan to resonantly excite plasma wakefields using a train of electron bunches separated by an integer number of plasma wavelengths. The multiple electron bunches are generated by a photocathode based RF gun by splitting the laser beam into temporally separated pulses. The amplitude of the wakefields generated by the sequence of bunches is expected to be higher than that generated if all charge had been in only one bunch, because this single bunch would be considerably longer than the individual sub-bunches due to space charge effects in our gun.

## INTRODUCTION

High charge short electron bunches are needed to achieve high accelerating gradients in electron beam driven plasma wakefield accelerators. Electron bunches generated by photocathode based RF guns in general become longer as the bunch charge is increased. An attractive alternative to generating high charge single electron bunches is to generate a train of bunches separated by a plasma wavelength (1,2). In this way the plasma wakefield is resonantly excited by the bunch train and each bunch can be kept short because the total charge is distributed over many bunches.

## GENERATION OF MULTIPLE BUNCHES

We can generate multiple electron bunches from our photocathode based RF gun simply by splitting the laser beam into a sequence of temporally separated pulses. The delay between the pulses needs to be equal to the plasma wavelength, which in our case is about 6 mm (assuming a plasma density of $3 \times 10^{13}$ cm$^{-3}$). This delay between the laser pulses will introduce an energy difference between the various electron bunches in the train, as they will be emitted from the photocathode at different phases of the RF power that is present  in the gun. However, since our RF gun operates at 1.3

CP472, *Advanced Accelerator Concepts: Eighth Workshop,*
edited by W. Lawson, C. Bellamy, and D. Brosius

GHz, we can generate three or four electron bunches separated by the required 6 mm without introducing an excessive energy spread between the bunches; electron bunches separated by 20 ps would have a launching phase difference of 9.4°.

Figure 1 illustrates how we plan to split the laser beam into four pulses. The laser beam would be directed onto two beam-splitters at near-normal incidence. Each beam-splitter has the appropriate dielectric coating on each surface, generating two pulses that are delayed by twice the optical thickness of the substrate. Thus, to generate pulses spaced by 6 mm we need a substrate approximately 2 mm thick with an index of refraction of 1.5. The two beam-splitters need to be separated by 3 mm. Table 1 shows the values of reflectivity for each of the four surfaces if four pulses of equal intensity are desired. Other charge profiles of the bunch train can be easily achieved by adjusting the reflectivity of the four surfaces.

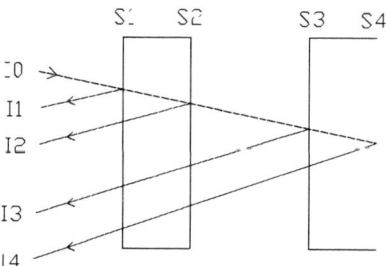

**FIGURE 1.** Laser beam splitting scheme. The incoming beam I0 reflects off of surfaces S1, S2, S3, S4, generating the beams I1 through I4.

TABLE 1. Beam Splitter Parameters

| Surface | Reflectivity |
|---------|--------------|
| S1 | 0.15 |
| S2 | 0.21 |
| S3 | 0.33 |
| S4 | 0.74 |

# THE AWA ELECTRON GUN PERFORMANCE

It is obviously easier to generate shorter electron bunches when the bunch charge is smaller. As we increase the bunch charge, space charge forces become more intense, in general making the bunch longer. We have measured the dependence of bunch length on charge for the AWA drive gun (3). Figure 2 shows the result of this measurement. Bunch length was measured with an aerogel Cerenkov radiator and a streak camera (Hamamatsu M1952/C1587). Bunch charge was measured with an integrating current transformer (Bergoz ICT-082-070-20:1). The plot shows the FWHM of the bunch length and also the 95% RMS values (i.e. the RMS calculated using only the section of the pulse profile with intensity within 95% of the peak value, with the purpose of discarding the effect of the small background noise at the wings of the distribution). The ratio between the 95% RMS and the FWHM values shows that the pulses are not gaussian. Each point on the graph corresponds to the average of three pulses. The large fluctuation in the FWHM of the pulses also shows that the detailed shape of the temporal profile varies considerably from pulse to pulse.

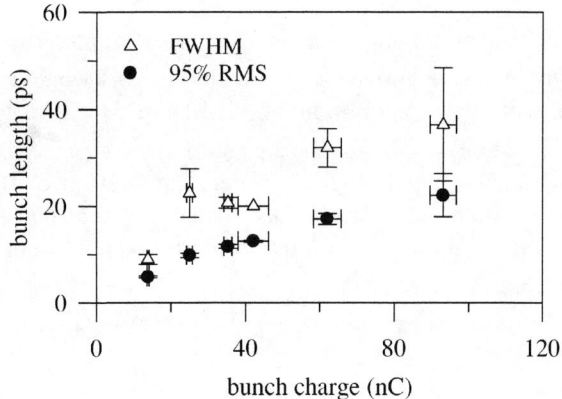

**FIGURE 2.** Measurements of electron bunch length as a function of bunch charge. The error bars indicate the standard deviation of the average of three pulses.

# ESTIMATES OF WAKEFIELD AMPLITUDE

The amplitude of the plasma wakefield is approximately given by:

$$W_z(z) = \sum_m -\frac{E_0 F_m}{\sqrt{2\pi}\,\sigma_z} \int_{-\infty}^{z} \cos\left[k_p(z_0 - z)\right] \exp\left[-\frac{(z_0 - m\lambda_p)^2}{2\sigma_z^2}\right] dz_0 \qquad (1)$$

where $\sigma_z$ is the RMS value of the bunch length, $k_p = 2\pi/\lambda_p$ is the plasma wavenumber, $E_0$ is the maximum amplitude of the field in V/m given by $E_0 = 192\, n_0\, Q$ with $n_0$ as the plasma density in cm$^{-3}$ and $Q$ as the bunch charge in nC; $F_m$ is the coefficient associated with the bunch number $m$, such that the distribution of charge along the $z$ axis is given by:

$$\rho(z) = \sum_m \frac{Q F_m}{\sqrt{2\pi}\,\sigma_z} \exp\left[\frac{(z - m\lambda_p)^2}{2\sigma_z^2}\right] \qquad (2)$$

Figure 3 shows the wakefield amplitude as calculated from equation (1) for two different cases. The first plot (Fig. 3a) shows the wakefield that would be excited by a single bunch of 60 nC with $\sigma_z = 3.5$ mm. In this case the plasma density is equal to 6 $\times 10^{12}$ cm$^{-3}$, and the wakefield amplitude has a peak of approximately 20 MV/m. The second plot shows that a considerably higher accelerating gradient of 100 MV/m can be achieved if the same amount of charge is distributed into four bunches of 15 nC, each one with $\sigma_z = 1.5$ mm. In this case, with this shorter bunch length, the optimum plasma density is about $3 \times 10^{13}$ cm$^{-3}$.

# ACKNOWLEDGMENTS

We thank N. Barov for useful discussions. This work was supported by the Department of Energy, Division of High Energy Physics, under contract W-31-109-ENG-38.

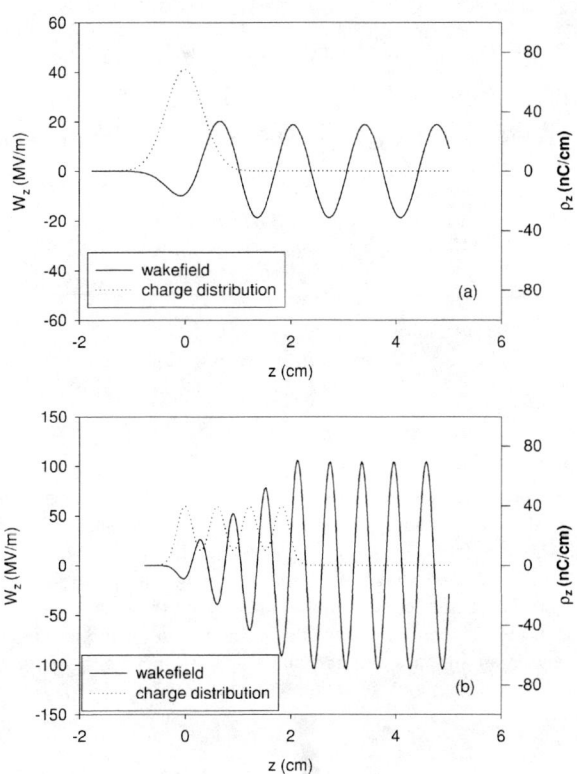

**Figure 3.** Wakefield amplitude and bunch charge distribution: (a) single 60 nC drive bunch with $\sigma_z = 3.5$ mm; plasma density is $6 \times 10^{12}$ cm$^{-3}$; (b) train of four 15 nC bunches with $\sigma_z = 1.5$ mm; plasma density is $3 \times 10^{13}$ cm$^{-3}$.

# REFERENCES

1. A.M. Kudryavtsev, K.V. Lotov, and A.N. Skrinsky, *Nucl. Instr. Meth.* **A 410**, 388 (1998).
2. K.V. Lotov, *Nucl. Instr. Meth.* **A 410**, 461 (1998).
3. M.E. Conde, W. Gai, R. Konecny, X. Li, J. Power, P. Schoessow, and N. Barov, *Phys. Rev. ST-AB* **1**, 041302 (1998).

# High Energy Electrons and Nuclear Phenomena in Petawatt Laser-Solid Experiments

T.E. Cowan[1], B. Dong[2], T. Ditmire[1], W. Fountain[3], S. Hatchett[1], A.W. Hunt[4], J. Johnson[5], T. Kuehl[6], T. Parnell[3], D.M. Pennington[1], M.D. Perry[1], T.W. Phillips[1], Y. Takahashi[2], S.C. Wilks[1] and P.E. Young[1]

[1]Lawrence Livermore National Laboratory, Livermore, CA, USA
[2]University of Alabama, Huntsville, AL, USA
[3]Marshall Space Flight Center, Huntsville, AL, USA
[4]Harvard University, Cambridge, MA, USA
[5]University Space Research Association, Huntsville, AL, USA
[6]Gesellschaft für Schwerionenforschung, Darmstadt, Germany

**Abstract.** The Petawatt laser at LLNL has opened a new regime of laser-matter interactions in which the quiver motion of plasma electrons is fully relativistic with energies extending well above the threshold for nuclear processes. We have developed broad-band magnetic spectrometers to measure the spectrum of high-energy electrons produced in laser-solid target experiments at the Petawatt, and have found that in addition to the expected flux of ~few MeV electrons characteristic of the ponderomotive potential, there is a high energy component extending to ~100 MeV apparently from plasma acceleration in the underdense pre-formed plasma. The generation of hard bremsstrahlung, photo-nuclear reactions, and preliminary evidence for positron-electron pair production will be discussed.

## INTRODUCTION

Much of this Workshop has addressed the potential of plasma acceleration concepts for future high energy collider technology, and substantial progress has been reported on plasma channel formation, optical guiding and initial acceleration experiments with injected electron beams. Important next steps include developing the channel-guided laser wakefield into a suitable accelerating structure and understanding the electron beam quality through a single structure, before eventually demonstrating acceleration through multiple stages. The feasibility of plasma acceleration for practical high energy accelerators will ultimately hinge on the system-level issues such as plasma reproducibility and control, beam emittance growth through multiple stages and total system efficiency. However, it may be possible to utilize laser acceleration concepts, which have already been demonstrated, for less demanding applications such as for compact single-stage accelerators or as radiation sources. For example, Self-Modulated

CP472, *Advanced Accelerator Concepts: Eighth Workshop,*
edited by W. Lawson, C. Bellamy, and D. Brosius
© 1999 The American Institute of Physics 1-56396-889-4/99/$15.00

Laser Wakefield Acceleration has desirable properties for use as bremsstrahlung x-ray sources for which the electron beam quality is not particularly important -- the large bunch charge and short pulse duration may be sufficient to create intense pulsed x-ray sources for studies of dynamic properties of materials under extreme conditions. At LLNL, we are both developing a laser/linac facility (1,2) for the long range pursuit of stageable laser acceleration concepts and advanced x-ray source designs, and we are using existing lasers to evaluate single-stage accelerators for intense short-pulse radiation generation. This contribution will describe first radiation generation experiments on the Petawatt laser in which we observed 100 MeV electrons, photo-nuclear reactions and laser-assisted nuclear transmutation in ultra-intense laser-solid interactions (3).

## ELECTRON SPECTROSCOPY AT THE PETAWATT

The Petawatt laser (4) at the LLNL NOVA facility is presently the world's highest power chirped pulse amplification laser. Developed primarily for ICF research (5), it uses one arm of the NOVA Nd:glass laser to amplify a frequency-chirped pulse to kJ energies before temporal compression to ~450 fs in a vacuum compressor using very large (94 cm diameter) gratings. The peak power is well in excess of 1000 TW, and the 58 cm diameter beam can be focused with an 80 cm diameter f/3 on-axis parabola to a <20 μm spot, with intensity exceeding $10^{20}$ W/cm$^2$. At these extreme intensities the electromagnetic fields at the laser focus are enormous (E > $10^{13}$ V/m, and B > $10^5$ Tesla) and the motion of the electrons in the target plasma is fully relativistic. The cycle-averaged oscillation or "quiver" energy of the electrons,

$$\overline{E} = mc^2 \left[ 1 + 2U_p / mc^2 \right]^{1/2}, \tag{1}$$

can exceed several MeV, where $U_p = 9.33 \times 10^{-14} I \ (W/cm^2) \lambda^2 (\mu m)$ is the non-relativistic ponderomotive potential. The resulting distribution of electron energies in the target resembles a Maxwellian (6), with mean energy given by Eq. 1, and it extends far beyond the threshold for which nuclear effects become important. An adjustable pre-pulse, at 2 or 10 ns before the main pulse and tunable over $10^{-4}$ to $10^{-1}$ of the peak energy, is used to pre-form an underdense plasma on the face of a solid target in order to produce beam filamentation, self-focusing, and self-modulated wakefield acceleration of electrons which then produce bremsstrahlung cascades, positron-electron pairs, and photonuclear reactions in the solid portion of the target.

To characterize the entire electron energy range of interest, from few MeV ponderomotive electrons to several hundred MeV or GeV-scale plasma-accelerated electrons, we have fielded two types of compact, permanent magnet electron spectrometers (3,7) as shown in Fig. 1. The "low-energy" spectrometers, covering the electron energy range of 0.2 to 100 MeV, are mounted at 30° and 95° with respect to the laser beam incident on the target. The neodymium-iron-boron permanent magnet material is sufficient to attain a field strength of 5.5 kG, over a 10 cm × 15 cm pole face dimension with a 3.8 cm gap. The electrons are recorded in nuclear emulsion track detectors which are positioned in the gap of the magnet such that the electron angle of incidence is a constant value of $10°$ throughout the dispersion plane. The emulsions consist of two layers of 50 μm thick, fine-grained silver bromide emulsion coated on the front and back surface of a 500 μm polystyrene strip. Microscopic examination of the developed emulsion strips allows clear identification of charged particle tracks and distinguishes electrons emitted from the target by virtue of the density of exposed grains along the track, the angle of incidence, and the transverse position along the emulsion strip (which must be within the image of the entrance aperture). A second emulsion strip in each spectrometer is oriented to detect positrons within the energy range of 0.2 to 40 MeV. Below about 2 MeV, however, both electrons and positrons undergo sufficient multiple scattering in the emulsion that the incidence angle and track length cannot be unambiguously determined and therefore cannot be distinguished from Compton electrons generated in the emulsions by the large flux of hard x-ray bremsstrahlung present in the solid target experiments.

The "high energy" spectrometer, covering an electron energy range of 100 MeV to 2 GeV, is a hybrid system which uses an 8 kG NdFeB permanent magnet as a dispersive element, followed by a second, 5 kG NdFeB magnet to provide redundant energy

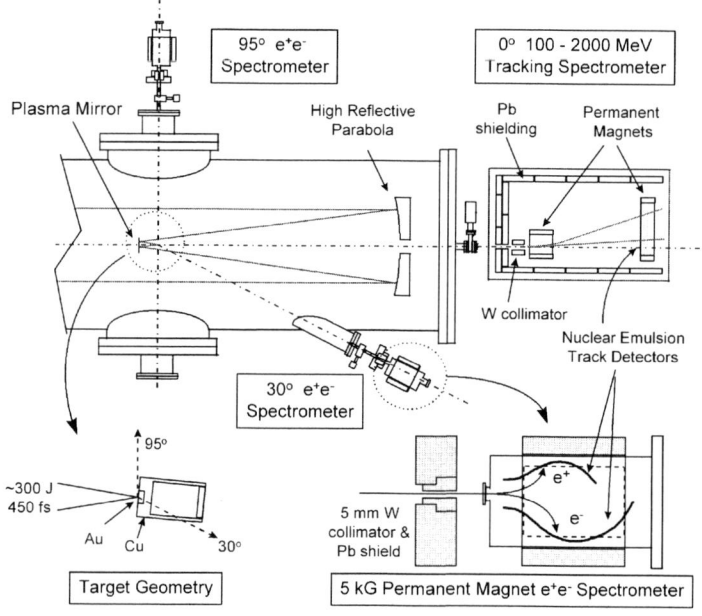

**FIGURE 1.** Schematic of the Petawatt target chamber and electron spectrometers. The compressed laser beam enters from the left, is reflected from the f=180 cm parabola to a secondary plasma mirror, and reflected onto the target at chamber center. The high-energy spectrometer (at right) measures electrons (~100 - 2000 MeV) emitted near $0°$, which pass through a hole in the parabola. Low energy spectrometers measure electrons (0.2-140 MeV) and positrons (0.2-40 MeV) emitted at $30°$ and ~$95°$ from the target (lower right).

determination by measuring the electron's track curvature through multiple detector planes. The spatial resolution for locating a track's coordinates at each emulsion plane is <2 μm, and the eventual resolution of the reconstructed energy is ~20% at 1 GeV, limited by multiple scattering in the emulsion planes. The spectrum of electron energies is obtained by counting the number of individual tracks in a specified microscopic field of view at several locations along the dispersion plane, and converting the track densities to absolute differential cross-sections. The energy dispersion (i.e., position versus energy) was calculated by ray-tracing and confirmed by calibration measurements performed at the LLNL 100 MeV electron Linac.

The first high energy electron spectra measured on the Petawatt are shown in Fig. 2, from an experiment in September 1997 to characterize the high-energy bremsstrahlung x-ray yield in short-pulse solid target interactions (8). The target was a 0.5 mm thick × 2 mm diameter disk of gold, mounted in the end plate (~0.5 mm thick) of a cylindrical copper can (subsequent targets in this shot series were thicker). As shown in Fig. 1, the laser light was focused using a secondary plasma mirror in a Cassegranian geometry. The total laser energy incident on the Au target was ~280 J, with a pulse duration of 450 fs. The energy enclosed in the central focal spot was sufficient to achieve a focused intensity of ~8×10$^{19}$ W/cm$^2$.

Two important features of these data are the large excess of 5-15 MeV electrons observed in the forward direction, and the presence of a very high energy tail extending to above 80 MeV in both spectra. The forward enhancement of the electron angular

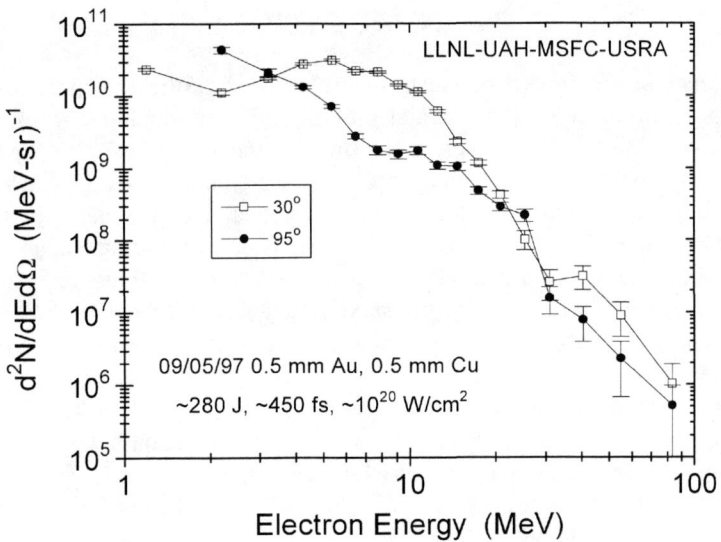

**FIGURE 2.** Distribution of electrons measured in magnetic spectrometers at 30° (open) and 95° (solid), from the first short-pulse (0.5 ps) Petawatt shot series on September 5, 1997. The 0.5 mm Au target was backed by 0.5 mm Cu. The estimated incident laser energy was 280 J, delivered in 450 ± 50 fs, with an estimated focused intensity of 8×10$^{19}$ W/cm$^2$.

distribution is even larger than it appears because those electrons detected in the 30° spectrometer had to penetrate the target and in so doing lost substantial energy ($\Delta E \sim$ 2.5 MeV for minimum ionizing particles) and were multiply scattered through rather large angles. Monte-Carlo electron transport calculations suggest that the forward electron flux between 2-15 MeV may be an order of magnitude or more larger than the factor of ~10 enhancement already apparent from Fig. 2. The low energy portion of the electron spectrum at both angles is also suppressed, compared to the actual energy distribution of electrons present in the target plasma, due to a large, ~few MV space charge potential which develops as electrons are expelled from the focus during the laser pulse. Because the laser pulse duration is so short, the return current of electrons flowing from the bulk of the target material does not have time to fully neutralize the positive space charge potential formed by the current of ejected relativistic electrons.

Including reasonable estimates of the above effects, the lower energy portion of these spectra, up to ~20 MeV, appear reasonably consistent with the expected Maxwellian distribution having a mean energy of ~3 MeV, predicted from Eq. 1, at the nominal intensity of ~$10^{20}$ W/cm$^2$. However, the presence of the high energy tail in each spectrum out to nearly 100 MeV, indicates a more complicated laser-target interaction. The laser pre-pulse in this experiment is estimated to have pre-formed a ~50 μm scale-length plasma at the surface of the gold target, sufficient to cause either self-focusing of the laser light to much higher intensity, or self-modulated laser plasma acceleration.

## PHOTO-NEUTRON REACTIONS AND LASER-ASSISTED TRANSMUTATION

The presence of very energetic electrons in the forward directed spectrum in Fig. 2 implies also the presence of a substantial flux of hard bremsstrahlung x-rays produced by the passage of the relativistic electrons through the Au target. The portion of the bremsstrahlung spectrum above the threshold for photonuclear reactions should contribute to photoneutron emission from the gold and copper and, if sufficiently intense, produce measurable long-lived activities in the target. Immediately following each laser shot, the target assemblies were measured with a high resolution HPGe gamma-ray spectrometer to identify the specific daughter nuclides present. A typical gamma-ray energy spectrum for an activated target is shown in Fig. 3. The production of $^{196}$Au nuclei by $^{197}$Au($\gamma$,n)$^{196}$Au is clearly identified by the appearance of the nuclear de-excitation gamma-rays in the $^{196}$Pt daughter nuclide at 356 keV and 333 keV. In addition, the less probable $\beta^-$ decay of $^{196}$Au to $^{196}$Hg is also identified by the line at 426 keV. The intensities of the 333, 356 and 426 keV gamma-ray lines were observed to exhibit a decay rate consistent with the 6.18 day half-life of the $^{196}$Au parent, further confirming their identification.

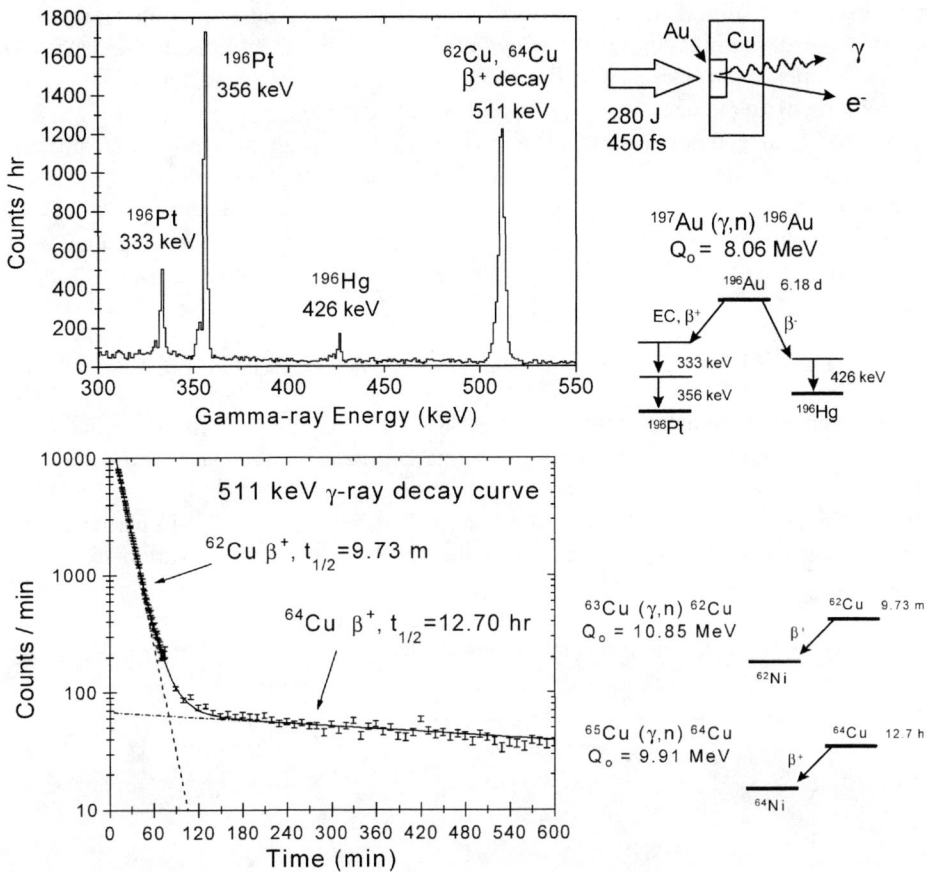

**FIGURE 3**. Photoneutron activation of Au and Cu target material in an intense Petawatt laser shot. (Upper Left) Gamma-ray energy spectrum measured with a high purity Ge detector. Peaks correspond to nuclear de-excitation gamma-rays following decay of $^{196}$Au to $^{196}$Pt and $^{196}$Hg (see decay scheme at Right). (Lower Left) Count-rate vs. Time for the 511 keV gamma-ray peak from positron annihilation following β+ decay of $^{62,64}$Cu to $^{62,64}$Ni. Curves show two-component fit of exponential decay corresponding to the half-lives for the β+ decay of $^{62}$Cu and $^{64}$Cu.

The measurable yield of $^{197}$Au(γ,n)$^{196}$Au reactions indicates a large flux of bremsstrahlung photons above the threshold energy for this reaction of $Q_o$=8.06 MeV. Photoneutron activation of the principal copper isotopes by the reactions $^{65}$Cu(γ,n)$^{64}$Cu [$Q_o$=9.91 MeV], and $^{63}$Cu(γ,n)$^{62}$Cu [$Q_o$=10.85 MeV] are also observed by the beta decay of their neutron deficient daughters to $^{64}$Ni and $^{62}$Ni respectively. These were identified in this experiment by the 511 keV positron-annihilation gamma-ray line, whose decay curve shown in Fig. 3 reveals two principal components whose decay constants match the 9.7 min and 12.7 hr half-lives of $^{62}$Cu and $^{64}$Cu, respectively. We have therefore identified three photo-neutron reactions in gold and copper produced from the laser-target interaction. We believe that their subsequent radioactive decays to isotopes of platinum, mercury and nickel represent the first observations of the transmutation of an element from a laser-produced plasma (3).

From the total activation yield of each nuclide, and using the known $\sigma_{\gamma,n}$ nuclear cross sections, one may estimate the total flux of bremsstrahlung photons above the threshold energy $Q_0$ emitted into the forward hemisphere (i.e., intercepting the bulk of the target material) (3,8,9). For example, the $^{63}Cu(\gamma,n)^{62}Cu$ activation yield implies a total forward-going flux of $\sim 10^{11}$ photons above 10.85 MeV, for the data of Fig. 2.

## REPRODUCIBILITY AS A RADIATION SOURCE

The eventual usefulness of ultra-intense laser plasma interactions as single-stage accelerators and radiation sources requires some degree of control and reproducibility. In this experimental campaign, seven additional target shots were accumulated and for the most the basic features of the electron energy distributions in Fig. 2, in particular the presence of the high energy electron tail, and the photoneutron activation discussed above, were reproduced. The overall electron and activation yields are summarized in Fig. 4. It should be noted that the seven later targets were somewhat thicker (2 mm Au, 2 mm Cu) than the first, which further attenuated the low energy portion of the electron

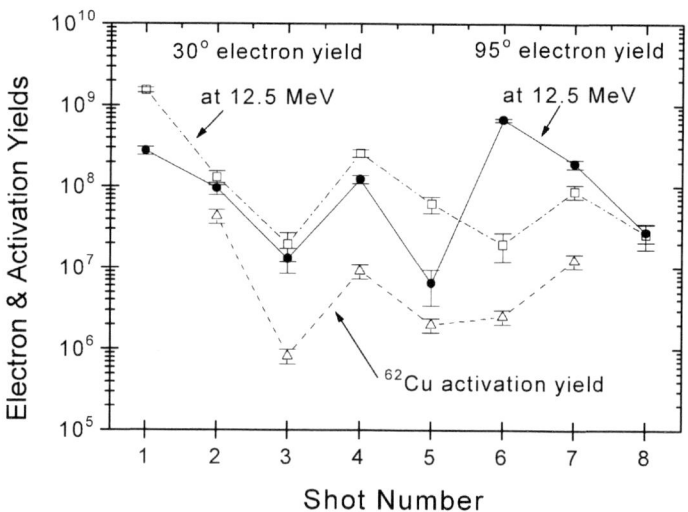

**FIGURE 4.** Comparison of electron and activation yields for the eight Petawatt target shots in this experiment. The curves plot the electron differential cross section (MeV$^{-1}$ sr$^{-1}$), measured in the 30° and 95° spectrometers, at a fixed electron energy of 12.5 MeV. The activation curve plots the number of $^{62}Cu$ nuclei produced in the target assembly, calculated from the time=0 intercept of the fitted $^{62}Cu$ decay curve from Fig. 3, corrected for the 511 keV full-energy-peak detection efficiency in the high-purity Ge gamma-ray spectrometer.

spectrum measured in the 30° spectrometer (cf Fig 5). Apart from this systematic difference, and a marked reversal of the electron angular distribution on the 6th shot, the yield data are well correlated, particularly the forward electron and the activation yields which are, respectively, the source for, and the result of, the bremsstrahlung produced in the target.

Most of the order-of-magnitude shot-to-shot variation in the yields is attributable to variations in the laser pulse energy and focal aberrations caused by pump-induced thermal distortions in the NOVA Nd:glass amplifiers. This has been solved in recent experiments by an adaptive optics system which successfully corrects for these distortions on a given shot by preconfiguring a deformable mirror based on the wavefront data in a Hartmann sensor accumulated on prior shots (10). There remains, however, residual shot-to-shot variations particularly in the directionality of the radiation pattern. Recent measurements indicate that the maximum position of the angular distribution of the ~MeV x-rays can shift dramatically from shot to shot (10). Similar effects have been reported shifts dramatically from ICF experiments on the GEKKO 100 TW laser at Osaka (11). Several groups have noted a dependence of the yield of hard X-ray generation on laser pre-pulse and pre-formed plasma conditions in lower energy laser experiments (12). Most probably, differences in the pre-formed plasma (presently generated with a focused pre-pulse instead of a fully controlled independent pre-form beam) lead to laser filamentation and/or self-focusing with some degree of directional instability. The presence of exceptionally strong magnetic fields (~10 MG) generated by the ponderomotive electron current in the target might induce additional beaming.

More complicated laser-plasma effects appear also to be important in these experiments as shown by the extreme deviation of the electron distribution on the 6th shot. As shown in Fig. 5, in addition to a general shifting of electron intensity from the forward to transverse directions, a remarkably narrow enhancement in the energy

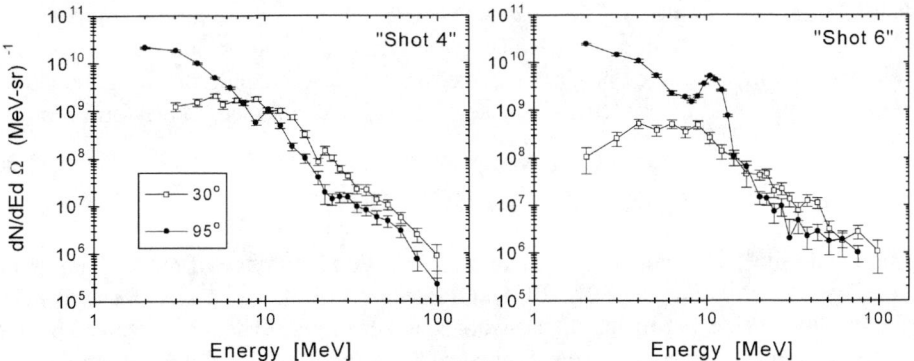

**FIGURE 5.** Electron spectra for Shots 4 and 6. Note the thicker target in shots 2-8 further attenuates the measured electron distribution in the forward, 30° spectrometer (compare Fig. 2). Electron spectra of Shot 4 are typical of most other shots. Shot 6 exhibits a decrease in forward directed electron flux below 20 MeV, and a well defined peak in the transverse directed electron distribution at 10.75 MeV.

spectrum of electrons at 95° is observed at 10.75 MeV, having a FWHM of ~2.5 MeV. Repeated, independent scans of the emulsion strip, at different positions within the projected image of the entrance aperture, verify the statistical reproducibility and significance of the peak-like feature. Although the origin of this jet-like feature in energy is at present unknown, it is presumably associated with some aspect of the evolution of the plasma at the target vacuum interface. If controllable, it could have future usefulness as a high-current, short-pulse electron beam source for radiation generation.

## SUMMARY AND FUTURE DIRECTIONS

In summary, we have begun to explore a new regime of laser-matter interactions where the electron quiver energies are well above the threshold for causing nuclear reactions. In first petawatt-class laser-solid target experiments, we have observed a high energy component in the electron spectrum, up to ~100 MeV, in addition to a forward-directed lower-energy distribution which appears to be consistent with the relativistic dynamics of the electrons in the laser focus at the nominal focused intensities of ~$10^{20}$ W/cm$^2$. High energy bremsstrahlung produced in the Au targets was observed to produce photoneutron reactions in the Au target and Cu backing material. The production of radioactive isotope daughters, and their subsequent decay to stable isotopes of Pt, Hg, and Ni, represents to our knowledge the first observation of laser-assisted transmutation of elements. Subsequent experiments have revealed evidence for higher energy photoneutron reactions, up to ($\gamma$,5n), indicating bremsstrahlung photon energies above 40 MeV (10). Evidence for positron production by external conversion of the high energy bremsstrahlung in the solid target is presently being examined (12). Residual shot-to-shot variations in the energy and angular distributions of electron and bremsstrahlung yields may be associated with variable pre-formed plasma conditions which are not yet under strict experimental control. The intensity and spectrum of radiation from this single-stage device appears to be interesting for potential future applications as an intense short-pulse bremsstrahlung radiation source. Future developments will include controlling the pre-formed plasma conditions by eliminating the laser pre-pulse and by implementing a separately timed and focused pre-form beam.

## ACKOWLEDGEMENTS

We would like to acknowledge valuable contributions by many of our collaborators on the Petawatt program (see Ref. 10), and the support of the Nova staff for their help in fielding of these experiments. This work was performed under the auspices of the U.S. Department of Energy by the Lawrence Livermore National Laboratory under Contract W-7405-Eng-48.

# REFERENCES

1. Cowan, T.E., *et al.,* "Laser Acceleration Research at LLNL," in *Applications of High Field and Short Wavelength Sources VII* (OSA Technical Digest Series Vol 7, 1997), p. 243.

2. Cowan, T.E., T. Ditmire, G. Le Sage, *et al.* The FALCON Laser/Linac Facility combines a 100 TW, 30 fs Ti:sapphire CPA laser synchronized to a 100 MeV S-band Linac with an RF photoinjector. Research on laser acceleration, laser-plasma x-ray sources, advanced accelerator techniques and ultra-short pulse high-brightness Thomson scattering x-ray source development experiments are planned.

3. Cowan, T.E., A.W. Hunt, M.D. Perry, *et al.,* "High Energy Electron Emission and Laser-Assisted Nuclear Transmutation in Petawatt Laser-Solid Interactions," in *Proceedings of the International Conference on Lasers '97,* J.J. Carroll & T.A. Goldman, eds., (1998, STS Press, McLean Va), pp. 882-889.

4. Perry, M.D., B.C. Stuart, D. Pennington, G. Tietbohl, J.A. Britten, C. Brown, S. Herman, J. Miller, H.T. Powell, M. Vergino and V. Yanovsky, "Petawatt Laser Pulses," submitted to Optics Letters.

5. Key, M.H., M.D. Cable, T.E. Cowan *et al.,* "Hot electron production and heating by hot electrons in fast ignitor research,"*Phys. Plasmas* 5, 1966 (1998).

6. Wilks, S.C., W.L. Kruer, M. Tabak and A.B. Langdon, Phys. Rev. Lett. 69, 1383 (1992).

7. Cowan, T.E., Y. Takahashi, W. Fountain, J. Johnson, W.S. Patterson, and T. Parnell, "Design and performance of a magnetic electron and positron spectrometer using nuclear emulsion track detection," to be published.

8. Perry, M.D., J.A. Sefcik, T.E. Cowan, S. Hatchett, A. Hunt, M. Moran, D. Pennington, R. Snavely, and S.C. Wilks, "Hard x-ray production from high intensity laser solid interactions," to be published in: *Proceedings of Workshop on Diagnostics for High Temperature Plasmas* (Monterey, May 1998).

9. Phillips, T.W., M.D. Cable, T.E. Cowan *et al.,* "Diagnosing hot electron production by short pulse, high intensity lasers using photonuclear reactions," in: *Proceedings of Workshop on Diagnostics for High Temperature Plasmas* (Monterey, May 1998).

10. Perry, M.D., T.E. Cowan, T. Ditmire, B. Dong, W. Fountain, S. Hatchett, E.A. Henry, A.W. Hunt, J. Johnson, M. Key, T. Kuehl, J. Moody, M. Moran, T. Parnell, D.M. Pennington, T.W. Phillips, T.C. Sangster, J.A. Sefcik, M. Singh, R.A. Snavely, M.A. Stoyer, Y. Takahashi, S.C. Wilks, to be published.

11. Tanaka *et al.,* to be published in *Proceedings of XXVth European Conference on Laser Interactions with Matter,* Formia Italy, May 2-8, 1998.

12. Malka, G. and J.L. Miquel, Phys. Rev. Lett. 77, 75 (1996).

13. Cowan, T.E., Y. Takahashi, T.W. Phillips, *et al.,* "Relativistic Electron and Positron Production and Nuclear Phenomena in Petawatt Laser-Solid Interactions," to be published.

# Recent Developements for the 2nd Generation LWFA Experiments

H. Dewa*. T. Hosokai*. M. Kando*, S. Kondo*, H. Kotaki*, F. Sakai*
K. Nakajima*†, H. Nakanishi†, A. Ogata†
H. Harano‡, T. Ueda‡, M. Uesaka‡. T. Watanabe‡, K. Yoshii‡
Y. Kotaka**, K. Niwa**

*Advanced Photon Research Center, Japan Atomic Energy Research Institute
2-4 Shirakatashirane, Tokai, Naka, Ibaraki 319-1195, Japan
†High Energy Accelerator Research Organization (KEK)
1-1 Oho, Tsukuba, Ibaraki 305-0801, Japan
‡Nuclear Engineering Research Laboratory. The University of Tokyo
2-22 Shirakatashirane, Tokai, Naka, Ibaraki 319-1106, Japan
**Graduate School of Science. Nagoya University
1 Furo, Chikusa, Nagoya, Aichi 464-8601, Japan

**Abstract.** The laser wakefield acceleration (LWFA) experiments has been conducted by the use of 2 TW. 90 fs laser pulses synchronized with an electron beam injected from a 17 MeV RF linac. Preliminary experiments demonstrated the energy gain more than 200 MeV attributed to ionization-induced self-modulation and self-guiding of intense ultrashort laser pulses in plasmas. Measurements of the plasma density oscillation and laser-plasma interactions support the enhanced acceleration. We are also developing an RF photo-injector producing low emittance beams, a capillary plasma wave guide generating long stable wakefields and a energy measurement system with emulsion technique.

## INTRODUCTION

The present high energy accelerators are based on the RF acceleration, in which the accelerating field is limited to approximately 100 MV/m due to electrical break-down. The laser-plasma accelerators have been conceived to be the next-generation particle accelerators, promising ultra-high field particle acceleration [1]. It has been experimentally demonstrated that the laser wakefield acceleration (LWFA) has great potential to produce ultra high field gradients of plasma waves [2] [5]. Recently a wakefield of the order of 10 GV/m in a plasma has been directly observed by the use of a compact terawatt laser system so called T³ lasers [6,7].

In a homogeneous plasma, however, diffraction of the laser propagation limits

CP472, *Advanced Accelerator Concepts: Eighth Workshop,*
edited by W. Lawson, C. Bellamy, and D. Brosius

the laser-plasma interaction distance to the extent of the vacuum Rayleigh length. This deducts the advantage of ultrahigh gradient acceleration from laser-driven accelerators. It would be of importance for practical application of the laser wakefield accelerator concept to generate a high energy gain as well as high gradient acceleration. Therefore it is essential for laser wakefield acceleration to achieve a long interaction of an intense ultrashort laser pulse with an underdense plasma. We made an attempt to demonstrate laser wakefield acceleration by intense ultrashort laser pulses propagating in a plasma. This paper reports the results of LWFA experiments and developements of an RF photo-injector, a plasma guiding and a energy measurent system with emulsion.

## LWFA EXPERIMENTS

The laser wakefield results from the ponderomotive force exciting the density oscillation of plasma with the frequency $\omega_p = \sqrt{4\pi e^2 n_e/m_e}$ for the ambient electron density $n_e$ in plasma. Assuming a Gaussian laser pulse of a temporal $1/e$ half-width $\sigma_z$ with the peak power $P$, a peak amplitude of the accelerating wakefield is

$$eE_z = \frac{\Omega_0 P}{\sqrt{\pi} m_e c^2}\left(\frac{\lambda_0}{\lambda_p}\right)\left(\frac{k_p \sigma_z}{2Z_R}\right)\exp\left(-\frac{k_p^2\sigma_z^2}{4}\right). \tag{1}$$

where $\Omega_0$ is the vacuum resistivity ($377\Omega$), $\lambda_0$ is the laser wavelength, $\lambda_p$ is the plasma wavelength, $k_p = 2\pi/\lambda_p$ and $Z_R$ is the vacuum Rayleigh length, i.e. $Z_R = \pi R_0^2/\lambda_0$, where $R_0$ is the spot radius at the focus. The maximum amplitude is achieved at $\lambda_p = \pi\sigma_z$. The maximum energy gain of relativistic electrons is given by $\Delta W = eE_z L_{ac}$ with the acceleration length $L_{ac}$. Diffraction limits the acceleration length to $L_{ac} \simeq \pi Z_{eff}$, where $Z_{eff}$ is the effective Rayleigh length for the laser propagation in plasma. For the resonant plasma density, $n_e = 1/(\pi r_e \sigma_z^2)$ where $r_e$ is the classical electron radius,

$$\Delta W_{max}|\text{MeV}| \simeq 850P[\text{TW}]\lambda_0[\mu m]/\tau_0[\text{fs}] \times (Z_{eff}/Z_R). \tag{2}$$

where $\tau_0$ is the FWHM pulse duration, $c\tau_0 = 2\sqrt{\ln 2}\sigma_z$. Note that the maximum energy gain is independent of a focal spot size of laser pulses for Gaussian propagation in homogeneous plasma where $Z_{eff} = Z_R$.

## Laser Wakefield Acceleration Test Facility

We have constructed the test facility for the LWFA experiments, consisting of a $T^3$ laser system and an electron beam injector [8]. The Ti:sapphire $T^3$ laser system produces output pulses compressed by a grating compressor to 90 fs with an energy greater than 200 mJ at the repetition rate of 10 Hz. We used the 2856 MHz RF linac as an electron injector to produce a 17 MeV single bunch beam with a 10 ps FWHM pulse duration containing $\sim 1$ nC at the repetition rate of 10 Hz.

The setup for LWFA experiments is shown in Fig. 1. Focusing optics and magnets were installed in the acceleration chamber filled with He gas. Laser pulses were focused by a f/10 off-axis parabolic (OAP) mirror with a focal length of 480 mm. The measured focal spot radius was 13 μm. An electron beam from the injector is brought to a focus with the FWHM beam size of 0.8 mm through a triple focusing magnet and a permanent magnet quadrupole (PMQ) triplet. The beamline and the acceleration chamber were separated by a 50μm thick titanium window. An electron pulse was synchronized to laser pulses within the rms jitter of 3.7 ps by a phase locked control of a mode-locked oscillator.

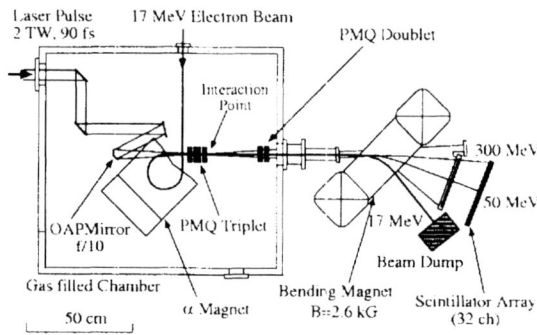

**FIGURE 1.** Schematic of the experimental setup for the laser wakefield acceleration.

## Experiments

The energy of accelerated electrons was measured with the magnetic spectrometer consisting of a dipole magnet and an array of 32 scintillation detectors. Injected electrons undergoing no acceleration were swept out of the detectors by the spectrometer magnet. Fine adjustments of overlapping two spots of laser and electron beams was carried out within 50 μm. Two sets of pulse height data from the scintillator array were taken with pump laser pulses and without them as a background. The pulse height was averaged over 500 to 1000 shots to reduce fluctuations. A net pulse height proportional to the number of electrons was obtained by subtracting the data without the pump pulses from the data with them.

In the acceleration experiments the gas pressure of He was scanned from 0 Torr to 300 Torr. Figure 2 shows energy gain spectra of electrons accelerated at the laser peak power of 0.9 TW and 1.8 TW. The maximum energy gain up to 300 MeV was obtained from these data. There was no acceleration of injected electrons at 4.3 mTorr. When the electron pulse preceded the pump laser pulse, no acceleration was observed. Accelerated electrons visibly appeared as the electron pulse was delayed. We have investigated the self-trapping of plasma electrons without the electron beam injection, but could not observe any electron accelerated higher than 1 MeV.

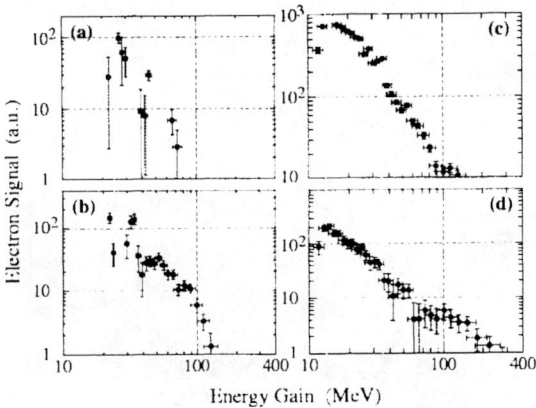

**FIGURE 2.** Energy gain spectra of accelerated electrons for a) 3.4 Torr, P 0.9 TW, b) 20 Torr, P 0.9 TW, c) 2 Torr, P 1.8 TW, and d) 20 Torr, P 1.8 TW.

In order to make confirmation of wakefield excitation by an ultrashort laser pulse in an underdense plasma, we measured the plasma wave oscillation with the frequency domain interferometer [3]. Figure 3 shows the electron plasma wave measured at 2 Torr for the pump peak power of 1 TW. The plasma electron density is deduced to be $9.6 \times 10^{16}$ cm$^{-3}$ from the measured period of the density oscillation, while the electron density is $1.4 \times 10^{17}$ cm$^{-3}$ for a fully ionized plasma at 2 Torr. The measured density perturbation was $\langle \delta n/n_e \rangle \sim 15\%$ corresponding to the longitudinal wakefield of $\sim 2$ GeV/m. This measured amplitude is in good agreement with the accelerating wakefield of 2.2 GeV/m theoretically expected for 1 TW.

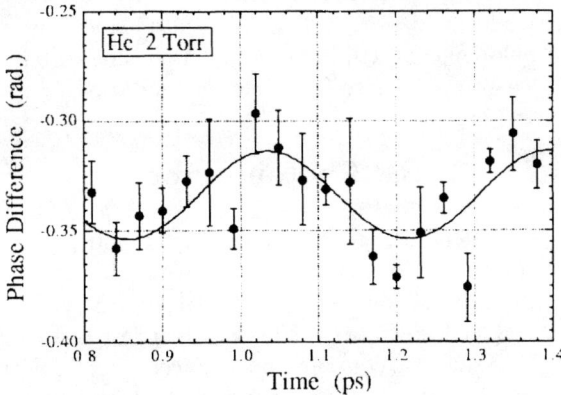

**FIGURE 3.** Measurement of the plasma density oscillation excited by a 1 TW pump power in a He gas of 2 Torr. The solid curve shows a fit of the plasma wave with oscillation period of 360 fs.

As shown in Figure 4, we found several irregular bright spots in the Thomson scattering image of the 1.8 TW pump laser in He gas plasma at 20 Torr. It indicates that strong self-focusing of the laser occurs in plasma, which would relate to the generation of ultra-high wakefield that accelerate electrons to higher energy than expected from the linear fluid plasma theory.

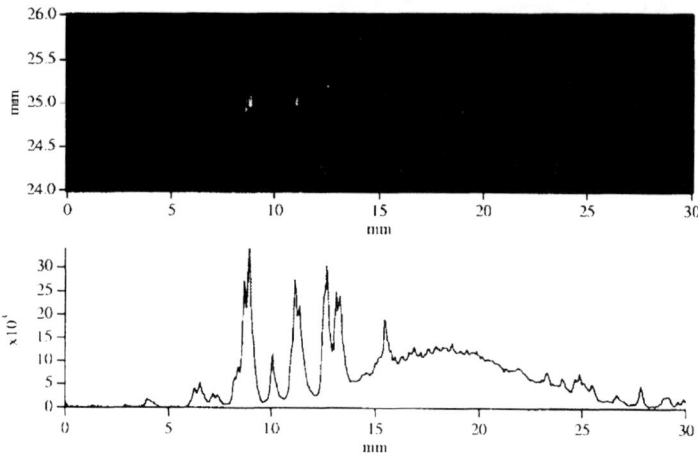

**FIGURE 4.** Thomson scattering image of the 1.8 TW pump laser in He gas plasma at 20 Torr.

# LWFA DEVELOPEMENTS

For the 2nd generation LWFA experiments, we have developed an RF photo-cathode, a capillary plasma wave guide and a energy measurement system with emulsion. These developements will increase possible accelerated electrons ,energy gains and energy measurements free from background noise.

## RF Photoinjector

An RF photoinjector generating the low emittance and short bunch beam was installed as an electron source for LWFA to increase acceleration efficiencies. It was developed by an international collaboration between BNL(Brookhaven National Laboratory), KEK and SHI (Sumitomo Heavy Industries, Ltd.) for the heavy duty operation (50Hz). The system is consisted on a 1.6 cell RF gun with the Cu cathode , a solenoidal magnet for the emittance compensation and the diagnostic system with a beam profiler and a Faraday-cup.

A prototype laser illuminating a photocathode is a laser diode pumped Nd:YLF regenerative amplifier to stable the system and make the laser system compact.

The laser pulse energy is about 200 μjoule for the 4th-harmonics (263nm) at the maximum with 100Hz repetition rate. The pulse width of the 4th harmonics is about 20ps FWHM by the measurement with a femto second streak camera.

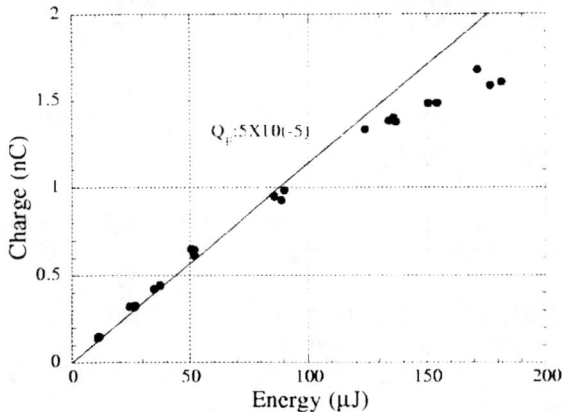

**FIGURE 5.** Emitted Charge vs. Laser Energy (263nm)

As experimental results, the relation between the maximum electron charge and laser pulse energy is shown in in Fig. 5. Quantum efficeincy is to be about $5 \times 10^{-5}$ at the range of less than 1 nC. The normalized emittance versus the current of the solenoid magnet was measured for about 200pC electron bunch charge (in Fig. 6).

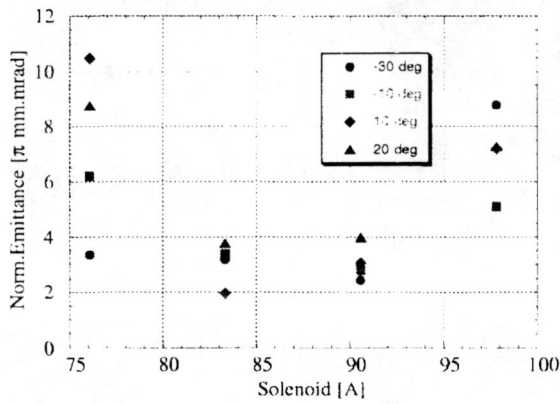

**FIGURE 6.** Horizontal emittance vs. Current of solenoid coil Linac RF phase was changed within 50 degree.

# Optical Wave Guiding by Capillary Discharge

Propagating a laser pulse over distances larger than the vacuum diffraction length is critical issue of LWFA. Our approach is that of guiding in a preformed plasma column with concave electron density profiles. Recently it has been demonstrated by Ehrlich *et al.* that capillary discharges can guide intense optical laser pulse (up to $10^{16}$ W/cm$^2$) over length of 1cm [9].

We propose alternative scheme of capillary discharges to guide intense laser pulse over longer length. In this scheme plasma column is generated by fast Z discharges through prefilled gas in a capillary. Stably imploded Z pinch plasma columns exhibit good coaxial symmetry, and the core of the compressed column has a internal structure due to fast imploding plasma sheet and shock wave [10]. We expect that the laser pulse can be guided in the compressed core with concave electron density profiles.

Since density profiles in the core plasma strongly depend on discharge dynamics, as a first step of the study, we started out to investigate the discharge dynamics experimentally. Experiment were conducted using polyethylene capillaries of 1 mm inner diameter with length of 20 mm which was filled with 1 Torr helium gas. The discharge was driven by current of up to 2.5kA. The discharge dynamics was observed with a streak camera (HAMAMATSU C-2830) from the capillary axis direction. As shown in Fig. 7, recent results were indicate that concave luminous profiles were observed on the axis of the capillary at second implosion phase.

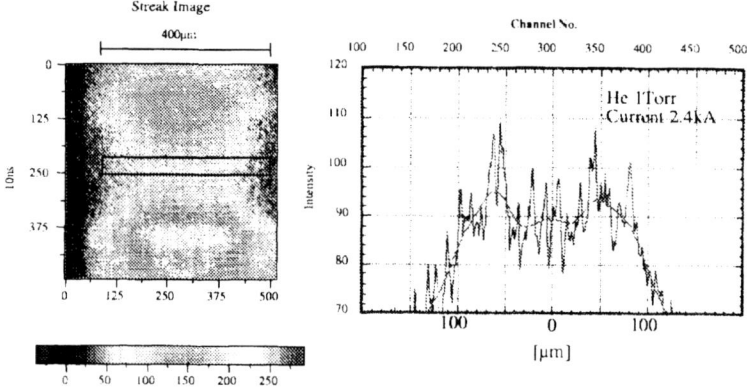

**FIGURE 7.** Convave luminous profiles in capillary at second implosion phase measured with a streak camera

# Energy Measurement Using Emulsion

In order to improve the energy spectrum measurement of the accelerated electerons, we are developing an energy measurement system, consisting of a

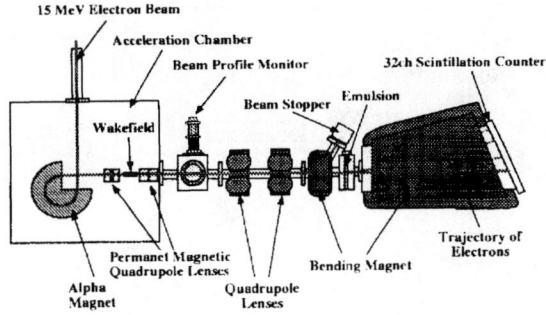

FIGURE 8. The setup for the energy measurement of the electrons accelerated by LWFA

quadrupole doublet, two bending magnets and a emulsion track detector. It is important to distingish the accelerated electrons by laser acceleration from other background noise of the scattered electrons and bremsstrahlung x-ray from the vacuum chamber and the beam dump.

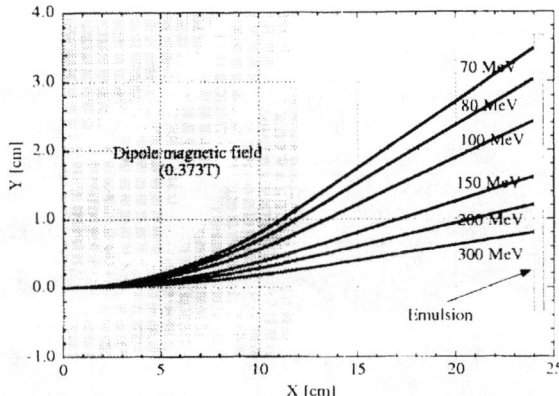

FIGURE 9. Trajectries of electrons with various energies from 70 MeV to 300 MeV

The new energy measurement system is shown in Fig. 8. The electron beams from the acceleration chamber are focused by a quadrupole doublet. The electrons that are not accelerated by the wakefield are bent by 60 degrees in a dipole magnet at the curvature radius of 13.9 cm and then go to a beam stopper. The length of the dipole magnet is 12.0 cm and the magnetic field density is 0.373 T. After the first dipole magnet, we install a chamber for a recutangular emulsion film of $60 \times 40$ mm$^2$. The emulsion 100 $\mu$m thick is applied on the both sides of an acrylic film 300 $\mu$m thick and is enveloped in a shading paper bag. The effective measurement area of the emulsion is $40 \times 20$ mm$^2$. The accelerated electrons more than 70 MeV can go through an emulsion sheet after the dipole magnet. The trajectries of the electrons from 70 MeV to 300 MeV are shown in Fig. 9.

375

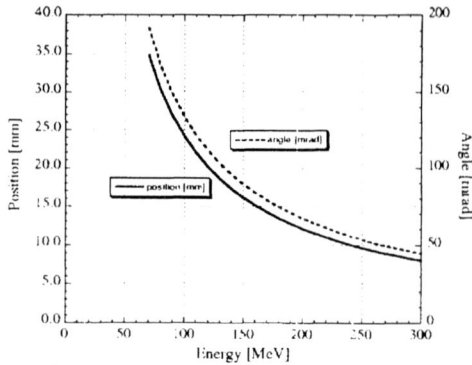

**FIGURE 10.** The relation between posiotions and angles of tracks on a emulsion for electrons with various energies

Thousands number of electrons are expected to be accelerated and leave thier tracks in the emulsion film. Each energy of the electrons is given by the position of the track in the emulsion. The position and the angle of the electron tracks for different energies are ploted in Fig. 10. Most of the angles of the tracks for the scattered low energy electrons are different from those of accelerated ones. Thus it is possible to distingish the tracks of the accelerated electrons from the background electrons. The x-rays are also identified easily from the Compton scattering in the emulsion. We expect to get energy spectra of accelerated electrons at lower background noise than previous measaurement system. In addition, as the emulsion spectrometer is compact, it is also suitable for measuring the energy of electrons higher than 1 GeV in the next generation LWFA experiments.

# REFERENCES

1. T. Tajima and J. M. Dawson, Phys. Rev. Lett., **43**, 267 (1979).
2. K. Nakajima et al., Phys. Rev. Lett., **74**, 4428 (1995)
3. A. Modena et al., Nature (London) **377**, 606 (1995)
4. D. Umstadter et al., Science **273**, 472 (1996)
5. S. P. Le Blanc et al., Phys. Rev. Lett. **77**, 5381 (1996)
6. J. R. Marquès et al., Phys. Rev. Lett., **76**, 3566 (1996)
7. C. W. Siders et al., Phys. Rev. Lett., **76**, 3570 (1996)
8. K. Nakajima, Phys. Plasmas, **3**, 2169 (1996)
9. Y. Ehrlich et al., Phys. Rev. Lett.,**77**,4186(1996)
10. T. Hosokai et al., Jap. J. Appl. Phys., **36**, No4. Part1, 2327(1997)

# Efficient excitation and measurement of plasma channels

E. W. Gaul*, S. P. Le Blanc, and M. C. Downer

*University of Texas at Austin, Department of Physics, MC C1600, Austin, Tx 78712*
*\* email:gaul@physics.utexas.edu*

**Abstract.** We report formation of a guiding channel for high intensity laser pulses in pure, preionized helium using an axicon focused laser pulse, and demonstrate a single shot technique to measure the temporal and spatial evolution of the plasma channel.

## I INTRODUCTION

Laser wakefield acceleration (LWFA) can be used to accelerate particles to high energies in a short distance [1]. To create wakefields over lengths much longer than the diffraction limit of a laser pulse focused to $10^{18}$ $W/cm^2$, optical guiding of the laser is necessary [2]. Today, there are two primary methods for guiding high intensity laser pulses though plasmas. The first is based on relativistic self-focusing and requires that the power of the laser pulse $P$ exceeds the critical power for relativistic self-focusing $P_c$. Relativistic self channeling has been observed [3–5] in self-modulated LWFA experiments at high densities $n_e \geq 10^{19} cm^{-3}$ where the dephasing length is comparable to the Rayleigh length of the focused laser beam. Evidence of acceleration beyond the dephasing limit has also been reported for self-modulated LWFA [5]. In all these cases, channeling requires high densities, and, therefore, acceleration is based on self-modulated LWFA which relies on a plasma instability. The second guiding method relies on propagation though a pre-formed plasma channel in which the electron density radial profile has a minimum on axis. Under these conditions, linear guiding of the laser pulse occurs and a plasma waveguide can be produced in a regime in which the standard LWFA ($\omega_p\tau \sim 1$) can be realized. The channeled standard LWFA provides the best opportunity for a well controlled, efficient accelerating structure, but has not been realized in the laboratory. Several pre-formed channel generation methods for the standard LWFA are currently being explored, including those based on hydrodynamic expansion of a laser heated plasma column [6], and propagation through hollow core fibers [7]. The challenges facing channeled LWFA include: the generation of plasma channels of several cm to 1 meter in length with on axis electron densities near $10^{18}$ $cm^{-3}$;

CP472, *Advanced Accelerator Concepts: Eighth Workshop,*
edited by W. Lawson, C. Bellamy, and D. Brosius
© 1999 The American Institute of Physics 1-56396-889-4/99/$15.00

the development of efficient coupling into the channel [8] and maintenance of robust radial electron density profiles; the realization of hollow core plasma fibers to reduce wakefield decay [9–11]; and the development of low cost channeling techniques. Equally important will be the development of accurate single-shot optical diagnostics for the plasma channel itself as well as for the properties of channeled wakefields [12].

## II  ELECTRICAL PREIONIZATION FOR CHANNEL GENERATION

For preformed channels generated by an axicon focused laser pulse, a long pulse ionizes a gas by multi-photon ionization and heats the electrons by inverse bremsstrahlung absorption [6,13]. The hydrodynamic evolution of the resulting cylindrical blastwave forms an electron density profile suitable as an optical guide. Guiding of high intensity pulses up to $10^{17}$ $W/cm^2$ has been demonstrated [14]. In this technique, it has been recognized that channel formation is a two step process. The first is the ionization of the gas to provide the necessary seed electrons for inverse bremsstrahlung absorption, and the second is the heating of the gas on axis which then drives a radial shock wave. Both steps can be done by one laser pulse, but its parameters can not be optimized for both processes. The heating rate is given by the collision frequency $\nu_c \propto n_e/(U_q + T_e)^{3/2}$, where $n_e$ is the electron density, $U_q$ the electron quiver energy in the electric field of the laser, and $T_e$ is the electron temperature [15]. To have the highest heating rates, intensities of about $10^{13}$ $W/cm^2$ are required. In contrast, the first electron of Helium is stripped at intensities of $> 10^{15}$ $W/cm^2$. For the guiding channel, it is preferred to have a degree of ionization such that no further electrons are removed from the ions via multi-photon ionization by the guided high intensity ($\sim 10^{17} - 10^{19}$ $W/cm^2$) laser pulse. Additional electrons in the center of the guide would alter the electron density profile and destroy the guide. Suitable plasmas are fully ionized hydrogen or helium, and fully helium-like nitrogen, oxygen, or neon.

To generate seed electrons for the first step of the channel formation process, we preionize $100 - 700$ Torr helium in a backfilled vacuum chamber with an electrical discharge (Fig. 1). For this a 0.5 $nF$ storage capacitor is charged to $\sim 30$ $kV$ by a high voltage power supply though a current limiting resistor. The charge on the capacitor is held off by a spark gap, pressurized to $2 - 3$ $atm$ nitrogen, which is triggered by a Q-switched laser pulse (532 nm, 5 ns, 20 mJ) synchronized to the channel forming pulse. The spark gap is enclosed in a pressurized plastic cell in order to isolate its contents from the helium discharge and to increase the hold off voltage of the gap. When the spark gap conducts, charge flows onto a second capacitor in parallel with discharge electrodes separated by 15 mm in a helium backfilled chamber. The compact design has a low wire inductance which leads to a fast overvoltage built up across the electrodes, and hence a rapid breakdown of the helium gas and strong ionization. About 100 $mJ$ energy is deposited into

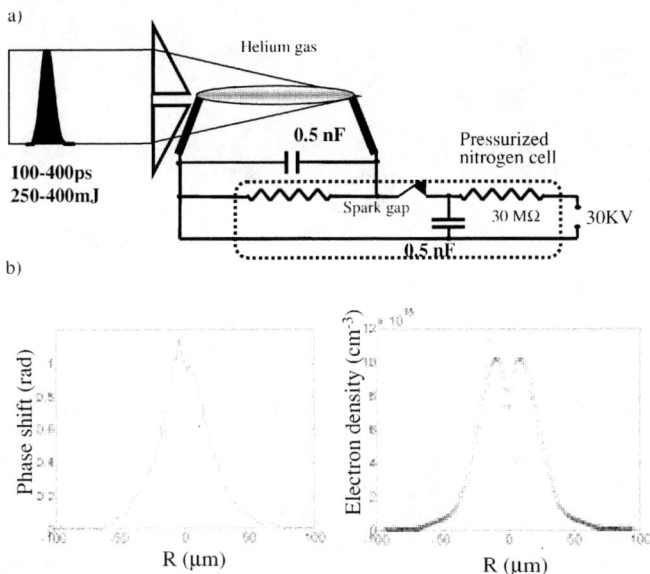

**FIGURE 1.** a)Sketch of the electrical preionization setup. The pressurized nitrogen cell is made of plastic in order electrically isolate its contents from the helium discharge. The BK7 axicon has a base angle of 30 degrees. b) Measured phase shift of a transverse probe pulse through the laser spark in 300 $Torr$ helium. The electron density is obtained by Abel inversion.

the helium gas in $\sim 30\ ns$. From the current density and the size of the discharge we estimate a peak preionized electron density $n_e$ of $10^{16}\ cm^{-3}$ and an electron temperature of 1 eV. This is sufficient to absorb about 1% of the energy of the axicon focused Nd:YAG laser pulse ($\tau = 100\ ps, E = 250\ mJ$) at intensities of $10^{13}\ W/cm^2$ in 300 $Torr$ helium. The axicon focused laser pulse generates a high density ($n_e \sim 10^{19}\ cm^{-3}$) plasma column over 10 mm of the 15 mm long pre-ionized region. The laser spark is easily distinguished from the discharge as a thin, straight spark with much smaller radius. The laser spark was most consistent when it arrived within $\simeq 10\ ns$ of the peak discharge current. At later times diffusion of the electrons decreases the peak current density and hence, the electron density. Without pre-ionization by the electrical discharge, the axicon focused laser pulse has insufficient intensity to produce a plasma in pure helium.

Fig. 1b shows the interferometric measured phase shift of a transverse probe pulse and the electron density profile obtained by Abel inversion. For a delay of $\sim 1.5\ ns$ between the axicon focused laser pulse and the probe laser pulse, Fig. 1b shows that the electron density profile has a depression on axis indicating the formation of a plasma channel. Numerical simulations suggest that a $300 - 500\ ps$ pulse at the same intensity level will be absorbed more strongly, which then leads to almost fully ionized helium. In the first $50 - 100\ ps$ of the pulse the ionization is increased from the preionization level of $n_e \simeq 10^{16}\ cm^{-3}$ to complete single

379

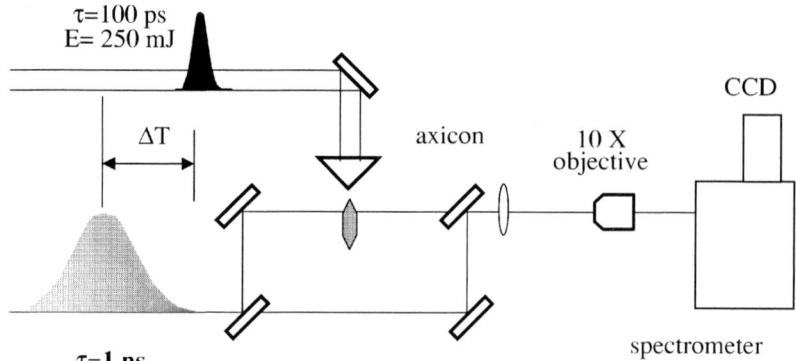

**FIGURE 2.** Setup of chirped pulse interferometry. The 1 $ns$ probe pulse has a linear frequency chirp.

ionization $n_e \simeq 10^{19}$ $cm^{-3}$. Therefore, for 100 $ps$ pulses the energy gets absorbed only in the tail of the pulse.

The seed electrons for the first step of channel formation can also be obtained by using a high intensity, femtosecond laser pre-pulse [16] or by multi-photon ionization of a high-Z seed gas with low ionization threshold [6]. Use of a high-Z seed gas has the disadvantage that additional ionization stages of the seed gas may be ionized by a guided high intensity laser pulse. Pre-ionization by use of an electrical discharge is an inexpensive alternative to a femtosecond ionizing pre pulse, especially if we scale the channel to lengths $\geq 10$ $cm$ and repetition rates $\gg 10$ $Hz$.

## III   CHIRPED PULSE INTERFEROMETRY

To probe the temporal and spatial evolution of a plasma channel, we demonstrate a new single-shot technique based on chirped pulse interferometry. In this experiment, a Nd:YAG laser pulse ($\tau = 100$ $ps$, $E = 250$ $mJ$) is focused by an axicon to intensities of $10^{13}$ $W/cm^2$ into 400 $Torr$ argon to form a plasma channel. A 20 $nm$ bandwidth 35 $fs$ Ti:Sapphire laser pulse ($\lambda = 800$ $nm$) is linearly chirped to 1 $ns$, regeneratively amplified, and used to interferometrically probe the spark (Fig. 2). The probe pulse is synchronized to the Nd:YAG pulse with less than 100 $ps$ jitter. The interferogram at the output of the Mach-Zehnder interferometer is imaged by a lens and magnified tenfold by a microscope objective on to a 100 $\mu m$ wide slit of a spectrometer. Because the probe pulse is linearly chirped, the temporal behavior of the interferogram is mapped onto the horizontal (wavelength) axis of the CCD camera. The vertical axis, which is parallel to the entrance slit of the spectrometer, records the transverse (radial) dependence of the interferogram.

In a simulation of the experiment a 35 $fs$ pulse, linearly chirped ($\omega = \omega_o + at, a = .01$ $ps^{-2}$) to 1 $ns$, is passed though a cylindrical electron density profile (Fig. 3a right), which was assumed to have a Gaussian profile moving out at a speed of

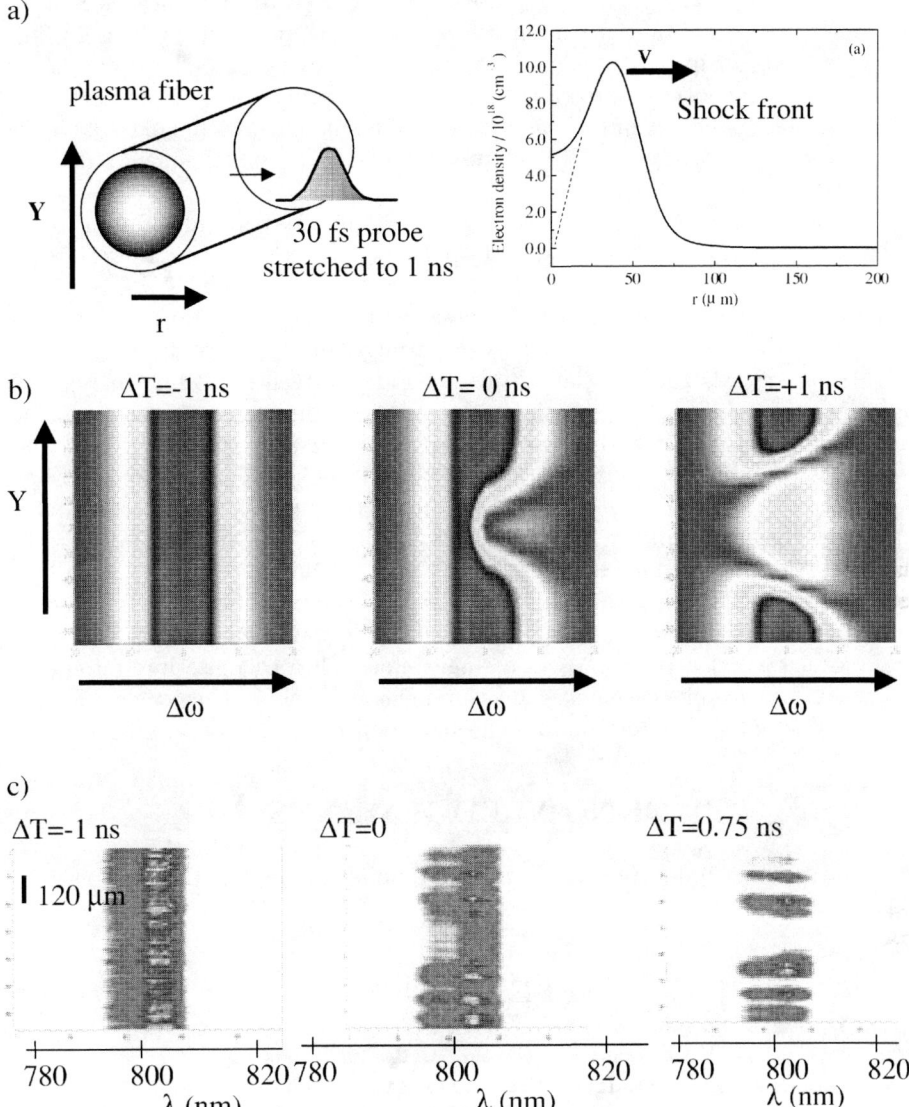

a) plasma fiber

Y

r

30 fs probe
stretched to 1 ns

Shock front

Electron density / $10^{18}$ (cm$^{-3}$)

r ($\mu$ m)

b) $\Delta T=-1$ ns     $\Delta T= 0$ ns     $\Delta T=+1$ ns

Y

$\Delta\omega$     $\Delta\omega$     $\Delta\omega$

c) $\Delta T=-1$ ns     $\Delta T=0$     $\Delta T=0.75$ ns

120 $\mu$m

780     800     820 780     800     820     780     800     820

$\lambda$ (nm)     $\lambda$ (nm)     $\lambda$ (nm)

**FIGURE 3.** a) Sketch of a chirped pulse probing the plasma channel and electron density profile used in the simulation. b) Simulated intensities versus radius and frequency for various delays between spark and probe pulse. c) Measured interferograms for the same delay times as in b). The structure in the vertical direction is due to probe ray refraction caused by transverse electron density gradients.

$4 \times 10^6$ $cm/s$ (Fig. 3a right). The pulse interferes with its reference, is Fourier transformed, and we obtain the intensity versus radius and frequency (Fig. 3b). For a shock front moving out with constant velocity, we expect a $\subset$-shaped intensity contour since frequency maps to time for the chirped probe pulse. Note that both paths in the Mach-Zehnder interferometer have to be matched within one coherence length of the unchirped pulse (35 $fs$) to minimize additional frequency domain interference fringes.

Fig. 3c shows experimental data at three different delay settings for the 1 $ns$ probe pulse. At time $t = t_0$ the Nd:YAG pulse arrives. In Fig. 3c, the center of the probe pulse arrives at times $t = t_0 - 1$ $ns$, $t = t_0 + .25$ $ns$, and $t = t_0 + .75$ $ns$ from left to right, respectively. First, the interferogram is uniform, since the plasma develops after the probe pulse arrives. We get increasing phase shifts from about $\lambda = 800$ $nm$ towards the left (later time) of Fig. 3c (center), while in Fig. 3c (right) the phase shift is through the center of the interferogram. The evolution of the shock front radius with delay time agrees with previous results [13] and agrees qualitatively with the expansion of a cylindrical explosion with an absorbed energy of $/sim$10 mJ/cm. The temporal resolution of chirped pulse interferometry is limited by the spectrometer resolution and the pixel size of the CCD camera to about 3 $ps$, while the spatial resolution of the recovered electron density depends on the imaging onto the camera and is limited by the accuracy of the Abel inversion. A wider bandwidth and more strongly chirped probe would enable this technique to probe over several nanoseconds in a single shot. Chirped pulse interferometry can be used for any interferometric measurement on a short time scale where a spatial dimension of the interferogram can be sacrificed for temporal information.

## ACKNOWLEDGEMENTS

This research was supported by the US DOE under Grant No. DEFG03-96-ER-40954.

## REFERENCES

1. E. Esarey et. al., IEEE Trans. on Plasma Sci. **24**, 252 (1996).
2. W. Leemans et. al., IEEE Trans. Plasma Sci. **24**, 331 (1996).
3. S.-Y.Chen et. al., Phy. Rev. Lett. **80**, 2610 (1998).
4. K. Krushelnick et. al., Phys. Rev. Lett **89**, 4047 (1997).
5. D. Gordon et. al., Phys. Rev. Lett. **80**, 2133 (1998).
6. C. G. Durfee III and H. M. Milchberg, Phys. Rev. Lett. **78**, 2409 (1993).
7. Y. Ehrlich et. al., Phys. Rev. Lett. **77**, 4186 (1996).
8. S. P. Nikitin et. al., Opt. Lett. **22**, 1787 (1997).
9. T. C. Chiou et. al. Phys. of Plasmas **2**, 310 (1995).
10. N. E. Andreev et. al. Phys. of Plasmas **4**. 1145 (1997).
11. B. Shadwick et. al. Bull. Am Phys. Soc. **42**, 1935 (1997).

12. S. P. Le Blanc, E. W. Gaul, and M. C. Downer, these proceedings.
13. T. R. Clark and H. M. Milchberg, Phys. Rev. Lett. **78**, 2374 (1997).
14. S. P. Nikitin, I. Alexeev, T. Antonsen, and H. M. Milchberg, these proceedings.
15. Yu. P. Raizer, Laser induced discharge phenomena, [Consultants Bureau, New York] (1977).
16. P. Volfbeyn and W. P. Leemans, these proceedings.

# Experimental Observation of Return Current Effects in Passive Plasma Lenses

R. Govil and W. P. Leemans

*Lawrence Berkeley National Laboratory, One Cyclotron Road, Berkeley, CA 94720*

**Abstract.** An overdense plasma lens experiment was conducted at the Beam Test Facility at LBNL [1]. Measurements showing focusing of 50 MeV electron bunches by laser produced overdense plasmas have been reported previously [2]. Here we present experimental observation of plasma focusing in the return current regime, where the electron beam is both charge and (partially) current neutralized. Theoretical and numerical analysis of plasma return current is presented and shown to agree with measurements.

## INTRODUCTION

The next generation colliders require strongly focused beams with high luminosity. To focus charged particle beams for such applications, a plasma focusing scheme has been proposed [3]. Overdense plasma lenses (plasma density $n_p$ much greater than electron beam density $n_b$), rely on beam charge neutralization by plasma. When the beam charge is neutralized, the self-magnetic field of the bunch can pinch it into a small size. In addition, sufficiently dense plasmas can also neutralize the current of a beam. Due to the changing magnetic flux of a bunch propagating in plasma, current is induced in the plasma in the opposite direction (Lenz' law). The plasma return current flows within a radius of the order of $c/\omega_p$, where $c$ is the speed of light and $\omega_p$ is the plasma frequency (Fig. 1). The fractional current flowing within the beam volume reduces the net current and, therefore, the self-focusing.

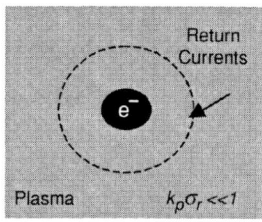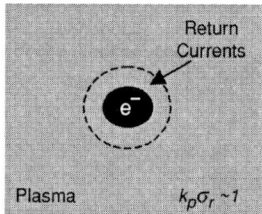

**FIGURE 1**. Return current distribution in plasma. For low plasma density, the return current occupies a larger volume. For high density, the current flows predominantly within the bunch and cancels the bunch current and self-focusing.

CP472, *Advanced Accelerator Concepts: Eighth Workshop,*
edited by W. Lawson, C. Bellamy, and D. Brosius
© 1999 The American Institute of Physics 1-56396-889-4/99/$15.00

In this paper, we present results from the experimental study of return current in plasmas and its effect on beam focusing. We start with a physical derivation of plasma return current. Then we describe the experimental setup and beam diagnostics. An analysis of return current in plasma lenses is presented and compared to experimental measurements. The effect of finite plasma size on return current distribution and beam focusing is described.

## THEORY

Return current in plasmas was first calculated by Roberts and Bennett [4]. A physical derivation following Whittum et al. [5] is presented below. Results from a rigorous treatment are presented later in the analysis section.

The plasma current within the beam volume can be calculated from the definition,

$$\frac{d}{dt} I_p = \frac{d}{dt}(J_p \, 2\pi\sigma_r^2) \, , \tag{1}$$

where $\sigma_r$ is the beam radius and $J_p$ is the plasma current density. Using a linearized cold fluid plasma model, $J_p = -en_0 v_1$ and $dv_1/dt = -eE/m$, we have,

$$\frac{d}{dt} I_p = \frac{\omega_p^2 \sigma_r^2}{2} E \, , \tag{2}$$

where $v_1$ is the velocity of the plasma fluid element, $\omega_p^2 = 4\pi n_0 e^2 / m$ is the plasma oscillation frequency, $m$ is the electron mass and $E$ is the self-consistent electric field in the plasma.

The electric field can be calculated using the vector potential:

$$|E| = -\left|\frac{\partial A}{\partial t}\right| = -\frac{1}{c}\frac{\partial I_{tot}}{\partial t} \, , \tag{3}$$

where we have dropped geometrical factors of order unity. Substituting eq. (3) into eq. (2) yields,

$$\frac{d}{dt} I_p = \frac{\omega_p^2 \sigma_r^2}{2}|E| = -\frac{k_p^2 \sigma_r^2}{2}\frac{dI_{tot}}{dt} \, . \tag{4}$$

Here $k_p = \omega_p / c$ is the plasma wavenumber. Using $i_p = i_{tot} - i_b$ and integrating the above equation, we obtain,

$$\frac{I_{tot}}{I_b} = \frac{1}{1 + k_p^2 \sigma_r^2 / 2} \, . \tag{5}$$

385

As expected, the effect of the return current is to reduce the total current. Overdense plasma lens focusing strength is defined as, $K \equiv |F_r / r| / \gamma m_e c^2$, where $F_r$ is the radial force on the beam and $\gamma m_e c^2$ is the beam energy. For the overdense plasma $|F_r| = 2\pi e^2 n_b r$, corresponding to a focusing strength of,

$$K = \frac{2\pi r_e}{\gamma} n_b, \qquad (6)$$

where, $r_e$ is the classical electron radius. The plasma return current effectively reduces the focusing strength to,

$$K = \left( \frac{1}{1 + k_p^2 \sigma_r^2 / 2} \right) \frac{2\pi r_e}{\gamma} n_b. \qquad (7)$$

Very high density plasmas, with $k_p \sigma_r >> 1$, would completely neutralize the beam's charge and current. Although such extremely overdense plasmas are not useful as focusing devices, they can be used to reduce beam-beam disruption at the interaction point in colliders [5]. At the interaction point, the colliding beams exert very strong forces on each other. If a very-overdense plasma were placed at the interaction point, it would neutralize the beams' fields, allowing them to collide and pass through each other without disruption.

The goal of the LBNL experiment was to study the plasma return current and its impact on beam focusing in plasma lenses. Since return current is significant only for large values of $k_p \sigma_r$, high plasma densities and large beam sizes were needed to study their effect. In addition, $k_p \sigma_z > 2$, so that the plasma can respond within the beam duration. The production and characterization of plasmas meeting these requirements has been reported previously [6,7]. Plasmas with density 2 x $10^{13}$- 4 x $10^{14}$ cm$^{-3}$ are produced through two-photon ionization of Tripropylamine (TPA) vapor. The plasma density is controlled through TPA fill pressure and laser intensity. The plasma is produced in a slab shape, with one transverse dimension (height) comparable to beam dimensions. This allows for studying the effect of finite plasma sizes on return current distribution and beam focusing. For transverse beam sizes of about 200 - 400 microns, $k_p \sigma_r$ is in the range of 0.3 - 1.1 and $k_p \sigma_z$ is 4 - 14.

To measure the beam envelope evolution as a function of propagation distance over the entire 1 meter length of the plasma chamber, an optical transition radiation (OTR) foil on a moveable stage was used [8]. Production of plasma through laser ionization allowed access to plasma volume for beam size measurements inside and at close proximity to the plasma. In addition, bunch charge is measured at two fixed locations along the beam line and at every location inside the plasma chamber using Integrating Current Transformers (ICTs).

In the next section, we describe the experimental setup and beam diagnostics. We present beam envelope measurements in vacuum and plasma lenses.

# EXPERIMENTAL RESULTS

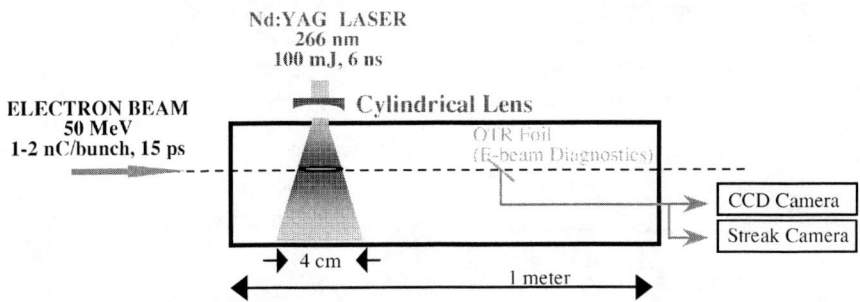

**FIGURE 2.** Schematic of plasma lens experiment at LBNL

The electron beam for the plasma lens experiment is provided by the injector of the Advanced Light Source (ALS) at LBNL. The 50 MeV beam carries 1 nC charge in 15 ps (rms) short bunches. The electron beam was transported through a dedicated beamline to the experimental chamber, separated from the beamline by a 7.6 μm thick Kapton window. Inside the experimental chamber, plasma was produced through two-photon ionization of Tripropylamine (TPA) vapor using a 266nm laser beam. The plasma is produced in slab-shape, by modifying the laser beam profile through a cylindrical lens. The plasma density can be controlled through neutral gas fill pressure and laser beam intensity profile. Plasma densities up to $2 \times 10^{14}$ cm$^{-3}$ were measured using microwave interferometry at 94.3 GHz. A microwave beam tunneling technique was developed to measure the plasma density greater than the cutoff density for 94 GHz radiation [7].

As mentioned earlier, the electron beam sizes were measured using OTR images of the bunch. Figure 3 shows transverse bunch images in false color measured a few centimeters downstream of the plasma location inside the chamber. The images show a dramatic increase in intensity and a corresponding decrease in transverse bunch size when plasma is produced. As reported previously [2], we also observed focusing of the electron beam by the background TPA vapor, in the absence of the ionizing laser beam. An analysis of the gas focusing is under study. Here, we discuss measurements and analysis of beam focusing due to laser-ionized plasma lenses in the return current regime.

A full beam envelope is shown in Figure 4. The importance of detailed diagnostics can be seen from this data. Note that the beam sizes before the plasma location (z<2cm) are the same for both curves.

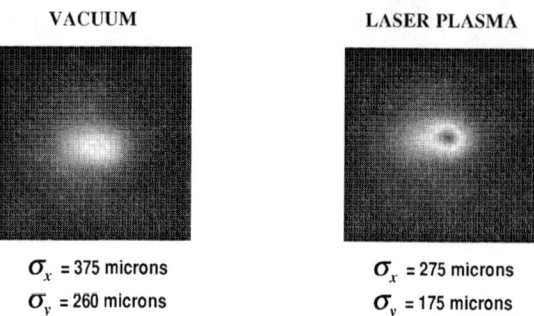

VACUUM                           LASER PLASMA

$\sigma_x$ = 375 microns          $\sigma_x$ = 275 microns
$\sigma_y$ = 260 microns          $\sigma_y$ = 175 microns

**FIGURE 3**. Single shot OTR images of electron beam, with false color representing intensity.

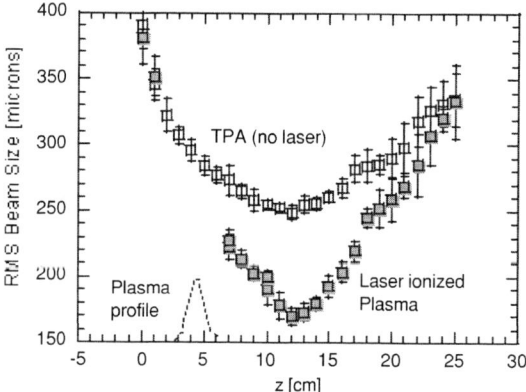

**FIGURE 4**. Measured electron beam envelope in the y-axis in TPA (hollow dots) and with laser-produced plasma lens (solid dots) as a function of propagation distance, $z$. Each dot is an average of 10-20 shots, and the error bars represent the statistical standard deviation. Plasma with density 4.1 x 10$^{13}$ cm$^{-3}$ is centered at z=4.4 cm.

Since the goal of the experiment was to study the return current, we varied the plasma density and beam sizes to produce lenses with large and small values of $k_p\sigma_r$. Figure 5 shows the beam envelope measured for vacuum and two plasma densities. The round dots represent the beam sizes measured in vacuum. When laser-plasma of density 4.9 x 10$^{13}$ cm$^{-3}$ is produced along the its path at z=4.4cm, the electron beam is observed to focus to a smaller beam size (triangular dots). When the plasma density is increased to 1.5 x 10$^{14}$ cm$^{-3}$, the measured beam sizes, though still smaller than in vacuum, show less focusing (squares). The value of $k_p\sigma_r$ for the two cases was 0.38 and 0.65 respectively. As expected, the plasma lens with higher $k_p\sigma_r$ produces less beam focusing.

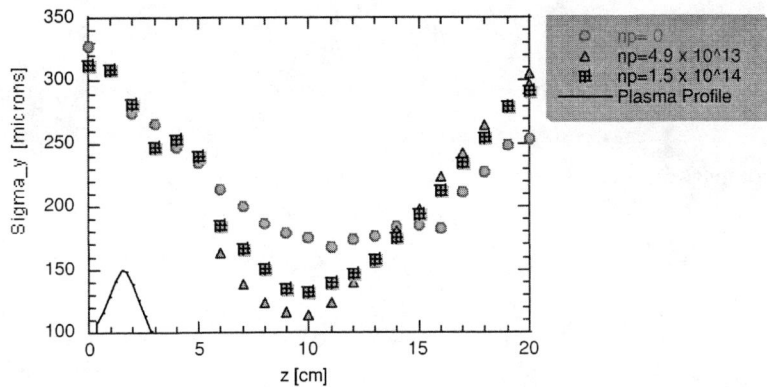

**FIGURE 5.** Beam envelope measured for various plasma densities.

Next, we carefully examine the affect of the return current on beam focusing. We compare beam evolution in plasma lenses with different values of $k_p \sigma_r$, and model the data with simulation.

## PLASMA FOCUSING IN RETURN CURRENT REGIME

In this section we examine two plasma lenses, one with low return current ($k_p \sigma_r = 0.33$), the other with high current ($k_p \sigma_r = 1.1$). The data is analyzed using a simulation code by comparing two wakefield models - one that includes plasma return current ("on") and one that doesn't ("off").

The plasma current for a gaussian bunch in an overdense lens is given by,

$$J_{pz}(r,z) =$$
$$ecn_{b0}\frac{R_0^2}{R^2}k_p^2\left\{K_0(k_p r)\int_0^r dr' r' I_0(k_p r')e^{-r'^2/2R^2} + I_0(k_p r)\int_r^\infty dr' r' K_0(k_p r')e^{-r'^2/2R^2}\right\}, \quad (8)$$

where a cold fluid model for the plasma, and a gaussian radial bunch profile with R as the rms beam size are assumed. Here $n_{b0}$ is the initial peak beam density, $R_0$ the initial bunch size and $R$ the bunch size as a function of z.. The wakefield $W_\perp$ can be calculated under the thin lens approximation [9]:

$$W_\perp(r,z,s) = 4\pi e n_{b0}\frac{\partial G(r,z,s)}{\partial r}\left[k_p\int_{-\infty}^s ds' \sin k_p(s-s')h(s') - \frac{h(s)}{\gamma^2}\right], \quad (9)$$

where,

389

$$G(r,z,s) = \int_0^r dr' \, r' \, g(r',z,s) I_0(k_p r') K_0(k_p r) + \int_r^\infty dr' \, r' \, g(r',z,s) I_0(k_p r) K_0(k_p r'), \quad (10)$$

and, $g(r,z,s)$ and $h(s)$ are the radial and longitudinal beam profiles, respectively.

For comparison, in a plasma response model ignoring return current, the wakefield with return current "off" can be calculated to be [10],

$$W_\perp(r,z,s) = -4\pi e n_{b0} k_p \int_{-\infty}^s ds' \sin k_p (s-s') h(s') \int_0^r dr' \, r' \, g(r,z,s'), \quad (11)$$

The electron beam evolution is modeled through the beam envelope [10],

$$\frac{d^2 R(z,s)}{dz^2} = \frac{\varepsilon^2}{R^3(z,s)} - \frac{e}{\gamma m c^2} \frac{\langle r \cdot W_\perp(r,z,s)\rangle_r}{2R(z,s)},$$

where

$$\frac{\langle r \cdot W_\perp(r,z,s)\rangle_r}{R(z,s)} = \frac{1}{R(z,s)} \frac{\int_0^\infty dr \, r \, e^{-r^2/2R^2(z,s)} \left[ r W_\perp(r,z,s) \right]}{\int_0^\infty dr \, r \, e^{-r^2/2R^2(z,s)}}. \quad (12)$$

Here, $R(z,s)$ is the rms beam size as a function of propagation distance and $s=ct-z$. A simulation is developed to solve the envelope equation numerically for specified wakefields. For each set of data, the initial beam envelope $R(0)$ and slope $R'(0)$ are obtained from the TPA vapor scans, and used as initial conditions for the plasma scans. The code calculates the additional focusing due to laser-ionized overdense plasma.

First we examine a plasma lens with a low value of $k_p \sigma_r = 0.33$. For this lens, peak $n_p = 2.3 \times 10^{13}$ cm$^{-3}$, corresponding to $k_p \sigma_z = 4.1$. The calculated return current distribution is plotted in Fig. 6a. Only 12% of the plasma current flows within the rms beam area. Therefore, we expect the reduction in focusing to be minimal. Figure 6b shows the measured beam evolution in the plasma lens. The solid (dashed) curves are the calculated beam envelope in TPA and overdense plasma with (without) return current. The calculated beam evolution is in reasonable agreement with measurements. The simulated beam evolution without return current shows slightly stronger focusing than the solid curve with return current. As expected, the focusing strength reduction due to current cancellation is only minimal.

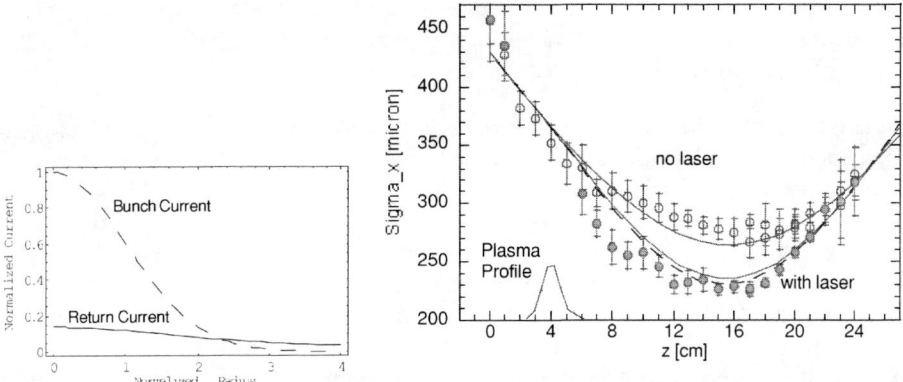

**FIGURE 6.** a) Comparison of plasma and beam current profiles for $k_p\sigma_x$ =0.33, plotted against radius normalized to bunch size. b) Corresponding beam envelope measurement (dots) and simulation (lines). Solid curves are obtained using envelope equation for TPA (no laser) and overdense plasma. Dashed curve is calculated beam evolution in plasma with return current off. Data symbols have the same meaning as in Figure 4.

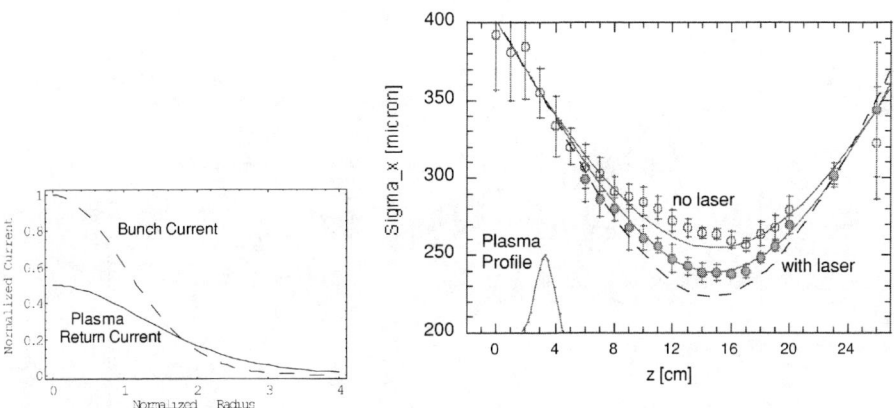

**FIGURE 7.** a) Comparison of plasma and beam current profiles for $k_p\sigma_x$ =1.1, plotted against radius normalized to bunch size. b) Corresponding beam envelope measurement (dots) and simulation (lines). Solid curves are obtained using envelope equation for TPA (no laser) and overdense plasma. Dashed curve is calculated beam evolution in plasma with return current off. Data symbols have the same meaning as in Figure 4.

For comparison, a plasma with $n_p$= 2.9 x $10^{14}$ cm$^{-3}$, corresponding to $k_p\sigma_r$=1.1 was produced. For this lens, about 25% of the return current flows inside the electron bunch rms area, as shown in Fig. 7a. Therefore, the return current cancels the beam current strongly. Figure 7b shows the measured beam evolution for this plasma lens. Simulation results are shown as solid lines. The beam evolution calculated ignoring current cancellation (dashed line in Fig. 7b) shows how much stronger the beam focusing would have been if there were no plasma return current. Thus, in agreement with calculations, the return current reduces the focusing strength of the plasma lens significantly for large $k_p\sigma_r$.

Finally, we examine the return current in finite-sized plasmas. As mentioned earlier, the laser-produced plasma has a slab shape, with one transverse dimension (height) comparable to beam size and plasma wavelength. Consequently, less return current flows in the vertical axis. To gain a qualitative understanding of the return current reduction in the vertical axis, we compare the measured envelope with calculations from the two models -- with return current "on" and "off". As seen from Fig. 8, the measured beam envelope is in between the calculated envelopes with return current turned on or off, as expected. By terminating plasma current integration (eq. (8)) at the plasma edge instead of at infinity, it is found that only 48% of the return current is able to flow in the finite size. The beam envelope calculated for this case is in reasonable agreement with measurements.

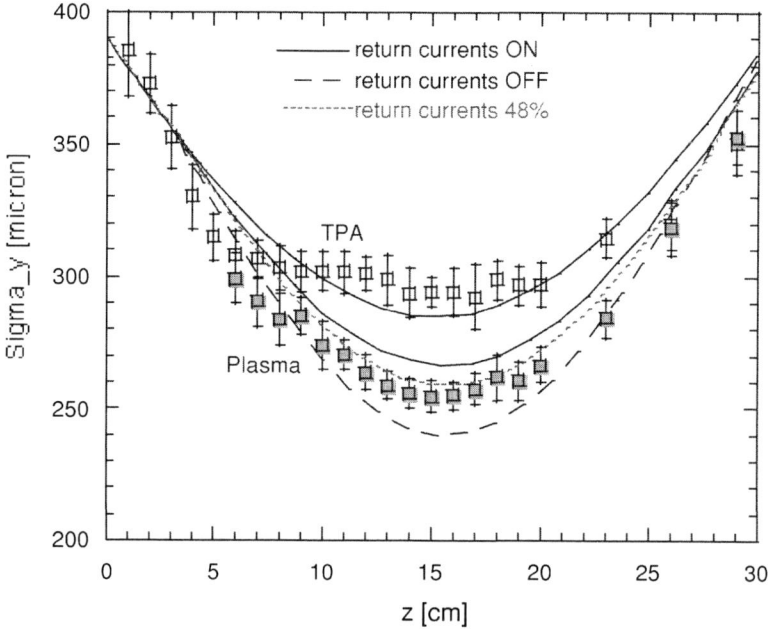

**FIGURE 8.** Measured electron beam vertical envelope is bracketed by calculations in TPA (upper solid line), with return current on (solid line) and off (dashed line). Data symbols have the same meaning as in Figure 4. Short-dashed line is beam envelope calculate with 48% return current.

# SUMMARY

An overdense plasma lens experiment was conducted at the Beam Test Facility at LBNL. Short (15 ps rms) 50 MeV electron bunch were focused by laser produced overdense plasmas. Simulations based on beam envelope models were developed and shown to agree with measurements.

In the return current regime, reduction in focusing strength due to current cancellation was experimentally measured and modeled. Return current was shown to reduce the focusing strength of the plasma lens significantly for large values of $k_p \sigma_r$.

Beam evolution calculations using return current models were shown to be in agreement with data. A simple model for return current in finite-sized plasmas was developed and shown to be in reasonable agreement with measurements. Further analysis is being conducted using PIC simulations [11]. In addition, beam focusing in TPA will be examined using PIC simulations.

# ACKNOWLEDGMENTS

This work would not have been possible without the help of S. Wheeler in running the experiment. Engineering support was provided by J. Dougherty and L. Archambault, and operations support by W. Byrne. We would like to thank K. Backhaus and J. Wurtele for many enlightening discussions on envelope models.

# REFERENCES

1. Leemans, W. P. et al., "The 50 MeV Beam Test Facility at LBL" in *Proceedings of the Particle Accelerator Conference*, 1993, pp.83-85.
2. Govil, R. et al., "Experimental Observation of Electron Beam Focusing through Plasma Lenses" in *Proceedings of the Particle Accelerator Conference*, 1997, to be published.
3. Chen, P., *Particle Accelerators*, **20**, 171 (1987).
4. Roberts, T. G., and Bennett, W. H., *Plasma Physics*, **10**, 381-389 (1968).
5. Whittum, D. H., et al., *Particle Accelerators*, **34**, 89 (1990).
6. Leemans, W.P. et al., "Plasma Production for the 50 MeV Plasma Lens Experiment at LBL" in *Proceedings of the European Particle Accelerator Conference*, pp. 831-833, 1994.
7. Govil, R., et al., "UV Laser Ionization and Electron Beam Diagnostics for Plasma Lenses" in *Proceedings of the Particle Accelerator Conference*, pp.776-778, 1995.
8. de Loos, M. J., et al., "Characterization of the 50 MeV ALS Linac Beam with Optical Transition Radiation" in *Proceedings of European Particle Accelerator Conference*, pp. 1679-1681, 1994.
9. Keinigs, R., and Jones, M.E., *Physics of Fluids*, **30**, 252-263 (1987).
10. Govil, R., Ph.D. Dissertation, University of California, Berkeley (1998), to be published.
11. Backhaus, E., et al., "Envelope Equation for a Magnetically Self-Focused Beam in a Plasma", presented at Advanced Accelerator Concepts Workshop, Baltimore, MD, July 1998.

# Intense Laser Pulse Propagation In Capillary Discharge Plasma Channels

R. F. Hubbard,[1] Y. Ehrlich,[2] D. Kaganovich,[2] C. Cohen,[2]
C. I. Moore,[1] P. Sprangle,[1] A. Ting,[1] and A. Zigler[2]

[1]*Beam Physics Branch, Plasma Physics Division*
*Naval Research Laboratory, Washington, DC 20375*

[2] *Racah Institute of Physics, Hebrew University, Jerusalem, Israel 91904*

**Abstract.** Optical guiding of intense laser pulses is required for plasma-based accelerator concepts such as the laser wakefield accelerator. Reported experiments have successfully transported intense laser pulses in the hollow plasma column produced by a capillary discharge. The hollow plasma has an index of refraction which peaks on-axis, thus providing optical guiding which overcomes beam expansion due to diffraction. In more recent experiments at Hebrew University, 800 nm wavelength, 0.1 mJ, 100 fs pulses have been guided in ~300 micron radius capillaries over distances as long as 6.6 cm. Simulations of these experiments using a 2-D nonlinear laser propagation model produce the expected optical guiding, with the laser pulse radius $r_L$ exhibiting oscillations about the equilibrium value predicted by an analytical envelope equation model. The oscillations are damped at the front of the pulse and grow in amplitude in the back of the pulse. This growth and damping is attributed to finite pulse length effects. Simulations also show that further ionization of the discharge plasma by the laser pulse may hollow the laser pulse and introduce modulations in the spot size. This ionization-defocusing effect is expected to be significant at the high intensities required for accelerator application. Capillary discharge experiments at much higher intensities are in progress on the Naval Research Laboratory T$^3$ laser, and preliminary results are reported.

## INTRODUCTION

Plasma-based accelerators such as the laser wakefield accelerator (LWFA) have demonstrated very large accelerating gradients and acceleration of electrons (1-6) to energies above 100 MeV. Unless the laser beam can be optically guided, the beam will expand quickly due to diffraction. This results in a short interaction length which severely limits the energy gain. Although the laser pulse may be self-guided, as in the self-modulated LWFA, such methods are usually difficult to control and are likely to be unsuitable for a practical accelerator. Preformed plasma channels, such as those

CP472, *Advanced Accelerator Concepts: Eighth Workshop,*
edited by W. Lawson, C. Bellamy, and D. Brosius

produced by an axicon-focused laser (7,8) or capillary discharge (9-11), may be better suited for applications.

The characteristic distance for the beam to expand due to diffraction is the Rayleigh length $Z_R$, given by

$$Z_R = \pi r_0^2 / \lambda, \tag{1}$$

where $r_0$ is the radius of the laser at the focus, and $\lambda$ is the laser wavelength. Since accelerator applications require the laser to be focused to ~10 μm, the Rayleigh length is typically less than 1 mm.

In general, refractive guiding of optical pulses can occur when the index of refraction η peaks on-axis. The index of refraction may include contributions from a number of sources, including relativistic (1-6,12,13) and nonlinear (14,15) self-focusing, preexisting plasma, and laser-produced plasma (15,16). A plasma with a density variation $n(r)$ introduces a refractive index change $\Delta\eta(r) = -\omega_p^2(r)/2\omega_0^2$ where $\omega_p = (4\pi n e^2/m)^{1/2}$ is the plasma frequency, e and m are the electron charge and mass, and $\omega_0 = 2\pi c/\lambda$ is the laser angular frequency. A hollow plasma column or channel with a density minimum on-axis produces the desired refractive index profile. Optical pulses have been guided in preformed plasma channels produced by an axicon focus pump laser pulse (7,8), an intense self-guided laser pulse (6,17), and capillary discharges (9-11).

This paper presents both theoretical and experimental results on intense laser pulse propagation in hollow plasma channels produced by a capillary discharge. We have previously reported laser guiding experiments (9-11) employing a linearly-polarized Ti-sapphire laser generating pulses with duration $\tau_L$ = 110 fs, wavelength $\lambda$ = 0.8 μm, and energy $E_L$ ~ 10 mJ. The laser is focused to a spot size $r_0$ = 15 μm. The intensity (~$10^{16}$ W/cm$^3$ at focus) is high by most standards, but is well below the intensity required for most accelerator applications. In addition, preliminary results from the more powerful $T^3$ laser at the Naval Research Laboratory (NRL) are described. This laser has $\tau_L$ = 400 fs, $\lambda$ = 1.06 μm, and $E_L$ ~ 1 J.

## REVIEW OF CAPILLARY DISCHARGE GUIDING

Capillary discharges provide a simple method for creating a narrow, hollow plasma column. A typical experimental configuration is shown in Fig. 1. The capillary is constructed from a thin cylinder of plastic in an insulating medium with annular high voltage electrodes at the ends. In the single capillary configuration, a high voltage is applied to the electrodes, initiating a surface discharge on the capillary walls. A plasma forms from material ablated from the surface. There is generally a period of time lasting several hundred nsec during which plasma temperature peaks on-axis, resulting in an on-axis density minimum. If the intense laser pulse is triggered during this period and injected into the capillary, the pulse may be guided over many Rayleigh lengths. Experimental results are described in Section III.

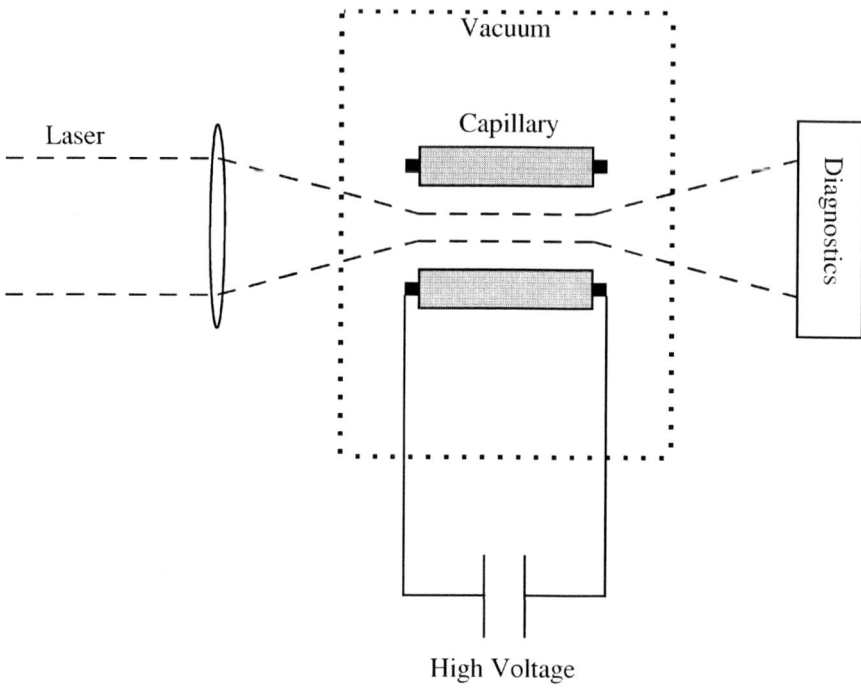

**FIGURE 1.** Generic experimental setup for capillary discharge guiding experiments. The laser pulse is focused on a point near the capillary entrance.

Laser propagation in a refractive medium can often be described theoretically using an envelope equation of the form

$$\frac{\partial^2 r_L}{\partial z^2} - \frac{r_0^4}{Z_R^2 r_L^3} + \frac{2}{r_L}\langle r\partial\eta / \partial r\rangle = 0, \tag{2}$$

where the brackets denote the intensity-weighted radial average. The second and third terms in Eq. (2) describe vacuum diffraction and refraction respectively. A matched (constant spot-size) optical beam can occur if the diffraction term is balanced by the refraction term.

If one includes only the contribution from a parabolic density channel profile $n(r) = n_0 + \Delta n r^2 / r_{ch}^2$, where $r_{ch}$ is the channel radius, then the equilibrium or matched beam radius is (12)

$$r_M = (r_{ch}^2 / \pi r_e \Delta n)^{1/4},\tag{3}$$

where $r_e = e^2 / mc^2$ is the classical electron radius. If the focused radius $r_0$ is near the matched beam radius $r_M$, so that $|r_0 - r_M| \ll r_M$, then the spot size undergoes envelope oscillations with a period

$$\lambda_e = \pi Z_M = \pi^2 r_M^2 / \lambda.\tag{4}$$

Here $Z_M = \pi r_M^2 / \lambda$ is the Rayleigh length associated with the matched beam radius.

The capillary discharge plasmas used in the experiments described in this paper have channel depth $\Delta n \sim 10^{18}\text{-}10^{19}$ cm$^{-3}$, peaking at a distance $r_{ch} \sim$ 100-150 μm from the channel axis. The resulting matched beam radius from Eq. (3) is thus typically 30 μm, independent of the laser pulse parameters.

The envelope equation assumes that the radial profile does not change its Gaussian shape as the beam propagates. This assumption is relaxed in the 2-D axisymmetric LEM simulation model (10,11). This code solves a nonlinear wave equation for the normalized vector potential a(r,ζ,τ) in a frame moving with the pulse at the speed of light with independent variables r, ζ = z - ct, and τ = t. The plasma response is treated using a fluid model. This code has been used extensively to study intense laser pulse propagation in uniform and hollow plasma channels.

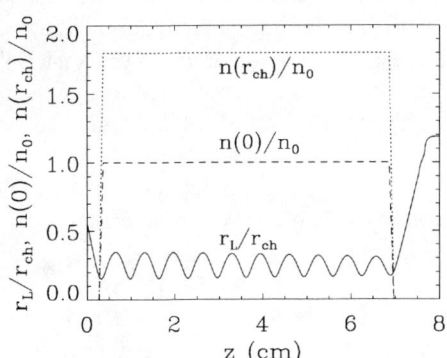

FIGURE 2. Simulation results for the normalized spot size $r_L/r_{ch}$ (solid curve) in a 6.6 cm long channel, where the channel radius $r_{ch}$ = 100 μm. Also shown are the on-axis and channel edge densities $n(0)/n_0$ and $n(r_{ch})/n_0$, normalized to $n_0$ = 5x10$^{18}$ cm$^{-3}$.

Figure 2 gives a plot of a typical simulation result. The laser pulse is injected in vacuum at z = 0 and focused onto the front of the channel. The laser pulse had $\lambda$ = 0.8 μm, pulse length $\tau_p$ = 100 fs, focal spot size $r_0$ = 15 μm, and the peak power $P_0$ = 0.15 TW. The plasma channel is 6.6 cm long, with channel radius $r_{ch}$ = 100 μm, peak on-axis density $n_0$ =5x10$^{18}$ cm$^{-3}$, and channel depth $\Delta n$ =

$4 \times 10^{18}$ cm$^{-3}$. The normalized on-axis density $n(r = 0, z)/n_0$ and channel edge density $n(r_{ch}, z)/n_0$ are plotted as the dashed and dotted lines in the figure. Also plotted is the laser spot size $r_L(z)$ at the reference point $\zeta^*(z)$ from the simulation. This point slips back in the pulse at the analytical group velocity which is $\beta_{g0} = \omega_p^2(r = 0)/2\omega_0^2$ in this frame. As expected, $r_L$ oscillates about an equilibrium value in the channel and expands rapidly once the laser pulse exits the channel. The equilibrium radius $r_{MS}$ in the simulation, based on an average of $r_L$ over many oscillations, is 25 μm, while the analytical value for $r_M$ based on Eq. (3) is 23 μm. Both the analytical oscillation period $\lambda_e(r_M)$ given by Eq. (4) and simulation oscillation period $\lambda_{eS}$ are 0.66 cm.

## EXPERIMENTAL APPROACH AND RESULTS

Ehrlich, *et al.* (10,11) and Zigler, *et al.* (9) have reported successful transport of short-pulse lasers in both slotted (9) and cylindrical (10,11) capillary discharge channels. The typical experimental setup is shown in Fig. 1. The major components include the short pulse laser, focusing optics, the capillary discharge and its driving circuitry, and diagnostics for measuring capillary plasma parameters, transmitted laser power, and laser beam expansion beyond the capillary.

The capillary discharge provides a simple, controllable method for generating a narrow, hollow plasma column. The device consists of a thin cylinder of easily-ablated insulating material such as a plastic with high voltage electrodes at the ends. The capillary is placed under vacuum, so the plasma is generated from material ablated from the inner wall of the capillary. The electrodes are connected to a capacitor which is typically charged to $V_D = 10$ kV in order to drive the discharge current. Additional inductance may be added to the driving circuit to slow the time scale of the discharge. In the double capillary configuration described by Kaganovich, *et al.* (18), high voltage is applied first to a short trigger section. This produces a source plasma which can trigger the main capillary at a lower charging voltage (1-2 kV) which allows more control over plasma parameters.

Simulations (18) indicate that there is a period of time where the pressure is approximately constant in space, and the plasma temperature is higher in the center of the capillary than at the edges. This results in a density minimum on-axis. The peak temperature is estimated to be 3 eV. Since the walls of the capillary are polyethylene (CH$_2$), the plasma at this temperature should be nearly fully dissociated and consist of protons and low charge states of carbon. The existence of the density minimum has been confirmed experimentally from Stark broadening measurements (11). The capillaries are 1-6 cm long and several hundred microns in radius. The radius $r_{ch}$ where the density is maximum varies with time but is typically 50-70% of the wall radius $r_w$.

These experiments were carried out at Hebrew University and utilized a linearly-polarized Ti-sapphire laser with a 100 fs long pulse at 0.8 micron wavelength and a pulse energy of up to 10 mJ. The pulses were focused onto the entrance of the capillary with an f/11 lens, producing a focused spot size $r_0$ of 15 μm. This spot size was approximately 1.7 times the diffraction limit. The external circuit parameters for the capillary and the time delay $\Delta t_D$ between the initiation of the capillary and the firing of the laser were varied in order to optimize laser transport. The first experiments with cylindrical capillary discharges (10) demonstrated transport efficiencies of up to 85% in both straight and curved configurations. CCD camera images of the transmitted laser light indicated a spot size $r_L$ of 30-50 μm, which is consistent with the analytical prediction from Eq. (3). The capillaries were 1 cm long with an inner radius of 175 μm. When the discharge was not triggered, transport efficiencies dropped dramatically, and the transported laser spot appeared to fill the capillary.

Recent experiments (11) using the same laser employed capillaries with lengths up to 6.6 cm. Post-capillary expansion of the beam was measured by placing four beam splitters at several locations after the capillary exit and deflecting the resulting images to different areas of a CCD camera. A series of experiments in 2 cm long channels using this technique revealed substantial qualitative shot-to-shot differences in $r_L(z)$. This is believed to be due to small variations in the matched radius $r_M$ and envelope oscillation phase at the exit of the capillary. Simulation results (11) support this interpretation. Laser guiding experiments were also performed in 3 and 6.6 cm long capillaries. In the latter case, the pulse radius at capillary exit was <25 μm and did not vary greatly with variations of the discharge parameters. In terms of the Rayleigh lengths for the vacuum focus $Z_R$ ($r_0$ = 15 μm) and the matched beam spot size $Z_M$ ($r_M \cong 24$ μm), the total propagation length is $75 Z_R \cong 30 Z_M \cong 6.6$ cm. Light transmission was ~10% in comparison to the ~1% without the discharge. The origin of this substantial loss is discussed in Section IV.

Accelerator applications require substantially higher intensities than are produced by this laser. To investigate the high intensity regime, capillary discharge experiments have recently been carried out using the $T^3$ laser at the Naval Research Laboratory. The laser has a 1.06 μm wavelength, a 400 fs pulse length, and peak power exceeding 2 TW. A 2 cm long double capillary was used. The focused intensities in these experiments exceeded $10^{17}$ W/cm$^2$. Transport efficiencies were generally higher than at lower intensities. The interpretation and implication of this result are discussed in the next section.

## LASER GUIDING SIMULATIONS

The LEM simulation code described in Section II has been used to interpret the discharge guiding experiments and to analyze various physical processes. If laser-induced modifications to the plasma density are small, one would expect the analytical model (Section II) to be reasonably accurate.

Changes in the channel depth $\Delta n$ or radius $r_{ch}$ should modify the equilibrium radius $r_M$ and envelope wavelength $\lambda_e$ according to the scalings in Eqs. (3) and (4). LEM simulations for 2 cm long capillaries were carried out with different values of $\Delta n$ with other parameters held fixed. In these simulations, the channel had $n_0 = 5 \times 10^{18}$ cm$^{-3}$, $r_{ch} = 100$ μm, and a 0.1 cm long density ramp at each end. The laser parameters were $\lambda_0 = 0.8$ μm, $r_0 = 15$ μm, pulse length $L_p = 60$ μm, and peak power $P_0 = 0.1$ TW.

Figure 3 illustrates the scaling of $r_M$ and $\lambda_e$ with $\Delta n$. The asterisks in the left frame are from the series of LEM simulation described above, using the average of the minimum and maximum values of $r_L$ over several oscillations to estimate the matched beam radius. The solid line is the analytical result from Eq. (3). The simulation results become increasingly closer to the analytical result as the channel depth $\Delta n$ is increased. This is because the mismatch between $r_M$ and $r_0$ and the resulting amplitude of the envelope oscillations are reduced. The right frame provides a similar plot for the envelope wavelength $\lambda_e$. The asterisks are again from simulations, and the solid line is the analytical estimate from Eq. (4). The agreement between the simulations and the analytical model is excellent.

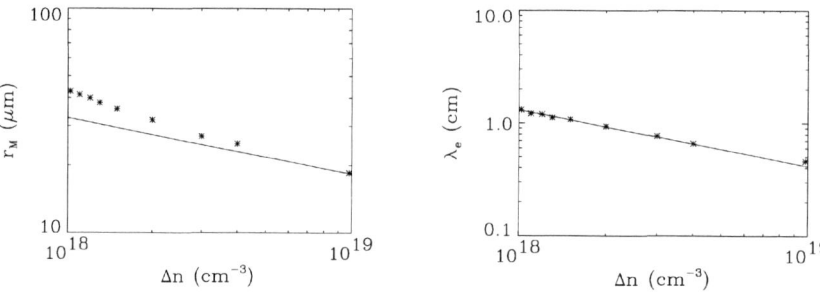

**FIGURE 3.** Comparison of simulation results with analytical model predictions for matched beam radius $r_M$ (left frame) and envelope wavelength $\lambda_e$ (right frame). The asterisks are the simulation results, and the solid lines are the analytical results from Eqs. (3) and (4).

Even at low laser intensities, the simulations reveal higher order structure and pulse distortion not contained in the simple envelope model. The plot of $r_L(z)$ in Fig. 2 reveals a modest damping of the envelope oscillations. The apparent damping becomes stronger as the pulse length or matched radius decreases. Plots of $r_L(\zeta)$ at constant $\tau \approx z/c$ show systematic variations within the pulse which become more pronounced as the pulse propagates. This behavior

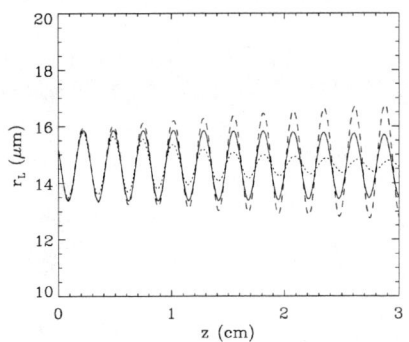

**FIGURE 4.** Plot of laser spot size versus z at 3 different locations $\zeta^*$ in the pulse. The (dotted, solid, dashed) curves are at $\zeta^* = (-6, -30, -48)$ μm which are in the (front, center, rear) of the pulse.

is believed to be due to finite pulse length effects contained in the $\partial^2 a / \partial \zeta \partial \tau$ term in the wave equation which become important when the number of optical periods contained in the laser pulse is not large. Sprangle, *et al.* (19) show analytically that for a short pulse propagating in an ideal channel, these effects cause the envelope oscillations in $r_L(z)$ to be damped in the front of the pulse and to grow in the back. This growth and damping is illustrated in the simulation results shown in Fig. 4, which plot $r_L(z)$ at 3 different locations $\zeta^*$ in the pulse. The pulse length $L_p = 60$ μm (half sine wave base) as in previous simulations. The channel size $r_{ch}$ was reduced to 40 μm, and $\Delta n = 4 \times 10^{18}$ cm$^{-3}$, so the beam is nearly matched with $r_M \approx r_0 = 15$ μm. The reference position $\zeta^*$ moves at the instantaneous simulation group velocity $\beta_{gs}$ rather than the analytical value used in previous figures. At $\zeta^* = -6$ μm, which is near the front of the beam, the oscillations are damped (dotted curve). The oscillations are very weakly damped at the beam center ($\zeta^* = -30$ μm, solid curve) and grow in amplitude at the rear of the beam ($\zeta^* = -48$ μm, dashed curve).

These simulations do not include several effects which may degrade the pulse, including collisional absorption (inverse bremsstrahlung) and photo-ionization. Inverse bremsstrahlung is probably the dominant loss mechanism in the long channel experiments described in Ref. 11. Although photo-ionization causes direct depletion of the pulse energy, the most serious potential concern is loss of confinement. This may occur because the laser generates additional plasma in high intensity regions. Initially, the highest intensity will be near the axis, so ionization will eliminate part of the desired hollow channel profile and may cause the beam to hollow and thus increase the beam spot size. The effect is often termed ionization defocusing (16). Loss of confinement due to ionization defocusing has been investigated by adding a simple ionization model to the LEM code. It assumed for simplicity that the discharge plasma consists entirely of H$^{+1}$ and C$^{+1}$ ions. The barrier suppression ionization (BSI) model (20) is used to estimate a threshold value of intensity for ionization to the next charge state.

For $C^{+2}$, this threshold value is $3.3 \times 10^{14}$ W/cm$^2$ which is significantly below the peak intensities produced in the experiment. Thus, laser-generated ionization should modify the laser pulse significantly.

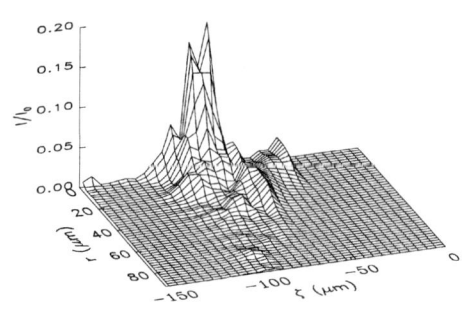

**FIGURE 5.** Surface plot of normalized intensity $I/I_0$ versus r and $\zeta$ for a simulation which includes ionization of carbon in the capillary plasma  The pulse has propagated to z = 3 cm.

Figure 5 is a surface plot of the normalized intensity $I/I_0$ versus r and $\zeta$ at z = 3 cm in a simulation which includes the ionization model described above. The channel and beam parameters are the same as in Fig.2 except that the peak intensity $I_0$ at the 15 μm focus was $7 \times 10^{15}$ W/cm$^2$. The ionization introduces radial and longitudinal structure into the pulse, with different portions undergoing hollowing and recollapse. Plots of the laser envelope radius $r_L(z)$ no longer have the regular oscillations seen in Figs. 2 and 4, and a significant fraction of the beam energy may be lost.

## SUMMARY AND DISCUSSION

Guiding of high power (0.1 TW) laser pulses in hollow plasma channels produced by a capillary discharge has been demonstrated experimentally over distances as long as 6.6 cm (10,11). Recent experiments in 2 cm long channels with the NRL T$^3$ laser have extended the technique to terawatt power pulses with focused intensities of $10^{17}$ W/cm$^2$. These experiments approach the regime necessary for laser wakefield accelerator applications. Although ionization defocusing by these higher power pulses remains a concern, high transport efficiencies were observed in the T$^3$ laser experiments than at lower powers.

The LEM simulation code has been used to interpret the experiments and to study fundamental processes in plasma channel guiding of laser pulses. At moderate beam powers, the simulations agree well with analytical models based on an envelope equation. Finite pulse length effects may distort the pulse, causing envelope oscillations to be damped in the beam front and grow in the tail. Simulations with a simple model for laser-generated ionization often produce longitudinal and radial structure in the beam which could affect its usefulness for accelerator applications.

# ACKNOWLEDGMENTS

The authors acknowledge useful conversations with E. Esarey, J. Krall, P. Sasorov, K. Krushelnick, H. Milchberg, T. Antonsen, B. Hafizi, and H.-J. Kunze. This work was supported by the U.S.-Israeli Binational Science Foundation, the U.S. Office of Naval Research, and the U.S. Department of Energy.

# REFERENCES

1.  Sprangle, P., Esarey, E., Ting, A., and Joyce, G., Appl. Phys. Lett. **53**, 2146-2148 (1988).
2.  Esarey, E., Sprangle, P., Krall, J., and Ting, A., IEEE Trans. Plasma Sci. **24**, 252-288 (1996).
3.  Nakajima, K., Fisher, D., Kawakubo, T., Hashaniki, H., Ogata, A., Kato, Y., Kitagawa, Y., Kodama, R., Mima, K., Shiraga, H., Suzuki, K., Yamakawa, K., Zhang, T., Sakawa, Y., Shoji, T., Nishida, Y., Yagami, N. M., Downer, M.,and Tajima, T. Phys. Rev. Lett. **74**, 4659-4662 (1995).
4.  Modena, A., Najmudin, Z., Dangor, A. E., Clayton, C. E., Marsh, K. A., Joshi, C., Malka, V., Darrow, C. B., and Danson, C., IEEE Trans. Plasma Sci. **PS-24**, 289-295 (1996).
5.  Umstadter, D., Chen, S. Y., Maksimchuk, A., Mourou, G., and Wagner, R., Science **273**, 472-475 (1996).
6.  Ting, A., Moore, C. I., Krushelnick, K., Manka, C., Esarey, E., Sprangle, P., Hubbard, R., Burris, H. R. and Baine, M., Phys. Plasmas **4**, 1889-1899 (1997).
7.  Durfee, III, C. G., and Milchberg, H. M., Phys. Rev. Lett. **71**, 2409-2411 (1993).
8.  Milchberg, H. M., Clark, T. R., Durfee, III, C. G., Antonsen, T. M., and Mora, P., Phys. Plasmas **3**, 2149-2155 (1996).
9.  Zigler, A., Ehrlich, Y., Cohen, C., Krall, J., and Sprangle, P., J. Opt. Soc. Am. B **13**, 68-71 (1996).
10. Ehrlich, Y., Cohen, C., Zigler, A., Krall, J., Sprangle, P., and Esarey, E., Phys. Rev. Lett. **77**, 4186-4189 (1996).
11. Ehrlich, Y., Cohen, C., Kaganovich, D., Zigler, A., Hubbard, R. F., .Sprangle, P., and Esarey, E., "Guiding and damping of high intensity laser pulses in long plasma channels," to appear in J. Opt. Soc. Am. B (1998).
12. Sprangle, P., and Esarey, E., Phys. Fluids B **4**, 2241-2248 (1992);
13. Borghesi, M., McKinnon, A. J., Barringer, L., Gaillard, R., Gizzi, L. A., Meyer, C., Willi, O., Pukhov, A., and Meyer-ter-Vehn, J., Phys. Rev. Lett. **78**, 379-382 (1996).
14. Braun, A., Korn, G., Liu, X., Du, D., Squier, J., and Mourou, G., Opt. Lett. **19**, 1544-1546 (1995).
15. Sprangle, P., Esarey, E., and Krall, J., Phys. Rev. E 54, 4211-4232 (1996).
16. McKinnon, A. J., Borghesi, M., Iwase, A., Jones, M. W., Pert, G. J., Rae, S., Burnett, K., and Willi, O., Phys. Rev. Lett. **76**, 1473-1476 (1996).
17. Krushelnick, K., Ting, A., Moore, C. I., Burris, H. R., Esarey, E., Sprangle, P., and Baine, M., Phys. Rev. Lett. **78**, 4047-4050 (1997).
18. Kaganovich, D., Sasarov, P. V., Ehrlich, Y., Cohen, C., and Zigler, A., Appl. Phys. Lett. **71**, 2295-2297 (1997).
19. Sprangle, P., Hafizi, B. and Serafim, P., submitted to Phys. Rev. Lett.
20. Augst, S., Meyerhoffer, D. D., Strickland, D., and Chin, S. L., J. Opt. Soc. Am. B **8**, 858-867 (1991).

# Cold Relativistic Wavebreaking Threshold of Two-Dimensional Plasma Waves

J. K. Kim, and D. Umstadter

*Center for Ultrafast Optical Science, University of Michigan, Ann Arbor, MI 48109*
July 28, 1998

**Abstract.** The two-dimensional wave-breaking of relativistic plasma waves driven by a ultrashort high-power lasers, is described within a framework of cold 2-D fluid theory. It is shown that the transverse nonlinearity of the plasma wave results in temporally increasing transverse plasma oscillation in the wake of the laser pulse, inevitably inducing wave-breaking below the 1-D threshold. A condition for wave-breaking is obtained and evaluated. A preformed density channel is found to partially cancel the effect and increase the length of wakefield that survives before wavebreaking occurs.

## I  INTRODUCTION

There has been recently much interest in the propagation of ultrashort high intensity laser pulse propagation through a plasma thanks to the proliferation of compact tabletop terawatt laser systems [1], and applications to laser accelerator of electrons, fast ignitor fusion, X-ray lasers and laboratory astrophysics. When a terawatt laser pulse propagates through an underdense plasma, it generates a relativistic longitudinal plasma wave, so called wakefield, trailing behind the laser pulse. Considering that the field gradient obtainable from the wakefield ($>1$ GeV/cm) is enormous in amplitude and the wakefield itself is traveling with the speed very close to $c$, the wakefield can serve as an extraordinary accelerating structure, as was first suggested by Tajima and Dawson [2]. Charged particles, if trapped, can be accelerated until they are eventually dephased with respect to the accelerating field gradient by outrunning the wave.

However, it is a well-known fact that the plasma wave breaks if the wave amplitude is very large. At this electron plasma wave-breaking(epwb) amplitude, the excursion distance of electrons in the wave becomes comparable to the plasma wavelength, thus enabling those electrons to penetrate the acceleration phase of the wakefield, called the separatrix. Once in the separatrix, the electrons gain a net longitudinal momentum consuming all the wave energy and destroying the

CP472, *Advanced Accelerator Concepts: Eighth Workshop,*
edited by W. Lawson, C. Bellamy, and D. Brosius

wake. Thus, epwb leads to bulk production of energetic electrons and, if controlled [3], could serve as a useful technique for an efficient plasma cathode. However, due to the inherently catastrophic nature of uncontrolled epwb, it cannot provide the mono-energetic beam, required for FELs and colliders. For applications such as laser-plasma accelerators, which require quasi-static plasma wave oscillations, epwb sets an upper limit on the largest field gradient extractable from the wake and thus should be avoided in order to maintain a constant peak-amplitude wake structure rather than exponentially decaying one. If, as we will show, epwb inevitably occurs at some distance behind the laser pulse, then it is crucial to know where, since it may be possible to drive the wave back down before epwb occurs. This information is also important because the maximum length of the plasma wave that survives before the onset of epwb also determines the maximum electron current that can be usefully produced by a wakefield.

Much theoretical work has been performed on the subject of epwb of a plasma wave in 1-D geometry. Work on a nonrelativistic cold epwb amplitude was first carried out by Dawson [4], which can be written as $E_{wb} = mc\omega_p/e$, where $\omega_p = (4\pi e^2 n_{e0}/m_e)^{1/2}$ is electron plasma frequency and $n_{e0}$ is the ambient electron density. The correction on the amplitude including a finite temparature was given by Coffey [5]. It was followed by the 1-D cold relativistic epwb amplitude as $E_{wb} = (m_e c\omega_p/e)\sqrt{2(\gamma_\phi - 1)}$ [6], where $\gamma_\phi$ is the relativistic factor of the plasma wave. Katsouleas $et.$ $al.$ integrated the previous results into a single theory of relativistc warm epwb amplitude in 1-D geometry [7]. Rosenzweig claimed a different answer from Katsouleas' solution, which was eventually resolved, to be unified into a single description by Sheng [8]. All these expressions can be obtained either by evaluating the break point of 1-D relativistic fluid theory [6,9], or by analysis of plasma oscillations of a single electron around an equilibrium point [10]. In as much as it is solely determined by $\gamma_\phi$ associated only with longitudinal characteristics of the wakefield, we call it $longitudinal$ $wave\text{-}breaking$. However, all of these works concern only 1-D plasmas. Accordingly, it is inadequate to apply these conditions to 2-D epwb phenomena.

Wave-breaking in the multi-dimensional case is much more complicated than in 1-D, due to the coupling between the transverse and longitudinal plasma oscillation [4]. Transverse variation of the 1-D nonlinearity of a plasma wave results in many interesting aspects of 2-D wakefield such as horse-shoe-shaped wakefield phase front [11]. Due to the nonlinear plasma freqency $\Omega_p$, which is smaller than $\omega_p$ on the propagation axis of the pump, the phase front of the wakefield experiences increasing distortion and the wake is shown to break after a number of plasma periods behind the laser pulse [12]. This new type of epwb is called $transverse$ $wave\text{-}breaking$. Bulanov $et.$ $al.$ first presented an estimate for the number of unbroken wakefield "buckets" by two-dimensional analysis on the self-intersection of electron trajectories [12]. Their analysis was based on the study of electron fluid oscillation perpendicular to the phase front of the wakefield by means of catastrophic theory. It however explains transverse epwb only qualitatively with the fixed ex-

405

cursion distance of an electron fluid element perpendicular to the phase front. For this reason, their expressions of the amplitude and the bucket where epwb takes place for a given plasma channel may not be accurate. The fluid description of the wakefield provides a more solid and quantitative analysis of the epwb condition for a 2-D plasma wave than does a single particle analysis. Since the 1-D epwb amplitude was obtained from both fluid and single particle analysis, it is expected that a generalized 2-D epwb condition can be arrived at with a set of relativistic fluid equations.

We present here such a fluid analysis (with finite $\gamma_\phi$) of the 2-D epwb process in order to derive a criterion for the epwb condition. We then use our epwb condition to determine how long a 2-D wave will survive without breaking for a given wave amplitude. For the sake of simplicity, we neglect all competing mechanisms, such as the modulational instability, which can damp the wave on the timescale of several ion periods. Also for the sake of simplicity, the plasma wave is assumed to be excited by a single short-pulse laser (called pump). If the pulse width is on the order of the plasma wavelength, the wake is driven resonantly to very large amplitude prescribed by $E_{max}(\mathrm{GeV/m}) \approx 3.8 \times 10^{-8} n_0^{1/2}(\mathrm{cm}^{-3})a_0^2/(1+a_0^2/2)^{1/2}$ [13], while the transverse extent of the channel linearly scales with the laser spot radius $r_0$. $a_0$ is the amplitude of the normalized vector potential of the pump.

For a more quantitative characterization of epwb in 2-D fluid theory, it is crucial to understand the transverse nonlinearity of the relativistic plasma wave as well as the longitudinal one. It has been pointed out [12] that for an unguided wakefield accelerator, the change of $\Omega_p$ across the channel causes the reduction of the curvature radius of the phase front eventually breaking the wave when the electron displacement becomes larger than the curvature radius. In the longitudinal direction appears the nonlinearity of the sawtooth-shaped oscillation of the wakefield gradient, in which the maximum acceleration and deceleration phase are so spatially close to each other that electrons in the deceleration phase are easily accessible by the acceleration phase via thermal velocity fluctuations. If $r_0 \gtrsim \lambda_p$ where $\lambda_p$ is the plasma wavelength, the excursion distance of the transverse plasma oscillation is small compared to the longitudinal one, and thus can be neglected. With this assumption, the approximation that the transverse shape of the plasma wave follows the transverse intensity profile of the pump can be justified.

## II   WAVEBREAKING THRESHOLD

We start from considering the self-consistent set of 2-D relativistic cold fluid equations coupled to Maxwell's equations with the choice of Gaussian gauge $\nabla \cdot \mathbf{a} = 0$,

$$\left(\nabla_\perp^2 + \frac{1}{\gamma_g^2}\frac{\partial^2}{\partial\zeta^2}\right)\mathbf{a} = k_p^2\eta_e\mathbf{u} - \beta_g\nabla\frac{\partial\phi}{\partial\zeta}, \tag{1}$$

$$\left(\nabla_\perp^2 + \frac{\partial^2}{\partial\zeta^2}\right)\phi = k_p^2\left(\gamma\eta_e - \eta_i\right), \tag{2}$$

$$-\beta_g\frac{\partial}{\partial\zeta}(\mathbf{u} - \mathbf{a}) = \nabla(\phi - \gamma), \tag{3}$$

$$-\beta_g\frac{\partial}{\partial\zeta}(\eta_e\gamma) + \nabla\cdot(\eta_e\mathbf{u}) = 0. \tag{4}$$

For our convenience, we denote the perpendicular direction by $r$ and the longitudinal direction by $\zeta$. Therefore, $\nabla_r \equiv \nabla_\perp$. Normalized scalar and vector potential are, $\mathbf{a} = (m_ec^2/e)\mathbf{A}$ and $\phi = (m_ec^2/e)\Phi$. The group velocity of the laser pulse is $\gamma_g = (1 - \beta_g^2)^{-1/2} \sim \gamma_\phi$, $k_p = \omega_p/c$, and $\gamma m_ec^2$ the electron energy. $\mathbf{u} = \gamma\beta$ is the normalized momentum of the electron, $\eta_e = n_e/\gamma n_{e0}$ normalized electron density, and $\eta_i = n_i/n_{i0}$ the normailized ion density of the channel where $n_{i0} = n_{e0}$ for a singly-ionized plasma. In order to obtain Eqs. 1-4, the slowly varying envelope approximation, the two-variable expansion, and the quasistatic approximation are used [14]. Cylindrical geometry is also implicitly assumed. With the coordinate transformation $\zeta = z - \beta_gct$, Eqs. 1-4 become

$$u_\zeta = \left[\gamma - (1 + \psi)\right]/\beta_g, \tag{5}$$

$$\mathbf{u}_\perp = \frac{1}{k_p^2\eta_e}\left[\beta_g\nabla_\perp\frac{\partial\psi}{\partial\zeta} + \frac{1}{\gamma_g^2}\left(\frac{\partial^2}{\partial\zeta^2} + \nabla_\perp\nabla_\perp\cdot\right)\mathbf{a}_\perp\right], \tag{6}$$

where $\gamma = \sqrt{1 + u_\zeta^2 + u_\perp^2 + \mathbf{a}_L^*\cdot\mathbf{a}_L/2}$, $\psi = \phi - \beta_ga_z$, and $\delta = (k_p^2\beta_g\gamma_g^2)^{-1}(\nabla_\perp^2 + \partial^2/\partial\zeta^2)a_z$. $u_\zeta$ and $\mathbf{u}_\perp$ are the longitudinal and transverse momentum of the electron fluid element respectively. $\mathbf{a}_L$ is the oscillating component of $\mathbf{a}$ at the laser frequency. Plugging Eq. 5 and Eq. 6 in $\gamma$ yields $\gamma = (1 + \psi)\gamma_g^2\left\{1 - \sqrt{1 - G}\right\}$, where

$$G = \frac{(1 + \psi)^2\gamma_g^2 + (1 + u_\perp^2 + \frac{1}{2}\mathbf{a}_L^*\cdot\mathbf{a}_L)(\gamma_g^2 - 1)}{(1 + \psi)^2\gamma_g^4}.$$

It is noted that all of the above equations converge to the well-known 2-D relativistic fluid equations obtained with the assumption $\beta_g = 1$ [14]. If $G > 1$, the energy of the electron fluid, $\gamma$, becomes imaginary and the fluid theory breaks, i.e., epwb occurs. Physically, the condition $G = 1$, where $\gamma = (1 + \psi)\gamma_g^2$, corresponds to infinite electron density in Eq. 7. The description of 2-D epwb is thus very similar to the procedures used in the 1-D case. Inside the wakefield after the pump, $\mathbf{a}_L = 0$. Therefore, the generalized epwb condition of a relativistic plasma wave in 2-D can be written as $G = 1$, which is equivalent to

$$1 + \psi = \frac{\sqrt{1 + u_{\perp0}^2}}{\gamma_g}, \tag{7}$$

407

where $\mathbf{u}_{\perp 0} = (1 + \psi)\nabla_{\perp}(\partial\psi/\partial\zeta)/k_p^2(\eta_i + \nabla_{\perp}^2\psi/k_p^2)$. The terms in the order of $1/\gamma_g^2$ are neglected in the calculation of $u_{\perp 0}$ whereas $\gamma_g \gg 1$ is assumed.

Eq. 8 takes a rather simple form to elucidate both longitudinal and transverse epwb in terms of $\psi$ and $u_{\perp 0}$. As $\gamma_g$ goes to infinity, $\gamma$ asymptotes to $\left\{1 + u_{\perp}^2 + \frac{1}{2}\mathbf{a}_L^* \cdot \mathbf{a}_L + (1 + \psi)^2\right\}/2(1 + \psi)$, which is identical to the previous calculation by Feit [14]. Also, if $u_{\perp 0} = 0$, which represents a 1-D plasma wave, Eq. 8 states $\psi = \gamma_g^{-1} - 1 \approx -1$ for $\gamma_g \gg 1$. On account that epwb takes place where $\psi$ is near the minimum value $\psi_{min}$, the expression properly describes longitudinal epwb by a large-amplitude plasma wave. In addition to this, $u_{\perp 0}^2$ in the numerator of $G$ is related to transverse epwb.

Since $u_{\perp 0}$ involves the cross derivative with respect to $r$ and $\zeta$, it is proportional to the increase of the curvature of the phase front. As $\psi$ becomes close to $\psi_{min}$, the increase of $u_{\perp 0}^2$ as well as $(1 + \psi)^{-2}$ stimulates the break of the wave. It is also remarked that the epwb threshold in 2-D should be smaller than that of the 1-D case; In a 1-D plasma, the threshold is given by $\psi_{max,wb} = (2\gamma_g^2 - 1)/\gamma_g - 1$, which is equal to $\psi_{min,wb} = \gamma_g^{-1} - 1$. As $\gamma_g^{-1} \longrightarrow \gamma_g^{-1}\sqrt{1 + u_{\perp 0}^2}$, the expression for the threshold for 2-D plasma wave is turned out to be $(\psi_{max,wb})_{2d} = (2\gamma_g^2 - 1 - u_{\perp 0}^2)/\gamma_g\sqrt{1 + u_{\perp 0}^2} - 1$. For example, if $u_{\perp 0} \sim 1$, then the threshold is decreased by a factor of $1/\sqrt{2}$. This suggests that a 2-D plasma wave has the longitudinally varying threshold, which decreases as $(1 + u_{\perp 0}^2)^{-1/2}$ along the channel axis.

# III   CHANNEL LENGTH CALCULATION

It will take several or more plasma periods for $u_{\perp 0}$ to grow until the wave breaks by this type of epwb. Estimation of the number of plasma wave oscillations behind the pump, $N_{wb}$, is of particular importance to many applications such as laser plasma accelerators considering that the length of the plasma wave train driven by a laser pulse cannot be greater than $N_{wb}$. To acquire a simple expression for $N_{wb}$, only the case of $r_0 \gg \lambda_p$ is considered, allowing us to depict the 2-D wakefield by a solution to $\partial^2 x/\partial\zeta^2 = (k_p^2/2)(\gamma_{\perp}^2/x^2 - 1)$ [15], where $\gamma_{\perp}^2 = 1 + a^2$ and $x = 1 + \psi(\zeta, r)$. The formal solution [15] of the differential equation by Esarey $et.$ $al.$ is $k_p(\zeta - L) = 2(x_0)^{1/2}E(\alpha, \rho)$, where $\alpha = \sin^{-1}\left[(x_0 - x)/(x_0 - x_0^{-1})\right]^{1/2}$, $\rho = (1 - 1/x_0)^{1/2}$, and $E(\alpha, \rho)$ is the incomplete Legendre elliptic integral of second kind. $L$ denotes the pulse length of the pump. In this equation, the transverse profile of the maximum potential, $x_0$, is assumed to be small compared to $\gamma_g$ but larger than unity. By defining a transverse shape function $S(r)$ of the pump, it is further assumed that $x_0(r) = 1 + \psi_{max}S(r)$, where $\psi_{max}$ is the maximum value of the potential on the propagation axis and a constant. For example, if the pump is Gaussian in the transverse dimension, the perturbation on the wake potential approximately follows the shape of the pump, i.e., $S(r) = \exp(-r^2/r_0^2)$. To validify this premise, 1-D and 2-D particle-in-cell (PIC) simulations (TRISTAN) [16] are used. Both

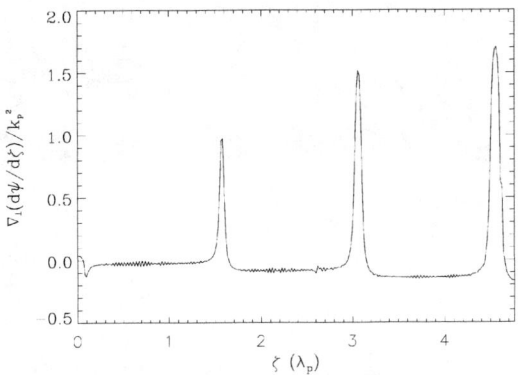

**FIGURE 1.** Example plot of $\nabla_\perp(\partial\psi/\partial\zeta)$ versus $\zeta$ in units of $\lambda_p$. At $x \sim x_0^{-1}$, where the plasma density has reached maximum, the transverse oscillation also peaks, with an amplitude that increases with $\zeta$. This growth contributes to epwb below the 1-D threshold.

simulations were previously benchmarked with the analytic expression for the 1-D wakefield amplitude and proven to work properly. With $a_0 \sim 2$ and $r_0 \gtrsim 2\lambda_p$, a very good agreement on the amplitude of the wakefield is achieved betwen 1-D and 2-D PIC simulations, which justifies the use of the solution to model a moderately two-dimensional plasma wave.

As can be seen from Eq. 8, 2-D epwb critically depends on $u_{\perp 0}$ which is the cause of transverse break of the wave. Hence, for a given value of $\psi$ and a wave-breaking coordinate $P \equiv (\zeta, r)$, $\nabla_\perp(\partial\psi/\partial\zeta)$ can be expressed as an explicit function of $\zeta$ and $r$, to be plugged into $u_{\perp 0}$. Denoting $x' = \partial x/\partial\zeta$, differentiation of $k_p(\zeta - L) = 2(x_0)^{1/2}E(\alpha, \rho)$ with respect to $r$ gives $\nabla_\perp x = (F - (\zeta - L)x'/2x_0)\nabla_\perp x_0$, where $F$ is a polynomial funtion of $x$ and $x_0$. Integrating the differential equation for the wake potential with respect to $\zeta$ and then differentiating it with respect to $r$, $\nabla_\perp x'(\equiv \nabla_\perp(\partial\psi/\partial\zeta))$ is calculated to be

$$\nabla_\perp x' = \frac{k_p^2}{2x'}\left[(1 - 1/x_0^2) - (1 - 1/x^2)F\right]\nabla_\perp x_0$$
$$+\frac{k_p^2}{4x_0}(1 - 1/x^2)(\zeta - L)\nabla_\perp x_0. \tag{8}$$

Wave-breaking occurs near $x \sim x_0^{-1}$ where the electron density is maximum. Hence, the evaluation of Eq. 9 at $x = x_0^{-1}$ yields $\nabla_\perp x' = -(k_p^2\psi_{max}/4)(x_0-x_0^{-1})(\zeta-L)\nabla_\perp S$. In Fig. 1, $\nabla_\perp(\partial\psi/\partial\zeta)$ is plotted and the peaks where $x = x_0^{-1}$ are shown to linearly increase with $\zeta$.

Plugging Eq. 9 into $u_{\perp 0}$, we rewrite Eq. 8 into

$$\gamma_g^2 = x_0^2\left[1 + \frac{(\zeta_{wb} - L)^2|\nabla_\perp x_0|^2}{16\eta_i^2}\right], \tag{9}$$

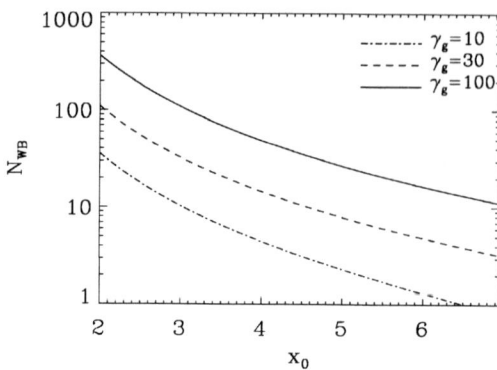

**FIGURE 2.** $N_{wb}$ in Eq. 11 is plotted versus $x_0$, varying $\gamma_g$, where $r_0 = 2\lambda_p$ and $\eta_i = 1$ are used. Decreasing $\gamma_g$ reduces $N_{wb}$, effectively limiting the plasma channel length.

where $\zeta_{wb}$ is the $\zeta$-coordinate of $P$. In Eq. 10, the contribution from $\nabla_\perp^2 x$ is neglected since $r_0 \gg \lambda_p$ and $\eta_i \gg \nabla_\perp^2 x / k_p^2$ are assumed. Solving Eq. 10 for $\zeta_{wb}$, we obtain an expression for $N_{wb}$ as

$$N_{wb} \sim (\zeta_{wb} - L)/\lambda - \frac{k_p \sqrt{\eta_i}}{x_0^{3/2} E_c(\rho)} \frac{\sqrt{\gamma_g^2 - x_0^2}}{|\nabla_\perp x_0|}, \tag{10}$$

where $k_p \lambda = 4\sqrt{x_0} E_c(\rho)$ and $E_c(\rho)$ is the complete elliptic integral of the second kind. Since it is observed that the maximization of $u_{\perp 0}$ occurs at $r \sim r_0/2$, the estimation of $|\nabla_\perp x_0|$ as $x_0/r_0$ states that $N_{wb}$ is approximately proportional to $\eta_i^{1/2} r_0 x_0^{-5/2}$. For instance, the parameters $\psi_{max} = 3$, $\eta_i = 1$, $r_0 = 6\lambda_p$ and $\gamma_g = 10$ yield $N_{wb} \sim 7.5$. However, the same parameters with $r_0 = 2\lambda_p$ will give $N_{wb} \sim 2.5$, which represents a case of more tightly focused pump. Eq. 11 is plotted in Fig.2, with $r_0 = 2\lambda_p$. Since $N_{wb}$ scales with $x_0^{-5/2}$, the drastic decrease of $N_{wb}$ is an unavoidable consequence as $x_0$ increases.

Note that $\eta_i = \eta_i(r)$. Variation of $\eta_i$ as a function of $r$ in a preformed density channel used for channel-guiding of the pump, could alter the value of $N_{wb}$, which suggests that the channel guiding can control the number of surviving buckets against epwb. As can be seen in $u_{\perp 0}$ and $N_{wb}$, a strong ion density depression across the plasma channel can be a preventive measure to suppress the transverse break of the wake and extend the length of the plasma wave significantly. The increase of $N_{wb}$ with $\sqrt{\eta_i}$ can be interpreted in terms of density-dependent $\gamma_\phi$ of the wave. $\gamma_\phi = \omega/\omega_p$ and $\omega_p \propto \sqrt{n_e} = \sqrt{n_i}$. Subsequently, $\gamma_\phi$ is proportional to $\eta_i^{-1/2}$. It means that the plasma wave propagates faster on the axis than the peripheral region of the channel, thus effectively reducing the phase front distortion of the plasma wave for large $\zeta$. In Fig. 3, $N_{wb}$ is shown as a function of the depth of density depression $\Delta\eta_d$ of a preformed plasma density channel. The transverse

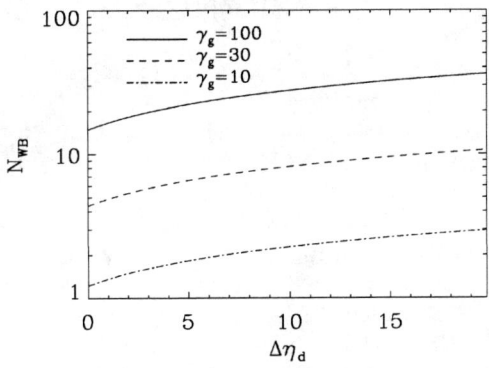

**FIGURE 3.** $N_{wb}$ is plotted versus $\Delta\eta_d$, varying $\gamma_g$, where $r_0 = 2\lambda_p$ and $x_0 = 7$. It is shown that for larger $\Delta\eta_d$ and $\gamma_g$, the length of a plasma wave train increases.

ion density profile $\eta_i = 1 + \Delta\eta_d(r^2/r_0^2)$ is assumed for $0 < r < r_0$.

Because $N_{wb}$ scales with $\sqrt{\eta_i}$, larger values of $\Delta\eta_d$ are preferred in order to increase $N_{wb}$. Conversely, for a given index $j$ for the epwb bucket, the minimum depth of density depression $\Delta\eta_{wb}$ can be calculated to be $4\left\{j^2 x_0^5 E_c^2/r_0^2 k_p^2(\gamma_g^2 - x_0^2) - 1\right\}$, assuming $r = r_0/2$. Furthermore, a certain level of depression depth is also required for a channel-guiding and it is found to be $\Delta\eta_{ch} = m_e c^2/\pi e^2 r_0^2 n_{e0}$ [17]. Therefore, if $\Delta\eta_d > \max(\Delta\eta_{wb}, \Delta\eta_{ch})$, it is possible to simultaneously guide the pump in the channel and acquire a plasma wave of which length is approximately $j\lambda_p$.

## IV  CONCLUSION

We showed that transverse wave-breaking can be analyzed by standard 2-D cold relativistic fluid theory. It is shown that the transverse nonlinearity of the plasma wave channel results in transverse wave-breaking, which cooperates with longitudinal wave-breaking to break the wave, lowering the break threshold below the 1-D limit. The effect can be at least partially mitigated by use of a density channel. Complete cancellation may be possible, which will be the subject of a future publication.

We gratefully acknowledge support by the Division of High Energy Physics, Office of Energy Research, U. S. Department of Energy, DE-FG02-98ER41071; and the National Science Foundation, STC PHY 8920108. Computing services were provided by the National Partnership for Advanced Computing Infrastructure, PHY 980027 S.

# REFERENCES

1. M. D. Perry and G. Mourou, Science **264**, 917 (1994).
2. T. Tajima and J. M. Dawson, Phys. Rev. Lett. **43**, 267 (1979).
3. D. Umstadter, J. K. Kim, and E. Dodd, Phys. Rev. Lett. **76**, 2073 (1996).
4. J. M. Dawson, Phys. Rev. **113**, 383 (1959).
5. T. P. Coffey, Phys. Fluids **14**, 1402 (1971).
6. A. I. Akhiezer and R. V. Polovin, Sov. Phys. JETP **3**, 696 (1956).
7. T. Katsouleas and W. B. Mori, Phys. Rev. Lett. **61**, 90 (1988).
8. Z. M. Sheng and J. M. ter Vehn, Phys. Plasmas **4**, 493 (1997).
9. E. Esarey and M. Pilloff, Phys. Plasmas **2**, 1432 (1995).
10. J. Albritton and P. Koch, Phys. Fluids **18**, 1136 (1975).
11. S. V. Bulanov, F. Pegoraro, and A. M. Pukhov, Phys. Rev. Lett. **74**, 710 (1995).
12. S. V. Bulanov, F. Pegoraro, A. M. Pukhov, and A. S. Sakharov, Phys. Rev. Lett. **78**, 4205 (1997).
13. P. Sprangle and E. Esarey, Phys. Fluid. B **4**, 2241 (1992).
14. M. D. Feit, J. C. Garrison, and A. M. Rubenchik, Phys. Rev. E **53**, 1068 (1996).
15. E. Esarey, P. Sprangle, J. Krall, and A. Ting, IEEE Trans. Plasma Sci. **24**, 252 (1996).
16. O. Buneman, T. Neubert, and K. Nishikawa, IEEE Trans. Plasma Sci. **20**, 810 (1992).
17. E. Esarey, P. Sprangle, J. Krall, and A. Ting, IEEE J. Quantum Electron. **33**, 1879 (1997).

# Excitation and measurement of laser induced wakefields

S. P. Le Blanc *, E. W. Gaul, and M. C. Downer

University of Texas at Austin, Department of Physics, MC C1600, Austin, Tx 78712
* email:spl@physics.utexas.edu

**Abstract.** The amplitudes of wakefields excited by the ponderomotive force of an intense laser pulse are dependent on laser pulse shape and amplitude. Pulses with steep edges can drive wakefields in the standard LWFA when $\omega_p \Delta \tau_L \gg 1$. The temporal and spatial properties of the wakefield micro-structure can be measured by a single shot optical diagnostic based on frequency domain holography. In a proof of principle demonstration of this technique, cross-phase modulation of a weak probe pulse by two strong pump pulses is measured using the the nonlinear index of refraction of fused silica.

## I  INTRODUCTION

Of the several methods for driving large amplitude plasma waves, the laser wakefield accelerator (LWFA) and the self-modulated LWFA have recently received considerable attention because of the reduction in size of terawatt class laser systems necessary to drive them [1]. In the LWFA, the amplitude $\delta n_e/n_e$ of the plasma wave can be optimaly excited by the ponderomotive force of the laser pulse if the laser pulse duration $\Delta \tau_L$ is approximately half of the plasma wave period, or $\Delta \tau_L \sim \pi/\omega_p$. Terawatt class femtosecond laser systems are necessary both to achieve this resonance condition in near atmospheric density plasmas and to generate laser intensities near $10^{18}\ W/cm^2$. For the self-modulated LWFA, the plasma density is chosen to be much larger than for the standard LWFA so that $\Delta \tau_L \gg \pi/\omega_p$ and the forward Raman scattering (FRS) instability can grow. The FRS instability is the conversion of an electromagnetic wave $(\omega_o, k_o)$ into a plasma wave $(\omega_p, k_p)$ and Stokes $(\omega_o - \omega_p, k_o - k_p)$ and anti-Stokes$(\omega_o + \omega_p, k_o + k_p)$ electromagnetic side bands.

Laser wakefield accelerators can excite large amplitude electrostatic fields (E $\geq$ 100 GV/m) which are potentially suitable for compact accelerators and advanced high energy colliders. Just as cavity geometry is important for conventional RF accelerators, plasma wave micro-structure is similarly critical for determining the accelerating properties of the LWFA. In previous work, we have demonstrated that

CP472, *Advanced Accelerator Concepts: Eighth Workshop,*
edited by W. Lawson, C. Bellamy, and D. Brosius
© 1999 The American Institute of Physics 1-56396-889-4/99/$15.00

ultrafast optical techniques based on frequency domain interferometry [3,4] and collective Thomson scattering [5,6] can be used to diagnose critical wakefield parameters such as amplitude, decay time, and radial extension. Optical diagnostics with femtosecond resolution are important for measuring wakefield growth and decay, establishing feedback control on the wakefield, and synchronizing the injection of particles at the appropriate wakefield phase. In this paper, we discuss the effect of laser pulse shape on wakefield resonance bandwidth. Additionally, we report the development of a 2D frequency domain holographic technique capable of the radial and temporal properties of the wakefield in a single laser shot.

## II  INFLUENCE OF PULSE SHAPE

Excitation of a laser induced wakefield is not only influenced by the temporal duration of the laser pulse, but also by its shape. To explore the influence of pulse shape on wakefield amplitude, we begin by considering the cold fluid equations describing plasma wave excitation for the linear regime of the standard LFWA [1]:

$$\left(\frac{\partial^2}{\partial \zeta^2} + \omega_p^2\right)\frac{\delta n_e}{n_e} = c^2 \nabla^2 \frac{a_L^2}{2} \tag{1}$$

where $\delta n_e/n_e \ll 1$ is the perturbed density of the plasma wave, $a_L^2 \ll 1$ is the normalized vector potential of the driving pulse. $\omega_p$ is the plasma frequency, and $\zeta$ is the longitudinal coordinate in the speed of light reference frame. Assuming a pre-ionized plasma, and a laser pulse with longitudinal profile $f(\zeta)$ and Gaussian transverse profile,

$$a_L^2(\zeta, r) = a_o^2 f(\zeta) e^{-(r/\sigma_r)^2}, \tag{2}$$

solution of Eqn. 1 yields a wakefield amplitude given by [7,8]:

$$\frac{\delta n_e}{n_e} = \frac{a_L^2}{2} + k_p \frac{a_o^2}{2} e^{-(r/\sigma_r)^2} \left[1 + \frac{4}{(k_p\sigma_r)^2}\left\{1 - \frac{r^2}{\sigma_r^2}\right\}\right] \int_\zeta^\infty f(\zeta') \sin[k_p(\zeta - \zeta')]d\zeta'. \tag{3}$$

Far behind the laser pulse ($\zeta \to -\infty$), the wakefield amplitude reduces to:

$$\frac{\delta n_e}{n_e} = \frac{a_o^2}{2} e^{-(r/\sigma_r)^2} \left[1 + \frac{4}{(k_p\sigma_r)^2}\left\{1 - \frac{r^2}{\sigma_r^2}\right\}\right] A \sin[\omega_p t + B], \tag{4}$$

where

$$A = \sqrt{2\pi}[|\omega F.T.\{f(t)\}|^2_{\omega=\omega_p}]^{1/2}, \tag{5}$$

$$B = arg[F.T.\{f(t)\}]. \tag{6}$$

$k_p$ is the plasma wavenumber, and F.T. indicates a Fourier transform. Within the assumptions of small density perturbations and the quasi-static approximation,

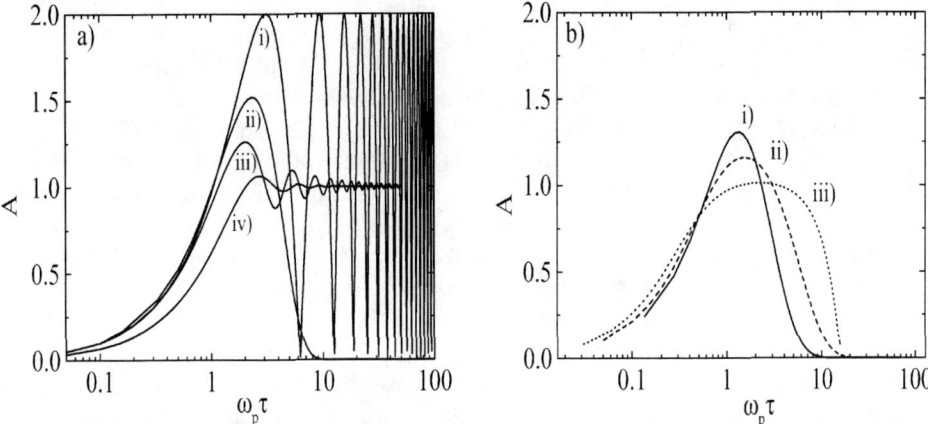

**FIGURE 1.** Dependence of wakefield resonance spectrum on pulse shape. a) LWF resonance spectrum for various pulse shapes assuming $a_o^2 \ll 1$ and constant peak intensity : i) Square pulse, ii) Gaussian pulse; iii) triangle pulse, and iv) triangle pulse with parabolic rising edge. $\tau$=FWHM of pulse. b) Resonance spectrum for a Gaussian pulse with different rising edges $t_r$: i) $t_r = 1$, ii) $t_r = 0.1$, and iii) $t_r = 0.01$.

Equations 4-6 provide incite into several aspects of wakefield excitation: 1) The resonance spectrum ($A$) and phase ($B$) of the wakefield are dependent on the pulse shape $f(t)$. 2) The optimal pulse duration is dependent on pulse shape. For example, optimal excitation for a square or Gaussian pulse occurs for $\omega_p \Delta \tau_L = \pi$ or $\omega_p \Delta \tau_L = 0.75\pi$, respectively [7]. 3) Since the 1D resonance factor $A$ is proportional to the autocorrelation function of $f$, $A$ is independent of the asymmetry of the pulse shape. Hence, the wakefield amplitude excited by a pulse with a sharp leading edge or a sharp trailing edge will be equal since their autocorrelation functions are equal.

Figure 1 shows the dependence of $A$ as a function of $\omega_p \Delta \tau_L$ for various pulse shapes. For a Gaussian pulse shape, Fig. 1a shows that the wakefield amplitude is maximum for $\omega_p \Delta \tau_L = 2.36$ and falls off for smaller and larger values of this product. In contrast, a pulse with an infinitely sharp edge - a square pulse, triangle pulse, or parabolic triangle pulse - can continue to excite a wakefield for $\omega_p \Delta \tau_L \gg 1$. This trend points to the fact that excitation of the wakefield occurs whenever an edge of the pulse has a rise (or fall) time smaller than, or comparable to, the plasma wavelength. For the square pulse, wakeless regimes [8 10] exist whenever $\omega_p \Delta \tau_L = 2\pi n$, (n=1,2,3....). In this case, the wakefield generated by the front of the pulse destructively interferes with the wake generated by the back. For a triangle pulse or a triangle pulse with a parabolic slope, these wakeless regimes do not exist due to the asymmetry of the pulse. Instead, the amplitude of the wakefield saturates for $\omega_p \Delta \tau_L \gg 1$. For the square pulse, whenever $\omega_p \Delta \tau_L = \pi m$, (m=1,3,5,...), the wake excited by the front and back of the pulse add constructively to give the maximum wakefield amplitude. Fig. 1b shows the dependence of the resonance spectrum for Gaussian pulses with finite rise times. As the rise time becomes smaller, the resonance bandwidth broadens and extends to larger values of $\omega_p \Delta \tau_L$. When ionization effects are included in LWFA simulations, it has been shown that the rising edge of an initially smooth Gaussian laser pulse can be steepened by the ionization front due to differences in electron density and group velocity [11]. Hence, symmetric Gaussian pulses satisfying $\omega_p \Delta \tau_L \gg 1$ can eventually excite a large amplitude wakefield if the laser pulse propagates far enough so that the ionization front can steepen the rising edge of the laser pulse.

Similar conclusions about broadening of the resonance bandwidth for the standard LWFA have been obtained for intense pulses ($a_o \geq 1$) much longer than $\lambda_p/2$ [12]. In Ref. [12], wakefield enhancement for $\omega_p \Delta \tau_L \gg 1$ was not due to changes in pulse shape, but rather due to nonlinearities in the response of the electrons to the ponderomotive force of the laser. When $a_o \geq 1$, the nonlinearities in the plasma wave effectively increase the resonance bandwidth by adding additional Fourier terms to Eqn. 5. The broadening of the resonance bandwidth for the standard LWFA is more than of academic interest for exciting wakefields in plasma fibers. With current techniques, pre-formed plasma channels can only generate electron density radial profiles which have on axis electron densities greater than $n_e \geq 3 \times 10^{18}$ cm$^{-3}$ [13]. From linear theory, such a high density implies that ~20 fs pulses are required to excite the wakefield in the plasma channel. However, longer laser pulse with $a_o \geq 1$ or with a steep rising edge may also be used to drive a wakefield in the plasma fiber if one considers nonlinearities which broaden the resonance bandwidth.

In the linear regime, the resonance factor $A$ is independent of the asymmetry of the pulse shape. Figs. 2a and 2b show this to be the case for a triangle pulse with a positive or negative slope and $a_o$=0.2. For each case, the wakefield trailing the laser pulse has the same magnitude. However, whether the triangle pulse has positive or negative slope influences the magnitude of the wakefield during the laser pulse. Since the triangle pulse with positive slope generates a smaller amplitude wakefield during the pulse, it will suffer less distortion from modulational instabilities during propagation [14,15]. Similar results are obtained for asymmetric pulses with finite rise and fall times. In the nonlinear regime, the relativistic cold fluid equation describing the plasma motion is [1]:

$$\frac{\partial^2 \phi}{\partial \zeta^2} = \frac{k_p^2}{2} \left( \frac{1 + a_L^2}{(1 + \phi)^2} - 1 \right) \tag{7}$$

where $\phi = e\Phi/mc^2$ is the dimensionless electrostatic potential of the plasma. Figs. 2c and 2d demonstrate that the magnitude of the wakefield behind a positive or negative triangle pulse with $a_o = 1$ depends on the sign of the pulse asymmetry. The triangle pulse with negative slope generates a larger wakefield (by 18 % for the present example) than the positively sloped laser pulse [10]. This difference occurs

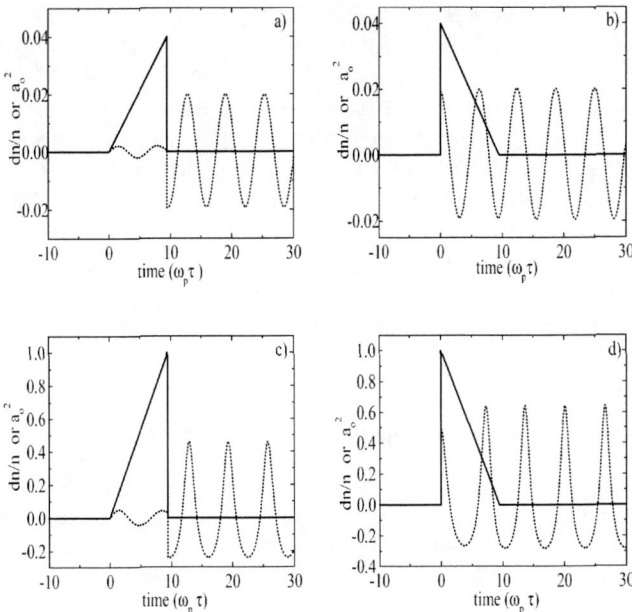

**FIGURE 2.** Wakefield (dotted line) excited by a positive triangle pulse (solid line) with a) $a_o = 0.2$, and c) $a_o = 1$. Wakefield (dotted line) excited by a negative triangle pulse (solid line) with b) $a_o = 0.2$, and d) $a_o = 1$. For each case, the wakefield is calculated by solving Eqn. 7.

in part because of the change in wakefield wavelength with increasing wakefield amplitude. Since the wakefield amplitude during the negative triangle pulse is larger and slightly shifted in frequency compared to the positive triangle pulse, interference between the wakes from the front and back of the pulse is different for the two pulse shapes.

Laser pulses [16,17] (or electron bunches for the plasma wakefield accelerator [14,15]) can be shaped for optimal excitation of the wakefield. According to Ref. [17], the optimal shape corresponds to that which maximizes the wakfield amplitude while achieving a constant rate of photon deceleration, or frequency red shifting, of the laser pulse. In the linear regime, the optimal shape is close to that of the triangle pulse with positive slope.

## III 2D SINGLE SHOT WAKEFIELD DIAGNOSTIC

Femtosecond time resolved measurements of the longitudinal and radial structure of laser-induced wakefield oscillations can be realized using all optical techniques known as photon acceleration [18–20] and frequency domain interferometry (FDI)

[21,22]. In the simplest version of the photon acceleration technique, a single short laser pulse co-propagating with the pump pulse over an interaction length L will have impressed upon it a time-dependent phase change given by $e^{i\phi(t)}$, where $\phi(t) = \eta(t)\omega_o L/c$, $\omega_o$ is the center frequency of the probe pulse, $\eta = (1 - n_e(r, z, t)/n_{crit})^{1/2}$ is the index of refraction of the plasma, and $n_{crit}$ is the critical electron density. If $n_e$ increases during the time of the probe pulse (either at the ionization front or in the longitudinal gradients of the wakefield in which negatively charged particles would be accelerated), then $\delta\eta/\delta t \leq 0$ and the spectral shift $\Delta\omega$ becomes positive (blueshift). The correlation of a blueshift (photon energy increase) with particle acceleration motivates the term "photon acceleration". Conversely, red shifts occur in regions where $n_e$ decreases. For the standard LWFA, the frequency shift from the wakefield scales as [23]:

$$\frac{\Delta\omega}{\omega} = 10^5 \left(\frac{\lambda}{\tau}\right)^4 \left(\frac{L}{A}\right) W_p \qquad (8)$$

where $W_p$ is the laser pulse energy in Joules; $\tau$ is the pulse width in femtoseconds; $\lambda, L, and A$ are the laser wavelength, interaction length, and cross-sectional area in $\mu m$ and $\mu m^2$. Centroid based methods are often used to determine the frequency shift of ultrashort laser pules for $\Delta\omega/\omega \sim 10^{-3} - 10^{-4}$.

To detect smaller changes in frequency and phase on a femtosecond time scale, frequency domain interferometry may be utilized. In frequency domain interferometry, a probe pulse co-propagates in front of and behind the intense pump pulse which excites the wakefield. To measure small changes in phase and frequency caused by the wakefield, the spectrum of the probe pulse sequence is recorded by a spectrometer. The total electric field of the two probe pulses can be written as: $E_{tot} = E_1 + E_2$, where $E_1 = E(t)e^{i\omega t}$, $E_2 = E(t-T)e^{(i(\omega(t-T)+\phi))}$, and $\phi$ is the phase shift of the second pulse relative to the first. The power spectrum of this probe pulse sequence is then: $I(\omega) = I_o(\omega)(1 + cos(\omega T + \phi))$. Consequently, two pulses separated in time generate frequency domain fringes whose modulation depends on the pulse separation $T$ and whose phase depends on the phase offset $\phi$ between the two pulses. By Fourier transforming the frequency domain interferogram, the original phase shift $\phi$ can be recovered.

Using FDI, 2D wakefield oscillations with amplitudes near unity ($\delta n_e/n_e \sim 1$) have been measured for intense laser pulses ($I \geq 10^{17}$ $W/cm^2$) propagating in helium plasmas ($n_e \sim 10^{17}$ cm$^{-3}$) [3,4,24]. Frequency domain interferometry (FDI) and photon acceleration are complementary optical techniques for wakefield characterization. In the regime of small frequency and phase shifts ($\Delta\omega/\omega < 10^{-3}$), FDI provides sensitive characterization methods of the wakefield by placing the phase shift information on a carrier wave that is detected by phase-sensitive lock-in detection. In the regime of large frequency shifts, centroid based techniques for measuring frequency shifts due to photon acceleration are simpler.

When using conventional FDI to map plasma density oscillations, the delay between the probe pulse sequence and the intense pump pulse must be changed many times in order to record the wakefield oscillations over many plasma cycles. To record several wakfield oscillation cycles in a single measurement, we introduce a new technique based on the extension of frequency domain interferometry to the regime of frequency domain holography (FDH). In single shot frequency domain holography (SS-FDH), 1 short and 1 long probe pulse are used to measure the wakefield excited by an intense pump pulse. The broad probe pulse is made sufficiently long so that it samples several cycles of the wakefield. Spectral interference of the two probe pulses produces a frequency domain interferogram sensitive to temporal phase changes (caused by the wakefield or any other type of phase modulator) over the duration of the long pulse. The temporal resolution is approximately equal to the short probe pulse duration.

In a proof of principle experiment, we have used SS-FDH to measure phase changes caused by the non-linear susceptibility ($\chi^{(3)}$) of a 150 $\mu m$ thick fused silica plate in which 1 or 2 strong pump pulses co-propagate with and cross-phase modulate a 1 ps probe pulse. In this experiment, the short and long probe pulses where produced by second harmonic generation in a nonlinear crystal. The short pulse was generated by type I phase matching in a thin (100 $\mu m$) LiIO$_3$ crystal, while the long probe pulse was generated in a 2 mm thick LiIO$_3$ crystal. Such a thick crystal has a small phase matching bandwidth ($\Delta\lambda = 0.45$ nm), resulting in a temporally long second harmonic pulse. Fig 4a shows a 1D lineout from a single shot 2D frequency domain hologram recorded by a CCD camera at the output of the spectrometer. The broad pedestal from the short (75 fs) probe, the sharp central feature from the long (1ps) probe pulse, and their frequency domain interference fringes are evident. In this example, the short and long probe pulses are separated by 3 ps. Fig. 4b shows the temporal phase changes extracted from such a hologram during a single shot experiment in which a sequence of two 75 fs pump pulses ($I = 10^{11}$ W/$cm^2$) separated by 333 fs overlapped the 1 ps probe pulse as they co-propagated through the fused silica plate. Two 0.1 rad phase-modulation "peaks" separated by 333 fs (corresponding to the two 75 fs pump pulses) are recovered. On each laser shot, the phase shift on the second probe pulse is probed over a range of 1 ps with a resolution of ~75 fs. As the temporal separation $\Delta T1$ between the pump pulse sequence and the probe pulse sequence is varied, Fig 4b shows the expected shifts in the phase modulation peaks. The magnitude of the cross phase modulation peak (0.1 rad) is in good agreement with the expected value based on the nonlinear index of refraction of fused silica [25]. When the phase mulation on the second probe pulse is caused by wakefield plasma oscillations, SS-FDH will enable measurement of the wakefield structure over several plasma oscillations.

## IV   DIAGNOSTIC LIMITATIONS

In the limit of large amplitude plasma waves and long interaction lengths ($L \geq 1$ mm), large frequency shifts of the probe pulses occur, and in FDI based methods, the visibility of the frequency domain interferogram can be significantly reduced

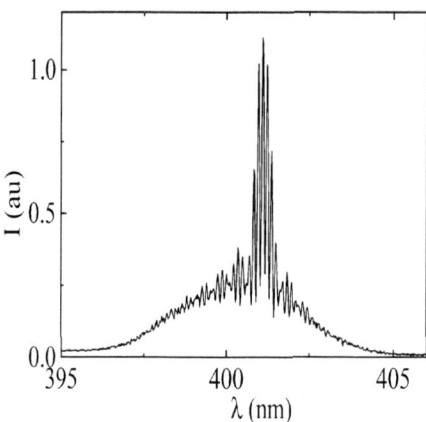 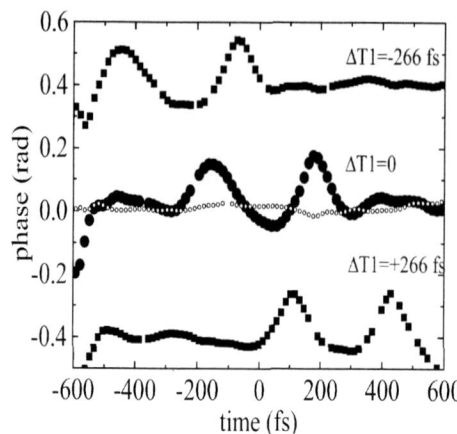

**FIGURE 3.** a) Frequency domain hologram generated by the interference of the long and short probe pulses. b) Single shot temporal phase change measured during the 1 ps probe pulse caused by cross-phase modulation from a sequence of two pump pulses copropagating with the probe pulses in a glass plate . open circles, phase change without 2 pump pulses; filled points, phase change with 2 pump pulses separated by 333 fs for different delays $\Delta T1$ between the pump and probe pulse sequences.

due to the incomplete interference between probe pulses with different central frequencies. Fig. 4 shows the change in spectral shape of a conventional frequency domain interferogram as the amount of phase modulation impressed on the probe pulse begins to produce large frequency shifts. In Fig 4a, the frequency domain fringes extend over the full bandwidth of the probe spectrum. As the frequency shift on one probe pulse increases (due to an increase in propagation length and/or wakefield amplitude), Fig. 4b shows that the bandwidth and contrast of the spectral fringes decreases. Note that when the probe overlaps the portion of the plasma wave between the crest and trough, the phase modulation is appriximately linear in time and causes the probe pulse spectrum to have an Airy function shape [26].

Plasma dispersion can also limit ultrashort pulse optical diagnostics of the plasma wave. For example, dispersive broadening $\Delta T_B$ for a transform limited sech$^2$ pulse amounts to:

$$\frac{\Delta T_B}{L} = \frac{0.315\lambda^3\omega_p^2}{4\pi^2c^4\Delta T_L}$$

$$\frac{\Delta T_B}{L}\left[\frac{fs}{mm}\right] = 3.1 \times 10^{-18}\frac{n_e[cm^{-3}]}{\Delta T_L[fs]}\lambda^3[\mu m], \tag{9}$$

where $L$ is the propagation length. For typical conditions, though, this broadening is small: for example, a 30 fs Ti:Sapphire laser pulse would experience a broadening

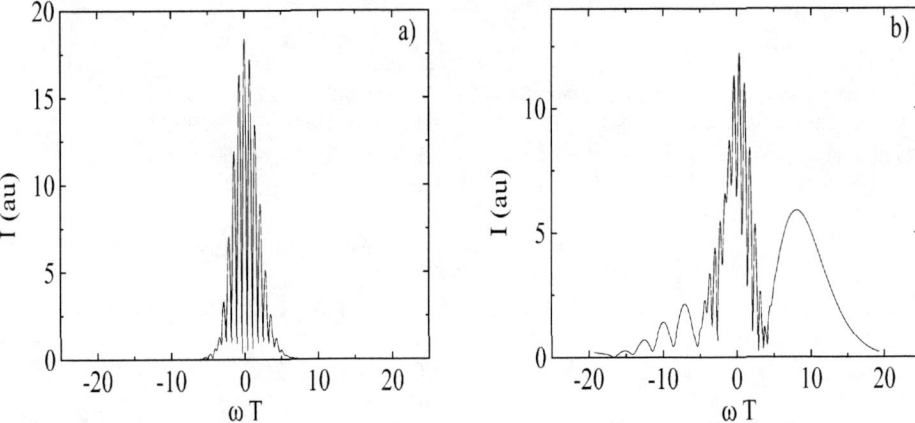

**FIGURE 4.** Effect of strong frequency shifts on frequency domain interferograms. Probe pulse frequency domain interferogram generated when the second probe pulse propagated though a plasma with density $n_e = 10^{18}\ cm^{-3}$ , plasma wave amplitude $\delta n_e/n_e = 0.1$, and interaction length a) 0.5 mm, and b) 10 mm. For L=10 mm, phase modulation by the wakefield causes spectral broadening of the second probe pulse. This broadening is in the shape of a shifted Airy function since the phase modulation caused by the wakefield is linear in time between the crest and trough of the plasma wave.

of only 0.1 fs/mm at $n_e = 10^{19}\ cm^{-3}$. For 2 color pump-probe experiments, optical diagnostic techniques must also take into consideration temporal walk-off effects which can result from differences in group velocity. The temporal walkoff $\Delta T_W$ between pump and probe is given by:

$$\frac{\Delta T_W}{L} = \frac{\omega_p^2}{8\pi^2 c^3}\left|\lambda_{probe}^2 - \lambda_{pump}^2\right| \qquad (10)$$

For $n_e = 10^{19}\ cm^{-3}$, the walkoff between $\lambda_{probe} = 400\ nm$ and $\lambda_{pump} = 800\ nm$ is 7 fs/mm. Such a temporal walkoff will introduce broadening and smearing of the measured wakefield micro-structure. Measured phase shifts can also be modified when the plasma wave phase velocity is different from the group velocity of the laser pulse which excites or probes the wakefield [27].

## ACKNOWLEDGEMENTS

This research was supported by the US DOE under Grant No. DEFG03-96-ER-40954.

# REFERENCES

1. E. Esarey *et. al.*, IEEE Trans. on Plasma Sci. **24**, 252 (1996).
2. E. Esarey *et. al.*, Comm. Plasma Phys. Controlled Fusion **12**, 191 (1989).
3. C. W. Siders *et. al.*, Phys. Rev. Lett. **76**, 3570 (1995).
4. J. R. Marques *et. al.*, Phys. Rev. Lett. **76**, 3566 (1995).
5. S. P. Le Blanc *et. al.*, Phys. Rev. Lett. **77**, 5381 (1996).
6. A. Ting *et. al.*, Phys. Rev. Lett. **77**, 5377 (1996).
7. C. W. Siders *et. al.*, IEEE Trans. on Plasma Science, **24**, 301 (1996).
8. L. M.Gorbunov and V. I. Kirsanov, Sov. Phys. JETP **66**, 290 (1987).
9. S. V. Bulanov, V. I. Kisanov, and A. S. Sakharov, JETP Lett. **50**, 198 (1989).
10. V. I. Berezhiani and I. G. Murusidze, Physica Scripta **45**, 87 (1992).
11. S. Kato, Y. Kishimoto, and J. Koga, Phys. of Plasmas **5**, 292 (1998).
12. R. J. Kingham and A. R. Bell, Phys. Rev. Lett. **79**, 4810 (1997).
13. T. R. Clark and H. M. Milchberg, Phys. Rev. Lett. **78**, 2374 (1997).
14. K. L. F. Bane, P. Chen, and P. B. Wilson, IEEE Trans. on Nuclear Sci. **32**, 3524 (1985).
15. T. Katsouleas, Phys. Rev. A **33**, 2056 (1986).
16. S. V. Bulanov *et. al.*, IEEE Trans. on Plasma Sci. **24**, 393 (1996).
17. P. Chen, A. Spitkovsky, and W. B. Mori, Nucl. Instrum. Meth. A, in press.
18. S. C. Wilks *et. al.* , Phys. Rev. Lett. **62**, 2600 (1989).
19. W. M. Wood, C. W. Siders, and M. C. Downer, Phys. Rev. Lett. **67**, 3523 (1991).
20. S. P. Le Blanc, R. Sauerbrey, S. C Rae, and K. Burnett, J. Opt. Soc. Am. B **10**, 1801 (1993).
21. C. Frochly, J. Optics (Paris) **4**, 183 (1973).
22. J. P. Geindre *et. al.*, Opt. Lett. **19**, 1997 (1994).
23. W. Leemans *et. al.*, IEEE Trans. Plasma Sci. **24**, 331 (1996).
24. H. Dewa *et. al.*, Nucl. Instrum. Meth. A, in press.
25. G. Rodrigues and A. J. Taylor, Opt. Lett. **23**, 858 (1998).
26. S. P. Le Blanc and R. Sauerbrey, J. Opt. Soc. Am. B **13**, 72 (1996).
27. J. M. Diaz, L. Oliveira e Silva, and J. T. Mendonca, Proceedings of the $13^{th}$ Advanced ICFA Beam Dynamics Workshop, Kyoto, Japan, 1997.

# Simulations of the SLAC Plasma Lens Experiment

Shinichi Masuda* and Pisin Chen†

*Department of Physics and Astronomy, University of California,
Los Angeles, CA 90024-1547
† Stanford Linear Accelerator Center, Stanford University, Stanford, CA 94309

**Abstract.** A plasma lens experiment is planned at SLAC with high energy electron beams and high plasma density. We have developed a PIC code, SPLASH (Simulation of Plasma Lens And Synchrotron-radiation in Hybrid), for computer simulations of this experiment. The plasma lens focusing effects are simulated for the beam energy at 30 and 50 GeV and the plasma density between $n_p/n_b = 0.1$ and 100. The original bunch size is 3 and 5 $\mu$m in radius. Due to plasma focusing, the beam size is reduced, which is progressively more effective when the relative plasma density increases from the "underdense"regime to the "overdense"one. However, field compensation due to the plasma return current renders the plasma focusing effect suppressed when the plasma density further increases to the "superdense"regime. The beam waist is located between ±20 mm around the original focal point. In the most effective case, the beam is focused from 5 $\mu$m to 2 $\mu$m and from 3 $\mu$m to 1 $\mu$m.

## INTRODUCTION

The effect of the plasma on the charged particle beam passing through it has been studied in recent years. The beam excites the electromagnetic waves in the plasma which affects the beam itself. The effects of the excited wave acting on the beam appear as the focusing and the acceleration. The plasma lens as a final focus device for future linear accelerators has been developed theoretically [1–6] and experimentally [7–9]. The experiments conducted in past demonstrated the plasma lens effect for the electron beam with the energy of several MeV and the plasma density of the order of $10^{10}$ to $10^{12} \mathrm{cm}^{-3}$.

For future linear colliders, there is need to investigate the plasma lens with the energy and the plasma density many order of magnitude higher than that in these proof of principle experiments. The E-150 SLAC plasma lens experiment is planned with the electron energy up to 50 GeV, the wide range of the plasma density and the beam density around $10^{18}$ cm$^{-3}$ [10].

A PIC (Particle-In-Cell) code, *SPLASH* (Simulation of Plasma Lens And Synchrotron-radiation in Hybrid), was developed for computer simulations of this

CP472, *Advanced Accelerator Concepts: Eighth Workshop,*
edited by W. Lawson, C. Bellamy, and D. Brosius

experiment. In this paper we report on the simulation model and the simulation results.

## SIMULATION MODEL AND PARAMETERS

Our simulation code treats two dimensional axially symmetric system. The electromagnetic field and the current density on the $r$-$z$ plane are calculated on each grid point by the axially symmetric Maxwell equations. Particle motion is calculated by the Lorentz force equation in the three dimensional Cartesian coordinates. Each macro particle has the shape like a ring. The current densities on each grid point are calculated by the volume weighting scheme in the $r$-$z$ around the $z$-axis for the position and velocity of each particle projected on the $r$-$z$ plane. The electromagnetic field acting on each macro particle is interpolated by the same weighting scheme. The quantities such as the electromagnetic field and the position and velocity of the particle are advanced in time by the leap-frog method.

Figure 1 shows the simulation model. The system is enclosed by a cylinderical conducting wall. A plasma column is located at the center of the system. An initially given electron bunch goes into the system as shown in Fig.1.

Following the convention in accelerator physics, we take a $s$-axis along the $z$-axis so that an original waist of the beam without plasma lens is located at $s = 0$. Position of the focused beam waist is represented by the $s$-axis in this paper.

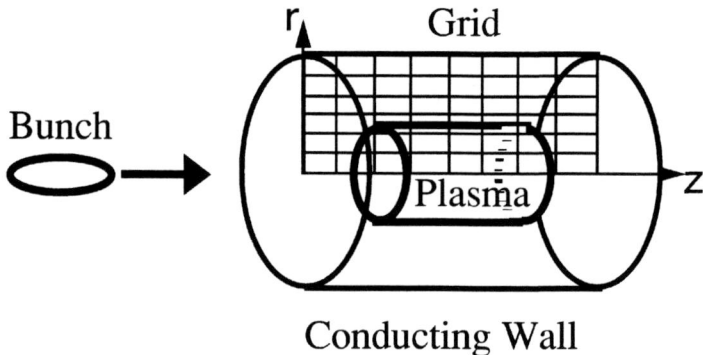

**FIGURE 1.** The simulation model for the plasma lens.

**TABLE 1.** Beam parameters in the SLAC plasma lens experiment.

| Quantities | Values |
|---|---|
| Energy $E$ | 30 and 50 GeV |
| Population $N$ | $1 \sim 4 \times 10^{10}$ |
| Emittance $\epsilon_n$ | 2 or $3 \times 10^{-5}$ m rad |
| Bunch size $\sigma_{r0}^*$ | $3 \sim 5\,\mu$m |
| Bunch length $\sigma_z$ | $0.45 \sim 0.7$ mm |
| Density ratio $n_p/n_{b0}$ [a] | $0.1 \sim 100$ |

[a] $n_p$ and $n_{b0}$ are the plasma and beam density at the lens entrance, respectively

Spatial distribution of the plasma column, in our case 3 mm in length and 30 $\mu$m in radius, is assumed to be uniform. The plasma lens entrance is located at $s = -20$ mm. Cold plasma is supposed, i.e., initial velocity of each plasma particle is zero because the thermal effect of the plasma is negligible due to small thermal velocity in our experiment.

The bunch comes into the system along the $s$-axis as shown in Fig.1. Initial distribution of the beam particles is assumed to be Gaussian for both of spatial and velocity distributions. The Gaussian distribution at the initial position, $s = s_0$, is

$$n_b(r,s) = n_{b0} \exp[-\frac{r^2}{2\sigma_{r0}^{*2}(1 + (s - s_0)^2/\beta_0^{*2})}] \exp[-\frac{(s - s_0)^2}{2\sigma_z^2}] \qquad (1)$$

where $n_{b0}$ is the density at center of the beam, $\beta_0^*$ and $\sigma_{r0}^*$ is the beta function and the rms size at the original beam waist, respectively, $\sigma_z$ is the rms bunch length. Beam parameters are shown in Table 1.

In this simulation changes in the beam size along the beam line are calculated to obtain some information for real measurements. The beam particles are tracked after they pass through the plasma lens. Intersections of each particle orbit on the $x$-$y$ planes, which are located on the $s$-axis along the beam line, are collected. Then the *rms* and the *hwhm* (half with half maximum) sizes are calculated from the time integrated distributions of the points of intersections on each plane. We will discuss the simulation results in next section.

# SIMULATION RESULTS AND DISCUSSION

## Characteristics of Plasma Focused Beam

The plasma lens works differently from the conventional lens using magnetic fields because the intensity of the excited wakefield changes along the bunch. Then, the

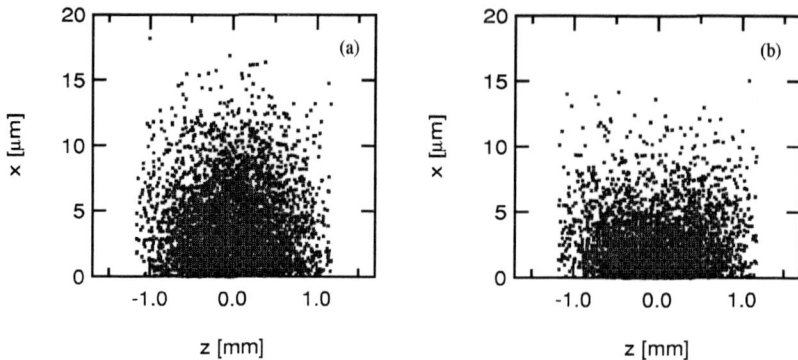

**FIGURE 2.** Snap shots of the bunch. Positions of each particle in the bunch projected on $x$-$z$ plane are plotted. Only upper half of $x$-$z$ distribution is shown because lower half of the distribution is symmetric. The beam parameters are $E = 46\,\mathrm{GeV}$, $N = 2 \times 10^{10}$, $\epsilon_n = 3 \times 10^{-5}\,\mathrm{mrad}$, $\sigma_{r0}^* = 4.74\,\mu\mathrm{m}$ and $\sigma_z = 0.47\,\mathrm{mm}$, $n_p = 20 \times 10^{16}\,\mathrm{cm}^{-3}$. (a) shows plots of the particles in the bunch at the plasma lens entrance. (b) Snap shot for the bunch near the focal point.

focusing strength is different for different $z$-position in the bunch. Snap shots of the beam at different time are shown in Fig.2. The initial profile of the beam at the lens entrance is Gaussian (see Fig.2(a)). Figure 2(b) shows the bunch near the focal point. It is seen that the focusing effect is strongest around the center of the bunch. Because of this, we expect that the radial profile of the bunch at the focal point deviates from the Gaussian distribution.

The evolution of the beam envelope along the beam line downstream of the plasma lens is shown in Fig.3. The beam size is determined by two different ways, the rms and the hwhm. When without plasma lens, the envelopes calculated by both rms and hwhm sizes evolve much the same way. This indicates that the radial profile does not much change along the beam line in this case. On the contrary, when the plasma is turned on the beam rms and hwhm sizes evolve differently. The focal length deduced from the hwhm envelope tends to the much shorter than that calculated by the rms size.

Changes of the radial profile along the beam line for the same beam parameters as that in Fig.3 are shown in Fig.4. Figure 5 shows phase space plots along the beam line with the same parameters. The divergence angle of the focused beam changes from $85\,\mu\mathrm{rad}$ to $200\,\mu\mathrm{rad}$ in this case. It is seen from Fig.5(a) that the focusing effect is strongest at $r(=\sqrt{x^2+y^2}) \sim 5\,\mu\mathrm{m}$ and is small at peripheral of the bunch. This is because the self-focusing plasma wakefield increases linearly with $r$ for $r \ll \sigma_r$ and reaches a maximum at $r \approx 1.6 \times \sigma_r$. Then it decreases for $r > 1.6 \times \sigma_r$ as described by the equation for the overdense regime [1],

$$F_r(r) = k_p^2 \left\{ \int_{-\infty}^{r} r' dr' \exp\left(-\frac{r'^2}{2\sigma_r^2}\right) K_1(k_p r) I_0(k_p r') \right.$$
$$\left. - \int_r^{\infty} r' dr' \exp\left(-\frac{r'^2}{2\sigma_r^2}\right) I_1(k_p r) K_0(k_p r') \right\}, \qquad (2)$$

where $k_p$ is the wave number of the plasma oscillation, $K_n$ and $I_m$ are the modified Bessel's functions.

Deviation of the radial profile from the Gaussian profile is small for the bunch just leaving the plasma lens (see Fig.4(a)). However, the profile becomes spiky near the beam waist further as shown in Fig.4(b) and (c). Figure 5 shows that the phase space ellipse rotates clockwise along the beam line. Strongly accelerated particles at $r \sim 1.6 \times \sigma_r$ (Fig.5(a)) concentrate to the core of the bunch when the bunch reaches the focal point as shown in Fig.5(b).

Such spiky distribution can be attributed to the dynamical nature of the plasma focusing in the overdense regime, where particles near the head and tail of the beam tend to receive a weaker focusing. In addition, nonlinear $r$-dependence of self-focusing field also contributes to this spiky distribution as mentioned above.

## Dependence on Beam Energy and Density

SLAC E-150 experiment will be conducted with a wide range of the plasma density between $n_p/n_b = 0.1 \sim 100$. We obtained results from simulations for several plasma densities. Figure 6 shows the beam envelopes calculated by the hwhm for three different ratios of the plasma density to the beam density. For the

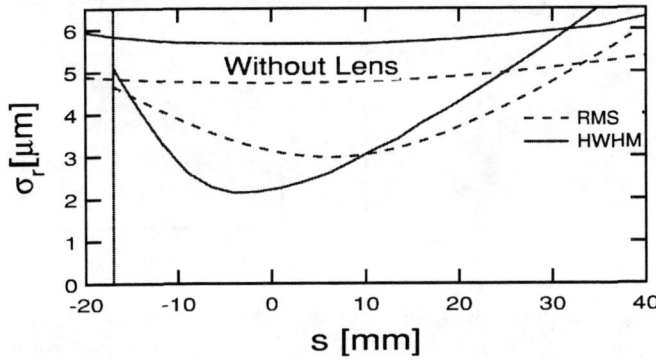

**FIGURE 3.** Comparison between the rms size and hwhm size of the bunch along the beam line. Original waist of the beam without lens is located at $s = 0$. The plasma lens entrance and exit are at $s = -20$ mm and $s = -17$ mm, respectively. The beam parameters are $E = 46$ GeV, $N = 2 \times 10^{10}$, $\epsilon_n = 3 \times 10^{-5}$ mrad, $\sigma_{r0}^* = 4.74\,\mu$m and $\sigma_z = 0.47$ mm, $n_p = 20 \times 10^{16}\,\mathrm{cm}^{-3}$.

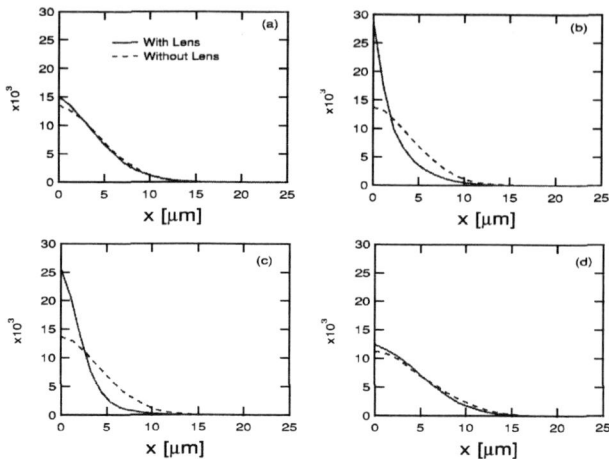

**FIGURE 4.** Radial profile of the bunch. Dashed curve indicates the radial profile of the original beam (without lens case). Solid curve is the radial profile when the plasma is turned on. The beam parameters are the same with those in Fig.3. (a) At the plasma lens exit, $s = -17$ mm. (b) At the beam waist calculated by the hwhm size, $s = -4$ mm. (c) At the beam waist calculated by the rms size, $s = 8$ mm. (d) On downstream away from the plasma lens, $s = 40$ mm.

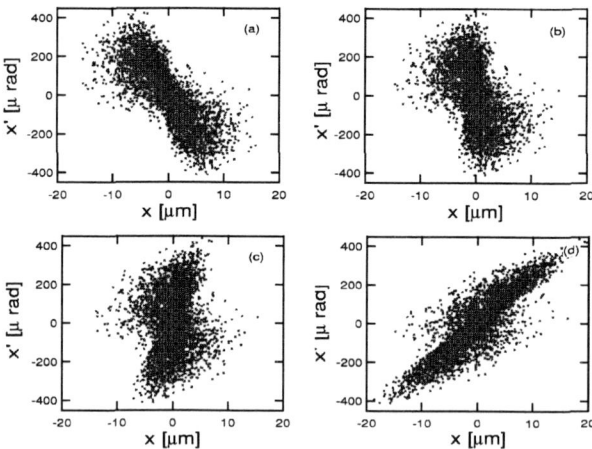

**FIGURE 5.** Phase space plots along the beam line. The beam parameters are the same with those in Fig.3. (a) At $s = -17$ mm. (b) At $s = -4$ mm. (c) At $s = 8$ mm. (d) At $s = 40$ mm.

underdense regime of $n_p/n_b = 0.3$, the beam waist moves to downstream of the original waist. On the contrary, the beam waist for the overdense case is located on upstream of the original waist. We found that the focusing effect is strongest

428

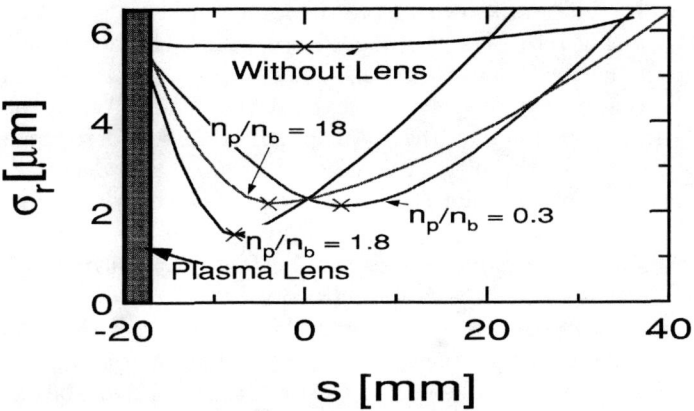

**FIGURE 6.** Beam envelopes for three different plasma densities, underdense ($n_p/n_{b0} = 0.3$), overdense ($n_p/n_{b0} = 1.8$) and superdense ($n_p/n_{b0} = 18$). Beam parameters are $E = 30\,\mathrm{GeV}$, $N = 2 \times 10^{10}$, $\epsilon_n = 3 \times 10^{-5}\,\mathrm{mrad}$, $\sigma_{r0}^* = 4.74\,\mu\mathrm{m}$ and $\sigma_z = 0.47\,\mathrm{mm}$. Cross symbols mark the focal point for each beams.

when $n_p/n_b = 1.8$. An incoming beam with $\sigma_{r0}^* \sim 5\,\mu\mathrm{m}$ is focused to $1.5\,\mu\mathrm{m}$ at the beam waist as shown in Fig.6.

Dependence of the focusing effect on the plasma density is shown in Fig.7. Po-

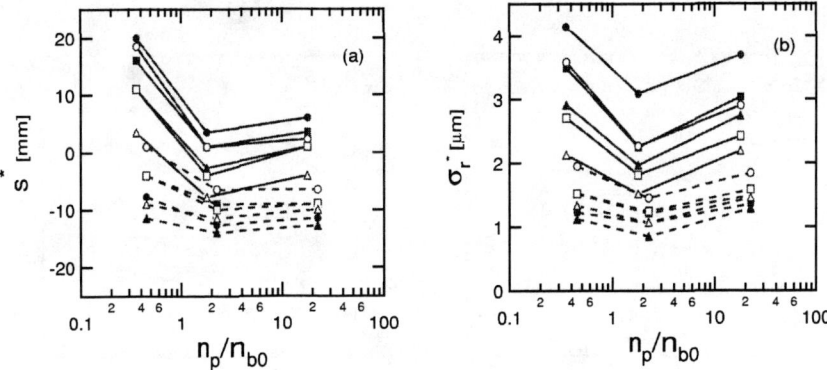

**FIGURE 7.** Dependency on density ratio. (a) Positions of the beam waist and (b) the beam sizes (hwhm) at the focal points are plotted versus the ratio of the plasma density to the density of the beam bunch. The beam parameters are $E = 30\,\mathrm{GeV}$ (open symbols), $E = 46\,\mathrm{GeV}$ (solid symbols), $N = 1 \times 10^{10}$ (circles), $N = 1.5 \times 10^{10}$ (squares), $N = 2 \times 10^{10}$ (triangles), $\sigma_{r0}^* = 4.74\,\mu\mathrm{m}$ (solid curves) $\sigma_{r0}^* = 3\,\mu\mathrm{m}$ (dashed curves) and $\sigma_z = 0.47\,\mathrm{mm}$.

sitions of the focal points are shown in Fig.7(a) and the beam sizes at the focal points are shown in Fig.7(b). For the underdense regime the beam waist moves closer to the plasma lens with an increase in the density ratio of the plasma to the beam. Then the focal length decreases with the density ratio. When the plasma density exceeds the beam density, the focal length increases for $5\,\mu$m incoming beam with the increase in the plasma density. The beam size decreases for the underdense regime and increases for the overdense regime with the increase of the plasma density.

Figure 7 shows also the energy dependence indicated by open symbols ($30\,$GeV) and solid symbols ($46\,$GeV). For the incoming beam of $5\,\mu$m in radius (solid curves) the focal length and the waist size at the energy of $46\,$GeV larger than that at the energy of $30\,$GeV. The focusing strength decreases with an increase in the energy in this case. However this dependence is opposite for case of $3\,\mu$m beam (dashed curves).

Circles, squares and triangles in Fig.7 indicate the results for the particle populations within the bunch, $1.0 \times 10^{10}$, $1.5 \times 10^{10}$ and $2.0 \times 10^{10}$, respectively. It is seen that the focusing strength increases with an increase in the particle population for both of $3\,\mu$m and $5\,\mu$m beams.

## Focusing of Positron Beams

For an eventual application of the plasma lens to high energy linear colliders, we are also interested in the plasma focusing of a positron beam. The theory [1]

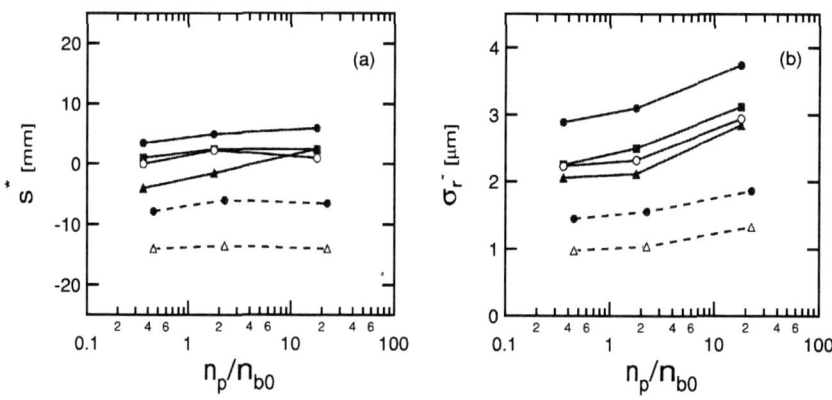

**FIGURE 8.** Focusing effect of the plasma lens for the positron beam case. (a) Position of the beam waist and (b) the beam size at the waist. The beam parameters are $E = 30\,$GeV (open symbols), $E = 46\,$GeV (solid symbols), $N = 1 \times 10^{10}$ (circles), $N = 1.5 \times 10^{10}$ (squares), $N = 2 \times 10^{10}$ (triangles), $\sigma_{r0}^{*} = 4.74\,\mu$m (solid curves) $\sigma_{r0}^{*} = 3\,\mu$m (dashed curves) and $\sigma_z = 0.47\,$mm.

430

predicts that the focusing effect on the positron beam is the same as the electron case in the overdense regime.

We have performed the plasma lens simulations for the positron beams. The beam parameters are the same as those for the electron bunch. Figure 8 shows the focusing effect for the positron case. Dependency on the plasma density for the overdense regime shows good agreement with the electron case (see Fig.7 also) as expected by the theory. In the underdense regime, contrary to the electron case, the waist size increases with the increase of the plasma density. In addition, it is seen that the focusing strength of the underdense regime is stronger for the positron bunch than for the electron bunch.

Other characteristics of positron focusing, such as the dependencies on the energy and the population of the particle, are the same as those for the electron beam. The focusing strength decreases with the beam energy. The focused beam size as well as the focal length is inversely proportional to the beam density.

## Simulation of Synchrotron Radiation

The plasma lens strongly deflects the particle orbit in the beam and would induce *SR* (synchrotron radiation). Experimentally, these SR photons induced by the plasma focusing provide a direct information on the plasma focusing strength. The E-150 experiment intends to construct a SR monitor for this purpose.

Computer simulation on this process is therefore important for the design of such an apparatus. The synchrotron radiation based on the quantum mechanical formulation [11] is included in our simulation code. An algorithm to handle dis-

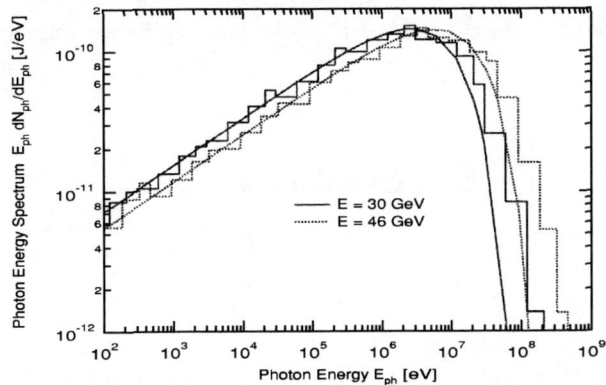

**FIGURE 9.** Energy spectra of photons emitted from electron bunch with energy of 30 GeV (solid) and of 46 GeV (dashed) during it passing through the plasma lens. Histograms show results from the simulations. Theoretical curves are also plotted. Beam parameters are $N = 2 \times 10^{10}$, $\sigma_{r0}^* = 5\,\mu\text{m}$ and $\sigma_z = 0.65\,\text{mm}$. The plasma density is $20 \times 10^{16}\text{cm}^{-3}$ $(n_p/n_b \sim 2)$.

crete photon emission is implemented following the technique in the code, DRNO [12]. Number of photons emitted from an electron during one simulation cycle can be calculated by $N_{ph}\Delta t$, where $N_{ph} = \int_0^{\hbar\omega_{\max}}[dN_{ph}(\hbar\omega)/d(\hbar\omega)]d(\hbar\omega)$ is the photon emission rate from an electron and $\Delta t$ is the time step in the simulation. $N_{ph}\Delta t$, which is much smaller than unity, is proportional to the probability of the photon emission during the $\Delta t$. To simulate the discrete photon emission, a number provided by the uniform random number generator, which has value between $0 \sim 1$, is compared with $N_{ph}\Delta t$. When the number is smaller than $N_{ph}\Delta t$, the algorithm decides that a photon is emitted. The emitted photon energy is determined randomly according to the photon number spectrum, $dN_{ph}(\hbar\omega)/d(\hbar\omega)$; A uniform random number $(0 \sim 1)$ is provided as the argument of the inverse function of the accumulated spectrum, $\int_0^{\hbar\omega}[dN_{ph}(\hbar\omega')/d(\hbar\omega)']d(\hbar\omega)'/N_{ph}$. Then the photon energy, $\hbar\omega$, can be calculated from the inverse function. This inverse function works as a random number generator providing the photon number spectrum.

Figure 9 shows energy spectra of the photon emitted from the electron bunch. Histograms of energy spectra are plotted for the beams with energy of 30 GeV (solid) and of 46 GeV (dashed). Shapes of both spectra are similar as expected. Since the initial beam sizes of the 30 GeV and 46 GeV beams are the same and the plasma density was also the same, we expect that the SR critical frequencies in the two cases differ by the ratio of the beam energies.

To obtain the magnitude of the equivalent magnetic field for the plasma focusing, a theoretical intensity, $\hbar\omega[dN_{ph}/d(\hbar\omega)]N\Delta t_f$, was calculated, where $\Delta t_f \sim 10^{-11}$ sec is the flight time of the electron bunch passing through the lens, to fit the simulation results. The equivalent focusing field gradient was found to be $B/\sigma_r \sim 2.6 \times 10^8$ Gauss/cm for both 30 GeV and 46 GeV cases. This is to be compared with the theoretical prediction from the focusing strength of plasma lens [10], $K = 2\pi r_e n_b/\gamma$, where $r_e$ is the classical electron radius and $\gamma$ is the relativistic kinetic energy factor of the beam. The equivalent focusing field gradient is calculated as $B/\sigma_r \approx KE/e \sim 1.6 \times 10^8$ Gauss/cm. The simulation result and the theoretical prediction are reasonably agreed.

# SUMMARY

We reported the results from the PIC simulations for the plasma lens experiment. The results indicate that the electron bunches of $3\,\mu$m and $5\,\mu$m are focused down to $1\,\mu$m and $2\,\mu$m, respectively. We confirm that the focusing strength varies from the underdense regime to the overdense regime, and the superdense regime, in agreement with the theoretical prediction. The electron bunch is strongly focused around $n_p/n_b \approx 1 \sim 2$. Table 2 summaries our simulation results relevant to the E-150 experiment.

We have also shown that the synchrotron radiation algorithm in our code appears to produce the correct physics, although further studies is needed before we can fully utilize it to assist the design of a synchrotron radiation monitor.

**TABLE 2.** Simulation results. ($\sigma_{r0}^* = 5\,\mu\text{m}$, $\sigma_z = 0.65\,\text{mm}$.)

| Population $N$ | Density $n_p/n_b$ | Waist $s^*$ [mm] | Size $\sigma_r^*$ [$\mu$m] | Divergence [$\mu$rad] |
|---|---|---|---|---|
| $E = 29.6\,\text{GeV}$ | | | | |
| | 0.3 | 9.35 | 3.42 | 125 |
| $1.0 \times 10^{10}$ | 1.8 | 4.30 | 2.93 | 135 |
| | 18 | 5.98 | 3.39 | 108 |
| | 0.3 | 2.62 | 2.27 | 205 |
| $2.0 \times 10^{10}$ | 1.8 | -1.86 | 2.05 | 220 |
| | 18 | -0.18 | 2.70 | 143 |
| | 0.3 | -6.35 | 1.48 | 374 |
| $4.0 \times 10^{10}$ | 1.8 | -10.3 | 1.51 | 382 |
| | 18 | -7.47 | 2.08 | 207 |
| $E = 46.0\,\text{GeV}$ | | | | |
| | 0.3 | 13.8 | 4.19 | 104 |
| $1.0 \times 10^{10}$ | 1.8 | 12.2 | 3.45 | 107 |
| | 18 | 9.34 | 4.04 | 91 |
| | 0.3 | 8.79 | 2.99 | 148 |
| $2.0 \times 10^{10}$ | 1.8 | 2.62 | 2.53 | 156 |
| | 18 | 2.06 | 3.38 | 110 |
| | 0.3 | -1.30 | 1.95 | 259 |
| $4.0 \times 10^{10}$ | 1.8 | -6.91 | 1.84 | 267 |
| | 18 | -2.98 | 2.49 | 146 |

# ACKNOWLEDGMENTS

We deeply appreciate the helpful discussion and kind assistance from R. J. Noble, who lent us the SR code, DRNO.

# REFERENCES

1. Chen P., *Part. Accel.* **20**, 171 (1987).
2. Chen P., *et. al.*, *IEEE Trans. Plasma Sci.* **15**, 218 (1987).
3. Rosenzweig J. B., *et. al*, *Phys. Rev. D* **39**, 2039 (1989).
4. Chen P., *et. al.*, *Phys. Rev. D* **40**, 923 (1989).
5. Su J.J., *et. al.*, *Phys. Rev. A* **41**, 3321 (1990).
6. Chen P., *Phys. Rev. A* **45**, R3398 (1992)
7. Rosenzweig J. B., *et. al.*, *Phys. Rev. Lett.* **61**, 98 (1988).
8. Nakanishi H., *et. al.*, *Phys. Rev. Lett.* **66**, 1870 (1991).
9. Hairapetian G., *et. al.*, *Phys. Rev. Lett.* **72** 2403 (1994).
10. Chen P., *et. al.*, "SLAC Proposal E-150", SLAC (1997);
    Chattopadhyay S., *et. al.*, Wkshp. Adv. Accel. Concepts, 1998.
11. Sokolov A. A, and Ternov I. M., *Synchrotron Radiation*, Pergamon, Berlin, 1968.
12. Noble R. J., *Nucl. Inst. Meth.* **A256**, 427 (1987).

# High efficiency coupling and guiding of intense femtosecond laser pulses in preformed plasma channels in an elongated gas jet

S.P. Nikitin, I. Alexeev, J. Fan, and H.M. Milchberg

*Institute for Physical Science and Technology*
*University of Maryland*
*College Park, MD 20742*

**Abstract.** We report coupling and guiding of pulses of peak power up to 0.3 TW in 1.5 cm long preformed plasma waveguides generated in a high repetition rate argon gas jet. Coupling of up to 52% was measured for 50 mJ, ~110 fs pulses injected at times longer than 20 ns, giving guided intensities up to ~$5 \times 10^{16}$ W/cm$^2$. It was found that for short delays between waveguide generation and pulse injection, pulse shortening occurred, with this effect reduced as delay was increased. Injection into the waveguide of two consecutive pulses separated by a few nanoseconds resulted in the reduction of shortening of the second pulse at all delays. Femtosecond time-resolved shadowgrams of the coupling of injected pulses into the waveguide show that there is ~0.5 mm of neutral gas remaining at the waveguide entrance after waveguide generation.

## INTRODUCTION

Optical guiding of intense laser pulses in plasmas has several important applications including laser-driven electron acceleration [1] and has been demonstrated both in the self-guiding regime [2] and using prepared plasma channels. The latter includes guiding in high voltage capillary discharges [3] and in thermally driven laser-induced plasma waveguides [4], the approach developed by our group. In our previous experiments, these laser-driven channels were produced in the focus of an axicon in an ambient background gas, with neutral gas extending in front of the channel entrance and beyond the exit. Maximum injected pulse coupling efficiencies of ~70% were measured in those experiments, which were limited to moderate intensity ($<10^{15}$ W/cm$^2$) pulse injection. At higher intensity there is sufficient optical field ionization of neutral gas in advance of the channel entrance to cause refraction of most of the injected pulse before it enters the channel. Recently, we measured reduced coupling

CP472, *Advanced Accelerator Concepts: Eighth Workshop,*
edited by W. Lawson, C. Bellamy, and D. Brosius
© 1999 The American Institute of Physics 1-56396-889-4/99/$15.00

efficiency of less than 30% and refraction-induced pulse shortening in the injection of $\sim5\times10^{15}$ W/cm$^2$ pulses into plasma channels produced in an ambient gas [5].

## ELONGATED GAS JET

The absence of gas from the region in advance of the waveguide entrance would be expected to reduce the refraction problem, and to this end we designed a high repetition rate gas jet (see Fig. 1) in which to produce the plasma waveguide [6]. In the experiments reported here, the nozzle orifice is an elongated slot 1.5 cm long and 0.75 mm wide. The nozzle is fed through a 0.75 mm diameter throat by a pulsed solenoid valve. The gas flow pattern beyond the orifice is controlled with adjustable teflon flow guides inside the nozzle. The nozzle end face is tapered to minimize its obstruction of the axicon beam used to generate the plasma waveguide. The presence of this obstruction had minimal effect on the gas breakdown [6], and this observation was supported by calculations which show that the main effect of the obstruction is to reduce the peak intensity by an amount equal to the fraction of the obstructed beam energy, here <10%, with little disruption to the Bessel field profile in the focus [7].

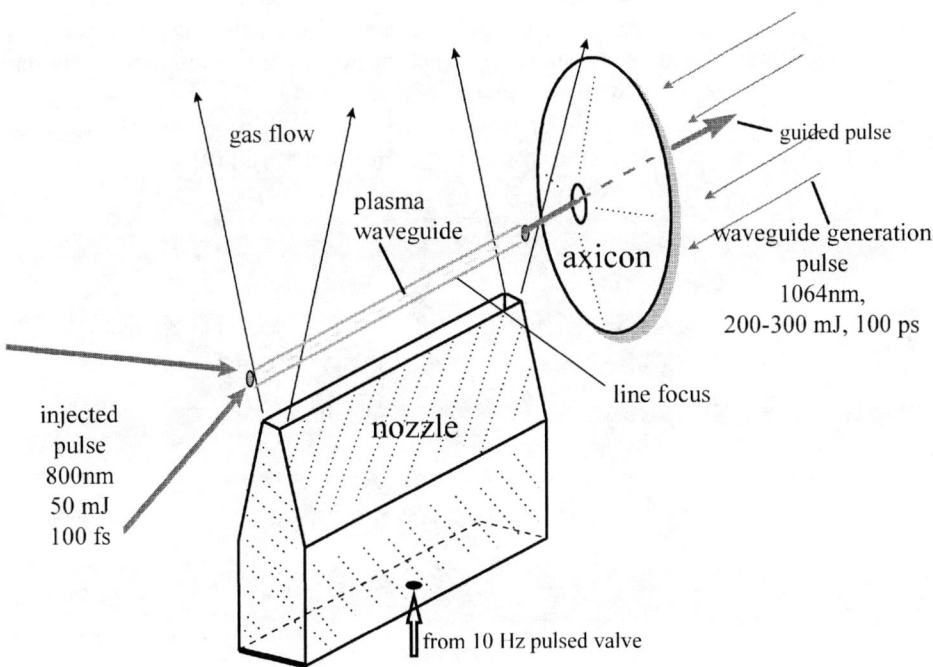

**FIGURE 1.** Elongated nozzle and laser focusing geometry.

The plasma waveguides in the experiments reported here were produced in the 1.7 cm long line focus of a 25° base angle axicon using 200-300 mJ, 100 ps, λ=1.064μm pulses (peak intensity ~4x10^{13} W/cm²) from a Nd:YAG regenerative amplifier/power amplifier system [8]. Argon gas was used in the jet, with a valve backing pressure of 33 atm and a gas puff FWHM of 550μs, with the laser pulse arriving at the temporal peak of the puff. The axicon line focus was aligned 2mm from the nozzle orifice and overfilled the length of the gas region.

Folded wavefront interferometry was used to measure the spatial profiles of neutral gas density and plasma waveguide electron density from the jet, using λ=532 nm laser pulses obtained from frequency doubling λ=1.064 nm pulses from the Nd:YAG system. The interferometer probe pulse width was 70 ps , sufficient to capture the rapid hydrodynamic evolution of the plasma waveguide [9]. The neutral density profiles, shown in ref. [6], were quite uniform along the jet, as evidenced by the uniform waveguide shadowgram images shown in Figs. 8 and 9 below. A sample interferogram from a laser-induced plasma waveguide produced under these conditions with 225mJ laser energy is shown in Fig. 2, and a sequence of electron density profiles, obtained from Abel inversion of the phase plots extracted from the interferograms [9], is shown in Fig. 3. The time scales for channel evolution are similar to our earlier backfill experiments, and the channels are as axially uniform as in the backfill case [9]. There are two notable differences, however. First, we find that with the jet, lower laser energy is sufficient to generate comparable electron density waveguides from similar initial neutral densities as in the backfill case, without the use of avalanche seed gases such as $N_2O$ [9]. Second, the ~100 μm radial extent of the electron density profile at $t=0$ (see Fig. 3) is much wider in the case of the jet than in backfill [9].

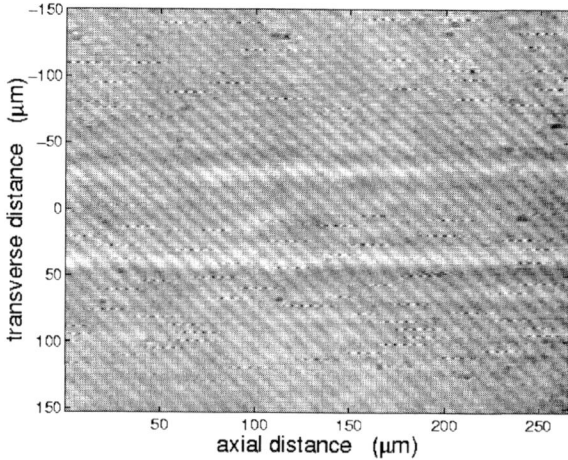

**FIGURE 2.** Interferogram of typical section (away from ends) of plasma waveguide in argon, produced 2 mm from the orifice, with gas jet backing pressure 33 atm and gas pulsewidth 550μs. The waveguide was generated with a 225 mJ, 100ps, λ=1064 nm pump pulse and the pump- interferometer probe delay was 1.8 ns.

A likely explanation of both of these observations is that clustering [10] of argon occurs in the nozzle flow. By reducing the threshold for breakdown [11], clusters would reduce the need for a seed gas and also would result in initial gas breakdown by the Bessel beam farther radially from the peak intensity on axis, both consistent with our observations.

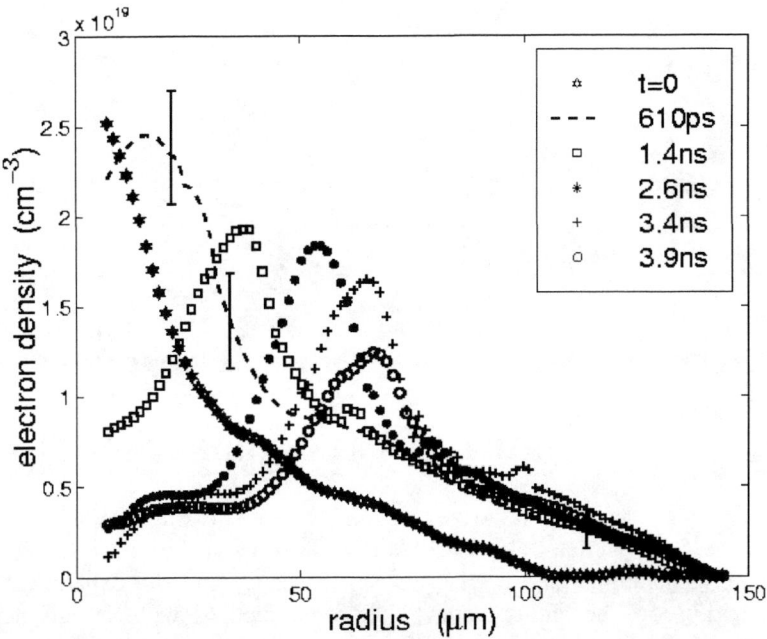

**FIGURE 3.** Time resolved electron density profiles of evolving plasma waveguide for conditions of Fig. 2, where $t=0$ refers to the center of the pump pulse.

The degree of taper of the end of the plasma waveguide is an important consideration for coupling. Figure 4 shows electron density profiles as a function of axial distance at the end of the waveguide. The electron density profile fall-off to the end of the guide takes place over ~300 μm, longer than ~150μm in the case of gas backfill [9]. This likely results from reduced laser heating as the gas density drops off at the end of the jet, and is consistent in extent with the estimated gas density gradient at the end. Since the laser-plasma heating rate scales as $N^2$, the temperature scales as $N$, and the collisional ionization rate scales as $N^2 \exp(-\chi/kT)$, where $N$ is the density and $\chi$ is the ionization potential, we would expect collisional ionization to be reduced in the edge region of the jet. An end density fall-off distance of ~0.5 mm was estimated from both the shadowgrams (see Figs. 8 and 9 below) and by focusing a high intensity Ti:Sapphire laser pulse (100fs, 20 mJ) into the end of the jet and imaging the resulting visible recombination radiation, which is proportional to the gas density for such a short pulse width. Our hydrocode simulations of laser heating in such a density gradient show the rapid fall-off in efficiency of collisional ionization.

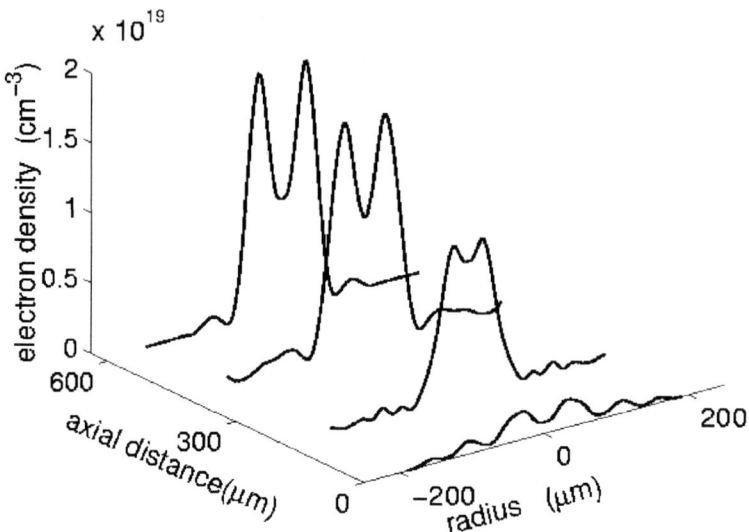

**FIGURE 4.** End region of plasma waveguide, showing tapering of electron density profile. Probe delay = 3.0 ns.

## HIGH INTENSITY GUIDING

Guiding experiments were performed with a high power femtosecond Ti:Sapphire laser system (10 Hz, 50mJ, 100 fs) synchronized to the Nd:YAG waveguide generating laser as described in reference [5]. The experimental setup is shown in Fig. 5. The femtosecond pulses were focused with a MgF$_2$ lens at f/15 through a MgF$_2$ window in the experimental chamber. The vacuum focus intensity FWHM was 30 μm (see Fig. 7), approximately twice diffraction limited. A small portion of the main pulse (~5%) was split off with a pellicle (BS1) and directed to a delay line for femtosecond time-resolved shadowgrams, captured with camera CCD-1. After the jet interaction region, the main pulse was reflected to the chamber exit by a dichroic beam splitter (DS), after which it was split by another pellicle (BS2) and directed to a FROG (frequency-resolved optical gating [12]) diagnostic, and to a microscope objective and camera CCD-2 to image the channel exit plane.

Figure 6 shows waveguide output pulsewidth (derived from FROG) vs. injection delay (with respect to waveguide generation) for injection with single and double pulses (first to second pulse energy ratio 1:5) of total energy 50 mJ. The double pulse was produced by injecting the Ti:Sapphire regenerative amplifier with two variably separated seed pulses from the oscillator. The waveguides were produced in argon under the conditions of Figs. 2 and 3. For the double pulses, which were separated by 4ns, the FROG traces effectively register the second pulse alone since the signal level is proportional to the cube of intensity [12]. It is seen that the measured pulse shortening is reduced for longer delays (for both single and double pulses), and at all delays the shortening is less for the double pulse case.

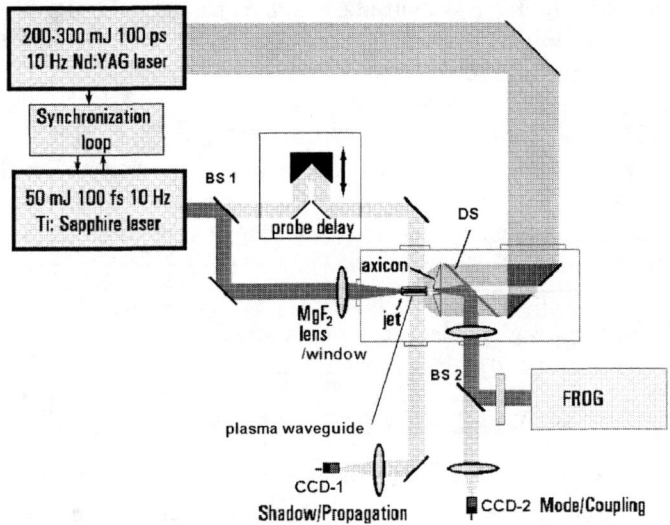

**FIGURE 5.** Experimental setup for femtosecond pulse guiding experiments.

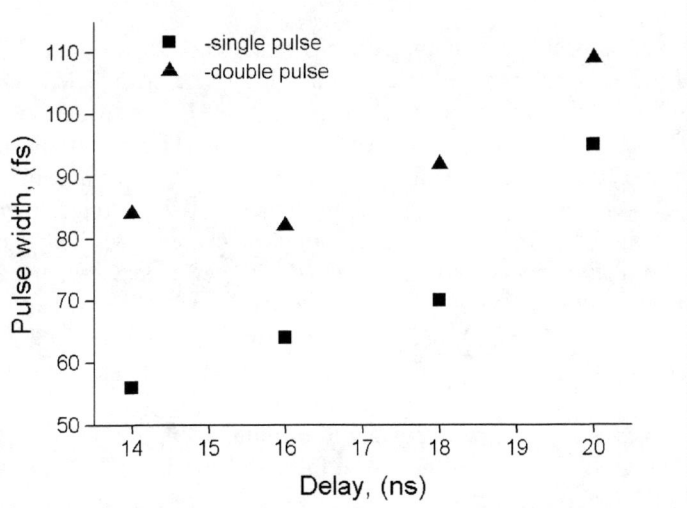

**FIGURE 6.** FROG measurement of waveguide output pulsewidth for injection with single or double (first to second in 1:5 ratio) pulses. Delay is time between channel generation and main pulse injection.

The reduction in shortening for the double pulse case reinforces our earlier conclusion [5] that the origin of the pulse shortening is refraction at the channel entrance and indicates that the 20% level prepulse pre-ionizes residual gas or incompletely ionized ions at the channel entrance, reducing the effect of ionization on the main pulse. The reduction in shortening with delay for the single pulse is consistent with increased channel width at the entrance, resulting in better capture of refracted light and decreased interaction with the less ionized gas at the channel radial periphery. We note that the waveguides of this experiment are too short (at the measured electron density) for the negative group velocity dispersion of plasma to be responsible for the maximum 30-40% shortening observed.

The FROG traces show only a small level of blue shifting for the guided pulses, even in the presence of shortening. The slight increase in blue shift with delay (not shown) suggests that at early delay, light refracted at the channel entrance is diverted outside the channel acceptance and at later delay more of this light is captured by the channel. The contribution of waveguide ionization during guiding to blue shifting of the guided pulse is negligible. An estimate of the average degree of ionization $\overline{Z}$ induced by the 100 ps waveguide generation pulse alone is found by dividing the central electron density at early times in the electron density profiles by the known gas density, $\overline{Z} = \dfrac{N_e}{N_0}$, where $N_0 = \sum_i N_i$ is the original neutral gas density and $N_i$ is the density of ion species $i$ [6,9]. For the waveguide conditions of this experiment shown in Fig. 2, $\overline{Z} \approx 8$, which indicates that Ne-like argon is the dominant species in the waveguide. The approximate intensity threshold for optical field ionization from Ne-like to F-like argon is $\sim 10^{18}$ W/cm$^2$ [13], so that further ionization by the guided pulse would not be expected at our peak guided intensity of less than $\sim 10^{17}$ W/cm$^2$ (see below).

The vacuum focal spot and guided mode images are shown in Fig. 7. In contrast to the case of waveguides produced in a gas backfill, where the injected pulse delay for efficient coupling at this f/# is no more than 5-6 ns [4], typical delays for efficient coupling with the jet channel are in excess of 10-15 ns. This results from the longer taper at the end of the gas jet waveguides, which requires longer expansion times for efficient pulse injection. Typical coupling efficiency at delays longer than ~20 ns is slightly better than 50%, measured by integrating the channel exit image and dividing by the integral of the vacuum spot image. The coupling efficiency remains at ~50% out to ~ 40 ns, where it begins to slowly decrease. At 24 ns delay, typical single mode output is shown (FWHM 35µm), and it is larger than the vacuum spot, owing to the channel expansion. At even longer delay of 36 ns, $m=1$ mode structure appears with the onset of 2 lobes. Depending on delay, pulse shortening down to ~70fs occurs (as shown in Fig. 6), resulting in peak powers transmitted of up to ~0.3 TW, and guided intensity up to ~5x10$^{16}$ W/cm$^2$. However, at these rather large spot diameters, there is only 8-9 Rayleigh lengths of guiding.

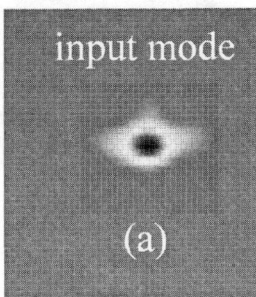 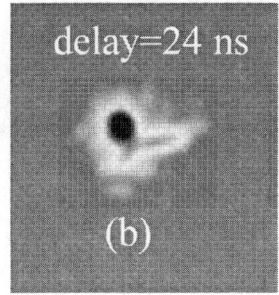 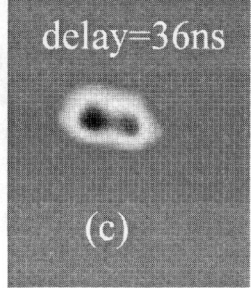

**FIGURE 7.** Mode images for (a) vacuum case (FWHM 30μm); guided mode for (b) 24 ns delay (FWHM 35μm) and (c) 36 ns delay.

Femtosecond time-resolved shadowgrams were used to examine the propagation and coupling of injected pulses near the waveguide entrance. The left side of Fig. 8 shows a sequence of shadowgrams for the injected pulse propagating into the gas jet, with the waveguide off.

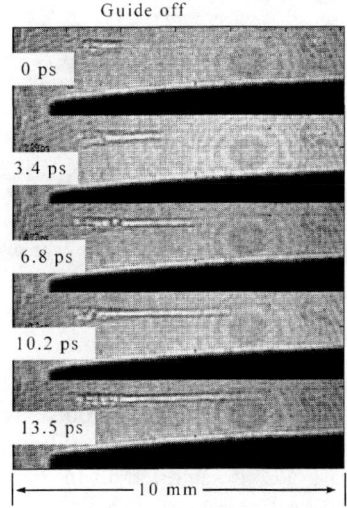 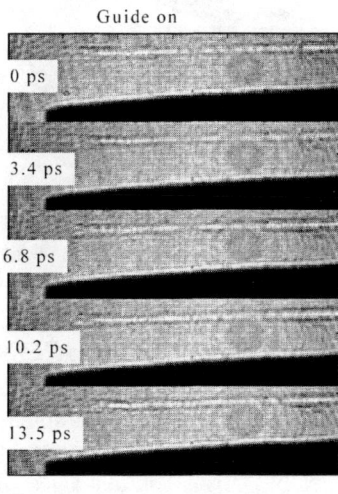

**FIGURE 8.** Sequence of shadowgrams showing coupling of intense femtosecond pulses to the jet alone (guide off) and to the plasma waveguide (guide on). Injection delay is 24 ns.

The abrupt onset of ionization (within ~0.5 mm) is seen at the jet entrance, with the ionization front propagating from left to right at the speed of light. The relatively collimated looking ionization region results from the on-axis intensity of the pulse remaining sufficiently high for ionization of neutral Ar. With the waveguide on (right side of Fig. 8), the region of strong refraction near the jet entrance (associated with the

441

turbulent-looking region there) is less evident, and there is little apparent effect on the shadowgram induced by the guided pulse, as is consistent with the FROG measurements, which indicated negligible additional ionization in the waveguide. Figure 9 is a shadowgram of vertically misaligned coupling between the injected pulse and the waveguide, showing that ~0.5 mm of gas at the entrance to the jet is insufficiently ionized by the 100 ps waveguide generation pulse. This is likely the source of the residual refraction-induced shortening seen in the FROG measurements of the waveguide output.

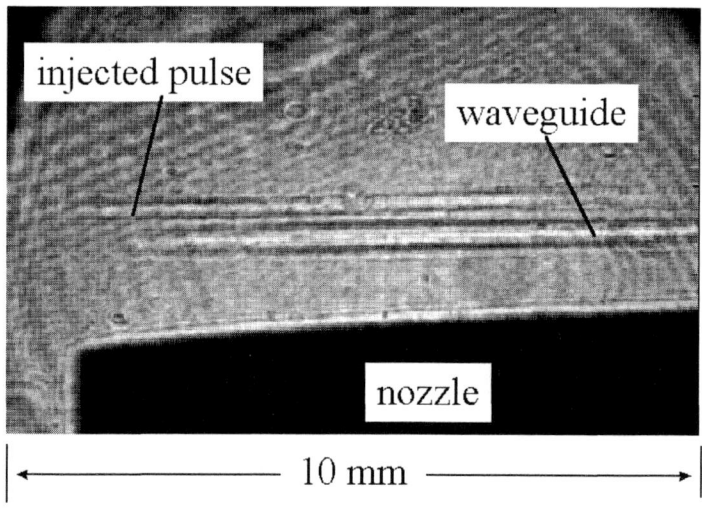

**FIGURE 9.** Shadowgram showing misaligned coupling between the injected pulse and the plasma waveguide. Probe delay after femtosecond pulse enters jet is 17 ps.

## CONCLUSIONS

We have demonstrated optical guiding of up to 0.3 TW pulses in plasma waveguides generated in a high repetition rate pulsed gas jet, achieving high coupling efficiencies of ~50%. The existence of a finite gradient scale length in gas density of ~0.5 mm at the edge of the jet results in reduced efficiency of inverse bremsstrahlung heating and ionization at the jet edge by the 100 ps waveguide generation pulse. This has two consequences. First, in the jet edge region the waveguide end is more tapered than in the backfill case due to the reduced heating driving the expansion, so longer delays before injection are required for efficient coupling. Second, approximately 0.5 mm of gas at the edge is less ionized than in the fully developed region of the waveguide, where $\overline{Z} \approx 8$. With single injected pulses and short delay, this results in an observed pulse shortening from ionization induced refraction at the entrance. The shortening is reduced with double pulse injection or single pulse injection at longer delay. For efficient injection and guiding of the fundamental mode at a small spot size over a large number of Rayleigh lengths, future work will require increasing the heating and ionization at the jet edge.

## ACKNOWLEDGEMENTS

The authors thank T. Antonsen for useful discussions. This work was supported by the Department of Energy (DEF G0297 ER 41039).

## REFERENCES

1.  C. J. Joshi and P. B. Corkum, Phys. Today **48**, 36 (1995).
2.  A. B. Borisov et al., Phys. Rev. Lett. 68, 2309 (1992); P. Monot et al., Phys. Rev. Lett. **74**, 2953 (1995); K. Krushelnick et al., Phys. Rev. Lett. **78**, 4047 (1997); R. Wagner et al., Phys. Rev. Lett. 78, 3125 (1997); J. Fuchs et al., Phys. Rev. Lett. **80**, 1658 (1998).
3.  Y. Erlich, C. Cohen, A. Zigler, J. Krall, P. Sprangle, and E. Esarey, Phys. Rev. Lett. **77**, 4186 (1996).
4.  C.G. Durfee, J. Lynch, and H.M. Milchberg, Phys. Rev. E **51**, 2368 (1995); C.G. Durfee and H.M. Milchberg, Opt. Lett. **19**, 1937 (1994); C. G. Durfee and H. M. Milchberg, Phys. Rev. Lett. **71**, 2409 (1993).
5.  S. P. Nikitin, T. M. Antonsen, T.R. Clark, Yuelin Li, and H.M. Milchberg, Opt. Lett. **22**, 1787 (1997).
6.  J. Fan, T.R. Clark, and H.M. Milchberg, Appl. Phys. Lett. **73**, 3064 (1998).
7.  Z. Bien and T. M. Antonsen, unpublished.
8.  T.R. Clark, Ph.D. Thesis, University of Maryland, 1998.
9.  T.R. Clark and H.M. Milchberg, Phys. Rev. Lett. **78**, 2373 (1997).
10. O. F. Hagena, Surf. Sci. **106**, 101 (1981).
11. T. Ditmire, T. Donelly, A.M. Rubenchik, R.W. Falcone, and M.D. Perry, Phys. Rev. A **53**, 3379 (1996).
12. D.J. Kane and R. Trebino, Opt. Lett. **18**, 823 (1993).
13. S. Augst, D. Strickland, D. Meyerhofer, S.L. Chin, and J.H. Eberly, Phys. Rev. Lett. **63**, 2212 (1989).

# Practical Approach to Monochromatic LWFA

I.V. Pogorelsky

*Accelerator Test Facility, BNL, Upton, NY 11973, USA*

N.E. Andreev and S.V. Kuznetsov

*High Energy Density Research Center, IVTAN, Moscow, 127412, Russia*

**Abstract.** Dependence of the LWFA performance upon the laser wavelength is applied to optimization of the plasma-channeled standard LWFA operating in a linear regime. Electron beam energy spread, emittance and luminosity depend upon the proportion of the electron bunch size to the plasma wavelength. This proportion tends to improve with the laser wavelength increase. We propose the two-stage ~1 GeV LWFA with the controlled energy spread and emittance based on realistic capabilities of the BNL ATF that features: picosecond terawatt $CO_2$ laser and a high-brightness electron gun.

## 1. INTRODUCTION

After the successful experimental demonstration of the ultra-high gradient (up to 100 GeV/m) laser electron acceleration [1-3], the next goal in the laser accelerator development is to produce extended quasi-monoenergetic acceleration for appreciably bulk electron charges (~0.1 nC). Similar to conventional RF linacs, injection of electron bunches small to compare with a period of the oscillatory driver field is a condition for producing highly monochromatic, low emittance electron beams in laser accelerators.

In the case of RF linacs, with their ~100 MHz driving fields, the required electron bunches are generated using photocathode RF guns driven by picosecond lasers. Orders of magnitude higher frequencies of the drive field in laser accelerators require proportionally short electron bunches. Proposals to generate such bunches for plasma-based laser accelerators look as intricate as the laser accelerators themselves. One of such accelerator schemes, called LILAC [4], is based on using focused femtosecond laser pulses that inject plasma electrons precisely into the maximum of the accelerating wake field. According to another scheme, the quasi-relativistic electron bunches are formed due to frequency beating of two laser beams counter-propagating in plasma [5].

Practicalities of the mentioned above electron microbunch injection methods are subject to further analysis which we do not attempt here. In the present paper, we approach monochromatic laser acceleration from a different position. Our concept is based on a combination of the emerging ps-TW $CO_2$ lasers with a high-brightness conventional electron injectors.

The first ps-TW $CO_2$ laser is being commissioned at the BNL ATF [6-8]. Because the 10 times longer wavelength, $\lambda$, than the conventional $T^3$ solid state lasers, $CO_2$ laser may open new opportunities for the advance laser accelerator development. For processes based on electron quiver motion, like in the LWFA, the advantage of $CO_2$ lasers is primarily due to a gain of two orders of magnitude in the energy acquired

CP472, *Advanced Accelerator Concepts: Eighth Workshop,*
edited by W. Lawson, C. Bellamy, and D. Brosius

by the electron oscillating in EM laser field. The enhancement of the plasma wake formation with the $CO_2$ laser driver allows to reduce the plasma density in the LWFA without jeopardizing the net acceleration. Then, a conventional photocathode electron gun may be considered as a suitable injector for the LWFA with the controlled beam properties. The BNL/ATF, with its high-brightness RF gun and the world first ps-TW $CO_2$ laser, is equipped for the proof-of principle experimental test of such laser accelerator approach.

The ATF's electron linac delivers up to 3 nC, 10-0.3 ps, 50 MeV electron bunches with a peak current of 100-300 A, energy spread of 0.2%, and a normalized emittance $\varepsilon_n$=0.5 mm.mrad. Note, that with the present 10-ps UV photocathode laser driver, the electron bunch compression to 370 fs is achieved by the proper electron phasing to the RF field in the gun [9]. A shorter, down to 100 fs, electron bunch duration may be attained with the faster, ~1 ps, laser driver.

In Section 2, we discuss the dependence of the LWFA performance upon the laser wavelength and apply our findings to improving the energy spread of the plasma-channel LWFA operating in a linear regime.

In Section 3, we describe the 2D PIC simulations of the single-stage LWFA driven by the prospective 50 TW, 1-ps $CO_2$ laser and 50 MeV linac used as an injector. As is shown in Section 4, the performance of the GeV LWFA may be improved and as small as 1% energy spread obtained when the LWFA buncher is used between the conventional 5-MeV electron injector and the LWFA acceleration stage.

## 2. DEPENDENCE OF THE LWFA PERFORMANCE UPON THE LASER WAVELENGTH

The LWFA [10] operating in the linear regime is generally considered as the preferable scheme for the advanced high-gradient electron laser accelerators. The interest to this regime is due to a relatively high stability and regularity of the plasma wake. This offers an opportunity to achieve a reasonably good quality (emittance, monochromaticity) of the accelerated beam while maintaining the acceleration gradient ~100 times higher than with conventional RF accelerators.

In the standard LWFA scheme, plasma wave is initiated by a short laser pulse equal in duration to the half-period of the plasma wave, $\tau_L \approx \lambda_p/2c$, that depends upon the plasma density, $n_e$,

$$\lambda_p[\mu m] = 3 \times 10^{10} n_e^{-1/2}[cm^{-3}],$$ (1)

An important parameter that characterizes how strong is the laser effect on the plasma electrons is a normalized laser vector-potential,

$$a = eE_L / mc\omega = 0.3\, E_L\,[TV/m]\lambda[\mu m],$$ (2)

where $$E_L[TV/m] = 15 P_L^{1/2}[TW] / r_L[\mu m],$$ (3)

$P_L$ is the laser power, and $r_L$ is the radius of the laser focus spot.

The amplitude of the accelerating field, $E_a$, due to the charge separation in a plasma wave is [11,12]

$$E_a[V/cm] = 0.37a^2\sqrt{n_e}\left[cm^{-3}\right],$$ (4)

or using Eq.(1),

$$E_a^{max}[GV/m] = 10^3 a^2 / \lambda_p[\mu m].$$ (5)

Thus, increase of plasma density favors attaining ultra-high acceleration fields much above the conventional RF linacs. However, energy is not the only parameter important for a practical particle accelerator, which shall provide also a reasonable monochromaticity ($\Delta E/E \leq 1\%$) and an appreciably high charge ($N_e \geq 10^9$ electrons/bunch).

The maximum number of the electrons per bunch is defined by the condition that the self-field of the bunch does not alter the plasma wakefield structure,

$$N_e \leq n_e\left(c/\omega_p\right)^3 = 4\times10^6 \lambda_p[\mu m].$$ (6)

The condition for a small energy spread requires the electron bunch to be much shorter than the plasma wake period, $\tau_b \ll \lambda_p/c$. This again favors bigger $\lambda_p$. Thus, a certain trade-off needs to be resolved during the design of the high-gradient laser accelerator capable to deliver a good quality electron beam.

It is shown [13,14] that due to the enhanced ponderomotive action of the long wavelength radiation, $CO_2$ laser may help in designing the GeV compact electron accelerator with the appreciably high, ~0.1 nC, bunch charge. Below, we discuss how to do it with a low energy spread and a controlled emittance.

## 3. LIMITATIONS OF THE SINGLE STAGE LWFA DUE TO THE ELECTRON BUNCH DURATION

The approach to the mono-energetic LWFA is based on using the seed electron bunches small to compare with $\lambda_p$. The initial bunches with the required quality and charge may be produced with the conventional technique using a photo-cathode RF gun.

2-D PIC simulations [15,16] for electron acceleration by plasma wake excited in the matched parabolic plasma channel help to verify the conditions for practically achievable monochromatic LWFA. The following input parameters have been used in simulations: $CO_2$ laser with $\tau_L=1$ ps, $P_L=50$ TW, $a^2=0.5$; plasma parameters $k_p r_L=3.8$, $k_p R_{ch}=14.3$, $\lambda_p=800$ $\mu m$. Note that $\lambda_p$ and $k_p$ are defined at the axis of the plasma channel that has the profile $n_e(r) = n_e^0\left(1+\left(\dfrac{r}{R_{ch}}\right)^2\right)$.

In the preliminary simulations discussed in this section, we considered the 50 MeV injected electron bunch with the energy spread 0.2%, geometric emittance of 3 mm.mrad, and the 30 μm (100 fs) length. These experimental conditions can be practically realized at the BNL ATF using a combination of the prospective ps-TW $CO_2$ laser, a compact high-brightness linac, and a plasma channel produced by the high-current capillary discharge in vacuum [17].

As is demonstrated in Fig.1, the above plasma and laser parameters lead to excitation of fairly regular wakefield. For the electron bunch injection, we choose the moment of the maximum accelerating field at the beginning of the focusing phase.

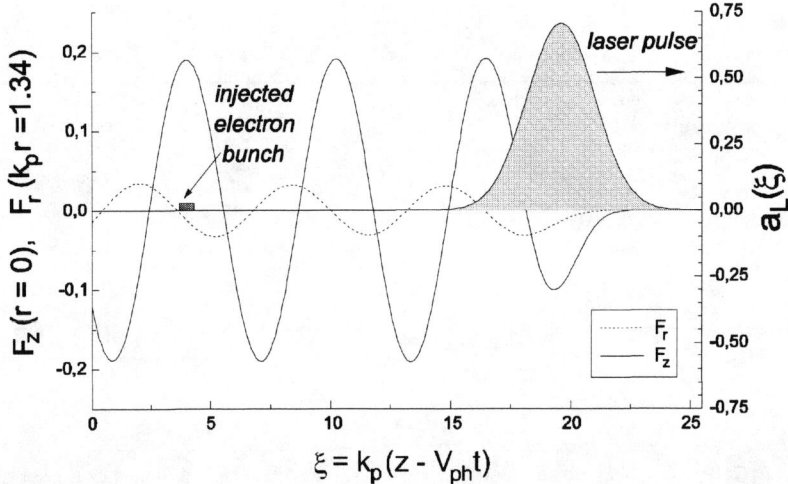

**FIGURE 1.** Accelerating $F_z$ and focusing $F_r$ wakefield forces in a matched parabolic plasma channel; $CO_2$ laser $\tau_L=1$ ps, $P_L=50$ TW, $a^2=0.5$; plasma parameters $k_p r_L=3.8$, $k_p R_{ch}=14.3$, $\lambda_p=800$ μm.

Simulations of the electron bunch acceleration in the plasma wake illustrated by Fig.2 demonstrate a significant, up to 20%, energy spread to the end of the acceleration cycle when up to 1.7 GeV net energy gain is expected. More detailed dependencies of the electron beam quality upon the initial emittance and the bunch size are shown in Fig.3. The simulations demonstrate the importance of using possibly small (longitudinally and transversely) electron bunches in order to control the beam quality in the course of acceleration. Strong bunch focusing in the wakefield allows to maintain a low emittance. However, the selected initial bunch duration and a highly relativistic initial energy do not permit to achieve a low energy spread.

# 4. MONO-ENERGETIC LWFA WITH ps-TW CO₂ LASER DRIVER

Subsequently, the LWFA scheme evolved into the two-stage design where the first stage serves for bunch compression and second for monochromatic acceleration. In this section, we consider the 5 MeV injected electron bunch with the energy spread 1.5%, geometric emittance between 0.3-0.6 mm.mrad, the 30 μm (100 fs) length, and 50-100 μm radius at the entrance to the plasma channel. Laser and plasma parameters are taken similar for both stages and the same as in the simulations discussed in Section 3.

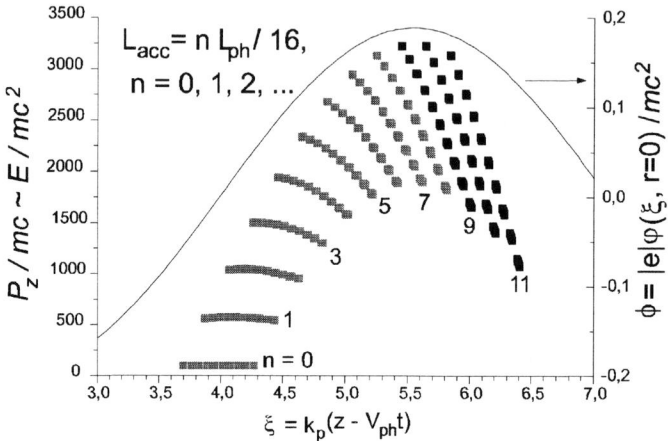

**FIGURE 2.** Wakefield acceleration of electron bunch by channel guided $CO_2$ laser pulse; $a=0.71$, $k_p r_L=3.8$, $k_p R_{ch}=14.3$, $L_{ph}=512$ cm, $\gamma_0=100$, $L_h=r_h=0.1\ \lambda_p$.

Simulations of bunch dynamics through the bunching stage are illustrated in Fig.4 and 5. We see that by choosing the proper injection phase, at the negative slope of the wake, significant compression of the bunch (up to 6 times) may be attained over the short acceleration distance. Simulations show again the importance of maintaining a small radius of the electron bunch at the entrance to the bunching stage for both, bunch compression and a low emittance.

Extracted from the bunching stage at the 40 MeV energy, the compressed to 20 fs electron bunch is injected into the accelerating stage at the moment of the maximum accelerating field at the beginning of the focusing phase. The quasi-monochromatic energy gain up to 1.5 GeV with the energy spread of <2% is observed over the rest of the accelerating phase (see Fig.6). An important result is the preservation of the normalized emittance in the course of the acceleration (see Fig.7).

The presented simulations shall be considered as preliminary, and further optimization is under way. For example, the parameters of the bunching stage are assumed similar to the acceleration stage. Just the injection phase and the channel length are different. The bunching stage may probably be optimized to consume less of the laser energy. We can also extend the acceleration stage over the entire $L_{ph}$ distance that may double the net acceleration.

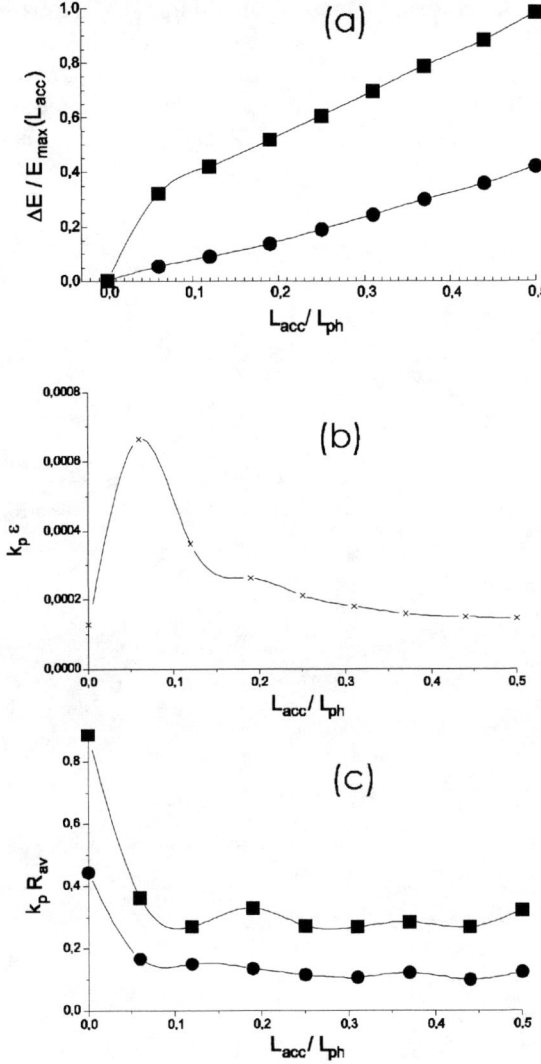

**FIGURE 3.** Quality of accelerated electron bunch in channel guided LWFA; $a$=0.71, $k_p r_L$=3.8, $k_p R_{ch}$=14.3, $\gamma_0$=100, $L_{ph}$=512 cm, (a), energy spread; (b) dynamics of emittance with the initial bunch sizes $L_b$=$r_b$=0.1 $\lambda_p$.; (c) average e-beam radius; circles- $L_b$=$r_b$=0.1 $\lambda_p$ , boxes-$L_b$=$r_b$=0.2 $\lambda_p$

Summarizing, we see that a small proportion of the electron bunch dimensions to the plasma wavelength is a prerequisite for the reasonably monochromatic acceleration. As far as the used initial electron bunch parameters represent state of the art in the conventional photocathode RF gun technology, then the chosen plasma wavelength may also be considered being close to the minimum allowed for the proposed monochromatic LWFA scheme. Under these conditions, replacement of the $CO_2$ laser with the solid state laser of the equivalent peak power results, according to Eqs.(2) and (4), in approximately two orders of magnitude reduction in the acceleration gradient. Thus the proposed monochromatic GeV LWFA can be realized only with the $CO_2$ laser driver.

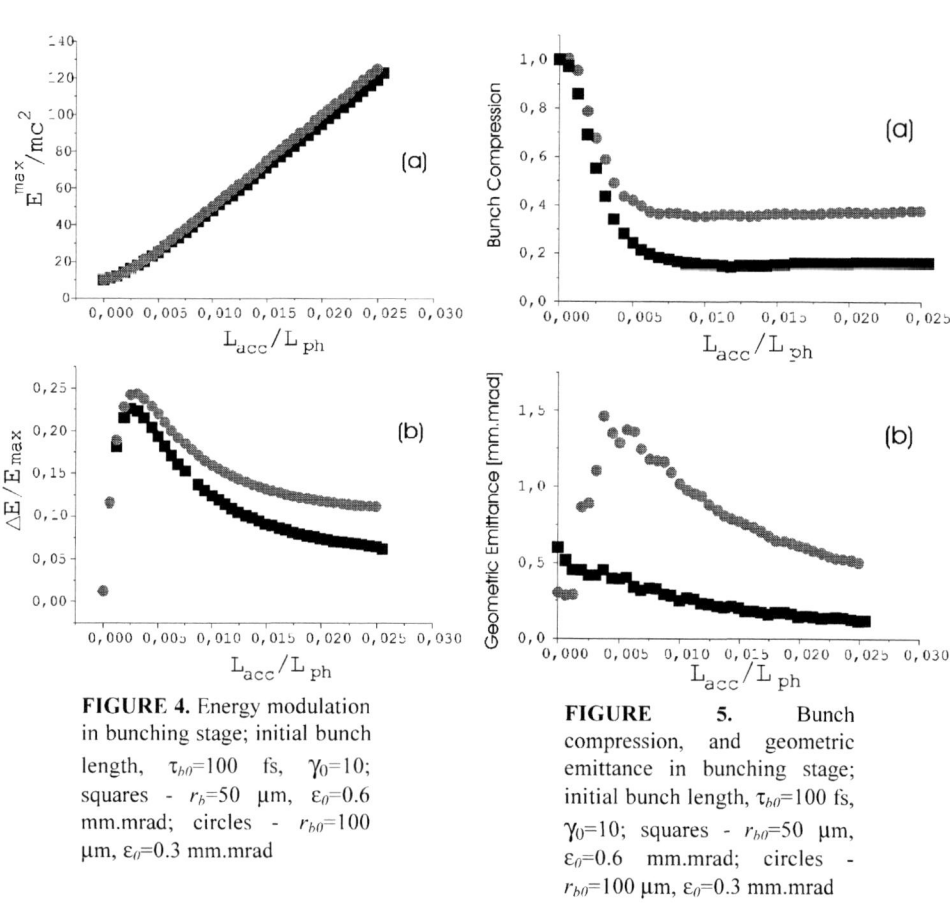

**FIGURE 4.** Energy modulation in bunching stage; initial bunch length, $\tau_{b0}$=100 fs, $\gamma_0$=10; squares - $r_b$=50 μm, $\varepsilon_0$=0.6 mm.mrad; circles - $r_{b0}$=100 μm, $\varepsilon_0$=0.3 mm.mrad

**FIGURE 5.** Bunch compression, and geometric emittance in bunching stage; initial bunch length, $\tau_{b0}$=100 fs, $\gamma_0$=10; squares - $r_{b0}$=50 μm, $\varepsilon_0$=0.6 mm.mrad; circles - $r_{b0}$=100 μm, $\varepsilon_0$=0.3 mm.mrad

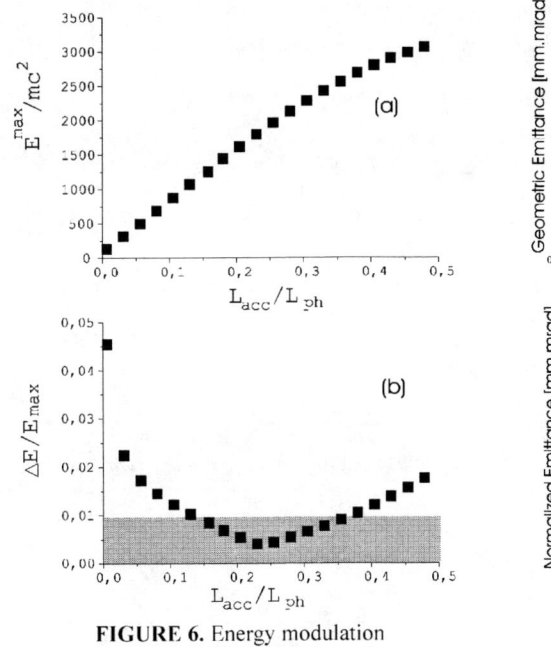

**FIGURE 6.** Energy modulation in the first acceleration stage; initial electron bunch parameters (before bunching stage): $\tau_{b0}=100$ fs; $r_{b0}=50$ $\mu$m, $\varepsilon_0=0.6$ mm.mrad

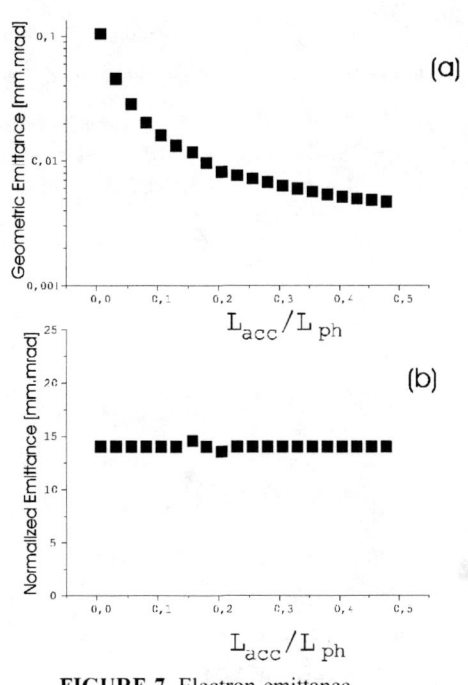

**FIGURE 7.** Electron emittance in the first acceleration stage; initial electron bunch parameters (before bunching stage): $\tau_{b0}=100$ fs; $r_{b0}=50$ $\mu$m, $\varepsilon_0=0.6$ mm.mrad

# 5. ACKNOWLEDGMENTS

The authors wish to thank I. Ben-Zvi and X.J. Wang for valuable input and discussions.

This work was supported by the U.S. Dept. Of Energy under the Contract DE-AC02-76CH00016 and by Russian Foundation for Basic Research under the Grant 98-02-16263.

# 6. REFERENCES

1.  A. Madena, Z. Najmudin, A.E. Dangor, C.E. Clayton, K.A. Marsh, C. Joshi, V. Malka, C.B. Darrow, C. Danson, D. Neely, and F.N. Walsh, *Nature*, **377**, 606-608 (1995)

2.  K. Nakajima, D. Fisher, T. Kawakubo, H. Nakanishi, A. Ogata, Y. Kato, Y. Kitagawa, R. Kodama, K. Mima, K. Shiraga, K. Suzuki, K. Yamakawa, T. Zhang, Y. Sakawa, T. Shoji, Y. Nishida, N. Yugami, M. Downer, and T. Tajima, *Phys. Rev. Lett.*, **74**, 4428-4431 (1995)

3. M. Everett, A. Lal, D. Gordon, C. Clayton, K. Marsh, C Joshi, *Nature*, **368**, 527-529 (1994)

4. D. Umstadter, J.K. Kim, and E. Dodd, *Phys. Rev. Lett.* **76**, 2073-2076 (1996)

5. E. Esarey, R.F. Hubbard, W.P. Leemans, A. Ting, and P. Sprangle, *Phys. Rev. Lett.* **79**, 2682-2685 (1997)

6. I.V.Pogorelsky, I. Ben-Zvi, J. Skaritka, Z. Segalov, M. Babzien, K. Kusche, I.K. Meshkovsky, V.A. Lekomtsev, A.A. Dublov, Yu.A. Boloshin, G.A. Baranov, 7th Workshop on Advanced Accelerator Concepts, October 12-18, 1996, Lake Tahoe, CA, *AIP Conference Proceedings* **398**, 937-950 (1997).

7. I.V. Pogorelsky, I. Ben-Zvi, M. Babzien, K. Kusche, J. Skaritka, I. Meshkovsky, A Dublov, V. Lekomtsev, I. Pavlishin, and A. Tsunemi, "The first picosecond terawatt $CO_2$ laser at the Brookhaven Accelerator test facility", *to be published in Conference Proceedings of LASERS '97*, New-Orleans, LA, December 15-19, 1997

8. I.V. Pogorelsky, I. Ben-Zvi, M. Babzien, K. Kusche, J. Skaritka, I.K. Meshkovsky, A.A. Dublov, V.A. Lekomtsev, I.V. Pavlishin, Y.A. Boloshin, G.B. Deineko and A. Tsunemi, " The first picosecond terawatt $CO_2$ laser", *to be published in Conference Proceedings of Laser Optics '98*, St. Petersburg, June 22-26, 1998

9. X.J. Wang, X. Qiu, and I. Ben-Zvi, *Phys. Rev.* **E54**, R3121-R3124 (1996)

10. T. Tajima & J.M. Dawson, *Phys. Rev. Lett.* **43**, 267-270 (1979)

11. E. Esarey, P. Sprangle, J. Krall, and A. Ting, *IEEE Trans. on Plasma Sci.* **24**, 252-288 (1996)

12. W. Leemans, C.W. Siders, E. Esarey, N. Andreev, G. Shvets, and W.B. Mori, *IEEE Trans. on Plasma Sci.* **24**, 331-342 (1996)

13. I.V. Pogorelsky, "Terawatt picosecond $CO_2$ laser technology for high energy physics applications", *these Proceedings*

14. I.V. Pogorelsky, "Optimization of laser wakefield accelerator parameters", *to be published in Conference Proceedings of LASERS '97*, New-Orleans, LA December 15-19, 1997

15. N.E. Andreev, L.M. Gorbunov, V.I. Kirsanov, K. Nakajima, and A. Ogata, "Structure of the wakefield in plasma channels", *Phys. Plasmas* **4** 1145-1153 (1997);

16. N.E. Andreev, A.A. Frolov, S.V. Kuznetsov, E.V. Chizhonkov, and L.M. Gorbunov, "The laser wakefield electron acceleration in homogeneous plasma and plasma channels", *Proceedings of LASERS'97*, New-Orleans, LA, December 15-19, 1997 (to be published)

17. Y. Ehrlich, C. Cohen, A. Zigler, J. Krall, P. Sprangle, and E. Esarey, *Phys. Rev. Lett.* **77**, 4186-4189 (1996)

452

# Multimode Analysis of the Hollow Plasma Channel Accelerator

C. B. Schroeder[1], J. S. Wurtele[1], D. H. Whittum[2]

[1]Department of Physics, University of California at Berkeley,
Berkeley, California 94720
[2]Stanford Linear Accelerator Center, Stanford University,
Stanford, California 94309

**Abstract.** The hollow plasma channel is analyzed as an accelerating structure. The loss factors and mode frequencies are calculated for all azimuthal modes. Implications for plasma-based accelerator design are discussed, including beam loading, beam breakup, and tuning tolerances. The asymptotic growth of the transverse beam displacement is computed. Methods for beam breakup reduction are considered.

## INTRODUCTION

The application of plasma-based accelerators to high-energy physics has received much attention (1-3). Yet their characterization as accelerators is largely incomplete. The hollow plasma channel appears to be the most promising plasma-based structure, decoupling the transverse profile of the driver and the transverse profile of the accelerating mode. Such a structure has been studied for laser wake field acceleration (4), bringing with it the advantages of optical guiding and extended interaction length. In this work, we characterize an externally formed hollow plasma channel as an accelerating structure independent of the structure excitation mechanism (beam or laser driven). Our results allow us to set down the basic scalings for the plasma channel accelerator, including current limiting higher order mode couplings.

## CHANNEL MODES

We consider an equilibrium plasma density $n_e = nH(r - b)$, where $H$ is the step function and $b$ is the channel radius. We assume that the driver pulse length is much shorter than the response time of the ions, which are therefore motionless. The mode structure of this hollow plasma-enclosed channel can be derived

CP472, *Advanced Accelerator Concepts: Eighth Workshop,*
edited by W. Lawson, C. Bellamy, and D. Brosius

from Maxwell's equations and the linearized fluid equations for a cold collisionless plasma. Considering only transverse surface modes ($\nabla \cdot \vec{E} = 0$) the equation for the plasma wave electric field after the driver is

$$\left[ c^2 \nabla^2 - \frac{\partial^2}{\partial t^2} - \omega_p^2 \right] \vec{E} = 0. \tag{1}$$

We assume the driver is nonevolving and propagating axially at near the speed of light, $c$, such that we may make the "frozen-field" approximation. Axial variation at a fixed position is small and the modes are functions of the driver coordinate $\tau = t - z/c$. The fields are further decomposed into discrete modes by a Fourier transform in azimuth $\theta$ with mode index $m$. With these assumptions we look for solutions of Eq. (1) of the form $\exp[-i\omega\tau + im\theta]$ with the appropriate boundary conditions across the channel wall: continuity of the electric and magnetic field components $\epsilon \vec{E} \cdot \hat{r}$, $\vec{E} \times \hat{r}$, and $\vec{B}$. The dielectric function is $\epsilon = 1$ in the channel and $\epsilon = 1 - \omega_p^2/\omega^2$ in the plasma, where $\omega_p^2 = 4\pi n_e e^2/m_e$ is the electron plasma frequency, with $m_e$ the electron rest mass and $-e$ the electron charge.

## Fundamental Mode

In the channel ($\omega_p = 0$) the field components of the fundamental mode ($m = 0$) solutions are

$$
\begin{aligned}
E_z &= G e^{i\Psi} \\
E_r &= B_\theta = \frac{1}{2} G \Omega_0 x i e^{i\Psi},
\end{aligned}
\tag{2}
$$

where the gradient $G$ is a constant determined by the excitation mechanism, $x = \omega_p r/c$ is the normalized radial position, $\Omega_0 = \omega_0/\omega_p$ is the normalized fundamental mode frequency, and $\Psi = \Omega_0 \omega_p \tau + \varphi$ is the wake phase with $\varphi$ a constant phase factor. In the plasma ($r > b$) the fundamental mode field components are

$$
\begin{aligned}
E_z &= G \frac{K_0(x)}{K_0(B)} e^{i\Psi} \\
E_r &= -G \Omega_0 \frac{K_1(x)}{K_0(B)} i e^{i\Psi} \\
B_\theta &= (1 - \Omega_0^{-2}) E_r,
\end{aligned}
\tag{3}
$$

where $B = \omega_p b/c$ is the normalized channel radius (fiber parameter), and $K_m$ are $m^{\text{th}}$-order modified Bessel functions of the second kind. The fundamental mode frequency is

$$\omega_0 = \omega_p \Omega_0 = \omega_p \left[ 1 + \frac{B K_0(B)}{2 K_1(B)} \right]^{-\frac{1}{2}}. \tag{4}$$

454

The excited fundamental mode fields in the hollow plasma channel structure, Eq. (2), are particularly favorable for the propagation of high quality electron beams. In the channel, the accelerating field is uniform across the channel, so that particles gain energy at the same rate for all radii, thereby minimizing their energy spread. The focusing force due to the fundamental mode fields in the channel is linear with respect to transverse position, therefore the root-mean-squared transverse normalized emittance will be conserved in such a structure.

## Energetics

The invariant energy density, averaged over wake phase $\Psi$, stored in the excited fields is

$$u_{field} = \frac{1}{16\pi} \left[ (E_r - B_\theta)^2 + (E_\theta + B_r)^2 + E_z^2 + B_z^2 \right],$$ (5)

and that stored in the plasma fluid motion is

$$u_{fluid} = \frac{1}{16\pi} \Omega^{-2} \left[ E_r^2 + E_\theta^2 + E_z^2 \right].$$ (6)

Performing the integral over the transverse coordinates of the field and fluid energy densities of the fundamental mode fields yields the total energy per unit length stored in the structure due to the excitation of the fundamental mode fields,

$$U = \int_0^\infty d^2\vec{r}_\perp (u_{field} + u_{fluid}) = \frac{G^2 c^2}{\omega_p^2} \left[ \Omega_0^{-2} \frac{B K_1(B)}{4 K_0(B)} \right].$$ (7)

This is a lower bound on the amount of energy per unit length that must be deposited in such an accelerator to produce a specific gradient $G$.

## Loss Factor

As a beam travels through the structure, it will interact with the accelerator environment. This interaction can be quantified by the loss factor. The loss factor per unit length $\kappa$ determines the energy loss per unit length into an excited mode $\Delta\varepsilon$ of a short bunch of charge $q$ passing through an unpowered structure by the relation $\Delta\varepsilon = \kappa q^2$, and relates the accelerating gradient to the energy stored per unit length in the structure by $\kappa = G^2/4U$. It is a purely geometrical factor of the structure independent of excitation mechanism. The loss factor $\kappa$ is related to the more familiar quantity $[R/Q]$ by $\kappa = \omega[R/Q]/4$. Using Eq. (7), the loss factor per unit length for the fundamental mode is

$$\kappa_0 = \frac{\omega_p^2}{c^2} \Omega_0^2 \left[ \frac{K_0(B)}{B K_1(B)} \right].$$ (8)

455

This loss factor is somewhat lower than conventional resonantly excited conducting structures (5) which implies lighter beam loading and higher stored energy per unit length for a given gradient.

To appreciate the implications of Eq. (8) to accelerator design, consider an example. For a fundamental mode wavelength $\lambda_0 = \Omega_0^{-1} 2\pi c/\omega_p \approx 100$ $\mu$m and channel radius $b \approx 50$ $\mu$m, the fiber parameter is $B \approx 6.2$ and the loss factor is $\kappa_0 \approx 5.3$ MV/(pC m). If a 10 GV/m gradient is desired, the energy stored in the structure is $U = G^2/4\kappa_0 \approx 5$ J/m. If laser excitation is employed and the laser pulse energy is fed once per meter one can see that the pulse energy must be in excess of 5 J, accounting for launching losses and coupling to the accelerating mode. If an energy spread of order 0.1% is required then the beam-induced voltage should be held to $2\kappa_0 q \approx 20$ MV/m and thus the single-bunch charge is limited to 4 pC or $2 \times 10^7$ particles.

# Higher Order Modes

If higher order moments of the driver exist, due to driver shape or misalignment, they will excite higher order modes in addition to the fundamental (accelerating) mode. These higher order modes can cause beam breakup instabilities, which will provide another limit on beam current. The higher order modes can be found by solving Eq. (1) for $(m \geq 1)$. This calculation yields the higher order mode frequencies,

$$\omega_m = \omega_p \Omega_m = \omega_p \left[ 2 + \frac{BK_m(B)}{(m+1)K_{m+1}(B)} \right]^{-\frac{1}{2}}. \tag{9}$$

As with the fundamental mode, the loss factor per unit length for higher order modes can be calculated by integrating the field and fluid energy densities, Eqs. (5) and (6), for the excited mode fields. The loss factor per unit length for mode $m$ has the form

$$\kappa_m = \frac{\omega_p^2}{c^2} \left[ \frac{K_m(B)}{BK_{m+1}(B)} \right] \left[ 1 + \frac{BK_m(B)}{2(m+1)K_{m+1}(B)} \right]^{-1}, \tag{10}$$

with the axial electric field evaluated at the channel radius. The loss factors, Eq. (10), and mode frequencies, Eq. (9), for all modes constitute the main results of this work.

The energy deposited in the plasma structure by a short ultra-relativistic bunch of charge $q$ at a radius $a$ is $\Delta \varepsilon_{total} = \sum_m \Delta \varepsilon_m = \sum_m \kappa_m (a/b)^{2m} q^2$. This is the total energy loss in the sense that there are no higher synchronous modes other than those calculated. This can be understood by noting that the relation $\beta^2 \epsilon < 1$ will always be satisfied for the plasma-vacuum structure since $\epsilon = 1 - \omega_p^2(r)/\omega^2 \leq 1$. Therefore no energy will be lost through (Cherenkov) radiation, and the loss factors calculated Eq. (10) provide a complete description of the energetics of the hollow plasma channel.

456

# WAKE FIELDS

With the loss factors and mode frequencies calculated, the wakefields are determined. The wakefields $\vec{W} = \vec{E} + \hat{z} \times \vec{B}$ excited in the channel by a ultra-relativistic point charge $q$, passing through the channel at radius $a$ (with $a < b$) and azimuthal angle $\theta = 0$, are given by

$$
\begin{aligned}
\vec{W}_{\parallel} &= -q\hat{W}_{\parallel m}(\tau) r^m a^m \cos(m\theta)\hat{z} \\
\vec{W}_{\perp} &= q\hat{W}_{\perp m}(\tau) r^{m-1} a^m [\hat{r}\cos(m\theta) - \hat{\theta}\sin(m\theta)],
\end{aligned}
\tag{11}
$$

where the wake functions $\hat{W}_{\parallel m}$ and $\hat{W}_{\perp m}$ are

$$
\begin{aligned}
\hat{W}_{\parallel m}(\tau) &= \frac{2\kappa_m}{b^{2m}}\cos[\Omega_m \omega_p \tau] \\
\hat{W}_{\perp m}(\tau) &= \frac{2m\kappa_m c}{b^{2m}\Omega_m \omega_p}\sin[\Omega_m \omega_p \tau].
\end{aligned}
\tag{12}
$$

These wake functions can be used as Green functions to compute the longitudinal and transverse wakefield response for an arbitrary charge distribution through a convolution of the distribution with the point charge response.

## Single-Bunch Beam Breakup

With the wakefields excited in the plasma accelerator known, the wakefield induced transverse instabilities can be analyzed. From the Lorentz force equation the evolution of the transverse displacement of a beam $X(z, \tau)$ is given by,

$$
\left[\frac{\partial}{\partial z}\gamma(z)\frac{\partial}{\partial z} + \gamma(z)k_\beta^2(z)\right] X(z, \tau) = \int_0^\tau c\,d\tau' \frac{I(\tau')}{I_o}\hat{W}_{\perp 1}(\tau - \tau')X(z, \tau'),
\tag{13}
$$

where $I(\tau)$ is the beam current, $I_o = mc^3/e \approx 17$ kA is the Alfvén constant, $k_\beta$ is the betatron wavenumber due to transverse focusing in the channel, and $\hat{W}_{\perp 1}$ is the transverse dipole ($m = 1$) wake function. Eq. (13) can be used to study the single-bunch beam breakup instability (6).

We consider the case of a bunch much shorter than the natural periods of the wakefield. The asymptotic growth of the beam displacement can be analyzed following the method in Ref. (7). Assuming weak focusing, the asymptotic growth of the transverse beam displacement of a short bunch is $X(z, \tau)/X_0 \approx (\gamma_0/\gamma)^{1/4}(8\pi A)^{-1/2}\exp[A]$, where $X_0$ is the initial transverse displacement of the beam and $\gamma_0$ is the initial injection beam energy. The exponent has the form

$$
A = 2^{7/4}\left(\frac{I}{I_o}\frac{\kappa_1(\omega_p \tau)^2}{q^2 B^2}\right)^{1/4}\left[\gamma^{1/2} - \gamma_0^{1/2}\right]^{1/2}.
\tag{14}
$$

Here we have assumed the beam energy is $\gamma = \gamma_0 + gz$, where $g$ is the gradient in beam energy and is assumed constant. Asymptotically, $A \to (z/L_g)^{1/4}$, with the instability growth length,

$$L_g = 2^{-7} \frac{I_o}{I} \frac{gB^2}{\kappa_1(\omega_p\tau)^2}. \tag{15}$$

For illustration, consider a numerical example. The instability growth length, Eq. (15), for a 3 fs beam with a charge of 1 pC travelling through a plasma channel with plasma wavelength of 50 $\mu$m, channel radius of 50 $\mu$m, and accelerating gradient of 10 GV/m is $L_g \approx 0.02$ m.

## Beam Breakup with External Focusing

For high-energy applications one would prefer not to operate in the weak focusing regime, $k_\beta L_g \ll 1$; yet the intrinsic focusing in the plasma channel is typically weak. Using Eq. (2), the transverse force from the fundamental mode fields will produce a betatron wavenumber which scales as $k_\beta \propto \gamma^{-3/2}$ to leading order in beam energy. For example, using the structure parameters above, a 10 GeV beam will have a betatron wavenumber due to the fundamental mode fields of $k_\beta \approx 6 \times 10^{-3}$ m$^{-1}$, and intrinsic focusing will be ineffective in controlling beam breakup. This is a consequence of our dual assumptions of an ultra-relativistic driver and a hollow channel. A laser driver propagating at $v \sim v_g$ would result in increased focusing and a betatron wavenumber which scales as $k_\beta \propto \gamma^{-1/2}\gamma_g^{-1} \sim \gamma^{-1/2}(\omega_p/\omega)$.

In contrast, if external focusing is applied in the plasma structure, the asymptotic growth of the transverse beam displacement is much reduced. Assuming the external focusing has a dependence on beam energy such that $k_\beta \propto \gamma^{-\alpha}$, the asymptotic form of the transverse beam displacement of a short bunch is

$$\frac{X(z,\tau)}{X_0} \approx \frac{3^{1/4}}{2^{3/2}\pi^{1/2}} \left(\frac{\gamma_0}{\gamma}\right)^{(1-\alpha)/2} \frac{\exp[A_e]}{A_e^{1/2}} \cos\left[\theta - \frac{A_e}{3^{1/2}} + \frac{\pi}{12}\right], \tag{16}$$

with the betatron phase $\theta = (g^{-1}\gamma_0^\alpha k_0/(1-\alpha))[\gamma^{1-\alpha} - \gamma_0^{1-\alpha}]$ and exponent

$$A_e = \frac{3^{3/2}}{2^{5/3}} \left(\frac{I}{I_o} \frac{\kappa_1(\omega_p\tau)^2}{\alpha g \gamma_0^\alpha k_0 B^2} [\gamma^\alpha - \gamma_0^\alpha]\right)^{1/3}, \tag{17}$$

where $k_0$ is the initial betatron wavenumber at injection. Asymptotically, $A_e \to (z/L_e)^{\alpha/3}$, with the instability growth length,

$$L_e = \frac{2^{5/\alpha}}{3^{9/2\alpha}} \left(\frac{I_o}{I}\right)^{1/\alpha} \left(\frac{\alpha g^{1-\alpha}\gamma_0^\alpha k_0 B^2}{\kappa_1(\omega_p\tau)^2}\right)^{1/\alpha}. \tag{18}$$

For example, for $\alpha = 1/2$, the growth rate scales as $L_e \propto (I/I_o)^{-2}(\omega_p\tau)^{-4}$, a more favorable scaling than Eq. (15).

# TOLERANCES

The promise of the hollow plasma channel examined in this work depends on the ability to form such a structure. Channel creation is currently being explored experimentally (8). Possible methods of forming a hollow plasma channel include: ionization of a preformed capillary tube, or inverse bremsstrahlung heating of the plasma by a pre-cursor laser pulse resulting in hydrodynamic expansion and channel formation (9). Errors in the channel creation will produce errors in the mode frequencies and loss factors.

Eq. (9) implies that any fractional error in the plasma frequency will produce the same fractional error in the mode frequency. Errors in the channel radius will also produce errors in the mode frequency. The ratio of the fractional error in frequency to the fractional error in channel radius (assuming no fluctuations in plasma frequency) is

$$\left[\frac{\delta\omega_m}{\omega_m}\right]\left[\frac{\delta B}{B}\right]^{-1} = \frac{B^2 K_{m+1}(B)}{4(m+1)K_{m+1}(B) + 2BK_m(B)} - \frac{BK_m(B)}{2K_{m+1}(B)}. \tag{19}$$

Operating at a smaller channel radius will yield greater tolerances of frequency to errors in channel radius.

If a finite plasma density exists inside the channel $n_{in}$, then the shift in mode frequencies $\Delta\omega_m = \omega_m(n_{in} > 0) - \omega_m(n_{in} = 0)$ to lowest order in the ratio of plasma density inside the channel to plasma density outside the channel is

$$\frac{\Delta\omega_m}{\omega_m} = \frac{n_{in}}{n}\left[\frac{(1 - \delta_{m0})(m+1)K_{m+1}(B) + BK_m(B)}{2(1 + \delta_{m0})(m+1)K_{m+1}(B)}\right], \tag{20}$$

where $\delta_{m0} = 1$ for $m = 0$ and zero otherwise. An electrostatic mode will also be excited by the driver in a partially filled channel. The ratio of energy deposited into the elecrostatic mode to the ratio of energy deposited into the electromagnetic modes will be of the order $\sim n_{in}/n$.

# DISCUSSION

In summary, we have characterized the hollow plasma channel in terms of the fundamental accelerator parameters: mode frequencies and loss factors. The monopole and dipole results provide the limits due to beam loading and single-bunch beam breakup. With these results one can consider for the first time designing a high-energy machine based on this structure.

For example, such a machine requires high charge for high luminosity. Beam loading and beam breakup constrain the charge in a single bunch, and thus one is interested in multiple bunches. For multibunch operation, control of beam breakup requires stagger-tuning (10.11). Our results permit such a stagger-tuned design for the plasma channel accelerator. The mode frequencies are functions of two independent design parameters: the channel radius and plasma frequency. Therefore these two parameters can be varied such that the higher order mode frequencies vary over the length of the accelerator while maintaining a constant fundamental (accelerating) mode frequency.

One problem not addressed in this work is the quality factor Q of the plasma structure. In our model this figure is infinite since we have neglected collisions, nonlinearities, and finite wall thickness. Linear analysis including finite wall effects (12) and numerical simulations (13) suggests that the Q may not be large. Systematic studies of Q, and design of a realistic channel with high Q is in our opinion the foremost problem for plasma-based accelerators.

## ACKNOWLEDGMENTS

The authors acknowledge useful conversations with B. A. Shadwick. This work was supported by the U. S. Department of Energy Division of High Energy and Nuclear Physics contract No. PDDEFG-03-95ER-40936. Work at SLAC was supported by the U. S. Department of Energy contract No. AC03-76SF-00515.

## REFERENCES

1. Tajima, T. and Dawson, J., *Phys. Rev. Lett.* **43**, 267 (1979).
2. Wurtele, J. S., *Phys. Fluids B* **5**, 2363 (1993).
3. Esarey, E., Sprangle, P., Krall, J., and Ting, A., *IEEE Trans. Plasma Sci.* **24**, 252 (1996).
4. Chiou, T. C., Katsouleas, T., Decker, C., Mori, W. B., Wurtele, J. S., Shvets, G., and Su, J. J., *Phys. Plasmas* **2**, 310 (1995).
5. Wilson, P., *Laser Acceleration of Paricles*, edited by C. Joshi and T. Katsouleas, AIP Conf. Proc. 130 (AIP, New York, 1985), p. 560.
6. Lau, Y. Y., *Phys. Rev. Lett.* **63**, 1141 (1989).
7. Whittum, D. H., Lampe, M., Joyce, G., Slinker, S. P., Yu, S. S., and Sharp, W. M., *Phys. Rev. A* **46**, 6684 (1992).
8. Durfee, III, C. G., Lynch, J., and Milchberg, H. M., *Phys. Rev. E* **51** 2368 (1995).
9. Volfbeyn, P. (private communication).
10. Colombant, D. G. and Lau, Y. Y., *Appl. Phys. Lett.* **55**, 27 (1989).
11. Kroll, N. M. *et al.*, *Advanced Accelerator Concepts*, edited by S. Chattopadhyay, J. McCullough and P. Dahl (AIP, New York, 1997), p. 455.
12. Shvets, G., Wurtele, J. S., Chiou, T. C., and Katsouleas, T., *IEEE Trans. Plasma Sci.* **24**, 351 (1996).
13. Shadwick B. A., these proceedings.

# Generation of Ultrashort Electron Bunches by Colliding Laser Pulses

C. B. Schroeder[1], P. B. Lee[1], J. S. Wurtele[1], E. Esarey[2], W. P. Leemans[3]

[1]Department of Physics, University of California at Berkeley, Berkeley, California 94720
[2]Beam Physics Branch, Plasma Physics Division, Navel Research Laboratory, Washington D.C. 20375
[3]Center for Beam Physics, Ernest Orlando Lawrence Berkeley National Laboratory, Berkeley, California, 94720

**Abstract.** A proposed laser-plasma based relativistic electron source [E. Esarey *et al.*, Phys. Rev. Lett. **79**, 2682 (1997)] using laser triggered injection of electrons is investigated. The source generates ultrashort electron bunches by dephasing and trapping background plasma electrons undergoing fluid oscillations in an excited plasma wake. The plasma electrons are dephased by colliding two counter-propagating laser pulses which generate a slow phase velocity beat wave. Laser pulse intensity thresholds for trapping and the optimal wake phase for injection are calculated. Numerical simulations of test particles, with prescribed plasma and laser fields, are used to verify analytic predictions and to study the longitudinal and transverse dynamics of the trapped plasma electrons. Simulations indicate that the colliding laser pulse injection scheme has the capability to produce relativistic femtosecond electron bunches with fractional energy spread of order a few percent and normalized transverse emittance less than 1 mm mrad using 1 TW injection laser pulses.

## INTRODUCTION

The characteristic scale length of the accelerating field (plasma wake) in a plasma-based accelerator (1) is the plasma wavelength $\lambda_p[\text{m}] = 2\pi c/\omega_p \simeq 3.3 \times 10^4 n_e^{-1/2}[\text{cm}^{-3}]$ where $n_e$ is the plasma density, $\omega_p$ is the plasma frequency, and $c$ is the speed of light. In such short wavelength accelerators (typically $\lambda_p < 100~\mu\text{m}$), production of electron beams with low momentum spread and good pulse-to-pulse energy stability requires femtosecond electron bunches to be injected with femtosecond synchronization with respect to the plasma wake. Although conventional electron sources (photocathode or thermionic RF guns) have achieved sub-picosecond

CP472, *Advanced Accelerator Concepts: Eighth Workshop,*
edited by W. Lawson, C. Bellamy, and D. Brosius
© 1999 The American Institute of Physics 1-56396-889-4/99/$15.00

electron bunches (2), the requirements for injection into plasma-based accelerators are presently beyond the performance of these conventional electron sources. Novel schemes which rely on laser triggered injection of plasma electrons into their own plasma wake have been proposed (3) to generate the required femtosecond electron bunches.

Recently a new optical injection scheme was proposed (4) which uses two relatively low intensity laser pulses in addition to a pump laser pulse for plasma wake excitation. In this scheme, the pump laser pulse generates a plasma wake through its pondermotive force, as in the standard laser wakefield accelerator (1). The two low intensity laser pulses, one pulse propagating in the forward direction behind the pump laser pulse and the other pulse counter-propagating to the pump laser pulse, collide at a predetermined phase of the plasma wake. The beating of these injection laser pulses provides a pondermotive force that dephases and boosts background plasma electrons undergoing fluid oscillations. Under appropriate conditions, described in this paper, some of the background plasma electrons will attain sufficient momentum and phase shift to be trapped by the plasma wake.

This colliding pulse scheme has the capability to produce femtosecond electron bunches with low fractional energy spreads using relatively low injection laser pulse intensities compared to the pump laser pulse $a^2_{inj} \ll a^2_{pump} \sim 1$ where $a = eA/mc^2 \simeq 8.5 \times 10^{-10}\lambda[\mu m]I^{1/2}[\text{W/cm}^2]$ is the normalized vector potential, $I$ is the laser pulse intensity and $\lambda$ is the laser wavelength. The colliding pulse concept also offers detailed control of the injection process. The injection phase is determined by the relative timing between the forward propagating injection laser pulse and the pump laser pulse. The beat wave velocity is adjusted by varying the frequency detuning between the injection laser pulses, and the number of trapped electrons can be controlled by the injection laser pulse intensities.

## PHASE SPACE ANALYSIS

The colliding laser pulse optical injection scheme employs three short laser pulses as shown in Fig. (1). An intense ($a^2_0 \sim 1$) laser pulse (denoted by subscript 0) for plasma wake generation, a forward propagating injection laser pulse (subscript 1), and a backward propagating injection laser pulse (subscript 2). The pump laser pulse generates a plasma wake with phase velocity near the speed of light $v_\varphi \simeq c$. The injection laser pulses collide some distance behind the pump laser pulse. When the injection laser pulses collide, they generate a beat wave with a phase velocity $v_b = \Delta\omega/\Delta k \simeq \Delta\omega/2k_0$, where the frequency difference of the injection laser pulses is $\Delta\omega = \omega_1 - \omega_2$ and the wavenumber difference is $\Delta k = k_1 - k_2$ with $k_1 \simeq |k_2| \simeq k_0$. During the time when the two injection laser pulses overlap, the slow beat wave injects plasma electrons into the fast plasma wake for acceleration to high energies.

The evolution in longitudinal phase space of a uniform distribution of electrons as they interact with the plasma wake and the beating injection laser pulses is shown in Fig. (2). This figure was generated using a particle transport code

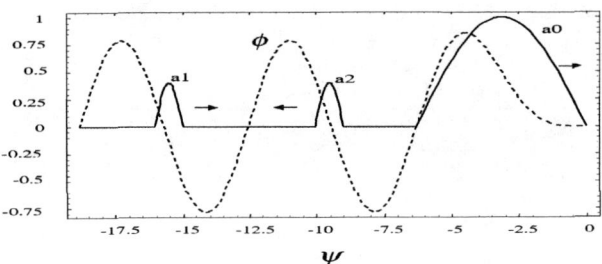

**FIGURE 1.** Normalized potential profiles of the pump laser pulse $a_0$, the plasma wake $\phi$, forward injection laser pulse $a_1$, and the backward injection laser pulse $a_2$.

described in the next section. Also shown is the separatrix (dotted line) between the trapped and untrapped orbits of the plasma wake Hamiltonian. Fig. (2) shows the electron distribution (a) before the collision of the injection laser pulses (in an untrapped orbit of the plasma wake), (b) during the collision of the injection laser pulses (crossing the wake separatrix), (c) just after the collision, and (d) the resulting energetic electron bunch.

The colliding laser pulse injection mechanism can be studied using a Hamiltonian approach. The electron motion in the plasma wake is described by the Hamiltonian

$$H(u_z, \psi) = \left[1 + u_z^2\right]^{1/2} - \beta_\varphi u_z - \phi(\psi), \tag{1}$$

where $u_z m_e c$ is the electron momentum and $\beta_\varphi = (1 - \omega_p^2/\omega_0^2)^{1/2}$ is the plasma wake phase velocity (group velocity of the pump laser pulse). Here $\omega_p = k_p c = (4\pi e^2 n_e/m_e)^{1/2}$ is the plasma frequency with $-e$ the electron charge and $m_e$ the electron mass. The scalar potential of the plasma wake is assumed to have the form $\phi(\psi) = \phi_o \cos \psi$ where the wake phase is $\psi = k_p z - \omega_p t$ and the normalized wake potential amplitude is $\phi_o = e\Phi_o/m_e c^2$. The amplitude of the wake potential is determined by the pump laser pulse amplitude and shape. The normalized axial momentum of the electron in an orbit $H$ of the plasma wake is

$$u_z = \beta_\varphi \gamma_\varphi^2 [H + \phi(\psi)] \pm \gamma_\varphi \left(\gamma_\varphi^2 [H + \phi(\psi)]^2 - 1\right)^{1/2}, \tag{2}$$

where $\gamma_\varphi = (1 - \beta_\varphi^2)^{-1/2} = \omega_0/\omega_p$. The boundary between trapped and untrapped orbits is given by the separatrix orbit $H = H(u_z = \gamma_\varphi \beta_\varphi, \psi = \pi) = \gamma_\varphi^{-1} + \phi_o$. Assuming the plasma is initially cold, the background electron fluid motion in the plasma wake is given by the orbit $H = 1$.

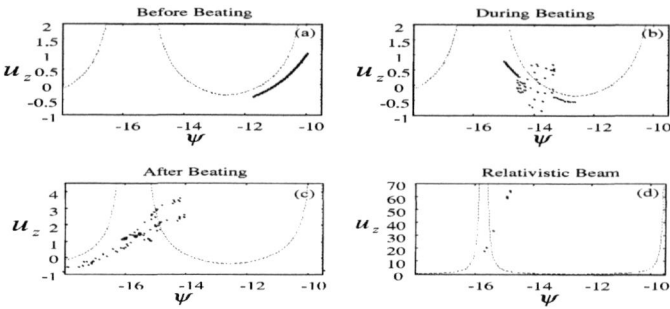

**FIGURE 2.** Electron distribution in longitudinal phase space $(\psi, u_z)$ (a) before the collision of injection laser pulses (in untrapped orbit of the plasma wake), (b) during the collision of injection laser pulses, (c) just after the collision, and (d) much after the collision (an energetic electron beam). The separatrix between trapped and untrapped plasma wake orbits (dotted line) is shown.

The colliding injection laser pulses lead to formation of a beat wave with phase space buckets (separatrices) of width $2\pi/\Delta k \simeq \lambda_0/2$ (much shorter than those of the wake field $\lambda_p$). The motion of the electron in the beat wave is described by the beat wave Hamiltonian

$$H_b(u_z, \psi_b) = \left[\gamma_\perp^2(\psi_b) + u_z^2\right]^{1/2} - \beta_b u_z, \tag{3}$$

where $\beta_b = \Delta\omega/c\Delta k \simeq (\lambda_2 - \lambda_1)/(\lambda_2 + \lambda_1)$ is the beat wave phase velocity, $\psi_b = \Delta k(z - \beta_b ct)$ is the beat wave phase, and $\gamma_\perp^2(\psi_b) = 1 + \hat{a}_1^2 + \hat{a}_2^2 + 2\hat{a}_1\hat{a}_2 \cos\psi_b$ with $\hat{a}_1$ and $\hat{a}_2$ the slowly varying (compared with the phase oscillation) amplitudes of the forward and backward injection laser pulses respectively. The separatrix orbit in phase space of the beat wave Hamiltonian has the value $H_b = H_b(u_z = \gamma_b\beta_b\gamma_\perp(0), \psi_b = 0) = \gamma_\perp(0)\gamma_b^{-1}$ where $\gamma_b = (1 - \beta_b^2)^{-1/2}$. The maximum and minimum normalized axial momentum of an electron in a beat wave orbit (extrema of the separatrix) are

$$[(u_z)_{beat}]_{max/min} = \gamma_b\beta_b\gamma_\perp(0) \pm 2\gamma_b(\hat{a}_1\hat{a}_2)^{1/2}. \tag{4}$$

As we will show, the beat wave amplitude parameter $(\hat{a}_1\hat{a}_2)^{1/2}$ is a critical parameter in the injection process. For $\hat{a}_1 = \hat{a}_2 = \hat{a}_{inj}$ the beat wave amplitude parameter $(\hat{a}_1\hat{a}_2)^{1/2} = \hat{a}_{inj} = eE_{inj}/m_e c\omega_{inj}$ is the normalized root-mean-squared (rms) electric field amplitude of the injection laser pulses.

The threshold injection laser pulse intensities required for trapping of background plasma electrons into the plasma wake can be estimated by considering the

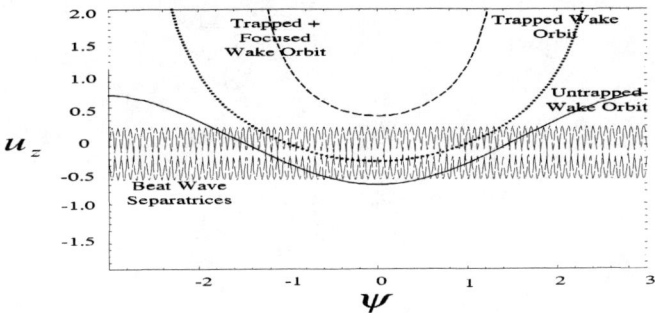

**FIGURE 3**. Phase space $(\psi, u_z)$ showing the beat wave separatrices, an untrapped plasma wake orbit (solid line), a trapped plasma waked orbit (dotted line), and a trapped and focused plasma wake orbit (dashed line).

effects of the plasma wake and the beat wave individually and requiring resonance overlap (shown in Fig. (3)), namely that the maximum momentum of the beat wave separatrix exceed the minimum momentum of the plasma wake separatrix and the minimum momentum of the beat wave separatrix be less than the fluid momentum of electrons in the plasma wake. With these requirements, the necessary conditions for trapping are

$$[(u_z)_{beat}]_{max} \geq (u_z)_{trapped} \tag{5}$$

$$[(u_z)_{beat}]_{min} \leq (u_z)_{untrapped}. \tag{6}$$

The maximum and minimum momentum of an electron in a beat wave orbit are given in Eq. (4). The momentum of an electon in an untrapped orbit of the plasma wake $(u_z)_{untrapped}$ is given by Eq. (2) with $H = 1$. The momentum of an electron in a trapped orbit of the plasma wake $(u_z)_{trapped}$ is given by Eq. (2) with $H \leq H(u_z = \gamma_\varphi \beta_\varphi, \psi = \pi) = \gamma_\varphi^{-1} + \phi_o$. For an electron in a trapped and focused orbit, $(u_z)_{trapped}$ is given by Eq. (2) with $H \leq H(u_z = \gamma_\varphi \beta_\varphi, \psi = \pi/2) = \gamma_\varphi^{-1}$. A trapped and focused electron orbit is one where the wake phase remains in a range such that the transverse electric field due to the plasma wake is always providing a focusing force on the electron.

Solving for the minimum $(\hat{a}_1 \hat{a}_2)^{1/2}$ which satisfies the conditions Eqs. (5) and (6) yields the threshold beat wave amplitude parameter for trapping plasma electrons

$$(\hat{a}_1 \hat{a}_2)_{th}^{1/2} = \frac{1 - H}{4\gamma_b(\beta_\varphi - \beta_b)}, \tag{7}$$

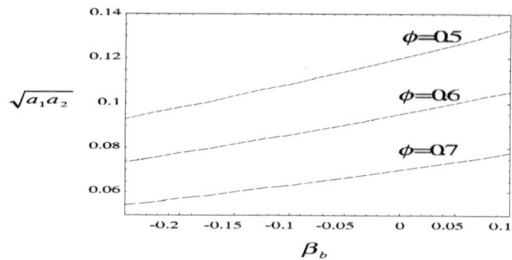

**FIGURE 4.** The threshold beat wave amplitude parameter $(\hat{a}_1\hat{a}_2)^{1/2}$ for trapping at the optimal injection phase (Eq. (7)) versus beat wave phase velocity (for $\phi_o = 0.5$, 0.6, and 0.7).

and the optimal wake phase for injection

$$\cos\psi_{opt} = \phi_o^{-1}\left[\gamma_b\left(1 - \beta_\varphi\beta_b\right)\gamma_\perp(0) - \frac{1}{2}\left(1 + H\right)\right]. \tag{8}$$

Here $H = \gamma_\varphi^{-1} + \phi_o$ for injection into a trapped plasma wake orbit and $H = \gamma_\varphi^{-1}$ for injection into a trapped and focused plasma wake orbit. In the limit $\gamma_\varphi^2 \gg 1$, $\beta_b \ll 1$ and $\hat{a}_{1,2}^2 \ll 1$, Eqs. (7) and (8) become $4(\hat{a}_1\hat{a}_2)_{th}^{1/2} \simeq (1 + \beta_b)(1 - H)$ and $2\phi_o\cos\psi_{opt} \simeq 1 - 2\beta_b - H$ with $H \simeq \phi_o$ for a trapped orbit and $H \simeq 0$ for a trapped and focused orbit.

Fig. (4) shows the threshold beat wave amplitude parameter for trapping at the optimal injection phase (Eq. (7) with $H = \gamma_\varphi^{-1} + \phi_o$) versus the beat wave phase velocity $\beta_b$ for several wake potential amplitudes. From Fig. (4) one can see that, the larger the plasma wake, the smaller the injection laser pulse intensity required for trapping plasma electrons. The threshold for injection into a trapped and focused orbit is independent of plasma wake amplitude as indicated by Eq. (7) with $H = \gamma_\varphi^{-1}$. Fig. (4) also shows the threshold slightly decreases for decreasing $\beta_b$. Note that the estimate of the trapping threshold Eq. (7) will not be valid when $|\beta_b|$ is large enough such that the bounce time in a beat wave orbit $\sim \lambda_p/c$ (i.e. when a separation of time scales no longer applies).

Minimizing the injection pulse amplitudes (operating near the threshold amplitude given by Eq. (7)) will minimize the laser power $P[GW] \simeq 21.5(a_i r_i/\lambda_i)^2$ required for trapping and is therefore important for the experimental realization of this injection scheme. For illustration, if the injection laser pulses have a wavelength of 0.8 $\mu$m and a spot size of 15 $\mu$m, then the injection laser pulse power required for trapping is $P < 1$ TW.

# NUMERICAL STUDIES

To further evaluate the colliding laser pulse scheme and to test the analytic predictions for the trapping thresholds, the motion of test particles in the combined plasma wake and laser fields was simulated by numerically solving the equations of motion for the electrons. In the numerical simulations, we assume the laser pulses are linear polarized fundamental Gaussian beams with half-period cosine longitudinal envelopes.

The polarizations of the laser pulses are chosen to be orthogonal such that $\vec{a}_0 \cdot \vec{a}_2 = 0$ and thus there is no beating (no wake generation) from the interaction of the pump laser pulse and the counter-propagating injection laser pulse. The pondermotive potential due to the beating of the injection laser pulses (averaged over the fast oscillations) is $\vec{a}_1 \cdot \vec{a}_2 = \hat{a}_1^2 + \hat{a}_2^2 + 2\hat{a}_1\hat{a}_2 \cos(\psi_b)$. The plasma wakefields produced by the injection laser pulses can be neglected ($\phi_1 \sim \phi_2 \ll \phi_0$) since the injection laser pulse amplitudes required for trapping are much less than the pump laser pulse amplitude and the pulse lengths of the injection laser pulses can be chosen to provide poor coupling between the plasma response and the injection laser pulses.

Assuming $a_0^2 < 1$, the plasma wake potential $\phi$ excited by the pondermotive force generated by the pump laser pulse (to lowest order in pump pulse amplitude) near the waist of the pump laser pulse ($z \ll Z_{R0}$) satisfies

$$\phi = \frac{a_0^2}{4} \exp\left[-2r^2/r_{s0}^2\right] \left[1 + \sin\psi + \left(\frac{3\pi}{4} - \frac{\psi}{2}\right)\cos\psi\right] \tag{9}$$

inside the pump laser pulse, and

$$\phi = \frac{\pi a_0^2}{4} \exp\left[-2r^2/r_{s0}^2\right] \cos\psi \tag{10}$$

after the pump laser pulse. Here $r_{s0}$ is the minimum spot size (waist) of the pump laser pulse and $Z_{R0} = k_0 r_{s0}^2/2$ is the Rayleigh length. The axial and radial components of the electric field due to the plasma wake potential after the pump laser pulse are

$$E_z = \frac{m_e c^2}{e} k_p \phi_0 \exp\left[-2r^2/r_{s0}^2\right] \sin\psi \tag{11}$$

$$E_r = \frac{m_e c^2}{e} \frac{4r}{r_{s0}^2} \phi_0 \exp\left[-2r^2/r_{s0}^2\right] \cos\psi, \tag{12}$$

where $\phi_o = \pi a_0^2/4$. The radial electric field will provide a focusing force for an electron at a plasma wake phase of $\cos\psi > 0$ and a defocusing force for $\cos\psi < 0$.

The equations of motion for the plasma electrons in the combined fields of the three lasers and the plasma wake were numerically integrated using an adaptive

**TABLE 1.** Simulation Parameters

| | |
|---|---|
| Plasma wavelength $\lambda_p$ | 40 $\mu$m |
| Pump laser strength $a_0$ | 0.94 |
| Plasma wake potential $\phi_o$ | 0.7 |
| Pump pulse length $L_0 = \lambda_p$ | 40 $\mu$m |
| Pump pulse wavelength $\lambda_0$ | 0.8 $\mu$m |
| Laser spot size $r_{s0} = r_{s1} = r_{s2}$ | 15 $\mu$m |
| Injection laser pulse strength $a_1 = a_2$ | 0.3 |
| Injection pulse length $L_1 = L_2 = \lambda_p/2$ | 20 $\mu$m |
| Injection pulse (forward) wavelength $\lambda_1$ | 0.83 $\mu$m |
| Injection pulse (backward) wavelength $\lambda_2$ | 0.80 $\mu$m |

stepsize Runge-Kutta method. The plasma was assumed to be initially homogeneous and cold such that the test particles were loaded uniformly with no initial momentum. Unless otherwise stated, the parameters used in the numerical simulations are listed in Table 1.

The minimum injection laser pulse amplitude for injection of plasma electrons into a trapped orbit of the plasma wake is shown in Fig. (5)(a). The solid line is the analytic estimation, Eq. (7) with $H = \gamma_\varphi^{-1} + \phi_o$ and $\beta_\varphi = -0.2$, and the points correspond to simulation results.

To determine the optimal injection wake phase which minimizes the injection laser pulse amplitude required for trapping of background plasma electrons, the fraction of loaded test electrons which become trapped as a result of the colliding injection laser pulses was examined as a function of the injection wake phase. Fig. (5)(b) shows the fraction of trapped electrons versus the injection wake phase (the plasma wake phase where the maxima of the injection laser pulses collide). The curve is peaked at $\psi_{opt} \simeq \pm 1.0$ which agrees well with the analytic predictions (Eq. (8) with $\beta_b = -0.2$ and $\phi_o = 0.7$). Significant trapping of electrons occurs for an injection wake phase region of $-1.5 < \psi_{inj} < 1.5$. This indicates that the two colliding injection laser pulses must be synchronized to the plasma wake with an accuracy of $\sim 10$ fs, which is not a serious timing constraint for present laser technology.

The quality of the electron bunch can be examined as the beat wave amplitude parameter $(\hat{a}_1 \hat{a}_2)^{1/2}$ is increased beyond the threshold value for injection into a trapped and focused orbit, Eq. (7) with $H \leq \gamma_\varphi^{-1}$. Fig. (6)(a) shows the fraction of loaded test electrons which become trapped and focused versus the beat wave amplitude parameter. The maximum value shown on Fig. (6)(a) corresponds to a bunch number of $N_b \simeq 0.5 \times 10^7$ electrons for a plasma density of $n_e = 7 \times 10^{17} \text{cm}^{-3}$. Note that the bunch number can be increased by increasing the injection laser spot size (i.e. increasing the injection laser pulse power).

The rms phase spread (bunch duration) is constant for a highly relativistic

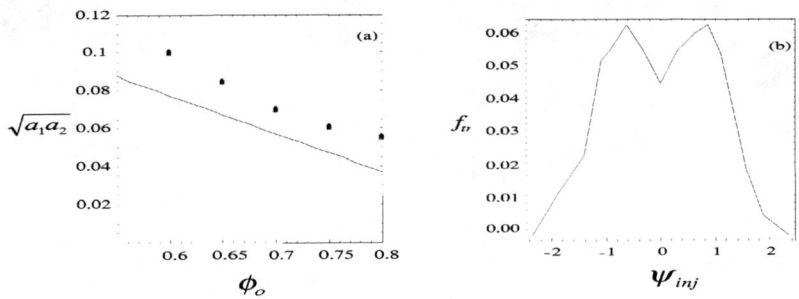

**FIGURE 5.** (a) Threshold beat wave amplitude parameter $(\hat{a}_1\hat{a}_2)^{1/2}$ versus plasma wake potential amplitude. Solid line is Eq. (7) with $\phi_o = 0.7$ and $\beta_b = -0.2$. Points are numerical simulation results. (b) Fraction of loaded test electrons which become trapped after the colliding laser pulses versus injection wake phase $\psi_{inj}$.

bunch ($\delta\psi = (\beta_z - \beta_\varphi)k_p\delta z \simeq 0$ for the interaction lengths considered in these simulations), the fractional energy spread $\sigma_\gamma/\langle\gamma\rangle$ is asymptotic, and the transverse normalized rms emittance is conserved for large pump laser spot size (to order $\mathcal{O}((k_p r_{s0})^{-2})$). Therefore we examined these three measures of bunch quality versus increasing beat wave amplitude parameter. Fig. (6)(a) shows the bunch duration of the trapped electron bunch versus the beat wave amplitude parameter. The asymptotic fractional energy spread $\sigma_\gamma/\langle\gamma\rangle$ and the transverse normalized rms emittance versus the beat wave amplitude parameter are shown in Fig. (6)(b). These figures indicate the production of $\sim 1$ fs election bunches with $< 10\%$ fractional energy spread and $\sim 1$ mm mrad normalized transverse rms emittance.

## SUMMARY

We have explored the generation of ultrashort electron bunches by using colliding laser pulses to dephase background plasma electrons undergoing fluid oscillations in a plasma wake. This scheme has the ability to produce femtosecond electron bunches with low fractional energy spread ($< 10\%$) and low normalized transverse emittance ($\sim 1$ mm mrad). The colliding pulse scheme requires relatively low laser power compared to the pump pulse $a_1^2 \sim a_2^2 \ll a_0^2$, and allows for detailed control of injection process through the injection phase (position of the forward injection laser pulse), beat wave velocity (frequencies of the injection laser pulses), and the beat wave amplitude parameter (injection laser pulse intensities). We believe these capabilities are critical for the experimental realization

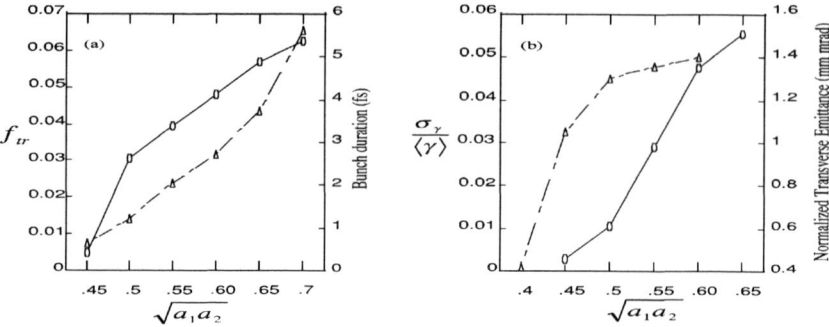

**FIGURE 6.** (a) Fraction of loaded test electrons which become trapped and focused after the colliding laser pulses (solid line) and bunch duration (fs) of trapped electron bunch (dashed line) versus beat wave amplitude parameter. (b) Asymptotic fractional energy spread $\sigma_\gamma / \langle \gamma \rangle$ (solid line) and normalized transverse rms emittance $\varepsilon_\perp$ [mm mrad] (dashed line) of trapped electron bunch versus beat wave amplitude parameter.

of laser triggered injection and subsequently compact laser-plasma based particle accelerators.

## ACKNOWLEDGMENTS

We acknowledge useful conversations with P. Volfbeyn and M. Zolotorev. The research at UCB was supported by the U. S. Department of Energy Division of High Energy and Nuclear Physics contract No. PDDEFG-03-95ER-40936. The work at NRL was supported by the U. S. Department of Energy and the Office of Naval Research, and the work at LBNL was supported by the U. S. Department of Energy under contract No. DE-AC-03-76SF0098.

## REFERENCES

1. Esarey, E., Sprangle, P., Krall, J., and Ting, A., *IEEE Trans. Plasma Sci.* **24**, 252 (1996).
2. Kung, P., Lihn, H., Wiedemann, H. and Bocek, D., *Phys. Rev. Lett.* **73**, 967 (1994).
3. Umstadter, D., Kim J. K., and Dodd E., *Phys. Rev. Lett.* **76**, 2073 (1996).
4. Esarey, E., Hubbard, R. F., Leemans, W. P., Ting, A., and Sprangle, P., *Phys. Rev. Lett.* **79**, 2682 (1997).

# Electromagnetically Induced Guiding and Superradiant Amplification of Counter-Propagating Lasers in Plasma

G. Shvets and N. J. Fisch

*Princeton Plasma Physics Laboratory, Princeton University, Princeton, NJ 08543*

A. Pukhov

*Max-Plank-Institut fur for Quantenoptik, D-85748 Garching, Germany*

**Abstract.** The interaction of counter-propagating laser pulses in a plasma is considered. When the frequencies of the two lasers are close, nonlinear modification of the refraction index results in the mutual focusing of the two beams. A short (of order the plasma period) laser pulse can be nonlinearly focused by a long counter-propagating beam which extends over the entire guiding length. It is also demonstrated that a short ($< 1/\omega_p$) laser pulse can be superradiantly amplified by a counter-propagating long low-intensity pump while remaining ultra-short. Particle-in-Cell simulations indicate that pump depletion can be as high as 40%. This implies that the long pump is efficiently compressed in time without frequency chirping and pulse stretching, making the superradiant amplification an interesting alternative to the conventional method of producing ultra-intense pulses by the chirped-pulse amplification.

## I  INTRODUCTION

One of the important challenges to laser-plasma accelerators is to propagate an intense laser pulses over a long distance. A laser will, in free space, remain focused over a Rayleigh range $Z_R = \pi\sigma^2/\lambda_0$, where $\lambda_0$ is the laser wavelength, and $\sigma$ is the laser spot size at the focus. Relativistic self-focusing, which can overcome diffraction, is ineffective for short laser pulses and requires a very high laser power $P_{c1} = 17\omega_0^2/\omega_p^2$ GW, where $\omega_p^2 = 4\pi n_0 e^2/m$ is the plasma frequency, $n_0$ is the background electron density, and $-e$ and $m$ are the electron density and mass, respectively.

In this paper we demonstrate how a second counter-propagating pulse can be used to create an "electromagnetic channel". Such electromagnetically induced guiding (EIG) occurs when the intensity of the guiding laser is well below the relativistic threshold $P_{c1}$. The guiding and guided beams interfere, forming a periodic intensity

CP472, *Advanced Accelerator Concepts: Eighth Workshop,*
edited by W. Lawson, C. Bellamy, and D. Brosius
© 1999 The American Institute of Physics 1-56396-889-4/99/$15.00

pattern which ponderomotively drives the plasma wave. The guided beam undergoes a Bragg reflection off this periodic density perturbation, scattering into the guided pulse. Basic equations for the plasma wave excitation and laser scattering are given in Section II

When laser intensities are small, plasma wave remains linear, and the complex index of mutual refraction $\chi_0$ can be introduced. Depending on the frequency detuning between the lasers, duration of the guided pulse, and the transverse intensity profile of the guiding beam, the real part of $\chi_0$ can have a maximum on axis, resulting in the focusing of the pulse. It turns out that the most power-efficient way of guiding a short (of order plasma period) pulse is to employ a long (twice the desired propagation distance) low-intensity Bessel beam. The theory and numerical simulations of the EIG of long and short pulses is presented in Section III.

The imaginary part of the mutual refraction index describes the energy exchange between the two beams. If the frequency detuning between the beams is close to $\omega_p$, the lower-frequency (signal) beam can be amplified at the expense of the higher-frequency (pumping) beam. This linear amplification has an important limitation: pulses shorter than a plasma period broaden because of the narrow bandwidth of the Raman backscattering instability. Building up on our earlier work [1] on superradiance in Raman backscattering, we found an alternative technique for nonlinearly amplifying ultra-short laser pulses, which is described in Section IV.

## II  BASIC EQUATIONS

Consider two circularly polarized counter-propagating laser pulses $\vec{a}_0$ and $\vec{a}_1$, where $\vec{a}_{0,1} = a_{0,1}/2(\vec{e}_x \pm i\vec{e}_y)e^{i\theta_{0,1}} + $ c.c. are the normalized vector potentials: $a_{0,1}(\vec{r}_\perp, z, t) = eA_{0,1}/mc^2$. Depending on the problem at hand (guiding or amplification), the subscripts 0 and 1 distinguish the right-moving *guided* (or *amplified*) pulse and the left-moving *guiding* (or *pumping*) pulses, respectively. The phases of the waves are $\theta_0 = (k_0 z - \omega_0 t)$ and $\theta_1 = (k_1 z + \omega_1 t)$, and we choose $|\Delta\omega| = |\omega_0 - \omega_1| \ll \omega_0$. We also assume a tenuous plasma $\omega_p \ll \omega_{0,1}$, so that $k_0 \approx k_1 \approx \omega_0/c$.

Assuming that both lasers are of nonrelativistic intensity, $a_{0,1} \ll 1$, the dominant nonlinear force experienced by the plasma electrons is the ponderomotive $\vec{v}_1 \times \vec{B}_0 + \vec{v}_0 \times \vec{B}_1$ force, where $\vec{v}_{0,1} = c\vec{a}_{0,1}$. On a time-scale much longer than $1/\omega_0$ plasma electrons are pushed by the intensity gradient of the "optical lattice" which is produced by the interference of the two beams and has a spatial periodicity $k_0 + k_1 \approx 2k_0$. Therefore, the nonlinear force experienced by an electron $j$ is a function of its position in the optical lattice, which is characterized by its ponderomotive phase $\psi_j = (k_0 + k_1)z_j - \Delta\omega t_j$. As was shown earlier [1], the equations of motion take the form

$$\ddot{\psi}_j + \omega_B^2 \sin\psi_j = -2\omega_p^2 \sum_{l=1}^{\infty} \frac{\cos\psi_j <\cos\psi_j> + \sin\psi_j <\sin\psi_j>}{l} - \frac{2\omega_0 e E_z}{mc^2}, \quad (1)$$

where $\omega_B^2 = 4\omega_0^2 a_0 a_1$ is the bounce frequency of an electron in the optical lattice, and averaging is over one lattice period.

The second term in the LHS of Eq. (1) is the ponderomotive force, the sum in the RHS is over the harmonics of the rapidly spatially varying space-charge force, and $E_z$ is the global electric field. The average of $E_z$ over the lattice period does not vanish. The origin of $E_z$ was explained in Ref. [1]: as the photons are exchanged between the counter-propagating beams, electrons receive the recoil momentum. However, the average current in 1-D is equal to zero – electrons cannot run away from the immobile ions. To ensure this an electric field $E_z$ is produced, satisfying $\partial E_z / \partial t = 4\pi e n_0 < v_{zj} >$.

The periodic density variation, induced in the plasma, serves as an index grating which causes the $a_1$ to undergo Bragg backscattering into $a_0$, and *vice versa*. The nonlinear interaction between the beams is calculated by substituting the modified plasma density into the respective paraxial wave equations. For example, for the guided pulse obtain $2ik_0 \left( \dfrac{\partial}{\partial z} + \dfrac{1}{c}\dfrac{\partial}{\partial t} \right) a_0 + \nabla_\perp^2 a_0 = -k_p^2 \chi_0 a_0$, where $\chi_0 = - < e^{-\psi_j} > a_1^*/a_0$ is the mutual refraction index witnessed by the beam 0.

Refraction index can be evaluated in closed forms if plasma motion is linear, which is the case if $\omega_B^2 \ll \omega_p^2$:

$$\chi_0(\tau) = \frac{2\omega_0^2 |a_1|^2}{\omega_p} \int_0^\infty du \sin(\omega_p u) e^{i\Delta\omega u} \frac{a_0(\tau - u)}{a_0(\tau)}, \qquad (2)$$

where $\tau = t - z/c$ is a co-moving with the pulse 0 coordinate, and the guiding (pumping) pulse is assumed very long. As apparent from Eq. (2), refraction index scales as $\chi_0 \propto |a|^2 \omega_0^2/\omega_p^2$. For co-propagating beams (or a single relativistic beam) this scaling would change to $\chi_R \propto |a|^2$. The enhancement factor $\omega_0^2/\omega_p^2$ is due to the fact that the ponderomotive force is proportional to the gradient of the laser intensity, which is the largest for the small-scale $\approx \lambda_0/2$ interference pattern of the counter-propagating beams.

## III   LINEAR THEORY OF EIG

The mutual refraction index $\chi_0$ is, in general, complex. The real part of $\chi_0$ is much larger than the imaginary part when the beams are detuned far from the plasma resonance, $|\Delta\omega^2 - \omega_p^2| \geq \omega_p^2$. For such detunings there is no energy exchange between beams, and their interaction is purely refractive. This is assumed for the rest of this section. The nonlinear interaction between $a_0$ and $a_1$ can then lead to electromagnetically-induced guiding (of one or both beams) if $\chi_0$ is peaked on axis. The refraction index depends on the spatial and temporal profiles of the pulses and their frequency detuning. For long pulses $\tau_L \gg \min(1/\Delta\omega, 1/\omega_p)$ the mutual refraction index is independent of $\tau$: $\chi_0 = 2|a_1|^2 \omega_0^2/(\omega_p^2 - \Delta\omega^2)$. Thus, two long transversely Gaussian beams focus each other if $\Delta\omega < \omega_p$. Since this type

473

of nonlinear guiding relies on a rather delicate mechanism of generating an index grating with a very short wavelength, and then backscattering off this grating, direct Particle in Cell (PIC) simulation in a slab geometry are used to demonstrate the existence of the EIG.

A 2D version of the relativistic electromagnetic PIC code VLPL (Virtual Laser Plasma Lab) [2] running on a single processor workstation was used to simulate the problem. The grid size was $4000 \times 120$ with $4 \times 10^6$ electrons on it. The two counter-propagating laser pulses with wavelengths $\lambda_1 = \lambda_0 = 1\mu m$ are focused to the spotsizes $\sigma_0 = \sigma_1 = 3\mu m$ at their corresponding entrances into a $400\mu m \times 40\mu$ slab of $n_0 = 10^{19}\text{cm}^{-3}$ plasma. The normalized vector potentials of both lasers at their respective foci are equal to $\alpha_0 = 0.07$. The intensity gray scale plots for the right-moving pulse *in vacuo* and in the plasma are shown in Figs. 1(b) and (c), respectively. The snapshots at $t = 500\lambda_0/c$ of the on-axis ($x = 0$) laser intensities for both cases are plotted versus the propagation distance in Fig. 1(a). As Fig. 1(a) demonstrates, in the presence of the plasma the intensity of the right-going laser $|a_0|^2(z = L_g)$ is increased by a factor 2. Thus, the nonlinear interaction between the lasers strongly reduces laser spreading, confirming the electromagnetically induced guiding. Note that in this example EIG occurs at a very modest laser power $P/P_{cl} = 0.0016$.

A simplified analytical treatment of the EIG can be developed by assuming that the guided laser pulse has a Gaussian transverse profile when it enters the plasma, and that it remains such in the plasma. Guided beam is assumed to have an intensity profile $|a|_0^2 = (1/R_0^2)a_0^2 \exp\left[-r^2/R_0^2\sigma^2\right]$ in cylindrical geometry, where $R_0$ is the dimensionless spotsize. Applying the source-dependent expansion to the envelope equations for $a_0$ similarly to the way it was done in Ref. [3], the equation

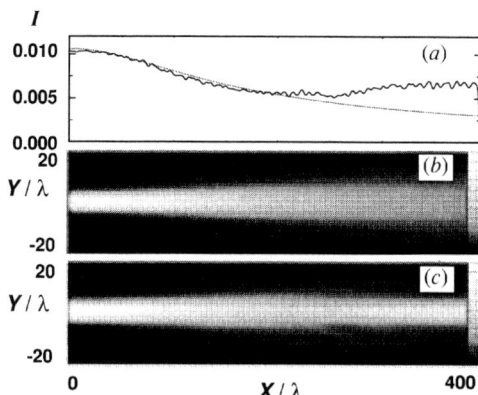

**FIGURE 1.** (a) On-axis intensity of a right-going laser focused at $z = 0$ in vacuum (dashed line) and in the plasma (solid line). Contour plot of the intensity of the right-going laser in vacuum (b) and in the plasma (c).

is derived for the normalized beam radius:

$$\frac{d^2 R_0}{d\bar{z}^2} = \frac{4}{R_0^3} + \frac{4k_p^2\sigma^2}{R_0^3}\int_0^{+\infty} d\rho\rho^2 e^{-\rho^2/R_0^2}\frac{\partial\chi_0}{\partial\rho}$$ (3)

where $\rho = r/\sigma$, and $d/d\bar{z} = 2k_0\sigma^2(\partial/\partial z + \partial/c\partial t)$. In Ref. [4] a similar equation was derived for flat beams in order to make a comparison with the PIC simulation.

Guiding of ultra-short (of order $1/\omega_p$) pulses is particularly interesting for laser-wakefield acceleration since such pulses are optimal for generating plasma wakes. For short guided pulses $\chi_0$ becomes a function of $\tau$. This is because, as Eq. (2) indicates, $\chi_0(\tau)$ of a given longitudinal laser slice $\tau$ is determined by the laser intensity at all earlier instances $\tau' < \tau$. Therefore, the nonlinear focusing experienced by the short pulse is not uniform along the pulse and determined by its longitudinal profile. One finds that, if the pulse intensity drops off faster than exponentially in $|\tau|$, the ratio $a_0(\tau - u)/a_0(\tau)$ decreases with $|\tau|$. For example, for a longitudinally Gaussian profile $a_0 = \alpha_0\exp(-\tau^2/2\tau_L^2)$ the leading edge $\tau < -\tau_L$ slowly erodes because $\chi_0(\tau) \propto \tau_L^2/\tau^2$. However, for $a_0 = \alpha_0\sec(\tau/\tau_L)$ (exponential decay in $|\tau|$) the leading edge is uniformly focused: $\chi_0 = 2|a_1|^2\omega_0^2\tau_L^2/(1 + \omega_p^2\tau_L^2)$ for $\Delta\omega = 0$. Note that guiding is reduced for ultra-short pulses because the plasma does not respond faster than on a $1/\omega_p$ time scale.

The ultra-short pulse is guided by the long guiding pulse along its entire passage through the plasma. In contrast, the guiding pulse is only briefly affected by the guided pulse. Therefore, it has to propagate through the entire plasma region without any additional nonlinear focusing by the short pulse. We found that the most power-efficient choice of the guiding beam is a Bessel beam.

Apertured Bessel beams [5] with sharply peaked radial profiles propagate without diffraction over the distances much exceeding the Rayleigh length (as defined using the width of the central maximum for a spotsize). Such beams transport energy within a narrow spot $\sigma_1 \ll W$ over the distance of order $L_g = 2\pi W\sigma_1/\lambda_1$, where the laser intensity profile is given by $|a_1|^2 = \alpha_1^2 J_p^2(r/\sigma_1)$, and the beam is apertured at the radius $W \gg \sigma_1$ [5]. The total power of a Bessel beam is given by $P_{Bp}/P_0 = \pi W\sigma_1|\alpha_1|^2/\lambda_0^2$, where $P_0 = mc^3/r_e = 8.0$ GW and $r_e = e^2/mc^2$ is the classical electron radius. Therefore, the propagation distance is proportional to the beam power.

A large spotsize Gaussian beam, such that its Rayleigh length is equal to the required guiding distance $L_g$, may appear to be an alternative to the Bessel beam. At first glance, the on-axis mutual refraction index $\chi_0$, which is proportional to the peak intensity of the guiding beam, would be, roughly, the same for the equal power Gaussian and Bessel beams. This is because the product of the peak on-axis intensity and the propagation distance is, approximately, the same for the equal power Gaussian and Bessel beams [6]. However, following Eq. (3), the focusing strength of the guiding beam is proportional to the *curvature* of $\chi_0$, and is, therefore, larger for a Bessel beam.

The choice of $p$, the order of the Bessel beam, depends on the frequency detuning of the lasers. Consistently with our assumption of purely refractive interaction between beams, we consider two cases (i) zero detuning $\Delta\omega = 0$, $p = 0$, and (ii) $\Delta\omega \gg \omega_p$, $p = 1$. For a zero-detuning case assume $a_0 = \alpha_0 \sec(\tau/\tau_L)$ to facilitate the calculation of $\chi_0$ at the leading edge: $\chi_0 = 2|a_1|^2\omega_0^2\tau_L^2/(1 + \omega_p^2\tau_L^2)$. Clearly, $\chi_0$ peaks on axis for a zeroth order Bessel beam.

The guiding power threshold $P_{B0}$ is estimated by substituting $|a_1|^2 = \alpha_1^2 J_0^2(r/\sigma_1)$ into Eq. (3) and assuming $R_0 = 1$:

$$k_0^2\sigma^2|a_1|^2\frac{2\omega_p^2\tau_L^2}{1 + \omega_p^2\tau_L^2}\left[t\frac{d}{dt}\left(e^{-t}I_0(t)\right)\right] = -1, \tag{4}$$

where $t = \sigma^2/2\sigma_1^2$ and $I_0(t)$ is a modified Bessel function. The strongest focusing occurs for $\sigma_1 \approx 0.8\sigma$. Since the guiding distance $L_g$ is related to the beam power $P_{B0}$, Eq. (4) can be rewritten as a power threshold for the EIG:

$$P_{B0} = 1.27\text{GW}\frac{1 + \omega_p^2\tau_L^2}{\omega_p^2\tau_L^2}\frac{L_g}{k_0\sigma^2}. \tag{5}$$

The EIG threshold $P_{B0}$ differs from the relativistic guiding threshold $P_{c1}$ in two respects: first, it is *independent* of the plasma density, enabling electromagnetically-induced channeling in very tenuous plasmas, and, second, it depends on the propagation distance. In deriving Eq. (2) linear plasma response was assumed, which is only valid when $|\chi_0| < 1$. Combining this restriction with the guiding condition, given by Eq. (4), imposes an upper limit on the intensity of the guided pulse:

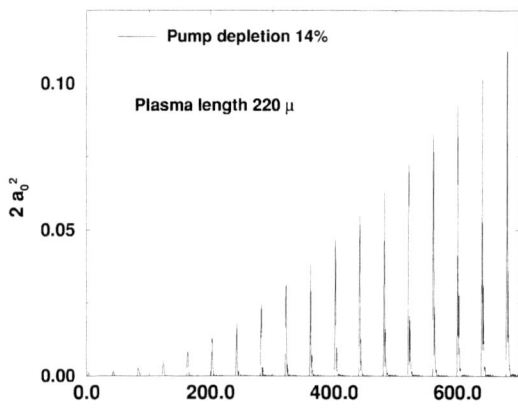

**FIGURE 2.** Superradiant amplification by factor 100, results of the time-averaged code. Pump $a_1 = 0.025$, initial signal $a_0 = 0.025$ and $\tau_L = 1/\omega_p$, plasma $n_0 = 2.5 \times 10^{18}$ cm$^{-3}$.

$\alpha_0 \leq \lambda_0 \sigma / \lambda_p^2$. More extensive numerical study is needed to fully understand how guiding is affected when the linear theory breaks down, but it is reasonable to expect that guiding saturates and weakens at higher intensities of the guided pulse.

Since the focusing strength is determined by the curvature of the $\chi_0$ rather than by its absolute value, a possible solution to the nonlinear saturation of the EIG is to use a first-order Bessel beam, which has an intensity node on axis. The density perturbation on axis vanishes while its curvature does not. To utilize the $J_1$ Bessel beam frequency detuning has to be chosen $\Delta \omega > \omega_p, \tau_L^{-1}$. For a guided beam of arbitrary longitudinal shape $\chi_0 \approx 2|a_1|^2 \omega_0^2 / (\omega_p^2 - \Delta \omega^2)$, enabling focusing by a beam with intensity minimum on axis. The guiding condition similar to Eq. (4) can be derived for the $J_1(r/\sigma_1)$ beam. The strongest focusing occurs for $\sigma_1 = 1.04\sigma$, yielding $k_0^2 \sigma^2 \alpha_1^2 = 5[(\Delta \omega / \omega_p)^2 - 1]$, from which the power threshold condition is be derived:

$$P_{B1} = 3.6 \text{GW} (\Delta \omega^2 / \omega_p^2 - 1) \frac{L_g}{k_0 \sigma^2}. \qquad (6)$$

The linear theory breaks down at $r \approx \sigma$ if $\alpha_0 > 2.7 \lambda_0 \sigma / \lambda_p^2$. In practice, this limitation is likely to be over-stringent since the nonlinear saturation of the EIG in a limited spatial region may not strongly influence the overall focusing. As a numerical example, consider guiding a $\tau_L = 10$ fs long, $\sigma = 18\mu$m wide laser pulse through 1 cm of $n_0 = 10^{19} \text{cm}^{-3}$ plasma. For $\Delta \omega / \omega_p = 1.5$ the threshold power of the Bessel beam $P_{B1} \approx 16$ GW. This beam can guide a pulse with a normalized vector potential up to at least $\alpha_0 = 0.5$, or 2.2TW.

# IV   SUPERRADIANT AMPLIFICATION OF ULTRA-SHORT PULSES

In this Section we consider the possibility of amplifying an ultra-short laser pulse by colliding it with a long counter-propagating pump. If the frequency of the short pulse $\omega_0$ is smaller than the pump frequency $\omega_1$, Manley-Rowe relations dictate the direction of the energy flow – from the pump to the pulse. The difficulty is to sustain the short duration of the signal $\tau_L \approx \omega_p^{-1}$ as its intensity grows by several orders of magnitude. In the linear (exponential) regime of amplification the signal inevitably broadens due to the narrow amplification bandwidth of the Raman backscattering (RBS). It turns out that in the nonlinear regime, defined by $\omega_B > \omega_p$, superradiant amplification (SRA) [1] is possible. In the course of the SRA the width of the signal's leading spike is about $\tau_L \approx \pi / \omega_B$. Since $\omega_B \propto \sqrt{a_0}$, the signal slowly narrows down as it is amplified.

We simulated a collision between a short $\tau_L = \omega_p^{-1}$ signal of initial amplitude $a_0 = 0.025$ and a long pump with $a_1 = 0.025$ using the one-dimensional time-averaged particle code, specifically developed for the study of the RBS instability [1]. This code solves the plasma equation of motion (1) and times-averaged (eikonal) wave equations for the forward and backwards radiation. Simulation results are presented in Fig. 2. Plasma density was chosen $n_0 = 2.5 \times 10^{18} \text{cm}^{-3}$, and the frequency detuning between the signal and the pump is $\Delta\omega = -1.7\omega_p$. As shown in Fig. 2, the signal is amplified by a factor 100 in intensity to a mildly relativistic $I = 2 \times 10^{17}$ W/cm$^2$. It's final FWHM is about one-third of the original.

The initial signal amplitude $a_0$ was chosen to ensure that $\omega_B \approx \omega_p$ at the entrance into the plasma, so that the nonlinear regime is accessed from the start. Taking a smaller initial $a_0$ was resulting in the initial spreading of the pulse. This is consistent with the earlier findings [1] which have shown that an amplified RBS signal spreads during the linear stage of the instability and shrinks during the nonlinear stage.

The physical explanation of the nonlinear SRA is fairly intuitive. When $\omega_B$ exceeds $\omega_p$, all the terms in the RHS of Eq. (1) become much smaller than the ponderomotive term in the LHS. Hence, the particle motion is described by the nonlinear oscillator equation $\ddot{\psi}_j + \omega_B^2 \sin\psi_j = 0$. In the reference frame moving with the short pulse electrons enter the ponderomotive bucket with the initial "speed" $\dot{\psi} = -\Delta\omega$. If this speed is smaller than the bucket height $\dot{\psi}_{max} = 2\omega_B$, electrons become trapped and execute a synchrotron oscillation in the bucket. For example, in the simulated case of $\Delta\omega = 1.7\omega_B$ electrons were initially entering the pulse near the maximum of the ponderomotive bucket. In a set of numerical simulations we varied the laser detuning between $-2.0\omega_p < \Delta\omega < -1.5\omega_p$ and did not see any qualitative differences in the SRA.

In the nonlinear regime signal duration adjusts itself, so that the trapped electrons make exactly half a bounce in the ponderomotive potential of the signal and the pump. Since the amount of energy transferred from pump to signal is proportional to the momentum deposited into the plasma electrons, $\pi/\omega_B$ pulse duration corresponds to the most efficient amplification of the pulse. As the signal

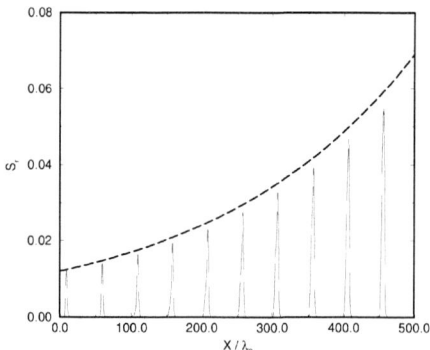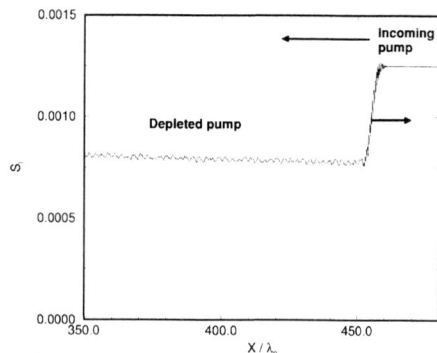

**FIGURE 3.** Superradiant amplification in the large pump depletion regime, results of the PIC simulation with 1-D VLPL code. Pump $a_1 = 0.025$, initial signal $a_1 = 0.07$ and $\tau_L = 1/\omega_p$, plasma $n_0 = 10^{19}$ cm$-3$.

is amplified, its duration decreases. In fact, a self-similar superradiant solution has been found by Bonifacio *et. al.* [7] in the context of free-electron lasers. The peak intensity of this solution increases quadratically (not exponentially!) with the propagation distance. For the RBS the growth of the superradiant spike was estimated earlier [1] as

$$|a_0| \approx \frac{4\omega_p}{3\pi\omega_0} k_p z |a_1|. \tag{7}$$

Eq. (7) predicts that $|a_0/a_1| = 10$ after $k_p z = 500$. From Fig. 3 the ten-fold increase in amplitude occurs for $k_p z = 700$, which is in qualitative agreement with the analytic estimate.

Pump depletion is very modest for tenuous plasmas. In the example shown in Fig. 3, with $\omega_0/\omega_p = 20$, pump depletion is only 14%. This translates into the low efficiency of short-pulse amplification. To demonstrate that much higher extraction efficiency is possible, we simulated a higher density case $\omega_0/\omega_p = 10$ using the one-dimensional version of the VLPL PIC simulation code. As before, the pump amplitude is $a_1 = 0.025$, the initial signal intensity is $a_0 = 0.07$. The results are shown in Fig. 3. By the time the signal intensity grows by factor 5 pump depletion reaches 40%.

Our quantitative understanding of the superradiant amplification is far from complete. However, the numerical and analytical results presented here indicate that both the efficient and manifold amplification of an ultra-short pulse is possible. It appears that SRA has the potential for becoming an efficient and simple pulse-compression technique.

## CONCLUSIONS

In conclusion, we demonstrated analytically and numerically the electromagnetically-induced guiding of two counter-propagating lasers in the plasma at intensities much below the threshold for relativistic guiding. We also numerically demonstrated the possibility of superradiant amplification of ultra-short laser pulses by a properly detuned counter-propagating laser pump.

## ACKNOWLEDGMENTS

The authors gratefully acknowledges helpful discussions with J. S. Wurtele. This work was supported by the DOE contract DE-FG030-98DP00210 and the United States Department of Energy (US DoE) Division of High Energy Physics.

## REFERENCES

1. G. Shvets, J. S. Wurtele, and B. A. Shadwick, Phys. Plasmas, **4**, 1872 (1997).

2. A. Pukhov and J. Meyer-ter-Vehn, APS Bulletin **41**, 1502 (1996).

3. E. Esarey *et. al.,* Phys. Rev. Lett. **72,** 2887 (1993).

4. G. Shvets and A. Pukhov, "Electromagnetically Induced Guiding of Counter-Propagating Lasers in Plasmas", internal report PPPL-3297, May 1998.

5. J. Durnin *et. al.,* Phys. Rev. Lett. **58**, 1499 (1987).

6. B. Hafizi, E. Esarey, and P. Sprangle, Phys. Rev. E **55,** 3539 (1997).

7. R. Bonifacio, L. De Salvo, P. Pierini, N. Piovella, and C. Pellegrini, Phys. Rev. Lett. **73**, 70 (1994), and references therein.

# Laser-Driven Undulator Radiation

G. Shvets and N. J. Fisch

*Princeton Plasma Physics Laboratory, Princeton University, Princeton, NJ 08543*

A. Pukhov

*Max Plank Institute for Quantum Optics, Garching-bei-Munchen, Germany*

**Abstract.** An electromagnetic wake of infra-red radiation can be generated by an intense laser pulse, propagating through an underdense plasma in the presence of a magnetostatic undulator. As opposed to undulator radiation from a charged bunch propagating in a periodic magnetic field, here radiation comes from almost stationary plasma electrons, ponderomotively pushed by the laser pulse. Such laser-driven undulator radiation can be produced by either a single ultra-short pulse, or by two frequency-detuned long pulses. In the latter case the difference frequency is efficiently produced by quasi phase-matched optical heterodyning.

## I  INTRODUCTION AND MOTIVATION

This work is motivated by a search for the new ways of generating high-power tunable far infra-red (FIR) radiation, which is an important research tool in various scientific applications. One such application is to use the FIR radiation as a seed for a free-electron laser (FEL) amplifier. The initially uniform electron beam can emerge from the amplifier, tightly bunched at the right frequency for injecting it into a plasma beatwave accelerator [1]. If the FIR radiation is generated by the same laser beatwave which drives the accelerating plasma beatwave, the beam bunching is naturally synchronized with the plasma beatwave.

We describe a novel Laser-driven Undulator Radiation in the Infrared (LURI), which is produced when a short laser pulse (or laser beatwave) propagates in a plasma-filled undulator. In contrast with the *electrostatic* wake at $\omega_p$ in unmagnetized plasma, or Cerenkov radiation at a frequency close to $\omega_p$ in uniformly magnetized plasma [2], the frequency of the LURI radiation is given by $\omega_s = \omega_p^2/2k_w c$, where $\omega_p = \sqrt{4\pi n e^2/m}$ is the plasma frequency, $\lambda_w = 2\pi/k_w$ is the undulator periodicity, $n$ is the plasma density, $-e$ and $m$ are electron charge and mass, respectively.

CP472, *Advanced Accelerator Concepts: Eighth Workshop,*
edited by W. Lawson, C. Bellamy, and D. Brosius

Conventionally, one thinks of undulator radiation as being produced by a relativistic electron beam. Magnetostatic field is Doppler-upshifted into the infrared photons, with the amount of upshift determined by the electron energy, providing the tunability. In a LURI, however, the electron beam is substituted by the stationary plasma and laser pulse. Plasma is the radiating medium, and the laser pulse provides the Doppler shift. The schematic c of the LURI source, recently described by us [3], is shown in Fig. 1. Plasma electrons experience a ponderomotive kick from the laser, giving them axial momentum. The undulating magnetic field couples with the longitudinal motion of the electrons to produce transverse acceleration, causing them to radiate in the axial direction.

As in undulator radiation, a "radiation zone", roughly defined as the extent of the laser pulse, propagates with relativistic speed in the direction of the undulator periodicity. Here, however, the Doppler upshift is caused by the group velocity of the laser pulse instead of the, typically small, axial velocity of the plasma electrons. LURI is emitted by mostly stationary (although transversely accelerated) plasma electrons.

In this paper we present a detailed treatment of LURI, expanding upon earlier one-dimensional calculation [3](Section II). While the original calculation emphasized the LURI excitation by an ultra-short pulse, here we emphasize an alternative excitation method by a beatwave of two long, suitably detuned in frequency, laser pulses (Section III). This method of producing the difference frequency radiation is compared with the Cerenkov source (Section IV). In Section III we also present the results of the two-dimensional Particle-in-Cell (PIC) simulation of the beatwave excitation of the LURI and discuss the applications of this radiation scheme to beam injection into a plasma beatwave accelerator.

## II  GENERAL DESCRIPTION

We start with simple 1-D geometry, assuming that the laser is infinite in $x - y$ plane. Linearly polarized undulator $\vec{B}_w = B_w \cos{(k_w z)}\vec{e}_x$ laser pulse $\vec{A}_0 = \vec{e}_x a_0(\zeta)mc^2/e \cos{(\omega_0 t - k_0 z)}$ are assumed, where $\zeta = v_g t - z$, and $\omega_0 \gg \omega_p$,

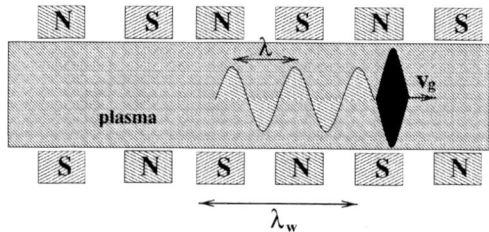

**FIGURE 1.** Schematic for generation of a laser-driven undulator radiation. The resonant wavelength $\lambda = 2\lambda_p^2/\lambda_w$.

so that $v_g \approx c$. Plasma electrons experience several forces: ponderomotive force $\vec{F}_p = \vec{e}_z mc^2 \partial (a_0^2/2) \partial \zeta$, restoring force of the fixed ions, $\vec{v} \times \vec{B}_w$ force of an undulator, and the electric fields of the LURI. Linearized equations of motion are

$$\ddot{\xi}_z = -\omega_p^2 \xi_z - \omega_B \cos(k_w z)\dot{\xi}_y + \frac{c^2}{4}\frac{\partial a_0^2}{\partial \zeta} \tag{1}$$

$$\ddot{\xi}_y = \omega_B \cos(k_w z)\dot{\xi}_z - eE_s/m, \tag{2}$$

where $\xi_y$, $\xi_z$ are Lagrangian electron displacements, $\omega_B = eB_w/mc$, $E_s$ is the LURI electric field, and $\dot{\xi} = \partial\xi/\partial\zeta$.

We further assume that (a) LURI is significantly above the plasma cutoff (a true electromagnetic mode!) $\omega_s \gg \omega_p$, and (b) plasma is weakly magnetized $\omega_B \ll \omega_s$. Hence, all forces except the ponderomotive can be neglected in the RHS of the longitudinal equation of motion (1), yielding $\dot{\xi}_z = ca_0^2(\zeta)/4$. Therefore, only the electrons overlapping with the laser pulse are moving axially, thereby experiencing a $z$-dependent $\vec{v} \times \vec{B}_w$ undulator acceleration. Thus, periodic magnetic field has two functions: (i) enabling plasma electrons to radiate in the forward direction by imparting a finite transverse acceleration $\dot{v}_y$; and, (ii) ensuring constructive interference of the radiation emitted from different regions along the undulator. The combination of a laser pulse, moving at a speed approaching $c$, and a periodic magnetic field introduces the necessary spatio-temporal correlation between different regions of the plasma, resulting in a Doppler-upshifted radiation.

In 1-D condition (ii) is equivalent to requiring that the phase of the LURI changes by $-2\pi$ as the laser pulse moves by one wiggler period: $k_{sz}\Delta z - \omega_s \Delta t = -2\pi$, where $\Delta z = \lambda_w$ and $\Delta t = \lambda_w/v_g$. This can be recast in a form familiar from FEL physics, where the group velocity of the laser $v_g = c\sqrt{1 - \omega_p^2/\omega_0^2}$ assumes the role of the beam velocity:

$$(k_{sz} + k_w)v_g = \omega_s, \tag{3}$$

where $\omega_s$ and $k_{sz}$ are related through the dispersion relation in unmagnetized plasma $\omega_s^2 = k_{sz}^2 c^2 + \omega_p^2$. Equation (3) is only meaningful when the frequency of the LURI $\omega_s$ is much smaller than the frequency of the driver laser $\omega_0$.

Equation (3) determines the frequency of the emitted radiation, and, depending on $\omega_0$, $\omega_p$, and $k_w$, can have two or no solutions. The two solutions coalesce when $\omega_p^2/c^2 = k_w k_0$. For $\omega_p < k_w k_0$ two solutions exist: $\omega_s < \omega_0$ and $\omega_U > \omega_0$. The most interesting case is

$$k_w \ll \omega_p/c \ll \sqrt{k_0 k_w}, \tag{4}$$

when the two solutions are, approximately, given by

$$\omega_s = \frac{\omega_p^2 + k_w^2 c^2}{2k_w c} \quad \text{and} \quad \omega_U = 2k_w c\omega_0^2/\omega_p^2. \tag{5}$$

When condition (4) is satisfied, it turns out that $\omega_s \ll \omega_0 \ll \omega_U$. For example, for $n_0 = 4.0 \cdot 10^{15} \text{cm}^{-3}$, $\lambda_w = 1\text{cm}$, $\lambda_0 = 1.0\mu\text{m}$ we find $\lambda_s = 50\mu\text{m}$, $\lambda_U = 200\text{Å}$. Another important distinction between the low-frequency and high-frequency solutions is that the former is almost independent of the group velocity of the laser (as long as $v_g \approx c$), while the latter is proportional to $\gamma_g^2$, where $\gamma_g = 1/\sqrt{1 - v_g^2/c^2}$ is the relativistic factor of the laser pulse. The high-frequency solution is more reminiscent of the undulator radiation from a relativistic beam, upshifted by $\gamma_\parallel^2$. As we explain below, the rate of emission at $\lambda_U$ is negligible. However, by appropriately shaping the laser pulse, or by utilizing a beatwave of two long frequency-detuned lasers, a considerable amount of far infra-red/THz radiation at $\lambda_s$ can be obtained.

Geometry of the problem dictates that the electric and magnetic fields of the LURI point in $y$ in $x$ directions, respectively. Neglecting the end effects near the entrance and exit of an undulator, the electric field of the LURI is assumed as $\vec{E} = \vec{e}_y E_y(\zeta) \exp(-i k_w z)/2 + \text{c. c.}$ We substitute this expression into the wave equation for the transverse electric field $(-c^2 \nabla^2 + \partial^2/\partial t^2) E_y = 4\pi e n_0 \ddot{\zeta}_y$, where $\ddot{\zeta}_y$ is given by Eq. (2). After some algebra, obtain

$$-i \frac{v_g}{\omega_s + \omega_U} \left( \frac{\partial}{\partial \zeta} + i \frac{\omega_s}{v_g} \right) \left( \frac{\partial}{\partial \zeta} + i \frac{\omega_U}{v_g} \right) E_y = i \frac{k_p^2}{8 k_w} B_w \, a_0^2(\zeta) \tag{6}$$

Equation (6) is integrated, yielding $E_y = E_{\omega_s} + E_{\omega_U}$, where

$$E_{\omega_s} = i \frac{k_p^2 B_w}{8 k_w} \frac{\omega_s + \omega_U}{\omega_U - \omega_s} e^{-i\omega_s \zeta/c} \int_{-\infty}^{\zeta} d\zeta' \, e^{i\omega_s \zeta'/c} \, a_0^2(\zeta') \tag{7}$$

and

$$E_{\omega_U} = i \frac{k_p^2 B_w}{8 k_w} \frac{\omega_s + \omega_U}{\omega_s - \omega_U} e^{-i\omega_U \zeta/c} \int_{-\infty}^{\zeta} d\zeta' \, e^{i\omega_U \zeta'/c} \, a_0^2(\zeta'). \tag{8}$$

Equations (7,8) indicate that a finite duration laser pulse leaves behind a wake of electromagnetic radiation which contains the two resonant frequencies $\omega_s$ and $\omega_U$, and the amplitude of each of the frequency components is proportional to the Fourier transform of the laser intensity profile. When Eq. (4) is satisfied, $\omega_U \gg \omega_0$, and the Fourier transform of the time-averaged laser intensity, evaluated at $\omega_U$, vanishes. Therefore, there is no laser-driven undulator radiation at the higher frequency, upshifted by $\gamma_g^2$, and the frequency of the LURI is equal to $\omega_s$.

Mathematically, Eq. (6) has two solutions with different frequencies because it is a second order differential equation. Since the high-frequency contribution is negligible, Eq. (6) can be turned into a first-order equation by noting that $\partial/\partial \zeta \approx -i\omega_s/c$ and $\omega_U \gg \omega_s$:

$$\left( \frac{\partial}{\partial \zeta} + i \frac{\omega_s}{c} \right) E_y = i \frac{\omega_s}{4c} B_w \, a_0^2(\zeta) \tag{9}$$

The solution to Eq. (9) is given by $E_{\omega_s}$. It was was studied in Ref. [3] with an emphasis on the LURI excitation by a "matched pulse" flat-top laser pulse of duration $\tau_L = \pi/\omega_s$, resulting in $|E_y| = a_0^2 B_w/2$. The emphasis of this paper is on the laser beatwave excitation of the LURI. We briefly note that a wake of stationary periodic magnetic field is left behind a laser pulse, as demonstrated earlier [3]. This wake vanishes in the regime of the beatwave excitation, and will not be discussed in this paper.

## III  BEATWAVE EXCITATION OF LURI: QUASI-MATCHED OPTICAL HETERODYNE

The matched pulse excitation of the LURI invokes an analogy with the plasma wake generation in unmagnetized plasma. This analogy suggests an alternative approach: to use the beat of two lasers with frequencies $\omega_1$ and $\omega_2$, suitably detuned by $\omega_1 - \omega_2 = \omega_s$, to resonantly excite the LURI. This makes the LURI a promising method of difference-frequency generation.

The periodically magnetized plasma is, therefore, a quasi phase-matched optical heterodyne, where the plasma is the nonlinear medium, and the periodicity of the magnetic field creates the quasi-phase matching of the beatwave and the LURI. Using the quantum mechanical language, the high-frequency photon 1 scatters into a lower-frequency photon 2 and a LURI photon. By energy conservation, $\omega_s = \omega_1 - \omega_2$, and by the conservation of the quasi-momentum $k_s = k_1 - k_2 - k_w$. Combining these two conservation laws and assuming that $v_g \approx (\omega_1 - \omega_2)/(k_1 - k_2)$ yields Eq. (3). This is another way of deriving the frequency of the LURI, which demonstrates the phase-matching role of the undulator – it takes the excess momentum out of the beatwave driven LURI photon, enabling the resonant excitation. Thus, the generation of the LURI is a process of downshifting the high-frequency photons in a periodically magnetized plasma.

To describe the beatwave excitation of the LURI assume two overlapping Gaussian pulses:

$$\vec{A}_{1,2} = \vec{e}_x a_{1,2} mc^2/e \cdot \cos\left(\omega_{1,2}t - k_{1,2}z\right)e^{-\zeta^2/2\tau_L^2}.$$

Assuming $\omega_1 - \omega_2 = \omega_s$, we replace $a_0^2/2$ by $a_1 a_2 \cos\left(\omega_s\zeta/c\right)$ in Eq. (7) to obtain:

$$|E|_y^{max} = \frac{\pi^{1/2}\omega_s\tau_L}{4}a_1 a_2 B_w \tag{10}$$

It is instructive to understand why the LURI amplitude grows linearly with the pulse length. physical mechanism behind this dependence. Consider a plasma electron at a given location $z = \mathcal{N}\lambda_w$ along the undulator, executing the steady-state longitudinal oscillation in the electromagnetic beatwave (EMBW) of the two lasers and continuously radiating the LURI. After the second beat of the EMBW this radiation combines *in phase* with the radiation emitted by an electron at the location $z = (\mathcal{N}-1)\lambda_w$ after the *first* beat of the EMBW. This is because the LURI falls behind the driving laser by one wavelength per wiggler period. Thus combined contributions to the LURI travel forward and, once again, add in phase with the radiation emitted after the *third* beat by an electron at the location $z = (\mathcal{N}+1)\lambda_w$, and so on.

In a resonantly driven LURI we observe a phenomenon of constructive interference of the radiation emitted from different places along the undulator. The resonant excitation of the LURI is, thus, of an entirely different origin than, for example, the resonant plasma-wave excitation [4]. The latter results in an increasingly vigorous motion of the plasma electrons and saturates as soon as the wave breaking limit is reached, when the particle displacement becomes comparable with the wavelength of the plasma wave. Increasingly relativistic motion of the plasma electrons is also responsible for the saturation of the plasma beatwave [4]. To the contrary, individual electrons in the LURI source are not resonantly driven; they execute a steady-state longitudinal oscillation in the EMBW with frequency $\omega_s$ and amplitude $v_z/c = a_1 a_2/2$.

We simulated the beatwave excitation of the LURI using a 2-D Particle in Cell (PIC) simulation. Since both the LURI and the excitation pulse are, typically, shorter than the undulator length, we've chosen a time-saving "moving window" simulation, where only the plasma electrons inside a moving simulation box are followed. In Fig. 2 the LURI with $\lambda_s = 3\lambda_1$ is driven by the beatwave of the $\lambda_1$ and $\lambda_2 = 1.5\lambda_1$ lasers. The plasma wavelength $\lambda_p = 10\lambda_1$ is assumed, and the undulator period, corresponding to the exact one-dimensional phase-matching, is $\lambda_w = 56\lambda_1$. The transverse profiles of the laser pulses in both simulations were Gaussian, with $\sigma = 14\lambda_1$, and the width of the simulation box is $200\lambda_1$.

To distinguish between the much more intense laser radiation and the much weaker LURI we chose the laser electric field and the undulator magnetic field to be linearly polarized in $z$ direction. The laser magnetic field is then in $y$ direction, and we can plot the magnetic field of the LURI (which points in $z$ direction) without interference from the laser magnetic field. The undulator $B_w = 10^{-3} mc\omega_1/e$, and the two lasers have identical normalized vector potentials $a_1 = a_2 = 0.1$, durations $\tau_L = 18\lambda_1$, and spotsizes $\sigma = 14\lambda_1$. For $\lambda_1 = 10\mu$ the undulator and laser parameters correspond to $B_w = 1$ Tesla, $P_1 = 83$GW, and $P_2 = 37$GW. The peak power of the LURI is then about 100 kW.

One application of the difference frequency generation is beam injection into a plasma beat wave accelerator [1,5]. Efficient operation of a plasma beatwave accelerator requires an injector which supplies an electron beam consisting of microbunches separated by a plasma wave. Each of the microbunches should be significantly shorter than the inter-bunch distance to ensure that all the particles are injected into a plasma wave at, roughly, the same accelerating phase. A typical plasma wavelength $\lambda_p = 100\mu$m [1] requires $\tau_b = 40$ fs long electron bunches (arbitrarily, $c\tau_b/\lambda_p = 1/8$) – too short for the modern photoinjectors. Moreover, the injected electrons have to be synchronized with the EMBW of the two lasers.

Since free-electron lasers (FELs) are known to produce a highly bunched electron beam [5], one possibility is to use a high-gain single-pass FEL, injected with a seed electromagnetic wave to pre-bunch the beam before injecting it into a plasma beat wave. The challenge is to produce the electromagnetic seed which is synchronized with the electromagnetic beat wave (and, therefore, with the plasma beat wave). The LURI, generated by the electromagnetic beatwave, could be used as a seed. Below we present an examples of a multi-kilowatt $\lambda_s = 100\mu$m LURI source driven by a beatwave of two $CO_2$ laser pulses with $\lambda_1 = 9.6\mu$m and $\lambda_2 = 10.6\mu$m.

We'll choose the pulse durations $\tau_L$ and the laser radii $\sigma$ using the following qualitative argument. Peak power of the LURI is maximized when the laser pulse duration is at least $N_w$ beats long, so that $c\tau_L = N_w\lambda_s$. For the one-dimensional

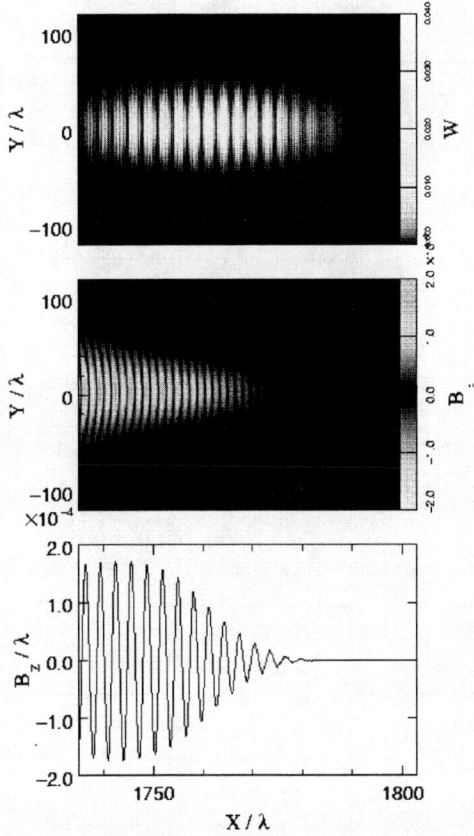

**FIGURE 2.** Particle in Cell simulation of the beatwave LURI excitation. Laser parameters: $a_1 = a_2 = 0.1$, $\sigma = 14\lambda_1$, $\tau_L = 18\lambda_1$; plasma: $\lambda_p = 10\lambda_1$; undulator: $\lambda_w = 56\lambda_1$. (a)–(c) are the snapshots, at $t = 1000\lambda_1/c$, of (a) excitation pulses; (b) magnetic field of the LURI (short wavelength $\lambda_s = 3\lambda_1$) and Cerenkov wake (long wavelength $\lambda_p$), normalized to $mc\omega_1/e$; (c) same as (b), on axis.

theory to remain valid, the LURI has to propagate without diffraction over the entire length of the undulator, so that $N_w\lambda_w = 2\pi\sigma^2/\lambda_s$. The peak power of the LURI is then given by

$$P_s = \frac{N_w P_1 P_2}{4P_0} \frac{\omega_B^2 \lambda_1^2 \lambda_2^2}{c^2 \lambda_w \lambda_s} \tag{11}$$

where, for this example, $\lambda_s = 100\mu$m and $\lambda_0 \approx 10\mu$m.

Using the laser and undulator parameters similar to the ones available at the UCLA Saturnus laboratory [1], we assume $P_1 = P_2 = 400$GW, $B_w = 1$Tesla, $\lambda_w = 1.5$cm, and $N_w = 38$. This requires plasma density of $n_p = 1.3 \times 10^{15}$cm$^{-3}$, laser spotsize $\sigma = 1$ mm, and $\tau_L = 15$ps. From Eq. (11), $P_s \approx 400$kWatt of 100 micron is generated, which, according to Ref. [1], is more than sufficient to inject an FEL amplifier.

# IV COMPARISON OF THE LURI AND CERENKOV WAKES

An alternative method of the difference-frequency generation by exciting a Cerenkov wake in a uniformly magnetized plasma was recently suggested [2]. Since the two methods are somewhat similar in that they involve magnetized plasma, in this Section we compare the underlying principles of the two approaches.

An important distinction between the resonant excitations of the CW and the LURI is that the former involves *phase-matching* between the vacuum beatwave and the CW, while the latter involves the *quasi phase-matching*. The Cerenkov wake (CW) is an extraordinary plasma mode in a *uniformly* magnetized plasma, which has the frequency $\omega^{(ch)} \approx \omega_p$. The exact phase-matching between the beatwave and the CW, $\omega^{(ch)} = \omega_1 - \omega_2$ and $k^{(ch)} = k_1 - k_2$, is achieved in magnetized plasma. Note that such a 3-wave phase-matching is impossible in unmagnetized plasma. It is the presence of the static magnetic field, which strongly modifies the propagation of the low-frequency Cerenkov wave without significantly affecting the higher frequency drive lasers, that makes phase-matching possible.

The electric field of the CW has both the transverse($E_y$) and the longitudinal ($E_z$) components, connected by the relationship $E_y/E_z = eB_0/mc\omega_p$. For example, when a 1 Tesla magnetic field is used to generate 100 micron radiation, the longitudinal field is 100 times larger than the transverse field. Therefore, if high power output (large $E_y$) is needed, the longitudinal field $E_z$ is going to be even larger. This implies large-amplitude longitudinal oscillations of the plasma. Nonlinear effects, such as the relativistic detuning and, possibly, the wave breaking, are likely to become important.

In contrast, the dispersion relation of the LURI is that of a transverse electromagnetic wave since $\omega_s \gg \omega_p$. Quasi phase-matching is made possible by the *periodic* undulator field, which modifies the resonance condition by absorbing momentum $k_w$: $\omega^s = \omega_1 - \omega_2$ and $k^s = k_1 - k_2 - k_w$. Resonantly driven LURI builds up to large power not because of the excitation of the large-amplitude plasma oscillations, but, instead, because of the constructive interference between the radiation emitted at different locations along the plasma. Therefore, plasma nonlinearities will not affect generation of the LURI.

The group velocity of the CW is $v_{ch}/c = (eB_0/mc\omega_p)^2 \ll 1$, so that there is a substantial group velocity mismatch between the vacuum and the plasma. As was pointed out by the authors [2], a very sharp vacuum/plasma interface (of order $c/\omega_p$) is needed to overcome the radiation absorption at the exit from the plasma. Small group velocity can also be viewed as an advantage: the $L_{ch} = c^2 T_{ch}/v_{ch}$ long plasma region is required to produce a radiation pulse of duration $T_{ch}$. In contrast, a much longer plasma region $L = cT_{ch}\lambda_w/\lambda_s$ is needed to produce a LURI pulse of the same duration.

Interestingly enough, for identical flat-top laser drivers and magnetic field magnitudes the amplitudes of the CW and the LURI are identical: $E_y^{ch} = (\omega_{ch}\tau_L/4)a_1a_2B_0$ for the CW and $E_y^s = (\omega_{ch}\tau_L/4)a_1a_2B_w$ for the LURI. The important difference, however, is that the longitudinal velocity of the plasma electrons in the Cerenkov wake is much larger than in the LURI source: neglecting the relativistic mass increase, $v_z/c = a_1a_2(\omega_p\tau/4)$.

When relativistic mass increase is taken into account, the peak electron velocity is $v_z^m/c = (16a_1a_2/3)^{1/3}$ [4], corresponding to the peak transverse field $E_y^{(ch)} = v_z^m B_0/c$. For example, for $a_1 = a_2 = 0.2$ we find $E_y^{(ch)} = 0.6B_0$. The same LURI intensity is obtained by using identical laser pulses of only $10\lambda_s$ long. Since there is no relativistic detuning, longer pulses result in higher power. The peak intensity stops increasing when the number of beats in the pulse exceeds the number of the undulator periods: $\omega_s\tau_L > 2\pi N_w$. Hence, the limiting peak amplitude of the LURI, generated by the EMBW, is $E_y = \pi N_w a_1a_2B_w/2$.

Interestingly, the Cerenkov wake is present even in a periodically magnetized plasma. The undulator field with $k_w \ll \omega_p/c$ is perceived by the plasma electrons as an almost homogeneous field. Hence, the usual plasma wake, electrostatic in the absence of the external magnetic field, turns into a Cerenkov wake. The magnitude of the wake is slowly modulated with periodicity $\lambda_w$. This wake, however, does not escape from the plasma – the nodes of the magnetic field serve as bottlenecks for the field energy since the group velocity vanishes when magnetic field goes to zero. Although the Cerenkov wake appears as a necessary bi-product whenever the LURI is excited, its magnitude can be controlled by adjusting the duration of the laser pulse. Beatwaves tuned to $\omega_s$, which are many plasma periods long, do not excite the CW while exciting the LURI. This is because, for a given plasma density, the LURI has a much higher frequency than the CW.

# V  CONCLUSIONS

In conclusion, a new method for generating coherent tunable radiation in the plasma is described. Radiation is emitted when an laser pulse propagates through the plasma in the presence of a static periodic magnetic field. The resulting laser-driven undulator radiation in the infrared (LURI) can be viewed as an electromagnetic wake, propagating along with the pulse, and can be efficiently excited by two methods: using an ultra-short laser pulse, which matches in duration the period of the LURI, or using a long beatwave of two lasers, detuned by the LURI frequency. The latter excitation scenario was numerically modeled using direct Particle-in-Cell (PIC) simulations. The beatwave generation of the LURI is, to our knowledge, the first example of a quasi phase-matched optical difference frequency generation of high-intensity FIR radiation in the plasma. This easily tunable radiation can be used for injecting a free-electron laser with the purpose of pre-bunching an electron beam. Since the beam bunching would be naturally synchronized with the laser beatwave, this scheme may be used as an injector into a plasma beatwave accelerator. Comparisons between this and an alternative method of difference-frequency generation, the beatwave excitation of the Cerenkov wake, are also presented.

## VI ACKNOWLEDGMENTS

This work was supported by the United States Department of Energy, Division of High-Energy Physics. We gratefully acknowledge discussions with Dr. T. Katsouleas.

## REFERENCES

1. M. Lampel, C. Pellegrini, R. Zhang, C. Joshi, and W. M. Fawley, Proceedings of the 1995 Particle Accelerator Conference, v. 2, 764 (IEEE, New York, NY, 1995).
2. J. Yoshii, C. H. Lai, T. Katsouleas, C. Joshi, and W. B. Mori, Phys. Rev. Lett. **79**, 4194 (1997).
3. G. Shvets, N. J. Fisch, and J.-M. Rax, Phys. Rev. Lett. **80,** (1998).
4. M. N. Rosenbluth and C. S. Liu, Phys. Rev. Lett. **29**, 701 (1972).
5. C. E. Clayton and L. Serafini, IEEE Trans. Plasma Sci. **24, 400** (1996).

# Collisionless Damping of Laser Wakes in Plasma Channels

G. Shvets and X. Li

*Princeton Plasma Physics Laboratory, Princeton NJ 08543*

**Abstract.** Excitation of accelerating modes in transversely inhomogeneous plasma channels is considered as an initial value problem. Discrete eigenmodes are supported by plasma channels with sharp density gradients. These eigenmodes are collisionlessly damped as the gradients are smoothed. Using collisionless Landau damping as the analogy, the existence and damping of these "quasi-modes" is studied by constructing and analytically continuing the causal Green's function of wake excitation into the lower half of the complex frequency plane.

Electromagnetic nature of the plasma wakes in the channel makes their excitation nonlocal. This results in the algebraic decay of the fields with time due to phase-mixing of plasma oscillations with spatially-varying frequencies. Characteristic decay rate is given by the mixing time $\tau_m$, which corresponds to the dephasing of two plasma fluid elements separated by the collisionless skin depth. For wide channels the exact expressions for the field evolution are derived. Implications for electron acceleration in plasma channels are discussed.

## I  INTRODUCTION

Plasma channel is an important tool for a variety of laser-plasma applications, such as laser-driven particle accelerators, x-ray lasers, harmonic generation, and inertial confinement fusion. The primary objective of creating a plasma channel is to guide radiation (e.g. an intense laser pulse in a laser wakefield accelerator) over many Rayleigh lengths. Plasma channels also support low-frequency waves which can be used for particle acceleration. For example, the accelerating wake in the hollow plasma channel is transversely homogeneous and provides a weak linear focusing [1,2].

Theoretical analysis of laser wake excitation in more realistic plasma channels with continuous density profiles is far more challenging, and was first undertaken by Shvets *et. al.* [2], with the emphasis on the hollow channels with a sharp (although not infinitely sharp) plasma-vacuum interface. In an ideal hollow channel of width $b$ only two plasma modes of oscillation are excited by a short laser pulse: (a) global surface mode with frequency $\omega_{ch} = \omega_p / \sqrt{1 + k_p b}$, which exists in the channel and inside the plasma, and is excited at the plasma-vacuum interface, and (b)

CP472, *Advanced Accelerator Concepts: Eighth Workshop,*
edited by W. Lawson, C. Bellamy, and D. Brosius

electrostatic plasma mode with frequency $\omega_p$, which is locally excited in the bulk of the plasma.

In a realistic plasma channel the surface mode turns into a damped quasi-mode, which decays exponentially in time. Qualitatively, a quasi-mode, whose frequency is close to the frequency of the surface mode in a hollow channel [1], is absorbed at the resonant location $x_r$, defined by $\omega_p(x_r) = \omega_{ch}$. The locally excited plasma wave also decays, but much slower, according to the power law. This occurs because, due to the electromagnetic nature of the plasma wave in a channel, fields at a given spatial location $x$ are affected by the plasma currents within a collisionless skin depth $c/\omega_p$ from $x$. Therefore, wakes damp when, roughly, two fluid elements, separated by $c/\omega_p$ get out of phase. The detailed analysis of this collisionless damping will appear in an upcoming issue of Physics of Plasmas [3]. Here we outline the general theoretical approach (Section II), present the results of the numerical study of quasi-modes (Section III), and present the analytical results of wake damping in wide plasma channels (Section IV).

## II  THEORETICAL MODEL

A number of simplifying assumptions make the problem at hand analytically tractable. Slab geometry is assumed, so that the plasma density $n_0(x)$ is only a function of the transverse coordinate $x$. Ions are assumed immobile, providing a neutralizing background. Wakes are driven by a non-evolving nonrelativistic laser pulse with $a = e|A|/mc^2 < 1$, so that all calculations are performed to order $a^2$. The group velocity of the pulse is assumed close to $c$ (tenuous plasma), and all the plasma quantities are functions of $\zeta = t - z/c$.

In a recent work [2] we demonstrated that the Laplace transformed electric field of the wake is

$$c\vec{E} = \frac{\omega_{p0}^2(x)}{\omega^2}\vec{\nabla}\tilde{f} + i\frac{c}{\omega}\nabla \times \vec{\tilde{B}},\tag{1}$$

where $f = -mc^2|\vec{a}|^2/4e$ is the ponderomotive potential and $\epsilon(x,\omega) = 1 - \omega_{p0}^2(x)/\omega^2$. All Laplace transformed variables are "tilded". Magnetic field is calculated by substituting Eq. (1) into Faraday's law: $\mathcal{L}_S(\omega)\tilde{B}_y = \omega^2\epsilon'\tilde{f}/c^2\epsilon^2$, where $\mathcal{L}_S$ is a Sturm-Louville operator, parametrically dependent on the complex frequency $\omega$:

$$\mathcal{L}_S(\omega) = \frac{d}{dx}\left(\frac{1}{\epsilon(x,\omega)}\frac{d}{dx}\right) - \frac{\omega_p^2(x)}{c^2\epsilon(x,\omega)}.\tag{2}$$

Magnetic field in $\zeta-$ domain is then obtained by inverse Laplace transforming $\tilde{B}_y(x,\omega)$ along any contour $\Gamma$ in the upper half plane of complex $\omega$, as shown in Fig. 1a. Since we only consider accelerating TM modes, boundary conditions $\tilde{B}_y(x = 0) = \tilde{B}_y(x = \infty) = 0$ must be satisfied. Assuming a monotonic density plasma density profile, which asymptotes to $n_0$ at infinity, one can choose a transverse location $x_{max}$ (finite or infinite), such that $\omega_p(x) \equiv \omega_{p0}$ for $x > x_{max}$. Therefore, the second boundary condition can be chosen at $x_{max}$ as $\tilde{B}' + k_{p0}B = 0$.

Magnetic field $\tilde{B}_y$ is computed by constructing the Green's function $G(x, x', \omega)$ of the operator $\mathcal{L}_S$, satisfying the boundary conditions at $x = 0$ and $x = x_{max}$, and a differential equation $\mathcal{L}_S\, G(x, x', \omega) = \delta(x - x')$. Such approach was originally used by Briggs et. al. to study the eigenmodes of a strongly magnetized noncutral plasma [4]. By causality, we require that $G(x, x', \omega)$ has no singularities in the upper half $\omega-$ plane.

The Green's function of $\mathcal{L}_S$ is

$$G(x, x', \omega) = \frac{1}{D(\omega)} \begin{cases} \phi_+(x, \omega)\phi_-(x', \omega) & \text{for } x' < x \\ \phi_-(x, \omega)\phi_+(x', \omega) & \text{for } x' > x \end{cases} \tag{3}$$

where $\phi_-$ and $\phi_+$ are linearly independent solutions of $\mathcal{L}_S\phi_\pm = 0$, satisfying their respective boundary conditions: $\phi_-(x = 0, \omega) = 0$ and $\phi_+'(x = x_{max}, \omega)/\phi_+(x = x_{max}, \omega) = -\omega_{p2}/c$. Constant $D(\omega)$ is the Wronskian of $\phi_\pm$, given by $D(\omega) = [\phi_+'\phi_- - \phi_-'\phi_+]/c$. $D(\omega)$ will be referred to as a dispersion function because its zeros correspond to the eigenmodes of a plasma channel. For example, for an ideal hollow channel $D(\omega_{ch}) = 0$. Incidentally, for a plasma channel with continuous density profile dispersion function never vanishes. Therefore, realistic channels do not support any continuous eigenmodes. A rigorous proof is given in [3].

We can now formally solve for the magnetic field:

$$\tilde{B}_y(x, \omega) = \int_0^{x_{max}} dx' G(x, x', \omega) \left( \frac{\omega^2}{c^2} \frac{c'}{c^2} \tilde{f}(x', \omega) \right) \tag{4}$$

Similar Green's functions can be derived for the electric field.

For $\omega_i > 0$ the eigenfunctions $\phi_-$ and $\phi_+$ can be expressed in terms of any two linearly independent solutions of the homogeneous equation $\mathcal{L}_S\phi = 0$. One

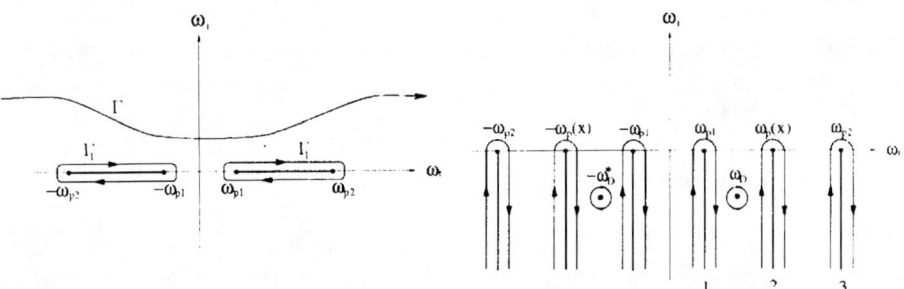

**FIGURE 1.** (a) $\Gamma$ is integration path for the inverse Laplace transform). Integral along the contour $\Gamma$ can be reduced to a sum of two integrals along branch cuts (contours $\Gamma_i$). (b) A different cut which recovers the quasi-modes $\omega_D$ and $\omega_D^*$ zeros of the analytic continuation of the dispersion function $D_*(\omega)$.

493

example of such functions is a set $(\phi_0, \phi_1)$, satisfying $\phi_0(0) = 0$, $\phi_0'(0) = 1$, and $\phi_1(0) = \epsilon(x = 0, \omega)$, $\phi_1'(0) = 0$. Integrating these solutions according to $\mathcal{L}_S \phi_{0,1} = 0$ between $x = 0$ to $x = x_{max}$ yields the dispersion function $D(\omega)$, which is given by

$$D = \frac{\phi_0'(x_{max}) + k_{p2}\phi_0(x_{max})}{\phi_1'(x_{max}) + k_{p2}\phi_1(x_{max})}. \tag{5}$$

and the $\phi_\pm$ basis functions: $\phi_- = \phi_0$ and $\phi_- = \phi_0 - D(\omega)\phi_1$. Integration of the homogeneous equation along the real axis is unambiguous for $\omega_i > 0$ since the singularities of $\mathcal{L}_S$ lie above the real $x$–axis.

Singularities of $\mathcal{L}_S$ are encountered when $\omega$ assumes a real value from the interval $[\omega_{p1}, \omega_{p2}]$. The dispersion function $D(\omega)$ thus has branching points at $\omega_{p1}$ and $\omega_{p2}$. Basis functions $\phi_\pm(x, \omega)$ have another branching point at $\omega = \omega_p(x)$. One can insure that $B_y(x, \omega)$ is single-valued by making a single cut between $\omega_{p1}$ and $\omega_{p2}$.

To recover the damped quasi-modes a different cut can be used (following Ref. [4]), as shown in Fig. 1b, where the integration contour is pushed into the lower half plane. To do that, $D(\omega)$, $\phi_-(x, \omega)$, and $\phi_+(x, \omega)$ have to be analytically continued for $\omega_i < 0$. Although $D(\omega)$ has no zeros above or below the real $\omega$ axis, its analytic continuation $D_*(\omega)$ may have zeros $\omega_D$ and $-\omega_D^*$ in the lower half-plane, as shown in Fig. 1b. This phenomenon is analogous to the collisionless Landau damping of electrostatic plasma waves, where the dielectric constant does not vanish for any complex frequency while its analytic continuation does vanish, resulting in Landau-damped quasi-modes.

With $\omega_i > 0$, the complex resonant point $x_r(\omega)$, defined by $\epsilon(\omega, x_r) = 0$, is located in the upper half of the $x-$ plane. Basis functions $\phi_{0,1}$ can be integrated along any contour which connects the origin and $x$ (including the straight line, of course), provided that the contour stays below the resonant point $x_r(\omega)$, as shown in Fig. 2a. When $\omega$ is moved into the lower half-plane, the resonant point $x = x_r(\omega)$ is also lowered below the real axis. Therefore, to ensure analyticity of $\phi_{0,1}(x, \omega)$ as a function of $\omega$ the integration contour in the complex $x-$ plane must be deformed to stay *below* the branching point $x = x_r(\omega)$, as shown in Fig. 2b.

For a given $\omega$ we have to choose the branch cut in the $x-$ plane, corresponding to the cuts in the $\omega$ plane, shown in Fig. 1a. It can be shown [3] that this cut intersects the real $x-$ axis at a point $x_d$, such that $\omega_p(x_d) = \text{Re}(\omega)$. Basis functions $\phi_\pm(x, \omega)$ are discontinuous at $x_d$. As was pointed out by Briggs [4], discontinuity of the field is not "real" in the sense that it only exists in frequency domain and disappears after integration over $\omega$.

With the procedures for the analytic continuation of the basis functions and the dispersion function $D(\omega)$ rigorously established, the inverse Laplace transform integration can be carried out along the branch cuts and around the poles of the analytically continued Green's function, as shown in Fig. 1b. Zeros of the analytically continued dispersion function $(\omega_D, -\omega_D^*)$ define the collisionlessly damped quasi-modes of the plasma channel. Note that since the quasi-modes are discontinuous at $x_d(\omega_D)$, their existence does not contradict the earlier statement that realistic channels do not support *continuous* eigenmodes. Contributions of the quasi-modes are important for the short-time evolution of the plasma wakes, $\zeta < 1/\mathrm{Im}(\omega_D)$. For longer times quasi-modes damp out, and branch cut contributions dominate. This is because the fields associated with the branch cuts *algebraically* decay in $\zeta$. Moreover, for very wide plasma channels quasi-modes are important in a very small part of the channel, and the branch cut 2 dominates throughout the rest of the channel. In the rest of the paper we separately consider the global quasi-modes and the local plasma excitations, associated with the branch cut 2.

## III    WEAKLY DAMPED QUASI-MODES

Undamped electromagnetic surface mode, supported by an ideal hollow channel with an infinitely sharp vacuum-plasma interface, becomes a weakly damped quasi-mode as the plasma-vacuum interface is smoothed out. For a channel with continuous plasma density a resonant location $x_d$ exists, where the surface mode frequency is equal to the local plasma frequency. As a result of the coupling between the surface mode and the electrostatic plasma wave, the former damps. Quasi-mode is peaked at and exponentially decays away from the resonance point $x_d$. Quasi-mode is distinct from the localized electrostatic plasma waves in that it is global (i. e. exists throughout the channel), yet possess a well-defined frequency which is independent of the transverse location $x$. It is not, however, a true eigenmode: it belongs to the continuum of the plasma waves. Weakly damped quasi-modes are important only for short times of order the damping time, after which the coherent.

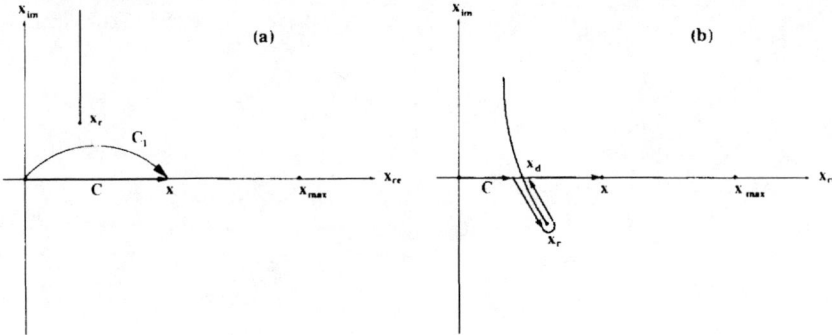

**FIGURE 2.** Integration paths in the complex $x$ plane of the differential equation $\mathcal{L}_S(\omega)\phi_{0,1} = 0$ for the basis functions $\phi_0$ and $\phi_1$, whose boundary conditions are set at $x = 0$. $x_r$ is the resonance point, defined by $\epsilon(x_r,\omega) = 0$. (a) $\omega$ in the upper half plane, $\mathrm{Im}\,\omega > 0$. $C_1$ and $C$ possible integration paths. Basis functions are continuous on the real axis. (b) $\omega$ in the lower half plane, $\mathrm{Im}\,\omega < 0$. Integration contour remains below the resonance point $x_r$ to ensure analytic continuation of the Green's function. Basis functions are discontinuous on the real axis at $x = x_d$, where $\epsilon(x_d, \mathrm{Re}\,\omega) = 0$.

single-frequency motion of the plasma electrons is destroyed, and the local fields, oscillating with the local plasma frequencies $\omega_p(x)$, prevail.

As an example, consider the parabolic plasma channels

$$\omega_p^2(x) = \omega_{p1}^2 + \frac{(\omega_{p0}^2 - \omega_{p1}^2)x^2}{x^2 + b^2},\qquad (6)$$

of two types: (a) hollow on axis ($\omega_{p1} = 0$), or (b) with finite on-axis density ($\omega_{p1} \neq 0$). For such chanels $D_*(\omega)$ is found from Eq. (5), where the basis functions $\phi_{0,1}$ are numerically integrated from $x = 0$ to $x = x_{max}$ along the contour shown in Fig. 2b. In practice, the integration contour was taken as a sum of two straight lines in the complex plane: one, connecting $x = 0$ and $x = x_r(\omega) - i\nu$ (where $\nu$ is an arbitrary positive number) and the other, connecting $x = x_r(\omega) - i\nu$ and $x_{max}$ (chosen at $x_{max} = 3b$ ). The real frequency and the damping rate of the quasi-modes of are plotted as functions of the dimensionless channel width $k_{p0}b$ in Figs. 3a,b.

For a hollow channel the only contribution to the accelerating gradient $E_z(r = 0)$ comes from the quasi-mode. A fluid code, which calculates all the electromagnetic fields and plasma fluid quantities in the $\zeta$ domain was recently developed [5]. To extract the damping coefficient and the frequency from the fluid code, we fitted $E_z(r = 0, \zeta)$ by $E_0 \cos(\omega_r\zeta + \phi_0) \exp(-\gamma\zeta)$, where $E_0$ and $\phi_0$ are constants. The results of the fit are shown in Fig. 4, where a narrow plasma channel $k_{p0}b = 1$ was simulated. Both the damping rate and the frequency exactly correspond to those of the eigenmode ($k_{p0}b = 1$ point of the plots in Fig. 3a).

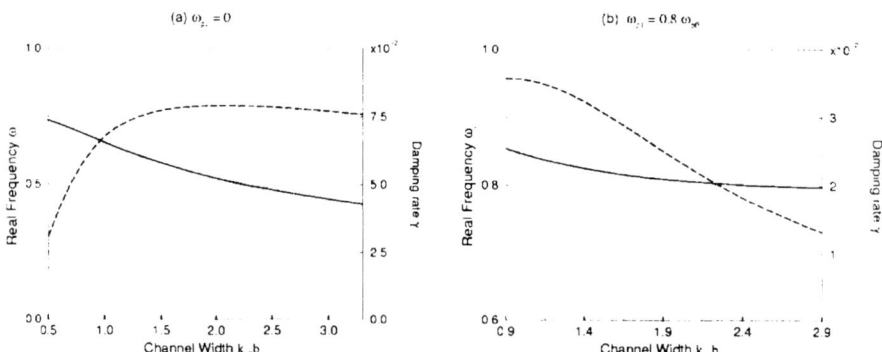

**FIGURE 3.** The quasi-mode's real frequency $\mathrm{Re}(\omega_p)$ (solid line) and damping rate $\mathrm{Im}(\omega_p)$ (dashed line) as a function of the channel width $k_{p0}b$ for two types of plasma channels: (a) zero plasma density on axis and (b) finite plasma density on axis. Plasma frequency satisfies $\omega_p^2(x) = \omega_{p1}^2 + (\omega_{p0}^2 - \omega_{p1}^2)x^2/x^2 + b^2$.

# IV  WAKE DAMPING IN WIDE CHANNELS

Contributions of the quasi-modes are not sufficient to describe the electromagnetic wakes for all times and at all transverse locations inside the plasma channel. The quasi-mode contribution to the total electromagnetic fields at a given transverse point can be negligible if the quasi-mode is localized far away from that location. For example, the quasi-mode of a wide channel becomes exponentially small sufficiently far away from the resonance point. Figure 5 shows the spatial profiles of the accelerating and focusing electric fields, corresponding to the quasi-mode of a channel, described by Eq. (6) with $k_{p0}b = 3$ and $\omega_{p1} = 0$. Both fields peak at $x_d(\omega_D) \approx 1.5/k_{p0}$, where $\omega_D = \omega_{p0}(0.45 - 0.077i)$ is the complex frequency of the quasi-mode. Note that the accelerating field, corresponding to the quasi-mode, is discontinuous at $x_d$. As explained earlier, this does not correspond to a physical discontinuity. The values of both fields at the edge of the channel ($x = 2b$) are negligible in comparison with their peak values at $x = x_d$. Therefore, EM fields far from $x_d$ are dominated by local excitations. The evolution of these excitations, described by the branch cut 2, is computed below for wide channels.

We assume that the scale of the plasma inhomogeneity is much longer than the skin depth: $k_p L \gg 1$, where $L = k_p/k_p'$. The laser profile is also assumed only weakly nonuniform on a scale $1/k_p$: $\partial \ln|f|/\partial x \ll k_p$. The exact analytic expressions for the $\phi_+$ and $\phi_-$ can be obtained, enabling the exact integration of the Green's function in $x'$ [see Eq. (4)] and $\omega$ (inverse Laplace transform).

In the vicinity of the resonant point $|x - x_r| \ll L$ the lowest-order Taylor expansion for the plasma dielectric function can be used: $\epsilon = \epsilon'(x - x_r)$. Therefore, in the vicinity of $x_r$ basis functions $\phi_\pm$ satisfy $\phi'' - \phi'/(x - x_r) - k_p^2\phi = 0$. We further assume that $x$ is at least $1/k_p$ away from both the origin and the discontinuity point $x_d(\omega_D)$. The origin can now be treated as being at $-\infty$. The basis functions are $\phi_- = tK_1(t) + i\pi t I_1(t)$ and $\phi_+ = tK_1(t)$, where $t = k_p(x - x_r)$, and $I_1(t)$ and $K_1(t)$ are the modified Bessel functions.

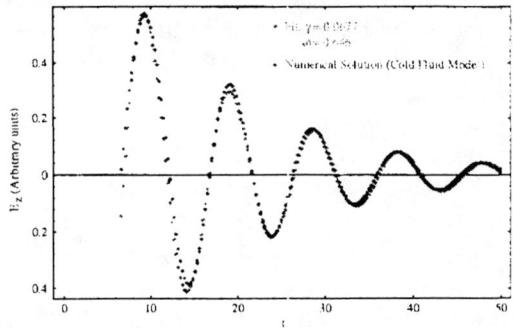

**FIGURE 4.** Numerical fit of the on-axis accelerating field from a fluid simulation by the formula $E_z(x = 0, \zeta) = E_0 \cos(\omega_r\zeta + \phi_0) \exp(-\gamma\zeta)$.

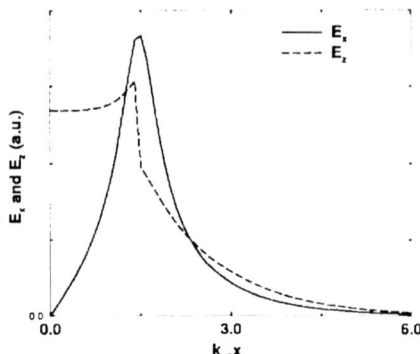

**FIGURE 5.** Spatial profiles of the focusing (solid line) and accelerating (dashed line) electric fields of the collisionlessly damped quasi-mode of a wide plasma channel. Plasma frequency satisfies $\omega_p^2(x) = \omega_{p0}^2 x^2/x^2 + b^2$; channel width $k_{p0}b = 3$. Accelerating field $E_z$ of the quasi-mode is discontinuous at $x_d \approx 1.5 k_{p0}^{-1}$.

The details of the integrations in $x'$ and in $\nu$ (which parametrizes the branch cut 2: $\omega = \omega_p(x) - i\nu$) are given elsewhere [3]. Here we only present the results:

$$E_x(x,\zeta) = -\sin(\omega_p\zeta)\,\omega_p\frac{\partial \tilde{f}(x,\omega_p)}{\partial x}\left[1+\frac{\zeta^2}{\tau_m^2}\right]^{-\frac{1}{2}} - \cos(\omega_p\zeta)\frac{\omega_p^2}{c}\tilde{f}(x,\omega_p)\frac{\zeta/\tau_m}{\sqrt{1+\zeta^2/\tau_m^2}} \quad (7)$$

$$E_z(x,\zeta) = \cos(\omega_p\zeta)\frac{\omega_p^2}{c}\tilde{f}(x,\omega_p)\frac{1}{\sqrt{1+\zeta^2/\tau_m^2}} \quad (8)$$

$$B_y(x,\zeta) = -2\sin(\omega_p\zeta)\frac{k_p}{\tau_m}\tilde{f}(x,\omega_p)\left[1+\frac{\zeta^2}{\tau_m^2}\right]^{-\frac{1}{2}}. \quad (9)$$

where $\omega_p \equiv \omega_p(x)$ is the *local* plasma frequency at the transverse position $x$. In deriving Eqs. (7,8,9) the second and higher derivatives of $f$ are neglected.

In a homogeneous plasma the second term in Eq. (7) identically vanishes. Interestingly, in a channel it comes to dominate the locally excited transverse electric field (described by the first term) for $\zeta \gg \tau_m$. Also note that this non-decaying component of the focusing electric field is *in phase* with the accelerating field. This is potentially important for plasma-based particle accelerators. In a standard laser wakefield accelerator, based on a transversely uniform plasma, the accelerating and focusing electric fields are 90 degrees out of phase. Therefore, the phase region, where an injected particle is both accelerated and focused, is only $\lambda_p/4$ long. In a channel this region can be $\lambda_p/2$ long, as seen from Eqs. (7,8).

Equations (7,8,9) constitute an important result of this paper. For the first time, the exact expressions for the wake evolution in wide plasma channels, valid for arbitrary times, have been derived, and the phase-mixing time introduced. The physical meaning of the phase-mixing in plasma channels is explained below.

Damping of the electromagnetic fields in plasma channels is reminiscent of the decay of two-dimensional plasma oscillations due to the trajectories crossing, as described by Dawson [6]. To appreciate the differences between these two damping mechanisms, consider the underlying basic physics. Local plasma oscillations (and the corresponding electric field) decay when different plasma fluid elements, oscillating with their local plasma frequencies, get out of phase. Consider two fluid elements, initially located at $x = x_0$ and $x = x_0 + \Delta x$ and oscillating with their local plasma frequencies $\omega_p(x_0)$ and $\omega_p(x_0 + \Delta x)$, respectively. These two elements get out of phase after $\tau = |\omega_p' \Delta x|^{-1}$.

Electrostatic cold plasma oscillations, considered by Dawson, collapse when adjacent fluid elements collide with each other. This is because there is no Pointing flux associated with such oscillations, so the energy can only be physically carried by the plasma electrons. This trajectory crossing can be interpreted as dephasing between the electrons separated by the distance equal to the amplitude of their oscillation $A$. Therefore, the nonlinear wave breaking time $\tau_{WB} = |A\omega_p'|^{-1}$ is calculated by taking $\Delta x = A$. For very small oscillation amplitudes this transverse wave breaking takes place several plasma periods behind the laser pulse [7].

On the other hand, wakes in plasma channels have a non-vanishing Pointing flux, enabling communication between different transverse positions. As a result, electromagnetic wakes in channels become nonlocal: fields at a given spatial location $x$ are affected by the plasma currents within a collisionless skin depth $c/\omega_p$ from $x$. Therefore, wakes damp when, roughly, two fluid elements, separated by $c/\omega_p$ get out of phase, that is, $\Delta x = c/\omega_p$, or $\tau_m = \omega_p/c\omega_p'$. Below we demonstrate how this estimate can be obtained in a more formal way.

As Eqs. (7,8,9) indicate, magnetic field decays faster than both components of the electric field. Due to the conservation of the total vorticity ($\nabla \times \vec{v} = e\vec{B}/mc$), plasma flow becomes almost curl-free for large times: $\partial v_z/\partial x = \partial v_x/\partial z$, or $v_z = -v_x \tau_m/\zeta$. Combining Maxwell's equations with the vorticity conservation yields

$$\left[\frac{\partial^2}{\partial x^2} - k_p^2(x)\right] B_y \approx \frac{4\pi e n_0}{c^2 \zeta} \hat{v}_x e^{-i\omega_p(x)\zeta} + \text{c. c.,} \tag{10}$$

where we have used the curl-free flow assumption and the fact that, for large $\zeta$, $v_x$ oscillates with a constant amplitude. Equation (10) can be solved in closed forms by expanding $\omega_p(x') \approx \omega_p(x) + \omega_p'(x' - x)$:

$$B_y(x, \zeta) = -\frac{2\pi e n_0 \hat{v}_x}{k_p c^2 \zeta} e^{-i\omega_p(x)\zeta} \int_{-\infty}^{+\infty} dx' e^{-k_p|x-x'|} e^{i\omega_p'\zeta(x-x')} + \text{c. c.} \tag{11}$$

As illustrated by the integrand of Eq. (11), fluid elements within one collisionless skin depth from $x$ contribute to $B_y(x)$. For large $\zeta$ these contributions get out of phase, and the integral in Eq. (11) decays as $(1 + \zeta^2/\tau_m^2)^{-1}$. Therefore, magnetic

499

field decays as $\zeta^{-3}$, in agreement with Eq. (9). This calculation illustrates how the electromagnetic nature of the channel wake results in the nonlocal excitation on a $c/\omega_p$ scale, leading to phase-mixing of the plasma fluid and algebraic decay of the wake.

If the plasma oscillation amplitude $A$ is much smaller than the collisionless skin depth, phase mixing occurs before the wave breaking. Transverse wave breaking can still take place after $\tau_{WB} \gg \tau_m$, which can be estimated by requiring that $|\partial_r E_x| = 4\pi e n_0$. Assuming a short laser driver with duration of order $1/\omega_p$ and nonrelativistic normalized vector potential $|\vec{a}^2| \ll 1$, estimate that $\tau_{WB} \approx 2\tau_m/|\vec{a}^2|$.

# V    CONCLUSIONS

In summary, the highlights of the theory of collisionless damping of laser wakes in plasma channels is presented. Two types of decaying solutions are identified: exponentially decaying global quasi-modes and algebraically decaying almost-local excitations. Frequencies and damping rates of the quasi-modes for various channels are found, in excellent agreement with the results of the explicit fluid simulations. Almost-local wake fields decay because of the phase-mixing of the plasma currents at different transverse locations inside the channel.

# ACKNOWLEDGMENTS

This work was supported by the US DOE Division of High Energy Physics. The authors acknowledge useful conversations with N. J. Fisch, B. A. Shadwick, and J. S. Wurtele.

# REFERENCES

1. T. C. Chiou et. al., Phys. Plasmas **2**, 310 (1995).
2. G. Shvets et. al., IEEE Trans. Plasma Sci. **24**, 351 (1996).
3. G. Shvets and X. Li, "Theory of Laser Wakes in Plasma Channels", submitted to Phys. Plasmas (1998).
4. R. J. Briggs, J. D. Daugherty, and R. H. Levy, Phys. Fluids **13**, 421 (1970).
5. B. A. Shadwick and J. S. Wurtele, *Proceedings of the Sixth European Particle Accelerator Conference, Stockholm, 1998*, edited by S. Myers, L. Liljeby, Ch. Petit-Jean-Genaz, J. Poole, and K.-G. Rensfelt (Institute of Physics Publishing, Philadelphia, 1998), p. 827.
6. J. M. Dawson, Phys. Rev. , **113**, 383 (1959).
7. S. V. Bulanov et. al., Phys. Rev. Lett., **78**, 1205 (1997).

# Test Results of the Plasma Source for Underdense Plasma Lens Experiments at the UCLA Neptune Lab

H. Suk, C. E. Clayton[†], R. Narang[†], P. Muggli[†],
J. Rosenzweig, C. Pellegrini, and C. Joshi[†]
Department of Physics and Astronomy
University of California, Los Angeles, CA 90095
[†]Department of Electrical Engineering

## Abstract

A plasma source was developed at UCLA for planned underdense plasma lens experiments, where the plasma density is less than the electron beam density. The argon plasma, produced by a discharge between a $LaB_6$ cathode at 1330 °C and a tantalum anode, is confined by a solenoidal magnetic field and flows transversely across the electron beam path. Extensive test of the plasma source is under way for various conditions before it is assembled with the UCLA photocathode-based electron linac. In particular, different longitudinal (with respect to the electron beam) plasma profiles and effective plasma lengths can be obtained by adjusting the moveable sliding door between the plasma source and the transverse beamline. Test results of the plasma source are presented.

# 1 Introduction

Plasma lenses have potential applications to next generation linear colliders because plasma focusing can be several orders of magnitude stronger than that of conventional magnets [1, 2]. For this purpose, numerous studies have been done both in theory and experiment [3, 4, 5, 6, 7]. In the experimental area, the plasma lens research has been focused on the overdense plasma regime where the plasma density is larger than the beam density. In the case of overdense plasma lenses, however, a plasma return current flows inside the beam region when the beam propagates in the plasma. Hence, overdense plasma lenses lead to rather large spatial and temporal aberrations. In contrast, in underdense plasma lenses most of the plasma electrons are radially expelled from the beam

CP472, *Advanced Accelerator Concepts: Eighth Workshop,*
edited by W. Lawson, C. Bellamy, and D. Brosius

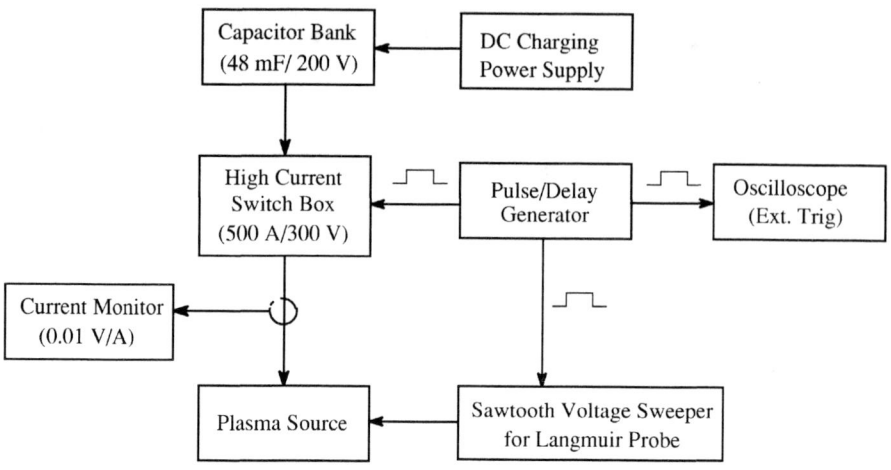

Figure 1: Block diagram of the plasma source test.

path when the beam propagates in the plasmas. Therefore, a nearly uniform electron-free ion channel is produced and this ion channel provides a high quality focusing. In other words, the focusing force is linear in radius and independent of the longitudinal position so that aberrations are avoided in underdense plasma lenses. This focusing effect was observed in UCLA/ANL experiments where the plasma was thick, compared to the focal length [8]. For practical underdense plasma lenses, however, the plasma thickness should be less than the focal length and this kind of thin underdense plasma lenses have not been explored experimentally so far. Hence, experiments in this regime are planned at the UCLA Neptune laboratory [9] which is based on the 1.625-cell photoinjector RF gun and the PWT linac. For this underdense plasma lens experiment, a discharge plasma source has been developed and tested at UCLA. In this paper, details of the plasma source and test results are presented.

## 2   Description of the Plasma Source

Figure 1 shows a block diagram of the setup for testing the plasma source. The capacitor bank consists of four 12 mF capacitors connected in parallel.

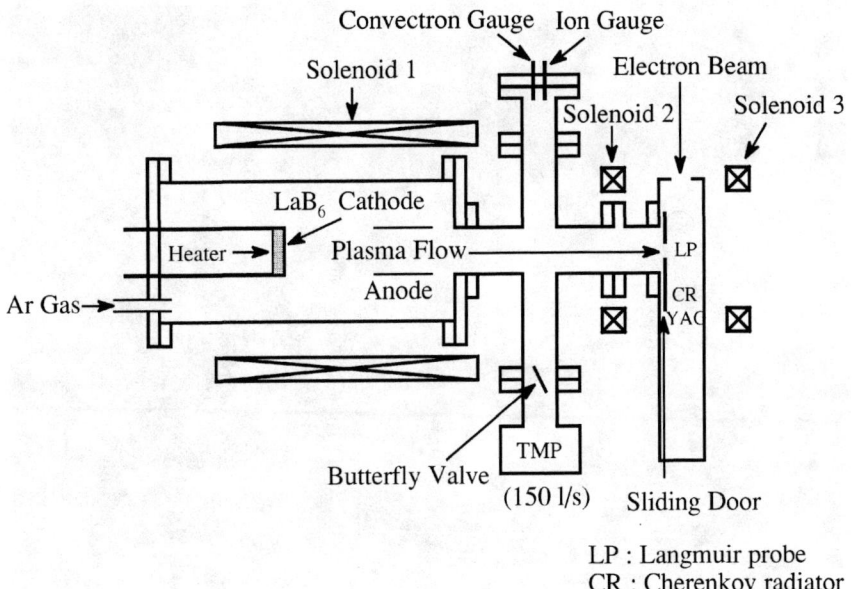

Figure 2: Schematic drawing of the plasma source.

It is charged by the DC charging power supply through a 33.5 Ω resistor and the charging RC time constant for this capacitor bank is 1.6 sec. For plasma generation, the high-current switch box (up to 500A/300V) is switched on by a trigger signal from the pulse/delay generator and the charged capacitor bank is discharged. This discharge current is measured with a current monitor with a sensitivity of 0.01 V/A. At the same time, a voltage sweeper provides a sawtooth voltage to the Langmuir probe for plasma diagnostics. The sawtooth voltage signal is adjustable between -120 V and 120 V. Figure 2 shows a detailed schematic drawing of the plasma source. As shown in the figure, the LaB$_6$ cathode with a diameter of 7 cm is indirectly heated by a tungsten wire heater up to 1330 °C and the discharge pulse from the capacitor bank is applied between the tantalum hollow anode and the grounded cathode, where the separation is 6.7 cm. As a result, an electric discharge occurs and a plasma is generated because the plasma source is filled with an argon gas in mTorr range.

Figure 3 shows a typical discharge current pulse (bottom trace) and a saw-

503

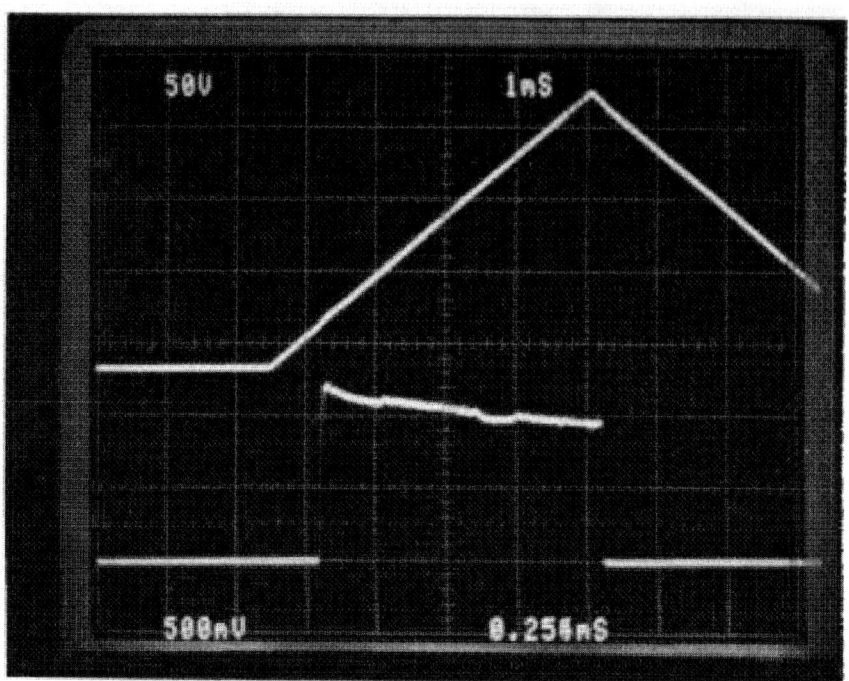

Figure 3: Typical discharge current pulse (bottom trace) and sawtooth voltage signal from the voltage sweeper (top trace). The discharge current pulse was obtained for P=2 mTorr, $V_d$=90 V and $T_c$=1280 °C.

tooth voltage (top trace) applied to the Langmuir probe. The current is about 100 A and its duration is 4 ms. For the current measurement, a discharge voltage of 90 V and a pressure of 2 mTorr were used when the cathode temperature was 1280 °C. Figure 3 also shows that the sawtooth voltage applied to the Langmuir probe is synchronized with the discharge current and it sweeps from -120 V to 75 V in 4.5 ms. The generated plasma is guided by a solenoidal magnetic field to flow towards the long diagnostic chamber. As shown in Fig. 2, a sliding door is located at the entrance of the long diagnostic chamber and

the plasma thickness traversed by the electron beam can be controlled by adjusting the sliding door gap. The gap can be changed from 0 to 7.5 cm. Hence, different longitudinal plasma profiles can be obtained in the long diagnostic chamber, as will be shown below. In the beam-plasma interaction region, a Langmuir probe, which consists of a ceramic tube and a tungsten wire with a 1.43 mm$^2$ surface area, is installed to measure the plasma density and the electron temperature. The Langmuir probe is moveable in the direction of the high energy electron beam path so that it can be used to measure longitudinal plasma density profiles for various conditions. When a high energy electron beam passes through the plasma lens, it is focused downstream and the focal point is located outside the plasma in the case of a thin lens. To get information about time-resolved beam dynamics a Cherenkov radiator is installed in the long diagnostic chamber. In addition, a YAG crystal is located in the chamber to measure the beam size.

# 3   Test Results

The plasma source has been tested for various conditions. As expected, the plasma density and its longitudinal profile in the diagnostic chamber are determined by many experimental parameters, such as gas pressure, discharge current/voltage, cathode temperature, magnetic fields of the solenoids, and sliding door gap. These variables were changed consecutively to find an optimum condition for plasma lens experiments.

First, the magnetic fields of the three solenoids were changed up to 70 G to study the effect of magnetic fields. It was found that the main solenoid in cathode-anode region, denoted by Solenoid 1 in Fig. 2, does not make a significant difference in discharge current and plasma density, while the magnetic field by the two small solenoids, denoted by Solenoid 2 and Solenoid 3 in Fig. 2, significantly affects the plasma density. However, the magnetic field strength in the long diagnostic chamber can not be increased too much because it bends the electron beam path transversely when the beam propagates in the magnetic field region. Hence, there is a certain limit in magnetic field strength. For this reason, the magnetic field in the plasma lens region was set between 50 to 70 G. In this magnetic field region, the plasma density turned out to be large enough for plasma lens experiments. For a fixed magnetic field at the location of the plasma lens, the magnetic field of Solenoid 1 was adjusted to obtain a maximum plasma density in the plasma lens region. It was observed that a maximum plasma density is achieved when the magnetic field is almost uniform from the cathode-anode region to the plasma lens region.

To investigate the dependence of the discharge current on the discharge voltage, pressure and cathode temperature, these parameters were varied. Figure 4(a) shows the discharge current vs. discharge voltage in the pressure range of 0.5 to 8 mTorr when the cathode temperature $T_c$ and the magnetic field are 1280 °C and 53 G, respectively. The discharge current increases as the voltage

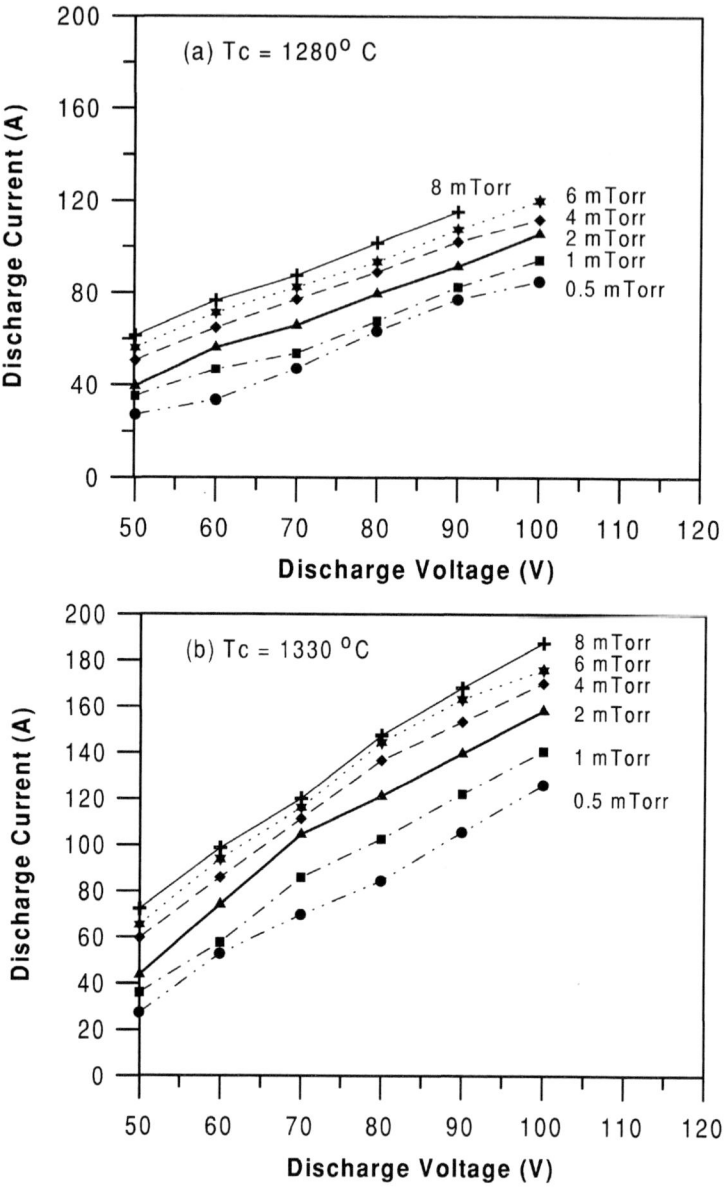

Figure 4: Discharge current vs. discharge voltage for different pressures and different cathode temperatures. (a) $T_c$=1280 °C, (b) $T_c$=1330 °C.

increases for each pressure, and the discharge current also increases when the pressure is increased. However, it should be noted that the effect of pressure on the discharge current is small, compared to that of the discharge voltage. This is because the discharge current is dominated by seed thermal electrons emitted from the cathode. When the discharge voltage was increased beyond 110 V, an arc began to occur and it was more easily produced in the higher pressure range. During arcing, the discharge current increased a lot, but the plasma density in the long diagnostic chamber was observed to remain very low. In order to investigate the effect of the cathode temperature, it was raised to 1330 °C. The result for this case is shown in Fig. 4(b). It shows that the discharge current increases significantly, compared to Fig. 4(a). For 2 mTorr and 100 V, for example, the discharge current increases from 105 A to 158 A as the cathode temperature is raised (i.e., a 50 % current increase was observed). Furthermore, arcing was suppressed noticeably when the cathode temperature was higher.

Plasma densities were measured along diagnostic chamber for different discharge voltages and pressures. Figure 5(a) shows the density measurement when the cathode temperature and the sliding door gap were 1330 °C and 2.5 cm, respectively. The figure shows that the plasma density increases for higher discharge voltages, as expected. It also shows that the three density curves have a maximum around 2 mTorr. This is due to a larger diffusion rate for a higher pressure, i.e., a diffusion rate is proportional to the pressure during transport from the cathode-anode region to the diagnostic chamber, while a plasma production is not proportional to the pressure for a given discharge voltage, as shown in Fig. 4(b). Therefore, the density will decrease beyond a certain pressure. Electron temperature was also measured for different discharge voltages and pressures. Figure 5(b) shows a graph of the electron temperature vs. pressure in the case of $V_d$=100 V. The figure shows that the electron temperature decreases from 7.1 eV to 3.5 eV as the pressure increases.

Plasma density profiles along the beam direction were measured for different conditions to find an optimum condition for the plasma lens experiment. Two cases are shown in Fig. 6(a) and Fig. 6(b). In Fig. 6(a), the sliding door gap is 1.5 cm and the discharge voltage is 100 V. In this case, the peak plasma density and the plasma thickness are too large for underdense plasma lens experiments, as shown in the next section. Hence, the gap was changed to 1 cm to reduce the peak plasma density and the plasma thickness. In this case, however, the peak density was observed to decrease too much so that the discharge voltage was increased to 110 V to compensate the density decrease.

507

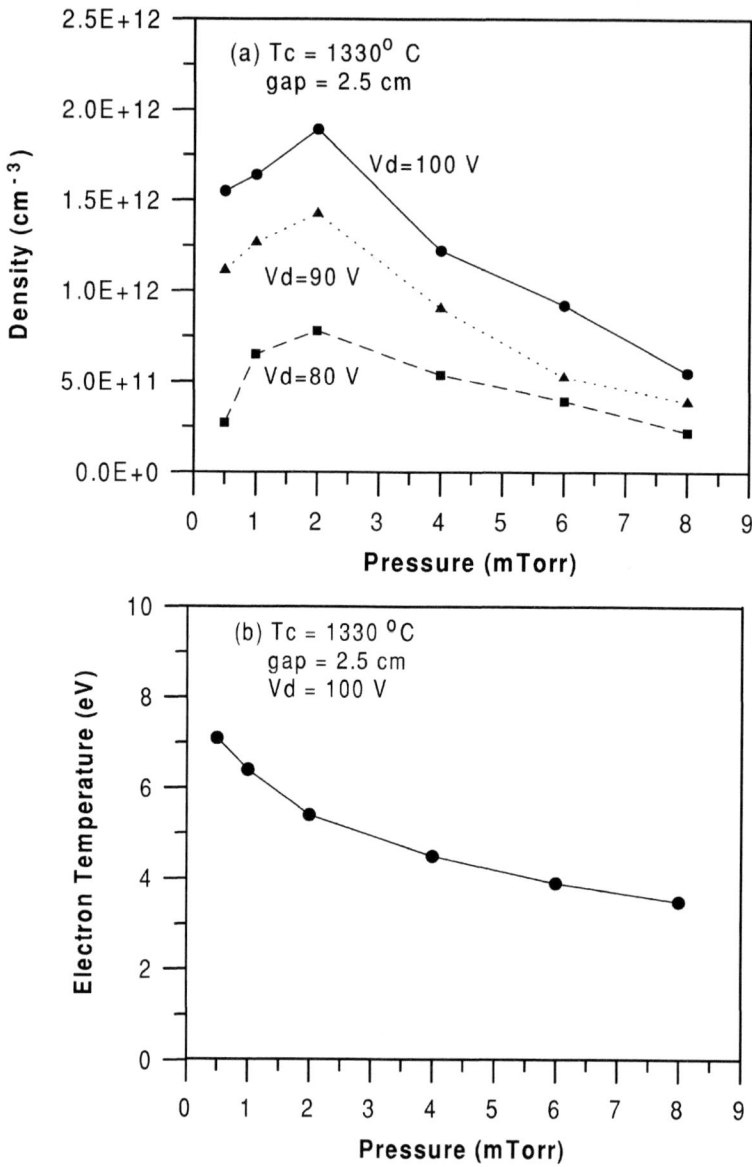

Figure 5: (a)Plasma density vs. pressure for different discharge voltages. The plasma density is measured at the long diagnostic chamber. (b) Electron temperature vs. pressure for $V_d$=100 V.

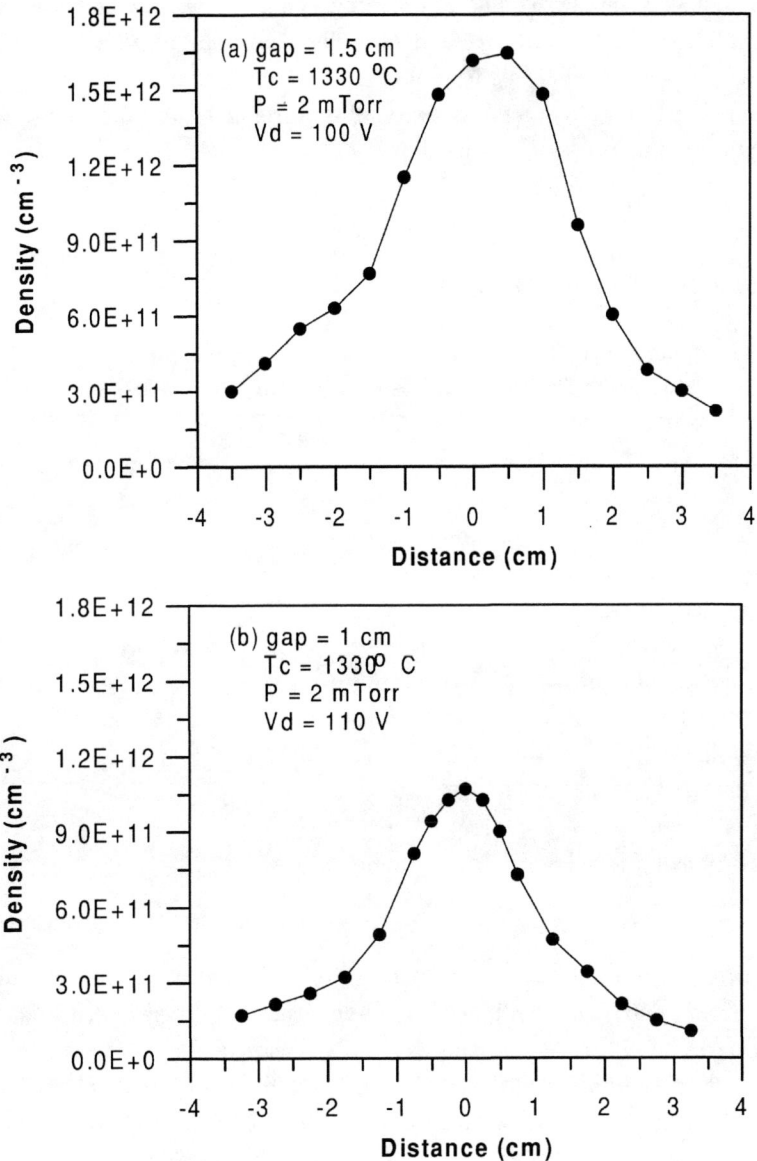

Figure 6: Plasma density profiles along the beam direction. (a) gap=1.5 cm, $V_d$=100 V, (b) gap=1 cm, $V_d$=110 V.

Fig. 6(b) shows this case. The graph shows that the peak plasma density $n_p$ is about $1.1 \times 10^{12}$ cm$^{-3}$ and the effective length of the plasma lens, which is defined by $L_p = \int f(z)dz/n_p$, is about 2.9 cm. Here, $f(z)$ is the density profile function. For these plasma parameters all requirements for underdense plasma lens experiments are satisfied, as shown in the next section.

# 4   Experimental Parameters

Experiments on the underdense plasma lens will be conducted at the UCLA Neptune laboratory. The basic parameters for the planned plasma lens experiment are summarized in Table 1, where the plasma parameters are based on the data shown in Fig. 6(b).

Table 1 : Experimental parameters.
========================================
Electron beam energy $E = 16$ MeV
Charge per bunch $Q = 5$ nC
Normalized beam emittance $\epsilon_n = 20$ mm-mrad
Beam duration $\sigma_t = 25$ ps
Beam radius (rms) at plasma entrance $\sigma_i = 400$ $\mu$m
Beam density $n_b = 3.2 \times 10^{12}$ cm$^{-3}$
Peak plasma density $n_p = 1.1 \times 10^{12}$ cm$^{-3}$
Effective plasma lens length $L_p = 2.9$ cm
========================================

As shown in the table, the beam density $n_b$ is larger than the peak density of the plasma lens, i.e., the plasma is underdense. Furthermore, the quantity $k_p \sigma_z \sqrt{n_b/n_p}$ [9], where $k_p$ is given by $k_p = \sqrt{4\pi r_e n_p}$, is calculated to be 2.5 so that the electron bunch length $\sigma_z$ is long enough to expel plasma electrons from the beam path. Here, the parameter $r_e$ is the classical radius of an electron. The focal length $f$ of this plasma lens can be calculated by $f = 2\gamma c^2 / L_p \omega_p^2$ if the thin lens approximation, i.e., $L_p < f$, is satisfied [10]. In this equation, $\gamma$ is the Lorentz factor and $c$ is the velocity of light. For the parameters of Table 1, the focal length $f$ is calculated as 5.7 cm and is larger than the effective plasma length. Hence, the thin lens approximation is satisfied. For a plasma lens to be useful, its length should be less than $\lambda_\beta/4$, where $\lambda_\beta$ is the betatron wavelength given by $(2\pi\gamma/r_e n_p)^{1/2}$. Otherwise, the beam emittance will increase significantly. In the case of Table 1, $\lambda_\beta/4$ is calculated as 6.4 cm and this is larger than $L_p$. Hence, the requirement of $L_p < \lambda_\beta/4$ is satisfied.

The beam radius at the focal point, $\sigma_f$, can be estimated for the parameters given in this section. The $\beta$-function at the entrance of the plasma lens, $\beta_0$, is calculated as 25.1 cm, and $\beta_0^2$ turns out to be much larger than $f^2$. In this case, the beam radius at the focal point is approximately given by $\sigma_f \simeq f\epsilon_n/\gamma\sigma_i$ [10], where $\sigma_i$ is the initial beam radius at the entrance of the plasma lens, and $\sigma_f$ is calculated to be 91 $\mu$m. Therefore, the beam size will be reduced by a factor of 4.4, as a result of plasma focusing. The beam size and the time-resolved beam profile will be measured at different positions along the beam path by using the YAG crystal, the Cherenkov radiator, and a psec-resolution streak camera.

In this plasma lens experiment, there is one potential problem related with bending of the electron beam path. When the beam propagates in the plasma lens region, its path will be bent because of the transverse magnetic field from the solenoids. To investigate this issue, the transverse magnetic field was measured along the beam path and this result was used to calculate the beam trajectory. In Fig. 7, the measured magnetic field at different positions is represented by the round dots and a Gaussian profile is used for best fitting. The resulting Gaussian function was used for calculation of the beam trajectory. According to this calculation, the beam is deflected in the magnetic field region, as represented by the dotted line in the figure. The calculation shows that the beam is deflected by 1.3 mm at the position of the focal point ($f=5.7$ cm) and deflected by 4 mm at a position of 20 cm downstream the plasma lens. This will be a problem for Cherenkov radiation measurements. Hence, a couple of methods, such as adding a special iron structure to reconfigure the magnetic field shape, using a small Helmholtz coil to compensate for the magnetic field, etc., will be investigated to solve this problem.

# 5 Summary

A LaB$_6$-based discharge plasma source was developed for thin underdense plasma lens experiments at UCLA and it was tested extensively for various experimental conditions. The results show that the plasma source can provide satisfactory plasma parameters for thin underdense plasma-lens experiments, i.e., plasma densities in low $10^{12}$ cm$^{-3}$ range and an effective length of a few cm. The plasma source is expected to be assembled with the Neptune Laboratory beamline by the end of this year, after beam characterizations of the new

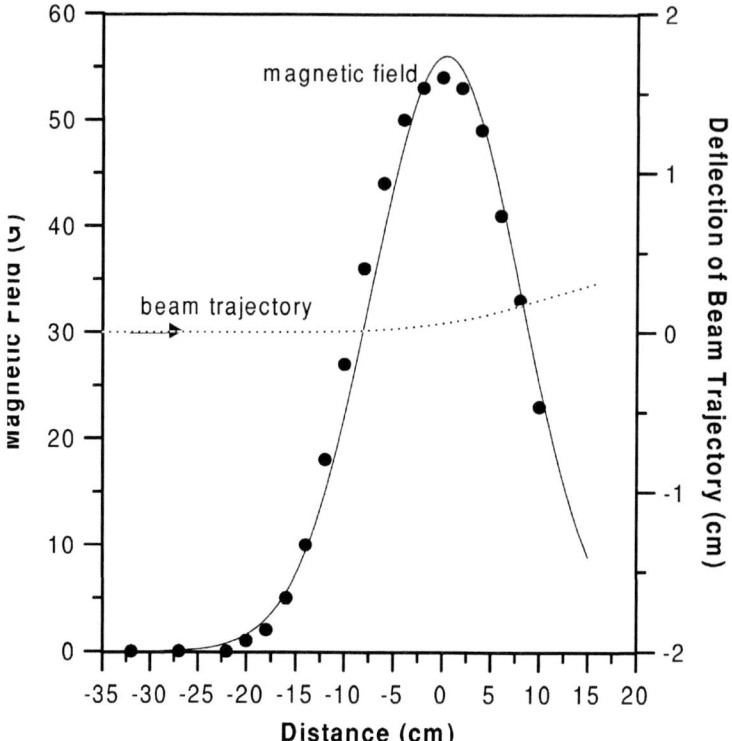

Figure 7: Measured magnetic field profile along the beam direction (round dots) and calculated deflection of the beam trajectory (dotted line) for E=16 MeV. The solid line is a Gaussian fit to the measured magnetic field.

1.625-cell photoinjector gun.

## Acknowledgements

The authors of the paper would like to thank Dr. G. Hairapetian for construction and initial measurements of the prototype for the plasma source.

# References

[1] P. Chen, Part. Accel. **20**, 171 (1987).

[2] T. Katsouleas *et al.*, Phys. Fluids B **2**, 1384 (1990).

[3] W. E. Martin *et al.*, Phys. Rev. Lett. **54** 685 (1985).

[4] N. Barov and J. Rosenzweig, Phys. Rev. E **49**, 4407 (1994).

[5] H. Nakanishi *et al.*, Phys. Rev. Lett. **66**, 1870 (1991).

[6] J. J. Su *et al.*, Phys. Rev. A **41**, 3321 (1990).

[7] G. Hairapetian *et al.*, Phys. Rev. Lett. **72**, 2403 (1994).

[8] N. Barov *et al.*, Phys. Rev. Lett. **80**, 81 (1998).

[9] J. Rosenzweig *et al.*, Nucl. Intrs. Meth. A **410**, 437 (1998).

[10] J. Davis, Ph.D. Dissertation, Dept. of Electrical Engineering, UCLA (1996).

# Guiding of High Intensity Ultrashort Laser Pulses in Plasma Channels Produced with the Dual Laser Pulse Ignitor-Heater Technique.

P. Volfbeyn and W. P. Leemans,

*LBNL, Berkeley, CA 94720*

**Abstract.** We present results of experimental investigations of laser guiding in plasma channels. A new technique for plasma channel creation, the Ignitor – Heater scheme is proposed and experimentally tested in hydrogen and nitrogen. It makes use of two laser pulses. The Ignitor, an ultrashort (<100 fs) laser pulse, is brought to a line focus using a cylindrical lens to ionize the gas. The Heater pulse (160 ps long) is used subsequently to heat the existing spark via inverse Bremsstrahlung. The hydrodynamic shock expansion creates a partially evacuated plasma channel with a density minimum on axis. Such a channel has properties of an optical waveguide. This technique allows creation of plasma channels in low atomic number gases, such as hydrogen, which is of importance for guiding of highly intense laser pulses. The channel density was diagnosed with time resolved longitudinal interferometry. From these measurements the plasma temperature was inferred. The guiding properties of the channels were tested by injecting a $>5 \times 10^{17}$ W/cm$^2$, 75 fs laser pulse.

## 1 INTRODUCTION

Since the original proposal of laser driven plasma accelerators [1], where it was proposed that plasma waves with extremely high longitudinal electric fields and with phase velocity close to the speed of light can be excited by ponderomotive pressure of a laser pulse, the field has progressed past the proof-of-principle stage (for a recent review see E. Esarey et al. [2]).

The most severe limit on the energy gain in a homogeneous plasma LWFA is laser diffraction. To reach the high intensities needed for plasma wave excitations, on the order of $10^{18}$ W/cm$^2$, the laser pulse (or pulses) must be focused to spotsizes on the order of several laser wavelengths. In order for laser acceleration to achieve higher energies, the laser beam must remain

CP472, *Advanced Accelerator Concepts: Eighth Workshop,*
edited by W. Lawson, C. Bellamy, and D. Brosius
© 1999 The American Institute of Physics 1-56396-889-4/99/$15.00

tightly focused over distances of many Rayleigh ranges (diffraction distance in vacuum).

Laser guiding in plasma channels has been proposed [3] as a means to extend the distance over which the laser remains intense. The index of refraction in a plasma of density n can be approximated by $\eta_R \approx 1 - \omega_p^2/2\omega^2$. As in an optical fiber, a plasma channel can provide optical guiding if the index of refraction peaks on axis. This requires a plasma density profile that has a local minimum on axis.

Experimentally, laser pulses have been guided in plasma channels [4]. In these experiments one laser pulse (~100 ps, ~100 mJ) was brought to a line focus in a mixture of high atomic number (Z) gases with an axicon lens to produce a few cm long plasma, and subsequently heat it via inverse Bremsstrahlung. The resulting hydrodynamic expansion led to a time-dependent density profile with a minimum on-axis. Pulse propagation over distances of up to 70 Rayleigh lengths (about 2.2 cm) of moderately intense laser pulses ($<5 \times 10^{14}$ W/cm$^2$), with pulse lengths much larger than the plasma period, was demonstrated in these experiments. The intensity of the channel creation laser pulse, achieved in these experiments was not sufficient for ionization of low Z gases, but required instead the use of high Z gases. Unfortunately, in channels produced with high Z gases, an ultra-intense pulse would further ionize the gas on the channel axis, thereby negating the guiding. While the 100ps long laser pulse was energetic enough to cause significant plasma heating, ionization of low Z atoms requires an order of magnitude higher laser intensity. The intensity of the laser pulse, needed for channel creation in hydrogen, for instance, has to be ~$1.5 \times 10^{14}$ W/cm$^2$ [5]. It is important to note that once partial ionization has occurred, and plasma heating through inverse Bremsstrahlung takes place, collisional ionization plays a significant role as an additional mechanism for plasma creation. For low laser intensities the inverse Bremsstrahlung rate is independent of the laser intensity [6]. When the electron quiver velocity due to the laser field exceeds the thermal velocity, the heating crossection drops precipitously [7]. The optimum intensity for plasma heating is on the order of $1-2 \times 10^{13}$ W/cm$^2$ [6]. Ionizing the gas and subsequently heating it with one pulse is inefficient.

To allow the use of low Z atoms and to demonstrate the feasibility of guiding of the highly intense laser pulses over many Rayleigh ranges we have developed a novel method for channel production: the Ignitor - Heater technique. Rather than utilizing a single laser pulse for ionization and heating, this scheme makes use of two laser pulses. A femtosecond "Ignitor" pulse is used to create the initial spark. A longer, ~160 ps, perfectly time synchronized "Heater" pulse is then introduced to heat the plasma. Results of Ignitor –

Heater channel production experiments and measurements of the channel transverse plasma density profile with femtosecond Mach-Zehnder interferometry will be presented in Section 2. Results of guiding studies are reported in Section 3.

## 2 CHANNEL PRODUCTION

To implement the Ignitor-Heater channel creation scheme, the two laser pulses were combined in a line-focus by means of cylindrical optics onto a gas jet, Figure 1.

A gas jet was used to avoid ionization induced refraction [8] in a statically filled experimental chamber. The femtosecond intense (~5 $10^{14}$ W/cm$^2$) Ignitor pulse was focused to a line by reflecting off a cylindrical reflector. The cylindrical reflector is a plano-concave (R=38 mm) cylindrical lens, coated with a dielectric high reflection coating for 45° angle of incidence 800 nm radiation. By using a reflective optic we have avoided beam filamentation, self-focusing, and other undesirable nonlinear effects that would prevent from obtaining a well focused, near diffraction limited beam spot. The Heater pulse was focused with an F/5 refractive cylindrical lens (focal length fl=50mm) at the exact location of the ignitor focus.

In addition to the fact that the channel forming beams propagate perpendicularly to the guided pulse, the use of two independent cylindrical optics provides precise independent adjustment of both the positions, angles of incidence, and sizes of the line foci.

A Mach-Zehnder type interferometer with a measured spatial resolution of 4 μm was built to measure line integrated plasma density. This interferometer measures the relative spatial phase shift between two blue (400 nm) 50 fs pulses, one propagating through plasma and one through air. These pulses are produced by frequency doubling and are perfectly synchronized with the high power beams used in plasma production. The evolution of the 2-D transverse plasma density profile can be measured with a temporal resolution determined by the duration of the blue pulse.

The Ignitor-Heater scheme was implemented with 20-40mJ, 75 fs Ignitor pulse and ~270 mJ, 160 ps Heater pulse in nitrogen and hydrogen backed gas jet. Figure 2 a) shows an interferogram taken with the interferometer pulse delayed by 1 ns with respect to the Heater pulse. The X-size of the channels roughly corresponds to the Rayleigh range of the Ignitor pulse. From the inferred plasma density lineouts of Figure 2 b), it is seen that a plasma channel is created only in the vertical direction. These channels are expected to guide in Y-direction only.

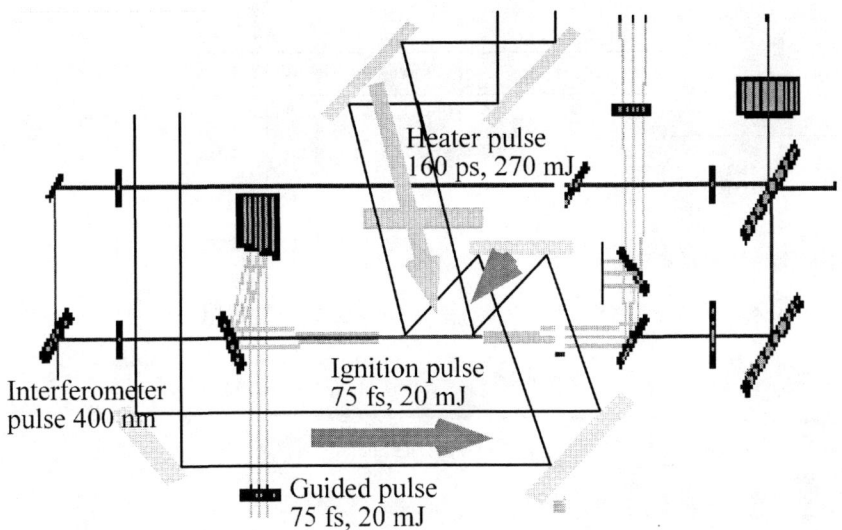

Heater pulse
160 ps, 270 mJ

Interferometer
pulse 400 nm

Ignition pulse
75 fs, 20 mJ

Guided pulse
75 fs, 20 mJ

**FIGURE 1.** Experimental Setup.

From the interferograms of the plasma channels, the shock front diameter, D, is found by measuring the separation (in Y) between the points of commencement of the fringe shifts in the middle section of the channel. From the channel size and density dynamics, the initial temperature of the spark is inferred in two ways. By equating the shock speed to ion acoustic speed, the electron temperature can be found. From Sedov's solution of strong explosion in a homogeneous atmosphere [9], a theoretical calculation that relates the energy per unit length in the initial spark to the form of the expansion curve, the temperature can be calculated once again. From the shock speed, the initial temperatures (right after the Heater pulse) are calculated to be ~20 eV and ~120 eV in hydrogen and nitrogen respectively. This agrees well with the deposited energy calculation from Sedov' solution. From the inverse Bremsstrahlung theory [7], in hydrogen, with $n_i=2\times10^{18}$ cm$^{-3}$, laser $I=7\times10^{12}$ W/cm$^2$, and laser pulse duration w=150 ps, the temperature was calculated to be $T_e=19$ eV. In nitrogen, with $n_i=1.6\times10^{18}$ cm$^{-3}$, $<Z>=3.5$, and the same laser parameters, the temperature was calculated to be Te=118 eV.

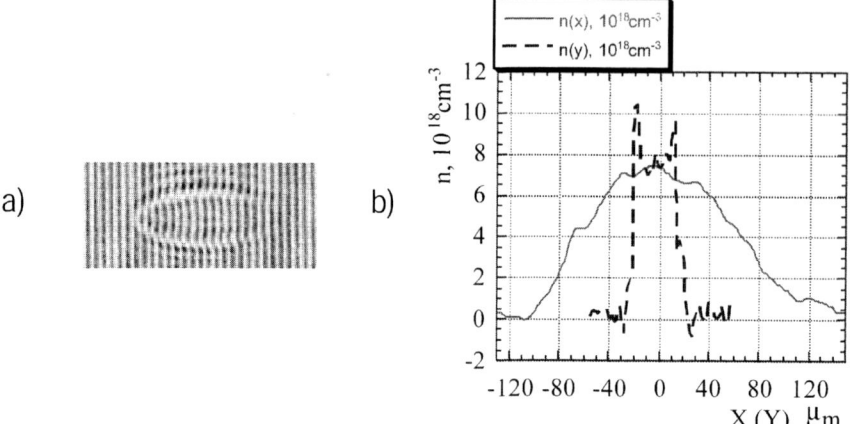

**FIGURE 2.** a) Channel interferogram at 1ns after the heater pulse; b) Inferred plasma density lineouts.

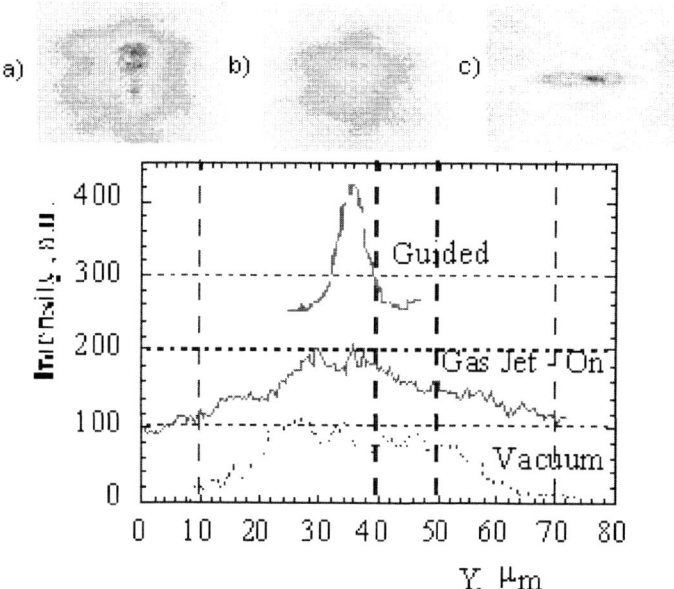

**FIGURE 3.** Laser beam images and vertical lineouts. pulse a) gas jet turned off, b) gas jet - on, without Heater pulse, and c) guided by the channel, gas jet backed with nitrogen at 1000 psi.

# 3 GUIDING

This Section describes the results of experiments on guiding high intensity laser pulses in the plasma channel. The laser pulse (injection pulse) was focused near the entrance of the channel using an off-axis parabola. The time delay between the Ignitor pulse and the injection pulse was fixed to 600 ps. (This constraint arose from physical limitations in the available vacuum chamber.)

To diagnose the guiding, the laser beam was imaged onto a CCD camera with a $MgF_2$ lens of 1 inch diameter and focal length of fl=68.3 mm at 800 nm. The CCD camera was mounted on an optical rail so that it could be moved over about 50 cm range, thus changing the position of the imaging plane. The resolution and magnification of the imaging system was calibrated for different CCD camera locations. By comparing the laser beam images with and without the gas flowing out of the gas jet (valve pulsing or not), it was possible to clearly observe the effect of guiding on the laser beam.

Figure 3 shows images of the injection laser pulse a) propagating through vacuum (gas jet turned off), b) after undergoing ionization induced refraction in the gas jet plume without the Heater pulse being present, hence no channel formed, and c) guided by the channel, for a gas jet backed with nitrogen at 1000 psi. Vertical lineouts of images of Figure 3 clearly demonstrate the changes induced by the plasma channel on the guided laser pulse. The change in size of ~8 times is consistent with a laser beam of $Z_R$~0.1 mm propagating a distance of 0.8 mm (the width of the jet).

As seen in Figure 2, for the specific Ignitor and Heater pulse parameters, plasma channels were created in an elongated, elliptical shape. In turn, the guided beam images (Figure 3) had a similar elongated shape. Through control of the Ignitor pulse intensity, channels with circular cross-sections, possessing guiding properties in X as well as in Y direction, can be created [6].

A study of the guided beam image vs. the CCD camera position is used to find the guiding length and to prove that, as the gas jet is displaced, the guided beam waist is shifted accordingly. The guided laser beam size was measured vs. z by moving the CCD camera. Measurements were performed at two different gas jet z-positions, z=0.59 mm and z=1.02 mm. The positions of the vertical waist (smallest vertical spotsizes) in the two cases are ~0.9 mm and 1.25 mm respectively (while the waist position of the injection beam in vacuum is at z=0 mm). The change in the waist position is in agreement with the measured gas jet size of FWHM~0.8 mm.

A set of data was taken for several different gas jet z-positions. The CCD camera was set at the point of smallest vertical spotsize for each gas jet z-position. The corresponding object plane locations were then plotted vs. the gas jet z-positions. The resultant graph is shown in Figure 4. This clearly shows that moving the gas jet moves the beam waist.

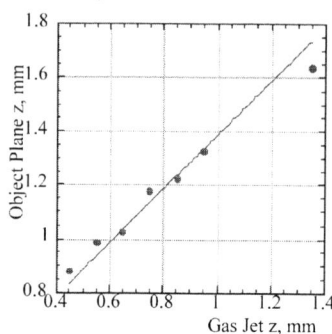

**FIGURE 4.** Changes in best camera position for imaging the guided lobe (vertical spotsize at minimum) vs. Gas Jet position. The line in the plot is drawn at 45° to the axes.

# 3.1 Mode Coupling and Propagation in Realistic Plasma Channels

A plasma channel with a parabolic density profile with infinite radial extent, $\Delta n_p = \Delta n r^2 / r_0^2$, supports a Gaussian guided mode $a^2 \propto \exp(-2r^2 / r_0^2)$, provided that the channel depth satisfies $\Delta n = \Delta n_c$, where $\Delta n_c = (\pi r_e r_0^2)^{-1}$. If the injection is done with a perfectly Gaussian laser beam focused to a waist size of $r_0$ at the channels entrance then 100% of the injection laser energy will couple into the guided mode. If a perfectly Gaussian laser beam is not at its waist at the channel entrance, or if the waist size is different from the guided mode size for that particular channel, a corresponding fraction of the laser energy will couple to higher order modes in the channel. This loss mechanism we will refer to as the coupling loss.

Another loss mechanism in a realistic channel results from the laser tunneling through the channel walls. Experimentally created channels do not have infinitely high walls. Rather, as the profile in Figure 2 b), the channel walls will reach a peak height and then fall off rapidly as shown in Figure 5.

520

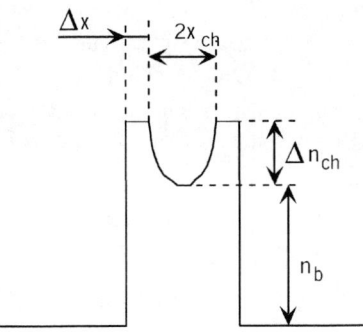

**FIGURE 5.** Schematic of a realistic plasma channel profile.

In our experiments, the channel depth, width, and wall thickness at 600 ps after ignition were $\Delta n_{ch} \approx 1.5 \times 10^{18} \, cm^{-3}$, $n_b \approx 7 \times 10^{18} \, cm^{-3}$, $x_{ch} \approx 8 \, \mu m$, and $\Delta x \approx 3 \, \mu m$ respectively. In such a channel there are no bound modes. The laser can leak out through the finite thickness walls, coupling to the continuum outside the channel. The degree of this leakiness depends on the channel depth, $\Delta n_{ch}$, width, 2xch, and the wall thickness, $\Delta x$. The leakage loss is much larger for higher order modes, and any power not coupled into the fundamental mode of the channel is rapidly lost. Because of the finite transverse extent of the channel, peripheral parts of the laser pulse may miss the plasma all-together, spilling outside the channel, the spillage.

The loss due to leakage can be estimated [6], assuming that the channel mode is close to the fundamental mode of the infinite wall channel. By calculating the leakage exponent of the evanescent mode within the channel wall from the paraxial wave equation, the energy loss from the guided mode through tunneling is estimated to be ~ 35%.

As the gas jet is moved further away from the laser focus, the mode coupling into the channel worsens, i.e. larger portions of the injected laser energy are not coupled into the channel, but rather spilled outside the channel. Quantitatively, the image intensity integral is a measure of the laser energy. By taking intensity integrals of the CCD images, we will compare the amount of the laser energy guided by the channel to the total intensity integral of the CCD image (full beam energy, $E_{full}$) and the image intensity integral of the vacuum propagated beam ($E_{vacuum}$). The intensity integral of the isolated central lobe of the images is taken to be the guided energy.

The ratio of the guided energy, $E_g$, to the total image intensity integral, $E_{full}$ is shown in Figure 6. As expected, the fraction of the laser beam that is coupled into the channel is larger as the injection pulse focus is moved closer to the jet's edge than when the gas jet is moved further away from the laser focus position in vacuum. It should be noted that the F-numbers of the off-axis parabola and the imaging $MgF_2$ lens are rather close and the collection angle is limited. Nevertheless, the leakage of the fundamental mode of the plasma channel is fully collected.

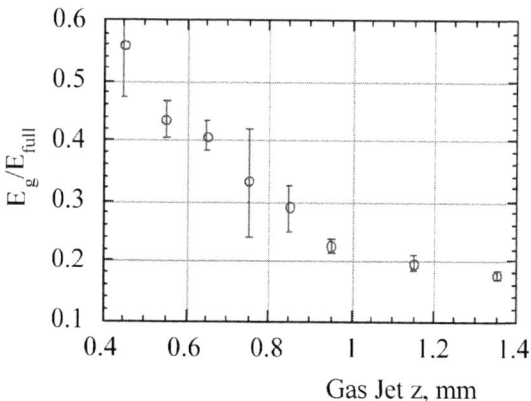

**FIGURE 6.** Ratio of energy in the central lobe or guided, $E_g$, to that of the full beam, $E_{full}$, plotted vs. the gas jet position.

As the laser waist is moved closer to the channel entrance, the laser spotsize becomes comparable with the channel size and most of the power is coupled to the modes of the channel. The ratio of $E_g$ to the $E_{full}$ approaches ~60 %. This number is in good agreement with previous results where the leakage fraction was calculated to be ~35% (i.e. $E_g/E_{full}$~ 65%).

## 4 CONCLUSIONS

To overcome the laser diffraction length limit, a novel method of plasma channel production for laser guiding, the Ignitor - Heater technique, was proposed and tested experimentally. This scheme made it possible, to create preformed guiding plasma channels in hydrogen and deeply ionized nitrogen without high atomic number additives, thereby allowing high intensity laser pulse guiding. To avoid the ionization induced refraction of the guided laser pulse, the plasma channels were formed in a plume of a pulsed gas jet. It should be also noted that the Ignitor – Heater scheme employs cylindrical optics that could be kept out of the path of the accelerator beam and, potentially, allow the recycling of the laser beams. The channel formation process was fully characterized with time resolved 2-D longitudinal interferometry diagnostic using a femtosecond probe pulse. From the measured dynamics of the radial shock expansion, the temperature and energy of the heated plasma were calculated. The ability to independently control the intensity of the Ignitor pulse allowed us to control the transverse extent of the initial ionization spark. The length of the initial spark affected the shape of the plasma channel. In this fashion channel transverse aspect ratio was controlled from ~ 3 to ~10. Future work will concentrate on further improving this aspect ratio.

Laser pulses at record high intensity ($\sim 5 \times 10^{17}$ W/cm$^2$) were guided in these channels over ~ 10 Rayleigh lengths. Control over the channel shape allowed us to observe guiding in one transverse dimension, for channels with high aspect ratio, or guiding in both X and Y, if a round channel was formed.

Insertion losses were measured as functions of the gas jet position with respect to the vacuum focus position of the injection laser pulse. Spillage and leakage mechanisms were found to agree well with the theoretical predictions.

## REFERENCES

1. Tajima, T. and J.M. Dawson, Phys. Rev. Lett. **43**, pp. 267-270 (1979).
2. Esarey, E. et al., IEEE Trans. Plasma Sci., vol. PS-24, pp. 252-288 (1996); E. Esarey et al., IEEE J. Quant. Electron. 33, pp. 1879-1914 (1997).
3. Tajima,T., Laser Part. Beams, vol. 3, pp. 351-413, 1985
4. Durfee, C.G. III and H. M. Milchberg, Phys. Rev. Lett., vol. 71, pp. 2409, 1993
5. Augst, S. et al., PRL, vol. 63, no. 20, 2212 (1989)
6. Volfbeyn, P., Ph.D. Thesis, MIT, June 1998
7. Shvets, G. and N. J. Fisch, Phys. Plasmas, Vol. 4, No. 2, 428 (1997), C. D. Decker et al., Phys. Plasmas 1 (12), 4043 (1994)
8. Leemans, W.P. et al., Phys. Rev. A, 46, 1091 (1992)
9. Sedov, L. I. "Propagation of strong blast waves", Prikl. Mat. I Mekh. v. 10, 241-250 (1946).

# Wakefield Accelerators In The Blowout Regime With Mobile Ions

S. Lee and T. Katsouleas

*Department of Electrical Engineering/Electrophysics,
University of Southern California, Los Angeles, CA 90089*

**Abstract.** In the Plasma Wakefield Accelerator a high current drive-beam excites a large wake that can accelerate trailing particles. The wake is created when the space charge of the drive beam displaces plasma electrons. The plasma ions provide the restoring force on the displaced electrons. For symmetric bunches, the peak accelerating gradient is proportional to the current over a pulse length. For example, for a Gaussian bunch with 6 $n$C of charge and bunch length $\sigma_z \approx 0.6$mm, a gradient of 1GeV/m can be obtained. For the case of dense (beam density greater than plasma density), narrow (beam spot size $\sigma_r$ smaller than $c/\omega_p$ ) beams the plasma response is non-linear and is dominated by the radial blow out of all the plasma electrons. However, such dense beams are strongly focused by the plasma lens effect. As a result they become so dense that ion motion should become important even on the electron plasma frequency time-scale. We will present analytic and 2-D particle-in-cell (PIC) models of wake excitation including mobile ions. The effect of the ion motion on the accelerating and focusing wake and the dynamics of the drive beam are discussed.

## INTRODUCTION

In the usual description of the Plasma Wakefield Accelerator [4], a high-current drive-beam excites a large wake that can accelerate trailing particles. The wake is created as the space charge of the drive-beam displaces plasma electrons. The plasma ions provide the restoring force on the displaced electrons. For symmetric drive-beams, the wake amplitude is proportional to the current over the pulse length. For example, a Gaussian electron beam with bunch length $\sigma_z=0.6$ mm and 6 nCoulombs of charge can generate a wake amplitude (i.e., peak accelerating gradient) of 1 GeV/m [4]. For the case of dense (beam density $n_b$ greater than plasma density $n_o$ ) and narrow (spot size less than plasma skin depth $c/\omega_p$ , $\omega_p = \sqrt{4\pi\,n_o e^2/m}$ ) beams, the plasma response is nonlinear and is dominated by the radial blowout of all the plasma electrons [5]. However, such dense beams are strongly focused by the plasma lens effect. As a result they become so dense that ion motion can become important even on the electron plasma frequency time scale. In this paper we present analytic models and 2-D particle-in-cell (PIC) simulations of the PWFA with mobile ions. We model the effect of ion motion on the longitudinal (accelerating) and transverse (focusing)

CP472, *Advanced Accelerator Concepts: Eighth Workshop,*
edited by W. Lawson, C. Bellamy, and D. Brosius

wakefield. We then use these results to assess the impact of ion motion on the beam dynamics of the driving electron beam.

The remainder of the paper is organized as follows; In Sec. II we review the dynamics of dense electron beams in plasmas and use envelope equations to estimate the conditions for which ion motion becomes important in the PWFA. In Sec. III, we present analytic models for the ion dynamics in a PWFA. In Sec. IV 2-D self-consistent PIC simulations are presented and compared to the analytic predictions and to simulations in which the ions are immobile. In Sec. V the effects of the ion motion on drive-beam propagation are assessed by modifying the formalism of Sec. II.

## REVIEW OF DRIVE-BEAM DYNAMICS AND CONDITIONS FOR SIGNIFICANT ION MOTION

As described in Sec. I intense beams of interest for generating large wakes are also subject to strong focusing by the plasma. This is essentially the mechanism of an underdense plasma lens [6]. In this section we review the envelope equation describing the evolution of such beams and use this to estimate the conditions for which ion motion can no longer be neglected.

The equation describing the envelope of a relativistic particle beam traveling in the z-direction in a focusing channel is well known [7];

$$\frac{d^2\sigma}{dz^2} + K\sigma - \frac{\varepsilon_N^2}{\sigma^3} - \frac{\rho}{\sigma} = 0 ,$$

(1)

Where $\sigma$ is the transverse spot size of the beam (in the x or y direction), $\varepsilon_N$ is the rms normalized emittance of the beam (=area in transverse $p_x - x$ or $p_y - y$ space/$\pi$) and $\rho$ is the space charge parameter (=$(2I/I_A)\sigma_o^2/\varepsilon_N^2$, $I_A$=17,000$\beta\gamma$ Amps). For relativistic drive-beams, the space charge term is generally very small compared to the emittance term (i.e. emittance-dominated beams) and can be neglected.

Here $K$ is the focusing strength ($K = (F_x/x)/\gamma mc^2$). For dense drive-beams ($n_b > n_o$), plasma electrons are rapidly expelled by the beam's space charge leaving a focusing column of ions of density $n_o$. The focusing strength of the ion column is easily estimated from Gauss' law to be $K = 2\pi n_o e^2/\gamma mc^2 = (1/2\gamma)\omega_p^2/c^2$. Equation (1) describes the periodic focusing (i.e., betatron oscillations) of the beam. We can easily find the peak density of the beam by multiplying equation (1) by $d\sigma/dz$ and integrating. This gives

$$\sigma'^2 + K\sigma^2 + \varepsilon^2/\sigma^2 = \text{constant} ,$$

(2)

, where we have neglected the space charge term.

525

For the purpose of illustration, consider a beam entering the plasma at a waist $(\sigma'=a)$ of size $\sigma_o$ that is significantly larger than the matched beam radius $\sigma_{eq}$ $(\sigma_{eq}=[\varepsilon^2/K]^{1/4}$ from equation (1)). From Eq. (2) the beam focuses to a minimum $(\sigma'=0)$ at $K\sigma^2+\varepsilon^2/\sigma^2 = K\sigma_o^2+\varepsilon^2/\sigma_o^2$ or $\varepsilon^2/\sigma^2 \approx K\sigma_o^2$. The density of the beam at this focus is approximately

$$n_b = \frac{N}{2\pi\sqrt{2\pi}\sigma^2\sigma_z} \approx \frac{I}{2\pi\sigma^2 e}c^{-1} \,, \tag{3}$$

Substituting for $\sigma^2$ and K from above gives the peak beam density normalized to the plasma density;

$$\frac{n_b}{n_o} = I\frac{\sigma_o^2}{\varepsilon^2}\frac{e}{\gamma mc^2} = \frac{I}{I_A}\frac{\sigma^2}{\varepsilon^2} \,, \tag{4}$$

As we shall verify in the next section, ions respond on a time scale of the ion beam-plasma frequency $\omega_{bi}^{-1} = \sqrt{M/4\pi\, n_b e^2}$; thus ion motion becomes important on an electron plasma frequency time scale when $M/n_b$ is comparable to $m/n_o$ or $n_b/n_0 > M/m_0$. From (4) we see that ion motion can be significant when

$$\frac{I}{I_A}\frac{\sigma_o^2}{\varepsilon^2}(=\frac{\rho}{2}\gamma^2) \;>\; \frac{M}{m} \,, \tag{5}$$

As an example, consider the parameters of the SLC proposed experiment [3]; I=1.3 k Amps, $\sigma_o = 0.63\text{mm}$, $\varepsilon = 0.6\text{mm}-\text{mrad}$. Then the left-hand side is approximately 20,000 which is large compared to $M/m \approx 13{,}000$ for the Lithium plasma (Z=3). Thus ion motion will be significant. We comment that ion motion can be prevented by "spoiling the emittance of the beam" (e.g., by scattering of the beam upon passage through a foil) so that it cannot pinch to too small radius.

## ANALYTIC MODELS

In this section we describe analytic models for the motion of plasma ions in the blowout regime of the PWFA. In this regime the head of the electron driver is assumed to have expelled all of the plasma electrons from the region occupied by the main body of the drive-beam. The ions are attracted toward the axis by the negative space charge of the drive-beam and eventually repelled by their own space charge. To analyze the ion dynamics we consider a sheet or ring model (Fig. 1). Identical results can be obtained by treating the ions as a "stationary beam" and solving their beam envelope equation (i.e., using equation (1) with $\varepsilon=0$, $\gamma=1$ and the appropriate $K$).

For simplicity we consider initially a cylindrical uniform electron beam of radius $\sigma \approx r_b$ and density $n_b$. We treat the plasma ions as a group of concentric rings as illustrated in Fig. 1.

Consider an ion ring of initial radius $r_o \leq r_b$ pulled in by the electron beam by amount $\xi$ to a new position $r = r_o - \xi$.

The equation of motion for the ion is

$$F_i = M\frac{d^2\xi}{dt^2} \qquad (6)$$

From Gauss' law the force on the ion is

$$F_i = -2\pi \, n_b e^2 r + 2\pi \, n_i e^2 r \, , \qquad (7)$$

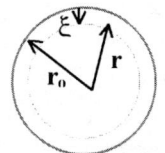

**FIGURE 1.** Model of electron-ion ring sheet.

Where $n_i$ is the density of ions. $n_i$ is initially $n_0 (<< n_b)$ but increases as the rings collapse. If the sheets don't cross (an assumption to be checked later), the force on the collapsing ion ring due to the other ions increases as $n_i r = (n_o r_o^2 / r^2) r = n_o r_o^2 / r$.

The equation of motion may then be written in terms of r or $\xi$. In terms of r, it is written

$$M\frac{d^2 r}{dt^2} + 2\pi \, n_b e^2 (r - \delta/r) = 0 \quad , \qquad (8)$$

where $\delta = n_o r_o^2 / n_b$. Note that this has the form of the envelope equation (equation (1) with $\varepsilon=0$). The solution is "betatron" oscillations of the ions at a frequency $\omega \approx \sqrt{2\pi \, n_b e^2 / M} \equiv \omega_{bi} / \sqrt{2}$.

All of the rings within the electron beam initially collapse according to $r = r_o \cos(\omega_{bi} t / \sqrt{2})$. Thus rings which start within the electron beam radius never cross. The collapse stops at a turning point for the ring at a minimum radius $r_o e^{-n_b / 2 n_o}$. The last expression follows from an integration of equation (8) (as was done in Sec. II). From this result we can describe the behavior of the focusing wake in the PWFA including ion dynamics. When $n_b >> n_o$, the focusing wake rises rapidly from zero at the head of the beam to a value of $F_r = 2\pi \, n_o e^2 r$ due to the electron

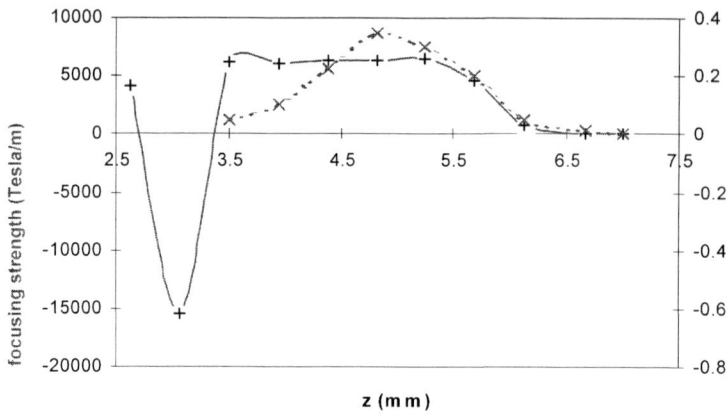

**FIGURE 2.** Focusing strength and axial current $J_z$ vs. longitudinal position z in a PIC simulation, with immobile ions. $\sigma_r$=75 μm, $n_o$=2.1e14 cm$^{-3}$

blow out. This behavior is illustrated in Fig. 2 (solid curve) which shows the focusing wake in a PIC simulation with immobile ions.

Later in the bunch the wake can increase due to the increasing ion density. The ion density rises initially as, $n_i \approx n_o r_o^2/r^2 \approx n_o/\cos^2\left(\omega_{bi}t/\sqrt{2}\right)$, causing the wake to increase as $F_r = 2\pi n_o e^2 r/\cos^2\left(\omega_{bi}\xi/\sqrt{2}c\right)$, where $\xi$ is distance measured from the head of the beam. After a distance $\xi \approx \pi c/\sqrt{2}\omega_{bi}$, the focusing force peaks near a value $2\pi n_o e^2 \exp(n_b/n_o)r$ and oscillates between this value and $2\pi n_o e^2 r$ with a wavelength

**TABLE 1.** Particle-in-Cell (PIC) simulation parameters for beam and plasma

| Beam | number(N) | $4 \times 10^{10}$ |
|---|---|---|
| | length($\sigma_z$) | .6 mm |
| | radius($\sigma_r$) | 5-75 μm |
| | emittance | $\gamma\varepsilon <$ 60 mm-mrad |
| **Plasma** | density($n_o$) | $2$-$10\times10^{15}$ cm$^{-3}$ |
| | Ion/e$^-$ mass ratio | $3\times1800$ |
| | Width | 50-500 μm |
| | cell size | 1-10 μm |
| | number of particles | $10^5$ |

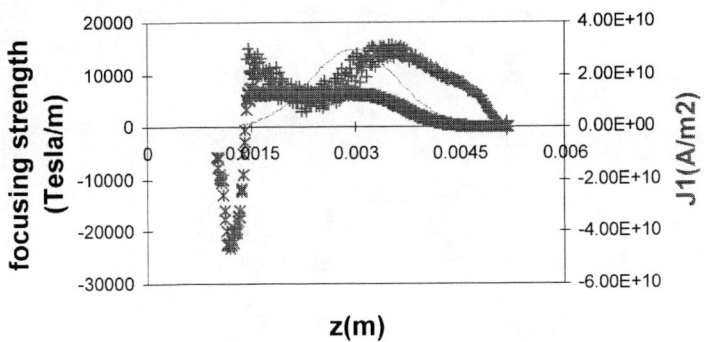

**FIGURE 3.** Focusing strength vs. z (lower curve $\sigma_r = 75\ \mu m$; upper; $\sigma_r = 5\ \mu m$) with mobile ions. Beam current is also plotted(dashed).

$\lambda_i = \pi \sqrt{2}c/\omega_{bi}$ . When this wake oscillation is calculated for a Gaussian beam profile, the wavelength $\lambda_i$ is modified by the factor $1/\sqrt{\pi}$ . The predicted wavelength $\lambda_i$ is approximately 2 *mm* for the SLC proposed experiment. The oscillatory behavior of the ions can be inferred from the focusing wake in the self-consistent PIC simulations of Figs. 2 and 3.

## PIC SIMULATIONS

We now model the effect of ion motion on the plasma wakefield using fully self-consistent particle-in-cell simulations. We use the code MAGIC in 2-D cylindrical geometry (i.e., coordinates r, z, $P_r, P_z, P_\theta$ ). The simulation models $10^5$ electrons on a $52 \times 140\,(r \times z)$ mesh. The cells are $.06\,c/\omega_p$ (20μm) in the z-direction and vary from 1-10 μm in the r-direction (smaller near the axis). The ions have a mass of 3 x 1800 $m_e$. The plasma and beam parameters are as shown in table [1];

First we present in Fig. 2 –a simulation without ion motion; the beam radius is 75 μm. One can see the transverse wakefield $W_r \approx E_r - v_z B_\theta\,/c \approx E_r - B_\theta$ plotted at r=75 μm.

$W_r$ rises quickly to a constant value. The value of $W_r/r$ is ~ 6000Tesla/m This is in good agreement with that expected from a uniform positive column of charge density $n_o$ ;

$$W_r\,/r = 2\pi\,n_o e \approx 960\pi\ \text{Tesla}\,/\,m \times (n_o\,/10^{14}\,\text{cm}^{-3}) \approx 6{,}400\ \text{Tesla}\,/\,m$$

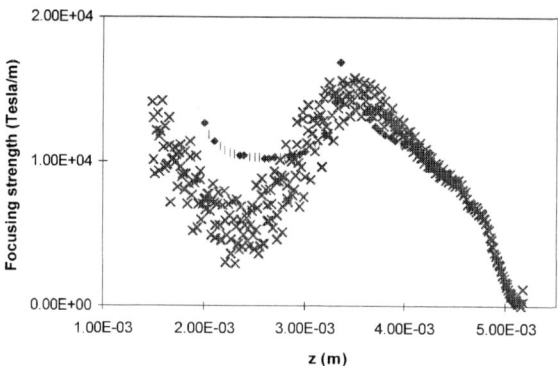

**FIGURE 4.** Focusing force vs. z (mobile ions) with $\sigma_r = 5 \ \mu m$, analytic mobile ion model is overlaid.

For this case of relatively large beam radius, including ion motion has no noticeable effect on the simulation results.

In the second set of simulations (Fig. 3), the transverse wakefield $W_r$ is plotted for two mobile ion cases (one with $\sigma_r = 75 \ \mu m$, another with $\sigma_r = 5 \ \mu m$). Overlaid in Fig. 4 is the result of the analytic model from Sec. III. We see the expected qualitative behavior of a rapidly rising wake at the head due to electron blowout. Once the electrons blowout there is a knee in the curve which then increases at a lower slope, as the ions collapse toward the beam axis, followed by oscillations. The wavelength of the wakefield oscillation in Fig. 4 is ~2 mm in agreement with the prediction in Sec. III. Thus we see that ion dynamics significantly alter the focusing wakefield in the second case.

In Fig. 5 we show the accelerating wakefield $E_z$ with and without mobile ions for the case of Fig. 3($\sigma_r = 5\mu$m). As expected, the ion motion has less effect on $W_z$ than $W_r$.

## DRIVE-BEAM DYNAMICS WITH MOBILE IONS

Knowledge of the betatron oscillations of the drive-beam in the plasma is important to the diagnostics of the proposed PWFA experiment at SLAC. Ion motion appears to complicate the betatron behavior. To assess the effect of the ion motion we return to Eq. (1) modifying the focusing strength K.

We adopt the following simplified model for the beam envelope $R$.

$$R'' + K(R(z))R = \varepsilon^2 / R^3 \ , \tag{9}$$

where $K = (\omega_p^2/2\gamma c^2)(n_i(R)/n_o)$ and $n_i(R) \approx n_o R_o^2 / R^2$. Before solving equation (9), we introduce a phenomenological constant, $a^2$ which causes the rise in $n_i$ to peak at a value of 10 times its initial value (see Fig. 6). Then the ion density term $n_i(R)$ has the form of $n_i(R) \approx n_o(R_o^2 + a^2)/(R^2 + a^2)$. The choice of $a \approx .1R_o$ is based on previous simulations which showed that non-linearity in the focusing of the ions limits the peak density they reach (this is analogous to spot size limits due to spherical aberrations of a plasma lens [3]).

Next we solve equation (9) numerically with and without mobile ions. The results for K and $\sigma$ are shown in Fig. 6. We see that although $W_r$ and K increase dramatically at a focus of the electron beam, the effect on the betatron phase advance is rather small. This is because the increase in focusing strength is only in a limited region and is greatest at the point that the beam is smallest and least sensitive to focusing.

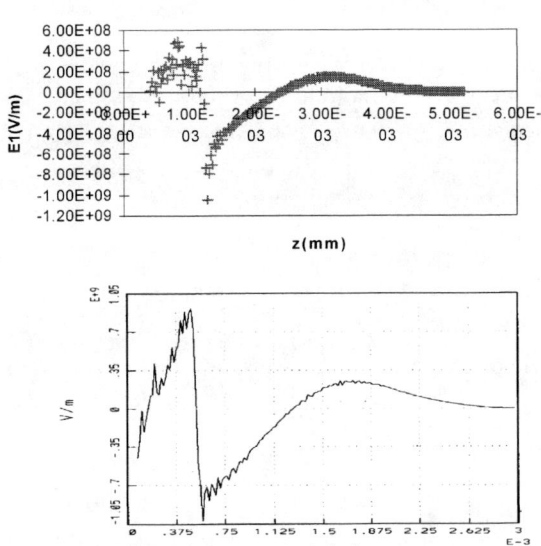

**FIGURE 5.** Longitudinal accelerating field $E_z$ vs. longitudinal position $z$ with $\sigma_z = 75\mu\, m$ (top :with mobile ions, below: without mobile ions)

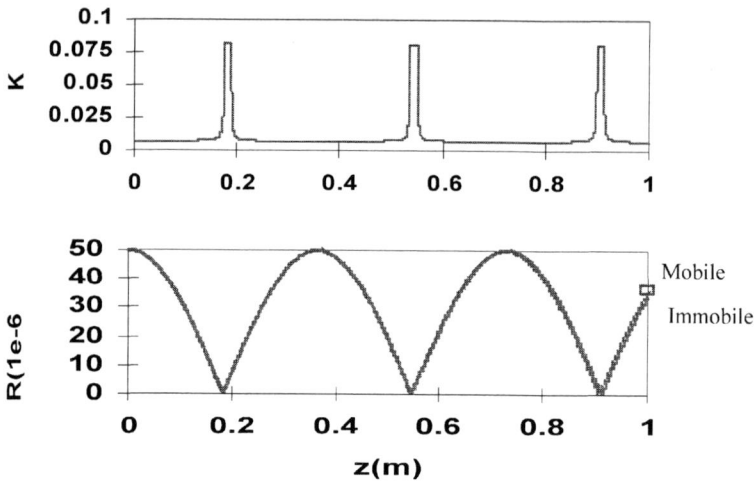

**FIGURE 6.** Focusing strength $K$ (above) and beam envelop $R$ over one meter of plasma with and without ions.

## CONCLUSION

We have modeled ion motion in the PWFA. Ion motion can be significant in the blowout regime. The effect of ion motion is primarily to increase the focusing wakefield at points when $n_b > (M/m)n_0$. There is less effect on the longitudinal wakes. Moreover, the drive-beam dynamics are only slightly effected by the increased focusing wakes for the parameters of an upcoming experiment.

## ACKNOWLEDGEMENTS

The authors wish to acknowledge J. Rosenzweig for raising this question. Useful discussions with Warren Mori, and Chan Joshi are appreciated. We thank T. C. Chiou for assistance with numerical and computational modeling. Work supported by US DOE, AFOSR, and NSF

# REFERENCES

1. Tajima T. and Dawson J. M., *Phys. Rev. Lett.* **43**, 267-270 (1979).
2. Clayton C. E. et al, K. Nakajima et al, *Proc. Adv. Accelerator Concepts Workshop*, Lake Tahoe, 1996(AIP, NY 1997).
3. Katsouleas T., *American Journal of Physics* **60**, 568-569 (1992)
4. Katsouleas T., *The American Physical Society* **33**, 2056-2064 (1986)
5. Rosenzweig J. B. *Phys. Rev. Lett.* **58**, 555-558 (1987).
6. Chen P., Su J. J., Dawson J. M., Bane K. L. F., and Wilson P. B., *Phys. Rev. Lett.* 56, 1252-1255 (1986)
7. Envelope equation of relativistic particle beam in focusing channel from the Lecture of Katsouleas T.
8. Su J. J., Ph.D. Dissertation 1998 "Plasma Wake Field Excitation by Charge Particle Beams"
9. Clayton C. E., Joshi C., Darrow C., and Umstadter D., *Phys. Rev. Lett.* 54 2343-2346 (1985).

# Meter Long, Homogeneous Plasma Source for Advanced Accelerator Applications

P. Muggli,[†] K.A. Marsh,[*] S. Wang,[*] C.E. Clayton,[*] T.C. Katsouleas,[†] and C. Joshi[*]

[†]*University of Southern California*
*Los Angeles, CA 90089*

[*]*University of California, Los Angeles*
*Los Angeles, CA 90095*

**Abstract.** A prototype for a 1-meter long plasma source is developed for plasma acceleration experiments. The lithium neutral vapor with a density of $2\times10^{15}$ cm$^{-3}$ is ionized by a uv laser pulse and produces a plasma density of $2.6\times10^{14}$ cm$^{-3}$. The plasma density is limited by the available uv energy. In this prototype, the length of the neutral vapor and plasma column is 25 cm. After ionization the plasma density decreases by a factor of two in about 12 µs. Interferometry in the visible on the lithium neutrals as well as $CO_2$ laser interferometry on the plasma electrons are used as diagnostics for the plasma density.

## 1. INTRODUCTION

The latest experimental results on electron acceleration in relativistic plasma waves demonstrate energy gains up to 100 MeV.[1,2] The maximum energy observed in these experiments is limited by the length over which the large amplitude coherent plasma waves (wakes) can be excited. In these experiments the observed energy spread is 100% because the electrons are either self-trapped or injected in a bunch longer than a plasma period. In the LWFA scheme, gradients up to 100 GeV/m are excited over a few millimeters with a plasma density $n_e$ in the $10^{18}$ cm$^{-3}$ range. In the PBWA scheme,[2] gradients of the order of 3 GeV/m are excited over 1 cm ($n_e \approx 10^{16}$ cm$^{-3}$). In all these experiments the energy gain $\Delta W$ is given by:

$$W \propto \varepsilon \cdot \sqrt{n_e} \cdot L$$

where $\varepsilon = \delta n_e / n_e$ is the wake amplitude ($0.3 < \varepsilon < 0.5$), and $L$ is the length over which the wake is excited. Future experiments will aim at demonstrating energy gains of the order of 1 GeV by operating at lower plasma densities ($n_e \approx 10^{14}$ cm$^{-3}$) but over a longer length ($L \approx 1$ m).

Recently a plasma-wakefield acceleration (PWFA) experiment has been proposed[3] in which the goal is to demonstrate electron acceleration by 1 GeV in a 1-meter long plasma. In that experiment known as E-157, the bulk of a 30 GeV SLAC-FFTB electron bunch generates a 1 GeV/m plasma wake, and the trailing electrons of the same bunch experience the acceleration. An energy gain of 1 GeV is in principle possible with a wake excited over a distance of 1 meter in a $n_e \approx 2\text{-}4\times10^{14}$ cm$^{-3}$ plasma. A prototype for a 1-meter plasma source that should fulfill the requirements for the proposed experiment is described here.

CP472, *Advanced Accelerator Concepts: Eighth Workshop,*
edited by W. Lawson, C. Bellamy, and D. Brosius
© 1999 The American Institute of Physics 1-56396-889-4/99/$15.00

The plasma source consists of a lithium (Li) heat-pipe oven which produces a Li vapor density in the $10^{15}$ cm$^{-3}$ range over 25 cm. The vapor is ionized by a uv laser pulse to a plasma density in the $10^{14}$ cm$^{-3}$ range.

## 2. PLASMA REQUIREMENTS

The plasma parameters required for the E-157 experiments, listed in Table 1, can be deduced for the SLAC-FFTB beam parameters, and from numerical PIC simulations.[3] They are as follows (Table 2). a) An electron plasma density $n_e$ adjustable around $4\times10^{14}$ cm$^{-3}$ over 1 meter to optimize the energy gain and the number of accelerated particles. Larger energy gains can be achieved at lower plasma density, but in this case less particles would be accelerated. b) A density variation $\delta n_e/n_e < 25\%$ over 1 m to avoid dephasing between the accelerated particles and the plasma wake. c) A plasma diameter larger than 600 μm to avoid the degradation of the wake amplitude caused by the excursion of the plasma electrons to outside of the plasma when blown out by the electron bunch. d) An adjustable plasma length $L$, and sharp plasma/vacuum boundaries to match the electron bunch betatron wavelength ($\approx 40$ cm with $n_e = 2\times10^{14}$ cm$^{-3}$). e) A low atomic number $Z$ for the gas/plasma nuclei, and a large fractional ionization ($n_e/n_0 > 15\%$), to minimize the influence of impact ionization of the neutrals by the driving electron bunch. Estimates[4] show that with $N_e = 3\times10^{10}$ electrons about 1.6% ionization for $Z=2$ (He) and 4.5% for $Z=10$ (Ne) could be generated by this process at the locations where the beam pinches to its minimum spotsize ($\approx 0.5$ μm).

Lithium is chosen (Table 3) because it can be photo-ionized by uv light (1-photon process) and has $Z=3$. Vapor pressures corresponding to neutral densities in the $10^{15}$ cm$^{-3}$ range can be obtained at temperatures around 750°C in a heat-pipe oven.

| Number of Electrons | $N_e$ | 3.5-4.0×$10^{10}$ |
|---|---|---|
| Initial Energy | $E_0$ | 30 GeV |
| Bunch Length | $\sigma_z$ | 0.6 mm |
| Bunch Size | $\sigma_x$ | 23 μm |
| (@ 1×$10^{10}$ e$^-$) | $\sigma_y$ | 37 μm |

**Table 1**: Parameters for the SLAC-FFTB driving electron beam.[3]

| Plasma Density | $n_e$ | 4 ×$10^{14}$ cm$^{-3}$ |
|---|---|---|
| Length | $L$ | 1 m |
| Density Uniformity | $\delta n_e/n_e$ | <25% |
| Ionization Fraction | $n_e/n_0$ | >15% |
| Diameter | $d$ | >600 μm |

**Table 2**: Required plasma parameters.[3]

| Li Atomic Number | 3 |
|---|---|
| Atomic Weight | 6.94 |
| Density (@ 25 °C) | 0.534 |
| Thermal Conductivity | 0.847 W/cm K |
| Heat Capacity @ 25°C | 3.582 J/g K |
| Melting Point | 180.54 °C |
| Boiling Point @ 1 atm. | 1342 °C |
| Fusion Enthalpy | 432 J/g |
| Vaporization Enthalpy | 19 J/g |
| Ionization Potential: Li$^I$ | 5.392 eV |
| : Li$^{II}$ | 75.638 eV |

**Table 3**: Lithium physical characteristics.[5]

# 3. HEAT-PIPE OVEN

The heat-pipe oven (Fig. 1) consists of a stainless steel tube heated along its central part, cooled at both ends, and containing a stainless steel mesh (wick).[6] The cold oven is filled with a given pressure of helium (buffer gas, $P_{buffer} \approx 200$ mT), and contains an ingot of Li. When heated, the Li melts and its vapor pressure increases. At some heating power, the oven reaches a temperature such that the vapor pressure of the Li is equal to $P_{buffer}$. Above this heating power, the temperature of the Li does not increase. In this regime, the buffer gas at both ends of the oven contains the pure Li vapor to the central region of the oven. The Li pressure and temperature (i.e., density $n_0$) are fixed by the buffer pressure ($P_{Li} = P_{buffer}$ in absence of flow), while the Li column length is proportional to the heating power. The Li evaporates in the heated zone of the oven, and condenses at the ends. The wet wick returns the liquid Li toward the evaporation zone by capillary action. The prototype described here is heated over 40 cm, the insulation is 46 cm long, while the distance between the cooling jackets is 52 cm. Optical windows (quartz or $BaF_2$) are located at the ends of the oven, and are in contact only with the room temperature buffer gas. They provide access for the ionizing laser pulse and for optical diagnostics of the Li vapor and plasma.

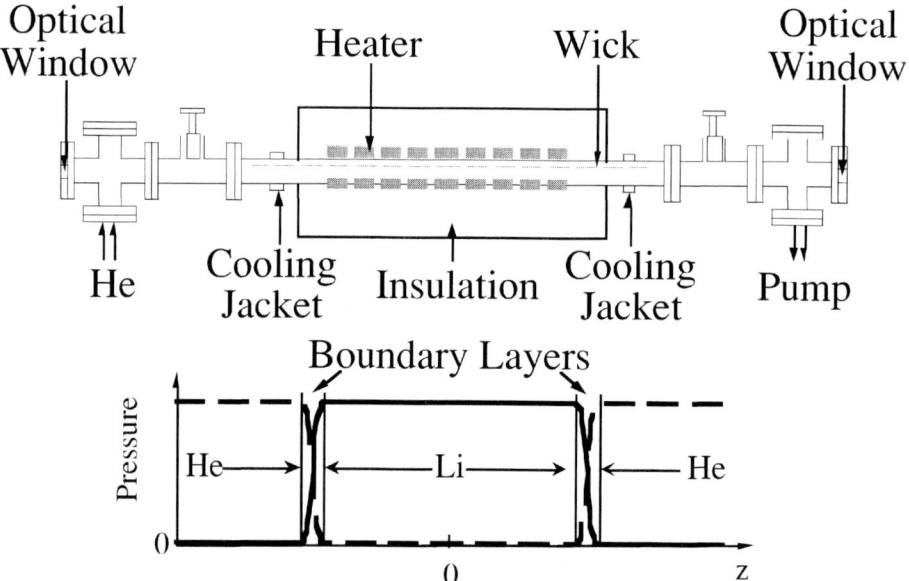

**Figure 1:** Schematic of the heat-pipe oven (not to scale), and expected pressure profiles for the buffer gas (He) and the Li vapor (assuming no significant flow in the oven). The length of the Li column and width of the boundary layers are arbitrary on this figure.

# 4. TEMPERATURE PROFILE MEASUREMENTS

The length over which the Li vapor extends is estimated from temperature profiles measured with a thermocouple probe along the axis of the heat-pipe oven with and without Li. The profile without Li in the oven (Fig. 2) shows that the temperature is decreasing away from the oven center, due to heat conduction along the oven wall and along the stainless steel probe housing itself. In contrast, the profiles measured with Li in the oven and with comparable heating power exhibit a relatively constant temperature in the center of the oven, followed by a rapid drop toward the end of the oven. The Li takes the heat from the center of the oven upon evaporation (Li enthalpy of evaporation: 19 J/g), the vapor transports it towards the end of the oven where it is released upon condensation of the vapor on the wick. In this oven, the length over which the vapor density drops to 80% of the center density is estimated from the temperature profiles and the Li vapor pressure curve to be $L$=25 cm ($P_{heat}$=307 W). Note that the temperature of the oven wall measured by a thermocouple placed outside of the oven, and referred to as $T_{ext}$ hereafter, is about 60°C higher than that measured inside the oven.

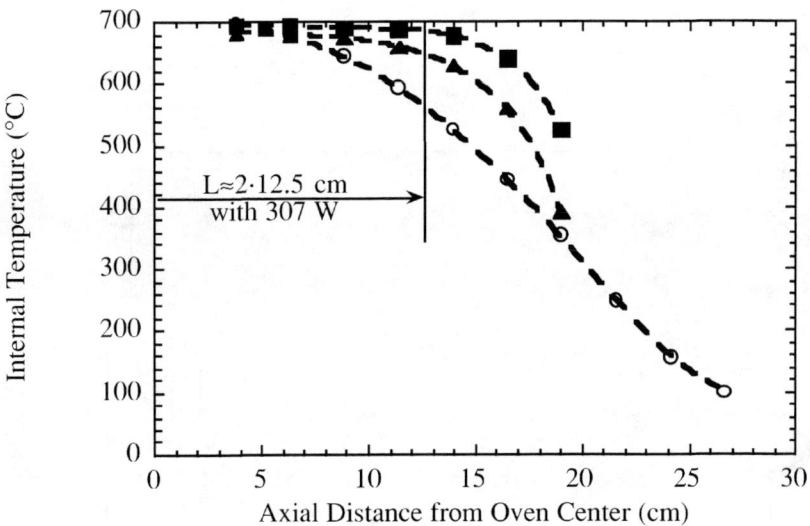

**Figure 2**: Temperature profiles along the oven axis without Li in the oven and $P_{heat}$=250 W (open circles), and with Li in the oven and $P_{heat}$=265 W (filled triangles) and $P_{heat}$=307 W (filled squares). The lines are drawn to guide the eye. The Li column length defined as the length over which the density $n_0$ drops by 20% (according to the Li vapor pressure curve) is $L$=25 cm with $P_{heat}$=307 W.

537

# 5. NEUTRAL DENSITY MEASUREMENTS

In this section two diagnostics for measuring the line integrated Li neutral density $n_0L$ are presented. Assuming a length $L$ of 25 cm, estimates for the average neutral density are obtained.

## 5.1 uv absorption

A low energy (<1 mJ) laser pulse of 5.83 eV (uv) photons (frequency quintupled Nd:YAG, nanosecond laser pulse) is sent along the heat-pipe oven axis. It ionizes the Li vapor through a single-photon process (Li ionization potential: 5.392 eV). The product $n_0L$ is obtained by measuring the ratio of the transmitted to incident uv energy:

$$n_0 \cdot L = -\frac{1}{\sigma} \cdot \ln\left(\frac{E_{transmitted}}{E_{incident}}\right)$$

where $\sigma = 1.8 \times 10^{-18}$ cm$^{-2}$ is the ionization cross-section.[7] For this measurement, the incident and transmitted energies are monitored by photodiodes, and their ratio calculated for each shot. Figure 3 shows the values of $n_0L$ measured as function of the oven external temperature. A value of $5.29 \times 10^{16}$ cm$^{-2}$ is obtained with $T_{ext}$=749°C, corresponding to $n_0$=2.12×10$^{15}$ cm$^{-3}$ for $L$=25 cm. At this temperature (or density) the fractional uv energy absorbed is only 9.5%.

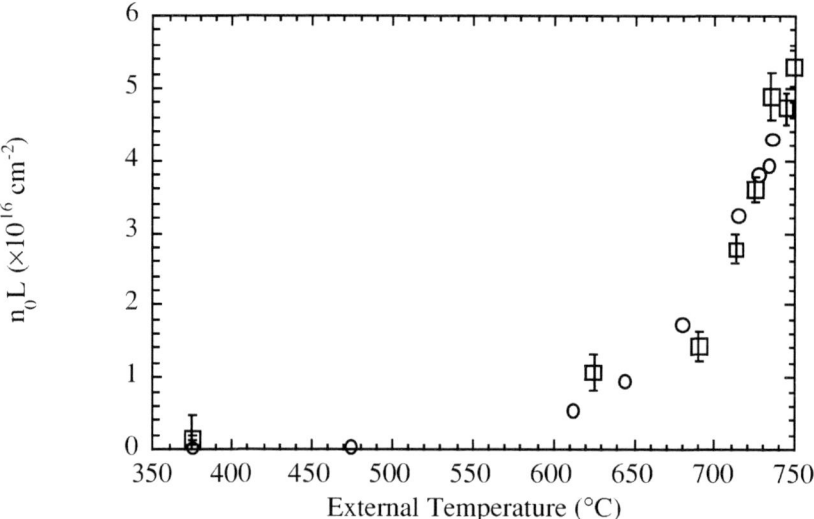

**Figure 3**: Neutral density measurement. Values of $n_0L$ as measured by the uv absorption method (squares) and by the hook method (circles). Assuming $L$=25 cm (see Fig. 2), both method show that $n_0 \approx 2 \times 10^{15}$ cm$^{-3}$ with $T_{ext} \approx 750°C$.

## 5.2 Hook method

Lithium, as do other alkali metals, exhibits a strong transition line in the visible frequency range from its ground state to the first excited state (2s to 2p state, $J=1/2 \rightarrow J'=3/2$ at $\lambda=670.78$ nm, and $J=1/2 \rightarrow J'=1/2$ at $\lambda=670.79$ nm). However for the experiments described here, the two Li transitions can be considered as one at $\lambda_i=670.785$ nm. The susceptibility $\chi$ associated with this atomic transition significantly modifies the index of refraction of the Li vapor in the vicinity of $\lambda_{ij}$, even at relatively low density ($n_0 \approx 10^{15}$ cm$^{-3}$). The corresponding dispersion is used to measure the neutral vapor density and as a diagnostic for the plasma density (see Sect. 6).

The oven is placed in one arm of a Mach-Zehnder white light interferometer. The interferometer light is sent to a stigmatic, f/4, 27 cm spectrograph to observe the dispersion ($n(\lambda)=(1+\chi(\lambda))^{1/2}$) in the vicinity of the neutral Li line at 670.785 nm (hook method,[8] see Fig. 4). The evaluation of the hook interferograms uses the following formula:

$$N_i \cdot L = \frac{\pi K}{r_0 \lambda_{ij}^3 f_{ij}} \Delta_{ij}^2$$

where $N_i$ is the population of the lower level $i$, $K=p\lambda'/\Delta\lambda$ is the hook-interferogram constant, $r_0= 2.82 \times 10^{-15}$ m is the classical electron radius, and $f_{ij}$, and $\Delta_{ij}$ are the oscillator strength ($f_{ij}=0.75$), and the hook separation of the $i \rightarrow j$ transition. The transition originates from the ground state of the Li atom. The thermal energy of the Li atoms is low ($k_B T_{Li} \ll h \, v_{ij}$), and the population of the ground state is therefore a very good approximation for the vapor density $n_0=N_i$. The measured values of $n_0 L$ obtained by this method are shown on Fig. 3 as a function of the oven external temperature. A value of $4.27 \times 10^{16}$ cm$^{-2}$ is obtained with $T_{ext}=736°$C, corresponding to $n_0=1.71 \times 10^{15}$ cm$^{-3}$ for $L=25$ cm. The values of $n_0 L$ obtained by the hook method are in very good agreement with the value obtained by the uv absorption method.

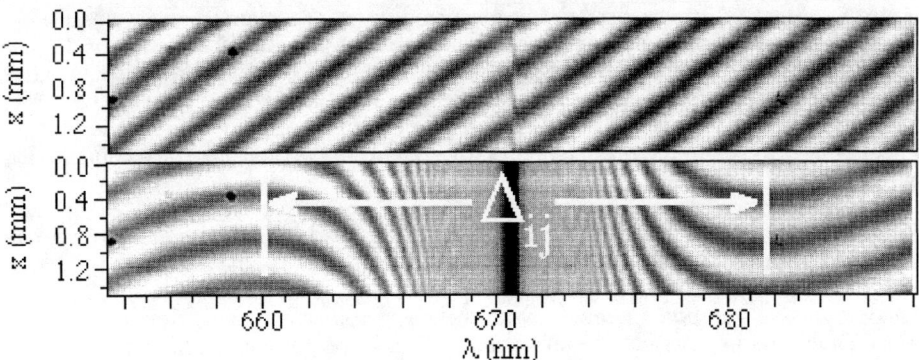

**Figure 4**: Interferogram (hook method[8]), observed in the plane of the imaging spectrograph, showing the hooks created by the dispersion associated with the $2s \rightarrow 2p$ transition of the Li atom ($\lambda_{ij}=670.785$ nm). Top picture: $T_{ext}=376°$C, there is a negligible density of Li in the oven. Bottom picture: $T_{ext}=715°$C, the measure of $\Delta_{ij}$ (distance between the hooks, in nm) yields $n_0 L=3.2 \times 10^{16}$ cm$^{-2}$, which corresponds to $n_0=1.3 \times 10^{15}$ cm$^{-3}$, assuming $L=25$ cm.

# 6. LITHIUM PLASMA

Changes in neutral and/or plasma density modify the dielectric constant n of the Li vapor and are measured by interferometry. The relative dielectric constant $\varepsilon_r$ at frequency $\omega_0$ near an atomic transition at frequency $\omega_{ij}$ of a partially ionized gas is given by:

$$\varepsilon_r = 1 + \frac{N_i e^2}{\varepsilon_0 m_e} \sum_j f_{ij} \left(\omega_{ij}^2 - \omega_0^2 - i\omega_0/\tau_{ij}\right)^{-1} - \frac{\omega_{pe}^2}{\omega_0^2}$$

where $\tau_{ij}$ is the lifetime of the upper state $j$ of the $i\rightarrow j$ transition. It includes a contribution from the resonant transitions (change in $N_i$) from the lower state $i$ of the neutral atoms to the possible upper states $j$ of the transition (see Section 5.2), and a contribution from the plasma (change in $n_e$). Depending on $\omega_0$, one or both of the contributions need to be retained. The phase difference $\Delta\Phi$ between light from the two arms of an interferometer arising from the change in index of refraction is given by:

$$\Delta\Phi = 2\pi(n-1)\frac{L}{\lambda_0}$$

where $n=(\varepsilon_r)^{1/2}$.

The Li vapor is ionized by the 20 ns, uv light pulse of an ArF excimer laser at 193 nm ($h\nu=6.43$ eV). The line integrated plasma density $n_e L$ is directly measured by $CO_2$ interferometry ($\lambda_0=10.6$ $\mu$m). At this wavelength, only the plasma contribution to $\varepsilon_r$ needs to be retained. The plasma source is placed in one arm of a Mach-Zehnder interferometer. The product $n_e L$ is obtained by measuring the phase shift on the interferometer signal resulting from the change in index of refraction of the recombining plasma:

$$n_e \cdot L \cong -\frac{\Delta\Phi}{\pi} \cdot \lambda_0 \cdot n_{crit} \quad \text{for} \quad n_e \ll n_{crit}$$

where $n_{crit}$ is the critical plasma density for $\lambda_0$: $n_{crit}=(2\pi c/\lambda_0)^2 \cdot \varepsilon_0 m_e/e^2=9.9\times10^{18}$ cm$^{-3}$. Figure 5 shows the plasma density measured by $CO_2$ interferometry (assuming $L=25$ cm) as a function of the uv energy incident on the plasma. The maximum density of $1.27\times10^{14}$ cm$^{-3}$ is limited by the uv energy reaching the Li vapor. The $CO_2$ laser pulse is coupled into the oven by grazing reflection ($\approx70$ deg.) upon a quartz window. When traversing this window about half of the ionizing laser pulse energy is lost by reflection. The time for the plasma to drop by a factor of two, because of recombination and possible diffusion, is about 12 $\mu$s.

The average plasma density $n_e$ is estimated from the absorbed uv energy and from the illuminated volume since every photon absorbed by an atom creates one free electron. About 6.6% of uv energy is absorbed over a 0.67 cm$^2$ beam cross section and $L=25$ cm. The average plasma density calculated by this method is also plotted on Fig. 5, and shows a good agreement with the interferometry results. Note that for this measurement the uv energies are averaged over a large number of laser pulses by a calorimeter. Shot-to-shot variation of the uv energy ($\pm10\%$) is responsible for the variations in the plasma density observed by interferometry. The neutral density necessary for a 6.6% uv absorption is $1.5\times10^{15}$ cm$^{-3}$; the fractional ionization is therefore about 12%.

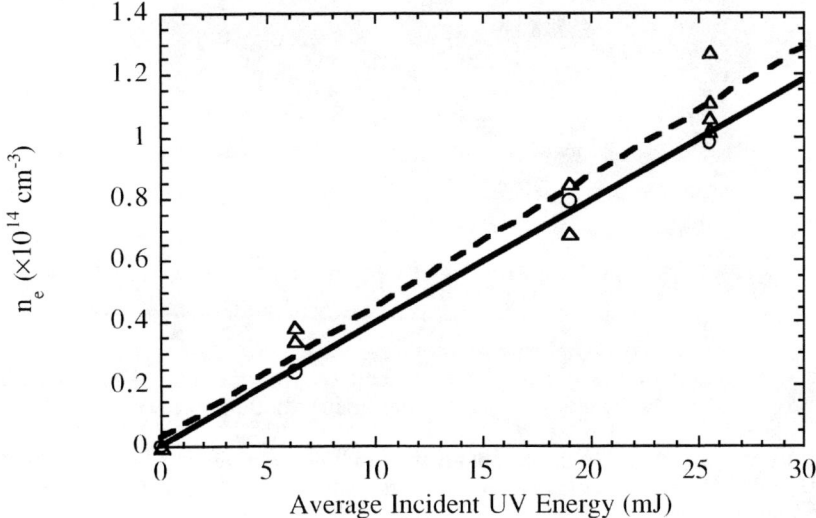

**Figure 5**: Plasma density as measured by $CO_2$ interferometry (triangles, dashes line), and by absorption of the uv ionizing laser pulse energy (circles, continuous line) as a function of the energy incident upon the Li vapor. The neutral density $n_0$ is $1.52\times10^{15}$ cm$^{-3}$, and the uv absorption is 6.6%.

Since the plasma is obtained by ionization of the neutrals, the plasma density can be inferred from the variation of the neutral Li density, i.e., by interferometry on the Li neutrals near $\lambda_{ij}=670.785$ nm (see section 5.2). The wavelength at which interferometry is performed can be chosen to observe a suitable number of fringes. An example of the fringe shift observed simultaneously by $CO_2$ ($\lambda_0=10.6$ μm) and by helium neon ($\lambda_0=632.8$ nm) interferometry is shown on Fig. 6. At $\lambda_0=10.6$ μm, only the plasma contribution to $\varepsilon_r$ needs to be retained, whereas at $\lambda_0=632.8$ nm both the neutrals and the plasma contribute to $\varepsilon_r$. The average plasma density obtained at these two wavelengths are $0.42\times10^{14}$ cm$^{-3}$, and $1.0\times10^{14}$ cm$^{-3}$ respectively, for a neutral density of $7.6\times10^{14}$ cm$^{-3}$ and an incident uv fluence of 30 mJ/cm$^{-2}$. The value of $n_e$ obtained from the uv absorption (3.1%) is $0.37\times10^{14}$ cm$^{-3}$ and is in good agreement with the $CO_2$ interferometry value. Note that the uv pulse spot shape is not uniform and may account for the discrepancy between the interferometry results. The uv absorption and the $CO_2$ interferometry yield volume integrated values of the plasma density, whereas the helium-neon pencil beam only samples a cord along the oven axis. Interferometry on the neutrals at the helium-neon laser wavelength is suitable for the 1-meter plasma source ($n_e\approx2$-$4\times10^{14}$ cm$^{-3}$) since the fringe shift is expected to be about ten times larger than the one observed with the prototype.

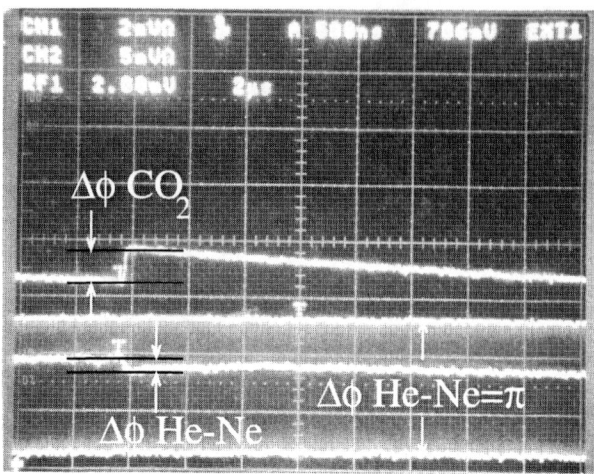

**Figure 6**: Oscilloscope traces obtained simultaneously from the $CO_2$ interferometry (top trace) and the helium-neon interferometry (bottom trace).

The maximum plasma density obtained with this prototype is at $T_{ext}=734°C$ ($n_0 \cong 2.0 \times 10^{15}$ cm$^{-2}$ over $L=25$ cm, Fig. 3) with a uv fluence of 35 mJ/cm$^{-2}$, leading to an average plasma density of $n_e=2.6\times10^{14}$ cm$^{-3}$, and a fractional ionization $n_e/n_0$ of 13%. This value is limited only by the available uv fluence.

# 7. SUMMARY

A lithium vapor density in the $2\times10^{15}$ cm$^{-3}$ range is obtained in a heat-pipe oven. The length of the neutral Li column is about 25 cm. The values for the neutral density obtained by uv absorption and by the hook method are in very good agreement. The vapor is ionized by uv light (1-photon process). The line integrated plasma density is measured by $CO_2$ laser interferometry on the plasma electrons, by visible interferometry on the Li neutrals (near the $\lambda_{ij}=670.785$ nm transition from the Li ground state), and by uv absorption. The maximum plasma density is $2.6\times10^{14}$ cm$^{-3}$, limited only by the maximum uv fluence available. The time for the plasma density to drop by a factor of two is about 12 μs. Visible interferometry on the neutrals (helium-neon laser, $\lambda_{ij}=632.8$ nm) provides a simple diagnostic for $n_e$ and will be implemented in the SLAC E-157 experiment. The plasma density will be optimized for the acceleration experiment by changing the delay between the ionizing laser pulse and the electron bunch. The plasma column length will be adjusted by changing the oven heating power. The 1-meter long plasma source is presently being assembled. This source differs from the prototype only by its length, therefore similar neutral/plasma densities should be obtained. A uv fluence of 200 mJ/cm$^{-2}$ incident upon a vapor density of $6\times10^{14}$ cm$^{-3}$ is required to produce a $2\times10^{14}$ cm$^{-3}$ plasma density with a 10% drop along the 1-meter plasma. The variation of the plasma density over the length of the oven due to the depletion of the ionizing pulse energy can be minimized by sending the pulse back through the oven. Double passing the oven with the uv beam would also almost double the plasma density.

# ACKNOWLEDGMENTS

The authors would like to thank Prof. P. Muntz from USC for lending the excimer laser. This work is supported by the US Department of Energy under Grants No. DE-FG03-92ER40727 and No. DE-FG03-92ER40745, and by the National Science Foundation under Grants No. ECS-9632735 and No. ECS-9617089.

# REFERENCES

[1] D. Gordon, K.C. Tzeng, C.E. Clayton, A.E. Dangor, V. Malka, K.A. Marsh, A. Modena, W.B. Mori, P. Muggli, Z. Najmudin, D. Neely, C. Danson, and C. Joshi, "Observation of electron energies beyond the linear dephasing limit from a laser-excited relativistic plasma wave," Phys. Rev. 80(**10**), 2133 (1998), and papers in this working group.

[2] C.E. Clayton, K.A. Marsh, A. Dyson, M. Everett, A. Lal, W.P. Leemans, R. Williams, and C. Joshi, "Ultrahigh-Gradient Acceleration of injected electrons by Laser-excited relativistic electron plasma wave," Phys. Rev. Lett. 70(**1**), 37 (1993).

[3] T. Katsouleas, S. Lee, S. Chattopadhyay, W.L. Leemans, R. Assmann, P. Chen, F.J. Decker, R. Iverson, T. Kotseroglou, P. Raimondi, T. Raubenheimer, S. Ronki, R.H. Siemann, D. Walz, D. Whittum, C.E. Clayton, C. Joshi, K. Marsh, W.B. Mori, and G. Wang, "A proposal for a 1 GeV plasma-wakefield acceleration at SLAC," Proceedings of the Particle Accelerator Conference, Vancouver, Canada, May 1997.

[4] D. Gordon, C.E. Clayton, and C. Joshi, "Collisional ionization of gases by GeV beams," presented at the Seventh Advanced Accelerator Concept Workshop, Lake Tahoe CA, (1996).

[5] CRC Handbook of Chemistry and Physics, 72nd Ed., CRC Press Inc. (1991).

[6] C.R. Vidal and J. Cooper, "Heat-pipe oven: a new, well-defined metal vapor device for spectroscopic measurements," J. Appl. Phys. 40(**8**), 3370 (1969).

[7] G.V. Marr, "Photoionization process in gases," Academic Press, (1967).

[8] W.C. Marlow, "Hakenmethode," Applied Optics 6(**10**), 1715 (1967).

# Photon Acceleration-Based Radiation Sources

J. R. Hoffman, P. Muggli, T. Katsouleas, W. B. Mori*, and C. Joshi*

*University of Southern California, Los Angeles, CA 90089-0484*
*\*University of California at Los Angeles, Los Angeles, CA 90095*

**Abstract.** The acceleration and deceleration of photons in a plasma provides the means for a series of new radiation sources. Previous work on a DC to AC Radiation Converter (DARC source) has shown variable acceleration of photons having zero frequency (i.e., an electrostatic field) to between 6 and 100 GHz (1-3). These sources all had poor guiding characteristics resulting in poor power coupling from the source to the load. Continuing research has identified a novel way to integrate the DARC source into a waveguide. The so called "pin structure" uses stainless steel pins inserted through the narrow side of an X band waveguide to form the electrostatic field pattern ($k \neq 0$, $\omega = 0$). The pins are spaced such that the absorption band resulting from this additional periodic structure is outside of the X band range (8-12GHz), in which the normal waveguide characteristics are left unchanged. The power of this X band source is predicted theoretically to scale quadratically with the pin bias voltage as $\sim 800 \text{W}/(\text{kV})^2$ and have a pulse width of $\sim 1$ns. Cold tests and experimental results are presented. Applications for a high power, short pulse radiation source extends to the areas of landmine detection, improved radar resolution, and experimental investigations of molecular systems.

## INTRODUCTION

Three methods of photon acceleration have been proposed in the past: the photon accelerator, which uses a plasma wakefield to accelerate a photon bunch; frequency upshifting by over and underdense ionization fronts; and DC to AC radiation converters (see figure 1). In the first scheme called the photon accelerator, an initial laser pulse moves through a plasma producing a plasma wake which accelerates a trailing laser pulse. The amount of acceleration, or equivalently, the frequency upshift of trailing photons is given by

$$\omega_1 = \omega_I [ 1 + 2(2\varepsilon^{1/2} )+ 4\varepsilon ].$$

where $\omega_1$ is the frequency of the accelerated photon, $\omega_I$ is the frequency of the incident photon, and $\varepsilon$ is the plasma wave amplitude (typically, 0.1 - 0.5). For typical values of $\varepsilon$, an upshift factor of about 5 can be achieved(4).

In the second scheme, a traveling electromagnetic wave is incident upon a moving ionization front (plasma/gas boundary usually produced by a laser), and is either reflected, if the plasma density is such that the corresponding plasma frequency is larger than the double Doppler shifted frequency (overdense), or is transmitted, if the plasma density is below the transmitted frequency (underdense). The frequency

CP472, *Advanced Accelerator Concepts: Eighth Workshop,*
edited by W. Lawson, C. Bellamy, and D. Brosius

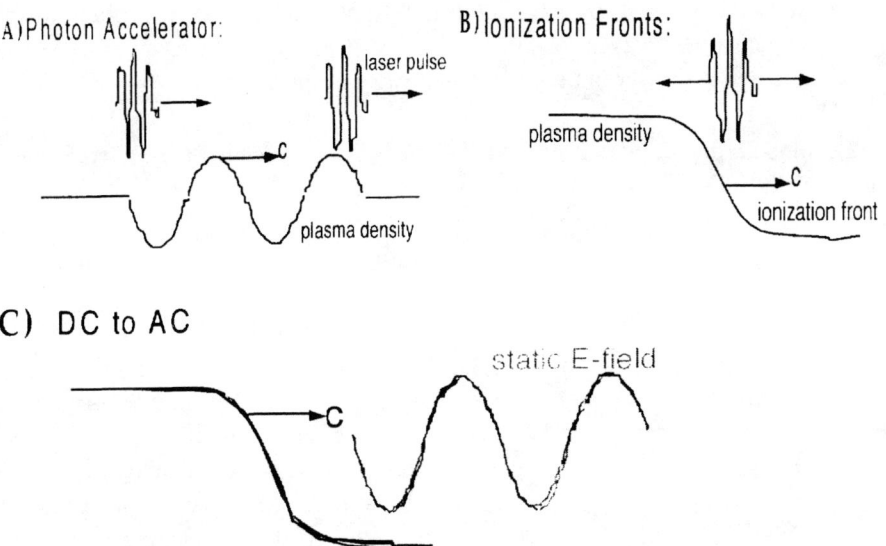

A) Photon Accelerator:

laser pulse

plasma density

B) Ionization Fronts:

plasma density

ionization front

C) DC to AC

static E-field

ionization front

FIGURE 1. Three schemes for accelerating photons. a) the photon accelerator; b) the moving ionization front; and c) the DC to AC converter.

upshift of the transmitted wave in the underdense case is given by

$$\omega_t = \omega_1 [ 1 + (\omega_p/2\omega_1)^2 ],$$

where $\omega_p$ is the plasma frequency. For a plasma density of $4 \times 10^{15}$ cm$^{-3}$ and an incident frequency of 33 GHz, the predicted upshifted frequency is 3 THz, (i.e., an upshift factor of a 100)(5).

In the last scheme, a static electric field is frequency upshifted from zero frequency to

$$\omega_t = k_0 v_f/2 + \omega_p^2/2k_0 v_f.$$

where $k_0$ is the wavenumber of the static field ($k_0 = \pi/d$), where d is the distance between the capacitor plates, and $v_f$ is the velocity of the front. For a static wavelength of 2 cm and a plasma density of $6 \times 10^{12}$ cm$^{-3}$, the predicted transmitted wave frequency is 15 GHz(6). Of the three schemes, the last is the easiest to realize in the laboratory, since only one laser and DC power supply are needed. In the DARC source or, DC to AC converter, alternately biased capacitors create a position dependent ($k_0 \neq 0$) static electric field. In the front frame, this static field approximates an electromagnetic wave with a frequency $\gamma k_0 v_f$, where $\gamma$ is the Lorentz factor $1/(1 - (v_f/c)^2 )^{1/2}$. For the case of a sharp front, the amplitude of the transmission coefficient is approximately 1(6). (Note, in the ionization front case, there are four transmitted and one reflected wave, all with different frequencies). Of practical interest, the laser energy is just used to make the moving ionization front, and does not contribute to the energy of the transmitted waves.

The theoretical power in the transmitted wave of interest is:

$$P = E_o^2 V_{group} A/8\pi. \qquad (1)$$

where $E_o$ is the amplitude of the first fundamental y component of the electric field between the plates. $V_{group}$ is the group velocity of the transmitted wave, and A is the transverse area of the plasma. For parallel plate capacitors, $E_o \propto V_o/b$. where $V_o$ is the voltage applied to the capacitors, b is the separation between the plates, and thus, the power is proportional to

$$P \propto (V_o/b)^2 A V_{group}. \qquad (2)$$

For the DARC structure shown in figure 3a, the theoretically predicted power scaling is $250W/kV^2$, whereas, experimental measurements(2) showed a power scaling of $100mW/kV^2$. This discrepancy will be discussed further in a later section.

Consider some different examples of DARC sources shown in figure 2. The first picture, 2a shows the proof-of-principle structure(2), that is a series of alternatively biased parallel plate capacitors. The second picture, 2b shows a Ka band waveguide structure with continuous sidewalls and sectioned top and bottom walls(3). The third picture, 2c is of a coaxial structure sectioned perpendicularly to the axis. Common to all of these structures are transverse cuts relative to the direction of propagation of the transmitted wave. In order to support the well-known TE, TM, or TEM modes of these structures, the structure has to be able to support the wall currents. A cold test of the Ka band structure (figure 2b) showed a 30db loss in transmission. The conclusion to be drawn

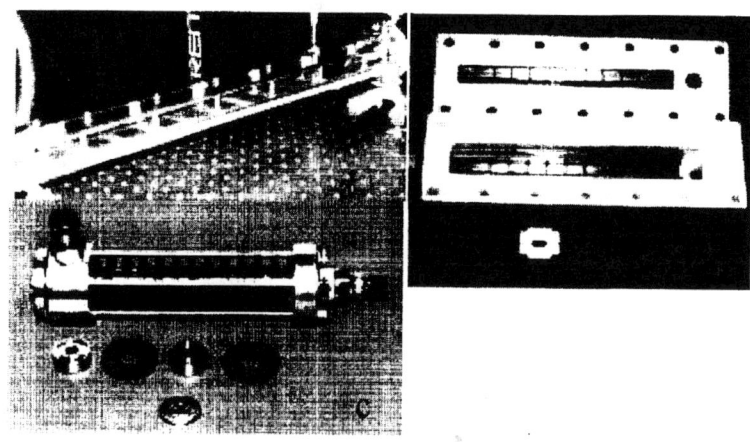

FIGURE 2. a) Proof of principle DARC source, 12 sets of parallel plate; b) Ka band DARC source, a transversely sectioned Ka waveguide; c) Coaxial DARC source, a transversely sectioned coax line.

from this is that sectioning of the waveguide is not an acceptable way to produce the static electric field.

A new idea is proposed here to produce the static field: insert pins through the short side of the waveguide, thus minimizing the perturbation to the continuity of the waveguide walls (figure 3). The pins form a periodic structure inside of the waveguide with stopband frequencies given by :

$$\omega_{sb} = c \left[ (\pi/h)^2 + (l\pi/d)^2 \right]^{1/2},$$

where c is the speed of light, h is the vertical spacing between the pins, d is the longitudinal spacing between the pins, and l is the mode number. By judicious selection of the spacing (d) between the pins, these stopband frequencies can be placed

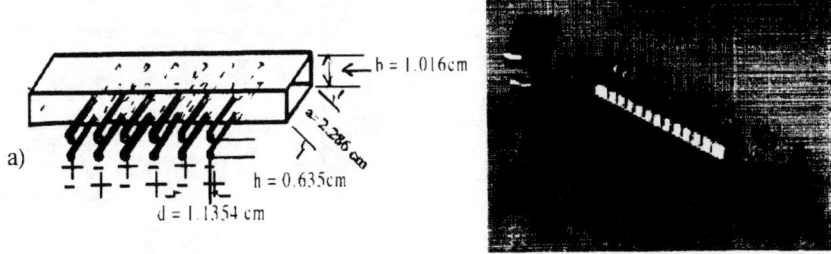

FIGURE 3. a) a dimensional cartoon of an X-band waveguide showing where the pin electrodes are inserted. b) the actual implementation of the X-band pin structure.

outside of the X band frequency range. For an X-band waveguide, with h = 0.635cm, and d = 1.1354cm, the first stopband $\omega_{sb}(l=1)$ corresponds to a frequency of 14.6 GHz, and the transmission characteristics of the fundamental $TE_{10}$ mode of the waveguide should remain unchanged between 8 and 12 GHz. The transmission characteristics of the pin structure was measured with an HP network analyzer, model 8720A, and compared to an ordinary waveguide. These results are shown in figure 4.

## PIN POWER AND PULSEWIDTH CALCULATIONS

Figure 5 shows a comparison of the potential and y component of the electric fields of the pins and parallel plates at the same separation, h = 0.663cm, and period d = 1.135cm. The y component of the electric field of the pins is well concentrated between the pins and falls off rapidly to zero in the void separating the pins. Whereas, in the parallel plate case, the electric field fills most of the space between the plates and has a fairly constant amplitude. The reduction factor in the expected power due to the different electric field configurations is currently under investigation, and will be reported on in a later paper.

The expected pulse width can be approximated by

$$\tau = Nc/\lambda = N/f,$$

where N is the number of cycles (i.e., the number of pairs of pins divided by two), c is the speed of light, $\lambda$ is the wavelength of the transmitted wave. Assuming an output frequency of 10GHz, the expected pulse width is 1.2 ns for N=6.

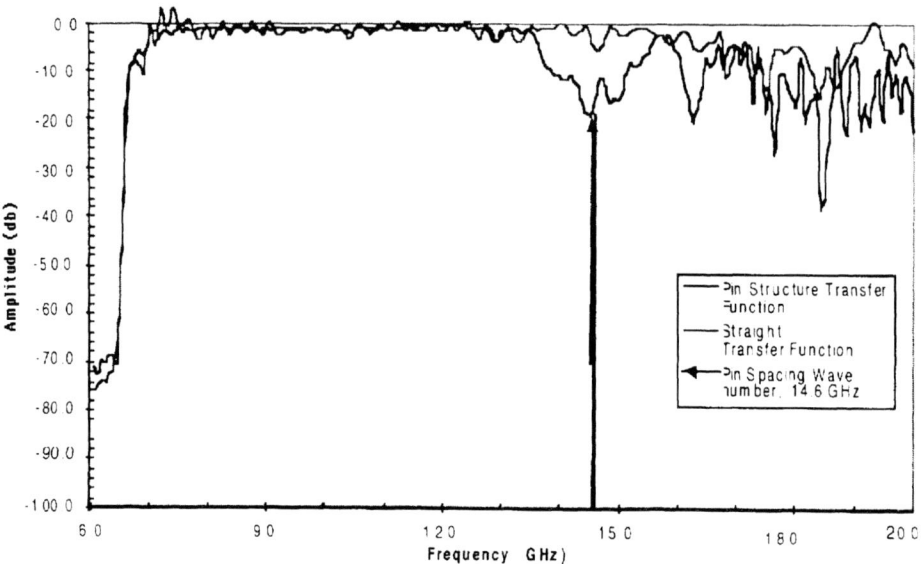

**FIGURE 4. The $S_{12}$ transmission characteristics of the X-band pin structure compared to a standard X-band waveguide. The vertical arrow points to the theoretical position of the periodic reflection due to the inserted pin.**

## EXPERIMENT

The experiment was performed at the UCLA Neptune lab facility using a frequency quadrupled YAG laser beam with a pulsewidth of 50ps, a wavelength of 266nm, and an energy of ~27mJ. The ionizable gas was a volatile organic compound called TMAE, and the pressure inside of the pin structure was maintained at ~50mT. During the experiment the shot to shot variation of the laser energy was observed to be ± 7 percent. The pressure drift was no more than ± 5%.

The microwave detection diode was calibrated using a microwave power meter and source (figure 6). As can be seen from the insert on the lower right side of figure 6, the diode saturates at relatively low power levels, approximately 0.4W. The detection diode was calibrated up to 1W of power, and a power law fit was made to this data. The power produced in the pin structure was inferred from the detection diode signal

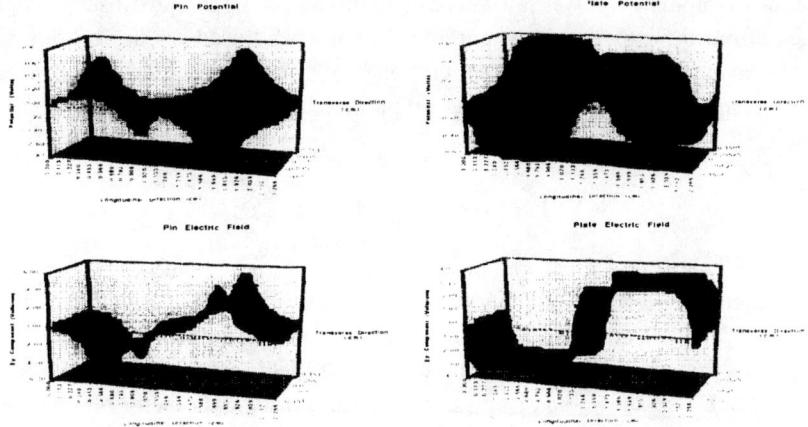

**FIGURE 5.** Plots of the analytical solutions of the potential and electric fields: a) 3D contour plot of the scalar potential between two sets of pins. b) 3D contour plot of the y component electric field between two sets of pins. c) 3D contour plot of the scalar potential between two sets of plates. d) 3D contour plot of the y component of the electric field between two sets of plates.

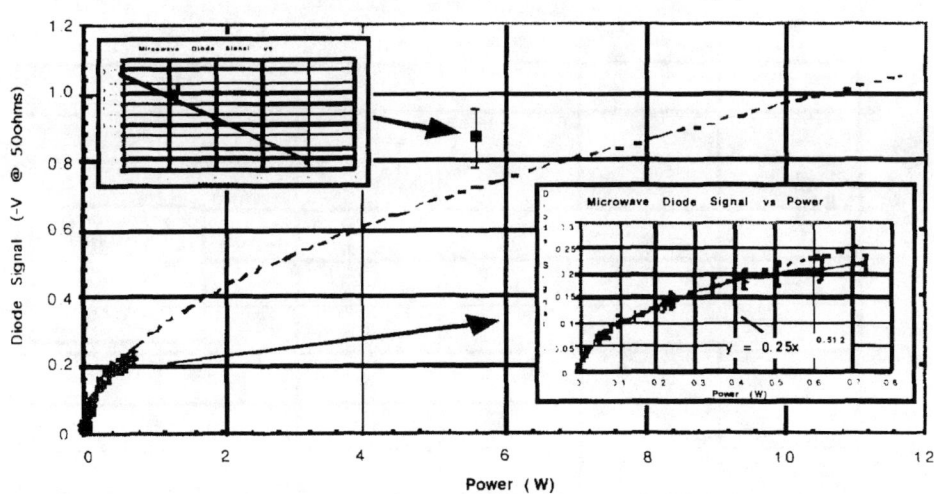

**FIGURE 6.** The microwave detection diode calibration results. The bottom right insert shows the measured diode voltage as a function of input power. The dotted line curve in the insert and in the main figure is a power curve fit to the measured data. The top left insert shows data points taken using the pin structure as a power source and placing various attenuators between the pin structure and the detection diode.

by using the extrapolated power law curve of figure 6. The upper left insert in figure 6 shows data from the experiment where an attenuator was used on the output signal. This was done in an attempt to get a calibration of the detection diode at higher power levels. The result is the point on the main graph indicated by the arrow from the upper insert.

## EXPERIMENTAL RESULTS

Figure 7 shows the measured microwave diode signal plotted versus applied voltage squared. The unsaturated data points (power less than 5W) are indicated by the diamonds, and the saturated data points are indicated by the squares. A least squared fit line drawn through the unsaturated points yields a slope of $\sim 1\,W/kV^2$. The microwave diode also showed a response time of $\sim$4ns, roughly a factor of 4 above the predicted pulse width (again, assuming an output frequency of 10GHz). Since the diode acts like an integrator, we estimate that the actual peak power for a 1 ns pulse is a factor of four greater, or $\sim 4\,W/kV^2$. For our peak applied voltage of 7kV, we infer the peak power to be $\sim$200W. This is much larger than the powers observed in either of the previous DARC structures. Further work is planned to improve the power diagnostic and to obtain theoretical power predictions for the pin structure.

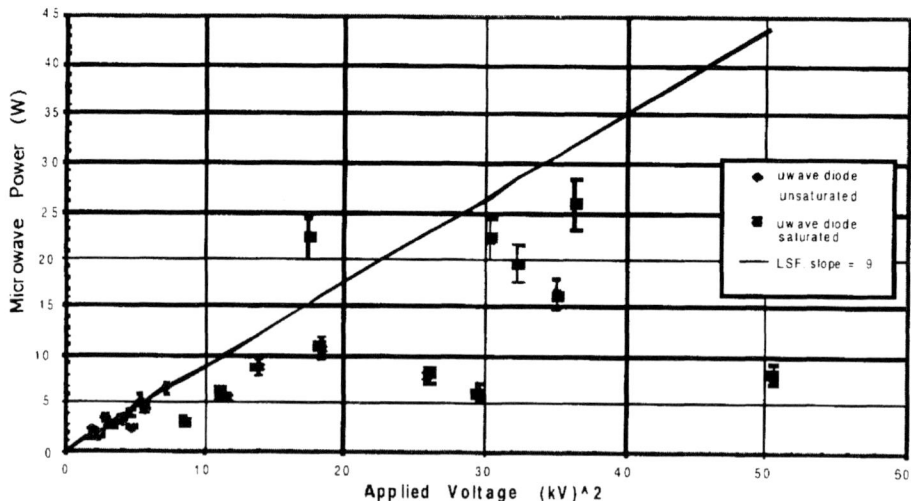

FIGURE 7. The measured microwave power versus applied voltage squared. The square data points are the data points we believe the microwave diode was saturated, so a least squared fit (LSF) was calculated only through the diamond points.

# CONCLUSION

In conclusion, the novel pin structure produced power that is inferred to be in the hundred Watts range, a large improvement with respect to other DARC sources that reached power levels of milliwatts. Future experiments will try to measure the frequency spectrum, and the feasibility of detecting buried objects (e.g., landmines) inthe ground with this ultrashort pulse microwave source.

# ACKNOWLEDGEMENTS

This work is supported by the National Science Foundation, Grant No. ECS-9632735; the Air Force Office for Scientific Research, Grant No. F49620-95-1-0248; and the US Department of Energy, Grant No. DE-FG03-92ER-40745.

# REFERENCES

1. Savage, Jr., R. L., Brogle, R. P., Mori et al., "Frequency upshifting and pulse compression via underdense relativistic ionization fronts," IEEE Transactions on Plasma Science 21 (1), 5-19 (1993).
2. Lai, C. H., Liou, R., Katsouleas, T. C., et al., "Demonstration of microwave generation from a static field by a relativistic ionization front in a capacitor array," Physical Review Letters 77 (23), 4764-7 (1996).
3. Muggli, P., Liou, R., Lai, C. H., et al., "Generation of microwave pulses from the static electric field of a capacitor array by an underdense, relativistic ionization front," Physics of Plasmas 5 (5), 2112-19 (1997).
4. Wilks, S. C., Dawson, J. M., Mori, W. B., et al., "Photon accelerator," Physical Review Letters 62 (22), 2600-3 (1989).
5. Mori, W. B., "Generation of tunable radiation using an underdense ionization front," Physical Review A 44 (8), 5118-21 (1991).
6. Mori, W. B., Katsouleas, T., Dawson, J. M., et al., "Conversion of dc fields in a capacitor array to radiation by a relativistic ionization front," Physical Review Letters 74 (4), 542-5 (1995)

# Proton Acceleration in Plasma Waves Produced by Backward Raman Scattering

A. Ogata* and T. Katsouleas†

*High Energy Accelerator Research Organization
Oho, Tsukuba, 305-0801 Japan
†University of Southern California
Los Angeles, CA 90089-0484 USA

**Abstract.** A schemes of proton acceleration is proposed. It uses plasma waves with a slow phase velocity produced by backward Raman scattering, making a density gradient in a plasma so that the wave phase velocity increases as the proton test beam is accelerated.

## I INTRODUCTION

Numerous experiments have demonstrated proof of principle of high-gradient laser- or electron-beam-driven plasma accelerators in recent years [1]. The test particles are limited to electrons in these experiments. Plasma-based electron accelerators are, however, still too premature to be used as a linear collider, and it is hard for them to find applications in other fields. This is because the technology of rf-based electron accelerators with energies below 100MeV is well established and their size is small enough. Certainly the acceleration distance of plasma-based accelerators is shorter than that of rf-based ones by two or three orders of magnitude or more, but if we take into account the size of drivers, i.e., lasers or particle accelerators of the plasma wakefield accelerators, the size reduction is not at all impressive. On the other hand, existing proton and ion accelerators with energies in the 100MeV-1GeV range are quite large. They could be applied more widely than the electron beams, if their sizes were reduced, for medical use in cancer therapy, and for studies of physics, chemistry and biology.

This paper proposes proton acceleration using plasma waves. Among the prior efforts to accelerate protons and ions based on novel principles, the most famous is the electron-ring accelerator (ERA). The aim of this scheme was to give protons the same speed as the electrons, mixing a small amount of protons in a large amount of electrons [2]. Several ERA projects were abandoned more than ten years

CP472, *Advanced Accelerator Concepts: Eighth Workshop,*
edited by W. Lawson, C. Bellamy, and D. Brosius

ago, because it became theoretically clear that it cannot increase the acceleration gradient dramatically from that of conventional accelerators [3].

The present proposal is based on a principle different from the ERA. It is based instead on the same principle as laser wakefield acceleration which has successfully accelerated electrons with a large acceleration gradient. It uses the electric field of a plasma wave excited by a laser pulse. The laser is scattered into either forward or backward directions in a plasma. The forward scattering excites a plasma wave whose phase velocity is almost equal to the group velocity of the pump laser. This fast plasma wave associated with forward scattering has been used in electron acceleration. It is called self-modulated laser wakefield acceleration, because the laser pulse is modulated at the plasma frequency by the forward instability. The backward scattering excites a slow plasma wave, and it is this slow wave that we are going to use for proton acceleration.

This paper consists of four sections. Following this introductory section, the second section briefly describes backward Raman scattering. The third section proposes acceleration in a plasma with tapered density, in which the phase velocity of the plasma wave increases as the test beam is accelerated. The last section gives conclusions.

## II BACKWARD RAMAN SCATTERING

Figure 1 shows dispersion relations of Raman scatterings, ignoring the plasma temperature. In the forward scattering shown in (a), both the plasma wave (shown by suffix 'p') and the scattered radiation (shown by suffix 'F') propagate in the same direction as the pump radiation (shown by suffix 'L'). In the backward scattering, the scattered radiation with suffix 'B' propagates in the opposite direction, though the plasma wave has the same direction as the pump radiation. As shown, the wavenumber of the plasma wave in the backward direction becomes large and its phase velocity becomes small.

The dispersion relations of the electromagnetc waves are given by the following:

$$\omega_x^2 = \omega_p^2 + c^2 k_x^2, \quad x = L, F, B, \tag{1}$$

where $\omega_p = [e^2 n/(\epsilon_0 m_e)]^{1/2}$ is the electron plasma frequency. The conditions for resonance are

$$\omega_L = \omega_p + \omega_F,$$
$$\omega_L = \omega_p + \omega_B,$$
$$k_L = k_p + k_F,$$
$$k_L = k_p - k_B. \tag{2}$$

The wave number of the plasma wave then becomes

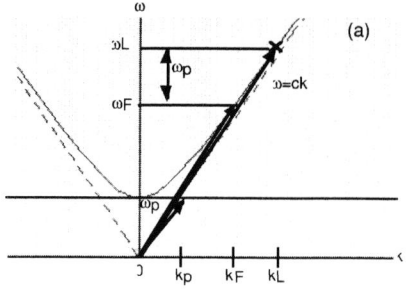

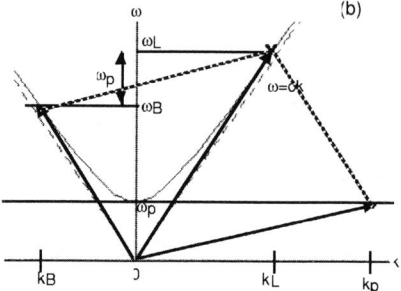

**FIGURE 1.** Dispersion relations of Raman scattering; (a) forward scattering, (b) backward scattering.

$$k_p = \frac{\omega_L}{c} \left[ \sqrt{1 - \left(\frac{\omega_p}{\omega_L}\right)^2} \mp \sqrt{1 - \frac{2\omega_p}{\omega_L}} \right], \tag{3}$$

where $-$ and $+$ signs correspond to forward and backward scatterings, respectively.

The phase velocity of the plasma wave is $v_p = c\beta_p$, where

$$\beta_p = \frac{\omega_p}{\omega_L} \frac{1}{\sqrt{1 - (\omega_p/\omega_L)^2} \mp \sqrt{1 - (2\omega_p/\omega_L)}}. \tag{4}$$

It is shown in Fig.2 as a function of $\omega_p/\omega_L$. Once the pump radiation is given, the phase velocity is thus determined by the plasma density. The broken line in the figure shows the phase velocity of the pump radiation

$$\beta_p = \sqrt{1 - \left(\frac{\omega_p}{\omega_L}\right)^2}, \tag{5}$$

which is usually used as an approximation of the phase velocity of the plasma wave produced in the forward scattering. As is clear in the figure, the curves showing phase velocities of forward and backward radiation merge at the point where $\omega_L = 2\omega_p$, where $\beta_p$ takes the value $1/\sqrt{3}$; i.e.,

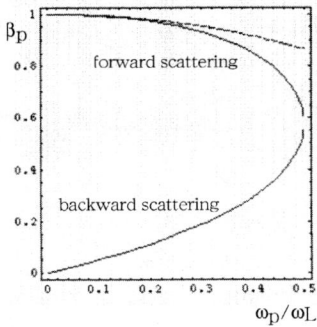

**FIGURE 2.** Dependence of phase velocity of the plasma wave produced in the Raman scattering on $\omega_p/\omega_L$. The broken line shows the phase velocity of the pump radiation.

$$\beta_p \xrightarrow{\omega_p = \omega_L/2} \frac{1}{\sqrt{3}} \sim 0.577. \tag{6}$$

The Raman backward scattering had been studied vigorously in the 1970's in connection with laser fusion. We summarize here the main features obtained then [4]. The growth rate of the backward Raman instability is given by

$$\gamma_B = \left(\frac{v_q}{c}\right)^2 \sqrt{\omega_L \omega_p}, \tag{7}$$

where $v_q = eE_L/(m_e\omega_L)$ is the quiver velocity of electrons in the laser field. That of the forward Raman instability is [5]

$$\gamma_F = \left(\frac{v_q}{c}\right) \frac{\omega_p^2}{\sqrt{8}\omega_L}. \tag{8}$$

We thus have $\gamma_B/\gamma_F = \sqrt{8}(v_q/c)(\omega_L/\omega_p)^{3/2}$. The backward instability grows faster than the forward instability if the laser field is large.

Landau damping suppresses the backward Raman scattering, unless [6]

$$\frac{\omega_p}{\omega_L} > \frac{2v_{th}}{c}, \tag{9}$$

where $v_{th} = (k_BT_e/m_e)^{1/2}$. This requirement is most difficult to fulfill at beam injection. First of all, the phase velocity of the plasma wave should match the injection velocity of the proton beams. In order to give a slow velocity, the plasma density has to be low, which makes the lefthand side of eq.(9) small. Approximating $k_p = 2k_L$ and $\omega_L = ck_L$, we have $\omega_p/\omega_L = 2\beta_p$. The $\beta$ value of the injected protons should be equal to the $\beta_p$ value of the plasma wave. The condition of eq.(9) then becomes $\beta > v_{th}/c$. Electron temperatures 10eV, 100eV, 1keV ... give

$v_{th}/c$ values of 0.006, 0.02, 0.06,...., respectively. If we want to use a proton source of $\beta = 0.02$ (with ~200keV kinetic energy) such as from a duoplasmatron, the electron temperature should be lower than 100eV. This is not a severe restriction.

The maximum electric field associated with a fast plasma wave is approximately $eE_{max} = m_e\omega_p c$. This cannot be applied to our case of a slow plasma wave. We have instead

$$eE_{max} = m_e\omega_p v_p. \tag{10}$$

In the following section, we propose a method to use a plasma with tapered density. In such an inhomogeneous plasma, the resonant condition can hold only locally. The propagation of the plasma wave out of this resonant region then provides an additional threshold, which is [4],

$$\left(\frac{v_q}{c}\right)^2 k_L L > 1, \quad (2\omega_p < \omega_L), \tag{11}$$

where

$$L = \left(\frac{1}{n}\frac{dn}{ds}\right)^{-1}, \tag{12}$$

is a length characterizing the density gradient, with $s$, the distance along the density gradient.

If the laser pulse width is short, a plasma wave must survive during the acceleration time. The plasma wave decays due to processes such as collisional damping and modulational instability. The collisional damping is eased by quiver motion of electrons in a high laser field [7]. So we consider here only the modulational damping.

Under the electric field satisfying the condition $v_q/v_{te} > (\omega_{pe}/\omega_{pi})^{1/3}$, the decay constant of the plasma wave due to modulational instability is given by [8],

$$\gamma_{mod} = \left(\frac{3}{2}\right)^{1/4} \omega_{pi} \left(\frac{\omega_{pe} v_{th}}{\omega_{pi} v_p}\right)^{1/2}, \tag{13}$$

where $\omega_{pi} = [Z^2 n_i e^2/(m_i \epsilon_0)]^{1/2}$ is the ion plasma frequency.

Figure 3 shows temperature and density dependencies of the decay constant due to the modulational instability at $v_p = 0.1c$ and $Z = 1$. This decay constant is larger than the case of plasma acceleration of electrons because of the smaller phase velocity.

The plasma wave decay is certainly slower than the growth of the backward instability given in eq.(7). However, once the pump laser fades out, the plasma wave decays in less than a ps. We cannot expect that a short laser pulse leaves a wake which survives until a slow proton beam arrives. In other words, the laser pulsewidth has to be longer than the acceleration time. This situation differs from the case of electron acceleration where "wakefield acceleration" is possible.

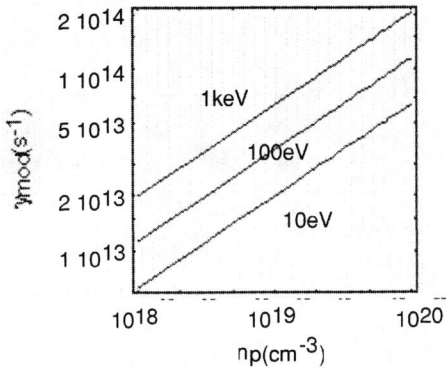

**FIGURE 3.** Decay constant due to modulational instability. $v_p = 0.1c, Z = 1$.

It is an open question whether an instability-triggered mechanism such as the Raman scattering can be used in an accelerator in real use. Because of this problem, many people are skeptical about the real use of the self-modulated laser wakefield acceleration mechanism in electron accelerators, in spite of the fact that these have attained the highest acceleration gradients [1]. Unfortunately, our slow plasma wave is also a product of an instability. One solution to control this instability is the use of a second laser which acts as a seed for the instability [9]. This seed laser should be frequency-shifted by $\omega_p$ from the pump laser and injected from the counter direction of the pump laser. In other words, the seed laser should have the same frequency and direction of the backscattered radiation. Without the seed laser, the laser-driven instability growth must just balances the plasma wave decay. It remains an open question whether such a balance is achievable.

## III   TAPERED DENSITY METHOD

Traveling wave can accelerate slow particles such as protons and ions if the wave phase velocity increases synchronously with the velocity of the particles accelerated. Figure 1 suggests this is possible if we control the plasma frequency with the plasma density. Figure 1 has a folding point at $\omega_p/\omega_L = 1/2$, where $\beta_p = 1/\sqrt{3}$. It suggests that if we use the backward scattering for acceleration up to $\beta_p = 1/\sqrt{3}$, and then use the forward scattering, decreasing the plasma density, we can give infinite energy to protons. The instability growth around $\beta_p = 1/\sqrt{3}$ is however absolute, while it is convective below $1/\sqrt{3}$ [4]. Because the mechanism changes at around $\beta_p = 1/\sqrt{3}$, we refrain from being too ambitious. We try here to increase the phase velocity of the plasma wave produced in the backward scattering up to $\sim 1/\sqrt{3}$, by giving a positive gradient to the plasma density.

We now try to carry out a simple simulation. For the reason above, we set the final velocity below $1/\sqrt{3}$, aiming at $\beta = 0.5$ ($\gamma = 1.154, W_K = 144MeV$). We assume the initial velocity of protons as $\beta = 0.02$ ($\gamma = 1.0002, W_K = 200keV$), which is the value obtainable by a small proton source. We assume use of a Ti:sapphire laser with $\lambda_L = 800nm$. We assume the acceleration gradient as $eE(s) = m_e c \beta_p(s) \omega_p(s)$. Of course this is the maximum possible value and the actual value is smaller, which will be determined by more realistic simulations. The calculation takes the following steps:

1)Assume initial value $\beta(0) = \beta_p(0)$, where $\beta$ and $\beta_p$ denote particle velocity and wave velocity, respectively.

2)Given $\beta(s)$ ($\beta(0)$ at the first step) calculate $\omega_p(s)$ and $n(s)$ ($\omega_p(0)$ and $n(0)$ at the first step) solving eq.(4).

3)Calculate the energy gain $\Delta\gamma(s)$ in a distance $\Delta s$, assuming the acceleration gradient as $eE(s) = m_e c \beta_p(s) \omega_p(s)$;

$$\frac{\Delta\gamma(s)}{\Delta s} = \frac{m_e c \beta_p(s) \omega_p(s)}{m_p c^2}. \tag{14}$$

Also calculate the density gradient length

$$L(s) = \left[ \frac{1}{n(s)} \frac{\Delta n(s)}{\Delta s} \right]^{-1}, \tag{15}$$

4)Calculate the new $\beta(s+\Delta s)$ corresponding to $\gamma(s+\Delta s)$ and return to step 2. If $\beta(s+\Delta s) > 0.5$, terminate the calculation.

Figure 4 shows results of the calculation. Figure 4 (a) and (b) show that, in order to attain $\sim$150MeV, the acceleration time is around 20ps and acceleration distance is around 0.5mm. Figure 4 (c) tells us that the plasma density has to be tapered from $\sim 3 \times 10^{18}cm^{-3}$ to $\sim 4 \times 10^{20}cm^{-3}$. The length characterizing the density gradient shown in Fig. 4(d) is far larger than the laser wavelength, so we need not worry about the condition given in eq. (11).

We meet a technical difficulty combining the seed laser technique introduced at the end of the previous section with this tapered density method. The seed laser has to be frequency-shifted by $\omega_p$, and $\omega_p$ has to be increased as the proton beam is accelerated. This means that the seed laser has to be chirped in quite a wide frequency range; from $\omega_{sL} \sim 0.99\omega_L$ at the injection to $\omega_{sL} = (1/2)\omega_L$ at the exit where the test beam gets the final energy. One possibility is use of a white laser. Light ranging from IR to 150 nm in flat continuum has been generated by a self-trapped femtosecond TW Ti:sapphire laser pulse in atomosperic-pressure rare gases [10].

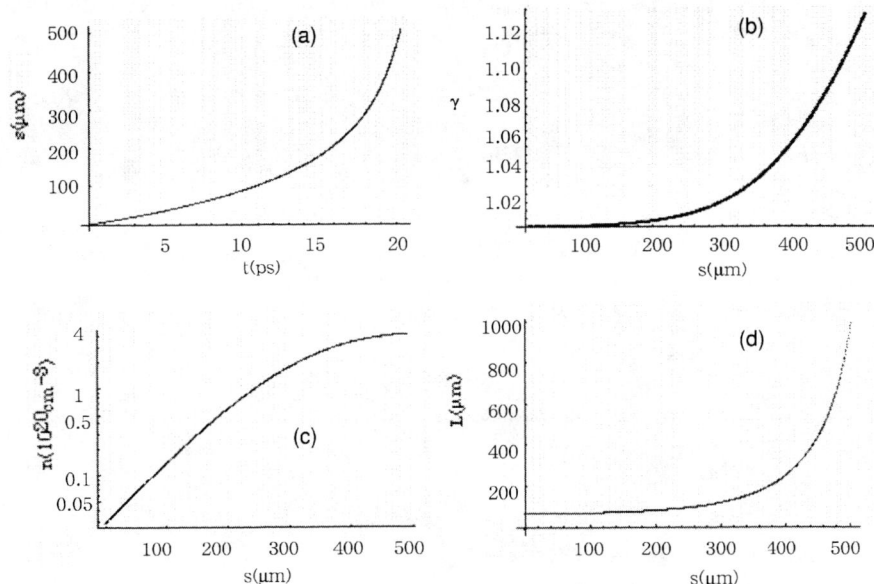

**FIGURE 4.** Results of the simulation. (a) propagation distance vs time, (b) beam energy ($\gamma$) vs. distance, (c) required plasma density in logarithmic scale vs distance, (d) gradient length of the plasma density vs distance.

## IV   DISCUSSION

In summary, a method is proposed to accelerate protons by a slow plasma wave produced in Raman backward scattering. It adjusts the plasma density to match the phase velocity of the plasma wave to the velocity of the particles accelerated.

There are some problems In this method. The phase continuity of the plasma wave in the density gradient is a priori assumed. Though it is plausible, prior analyses [4] have not addressed this issue. This is the first thing to be verified by simulation studies. Technical problems of the tapered-density methods are creation of the plasma density gradient, and chirping of the seed laser. Even if the two techniques were independently established, there would remain the problem of synchronization among the acceleration gradient, the plasma-density gradient and the laser-frequency chirping. Some good diagnostics and feedback control are essential to perform the experiment.

However, proton acceleration does not seem impossible. This method can be applied also to the acceleration of ions.

## REFERENCES

1. Esarey, E. et al., *IEEE Trans. Plasma Sci.* **24**, 252 (1996).

2. Veksler, V. I. et al., *Proc. VI Int. Conf. High Energy Accelerators*, Cambridge Electron Accelerator CEAL-2000, (1967), p289.
3. Mohl, D. et al., *Part. Accel.* **4**, 159 (1973).
4. Liu, C. S. et al., *Phys. Fluids* **17**, 1211 (1974); Liu, C. S. et al., *Advances in Plasma Physics*, **6** Kaw, P. K. et al., ed., Wiley (1976), p121.
5. Estabrook, K. and Kruer, W. L., *Phys. Fluids* **26**, 1892 (1983).
6. Tripathi, Y. K. and Liu, C. S., *Phys. Fluids* **B3**, 468 (1991).
7. Matte, J. P. and Martin, F, *Plasma Phys. Conr. Fusion* **30**, 395 (1988).
8. Mora, P. et al., *Phys. Rev. Lett.* **361**, 1611 (1988).
9. Andreev, N. E. et al., *JETP Lett.* **60**, 713 (1994).
10. Nishioka, H. et al., *Opt Lett.* **20**, 2505 (1995).

# Working Group 3:

# Novel Structure-Based Acceleration Concepts

# STELLA Experiment:
# Design and Model Predictions

W. D. Kimura,* M. Babzien,[†] I. Ben-Zvi,[†] L. P. Campbell,* D. B. Cline,[‡]
R. B. Fiorito,[§] J. C. Gallardo,[†] S. C. Gottschalk,* P. He,[‡] K. P. Kusche,*[,†]
Y. Liu,[‡] R. H. Pantell,[□] I. V. Pogorelsky,[†] D. C. Quimby,*
K. E. Robinson,* D. W. Rule,[#] J. Sandweiss,[∀] J. Skaritka,[†]
A. van Steenbergen,[†] L.C. Steinhauer,* and V. Yakimenko[†]

*STI Optronics, Inc., Bellevue, WA 98004
[†]Brookhaven National Laboratory, Upton, NY 11973
[‡]UCLA, Los Angeles, CA 90024
[§]Catholic University of America, Washington DC 20064
[□]Stanford University, Stanford, CA 94305
[#]Naval Surface Warfare Center, West Bethesda, MD 20817
[∀]Yale University, New Haven, CT 06520

**Abstract.**    The STaged ELectron Laser Acceleration (STELLA) experiment will be one of
the first to examine the critical issue of staging the laser acceleration process.  The BNL inverse
free electron laser (IFEL) will serve as a prebuncher to generate ~1-μm long microbunches.
These microbunches will be accelerated by an inverse Cerenkov acceleration (ICA) stage.  A
comprehensive model of the STELLA experiment is described.  This model includes the IFEL
prebunching, drift and focusing of the microbunches into the ICA stage, and their subsequent
acceleration.  The model predictions will be presented, including the results of a system error
study to determine the sensitivity to uncertainties in various system parameters.

## INTRODUCTION

Exciting progress has been made in the last several years on laser accelerators
including energy gains of 100 MeV and GeV/m acceleration gradients (1). At the BNL
Accelerator Test Facility (ATF), there are two laser acceleration experiments — an
inverse Cerenkov accelerator (ICA) (2) and an inverse free electron laser (IFEL) (3).
Both of these experiments have been routinely accelerating electrons.  Thus, laser
acceleration is becoming more viable as an advanced acceleration technique.

CP472, *Advanced Accelerator Concepts: Eighth Workshop,*
edited by W. Lawson, C. Bellamy, and D. Brosius
© 1999 The American Institute of Physics 1-56396-889-4/99/$15.00

Now the next logical step in its evolution is to address the issue of multi-accelerator module staging and acceleration of the microbunches produced during the laser acceleration process. To efficiently accelerate the electrons throughout these stages it is necessary to prebunch the electrons into a microbunch whose longitudinal length is a small fraction of the accelerating wave. In laser accelerators this accelerating wave can be of order 10 μm. Evidence of microbunching at optical wavelengths has already been detected from the IFEL using a coherent transition radiation (CTR) diagnostic (4).

Staging requires rephasing the microbunches with the accelerating light wave. Efficient acceleration of the electrons contained within a microbunch also requires trapping the electrons within the acceleration potential well. This implies the need to minimize effects that can lead to bunch smearing where the electrons no longer stay within the main bunch distribution. This may be due to trajectory differences of the electrons within the microbunch, gas scattering effects in the case of ICA, and space-charge spreading. While these effects are analogous to those encountered in microwave linacs, they can have a much greater impact in laser acceleration because of the orders of magnitude shorter wavelengths that are involved.

Therefore, the primary goal of the Staged Electron Laser Acceleration (STELLA) experiment is to demonstrate effective trapping and acceleration within an ICA acceleration stage of microbunches generated by an IFEL. During the course of this experiment, we will be addressing the many issues related to generation and preservation of the microbunch, and effective control of the rephasing process. These results will directly benefit other laser acceleration research by demonstrating that efficient staging is possible. This accomplishment will open the door to the next step of adding multiple laser acceleration stages in series to achieve high net energy gain.

## DESCRIPTION OF EXPERIMENT

Figure 1 is a conceptual layout of the STELLA experiment. The $e$-beam from the ATF linac, which consists of a single 10-ps macropulse, enters the IFEL prebuncher from the left. A beam splitter sends a relatively small amount of laser power to the IFEL for modulating the electron energy. The objective is to modulate the energy just enough so that maximum bunching occurs at the end of the drift region just before the entrance to the ICA laser acceleration stage. The rest of the laser power passes through a trombone delay line and to the ICA stage. This trombone delay line will be used to adjust the relative phase of the laser light in the ICA cell with the microbunches created by the IFEL.

For a given wiggler configuration (i.e., spacing and magnetic field strength), $e$-beam energy, and laser wavelength, the optimum bunching distance for the IFEL is controlled by the amount of energy modulation imparted by the laser beam. Therefore,

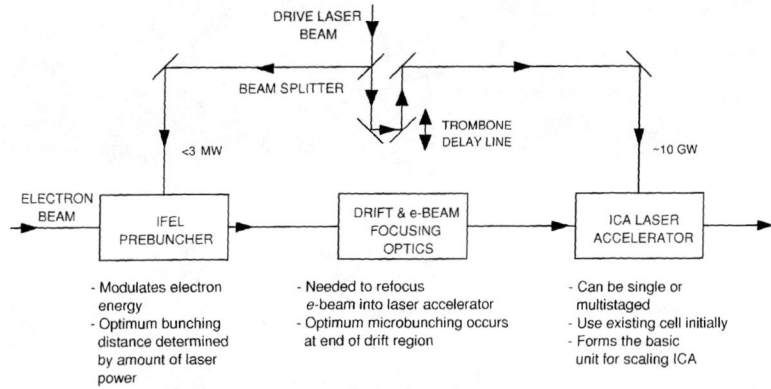

FIGURE 1. Conceptual layout for STELLA experiment.

during the experiment the IFEL prebunching characteristics can be "tuned" by adjusting the amount of laser power sent into the IFEL.

Between the IFEL and the ICA gas cell is a drift space in which quadrupoles will be located to focus the $e$-beam into the ICA gas cell. Next comes the ICA laser accelerator stage located at the end of the drift distance. The primary technical issue here is properly rephasing the optically bunched beam with the laser light wave in the ICA cell so that the bunch is trapped and accelerated.

## IFEL, ICA, and ATF System Parameters for STELLA

A schematic layout of the BNL IFEL (3) is given in Fig. 2. The ATF $CO_2$ laser beam is focused into a 2.8 mm ID 60-cm long sapphire circular waveguide located inside the 47-cm long wiggler. The guide extends beyond the front of the wiggler to permit the proper laser mode ($HE_{11}$) to form within the guide and eliminate other unwanted modes.

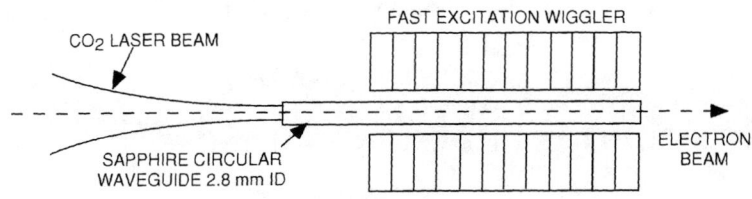

FIGURE 2. Schematic layout for BNL IFEL experiment.

565

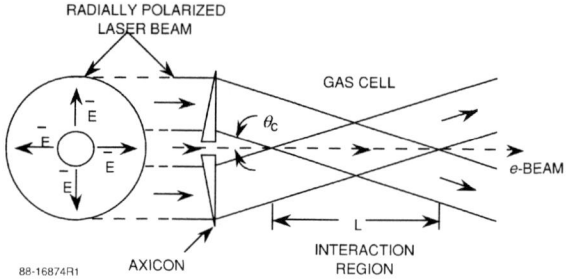

RADIALLY POLARIZED
LASER BEAM

GAS CELL

$\theta_c$

e-BEAM

L

INTERACTION
REGION

88-16874R1   AXICON

**FIGURE 3**. Basic geometry for inverse Cerenkov accelerator. From Ref. (6).

From earlier analysis (5) the IFEL wiggler parameters for STELLA have been selected to be: wiggler length $L_w$ = 47 cm, $\lambda_w$ = 3.3 cm (untapered), gap = 4 mm, $K$ parameter = 2.86, and on-axis peak field = 0.93 T. An untapered wiggler was found to provide the needed amount of energy modulation while at the same time being more tolerant of variations in the $e$-beam and laser beam parameters.

Figure 3 shows the basic scheme for the inverse Cerenkov accelerator. A radially polarized laser beam is focused by an axicon onto the $e$-beam traveling through a gas cell filled with $H_2$ gas (6). The laser light intersects the $e$-beam at the Cerenkov angle for phase matching, i.e., $\theta_c = \cos^{-1}(1/n\beta)$, where $n$ is the index of refraction of the gas, and $\beta$ is the ratio of electron velocity to the velocity of light. The electrons enter and exit the gas cell through thin diamond windows.

Previous analysis (5) demonstrated the importance of minimizing electron scattering off the windows and gas molecules. Scattering degrades the ability to trap the electrons within the microbunch. During STELLA the gas cell is being modified to reduce scattering by using thinner diamond windows (1 μm thick instead of 2.1 μm) and shortening the interaction length to 6.5 cm. With these changes, the model predicts that the effects of scattering will be reduced to the point where other effects, such as emittance and intrinsic energy spread, become dominant factors. The other gas cell parameters remain the same, i.e., $\theta_c$ = 20 mrad and gas pressure ≈ 1.8 atm.

Table 1 lists the system requirements for the ATF linac and ATF psec $CO_2$ laser. These were determined from the integrated modeling analysis discussed below. We should note that most of these values have been already demonstrated by the ATF or are achievable with the present system.

## DESCRIPTION OF DESIGN ANALYSIS

The design of the STELLA experiment is made more complex by the need to create and maintain micron-sized microbunches throughout the system. To understand how the various physical parameters affect this process, an integrated model was

**TABLE 1.** ATF System Requirements for STELLA.

| Parameter | Value Needed for STELLA |
|---|---|
| Electron Beam | |
|    Mean energy, $E$ | 45 MeV |
|    Normalized emittance, $\varepsilon_n$ | $\leq 0.85\ \pi$ mm-mrad |
|    Intrinsic energy spread, $\Delta E$ | 0.15% rms FWHM |
|    Macropulse charge, $Q$ | $\geq 0.1$ nC |
| Laser Beam | |
|    Laser wavelength | 10.6 μm |
|    Power delivered to IFEL | ~2.6 MW (linearly polarized) |
|    Power delivered to ICA | $\geq 10$ GW (radially polarized) |

developed, which consists of several models combined together to simulate the entire STELLA experiment. It uses a 1-D FEL tracking code developed at STI Optronics, Inc. (STI) for the IFEL wiggler. A 3-D ray tracing code was written to track the electrons through the drift region between the IFEL and ICA gas cell. Finally, the STI Monte Carlo ICA model (7), which includes the effects of scattering, was used to predict the acceleration of the microbunches. The integrated model includes all important effects, except space charge. Separate analysis (8) of space charge effects indicate that for the anticipated STELLA conditions it will be at most a 10% effect.

The 1-D IFEL model tracks the electron dynamics in ($\gamma$, $\psi$) phase space. It assumes that the $e$-beam is matched and aligned within the wiggler, the optical field is uniform along the wiggler length, the optical pulse is long (~200 ps) compared to the electron pulse (~10 ps), and the energy spread distribution is Gaussian. Emittance effects within the wiggler have been neglected, which is reasonable since the $e$-beam diameter is much smaller than the optical beam.

The drift region model includes the capability to simulate bunch smearing. TRANSPORT is used to design the $e$-beam optics (i.e., triplet between the IFEL and ICA systems). Differences in the axial velocity, path length difference, and optical phase relative to the resonant on-axis electron are calculated for each Monte Carlo electron traveling from the IFEL through the drift region. Chromatic effects are included by accounting for the magnetic rigidity of individual particles. Misalignments are simulated by introducing mis-steering at the wiggler exit and applying correction kicks at the triplet exit and ICA cell entrance.

The ICA model has been used extensively in the past to simulate the ICA process. It incorporates all relevant physics except space charge effects, including axicon focusing of the laser beam, optical beam propagation with diffraction and interference effects of an ideal laser beam, Rutherford scattering caused by the gas and $e$-beam windows, straggling and other inelastic losses, and the full $e$-beam characteristics (e.g., emittance, energy spread, etc.). A new version of the model has recently been developed that can handle nonideal laser beams, e.g., nonaxisymmetric, nonuniform, and imperfect radial polarization. It has not been applied to STELLA yet.

# SENSITIVITY ANALYSIS RESULTS

The integrated model was run to understand how variations in the different system parameters affect the experiment. The objective is to trap and accelerate the microbunch within a relatively narrow energy range (goal: <5%) while maintaining the original microbunch length (goal: <1 μm). During the first phase of the experiment there will be less emphasis on demonstrating high amounts of acceleration; although, with 10 GW delivered to the accelerator, an energy gain of nearly 10 MeV is predicted, corresponding to an acceleration gradient of >150 MeV/m.

For the baseline conditions, the model predicts for 2.6 MW delivered to the IFEL prebuncher a sinusoidal energy modulation of ≈ ±0.55% (±0.25 MeV). At the end of the drift region just before the ICA cell, roughly 50% of the electrons are bunched together with a microbunch length of ≈ 0.84 μm FWHM.

There are various effects that can degrade the prebuncher. One major effect is the intrinsic energy spread of the incoming e-beam, which directly affects the bunch length as depicted in Fig. 4. A baseline value of 0.15% has been chosen for the experiment.

E-beam energy stability affects the phase jitter of the microbunches leaving the IFEL, which can cause problems later in the accelerator because of phase mismatch with the optical field in the ICA cell. The model predicts a detuning of 2.7 radians per % of energy detuning. This implies the need for an energy jitter of <±0.11%.

Low emittance appears to be less critical for the prebuncher than the accelerator (see below). Emittance degrades the wiggler interaction by reducing the overlap with the laser beam. It also leads to path length differences in the drift region causing bunch smearing.

The prebuncher is also less sensitive to laser power fluctuations. This is demonstrated in Fig. 5. Fluctuation of ±50% are tolerable.

Table 2 summarizes the major sensitivities of the prebuncher.

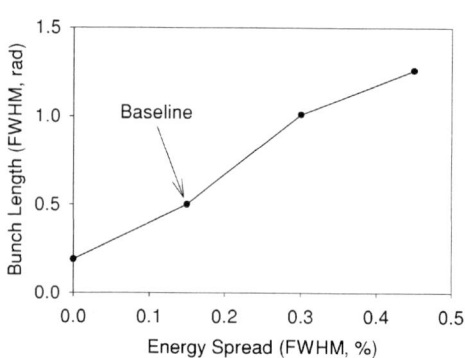

**FIGURE 4.** Scaling of bunch length with e-beam energy spread.

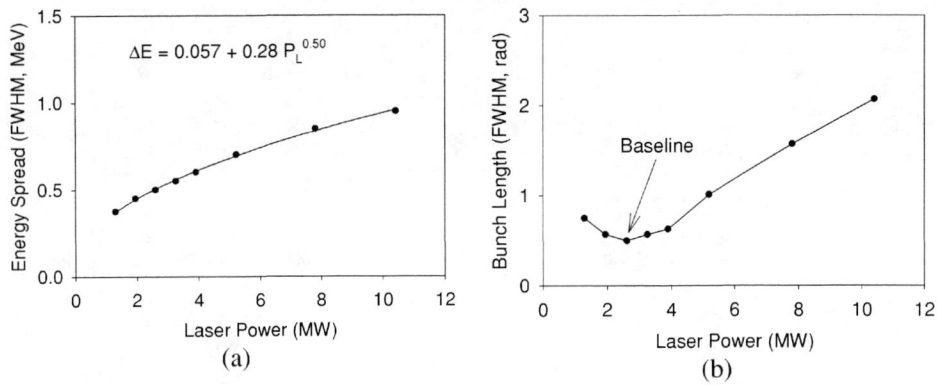

**FIGURE 5.** Scaling of induced energy spread and bunch length with laser power delivered to IFEL (a) Laser-induced energy versus laser power. (b) Bunch length versus laser power.

The baseline predictions for the output of the ICA accelerator are given in Fig. 6. This shows a net energy gain of $\approx 10$ MeV and bunch length of $<1$ μm. (The second small peak in Fig. 6(b) is a remnant from the initial prebuncher energy modulation.)

*E*-beam emittance strongly affects the ICA acceleration process. This is shown in Fig. 8. Although the peak energy gain does not change with emittance very much, the width of the energy peak and the bunch length tend to broaden with larger emittance.

In the ICA accelerator, small *e*-beam size and accurate alignment through the gas cell is critical. This is demonstrated in Fig. 8. Note, the laser beam size in the axicon focal region is $\approx 200$ μm FWHM. Hence, it is important that the *e*-beam diameter along the axicon focal region remain a fraction of this laser beam size in order for all the electrons within the microbunch to experience similar amounts of acceleration. Our goal is to have an *e*-beam diameter of $\approx 60$ μm.

**TABLE 2.** Summary of Major Prebuncher Sensitivities.

| Parameter | Tolerance | Deleterious Effect |
|---|---|---|
| Initial energy spread variance from baseline | $\Delta E/E < 0.3\%$ FWHM | Bunch smearing |
| *E*-beam energy jitter | $\Delta E < \pm 0.11\%$ | Causes bunch to appear at variable phase. Weakly affects bunching. |
| Emittance | $\varepsilon_n < 2\,\pi$ mm-mrad rms<br><br>(Note, ICA requires much smaller emittance.) | Bunch smearing in drift region. May also reduce IFEL bunching efficiency (not included yet in IFEL model). |

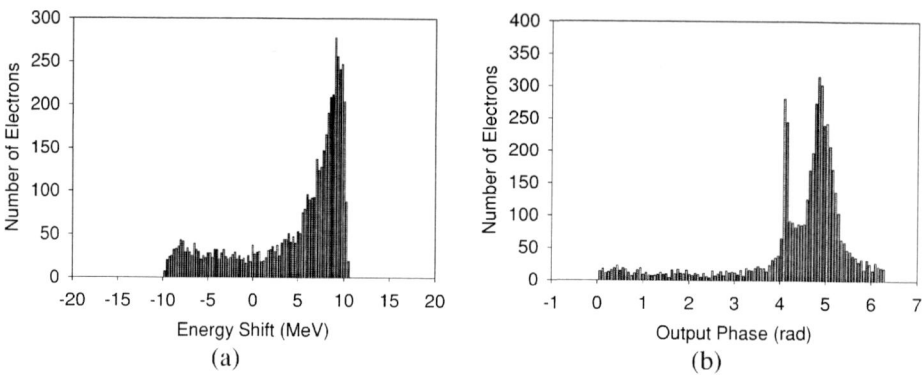

**FIGURE 6.** Output of ICA accelerator for $E = 45$ MeV, $\varepsilon_n = 0.5$ $\pi$ mm-mrad rms, $\Delta E/E = 0.15\%$ FWHM, 1-$\mu$m windows, with gas scattering. (a) Energy spectrum. (b) Longitudinal density.

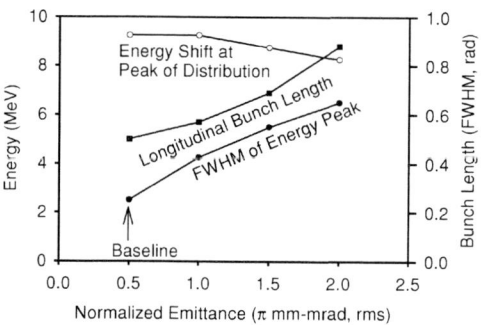

**FIGURE 7.** Scaling of ICA acceleration parameters with $e$-beam emittance.

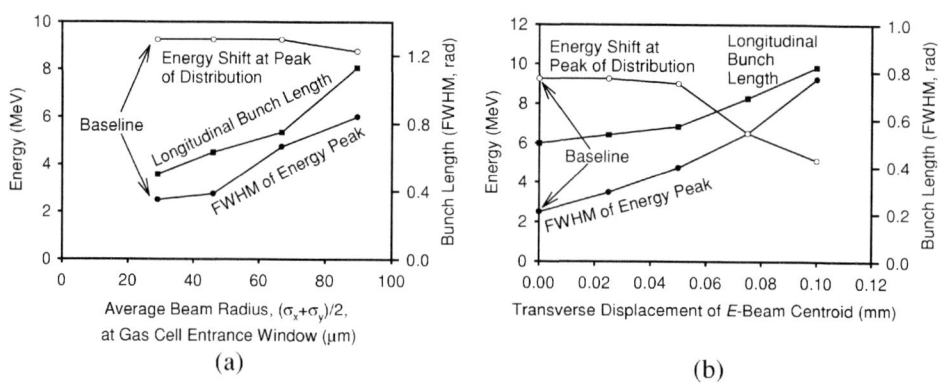

**FIGURE 8.** Scaling of ICA acceleration parameters with (a) $e$-beam size and (b) displacement.

570

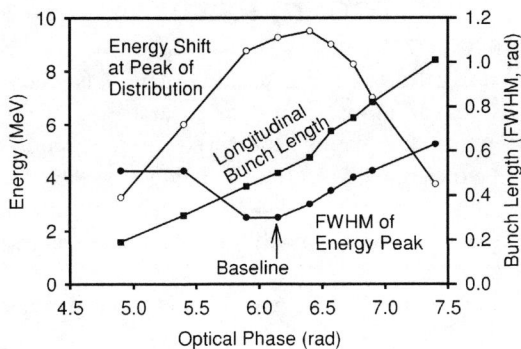

**FIGURE 9.** Variation of ICA acceleration parameters due to phase variation between electron microbunch and laser beam.

Not surprising, it is important to maintain tight control of the microbunch phase relative to the optical field inside the gas cell. Shifts in phase can be caused by fluctuations in the $e$-beam energy and the laser beam path length. Figure 9 shows the sensitivity to phase jitter. Note, in our simulations maximum acceleration and minimum energy spread of the accelerated bunch occurs at a phase value of 6.15 radians.

Based upon the simulations we have established an overall phase jitter tolerance of ±0.40 rad (±0.64 μm total path difference). This corresponds to an acceptable laser beam path length jitter of ≈±0.45 μm and $e$-beam energy jitter of ≈±0.11%., assuming these effects add in quadrature because they are uncorrelated.

The ICA process is more tolerant of laser power fluctuations with a tolerance of ±32% being chosen for STELLA.

Table 3 summarizes the major sensitivities of the ICA accelerator.

**TABLE 3.** Summary of Major ICA Accelerator Sensitivities.

| Parameter | Tolerance | Deleterious Effect |
|---|---|---|
| Emittance | $\varepsilon_n < 0.85\ \pi$ mm-mrad rms (Note, more stringent than IFEL requirement.) | Bunch smearing and loss of overlap with laser beam causes energy smearing. |
| $E$-beam energy jitter | $\Delta E < \pm 0.11\%$ | Causes jitter of initial bunch phasing. |
| Mis-steering of $e$-beam through gas cell | $\Delta r \leq 25$ μm | Reduces overlap with laser beam. |
| Phase jitter | $\Delta l < \pm 0.45$ μm | Reduces bunch acceleration and increases bunch energy spread. |

# CONCLUSION

An integrated model of the STELLA experiment has been developed, which predicts the sensitivity to variations in system parameters. This has identified key experimental issues, which are being addressed in the hardware design of the experiment (9). The experiment is currently being assembled and tested. Initial measurements will be conducted using 10 GW laser power to the ICA cell. Plans are to later modified the system to handle much higher laser power (e.g., several hundred gigawatts).

Additional information on the STELLA experiment can be found on the NSLS ATF Web Site at:

http://www.nsls.bnl.gov/AccTest/experiments/STELLA/STELLA.htm.

# ACKNOWLEDGMENTS

The authors wish to acknowledge Dr. J. R. Fontana for his technical advice during the initial planning of this experiment, and Dr. Xijie Wang for his on-going assistance and support. This work was supported by the U.S. Department of Energy, Grant Nos. DE-FG03-98ER41061, DE-AC02-76CH00016, and DE-FG03-92ER40695.

# REFERENCES

1.  See for example Proceedings of Workshop on Advanced Acceleration Concepts, Oct. 12-18, 1996, Lake Tahoe, CA.
2.  Kimura, W. D., Kim, G. H., Romea, R. D., Steinhauer, L. C., Pogorelsky, I. V., Kusche, K. P., Fernow, R. C., Wang, X. J., and Liu, Y., *Phys. Rev. Lett.* **74**, 546-549 (1995).
3.  van Steenbergen, A., Gallardo, J., Sandweiss, J., Fang, J.-M., Babzien, M., Qiu, X., Skaritka, J., and Wang, X.J., *Phys. Rev. Lett.* **77**, 2690 (1996).
4.  Liu, Y., Cline, D. B., Wang, X. J., Babzien, M., Fang, J. M., Kusche, K. P., 1997 Particle Accelerator Conference, May 12-16, 1997, Vancouver, B.C., Canada, Paper 3'C3.
5.  See Staged Electron Laser Acceleration (STELLA) Technical Proposal submitted to U.S. Department of Energy by STI Optronics, Inc., June 27, 1998.
6.  Fontana, J. R. and Pantell, R. H., *J. Appl. Phys.* **54**, 4285–4288 (1983).
7.  Romea, R. D. and Kimura, W. D., *Phys. Rev. D* **42**, 1807–1818 (1990).
8.  Steinhauer, L. C., and Kimura, W. D., "Space Charge Compensation in Laser Particle Accelerators," in these Proceedings of the 8th Workshop on Advanced Accelerator Concepts, Jul. 5-11, 1998, Baltimore, MD.
9.  Kusche, K. P., Ben-Zvi, I., Babizen, M., Campbell, L. P., Cline, D. B., Fiorito, R. B., Gallardo, J. C., Gottschalk, S. C., He, P., Kimura, W. D., Liu, Y., Pantell, R. H., Pogorelsky, I. V., Quimby, D. C., Robinson, K. E., Rule, D. W., Sandweiss, J., Skaritka, J., Steinhauer, L. C., van Steenbergen, A., and Yakimenko, V., "STELLA Experiment - Hardware Issues," in these Proceedings of the 8th Workshop on Advanced Accelerator Concepts, Jul. 5-11, 1998, Baltimore, MD.

# STELLA Experiment:
# Hardware Issues

K. P. Kusche,[*,†] M. Babzien,[†] I. Ben-Zvi,[†] L. P. Campbell,[*] D. B. Cline,[‡]
R. B. Fiorito,[§] J. C. Gallardo,[†] S. C. Gottschalk,[*] P. He,[‡] W. D. Kimura,[*]
Y. Liu,[‡] R. H. Pantell,[○] I. V. Pogorelsky,[†] D. C. Quimby,[*]
K. E. Robinson,[*] D. W. Rule,[#] J. Sandweiss,[∀] J. Skaritka,[†]
A. van Steenbergen,[†] L.C. Steinhauer,[*] and V. Yakimenko[†]

*STI Optronics, Inc., Bellevue, WA 98004
†Brookhaven National Laboratory, Upton, NY 11973
‡UCLA, Los Angeles, CA 90024
§Catholic University of America, Washington DC 20064
○Stanford University, Stanford, CA 94305
#Naval Surface Warfare Center, West Bethesda, MD 20817
∀Yale University, New Haven, CT 06520

**ABSTRACT.** The STaged ELectron Laser Acceleration (STELLA) experiment is currently being assembled and tested at the BNL Accelerator Test Facility (ATF). The existing BNL inverse free electron laser (IFEL) has been positioned upstream of the inverse Cerenkov acceleration (ICA) experiment on Beamline #1. This beamline also features new quadrupoles and a new spectrometer capable of a +/- 20% energy acceptance. A new laser beam transport system has been installed to permit accurate control of the laser phase for the laser beams sent to the IFEL and ICA devices. Detection of the microbunches will be performed using a coherent transition radiation (CTR) diagnostic similar to one already demonstrated at the ATF.

## INTRODUCTION

The acceleration of electrons via laser interaction has been demonstrated recently by various experimental schemes, including the inverse Cerenkov acceleration (ICA) (1) and inverse free electron laser (IFEL) (2) experiments. These schemes featured the modulation of electron macrobunches over a short distance at a single interaction point, where the macrobunch length (L ~ 10 psec) greatly exceeded the input laser

CP472, *Advanced Accelerator Concepts: Eighth Workshop,*
edited by W. Lawson, C. Bellamy, and D. Brosius

wavelength ($\lambda = 10.6\ \mu m$). Acceleration was inefficient, as expected, as evidenced by a wide energy spectrum. Thus, for monochromaticity, it is desired to have $L < \lambda$. This can be accomplished by converting the macrobunch into a train of microbunches, where the microbunch length, $\ell$, is shorter than $\lambda$.

The arrangement described here proposes to accomplish this by staging a microbunch-producing device in front of an accelerating section. Specifically, the IFEL wiggler, optimized for microbunching, will be utilized as a prebuncher for the ICA accelerating stage. Clearly, each of these sections has been proven to work independently of one other. But, such a combination presents a variety of engineering and hardware challenges, namely stabilizing and synchronizing each part of the experiment with respect to each other.

## DESIGN PHILOSOPHY

In the early preparations for STELLA, it was determined that major modifications to the ATF beamline #1 had to be performed; in fact, the entire beamline had to be rebuilt to meet the criteria of the project. This challenge presented certain optimizing opportunities that became practical due to the scope of the changes involved.

First, it was desirable to minimize the beamline assembly time in the experimental hall so as to maximize running time for all users of the facility. It was therefore decided to modularize the major sections of beamline #1. The sections were designed to be assembled and presurveyed at a separate facility, and then installed at the ATF with minimum impact to other experiments. Another reason for modularity comes from the future proposed uses of beamline #1, such as for the IFEL[2] and Compton Scattering experiments, where the section of beamline containing the ICA gas cell will be replaced by other interaction chambers.

A standardized low-profile rigid rail assembly was devised to locate and support all beamline components. This was done for several reasons. First, the rail system provides an extremely flexible component positioning method. The axial locations of many of the components, specifically the precise positions of the quadrupoles, had yet to be determined. The ability to reposition these elements was desirable. The rail system allows for the sliding of components into their desired axial locations with minimal need for survey and support modifications.

Standardized component clamping and the ability to presurvey the components to the rail and then to survey the rail onto the beamline was also desirable. This saved significant survey time. The geometry of beamline #1 presented certain challenges, as well. Specifically, the distances from the beamline axis to various horizontal surfaces varied greatly, and the mounting of Chinese quadrupoles required a special design of the interface between magnet and rail. It was found that, due to manufacturing tolerances of the support points on the quadrupoles and other beamline components, adjustable interface points were necessary.

It was, therefore, decided that all components which were sensitive to x,y alignment had to have an independently adjustable interface between the component and rail. Theoretically, once the component is presurveyed onto the beamline, it could be removed, serviced, replaced, and/or repositioned without the need for extensive resurvey. The rails were designed to kinematically locate onto the adjustable supports so that entire rail sections could be relocated with minimum survey.

The thermal stability of the laser (optical) and electron beamlines was evaluated for mechanical stabilization. The kinematic supports of the rails were configured so as to minimize the effects of random thermal excursions. The optical rails were located on the ceiling of the ATF experimental hall and anchored near where the bending mirrors are located. This philosophy was implemented to minimize the effect of thermal expansion on mirror alignment and stability, and took advantage of a unique feature of the hall. At the ATF, the entire overhead shielding structure of the experimental hall resides inside of a temperature-controlled building. Therefore, since the temperature is relatively stable on either side of the massive concrete roof beam, there is negligible flexing due to daily environmental changes. In short, the roof is as stable as the floor and can be safely used to support critical optical components and maintain precise alignment to components below.

The rail support system was analyzed and designed to be tolerant of thermal excursions. One example of this can be found in the supports for triplet-rail #3. Since the two supports had to be different lengths, with one mounted to a concrete pedastal and the other to the floor, it was decided to make them from different materials. For the longer support, steel was chosen; for the shorter, aluminum. In this way, the pitch of the rail would not be expected to vary more than about 2 $\mu$m/$\pm$1°C.

The rails and kinematic supports were analyzed for deflection and contact stresses. Beneath rails #3&4, the V-groove of the support system has two possible locations so as to minimize rail flexure in the event changes in the quadrupole size and/or position become necessary. All major electron beamline components have a 5-point support system incorporated into the rail interface. This provides ease of alignment as well as maximum rigidity.

Critical components such as the ICA gas cell, IFEL wiggler, and GPOP1 cross have axial anchoring such that they could theoretically be removed from the beamline, serviced, and replaced without the need for axial resurvey. The anchoring points are very near the cup/cone locator of the rail's kinematic support, so the thermal and mechanical effects on these critical components are minimized.

The two large crosses needed for laser beam incoupling (GPOP1 & GPOP3) presented significant engineering challenges. Large in-vacuum mirrors have to be repositioned in a precise manner to assure laser beam alignment stability. Each is mounted to two parallel ground stainless steel rods, with support provided by a spring-clamped vee and flat interface. Air-actuated linear feedthroughs are used to position the mirrors into and away from the electron beam axis. An ATF-designed pop-in flag assembly is mounted to the same flange, but is controlled by a separate actuator. The flags have the capability of mounting either a standard phosphor-coated screen or a

575

"YAG crystal" (3) in front of a feducialized 45° outcoupling mirror. Such flags are necessary in order to visualize the e-beam profile at a distance no more than a few centimeters away from the hole in the center of the large incoupling mirror. They also allow for precise presurvey and feducialization of the cross assembly.

## CRITICAL ISSUES

Ultimately, the ability to effectively synchronize the electron-laser interactions, upon which the success of the STELLA program relies, requires that several critical issues be addressed. Obviously, the ATF will be expected to deliver the 45 MeV and 1 GW $CO_2$ laser beam within specified parameters (4) that can be reasonably achieved. The STELLA project has taken responsibility for the creation and integration of systems to facilitate the transport of these beams to their respective interaction points (Figure 1). Comprehensive descriptions of the overall STELLA beamline functionalities (4,5), as well as specific subsystem details regarding electron beam microbunch formation and detection (4,6,7), are covered elsewhere. The critical issues discussed here include laser transport and phase stability, and spectrometry.

In order to enable high power (gigawatt regime) $CO_2$ laser beam delivery to the IFEL and ICA interaction regions, several requirements must be met. A zinc selenide (ZnSe) beamsplitter is necessary to divert <10% of the total laser power to the IFEL wiggler while allowing the remainder to be transmitted to the ICA gas cell. Due to the high peak power of each laser pulse (~100psec), the beam must be sufficiently expanded to avoid optical damage; <0.5J/cm$^2$ for ZnSe lenses and windows, <5J/ cm$^2$ for copper mirrors.

Delivery to the IFEL subsystem is achieved using axicon and spherical lens pairs to create an expanded-converging annular (central null) beam which is introduced into the vacuum beamline using the unique high-stability pop-in mirror (with central hole for e-beam passage) at the GPOP1 location. The annular beam is co-aligned with the electron beam as it forms a waist at the wiggler's sapphire waveguide, and is outcoupled at GPOP2 & 3 for diagnostic purposes.

Delivery to the ICA subsystem can be handled by one of two optical schemes (denoted "old" & "new") for converting the beam to the required radial polarization (RPC), after traversing the remotely-controllable delay rail path. Co-alignment of the axicon-focused laser beam to the triplet-focused e-beam at the gas cell has been previously facilitated via internal encoder-mike mirrors and pop-in screen. More stringent requirements for STELLA (<100µm size regime for both beams) have forced a redesign of the methods by which both beams can be visualized simultaneously at high resolution. A working group has been formed to investigate the possibilities and begin testing at least one promising scheme.

The stated stability requirement for the laser beam path length (one leg with respect to the other) is ±0.45µm (4). A stepper motor has been fitted to the delay rail to provide ~0.2µm steps over a total distance of >40cm, as confirmed using an LVDT

576

measurement system. Stability studies have been performed to determine partial-path jitter and drift, using a simple interferometer and IR camera/frame grabber combination to visualize fringes from 632nm HeNe and 10μm CW $CO_2$ lasers. Shot-to-shot (1 Hz) jitter was found to be $\leq\pm1$μm, but not better than 0.6μm, and drift over >0.5 hours was $\leq\pm2$μm. As this configuration used temporary optical mounts without beam path enclosures, these numbers represent upper limits on the expected full-path stability. Our goals are likely to be met with the installation of new optical mounts and full beam path enclosures. Total beam path interferometric studies will follow soon after installation of these upgrades, which will determine the need for active feedback via piezostage, etc. Nonetheless, falling short of the stated jitter tolerance will not totally prevent bunch acceleration, just make it more difficult to observe.

The observation of the total energy-modulated spectrum will be accommodated by a versatile spectrometer, designed to meet the needs of the various stages of the STELLA program. At first, resolution on the order of 1% is desired for IFEL wiggler recommissioning; later, successful acceleration using the ICA subsystem requires up to $\pm20\%$ acceptance. The ensuing design incorporated such variable energy acceptance via $\pm8°$ horizontal rotation of a large dipole magnet, with vertical and horizontal beam size control by three quadrupoles and energy compensation by a sextupole. For ease of survey, as well as flexibility with other experiments, the same type of rail support system was implemented as for the rest of beamline #1. A calculated energy resolution of up to 0.4% can easily accommodate all phases of STELLA. Imaging will initially be done with a standard phosphor screen and CCD/frame grabber system, as done in previous ICA experiments; an upgrade is planned and will be implemented if necessary.

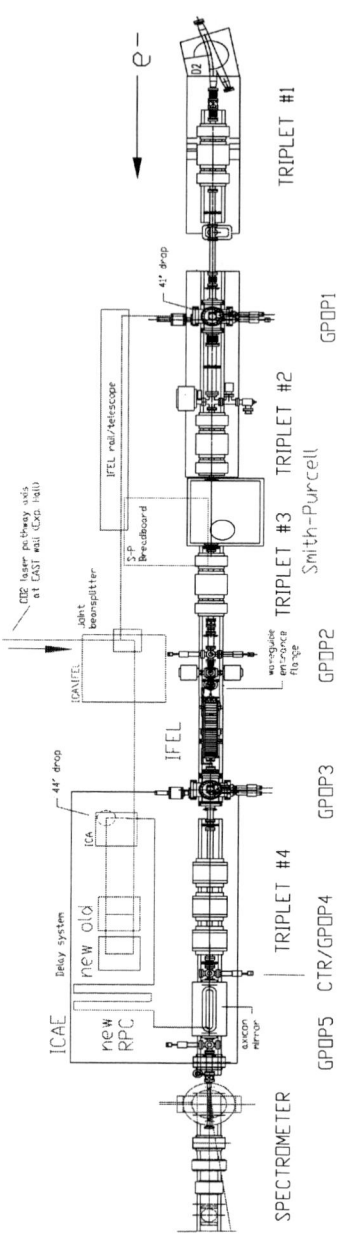

**FIGURE 1.** Actual layout for STELLA experiment on ATF beamline #1.

# CURRENT STATUS AND FUTURE PLANS

All four of the quadrupole triplet-rail assemblies have been installed and commissioned at the ATF. Safety and operational approval has been granted for these subsystems as well as for the entire experiment. The upgrade of the laser transport system is in progress, with the goal of meeting the aforementioned stability requirements. Upon completion, total beam path interferometric studies are planned to quantify laser beam path jitter and drift. A measurement scheme, similar to that described for initial tests, will consist of a $CO_2$ laser beam traversing both actual delivery paths terminating at a simple interferometer and IR camera/analyzer combination to visualize the motion of 10.6 micron-spaced fringes.

The installation and commissioning of the spectrometer assembly is expected shortly. Coupled with successful delivery of laser power on the order of a few MW to the prebuncher, the spectrometer will enable characterization of the wiggler and CTR diagnostic. We expect to observe negligible electron beam quality degradation as it exits the wiggler, as well as evidence of microbunching at the CTR location and about 1% energy modulation at the spectrometer.

Only after these steps have been passed will the ICA gas cell be installed and commissioned. Low laser power (<1GW) delivery to both interaction regions will enable the optimization of the laser synchronization system, and will culminate in attempts to accelerate microbunches. Preliminary results from the fully-implemented STELLA configuration are expected in time for presentation at PAC'99.

# CONCLUSION

Comprehensive plans for the STELLA experiment are being implemented at the BNL ATF. Given the size of the collaboration, the luxury of having several working groups address critical issues in parallel has been exploited. Installation of key hardware is nearing completion, and preliminary results from electron and laser beam studies thus far indicate an absence of major hurdles toward the success of the experimental program.

# ACKNOWLEDGMENTS

The authors wish to acknowledge the technical assistance of R. Harrington, L. Finch, X.J. Wang, and A. Doyuran, as well as the rest of the staff of the ATF, and the NSLS's survey, vacuum, and magnetic measurement groups. The long-standing support of the ATF and NSLS has been invaluable. This work was supported by the U.S. Department of Energy, Grant Nos. DE-FG03-98ER41061, DE-AC02-98CH10886, and DE-FG03-92ER40695.

# REFERENCES

1. Kimura, W. D., Kim, G. H., Romea, R. D., Steinhauer, L. C., Pogorelsky, I. V., Kusche, K. P., Fernow, R. C., Wang, X. J., and Liu, Y., *Phys. Rev. Lett*. **74**, 546-549 (1995).
2. van Steenbergen, A., Gallardo, J., Sandweiss, J., Fang, J.-M., Babzien, M., Qiu, X., Skaritka, J., and Wang, X.J., *Phys. Rev. Lett*. **77**, 2690 (1996).
3. Graves, W. S., Johnson, E. D., O'Shea, P. G., "A High Resolution Electron Beam Profile Monitor," Proceedings of the IEEE Particle Accelerator Conference, May 12-16, 1997, Vancouver, Canada; also, BNL Publication No. BNL-65006.
4. See Staged Electron Laser Acceleration (STELLA) Technical Proposal submitted to U.S. Department of Energy by STI Optronics, Inc., June 27, 1998.
5. Kimura, W. D., Babzien, M., Ben-Zvi, I., Campbell, L. J., Cline, D. B., Fiorito, R. B., Gallardo, J. C., Gottschalk, S. C., He, P., Kusche, K. P., Liu, Y., Pantell, R. H., Pogorelsky, I. V., Quimby, D. C., Robinson, K. E., Rule, D. W., Sandweiss, J., Skaritka, J., Steinhauer, L. C., van Steenbergen, A., and Yakimenko, V., "STELLA Experiment - Design and Model Predictions," in these Proceedings of the 8th Workshop on Advanced Accelerator Concepts, Jul. 5-11, 1998, Baltimore, MD.
6. Liu, Y., Cline, D. B., Wang, X. J., Babzien, M., Fang, J. M., Kusche, K. P., 1997 Particle Accelerator Conference, May 12-16, 1997, Vancouver, B.C., Canada, Paper 3'C3.
7. He, P., Babzien, M., Ben-Zvi, I., Campbell, L. J., Cline, D. B., Fiorito, R. B., Gallardo, J. C., Gottschalk, S. C., Kimura, W. D., Kusche, K. P., Liu, Y., Pantell, R. H., Pogorelsky, I. V., Quimby, D. C., Robinson, K. E., Rule, D. W., Sandweiss, J., Skaritka, J., Steinhauer, L. C., van Steenbergen, A., and Yakimenko, V., "STELLA Experiment - Microbunch Diagnostic," in these Proceedings of the 8th Workshop on Advanced Accelerator Concepts, Jul. 5-11, 1998, Baltimore, MD.

# The Proposed Interferometric-Type Laser-Driven Particle Accelerators

Y.C. Huang[1], Y.W. Lee[1], T. Plettner[2], and R.L.Byer[2]

[1]*Department of Atomic Science*
*National Tsinghua University, Hsinchu, Taiwan 30043*

[2]*Department of Applied Physics*
*Stanford University, Stanford, CA 94305-4085*

**Abstract** In a crossed-laser-beam accelerator, two properly phased laser beams, forming an interferometric configuration, may provide an adequate particle acceleration field over a phase matching distance. The two laser beams can be obtained by dividing a full, Gaussian laser beam equally in amplitude or in wavefront. We show in this paper that a wavefront-splitting laser-driven accelerator is relatively simple to set up and provides an acceleration gain comparable to that of an amplitude-splitting accelerator under the same laser damage fluence. We also present the noise characteristics measured from various interferometers, which may be useful for implementing interferometric-type accelerators.

## INTRODUCTION

In the past several decades, conventional RF accelerators have successfully generated GeV electrons with an average acceleration gradient of a few tens MeV per meter. The average gradient is limited by field-induced structure damage. The damage threshold is determined by the driving frequency, the pulse duration, and the structure material. To increase the acceleration gradient, recent research interests have moved toward higher operation frequencies including those in the millimeter-wavelength and optical regimes[1,2]. A high-frequency accelerator has the additional advantage of producing short electron bunches, which might be important in the future for generating coherent UV or x-ray[3].

To develop optical-frequency accelerators, efforts at Stanford University are being made in the experimental demonstration of crossed-laser-beam electron linear acceleration between two dielectric boundaries[4]. Although plasma-based laser-driven accelerators have demonstrated a ~100 GeV/m peak gradient over a submillimeter distance[5], theoretical studies indicate that a dielectric-based, laser-driven accelerator might achieve a ~1 GeV/m average gradient over a much longer distance[6].

According to the Lawson-Woodward theorem[7], first-order energy transfer from a photon to an electron can not occur in a vacuum. To obtain electron linear acceleration

CP472, *Advanced Accelerator Concepts: Eighth Workshop,*
edited by W. Lawson, C. Bellamy, and D. Brosius

from a laser field, Pantell *et al.*[8, 9] proposed to cross an electron beam with a linearly polarized laser beam at an angle and limit the interaction length to a phase slip less than $\pi$. To avoid any transverse force acting on the electron, a symmetric crossed-laser-beam acceleration scheme was proposed and analyzed by several authors[10,11,12]. The two laser beams in the proposed configurations are phased such that transverse-field components cancel and longitudinal-field components add along the electron axis. Figure 1 illustrates the symmetric, crossed-laser-beam accelerator configuration. The primed coordinate system defines the rotated laser beam axis, and the unprimed one includes the electron axis. Huang and Byer[13] extended the idea of the proposed accelerator to a configuration whose structure dimensions and the laser fields are constant in $y$. This novel design allows for the acceleration of more charge, reduced thermal loading, and decreased wake field generation.

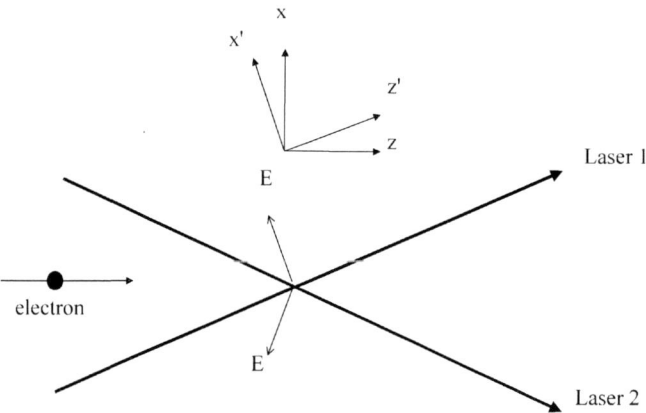

**FIGURE 1.** The proposed crossed-laser-beam accelerator configuration. Two linearly polarized laser pulses are crossed at an angle with respect to the electron trajectory. The two laser beams are phased such that their longitudinal electric fields add and their transverse electric fields cancel on the electron axis. The primed coordinate system defines the rotated laser beam axis, and the unprimed one includes the electron axis. The laser fields are constant in $y$.

Since the two laser beams form interference fringes in the transverse direction, in this paper we term the proposed structures interferometric-type accelerators. The distribution of the interference fringes indicates a proper longitudinal field for particle acceleration, as will be discussed.

## ACCELERATION FIELDS

In a linear accelerator, there ought to be a longitudinal electric field component in the acceleration direction or the $z$-direction defined in this paper. An observation from the divergent-free electric field vector $\nabla \cdot \vec{E} = 0$ indicates that the longitudinal field component $E_z$ is coupled to the transverse electric field $E_x$ through the approximation[14]

$$\tilde{E}_z \approx \frac{j}{k}\frac{\partial \tilde{E}_x}{\partial x}, \tag{1}$$

where $k \equiv \dfrac{2\pi}{\lambda}$ is the wave number, $j \equiv \sqrt{-1}$ is the imaginary unit, and the tilde designates phasor notations. In Eq. (1), the approximation comes from the assumption of a slowly varying laser field envelope along $z$. In an ideal accelerator it is desirable to have no transverse field on the electron axis or $E_x(x=0)=0$. To obtain a net $E_z$ along $z$, it is evident from (1) and the condition $E_x(x=0)=0$ that $E_x$ is an odd function in $x$. For example, the superposition of two crossed, equal-amplitude plane waves gives the following two electric field components:

$$E_x = 2E_0 \cos\theta \cdot \cos(kx\sin\theta + \varphi/2) \cdot \cos(\omega t - kz\cos\theta + \varphi/2), \tag{2}$$
$$E_z = -2E_0 \sin\theta \cdot \sin(kx\sin\theta + \varphi/2) \cdot \sin(\omega t - kz\cos\theta + \varphi/2), \tag{3}$$

where $E_0$ is the field amplitude of each plane wave, $\theta$ is the crossing angle relative to the electron axis, $\varphi$ is an arbitrary phase difference between the two waves, and $\omega$ is the angular frequency. To maximize the $z$-component field on the electron axis $\varphi$ is set to be $\pi$. Thus Eqs. (2, 3) become

$$E_x = 2E_0 \cos\theta \cdot \sin(kx\sin\theta) \cdot \sin(\omega t - kz\cos\theta), \tag{4}$$
$$E_z = -2E_0 \sin\theta \cdot \cos(kx\sin\theta) \cdot \cos(\omega t - kz\cos\theta), \tag{5}$$

which clearly show that $E_x$ is an odd function in $x$. In this crossed-laser-beam configuration, the laser power propagates primarily in the $z$-direction and thus the interference intensity along $x$ is mostly due to $E_x$, given by

$$I(x) = 2E_0^2 \cos^2\theta \cdot \cos^2(kx\sin\theta + \varphi/2). \tag{6}$$

For $\varphi = \pi$, a dark interference fringe appears at $x=0$ and therefore it indicates a $z$-component electric field for particle acceleration.

There are two types of interferometers, amplitude-splitting interferometers and wavefront-splitting interferometers[15]. Therefore the proposed accelerator configurations can be divided into two categories: amplitude-splitting accelerators and wavefront-splitting accelerators.

## AMPLITUDE-SPLITTING ACCELERATORS

Figure 2.a illustrates the proposed amplitude-splitting accelerator configuration where a full-Gaussian laser beam is split equally in amplitude to form two crossed Gaussian laser beams inside the accelerator cell. An external phase controller sets the relative phase $\varphi = \pi$ between the two laser beams. Figure 2.b shows the interference fringes in $x$ formed by two cylindrical Gaussian laser beams at the focal point. The dark fringe appearing at $x = 0$ is a signature of the longitudinal field for particle acceleration, as shown in the last section. Unlike the constant-amplitude fringes in Eq. (6) from two plane waves, Fig. 2.b shows an attenuating envelope in $|x|$ due to the Gaussian laser field distribution. In the plot we choose a laser crossing angle of 50 mrad, a laser waist size of 25 μm, and a laser wavelength of 1 μm.

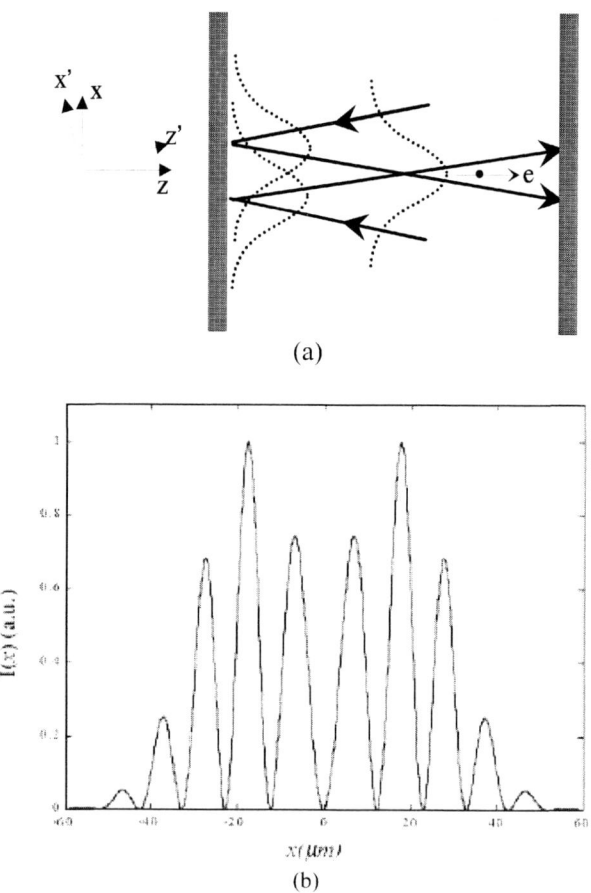

**FIGURE 2.** (a) The proposed amplitude-splitting accelerator configuration where a full-Gaussian laser beam is split equally in amplitude to form crossed laser beams inside the accelerator stage. An external optical phase controller sets the relative phase between the two laser beams. (b) The interference fringes in $x$ formed by two cylindrical Gaussian laser beams at the focal point. The dark fringe appearing at $x = 0$ is a signature of the longitudinal electric field for particle acceleration.

The electron energy gain of the proposed accelerator structure in Fig. 2.a can be calculated by integrating the $z$-component electric field of two superimposed, cylindrical, Gaussian laser beams. Structure damage by intense laser fields limits the acceleration gradient. At the dielectric damage fluence $2J/cm^2$ for ~100 fsec laser pulses[16], a 280 keV electron energy gain[13] is predicted for an interaction length of 340 μm.

## WAVEFRONT-SPLITTING ACCELERATORS

Figure 3 illustrates the proposed wavefront-splitting accelerator configuration where a full-Gaussian laser beam is split equally in wavefront to form two crossed, half-Gaussian beams at the coupling point. The $\phi = \pi$ phase shift in Fig. 3 is set by a phase step at the laser coupling point. The half-Gaussian laser fields in Fig. 3 can be numerically synthesized by superimposing many plane waves propagating at different angles and amplitudes[17]. Specifically an arbitrary vector-field component $\widetilde{E}_x(x,z)$ in $x$ can be expressed by

$$\widetilde{E}_x(x,z) = \int_{-\pi/2}^{\pi/2} a(\phi)\cos\phi \times \exp(-jkx\sin\phi - jkz\cos\phi)d\phi \qquad (7)$$

and the corresponding $\widetilde{E}_z(x,z)$ of the wave is

$$\widetilde{E}_z(x,z) = \int_{-\pi/2}^{\pi/2} a(\phi)\sin\phi \times \exp(-jkx\sin\phi - jkz\cos\phi)d\phi . \qquad (8)$$

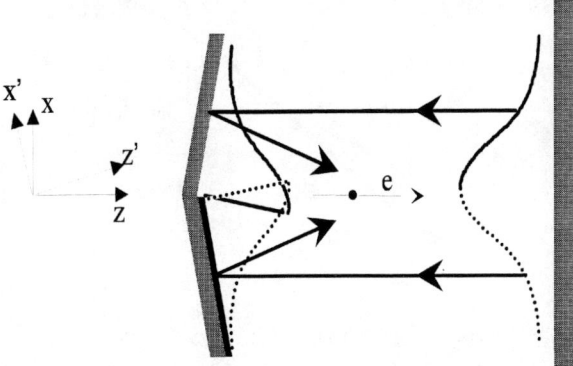

**FIGURE 3.** The proposed wavefront-splitting accelerator configuration where a full-Gaussian laser beam is split equally in wavefront to form two half-Gaussian beams at the coupling point. The $\phi = \pi$ phase shift between the two beams is set by a phase step on the corner reflector.

As long as the angular spectrum $a(\phi)$ is known, the vector fields $\widetilde{E}_x(x,z)$ and $\widetilde{E}_z(x,z)$ everywhere in space can be determined from Eqs. (7, 8). The angular spectrum $a(\phi)$ is evaluated numerically from Eq. (7) by knowing the boundary field $\widetilde{E}_x$ at the coupling point, which is essentially the transverse field sum of an angle-tilted Gaussian laser beam.

Integrating the thus-obtained $z$-component electric field gives the electron energy gain. Figure 4 illustrates the electron energy gain along a 340 μm interaction length for the amplitude-splitting accelerator in Fig. 2.a and the wavefront-splitting accelerator in Fig. 3. In our calculation the laser crossing angle is 50 mrad and the laser wavelength is 1 μm. We have assumed relativistic electrons, that is $\gamma \gg 1/\theta$, where $\gamma$ is the electron energy in units of the electron rest-mass energy and $\theta$ is the laser-electron crossing angle. Before split in wavefront, the full Gaussian beam in Fig. 3 has a waist size of 50 μm, twice that in Fig. 2.a. Thus the total laser power for acceleration and the laser damage intensity on the dielectric surface remain the same for both accelerator structures. The electron entrance phase for each case is adjusted to obtain the highest electron energy gain. For this particular set of parameters, the energy gain of the wavefront-splitting accelerator is about 15% lower due to the phase slip caused by large-angle plane-wave components at the coupling point. Numerical calculation shows that the single-stage energy gain of a wavefront-splitting accelerator need be only 10% lower than the amplitude-splitting interferometric accelerator when the laser crossing angle in Fig. 3 is reduced to 40 mrad.

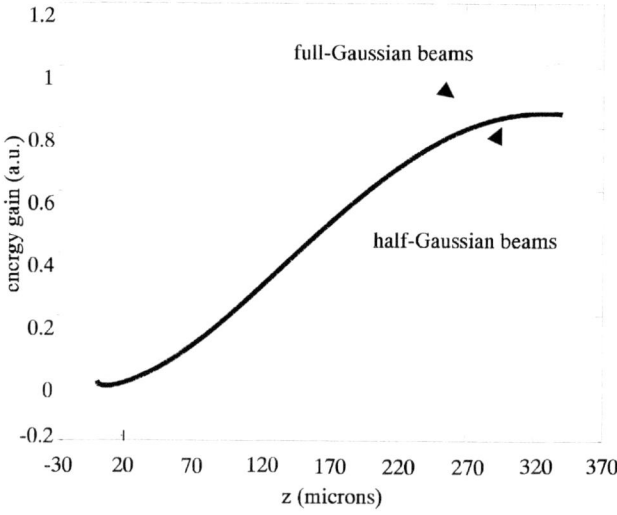

**FIGURE 4.** The electron energy gain along a 340 μm interaction length for the proposed amplitude-splitting accelerator and the wavefront-splitting accelerator. Both accelerators have the same laser damage intensity on the dielectric surfaces and the same laser power inside the accelerators. The laser crossing angle is 50 mrad and the laser wavelength is 1 μm. For this particular case, the electron energy gain of the wavefront splitting accelerator is lower by 15%.

# NOISE CHARACTERISTICS

Scaled to optical wavelengths, a laser-driven accelerator structure size is much smaller than a RF accelerator. The micron wavelength, together with the small structure size, imposes stringent requirements on optical phase control. In order to increase the phase coherence length and thus the accelerator size, the laser crossing angle in the proposed accelerators is typically of a few milliradians and the accelerator length is of a few hundred microns for a 1-μm laser wavelength[13]. Because of the small crossing angle and the large laser field $E_0 \approx 10$ GV/m, slight asymmetry in laser coupling or a small phase mismatch in ϕ could result in a large transverse field on the electron axis, as Eq. (2) indicates. Furthermore, recombining two sub-picosecond laser pulses within a small optical phase angle appears to be nontrivial. For example, overlapping the optical phase of two laser pulses within a one-degree angle requires a temporal precision of 10 atoseconds for a 1-μm laser wavelength. In view of the above, we argue that the proposed wavefront-splitting structure in Fig. 3 offers potential advantages over the amplitude splitting structure in Fig. 2.a. The reasons are:

1. In Fig. 3, the π relative phase between the two laser beams is set by a dielectric phase step on the corner reflector. The thickness of the dielectric step can be fabricated precisely by using optical coating or lithographic techniques.
2. In Fig. 3, since the source laser pulse is divided in wavefront at the accelerator coupling point, recombining the two wavefront-splitting laser pulses can be achieved with a great accuracy in time or in optical phase. The tilting angle between the two coupling mirrors determines the laser-crossing angle.

In addition, the wavefront-splitting structure is apparently immune from background acoustic noise. An acoustic wave in a solid typically has a wavelength of a few centimeters, which is much longer than a laser-driven particle accelerator structure. The accelerator stage occupies a very small phase of the acoustic wave. As a result, if the laser pulse splitting point is within the accelerator, as is the case for a wavefront-splitting accelerator, the relative optical phase of the two laser pulses remains fixed regardless of background vibration. An amplitude-splitting interferometer typically has more discrete elements, and requires a better technique in micro-fabrication or a noise-proof interferometer design.

As shown in Eq. (6), the stability of the interference fringes in $x$ manifests the noise characteristics of the proposed interferometric-type accelerators. In our noise measurements, the amplitude-splitting interferometer, mimicking the proposed structure in Fig. 2.a, is a Mach Zehnder interferometer and a Sagnac interferometer; whereas the wavefront-splitting interferometer, mimicking the proposed structure in Fig. 3, is a Fresnel two-mirror interferometer.

Figure 5.a illustrates the noise measurement setup for a Mach Zehnder interferometer and a Fresnel two-mirror interferometer; Fig. 5.b illustrates that for a Sagnac interferometer and a Fresnel two-mirror interferometer. A 3.5 mW, 670 nm-wavelength diode laser (LD) drives the two interferometers via a 50% beam splitter (BS1). Thus any noise from the diode laser and BS1 is common to both interferometers. Each of the two Fresnel mirrors is a 5 mm × 5 mm square silver reflector. The Mach Zehnder interferometer or the Sagnac interferometer occupies a 4 cm × 4 cm area, which is the smallest that we could build by using one-inch diameter optics (M1, M2, BS2, BS3). We adjusted the interferometers until we saw clear interference fringes on a screen. A 1 cm diameter PIN diode detector, covering the whole interference pattern, takes the laser signal into a Stanford Research SR780 spectrum analyzer for noise measurements.

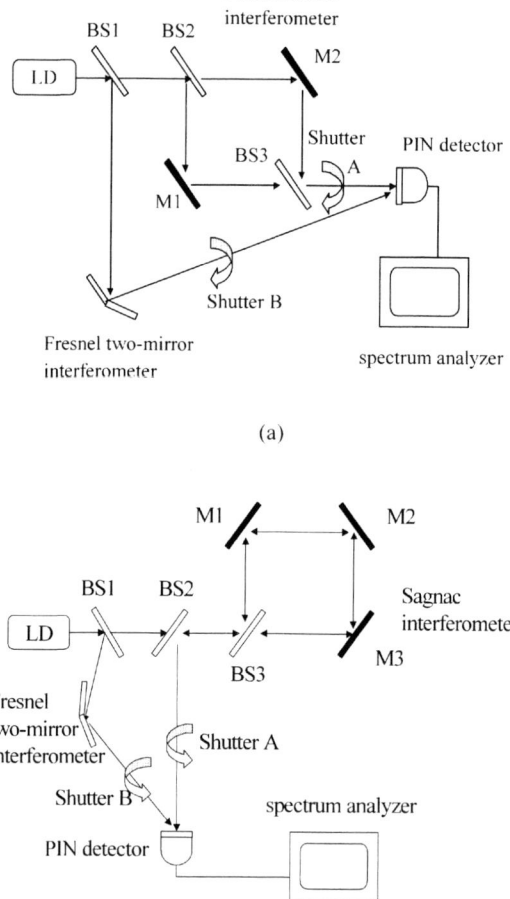

(a)

(b)

**FIGURE 5.** Noise measurement setups. (a) Mach Zehnder interferometer and Fresnel two-mirror interferometer. (b) Sagnac interferometer and Mach Zehnder interferometer.

We measured the root-mean-square noise spectra for the following three situations:

1.  Both Shutters A and B are closed. This measurement gives the electronic noise floor.
2.  Shutter A is closed and Shutter B is opened. This measurement gives the noise spectrum of the Fresnel two-mirror interferometer.
3.  Shutter B is closed and Shutter A is opened. This measurement gives the noise spectrum of the Mach Zehnder interferometer or the Sagnac interferometer.

Our measurements show no difference between the electronic noise floor and the noise spectrum of the Fresnel two-mirror interferometer. In fact, we visually noticed steady interference fringes from the Fresnel two-mirror interferometer, whereas blinking fringes from the Mach Zehnder interferometer. A Sagnac interferometer is a common-path interferometer, which rejects any noise symmetric to the two optical paths. In our experiments, the phase stability of a Sagnac interferometer is obviously better than a Mach Zehnder interferometer, but slightly worse than a Fresnel two-mirror interferometer. Figure 6.a shows the noise spectra of the Fresnel two-mirror interferometer and the Mach Zehnder interferometer between 10 Hz and 0.8 kHz with a 2 Hz resolution. Although the laser power of the Mach Zehnder interferometer is a factor of two lower due to the third beam splitter BS3, the noise amplitude of the 4 cm × 4 cm Mach Zehnder interferometer is apparently several orders of magnitude higher on the low frequency side. Figure 6.b shows the noise spectra of the Fresnel two-mirror interferometer and the Sagnac interferometer. The Sagnac interferometer is much less sensitive to acoustic noise when compared to the Mach Zehnder interferometer, but still slightly worse than the Fresnel two-mirror interferometer. Without miniaturizing the three interferometers, this preliminary result indicates

1.  A wavefront-splitting accelerator is simple to set up and insensitive to background acoustic noise.
2.  A Sagnac interferometer is less sensitive to noise and is a better choice for an amplitude-splitting accelerator.

## CONCLUSIONS

The particle acceleration field of an interferometric-type laser-driven accelerator can be obtained from an amplitude-splitting interferometric configuration or a wavefront-splitting configuration. We have showed numerically that a wavefront-splitting accelerator configuration gives a single-stage electron energy gain comparable to that of an amplitude-splitting accelerator configuration under the same laser damage fluence. Our noise measurements qualitatively confirm that the optical phase stability of a wavefront-splitting interferometric laser-driven particle accelerator is more stable due to its monolithic construction. The proposed wavefront-splitting accelerator configuration essentially alleviates the cumbersome phase and timing control on femto-second laser pulses.

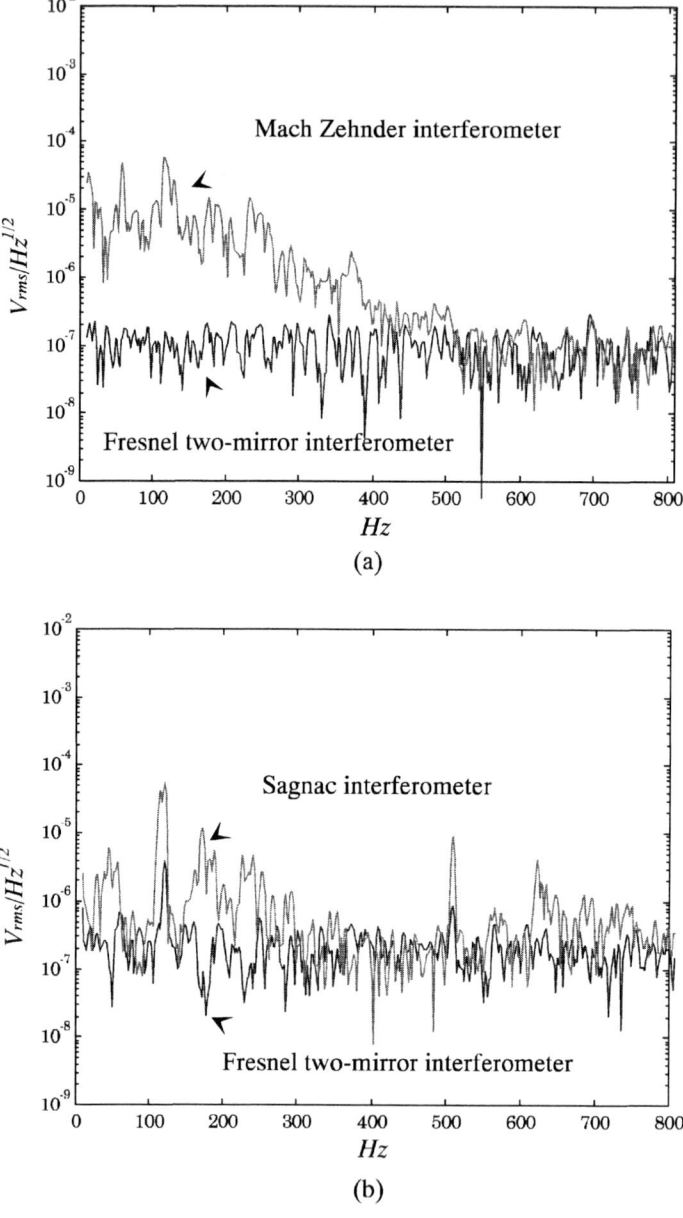

**FIGURE 6.** (a) The noise spectra of the Mach Zehnder interferometer and the Fresnel two-mirror interferometer. The noise amplitude of the Mach Zehnder interferometer is apparently much higher on the low-frequency side due to its discrete optical elements. (b) The noise spectrum of the Sagnac interferometer is not much worse than that of a Fresnel two-mirror interferometer.

## ACKNOWLEDGMENTS

The work at National Tsinghua University, Taiwan, is supported by the National Science Council under the contract number NSC-87-2112-M-007-024, and that at Stanford University, U.S.A., is supported by the Department of Energy under the contract number DE-AC03-76SF00515.

## REFERENCES

1.  Chou, P.J., Bowden, G.B., Copeland, M.R., Menegat, A., and Siemann, R.H., SLAC-PUB-7498, May 1997.
2.  Sprangle, P., Esarey, E., and Krall , J., *Phys. Plasmas* **3** (5), 2183-2188 (1996).
3.  Huang , Y.C. and Byer, R.L., *Nucl. Inst. Meth.* A393, I1-37 (1997).
4.  Huang, Y.C., et al. to be published in *Nucl. Inst. Meth. A* (1998).
5.  Clayton, C. E., *et al.* and Nakajima, K. *et al.*, Proceedings of the Advanced Accelerator Concepts Workshop, Lake Tahoe, 1996 (AIP, NY, 1997), S. Chattopadhay editor.
6.  Huang, Y.C. and Byer, R.L., *Appl. Phys. Lett.* **69** , 2175-2177 (1996).
7.  Palmer, R.B., Lecture Notes in Physics 296, Frontiers of Particle Beams p. 607 (Springer, Berlin, 1988).
8.  Edighoffer, J.A. and Pantell, R.H., *J. Appl. Phys.* **50** , p. 6120 (1979).
9.  Pantell, R.H. and Piestrup, *Appl. Phys. Lett.* **32**, 1 (1978).
10. Haaland , C.M., *Opt. Comm.* **114**, p. 280 (1995).
11. Sprangle, P., Esarey, E., Krall, J., Ting, A., *Opt. Comms.* 124, p. 69 (1996).
12. Huang, Y.C., Zheng, D., Tulloch, W.M., and Byer, R.L., *Appl. Phys. Lett.* **68** , p. 753 (1996).
13. Huang, Y.C. and Byer, R.L., *Appl. Phys. Lett.* **69** , p. 2175 (1996).
14. Scully, M.O., *Appl. Phys.* B **51**,238-241 (1990).
15. Hecht , Optics, , 3$^{rd}$ edition, Addison Wesley Inc. 1998.
16. Stuart, B.C., Feit, M.D., Rubenchik, A.M., Shore, B.W., and Perry, M.D., *Phys. Rev. Lett.* v **74**, 2248-2251 (1995).
17. Edighoffer, J.A. and Pantell, R.H., *J. Appl. Phys.* **50** , 6120 (1979).

# Laser Acceleration in Vacuum Using a Donut-Shaped Laser Beam

Y. Liu, D. Cline, and P. He

*Center for Advanced Accelerators*
*Department of Physics and Astronomy*
*University of California, Los Angeles, CA 90095*

**Abstract.** Utilizing the high power, radially polarized $CO_2$ laser and high quality electron beam at the Brookhaven Accelerator Test Facility, a laser acceleration in vacuum scheme is proposed. This scheme has several advantages: The configuration of the optics is simple, it is easy to produce a small beam, and optical damage is less important than in other schemes.

## INTRODUCTION

The high electric field associated with laser light has become an even more attractive possibility for building laser-driven particle accelerators. Laser acceleration in vacuum has been studied theoretically for many years and several schemes utilizing focused Gaussian laser beam(s) have been proposed [1–9]. In most of these schemes, the researchers attempt to confine the interaction to a finite length. However, none of the schemes has been successful until recently, when a group of French physicists [10] demonstrated that free electrons are accelerated in vacuum directly to MeV energies by a linearly polarized and intense subpicosecond laser pulse. It appears to directly violate the Lawson–Woodward (L–W) theorem [11]. In this paper, we propose a laser acceleration experiment based on the schemes discussed in Refs. [3] and [6]. In the proposed scheme, the interaction length of the electron beam with the laser beam is adjustable by remote control. Under the short interaction length limit which is less than two Rayleigh lengths, the scheme will violate one of the conditions of the L–W theorem and is expected to obtain a non-zero net energy gain. Under the limit of a long interaction length, which is much greater than the Rayleigh length, the scheme becomes a vacuum laser acceleration and it will be used to examine directly the results reported in Ref. [10].

CP472, *Advanced Accelerator Concepts: Eighth Workshop,*
edited by W. Lawson, C. Bellamy, and D. Brosius

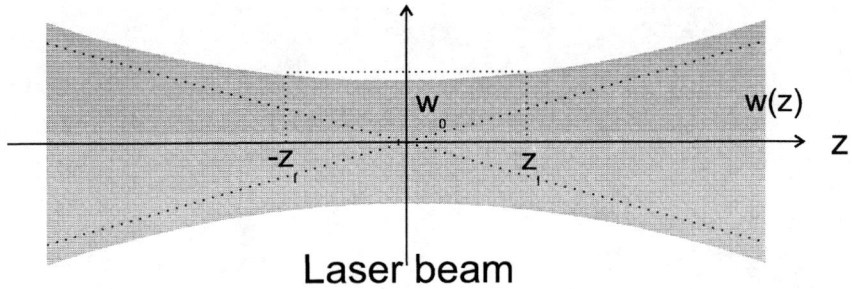

**FIGURE 1.** The focused Gaussian laser beam ($z_l$ is the Rayleigh length).

# ELECTRON ACCELERATION BY A GAUSSIAN LASER BEAM

Since Lax, Louisell, and McKnight [12] discovered that there is a longitudinal electric-field component in a focused Gaussian laser beam, analytical solutions of the electric-field for different modes of a laser beam have been well developed. A pair of solutions for a radially polarized, axially symmetric Gaussian laser beam near the focal region are given by [13,14]

$$E_r = E_0 \frac{rw_0}{w^2} \exp\left(-\frac{r^2}{w^2}\right) \sin\psi, \tag{1}$$

$$E_z = E_0 \frac{2w_0}{kw^2} \exp\left(-\frac{r^2}{w^2}\right) \times \left[\left(1 - \frac{r^2}{w^2}\right)\cos\psi - \frac{zr^2}{z_l w^2}\sin\psi\right], \tag{2}$$

where

$$\psi = kz - \omega t + \frac{zr^2}{z_l w^2} - 2\tan^{-1}\left(\frac{z}{z_l}\right) + \phi_0, \tag{3}$$

$w = w_0[1 + (z^2/z_l^2)]^{1/2}$ is the laser beam radius, $w_0$ is the beam radius at the focal waist, $z_l = \pi w_0^2/\lambda$ is the Rayleigh length, $\omega$ is the laser frequency, and $E_0$ and $\phi_0$ are constants (Figure 1). Now, we consider a relativistic electron ($\beta \sim 1$) traveling from the left side to the right side along the z-axis. It enters the laser focal region at position $z = z_i$, and exits at position $z = z_f$. The net energy gain is given by

$$\Delta E \approx -eE_0 \frac{2w_0}{k} \int_{z_i}^{z_f} \frac{1}{w^2} \cos\psi \, dz. \tag{4}$$

Where

$$\psi \approx -2\tan^{-1}\left(\frac{z}{z_l}\right) + \phi_1, \tag{5}$$

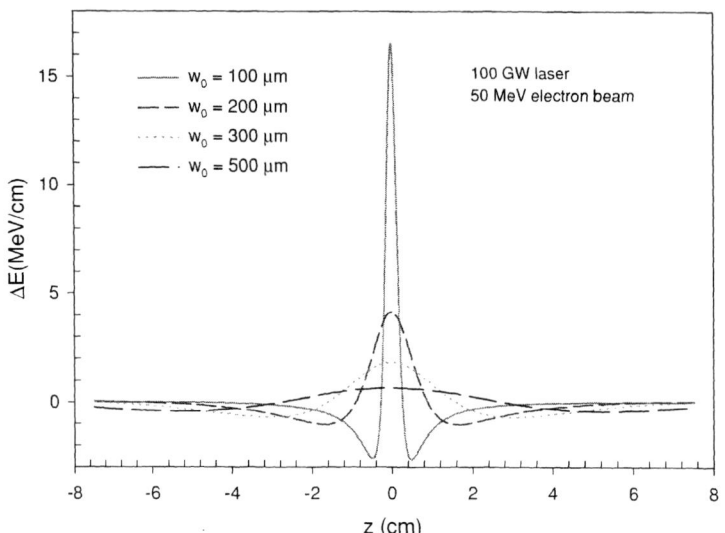

**FIGURE 2.** The initial electron energy is $E_e = 50$ MeV, laser power is $P = 100$ GW and total interaction length is 15 cm. Four different laser focal waist radii are 100 $\mu$m, 200 $\mu$m, 300 $\mu$m and 500 $\mu$m, respectively.

$\phi_1$ is determined by the initial phase and position of the injected electron. The acceleration gradients with optimized initial phase are shown in Figure 2. The initial electron energy is $E_e = 50$ MeV and laser power is $P = 100$ GW. The four curves represent the acceleration gradients calculated from different laser focal waist radii ($w_0 = 100$ $\mu$m, 200 $\mu$m, 300 $\mu$m and 500 $\mu$m, respectively). The total interaction length is 15 cm. Note that the curves are slightly asymmetric about the origin ($z = 0$).

The phase slippage is dominated by the first term in Eq. (5). The optimum interaction length ($l_0$) is equal to the phase slippage distance, defined as the distance required for the accelerated electron phase to the laser electric-field phase slip by $\pi$ [3,6]. From Eq. (5) we obtain

$$-\frac{\pi}{2} < 2\tan^{-1}\left(\frac{z}{z_l}\right) < \frac{\pi}{2}, \tag{6}$$

which yields

$$-z_l < z < z_l. \tag{7}$$

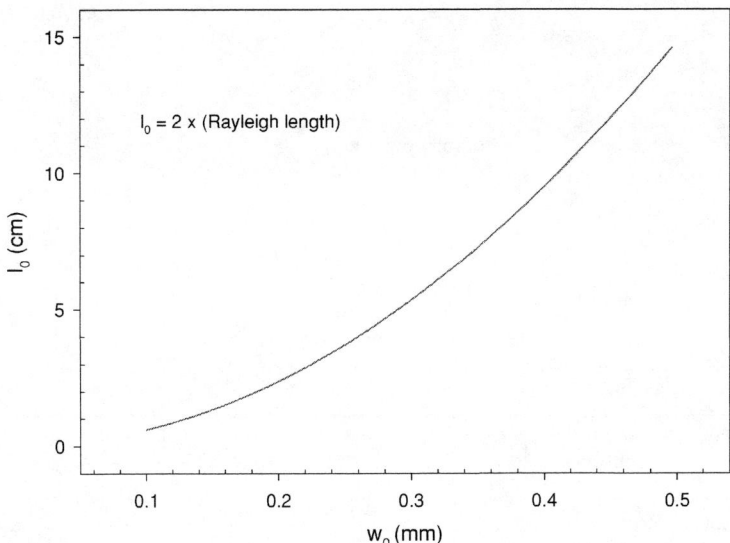

**FIGURE 3.** The optimum interaction length $l_0$ as a function of Gaussian beam radius $w_0$ at waist.

The optimized interaction length is $l_0 \approx 2z_l$. Integrating Eq. (4) over $l_0$ from $z_i = -z_l$ to $z_f = z_l$ we finally have

$$\Delta E = -eE_0 w_0 \cos \phi_1 \tag{8}$$

or

$$\Delta E_{max}[MeV] = 15.5(P[TW])^{1/2}, \tag{9}$$

where $\Delta E_{max}$ is the maximum energy gain with $\phi_1 = \pi$, and the total laser power $P = c\epsilon_0 \pi w_0^2 E_0^2/2$. Ideally, the maximum energy gain is proportional to the square root of the delivered laser power only over the optimized interaction length. The optimum interaction length is determined by the focused laser beam radius $w_0$ (Figure 3). For an upgraded 1-TW $CO_2$ laser at the Brookhaven Accelerator Test Facility [15], the maximum energy gain over the optimum interaction length is 15.5 MeV and the averaged gradient is 581 MeV/m with a radius of 300 $\mu$m at the laser beam waist.

## EXPERIMENT SETUP

A proposed optics configuration is shown in Figure 4. A radially polarized,

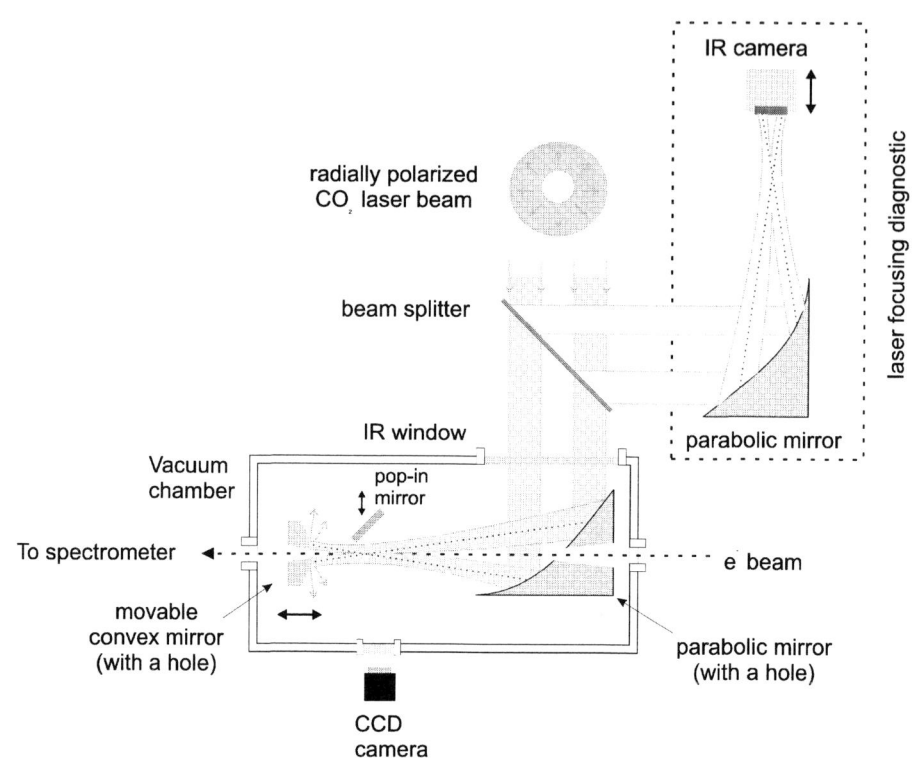

**FIGURE 4.** Schematic drawing of the configuration of the laser acceleration in vacuum experiment.

axisymmetric, annular (donut-shaped) Gaussian-distributed $CO_2$ laser beam is delivered to a vacuum chamber through a ZnSe window. A parabolic mirror with a hole at its surface is positioned along the electron beam path ($z$-axis) in the chamber. The hole located at the optical center of the mirror allows the electron laser beams to meet in the focal region. The 3-in diameter mirror, which is gold-coated and $90^0$ off-axis, has a 4-in focal length. After the focal point, there is a movable convex mirror which also has also a hole to allow the electron beam to exit from the interaction region. This convex mirror reflects the laser beam from its path and limits the interaction length.

A $45^0$ pop-in mirror located in the focal plane is used to diagnose the alignment of electron beam with laser beam. A laser-energy and focusing-position monitor could be located anywhere outside of the chamber. It consists of a parabolic mirror and an Infrared (IR) camera/detector. The parabolic mirror is identical to the one in the chamber so that the beam intensity pattern can be duplicated in the focal region. The IR camera is for image measurement and the IR detector (not shown

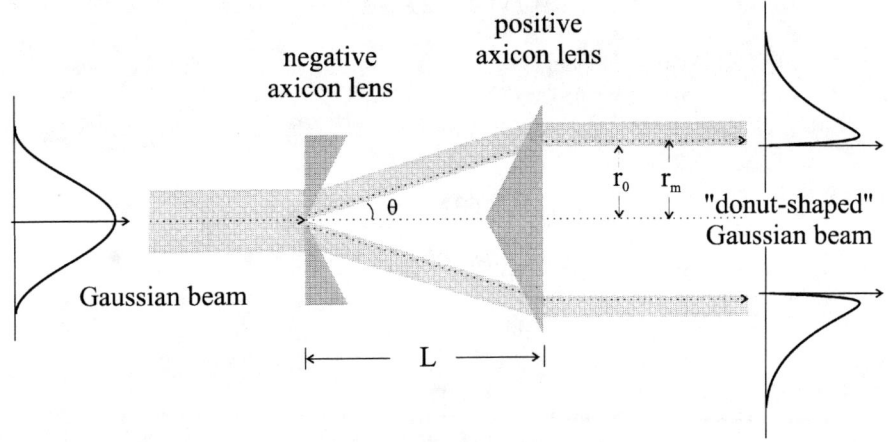

**FIGURE 5.** Schematic of axicon telescope for generating a donut-shaped beam.

in Figure 4) is for energy measurement. The laser beam used for diagnosis is split from the delivered beam by an IR beam splitter inserted $45^0$ to the laser beam path.

## DISCUSSIONS

Using a Gaussian beam that is donut-shaped automatically confined the interaction to a finite length. In addition, the interaction length is controllable by adjusting the position of the convex mirror. The energy gain as a function of the interaction length will be studied. When the interaction length is less than two Rayleigh lengths, significant energy gain is expected. As the interaction length increases and is much longer than the Rayleigh length, the scheme becomes a vacuum laser acceleration and will directly examine the results reported in Ref. [10].

In the optics setup a convex rather than a flat mirror is used, which reflects off the high intensity laser beam to avoid damaging other objects in the chamber. Most likely this convex mirror will be optically damaged by the tightly focused laser beam when it closes to the focal plane; but it will not limit the laser intensity threshold for electron acceleration.

The delivered laser power is only limited by the parabolic mirror optical damage threshold. However, by selecting a large diameter parabolic mirror, a threshold for optical damage can possibly be eliminated for the currently available laser power if the inner diameter of the laser beam is large enough (Figure 5). This allows the parabolic mirror to have a fairly large diameter hole, makes transporting and focusing the electron beam into the interaction region straightforward, and it is easy to manufacture as well.

# ACKNOWLEDGEMENTS

We thank X. J. Wang, I. Ben-zvi and I. V. Pogorelsky for helpful discussions and comments. This work is supported by the U.S. Department of Energy, Grant No. DE-FG03-92ER40695.

# REFERENCES

1. Scully M. O., *Appl. Phys. B*, **51**, 238 (1990).
2. Bochove E. J., Moore G. T., and Scully M. O., *Phys. Rev. A*, **46**(10), 6640 (1992).
3. Takeuchi S., Sugihara R., and Shimoda K., *J. Phys. Soci. Japan*, **63**(3), 1186 (1994).
4. Hauser T., Scheid W., and Hora H., *Phys. Lett. A*, **186**, 189 (1994).
5. Haaland C. M., *Opt. Commun.*, **114**, 280 (1995).
6. Esarey E., Sprangle P., and Krall J., *Phys. Rev. E*, **52**(5), 5443 (1995).
7. Sprangle P., Esarey E., Krall J., and Ting A., *Optics Comm.*, **124**, 69 (1996).
8. Hafizi B., Esarey E., and Sprangle P., *Phys. Rev. E*, **55**(3), 3539 (1997).
9. Hafizi B., Ting A., Esarey E., Sprangle P., and Knall J., *Phys. Rev. E*, **55**(5), 5924 (1997).
10. Malka G., Lefebvre E., and Miquel J.L., *Phys. Rev. Letts.*, **78**(17), 3314 (1997).
11. Lawson J. D., *IEEE Trans. Nucl. Sci.* **NS-26**, 4217 (1979); Woodward P. M., *J. Inst. Electr. Eng.* **93**, 1554 (1947); Palmer R. B., *Lecture Notes in Physics 296, Frontiers of Particle Beams* (Springer, Berlin, 1988) p. 607.
12. Lax M., Louisell W. H., and McKnight W. B., *Phys. Rev. A*, **11**(4), 1365 (1975).
13. Shimoda K., *J. Phys. Soci. Japan*, **60**(1), 141(1991).
14. Milonni P. W., and Eberly J. H., *Laser* (Wiley, New York, 1988). chap. 14; A. Yariv, *Quantum Electronics*, 3rd ed. (Wiley, New York, 1989), chap. 6.
15. Pogorelsky I. V., Ben-Zvi I., Skaritka J., et al., in *AIP Conf. Proc.* **398**, 937 (1997).

# Space Charge Compensation in Laser Particle Accelerators

L.C. Steinhauer and W.D. Kimura

*STI Optronics, 2755 Northup Way, Bellevue, WA 98004-1495*

**Abstract.** Laser particle acceleration (LPA) involves the acceleration of particle beams by electromagnetic waves with relatively short wavelength compared with conventional radio-frequency systems. These short length scales raise the question whether space charge effects may be a limiting factor in LPA performance. This is analyzed in two parts of an accelerator system, the acceleration sections and the drift region of the prebuncher. In the prebuncher, space charge can actually be converted to an advantage for minimizing the energy spread. In the accelerator sections, the laser fields can compensate for space charge forces, but the compensation becomes weaker for high beam energy.

## INTRODUCTION

Laser particle acceleration (LPA) concepts resemble radio-frequency (RF) linear accelerators in that a travelling wave is set up to move synchronously with the particle. Efficient acceleration in both LPA and RF systems requires that the spread of the particles along the wave be somewhat smaller than the wavelength, $\lambda = 2\pi c/\omega$ ($c$ is the speed of light in vacuum; $\omega$ is the frequency of the driving wave), or else that the particles be organized into narrow bunches separated by $\lambda$. However, LPA's differ from RF systems in that the wavelength corresponding to optical frequencies is on the order of $10^{-6}$ to $10^{-5}$ m rather than $10^{-2}$ m. Thus *microbunching* on a very fine scale is necessary. In RF accelerators, the small length scale is the transverse dimension of the particle beam. For a typical high-quality beam of moderate energy (50-100 MeV) the transverse dimension may be 200-500 μm, which is smaller than the needed *microbunching* scale. However, in an LPA with, e.g. $\lambda = 10$ μm, the microbunching scale (somewhat less than 10 μm) is the smaller length scale. The appearance of fine scales raises the issue of space charge effects.

The principal space charge effect in an RF linac is a defocusing tendency since the smallest length scale is the transverse dimension of the beam. However in an LPA the smallest length scale is longitudinal; hence the principal space charge effect is a de-bunching tendency. Moreover, the smaller length scale in an LPA makes space charge inherently more difficult. We consider two elements of an LPA system where space charge debunching may be important. One is in the accelerator sections themselves, and the other is in the drift regions between accelerator sections. The largest drift region is at the upstream end of the system where the initial microbunching is first created.

CP472, *Advanced Accelerator Concepts: Eighth Workshop,*
edited by W. Lawson, C. Bellamy, and D. Brosius

In this paper we analyze the de-bunching effect of space charge in (1) the prebuncher, and (2) in the acceleration sections. The dynamics of debunching are analyzed using quasi-one-dimensional models; the transverse beam dynamics are analyzed neglecting space charge. These results are applied to the practical case of the STaged ELectron Laser Acceleration (STELLA) experiment (1) on the BNL Accelerator Test Facility.

## SPACE CHARGE IN THE PREBUNCHER

The prebuncher may be the system element most vulnerable to space charge because there is no longitudinal force to oppose space charge effects. Therefore, the only thing opposing space charge forces is inertia. Fortunately, the effective "mass" for relative longitudinal motions is $\gamma^3 m_e$ ($\gamma$ is the relativistic factor and $m_e$ is the electron rest mass). A simple method of prebunching, shown in Fig. 1, is composed of an energy modulator (e.g. an inverse free-electron laser) followed by a drift section. A macrobunch with little energy spread enters the modulator and receives an energy modulation with longitudinal periodicity length $\sim\lambda$ (laser wavelength). Thus, within each segment of length $\lambda$, particles behind the segment center move slightly faster,

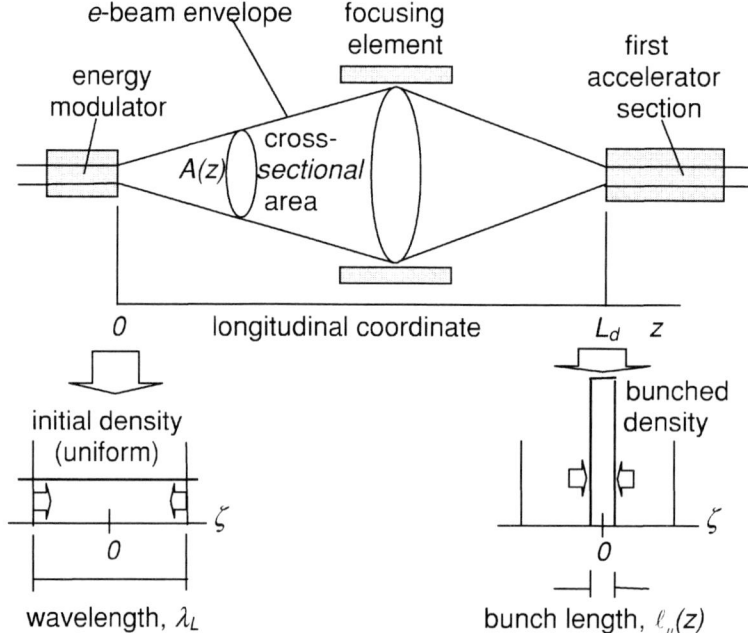

**FIGURE 1.** Idealized prebunching arrangement.

while those ahead move slightly slower. Then in the drift section the faster particles catch up while the slower slip back, gathering them into microbunches.

In this section we show that (1) space charge forces can be compensated by adjusting the strength of the energy modulator to allow effective microbunching, (2) a second level of compensation can reduce the energy spread. The latter is a positive byproduct of space charge: a coherent energy modulation is required to produce the initial bunching; space charge effects can reduce or eliminate this energy spread. It may be possible to both generate short microbunches and minimize the energy spread.

## Estimate of space charge debunching

It is helpful to begin with a simple scaling estimate. Longitudinal forces from space charge cause oscillations at the relativistically-corrected plasma frequency, $\omega_p = (\mu_0 c^2 e^2 n/m_e \gamma^3)^{1/2}$, where $\mu_0$ is the free-space permeability; $n$, $-e$ are the electron density and charge. Space charge effects become significant if the transit time through the drift region, $L_d/c$, ($L_d$ is the drift section length) is comparable to or exceeds the plasma oscillation time, $1/\omega_p$. i.e. if $\omega_p L_d/c \sim 1$. Accordingly, define the space charge parameter:

$$\sigma_{SC} \equiv \omega_p^2 L_d^2 / c^2 ; \tag{1}$$

if $\sigma_{SC} \sim 1$ or more, space charge will be important in the prebuncher. Since $\sigma_{SC} \propto n$, a practical expression for it is found once the density is identified. The macrobunch properties are charge $Q_M$, length $\ell_M$, and normalized emittance, $\varepsilon_N$. Note that in an LPA, the macrobunch length $\ell_M$ (usually in mm) is much longer than the microbunch length ($< \lambda/2$). The charge is related to the density by $Q_M = enA\ell_M$, where $A$ is the cross-sectional area of the beam. Suppose $A = \pi \varepsilon_N L_d / \gamma$, which is the characteristic area of a focused beam over a drift distance $L_d$: this area is proportional to the geometric mean of the waist area and the far-field area at a distance $L_d$ from the waist. Combining these gives the density $n = \gamma Q_M / \pi e \varepsilon_N L_d \ell_M$. Then

$$\sigma_{SC} = \frac{4 r_e L_d}{\gamma^2 \ell_M \varepsilon_N} \frac{Q_M}{e} = 7.04 \times 10^{-5} \frac{L_d(m) Q_M(nC)}{\gamma^2 \ell_M(m) \varepsilon_N(\pi m - rad)} \tag{2}$$

where $r_e = e^2 \mu_0 / 4\pi m_e$ is the classical electron radius.

# Quasi-one dimensional (1D) model

Since the thickness of each microbunch $\ell_\mu$ ($< \lambda/2$) is much smaller than the transverse beam dimension, the longitudinal motion in response to space charge effects is approximately 1D. We therefore adopt the 1D model, which has been checked elsewhere (2). Divide the beam into segments of length $\lambda$, and consider electrons within a single segment. Define a longitudinal coordinate in the moving frame of the segment, $\zeta \equiv z - z_0(t)$, where $z_0$ is the segment center and $|\zeta| \le \lambda/2$ within the segment. Identify $\zeta$ with individual electrons in the segment. The initial energy modulation is sinusoidal in $\zeta$. We approximate this crudely as a saw-tooth form, $\gamma = \bar{\gamma} - \Delta\gamma(2\zeta/\lambda)$, with corresponding velocity modulation $\Delta u_z = c(\Delta\gamma/\bar{\gamma}^3)\cdot 2\zeta/\lambda$, where $\bar{\gamma}$ denotes the mean value, and $\Delta\gamma$ is the amplitude of the energy spread. This is the 1D reduction of the spheroidal bunch model (3). The sawtooth model admits a self-similar solution. The drifting electrons gather into a microbunch, where $\pm\ell_\mu(t)/2$ is the position of the outermost electrons. Then within the bunch, $|\zeta| \le \ell_\mu/2$:

$$ n = n_i A_i \lambda / A \ell_\mu ; \quad u_z = (\zeta/\ell_\mu) d\ell_\mu/dt ; \quad E_z = -(m_e \lambda \omega_p^2 A_i/eA)\zeta/\ell_\mu . \quad \text{(3a, b, c)} $$

As in Fig. 1, $A(z)$ is the beam cross-sectional area; and $A_i$, $n_i$ denote initial values. Hereafter we adopt $\tilde{t} \equiv ct/L_d$ as the dimensionless time variable, with $0 \le \tilde{t} \le 1$ in the drift section.

The amount of microbunching (which varies with $\tilde{t}$) is expressed as a phase spread, $\psi_\mu \equiv 2\pi\ell_\mu/\lambda$. Before bunching begins, $\psi_\mu = 2\pi$, and the goal is to achieve as close to ideal microbunching ($\psi_\mu = 0$) as possible. Then the equation of motion for $\psi_\mu(\tilde{t})$ and its initial conditions are

$$ d^2\psi_\mu/d\tilde{t}^2 = 2\pi \sigma_{sc} A_i/A(\tilde{t}), $$

(4)

$$ \psi_\mu(0) = 2\pi, \qquad (\psi_\mu')_i \equiv (d\psi_\mu/d\tilde{t})_{\tilde{t}=0} = -4\pi L_d \Delta\gamma_i/\lambda\bar{\gamma}^3 . $$

Nominally only half the modulated electrons are in a phase range for which they gather into a bunch; electrons in the other half tend to spread out. Therefore, in the space charge parameter (Eqs. 1, 2) *half* the macrobunch charge should be used.

Consider the beam optics in the drift section. The prebuncher (Fig. 1) is divided into two drift regions: before "1" and after "2" the focuser ("triplet"). The width $w(z)$ of a drifting beam is $w^2 = W_{1,2}^2 + 4\theta_{1,2}^2(z - Z_{1,2})^2$. In each region, $W$, $Z$ are the waist width and position, and $\theta$ is the asymptotic divergence angle. The normalized emittance is $\varepsilon_N = \gamma W\theta/2$. For a circular beam, the area $A = \pi w^2/4$ is

$$A(\tilde{\tau})/A_w = 1 + (A_\varepsilon/A_w)^2(\tilde{\tau} - Z/L_d)^2 \tag{5}$$

where $A_w \equiv \pi W^2/4$, and $A_\varepsilon \equiv \pi \varepsilon_N L_d/\bar{\gamma}$. Regions 1 and 2 have the same $A_\varepsilon$ but different $A_w$, and areas match at the "idealized" triplet location, $\tilde{\tau} = z_m/L_d$.

These equations have an analytic solution of the form in terms of two factors $J$, $J'$ that depend on the beam optics,

$$\left(\psi_\mu'\right)_f = \left(\psi_\mu'\right)_i + 2\pi\,\sigma_{SC}J', \qquad \left(\psi_\mu\right)_f = 2\pi + \left(\psi_\mu'\right)_i + \pi\,\sigma_{SC}J \tag{6a, b}$$

For a constant area beam $A = A_i$: $J$, $J' = 1$. Focusing accentuates space charge effects so that $J$, $J' > 1$ and depend on the beam optics. Both are largest when the second focus lies *before* the end of the drift section because the second beam waist lies within the drift region. Their analytic forms are given in Ref. (2)

Space charge affects both bunching and energy spread. From Eq. 6b the bunching $(\psi_\mu)_f$ can made zero (ideal case) by making $(\psi_\mu')_i$ more negative. This is done by increasing the initial energy modulation. This also reduces the final energy spread, proportional to $(\psi_\mu')_f$ (Eq. 6a). By adjusting the focusing geometry (triplet and second focus locations), which modifies $J$, $J'$, one might simultaneously achieve maximal bunching $[(\psi_\mu)_f \to 0]$ *and* eliminate the energy spread introduced by the prebuncher $[(\psi_\mu')_f \to 0]$.

## Practical example

We apply this theory to estimate space charge effects in the STELLA experiment, which includes a prebuncher (inverse free electron laser) and accelerator section (inverse Cerenkov acceleration), both driven by a 10.6 μm wavelength laser. Table I lists the beam optics parameters in the prebuncher; for this example the space charge parameter is $\sigma_{SC} = 0.55$ (accounting for the factor of two effective reduction in the charge). Consider an example where the initial modulation and second focus location are chosen for dual compensation, i.e. maximal bunching and minimal final energy spread. Suppose the second waist is located at $Z_2/L_d = 0.985$, i.e. just inside the downstream end of the drift section. Then in this idealized model the final bunching and energy spread are perfect, i.e. $(\psi_\mu)_f = 0$ and $(\psi_\mu')_f = 0$, respectively. Dual compensation, (spatial *and* energy spread reduction) can only be achieved if the macrobunch charge is large enough.

This analysis is based on the 1D model. The reduction factor accounting for two-dimensional effects is $F_{2D} \approx 1/(1 + 1.1\,\gamma\ell_\mu/w)$. At the triplet ($z \approx z_m$, $\ell_\mu \approx 0.6\lambda$), $F_{2D} \approx 2/3$; and at the second waist ($z \approx 0.98L_d$, $\ell_\mu \approx \lambda/10$), $F_{2D} \approx 2/3$. Thus the 1D model *overestimates* the space charge effect by a factor of ~ 3/2. Accounting for 2D effects, this would corresponds to a macrobunch charge of $Q_M \approx 0.225$ nC.

**TABLE 1.** Beam optics parameters in STELLA

| | | | |
|---|---|---|---|
| Relativistic parameter | $\gamma = 80$ | Drift section length | $L_d = 2$ m |
| Normalized emittance | $\varepsilon_N = 1$ ($\pi$ mm-mrad) | First waist position | $Z_I/L_d = -0.125$ |
| Macrobunch charge | $Q_M = 0.15$ nC | Focuser position | $z_m/L_d = 0.6$ |
| Macrobunch length | $\ell_M = 3$ mm | | |

# SPACE CHARGE IN THE ACCELERATOR

In the acceleration sections, the electromagnetic fields that accelerate and focus the beam produce bunching forces. These can counteract space charge debunching effect. However, since in a practical LPA system there will be many acceleration sections, the inertial effect is unimportant. Thus the question is whether the laser-induced bunching is adequate to compensate for space-charge debunching. This question is again addressed using a quasi-1D, sharp-boundary bunch model.

## Generic laser fields and Lorentz force

Our interest is in the laser fields in the *e*-beam path , i.e. very near the axis. A radially polarized, axisymmetric laser can be expressed as a superposition of modes described by the Bessel functions $J_0$ and $J_1$. The propagation vector of each mode $\mathbf{k}$ has the same amplitude $|\mathbf{k}| = \omega N/c$, but different angle $\theta$ to the z axis, ($N$ is the index of refraction of the medium). Generally the dominant modes are clustered near a single angle $\theta_L$. Thus, near the axis the laser wave can be approximated as a single Bessel mode with $\theta \approx \theta_L$. Its nonzero components are

$$E_r = \frac{-E_0}{\tan\theta_L} J_1(k_r r)\sin\psi, \; E_z = E_0 J_0(k_r r)\cos\psi, \; B_\theta = \frac{-E_0 N}{c\sin\theta_L} J_1(k_r r)\sin\psi \quad \text{(7a, b, c)}$$

where $E_0 = const$; the components of $\mathbf{k}$ are $k_r = Nk_L \sin\theta_L$ and $k_z = Nk_L \cos\theta_L$ with $k_L = \omega/c = 2\pi/\lambda_L$ ($\lambda_L$ is the vacuum wavelength). The phase function is $\psi = \omega_L t - \kappa z = k_L(ct - N\cos\theta_L z)$. For later use we define the reference electron (subscript $R$) as one which runs along the *z*-axis and is perfectly synchronous ($\psi = const$). Since $z = \beta ct + const$ (for constant relativistic parameter $\beta$), the synchronism condition is $N\beta_R \cos\theta_L = 1$. If the *e*-beam path is small enough to lie well within the "Bessel spot" of the $J_0$ function, $k_r r << 2.405$, then the small argument approximations can be used:

$$E_r \approx -E_0 \frac{k_\perp r}{2} \sin\psi, \qquad E_z = E_0 \cdot \cos\psi, \qquad cB_\theta = -E_0 \cdot \frac{N^2 k_L r}{2} \sin\psi \qquad \text{(8a, b, c)}$$

## Equations of motion

The equations of motion for a relativistic electron are

$$\frac{d}{c\,dt}\left(m_e c^2 \gamma\boldsymbol\beta\right) = \mathbf{F}_L, \qquad\qquad \frac{d\mathbf{r}}{c\,dt} = \boldsymbol\beta \qquad\qquad \text{(9a, b)}$$

where $\mathbf{r}$ is its position and velocity, and $\boldsymbol\beta \equiv \mathbf{v}/c$, $\beta = (1-1/\gamma^2)^{1/2}$. The Lorentz force, $\mathbf{F}_L = -e(\mathbf{E}+\boldsymbol\beta\times c\mathbf{B})$, retaining first order components, is

$$F_{Lx} = -eE_0 \frac{Nk_L x}{2}\sin\psi\left(-\cos\theta_L + N\beta_z\right), \qquad F_{Lz} = -eE_0\cos\psi \qquad \text{(10a, b)}$$

where $F_{Ly}$ is the same as $F_{Lx}$ with $x$ replaced by $y$. The phase acts as the longitudinal coordinate of an electron. Its evolution is governed by

$$\frac{d\psi}{c\,dt} = k_L\left(1 - N\beta_z \cos\theta_L\right) \qquad\qquad \text{(11)}$$

In general an electron path will differ from the reference electron ($\beta_z = \beta_R$). We introduce perturbed quantities that represent the difference between an electron and the reference. Also consider the acceleration gradient $W' = dW/dz$, where $W = m_e c^2 \gamma$ is the electron energy. The acceleration gradient for the reference electron, using Eqs. 9a, 10b, and the identity $d(\beta\gamma) = d\gamma/\beta$, is

$$W_R' = eE_0 \cos\psi_R \qquad\qquad \text{(12)}$$

Consider the perturbed quantities $\beta_x$, $\beta_y$, $\tilde\beta_z = \beta - \beta_R$, $\tilde\psi = \psi - \psi_R$. The equation of motion in one transverse direction ($x$) is then governed by

$$\frac{1}{\gamma_R}\frac{d}{c\,dt}\left(\gamma_R \frac{dx}{c\,dt}\right) + k_\beta^2 x = 0 \qquad\qquad \text{(13)}$$

where the factor, $-\gamma_R^3 mc^2/k_L$, has been divided out. The "wave number" $k_\beta$,

$$k_\beta^2 = \frac{W_R'}{W} \frac{\theta_L^2 N k_L}{2} \tan\psi_R \tag{14}$$

is the length scale for the oscillatory "betatron" motion. If the change in $\gamma_R$ can be ignored, the transverse motion has the form $x = x_{max} \cos(k_\beta ct + \varphi)$, where $x_{max}$ and $\varphi$ are constants. Consider parameters relevant to the STELLA experiment: $W_R' = 100$ MeV/m; $W = 50$ MeV; $\theta_L = 20$ mrad; $N \approx 1$; $\lambda_L = 10$ μm; and $\psi_R = 0.85\pi$. Then the betatron "wave number" is $k_\beta = 11.3$ m$^{-1}$.

Although Eq. (13) is for a single particle, it traces a figure in phase space that coincides with that of a shell distribution. Therefore, an emittance can be found. The normalized emittance is $1/\pi$ times the area of the closed curve that the particle traces in $x$-$p_x$ phase space, where $p_x = \gamma dx/dz$. The semi-major axes of the phase-space ellipse are $x_{max}$ and $\gamma k_\beta x_{max}$: thus $\varepsilon_N = \gamma k_\beta x_{max}^2$. We are interested in the filled-in distribution for which the shell corresponding to $x_{max}$ is the phase space boundary. The emittance for this "top hat" distribution, averaging over all particles, is half that of the bounding shell: $\varepsilon_N = \gamma k_\beta x_{max}^2 / 2$. For axisymmetry (equal emittance in $x$ and $y$) the cross-sectional area of the beam is $A = \pi x_{max}^2 = 2\pi \varepsilon_N / \gamma k_\beta$.

Consider the longitudinal particle dynamics accounting only for the laser force. Expand Eq. (11) to separate reference and perturbed parts. If $\beta_R \approx 1$, then

$$\frac{1}{\gamma_R^3} \frac{d}{c\,dt}\left(\gamma_R^3 \frac{d\tilde\psi}{c\,dt}\right) + k_\parallel^2 \tilde\psi = 0 \tag{15}$$

where again the factor, $-\gamma_R^3 mc^2 / k_L$, has been divided out. The "wave number" $k_\parallel$,

$$k_\parallel^2 = \frac{W_R'}{W} \frac{k_L}{\gamma_R^2} \tan\psi_R \tag{16}$$

is the length scale for longitudinal oscillation, and represents the bunching effect of the laser. If $\theta_L \gg 1/\gamma_R$, the longitudinal oscillation is much slower than the transverse (betatron), i.e. $k_\parallel \ll k_\beta$.

Consider the effect of space charge on longitudinal motion. The space charge force is $F = -eE$, where the space charge field satisfies Poisson's equation, $\nabla \cdot E = -en/\varepsilon_0$. In the quasi-1D case $F_z = \gamma^3 mc^2 (\omega_p/c)^2 (z - z_R)$. Using the perturbed phase, $\tilde\psi = -k_L(z - z_R)$, the space-charge force to be included on the right side of Eq. (15) is

$$F_z = (\omega_p/c)^2 \tilde\psi \tag{17}$$

where again, $-\gamma_R^3 mc^2/k_L$, has been factored out. The plasma frequency can be expressed in terms of reference quantities. Given fixed cross-sectional area, continuity implies $n\ell_\mu = n_0\lambda_L$, where $n_0$ is the unbunched density and $\ell_\mu$ is the full width (longitudinal) of a microbunch; all electrons in the microbunch lie in the range $-\ell_\mu/2 \le z-z_R \le \ell_\mu/2$. The unbunched case has $\ell_\mu = \lambda_L$. Define $\pm\psi_\mu$ as the phase of the extremal electrons, $z-z_R = \pm\ell_\mu/2$: then $n = n_0 \cdot \pi/\psi_\mu$. With $Q_M = en_0 A\ell_M$, we have $\left(\omega_p/c\right)^2 = \pi k_{SC}^2/\psi_\mu$, where the constant associated with space charge oscillations is

$$k_{SC}^2 = \frac{2r_e k_\beta}{\gamma_R^2 \ell_M \varepsilon_N} \cdot \frac{Q_M}{e} \tag{18}$$

Adding the space-charge force in the equation of motion, Eq. (15),

$$\frac{1}{\gamma_R^3} \frac{d}{c\,dt}\left(\gamma_R^3 \frac{d\tilde{\psi}}{c\,dt}\right) + k_\parallel^2 \tilde{\psi} = \pi k_{SC}^2 \frac{\tilde{\psi}}{\psi_\mu} \tag{19}$$

Solve this for the outermost electrons $\tilde{\psi} = \pm\psi_\mu$:

$$\frac{1}{\gamma_R^3} \frac{d}{c\,dt}\left(\gamma_R^3 \frac{d\psi_\mu}{c\,dt}\right) + k_\parallel^2 \psi_\mu = \pm\pi k_{SC}^2 \tag{20}$$

The solution has two parts: oscillatory (homogeneous solution) and quasi-steady (inhomogeneous); our interest is in the latter. The quasi-steady phase spread is

$$\left|2\psi_\mu\right| = 2\pi \frac{k_{SC}^2}{k_\parallel^2} = 2\pi \frac{r_e N\theta_L^2}{\ell_M \varepsilon_N k_\beta} \cdot \frac{Q_M}{e} \tag{21}$$

## Practical example

Consider the parameters relevant to the STELLA experiment: $W_R' = 100$ Mev/m; $W = 50$ MeV; $\theta_L = 20$ mrad; $N \approx 1$; $\lambda_L = 10$ μm; $\psi_R = 0.85\pi$; $Q_M = 0.15$ nC; $\ell_M = 3$ mm; and $\varepsilon_N = 10^{-6}(\pi$ m-rad), the quasisteady phase of the outermost electrons is $|2\psi_\mu|/\pi = 0.06$. Thus, for these conditions the phase spread by space charge effects is relatively small. However, this becomes more difficult for high-energy accelerators. The scaling of phase spread with important parameters is

$$\left|2\psi_\mu\right| \propto \left(\frac{W\lambda_L}{W_R'}\right)^{1/2} \cdot \frac{\theta_L Q_M}{\ell_M \varepsilon_N}$$

This suggests that for high energy (W) the space charge phase spread will become significant. This is a consequence of the following: the bunching effect of the laser, $k_{\parallel}^2 \propto 1/W\gamma^2 \propto 1/\gamma^3$, decreases faster with increasing energy than the debunching effect of space charge, $k_{SC}^2 \propto k_\beta / \gamma^2 \propto 1/\gamma^{5/2}$. This tendency is mitigated if the acceleration gradient ($W'$) is also increased.

## ACKNOWLEDGMENTS

The authors acknowledge useful conversations with J. Fontana, R.H. Pantell, and J.C. Gallardo. This work is supported by the U.S. Department of Energy, Grant N. DE-FG03-98ER41061.

## REFERENCES

1. Kimura, W.D., Babzien, M., Ben-Zvi, I., Cline, D.B., Fiorito, R.B., Fontana, J.R., Gallardo, J.C., Gottschalk, S.C., He, P., Kusche, K.P., Liu, Y., Pantell, R.H., Pogorelsky, I.V., Quimby, D.C., Robinson, K.E., Rule, D.W., Sandweiss, J., Skaritka, J., van Steenbergen, A., and Yakimenko, V., in Proceedings of the 1997 Particle Accelerator Conference, May 12-16, 1997, Vancouver, BC.

2. Steinhauer, L.C., and Kimura, W.D., "*Space charge debunching and compensation in a laser particle acceleration system,*" (unpublished).

3. Haimson, J., in *Linear Accelerators*, P.M. Lapostolle and A.L. Septier, eds., North-Holland, Amsterdam, 1970, p. 435.

# Intensification of Harmonic Spontaneous Radiation with a Novel Undulator

T.C. Marshall and Yichen Shao

*Department of Applied Physics, Columbia University, New York City 10027*

Zohreh Parsa

*Physics Department 901A, Brookhaven National Laboratory, Upton, New York 11973*

**Abstract**. We have calculated the on-axis spectrum of spontaneous radiation emitted by an electron moving along a planar undulator that has a magnetic profile along the axis that approximates a square wave. (This could be obtained in practice by driving a ferromagnetic undulator into saturation by "excessive" current in the windings.) We find considerable enhancement of the harmonic radiation spectrum. We compare the harmonic power emitted by an electron moving through an undulator having a sine-wave field profile with the radiation emitted from an undulator having a "square-wave" profile; the latter is approximated by the first three Fourier components of the undulator magnetic field profile along the axial direction. Examples are computed for 40MeV electrons taking K < 1, for spontaneous radiation emitted along the axis of the system. The emission at harmonics f > 1 is greatly enhanced for the approximate square-wave magnetic profile: the ratio of the power emitted at f=5 by the "square-wave" undulator to that of the sine-wave undulator is about 15 (whereas the corresponding ratio at f=1 is only 1.5). While this enhancement might be expected because of the appreciable n=1 and n=5 Fourier components of the undulator field, higher odd harmonics are enhanced even more (e.g., x1000 at f= 11). FEL gain at the harmonics should be enhanced by similar factors.

## INTRODUCTION

In a typical free electron laser (FEL), the electron beam interacts with a "dipole" undulator that has a sinusoidal magnetic field variation; the electron motion is sinusoidal in the plane transverse to this field and emits odd-numbered harmonic radiations along the axis. However, one need not limit the choice of undulator field profile to the sinusoid, providing other profiles result in significant advantages. In connection with the IFEL accelerator, in the past we have pointed out [1] that the use of an undulator profile that approximates a "square wave" will result in an enhanced acceleration gradient, by as much as a factor of two (equivalent in effect to an increase of laser drive intensity by a factor of four), similar to the helical undulator. This improvement (essentially at the fundamental FEL resonance) results largely from the fact that, for a given peak undulator field amplitude, the rms electron acceleration obtained from the square wave undulator is larger than that from the sinusoid; furthermore, the electron orbit is stable as well. In this paper, we find additional advantages that should result particularly at harmonic numbers f >1 if the undulator field profile is nearly "square wave": namely, a large enhancement of the harmonic spontaneous power radiated, together with enhanced FEL gain. The modification of undulators to enhance FEL gain

CP472, *Advanced Accelerator Concepts: Eighth Workshop,*
edited by W. Lawson, C. Bellamy, and D. Brosius

has been examined in the past [2-5], usually with a particular design in mind, but with

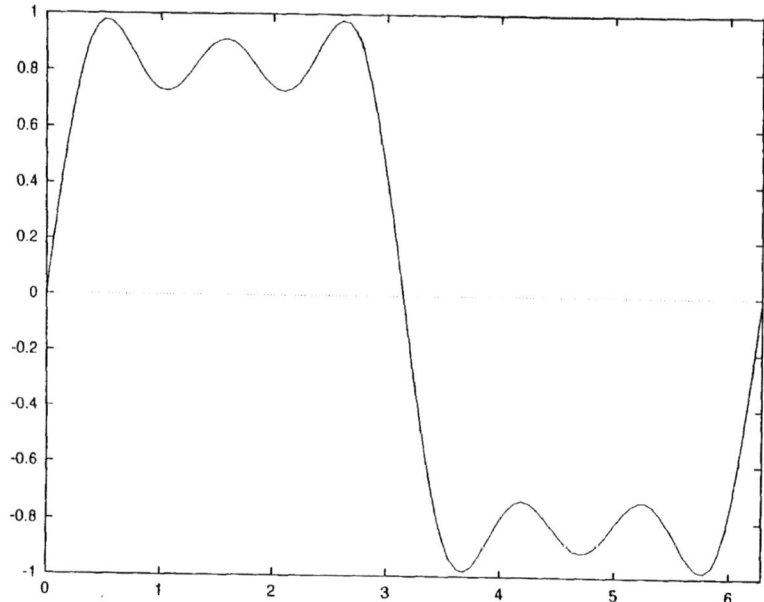

similar conclusions.

**FIGURE 1:** Representation of a "square-wave" undulator field by the first three Fourier components. Ordinate, normalized magnetic field; abscissa, axial distance measured in radians over one period ($2\pi$).

We retain the "undulator approximation", namely that not only is the amplitude of the motion $K/\gamma$ small but also $K < 1$: then the radiation will have sharp lines at the harmonics on the axis since the radiation cone (width ~ $1/\gamma$) overlaps the orbit. In a long undulator, the spectrum becomes sharply peaked at frequencies satisfying

$$\omega_{rf} = \omega_0\, f / (\,1\, -\, \beta_0 \cos\theta\,) = f\omega_{r1}\,, \qquad (1)$$

where $\omega_0 = k_0 c$ (the undulator wavenumber times the speed of light), $f$ is the harmonic number, $\theta$ is the angle from the axis of motion along the undulator, $\omega_{r1}$ is the FEL resonance frequency for $f=1$, and $\omega_{rf}$ is the resonance frequency for the $f^{th}$ harmonic.

The resonance line widths are all $\sim \omega_H/N$. K is the normalized magnetic vector potential, $eB_0/k_0mc^2$, and $B_0$ is the peak undulator field. The undulator is the planar dipole type, and we compute only radiation directed along the axis.

## THEORY

We begin by expanding the square-wave undulator field in a Fourier series:

$$B_y(z) = \Sigma F_n \exp(ink_0z), \tag{2}$$

where $F_n = 0$ for n= even integer, and $= 2B_0 / in\pi$ for n = odd integer. In order to keep the peak amplitude of the "square-wave" undulator comparable with the sinusoid, the coefficient is adjusted to be $3.3B_0/in\pi$, and we use a truncated Fourier series, retaining only the n=1,3,5 terms in the examples which follow, for simplicity. This representation is shown in Fig. 1; however, there are two matters to keep in mind. Going to higher n will "smooth" the top of the undulator field profile, but it also will increase the slope of the square-wave jump. The latter point is where the modelling of the actual saturated field will influence the decision of where to truncate.

As the spectrum requires the velocity, $\beta(t) = v(t)/c$, we integrate the equations of motion of a relativistic electron, finding:

$$\beta_x = \{3.3\ \omega_B/\pi\omega_0\} \{ \exp(i\omega_0t) + 1/9 \exp(3i\omega_0t) + 1/25 \exp(5i\omega_0t) \} \tag{3}$$

and $\beta_z = \beta_0 - \beta_x^2/4\beta_0$ , with $\omega_B = eB_0/\gamma_Bmc$. From this follows the orbits, z(t) and x(t), e.g.:

$$z = \beta_0ct - (3.3/\pi)^2(\omega_B^2c/4i\omega_0^3) \{ (1/2)\exp(2i\omega_0t) + (1/18)\exp(4i\omega_0t) + \tag{4}$$

$$+ (187/12150)\exp(6i\omega_0t) + (1/900)\exp(8i\omega_0t) + (1/6250)\exp(10i\omega_0t) \};$$

$$x = (3.3\omega_Bc/i\pi\omega_0^2) \{\exp(i\omega_0t) + (1/27)\exp(3i\omega_0t) + (1/125)\exp(5i\omega_0t) \}. \tag{5}$$

The terms introduced by the higher order Fourier components of the undulator are easily identified; they will cause frequencies up to $10\omega_0$ to appear in z and $\beta_z$. The technique of Fourier decomposition of the "square wave" undulator field profile permits one to retain a representation of the orbit as a sinusoid, thus the term $\exp[i\omega\ r(t)/c]$ in the radiation integral [7] can be replaced by a Bessel series expansion, a useful technique originally used for the sinusoid undulator [6]:

$$\begin{array}{c} n=\infty \\ \exp(iasinv) = \Sigma\ J_n(a)\ \exp(inv), \\ n=-\infty \end{array} \tag{6}$$

where the harmonic components of the orbit are represented by asinv.

Introducing this orbit into the formula for the energy spectrum emitted [6,7] we obtain the following including terms up to n=5, while setting $\theta = 0$:

$$dW/d\Omega d\omega =$$

$$(e^2/4\pi^2 c)[f^2/(1-\beta_0)]^2 [(\omega/\omega_0)(1-\beta_0)-f]^{-2} \sin^2 [N\pi\{(\omega/\omega_0)(1-\beta_0) - f\}]$$

$$\times \ (3.3/\pi)^2 \ (\omega_B/\omega_0)^2 \ \{A_1{}^2 + (1/81)A_3{}^2 + (1/625)A_5{}^2 + (2/9)A_1 A_3$$

$$+ (2/25)A_1 A_5 + (2/225)A_3 A_5\} \qquad (7)$$

where:

$$A_\alpha = \underset{n1,n2,........n8 = -\infty}{\overset{n1,n2,........ \ = +\infty}{\Sigma.............\Sigma}} J_{n1} \ J_{n2}........J_{n8} \ (\delta_{n1+ n2+...n8 + \alpha, \ f} + \delta_{n1+n2 +.....n8 - \alpha, \ f} ) \quad (8)$$

In eq. (8), the $\Sigma$'s denote summations over all the possible Bessel function of type $n_i$ permitted by the $\delta$'s, taking $\alpha = 1,3,5$; the arguments of these Bessel functions are (sequentially):

$$-3.3v; \ \ -(3.3/27)v; \ \ -(3.3/125)v; \ \ (3.3^2/8)\mu; \ \ (3.3^2/72)\mu;$$

$$[(3.3^2)(187)/48600]\mu; \ \ (3.3^2/3600)\mu; \ \ (3.3^2/25000)\mu,$$

{for example, we would have $J_{n3}(-[3.3/125]v)$},

where $\qquad\qquad\qquad v = (f\omega_B\sin\theta\cos\phi )/ \pi\omega_0[1- \beta\cos\theta] \qquad\qquad (9)$

and $\qquad\qquad\qquad \mu = (f\omega_B{}^2 \cos\theta )/ \pi^2\omega_0{}^2( 1- \beta\cos\theta ). \qquad\qquad (10)$

In these terms, $\phi$ is usually chosen to be zero, although we display the complete angular dependency in $v$ and $\mu$. This shows that along the axis, $v = 0$, and so the surviving Bessel functions connected with this argument are just the $J_0$ .

In eq. (7), the spectral dependence on $\omega$ is the same as it would be for the sinusoidal undulator. In the example of the sinusoidal undulator, only the term $A_1$ survives for radiation directed along the axis $\theta = 0$; but the new undulator introduces new sources of harmonic radiations.

## RESULTS

The spontaneous power was computed numerically, and in Fig. 2 we show a typical result where we have taken K = 0.64 (note that $\omega_B/\omega_0 = K/\gamma \sim \beta_x$) and $\gamma = 80$ (40MeV), N=16. Only the peak power emission data point is plotted at each harmonic, all intermediate points taken to be zero, and we compare the sinusoidal undulator with two approximations to the square wave undulator, where we include respectively only the n = 1 and 3 components, or the n= 1,3, and 5 components. The striking feature is the very substantial enhancement of

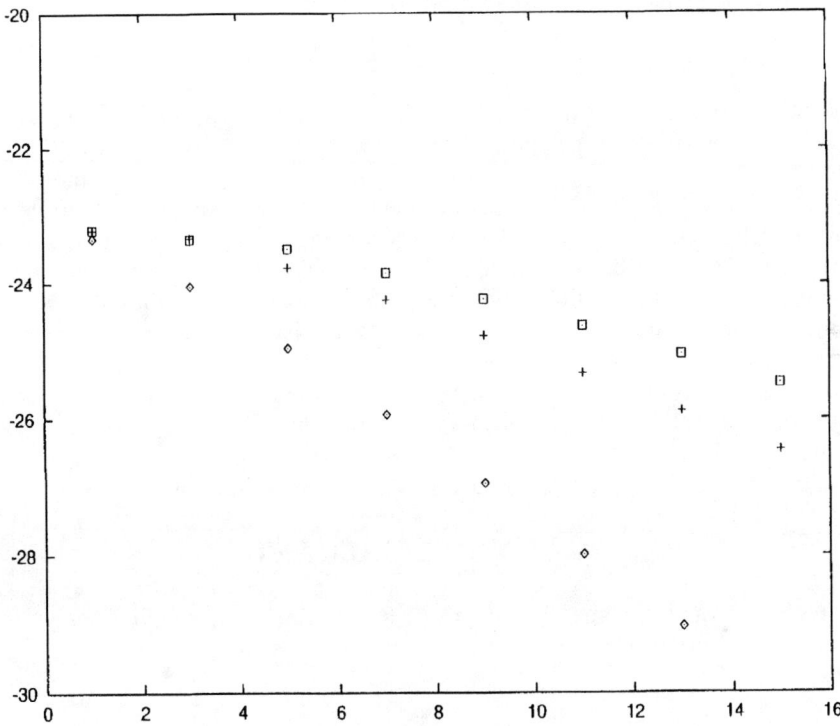

**FIGURE 2:** Logarithm to base 10 of {dW/dΩdω}, in {watt/ster.-sec⁻¹}, versus f, the harmonic number. The diamonds are the sinusoidal undulator; the crosses and squares are for the square-wave undulator approximated by the terms n =1,3; and by n = 1,3,5 respectively. The undulator parameter is K = 0.64.

spontaneous power emitted at the higher harmonics. That there should be *some* enhancement of radiation is apparant from the fact that electron radiation depends on the square of the electron speed, and the latter is proportional to the integral of the undulator field. The ratio of emission from the "square-wave" undulator to the "sinusoid" is 2.0 for f=1; this is the ratio of the mean square motion of the electron in these two different undulators that have equal peak field amplitudes. For the "approximate" square-wave undulator here, this ratio is about 1.5. However, discounting this factor of 1.5, there is still a remaining factor ~ 10 in enhanced radiation at the fifth harmonic, and larger enhancements at the higher harmonics, to be explained. A truly "square-wave" undulator field axial profile will cause a triangular -wave electron velocity response and a piecewise-quadratic orbit, which is very rich in higher-harmonic content, and this is the major source of this enhancement. This effect would be clearly identifiable in a simple experiment.

The enhancement of harmonic spontaneous emission using the "square-wave" undulator profile has further implications. From Madey's theorem [8], where $\gamma_f$ and $\gamma_i$ are the final and initial energy factors of an electron leaving and entering the undulator, and the brackets indicate an ensemble average:

$$<\gamma_f - \gamma_i> = 0.5 \; d/d\gamma_i \, [<(\gamma_f - \gamma_i)^2>]. \qquad (11)$$

Since the left-hand side is proportional to the gain, and the quantity within the square brackets is proportional to the spontaneous emission power, it follows that the gain is proportional to the frequency derivative of the spontaneous power spectrum. But, the width of this spectrum ( $\sim 1/N$ ) does not depend on the details of the profile of the undulator field, but rather the number of periods; hence the FEL gain (at small signal and small gain) should be proportional to the peak spontaneous power emitted at the various harmonics. In the illustration described above, the gain of a FEL should be enhanced by about 10 at the fifth harmonic. Notice that enhancements by even larger factors appear at the higher harmonics. Thus the harmonic enhancement effect we have found is not merely an artifact of our representation of the "square-wave" undulator, e.g., that the n=5 Fourier component of the undulator is driving just the n=5 FEL harmonic to high power. It almost goes without saying that for the higher gain indicated here to be realized, the customary conditions on beam quality apply.

## ACKNOWLEDGMENT

This research was supported by the Department of Energy, High Energy Physics Division. We (Z.P.) thank J. Lemmon for computing assistance.

## REFERENCES

[1] "Enhanced IFEL Performance Using a Novel Wiggler", Z. Parsa and T.C. Marshall, May 1997 Particle Accelerator Conference, Vancouver, Canada. BNL report #64491 and CAP – 168-ATF-97C; to be published, Conference Proceedings [IEEE].

[2] M.J. Schmitt and C.J. Elliott, *IEEE J. Quantum Electronics* **QE-23**, 1552 (1987)

[3] M. Asakawa et al, *Nucl. Instrum. Meth. in Phys. Res.* **A318**, 539 (1992)

[4] D. Iracane, D. Touati, and P. Chaix, *Nucl. Instrum. Meth. in Phys. Res.* **A341**, 220 (1994)

[5] M. Asakawa et al, *Nucl. Instrum. Meth. in Phys. Res.* **A375**, 416 (1996)

[6] W.B. Colson, *IEEE J. Quantum Electronics* **QE-17**, 1417 (1981)

[7] J.D. Jackson, *"Classical Electrodynamics"*, p. 670, John Wiley, NY, (1975)

[8] J.M.J. Madey, *Nuov. Cim.* **50B**, 64 (1979).

# A Laser Linear Collider Design[1]

## A. A. Mikhailichenko

*Wilson Laboratory, Cornell University, Ithaca, NY 14853-8001*

**Abstract.** A conceptual design of crucial elements of $2 \times 1$ *km* long linac is considered. This linac is driven by a laser radiation distributed within open accelerating structures with special sweeping devices. These devices deflect the laser radiation to the structures in accordance with instant position of accelerated particles. The power reduction and shortening the illumination time for every point on the structure equates to the number of resolved spots, associated with this sweeping device. A 300 *J* total, 100-*ps* laser flash could provide the final energy 30 *TeV* for $\lambda \cong 1\mu m$ and 3 *TeV* for $\lambda \cong 10\mu m$ on 1 *km* with the method described. This total power required could be generated with amplifiers distributed along the linac. For repetition rate 160 *Hz* the luminosity associated with colliding beams could reach $L \approx 10^{33} cm^{-2} s^{-1}$ per bunch with population $10^7$. Wall plug power required for operation of LLC is $\sim 2\ MW$.

## INTRODUCTION

The concept of a Linac, driven by a Traveling Laser Focus (TLF) method is described in [1]. TLF deals with accelerating structure, what scaled down to a laser wavelength. Each cell of the structure has an opening from one side (so called open acceleration structure). Laser radiation focused onto these openings. This spot of focused radiation has an instant size much smaller, than longitudinal dimension of the structure, thereby exciting the cells of the structure locally. Special sweeping (deflecting) device moves this focal spot in *longitudinal* direction so, that this spot is following the particle in its motion along the accelerating structure. Due to this arrangement, all impulse laser power is acting for generation of accelerating field at the instant particle's location only. It was shown [1], that the power reduction and shortening of illuminating time is equal numerically to the *number of resolved spots (pixels)*, associated with the sweeping device. On the basis of TLF method the scheme for Laser Linear Collider (LLC) was proposed, Fig. 1, [1c]. Parameters of this LLC are represented in a Table 1. As far as total laser flash energy (300 *J* in the Table 1), it could be generated either by a single amplifier unit or by *many amplifiers* distributed along the Linac. All the linac length is sectioned by modules having the length $L\sim3$ *cm*. Within each of these sections the laser flash having duration $\tau \cong 100$ *ps* becomes distributed along the length of accelerating structure. Energy required for excitation the field in the structure up to $3GeV/m$ is about 3 *mJ/cm* for the laser wavelength $\lambda \cong 10\mu m$. For $\lambda \cong 1\mu m$ this pulse will give $30GeV/m$. Interaction of beams at IP is going in deep quantum regime.

[1] Full version is available as CLNS 98/1562 at "www.lns.cornell.edu/public/CLNS/1998". *Supported by National Science Foundation.*

CP472, *Advanced Accelerator Concepts: Eighth Workshop,*
edited by W. Lawson, C. Bellamy, and D. Brosius

The beams of electrons and positrons can be polarized what gives the effective gain in luminosity and reduces the background [1c, 21]. We would like to stress here once again that TLF method does not exclude the problems, associated with breakdown limit of the structure, but *it allows drastically reduce the laser impulse power, required for excitation the structure* to this level[2]. This method also cuts down the illumination time for every point of the structure. The last brings hope that the break down threshold will be increased.

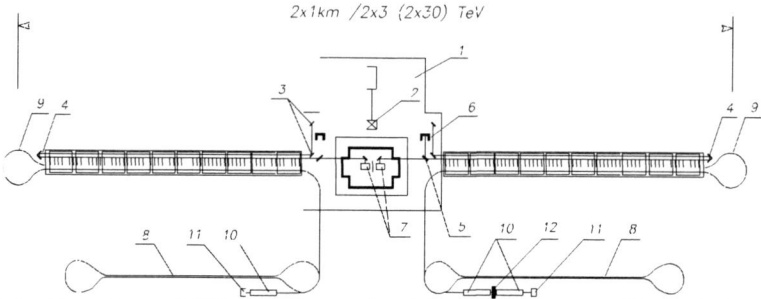

**FIGURE. 1.** Linear Laser Collider (LLC) complex. 1–is a laser master oscillator platform, 2 –is an optical splitter, 3,4–are the mirrors, 5–is a semi-transparent mirror, 6—is a light absorber. 7–are the Final Focus Systems. 8–are the damping systems for preparing particle's beams with small emittances, 9—are the bends for particle's beam. 10–are the accelerating X-band structures, 11–is an electron gun, 12 is a positron converter.

**TABLE 1.** Parameters of Laser Linear Collider.

| Parameter | $\lambda \cong 10\mu m$ | $\lambda \cong 1\mu m$ |
|---|---|---|
| Energy of $e^{\mp}$ beam | $3 \times 3\ TeV$ | $30 \times 30\ TeV$ |
| Total two-linac length | $2 \times 1\ km$ | $2 \times 1\ km$ |
| Main linac gradient | $3\ GeV/m$ | $30\ GeV/m$ |
| Luminosity/bunch | $\leq 1 \cdot 10^{33}\ cm^{-2}s^{-1}$ | $\leq 1 \cdot 10^{33}\ cm^{2}s^{-1}$ |
| No. of bunches/pulse | $3\ (\leq 100\ )*$ | $10\ (\leq 300\ )*$ |
| Laser flash energy/Linac | $300 J$ | $300 J$ |
| Repetition rate | $160\ Hz$ | $160\ Hz$ |
| Beam power/Linac | $2.3\ kW$ | $760 W$ |
| Bunch population | $3 \cdot 10^{7}$ | $10^{6}$ |
| Bunch length | $1\mu m$ | $0.1\mu m$ |
| $\gamma \varepsilon_x / \gamma \varepsilon_y$ | $10^{-8} / 4 \cdot 10^{-10} cm \cdot rad$ | $5 \cdot 10^{-9} / 1 \cdot 10^{-10} cm \cdot rad$ |
| Damping ring energy | $0.7\ GeV$ | $0.7\ GeV$ |
| Length of section/Module | $3 cm$ | $3 cm$ |
| Wall plug power** | $2 \times 0.5\ MW$ | $2 \times 0.5\ MW$ |

*–Maximal possible number. **–Laser efficiency $\approx 10\%$

---

[2]   *If electromagnetic field of some configuration accelerates particles, it has the electrical field maximal value on the surrounding boundaries* [1].

## TLF METHOD

Fig. 2 reminds the idea [1]. The pulse of laser radiation lasts for a time $\tau$, so it has the length $\approx c\tau$, pos. 1. The laser bunch 1 passed trough splitting devices 7 generated the secondary bunches 2-4. Beam of particles 5 is going inside this structure with velocity $V$. The laser bunch instant positions are numbered by 1-4. Structure 6 has a length $L$ in longitudinal direction. Device 9 is electrically driven by electrical pulse for *sweeping* the laser beam in longitudinal direction synchronously with the particle's passage. While arriving to the structure, the laser bunch has a slope $\alpha$. Angle $\Theta \cong L/R$, $\Omega \cong c/R$. These parameters connected as

$$\Theta \cdot \tan\alpha = cL/VR \cong \Omega\, c\tau/V \cong \Omega \cdot c\tau/V \cong c\Theta/V. \tag{1}$$

For $V \cong c$ this gives evidently $\alpha \cong 45°$. All deflecting elements are synchronized so that the bunch 5 has continuous acceleration in all sections.

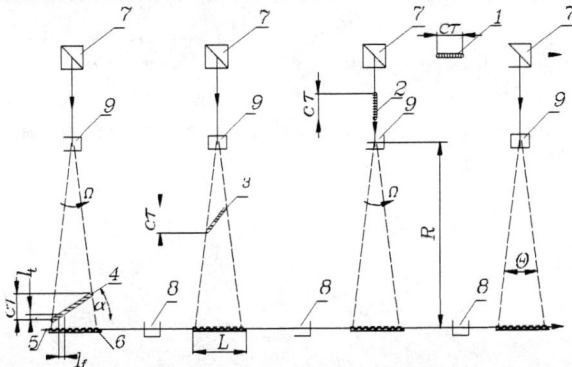

**FIGURE 2.** Accelerating complex scheme. 1–4–are instant laser beam positions; 1–is a primary laser bunch which is moving from the left side on the picture to the right. 5– is the beam of accelerated particles. 6–is the accelerating structure. 7–are the optical splitters. 8–are the particles' beam focusing elements (if no RF focusing in use). 9–are the deflecting devices.

The laser focus sweep is limited to the distance about $3cm$ for practical reasons (see lower), so the accelerating device looks like a sequence of 3 *cm* long accelerating structures. A shot focusing *cylindrical* lenses installed on the laser way close to accelerating structure reduce *the transverse size* of the spot to a minimal one pos. 13 in Fig.3. A natural way to increase the deflecting length $R$ is utilization of mirrors. Mirrors might be slightly curved to obtain some focusing properties. Each part of the grating structure is illuminated by duration, which is defined by $l_t/c$, Fig.2. For example, if we consider $l_t \cong L/N_R \cong 100\lambda$, ($N_R$–is the number of resolved spots, see lower), $\lambda = 1\mu m$, then $l_t/c \cong 3 \cdot 10^{-13} sec$. This time is less than the time between electron-electron collisions $\tau \approx l_{free}/v_F \cong 10^{-12} sec$, where $l_{free}$ – is the free path length, $v_F$ is the electron velocity at Fermi surface [15]. The time of illumination is, however longer, than the time, corresponding to reaction of electron plasma in a metal, what is

$\tau \cong 2\pi / \omega_p = 2\pi / \sqrt{4\pi n r_0 c^2} \approx 3 \cdot 10^{-16} s$, where $n$–is the density of electrons, $r_0 = e^2 / mc^2$.

## GENERAL DESCRIPTION OF LLC

Scaled view of regular part from Fig. 1 is represented in Fig. 3 below. To avoid the influence of fluctuations, it was suggested that both beams, the laser–1 and particle's–5 go first to the end of all linear system apart from central station. On this way the bunch's parameters (optical and particle's) picked up and processed with appropriate algorithms. After the bend in systems 2,6 on the *back way to IP* necessary voltages applied to correcting elements, distributed along the linac. The particle's beam goes further trough structures to next modules, 4. 7 and 8–are the focusing elements for the particle's and laser beam correspondingly. Optical platform 9 is standing on legs 10 with active damping system to minimize vibrations. Deflecting devices 11 sweep the laser radiation along the accelerating structures 14. 12–are the lenses for focusing the laser radiation in longitudinal direction. 13–are the short focusing cylindrical lenses. LLC could be located in a tunnel with 3 $m$ in diameter. For energy around $0.5 \times 0.5$ $GeV$ the LLC could be preferably located on the surface, so any other shape is acceptable.

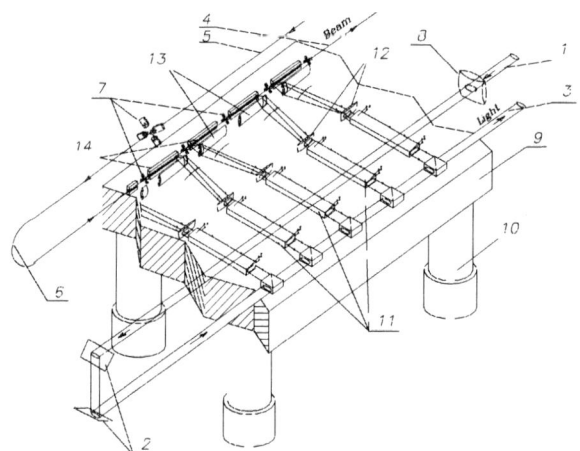

**FIGURE 3**. Table with a section of accelerator. Comments are in the text.

## ACCELERATING STRUCTURE

There are proposals for accelerating structures what could be scaled to match the wavelength of laser radiation [11-13]. We took the foxhole-type structure described in [12] as a basis, Fig. 4. This structure has an advantage in pumping possibilities. The most important feature, however, is that this structure gives a good positioning for

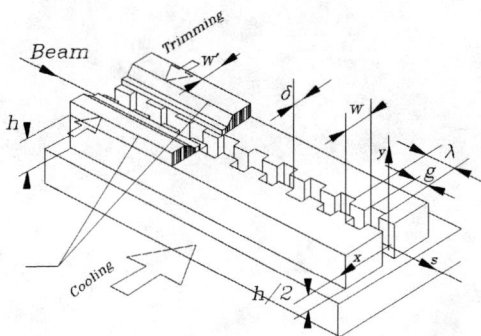

**FIGURE 4.** The foxhole type accelerating structure. Height $h$ is $h \cong \lambda_W / 2$, where $\lambda_W$—is a wavelength of laser radiation inside the cell. $g / \lambda \cong 1/2$, $W \cong 0.7\lambda$, $\delta \cong 0.2\lambda$. 1–are the masks along the structure. $W' \le W$. The masks are used for trimming the coupling ($Q_{RF}$-factor).

electrical field map [1c]. Covers 1 adjust the coupling between the groove and outer space. With these covers the height $h$ is about $h \cong \lambda_W / 2$ and the cells have *inductive* coupling with outer space. Structure developed in [13] is close to this type. In original foxhole structure [12] $h \cong 3\lambda_W / 4$, where $\lambda_W \cong \lambda / \sqrt{1 - (\lambda / 2W)^2}$. The $Q_{RF}$ factor of the order 5-10 could be expected here. Technological possibilities in lithography are advanced far beyond the requirements associated with this structure scaled down to $\lambda \cong 1\mu m$. Each structure is installed on a micro-table moved by a piezoelectric. This allows to align the structures for minimization the wakes. Structures are cooled down to keep the mechanical tolerances within the margins allowed. Calculations for this type of structure were done with of GdfidL code [14]. There were investigated different shapes of cell and transit slit between the cells. Typical longitudinal and transverse wakes *normalized to one cell* are $W_{\parallel} \cong -7 kV / pC$ and $W_{\perp} \cong 2.2 \cdot 10^2 V / pC / \mu m$ correspondingly for the accelerating structure with $\lambda \cong 10\mu m$, $\delta = 2\mu m$, $W = 7\mu m$ (see Fig. 4) and the bunch length $\sigma_l \cong 1\mu m$. Total charge is $eN \cong 0.16 pC$. An example of wake calculated is represented in Figs. 5, 6. The charge in the bunch here is $0.2 \, pC$. One can see, that the wake is slightly inductive.

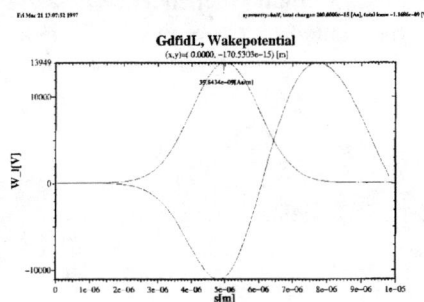

**FIGURE 5.** Example of the wake for 5 cells, Fig.6. Wake in the center of the bunch is about -10 $kV$.

619

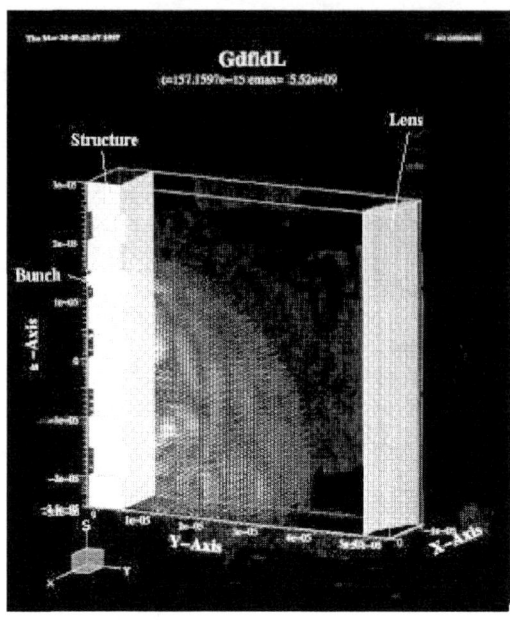

**FIGURE 6.** The bunch in the structure. A half of full picture is represented. The bunch is running through the structure on the left. A cylindrical lens (pos. 13 in Fig.3 is on the right side of the picture; ¼ is shown).

## DEFLECTING DEVICE

The ratio of deflection angle $\vartheta$ to diffraction angle $\vartheta_d \cong \lambda / a$, where $a$ is an aperture of the deflecting device, defines the *number of resolved spots (pixels)*, $N_R = \vartheta / \vartheta_d$. The deflection angle could be increased by some optics, but the number of resolved spots $N_R$ is *an invariant* under any transformations. As we mentioned above, $N_R$ value gives the number for the lowering the laser power for gradient desired and, also, the number for the duty time reduction.

*Electro-optical devices* use controllable dependence of refractive index on electrical field strength and direction applied to some crystals [5-10]. The change of reflecting index is equal to $\Delta n_i \cong (\partial n_i / \partial E_j) E^j(t)$. The refractive index in active media has a dependence like

$$1 / n_i^2 = 1 / n_{0i}^2 + \sum_j r_{ij} \cdot E^j , \qquad (2)$$

where $r_{ij}$–are $6 \times 3$ tensor[3]. This yields $\partial n_i / \partial E_j = -n_{0i}^3 r_{ij} / 2$ and the net change of refractive index becomes

---

[3] Index $i$ runs from 1 to 6. 1 stands for $xx$, 2–for $yy$, 3– $ss$, 4– for $ys$, 5–for $xs$, 6– for $xy$, $s$-longitudinal coordinate.

$$\Delta n_i \cong (\partial n_i / \partial E_j) E^j(t) \cong -n_{0j}^3 r_{ij} E^j / 2 \qquad (3)$$

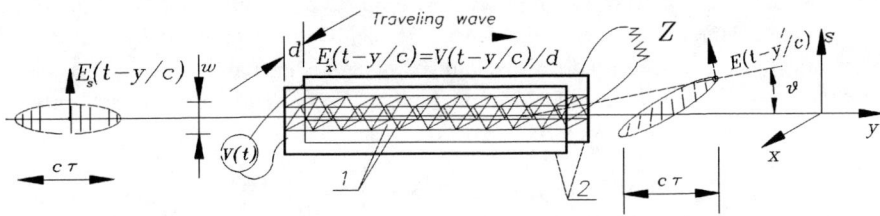

**FIGURE 7.** Prisms 1 with *oppositely directed optical axes* installed in series between two parallel strip–line electrodes 2, *d*–is the distance between them. *Z*– is a matching impedance. Lines across the laser bunch schematically show the wave fronts. $E_x(t - y/c)$–is a driving electrical field.

For *ZnTe* (transparent for $CO_2$ radiation) having $n_0 \cong 2.9$, $r_{41} \cong 4.4 \cdot 10^{-12} m/V$ what gives for 10 *kV/cm* the refraction index change as much as $\Delta n \cong 10^{-4}$. For *KTN* (*Potassium Tantalate Niobate*) $\Delta n \cong 7 \cdot 10^{-3}$ for $\lambda \cong 0.63\,\mu m$ is possible [8]. To increase the $N_R$, a *multiple-prism* deflectors were developed, see Fig.7. Here neighboring prismatic crystals have reversed optical axes. These crystals positioned between strip–line electrodes. To be able to deflect short light bunches, the voltage pulse $V(t)$ is propagating along this strip-line as a *traveling wave* together with the light to be deflected, as it was proposed in [1c]. The deflection angle and the number of resolved spots become now

$$|\Delta\vartheta| \cong \Delta n \cdot (2L_d / w) \cong L_d n_0^3 \cdot r_{ij} \cdot V / (w \cdot d), \qquad N_R \cong \Delta n \cdot 2L_d / \lambda \qquad (4)$$

Where $L_d$ stands for full length of deflecting device, *w*–is the light beam width (along direction of deflection). For $L=50cm$, one can expect for $w \cong 1cm$, that deflection angle is $\Delta\vartheta \cong 10^{-2}$ and $N_R \cong 10$ for $\lambda \cong 10\mu m$ and, respectively, $\Delta\vartheta \cong 10^{-2}$ and $N_R \cong 100$ for $\lambda \cong 1\mu m$. We estimated the field applied to the crystals as 10 *kV/cm*. This *variation is traveling with the light beam* along the deflecting device, Fig. 7. This strip-line deflecting device could be sectioned for voltage reduction. Special devices with avalanche diodes could be used here. Techical realization is represented below.

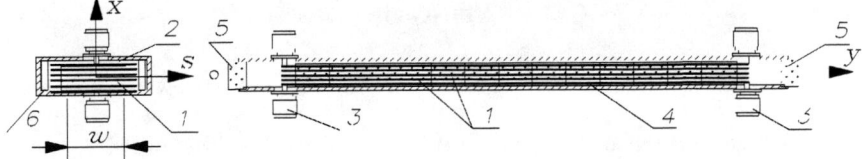

**FIGURE 8.** Arrangement of multi-prism traveling wave deflector. Electro-optical crystals 1 are positioned between strip-lines 2, attached at the end to the connectors 3. 4–is a cabinet with optical windows 5 from both sides. 6–is a matching dielectric.

The broad band traveling wave deflector could be obtained also if the same crystals placed in the middle of a waveguide [1c] shortened from both sides. Despite the materials transparent for longer wavelengths have lower value of $r_{ij}$–components, they have higher values of refracting index, so the variation of refractive index becomes about the same, (4) (see Table 2 below).

*Mechanical deflection system* and *Acousto-optical deflector* could be also used here[1c] (see references in there and [5-10]). Basically they give the same deflecting parameters. The power required for mechanical system is probably the smallest one.

*General conclusion for this section is that the light beam deflectors are available with parameters necessary for successful operation.*

## INJECTION SOURCE

Appropriate to a compact size of Laser Linear Collider complex a *kayak-paddle cooler*, Fig.1, pos. 8 [2,3]. This is basically a racetrack, which has a sequence of wigglers and RF cavities in a straight section. The straight sections narrowing at a center, so that they are geometrically congruent. Back and forward trajectories may be congruent also. The bends can be made to give a small input into cooling process. This yields the equilibrium emittances as for horizontal and vertical coordinate as the following[4] [2]

$$(\gamma \varepsilon_x) \cong 1 \cdot 10^{-8} \ cm \cdot rad \ , \quad (\gamma \varepsilon_y) \cong 4 \cdot 10^{-10} \ cm \cdot rad \ . \tag{5}$$

*Scrapping* extra particles remains a valid procedure for the emittance lowering at the expense of particle population [1b]. Optical Stochastic Cooling [17] could be used here. Intra-Beam Scattering remains the general problem.

## TRANSVERSE FOCUSING OF PARTICLE'S BEAM

A short wavelength of betatron oscillations helps against the resistive wall instability and wakefield influence reduction. The focusing system includes the quadrupole lenses of appropriate dimensions and a RF focusing [4], [19]. Quadrupole lenses could be placed between the accelerating sections, see Fig.2, 6, 7. This type of focusing was considered in [1c]. *Focusing with quadrupole lenses only is acceptable for the accelerating structure scaled to a laser wavelength* $\lambda \cong 10 \, \mu m$. RF focusing occurs if the particle is going out of the RF crest [18].

## ACCELERATING GRADIENT

For full energy carried by a laser bunch is $Q$ with time duration $\tau$ (see Fig. 2), the number $n$ of periods in the bunch is about $n = \tau / T = c\tau / \lambda$, where $T = \lambda / c-$ is a period of radiation. From he one can derive the maximal field strength $E_m$ as the following

$$E_m \cong \sqrt{Q / (\varepsilon_0 c \tau \lambda g)} \tag{6}$$

---

[4] With the wiggler's field oriented *vertically* and bending arcs in *horizontal* plane, the coordinates indicated here connected with the ones from Figs. 5, 7, 8 as $x \to y; y \to x$. This gives the minimum size of the beam for passing through the narrowings in the structure, marked with dimension $\delta$ in Fig. 5 if it is installed as in Fig. 3,4.

where $\varepsilon_o$ – is the dielectric permeability of the vacuum, $g$–is a dimension from Fig.4. For estimation let us take $Q = 0.01J$, $\tau \cong 0.1$ $ns$, $\lambda \cong 1\mu m$, $g \cong \lambda/2 = 0.5\mu m$. This gives the field strength $E_m \cong 270\,GeV/m$. This value must be reduced by the factor, taking into account the longitudinal length $l_f$ instead of $g$, see Fig.1. The longitudinal length could be estimated as a $l_f \approx L/N_R$ , where $L$ is the length of the structure. For $\lambda \cong 1\mu m$ one can expect $N_R \approx 100$, so the length of our interest is $l_f \approx c\tau/N_R$, what is $l_f \approx 0.03cm \approx 300\lambda$ . So the formula (4) must be rejected by factor $\sqrt{g/l_f} \approx 0.04$, what gives the field strength at the surface of the structure as high as $E_m \cdot \sqrt{g/l_f} \approx 10GeV/m$. The value of electrical field *inside* the structure could be found by taking into account RF quality factor $Q_{RF}$, see section about accelerating structure. For $Q_{RF} \approx 9$ it could reach again $E_m \cdot \sqrt{Q_{RF}} \cdot \sqrt{g/l_f} \approx 30GeV/m$. So we came to conclusion, that the laser flash with $Q \cong 10\,mJ$ at $\lambda \cong 1\mu m$ is able to feed the accelerating structure with the length about $c\tau \approx 3cm$ with $\geq 30$ $GeV/m$. A flash with $Q \approx 1J$ feeds $\sim 3$ $m$ of accelerating length, providing the energy gain $\approx 100GeV$ . $Q \approx 100J$ could feed 10 $TeV$ machine. For $\lambda \cong 10\mu m$ and the same $Q_{RF}$ factor accelerating gradient will be about $3GeV/m$ for laser flash $10mJ$ distributed along $3cm$ length.

The damaging level could depend on lot of factors, such as cleanness, vacuum, temperature of accelerating structure, technology of fabrication, heat treating and so on. The TLF method promises up to 30 $TeV/km$ or 300 $TeV$ on 10 $km$ with 0.3 $J/m$ or with 3 $kJ$ per pulse total. So the total output power of the laser must be within 0.5 $MW$ with repetition rate about 160 $Hz$ for $\lambda \cong 1\mu m$. Nd-Glass laser can be used here. $CO_2$ laser based system gives lower final energy. The power required $P_i \cong Q/\tau \approx 10^8\,W$ is within routinely obtained. As it was mentioned in Introduction, the laser amplifier could be sectioned with few structures or even each $3cm$ section could has it's own amplifier. For pumping the driving lasers the *diode laser arrays* could be used for the wavelengths indicated.

## BUNCH POPULATION

The energy, accepted from the field by $N$ particles is $W_a \cong eNE_m gI(g)$ where $e$ is the
charge of a particle, $I(g)$ is a function of the order of unity – an analog of the transit time factor. The share of the energy will be

$$\eta W \cong \eta \frac{1}{2} Q\lambda/(c\tau) \cong eNgI(g)\sqrt{Q/(\varepsilon_o c\tau\,\lambda\,g)} \qquad (7)$$

Equation (7) yields

$$N \cong \frac{\eta}{2eI(g)}\sqrt{\frac{\varepsilon_o\lambda^3 Q}{c\tau\,g}}\,. \qquad (8)$$

With $I(g) = 0.5$, $\eta \approx 0.05$ (5%), this gives $N \cong 1\cdot10^6$ for $\lambda \cong 1\mu m$. For $\lambda \cong 10\mu m$ this number will be $N \cong 3\cdot10^7$.

# FINAL FOCUS

We suggested to arrange a final focusing for our purposes with *a multiplet* of FODO structures having the number of the RF lenses in it of the order of few hundreds [1b]. The gradient in these lenses must vary from the very strong at the side closest to IP, to the weak one at the opposite side of the multiplet. Focusing properties of these RF lenses, discussed above can be used here. A laser radiation of general and *multiple* frequency can be used for such focusing. This is so called Adiabatic Final Focus. *Such a tiny lens, not sensitive to detector magnetic field could be easily installed inside the detector.*

# LUMINOSITY

For luminosity we have the formula

$$L \cong \frac{N^2 f H_B \gamma \cdot N_B}{4\pi \sqrt{(\gamma \varepsilon_x)(\gamma \varepsilon_y) \cdot \beta_x \beta_y}} \quad , \tag{9}$$

where $H_B$ is the enhancement parameter, $N_B$ – is the number of bunches per train. For emittances from (2, 3) one can obtain $L$ here for $\beta_x \approx \beta_y \approx 0.3\lambda$, $\lambda \cong 10\mu m$, $\gamma = 2 \cdot 10^6$ ($pc=1TeV$), $N \cong 2 \cdot 10^7$, $f \cong 160Hz$, $H_B \cong 2$, $N_B=1$ as $L \approx 1 \cdot 10^{33} cm^{-2} s^{-1}$. For $\lambda \cong 1\mu m$ result will be about the same as the number of the particles is lower here. The transverse dimensions will be $\sigma_x \cong 1.2 \cdot 10^{-9} cm \equiv 0.12 \overset{o}{A}$, $\sigma_y \cong 1.4 \cdot 10^{-10} cm \equiv 0.024 \overset{o}{A}$. Effective longitudinal distance between the particles has the order $l_b / N \approx \lambda / N$. The last number in our case is formally $\leq 10^{-11} cm$. As the distance $\approx \lambda_C \cong 3.8 \cdot 10^{-11} cm$ is absolute limit for such a spacing, we conclude that in each transverse slice $\approx \lambda_C$ the number of the particles $\approx 1-2$. The dimension as above, formally corresponds to $\approx 4$ particles per slice. As in each transverse slice $\approx \lambda_C$ the number of the particles $\approx 1-2$. As we mention in previous section 12, the transverse dimensions formally correspond to $\approx 4$ particles per slice. So the radiation is going in very strong quantum regime and models with coherent field are not applicable here. Strong reduction of radiation one can expect due to Landau-Pomeranchuk effect [22]. The radiation processes required more detailed considerations. *Using a train of bunches* (up to 100, see above) one can decrease the number of the particles in each bunch and/or to increase the cross section of the beam, thereby allow some reasonable losses. The possibility of operation with high repetition rate, up to *few kHz* is open.

# CONCLUSION

Any point of accelerating structure must be illuminated for the minimal time in order to avoid the damage associated with the overheating. Greater accelerating gradient requires higher power density of radiation at the structure. *Traveling Laser Focus,*

*TLF*- method solves both problems. Illuminating time and total laser power (or the flash energy) both defined by the number of resolved spots (pixels) associated with a deflecting device. Number of resolved spots $\approx 20 - 100$ achievable. Lasers for the TLF method need to have more power in intermediate time duration $\tau \approx 100 ps$ rather than in a shorter time interval. Equivalent time of illumination of accelerating structure with this pulse is $0.1 \div 1 ps$, nevertheless. For a $0.5 \times 0.5 \, TeV$ Collider the length becomes $2 \times 170 \, m$ for $\lambda \cong 10 \mu m$ and $2 \times 17 \, m$ for $\lambda \cong 1 \mu m$, respectively. One can compare these numbers with that from Linear Collider projects in consideration in many Laboratories around the World ($2 \times 10 \, km$ gives about $2 \times 500 \, GeV$)

*In conclusion we can say, that TLF method what brings a Laser driven Linac to a present day front technology, could be considered as a challenge for LLC.*

## REFERENCES

[1] Mikhailichenko, A.A., a) "The method of acceleration of charged particles", Author's certificate USSR N 1609423, Priority May 1989, Bulletin of Inventions (in Russian), N6, p.220, 1994. b) "A concept of a Linac Driven by a Traveling Laser Focus", 7th Advanced Accelerator Concepts Workshop, AIP 398 Proceedings, p.547. c) "Laser Acceleration: a Practical Approach", CLNS 97/1529, Cornell, 1997; also a Talk on LASERS'97, Fairmont Hotel, New Orleans LA, December 15-19, 1997.

[2] Mikhailichenko, A.A., "Injector for a Laser Linear Collider", this Workshop. Also CLNS 98/1668.

[3] Mikhailichenko, A.A., "On the physical limitations to the Lowest Emittance", Ref. [1b], p.294 and CLNS 96/1436, Cornell, 1996 .

[4] Mikhailichenko, A.A., "A Beam Focusing System for a Linac Driven by a Traveling Laser Focus", PAC97, Dallas, TX, Proceedings, p.784.

[5] C Ireland,.L. J., M. Ley, "Electrooptical Scanners" in "Optical scanning", Edited by G. Marshall, Marcel Dekker, Inc., 1991.

[6] *Selected Papers on Laser Scanning and Recording*, SPIE Vol. 378, Editor L. Beiser, 1985.

[7] Fowler, V.J., J.Schlafer, *Applied Optics,* Vol.5, N10, 1657(1966)

[8] Chen, F.S., et. al., Journ. *Applied Physics*, Vol.37, N1, p.388(1966).

[9] Lotspeich, J.F., *IEEE Spectrum*, 45 (Feb. 1968), see also [5].

[10] Bademain, L., "Acousto-Optical Laser Recording" in [6], p.255.

[11] Rosenzweig, J., A. Murokh, C. Pellegrini, *Phys. Rev. Let.* **74**, 2467(1995).

[12] Fernow, R.C., J.Claus, AIP Conference Proceedings, 279, 1992, p.212.

[13] Henke, H.,"mm Wave Linac and Wiggler structure", EPAC 94, London.

[14] Bruns, W., **GdfidL**, TU Berlin, 1997.

[15] Kittel, C., *Introduction to Solid State Physics*, Wiley, NY, 1976.

[17] Mikhailichenko, A.A., M.S.Zolotorev, "Optical Stochastic Cooling", *Phys. Rev. Let.* **71**, 4146(1993).

[18] W. Schnell, "Microwave quadrupoles for linear colliders", CLIC Note 34, 1987.

[19] Mills, F.E., A. Nassiri, Argonne National Laboratory, internal report ANL/APS/MMW-9, 1994.

[20] Baier, V.N., V.M. Katkov, BINP 97-70, Novosibirsk 1997.

[21] Mikhailichenko, A.A., SLAC-R-502, p.229.

[22] Davis, Ch. C., *Lasers and Electro-Optics. Fundamentals and Engineering*, Cambridge, 1996.

# RF POWER GENERATION AND COUPLING MEASUREMENTS FOR THE DIELECTRIC WAKEFIELD STEP-UP TRANSFORMER

M.E. Conde, W. Gai, R. Konecny, J. Power, P. Schoessow, P. Zou

*High Energy Physics Division, Argonne National Laboratory*
*9700 S Cass Ave, Argonne IL 60439*

**Abstract.** The dielectric wakefield transformer (DWT) is one route to practical high energy wakefield-based accelerators. Progress has been made in a number of areas relevant to the demonstration of this device. In this article we describe recent bench measurements and beam experiments using 7.8 and 15.6 GHz structures and discuss some remaining technical challenges in the development of the DWT.

## INTRODUCTION

Dielectric loaded structures driven by the wakefields of a high current electron beam have been under study for some time as high frequency high gradient accelerators [1]. The simplicity of this method, as well as the relative ease with which parasitic higher order modes can be damped [2] compared to conventional structures operating at a comparable frequency makes this technology an attractive option for future high energy $e^+e^-$ linear colliders.

A simple collinear drive-witness beam geometry suffers from inefficiencies due to the single bunch beam breakup instability of the drive beam unless unrealistic injection tolerances are imposed. Collinear devices are also limited to transformer ratios < 2, and beam staging is difficult. To circumvent these problems, a transformer geometry is used [3], where the drive and witness beams pass through separate structures (figure 1). The rf pulse generated by the drive beam is transferred via waveguide to a second structure which is adjusted to have the same fundamental frequency but smaller group velocity and transverse dimensions, thus providing an accelerating field step up by compressing the rf pulse. A drive structure can be designed with sufficiently low transverse impedance [4] to avoid beam breakup problems. Multiple drive bunches are used, spaced by an integral multiple of the rf period, to provide a long accelerating pulse.

CP472, *Advanced Accelerator Concepts: Eighth Workshop,*
edited by W. Lawson, C. Bellamy, and D. Brosius
1999 The American Institute of Physics 1-56396-889-4

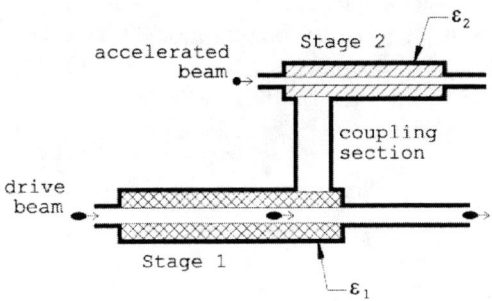

**FIGURE 1.** Schematic diagram of the dielectric wakefield transformer.

Initially it was useful to operate with a collinear drive and witness beam geometry in order to probe directly the fields generated in the drive structure. Measurement of the accelerating gradient also permitted indirect measurement of the rf power generated. Details of these measurements can be found in reference [6]. This paper is focussed on recent work involving coupling optimization between the drive and accelerating tubes and with direct measurements of the rf power generated using the intense electron source available at the Argonne Wakefield Accelerator [7].

## STRUCTURE DESIGN

The structures used for these experiments were designed to demonstrate the physics of the DWT while at the same time being compatible with the beam parameters currently available at the AWA. The device parameters are summarized in Table 1.

**TABLE 1.** Parameters for the dielectric structures considered here. a (b) is the inner (outer) radius, L is the structure length, $c\beta_g$ is the group velocity, and $E_z^{max}$ is the maximum wakefield accelerating gradient.

| f (GHz) | Stage | a (mm) | b (mm) | L (mm) | $\epsilon$ | $\beta_g$ | $E_z^{max}$ (MV/m) |
|---|---|---|---|---|---|---|---|
| 7.8 | I | 6 | 11.15 | 110 | 4.6 | .24 | 3.2 |
| | II | 3 | 5.41 | 160 | 20 | .05 | 8 |
| 15.6 | I | 5 | 7.22 | 140 | 4.6 | .31 | 8 |
| | II | 1.5 | 2.7 | 140 | 20 | .05 | 28 |

The dielectrics used are Cordierite which has $\epsilon = 4.6$ and MCT20, with $\epsilon =$

20. Both of these materials are low-loss ceramics which can be easily machined to the dimensions required [9]. Quality factors > 5000 have been measured for these structures. With optimized coupling, the 7.8 GHz structure has a transformer ratio of 2.5, while that of the 15.6 GHz device is 3.6.

# BENCH MEASUREMENTS AND RF COUPLING OPTIMIZATION

In the original concept for the DWT [3], the coupling between structures was accomplished by smoothly deforming the dielectric tubes and coupling the rf through a short unloaded section of waveguide which acted as a quarter wave transformer. This scheme involves no mode conversions and was found to be very efficient based on 2D numerical simulations [8]. The difficulty of deforming the ceramic tubes used as dielectrics led to an alternative method of rf coupling, using a section of rectangular waveguide as shown in figure 1.

The coupling of rf from the drive tube into the accelerating tube involves obtaining efficient broadband ($\Delta f / f \simeq 5 - 10\%$) power transfer with two mode conversions, from cylindrical $TM_{01}$ to rectangular $TE_{10}$ and back again. The problem of coupling optimization in the DWT was found to involve considerably more effort than a few hours work by a "halfway decent electrical engineer" as anticipated [5]. A lengthy trial and error procedure of coupling slot adjustments and network analyzer measurements was necessary to obtain reasonable coupling between the structures.

Figure 2 shows the bench measurement setup. A tapered launcher in combination with a tapered dielectric is used to achieve good coaxial to cylindrical waveguide coupling. A waveguide to coax adapter is located at the end of the rectangular waveguide. $S_{21}$ is measured from the tapered launcher to the coaxial coupler output.

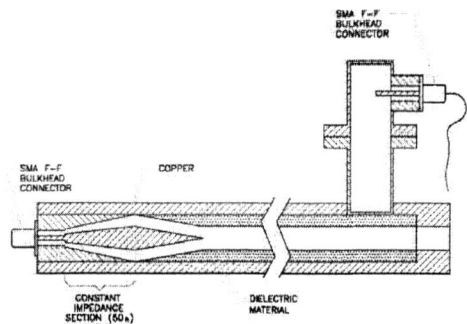

**FIGURE 2.** Bench measurement setup for dielectric structure to waveguide coupling measurements.

The coupling slot is located about $\lambda_{wake}/4$ from the end of the dielectric waveguide and is longer in the azimuthal direction than in the axial. The slot allows the azimuthal magnetic field from the dielectric structure to "leak out" into the rectangular transfer waveguide where it induces a transverse electric field. The coupling slot was gradually widened and lengthened to maximize $S_{21}$ and thus optimize the coupling into the transfer waveguide.

Figures 3-4 show the best results for the two stages of the 7.8 GHz structure. The best coupling obtained (from dielectric tube to wave guide) corresponds to $S_{21} \simeq 1.1$ db (stage I) and $S_{21} \simeq 1.8$ dB (stage II). Results for the 15.6 GHz device are similar. Work is currently underway to measure the coupling from stage I to stage II for the full transformer assembly.

# BEAM MEASUREMENTS

The stage I dielectric tube and waveguide assembly with the coupling optimized is installed in the test section of the AWA (fig.5). The rf from the wakefield of the beam in the stage I structure is coupled out through the waveguide to coax adapter to a calibrated rf diode detector via a -60 dB bidirectional coupler. The diode is placed in a lead shielding enclosure to avoid radiation damage to the diode. The diode signal is sent to the control room where it is read out using a digital oscilloscope.

Figure 6 shows a plot of rf power vs beam charge for single drive bunches in the 7.8 GHz structure. Although there is some scatter in the data due to shot to shot bunch length variations, a clear quadratic dependence of rf power on charge is observed, indicating that coherent generation of rf is occurring in the structure.

The multiple drive bunches essential for generating the long rf pulse necessary for the eventual operation of the DWT were generated by optically splitting and appropriately delaying the laser pulse to the drive gun. The present experiment used a bunch train of 4 bunches of 10 nC each (limited by available laser power), spaced by 3.07 ns ($= 4T_{linac} = 24\ T_{wake}$).

Figure 7 shows the envelope of the rf macropulse generated by the bunch train. Since the length $L$ of the dielectric structure is 11 cm, the rf micropulse length for a single drive bunch is $(L/c)((1 - \beta_g)/\beta_g) \simeq 1$ ns which is smaller than the bunch spacing, so that the individual micropulses are visible in the envelope. The difference in amplitudes of the peaks is due to slight misalignments in the beam splitting optics, resulting in different intensities for the micropulses.

Figure 8 shows the power output from the 15.6 GHz structure. Although the power output is low due both to poor beam transmission through the small aperture of the tube and longer than optimal drive bunch lengths, the observed rf coupling is good. Improved results are expected with planned improvements in the linac performance.

629

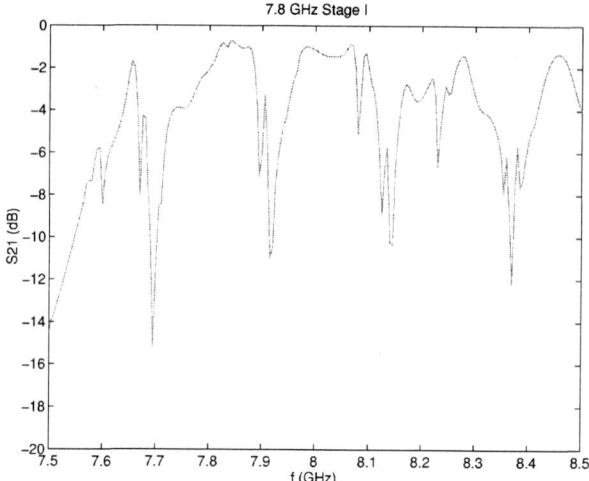

**FIGURE 3.** Measured $S_{21}$ for stage I of the 7.8 GHz DWT structure.

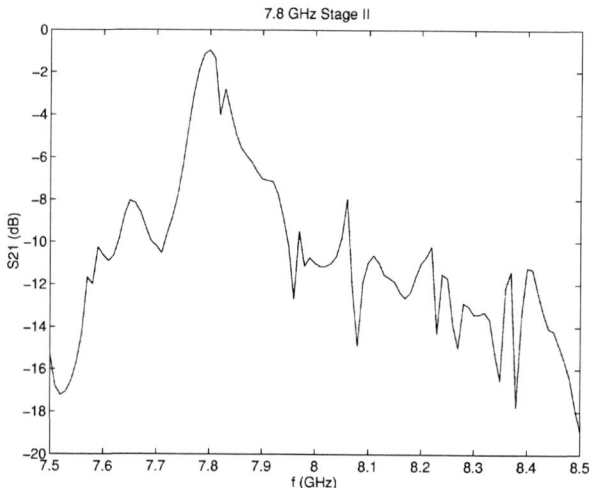

**FIGURE 4.** $S_{21}$ for stage II of the 7.8 GHz DWT structure.

**FIGURE 5.** Dielectric structure and transfer waveguide in place for beam measurements of power output.

# FUTURE EXPERIMENTS

The next task to be accomplished is the integration of the two stages and bench measurements of the coupling. A tapered pickup (analogous to launcher described above) for the stage II structure will be needed, but it is expected that little further coupling slot adjustment will be required.

Direct two beam acceleration in stage II will require some modifications to the AWA. The Y-chamber where the drive and witness beams are combined for collinear device wakefield measurements will be removed, and stage I will be mounted directly to the end of the drive linac. Stage II will be mounted to the end of the witness beam line and the transfer waveguide will be made sufficiently long to connect the two stages. The spectrometer magnet will be moved to a position immediately downstream of the stage II device. This arrangement avoids the drive and witness beam losses in the present chicane section, and also provides for easier tuning since the drive and witness beamlines are completely independent. Using a train of 8x20 nC drive bunches separated by $2T_{linac}$ (well within the capabilities of the existing drive linac), accelerating gradients of 10 (30) MeV/m can be obtained using the 7.8 (15.6) GHz structure parameters in Table 1. While the achievable gradients are modest, the experiment will demonstrate all the essential features of the DWT.

In the long term, improved beam quality would permit much higher gradients to be obtained. A design for a new photoinjector for the AWA drive linac has been developed [10] and with the improved emittance and bunch length available gradients > 100 MeV/m are attainable.

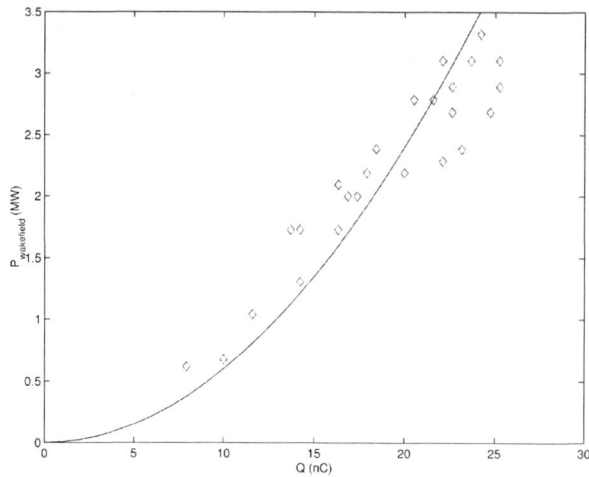

**FIGURE 6.** Rf power vs drive bunch charge, 7.8 GHz dielectric structure. The scatter in the data is due to bunch length fluctuations.

## SUMMARY

Considerable progress has been made towards a demonstration of the dielectric wakefield transformer. Some of the major steps on the way– optimizing the rf coupling from dielectric to transfer structure, multiple drive bunch generation in the AWA linac, and direct measurement of the rf generated by the beam– have been successfully achieved. We have some confidence that the remaining tasks can be accomplished without any major difficulties.

## ACKNOWLEDGEMENT

This work was supported by the US Department of Energy, Division of High Energy Physics, under contract W-31-109-ENG-38.

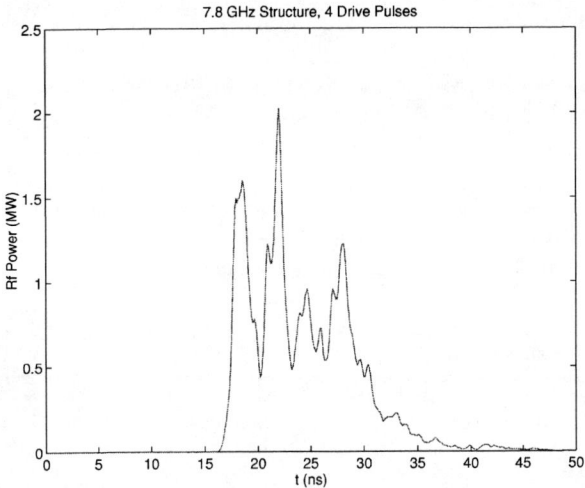

**FIGURE 7.** Rf macropulse envelope for train of 4 drive bunches, 7.8 GHz structure.

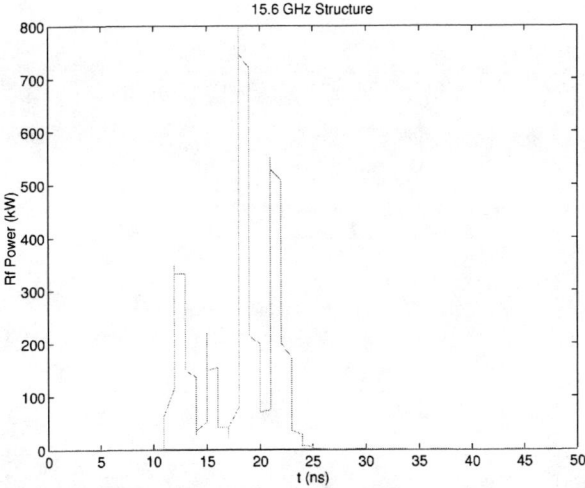

**FIGURE 8.** Measured rf macropulse envelope for a train of 4 drive bunches in the 15.6 GHz structure. The nonuniformities in peak power from bunch to bunch are due to imperfect alignment of the beam splitting optics.

# REFERENCES

1. W. Gai, P. Schoessow, B. Cole, R. Konecny, J. Norem, J. Rosenzweig, J. Simpson, Phys. Rev. Lett. **61** 2756 (1988)
2. E. Chojnacki, W. Gai, C. Ho, R. Konecny, S. Mtingwa, J. Norem, M. Rosing, P. Schoessow, J. Appl. Phys. **69** 6257 (1991)
3. E. Chojnacki, W. Gai, J. Simpson, P. Schoessow, Proc. 1991 Particle Accelerator Conf. p. 2557-2559
4. W. Gai, A. D. Kanareykin, A. L. Kustov, J. Simpson, Phys. Rev. **E 55** 3481 (1997)
5. J. Simpson, private communication
6. P. Schoessow, M. E. Conde, W. Gai, R. Konecny, J. Power, J. Simpson, to appear in J. Appl. Phys.
7. M. Conde et al., submitted to Phys. Rev. Special Topics– Accelerators and Beams
8. P. Schoessow, WF-169 (unpublished)
9. "Microwave Magnetic and Dielectric Materials Catalog", Trans-Tech Corp. n.d.
10. W. Gai, X. Li, M. E. Conde, J. Power, P. Schoessow, to appear in Nucl. Inst. Meth.

# Status of the Microwave Inverse FEL Experiment

R. B. Yoder,[a] T. C. Marshall,[b] Mei Wang,[a] and J. L. Hirshfield[a,c]

[a]*Physics Dept., Yale University, PO Box 208120, New Haven, CT 06520-8120*
[b]*Dept. of Applied Physics, Columbia University, New York, NY 10027*
[c]*Omega-P, Inc., 202008 Yale Station, New Haven, CT 06520-2008*

**Abstract.** A status report is presented on an inverse free-electron-laser accelerator experiment operating in the microwave regime (1). This proof-of-principle electron accelerator is powered by up to 15 MW of RF power at 2.86 GHz, which propagates in a smooth-walled circular waveguide surrounded by a pulsed bifilar helical undulator; solenoids provide an axial guiding magnetic field. Undulator pitch, which is initially 11.75 cm, is up-tapered to 13.5 cm over the 1-meter length of the structure to maintain acceleration gradient. Numerical computations predict an energy gain of 0.7 MeV using a 6 MeV injected beam from a 2-1/2 cell RF gun, with small energy spread and strong phase trapping. The maximum attainable acceleration gradient with such a design, using 150 MW of RF power at 34 GHz, is estimated to be at least 30 MV/m. Results from bench tests of the structure and undulator are presented, along with preliminary beam measurements.

## INTRODUCTION

Work on a prototype accelerator using the inverse FEL interaction at microwave wavelengths has been in progress for several years at Yale University and Omega-P, Inc. The device, known as the Microwave Inverse FEL Accelerator or MIFELA, is intended to demonstrate the IFEL principle with high efficiency and strong phase stability (2). At the 1996 Workshop, a proposed design was outlined for MIFELA (accompanied by simulation studies) which utilized output from the Yale gyroharmonic converter experiment at 11.424 GHz as an RF power source (1). However, since that time, the MIFELA experiment has been extensively redesigned to use instead RF power at 2.856 GHz which was already available in the laboratory; it was recognized that, while a frequency change will of course affect the output of the device, a lower-frequency MIFELA makes an equally effective proof-of-principle experiment. Because of these changes, details of the new design and simulation results will be described here, as well as the completed experimental layout, bench testing of components, and characterization of the electron beam to be injected.

In the device now nearing completion at the Yale Beam Physics Laboratory, the acceleration structure is a cylindrical waveguide into which RF power at 2.856 GHz is fed, creating a circularly polarized traveling wave. The undulator field is provided by a bifilar helical winding which is pulsed at high current and which is tapered in pitch for maximum acceleration gradient. This gives an undulator parameter $a_w = eB_w/k_w mc$

CP472, *Advanced Accelerator Concepts: Eighth Workshop,*
edited by W. Lawson, C. Bellamy, and D. Brosius

which varies from 2.4 to 2.75 at an initial period of 11.75 cm. A second (tapered) magnetic field, along the accelerator axis, is used for orbital stability and guiding.

The particle source for the experiment is a 2-1/2 cell RF gun, which produces 6 MeV electrons at low emittance. Focusing and energy selection are carried out by a 19-element beamline before injection into the accelerator. Numerical studies indicate that with 15 MW of RF power, such a beam will be accelerated to 6.7 MeV over 95 cm, and that this result can be scaled to higher energies; details of the calculations will be discussed below.

At the present, all elements of the system except for the acceleration structure itself are in place at Yale and have been fully tested, including measurements on the electron beam. The structure has undergone bench testing and field measurements, as reported below.


# EXPERIMENTAL OUTLINE

The details of the accelerator are outlined here; differences from the previously reported design are significant enough that most of the parameters in Table 1 of Ref. (1) have been altered. A new set of simulation studies (reported in a subsequent section) has been carried out, with a revised set of parameter values, as shown here in Table 1.

Up to 25 MW of RF power is available at 2.856 GHz from a SLAC XK-5 klystron, and is divided between the RF gun and the accelerator, so that at least 15 MW is usable for the RF drive in the MIFELA. In order to keep power levels high and the wiggler windings to a reasonable size, the RF structure is operated very near cutoff—with inner radius 3.14 cm at this frequency, we have waveguide refractive index $n = \omega/ck = 0.2$. A circularly polarized traveling wave in the $TE_{11}$ mode is generated by two waveguide inputs at 90° to each other, between which power is divided equally and phase matched. Unused RF power is absorbed into a ceramic matched load at the downstream end of the structure. Beam loading is not expected to be noticeable.

The magnetic fields in the MIFELA structure are shown in Figure 1. The field profiles make clear the three stages of beam movement through the MIFELA: beam injection (initial uptaper), acceleration (central region), and extraction (final downtaper). Each of these regions is discussed in more detail below.

The initial period of the wiggler is 11.75 cm, on the order of an RF wavelength. Since this gives rise to relatively large-diameter beam orbits, the beam travels first through an injection region in which the wiggler and guiding fields are gradually ramped up from zero to their acceleration values, so that the beam can be 'spun up' to its final gyration radius while ensuring that the orbits remain centered on the axis (3). This injection region occupies the first 5 wiggler periods, or 58.75 cm, and little energy change is seen there in computations. It was found in simulations that maximal orbit stability is obtained if the up-taper in the wiggler field follows a sine-squared shape— i.e., if the field profile obeys the relation $B(z) = B_0 \sin^2(2kz/\pi)$, $0<z<\pi/2k$, where $k$ is the wiggler wavenumber—rather than a linear taper. The axial guiding field is ramped linearly to its initial acceleration value.

**TABLE 1.** Simulation parameters for MIFELA.

**Entry region** (0 < z < 58.75 cm)

| | |
|---|---|
| Electron beam energy | $\gamma = 12.8$ |
| Electron beam radius | $r_b = 0.7$ mm |
| Wiggler field | $B_w = 0\text{--}2.2$ kG, sine-squared ramp |
| Axial magnetic field | $B_z = 0\text{--}1.7$ kG, linear ramp |
| Wiggler period | $\lambda_w = 11.75$ cm, constant |
| Wiggler radius | $r_w = 3.84$ cm |
| Peak electron beam current | $I_b = \leq 0.1$ A |
| Macropulse length | $\tau_p = 2$ µs |

**Acceleration region** (58.75 cm < z < 151.5 cm)

| | |
|---|---|
| Waveguide radius | $R = 3.14$ cm |
| Free-space RF wavelength | $\lambda_s = 10.5$ cm |
| Waveguide index | $n = 0.2$ |
| Input RF power level | $P_{in} = 15$ MW |
| Normalized RF field strength | $a_s = 0.14$, circularly polarized |
| Wiggler period | $\lambda_w = 11.75\text{--}12.32$ cm, linear ramp |
| Wiggler coil radius | $r_w = 3.84$ cm |
| Wiggler current | $I_w \leq 60$ kA |
| Wiggler field strength | $B_w = 2.2\text{--}2.3$ kG |
| Axial magnetic field | $B_z = 1.7\text{--}1.61$ kG |

**Exit region** (151.5 cm < z < 200 cm)

| | |
|---|---|
| Wiggler field strength | $B_w = 2.3\text{--}0$ kG, linear taper |
| Axial magnetic field | $B_z = 1.61\text{--}0$ kG, linear taper |

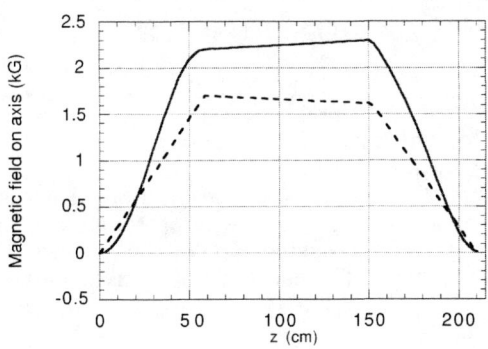

**FIGURE 1.** Variation of guide field (dashed line) and wiggler field (solid line) on the axis of MIFELA. The wigg

The beam itself, as provided by the RF gun, consists of 5 ps microbunches containing up to $10^9$ particles each, injected into each RF cycle during a macropulse of 2 μs duration. The gun design is optimized for a beam energy of 6 MeV (4). The bunch size is reduced during transport, since the beamline is designed to be achromatic and includes an energy selection slit which removes the low-energy tail from the beam, giving an energy spread of 1% or less at the exit of the beamline; in practice it is not anticipated to use more than about $10^7$ particles per bunch. The 5 ps bunch size is equivalent to a phase spread of about 0.1 radians, or $\pi/30$.

The acceleration region, which begins after the beam has been spun up, consists of linearly tapered wiggler and axial fields. The taper is due to the nature of the free-electron-laser interaction, which is approximately modeled by the following differential equation for the change in electron energy factor $\gamma$ as a function of axial distance $z$ along the structure:

$$\frac{d\gamma}{dz} = -\frac{\pi a_s a_w}{\lambda_w \gamma} \sin \varphi \tag{1}$$

where $a_s$ is the "pump parameter," the normalized vector potential of the RF driving field, $a_w$ is the normalized vector potential of the wiggler magnetic field, $\lambda_w$ is the wiggler period, $\gamma = (1-\beta^2)^{-1/2}$ is the relativistic energy factor, and $\varphi$ is the particle phase relative to the ponderomotive potential. The beat wave phase is controlled by the wiggler pitch and field amplitude; positive values of $\varphi$ correspond to particle deceleration (i.e., FEL radiation), and negative values to IFEL acceleration. As the relation shows, the acceleration gradient will decrease with increasing energy if the other parameters are fixed; hence, the wiggler field is up-tapered over the length of the acceleration region to maintain optimal acceleration gradient. This is accomplished by tapering the winding period of the bifilar helix from 11.75 to 12.32 cm over the 95-cm length of the interaction region, which produces a field tapered from 2.2 to 2.3 kG. If more power or a higher frequency enabled a sharper acceleration gradient, the taper would naturally become more important.

## SIMULATION RESULTS

Numerical simulations of acceleration in the MIFELA have been computed using a fully nonlinear, three-dimensional FEL simulation code written by Freund and Ganguly and known as ARACHNE (5), which has been benchmarked extensively against FEL experiments (6). The field and waveguide parameters have also been refined using this code. ARACHNE represents a slow-time-scale description of a steady-state FEL amplifier or IFEL accelerator with a single propagating frequency; the propagating electromagnetic field is assumed to be a superposition of vacuum modes, and the exact three-dimensional formula for the field of a helical wiggler is used, as derived by Park et al. (7).

The injected beam for these model calculations is assumed to be evenly spread in azimuth and radius, with an outer radius of 0.7 mm, and centered on axis with no energy spread. Equation 1 shows that the choice of a correct "phase window" for injection is

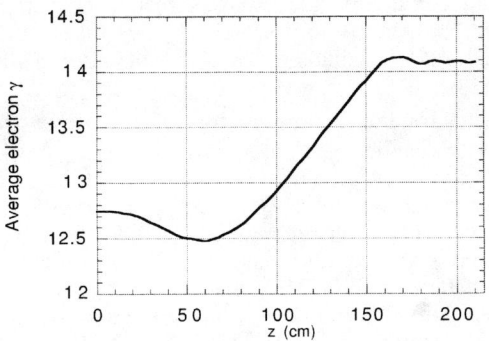

**FIGURE 2.** Average electron energy factor $\gamma$ as a function of axial distance in the accelerator. The "acceleration regime" is from 56 cm to 151 cm.

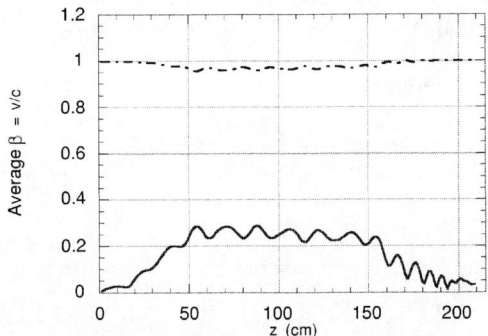

**FIGURE 3.** Average normalized axial (top, dashed) and transverse (bottom, solid) velocity components during acceleration in MIFELA. The injection and extraction regions are clearly visible.

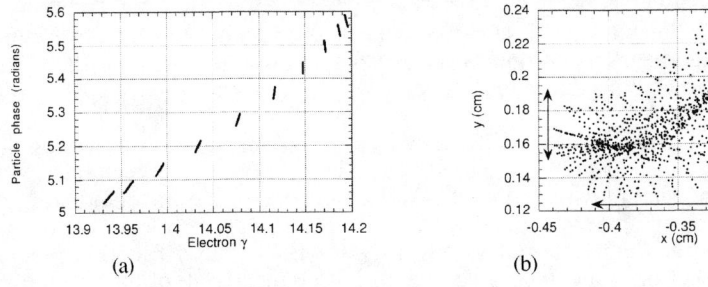

**FIGURE 4.** (a) Longitudinal phase space at the exit of MIFELA. Each "bunch" contains 100 particles. (b) Beam spot at the exit of MIFELA. The arrows on the axes include 80% of particles in each case. Note the different $x$ and $y$ scales.

vital for successful acceleration (2); however, with the long RF wavelength in use in this experiment and the consequential very small phase spread of the particle bunch, the phase window being used is almost negligibly small. For purposes of this simulation, a conservative phase spread of $\pi/10$ radians was used for injection, with good results.

The average value of $\gamma$ for the beam over the length of the structure is shown in Figure 2, where it is easy to see the injection and extraction regions on either end of the structure, during which little energy change takes place. During the injection region, energy is transferred from axial to azimuthal components of velocity; in the extraction region, the opposite occurs, as shown in Figure 3, which displays average axial and transverse velocity values as a function of distance. The gradient in the acceleration region is essentially constant at 0.81 MeV/m, which reflects the frequency of the experiment and the amount of available power. For comparison, the computed gradient for the previous MIFELA design at 11.424 GHz was 3.3 MeV/m.

In the extraction region, the particles pass out of resonance quickly and are essentially unchanged in energy. After extraction, the transverse velocity components of the beam have nearly vanished ($\beta_\perp = 0.03$) and the beam may be steered and focused into a spectrometer for energy analysis. The longitudinal phase space at exit, shown in Figure 4(a), shows that little energy spread has been introduced during acceleration, and the particles remain strongly phase bunched; in fact, it is clear that particles injected near each other in phase remain so throughout. The spread in $\gamma$ is 0.3, or 2%, for all particles in the simulation. The beam spot at the exit of MIFELA shows that the beam size is hardly increased by the acceleration process [Figure 4(b)]—it is extended to 1 mm in one direction and compressed to 0.5 mm in the other. For this case, with an initially cold beam, the computed average rms transverse emittance at exit is $0.43\pi$ mm mrad.

## EXPERIMENTAL CONSTRUCTION AND INITIAL TESTING

Construction of the prototype is nearly complete at present, with no major engineering tasks remaining. A schematic drawing of the structure, consisting of the acceleration waveguide and couplers for RF input and output, is shown in Figure 5. Initial testing of the accelerator systems has included diagnostics on the beam, performance tests of the beamline, and wiggler field measurements. The Yale 2-1/2 cell RF gun has been operated at beam energies ranging from 3 to 7 MeV and has produced bunches of up to $10^9$ electrons. Use of the beamline achromat to remove the initial energy spread on the beam has produced beam spot sizes on the order of the 0.7 mm radius used in the simulations discussed above.

The 15-period wiggler is powered by a capacitor bank capable of delivering up to about 50 kA with a pulse length of about 40 µsec. The windings of the wiggler, which are heavy-gauge copper, are wound directly onto the outside of the waveguide, which is also the vacuum vessel, and are separated from the waveguide by a layer of insulator. The waveguide walls were required to be fairly thick (~ 0.25") for high-precision machining of the interior; if a copper pipe had been used, nearly all of the wiggler field would have been prevented from entering the structure, since there is a finite diffusion time for the magnetic fields that is on the order of the pulse length. Because of this,

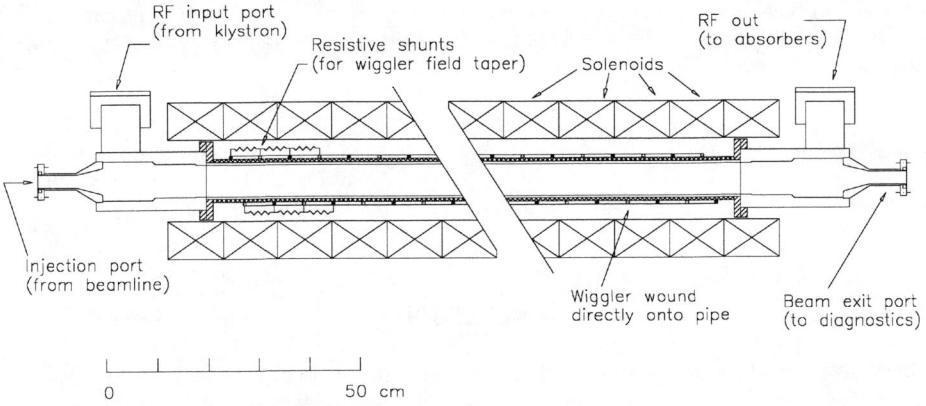

**FIGURE 5.** Schematic diagram of the MIFELA, showing input and output couplers for RF on either end of the acceleration waveguide itself. The wiggler helix is wound directly onto the beam pipe, and the whole is surrounded by the coils used to produce the axial field. The resistive shunts which are used to create the initial field up-taper are shown symbolically.

**FIGURE 6.** The completed acceleration tube, with wiggler wound on the outside. The winding pitch increases towards the near end.

stainless steel was used for the accelerator tube to reduce the conductivity, and both axial and azimuthal grooves were made in the outside surface of the tube to decrease the shielding effects of eddy currents. Preliminary tests indicate that wiggler fields of the strength required for acceleration can probably be achieved. An insulating spacer of varying thickness is used to control the winding period accurately as it is tapered along the structure. A photo of the acceleration structure, with wiggler windings present, is shown in Figure 6.

The axial guiding field is produced by a series of 18 coils, each independently computer-controlled so that any desired field profile may be created automatically. Such a system has been in place in the Beam Physics Laboratory's gyroharmonic conversion experiments and gives a high degree of flexibility in field configurations.

# SCALING

The low acceleration gradient of the present structure raises the question of the scalability of the MIFELA concept. In this case, the main limitations on the gradient which can be reached are the RF drive frequency and power level. Eqn. (1) can be written in the form $d\gamma^2/dz = k_w a_s a_w \sin\varphi$—i.e., for fixed phase, the rate of increase of energy squared is proportional to the RF field strength and wiggler field strength as well as the inverse of the wiggler period. This is of course assuming constant wiggler period as well, which is not the case in the present experiment; numerical computations would be necessary to scale the MIFELA accurately. Here, for convenience, we take the more simplistic assumption as a useful starting point. The squared rate of increase is proportional to the RF field and thus scales with the square root of RF power; if one assumes that the waveguide and wiggler dimensions scale inversely with the frequency, the overall increase in $d\gamma^2/dz$ is proportional to frequency as well. Hence, if an IFEL were to be built at 34 GHz, with a 150 MW power source, we would immediately see a factor of 36 gain in the squared rate of increase. The ultimate gradient depends on the length of the structure, since the increase is nonlinear; however, the initial gradient may be taken as an indication of the maximum *possible* gradient if tapered fields are used. With an initial beam energy of 6 MeV, as at Yale, the factor of 36 in the differential equation above leads to a factor of about 40 in the initial (and hence maximum) gradient attainable.

Such a device, which could conceivably thus have a gradient in the 30–40 MeV/m range, would be reasonably straightforward to construct, with a waveguide radius of perhaps 4 to 5 mm, and would be free from the fabrication difficulties posed by RF cavities at high frequencies. While a synchrotron-radiation limit of a few GeV must always apply to the final energy attainable by any undulating-beam device, the bunching capabilities of the IFEL mechanism could be an added benefit for accelerator applications, for example as a prebuncher for some higher-energy machine (8).

# CONCLUSION

The Yale/Omega-P MIFELA has been redesigned to operate at a lower frequency; it thus will have a smaller gradient but is equally effective as a proof of principle.

Accelerator construction is approaching completion, and the device is intended to be operated in upcoming months. Simulations predict a high degree of phase stability and low energy spread, which if verified will confirm the potential of the IFEL acceleration process to be efficient and easily achieved. Scaling of the accelerator to a more interesting energy regime appears straightforward, and would retain the phase trapping and stability properties of the prototype described here; at the same time, such an accelerator would be a relatively simple and easily manufactured device.

## ACKNOWLEDGMENTS

The authors acknowledge the collaboration of M. A. LaPointe (Omega-P), S. Y. Park (Yale and Postech, Korea), and M. Shapiro (MIT). This work was supported by the US Department of Energy, Division of High Energy Physics.

## REFERENCES

1.  Yoder, R. B., Zhang, T. B., Marshall, T. C., and Hirshfield, J. L., in Proceedings of the 1996 Advanced Accelerator Concepts Workshop, Lake Tahoe, CA, AIP Conf. Proc. 398, p. 629.

2.  Zhang, T. B., and Marshall, T. C., *Phys. Rev. E* **50**, 1491–1495 (1994).

3.  Zhang, T. B., and Marshall, T. C., *Nucl. Instr. Meth. Phys. Res. A* **375**, 515–518 (1996).

4.  Borland, M., unpublished report (1995).

5.  Ganguly, A. K., and Freund, H. P., *Phys. Rev. A* **32**, 2275–2286 (1985); Freund, H. P., and Ganguly, A. K., *Phys. Rev. A* **34**, 1242–1246 (1986).

6.  Ganguly, A. K., and Freund, H. P., Phys. Fluids **31**, 387–393 (1988); Freund, H. P., and Ganguly, A. K., IEEE Trans. Plasma Sci. **PS-20**, 245–255 (1992) and references therein.

7.  Park, S. Y., Baird, J. M., Smith, R. A., and Hirshfield, J. L., *J. Appl. Phys.* **53**, 1320–1325 (1982).

8.  Liu, Y., et al., *Phys. Rev. Lett.* **80**, 4418–4421 (1998).

# Phase Focusing for Finite Emittance Beams

C. Wang,[1] B. Hafizi,[1,2] and J. L. Hirshfield[1,2]

[1]*Department of Physics, Yale University, POB 8120, New Haven, CT 06520-8120*
[2]*Omega-P, Inc. 202008 Yale Station, New Haven, CT 06520-2008*

**Abstract.** An accelerator is a prime example of a physical system in which a beam of particles absorbs energy continuously from externally imposed electromagnetic fields only by remaining in, or very close to, synchronism with the fields. Traveling-wave amplifiers are another example. When the injected beam has finite emittance, deviations from exact synchronism can arise, usually limiting the ultimate energy exchange with the fields, and the ultimate emittance of an accelerated beam. A general theory is provided in this paper for a means of limiting deviations from exact synchronism. This can be achieved by providing a small space-varying detuning from synchronism for a particle near the center of the distribution, thus allowing otherwise non-synchronous particles brief opportunities to enjoy synchronism, and thus to limit their excursions in phase. To illustrate, an example is given of two-stage cyclotron autoresonance acceleration of a finite emittance beam, with and without detuning in the first stage. Space-varying detuning is shown to provide phase focusing in the first stage, lower phase spread at the entrance of the second stage, and thus higher ultimate energy.

## INTRODUCTION

Accelerators and traveling-wave tube amplifiers are two prime examples of physical systems in which a beam of particles must remain at, or very close to, phase synchronism with imposed electromagnetic fields in order to maintain cumulative energy exchange with the fields. When a finite-emittance beam is injected, only particles with a designated value of energy and momentum are exactly in phase synchronism. Particles with different energy and momentum only interact cumulatively if strong enough trapping is present to limit phase excursions, thereby imposing emittance growth for the beam. Detrapped particles can suffer significant phase slippage from synchronism, and thus may not experience energy gain (or loss) with the fields as great as for the trapped, synchronous particles. In traveling-wave amplifiers, initial beam velocity spread can reduce small-signal gain, and thereby delay or prevent the generation of large enough field amplitudes to effect strong trapping. Another instance in which initial beam emittance hampers tight synchronism between a beam and a wave is in the cyclotron autoresonance accelerator (CARA) [1-4].

Recently, a multi-stage CARA concept [4] has been described showing that much higher ultimate energy can be reached using several stages than for a single-stage CARA. For a sufficiently cold beam, it has also been shown that GeV energies could conceivably be reached in a multi-stage CARA, resulting in a novel "light source" with a synchrotron radiation spectrum comparable with traditional 3rd generation light sources. But strong trapping is not present in CARA, so a means is needed to permit nearly synchronous motion of all particles, when the beam has finite emittance. Such a means is suggested in this paper.

CP472, *Advanced Accelerator Concepts: Eighth Workshop,*
edited by W. Lawson, C. Bellamy, and D. Brosius

The paper is organized as follows. A general theory of phase focusing by introduction of spatially-varying detuning is developed. This general theory is specialized to the case of gyroresonant interaction in CARA. Numerical examples of the effect upon second stage performance from such detuning in the first stage are given. Finally, some general conclusions and comments are added.

# THEORY

For free particles interacting with electromagnetic fields, the general theory of phase focusing can be formulated in terms of the single-particle Hamiltonian $H(\mathbf{r},\mathbf{p}) = \gamma(\mathbf{r},\mathbf{p})mc^2$, where $m$ is the rest mass, $\mathbf{r}$ is the coordinate, $\mathbf{p}$ is the kinetic momentum, $c$ is the speed of light in vacuum, and $\gamma(\mathbf{r},\mathbf{p}) = \left[1 + \left(\mathbf{p}\cdot\mathbf{p}/m^2c^2\right)\right]^{1/2}$ is the Lorentz energy factor. We impose an external field (a strong static guide magnetic field, for example) represented by the vector potential $\mathbf{A}_o(\mathbf{r},t)$, and a small amplitude rf field represented by $\delta\mathbf{A}$ in the Coulomb gauge, where $\delta\mathbf{B} = \nabla \times \delta\mathbf{A}$ and $c\delta\mathbf{E} = -(\partial\delta\mathbf{A}/\partial t)$. To lowest order in the rf amplitude, and to lowest order in the ratio of gyro-radius to scale length, the effective Hamiltonian is given by

$$\overline{H}(\mathbf{r},\mathbf{p}) = mc^2\overline{\gamma} = mc^2\left[1 + \left\langle(\mathbf{P} - q\mathbf{A}_o/c)\cdot(\mathbf{P} - q\mathbf{A}_o/c)\right\rangle - 2\left\langle(q\delta\mathbf{A}/c)\cdot(\mathbf{P} - q\mathbf{A}_o/c)\right\rangle\right]^{1/2},$$
(1)

where $q$ is the charge, $\mathbf{P} = \mathbf{p} - q(\mathbf{A}_o + \delta\mathbf{A})/c$ is the canonical momentum, and $\langle\cdots\rangle$ indicates averaging over the synchrotron period. The last term in Eq. 1 represents the interaction between the particle and the rf fields.

For a given Fourier component of the electromagnetic field, the interaction phase $\psi$ can be written [5]

$$\psi = \mathbf{k}\cdot\mathbf{r} - \omega t + n\theta + \psi_o,$$
(2)

where $\mathbf{k}$ is the wave vector, $\omega$ is the wave frequency, $n$ is any integer, and $\psi_o$ is a constant. The quantity $\theta$ represents the particle contribution to the interaction phase; in a gyroresonance interaction it is the local gyrofrequency, while in a free electron laser it is the oscillation frequency due to the wiggler. In a resonant interaction, where the interaction phase remains nearly constant, one expects to find a given integer $N$ such that $|\mathbf{k}\cdot d\mathbf{r}/dt - \omega + N\,d\theta/dt| \ll |d\theta/dt|$. When this condition is satisfied, the $N$-th term in a Fourier expansion of the interaction is the dominant contribution to the effective Hamiltonian. Naturally, this assumes that the coefficient of the $N$-th term (often involving an $N$-th order Bessel function) is non-negligible.

During interaction of a particle beam with rf, there is a distribution of phase $\psi_i$, where the index labels the $i$-th particle. As the interaction proceeds, each phase $\psi_i$ evolves with, in many cases, a tendency for the phases to collapse towards a common value; this is referred to phase bunching or phase focusing and is responsible for the coherence of the interaction. However, phase bunching cannot be complete for a finite

645

emittance beam, and a spread in phases is inevitable. Phase spread interferes with ideal resonance and weakens the interaction. The following describes a means to control the phase spread.

Let the time-dependent phase for a given particle be written

$$\psi(t) = \int_0^t dt' \left[ \mathbf{k} \cdot d\mathbf{r}/dt' - \omega + N \, d\theta/dt' \right], \tag{3}$$

where the particle index has been deleted. Since the particle contribution to phase $\theta$ is generally a function of velocity $\mathbf{v}(t') = d\mathbf{r}(t')/t'$, the spread in phase may be written as

$$\delta\psi(t) = \int_0^t dt' \, \delta v \cdot \left( k + N \nabla_v \, d\theta/dt' \right). \tag{4}$$

Integration of Eq. 4 over the velocity distribution function yields the average phase spread

$$\langle \delta\psi \rangle \approx \left| \left( k + N \nabla_v \, d\theta/dt \right) \cdot \Delta v \right| \frac{L_{\text{int}}}{\bar{v}}, \tag{5}$$

where $\Delta v$ is the rms velocity spread, $\bar{v}$ is the average velocity, and $L_{\text{int}}$ is the interaction length. The maximum interaction length is given by the limit $\langle \delta\psi \rangle < \pi/2$.

In order to reduce phase spread below the value given by Eq. 5, a means is proposed to allow $\delta\psi(t)$ to pass through zero for most of the particles during the interaction. This can be accomplished by spatially varying one of the external system parameters, for example the imposed magnetic field or the wave group velocity, beyond the variation required to keep only a selected group of particles in resonance all along the interaction. Thus, nearly all particles in the distribution get to enjoy resonance at least some of the time, rather than limiting this luxury to only a few select particles, with the rest falling out of synchronism. As will be seen, this strategy can produce a beam with a smaller phase spread than for the case of ideal resonance for only a selected group of particles. This notion will be illustrated for the case of a gyroresonance interaction, such as that encountered in a multi-stage CARA.

## PHASE FOCUSING IN GYRORESONANT INTERACTIONS

Detailed examination of phase focusing will be carried out for a gyroresonance interaction of the type involved in CARA, where a laminar pencil beam is injected on the axis of a cylindrical waveguide in which a wave is propagating in the rotating $TE_{11}$-mode. An axial magnetic field profile is imposed that maintains gyroresonance as the beam interacts with the rf fields, gaining energy and transverse momentum. Thus we treat a system with a gently-varying static field with a vector potential $\mathbf{A}_o = -B_o(z)y\hat{\mathbf{e}}_x$, where $B_o(z)$ is the magnetic field, and $(\hat{\mathbf{e}}_x, \hat{\mathbf{e}}_y, \hat{\mathbf{e}}_z)$ are unit vectors. The rf vector potential for rotating $TE_{11}$-mode fields in a cylindrical waveguide of radius $a$ is

$$\delta \mathbf{A} = -\frac{1}{2k_c^2}\hat{B}\hat{\mathbf{e}}_z \times \nabla_\perp \left\{ J_1(k_c r)\exp\left[i(k_z z - \omega t + \varphi + \psi_o)\right]\right\} + c.c. , \qquad (6)$$

where $(r, \varphi, z)$ are cylindrical coordinates, $J_1$ is the first-order Bessel function of the first kind, $k_c = p'_{11}/a$ is the transverse wavenumber with $p'_{11}$ the first zero of $dJ_1(p)/dp$, and where (in the absence of beam loading) $\hat{B}$ is the constant rf amplitude. We introduce a canonical transformation from the variables $\left(x, P_x; y, p_y; z, p_z\right)$ to the variables $\left(\theta, \mu; Y, m\Omega_o X; z, p_z\right)$ using the relations

$$\mu = \frac{\left(P_x + m\Omega_o y\right)^2 + p_y^2}{2m\Omega_o}; \qquad \tan\theta = \frac{P_x + m\Omega_o y}{p_y};$$

$$X = x + \frac{p_y}{m\Omega_o}; \qquad Y = -\frac{P_x}{m\Omega_o}; \qquad (7)$$

where $\Omega_o = qB_o/mc$ is the non-relativistic gyrofrequency, $(X, Y)$ are the guiding center coordinates, and $\mu$ is the magnetic moment. In the limit $(X, Y) \to 0$, one finds

$$\bar{\gamma} = \bar{H}(\mathbf{r}, \mathbf{p})/mc^2 = \left[1 + \left(\frac{p_z}{mc}\right)^2 + \frac{2\Omega_o\mu}{mc^2} + \frac{4b}{mc}\sqrt{2m\Omega_o\mu}\, J'_1(k_c\rho)\sin\psi\right]^{1/2} , \qquad (8)$$

where $b = q\hat{B}/2mc^2k_c$, $\rho = (2\mu/m\Omega_o)^{1/2}$ is the gyration radius, and $\psi = k_z z - \omega t + \theta + \psi_o$ is the phase variable relevant for $N = 1$, the fundamental gyroharmonic interaction in the first stage of CARA. Prior theory of CARA can be derived from Eq. 8. Specifically, from $d\theta/dt = mc^2 \partial\bar{\gamma}/\partial\mu \approx \Omega_o/\bar{\gamma}$, the phase spread is

$$\delta\psi = \int_0^t dt' \left\{ \delta v_z \left( k_z - \frac{\Omega_o}{\bar{\gamma}^2}\frac{\partial\bar{\gamma}}{\partial v_z} \right) - \delta v_\perp \left( \frac{\Omega_o}{\bar{\gamma}^2}\frac{\partial\bar{\gamma}}{\partial v_\perp} \right) \right\}$$

$$= \int_0^t dt' \left\{ \delta v_z \left( k_z - \frac{\Omega_o \bar{\gamma} v_z}{c^2} \right) - \delta v_\perp \left( \frac{\Omega_o \bar{\gamma} v_\perp}{c^2} \right) \right\}, \qquad (9)$$

using $\bar{\gamma}^{-2} = 1 - v_z^2/c^2 - v_\perp^2/c^2$.

In general, the integrals in Eq. 9 cannot be performed analytically since the temporal variation of the particle quantities is not known *a priori*. Two limits of interest are the low efficiency case, where particle quantities change very little, and the high efficiency case. In the low efficiency case, particle quantities inside the integrals may be taken equal to their initial values. One observes, if the beam is initially monoenergetic, that terms proportional to $\Omega_o$ cancel one another and there can be no phase focusing.

647

For an initially monoenergetic beam, phase focusing can be obtained in the high efficiency case, after the beam develops an energy spread.

For simplicity, we consider the limit as $\delta v_\perp \to 0$, but $\delta v_z$ remains finite. We impose a slow, low amplitude spatial modulation on the guide magnetic field such that $\Omega_o \to \Omega_o[1 + \varepsilon \sin(2\pi z/\lambda_m)]$, where $\varepsilon$ is the amplitude, and $\lambda_m$ is the period of modulation. When this expression is inserted into Eq. 9, the integral is not straightforward to evaluate, since all the quantities vary with time. However, two general observations can be made. First, in the limit where $\lambda_m$ is much smaller than the interaction length, the sinusoidal term in Eq. 9 nearly integrates to zero and the modulation will have little effect. Secondly, in the limit where $\lambda_m$ is larger than the interaction length, and if the integrand does not vary much (i.e., low efficiency), an estimate for the integral in Eq. 9 can be obtained by simply integrating over the sinusoidal term. This gives an approximate condition on $\varepsilon$ for vanishing phase spread, namely

$$\varepsilon = \frac{(\pi L_{int}/\lambda_m)}{\sin^2(\pi L_{int}/\lambda_m)}\left(\frac{c^2 k_z}{\Omega_o \bar{\gamma} v_z} - 1\right). \tag{10}$$

The minimum value of $\varepsilon$ obtains when $\lambda_m = 2.7 L_{int}$.

The reasoning behind this means of reducing phase spread, or of providing phase focusing, can be gleaned from the equation of motion for the phase, namely

$$d\psi/dt = k_z dz/dt - \omega + d\theta/dt = k_z(\partial \bar{H}/\partial p_z) - \omega + (\partial \bar{H}/\partial \mu)$$
$$= \frac{\Omega_o}{\bar{\gamma}} - \omega + \frac{k_z p_z}{m\bar{\gamma}} + \frac{2bc}{\bar{\gamma}}k_c\left[(k_c\rho)^{-2} - 1\right]J_1(k_c\rho)\sin\psi \tag{11}$$

At injection into CARA, where $\rho$ is small, the last term proportional to rf amplitude dominates and phase focusing towards $\psi = \pi$ proceeds rapidly. But as $\bar{\gamma}$ increases and as $k_c\rho \to 1$, the focusing term becomes weaker than the detuning term, given by

$$\Delta = \frac{\Omega_o}{\bar{\gamma}} - \omega + \frac{k_z p_z}{m\bar{\gamma}}. \tag{12}$$

Since $\Delta$ can only be identically zero for a small portion of the distribution, the other particles will suffer defocusing. But when the sinusoidal variation of $\Omega_o$ is imposed, all particles get a chance to have $\Delta = 0$ at some point along the interaction path, so the last term in Eq. 11 can still provide some phase focusing. The net result can be a limitation to the growth in phase spread. This point will be illustrated for a two-stage CARA in the next section.

## SIMULATION ANALYSIS OF PHASE FOCUSING IN CARA

It has been shown by simulations for an ideal electron beam that the energy limit in a cyclotron autoresonance accelerator (CARA) can be overcome by operating a series of CARA stages [4]. An ideal beam has no initial velocity spread, and all electrons are injected into CARA with only axial momentum. Negligible phase and velocity spreads are produced when the electrons enter the interaction region, despite their different resonance phases and different rf forces: phase trapping is highly effective. But in practice electron beams have finite emittance, with a distribution of initial (small) transverse velocities. The rf field in CARA tends to phase focus the electrons, while the transverse velocity tends to defocus them. Finite initial transverse velocity can also lead to significant increases in the spreads in axial velocity and energy, even when phase trapping dominates. When beam loading due to finite beam current causes the field strength to diminish, the rf focusing force becomes weaker. Consequently, phase spread could be a serious problem for a beam coming from the first CARA stage, before injection to a second stage, and subsequent acceleration could be seriously hampered.

As mentioned previously, phase spread can be reduced by means of a space-varying detuning from synchronism for a central particle. This deviation may lower slightly the final acceleration energy in the first stage CARA, but can significantly reduce phase spread and increase acceleration energy of the beam in the next stage CARA. In this section, we will show by simulations for a two-stage CARA how the principle works.

We consider the first two stages of a multi-stage CARA, driven by a 75 MW, 11.424 GHz rf power source. A 250-kV, 15-A electron beam is injected from an rf gun with a bunch length of $0.1\pi$ rad, corresponding to 4.38 psec, with an initial axial velocity spread of 0.012%. The accelerating structure is made up of a $TE_{11}$-mode CARA (stage-I), and a $TE_{31}$ mode CARA (stage-II), separated by a 30-cm drift region where an ideal mode converter is inserted to convert leftover rf power from stage-I into the stage-II $TE_{31}$ mode. The beam is monochromatic, and is injected on axis without guiding center spread. The initial time-phase spread is $0.1\pi$ and the momentum angle spread is $2\pi$. Of the 256 computational particles, 8 values are selected for time-phase spread, 4 for momentum angle spread, and 8 for velocity spread. The parameters in the simulation for the two-stage CARA are given in table 1.

Table 1. Parameters in simulation for a two-stage CARA.

| Injection gun voltage | 250 kV |
|---|---|
| Injector gun perveance | $0.12 \times 10^{-6}\,A - V^{-3/2}$ |
| Beam current | 15 A |
| rms initial axial velocity spread | 0.012% |
| rf drive power | 75 MW |
| rf frequency | 11.424 GHz |
| Stage-I operating mode | $TE_{11}$ |
| Stage-I harmonic index | 1 |
| Stage-II operating mode | $TE_{31}$ |
| Stage-II harmonic index | 3 |

The detuned magnetic field profile we employ is given by

$$B_o(z) = B_{os}(z) + c_d(m_o\omega/e)\sin(2\pi z/\lambda_m), \tag{13}$$

where $B_{os}(z)$ is the synchronous magnetic field for the central particle without initial transverse velocity, $m_o$ and $e$ are the electron's rest mass and charge in magnitude, and $c_d$ is the detuning coefficient. From Eq. 13, the axial magnetic field is set to deviate

649

from exact resonance by adjusting the detuning coefficient and period. The detuning is only applied in the first stage while an exact synchronous magnetic field is employed in the second stage. Thus, for the examples shown below, the second stage can be considered as a diagnostic on the effects of phase spread in the first stage.

In the first stage, the waveguide has three segments: the first segment has a radius of 0.96 cm and a length of 5.0 cm; the third segment has a radius of 2.04 cm and a length of 4.2 cm; the second segment is tapered between them with a length of 10 cm. This arrangement of different waveguide radii is to allow a larger acceleration energy within a shorter acceleration length than for a constant radius. The 0.96-cm radius waveguide can only support the $TE_{11}$ mode; it has a larger accelerating gradient but a lower upper energy limit, as compared with 2.04 cm radius. Introduction of the taper is to avoid beam stalling and hence to increase the acceleration energy. For the second stage, the waveguide has two segments. The first segment with a length of 50 cm is gently tapered in radius to 2.44 cm from 2.47 cm for the case of $c_d = 0$, and 2.36 cm for the case of $c_d = -0.02$ with $\lambda_m = 38$ cm; the second segment is uniform, with radius of 2.44 cm and a length of 20 cm. Changing the detuning coefficient and period changes parameters of the output beam, and as a result the waveguide radius at the beginning of second stage changes in order to meet the resonance condition. Phase matching is acquired by adjusting the magnetic field profile in the drift section without changing axial magnetic fields at the two ends of drift [4].

Figure 1 shows the dependence of rms spread in resonance phase $\psi$ on axial distance $z$ for case-1, where $c_d = 0$; and for case-2 where $c_d = -0.02$ and $\lambda_m = 38$ cm. Case-1 means that an exact synchronous magnetic field for the central particle is applied; while case-2 means that the magnetic field satisfies the synchronous conditions only at the beginning and the end of the first stage CARA, but it is detuned in the rest. Figure 2 shows the dependence of averaged relativistic energy factor $< \gamma >$ and magnetic field $B_0$ on axial distance $z$ for the two cases. It can be seen from Fig. 1 and Fig. 2 that at the end of stage-I, $\psi$-spread is 0.1716 radians and $< \gamma >$ is 4.22 (1.6454 MV) for case-1, while $\psi$-spread is 0.1259 radians and $< \gamma >$ is 4.1772 (1.6235 MV) for case-2. The phase spread is reduced by 27% at a loss of 22 kV acceleration energy by detuning. When the beam passes through the drift, the magnetic field is decreased and the axial velocity is increased to satisfy the resonance condition for the third harmonic at $TE_{31}$ mode. At the beginning of stage-II, the case-1 $\psi$-spread becomes 0.8589 radians, but the case-2 $\psi$-spread is only 0.2379 radians. Because of so large phase spread for case-1, the beam acceleration energy only can reach 2.0398 MV ($< \gamma > =4.9918$) at $z = 86$ cm in the stage-II. After $z = 86$ cm, the beam is decelerated due to phase slippage, and the acceleration energy diminishes to 1.6990 MV ($< \gamma > = 4.3248$) at the end of stage-II. In contrast, for case-2 the beam energy is accelerated to 3.0693 MV ($< \gamma > = 7.0064$) at the end of stage-II, only reduced by about 9% compared with the acceleration energy 3.3773 MV ($< \gamma > = 7.6091$) for an ideal two-stage CARA [4]. Obviously, the magnetic field detuning for a warm beam has greatly improved the resonance phase spread and the beam of electrons resides longer in synchronism with the rf accelerating field. We have carried out simulations for a range of values of $c_d$ and $\lambda_m$. The plots in Figs. 1 and 2 are typical of the results obtained. In this case $\lambda_m \approx 2L_{int}$, in rough accord with the value indicated following Eq. 10.

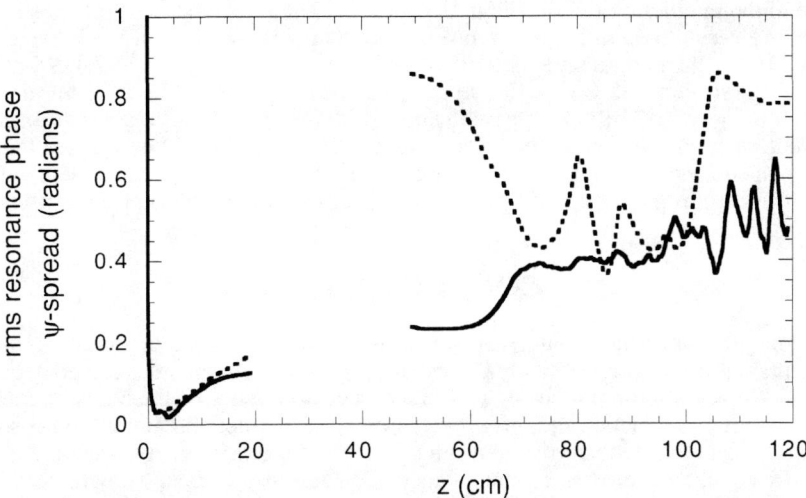

**Figure 1**. Dependence of rms resonance phase $\psi$-spread on axial distance $z$ for case-1 (dashed) where $c_d = 0$ and case-2 (solid) where $c_d = -0.02$ and $\lambda_m = 38$ cm. Stage-I: $z = 0 \sim 19.2$ cm; drift region: $z = 19.2 \sim 49.2$ cm; stage-II: $z = 49.2 \sim 119.2$ cm. In the drift region, the beam and the rf power are decoupled and the resonance phase spread is not defined.

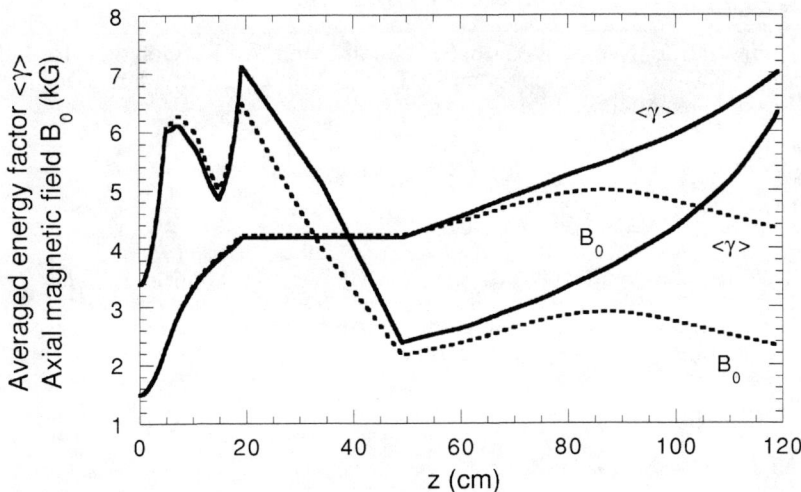

**Figure 2**. Dependence of averaged relativistic energy factor $< \gamma >$ and magnetic field $B_0$ on axial distance $z$. Dashed for case-1 without detuning; solid for case-2 with detuning.

It also can be seen from Fig. 2 that resonance phase spread seems a little sensitive to the magnetic field profile in the stage-I. The magnetic field profiles for the two cases are very close, with a difference of less than 200 G (out of 5-6 kG) except for near the end. At the end, the magnetic field is 6.4841 kG for case-1 and 7.1195 kG for case-2, a difference of 635 G or 10%. In stage-II, the magnetic field profile has a larger difference, namely 2.3214 kG for case-1 and 6.3250 kG for case-2 at the end because of different beam phase spread and acceleration energy. From this we can see that the detuning can considerably improve beam quality, but that sensitivity to precise magnetic field profile requirements could make the adjustment of magnetic field somewhat challenging.

## CONCLUSIONS

General theory has been given for a means to limit phase excursions when a finite-emittance beam interacts with a wave synchronous with only a selected group of particles in the beam. Phase focusing is achieved by varying one of the beam interaction parameters (guide magnetic field, group velocity, wiggler pitch) so that all the particles find themselves in exact synchronism during part of the interaction. To illustrate, phase orbits in a two-stage cyclotron autoresonance accelerator have been shown, both with and without a sinusoidal deviation of the guide magnetic field from exact resonance for a central group of particles. In this example, use of phase focusing was shown to allow acceleration in the second stage to 3.07 MV, rather than only 1.70 MV without the deviation in guide field. The 3.07 MV value compares favorably with the result for an ideal two-stage CARA with a cold beam, where the ultimate energy is 3.38 MV. The phase focusing principle described in this paper should find application in a wide range of synchronous interactions between particles and waves.

## ACKNOWLEDGMENTS

Appreciation is extended to S. Y. Park and A. K. Ganguly for helpful discussions. This work was supported by Department of Energy, Divisions of High Energy Physics and Advanced Energy Projects.

## REFERENCES

1. C. Chen, *Phys. Rev.* **A 46**, 6654-6661 (1992).
2. B. Hafizi, P. Sprangle, and J. L. Hirshfield, *Phys. Rev.* **E 50**, 3077-3086 (1994).
3. M. A. LaPointe, R. B. Yoder, C. Wang, A. K. Ganguly, and J. L. Hirshfield, *Phys. Rev. Lett.* **76**, 2718-2721 (1996); C. Wang and J. L. Hirshfield, *Phys. Rev.* **E 51**, 2456-2464 (1995).
4. C. Wang and J. L. Hirshfield, *Phys. Rev.* **E 57**, 7184-7191 (1998).
5. V. L. Bratman, N. S. Ginzburg, G. S. Nusinovich, M. I. Petelin, and P. S. Strelkov, *Int. J. Electronics* **52**, 541-567 (1981).

# A Field Symmetrized Dual Feed 2 MeV RF Gun for a 17 GHz Electron Linear Accelerator[*]

J. Haimson, B. Mecklenburg and G. Stowell

*Haimson Research Corporation*
*3350 Scott Blvd., Building 60, Santa Clara, CA  95054-3104*

**Abstract.** An undesirable feature that contributes to beam emittance growth, and prevents the full potential of laser driven, short wavelength RF guns from being realized, is the presence of field asymmetries introduced by the input power coupling aperture in the sidewall of the cavity. Means of avoiding these field asymmetries in a three cavity RF gun structure, by feeding RF power into the middle cavity through diametrically opposed coupling apertures in the sidewalls of the cavity and by using a racetrack (rather than a cylindrical) shaped coupling cavity, are described. An analysis of this $\pi$-mode structure indicated that, although E-field balancing of the three-cell circuit was very sensitive to small departures from the resonant condition, the proposed microtuner mechanisms on each end cavity could readily correct for these resonant errors. Furthermore, unlike the use of tuning probes in the outer wall of the cavities, the slight axial deformation produced by the end wall microtuners enables the symmetrized field distributions to be conserved.

Beam optics analyses indicate that a peak E-field of 200 MV/m and an energy gain of 2 MeV can be achieved with this 17 GHz RF gun using a peak input power of less than 3 MW. The design parameters of the three cavity structure are discussed; the results of beam optics studies and the racetrack coupler cavity field symmetrization data are presented; and details of the cavity microtuners and the high power rectangular waveguide network components, are described.

## INTRODUCTION

Theoretical investigations and the operation of laser driven RF guns over the past few years at laboratories around the world have confirmed that this type of photoinjector holds promise of satisfying the very low emittance, high current requirements of future linear collider and FEL applications. Early analytical studies [1], referencing the original work performed at the Los Alamos National Laboratory, investigated the beam emittance contributions due to the time varying RF and space charge fields. More recent studies

---

[*] Work supported by U.S. Department of Energy SBIR Grant No. DE-FG03-97ER82389.

CP472, *Advanced Accelerator Concepts: Eighth Workshop,*
edited by W. Lawson, C. Bellamy, and D. Brosius

have investigated means of improving the beam quality by attempting to reduce emittance growth attributed to RF and space charge effects [2], and have examined the beam emittance and brightness scaling dependence on the charge and wavelength of RF photoinjectors [3].

Examination of the beam brightness dependence on frequency scaling clearly favored the use of high frequency photoinjectors, and was a compelling argument for MIT to explore the 17 GHz domain as compared to the 3 GHz systems being investigated in other laboratories. The MIT 17 GHz photoinjector was based on a scaled down version of the Brookhaven National Laboratory 1.5-cell structure, and was driven by a relativistic TW klystron [4] and an RF phase synchronized picosecond UV laser system [5].

It should be noted that, until recently, the above photoinjector design studies were based on the assumption that the longitudinal electric field distribution was uniform in the transverse direction (i.e., axisymmetric), and no allowance had been made for the beam emittance growth caused by the field asymmetry associated with the commonly adopted photoinjector practice of using a single coupling aperture in the sidewall of the accelerating cavity. Early linac studies [6] of single feed 3 GHz coupling cavities had investigated the effects of these field asymmetries and had shown that the associated transverse gradient of the longitudinal electric field produced phase dependent transverse spreading [7] of the accelerated electron bunch.

More recently, in an effort to avoid beam emittance contributions caused by field asymmetry in coupling cavities, high gradient linac structures for linear collider applications have been fabricated with dual RF feed coupling cavities having diametrically opposed sidewall apertures. Influenced by this approach, in designing a new 3 GHz, 1.6-cell photoinjector, the BNL/SLAC/UCLA collaboration elected to compensate the asymmetry introduced by the high power RF coupling slot in the full cell by adding an identical slot (pumping port) on the opposite side of the cavity [8].

Numerical simulations of dual feed, cylindrical coupling cavities [9] using the 3D MAFIA code have shown that, although dual opposed coupling apertures can prevent bipolar field asymmetry, the technique introduces a significant quadrupolar asymmetry that produces an azimuthal variation of the longitudinal electric field. These quadrupolar transverse gradients of the longitudinal electric field produce bunch phase dependent, emittance degrading RF focusing and defocusing forces, and also cause electric field enhancement on the surfaces of the cavity end walls in the radial plane aligned with the sidewall coupling apertures. Recent investigations have shown that both the electric field enhancement effect and the quadrupolar field asymmetries can be essentially eliminated by the use of a racetrack shaped dual feed coupler cavity [10]. For high frequency photoinjectors, emittance growth contributions due to transverse gradients of the longitudinal electric field become a more critical issue when the laser beam diameter cannot be scaled down consistent with the operating wavelength of the RF system (e.g., 2.0 mm $\phi$ at 2856 MHz $\Rightarrow$ 0.33 mm $\phi$ at 17136 MHz).

The information presented herein describes a 17 GHz three cavity RF gun comprising a field symmetrized, dual feed, racetrack shaped coupling cavity positioned between cylindrical end cavities. The end cavities are fitted with microtuning mechanisms that enable the axial region of the end walls to be perturbed, while the axisymmetric field distributions in the structure are maintained. The structure is designed to operate with a peak input power of 2.5 MW to produce a 2 MeV electron beam for injection into an existing 17 GHz high gradient accelerator [11].

## MODE SEPARATION AND TUNING CHARACTERISTICS

Preliminary studies were performed on the three-cell $TM_{01}$ $\pi$-mode structure to establish the cavity geometry on either side of the coupling cavity, before embarking on the field symmetrization and optimization of the dual feed racetrack cavity. The URMEL and SUPERFISH codes were used to select a sufficiently large beam aperture and moderately thick disc to ensure a frequency separation of greater that 30 MHz between the zero, $\pi/2$ and $\pi$-mode resonances. Also, the SUPERFISH generated field distributions were used with a 2-1/2D PIC code to investigate the effect of adjusting the length of the first cell and to establish the lengths of the second and third cells so that a near-synchronous operating condition ($\beta_c\lambda_0/2$) would be obtained at a peak E-field of 200 MV/m.

Using a 4 mm diameter beam aperture, 2 mm thick discs, and successive cavity pitch lengths of 0.300 $\lambda_0$, 0.445 $\lambda_0$ and 0.474 $\lambda_0$, a near uniform $\pi$-mode peak E-field distribution was obtained in all three cells, as shown in Figure 1(a), by using radii of 0.2697″, 0.2722″ and 0.2713″ for the first, second and third cells respectively. The SUPERFISH computed E-field distributions and the values of $Q_0$ and power dissipation, for the corresponding $\pi/2$ and zero mode resonances are shown in Figures 1(b) and (c), respectively. This design gave a $\pi$-mode resonant frequency of 17132.82 MHz with mode separations of 33 and 50 MHz and was selected as the CONTROL for comparing resonance and field distribution variations caused by small geometric changes in this axisymmetric, (unloaded) structure. The $\pi$-mode 2.5 MV/m peak field achieved with 359 W of dissipation is based on a SUPERFISH normalization factor of 1 MV/m over the 0.0381 m (1.50″) length of the chosen baseline, and corresponds to a 200 MV/m peak gradient at an input power of 2.5 MW for a well matched structure having a practical external Q value of 5500.

Tuning of the three-cell structure is to be accomplished by first fabricating the cells to a high degree of accuracy (the resonant frequencies of the 94 cavities in the recently brazed 17 GHz linac structure were maintained within 4 MHz of the cold test values), and then fine tuning the coupler cavity (middle cell) resonance to the system operating frequency by adjusting the regulated temperature of the water flowing through cooling ducts machined in the cavity walls. Then, the precision microtuning mechanisms in the end walls of the structure will be used to independently align the resonant frequency of the first and last cavities with the coupler cavity resonant frequency.

Contra-rotation tuning and locking mechanisms attached to the end walls of the structure were chosen as the most favorable technique for providing both an adequate range of tuning and a desired tuning increment of approximately 1 MHz, while conserving

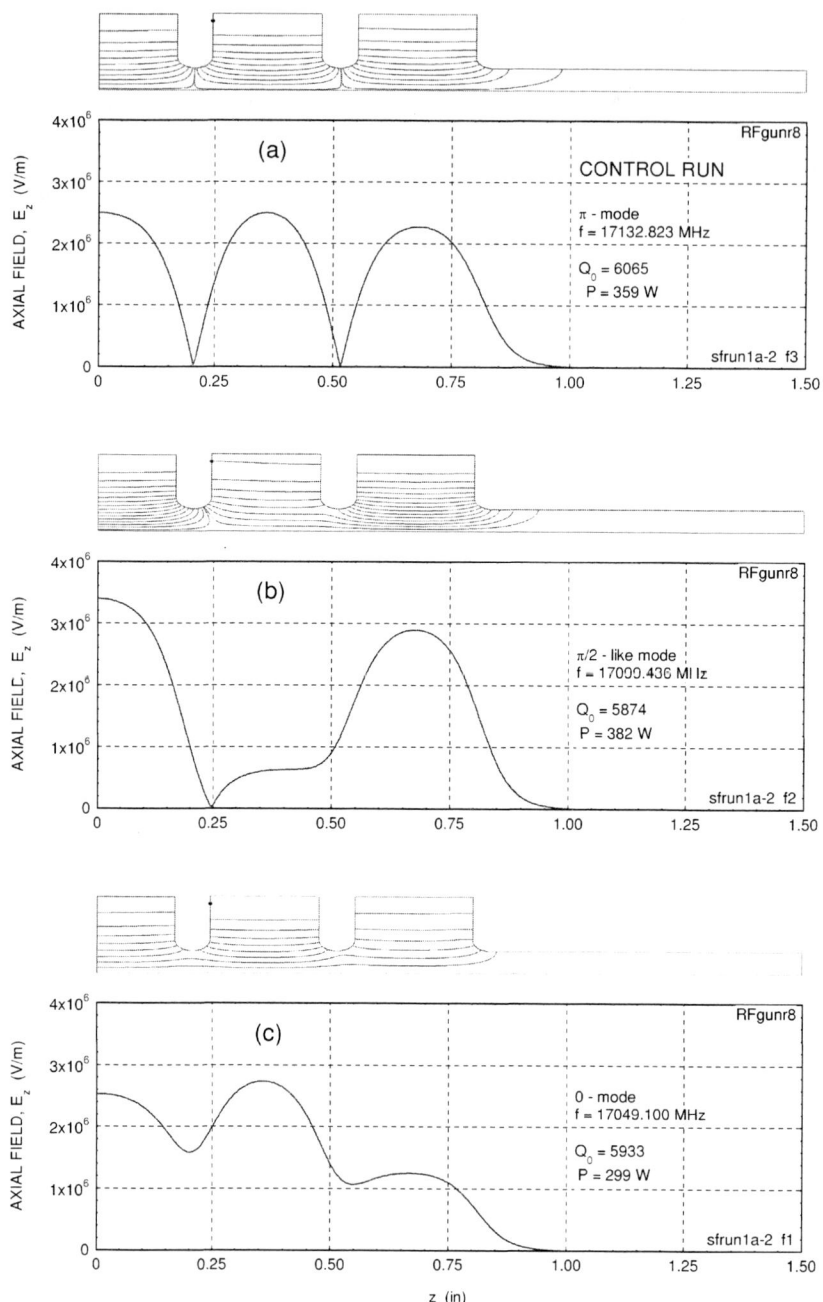

**FIGURE 1.** SUPERFISH E-field distributions for π, π/2-like and zero modes in the axisymmetric unloaded structure (i.e., in the absence of the sidewall coupling apertures).

the axisymmetric field distributions in the cylindrical end cavities. These tuning mechanisms incorporate precision, coaxial alignment guides and an annular shaped stainless steel plunger that is brazed to a flexible section of the copper cavity end wall to form a plunger and diaphragm assembly, as shown in Figure 2. Thus, a very precise and controllable cone-shaped axisymmetric distortion of the surface can be achieved with the displacement of the end wall decreasing from a maximum value near the axis to zero at the cylindrical wall of the cavity. With this technique, the tuning sensitivity is enhanced almost an order of magnitude compared to a uniform displacement of the full end wall because the positive tuning derivative in the axial E-field region is no longer strongly counteracted by the negative tuning derivative of the end wall outer annular surface in the H-field region of the cavity. For example, the end wall tuning sensitivities for the (shorter) first cavity are computed as 3.16, $-3.73$ and $-0.57$ MHz for a $0.001''$ longitudinal outward displacement of the axial (4 mm $\phi$) region, the outer annular region, and the entire end wall, respectively.

SUPERFISH simulations were performed to more accurately analyze the performance of the tuners and the sensitivity of the field balance, by modeling the three-cell structure

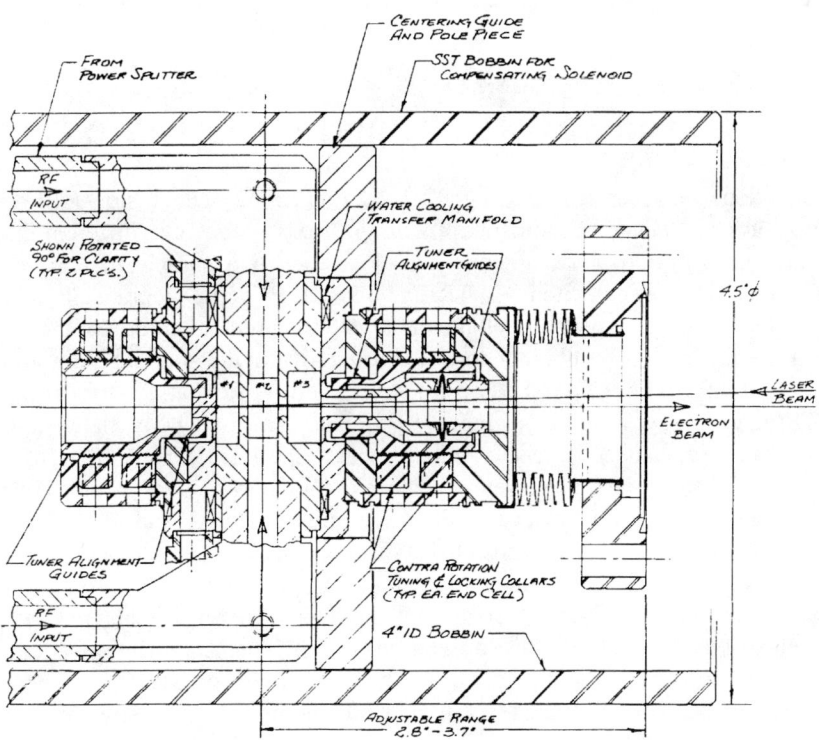

**FIGURE 2.** Three-cell, dual feed RF gun assembly showing microtuner mechanisms, water cooling and the compensating solenoid pole piece.

using oversize diameter "fabricated" cavities tuned into correct resonance by a cone-shaped deformation of the end wall. For example, Figure 3(a) compares the (solid line) field distribution of the correctly tuned CONTROL structure (f=17132.82 MHz) with the (dashed line) unbalanced field distribution that results from having a 0.0008" oversize diameter first cavity causing the resonant frequency of the structure to be reduced to 17128.46 MHz. Figure 3(b) shows that this detuned condition can be precisely realigned with the field distribution and frequency of the CONTROL run by a 0.0014" outward displacement of the axial region of the end wall. Similarly, Figures 4(a) and (b) illustrate that the substantial field imbalance and the 5.9 MHz error caused by the third cavity having a 0.0008" oversize diameter can be fully corrected by a 0.0022" axial extraction of the cavity exit nose.

These results confirmed earlier findings that, without tuning, a slightly oversize diameter first cavity (or an undersize diameter third cavity) can cause a very substantial reduction of the peak E-field at the cathode location. Simulation studies of this unloaded structure also indicated that with a 0.001" oversize diameter second cavity, the peak

E-field in this cavity would be reduced by 28 percent, with very little perturbation of the peak field in the first cavity and a 22 percent enhancement of the peak field in the third cavity. The results of a more rigorous analysis using MAFIA, to evaluate the loaded structure having dual feed coupling apertures in the second cavity, are given in a later section.

## PRELIMINARY PHASE ORBIT COMPUTATIONS

Simulations were performed using a 2-1/2D PIC code (PTP Version 1.17) to study energy gain and transverse momentum of the off-axis particles at different operating gradients and injection phase angles. This PIC code is a particle tracking program (without space charge) written in FORTRAN 90 for the PC. The program allows the user to input a cavity field array and track the radial and longitudinal motion of the electrons through these fields. A fourth order corrected Runge-Kutta integration algorithm is used for the field integration. The SUPERFISH $E_z$, $E_r$ and $H_\phi$ field arrays for the three-cell CONTROL geometry were used as input to PTP; and sets of single particle phase orbits were computed for an injection energy of 0.5 V and for peak E-field conditions of 150 and 200 MV/m. To study the transverse momentum imparted to off-axis particles by the RF fields, especially during traversal of the discs apertures and the RF cutoff nose at the end of the RF structure, particles were injected at different phase angles ($\phi_0$) and with starting radii of 0, 0.25 and 0.50 mm. An additional advantage of the 2-1/2D program was the ability to investigate the off-axis particle contributions to energy spread due to the radial dependence of the $E_z$ field. For the low charge (0.05 nC) present injection requirements of the 17 GHz linac, the photoinjector beam dynamics were considered to be dominated by the high gradient RF fields; and this initial study did not address space charge effects.

Phase orbit plots of particle energy ($\gamma$) versus longitudinal position during traversal of the 0.0381 m long CONTROL structure are shown in Figure 5 for a range of injection

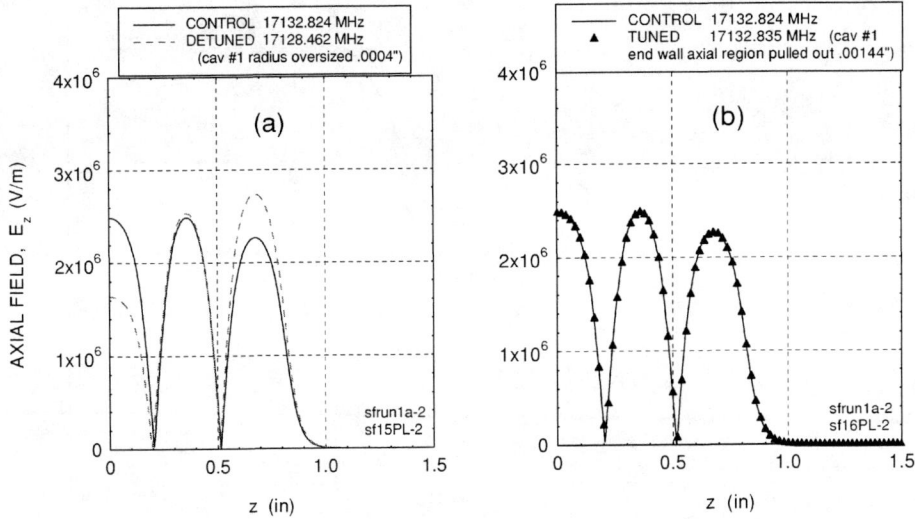

**FIGURE 3.** The detuning effects caused by cavity #1 having a 0.0008" oversize diameter and the correction achieved with a 0.00144" axial extraction of the end wall microtuner.

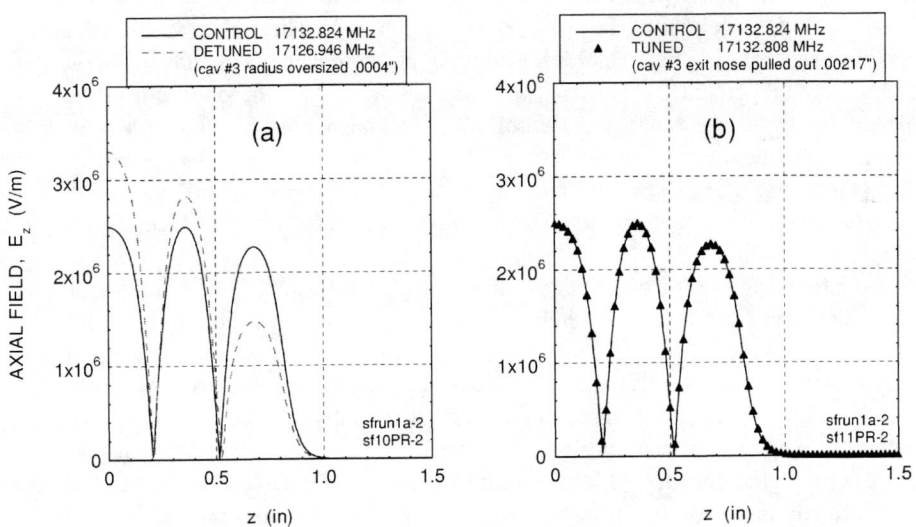

**FIGURE 4.** The detuning effects caused by cavity #3 having a 0.0008" oversize diameter and the correction achieved with a 0.00217" axial extraction of the cavity exit nose.

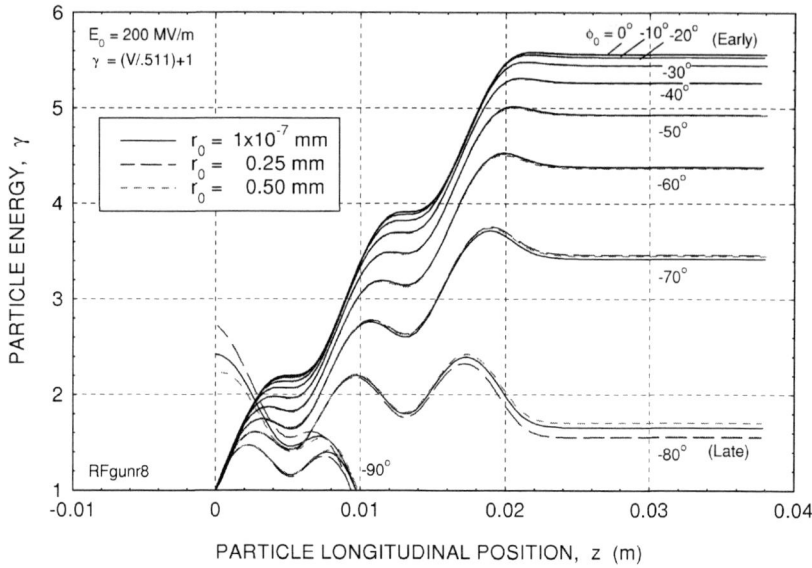

**FIGURE 5.** Particle total energy versus longitudinal position for different values of injection phase ($\phi_0$). Particle injection voltage = 0.5 V with starting radii ($r_0$) of 0, 0.25 and 0.50 mm used for each injection phase. Peak E-field ($E_0$) = 200 MV/m.

phase angles ($\phi_0$) and a peak E-field of 200 MV/m. (The linac phase convention of $-\pi/2$ at the peak of accelerating field has been used here.) These computations show the maximum energy gain to be 2.3 MeV with tight phase focusing and with an energy spread of less than 1 percent for particles injected within $20°$ of the optimum phase. With increasing departure from the optimum injection phase, not only is there a reduction in beam energy, there is also an increase in energy spread due to the off-axis particles. The dwells in energy gain, shown in Figure 5, correspond to the low intensity $E_z$ field regions (and the peak $E_r$ field regions) at the disc beam apertures (and at the RF cutoff nose at the end of the structure).

An interesting finding during these studies was that, for this 17 GHz regime, the laser pulse had to be fired at a moderately early phase to obtain the maximum electron beam energy. The phase orbit computations show that it is necessary to inject close to the zero crossing phase, well in advance of the peak accelerating E-field, to achieve the transit characteristics required for maximum energy gain. This early injection condition (also noted previously with standing wave, half-cavity capture of electrons injected into a 17 GHz high gradient linac [11,13]) compensates for the trajectory distance through the first cell, and provides the correct transit time to allow synchronization of the particles with the accelerating fields in the subsequently located cavities, i.e., the electron bunch crosses the midplane of each successive cavity at the peak of the accelerating field.

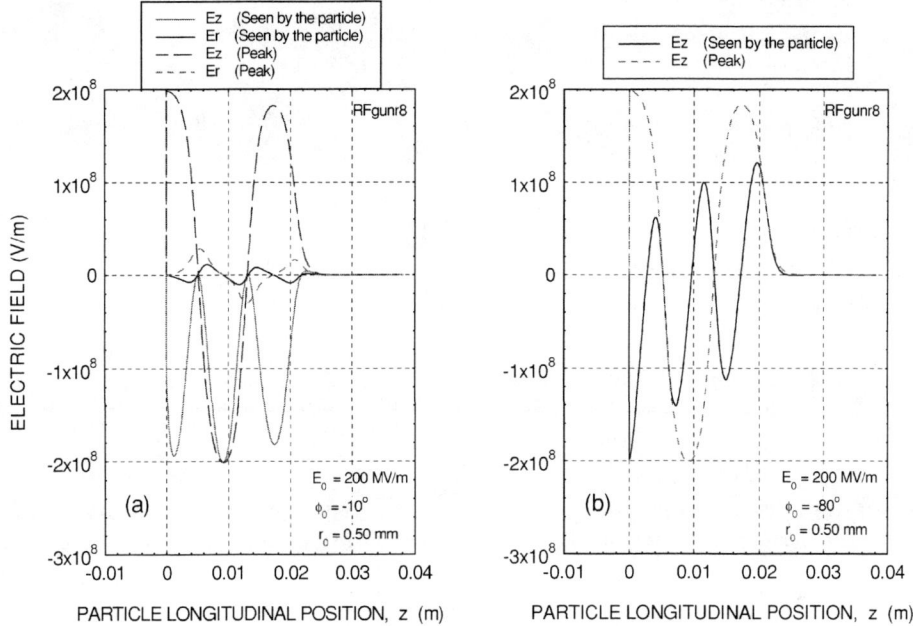

**FIGURE 6.** Comparison of the π-mode peak E-fields in the structure with the $E_z$ and $E_r$ fields experienced by an off-axis particle **(a)** injected at $\phi_0 = -10°$ and **(b)** injected at $\phi_0 = -80°$, with a starting radius of 0.50 mm. ($E_0 = 200$ MV/m.)

A good example of this synchronous operating condition is shown by the E-field plots in Figure 6(a). The dashed line curves show the $E_z$ and $E_r$ peak field distributions along the three-cell CONTROL structure, and the solid line curves show the $E_z$ and $E_r$ fields experienced by a particle launched 0.50 mm offset from the axis, with an energy of 0.5 V and at an injection phase angle of $\phi_0 = -10°$, i.e., $80°$ in advance of the peak accelerating field.. The high degree of synchronism achieved with this early phase injection, and the near optimum transfer of energy to the beam, can be noted by the absence of decelerating fields seen by the particle at the field reversal locations and by the close coincidence of the cavity peak fields with the peak fields seen by the particle during traversal of the structure. In contrast to this near optimum energy gain condition, Figure 6(b) shows the $E_z$ field seen by a particle injected at $\phi = -80°$, all other parameters remaining unchanged. The initial rapid acceleration causes loss of synchronism resulting in alternate acceleration and deceleration during the particle trajectory and a substantial energy loss, as illustrated by the $\phi_0 = -80°$ plot in Figure 5.

An overall matrix of r-z trajectories is shown plotted in Figure 7(a) for a peak field of $E_0 = 150$ MV/m, and in Figure 7(b) for $E_0 = 200$ MV/m. These plots show the r-z progression of particles after being launched at $5°$ injection phase intervals commencing at $0°$, and with starting radii of zero, 0.25 mm and 0.50 mm. Interesting features of the

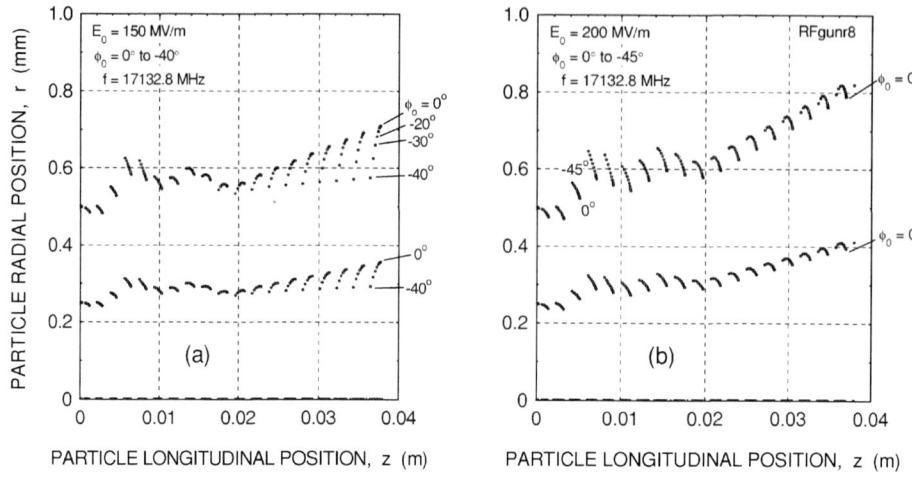

**FIGURE 7.** Particle r-z trajectory plots for a range of injection phase angles with starting radii of 0, 0.25 and 0.50 mm and for peak E-field values of 150 MV/m and 200 MV/m. (Length of first cell = 0.30 $\lambda_0$.)

Figure 7 plots are the radial displacements and corrections that occur at the first and second discs, located at z = 0.00525 m and 0.013 m, respectively, and the defocusing transverse momentum kicks that occur during traversal of the fringe field as the particles emerge from the RF environment at z = 0.022 m [refer also to z = 0.87″ on the CONTROL structure geometry in Figure 1(a)]. For the $E_0$ = 150 MV/m computations, it can be noted that the angular divergence of the emerging particles (in absence of space charge) ranges from 1 to 4 mradian for the $\phi_o$ = -40° to $\phi_o$ = 0° trajectories having starting radii of 0.25 mm; and from 2 to 8 mradian, over the same range of injection phases, for particles with starting radii of 0.50 mm. The phase dispersive characteristics of the defocused emergent trajectories are such that particles with different exit radii and the same injection phase angle have a common virtual object source, but the location of this source is subject to large excursions due to injection phase angle fluctuations.

Conversely, the Figure 7(b) plots show that by operating at a higher gradient ($E_0$ = 200 MV/m), although the angular divergence increases somewhat, the emergent trajectories have little phase dispersion, so that a common virtual source is maintained over a range of injection phase angles and exit radii.

The data presented above for this 17 GHz CONTROL structure, using a first cavity length of $0.30\lambda_0$, can be summarized as follows: (a) maximum beam energy is obtained by operating at an early phase close to the zero field crossing, but this results in a large angular divergence; (b) reducing the beam divergence requires operation at a later injection phase closer to the peak of the accelerating field (say, $\phi_0$ = -40°), but this will result in some loss of energy (<10%); (c) when operating at $\phi_0$ ≈ -40°, the beam energy spread will be sensitive to injection phase fluctuations due to the phase departure from the maximum energy position, refer Figure 5; and (d) operation at a high gradient can desensitize variations of the emergent beam angular divergence caused by laser pulse phase jitter (a critical issue at 17 GHz, since $10° = 1.62$ ps), and assist in maintaining a stable re-imaged focal source at entry to the linac.

An example of modifying the photoinjector operating characteristics by adjusting the length of the first cell is shown below in the Figure 8 plots of a field balanced 17 GHz three-cell structure, similar to the CONTROL structure discussed above, except the length of the first cell was reduced from $0.30\lambda_0$ to $0.18\lambda_0$. The energy gain plots in Figure 8(a) show that the maximum beam energy is now obtained at an injection phase of $\phi = -55°$ ($35°$ before the E-field peak); and for this operating condition, a laser pulse phase jitter of $\pm10°$ would result in an energy spread contribution of less than 1.5 percent. On the other hand, the Figure 8(b) r-z trajectories, computed for the same peak E-field as in Figure 7(b), show that, over a range of injection phase angles straddling the maximum beam energy region, the emergent beam diameter, the angular divergence and the phase dispersion are much greater than the values shown in Figure 7 for the CONTROL structure.

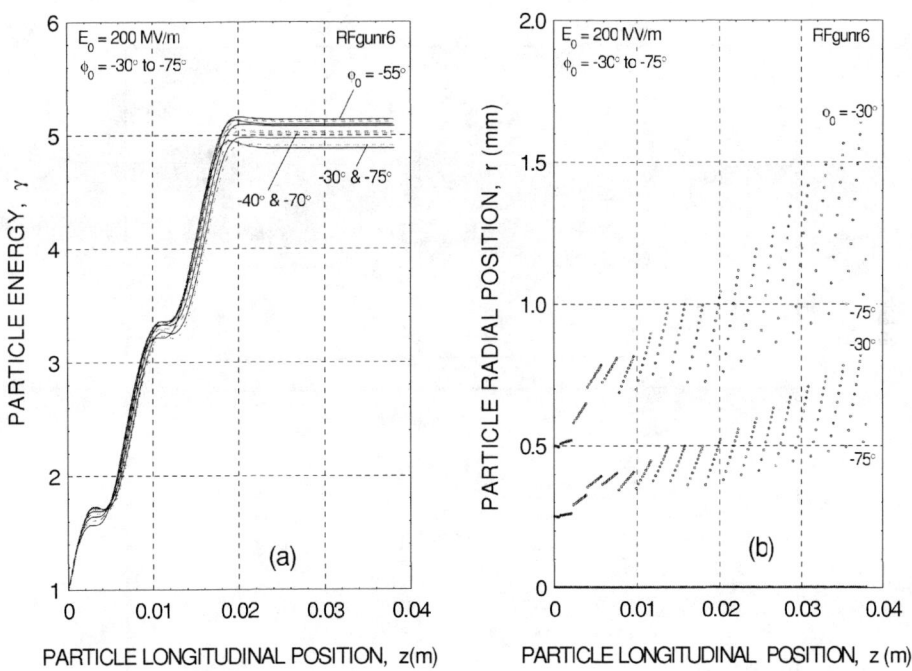

**FIGURE 8.** (a) Particle total energy versus longitudinal position and (b) particle r-z trajectory plots for a range of injection phase angles, using a three-cavity 17 GHz structure with a first cavity length of $0.18\lambda_0$.

Photoinjectors can be designed to emphasis the stabilization of specific beam parameters that best match the requirements of the linac beam optics, in order to satisfy a given application (sub-picosecond operation, high resolution, minimum emittance, etc.). For example, to achieve narrow bunch or high resolution performance with the single section high gradient 17 GHz linacs discussed in References 11 and 13, it has been shown that injection into the linac should occur over a limited phase interval ($\approx 25°$) and at specific phase positions. Operation of the three-cell CONTROL structure at, say, $\phi_0 = -35°$ with a phase jitter of $\pm 10°$, can produce injected beam energy variations of $+2.9\%$ to $-6.0\%$; and after traversing a 25 cm drift space, this transforms into linac entry phase variations of $+4.3°$ to $-9.4°$. Thus, excluding other phase variables, a laser pulse width of 1½ ps will satisfy the above injection conditions.

## DUAL FEED RACETRACK COUPLER CAVITY

A comparison of the geometric features of cylindrical and racetrack shaped, dual feed coupler cavities is shown in Figure 9. The racetrack cavity geometry is formed, in the first machining operation, by generating two semi-cylindrical shaped cavity walls with radii R having centers offset by a distance $\Omega$ on diametrically opposite sides of the centerline, as indicated in Figure 9(b); and the semi-cylindrical sections are connected with straight

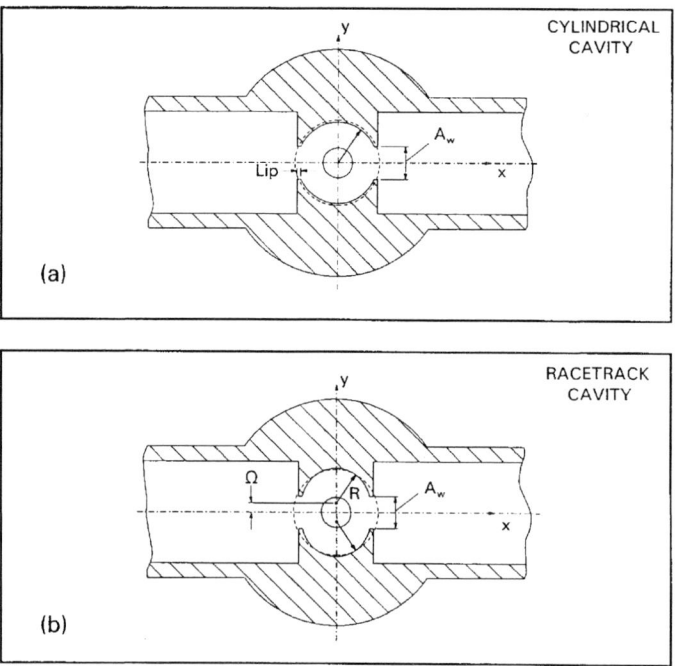

**FIGURE 9.** Comparison of cylindrical (conventional) and racetrack shaped dual feed coupler cavities.

parallel sidewalls of length $2\Omega$. The coupling apertures ($A_w$) for the rectangular waveguide feeds are then cut through the straight sidewalls of the cavity.

With the dual feed cylindrical cavity, the correct tune and match can be readily achieved by iterative adjustment of the cavity diameter and the coupling apertures ($A_w$); but the coupler fields, having different $E_z(x)$ and $E_z(y)$ distributions [10], will not be rotationally symmetric about the axis. The parameter, $\Omega$, shown in Figure 9(b) defines the racetrack geometry and provides a means of minimizing or eliminating the above field asymmetry because adjustment of the geometric ratio $R/(R + \Omega)$ enables the $E_z(x)$ and $E_z(y)$ distributions to be equalized (within the cylindrical volume bounded by the beam apertures) while still retaining the capability of readily matching and tuning the coupler by adjustment of $A_w$ and the transverse dimensions of the cavity; i.e., the cavity tune is dependent on the absolute values of R and $\Omega$, while the field symmetry, $E_z(x) = E_z(y)$, is dependent on the factor $R/(R + \Omega)$.

The MAFIA code was used initially to model the unloaded (axisymmetric) three-cell CONTROL structure to establish the mesh defining boundaries, especially around the contoured surfaces of the beam apertures, so that the computed resonant frequencies were accurately aligned with the SUPERFISH values. The loaded CONTROL structure was then modeled using a dual racetrack shaped second cell having diametrically opposed sidewall coupling apertures, and the external Q at the $\pi$-mode resonance was determined using the KY method [12].

Symmetrizing the coupler fields and achieving the correct resonant frequency required adjusting the racetrack parameters R and $\Omega$ to equalize the $E_z(x)$ and $E_z(y)$ fields; and matching the structure required adjusting the coupling aperture dimension ($A_w$) to give the desired external Q at the correct resonant frequency to achieve $\pi$-mode field balance. The computations required many iterations due to the interdependence and sensitivity of the cavity parameters, and Figure 10 shows the response of the E-field distribution in the three-cell structure for typical parameter variations of the racetrack coupler cavity.

The degree of field symmetrization achieved in the racetrack coupler cavity can be seen in the Figure 11 comparison of the orthogonal field distributions $E_z(x)$ and $E_z(y)$. These MAFIA computed fields are shown plotted at three parallel transverse planes through the racetrack cavity, one located parallel with and on mesh points near the surface of the upstream disc, one at the midplane of the coupler cavity and one near the surface of the downstream disc. By progressively adjusting the system parameters, it was possible to merge the $E_z(x)$ and $E_z(y)$ distributions so that the peak values were equalized to within a few tenths of a percent, while achieving a near matched condition at the resonant frequency (for a theoretical $Q_0 = 6050$).

The high power evacuated WR62 rectangular waveguide network for providing phase and amplitude adjustable 17 GHz input power for the new RF gun has been designed on the basis of a 10 percent loss in $Q_0$ for the fabricated and tuned three-cell structure, and a 30 percent loss in the transmission line. Power will be extracted from the linac high power input feed using a 6.5 dB high directivity directional coupler, and transmitted via remote controlled phase shifter and attenuator assemblies to a power splitter having dual, phase balanced waveguides connected to the RF gun racetrack coupler cavity. High

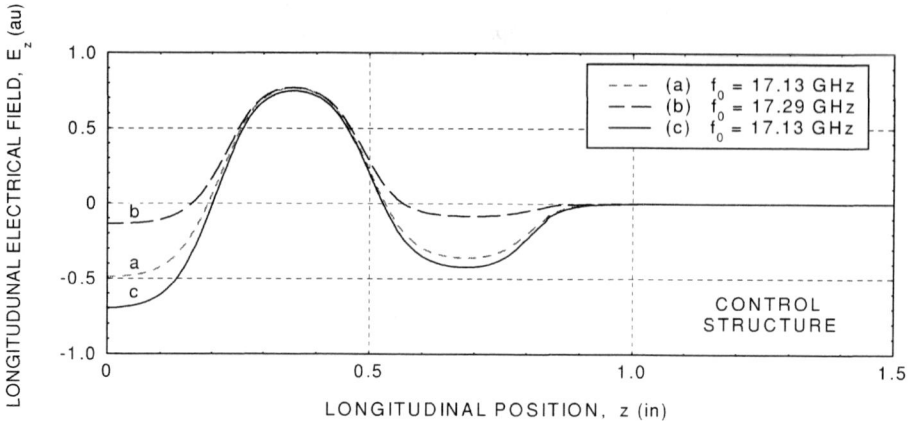

**FIGURE 10.** MAFIA generated longitudinal electric field distributions along the axis of the three-cell CONTROL structure, with the second cell comprising a racetrack shaped cavity with dual sidewall coupling apertures. Conditions: (a) cavity overcoupled ($\beta = 1.78$), $R/(R + \Omega) = 0.925$, E-field asymmetry 1.3%; (b) cavity overcoupled ($\beta = 1.77$), $R/(R + \Omega) = 0.945$, E-field asymmetry 1%; and (c) cavity undercoupled ($\beta = 1.08$), $R/(R + \Omega) = 0.97$, E-field asymmetry 0.3%.

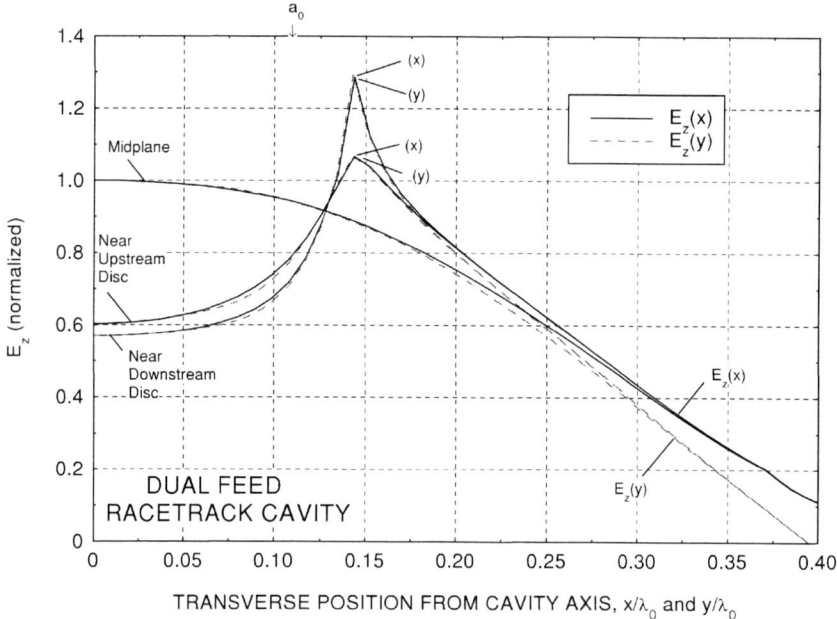

**FIGURE 11.** Longitudinal electric field distributions, $E_z(x)$ and $E_z(y)$, in three parallel transverse planes through the racetrack cavity dual feed coupler, showing the orthogonal distributions symmetrized at the locations of all three transverse planes. [$R/(R + \Omega) = 0.97$].

666

conductance, WR62 waveguide pumping ports shall be located at frequent intervals in the network to assist in power conditioning and maintaining a hard vacuum; and precision machined, thick stainless steel unisex flanges and metal gaskets will be used at the rectangular waveguide interfaces to simultaneously produce a self aligned high current RF joint and a high vacuum seal [13].

## ACKNOWLEDGMENTS

The authors wish to acknowledge the MAFIA code simulation contributions of E. L. Wright, and the collaboration of the MIT Plasma Science and Fusion Center staff in planning the interfacing of the existing UV laser system with this new RF gun.

## REFERENCES

1. Kim, K. J., "Rf and Space-Charge Effects in Laser-Driven Rf Electron Guns," *Nucl. Instr. Meth.*, **A275**, pp. 201–218, 1989, and references therein.
2. Serafini, L., "Improving the Beam Quality of RF Guns by Correction of RF and Space-Charge Effects,: in *Advanced Accelerator Concepts, AIP Conference Proceedings* **279**, New York: AIP Press, 1993, pp. 645–674.
3. Rosenzweig, J. and Colby, E., "Charge and Wavelength Scaling of RF Photoinjectors: A Design Tool," in *Proc. 1995 IEEE Particle Accelerator Conference* **95CH35843**, 1995, pp. 957–960.
4. Haimson, J., Mecklenburg, B. and Danly, B. G., "Initial Performance of a High Gain, High Efficiency 17 GHz Traveling Wave Relativistic Klystron for High Gradient Accelerator Research," in *Pulsed RF Sources for Linear Colliders, AIP Conference Proceedings* **337**, New York: AIP Press, 1995, pp. 146–159.
5. Trotz, S., *et al.*, "High Power Operation of a 17 GHz Photocathode RF Gun," in *Advanced Accelerator Concepts, AIP Conference Proceedings* **398**, New York: AIP Press, 1997, pp. 717–729.
6. Helm, R. H., *A Note on Coupler Asymmetries in Long Linear Accelerators,* Report No. M-167, Stanford Linear Accelerator Center, Stanford University, Stanford, California, 1960.
7. Loew, G. A. and Neal, R. B., *Linear Accelerators*, Amsterdam: North Holland Publishing Company, 1970, ch. B.1.1, pp. 97–99, 143–144.
8. Palmer, D.T., *et al.*, "Initial Commissioning Results of the Next Generation Photoinjector," in *Advanced Accelerator Concepts, AIP Conference Proceedings* **398**, New York: AIP Press, 1997, pp. 695–704.
9. Ng, C.-K. and Ko, K., "Numerical Simulation of Input and Output Couplers for Linear Accelerator Structures," presented at the *Computational Accelerator Conference (CAP 93)*, Pleasanton, California, February 1993.
10. Haimson, J., Mecklenburg, B. and Wright, E. L., "A Racetrack Geometry to Avoid Undesirable Azimuthal Variations of the Electric Field Gradient in High Power Coupling Cavities for TW Structures," in *Advanced Accelerator Concepts, AIP Conference Proceedings* **398**, New York: AIP Press, 1997, pp. 898–911.
11. Haimson, J. and Mecklenburg, B., "HV Injection Phase Orbit Characteristics for Sub-Picosecond Bunch Operation with a High Gradient 17 GHz Linac," in *Proc. 1995 IEEE Particle Accelerator Conference* **95CH35843**, 1995, pp. 755–757.
12. Kroll, N. M. and Yu, D. U. L., "Computer Determination of the External Q and Resonant Frequency of Waveguide Loaded Cavities," *Particle Accelerators* **34**, pp. 231–250, (1990).
13. Haimson, J. and Mecklenburg, B., "A Linear Accelerator Power Amplification System for High Gradient Structure Research," published in these Proceedings.

# Transverse Fields in
# Dielectric Wake Field Accelerators

J. L. Hirshfield,[1,2] S. Y. Park,[1,2,3] and T.-B. Zhang[1]

[1]*Omega-P, Inc. 202008 Yale Station, New Haven, CT 06520-2008*
[2]*Department of Physics, Yale University, POB 8120, New Haven, CT 06520-8120*
[3]*Permanent address: Department of Physics, POSTECH, Pohang, Korea*

**Abstract.** Theory is presented for excitation of hybrid electric/magnetic (*HEM*) wake-field modes by passage of an electron drive bunch in a dielectric-lined cylindrical waveguide. The drive bunch is moving parallel to the waveguide axis, but is displaced by a radial increment $r_o$. Knowledge of the amplitudes of all *HEM* modes allows calculation of the transverse forces on a bunch or bunches that follow the displaced drive bunch. Approximate formulae for the transverse forces on a trailing bunch are given, valid in the limit of small $r_o$. These transverse forces can lead to beam instability, if cumulative transverse motion is significant on the time-scale of passage of the bunch through the accelerator module. Constructive interference can be present amongst $TM_{0m}$ monopole modes that produce highly-peaked spatiotemporally-localized axial wake-fields, with peak fields at the locations of trailing bunches in a multi-bunch train. In this case, the spectrum of dipole (and higher order) *HEM* modes may not enjoy the same degree of constructive intereference, and may not have the same axial periodicity as the monopole modes. Further analytical and computational study is needed to determine the limitations that transverse wake fields may impose on a multi-bunch dielectric wake-field accelerator.

## INTRODUCTION

Elsewhere in this volume, Marshall, Zhang and Hirshfield (1) present analysis for the axisymmetric wake fields induced by a train of drive bunches moving along the axis of a dielectric-lined cylindrical waveguide. Formulation of wake field theory for this structure is carried out for cases where many $TM_{0m}$ waveguide modes can participate in wake-field formation. The waveguide is designed so that, with phase velocities equal to the bunch velocity and to one another, the excited modes have nearly equally-spaced eigenfrequencies. This leads to a strongly-peaked spatiotemporal superposition of many co-propagating waveguide modes, resulting in a net wake-field amplitude that can be much larger than the amplitudes of individual modes. When the nearly periodic highly-peaked wake-fields have the same period as the bunches, constructive interference occurs between fields from successive bunches, so that stronger wake-fields are stimulated than would be the case for a bunch not influenced by synchronous fields from bunches ahead of it. These considerations lead to a new approach to dielectric wake-field accelerators, in which a train of moderate-charge drive bunches is used to build up a strong cumulative wake field that can provide high-

CP472, *Advanced Accelerator Concepts: Eighth Workshop,*
edited by W. Lawson, C. Bellamy, and D. Brosius

gradient acceleration for a following test bunch. In an example discussed in (1), an unloaded acceleration gradient of 133 MV/m is predicted using three 2 nC drive bunches in an appropriately-designed dielectric-lined waveguide.

But one concern, as with any accelerating structure, is orbit instability driven by transverse forces that arise from non-axisymmetric excitations. That concern is addressed in this paper, in which theory is developed for excitation of hybrid *HEM* modes excited by a non-axisymmetric bunch, or train of such bunches. Theories along this line have been developed heretofore, including those by Rosing and Gai (2), and by Ng (3), but thes authors do not agree with one another on some important issues. These prior theories are not based on expansion of the wave equation in a sequence of orthonormal waveguide modes, so their treatment of multi-mode, multi-bunch effects may be more cumbersome than by use of the normal-mode expansion method that is expounded in this paper. Here we present an outline of the theory, leading to compact formulae for the transverse force on a test bunch that follows behind an non-axisymmetric drive bunch. This result can easily be generalized to a train of drive bunches, under conditions where the wake-fields interfere constructively. The purpose of this work is to provide the theoretical framework to allow scrutiny of beam instability driven by off-axis excursions of the drive bunches, especially for the case of a synchronous multi-bunch train.

## TRANSVERSE WAKEFIELD THEORY

The model analyzed here is a cylindrical waveguide, consisting of a dielectric cylinder with an axisymmetric hole, whose outer surface is coated with a low-loss conductor. The dielectric constant $\kappa$ is assumed to be independent of frequency. Radii of the central vacuum hole and the outer waveguide wall are $a$ and $R$, respectively. For axisymmetric excitations from a bunch centered on the waveguide axis, orthonormal wave functions can be found for the electromagnetic wake fields that separate into *TE* and *TM* classes. However, the general solution with a non-axisymmetric source, such as an off-axis bunch, does not so separate, and the orthonormal wave functions are combination of *TE* and *TM* modes, often called hybrid electromagnetic ( *HEM* ) modes. The field solutions and dispersion relation (but not the orthogonality relation) were obtained by Chang and Dawson (4). The eigenfunctions for the complete set of orthonormal *HEM* modes in the hole and dielectric regions are given by

$$E_{z,l}(r,\theta,k_z) = P_l(k_2 a)I_l(k_1 r)\cos l\theta \tag{1}$$

$$E_{r,l}(r,\theta,k_z) = i\frac{k_z}{k_1}\left[ P_l(k_2 a)I_l'(k_1 r)+\beta e_l\frac{l}{k_1 r}Q_l(k_2 a)I_l(k_1 r) \right]\cos l\theta \tag{2}$$

669

$$E_{\theta,l}(r,\theta,k_z) = -i\frac{k_z}{k_1}\left[\frac{l}{k_1 r}P_l(k_2 a)I_l(k_1 r) + \beta\, e_l\, Q_l(k_2 a)I_l'(k_1 r)\right]\sin l\theta \tag{3}$$

$$B_{z,l}(r,\theta,k_z) = e_l Q_l(k_2 a)I_l(k_1 r)\sin l\theta \tag{4}$$

$$B_{r,l}(r,\theta,k_z) = i\frac{k_z}{k_1}\left[\beta\frac{l}{k_1 r}P_l(k_2 a)I_l(k_1 r) + e_l Q_l(k_2 a)I_l'(k_1 r)\right]\sin l\theta \tag{5}$$

$$B_{\theta,l}(r,\theta,k_z) = i\frac{k_z}{k_1}\left[\beta P_l(k_2 a)I_l'(k_1 r) + e_l\frac{l}{k_1 r}Q_l(k_2 a)I_l(k_1 r)\right]\cos l\theta \tag{6}$$

for $0 \leq r \leq a$, and

$$E_{z,l}(r,\theta,k_z) = I_l(k_1 a)P_l(k_2 r)\cos l\theta \tag{7}$$

$$E_{r,l}(r,\theta,k_z) = -i\frac{k_z}{k_2}\left[I_l(k_1 a)P_l'(k_2 r) + \beta e_l\frac{l}{k_2 r}I_l(k_1 a)Q_l(k_2 r)\right]\cos l\theta \tag{8}$$

$$E_{\theta,l}(r,\theta,k_z) = i\frac{k_z}{k_2}\left[\frac{l}{k_2 r}I_l(k_1 a)P_l(k_2 r) + \beta e_l I_l(k_1 a)Q_l'(k_2 r)\right]\sin l\theta \tag{9}$$

$$B_{z,l}(r,\theta,k_z) = e_l I_l(k_1 a)Q_l(k_2 r)\sin l\theta \tag{10}$$

$$B_{r,l}(r,\theta,k_z) = -i\frac{k_z}{k_2}\left[\kappa\beta\frac{l}{k_2 r}I_l(k_1 a)P_l(k_2 r) + e_l I_l(k_1 a)Q_l'(k_2 r)\right]\sin l\theta \tag{11}$$

$$B_{\theta,l}(r,\theta,k_z) = -i\frac{k_z}{k_2}\left[\kappa\beta I_l(k_1 a)P_l'(k_2 r) + e_l\frac{l}{k_2 r}I_l(k_1 a)Q_l(k_2 r)\right]\cos l\theta \tag{12}$$

for $a \leq r \leq R$. All field components propagate as $e^{i(\omega t - k_z z)}$. The transverse wave number in the hole and the dielectric regions are $k_1^2 = k_z^2 - k_o^2$, and $k_2^2 = \kappa k_o^2 - k_z^2$, with $k_o = \omega/c$; $P_l(k_2 r) = J_l(k_2 R)N_l(k_2 r) - N_l(k_2 R)J_l(k_2 r)$; $J_l(x)$ and $N_l(x)$ are $l$-th order Bessel functions of the first and second kinds; $I_l(x)$ is the modified Bessel function; $Q_l(k_2 r) = N_l'(k_2 R)J_l(k_2 r) - J_l'(k_2 R)N_l(k_2 r)$; $\beta = k_o/k_z$; $e_l$ is a constant which relates the field components $E_z$ and $B_z$, i. e.,

$$e_l = -l\beta \left( \frac{1}{x_1^2} + \frac{\kappa}{x_2^2} \right) \frac{P_l(x_2)}{Q_l(x_2)} \left[ \frac{I_l'(x_1)}{x_1 I_l(x_1)} + \frac{Q_l'(x_2)}{x_2 Q_l(x_2)} \right]^{-1} \tag{13}$$

where $x_1 = k_1 a$ and $x_2 = k_2 a$.

The dispersion relation is

$$\left[ \frac{I_l'(x_1)}{x_1 I_l(x_1)} + \kappa \frac{P_l'(x_2)}{x_2 P_l(x_2)} \right] \left[ \frac{I'(x_1)}{x_1 I_l(x_1)} + \frac{Q_l'(x_2)}{x_2 Q_l(x_2)} \right] = l^2 \beta^2 \left( \frac{1}{x_1^2} + \frac{\kappa}{x_2^2} \right)^2 . \tag{14}$$

The first bracket represents *TM* modes, while the second represents *TE* modes. For the axial symmetric modes $l = 0$, and two decouple; otherwise, they are coupled hybrid *HEM* modes with, in general, six interdependent non-zero field components in each region. Eqs. 1-14 are all consistent with results obtained by Chang and Dawson (4).

We have confirmed by a detailed proof that a general orthogonality relation given by Collins (5) is applicable for any cylindrical waveguide with perfectly conducting walls and any number of annular dielectric layers, i. e.,

$$\int_0^R dr\, r \int_0^{2\pi} d\theta \left[ E_{r,lm} B_{\theta,l'n}^* - E_{\theta,lm} B_{r,l'n}^* \right] = \delta_{ll'} \delta_{mn} \alpha_{lm}^2 . \tag{15}$$

Eq. 15 states that power flow at each frequency occurs mode-by-mode, without mixing between field components of different modes. The normalization factor $\alpha_{lm}(\omega, k_{zm})$ is

$$\alpha_{lm}^2 = \pi a^2 \left\{ \left( \frac{k_{zm}}{k_{1m}} \right)^2 \left[ \beta \left( P_l^2(x_{2m}) + e_l^2 Q_l^2(x_{2m}) \right) C_{I,lm} + \left( 1 + \beta^2 \right) e_l P_l(x_{2m}) Q_l(x_{2m}) D_{I,lm} \right] \right.$$
$$\left. + \left( \frac{k_{zm}}{k_{2m}} \right)^2 I_l^2(x_{1m}) \left[ \beta \left( \kappa C_{P,lm} + e_l^2 C_{Q,lm} \right) + \left( 1 + \kappa \beta^2 \right) e_l D_{PQ,lm} \right] \right\} \tag{16}$$

where

$$C_{I,lm} = \frac{1}{x_{1m}^2} \int_0^{x_{1m}} dx\, x \left\{ I_l'^2(x) + \left[ \frac{l}{x} I_l(x) \right]^2 \right\}; \quad D_{I,lm} = \frac{l}{x_{1m}^2} I_l^2(x_{1m})$$

$$C_{P,lm} = \frac{1}{x_{2m}^2} \int_{x_{2m}}^{y_{2m}} dx\, x \left\{ P_l'^2(x) + \left[ \frac{l}{x} P_l(x) \right]^2 \right\};$$

671

$$C_{Q,lm} = \frac{1}{x_{2m}^2} \int_{x_{2m}}^{y_{2m}} dx\, x \left\{ Q_l'^2(x) + \left[\frac{1}{x} Q_l(x)\right]^2 \right\};$$

$$D_{PQ,lm} = \frac{1}{x_{2m}^2}\left[ Q_l(y_{2m})P_l(y_{2m}) - Q_l(x_{2m})P_l(x_{2m}) \right]; \quad x_{1m} = k_{1m}a, \quad x_{2m} = k_{2m}a \quad \text{and}$$

$y_{2m} = k_{2m}R$. The above integrals all reduce to standard forms, but writing out the full result would occupy more space than is justified. Prior analyses of dielectric-lined cylindrical waveguides do not give an orthogonality proof or a normalization constant.

To find wake fields induced by an electron bunch, one expands in orthonormal modes the solution of the inhomogeneous wave equations for $E_r$ and $E_\theta$, namely

$$\left[ \nabla^2 - \frac{\kappa(r)}{c^2} \frac{\partial^2}{\partial t^2} \right] E_r - \frac{2}{r^2} \frac{\partial E_\theta}{\partial \theta} - \frac{1}{r^2} E_r = S_r, \tag{17}$$

$$\left[ \nabla^2 - \frac{\kappa(r)}{c^2} \frac{\partial^2}{\partial t^2} \right] E_\theta + \frac{2}{r^2} \frac{\partial E_r}{\partial \theta} - \frac{1}{r^2} E_\theta = S_\theta \tag{18}$$

with the source functions $S_r(\mathbf{r},t) = 4\pi\, \partial\rho(\mathbf{r},t)/\partial r$ and $S_\theta(\mathbf{r},t) = 4\pi\, \partial\rho(\mathbf{r},t)/r\partial\theta$. These forms assume that the beam current is flowing in only in the $z$-direction; of course $\mathbf{S}(\mathbf{r},t) = 0$ for $r \ge a$. A complete solution can be constructed from fields as given in Eq.1-12, since these are solutions of Eq.17 and Eq.18 with $\mathbf{S}(\mathbf{r},t) = 0$ everywhere. To proceed, we expand the solutions of Eq.17 and Eq.18 in the interval $0 \le r \le R$ in a Fourier integral.

$$\begin{Bmatrix} E_r(\mathbf{r},t) \\ E_\theta(\mathbf{r},t) \end{Bmatrix} = \sum_{l=0}^{\infty} \sum_{m=1}^{\infty} \int_{-\infty}^{\infty} dk \int_{-\infty}^{\infty} d\omega\, A_{lm}(k,\omega) \begin{Bmatrix} E_{r,lm}(r,k,\omega)\cos l\theta \\ E_{\theta,lm}(r,k,\omega)\sin l\theta \end{Bmatrix} e^{-i(kz-\omega t)}. \tag{19}$$

Inserting Eq. 19 into Eqs. 17 and 18, multiplying Eq. 17 by $B_{\theta,l'n}^*(r,k',\omega')\cos l'\theta\, e^{i(k'z-\omega't)}$ and Eq. 18 by $B_{r,l'n}^*(r,k',\omega')\sin l'\theta\, e^{i(k'z-\omega't)}$, then combining the two equations, taking the integral over $d^3r\, dt$ and invoking the orthogonality relation Eq. 15, yields for the Fourier amplitude the result

$$A_{lm}(k,\omega) = \frac{-1}{4\pi^2 \, \alpha_{lm}^2 \left( k^2 - k_{zm}^2 \right)}$$

$$\times \iiint d^3 r' \int_{-\infty}^{\infty} dt' \left[ S_r(r',t') B_{\theta,lm}^*(r',k,\omega) \cos l\theta' - S_\theta(r',t') B_{r,lm}^*(r',k,\omega) \sin l\theta' \right] e^{i(kz'-\omega t')}$$

$$(20)$$

For the balance of this paper, we specialize to the case of a point charge displaced from the axis by an increment $r_0$, so that the charge density is given by $\rho(\mathbf{r},t) = Q_0(1/r)\delta(r-r_0)\delta(\theta)\delta(z-vt)$. In this instance, we find for the radial electric field

$$E_r(\mathbf{r},t) = \sum_{l=0}^{\infty} \sum_{m=1}^{\infty} A_{lm} \, E_{r,lm}(r,k_{zm}) \cos l\theta \, e^{-ik_{zm}z_0} . \qquad (21)$$

where $z_0 = z - vt$. The amplitude of all the field components (see Eqs. 1-12) is given by

$$A_{lm} = \frac{2\pi Q_0}{a^2} \left( \frac{a^2}{\alpha_{lm}^2} \right) b_{lm}(r_0, k_{zm}) \qquad (22)$$

where

$$b_{lm}(r_0, k_{zm})$$

$$= \frac{i}{k_{zm}} \left[ \frac{\partial}{\partial r} B_{\theta,lm}^*(r,k_{zm}) \bigg|_{r=r_0} + \frac{1}{r_0} B_{\theta,lm}^*(r_0,k_{zm}) - \frac{l}{r_0} B_{r,lm}^*(r_0,k_{zm}) \right] . \qquad (23)$$

This result, Eqs. 21-23, together with the other five field components, constitutes a complete formal solution for all the wake-field modes excited by a point charge moving with an axial velocity $v$, and displaced from the axis of the waveguide by a radial increment $r_0$.

For judging beam stability, one begins by calculating the transverse forces on a bunch with charge $q$ that follows a drive bunch whose wake-fields are given above. The results are

$$F_r(\mathbf{r},t) = q \left[ E_r(\mathbf{r},t) - \beta B_\theta(\mathbf{r},t) \right]$$

$$= \frac{q}{\gamma} \sum_{l=0}^{\infty} \sum_{m=1}^{\infty} A_{lm} P_l(k_{2m}a) I_l'(k_{1m}r) \cos l\theta \sin k_{zm} z_0 \qquad (24)$$

673

and

$$F_\theta(\mathbf{r},t) = q\big[E_\theta(\mathbf{r},t) + \beta B_r(\mathbf{r},t)\big]$$

$$= -\frac{q}{\gamma} \sum_{l=0}^{\infty} \sum_{m=1}^{\infty} A_{lm}\, P_l(k_{2\,m} a)\frac{l}{k_{1\,m} r}\, I_l(k_{1\,m} r)\sin l\theta \,\sin k_{zm} z_0 \tag{25}$$

The full significance of these results cannot be known without numerical computations using parameters for dielectric wake-field accelerators of interest. These computations are beyond the scope of this paper. However, some insight into the results can be inferred by considering the limit of a small radial displacement $r_0$ for the drive bunch, and small displacements $r$ for a following test bunch. If one inserts Eqs.5 and 6 for the eigenmodes $B_{r,lm}$ and $B_{\theta,lm}$ in Eq.23, and expands the result for small beam displacement , one has

$$b_{lm}(r_0, k_{zm}) \approx \frac{1}{2}\frac{1}{(l+1)!}\big[(l+2)\beta\, P_l(k_{2\,m} a) - l e_l\, Q_l(k_{2\,m} a)\big]\big(k_{1\,m} r_0/2\big)^l \tag{26}$$

This allows an estimate to be made of the forces associated with each azimuthal order of the HEM $_{lm}$ eigenmodes. Thus, for $l = 0$, where the eigenmode is actually TM $_{0m}$, one has to lowest order in $r_0$ the results

$$F_{r,0m}(\mathbf{r},t) = F_0 \sum_{m=1}^{\infty} c_{om}\big(k_{1\,m} r/2\big)\sin\big(k_{zm} z_0\big) \tag{27}$$

and
$$F_{\theta,0m}(\mathbf{r},t) = 0 \tag{28}$$
where,

$$F_0 = \frac{2\pi q Q_0}{a^2}, \text{ and } \quad c_{om} = \frac{1}{\gamma}\left(\frac{a^2}{\alpha_{om}^2}\right)\beta P_0^2(k_{2\,m} a).$$

One notes that Eq. 27 does not depend upon $r_0$, and that, mode-by-mode, the radial force is a quarter-cycle in $z_0$ out of phase with the axial electric field, which varies as $\cos k_{zm} z_0$. However, the train of drive bunches in the multi-bunch dielectric wake-field accelerator are spaced by a full cycle in $z_0$, and the bunch to be accelerated is spaced by a half-cycle from the last drive bunch. Thus, one can conclude that, to lowest order, the radial force from the TM $_{0m}$ modes of a slightly displaced driving bunch will not be felt by following bunches.

The radial and azimuthal forces for $l \geq 1$, for small displacement $r_0$ of the drive bunch, and for small deviations $r$ of the test bunch from the axis, are

$$F_r(\mathbf{r},t) = F_o \sum_{l=1}^{\infty} \sum_{m=1}^{\infty} c_{lm} \left( \frac{k_{1m} r_o}{2} \right)^l \left( \frac{k_{1m} r}{2} \right)^{l-1} \cos l\theta \sin k_{zm} z_o \ , \qquad (29)$$

and

$$F_\theta(\mathbf{r},t) = -F_o \sum_{l=1}^{\infty} \sum_{m=1}^{\infty} c_{lm} \left( \frac{k_{1m} r_o}{2} \right)^l \left( \frac{k_{1m} r}{2} \right)^{l-1} \sin l\theta \sin k_{zm} z_o \ , \qquad (30)$$

where

$$c_{lm} = \frac{1}{4\gamma} \left( \frac{a^2}{\alpha_{lm}^2} \right) \frac{P_l(k_{2m} a)}{(l-1)!(l+1)!} \left[ (l+2)\beta \ P_l(k_{2m} a) - l e_l \ Q_l(k_{2m} a) \right].$$

Clearly, the largest components are those for $l=1$, where both forces are independent of $r$ and linearly proportional to $r_o$. The degree to which the transverse dipole modes $HEM_{1m}$ propagate synchronously with one another depends upon the wavenumber spectrum for the $k_{zm}$ for $l=1$. The degree to which transverse dipole forces act at the locations of drive or test bunches depends upon the similarity of the wavenumber spectra of the $k_{zm}$ for $l=1$ to that for $l=0$. It would be unusual for these two spectra to not be dissimilar. These preliminary observations provide motivation for performing further analysis and detailed numerical computations of the full transverse forces obtained by summing over all significant $HEM_{lm}$ eigenmodes. This can lead to an accurate prediction of radial bunch trajectories and stability limits for the multi-bunch dielectric wake-field accelerator.

## ACKNOWLEDGMENT

This research was supported by the US Department of Energy, Division of High Energy Physics.

## REFERENCES

1. Marshall, T.C., Zhang, T-B., and Hirshfield⋅ J.L., this proceedings.
2. Rosing, M. and Gai, W., Phys. Rev. D**42**, 1829 (1990).
3. Ng, King-Yuen, Phys. Rev. D**42**, 1819 (1990).
4. Chang, C.T.M. and Dawson, J. W., J. Appl. Phys. **41**, 4493 (1970)
5. Collin, R. E., *Field Theory of Guided Waves*, ch. 6, (IEEE Press, New York, 1991).

# Diamond Coating in Accelerator Structure

Xintian E. Lin[1]

**Abstract.** The future accelerators with 1 GeV/m gradient will give rise to hundreds of degrees instantaneous temperature rise on the copper surface. Due to its extraordinary thermal and electric properties, diamond coating on the surface is suggested to remedy this problem. Multi-layer structure, with the promise of even more temperature reduction, is also discussed, and a proof of principle experiment is being carried out.

## I  INTRODUCTION

The quest for high accelerating gradient, on the order of 1 GeV/m, calls for the development at high frequency. While peak power requirement, breakdown voltage scaling with frequency is favorable at higher frequency, The pulsed temperature rise of a 1 GeV/m machine is still prohibitively high. Basically the heat generated by wall loss from microwave pulse diffuses slowly into the metal and local temperature increases with the one half power of the pulse duration. Diamond coating on the metal inner surface was suggested [1] to remove the heat away. The configuration is illustration in Fig. 1. Diamond being the hardest known material also has the

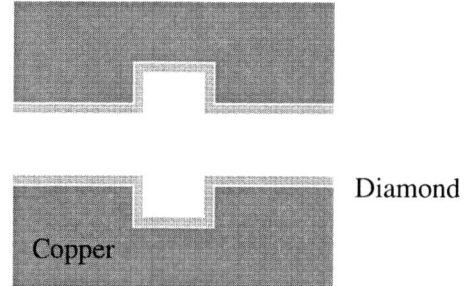

**FIGURE 1.** Surface coating in accelerator cavity.

highest thermal conductivity of 2000 W/m·K at room temperature, 5 times that of copper. The commercial success of Chemical Vapor Deposition (CVD) in artificial diamond synthesis makes it economically feasible to be used in many applications. One of them is heat sink in high power semiconductor device to lower its working

[1]  Work supported by the Department of Energy, contract DE-AC03-76SF00515.

temperature, and thus raise the reliability. In our application, the heat generated on the metal surface diffuses away through 2 thermal paths instead of one: one into metal and one into diamond. As long as diamond coating is substantially thicker than the diffusion length $\lambda = \sqrt{KT_p/C_v\rho}$ in diamond, the heat will not be reflected from diamond-vacuum interface until after the microwave pulse is over. The quantities $K$, $C_v$ and $\rho$ are the thermal conductivity, specific heat and density of diamond respectively. The pulse length is represented by $T_p$. With a 16 ns pulse, $\lambda = 4 \ \mu m$. The coating layer essentially acts as a fast heat diffuser and heat reservoir to temporarily store the energy.

## II  COATINGS

The following sections present a detail analysis of the thermal property of such layered system, and electric property will be mentioned briefly.

## A   thermal diffusion

Thermal diffusion in multi-layer system is common in applications ranging from heat spread and heat sinks. There are many papers on this very subject. But most of them calculate the steady-state equilibrium distribution, a solution of Helmholtz equation. This section provides the solution of the time dependent thermal diffusion equation in a multi-layered system.

The governing equation for heat diffusion is

$$\rho C_v \frac{\partial T(t, \vec{x})}{\partial t} - K\nabla^2 T(t, \vec{x}) = P(t, \vec{x}), \tag{1}$$

where $\rho$, $C_v$ and $K$ are the density, specific heat and thermal conductivity of the material respectively. The power density is denoted by $P$.

Laplace transforming Eq. 1, we have

$$\rho C_v(p\bar{T} - T(0+, \vec{x})) - K\nabla^2 \bar{T} = \bar{P}, \tag{2}$$

where the Laplace transform is defined as

$$\bar{T} \equiv L(T) = \int_0^\infty Te^{-pt} \, dt, \tag{3}$$

and the result

$$L(\frac{\partial T}{\partial t}) = pL(T) - T(0+) \tag{4}$$

is used to obtain Eq. 2.

677

To simplify the problem, we consider only the case of one spatial dimension. The operator $\nabla^2$, therefore reduces to $\frac{\partial^2}{\partial x^2}$. We assume the initial condition $T(0+, x) = 0$, which leads to

$$-K\frac{\partial^2 \bar{T}}{\partial x^2} + \rho C_v p\bar{T} = \bar{P}. \tag{5}$$

The homogeneous solution of Eq. 5 is then

$$\bar{T} = ae^{-\sqrt{p/D}x} + be^{\sqrt{p/D}x}, \tag{6}$$

where diffusivity is defined as $D \equiv K/\rho C_v$ and $a$ and $b$ are constants.

In a two medium system, illustrated in Fig. 2, the temperature distribution $T_1$

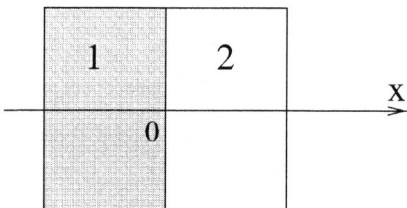

**FIGURE 2.** Heat diffusion in two medium with source at the interface.

and $T_2$ in the two regions has the form

$$\left.\begin{array}{rcl} \bar{T}_1 & = & a_1 e^{\sqrt{p/D_1}x} \\ \bar{T}_2 & = & a_2 e^{-\sqrt{p/D_2}x} \end{array}\right\} \tag{7}$$

as a result of the boundary condition at infinity. The heat source $\bar{P}$ is assumed to be a delta function $\delta(x)$. Continuity at $x = 0$ requires that $a_1 = a_2$. Integrating Eq. 5 from $x = 0^-$ to $x = 0^+$ yields

$$(-K_2)\frac{\partial T_2}{\partial x}\Big|_{0^+} - (-K_1)\frac{\partial T_1}{\partial x}\Big|_{0^-} = \int_{0^-}^{0^+} \bar{P} = 1. \tag{8}$$

Substituting Eq. 7 in Eq. 8, we obtain

$$a_1 = \frac{1}{1 + Z_2/Z_1}\frac{1}{Z_1\sqrt{p}}, \tag{9}$$

where we defined the heat wave impedance $Z_i \equiv \sqrt{\rho_i C_{vi} K_i}$. Using the property

$$L^{-1}\left(\frac{1}{\sqrt{p}}e^{-a\sqrt{p}}\right) = \frac{1}{\sqrt{\pi t}}e^{-a^2/4t}, \tag{10}$$

we obtain the temperature distribution in the medium

$$T_i = \frac{1}{1 + Z_2/Z_1} \frac{1}{Z_1 \sqrt{\pi t}} e^{-x^2/4D_i t}. \tag{11}$$

The solution is the Green's function of diffusion equation with source at $x = 0$. Eq. 11 can be understood in terms of heat generated at $x = 0$ being shared in the two medium according to their respective thermal wave impedances $Z_i$. When the two media are the same, Eq. 11 reduces to the well known result for the Green's function in a uniform medium

$$G(x, x', t, t') = \frac{1}{\sqrt{4\pi}} \frac{1}{\sqrt{K\rho C_v(t - t')}} e^{-(x-x')^2/4D(t-t')} \tag{12}$$

Since the skin depth at 91 GHz is only 0.22 $\mu$m in copper, the diffusion length of a 16 ns pulse $\sqrt{D_1 T_p} = 1.4\mu m$ is much larger. Thus to a good approximation, we may assume all the heat is generated at the interface $x = 0$. We will deal with finite skin depth effect later. The surface temperature rise is then

$$T = \frac{1}{1 + Z_2/Z_1} \frac{2P_0 T_p}{\sqrt{\pi K_1 \rho_1 C_{v1} T_p}} \tag{13}$$

after integrating Eq. 11 over a square pulse of length $T_p$. The surface power flux is represented by $P_0$. With diamond coating on copper, the peak temperature is $(1+Z_2/Z_1)^{-1} = 0.36$ times that of copper alone. The diamond thermal conductivity, density and specific heat are 1800 W/m·K, 3.54 kg/m³ and 674 J/kg·K respectively; and those of copper are 401 W/m·K, 8.96 kg/m³ and 385 J/kg·K.

## B    three-layer model

The above estimation assumes an infinite thick diamond layer. One way of modeling the finite diamond coating thickness is a three-layer configuration, illustrated in Fig. 3. Assuming the heat source is located at $x = x_1$, the temperature $\bar{T}_i$ can

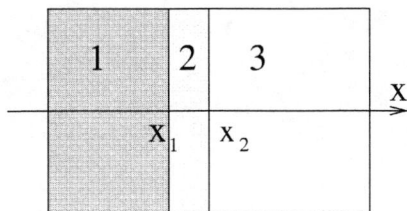

**FIGURE 3.** Three-layer coating model. The heat source is assumed to be located at $x = x_1$.

be expressed as

$$\left.\begin{aligned}
\bar{T}_1(x, p) &= a_1 e^{\sqrt{p/D_1}(x-x_1)} \\
\bar{T}_2(x, p) &= a_2 e^{-\sqrt{p/D_2}(x-x_1)} + b_2 e^{\sqrt{p/D_2}(x-x_1)} \\
\bar{T}_3(x, p) &= a_3 e^{\sqrt{p/D_3}(x-x_2)}
\end{aligned}\right\} \tag{14}$$

The continuity of temperature and heat flux at $x = x_2$ requires

$$
\left.
\begin{aligned}
a_2 e^{-\sqrt{p/D_2}(x_2-x_1)} + b_2 e^{\sqrt{p/D_2}(x_2-x_1)} &= a_3 \\
a_2 e^{-\sqrt{p/D_2}(x_2-x_1)} - b_2 e^{\sqrt{p/D_2}(x_2-x_1)} &= \frac{Z_3}{Z_2} a_3
\end{aligned}
\right\}
\tag{15}
$$

Solving Eq. 15, we obtain

$$
S_{22} \equiv \frac{b_2}{a_2} = \frac{1 - Z_3/Z_2}{1 + Z_3/Z_2} e^{-2\sqrt{p/D_2}(x_2-x_1)}.
\tag{16}
$$

Similarly, at $x = x_1$ we have

$$
\left.
\begin{aligned}
a_2 + b_2 &= a_1 \\
k_2\left(-a_2\sqrt{\tfrac{p}{D_2}} + b_2\sqrt{\tfrac{p}{D_2}}\right) - k_1 a_1 \sqrt{\tfrac{p}{D_1}} &= -1
\end{aligned}
\right\}
\tag{17}
$$

Combining Eq. 16 and 17, we can solve for $a_1, a_2, b_2$ and $a_3$:

$$
\left.
\begin{aligned}
a_1 &= \frac{1}{\sqrt{p} Z_1 (1+Z_2/Z_1)} \frac{1+S_{22}}{1+\alpha_{12} S_{22}} \\
a_2 &= \frac{1}{\sqrt{p} Z_1 (1+Z_2/Z_1)} \frac{1}{1+\alpha_{12} S_{22}} \\
b_2 &= \frac{1}{\sqrt{p} Z_1 (1+Z_2/Z_1)} \frac{S_{22}}{1+\alpha_{12} S_{22}} \\
a_3 &= \frac{1}{\sqrt{p} Z_1 (1+Z_2/Z_1)} \frac{e^{-\sqrt{p/D_2}(x_2-x_1)}}{1+\alpha_{12} S_{22}} \frac{2}{1+Z_3/Z_2}
\end{aligned}
\right\}
\tag{18}
$$

where

$$
\alpha_{12} = \frac{1 - Z_2/Z_1}{1 + Z_2/Z_1}.
\tag{19}
$$

The temperature at $x = x_1$ follows

$$
\bar{T}_1(x_1, p) = a_1 = \frac{1}{\sqrt{p} Z_1 (1 + Z_2/Z_1)} (1 + S_{22}) \sum_{n=0}^{\infty} (-\alpha_{12} S_{22})^n,
\tag{20}
$$

where we have expanded the denominator. Eq. 20 can be simplified further to read

$$
\bar{T}_1(x_1, p) = \frac{1}{(1 + Z_2/Z_1)} \frac{1}{Z_1\sqrt{p}} \left[ 1 + \sum_{n=1}^{\infty} (1 - \frac{1}{\alpha_{12}})(-\alpha_{12})^n \alpha_{23}^n e^{-2n\sqrt{p/D_2}\Delta x} \right],
\tag{21}
$$

where $\alpha_{23} = (1 - Z_3/Z_2)/(1 + Z_3/Z_2)$ and $\Delta x = x_2 - x_1$. Utilizing Eq. 10, we obtain the temperature at $x_1$

$$
T_1(x_1, t) = \frac{1}{1 + Z_2/Z_1} \frac{1}{\sqrt{K_1 \rho_1 C_{v1} \pi t}} \left[ 1 + \sum_{n=1}^{\infty} (1 - \frac{1}{\alpha_{12}})(-\alpha_{12})^n \alpha_{23}^n e^{-4n^2 \Delta x^2/4 D_2 t} \right].
\tag{22}
$$

Compared to Eq. 11 with $x = 0$, the finite thickness of medium 2 gives a correction factor; each term under the sum represents the multiple reflection of heat wave between 2 interfaces. Integrating Eq. 22 with a square pulse, we obtain the surface temperature

$$T = \frac{1}{1 + Z_2/Z_1} \frac{2P_0 T_p}{\sqrt{\pi K_1 \rho_1 C_{v1} T_p}} [1 + \alpha_c] \tag{23}$$

where the coating thickness correction factor

$$\alpha_c = \sum_{n=1}^{\infty} (1 - \frac{1}{\alpha_{12}})(-\alpha_{12})^n \alpha_{23}^n (e^{-n^2 \Delta x_s^2} - n\Delta x_s \sqrt{\pi} \mathrm{erfc}(n\Delta x_s)), \tag{24}$$

and the scaled coating thickness $\Delta x_s = \Delta x / \sqrt{D_2 T_p}$.

The value of $\alpha_c$, plotted in Fig. 4 for finite diamond thickness on copper, naturally

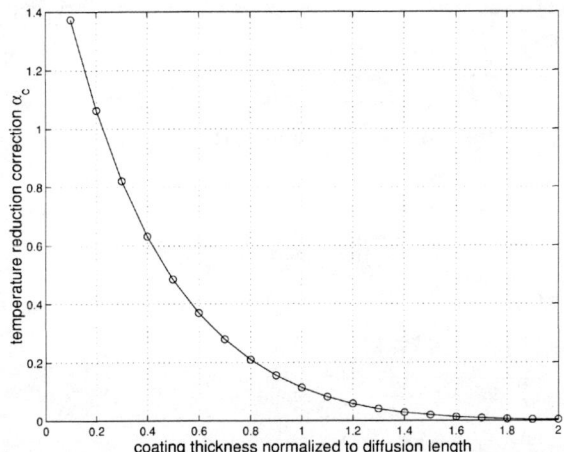

**FIGURE 4.** The correction factor $\alpha_c$ plotted as a function of coating thickness, which is normalized to diffusion length $\sqrt{D_2 T_p}$.

approaches zero as $\Delta x$ becomes large. But even with a coating thickness of 1.5 and 2 diffusion lengths, the correction is only 2% and 0.2% respectively.

## C   RF skin depth effect

The effect of RF skin depth can also be calculated from the three-layer model if we assume medium 1 and 2 are the same metal and medium 3 is diamond. From Eq. 18, we obtain

$$\bar{T}(x_2, p) = a_3 = \frac{1}{1 + Z_2/Z_1} \frac{1}{\sqrt{p} Z_1} e^{-\sqrt{p/D_1} \Delta x}, \tag{25}$$

where the subscript is renumbered to suit Fig. 5. Taking the inverse Laplace transform, we get

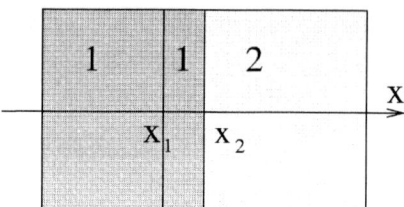

**FIGURE 5.** Model the effect of skin depth on temperature rise. Number 1 and 2 denote metal and diamond respectively.

$$T(x_2, t) = \frac{1}{1 + Z_2/Z_1} \frac{1}{\sqrt{\pi K_1 \rho_1 C_{v1} t}} e^{-\Delta x^2/4D_1 t}. \tag{26}$$

The RF loss decays exponentially into the metal, $e^{-2\Delta x/\delta}$, where $\delta$ is the skin depth. Considering an arbitrary pulse, the surface temperature at the end of the pulse is

$$T = \int_0^{T_p} \frac{P_0 f(t)}{1 + Z_2/Z_1} \frac{dt}{\sqrt{\pi K_1 \rho_1 C_{v1} t}} \int_0^\infty \frac{2d\Delta x}{\delta} e^{-2\Delta x/\delta} e^{-\Delta x^2/4D_1 t}, \tag{27}$$

where $f(t)$ represents the temporal profile of the pulse. After integrating $\Delta x$, we obtain

$$T = \frac{1}{1 + Z_2/Z_1} \frac{2P_0 T_p}{\sqrt{\pi K_1 \rho_1 C_{v1} T_p}} \frac{\sqrt{\pi}}{\delta_s} \int_0^1 e^{4t'/\delta_s^2} \mathrm{erfc}(2\sqrt{t'}/\delta_s) f(T_p t') dt'. \tag{28}$$

where the scaled skin depth $\delta_s = \delta/\sqrt{D_1 T_p}$. In the case of a square pulse, The integration gives

$$T = \frac{1}{1 + Z_2/Z_1} \frac{2P_0 T_p}{\sqrt{\pi K_1 \rho_1 C_{v1} T_p}} [1 + \alpha_s], \tag{29}$$

where the skin effect correction factor

$$\alpha_s = -\frac{1}{4}\sqrt{\pi}\delta_s + \frac{1}{4}\delta_s\sqrt{\pi}e^{4/\delta_s^2}\mathrm{erfc}(\frac{2}{\delta_s}) \tag{30}$$

$$= -\frac{1}{4}\sqrt{\pi}\delta_s + \frac{1}{8}\delta_s^2[1 + \sum_{m=1}^\infty (-)^m (2m-1)!!(\frac{\delta_s^2}{8})^m] \tag{31}$$

The first linear term of $\delta_s$ is the heat spread due to skin depth effect even without diffusion. It results in a lower surface temperature. The value of $\alpha_s$ is plotted in Fig. 6. For a skin depth of 0.22 $\mu$m and 16 ns pulse, $\alpha_s = -0.068$.

Taking another limit that skin depth is much greater than diffusion length, i.e. $\delta_s \gg 1$ or $T_p \ll 0.4$ ns at 91 GHz, Eq. 29 and 30 becomes

$$T = \frac{1}{1 + Z_2/Z_1} \frac{2P_0 T_p}{\rho_1 C_{v1} \delta} [\sqrt{\pi} - \frac{8}{3}\frac{1}{\delta_s} + O(\frac{1}{\delta_s^2})]. \tag{32}$$

The heat is concentrated in a layer roughly equal to the skin depth and diffusion is small. In this region the surface temperature is approximately linear with pulse length $T_p$ and the interior of the metal may be much hotter than the surface.

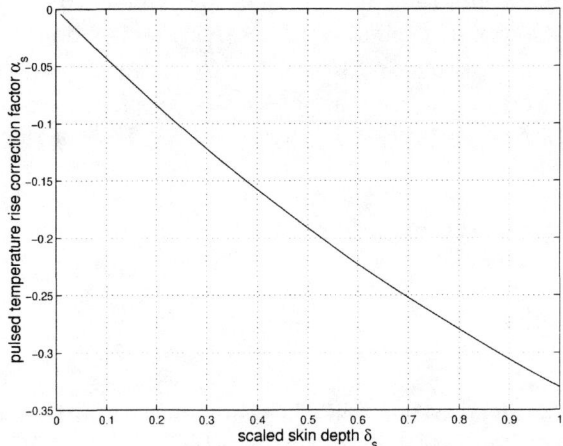

**FIGURE 6.** correction factor from skin effect.

# D  microwave property

Diamond not only has outstanding mechanical property, it is an excellent insulator with electric resistivity on the order of $10^{18}$ $\Omega$m and dielectric strength greater than 1 GeV/m [2] [3]. Its loss tangent $\tan \delta < 5 \times 10^{-4}$ is quite low. Thus for a pillbox cavity, the $Q_d$ due to dielectric loss is approximately $l\epsilon'/\Delta x \tan \delta$, where $l$ and $\Delta x$ are the cavity length and coating thickness respectively. With a 8 $\mu$m coating in a 250$\mu$m long cavity, $Q_d = 3 \times 10^5$. The dielectric loss is orders of magnitude smaller than metal wall loss.

# E  multi-layer structure

Taking advantage of the fact that diamond conducts heat better than copper. A three-layer system is proposed to reduce the pulsed temperature rise even further. Two advantages are realized here: First diamond conducts heat better than copper.

**FIGURE 7.** Three-layer configuration to reduce pulsed temperature rise.

Second heat generation is reduced by proper choice of metal thickness. The conduction current induced by the surface magnetic field decays exponentially inside metal; it also alternates direction with respect to the surface value because of the

683

phase lag. Choosing a film thickness of 0.345 $\mu$m, about 1.57 times skin depth at 91 GHz, the reverse current is eliminated and RF loss is reduced by 8% [4].

Thin metal film on the order of skin depth $\sqrt{\omega\mu/2\sigma}$ provides excellent RF shielding due to almost perfect reflection at the metal-dielectric interface. For a coating thickness about 1.57 times skin depth, less than $2 \times 10^{-4}$ of the total power loss leaks through metal film [4].

Taking the approximation that the metal film thickness is much smaller than the diffusion length, we can estimate the pulsed temperature rise from a point source in a medium of diamond, which gives

$$T = \frac{2P'_0 T_p}{\sqrt{4\pi K_2 \rho_2 C_{v2}}} = \frac{1}{2Z_2/Z_1} \frac{2P'_0 T_p}{\sqrt{\pi K_1 \rho_1 C_{v1} T_p}}, \tag{33}$$

where $P'_0 = 0.92 P_0$. The two diamond layer configuration further reduces the pulse temperature rise to 0.26 times that of the bare copper.

# III EXPERIMENT

We have chosen the microwave plasma enhanced CVD diamond process for its economy, deposition rate, lower temperature and good quality control [5]. Typically, methane and hydrogen mixture are introduced into the chamber. Under the microwave electric field, the gases are ionized into electrons and ions. The electrons, with their small mass, are quickly accelerated to high energy about a few thousands of degree or higher. The high energy electrons collide with gas molecules with resulting dissociation and generation of reactive chemical species and initiation of chemical reaction. The basic reaction is

$$CH_4 \rightarrow C + 2H_2$$

The substrate must be heated to 800-1000$^o$ C for diamond to form because at lower temperature graphite is the thermodynamically stable allotrope. The resulting polycrystalline diamond has almost the same property as single crystal diamond.

The vastly different thermal expansion coefficient between Copper ($17 \times 10^{-6}$/K) and diamond ($1 \times 10^{-6}$/K) makes it very difficult to grow adherent and continuous diamond film on copper substrate. The fact that copper does not form carbide also makes adhesion extremely poor. The diamond film can usually be pulled off after the substrate is cooled to room temperature.

The key to growing diamond on copper is to find an appropriate interlayer that bonds well with copper as well as diamond. We have tentatively identified Titanium Tungsten which has excellent adhesion with copper substrate and form strong covalent bond with carbon. A proof of principle experiment at X-band is being pursued.

There is program at SLAC to look at the maximum pulsed temperature rise that copper can withstand [6]. A TE$_{011}$ mode cavity with removable end plates is powered by approximately 1.5 $\mu$s 20 MW of X-band (11.424 GHz) pulse. Hundreds of degrees surface temperature rise can be achieved. A removable end plate coated with 70 $\mu$m of diamond, corresponding to about 2 diffusion lengths in diamond of a 1.5 $\mu$s pulse, should be able to reduce the pulsed temperature rise to 36% that of bare copper end plate.

## IV    CONCLUSION

Pulsed temperature rise is likely the most severe restriction on the gradient in a structure based accelerator. Diamond coating to reduce pulsed temperature rise by a factor of 2.7 to 3.8 is possible if appropriate interlayer can achieve good adhesion. A proof of principle experiment is being carried out. Diamond coating also has important application in high power vacuum device, especially in the output cavities.

## REFERENCES

1. S. Schultz, private communication.
2. L. S. Pan and D. R. Kania, "Diamond: electric properties and applications", pp 254.
3. K. E. Spear and J. P. Dismukes. "Synthetic Diamond: emerging CVD science and technology", pp 363.
4. Xintian E. Lin, "RF loss and leakage through thin metal film", submitted to IEEE transaction MTT.
5. H. O. Pierson, "Handbook of carbon, graphite, diamond and fullerenes", pp 302-321.
6. D. Pritzkau, et al. "Experimental study of pulsed heating of electromagnetic cavities", 1997 IEEE PAC.

# Coherent Multimoded Dielectric Wakefield Accelerators

J. Power, W. Gai and P. Schoessow

Argonne National Laboratory, High Energy Physics Division

9700 S Cass Ave, Argonne IL 60439

**Abstract.** There has recently been a study of the potential uses of multimode dielectric structures for wakefield acceleration [1]. This technique is based on adjusting the wakefield modes of the structure to constructively interfere at certain delays with respect to the drive bunch, thus providing an accelerating gradient enhancement over single mode devices. In this report we examine and attempt to clarify the issues raised by this work in the light of the present state of the art in wakefield acceleration.

## INTRODUCTION

The utility of dielectric loaded structures as wakefield accelerators has been investigated for the past decade [2–4]. Some progress has been made towards demonstrating the feasibility of this technique [5,6].

Previous work on dielectric devices has in general assumed that only a single $TM_{01}$ accelerating mode is present in the structure. Recently, a new scheme of wakefield acceleration using a dielectric loaded waveguide which makes use of many $TM_{0n}$ modes excited by a drive bunch train to enhance the accelerating gradient [1]. The structure is cleverly designed so that the coherent sum of the modes Fourier synthesizes a longitudinal wake which on the axis of the structure approximates a train of alternating sign delta functions.

In this report, we examine this interesting scheme, considering only the longitudinal wakefields. We attempt to clarify some of the confusion in terminology which has arisen from this work.

## WAKEFIELDS IN DIELECTRIC LOADED WAVEGUIDES

Consider a dielectric channel with an axial vacuum channel and conductor on the exterior. The dielectric channel has inner radius $a$ and outer radius $b$ with dielectric constant $\epsilon$. When a single charged particle beam (rms pulse length $\sigma_z$

CP472, *Advanced Accelerator Concepts: Eighth Workshop,*
edited by W. Lawson, C. Bellamy, and D. Brosius
1999 The American Institute of Physics 1-56396-889-4

and total charge $Q$) passes through the channel, it will produce longitudinal wake which can be expressed as

$$W_z(r, z, t) = Q \sum_n W_{zn}(a, b, \epsilon) \exp[-(\frac{2\pi\sigma_z}{\lambda_n})^2] \cos(\frac{2\pi z}{\lambda_n}) \tag{1}$$

where $\lambda_n$ is the wavelength of the nth mode, and the coefficients $W_{zn}$ depend only on the geometry of the structure.

If multiple bunches with longitudinal separation d are used then the beam distribution can be expressed as

$$f(z) = \frac{1}{\sqrt{2\pi}\sigma_z} \sum_l \exp[\frac{-(z - ld)^2}{2\sigma_z^2}]. \tag{2}$$

Then the wakefield can be expressed anywhere (including the region inside the bunch) as

$$W_z(z) = \sum_l \sum_n W_{zn}(a, b, \epsilon) \int_{-\infty}^{z} f(z') \cos[\frac{2\pi(z - z')}{\lambda_n}] dz'. \tag{3}$$

If $\sigma_z$ is comparable to $\lambda_n$, then modes up to the nth mode will be excited.

## TRANSFORMER RATIO

The conventional transformer ratio in a collinear wakefield accelerator is defined as [7]

$$R = -E_z^+ / E_z^- \tag{4}$$

Where $E_z^-$ is the *peak* deceleration field within the drive bunch and $E_z^+$ is the maximum accelerating field a test particle can see while travelling behind the drive bunch. In general, for a Gaussian drive beam profile, the transformer ratio cannot exceed 2.

(There are a number of circumstances where $R > 2$ can be attained through the use of noncollinear structures, nonlinearity, and multimoding. Reference [7] gives an example where a modest enhancement in transformer ratio beyond 2 is obtained in a multimode structure, but it is difficult to obtain significant enhancements to $R$ in this manner.)

However, if one uses an alternative definition of transformer ratio

$$R' = -E_z^+ / \overline{E_z^-} \tag{5}$$

where $\overline{E_z^-}$ is the average energy loss (or average deceleration field) in the drive beam (defined as the "drag field" in [1]), then by this alternative definition of the transformer ratio indeed can exceed 2. It is this definition which is used by the authors of [1].

# NUMERICAL EXAMPLES

In this section we will show some examples of wakefields in multimoded structures. The wakes are calculated using the analytic approach of reference [3]. The transcendental equation obtained for the dispersion relation is solved numerically for each mode and the mode amplitude $W_{zn}(a, b, \epsilon)$ is evaluated. The modes are then summed to obtain the wake potential according to Equation 3.

We first calculate the wakefield in a $\lambda = 21$ cm device designed by Hirshfield as a multimode device. This is a circular dielectric waveguide with $a = 0.15$ cm, $b = 1.97$ cm, $L = 60$ cm, and $\epsilon = 9.43$. The wake potential for a $\sigma_z = 0.1$ cm bunch (including the region inside the beam) is shown in figure 1. The transformer ratio $R = 2.384/1.307 = 1.824 < 2$.

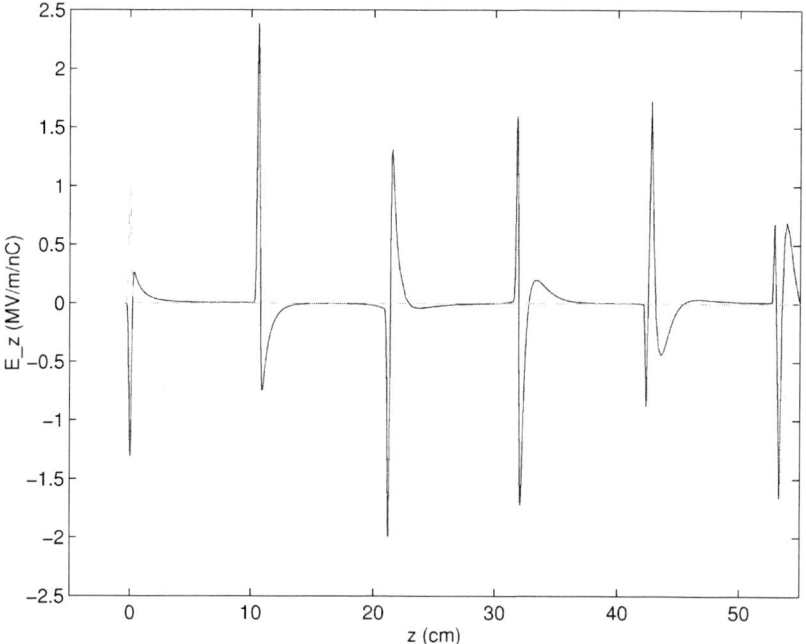

**FIGURE 1.** Wake potential for multimode 21 cm wavelength structure. The dashed curve shows the beam charge distribution.

Figures 2 shows the mode spectrum for this structure weighted by the current form factor (Fourier transform of $f(z)$) for 200 modes. It is interesting to note that in contrast to most structures considered for wakefield accelerators, the lowest lying modes do not dominate the spectrum.

Another aspect of the work in reference [1] is the suggestion that driving a multimode structure with a train of periodically spaced bunches would allow stimulated

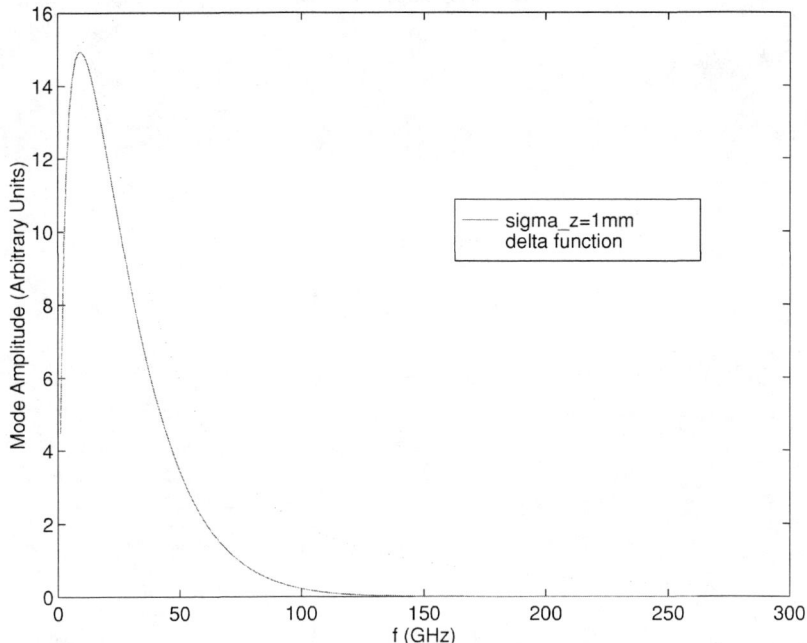

**FIGURE 2.** Current form factor weighted amplitude distribution for the first 200 modes of the 21 cm wavelength structure.

emission of wakefield energy to occur at a rate larger than the coherent radiation from a single bunch with the same total charge. Our calculations suggest that this is not the case. Figure 3 shows a comparison of the wake potentials of two 1 nC bunches spaced at 21 cm (the distance between the leading bunch and its first decelerating maximum) and that of a single 2 nC bunch. There is no apparent enhancement in the two bunch over the one bunch wake, and in fact this is to be expected by the principle of superposition.

We are planning to measure the wakes in a multimoded device optimized for the parameters of the AWA. For a structure with $a = 0.5$ cm, $b = 1.467$ cm, $\epsilon = 36$, the wavelength is 23.05 cm. The wake potential for this structure (taking $\sigma_z = 1$ mm) is shown in figure 4. Materials for this device have been ordered and the experiment is planned for later this year. The gradient expected for this device, using a train of 4x10 nC electron bunches is 16 MV/m.

## SUMMARY

We have studied the newly proposed multimode wakefield acceleration scheme. We find that using the usual definition of the transformer ratio, the wakefield

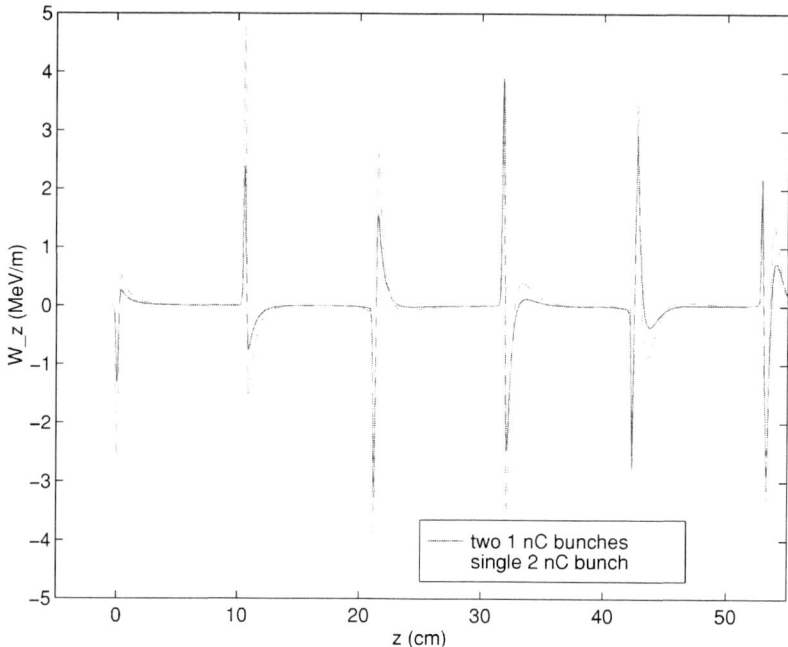

**FIGURE 3.** Comparison of wakes in multimode structure driven by two 1 nC bunches separated by 21 cm vs wake driven by a single 2 nC bunch.

theorem still holds, even though our numerical calculations agree with reference [1]. We also found the wakefield enhancement due to the use of multiple drive bunches simply scales with the total charge if one assumes all the beam parameters are the same for the individual bunches in the train. However, we believe that the multiple drive bunch scheme has an advantage considering both the capabilities of rf photocathode electron sources to produce bunch trains and the reduced sensitivity of multiple beams to parasitic wakefields in dielectric structures. We have designed an experiment at AWA to demonstrate this technique.

## ACKNOWLEDGEMENTS

We would like to thank J. Hirshfield and T. Zhang for stimulating discussions in this area. This work is supported by the Department of Energy, Division of High Energy Physics, under contract W-31-109-ENG-38.

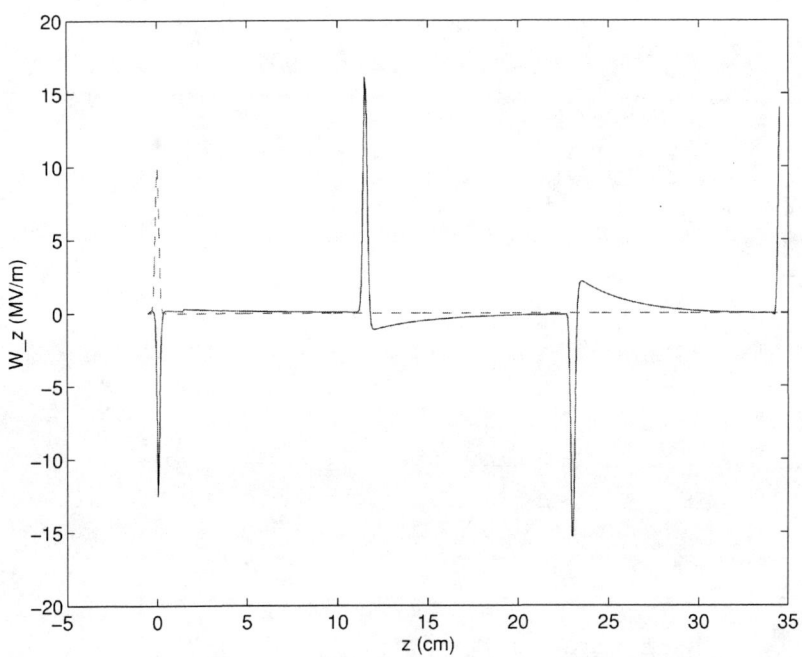

**FIGURE 4.** Wake potential for the planned AWA multimode structure experiment.

# REFERENCES

1. T-B. Zhang, J. L. Hirshfield, T. C. Marshall, B. Hafizi, Phys. Rev. **E56** 4647 (1997)
   J. L. Hirshfield et al., these Proceedings
2. R.Keinigs, M. Jones, W.Gai, Part. Accel. **24** 223 (1989)
3. M. Rosing, W. Gai, Phys. Rev. **D42** 1829 (1990)
4. W. Gai, P. Schoessow, B. Cole, R. Konecny, J. Norem, J. Rosenzweig, J. Simpson, Phys. Rev. Lett. **61** 2756 (1988)
5. P. Schoessow, M. E. Conde, W. Gai, R. Konecny, J. Power, J. Simpson, to appear in J. Appl. Phys.
6. M. E. Conde, W. Gai, R. Konecny, J. Power, P. Schoessow, P. Zou, these Proceedings
7. K. Bane, P. Chen, P. Wilson, IEEE Trans. Nucl. Sci. **NS-32** 3524 (1985)

# An Optimized Slab-Symmetric Dielectric-Based Laser Accelerator Structure[*]

## J.B. Rosenzweig

*UCLA Department of Physics and Astronomy*
*405 Hilgard Ave., Los Angeles, CA 90095-1547*

## P. V. Schoessow

*Argonne National Laboratory*
*9700 S. Cass Ave., Argonne, IL 60439*

**Abstract.** A slab-symmetric, partially dielectric filled, laser excited structure which may be used to accelerate charged particles is analyzed theoretically and computationally. The fields associated with the accelerating mode are calculated, as are aspects of the resonant filling and impedance matching of the structure to the exciting laser. It is shown through computer simulation that the accelerating mode in this structure can be excited resonantly and with large quality factor $Q$.. Practical aspects of implementing this structure as an accelerator are discussed.

The proposed use of lasers to accelerate charged particles has attracted much attention in recent years, spurred by the availability of increasingly large electromagnetic power densities from ever more sophisticated, yet cost-effective, laser sources. The proposed laser acceleration schemes range from mode conversion type devices, which the laser power is transformed into a different form of wave (e.g. plasma beat-wave accelerators[1]), to devices in which the laser is used to directly produce the accelerating fields. Direct laser accelerators can further be divided into two types of schemes — transverse-field dominated systems such as inverse Cerenkov accelerators[2], so-called vacuum accelerators, or other small-angle crossed laser beam devices[3,4], and systems in which the accelerating field profile is dominated by boundary conditions which are less than a vacuum wavelength away from the charged particle beam path[5]. The transverse-field dominated, or far-field, devices have the advantage of mitigating the fields on which induce breakdown on whatever structures are used to shape the field profile, but do so at a price — the longitudinal or accelerating fields are small compared to the transverse fields which may be experienced by the beam particles, leading to inefficient acceleration, and high sensitivity to asymmetries in the field profile. These asymmetries may lead to acceleration which is dependent on transverse position and concomitant strong transverse deflections of the beam particles.

On the other hand, in structures in which the field shaping boundary conditions are nearby, which are termed near-field structures, the material medium of the structure must support field levels a bit larger than the peak accelerating field experienced by the beam particles. For optical frequencies, the materials most resistant to field break down

---

[*] This work was supported by U.S. Dept. of Energy grants DE-FG03-93ER40796 and DE-FG03-92ER40693, and the Alfred P. Sloan Foundation grant BR-3225.

CP472, *Advanced Accelerator Concepts: Eighth Workshop,*
edited by W. Lawson, C. Bellamy, and D. Brosius

are dielectrics, which can support electric fields for short times (a few picoseconds), of a few GV/m[6], thus leading to a practical limit of about 1 GeV/m acceleration rate in a dielectric-based device. This limit being recognized, near field structures present considerable advantages over far-field devices[5]: they can be made resonant, allowing much smaller peak power lasers to be used; symmetry in the fields can be enforced more straightforwardly; the periodicity of the field can be controlled to keep the beam particles synchronous with the accelerating wave. In addition, use of structures with slab, as opposed to cylindrical symmetry, gives the additional advantage of allowing a larger current to be loaded into the device without inducing excessive beam loading, as the accelerating beam can be spread out in the "wide" transverse dimension. Furthermore, it has been shown that use of a beam which is wide compared to the structure gap greatly suppresses the excitation of dipole modes in the device, thus enhancing the current which can be confidently accelerated before onset of the beam breakup instability[7].

Given these advantages, it was proposed in Ref. 5 that a structure based on a Fabry-Perot highly reflective mirror pair, loaded with a dielectric whose permittivity is modulated in the longitudinal dimension, be used as a near-field accelerator. The purpose of the permittivity modulation in the device was two-fold: first, to enforce the proper periodicity of the resonant fields, giving a field profile which is optimized for acceleration, and second, to modulate the phase of the incoming light to couple the electromagnetic power into the accelerating mode. Unfortunately, simulations of this geometry showed that it was difficult to couple into the accelerating mode in the structure, and that a strong non-accelerating component was always present in the mode. This paper presents a new, simplified geometry, in which the functions of shaping of the resonant fields in the structure, and coupling of the laser power to the structure, are split. It will be shown that the performance of this system is quite good, leading to fields which are strongly reminiscent of those found in standard radio-frequency linear accelerators.

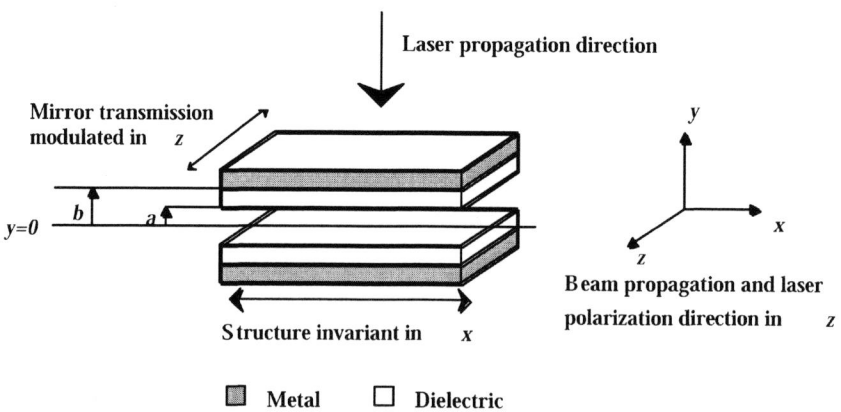

**FIGURE 1.** Geometry of slab-symmetric, laser-excited, dielectric-loaded resonant accelerator structure.

The structure geometry considered in this paper is shown in Figure 1. It consists of a pair of infinite (in $x$ and $z$) dielectric slabs of permittivity $\varepsilon$ and thickness $b - a$, separated by a gap of distance $2a$. The slabs are clad at $|y| = b$ by metallic mirrors whose transmission coefficient $T$ is modulated periodically in $z$. While in practice the reflecting surfaces which bound the structure may be made of dielectric layers, the use of the metallic boundary conditions in our analysis allows simple evaluation of the eigenmodes we wish to excite without loss of generality. In this structure the separation of functions is now clear cut; the dielectric, with metallic boundary conditions at $|y| = b$, specifies the electromagnetic mode structure, while the modulation of the mirror transmission coefficient controls the coupling of the incoming laser power to the trapped mode, which instead of being Fabry-Perot-like, is now a standing wave pattern capable of accelerating charged particles.

The electromagnetic characteristics of the resonant modes of the structure are analyzed using the usual approximation of perfectly conducting metallic boundaries. In this limit, the fundamental mode for the structure can be simply deduced from previous work on travelling waves in such a device found in Ref. 7. Inside of the gap ($|y| < a$), we have for the fundamental symmetric mode

$$E_z = E_0 \cos [kz] \cos [\omega t], \tag{1}$$

where $E_0$ is the field amplitude and $\omega = kc$. This speed-of-light phase velocity condition is enforced by the periodicity of the modulation in the transmissivity, which thus must be chosen with to be the free-space wavelength of the exciting laser. Without preferential coupling to the speed-of-light phase velocity components (spatial harmonics), it should be noted that the structure is essentially invariant under displacement in $z$. The fields which obey this symmetry belong to what is termed in accelerator physics a "zero-mode", meaning a mode in which the phase advance per period is zero. This fundamental spatial harmonic of this mode is the familiar Fabry-Perot field pattern; the accelerating fields we wish to excite are actually at the first higher harmonic of the Floquet expansion of fields[5].

In the dielectric ($a < |y| < b$) we have

$$E_z = AE_0 \cos [kz] \cos [\omega t] \sin \left[ k\sqrt{\varepsilon - 1}(y - b) \right], \tag{2}$$

Application of the boundary conditions at $y = a$ allows the eigenvalue of this mode to be determined through the transcendental relation

$$\cot \left( k\sqrt{\varepsilon - 1}(b - a) \right) = ka \frac{\sqrt{\varepsilon - 1}}{\varepsilon}, \tag{3}$$

and the relative amplitude of the longitudinal electric field within the dielectric to be obtained,

$$A = \csc\left[k\sqrt{\varepsilon - 1}\,(b - a)\right].\qquad(4)$$

For a gap much smaller than the free space wavelength $ka \ll 1$, the right hand side of Eq. 3 vanishes, and it is clear that $k \cong \pi/2\sqrt{\varepsilon - 1}\,(b - a)$, and that $A \cong 1$. On the other hand, for a large gap $(ka \gg 1)$, the right hand side of Eq. 3 is large, $k \cong \pi/\sqrt{\varepsilon - 1}\,(b - a)$, and $A \cong \sqrt{\varepsilon - 1}/\varepsilon ka$. Thus for a large gap the field in the dielectric becomes much larger than that in the gap, and for a dielectric-breakdown limited situation the achievable accelerating field is strongly diminished. In this limit one can show that $E_y$ also becomes quite large near the dielectric boundary. It will also be shown below that large gap structures suffer from introduction of Fabry-Perot components of the mode. For all of these reasons, and despite the fact that it implies stringent limits on the allowable accepted beam emittance, operation with the small gap is strongly preferred.

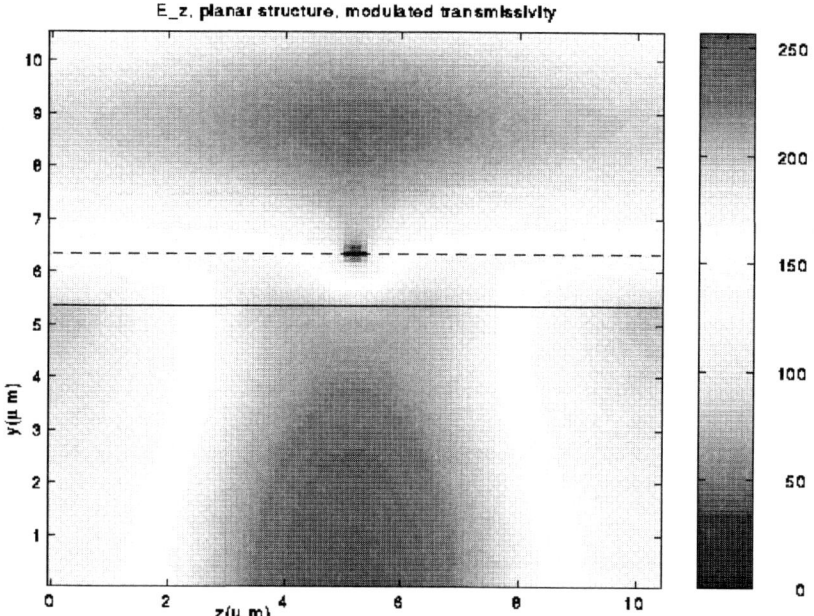

**FIGURE 2.** False-color contour map of $E_z$ in dielectric-loaded structure with modulated transmission mirrors, calculated by electromagnetic solver. The maximum transmission occurs at $z=5.3$ μm. The laser wavelength is $\lambda = 10.6$ μm; the structure parameters are $a=5.3$ μm, $b=6.27$ μm and $\varepsilon = 3.47$.

The model we have chosen for the structure neglects dissipative losses in the dielectric, but will allow dissipation in mirrors, as this is a standard way of computationally constructing a partially transmissive boundary. Thus the quality factor

$Q = \omega U / P$ is interpreted in the way familiar to designers of standing wave accelerators, with the power losses $P$ arising from ohmic losses in the cavity walls, as well as from fields radiated by the structure. Together, these effects combine to give an external $Q$ for the structure, with $1/Q = 1/Q_{ohmic} + 1/Q_{rad.}$. Further, as we are interested in accelerating fields near the breakdown limit, we are not free to choose this external $Q$ to be arbitrarily large, even though a high $Q$ device would lessen the required input laser power. The smallness of the assumed $Q$ (in the range of 100 to 1000) in practice may imply that $Q_{ohmic} >> Q_{rad.}$, which has implications for impedance matching of the structure to the input laser.

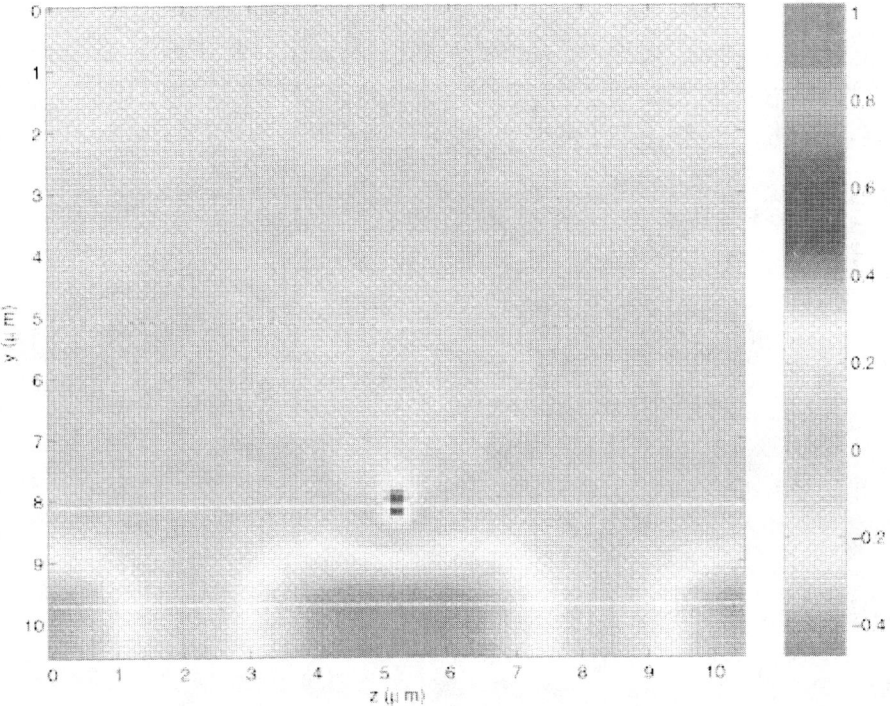

**FIGURE 3**. False-color contour map of $E_z$ in dielectric-loaded structure with modulated transmission mirrors, small gap case. The laser wavelength is $\lambda = 10.6$ µm; the structure parameters are $a=0.8$ µm, $b=2.4$ µm and $\varepsilon = 3.47$.

This structure has been studied by means of numerical simulations of the fields in a planar structure, performed using a custom 2-D (Cartesian) finite-difference time-domain electromagnetic solver code. Figure 2 shows a false-color contour map of $E_z$ after many fill structure times for a case in which the exciting laser ($\lambda = 10.6$ µm) impinges on a structure with $a=5.3$ µm, $b=6.27$ µm and $\varepsilon = 2.35$. Because of the computational symmetry, the laser is implicitly injected into both top and bottow mirrors of the structure. The mirror transmission coefficient is varied sinusoidally with variation between zero (z=0 and 10.6 μm) and one percent (z=5.3 μm). The gap size is large in this case, $ka = \pi$, and one can see that a pure accelerating mode, in which the phase fronts are normal to the beam (z) axis and $E_z$ is uniform in y, is not excited. The admixture of Fabry-Perot-like component, or spatial harmonic, in which $E_z$ is periodic in y at the free-space wavelength $\lambda = 2\pi / k$, is non-negligible in this case, in fact it is nearly equal to the accelerating component. This is because with large $ka$ the fundamental Fabry-Perot mode in this structure is nearly degenerate in frequency with the fundamental accelerating mode, and thus the fields excited display some characteristics of both modes.

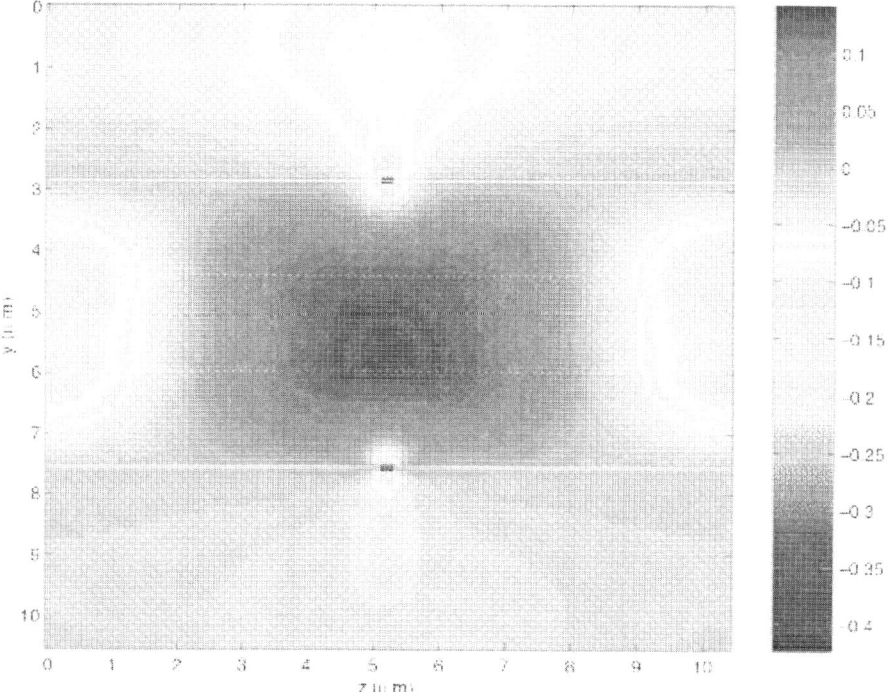

**FIGURE 4.** False-color contour map of $E_z$ in dielectric-loaded structure with modulated transmission mirrors, small gap case, without mirror symmetry enforced at z=0. The laser wavelength is $\lambda = 10.6$ μm; the structure parameters are a=0.8 μm, b=2.4 μm and $\varepsilon = 3.47$. Note that steady-state has not quite been reached, as the fields are still asymmetric within the structure.

This problem is overcome in the case of small $ka$, an example of which is displayed in Figure 3, where now a=0.8 μm, b=2.4 μm and $\varepsilon = 3.47$. The phase fronts $E_z$ are now much more uniform in y in the gap, and the ratio of the accelerating to Fabry-Perot spatial harmonics is roughly 6:1. In addition, the field in the device is better enhanced compared to the laser field than the case shown in Fig. 2.

With the structure's accelerating mode well understood by theory and the computational model, we turn to discuss some aspects of the structure coupling to the drive laser. In practice, one will not necessarily drive both sides of the structure, but illuminate one side, and use the output radiation emitted from the opposite side as a diagnostic tool. This case is shown in Figure 4, which shows the same dielectric-loaded structure as Fig. 3 without mirror symmetry enforced at $z=0$. In this case it can been seen that steady-state has not quite been reached, as the fields are still slightly asymmetric within the structure. In steady state with a lossless, impedance matched system, the field pattern should be symmetric with respect to the $z=0$ plane. The time dependence of the field amplitudes and stored (Figure 5) shows the approach to steady state typical of a high $Q$ resonator driven on its resonant frequency. In the case shown the $Q=80,000$ and the system has approached with characteristic fill time $\tau_f = Q/\omega =450$ psec. This is in fact a higher $Q$ than one needs in practice, as large fill times will imply more problems with avalanche breakdown of the structure.

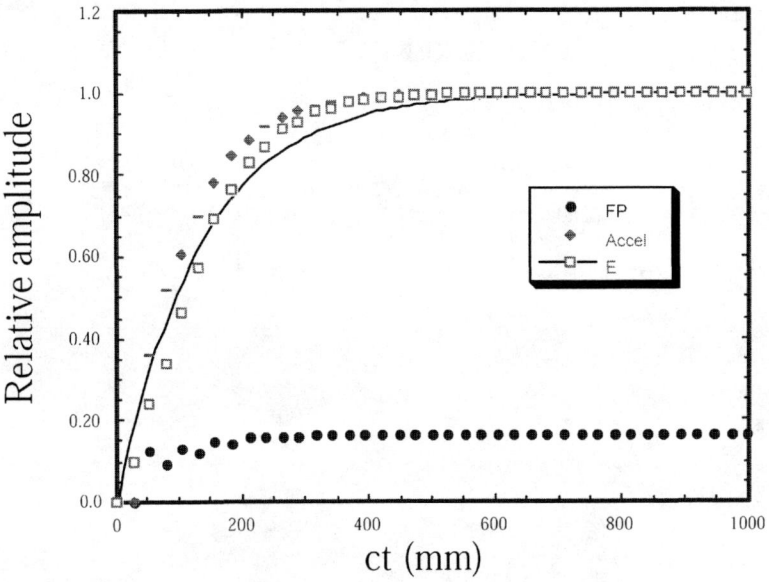

**FIGURE 5.** Time history of the accelerating (Accel), and Fabry-Perot (FP) components of the stored field in the structure for the case shown in Figure 4, with total stored energy (E).

In Figs. 3 and 4, it can be seen that the coupling system, while driving the correct mode pattern, does so at the price of introducing a high-field region near the mirror's highest transmissivity (1%) point. This is because, unlike a microwave cavity driven from a wave-guide, there is no region of free propagation linking the driving source and the resonant cavity. Thus the field must be large in the vicinity of the least poor transmissivity point. This effect in turn causes the introduction of higher spatial harmonics in the mode profile, which affect the field in the gap very little, as they exponentially attenuate in the vacuum region, but still serve to give a very high field region in the dielectric. This high field region undoubtedly would be the initiating point for structure breakdown.

In order to address this difficulty, it is sufficient to merely place small gaps in a uniform reflectance mirror at the periodicity of the free-space wavelength. This can be thought alternatively of as a simple deformation of the scheme shown in Figs. 3 and 4, or as a reversion of the coupling system to the microwave cavity model., in which a slot is used to couple the wave-guide to the cavity. Promising preliminary simulations of this coupling scheme are now underway. They will be published in an upcoming work.

In conclusion, we have shown that it is possible to construct a simple, high $Q$ resonant device which can be efficiently coupled to an external laser source. This coupling scheme gives good selectivity for the desired accelerating mode component. Near-term future work will be centered on lowering the field values in the vicinity of the coupling slot. When this task is completed, we hope to begin practical engineering design work on such a system in preparation for experimental testing.

## REFERENCES

1. C.E. Clayton, *et al., Phys. Rev. Lett.* **70**, 37 (1993).
2. J.R. Fontana and R.H. Pantell, *J. Appl. Phys.* **54**, 4285 (1983),
3. P. Sprangle, E. Esarey, A. Ting and G. Joyce, *Appl. Phys. Lett.* **53**, 2146 (1988)?
4. Y.C. Huang, and R.L. Byer, *Rev. Sci. Instrum.* **69**, 2629 (1998).
5. J.B. Rosenzweig, A. Murokh, and C. Pellegrini, *Phys. Rev. Letters* **74**, 2467 (1995).
6. D. Du, *et al., Appl. Phys. Lett.* **64**, 3073 (1994).
7. A. Tremaine, J. Rosenzweig, P. Schoessow, *Physical Review E* **56**, 7204 (1997).

# Laser Acceleration in Vacuum, Gases, and Plasmas with Capillary Waveguide

## Ming Xie

*Lawrence Berkeley National Laboratory, Berkeley, CA 94720, USA*

**Abstract.** I propose a new method for laser acceleration of relativistic electrons using the leaky modes of a hollow dielectric waveguide. The hollow core of the waveguide can be either in vacuum or filled with uniform gases or plasmas. In case of vacuum and gases, $_{01}$ mode is used for direct acceleration. In case of plasmas, $_{11}$ mode is used to drive longitudinal plasma wave for acceleration. Structure damage due to high power laser can be avoided by choosing a core radius sufficiently larger than laser wavelength. Effect of nonuniform plasma density on waveguide performance is also analyzed.

## MODE PROPERTIES

The capillary waveguide considered here is made of a hollow core with an index of refraction $\nu_1$ and radius $R$, embedded in a dielectric medium with an index of refraction $\nu_2$. We are interested only in oversized waveguide satisfying the condition $\lambda_1/R \ll 1$, where $\lambda_1 = \lambda/\nu_1$ and $\lambda$ is the wavelength in vacuum. As a result, EM wave in the core is dominantly transverse. Choosing appropriate dielectric medium such that $\sqrt{\nu^2 - 1} \gg \lambda_1/R$, where $\nu = \nu_2/\nu_1$, the eigenmodes of the waveguide can be solved following the same procedure by Marcatili and Schmeltzer [1].

Expressing the eigenmodes in the following form

$$\left\{ \begin{array}{c} \mathcal{E}(r,\phi,z,t) \\ \mathcal{H}(r,\phi,z,t) \end{array} \right\} = \left\{ \begin{array}{c} \mathbf{E}_{lm}(r,\phi) \\ \mathbf{H}_{lm}(r,\phi) \end{array} \right\} e^{i(\beta_{lm}z - \omega t) - \alpha_{lm}z} , \qquad (1)$$

the eigenvalues are given by

$$\beta_{lm} = k_1(1 - 1/2\gamma_g^2) , \quad \alpha_{lm} = \chi/\gamma_g^2 R . \qquad (2)$$

where $k_1 = \nu_1 k$, $k = 2\pi/\lambda$, $\gamma_g = 2\pi R/U_{lm}\lambda_1 \gg 1$, and $U_{lm}$ is the $m$th root of the equation, $J_{l-1}(U_{lm}) = 0$.

There are three types of modes, corresponding to

$$\chi = \left\{ \begin{array}{ll} \frac{1}{\sqrt{\nu^2 - 1}} & : \quad TE_{0m} \ (l = 0) \\[2mm] \frac{\nu^2}{\sqrt{\nu^2 - 1}} & : \quad TM_{0m} \ (l = 0) \\[2mm] \frac{\nu^2 + 1}{2\sqrt{\nu^2 - 1}} & : \quad EH_{lm} \ (l \neq 0) . \end{array} \right. \qquad (3)$$

CP472, *Advanced Accelerator Concepts: Eighth Workshop,*
edited by W. Lawson, C. Bellamy, and D. Brosius
1999 The American Institute of Physics 1-56396-889-4

For laser acceleration, we are interested primarily in two low-order modes: $TM_{01}$ mode for acceleration in vacuum and gases, and $EH_{11}$ mode for acceleration in plasmas. Consequently, we consider three cases: $\delta\nu_1 = 0$ when the core is in vacuum, $\delta\nu_1 > 0$ and $\delta\nu_1 < 0$ when the core is filled with uniform gases and plasmas, respectively, where $\delta\nu_1 = \nu_1 - 1$ and $|\delta\nu_1| \ll 1$. It is noted that $EH_{11}$ mode is often designated as $HE_{11}$ mode elsewhere in the literature.

The electric field within the core $r \leq R$ are given by

$$TM_{01} \quad : \quad \begin{cases} E_z = E_a J_0(k_{r1}r) \\ E_r = -i(\Gamma/k_{r1})E_a J_1(k_{r1}r) \ , \end{cases} \tag{4}$$

$$EH_{11} \quad : \quad \begin{cases} E_y = E_0 J_0(k_{r1}r) \\ E_z = -i(k_{r1}/\Gamma)E_0 J_1(k_{r1}r)\sin\varphi \ . \end{cases} \tag{5}$$

where $E_a$ is the peak acceleration field for $TM_{01}$ mode, $E_0$ is the peak transverse field for $EH_{11}$ mode, $\Gamma = \beta_{lm} + i\alpha_{lm}$, and

$$k_{r1} = (U_{lm} - i\chi/\gamma_g)/R \ . \tag{6}$$

To leading order, $\Gamma/k_{r1} = \gamma_g$, since by definition, $k_1^2 = \Gamma^2 + k_{r1}^2$. Given electric field, magnetic field of a mode can be determined by

$$\mathbf{H}_t = \hat{\mathbf{z}} \times (\Gamma\mathbf{E}_t + i\nabla_t E_z)/kZ_0$$
$$H_z = (i/\Gamma)\nabla_t \cdot \mathbf{H}_t \ , \tag{7}$$

where subscript $t$ denotes transverse component of a vector or operator, $\hat{\mathbf{z}}$ is a unit vector in z-direction, and $Z_0$ is the vacuum impedance.

As seen from Eq.(4) and Eq.(5), the transverse field dominates over the longitudinal one by a large factor, $\gamma_g$. For $TM_{01}$ mode, $E_r$ is peaked at $r = r_p$ with a maximum value

$$E_r^{max} = \gamma_g E_a J_1^{max}(U_{01}r_p/R) \ , \tag{8}$$

where $r_p/R = 0.481$ and $J_1^{max}(U_{01}r_p/R) = 0.582$. For $EH_{11}$ mode, $E_y$ is peaked on-axis.

A crucial factor of concern for using a waveguide for laser acceleration is structure damage due to high power laser. To evaluate surface field, $E_s$, at dielectric boundary we expand the Bessel function in the transverse field given by Eq.(4) and Eq.(5), using the expression for $k_{r1}$, Eq.(6), to obtain

$$\begin{aligned} E_s/E_a &\equiv |E_r(r = R)|/E_a = \chi|J_0(U_{01})| &: TM_{01} \\ E_s/E_0 &\equiv |E_y(r = R)|/E_0 = \chi|J_1(U_{11})|/\gamma_g &: EH_{11} \ . \end{aligned} \tag{9}$$

It is important to note that for $TM_{01}$ mode surface field is of the same order as the peak acceleration field, and for both modes surface field is much smaller than the peak transverse field.

Within the dielectric medium $r \geq R$, the maximum field intensities occur at the boundary $r = R$, and all fields have the radial dependence $\exp(ik_{r2}r)/\sqrt{r}$, where to leading order $k_{r2} = k_1\sqrt{\nu^2 - 1}$. A non-vanishing imaginary part of $\nu$ due to even slightly lossy dielectric medium could give rise to a rapid exponential decay of fields in radial direction. Thus the power carried in each mode is distributed dominantly within the core and can be expressed by

$$P(z) = P_0\, e^{-z/L_{attn}}, \tag{10}$$

where $L_{attn} = \gamma_g^2 R/2\lambda$ is power attenuation length due to refractive loss, and

$$P_0 = \int_0^R\int_0^{2\pi} |E_t|^2/2Z_0\; r\,dr\,d\phi = \begin{cases} \pi R^2\gamma_g^2 E_a^2 J_0(U_{01})^2/2Z_0 &:\quad TM_{01} \\ \pi R^2 E_0^2 J_1(U_{11})^2/2Z_0 &:\quad EH_{11}. \end{cases} \tag{11}$$

It is noted that $EH_{11}$ mode is linearly polarized, whereas $TM_{01}$ mode is radially polarized. However, when necessary, a linearly polarized mode can be formed by a proper mixing of $TM_{01}$ with $EH_{21}$ mode [2]. The electric fields for the mixed mode $TM_{01} + EH_{21}$ are given by

$$TM_{01} + EH_{21} \quad : \quad \begin{cases} E_z = E_a[J_0(k_{r1}r) + J_2(k_{r1}r)\cos 2\phi] \\ E_y = -2i(\Gamma/k_{r1})E_a J_1(k_{r1}r)\sin\phi. \end{cases} \tag{12}$$

To preserve the acceleration gradient on-axis, $E_a$, same as the $TM_{01}$ mode, the mixed mode requires a factor of two more laser power. For the three modes we have $U_{11} = 2.405$, $U_{01} = U_{21} = 3.832$, also, $J_0(U_{01}) = -0.403$ and $J_1(U_{11}) = 0.519$.

Coupling between the waveguide modes and free space Gauss-Laguerre modes (also known as $TEM$ modes [3]) can be very efficient. When focused at the waveguide input cross section, power coupling from a radially polarized $TEM_{01}$ mode to the $TM_{01}$ mode reaches a maximum of 97% at $w_0/R = 0.56$, this is true also for coupling from a lineally polarized $TEM_{01}$ mode to the mixed $TM_{01} + EH_{21}$ mode; and coupling from a $TEM_{00}$ mode to $EH_{11}$ mode is 98% at $w_0/R = 0.64$, where $w_0$ is the Gaussian beam waist. Despite the fact that the modes are leaky due to refractive loss, optical guiding is quite effective as the losses for low-order modes can be made very small by choosing $R$ sufficiently large relative to $\lambda$, as seen from Eq.(2).

## ACCELERATION IN VACUUM

According to Eq.(2), phase velocity, $v_p$, of the $TM_{01}$ mode is larger than the speed of light, $c$

$$v_p = \frac{\omega}{\beta_{01}} = \frac{c}{1 - 1/2\gamma_g^2}. \tag{13}$$

703

We define an acceleration phase slippage length over which a relativistic electron with energy $W_0 = \gamma mc^2$, while being accelerated, slips a full $\pi$ phase with respect to the fast acceleration wave

$$L_a = \frac{\lambda}{1/\gamma_g^2 + 1/\gamma^2} \ . \tag{14}$$

Over this distance, energy gain of the electron on-axis is

$$\Delta W_a = eE_a \int_0^{L_a} \sin(\pi z/L_a)dz = eE_a L_a T_a \ . \tag{15}$$

where $T_a = 2/\pi$ is a reduction factor due to a $\pi$ phase slippage during acceleration. Here we have neglected the small attenuation of the acceleration field due to waveguide loss over the distance $L_a$. In parallel, we may also define a deceleration phase slippage length, $L_d$, over which the electron slips another $\pi$ phase while losing energy amounted to $\Delta W_d = eE_a L_d T_d$, where $T_d$ can be different from $T_a$ if $L_d/L_a \neq 1$. The average acceleration gradient during a period of $2\pi$ phase slippage is then given by

$$G = \frac{\Delta W_a - \Delta W_d}{L_a + L_d} = G_a \frac{1 - (L_d/L_a)(T_d/T_a)}{1 + L_d/L_a} \ . \tag{16}$$

where $G_a = \Delta W_a/L_a = eE_a T_a$. To have net acceleration, the ratio $L_d/L_a$ should be made small. This can be done by introducing a static magnetic field during the half period of deceleration. The effect of the magnetic field is to reduce the longitudinal velocity of the electron such that it slips faster, thus taking shorter distance, $L_d$, in the field of deceleration.

For simplicity, we assume the magnetic field is sinusoidal as in a wiggler. $B_y = B_0 \cos(2\pi z/\lambda_w)$. with a period $\lambda_w$. Then $L_d$ is defined by

$$\left[ \frac{1}{\gamma_g^2} + \frac{1}{\gamma^2} + \frac{a_w^2}{\gamma^2} \right] \frac{\pi L_d}{\lambda} - \frac{a_w^2 \lambda_w}{2\gamma^2 \lambda} \sin\left[ \frac{4\pi L_d}{\lambda_w} \right] = \pi \ . \tag{17}$$

where $a_w = eB_0\lambda_w/2\pi\sqrt{2}mc$. If we set $\lambda_w = L_d$ then

$$L_d = \frac{\lambda}{1/\gamma_g^2 + 1/\gamma^2 + a_w^2/\gamma^2} \ , \tag{18}$$

and $a_w$ is now determined by

$$a_w = \sqrt[3]{Q_1 + \sqrt{Q_1^2 + Q_2^3}} + \sqrt[3]{Q_1 - \sqrt{Q_1^2 + Q_2^3}} \ . \tag{19}$$

where $Q_1 = eB_0\lambda\gamma^2/4\pi\sqrt{2}mc$ and $Q_2 = [1 + (\gamma/\gamma_g)^2]/3$. Due to longitudinal oscillation in electron orbit, $T_d$ is different from $T_a$ and given by

$$T_d = \frac{1}{\pi} \int_0^\pi \sin\left[\theta - \kappa \sin(4\theta)\right] d\theta \ . \tag{20}$$

where $\kappa = (1 - L_d/L_a)/4$. The value of $T_d$ varies in the range $\{1.84 \leftrightarrow 2\}/\pi$ for $L_d/L_a$ in the range $\{0 \leftrightarrow 1\}$.

We have assumed the electron is decelerated by the on-axis value of $E_z$, but as the electron is deflected off-axis, it will see a weaker longitudinal field and a stronger transverse EM fields. The maximum orbital offset in the x-direction due to the wiggler field is $\Delta X_{max} = \sqrt{2} a_w \lambda_w / \pi \gamma$.

Because of the magnetic deflection, the electron will radiate and lose energy. The radiative energy loss per wiggler period is

$$\Delta W_r = \frac{8\pi^2 mc^2}{3} \left(\frac{r_e}{\lambda_w}\right) a_w^2 \gamma^2 \ . \tag{21}$$

where $r_e$ is the classical radius of electron. The maximum possible energy that can be accelerated with this method can be determined by the condition: $\Delta W_a > \Delta W_d + \Delta W_r$.

Transverse force on a relativistic electron due to an EM wave does not vanish to order of $1/\gamma^2$ in a waveguide mode or when the index of refraction differs from unity. To see this, we note the magnetic field of the $TM_{01}$ mode can be expressed as $H_\varphi = (1 + 1/2\gamma_g^2 + \delta\nu_1)E_r/Z_0$. Correspondingly, the transverse force on a relativistic electron is

$$F_r = -e(1/2\gamma^2 - 1/2\gamma_g^2 - \delta\nu_1)E_r \ . \tag{22}$$

which can be either focusing or defocusing depending on acceleration phase, $\phi_a$, which varies constantly due to slippage. The beta function for an electron near the axis in vacuum is found to be

$$\beta_t = \gamma_g \lambda \sqrt{(\gamma mc^2/\pi e \lambda E_a \sin\phi_a)/[1 - (\gamma_g/\gamma)^2]} \ . \tag{23}$$

To avoid strong nonlinear transverse EM force when the electron is deflected off-axis by the wiggler field, $B_y$, during deceleration, the mixed mode $TM_{01} + EH_{11}$ may be used since it has zero transverse force along the x-axis, as seen from Eq.(12).

An example is given in Table 1 for acceleration in vacuum with the $TM_{01}$ mode. Here in calculating $\beta_t$ we set $\sin\phi_a = 1$, and $I_s$ is the laser intensity on the waveguide surface.

**Table 1**. Example for Laser Acceleration in Vacuum.

| $\lambda$ [$\mu$m] | 1 | $\gamma_g$ | 410 | $E_a$ [GV/m] | 3.7 |
|---|---|---|---|---|---|
| $R/\lambda$ | 250 | $L_a$ [cm] | 16 | $E_s$ [GV/m] | 3.0 |
| $\nu_2$ | 1.5 | $L_d$ [cm] | 6.2 | $I_s$ [TW/cm$^2$] | 1.2 |
| $W_0$ [GeV] | 1 | $L_{attn}$ [m] | 10 | $G$ [GV/m] | 1.1 |
| $P_0$ [TW] | 100 | $\beta_t$ [cm] | 12 | $\Delta W_a$ [GeV] | 0.38 |
| $B_0$ [T] | 1.5 | $\Delta X_{max}/R$ | 0.35 | $\Delta W_d$ [GeV] | 0.14 |
| $a_w$ | 6.2 | $T_d$ | 0.61 | $\Delta W_r$ [eV] | 88 |

# ACCELERATION IN GASES

The phase velocity of the $TM_{01}$ mode in uniform gases is given by

$$v_p = \frac{\omega}{\beta_{01}} = \frac{c}{1 - 1/2\gamma_g^2 + \delta\nu_1} , \tag{24}$$

which corresponds to an acceleration length or phase slippage length

$$L_{slip} = \frac{\lambda}{|1/\gamma_g^2 + 1/\gamma^2 - 2\delta\nu_1|} . \tag{25}$$

The phase matching condition is obtained by making the denominator zero

$$\delta\nu_1 = 1/2\gamma_g^2 + 1/2\gamma^2 . \tag{26}$$

This condition suggests an alternative way to maintain phase matching as $\gamma$ increases during acceleration: instead of varying $\delta\nu_1$ by adjusting gas pressure along the waveguide, one may change $\gamma_g$ by tapering waveguide radius. For highly relativistic electron satisfying the condition $\gamma \gg \gamma_g$, a steady state phase matching condition, $\delta\nu_1 = 1/2\gamma_g^2$, is approached. An example is given in Table 2 for laser acceleration in gases in the highly relativistic limit. The beta function in gases is smaller than that in vacuum, Eq.(23), by a factor of $\sqrt{2}$. The maximum acceleration gradient is limited by various processes occurring in gases in the field of high power laser, such as nonlinear self-focusing and gas breakdown. Here we assume the limit is set by $E_r^{max} \leq 10 \text{GV/m}$.

**Table 2.** Example for Laser Acceleration in Gases.

| $\lambda$ [$\mu$m] | 10 | $P_0$ [GW] | 50 | $E_a$ [GV/m] | 0.21 |
|---|---|---|---|---|---|
| $R/\lambda$ | 50 | $\gamma_g$ | 82 | $E_s$ [GV/m] | 0.17 |
| $\nu_2$ | 1.5 | $\delta\nu_1$ [$10^{-5}$] | 7.4 | $L_{attn}$ [m] | 0.84 |

# ACCELERATION IN PLASMAS

Wave equation for laser field propagation in weakly relativistic plasma under cold fluid condition is governed by [4]

$$\left[\nabla^2 - \frac{1}{c^2}\frac{\partial^2}{\partial t^2}\right]\mathbf{E}_l = \frac{\omega_p^2}{c^2}\left[1 + \frac{\delta n}{n_0} - \frac{a^2}{2}\right]\mathbf{E}_l . \tag{27}$$

where $\omega_p = \sqrt{e^2 n_0/\epsilon_0 m}$ is the electron plasma frequency, $n_0$ the ambient plasma density, and $\epsilon_0$ the dielectric constant in vacuum. The plasma density modulation, $\delta n/n_0$, driven by the ponderomotive potential of the laser field, $a^2 = <|e\mathbf{E}_l/mc\omega|^2>$, will generate a wakefield, $\mathbf{E}_w = -\nabla\Phi$, where the wake potential, $\Phi$, is determined by [5,6]

$$\left[\frac{\partial^2}{\partial t^2} + \omega_p^2\right]\Phi = \omega_p^2 \frac{mc^2}{c}\frac{a^2}{2} \ . \tag{28}$$

To close the loop, $\delta n/n_0$ is related to the wake potential by Poisson's equation

$$\nabla^2\Phi = (c/\epsilon_0)\delta n \ . \tag{29}$$

Under the condition $a^2 \ll 1$, we will have $\delta n/n_0 \ll 1$, as will be shown later. As a result, the second and third term on the right hand side of Eq.(27) can be dropped and the wave equation is then decoupled from the plasma equations, Eq.(28) and Eq.(29). The only effect of the plasma on laser propagation is through an index of refraction $\nu_1 = 1 - \omega_p^2/2\omega^2$.

We now consider laser wakefield acceleration [7] in a capillary waveguide filled with a uniform plasma. A laser pulse propagating through the waveguide will excite a wakefield with phase velocity equals the group velocity of the laser pulse. For $EH_{11}$ mode, the group velocity is given by

$$v_g = \frac{d\omega}{d\beta_{11}} = \frac{c}{1 + 1/2\gamma_g^2 + 1/2\gamma_p^2} \ , \tag{30}$$

where $\gamma_p = \omega/\omega_p \gg 1$. Correspondingly, the slippage length by definition is

$$L_{slip} = \frac{\lambda_p}{|1/\gamma_g^2 + 1/\gamma_p^2 - 1/\gamma^2|} \ . \tag{31}$$

To solve the plasma equations, we take the approach in parallel to the 1D linear analysis of laser wakefield by Gorbunov and Kirsanov [5], except here the solution we provide is in full 3D. Introducing a variable $\zeta = z - v_g t$, Eq.(28) can be solved as

$$\Phi = -(k_p mc^2/c)\int_\zeta^\infty d\zeta' \sin[k_p (\zeta - \zeta')]\frac{a^2}{2} \ . \tag{32}$$

where $k_p = \omega_p/v_g$. For a Gaussian pulse of $EH_{11}$ mode, we have from Eq.(5)

$$a^2(\rho, \zeta) = \frac{a_0^2}{2} J_0^2(U_{11}\rho)e^{-\zeta^2/2\sigma_\zeta^2 - z/L_{attn}} \ . \tag{33}$$

where $\rho = r/R$, the wake potential behind the laser pulse is

$$\Phi = -\Phi_0 J_0^2(U_{11}\rho)e^{-z/L_{attn}} \sin(k_p z - \omega_p t) \ . \tag{34}$$

where

$$\Phi_0 = (\sqrt{2\pi}mc^2/4c)\, a_0^2\, k_p\sigma_z\, e^{-(k_p\sigma_z)^2/2} \ . \tag{35}$$

The longitudinal wakefield is then given by

707

$$E_{wz} = E_a J_0^2(U_{11}\rho)e^{-z/L_{attn}}\cos\left(k_p z - \omega_p t\right) \ . \tag{36}$$

and the transverse wakefield by

$$E_{wr} = -2(\gamma_p/\gamma_g)E_a J_0(U_{11}\rho)J_1(U_{11}\rho)e^{-z/L_{attn}}\sin\left(k_p z - \omega_p t\right) \ . \tag{37}$$

where the peak acceleration field, $E_a = \Phi_0 k_p$, is maximized if the laser pulse length is chosen according to the condition, $k_p\sigma_z = 1$. From here on, we will use this optimal condition wherever relevant.

As wakefield is excited in the plasma channel, energy in the driver pulse, Eq.(33), will be depleted. To characterize this process, a pump depletion length, $L_{pump}$, can be defined by the condition, $W_l = W_w$, where $W_l$ is the initial energy of the laser pulse

$$W_l = \sqrt{\pi/8}J_1^2(U_{11})\epsilon_0\lambda_p R^2 E_0^2 \ , \tag{38}$$

and $W_w$ is the energy in the wakefield the laser pulse left behind as it propagates a distance $L_{pump}$

$$W_w = (\pi mc/4e)^2[\epsilon_0/\mathrm{Exp}(1)]a_0^4\omega_p^2 R^2 L_{pump}[I_z + (\gamma_p/\gamma_g)^2 I_r] \ . \tag{39}$$

The two terms above on the right hand side correspond to energy in the longitudinal and transverse wakefield, respectively, where $I_z = \int_0^1 d\rho\rho J_0^4(U_{11}\rho) = 7.62 \times 10^{-2}$ and $I_r = \frac{1}{4}\int_0^1 d\rho\rho J_0^2(U_{11}\rho)J_1^2(U_{11}\rho) = 6.35 \times 10^{-3}$. We then obtain

$$L_{pump} = [4\sqrt{2\pi}J_1^2(U_{11})\mathrm{Exp}(1)/\pi^2]\frac{\lambda_p}{a_0^2(I_z/\gamma_p^2 + I_r/\gamma_g^2)} \ . \tag{40}$$

In addition, we may define a characteristic pulse dispersion length, $L_{disp}$, over this propagation distance the driver pulse will double its length, $\sigma_z$. Given group velocity dispersion by Eq.(30), we have

$$L_{disp} = \frac{\sqrt{3}\,\gamma_p^2\,\lambda}{\pi(1/\gamma_g^2 + 1/\gamma_p^2)} \ . \tag{41}$$

There is yet another characteristic length, the beta function, due to the transverse wakefield. For electron near the axis the beta function is given by

$$\beta_t = (2/U_{11})[\mathrm{Exp}(1)/2\pi]^{1/4}\sqrt{\frac{\gamma}{\sin\phi_a}\frac{R}{a_0}} \ . \tag{42}$$

Finally, the plasma density modulation is

$$\frac{\delta n}{n_0} = \sqrt{\pi/8\mathrm{Exp}(1)}a_0^2\{1 + 2(\gamma_p/\gamma_g)^2[1 - J_1^2(U_{11}\rho)/J_0^2(U_{11}\rho)]\}$$

$$J_0^2(U_{11}\rho)e^{-z/L_{attn}}\sin\left(k_p z - \omega_p t\right) \ . \tag{43}$$

Indeed, we have $\delta n/n_0 \ll 1$, if $a_0^2 \ll 1$.

An example is given in Table 3 for laser acceleration in plasmas. Here the energy gain per stage is defined by, $\Delta W_a = eE_a L_{slip}T_a$, the energy gain per slippage length. In calculating $\beta_t$ we set $\sin\phi_a = 1$.

**Table 3.** Example for Laser Acceleration in Plasmas.

| $\lambda$ [$\mu$m] | 1 | $n_0$ [$10^{17}$/cm$^3$] | 1.1 | $E_a$ [GV/m] | 0.94 |
|---|---|---|---|---|---|
| $R/\lambda$ | 150 | $(\delta n/n_0)_{max}$ | 0.034 | $E_s$ [GV/m] | 1.7 |
| $\nu_2$ | 1.5 | $\gamma_p$ | 100 | $I_s$ [TW/cm$^2$] | 0.39 |
| $W_0$ [GeV] | 1 | $\gamma_g$ | 392 | $L_{slip}$ [m] | 0.94 |
| $P_0$ [TW] | 20 | $\beta_t$ [cm] | 1.6 | $L_{attn}$ [m] | 7.9 |
| $W_l$ [J] | 2.7 | $\sigma_z$ [$\mu$m] | 16 | $L_{disp}$ [m] | 52 |
| $a_0$ | 0.28 | $\Delta W_a$ [GeV] | 0.56 | $L_{pump}$ [m] | 126 |

We have analyzed the capillary waveguide when the core is filled with a plasma of uniform density, $n_0$. In reality, this can only be an approximation. To evaluate the effect of nonuniformity of the plasma, let's consider a special case when the uniform background is modified by a parabolic profile given by

$$n = n_0 + \Delta n[1 - (r/R)^2] . \tag{14}$$

Such a profile has a defocusing effect on the mode if $\Delta n > 0$, thus making the waveguide less effective. However, the attenuation length for the $EH_{11}$ mode is reduced by a factor of two at most, if the following criteria is satisfied

$$\frac{\Delta n}{n_0} \leq 4.6 \frac{\gamma_p^2}{\gamma_g^2} . \tag{15}$$

For the example in Table 3, this corresponds to $\Delta n/n_0 \leq 30\%$. Thus we may infer from here that the guiding provided by the capillary waveguide can also be rather stable, against either systematic or random variations in plasma density. The derivation of the criteria, Eq.(15), will be published elsewhere.

## CONCLUSIONS

I have introduced the concepts and techniques that are crucial for advancing the current development of laser acceleration into a new, more realistic stage. Of what being accomplished here, the most notable is the significant increase in acceleration distance by using the leaky modes of an oversized capillary waveguide. In vacuum, a new mechanism for energy transfer from laser to electron is proposed, and with which both limits on acceleration distance set by diffraction and phase slippage are overcome. In gases, a tuning technique is provided to maintain the phase matching over a longer acceleration distance. In plasmas, the approach taken here is radically different in concept from the current mainstream development. First of all, the prevailing notion on optical guiding in plasma is based on an analogy to optical fiber in which the index of refraction is maximal on axis, opposite to what proposed here. Secondly, it is commonly believed that no guiding structure could sustain the

709

laser power required for acceleration without being, at least partially, turned into plasma. Therefore the only method considered so far for optical guiding is to tailor the plasma itself in transverse density profile one way or the other, by either relativistic self-focusing [6], charge displacement [8], or capillary discharge [9]. By relieving the duty of optical guiding from the plasma, the medium for acceleration, to an external waveguide, our approach opens up a new avenue to more effective, stable and practical optical guiding and acceleration. In addition, the acceleration structure proposed here has the following advantages: dielectric damage [10] by high power laser is shown not to be a problem; the large waveguide cross section is favorable for achieving better electron beam quality; the efficient coupling between waveguide modes and free space modes eases mode handling such as mode injection, transport and recycling, thus leading to a better overall system efficiency; last but not least, all aforementioned desirable features are achieved without sacrificing a virtue of practical importance: the simplicity. This work was supported by the U.S. Department of Energy under contract No.DE-AC03-76SF00098.

# REFERENCES

1. E. Marcatili and R. Schmeltzer, *B.S.T.J.*, **43**(1964)1783.
2. J. Henningsen, M. Hammerich, and A. Olafsson, *Appl. Phys.*, **B51**(1990)272.
3. H. Kogelnik and T. Li, *Proc. of IEEE*, **54**(1966)1312.
4. T. Antonsen and P. Mora, *Phys. Fluids B.*, **5**(1993)1440.
5. L. Gorbunov and V. Kirsanov, *Sov. Phys. JETP*, **66**(1987)290.
6. E. Esarey, et al., *IEEE Tran. on Plasma Science*, **24**(1996)252.
7. T. Tajima and J. Dawson, *Phys. Rev. Lett.*, **43**(1979)267.
8. C. Durfee and H. Milchberg, *Phys Rev. Lett.*, **71**(1993)2409.
9. Y. Ehrlich, et al., *Phys Rev. Lett.*, **77**(1996)4186.
10. B. Stuart, et al., *Phys. Rev. B*, **53**(1996)1749.

# Working Group 4:

# Beam Monitoring, Conditioning, and Control at High Frequency and Ultrafast Timescales

# Precision Beam to Structure Alignment in Linear Accelerators

M. Seidel[a] and C. Adolphsen[b]

[a] *Deutsches Elektronen Synchrotron, DESY, 22603 Germany*

[b] *Stanford Linear Accelerator Center, Stanford University, CA 94309, USA*

## Abstract

Beam emittance dilution by self induced transverse fields (wakefields) is a key problem in linear accelerators, especially in small structures that operate at high frequencies. To minimize the wakefield effects the beam trajectory must be precisely centered within the reference frame of the structures. This paper describes two experimental attempts to align a high energy beam in accelerator structures (S-band and X-band) by detecting and minimizing beam induced dipole modes. In case of the X-band structure the achieved alignment precision has been verified by independent two beam deflection measurements.

# Introduction

When a high energy beam of charged particles passes off-center through an accelerator structure, it excites asymmetric electromagnetic fields, so called wakefields, that act back on the beam. To lowest order such fields are dipole modes, excited in proportion to the displacement of the beam from the electrical center of the structure. Wakefield effects are classified as either short range, which are important on the time scale of a single bunch length, or long range which couple the orbits of consecutive bunches in a train. Both effects can lead to intensity limiting instabilities and emittance blow up in the beam. An early example of this was the so called beam-breakup instability that was observed in the SLAC linac in 1966 [1].

The long range effects can be effectively suppressed by an appropriate design of the structure that employs damping and/or decoherence of the dipole modes (see for instance [2] [3]). On the other hand the short range effects, typically on a picosecond time scale, are more or less determined by the choice of accelerating frequency. The only way to reduce short range wakefields is to center the beam precisely in the structure. For times much shorter than the period of the accelerating mode the transverse wake can be approximated by [4]:

CP472, *Advanced Accelerator Concepts: Eighth Workshop,*
edited by W. Lawson, C. Bellamy, and D. Brosius
© 1999 The American Institute of Physics 1-56396-889-4/99/$15.00

$$W_\perp(z) \approx W_\perp'(z = 0 \ ) z \propto \left(\frac{a}{\lambda}\right)^{-7/2} \omega^4 z \qquad (1)$$

Here $a$ is the iris radius of the structure and $\lambda$ is the accelerating mode wavelength. The typical kick received by a test particle in the bunch is then:

$$\Delta x' = \frac{eQ_b L_s W_\perp(z = \sigma_z)}{E_b} \Delta x \qquad (2)$$

where $Q_b$ is the bunch charge, $L_s$ the structure length, $\sigma_z$ the RMS bunch length, $E_b$ the beam energy and $\Delta x$ the transverse bunch offset. The emittance blowup in the beam scales with the square of this kick angle. Equations (1), (2) suggest a rather strong scaling of the required beam-to-structure alignment with frequency: $\Delta x \propto 1/\omega^4$. In practice the scaling is not that strong since other parameters in (2) have implicit frequency dependencies that reduce this dependence. In particular, the bunchlength can be scaled in proportion to the wavelength of the accelerating mode. Also, the structures tend to operate at higher gradients at higher frequencies which reduces the accelerator length for the same energy gain. In addition, the iris radius to wavelength ratio can be slightly increased for higher gradients and the bunch charge is often reduced in the higher frequency designs. However, some of these parameter scalings are not desirable, and in fact are dictated by the inability to guarantee the necessary precision for the beam to structure alignment.

If one requires good alignment to reduce wakefields to manageable levels, then the best method is to use the structure itself as beam position monitor via detecting beam induced dipole modes. In the following sections we review the work that we have done in this regard.

# Experiences with S-Band Structures in the SLC Linac

The SLC linac consists of roughly 900 structures that are operated at 2.856 GHz (S-Band). The typical emittance blowup in the linac is in the range of 50% to 100% and is caused mainly by transverse wakefields. The emittance growth is minimized with BNS damping and by empirical optimization using orbit bumps. Otherwise one relies on beam steering using beam position monitor (BPM) information, which is difficult since the relative alignment between BPM's and structures is not known very precisely. The observed emittance growth suggests RMS beam-to-structure misalignments of $\approx$ 300 $\mu$m. This situation makes a more direct method of determining the beam orbit with respect to the electrical center of the structures very desirable. It would help reduce emittance growth and simplify the operation of the machine considerably. Although the SLC investigations did not ultimately lead to a scheme that was used for beam alignment in practice, it is useful to discuss the results and the observed difficulties.

The goal of the experiments was to identify a simple and inexpensive method to make dipole mode signals from the 900 SLC structures accessible for diagnostics and

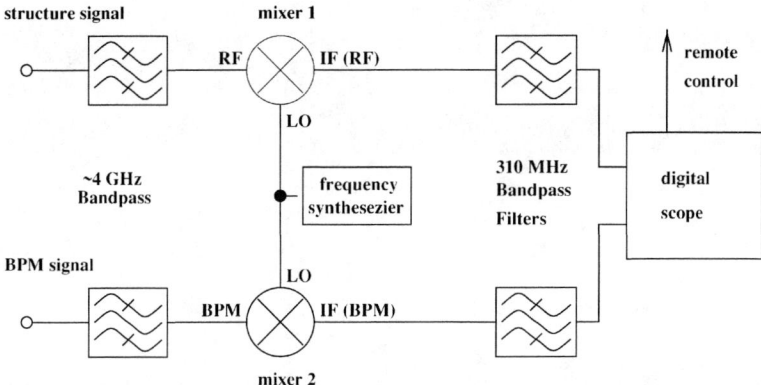

**Figure 1**: Setup used to detect phase and amplitude of dipole mode signals in the 4.0-4.25 GHz range.

beam steering. To be inexpensive, the scheme could not involve modifications of the structures such as the addition of higher-order-mode couplers. Each structure possesses fundamental mode input and output couplers. Beam induced dipole modes are backward waves and can be observed at the input coupler. The RF distribution system in the SLC feeds four structures with one klystron via a waveguide network. At a position close to the klystron the waveguide network is equipped with a -55 dB Bethe-Hole coupler. At this coupler it is possible to observe the $\approx 4.2$ GHz dipole mode signals. The disadvantage of this monitoring point is that it provides the sum signal from four structures, i.e., the effect of individual structures cannot be distinguished. In addition, both dipole polarizations (vertical and horizontal) contribute to the signal.

When the beam is centered precisely in the structures one expects the dipole mode signal to vanish. Since the signal power is a quadratic function of the offset it is technically difficult to determine the center position precisely for small offsets. On the other hand the phase of the signal should show a relatively fast shift by $180°$. This phase shift provides a more precise measure for the centered beam position. Furthermore the phase measurement allows one to determine the sign of the beam offset immediately. The phase and amplitude measurement were realized with the setup depicted in Fig.1.

The 4 GHz signal from the structure was bandpass-filtered to reject the fundamental mode and its harmonics, and then down-mixed to an IF frequency of 310 MHz. An RF mixer is basically a solid state device which performs a nonlinear transformation of the sum of two input signals. The output signal includes a contribution that oscillates at the difference frequency of the input signals:

$$
\begin{aligned}
U_{\text{out}} &= \sum_k c_k \left( U_1 \cos(\omega_1 t + \phi_1) + U_2 \cos(\omega_2 t + \phi_2) \right)^k \\
&= \frac{c_2}{2} U_1 U_2 \cos((\omega_1 - \omega_2)t + \phi_1 - \phi_2) + \text{other terms} \dots \quad (3)
\end{aligned}
$$

715

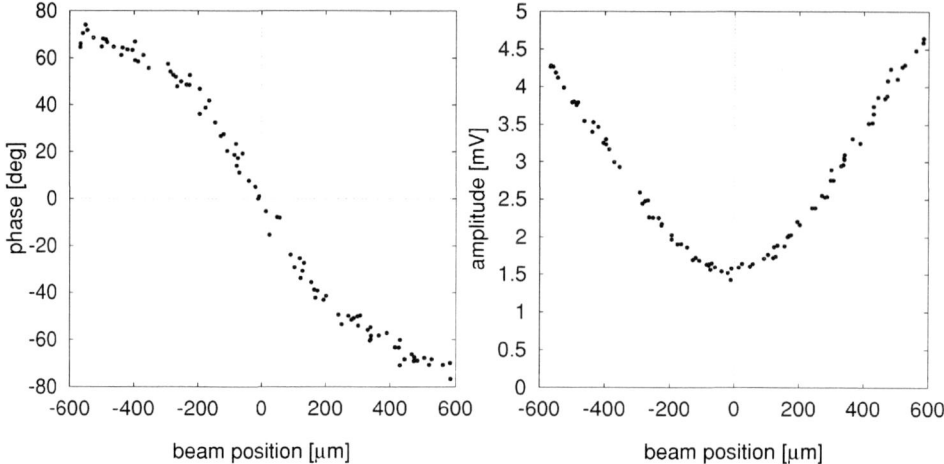

**Figure 2**: Measured phase and amplitude of the dipole mode signal as a function of the beam position. The zero beam position is arbitrarily defined.

Remarkably the phase information of the original high frequency signals is still contained in the low frequency output signal. The width of the 4 GHz bandpass has to be chosen such that only one sideband of the tunable local oscillator frequency is picked out. To provide a reference phase that was independent of beam position, a signal from a BPM antenna was bandpass-filtered and down-converted with the same LO-frequency in a second mixer arm. Both IF signals were measured with a digital scope and their relative phase determined by fitting cosine functions to the waveforms. Since the local oscillator phase subtracts out, one obtains the phase of the dipole mode signal relative to the reference phase. In an attempt to provide a signal that was easier to interpret, this setup was connected to a waveguide network in Sector 2 of the SLC linac where one klystron feeds only two structures instead of four.

During the experiment the beam was first steered in angle and position to minimize the dipole mode signal in both attached structures simultaneously. Then a parallel beam scan was performed that moved the beam over a range of 1.2 mm in the two structures. The relative beam position was measured using nearby BPM's. The result is shown in Fig.2.

The measurement shows that the minimum dipole mode amplitude is significantly larger than zero. This can be explained by a position independent out-of-phase component in the signal:

$$
\begin{aligned}
U(x) &= A(x - x_0)\cos(\omega t) + B\cos(\omega t + \Delta\phi) \\
&= \hat{U}(x)\sin(\omega t + \hat{\phi}(x)).
\end{aligned}
\tag{4}
$$

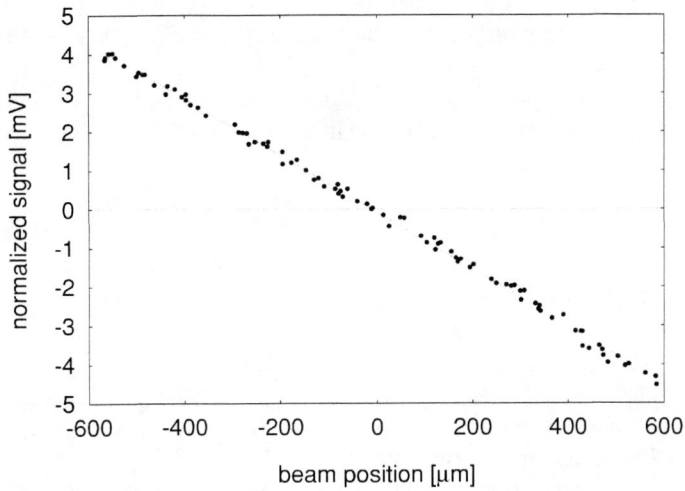

**Figure 3**: Linearized signal (data points) and a linear fit to these data (solid line).

Here the $B$ term represents the out-of-phase component with a phase difference of $\Delta\phi$ relative to the position dependent signal and $x_0$ is the electrical center of the structure in the horizontal plane. The origin of the $B$ component is not completely clear. Probably it is due to a residual offset of the beam in the vertical plane since the beam was not carefully centered in this dimension. It could also be due to a relative horizontal misalignment of the two accelerator structures. As a result of the out-of-phase component the minimum of the signal amplitude $\hat{U}$ does not occur at the center of the structure, but at some offset position:

$$x_{\min} = x_0 - \frac{B}{A}\cos(\Delta\phi) \tag{5}$$

The out-of-phase component also broadens the $180°$ phase transition:

$$\hat{\phi}(x) = \arctan\left(\frac{A(x - x_0)}{B\sin(\Delta\phi)} + \cot(\Delta\phi)\right) \tag{6}$$

In the limit where the $B$ component vanishes (either $B = 0$ or $\Delta\phi = 0$), the phase transition approaches a step function.

Using both amplitude and phase information the measurement can be linearized by computing:

$$A(x - x_{\min}) = \hat{U}\sin(\hat{\phi}(x)) \tag{7}$$

The result is shown in Fig. 3. Based on the goodness of the linear fit to the data, one obtains a resolution for this beam position measurement of better than $20\,\mu$m. This means that a beam position movement of $20\,\mu$m can be detected using the described method, however, it does not mean that the electrical center is be determined with this precision. The large out-of-phase component limits the applicability of the method for absolute beam position measurements.

In summary it was possible to detect beam induced dipole mode signals in SLC structures and to demonstrate beam position measurements with good relative precision. However, it was not possible to demonstrate reasonably good absolute positioning precision, which is actually the primary goal for successful wakefield control. The applied scheme suffers mainly from the fact that the SLC S-Band structures are not equipped with dedicated HOM couplers that would allow one to distinguish horizontal and vertical polarization of the modes as well as signals from individual structures.

Another potential problem is that modes which are coupled only weakly to the input coupler have possibly a preferred transverse polarization, determined by slight deviations of the cavity shape from the ideal round geometry. After being excited with a certain polarization such modes can rotate slowly into their preferred direction, thereby mixing information on horizontal and vertical offsets.

# Application to a Damped and Detuned X-Band Structure

In another experiment a prototype NLC structure which operates at 11.4 GHz (X-Band) was installed in a dedicated facility in the SLC beamline, called ASSET (Accelerator Structure Setup). The primary purpose of this experiment was to measure the long-range wakefield suppression in the structure by using two beams, one to drive the wakefields and one to witness them (see [5] [6] for a detailed description). The Damped and Detuned Structure (DDS) that was used for this experiment consists of 206 cells. By damping and detuning the dipole modes the long range wakefield is reduced by about a factor 100 by the time (1.4 ns) the next bunch passes through the structure in the nominal NLC design. The damping is achieved by four manifolds that run parallel along the structure and couple to the individual cells through slots. Horizontally and vertically polarized dipole modes can be detected separately in the vertical and horizontal manifolds, respectively. It has been verified experimentally that the crosstalk between the transverse planes is small. Furthermore, the manifolds allow one to detect dipole modes over the complete length of the structure whereas the SLC structure, for example, has trapped modes that are not accessible from the input coupler. Because of the cell detuning the frequency of the dipole mode signal is related to the source location of the signal along the structure. This relation can be used to perform an in-situ structure straightness measurement by sweeping the beam transversally across the structure and recording the complete dipole mode spectrum as a function of beam position. The position minima as a function of frequency are then converted into cell positions. Details of this method are given in references [7][8]. A typical dipole mode spectrum, observed with a parallel beam offset of about 1 mm, is shown in Fig. 4. The mid-frequency point of the spectrum for this X-Band structure is 15.1 GHz.

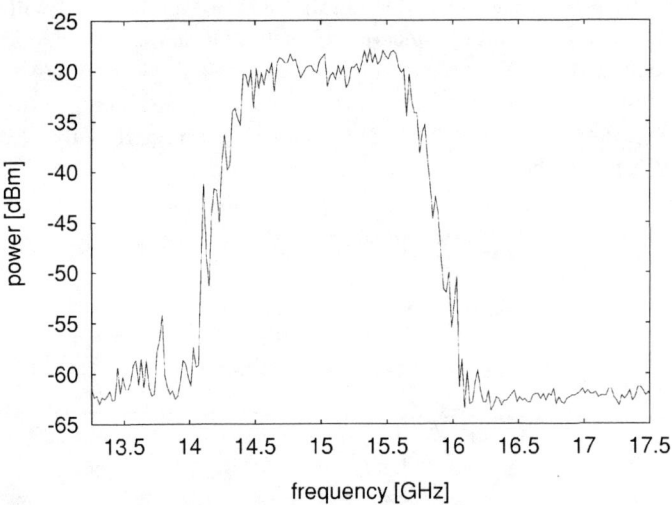

**Figure 4**: Typical dipole mode spectrum from the DDS for a parallel transverse beam offset.

The setup that was used to process the DDS signals was basically the same as that described above for the SLC structures (Fig. 1). The phase reference was also taken from a BPM antenna signal. Due to cable losses, however, the signal power in the 15 GHz region was too low to be used. Instead, the mixer in the dipole mode arm was operated at the fourth harmonic, which means that the fourth order contribution to equation (3) was selected, and the local oscillator frequency was tuned to $\omega_{LO} = (\omega_{RF} + \omega_{IF})/4$. In this case the lower side band was measured. The mixer in the BPM arm was operated at the first harmonic and picked out the upper sideband at a frequency of $\omega_{BPM} = \omega_{LO} + \omega_{IF}$. Consequently, for scanning dipole mode signals in the range 14.1 to 16.1 GHz the phase reference signal in the range 3.91 to 4.41 GHz was used. With these choices, the down-mixed signals in the two arms oscillate as:

$$
\begin{aligned}
U_{IF(RF)} &\propto \cos\left((4\omega_{LO} - \omega_{RF})(t - t_0) - \phi_{beam} + 4\phi_{LO}\right) \\
U_{IF(BPM)} &\propto \cos\left((\omega_{BPM} - \omega_{LO})(t - t_0) - \phi_{LO}\right)
\end{aligned}
\tag{8}
$$

Here $t_0$ is the trigger time of the digital scope in Fig. 1 and $\phi_{beam}$ is the phase of the dipole mode signal to be detected. After determining the phases of the two IF signals from the digitized waveforms the dipole mode phase is computed via:

$$
\begin{aligned}
-\phi_{IF(RF)} - 4\phi_{IF(BPM)} &= \phi_{beam} + (4\omega_{BPM} - \omega_{RF})\,t_0 \\
&= \phi_{beam} + 5\omega_{IF}\,t_0
\end{aligned}
\tag{9}
$$

Due to the operation of the two mixer arms on different harmonics the phase measurement depends now on the trigger time of the scope. The available trigger had a jitter of $\approx 20\,\mathrm{ps}$. According to (9) this number has to be multiplied with a frequency of $5f_{\mathrm{IF}} = 1.55\,\mathrm{GHz}$ to obtain the uncertainty of the phase measurement. The result is $11°$ which is adequate to determine the sign of the dipole mode signal. An example of a phase and amplitude measurement as a function of transverse beam position is shown in Fig. 5. For these data the relative beam position was determined by the steering magnet settings. The scatter of the data for a given setting was worsened by a $10\,\mu\mathrm{m}$ beam jitter. Nevertheless, the measurement shows a fast phase transition and a very small minimum signal, which indicates a negligible out-of-phase component.

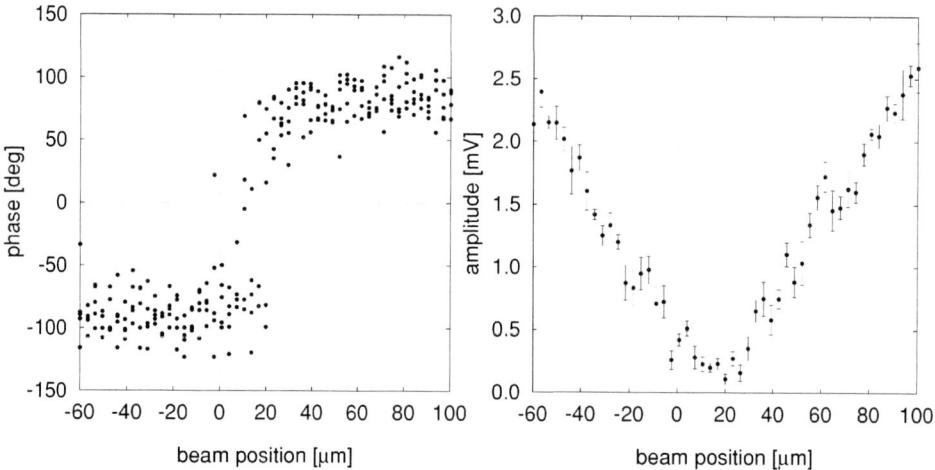

**Figure 5**: Dipole mode phase and amplitude versus beam position at a frequency of $14.97\,\mathrm{GHz}$. For each beam position 5 pulses were measured.

The experimental setup that we used in the SLC allowed us to go one step further and to verify the achieved absolute alignment by measuring the beam excited wakefield directly with a second "witness" bunch. The ASSET facility allows one to send two bunches, a positron bunch followed by an electron bunch, at a precisely defined time distance through the structure under test. The leading "drive" bunch is dumped just after passing through the structure whereas the witness bunch travels down the linac beamline (see Fig. 6). The deflection that the witness bunch received was measured from the differences in its trajectory in the linac for different drive bunch offsets. Both beams had energies of $1.2\,\mathrm{GeV}$ and bunch lengths of $550\,\mu\mathrm{m}$. The drive bunch had a charge of $3.2\ldots4.8\,\mathrm{nC}$ and the witness bunch had a charge of $-2.6\,\mathrm{nC}$.

To verify the positioning accuracy the witness bunch was first switched off and the drive bunch was centered as best as possible. For this purpose, steering magnets were used to adjust the position and angle of the drive bunch so as to minimize the dipole signals from two modes located at opposite ends of the structure. The witness bunch was then turned on and timed so it followed closely behind the drive bunch where the

720

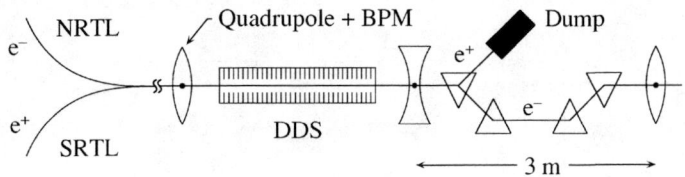

wakefield is strong and hence yields the largest witness bunch kick per unit drive bunch offset. The time difference between the two bunches was varied in three steps to allow a rough sampling of the 15 GHz wakefield waveform. The steps were chosen such that the variation of the longitudinal wakefield was minimal whereas the transverse kick varied strongly (see Fig. 7). After measuring the three kicks the amplitude of the oscillation was determined by fitting a sine function to the measured values. This result was then converted into an equivalent drive beam offset using the independently measured short-range wakefield amplitude of 65 V/pC/mm/m.

In practice, we found for small drive bunch offsets, that the relative ratios of the three measured kick amplitudes did not scale as the expected sine like curve in Fig. 7. In order to fit the data we had to allow another parameter to vary, either phase or frequency of the oscillation. In Fig. 8 the results are shown from several attempts at centering the beam in the structure using the procedure described above. The three measured kick amplitudes were fit to sine functions by varying frequency and amplitude of the fit-function and the results plotted as frequency vs. offset. One observes that the inferred frequency decreases as the beam comes closer to the optimum position from a positive offset, and then jumps suddenly to a frequency higher than the average mode frequency of 15.1 GHz.

This behavior is not not totally unexpected given that the structure is not perfectly straight. In fact the first 45 cells are offset transversely by about $60\,\mu$m step relative to the remaining cells as a result of slippage during the bonding of subsections of the structure. Thus the beam cannot be centered everywhere in the structure at one time and the excitation of wakefields cannot be avoided in general. At best the slope of the wakefield can be made to be zero near zero bunch separation, which would minimize single bunch effects. However, at later times a beating oscillation of the fields start because the initially cancelled dipole modes, excited in different regions of the structure, slip out of phase. This behavior is demonstrated in the simulation shown in Fig. 9.

The solid line in Fig. 8 is a simulation of how the fit frequency would vary with the fit drive bunch offset when the effects of the internal misalignments are included. One sees that it reproduces the general variation of the data but not the magnitude of the frequency jump. This disagreement may be due to an additional component of the wakefield that is independent of beam position. The presence of a 11.4 GHz sine-like component at the

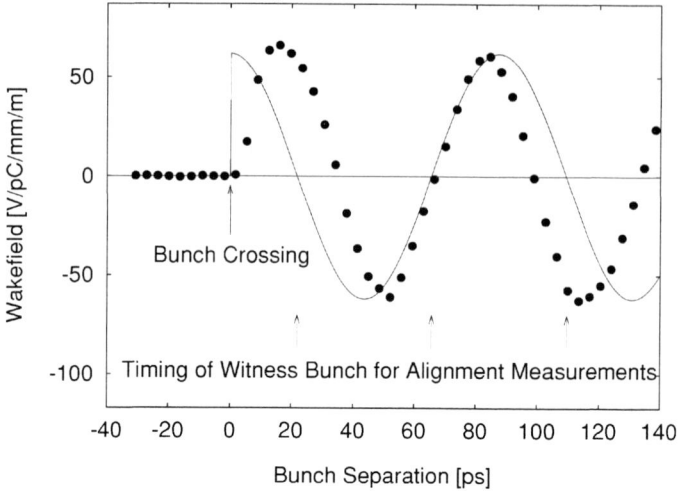

**Figure 7**: The dots show the transverse wakefield inferred from the witness bunch deflection for large parallel drive beam offsets, and the solid line is the expected variation of longitudinal wakefield. For the alignment measurements the witness bunch was positioned at the three indicated bunch separations.

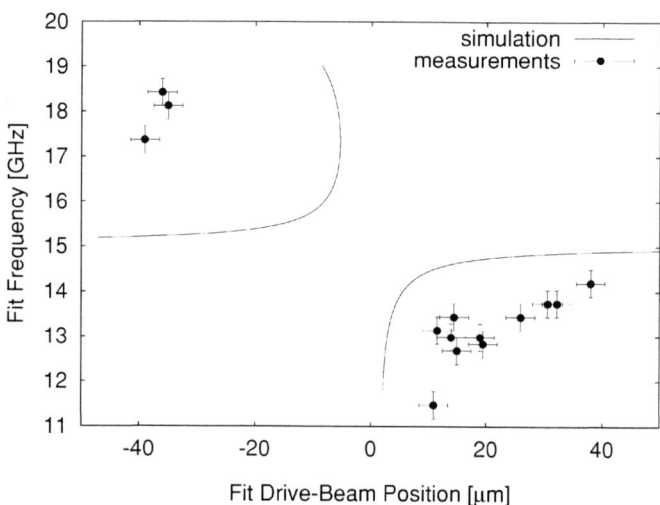

**Figure 8**: Results of several attempts at centering the drive beam in the DDS structure. The dots are the fit frequencies and amplitudes of the wakefield oscillation, which are converted into equivalent drive bunch offsets. The solid line is a prediction based on the mechanically measured internal misalignments of the structure.

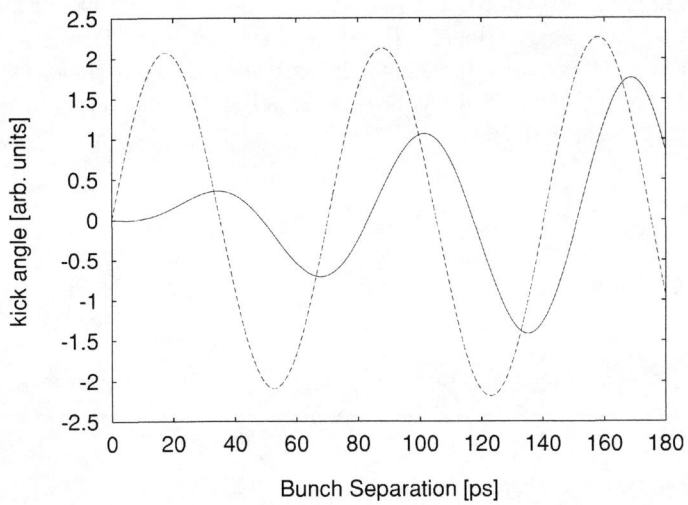

**Figure 9**: Simulated short range wakefield-kick for different beam positions in the DDS structure with internal misalignments. The solid line is the result for the optimum trajectory that zeroes the initial wakefield slope. The dotted line is the result when the beam is displaced by $10\,\mu$m with respect to the optimal trajectory.

level of 1/1000 of the beam loading would yield a curve that matches the data well. To be conservative, however, we conclude from these measurements that the optimum trajectory lies somewhere in between the two clusters of measured values and the achieved absolute positioning precision is at worse $40\,\mu$m. For a linear collider like the NLC the required beam to structure alignment is $15\,\mu$m RMS [9]. It can be expected that the described methods allow to achieve this accuracy with reasonably straight structures.

# Summary

As linear accelerator technology tends towards higher frequencies in order to achieve higher gradients, the strong wakefields generated by the smaller aperture structures will require precise absolute beam positioning within these structures. Beam steering based on detection and minimizing of beam induced dipole modes has the potential to provide the required precision.

We described two different experimental attempts to demonstrate these methods. The comparison of experiences with the SLC S-Band structures and the modern DDS structure shows that dedicated higher order mode couplers are essential to distinguish both transverse polarizations of the dipole modes. Furthermore the DDS manifolds allow one to access all modes along the structure for which the beam position information over the whole length can be obtained. Using phase and amplitude detection, an alignment precision better than $40\,\mu$m has been achieved and verified by measuring the wakefield deflection of a second test bunch. The resolution of the measurement was about $3\mu$m.

We thank R. Assmann, K. Bane, P. Emma, R. Jones, N. Kroll, R. Miller, C. Nantista, T.Slaton, J. Wang, D. Whittum for their help with the data taking and analysis of the experiments described here. This work was supported by the Department of Energy, contract DE-AC03-76SF00515.

# References

[1] G. Loew, R. Helm, H. Hoag, R. Koontz, R. Miller, Conf. on High Energy Part. Accel., Yerevan (1969).

[2] R.H.Miller *et al.* Linac 96, Geneva (1996).

[3] T. Shintake, Jpn. J. Appl. Phys. **31**, 1567 (1992).

[4] R. Palmer, Ann. Rev. Nucl. Part. Sci. **40**, 529 (1990).

[5] C. Adolphsen *et al.*, subm. to NIM (1997).

[6] M. Seidel, C. Adolphsen, K. Bane, R. Jones, N. Kroll, R.Miller, D.Whittum, NIM A 404 (1998) 231-236.

[7] M. Seidel, Proc. PAC'97, Vancouver (1997).

[8] N. Kroll, Proc. PAC'97, Vancouver (1997).

[9] T.O. Raubenheimer *et al.*, int. SLAC-Report 474 (1994).

# Noninvasive Beam Position, Size, Divergence and Energy Diagnostics Using Diffraction Radiation

R.B. Fiorito*, D.W. Rule[†] and W.D. Kimura[‡]

*The Catholic University of America, Washington, DC 20064
[†]Carderock Division Naval Surface Warfare Center, West Bethesda, MD 20817
[‡]STI Optronics Inc., Bellevue, WA 98004

**Abstract.** We have developed non-invasive methods to measure the size, position, divergence and energy of a relativistic charged particle beam using diffraction radiation (DR). The DR is produced by the interaction of the beam fields with single or multiple apertures inclined at an angle of $45^0$ with respect to the beam velocity, $v$. We propose to utilize the near field image, far field angular distribution (AD) and polarization of backward DR, observed at $90^0$ with respect to $v$, to diagnose the beam. Unlike transition radiation (TR), the AD of DR is generally a function of the beam's transverse size, $\rho$ and position of the centroid relative to the center of the aperture, $b$ as well as the divergence, $s$ and energy, $E$. We show how the effects of $\rho$ and $b$ on the AD can be minimized or maximized by proper choice of the ratio: $R = a/\gamma\lambdabar$., where $a$ is the distance from the edge of the aperture to the center of the beam, $\lambdabar = \lambda/2\pi$, $\lambda$ is the observed wavelength and $\gamma$ is the Lorentz factor. For example, when $R < 1$, the AD of DR is insensitive to $\rho$ and $b$; the AD can then be used to determine $s$ and $E$. For the case $R = 2.4$, the effect of $\rho$ for a beam centered in the aperture is maximized and the AD can be utilized to measure $\rho$. In addition, we present a preliminary analysis, which shows how the and polarization of DR can be employed to determine the beam centroid position $b$.

## INTRODUCTION

Linear accelerators presently used for high energy physics research, as well as those being developed for high power free electron lasers (FEL), self-amplified spontaneous emission devices (SASE), and to test advanced accelerator concepts require accurate measurements of beam parameters including position, size and divergence. The small beam size and high average power of modern accelerators used for these applications preclude the use of conventional interceptive diagnostics and make it very difficult to monitor beam parameters on line. Commonly used non-interceptive monitors such as RF wall beam position monitors provide limited information about the beam characteristics. Other types of non-interceptive monitors based on synchrotron or undulator radiation require large magnetic devices which are neither practical nor desirable in many instances. Few other general purpose, multi-parameter non destructive diagnostics are available.

CP472, *Advanced Accelerator Concepts: Eighth Workshop*,
edited by W. Lawson, C. Bellamy, and D. Brosius

We are developing non-interceptive diffraction radiation (DR) diagnostics to measure beam position, transverse size, divergence, energy, emittance and bunch length. These diagnostics are very compact, can be implemented at any position in the beam line, offer minimal perturbation to the beam, respond rapidly to changes in the beam and can diagnose a number of beam parameters simultaneously.

Diffraction radiation is produced when a charged particle passes through an aperture or near a discontinuity in the media in which the beam passes. DR has been studied theoretically since the early 1960's, (1-4) and is closely related to transition radiation (TR), which is produced when a charged particle traverses the boundary between media with different dielectric constants.

The potential of DR as a beam position, radius, divergence, emittance, and energy diagnostic has been analyzed by ourselves and others. (5-9) In particular, these analyses have examined the potential of using the angular distribution of DR observed from various kinds of apertures as a means of measuring divergence, position and size. A key issue in all these studies is the problem of separating out the competing effects of these parameters on the angular distribution of DR. We have solved this problem and present here new methods to unambiguously determine these beam parameters.

To our knowledge only two experimental studies (10,11) have been reported, which clearly identify DR and compare its theoretically predicted properties to real data. In one of these (10) the spectrum of *coherent* DR (CDR) from circular apertures at FIR and mm wavelengths was observed and actually used to determine a beam parameter, the bunch length of the beam.

## PROPERTIES OF DIFFRACTION RADIATION

The theory for DR generated from slits and circular apertures has been well developed (see Ref. 3 for a review). Hence we will only summarize and highlight the results.

When a charge particle or ensemble of particles passes through an aperture or near an edge at a distance $a$, appreciable DR is generated when the condition $a \lesssim \gamma \lambda$ is fulfilled. Relativistic charged particles traveling in a vacuum and passing through a hole or slit in a conducting screen, (see Fig. 1.), will emit a cone of diffraction radiation with peak intensity occurring at the angle $\theta \sim 1/\gamma$. This situation is similar to TR emitted by a particle passing through a continuous surface. The angle $\theta$ is measured from the electron trajectory for forward TR and DR and from the direction of specular reflection for backward TR and DR. Unlike TR, however, the detailed characteristics of the angular distribution of DR depends on the beam size and its position with respect to the center of the aperture.

When the observed DR wavelength is smaller than the size of the charged particle bunch dimensions, incoherent DR is produced independently from each particle and the total intensity $I \sim N$, the number of particles in the bunch. For wavelengths comparable to the bunch dimensions coherent DR is observed with an intensity $I \sim N^2$.

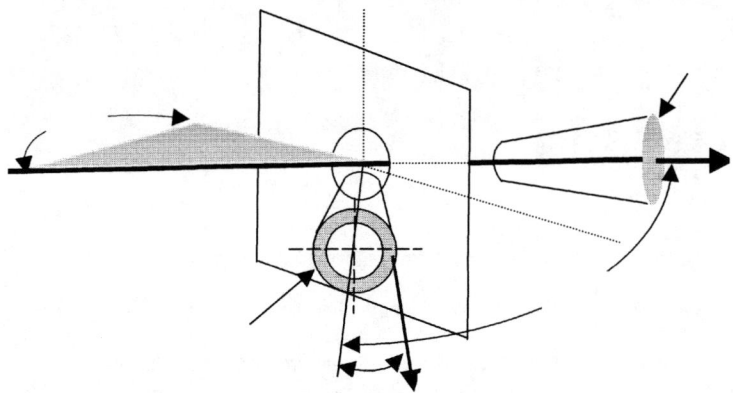

**FIGURE 1.** Forward and backward diffraction radiation of wave vector **k** produced by a particle with velocity **v** passing through an aperture of radius $a$ in a screen inclined at an angle $\Psi$ relative to the particle trajectory. The observation angle $\theta$ is measured from the direction of specular reflection.

The $N^2$ scaling is a key feature of coherent sources, since $N$ can be quite large (e.g. $10^8$). The diagnostics we are considering can utilize either incoherent or coherent DR from different types of apertures. However, we will focus our discussion here on incoherent DR emitted from a beam interacting with one or more circular apertures.

Ter-Mikaelian[3] shows that the number of incoherent forward directed DR photons of frequency $\omega$ radiated into $d\Omega$ for a single highly relativistic electron passing through a circular aperture of radius $a$ with an offset $b$ from the center of the aperture is given by

$$\frac{d^2 N}{d\omega\, d\Omega} = \frac{e^2}{\pi^2 c\hbar\omega} \frac{\theta^2}{\left(\gamma^{-2} + \theta^2\right)^2} \left[ J_o^2(qa) + (b/a)^2 J_1^2(qa) \right],\qquad(1)$$

where $e$ is the electron charge; $c$ is the vacuum light velocity; $q = k\sin\theta$, ($k = 2\pi/\lambda$); and it is assumed that $\gamma\lambdabar \gg a$ and $b \ll a$. The term in the squared brackets is the geometric form factor $G(q,a,b)$ which depends on the size and shape of the aperture. It is this term, which changes, for example, when a slit or other type of aperture is used. For a highly relativistic beam, the intensity of backward DR, like TR, will have the same angular distributions as forward DR but multiplied by the Fresnel reflection coefficient of the screen material. For moderate energy beams a more detailed analysis of the effect of the angle of incidence on the geometric form factor G(q,a,b) is required. For simplicity we will consider here only DR from highly relativistic beams ( $\gamma \gg 1$ ).

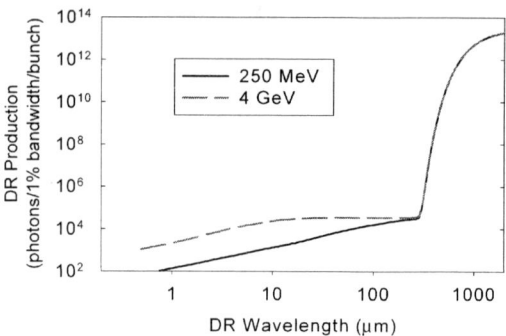

**FIGURE 2**. DR production for 250-MeV and 4-GeV *e*-beams from a 5-mm diameter aperture into 1% bandwidth. Note that the DR production is much higher at shorter wavelengths for the 4-GeV beam. The pronounced increase at 300 μm is directly related to the bunch length $\ell$ .

## REPRESENTATIVE INTENSITY CALCULATION

To illustrate the amount of potential DR production, which can be produced, we use as examples a 250-MeV and a 4-GeV electron beam. The other beam parameters assumed are: charge per bunch q = 135 pC ($8.4 \times 10^8$ electrons), bunch length $\ell = 300$ μm (1 ps), and beam rms radius $\rho_{rms} = 0.5$ mm., aperture radius $a = 2.5$ mm and a detector angle $\Theta = 10/\gamma$. The number of DR photons produced per 1% bandwidth per bunch as a function of DR wavelength is shown in Figure 2.

For $\lambda < 300$ μm the DR is incoherent; however, at larger $\lambda$ coherent DR production begins as evidenced by the orders of magnitude increase in the number of photons. Note for the 4-GeV case that roughly ~$10^4$ DR photons are generated in a 1% bandwidth per bunch for $\lambda < 10$ μm. Highly sensitive detectors within this wavelength range are readily available. Using a smaller aperture will generate more photons in the visible region of the spectrum making the observation of DR even easier.

The longitudinal length scale of most currently available beams is much longer than the transverse beam size. Therefore, it is most desirable to use short wavelength incoherent radiation to measure the position, radial size, and divergence, and coherent radiation, which is produced at longer wavelengths, to measure the beam's pulse length. This is the reason why bunch measurements for typical linacs ($\ell$ ~300μm) have been performed thus far at FIR and mm wavelengths. (10, 12)

## BEAM POSITION MEASUREMENTS USING DR

We have identified three possible DR observables which can be used to obtain beam position information: (1) the far-field angular distribution of DR, (2) near field image of DR at the screen surrounding the aperture and (3) the polarization measured at the aperture.

## Far field Angular Distribution

Unlike TR, the far field angular distribution of DR is affected by the beam's transverse size and position[**] with respect to the center of the aperture. The angular distributions of both TR and DR are affected by beam divergence. These effects must be separated for diagnostics purposes. As we will show below this can be done by properly choosing the aperture size and observed wavelength.

From Eq. 1 we see that the DR angular distribution is a function of beam particle position $b$ relative to the edge of the aperture. This effect has been previously considered for a slit in order to obtain beam position and size information[7] and it was shown that the DR intensity is a minimum when the beam is centered on a rectangular slit. A more recent (9) analysis has shown that the effect of beam size and offset on the angular distribution of DR from a rectangular slit are the same. Then the beam can centered in the slit by steering the beam to minimize the total DR intensity and the effect of offset on the angular distribution is nullified. The remaining effects on the angular distribution are then beam size and divergence only. A similar situation is to be expected in other geometries such circular apertures as well.

## Near Field Imaging

The electric field of the beam measured at position **x** from the center of the aperture, can be shown to be:

$$E_{x,y} = \frac{e\alpha}{\pi v} \{ \int d\varphi \int Q(\vec{x},z)I_0(\alpha x)xdx\} \exp(i\omega z/v) \frac{u_{x,y}}{u} K_1(\alpha, u) \tag{2}$$

where $\alpha = 2\pi/(\gamma\lambda)$, $Q(x,z)$ is the beam charge density distribution , $I_0$ is the modified Bessel function of the first kind, of order zero, $K_1$ is the modified Bessel function of the second kind, of order unity, $\mathbf{u} = \mathbf{x} - \mathbf{b}$ and **b** is along the line from the center of the aperture to the centroid of the beam,

The intensity of the DR at position **x**, $I(\mathbf{x}) = \{ |E_x|^2 + |E_y|^2 \}$, and is therefore a function of the offset $b$ and the beam distribution $Q$. It should be possible to use the asymmetry of the DR intensity measured around the aperture to locate the centroid of the beam, i.e. the optical analog to a RF beam position monitor (BPM). In order to use the intensity distribution to determine the beam's spatial distribution an inversion of Eq. 2. is required. We are presently investigating these possibilities.

---

[**]An important consideration for use of DR as a size or position monitor is the resolution of any such measurement. In Ref. 13 it was shown theoretically that the spatial resolution limit of TR (by analogy DR as well) is much less than the $\gamma\lambda$ limit thought to be imposed on TR by the uncertainty principle. More recent theoretical work (9,14) arrives at the same conclusion. Furthermore, OTR beam profile measurements (15) at a beam energy of 3.25 GeV have experimentally confirmed that beam sizes much smaller than $\gamma\lambda$ are actually observed. Therefore, there is very strong evidence that TR and similarly DR can be used to measure both the position and size of highly relativistic beams.

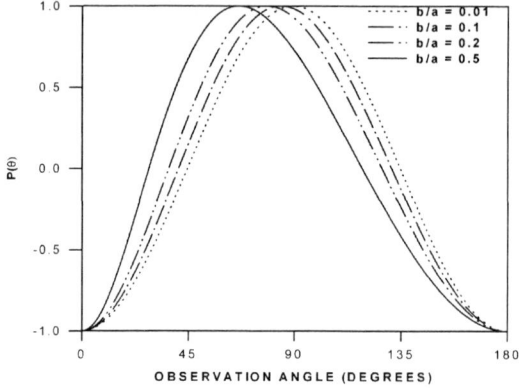

**FIGURE 3.** Polarization as a function of observation angle.

## *Polarization*

A finite size beam normally incident on and passing through the center of a circular aperture of radius $a$ produces DR which is purely radially polarized. This again is similar to the polarization of TR. We have calculated the effect on the polarization when the beam centroid is displaced from the center of the aperture by an amount $b$. The polarization measured around the aperture is:

$$P(\Phi) = \frac{\left|E_y\right|^2 - \left|E_x\right|^2}{\left|E_x\right|^2 + \left|E_y\right|^2} = \frac{a^2\left(\sin^2\Phi - \cos^2\Phi\right) + 2ab\cos\Phi - b^2}{a^2 + b^2 - 2ab\cos\Phi}, \qquad (3)$$

where **b** is along the x direction and the angle $\Phi$ is measured from the *x*-axis. Note that $P(\pi/4) \sim 1.414(b/a)$, i.e. the polarization is linearly dependent on b/a. The net polarization as a function of angle is shown in Figure 3. for various values of b/a.

## BEAM SIZE MEASUREMENT USING DR

We will now describe a new method of extracting the beam size from the far-field angular distribution pattern of DR. First of all we assume that the beam centroid has been centered in the aperture via one of the previous methods described above. Secondly we assume the particle position ρ within the finite beam spatial distribution has the same effect on the geometric form factor as the offset of a single particle, b. For simplicity we assume that the beam has a symmetric radial bunch distribution and integrate it over the geometric form factor for the circular aperture given in Eq. 1.

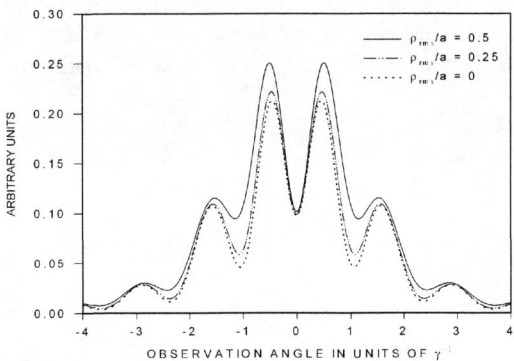

**Figure 4**. Angular distribution of DR for various values of the ratio $\rho_{rms}/a$.

The number of incoherent DR photons per electron per steradian per unit bandwidth is

$$\frac{d^2 N(x)}{d\omega d\Omega}\Delta\omega = \frac{1}{137\pi^2} \frac{4\gamma^2 x^2}{\left(1+x^2\right)^2}\left[J_0^2(Rx)+\left(< \rho^2 > / a^2 \right)J_1^2(Rx)\right]\frac{\Delta\omega}{\omega}, \qquad (4)$$

where $<\rho^2>$ is the averaged square beam radius and we have introduced the scaled variables $x = \gamma\theta$ (i.e. $\theta$ given in units of $\gamma^{-1}$) and $R = a/\gamma\lambda$. In terms of these variables the DR angular distributions of DR presented below can be interpreted for any value of beam energy.

To illustrate the use of the far-field DR pattern to obtain information on the beam's rms radius using Eq. 4, we choose values of aperture radius $a$, wavelength $\lambda$, and energy $\gamma$ such that $R = 2.4$. Then when $x = 1$, (i.e., $\theta = 1/\gamma$), $J_0(Rx) = 0$ and $J_1(Rx)$ is relatively large. The TR-like factor before the squared brackets in Eq. 4 is also a maximum when $x = 1$. Then the term proportional to $<\rho^2>/a^2$ in Eq. 4 is the main contributor at $x = 1$.

The patterns shown in Figure 4. were obtained by convolving Eq. 4 with a Gaussian distribution of beam particle angles characterized by an rms divergence $s = 0.2\gamma^{-1}$, and show that the rms beam size can be determined even in the presence of a relatively large beam divergence. Note that the intensity at $\theta = 0$ depends only on the value of $s$. This implies that both size and divergence can be obtained simultaneously .

## BEAM DIVERGENCE MEASUREMENT USING DR

We have developed several approaches for determining the beam divergence s, which have as their basis methods successfully developed with optical transition radiation. (16) The observables used to detemine the beam divergence are (1) the angular distribution of a single aperture and (2) the interferences of DR from two or more apertures, slits or edges. Under the proper experimental conditions, all these observables are insensitive to beam size and offset, and can be used to unambiguously determine the beam divergence.

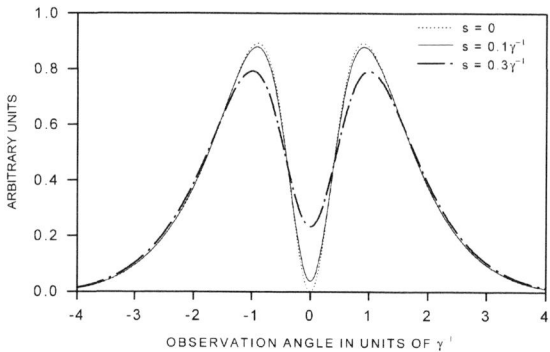

**FIGURE 5.** Single aperture DR angular distribution for divergences: $s = 0, 0.1\gamma^{-1}$, and $0.3\gamma^{-1}$; R=0.5.

## Single aperture method

Fiorito and Rule (16) have shown how $I_{TR}^{\parallel}(\theta)$ and $I_{TR}^{\perp}(\theta)$, the parallel and perpendicular components of TR, respectively, are influenced by $s_x$ and $s_y$, the rms divergences associated with the two orthogonal directions $(x, y)$ in a plane perpendicular to the velocity vector of the beam.

In order to estimate the sensitivity and effect of divergence on the DR angular distribution, Eq. 4. was numerically convolved with a Gaussian distribution of particle angles, characterized by a rms divergence $s$. The results for various values of $s$ are shown in Fig. 5., which is a scan of the horizontally polarized DR angular distribution pattern so that $s \equiv s_x$. Figure 5. was produced using $a = \gamma\lambda / 2$, i.e. $R = 0.5$ in Eq. 4, thus the zero of $J_0(Rx)$ occurs when $x = 4.8$. This choice of $a$ yields a DR pattern which is very similar to that of TR for angles $|\theta| \leq 2/\gamma$ or $|x| \leq 2$. This is just the region, which is most sensitive to beam divergence. Note that the observation angle and divergence shown in Fig. 5 are presented in units of $\gamma^{-1}$. Since both the divergence and the angle of peak intensity of DR scale with energy as $\gamma^{-1}$, the effect of divergence on the observable DR angular distribution pattern shown is valid for any value of beam energy.

The results shown in Fig. 5 are quite insensitive to the beam's radial distribution or position and therefore can be used to unambiguously measure the beam divergence.

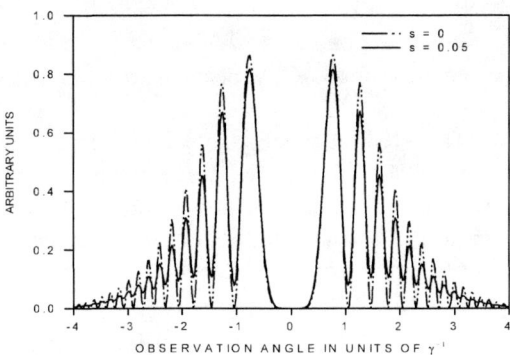

**FIGURE 6.** Angular distributions of interfering forward and backward DR from two apertures for two values of rms beam divergence: $s = 0$ and $s = 0.05\gamma^{-1}$; $R = 0.5$ and $L = 6L_c$.

## Multi-aperture interferences

When two or more apertures separated by a distance $L$ comparable to the coherence length (8) of TR or DR, interference fringes superimposed on the single-aperture angular distribution pattern are observed. This interference pattern is significantly more sensitive to divergence than the single aperture pattern. The fringe positions also provide a measure of the beam energy E.

Interference DR can be observed under the following conditions: (a) forward DR from the first aperture is reflected and interferes with backward DR produced by a second aperture, separated from the first by the *forward* coherence length, $L_c \sim \gamma^2\lambda$ ; (b) backward DR from the first aperture interferes with backward DR from the second, where the planes of two apertures are displaced by the *backward* coherence length, $L_c \sim \lambda$. The equation for the number of DR photons for two interfering sources is simply the single aperture DR function (Eq. 4) multiplied by the factor: $\{G_1 - G_2\,e^{-i\phi}\}$, where $G_{1,\,2}$ are the geometrical form factors for the first and second apertures, respectively, $\phi = L/L_c$, and $L_c$ is *either* the forward or backward coherence length.

Figure 6. shows horizontal scans of the DR angular distribution patterns theoretically predicted for interference of forward and backward DR from two apertures separated by a distance $L = 6L_c$ for two values of rms beam divergence, $s = 0$ and $s = 0.05/\gamma$. Figure 6. was obtained by again by convolving a Gaussian distribution of particle angles with the equation for a two aperture system (see Eq. 4 Ref. 8.) and assuming that the $G_{1,2}$ are equal and slowly varying functions of the observation angle $\theta$. The choice of R = 0.5 produces an interference pattern which is very similar to that observed for TR. Because of the slow variation of $G_{1,2}\,(\theta)$, the dominant effect on the fringe visibility is determined by the rapidly varying phase term $e^{-i\phi}$, just as for TR. A comparison of Fig. 6. with Fig. 5. clearly shows that the two aperture interferometer has a greater sensitivity to beam divergence than a single aperture. Such an interference pattern would be observed for example from a 20-GeV beam with an rms divergence of 1 μrad (SASE parameter regime), a 650-MeV beam with a divergence of 40 μrad (Argonne APS linac regime), or a 4-GeV beam with a divergence of 6.25 μrad (TJNAL regime).

As has been shown for OTR (16) by suitably selecting $L$ and $\lambda$, an interferometer optimally sensitive to a given value of divergence can be designed and used to measure $s$ by fitting the experimentally observed angular distribution pattern with theoretically generated patterns. Furthermore, by measuring both horizontally and vertically polarized DR interference patterns, one can obtain the two orthogonal values of beam divergence at corresponding $x$ and $y$ beam waist conditions, respectively. In addition, the position of each fringe provides a measure of the energy of the beam. (17) Using the multiple fringes shown here, it is possible to measure the energy to an accuracy of 1% or better, and simultaneously determine the divergence as well.

In order to observe measurable DR for certain beam energies, it may be necessary to use wavelengths that are comparable to the beam pulse length. In this case the observed DR will be coherent, and it becomes necessary to consider the effect of coherence on the angular distribution and geometrical form factors. These complications may affect the detailed form of the angular distributions, but should not substantially alter the results presented here. An analysis of the effect of coherence on DR measurements will be presented in a future study.

## REFERENCES

1. Bass, F.G. ,Yakovenko, V. M., Sov. Phys. Uspecki **8** (3), 420 (1965).
2. Keil, E., Nuc. Instrum. and Meth. **100**, 419 (1972).
3. Ter-Mikaelian, M.L. , *High Energy Electromagnetic Processes in Condensed Media*, J. Wiley-Interscience, New York, 1972.
4. Maresca, N.J. and Liboff, R.L. ,Can. J. Phys. **53**, 62 (1975).
5. Rule, D.W. and Fiorito, R.B., Naval Surf. Weap. Cent., Tech. Rep. No. TR 84-134 (1984).
6. Rule, D.W. and Ritchie, R.H. in Proc. of Werner Brandt Workshop on Penetration Phenomena, Oak Ridge National Laboratory, 1985 pp. 15-16.
7. Moran, M.J. and Chang, B., Nucl Instr. Meth. Phys. Res. **B40/41**, 970 (1989).
8. Rule, D.W. , Fiorito, R.B. and Kimura, W.D., "Non Interceptive Beam Diagnostics Based on Diffraction Radiation" in *Beam Instrumentation, Proc. of the Seventh Workshop, AIP Conf. Proc. No. 390*, AIP Press, NY, 1997, pp. 510-517.
9. Castellano, M., Nuc. Instr. Meth. Phys. Res. **A394**, 275 (1997).
10. Shibata, Y. *et al.*, Phys. Rev. E **52**, 6787 (1995).
11. Vnukov, I., et.al. , JETP Letters, **67**, 802 (1998).
12. Shibata, Y. *et al.*, Phys. Rev. E **49**, 785 (1994) and Phys. Rev. E **50**, 1479 (1994).
13. Rule, D.W. and Fiorito, R.B.,"Imaging Micron Sized Beams", in *AIP Conf..Proc. No. 229*, 1990, pp. 315-321.
14. Lebedev, V. A. , Nucl. Instr. Meth. Phys. Res. **A372**, 344 (1996).
15. Piot, P.,Denard, J.-C. *et. al.*, "High Current Beam Profile Monitors Using Transition Radiation at CEBAF" in *AIP Conf. Proc. No. 390*, AIP Press, NY, 1997, pp. 298-305.
16. See Fiorito, R. B. and Rule, D. W. "Optical Transition Radiation Emittance Diagnostics" in *AIP Conf. Proc. No. 319*, AIP Press, New York, 1994, pp. 21-35 and references therein.
17. Wartski, L. *et al.*, J. Appl. Phys. **46**, 3644 (1975).

# Sub-picosecond Electron Bunch Profile Measurement Using Magnetic Longitudinal Dispersion and Off-Phase RF Acceleration

K. N. Ricci, T. I. Smith, and E. R. Crosson

*Stanford Picosecond FEL Center*
*W. W. Hansen Experimental Physics Laboratory*
*Stanford University*
*Stanford, CA 94305-4085, USA*

**Abstract.** Using a longitudinally dispersive bending-magnet transport system followed by off-phase acceleration in a radio frequency electric field, it is possible to transform a highly relativistic electron bunch such that the relative time coordinate within the initial bunch is uniquely mapped to the relative energy coordinate of the final bunch. An energy spectrometer can be used to analyze the final energy distribution, revealing the longitudinal density profile of the initial electron bunch. We discuss factors which limit the resolution of this technique, and we present bunch profile measurements from the Stanford Superconducting Accelerator with a resolution better than 300 femtoseconds.

## INTRODUCTION

In recent years the production of picosecond and sub-picosecond relativistic electron bunches has become increasingly important to many experiments using free electron lasers (FELs) and other exotic light sources. In order to operate these light sources effectively it is necessary to develop electron beam diagnostics with sufficiently fine resolution to characterize such short bunches. Several diagnostic techniques are currently in use, including observation of coherent radiation from the bunch, streak cameras, and off-phase radio frequency (RF) acceleration. The spectra of coherent transition radiation (TR) or synchrotron radiation (SR) provide estimates of very short electron bunch lengths and bunch shapes through their Fourier relationship to the emitting electron distribution. However, the power spectrum of a waveform is not uniquely invertible to the original waveform, except for a limited class of well-behaved pulse shapes; and practical difficulties in measuring the entire spectrum of a broadband submillimeter light source reduce the accuracy of these sub-picosecond pulse-shape reconstructions (1,2). The streak camera measures the bunch profile unambiguously,

CP472, *Advanced Accelerator Concepts: Eighth Workshop,*
edited by W. Lawson, C. Bellamy, and D. Brosius
© 1999 The American Institute of Physics 1-56396-889-4/99/$15.00

but can achieve sub-picosecond resolution only with considerable effort and careful instrumentation which presents a significant investment for a typical accelerator facility (3). For this reason some labs have tried off-phase acceleration (OPA), in which some of the existing RF accelerator structures are run off-phase to impart a time-dependent energy slew to the electrons in the bunch. Electrons arriving later in the bunch receive a different energy increase than those arriving earlier, and the time coordinate is transformed in a known way to an energy coordinate, which is easily analyzed in an energy spectrometer. Unfortunately this technique does not map time to energy *uniquely*: since every real beam has some energy spread, the resolution of the final measurement is degraded by a factor proportional to the initial energy spread, as discussed below. For most RF accelerator systems, the degradation is significant enough to prohibit sub-picosecond bunch measurement (4).

Here we demonstrate a modification of the OPA technique, suggested three years ago (5), that can significantly improve temporal resolution. The beam passes first through a longitudinally dispersive bending magnet transport system, such as a chicane, and then into the OPA. The additional transformation from the chicane causes each time slice of the initial beam to map to a unique final energy regardless of the initial energy spread. This has allowed us to measure the beam profile with a resolution finer than 300 femtoseconds at the Stanford Superconducting Accelerator (SCA).

## THEORY OF THE TECHNIQUE

The distribution of electrons in a bunched, accelerated beam is often described in terms of phase-space coordinates, $(x, \dot{x}, y, \dot{y}, z, \dot{z})$ where $x$, $y$ are the transverse positions and $z$ is the longitudinal position (along the acceleration axis) relative to the center of an electron bunch; and the coordinates $\dot{x}, \dot{y}, \dot{z}$ are the relative momenta. In this section we will consider only the longitudinal coordinates $z$ and $\dot{z}$. For a highly relativistic beam, the position $z$ is equivalent to relative time $t$ within the bunch through the relation $z = ct$, and so $z$ will be described in units of time. Similarly, since the relativistic momentum is proportional to energy, we will treat $\dot{z}$ as the energy in the calculations that follow. The electron energy distribution is measured by bending the beam in a calibrated dipole magnetic field and observing the electron density or current as a function of bending radius, since the radius is proportional to momentum.

Measurement of the electron distribution in time is more difficult on the sub-picosecond time scale, but can be accomplished in the following way: consider an electron beam passing through a magnetic chicane and then into an RF accelerator section while the electric field is ramping up, off-phase from the maximal field. To first order, $z$ and $\dot{z}$ transform as shown in the transport matrices in equation (1).

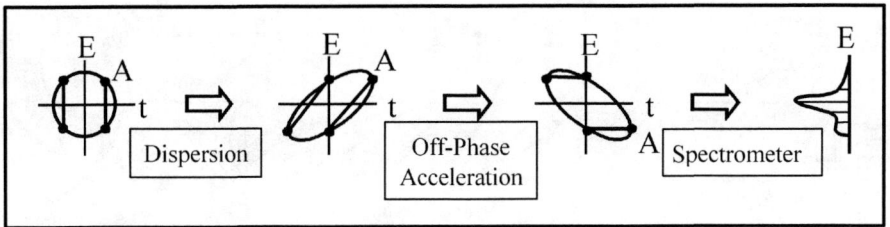

**FIGURE 1.** This linear transformation of phase space converts a temporal slice of the bunch, such as the segment containing point A, into a slice of the energy distribution which is imaged in the energy spectrometer. The stretching of the phase space in the dispersive section allows the OPA to remove the initial energy spread from the slice.

$$\underbrace{\begin{bmatrix} 1 & 0 \\ \alpha & 1 \end{bmatrix}}_{\text{OPA}} \underbrace{\begin{bmatrix} 1 & \delta \\ 0 & 1 \end{bmatrix}}_{\text{chicane}} \underbrace{\begin{bmatrix} z \\ \dot{z} \end{bmatrix}}_{\substack{\text{initial} \\ \text{coordinates}}} = \underbrace{\begin{bmatrix} z + \delta\dot{z} \\ \alpha z + (1 + \alpha\delta)\dot{z} \end{bmatrix}}_{\substack{\text{final} \\ \text{coordinates}}} \tag{1}$$

Using units of energy and time, $\delta$ is the longitudinal dispersion of the chicane in picoseconds/keV, and $\alpha$ is the rate of change of the OPA in keV/picosecond. Notice that the time coordinate $z$ has appeared in the first term of the final energy. This term can be isolated experimentally by adjusting the OPA and the chicane dispersion until $\alpha\delta = -1$, causing the second term to become zero. Then the energy will be $\dot{z}_{final} = \alpha z_{initial}$, and a measurement of the final energy profile will yield the time profile of the initial bunch. Figure 1 shows how the phase space evolves during these transformations. To first order, the temporal resolution of the bunch profile measurement is

$$\Delta t = \Delta E_{spec}/\alpha, \tag{2}$$

where $\Delta E_{spec}$ is the energy resolution of the spectrometer.

## EXAMPLE MEASUREMENTS

Using the apparatus shown in Figure 2, we measured the electron bunch profiles in Figure 3. The front end of the SCA accelerated a beam consisting of $4.2\times10^{7}$ electrons per bunch to 19 MeV, with an initial energy spread of 130 keV FWHM (full width at half-maximum) in the spectrometer. The beam from the front end is typically bunched to a few picoseconds or less in order to ensure high current density for the operation of the FEL.

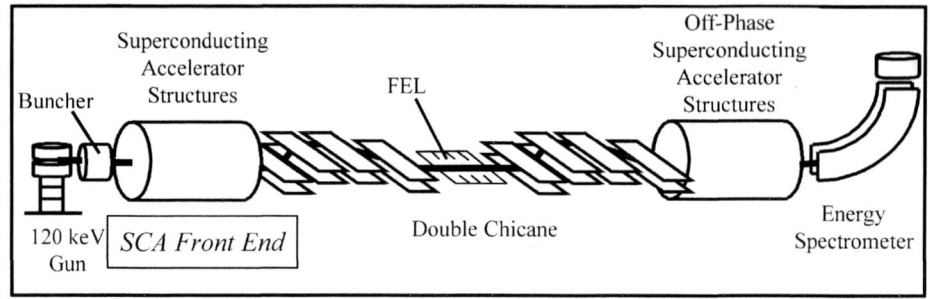

**FIGURE 2.** Schematic diagram of a part of the Stanford Superconducting Accelerator.

The double chicane, with dipole fields on the order of a kiloguass, was adjusted to provide about –0.012 ps/keV longitudinal dispersion at 19 MeV. The final sections of the 1.3 GHz RF accelerator, running at field amplitudes which normally provide 13 MeV total energy increase, were run off-phase to provide an acceleration increment that ramped up at a rate of 84 keV/ps.

It was easy to measure the value of $\alpha$ because the RF phases of the accelerator structures were locked with a relative stability of 200 fs over several minutes. It was more difficult to determine $\delta$ accurately because the chicane was not instrumented for this purpose, but the value of $\delta$ was inferred in the following way:

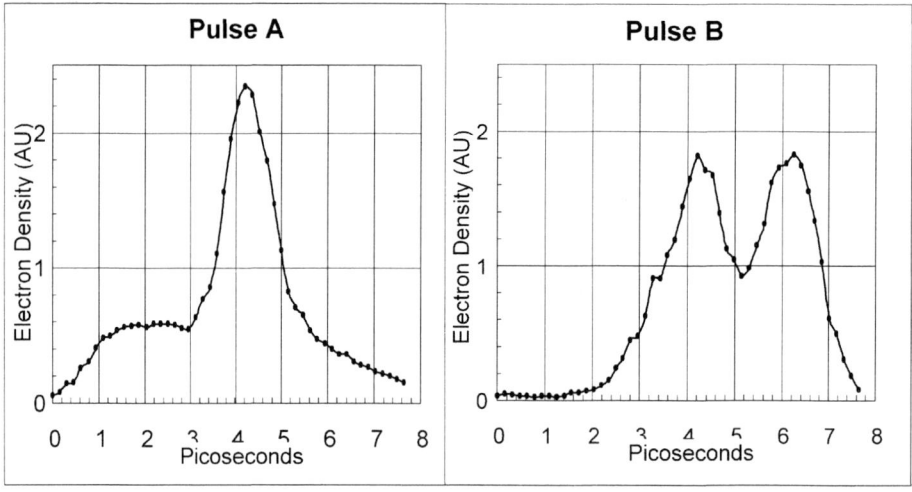

**FIGURE 3.** Longitudinal profiles of electron bunches from the SCA front end. Pulse A was generated with the SCA near its optimal configuration for FEL startup. Pulse B was produced by iterating the phases and amplitudes of buncher and accelerator structures specifically in an attempt to produce a structured bunch. The bunches can be visualized as travelling to the left, with time t=0 at their head.

according to the transport matrices in equation (1), energy fluctuations from the SCA front end are canceled by the OPA when $\alpha\delta = -1$. This prediction was verified by setting the chicane to a fixed field, gradually increasing the phase of the OPA from zero, and repeatedly incrementing the front end amplitude as the OPA approached the expected optimal phase. At the optimal phase, the front end energy increments ceased to affect the energy of the beam entering the spectrometer, verifying that $\alpha\delta = -1$ had been attained. At this point a 1.45° phase shift in the 1.3 GHz OPA produced an energy shift of 260 keV at the spectrometer. This provided the calibrations of $\alpha = 84$ keV/ps and $\delta \approx -0.012$ ps/keV, which agreed with our estimate based on the chicane geometry. The energy spectrometer was observed to have a precision of 0.08% or better at energies near 30 MeV for the beam emittance of this measurement. For this configuration, our temporal resolution was $\Delta E_{spec}/\alpha = (0.0008 \times 27000$ keV $/ 84$ keV ps$^{-1}) = 0.26$ ps or 260 femtoseconds.

Figure 3A shows a typical electron bunch when the front end is configured near-optimally for FEL startup. Most of the current is bunched tightly in a peak with FWHM of 1.4 ps, but a significant amount remains outside the peak in both front and back tails almost 3 ps long. The center of the main peak is extremely sharp—possibly too sharp to be resolved in this measurement. Figure 3B shows a double-peaked electron bunch created by adjusting various phases and energies at the SCA front end, including the microwave cavity buncher. A wide variety of interesting beam profiles can be reproducibly generated by monitoring the pulse shape while adjusting accelerator parameters in this way.

## DISCUSSION AND LIMITATIONS

Using a chicane with OPA has significantly improved our ability to understand the electron bunch shape at the SCA. Previous OPA-type experiments looked promising, but it was impossible determine whether the observed irregularities were from the temporal profile, the initial energy distribution, artifacts of the measurement technique, or some convolution of all three. Now that unambiguous pulse shape observations have been achieved at 300 fs resolution, other bunch diagnostics at the SCA will be easier to interpret, including attempts to map the longitudinal phase space of the beam using tomographic analysis (6). In developing this OPA technique as a practical beam diagnostic, it is important to consider the both low-resolution and high-resolution extremes.

### Utility at Low Resolution

As we have seen, the chicane makes high resolution temporal measurement possible by causing the OPA to eliminate initial beam energy spread at the same time that it is transforming the time coordinate to the energy coordinate. However, even if the

739

condition $\alpha\delta = -1$ cannot be reached, adding some longitudinal dispersion in front of an OPA can significantly improve bunch profile measurements. This may be of interest to accelerator facilities that do not want to invest in a streak camera or even build a new magnet, but would like to use existing apparatus to characterize short bunches. We may estimate the resolution of dispersion-improved OPA bunch measurements by considering again the final energy in equation (1): when the term $(1+\alpha\delta)\dot{z}$ does not go to zero it tends to dominate the random spread in the final energy distribution, which is equivalent to degrading the spectrometer resolution to the order of $\Delta E$ of the initial beam. For a beam with a gaussian energy distribution, very short temporal features will appear broadened by the initial energy spread to a resolution

$$\Delta t \approx \frac{1}{\alpha}\sqrt{\Delta E_{spec}^2 + ((1+\alpha\delta)\Delta E_{beam})^2}. \tag{3}$$

Obviously, the resolution improves significantly as $\alpha\delta$ gets closer to -1, for beams with an energy spread large compared to the spectrometer resolution; but even achieving $\alpha\delta$ = - 0.5 will generally improve temporal resolution by a factor of 2.

## High Resolution Limits

So far we have considered only the longitudinal coordinates of phase space and first-order beam transformations, and from these ideal predictions the only limits to temporal resolution are the available acceleration gradient and the resolution of the energy spectrometer. However, as we push these limits to very short time scales we expect that second-order terms and coupling to the transverse phase space will become important. We will briefly consider some of these limitations here.

Fortunately, the chicane is a transversely achromatic device, an important consideration when planning an experiment of such fine longitudinal resolution. There is no first-order coupling from any of the coordinates $x$, $\dot{x}$, $y$, or $\dot{y}$ to the longitudinal coordinates when transport through the chicane is taken as a whole. By symmetry, any transverse coupling introduced by the first two dipoles of a chicane is removed by the second pair. Since the rectangular dipoles are translationally invariant in the $x$ coordinate (in the bending plane), there is no coupling to any order in this dimension. However, there is some second-order coupling from the $\dot{x}$ momentum to the longitudinal coordinate $z$ due to increased path length in the dipoles, as we would expect; and there is also a second-order coupling from $y$ and $\dot{y}$ to $z$ due to focusing effects of the dipole fringing fields. We have considered these terms individually and they are comparable in magnitude to the second-order coupling of the quadrupole focusing discussed below.

The OPA provides no significant coupling between the transverse and the longitudinal coordinates because of the uniformity of the electric field across the accelerator aperture. Transverse emittance does play a significant role in limiting the resolution of the energy spectrometer, and this effect has been accounted for by calibrating the spectrometer resolution for the beam conditions of our measurement. Note that coupling from transverse to longitudinal phase space in general is not a problem after the OPA: since the initial longitudinal information has by this stage been encoded in the energy distribution, which is unaffected by magnetic transport, $z$ is of no further concern on the way to the spectrometer.

Important second order effects include the nonlinear longitudinal dispersion of the chicane, nonlinear velocity dispersion of the beam, and transverse focusing. The first two turn out to be negligible in the present case, but might be significant for later generation accelerators, and so we review them here. The path length through the chicane shown in Figure 4 is

$$S = 4r\theta + 2b/\cos\theta + w_2 . \tag{4}$$

The bending radius $r$ and angle $\theta$ vary as a function of the energy $\gamma$, and the path length varies accordingly. Expanding $S(\gamma)$ about the central energy $\gamma_0$ for the beam, we write

$$S(\gamma_0 + \Delta\gamma) = S(\gamma_0) + \Delta\gamma \, (dS/d\gamma) \,|\, \gamma_0 + \tfrac{1}{2} \, (\Delta\gamma)^2 \, (d^2S/d\gamma^2) \,|\, \gamma_0 + \dots , \tag{5}$$

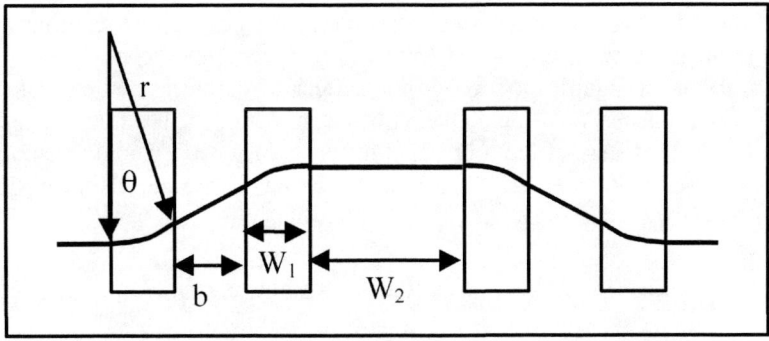

**FIGURE 4.** Geometry of a chicane and the electron beam path.

where $dS/d\gamma$ is the linear dispersion $\delta$ that we exploited in equation (1), and $d^2S/d\gamma^2$ is a troublesome quadratic dispersion. Proportional to $(\Delta\gamma)^2$, this dispersion cannot be eliminated by a linear transformation. It produces a phase-space distortion that may limit the ultimate resolution of this technique; although it conceivably could be partially canceled by the sinusoidal temporal nonlinearity of the RF field in the accelerator, or eliminated by a clever magnet design. Differentiating equation (4) with the help of some geometry and algebra, we find that

$$d^2S/d\gamma^2 = (4r/\gamma^3)\ tan^2\theta + (6\ b\ tan^2\theta)\ /\ (\gamma^3\ cos^3\theta). \tag{6}$$

For the parameters of our chicanes and beam, the quadratic chicane dispersion is less than 30 fs. A similar quadratic longitudinal dispersion results from the slight differences in drift velocities of highly relativistic electrons. Since the velocity

$$v = c\sqrt{1 - 1/\gamma^2}, \tag{7}$$

we can expand $v(\gamma)$ into linear and quadratic terms as we did for $S(\gamma)$ above. The linear term can be treated simply as an extension of the chicane, where linear dispersion is desirable. The quadratic term is of order

$$\tfrac{1}{2}\ (\Delta\gamma)^2\ d^2v/d\gamma^2 \approx -1.5\ (\Delta\gamma)^2\ c\ /\ \gamma^4 = -15 \text{ m/s} \tag{8}$$

for our beam. During the 50 nanoseconds it takes the beam to travel from the SCA front end to the OPA, the accumulated error is only a few femtoseconds.

Finally, the most significant second-order effect in our measurement comes from quadrupole and chicane focusing. Over a 6 meter section of beamline including the chicanes, our beam focuses from a one centimeter diameter to a 3 mm waist, and with typical transverse emittances of $10\pi$ mm mrad, $\dot{x}$ and $\dot{y}$ are approximately 3 mrad. The longitudinal error introduced by this angular spread is

$$\Delta S \approx \tfrac{1}{2}\ (\dot{x})^2\ L = (0.5) \text{ x } (0.003)^2 \text{ x } (6 \text{ meters}) = 27\mu\text{m}, \tag{9}$$

or 90 fs. So transverse emittance presents a serious lower limit to the possible resolution of this technique at the SCA. The problem can be reduced by using more quadrupoles to control beam size without tight focusing, and by placing the OPA as close after the magnetic dispersive section as possible, since beam divergence after the OPA does not couple to $\dot{z}$. With adjustments of this sort, and minor improvements to our spectrometer and RF accelerator controls, we hope to achieve 100 fs resolution at the SCA soon.

The next generation of linear accelerators, however, have improved transverse emittances by an order of magnitude, and this in turn reduces all of the transverse couplings by one or two orders of magnitude, putting femtosecond bunch profile measurements within reach. As an example, consider a 10 MeV section of a 10 GHz RF accelerator running 90° off-phase. The rate of change of acceleration amplitude will be $\alpha = 630$ keV/ps. Suppose a 100 MeV beam passes through a chicane with $\delta = -0.0016$ ps/keV (easily achievable with dipoles of a few kilogauss), through this OPA, and then into an energy spectrometer with 0.01% resolution. From equation (2) we estimate a beam profile resolution of

$$\Delta t = (0.0001 \times 10^5 \text{ keV}) / (630 \text{ keV ps}^{-1}) = 16 \text{ fs}. \qquad (10)$$

This result could be achieved in a machine with a transport geometry similar to ours, a transverse beam emittance of 1 $\pi$ mm mrad, and an energy spread of 0.5% or less. For an energy spread of 1%, the nonlinear longitudinal dispersion in the chicane would degrade the resolution to about 30 fs, unless corrective measures were taken.

# CONCLUSIONS

We have shown that, with the addition of magnetic longitudinal dispersion, the technique of off-phase RF acceleration can measure electron bunch profiles at sub-picosecond resolution. At the SCA we have measured profiles with a resolution better than 300 femtoseconds, and manipulated the bunch to produce a variety of specific shapes. We estimate that the latest generation of accelerators will be able to achieve resolutions of tens of femtoseconds using this method; the ultimate resolution is limited by the available acceleration gradient, the resolution of the energy spectrometer, second order coupling between longitudinal and transverse emittances, and nonlinear longitudinal dispersion at the femtosecond time scale. Conveniently, this technique may allow some accelerator facilities to improve their picosecond electron bunch characterization with little or no investment in new instruments and beamline modification. It requires only a longitudinally dispersive magnet system (preferably transversely achromatic), followed by an RF accelerator structure running off-phase from the beam source, followed by an energy spectrometer—standard components of many exotic light source laboratories.

In the near future, this bunch measurement technique will probably be compared favorably to the other techniques described in the introduction to this paper; we plan to use our OPA work to benchmark our coherent SR and TR diagnostics at the SCA. We expect that these different methods will complement each other: the coherent radiation can be used to observe higher resolutions, through high frequency radiation, than we can achieve with OPA at our facility. On the other hand, the shapes of the 3 ps long pulse tails observed with OPA probably cannot be recovered by coherent far-infrared radiation analysis, because the longer wavelength information is suppressed by beampipe aperturing, diffraction, and poor detector response.

Finally, there are many immediately interesting applications for bunch profile measurements at 100 to 300 fs resolution. The measured bunch shape can be used in electron gun and accelerator development—both to verify the proper operation of these components, and for detailed comparison to the predictions made by the accelerator codes used for future gun and accelerator design. A good bunch shape diagnostic will make it easier to tailor the bunch for purposes as such FEL startup or short pulse generation. It may help to explain why certain accelerator parameters produce unexpectedly high levels of coherent undulator radiation at wavelengths much shorter than the bunch FWHM. It may be possible to observe and to optimize electron beams micro-bunched by a terahertz FEL. The technology to produce the next generation of ultrashort electron bunches and advanced, modulated beams can only develop as fast as the technology develops to measure them.

## ACKNOWLEDGMENTS

This work was supported in part by the Office of Naval Research, Grant No. N000140-94-1-1024.

## REFERENCES

1. Lai R., Happek U., and Sievers A. J., *Physical Review E* **50**, R4294-7 (1994).
2. Schneider, Gi., et al, *Nuclear Instruments and Methods A* **396**, 283-292 (1997).
3. Takahashi, A., et al, "New Femtosecond Streak Camera," in *Proceedings of SPIE* **2116**, 275-84 (1994).
4. Wiedemann, H., "Summary of Discussions in the Radiation Working Group," in *Micro Bunches Workshop, AIP Conference Proceedings* **367**, 512-517 (1996).
5. Crosson, E. R., et al, "A Technique for Measuring an Electron Beam's Longitudinal Phase Space with Sub-picosecond Resolution," in *Micro Bunches Workshop, AIP Conference Proceedings* **367**, 397-405 (1996).
6. Crosson, E. R., et al, *Nuclear Instruments and Methods A* **375**, 212-219 (1995).

# The Effect of Detector Bandwidth on Microbunch Length Measurements Made with Coherent Transition Radiation

D. W. Rule [*], R. B.Fiorito [+], and W. D. Kimura [#]

[*]Carderock Division Naval Surface Warfare Center, W. Bethesda, MD 20817-5000
[+]The Catholic University of America, Washington, DC 20064
[#]STI Optronics Inc., Bellevue, WA 98004-1495

**Abstract.** We present an analysis on the effect of the finite bandwidth of a detector on the observed intensity of coherent transition radiation (CTR) produced by a long train of microbunches. The parameters of the planned CTR bunch length measurement for the Staged Electron Laser Acceleration (STELLA) experiment on the ATF linac at Brookhaven National Laboratory (BNL) are used as an illustrative example. We show how the effective number of bunches contributing to the observed coherent radiation depends on the band width of the detector. We conclude that meaningful CTR bunch length measurements on the STELLA experiment are possible using fairly wide band pass filters, e.g., ~5%. Micropulse widths of ~1 - 3 μm (3 - 10 fs) FWHM can easily be distinguished by observing up to 5 - 6 harmonics of the 10.6-μm bunch spacing. This analysis should also be applicable to other radiation produced by microbunches, such as coherent Cerenkov and synchrotron radiation.

## INTRODUCTION

Recently, Liu, *et al* (1) presented the results of a coherent transition radiation (CTR) measurement of the microbunching by the inverse free electron laser (IFEL) accelerator at BNL. They observed the near infrared CTR as a function of beam charge and found the expected quadratic dependence. By employing a 2.5-μm short wavelength pass filter with an InSb IR detector, they inferred a bunch length much shorter than 5 μm and closer to 2.5 μm, the fourth harmonic of the $CO_2$ laser. This same measurement technique will also be used to measure the microbunch length at the inverse Cerenkov accelerator (ICA) stage of the Staged Electron Laser Acceleration (STELLA) experiment (2 - 3). It is desirable to have a more precise measurement of the microbunch length in order to optimize the ICA efficiency. In the next phase of the bunch length diagnostic development, narrow band-pass filters, centered on the harmonics of the $CO_2$ laser, will be used to obtain more detailed information on the microbunch form factor. This will require at least one shot for each

CP472, *Advanced Accelerator Concepts: Eighth Workshop*,
edited by W. Lawson, C. Bellamy, and D. Brosius
© 1999 The American Institute of Physics 1-56396-889-4/99/$15.00

harmonic. Alternatively, a spectrometer operating in the 1 - 5 μm wavelength range could be used to obtain this information on a single shot.

Previous CTR bunch length measurements have been based on either autocorrelation of the CTR (4 - 5) or direct spectral observations (6 - 8). The former technique was used to measure bunches as short as 36 μm FWHM (120 fs) by averaging over $10^6$ microbunches (5). This technique requires a continuous train of bunches since it takes from ~30 seconds to several minutes to obtain the data. The direct spectral technique can, in principle, be used on a single shot; however, heretofore, it has been employed only on continuous linac beams to measure millimeter and sub-millimeter bunches (6 - 8).

The CTR microbunch measurement for the STELLA experiment poses additional challenges not previously encountered. The ~300 microbunches in a macropulse are closely enough spaced to produce coherent radiation from many microbunches, in addition to the intra-microbunch coherence. The interference of the radiation from the train of microbunches produces a spectrum of narrow lines about the harmonics of the $CO_2$ laser, so that the microbunch form factor, which is proportional to the Fourier transform of the envelope of the spectrum, can only be measured at these harmonics. Also, the repetition rate of the laser is so low that STELLA is essentially a single shot experiment. Furthermore, the microbunches are so short (1 - 3 μm or 3 - 10 fs) that observations must be made in the near IR.

Rosenzweig, et al., (9) have done an analysis of the CTR properties that are expected for STELLA-like microbunches. Pantell (10) has done a comparison of CTR and coherent Cerenkov radiation for the STELLA experimental conditions as well. In this paper, we focus on the effect of finite band-pass filters on the CTR observation. The same analysis applies to the effect of the resolution of spectrometers, if it is comparable to, or larger than the natural line width of the CTR.


# COHERENT RADIATION BY PARTICLE BUNCHES

In this section we review the standard treatment of coherent radiation by bunches of particles in order to highlight certain features of the process, which affect the use of coherent radiation for beam diagnostics purposes. The discussion of CTR by Vardanyan, et al., (11) will be followed [see also Wartski (12)]; however, the results are also applicable to coherent synchrotron, Cerenkov (13), and diffraction radiation. Next a heuristic discussion of the effect of long, finite trains of narrow microbunches on the longitudinal (temporal) coherence of the radiation is considered. The influence that a detector's angular acceptance and bandwidth has on the observations is discussed. Then the effects of the finite bandwidth is included in the analysis. Finally a particular model for the microbunch form factor is selected to illustrate what the experimental observations might yield.

# Review of coherent radiation production

Following Ref. 11, consider the amplitude for a single particle radiation process, $A = A(6,\phi)$, which can represent, for example, transition, Cerenkov, synchrotron, or diffraction radiation. $A(6,\phi)$ is a function of the wave vector $k$, which makes an angle $6$ with respect to the velocity vector $v$. The $z$-axis is taken parallel to the velocity vector, so $k_z = k\cos 6$ and $k_\perp = k\sin 6 \approx k6$, assuming $6 \ll 1$. The angle $\phi$ is measured in the plane perpendicular to the $z$-axis. For the special case of transition radiation, the single particle spectral angular radiation distribution is

$$\frac{d^2W}{d\omega d\Omega} = |A|^2 = \frac{e^2}{4\pi^2 c}\frac{\sin^2\theta}{(1-\beta\cos\theta)^2}. \tag{1}$$

Now, if we have $N$ particles, we can write the total amplitude $A_N$ as

$$A_N = Ae^{-ik\cdot r_0}\sum_{i=1}^{N}e^{-i(k_\perp\cdot\rho_i + k_z z_i)}, \tag{2}$$

where $r_i = \rho_i + z_i$ is the position of the $i^{th}$ particle relative to the center of the macrobunch at $r_0 = vt$. The vector $\rho_i$ lies in the plane perpendicular to the $z$-axis. Using this multiparticle amplitude, we can write the spectral angular distribution for the $N$-particle bunch as

$$\frac{d^2W_N}{d\omega d\Omega} = |A|^2\left\{N + \sum_{i\neq j}^{N}e^{-i\left[k_\perp\cdot\left(\rho_i-\rho_j\right)+k_z\left(z_i-z_j\right)\right]}\right\}. \tag{3}$$

The term $N$ in the large brackets above is the incoherent radiation from $N$ individual particles. The summation over $i \neq j$ gives the coherent part of the radiation. To proceed, we need to have some model for the probability $P(r\text{-}r')$ that the separation between any two particles, $i$ and $j$, is $r-r' = \rho - \rho' + (z-z')\bar{z}$. One of the simplest models used is to assume that the particles are completely uncorrelated so that

$$P(r\text{-}r') = p(r)p(r') \quad \text{and} \quad \iint P(r\text{-}r')drdr' \equiv 1, \tag{4}$$

with

$$p(r) = g(\rho)F(z), \tag{5}$$

i.e., the probability of radial position $\rho$ and longitudinal position $z$ are also uncorrelated. Using this probability model, we average Eq. (3) over $P(\mathbf{r}-\mathbf{r}') = p(\mathbf{r}')p(\mathbf{r}')$ in order to obtain the radiation from the bunch, the coherent part of which depends on how $P(\mathbf{r}-\mathbf{r}')$ weights the phase terms in Eq. (3),

$$\langle\ \frac{d^2W_N}{d\omega d\Omega}\ \rangle = |A|^2 \left\{ N + N(N-1)\iint p(\mathbf{r})p(\mathbf{r}')e^{-i\left[k_\perp\cdot(r_i-r_j)+k_z(z_i-z_j)\right]}d\mathbf{r}d\mathbf{r}'\right\}. \tag{6}$$

The separable form of $p(\mathbf{r})$ given by Eq. (5) yields the standard result for the integral in Eq. (6) which is

$$I_{coh.} = (2\pi)^3 |\tilde{g}(\mathbf{k}_\perp)|^2 |\tilde{F}(k_z)|^2, \tag{7}$$

where $\tilde{g}(\mathbf{k}_\perp)$ and $\tilde{F}(k_z)$ are the Fourier transforms of the radial and longitudinal form factors, respectively, for the macrobunch.

According to Eq. (7), if the radial beam profile is known, then a measurement of the coherent radiation will provide information on the longitudinal form factor squared, $|\tilde{F}(k_z)|^2$.

## Effects of angular aperture and bandwidth on observed coherence

We can use the discussion above as the starting point for the consideration of the effects of detector bandwidth and angular aperture size on the observed spectral angular distribution of a long pulse train of microbunches. Consider the $z$-dependent phase term $\phi = k_z(z-z')$ in Eq. (6) above. The change in phase with respect to bandwidth and angle while keeping the particle separation fixed is

$$d\phi = \frac{\partial\phi}{\partial k}dk + \frac{\partial\phi}{\partial\theta}d\theta = \cos\theta(z-z')dk - k\sin\theta(z-z')d\theta. \tag{8}$$

For coherent radiation from particles separated by the longitudinal distance $z-z'$, we need to have the total change in phase be such that $d\phi \leq \pi$.

With the help of Eq. (8), we can look at the effect of the variation of phase with $\theta$, while $k = k_n$ is held constant. Now the magnitude of the phase change is given by

$$|d\phi| = k_n(z-z')\theta d\theta \approx k_n(z-z')\frac{\theta^2}{2}, \quad \theta \ll 1, \tag{9}$$

748

**TABLE 1.** Maximum angular aperture $\theta$ for coherent radiation at various harmonics $n$ of $\lambda_0$=10.6 μm, assuming a particle spacing $(z - z')$=300$\lambda_0$.

| n | $\lambda$(μm) | Max $\theta$ *(mrad)* |
|---|---|---|
| 1 | 10.6 | 58 |
| 2 | 5.3 | 41 |
| 3 | 3.53 | 33 |
| 4 | 2.65 | 29 |
| 5 | 2.12 | 26 |

where $k_n = n(2\pi / \lambda_0)$, $n = 1,2,3....$, are the harmonics of the fundamental wave number. We impose the condition for coherent radiation,

$$k_n(z-z')\frac{\theta^2}{2} \leq \pi, \quad \text{or} \quad \theta \leq \left(\frac{\pi}{2k_n(z-z')}\right)^{\frac{1}{2}}. \tag{10}$$

If we take, for example, $(z - z') = 300\lambda_0$, we can illustrate the angular size of the detector aperture over which the radiation is coherent, as shown in Table 1. According to the table, the $n = 4$ harmonic would require an angular aperture which is one half that of the fundamental, i.e., the maximum aperture for coherence varies as $1/\sqrt{n}$, as can be seen from Eq. (10). If we have a beam energy of 40 MeV, then $\gamma^{-1} = 12.6$ mrad, and the maximum $\theta$ for $n = 1$ in Table 1 is $4.6\gamma^{-1}$; while for $n = 5$ it corresponds to $2\gamma^{-1}$. Thus, even for an infinitesimally narrow bandwidth detector, the observed coherent intensity will be reduced by an increasing amount as $n$ increases.

The $\theta$-dependence of the phase affects not only the detectability of coherence for a given aperture size, but it can also directly affect the coherence of the microbunch radiation production through the angular divergence of the beam. However, at least for the parameters of our example, the maximum allowed $\theta$ variation appears much larger than the expected beam divergence for the STELLA experiment. Nonetheless, in general, this $\theta$-dependent affect on coherence could be a concern.

If we keep $\theta$ constant in Eq. (8) and $6 \ll 1$, then the change in $\phi$ for a detector of bandwidth $\Delta k$ is $\Delta\phi = (z-z')\Delta k$. To illustrate bandwidth effects, we will take as an example the STELLA experimental parameters. In this case we expect 300 microbunches spaced at the $CO_2$ wavelength, $\lambda_0 = 10.6 \mu m$. Consider the harmonics $k_n$ of the fundamental wave number, and assume a detector bandwidth of 1% for each harmonic. The phase condition for coherent radiation then imposes a maximum particle separation of

**TABLE 2.** Maximum particle spacing $(z - z')$ for coherent radiation in a 1% bandwidth at the harmonics of $\lambda_0 = 10.6$ μm.

| n | $\lambda$(μm) | Max (z-z') |
|---|---|---|
| 1 | 10.6 | 50. $\lambda_0$ |
| 2 | 5.3 | 25. $\lambda_0$ |
| 3 | 3.53 | 16.7 $\lambda_0$ |
| 4 | 2.65 | 12.5 $\lambda_0$ |
| 5 | 2.12 | 10. $\lambda_0$ |

$$(z - z') \leq \pi / \Delta k = 100 \frac{\lambda_0}{2n}. \tag{11}$$

Note that the maximum particle separation for a fixed fractional bandwidth is proportional to $1/n$. In Table 2 we illustrate the maximum separation for the 1% bandwidth example. From Table 2 it can be seen that for a 1% bandwidth, the coherent addition of radiation from 50 microbunches can be observed at the fundamental wavelength, while coherent radiation from only 10 microbunches is observable for $n - 5$. Another way to look at this is to determine the bandwidth which is necessary to observe coherent radiation from all 300 microbunches. The first and last microbunch are separated by $300\lambda_0 = 3,180$ μm; this requires a bandwidth of $\Delta k / k = 1.67 \times 10^{-3}$.

## Analysis of observations made with a finite bandwidth detector

Now we return to Eq. (6) to investigate the $z$-dependent coherence integral when the macrobunch probability distribution for particles is given by Eq. (5). The $z$-dependent integral can be factored separately. If we have a detector band pass function $D(k, \Delta k)$, we can average the integral over it as follows

$$\langle I_z \rangle \equiv \iint dz dz' F(z) F(z') \int dk \, D(k, \Delta k) e^{-ik_z(z-z')}. \tag{12}$$

Let's take a simple square band pass filter $D(k, \Delta k) = 1$, $|k - k_n| \leq \Delta k / 2$, $D(k, \Delta k) = 0$, $|k - k_n| > \Delta k / 2$. Then we obtain

$$\langle I_z \rangle = \int_{-\infty}^{+\infty} dx e^{ik_z x} \frac{\sin(\Delta k x / 2)}{(\Delta k x / 2)} G(x), \tag{13}$$

where $G(x)$ is the autocorrelation of the macrobunch,

$$G(x) = \int_{-\infty}^{+\infty} dz' F(z') F(z'+x) = \frac{1}{(2\pi)^{1/2}} \int_{-\infty}^{+\infty} dk_z' \left| \tilde{F}(k_z') \right|^2 e^{ik_z' x}. \tag{14}$$

Using the last equality above for the autocorrelation, we can obtain the final expression for the average over the band pass function,

$$\langle I_z \rangle = \frac{1}{\Delta k} \int_{-\infty}^{+\infty} dk_z' \left| \tilde{F}(k_z') \right|^2 U(k_z - k_z' - \Delta k / 2), \tag{15}$$

where $U(k_z - k_z' - \Delta k / 2) = 1$ for $\left| k_z - k_z' \right| \leq \Delta k / 2$, and zero otherwise. Equation (15) is a convolution of the bunch form factor and the square function $U$, which is a function of both the spatial wave number of the bunch $k_z'$ and the radiation wave number $k_z$. Note that

$$\lim_{\Delta k \to 0} \langle I_z \rangle = \left| \tilde{F}(k_z) \right|^2, \tag{16}$$

which is the standard result, appearing in the last factor of Eq. (7).

## Model for the longitudinal bunch form factor

So far the discussion has been general. Now we need to choose a specific model for the macropulse and the microbunches. The $z$-dependent part of the bunch function will be taken to have the following form

$$F(z) = \sum_{m=0}^{M} f(z - m\ell), \tag{17}$$

where $\ell$ is the spacing between the $M + 1$ microbunches and $f(z - m\ell)$ is the microbunch shape function. For the STELLA experiment, $\ell = \lambda_0$, the $CO_2$ wavelength, but, in general, $\ell > \lambda_0$ downstream of the interaction region (1). The Fourier transform of $F(z)$ is

$$\tilde{F}(k_z') = f(k_z') \sum_{m=0}^{M} e^{-ik_z' m\ell} = f(k_z') \frac{\sin(M k_z' \ell / 2)}{\sin(k_z' \ell / 2)}, \tag{18}$$

which gives

$$|F(k_z')|^2 = \frac{\sin^2(Mk_z'\ell/2)}{\sin^2(k_z'\ell/2)}|f(k_z')|^2.$$  (19)

The first factor is the grating function which produces narrow lines at $k_n = n(2\pi/\ell)$, with full widths of $\Delta k_n = 2k_0/M$. Thus $\Delta k_n/k_n = 2/nM \propto 1/n$. The natural line widths are constant, but the fractional widths decrease with increasing harmonic number. Note that the effects of fluctuations in $\ell$ and beam divergence could be included by averaging Eq. (19) over the appropriate distributions in $\ell$ or $\theta$.

Next we specify the model for the microbunch charge distribution, which has total charge $Q = N_e e$, where $N_e$ is the number of electrons per microbunch. We take the charge distribution to be

$$q(r) = Qf(z)g(\rho),$$  (20)

where

$$f(z) = \frac{1}{W}\left[1 - \frac{|z|}{W}\right] \quad \text{and} \quad g(\rho) = \frac{1}{\pi\sigma^2}e^{-\rho^2/\sigma^2}.$$  (21)

The longitudinal shape is a triangular pulse of full width $2W$ and the radial shape is a Gaussian with rms radius $\sigma$. The Fourier transforms of these functions are

$$\tilde{f}(k_z) = \left(\frac{2}{\pi}\right)^{1/2}(k_zW)^{-2}[1 - \cos(k_zW)] \quad \text{and} \quad \tilde{g}(k_\perp) = e^{-(\sigma k_\perp/2)^2}.$$  (22)

The rms beam radius $\sigma$ has a very strong influence on the angular range over which there is coherence since, for $k_\perp \approx k_n\theta$, the intensity will be reduced by $1/e$ when $\theta_n = \sqrt{2}/(\sigma k_n)$. This angle falls off as $1/n$ for the higher harmonics.

## NUMERICAL RESULTS

In this section we will present the CTR spectra calculated with the microbunch model discussed above for microbunch FWHM values of $W = 1.5$, 2.0, and 3.0 μm (5 - 10 fs), assuming a 5% band pass filter. The beam energy is taken to be 40 MeV; the beam's rms radius $\sigma = 50$ μm; and the number of electrons per microbunch $N_e = 2 \times 10^6$, i.e., 0.3 pC in the microbunch and 90 pC in the macropulse.

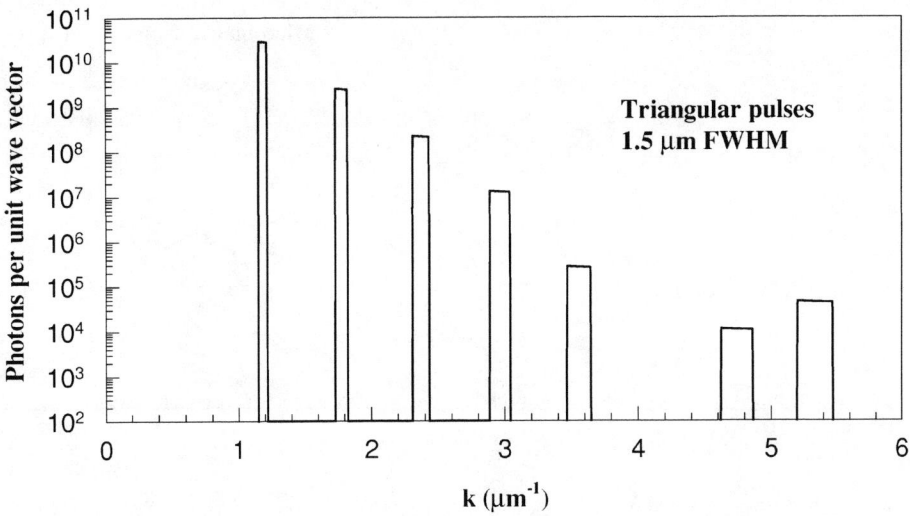

**FIGURE 1.** Calculated CTR spectrum observed by using a 5% band pass filter. The $n = 2 - 9$ harmonics of the 10.6-μm laser fundamental are shown. A triangular microbunch of 1.5 μm FWHM was used.

Using Eqs. (19) and (22), we evaluated $\langle I_z \rangle$, Eq. (15), numerically. This result was used in Eq. (6) to obtain the spectral angular distribution. Finally this is integrated over angles $\theta, \phi$ analytically by allowing the $\theta$-limit to go to infinity (9), which is reasonable because of the exponential fall off of $\left| \tilde{g}(k_\perp) \right|^2$ [see Eq. (22)].

Figure 1 is the spectrum obtained for a 1.5 μm FWHM microbunch case. The wave vectors correspond to the $n = 2 - 9$ harmonics of the 10.6 μm laser fundamental ($k = 0.593 \ \mu m^{-1}$). The missing line at 4.15 $\mu m^{-1}$ ($n = 7$) is a peculiarity caused by a zero of the form factor for the triangular microbunch $\tilde{f}(k_z)$, Eq. (22), for this particular value of $W$. Similar spectra also were calculated for the 2 and 3 μm FWHM microbunch cases.

Figure 2 is a plot of the integrated number of photons in each harmonic ($n = 2 - 9$) for the three different values of microbunch widths. The 2-μm case contains the $n = 7$ harmonic, unlike the other two cases. The InSb detector to be used on the STELLA experiment requires a minimum of $\sim 5 \times 10^4$ photons at a wavelength of 1.33 μm. Thus Figure 2 indicates that harmonics up to at least $n = 5 - 6$ can be observed. Note that, at each harmonic, the integrated number of photons for the spectra taken with band pass filters is identical to the integral over the natural line widths.

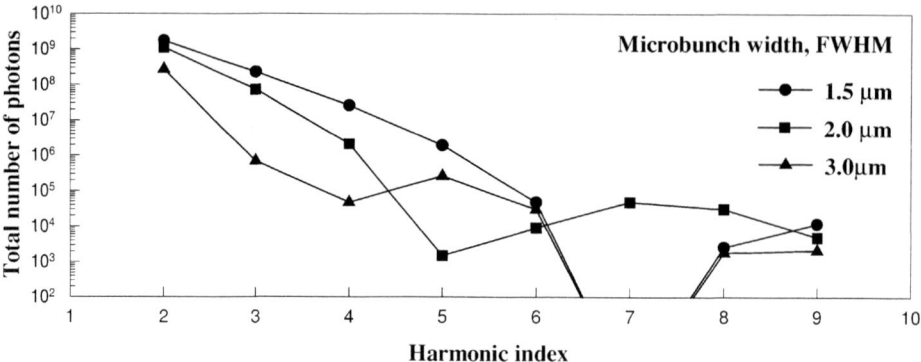

**FIGURE 2.** Integrated number of photons in each harmonic for triangular microbunches with widths of 1.5, 2.0, and 3.0 μm.

# CONCLUSIONS

We have calculated the CTR spectra which would be observed with 5% band pass filters for the case of a long, but finite train of microbunches. Bunch widths of ~1 - 3 μm FWHM (3 - 10 fs) can easily be distinguished by observing up to 5 - 6 harmonics of the 10.6-μm bunch spacing. Effects such as beam divergence, bunch spacing fluctuations, and bunch charge variations across a macropulse can be included within the framework presented. The analysis is also applicable to detection of CTR with spectrometers having a resolution on the order of the natural line width of the radiation.

# REFERENCES

1. Liu, Y. et al., Phys. Rev. Lett. **80**, 4418-4421 (1998).
2. Kimura, W.D. et al., Phys. Rev. Lett. **74**, 546 (1995).
3. Kimura, W.D. et al., "STELLA Experiment - Design and Model Predictions", these Proceedings.
4. Barry, Walter, "An Autocorrelation Technique for Measuring Sub-Picosecond Bunch Lengths Using Coherent TR", in *Proceedings of the Workshop on Advanced Beam Instrumentation, Vol. 1"* 22-24 April 1991, KEK, Tsukuba, Japan, A Ogata and J Kishiro, eds. (June 1991).
5. Kung, Pamela et al., Phys. Rev. Lett. **73**, 967 (1994).
6. Happek, U., Sievers, A.J., Blum, E.B., Phys. Rev. Lett. **67**, 2962-2965 (1991).
7. Lai, R., Happek, U. and Sievers, A.J. Phys. Rev. E **50**, R4294-4297 (1994).
8. Shibata, Yukio,et al., Phys. Rev. E **49**, 785-793 (1994).
9. Rosenzweig, J. et al., Nucl. Instr. and Meth. in Phys. Res. A **365**, 255-259 (1995).
10. Pantell, Richard, private communication, 1998.
11. Vardanyan, L.A., et al., Izvest. Acad. Nauk. Armyanskoy SSR. Fizika **10**, 350-360 (1975).
12. Wartski, L.W. et al., J. Appl. Phys. **46**, 3644 (1975).
13. Buskirk, F.R. and Neighbours, J.R., Phys. Rev. A **28,** 1531 (1983).

# STELLA Experiment - Microbunch Diagnostic

P. He, Y. Liu, and D.B. Cline

*Center for Advanced Accelerators*
*University of California, Los Angeles, CA 90095*

M. Babzien, J.C. Gallardo, K.P. Kusche, I.V. Pogorelsky, J. Skaritka,
A.van Steenbergen, and V. Yakimenko

*Brookhaven National Laboratory, Upton, NY 11973*

W.D. Kimura

*STI Optronics, Inc. Seattle, WA 98004*

**Abstract.** A microbunch diagnostic system is built at the Accelerator Test Facility (ATF) of Brookhaven National Laboratory for monitoring microbunches (10-fs bunch length) produced by the Inverse Free Electron Laser accelerator in Staged Electron Laser Acceleration experiment. It is similar to one already demonstrated at the ATF. With greatly improved beam optics conditions higher order harmonic coherent transition radiation will be measurable to determine the microbunch length and shape.

## 1 Introduction

With the success of the Inverse Cherenkov Acceleration (ICA) experiment[1] at the Acceleration Test Facility (ATF) of Brookhaven National Laboratory (BNL), a proposed 100-MeV acceleration demonstration experiment[2] is the next step towards developing the ICA scheme into a practical technique that may eventually lead to a high gradient linear accelerator. Therefore, an important new development in the ICA program is the joining of the ICA experiment with the BNL Inverse Free Electron Laser (IFEL) experiment[3, 4, 5, 6]. Since the IFEL operates in vacuum environment, it is possible to provide a high quality, pre-bunched $e^-$ beam for the next acceleration stage. Using the IFEL as the prebuncher for the ICA in the combined experiment, called Staged Electron Laser Acceleration (STELLA)[4], an

CP472, *Advanced Accelerator Concepts: Eighth Workshop,*
edited by W. Lawson, C. Bellamy, and D. Brosius
© 1999 The American Institute of Physics 1-56396-889-4/99/$15.00

important challenge is characterizing the microbunched $e^-$ beam entering the ICA interaction region.

Over the past year UCLA-BNL group has studied microbunching process using the IFEL accelerator[7] as a microbunch generator at the ATF. In the microbunching experiment the $e^-$ beam passing through the IFEL wiggler and interacting with the $CO_2$ laser beam in the existence of the magnetic field leaves the interaction region with a distinct velocity distribution pattern which, propagated over a certain distance in a free space, results in self-bunching of the $e^-$ beam. The distance (optimum bunching distance) resulting in a minimum longitudinal bunch length is controllable and characterized by input laser power. The experimental results show that the microbunch length is on a scale of several microns ($\sim 2 \ \mu$m)[8].

In the STELLA experiment electron beam optics elements located downstream of the IFEL wiggler have been greatly improved. A tightly focused electron beam with good profile geometry is achievable. A very intense coherent transition radiation (CTR) and its higher order harmonics will be measured to determine the microbunch length and shape.

## 2 Coherent Transition Radiation of Multiple Bunches

Coherent transition radiation is a collective effect of transition radiation produced by a large ensemble of particles being in phase with each other. The total number of photon radiated is highly enhanced when the observation wavelength is comparable to or greater than the bunch length. The total radiation distribution becomes[9]

$$\frac{d^2U}{d\omega d\Omega} = N\left[\ 1 + (N-1)F(\omega,\theta\ )\right]\frac{d^2u}{d\omega d\Omega},\tag{1}$$

where

$$F(\omega,\theta\ ) = |f(\omega,\theta\ )|^2 = \left|\int f(r,z)\exp(\ -i\vec{k}\cdot\ \vec{r})d^3x\right|^2\tag{2}$$

is a bunching factor, containing the electron distribution information. It transfers the electron distribution from a time domain to a frequency domain. The coherent effect scales like $N^2$ compare to the incoherent part, which scales linearly with the electron number, $N$. $k$ is the radiation wavenumber.

In multiple bunches the total coherent photon intensity depends on particle distribution in each bunch and number of bunches. In general the coherent effect contributed by transverse distribution of electrons is ignored while the transverse beam size is much smaller than its longitudinal length. However, if the bunch length (in several micron level) is shorter than or comparable with its transverse dimension, the electron phase difference caused by transverse distribution will reduce the CTR intensity. Furthermore, if a thin foil is inserted into the $e^-$ beam path and is $45^0$ to the $e^-$ beam direction, the individual electrons distributed at the surface of the foil will have an additional phase difference comparing to normal incidence. This additional phase difference will reduce the CTR intensity as well. Thus, these factors should be considered when one deals with microbunch coherent transition radiation.

For a two-foil configuration (Figure 1), the first foil is perpendicular to the electron beam direction, contributing a forward CTR, and the second foil is $45^0$ to the electron beam direction, contributing a backward CTR. The individual electrons distributed at the second foil will have an additional phase difference $\varphi(x,L)$ relative to the electrons at the first foil. The phase difference is given by

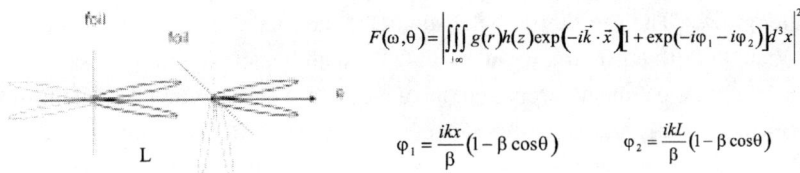

$$F(\omega,\theta)=\left|\iiint_{loc} g(r)h(z)\exp\left(-i\vec{k}\cdot\vec{x}\right)\left[1+\exp\left(-i\varphi_1-i\varphi_2\right)\right]d^3x\right|^2$$

$$\varphi_1=\frac{ikx}{\beta}(1-\beta\cos\theta) \qquad \varphi_2=\frac{ikL}{\beta}(1-\beta\cos\theta)$$

Figure 1: Two-foil configuration. First one generates a forward CTR and redirected by the second foil, the second foil generates a backward CTR.

$$
\begin{aligned}
\varphi(x,L) \quad & \varphi_1(x) + \varphi_2(L) \\
= \quad & \frac{kx}{\beta}(1 - \beta\cos\theta) + \frac{kL}{\beta}(1 - \beta\cos\theta),
\end{aligned}
\tag{3}
$$

where $L$ is the separated distance between the two foils. Following the same treatment for a single foil[10, 11], the two-foil CTR photon angular distribution is expressed as follow:

$$
\begin{aligned}
\frac{dN_{ph}}{d\theta} \approx & \frac{\alpha\beta^2}{8\sqrt{\pi^3}}\left(\frac{k_r\sigma_z}{\pi}\right)\left(\pi\frac{Nb_n}{k_r\sigma_z}\right)^2\left(\frac{1}{n}\right)\frac{\sin^3\theta}{(1-\beta\cos\theta)^2}\times \\
& \left\{\exp\left[-\left(\frac{nk_r\sigma_r}{\beta}\right)^2(\beta\sin\theta-\beta\cos\theta+1)^2\right]+\right. \\
& \exp\left[-(nk_r\sigma_r\sin\theta)^2\right]+ \\
& \left.2\exp\left[-\frac{1}{2}\left(\frac{nk_r\sigma_r}{\beta}\right)^2(\beta\sin\theta-\beta\cos\theta+1)^2\right]\times\right.
\end{aligned}
$$

$$\exp\left[-\frac{1}{2}\left(nk_r\sigma_r\sin\theta\right)^2\right]\times$$

$$\exp\left[-\left(\frac{L}{2\beta\sigma_z}\right)^2(\beta\cos\theta-1)^2\right]\times$$

$$\cos\left[\left(\frac{nk_rL}{\beta}\right)(\beta\cos\theta-1)\right]\bigg\}.\tag{4}$$

From the above equation we know that the total CTR intensity is produced by two individual CTR sources plus their interference at far region. The contribution of interference depends on the separation distance ($L$) of two foils.

The physical meaning of Eq.(4) is quite clear. In the first row of the equation the second term represents the length ($2\sigma_z$) of the macrobunch in unit of the modulation wavelength $\lambda_r$. The third term is the square of the total number of electrons which contribute the $n$th harmonic wavelength CTR within a single microbunch (the laser modulation 'chops' the macrobunch into a number of microbunches). The fifth term describes the contribution of single-electron transition radiation and $n$ is a harmonic number. The second row represents the contribution from the transverse electron distribution of the $45^0$ foil, the third row represents from the $90^0$ foil, the rest part is contributed by the CTR interfernce of two foils. The CTR intensity of higher harmonic wavelengths will be significantly weaker than that of the fundamental wavelength. By Fourier analysis of the CTR spectrum, the coefficient $b_n$ will be found. If the microbunch length is very short, the contribution from the higher harmonic components can be detectable.

## 3   Improved Beam Optics Conditions

In the microbunching experiment[8] there were no electron beam optics elements between the exit of the wiggler and the vacuum chamber which contains CTR target. The electron beam position and profile at the target surface were very difficult to optimize. These resulted in a relative small signal/noise (background) ratio. In the STELLA beamline has been refurbished and number of new beam optics elements have been added. A plan view drawing of the beamline is shown in Figure 2. The electron and $CO_2$ laser beams pass through the IFEL wiggler from right and travel towards the ICA cell. In the ICA cell the energy modulated electron beam will form microbunches and meet with the second $CO_2$ beam for high gradient acceleration. The ICA cell (center) is separated from the wiggler exit by 220cm. The CTR chamber attaches to the ICA cell entrance. The distance between target in the chamber to the wiggler exit is about 190cm. In between there is a quadruple triplet (triplet #4)

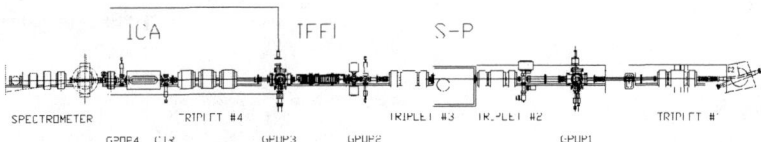

Figure 2: Plan view drawing of the beamline for the STELLA experiment at the ATF.

to refocus the microbunched beam and two steering magnets in front and behind the triplet to adjust the beam position shift. The expected transverse beam diameter at the target surface is between 50-100$\mu$m. In the previous microbunching experiment the minimum beam size was larger than 1mm. This tremdous improvement of the beam size and geometry will result in a great increase of CTR intensity. Thus, we expect that higher order harmonic wavelength CTR can be measured using our exist liquid Nitrogen cooled Indium Antimonide (InSb) detector.

# 4   System Layout

The diagnostic system consists of a transition radiation generator (CTR target), an optical transport line, a CCD camera and an INFRARED (IR) detector. The components are assembled on a breadboard above the beamline. A strip-line detector is placed the upstream of the chamber to provide the $e^-$ beam charge information. A schematic diagram drawing is shown in Figure 3. The basic features are similar to the previous one[8], except that the whole system including the target chamber is fixed. Because it is experimentally known that the optimum bunching distance is controllable and characterized by input laser power[8].

The CTR light is transported from the chamber through a ZnSe window and an IR lens, collected by a gold coated parabolic mirror and focused onto the IR detector. The IR lens helps reduce the CTR divergent angle, espectically, for a relative large beam spot size. The parabolic mirror is remote-controlled and is able to rotate in

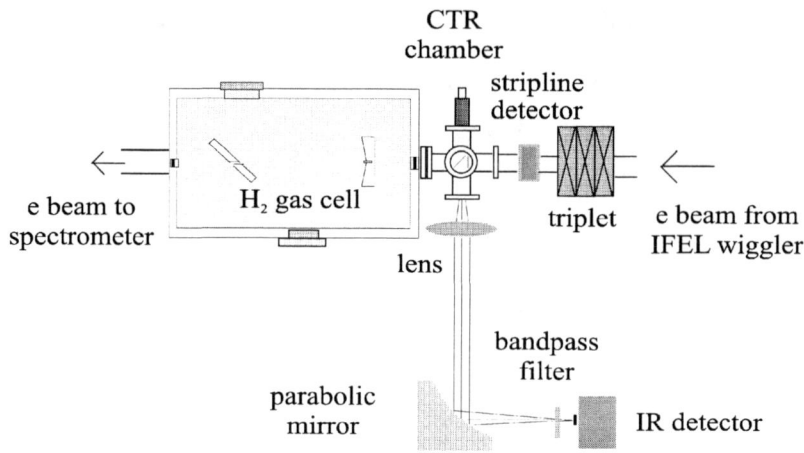

Figure 3: Schematic of the CTR diagnostic setup for the STELLA experiment.

two axes ($\alpha$, $\beta$) to bring the CTR to the IR detector. Different wavelength bandwidth pass filters or short wavelength pass filters can be placed in front of the detector to help measure either different single harmonic wavelength CTRs or total CTR up to the cut-off wavelength.

The detector active area is about 1.00 mm in diameter and its sensitive wavelength region is in 1.1 to 5.5 $\mu$m. Its quantum efficiency at 4.52 $\mu$m is 74.74%. Also, the sensitivity has been measured at the ATF using different wavelength diode lasers. The measured sensitivities are $5.7 \times 10^4$ photon and $4.8 \times 10^4$ photon at wavelength 1.55 $\mu$m and 1.33 $\mu$m, respectively. The detector surface is located in the parabolic mirror focal plane and well shielded by led bricks to reduce x-ray noise level. A HeNe laser is used to align the optics and to monitor parabolic mirror rotation trace.

# 5   CTR Target Chamber

The CTR target is mounted in a 6-way cross which serves as a small vacuum chamber. The chamber consists of two actuators placed in horizotal: one holds the CTR target, other holds a phosphor screen for electron and laser beam alignment. The target and phospher screen can be inserted and retracted by remote-controlled actuators. But only one is allowed to present in the beam path at the same time or none of them presents in the beam path if not in use. Both CTR and phosphor emittions are reflected out vertically through the same ZnSe window. Figure 4 shows the CTR target configuration only. The target is made up of a thin copper foil (2.5mil) and a gold-coated copper mirror in $45^0$ behind it, both mounted on the same actuator. A metal cone is placed inside beam pipe to interface with the foil frame through an

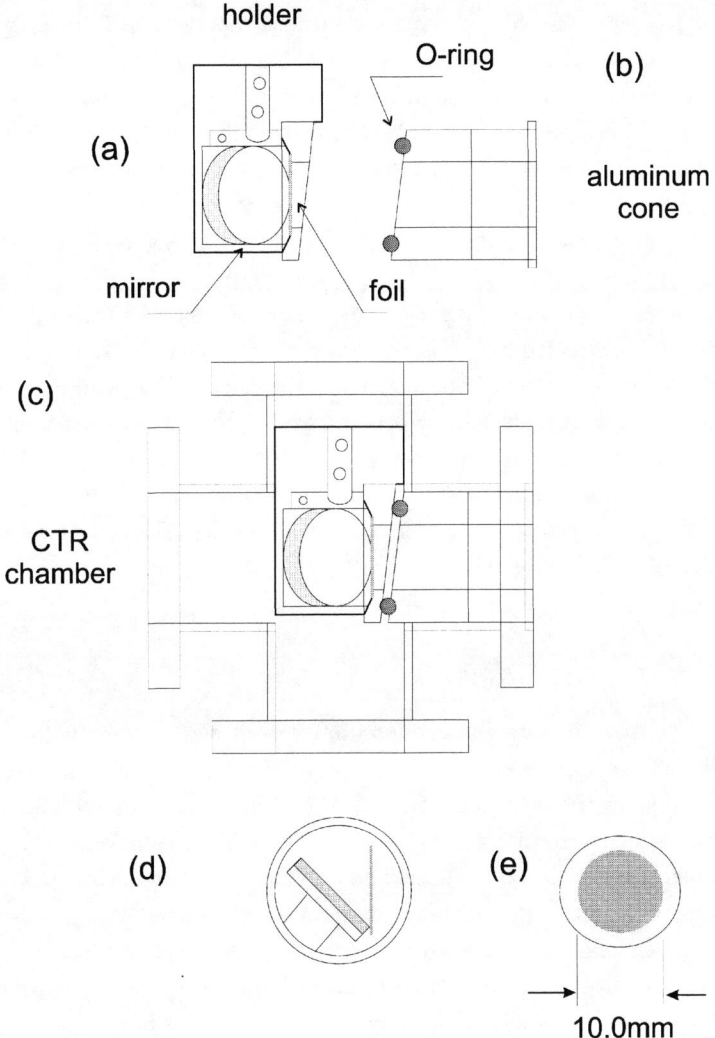

Figure 4: A CTR chamber. (a) A foil and a mirror, mounted on the same holder. (b) A cone, used inside of beam pipe to interface with (a). (c) A side view of CTR chamber with positioned target. $e^-$ beam enters from right. (d) Bottom view of positioned target in the chamber. (e) the size of a thin copper foil.

O-ring which isolates the chamber from the beam pipe when the target is inserted.

This particular target design is based on the following facts. Both the residual $CO_2$ laser light (usually very strong) and the $e^-$ beam will reach the target at the same time after they interact in the wiggler. In addition, while the $CO_2$ laser passes through the wiggler (Sapphire waveguide) it loses about 50% of its power in the wiggler and results in a very strong and broad bandwidth radiation. Such a broad bandwidth radiation will overwhelm the CTR if it leaves the chamber entering the detector. Thus, the foil before the mirror prevents this extraneous radiation from entering the CTR detector. It seals against the cone to ensure that no radiation makes it around the foil.

The $45^0$ mirror is used to direct the forward transition radiation away from the $e^-$ beam and out of the chamber through the ZnSe window to the detector. However, the passed $e^-$ beam hits the $45^0$ mirror and generates a backward CTR. The backward CTR is directed by the same mirror and following the same path through the ZnSe window to the outside of the chamber. The two CTRs will interfere in far region and enhance the detected CTR intensity. The foil, besides to use in measuring the CTR, has another use in measuring the $e^-$ beam profile by focusing a CCD camera on the foil surface to observe optical transition radiation (OTR). Since the foil diameter is 1mm, the $e^-$ beam spot size generated by OTR is easy to calculated according to the calibration of foil image.

# 6  Goals

The improved beamline optics configuration will provide a great help with optimizing the microbunched electron beam profile at the CTR target surface. As a use of microbunch monitor in the STELLA experiment and a continuation of the microbunch study, there are a number of goals could be achieved. The previous experimental results will be confirmed and a short wavelength (less than 2.5 $\mu$m) CTR intensity quadratic dependency curve on electron beam charge ($eN$) will be given. An inverse Fourier analysis of the harmonic components yields information about the microbunch characteristics. Obtaining this type of information requires sensing the magnitude of as many higher harmonics as possible.

The signal strength tends to decrease with harmonic number. Hence, detecting higher harmonics implies the need for a strong CTR signal. This signal strength is dependent on many factors, such as the degree of microbunching, the $e$-beam current, and the diameter of the $e$-beam on the CTR target. This latter factor is an important one since the photon yield inverses to the square of rms beam radius exponentially. We have utilized the TRANSPORT code to systematically calculate the electron beam size (transverse) for the use of the microbunching study with the aid of new beam optics between the IFEL wiggler and the ICA cell. Based on the simulation results it is possible to expect the beam size ($\sigma$) less than 50 $\mu$m at the ICA cell entrance (CTR chamber location). This could lead to the CTR intensity increase of 3-4 magnitudes comparing to previous experimental conditions. Higher harmonic wavelength CTR should be measurable using the exist InSb detector. The microbunch length and shape will be studied experimentally.

# 7  Acknowledgements

We thank the staff of the ATF at Brookhaven National Laboratory, especially I. Ben-Zvi and X.J. Wang for their enconverage and help, B. Cahill, B. Harrington and M. Montemagno for their technical support. We acknowledge prompt support from J. Kolonko, Physics & Astronamy Department of UCLA. This work is supported in part by DE-FG03-92ER40695, DE-FG06-93IR40803, and DE-FG03-90ER40545.

# References

[1] W. D. Kimura, G. H. Kim, R. D. Romea, L. C. Steinhauer, I. V. Pogorelsky, K. P. Kusche, R. C. Fernow, X. Wang, and Y. Liu, *Phys. Rev. Lett.* **74**, 546 (1995).

[2] STI Technical Proposal, *100-MeV Inverse Cherenkov Laser Acceleration Demonstration Experiment*, 20 June 1994.

[3] W. D. Kimura, I. V. Pogorelsky, Y. Liu, K. P. Kusche, A. van Steenbergen, J. Gallardo, J. Sandweiss, D. Cline, D.C. Quimby, and M. Babzien, "Inverse Cherenkov Acceleration Using an IFEL Prebuncher", in *Proceedings of Advanced Accelerator Concept Workshop*, Lake Tahoe, CA, 12-18 Oct. 1996.

[4] STI/BNL/UCLA, Scientific Plan *Staged Electron Laser Acceleration(STELLA)*, 29 Jan. 1998.

[5] W.D. Kimura, and STELLA Colleboration, "Design and Model Simulations of Inverse Cerenkov Acceleration Using Inverse Free Electron Laser Prebunch", PAC'97.

[6] W.D. Kimura, and STELLA Colleboration, "STELLA Experiment - Design and Model Predictions"; K.P. Kusche, and STELLA Colleboration, "STELLA Experiment - Hardware Issues", in this proceeding.

[7]  A. van Steenbergen, J. Gallardo, J. Sandweiss, J-M. Fang, M. Babzien, X. Qiu, J. Skaritka, and X.J. Wang, *Phys. Rev. Lett.* **77**, 2690 (1996).

[8]  Y. Liu, X.J. Wang, D.B. Cline, M. Babzien, J.M. Fang, J. Gallardo, K. P. Kusche, I. V. Pogorelsky, J. Skaritka, and A. van Steenbergen, *Phys. Rev. Lett.* **80**, 4418 (1998).

[9]  H.-C. Lihn, P. Hung, C. Settakorn, Helmut Wiedemann, and D. Bocek, Phys. Rev. E **53**, 6413 (1996).

[10]  J. Rosenzweig, G. Travish, and A. Tremaine, *Nucl. Instr. Methods in Phys. Res.* **A365**, 255 (1995).

[11]  Y. Liu, D.B. Cline, X.J. Wang, M. Babzien, J.M. Fang, and V. Yakimenko, "Micro-bunching Diagnostics for the IFEL by Coherent Transition Radiation" in *Proceedings of the Advanced Accelerator Workshop*, Lake Tahoe, CA, 12-18 October 1996.

# Microbunch Temporal Diagnostic by Compton Scattering in Interfering Laser Beams

A.Ts. Amatuni

*Yerevan Physics Institute, Alikhanian Brother's St. 2, Yerevan 375036, Republic of Armenia*

I.V. Pogorelsky

*Accelerator Test Facility, BNL, Upton, NY 11973, USA*

**Abstract.** The exact solution of the classical nonlinear equation of motion for a relativistic electron in the field of two electromagnetic (EM) waves is obtained. For the particular case of the linearly polarized standing EM wave in the planar optical cavity, the intensity of the nonlinear Compton scattering, the time of flight, and the momentum variation after the relativistic electron passes along the cavity axis are calculated in weak and strong field limits. The extent of these effects depends on the initial phase of the EM wave when the electron enters the cavity. This can be used for the production, diagnosis, and acceleration of relativistic electron (positron) microbunches.

## INTRODUCTION

The theory of the Compton effect in interfering EM waves, in particular, in two counter-propagating plane waves, has been addressed previously to describe the Kapitza-Dirac effect [1,2], Compton lasers [3,4], and inverse Compton laser acceleration [2,5]. The physical principle of the nanometer-resolution Shintake electron beam profile monitor [6-9] is also based upon the understanding of the Compton effect in a standing EM wave. The vacuum beat wave laser accelerator concept [10] relies on the ponderomotive acceleration resulting from the beat wave produced by the interaction of two copropagating laser beams

Temporal diagnostics of ultra-fine electron microbunches (sized to a portion of the laser wavelength) is another potential application for intense standing EM waves. Production and reliable characterization of such microbunches are essential for development of far-field and near-field laser accelerator schemes into practically meaningful monochromatic electron (positron) accelerators. The example of such a scheme is the staged electron laser acceleration experiment, STELLA, at the Brookhaven Accelerator Test Facility [11]. In this experiment, a train of the 1 $\mu$m (3 fs) thin electron microbunches, produced by the IFEL method and periodical to the $CO_2$ laser wavelength, $\lambda = 10$ $\mu$m, is phased to the inverse Cherenkov laser acceleration stage driven by the same $CO_2$ laser. Observation of Compton scattered radiation from the electron interacting with a standing EM wave, produced by two counter-propagating $CO_2$ laser beams, may permit a direct assessment of the microbunch temporal structure. The inverse process may provide also an alternative mechanism to generate microbunches starting with a quasi-continuous electron pulse.

CP472, *Advanced Accelerator Concepts: Eighth Workshop,*
edited by W. Lawson, C. Bellamy, and D. Brosius

Theoretical studies of the Compton effect in two interfering EM waves, comprehensively reviewed by M.V. Fedorov [2], generally capitalize on various aspects of the perturbation theory or other approximate approaches. As long as we are interested in processes nonlinear to the field, the approach based on exact solution of equations of motion is the most appropriate. The exact solutions of classical equations of the electron motion in a single plane EM wave have been obtained by many authors using different methods (see [12-16] and references therein). A natural extension of this approach is to apply the same methods to the case of two interfering plane waves.

In the recently submitted publication [17], this intent is partially accomplished using the approach developed in [13, 14]. In the present paper, we review results from [17] which are relevant to the problem of the microbunch diagnostics. In Section 2, we show the exact solution of classical nonlinear equations of electron motion in the field of two plane EM waves.

The particular case of a standing wave is considered in Section 3. We calculate the time of flight for the electron passing through the radiation-filled plane-parallel optical resonator and demonstrate that this parameter depends upon the phase of the standing EM wave at the moment when the electron enters the cavity.

In Section 4, we calculate the phase dependent intensity radiated due to Compton scattering when the relativistic electron passes along the axis of the plane-parallel optical cavity. The obtained results, discussed in Section 5, can be used for short relativistic electron bunch diagnostics.

## SOLUTION OF NONLINEAR EQUATION OF MOTION

Following the approach outlined in [13,14], we consider a classical equation of the electron motion in EM field:

$$\frac{d\pi_\mu}{d\tau} = \frac{e}{m} F_{\mu\nu}\pi_\nu, \tag{1}$$

where $\pi_\mu$ is the four-dimensional vector of the electron energy-momentum $\pi_\mu(\varepsilon, \vec{p})$ equal to $\pi_\mu^2 = \varepsilon^2 - p^2$, $\tau$ is time in the relativistic frame of the electron (local or proper time), and $F_{\mu\nu}$ is a tensor of EM field, expressed through the vector potentials $A_{\nu,\mu}$,

$$F_{\mu\nu} = \frac{\partial A_\nu}{\partial x_\mu} - \frac{\partial A_\mu}{\partial x_\nu}, \tag{2}$$

$$A_\nu = a_\nu^{(1)}(\varsigma_1) + a_\nu^{(2)}(\varsigma_2), \tag{3}$$

where $a_\nu^{(1)}(\varsigma_1)$, $a_\nu^{(2)}(\varsigma_2)$ are vector potentials of two linearly polarized plane waves with frequencies $\omega_1$ and $\omega_2$ , $\varsigma_1 = k_\mu^{(1)} x_\mu$ and $\varsigma_2 = k_\mu^{(2)} x_\mu$ are the phases of corresponding plane waves, and $k_\mu^{(1)}(\omega_1, \bar{\omega}_1)$ and $k_\mu^{(2)}(\omega_2, \bar{\omega}_2)$ are four-dimensional wave vectors. In Eqs. (1), (2) and below, the system of units with $c = 1, h = 1$ is used.

A four-dimensional electron vector $x_\mu$ is

$$x_\mu(\tau) = \frac{1}{m}\int_{\tau_i}^{\tau}\pi_\mu(\tau)d\tau + x_\mu(\tau_i). \tag{4}$$

Consider linearly polarized waves that satisfy generalized transverse conditions

$$\left(k^{(1)}a^{(2)}\right) = 0, \quad \left(k^{(2)}a^{(1)}\right) = 0 \quad . \tag{5}$$

We search for the solution of Eqs. (1)-(4) as a linear decomposition over four-vectors $p_\mu, a_\mu^{(1)}, a_\mu^{(2)}, k_\mu^{(1)}, k_\mu^{(2)}$,

$$\pi_\mu(\tau) = p_\mu - e\left[a_\mu^{(1)}(\varsigma_1) + a_\mu^{(2)}(\varsigma_2)\right] + k_\mu^{(1)}f_1(\tau) + k_\mu^{(2)}f_2(\tau), \tag{6}$$

where $p_\mu$ is the four-vector of the electron initial momentum before the electron enters the EM field. This brings us to the following result [17]:

$$f_{1,2} = \frac{\left(k^{(2,1)}p\right)}{k^{(1)}k^{(2)}}\left[\exp\left(\int_{\tau_i}^{\tau}F_{1,2}d\tau\right) - 1\right], \tag{7}$$

where quantities $F_1$ and $F_2$ are

$$F_{1,2} \equiv \frac{e\dfrac{d}{d\tau}\left[\left(a^{(1,2)}p\right) - \dfrac{e}{2}\left(a^{(1,2)}\right)^2\right] - e^2 a^{(2,1)}\dfrac{da^{(1,2)}}{d\tau}}{e\left(\left(a^{(1)} + a^{(2)}\right)p\right) - \dfrac{e^2}{2}\left(a^{(1)} + a^{(2)}\right)^2 + \dfrac{\left(k^{(1)}p\right)\left(k^{(2)}p\right)}{k^{(1)}k^{(2)}}}, \tag{8}$$

and the phases of the fields satisfy the equation

$$d\varsigma_{1,2} = \frac{\left(k^{(1,2)}p\right)}{m}\exp\left(\int_{\tau_i}^{\tau}F_{2,1}d\tau\right) \quad . \tag{9}$$

# ELECTRON IN PLANE-PARALLEL OPTICAL RESONATOR

Let us apply the solution obtained in Section 2 to the particular case of a standing EM wave formed by two linearly polarized plane waves of the same frequency $\omega$, counter-propagating along the $x$-axis and polarized along the $y$-axis with

$$a_y^{(1)}(\varsigma_1) = -(E_0/2\omega)\cos\varsigma_1 \ , \ \varsigma_1 = \omega(t - x) \ ,$$

$$a_y^{(2)}(\varsigma_2) = (E_0/2\omega)\cos\varsigma_2 \ , \ \varsigma_2 = \omega(t + x) \ , \tag{10}$$

$$A_y = a_y^{(1)} + a_y^{(2)} = -(E_0/\omega)\sin\omega t \sin\omega x \ ,$$

where $E_0$ is the amplitude of electric field. A standing wave is confined between two conducting surfaces (mirrors) placed at $x=0$ and $x=L=n\lambda/2 = n\pi/\omega$ , $n=1,2,3....$

The electron moving along the axis of the optical resonator enters the cavity at the moment $t = t_i(\tau_i)$ and leaves it at $t = t_f(\tau_f)$ ($t$ is time in the laboratory scale, $\tau$ is the local time of the moving electron). Time of flight in the lab system is

$$T = t_f - t_i = \int_{\tau_i}^{\tau_f} \gamma d\tau = \frac{1}{m}\int_{\tau_i}^{\tau_f} \pi_0 d\tau \ , \tag{11}$$

where $\gamma$ is a Lorentz factor of the electron.

## Weak Fields

Let us consider the case of the weak field, when the normalized field amplitude is $\eta^2 = e^2 a^2/m^2 \ll 1$, where $a = E_0/2\omega$ is the amplitude of the potentials in Eq. (10). In this case, Eq. (8) for $F_{1,2}$ can be expanded in series of $\eta^2$, and Eq. (7) for $f_{1,2}$ takes the form

$$f_{1,2} \approx \frac{e}{(k^{(1,2)}p)}\left[(a^{(1,2)}p) - \frac{e}{2}(a^{(1,2)})^2\right] -$$

$$\frac{e}{(k^{(1,2)}p)}\left\{\left[(a^{(1,2)}p) - \frac{e}{2}(a^{(1,2)})^2\right]_i - e\int_{\tau_i}^{\tau} a^{(2,1)}\frac{da^{(1,2)}}{d\tau}d\tau\right\}. \tag{12}$$

In the considered below ultrarelativistic case when $\vec{p}$ is collinear to $\vec{k}^{(1)}$, we obtain:
$$k^{(1)}p = \omega(\varepsilon - p) \approx \omega\varepsilon m^2/2p^2 \ , \quad k^{(2)}p = \omega(\varepsilon + p) \approx 2\omega\varepsilon, \quad \text{and } |f_2| \ll |f_1|.$$

By Eq. (12),

$$f_{1f} = -\frac{2p^2\eta^2}{\omega}\left\{\frac{1}{2}\left(\cos^2\varsigma_{1f} - \cos^2\varsigma_{1i}\right) - \int_{\varsigma_{1i}}^{\varsigma_{1f}}\cos\varsigma_2\sin\varsigma_1 d\varsigma_1\right\}, \tag{13}$$

where

$$\varsigma_{1i} = \omega t_i \ \ (t = t_i \ , \ x = 0), \ \ \varsigma_{1f} = \omega(t_i + T) - \omega L \ , \tag{14}$$

$L$ is the length of the resonator, and $T$ is the time of flight,

$$T \equiv t_f - t_i = \frac{1}{m} \int_{\tau_i}^{\tau_f} \pi_0 d\tau \approx \frac{p}{m}(\tau_f - \tau_i) + \frac{\omega}{m} \int_{\tau_i}^{\tau_f} f_1 d\tau ,$$

$$L = \frac{p}{m} \left[ 1 - \frac{p \eta^2}{\varepsilon} \left( \frac{1}{2} - \cos^2 \varsigma_{1i} \right) \right] (\tau_f - \tau_i) . \tag{15}$$

Neglecting in Eq. (13) terms proportional to $\left( k^{(1)} p \right) \big/ \left( k^{(2)} \pm k^{(1)}, p \right) \sim 1 \big/ 4\gamma^2$ ,

we arrive to

$$f_{1f} = -\frac{p^2 \eta^2}{\omega} \left( \cos^2 \varsigma_{1f} - \cos^2 \varsigma_{1i} \right) . \tag{16}$$

Expressing $T$ and $L$ through the local time interval $\left( \tau_f - \tau_i \right)$ we obtain:

For a short resonator, when $\quad k^{(1)} p \left( \tau_f - \tau_i \right) \big/ m \approx \omega L / 2\gamma^2 \lll 1, \quad (\omega L = n\pi),$

$$T = \varepsilon \left( \tau_f - \tau_i \right) / m, \quad L = p \left( \tau_f - \tau_i \right) / m, \quad T = \varepsilon L / p . \tag{17}$$

For a long resonator, when $\quad k^{(1)} p \left( \tau_f - \tau_i \right) \big/ m \approx \omega L / 2\gamma^2 \gg 1,$

$$T \approx \frac{\varepsilon}{m} \left[ 1 - \frac{p^2 \eta^2}{\varepsilon^2} \left( \frac{1}{2} - \cos^2 \varsigma_{1i} \right) \right] (\tau_f - \tau_i) ,$$

$$L = \frac{p}{m} \left[ 1 - \frac{p \eta^2}{\varepsilon} \left( \frac{1}{2} - \cos^2 \varsigma_{1i} \right) \right] (\tau_f - \tau_i) , \tag{18}$$

or

$$T = \frac{\varepsilon}{p} L \left[ 1 + \frac{m^2 \eta^2}{2\varepsilon p} \left( \frac{1}{2} - \cos^2 \varsigma_{1i} \right) \right] . \tag{19}$$

In any case, the final phase, $\varsigma_{1f}$, is

$$\varsigma_{1f} \approx \varsigma_{1i} + \omega L (\varepsilon - p) \big/ p \approx \varsigma_{1i} + \omega L / 2\gamma^2 . \tag{20}$$

We can define the coherent interaction distance, $d_c$, which corresponds to the distance where the electron phase slippage relative to the copropagating wave is $\Delta \varsigma_1 = \varsigma_1 - \varsigma_{1i} = \pi$ (compare with coherent radiation distance or radiation formation zone [18]). Then, according to Eq. (20), $d_c = \frac{2\pi}{\omega} \gamma^2 = \lambda \gamma^2$ and definitions for "short" and "long" resonators can be modified to $L \ll d_c$ for the short resonator, and $L \gg d_c$ for the long resonator.

769

# Strong Fields

When trying to consider the strong field case, $\eta^2 = e^2 a^2 / m^2 \gg 1$, we realize that straightforward calculation of the integrals in Eqs. (7) and (8) is difficult due to rapid variations of the expressions under the integrals. By transformation and integration of general expressions for $f_{1,2}$ the exact equation valid for arbitrary orientation of $\vec{p}$ and $\vec{k}^{1,2}$ and for an arbitrary field strength, $\eta$, can be obtained [17]. For counter-propagating plane waves, $k_\mu^{(1)}$ and $k_\mu^{(2)}$, linearly polarized along the y-axis, and with $\vec{p}$ directed along $k_1(\vec{k}_x, 0, 0)$, the exact equation takes the form

$$\frac{2\omega}{\varepsilon + p} f_1 + \frac{2\omega}{\varepsilon - p} f_2 + \frac{4\omega^2}{m^2} f_1 f_2 = \eta^2 X^2, \qquad (21)$$

where the condition $\varsigma_{1i} = \varsigma_{2i}$ is taken into account and

$$X^2 \equiv \left( \cos\varsigma_1 - \cos\varsigma_2 \right)^2. \qquad (22)$$

By applying Eq. (21) to the case of $\eta^2 \ll 1$, that corresponds to $E_0 \to 0$ (or $\omega \to \infty$), we obtain

$$f_{1,2} = m^2 \eta^2 y_{1,2} / \left( k^{(1,2)} p \right), \qquad (23)$$

where $y_{1,2}$ are functions of $\varsigma_{1,2}$ and $y_1 + y_2 = X^2$.

For the strong field case, $\eta^2 \gg 1$, we use similar arguments as for $\eta^2 \ll 1$ above. First, notice that the condition $\eta^2 \gg 1$ is fulfilled when $E_0 \to \infty$ (or $\omega \to 0$). Applying the limit $\eta^2 \to \infty$ to left and right sides of Eq. (21), we find the solution

$$f_{1,2} = m^2 \eta y_{1,2} / 2 \left( k^{(1,2)} p \right), \qquad (24)$$

where $y_{1,2}$ are functions of $\varsigma_{1,2}$ and $y_1 y_2 = X^2$. The simplest choice that provides the symmetry condition, $k^{(1)} \leftrightarrow k^{(2)}$, $f_1 \leftrightarrow f_2$, is $y_1 = y_2 = |X|$ as is adopted below.

# COMPTON SCATTERING IN STANDING EM WAVE

The total radiated energy during the passage of the electron through the optical resonator of the length $L$ is

$$\Delta E = \frac{2}{3}e^2 \int_{t_i}^{t_f} w^2 \, dt = \frac{2}{3}e^2 m \int_{\tau_i}^{\tau_f} w^2 \pi_0 \, d\tau, \tag{25}$$

where $w$ is the four- vector of acceleration,

$$w^2 = \frac{1}{m^2} \frac{d\pi_\mu}{d\tau} \frac{d\pi_\mu}{d\tau}. \tag{26}$$

For the weak field case, $\eta^2 \ll 1$, $\pi_\mu$ is given by Eqs. (6)-(8), and (23). For the strong field case, use Eq. (24) instead of Eqs. (23).

Consider the weak field approximation for the ultrarelativistic electron motion along the standing wave axis. Then, the expression under the integral in Eq. (25), up to the terms proportional to $\eta^4$, is

$$w^2 \pi_0 \approx \frac{e^2 \varepsilon}{m^4} \left( \frac{da^{(2)}}{d\varsigma_2} \right)^2 \left\{ \left(k^{(2)} p\right) + f_1 \left[ 2\left(k^{(2)} p\right)\left(k^{(1)} k^{(2)}\right) + \frac{\omega}{\varepsilon}\left(k^{(2)} p\right) \right] \right\}. \tag{27}$$

The integral in Eq. (25) looks slightly different for long and short cavities:

$$\Delta E = \frac{1}{3} \frac{e^2 \eta^2 \omega^2 \varepsilon(\varepsilon + p)^2}{m^3} \left(\tau_f - \tau_i\right)\left\{1 + 3\eta^2 \varphi(\varsigma_{1i})\right\}, \tag{28}$$

$$\varphi(\varsigma_{1i}) = \begin{cases} 1/2 - \cos^2 \varsigma_{1i}, & L\omega/2\gamma^2 \gg 1; \\ 1/2 - \cos^2 \varsigma_{1i} + \cos\varsigma_{1i}, & L\omega/2\gamma^2 \ll 1. \end{cases} \tag{29}$$

Using Eqs. (17)-(19) for the time of flight, $T$, we obtain the following expressions for the average intensity, $I = \Delta E / T$, of the scattered radiation:

for a long resonator

$$I_{long} \approx \left[ e^2 \eta^2 \omega^2 (\varepsilon + p)^2 / 3 m^2 \right] \left[ 1 + 4\eta^2 \left( 1/2 - \cos^2 \varsigma_{1i} \right) \right], \tag{30}$$

and for a short resonator

$$I_{short} \approx \left[ e^2 \eta^2 \omega^2 (\varepsilon + p)^2 / 3 m^2 \right] \left[ 1 + 3\eta^2 \left( 1/2 - \cos^2 \varsigma_{1i} + \frac{1}{2}\cos\varsigma_{1i} \right) \right]. \tag{31}$$

In the CGS system, the right hand side of Eqs. (28) and (29) need to be multiplied by the factor $c^{-7}$, the right hand side of Eqs. (30) and (31) - multiplied by the factor $c^{-5}$, and $p$ will be replaced by $pc$.

Eqs. (30) and (31) comply, to the precision of up to $\eta^2$, with the results obtained for a single plane wave in Ref. [13], [14] (if we take $\bar{p} = 0$). The difference is due to the terms proportional to $\eta^4$ which depend upon the initial phase. The dependence of the nonlinear Compton scattering on phase can be used for the electron microbunch diagnostics.

For the ultrarelativistic strong field case, when $\eta^2 \gg 1$, the total radiated energy during the electron passage through the optical resonator is

$$\Delta E \approx \left(16\varepsilon^4\eta^4 e^2\omega/3m^4\right)\int_{\varsigma_{1i}}^{\varsigma_{1j}}\sin^2\varsigma_2\left(\cos\varsigma_1 - \cos\varsigma_2\right)d\varsigma_1 \approx$$
$$\left(2\varepsilon^4\eta^4 e^2\omega/3m^4\right)\left[4\left(\varsigma_{1f} - \varsigma_{1i}\right) + \sin 2\varsigma_{1f} - \sin 2\varsigma_{1i}\right].$$
(32)

In Eq. (32), $\sin^2\varsigma_2$ and $\cos^2\varsigma_2$ are set equal to their average value 1/2, that is possible due to rapid variation of $\varsigma_2$.

For a short resonator ($\omega L/2\gamma^2 \ll 1$), Eq. (32) can be reduced to

$$\Delta E = 2\varepsilon^2\eta^4 e^2\omega^2 L\left[2 + \cos 2\varsigma_{1i}\right]/3m^2 ;$$
(33)

and for a long resonator ($\omega L/2\gamma^2 \gg 1$)

$$\Delta E \approx 8e^2\varepsilon^2\eta^4\omega^2 L/3m^2 .$$
(34)

In the CGS system, the right hand side of Eqs. (33) and (34) is multiplied by $c^{-6}$.

## DISCUSSION AND CONCLUSIONS

Following the classical approach developed in [13,14] for a single planar EM wave, the general solutions for electron motion in the field of two linearly polarized planar EM waves are obtained. Special consideration is given to the problem of radiation of the ultrarelativistic electrons, propagating along the standing EM wave axis.

The total radiated energy due to Compton scattering is defined primarily by the wave counter-propagating to the direction of the electron momentum. This feature is physically understandable, if we take into account that the relativistic electron experiences quickly oscillating force from the counter-propagating component of the standing wave that causes the electron to radiate, but moves practically in phase with the co-propagating component. Constant amplitudes of the electric and magnetic fields produced by the co-propagating wave define a drift of the electron trajectory in the cavity. As a result, the amplitude of the electron oscillation is modulated by the initial phase, and this is revealed through radiation.

We can expect that the overall angular and spectral spread of the radiated photons shall generally obey distributions obtained in [13,14,19,20] for backscattering of a single laser beam on the relativistic electron beam. However more detailed analysis may be necessary, especially for the spectral distribution of the phase dependent terms in Eqs. (19, 28-29, 33).

Phase dependence of the nonlinear Compton scattering in a standing laser wave may be used for microbunch characterization (bunch duration, longitudinal charge distribution). This is required for the advanced laser acceleration experiments, such as the STELLA experiment at the Brookhaven ATF [11].

Propagating along the axis of the standing EM wave, the electron can loose its energy via Compton scattering or acquire it in the inverse process.

Using general solutions obtained in Section 3 for the electron passing through the radiation filled plane-parallel cavity, we can address a problem of electron acceleration in vacuum. For weak field case, $\eta \ll 1$, the momentum of the electron at the exit from the cavity is

$$p_f = p_i - \frac{p^2\eta^2}{\varepsilon}\left(\cos^2\varsigma_{1f} - \cos^2\varsigma_{1i}\right) = p_i\left\{1 - \frac{p\eta^2}{\varepsilon}\left[\cos^2\left(\varsigma_{1i} + \frac{L\omega}{2\gamma^2}\right) - \cos^2\varsigma_{1i}\right]\right\}.(35)$$

For a short resonator ($\omega L/2\gamma^2 \ll 1$)

$$p_f = p_i\left[1 - \frac{p\eta^2}{\varepsilon}\left(\frac{1}{2} - \cos^2\varsigma_{1i} - \sin 2\varsigma_{1i}\right)\right], \tag{36}$$

and for a long resonator ($\omega L/2\gamma^2 \gg 1$), after averaging over small changes in frequency, $\frac{\Delta\omega}{\omega} \ll \frac{4\gamma^2}{\omega L} \ll 1$,

$$p_f = p_i\left[1 - \frac{p\eta^2}{\varepsilon}\left(\frac{1}{2} - \cos^2\varsigma_{1i}\right)\right]. \tag{37}$$

Thus, if $\cos^2\varsigma_{1i} \rangle 1/2$, the acceleration takes place. For example, if $\varsigma_{1i} = \omega t_i = 0$ and $\eta^2 = 0.25$, then, after passing the resonator, $p_f = 1.12\,p_i$.

Contrary to Compton scattering, electron acceleration is primarily due to the component of the standing wave collinear with the electron propagation. The effect of the second counter-propagating wave is dumped due to the phase averaging. In the classical approach considered here, the acceleration of electrons is due to the ponderomotive force related to the linear and nonlinear parts of the Lorentz force as it is typical for the ponderomotive acceleration processes [10, 15, 16].

The facts that the electron-laser interaction is localized within the finite optical cavity and the accelerating force is nonlinear to the laser field circumvent the Lawson-Woodward theorem [21], that otherwise forbids the residual electron energy gain from EM waves in vacuum.

The revealed dependence of the electron momentum at the exit of the optical resonator upon the initial phase of the standing EM wave may be potentially used for electron (positron) microbunch diagnostics as well.

# REFERENCES

1. P.L. Kapitza, P.A.M. Dirac, *Proc. Cambr. Phys. Soc.* **29**, 297 (1933).

2. M.V. Fedorov, "Electron in Strong Light Field", M. "Nauka", 1991 (in Russian).

3. R. Pantell, G. Soncini, and H.E. Puthoff, *IEEE J. Quantum Electron.* **QE-4**, 905 (1968).

4. B.G. Bagrov, Yu. I. Klimenko, V.R. Khalikov, *JETPh* **57**, 922 (1969) (in Russian).

5. V.V. Appolonov, Yu. L. Kalachev, A.M. Prokhorov, M.V. Fedorov, *Pis'ma JETPh* **46**, 61 (1986) (in Russian).

6. T. Shintake, *NIM in Phys. Res.* **A311**, 453 (1992).

7. T. Shintake et al. KEK-- preprint 94-129, Oct. 1994 .

8. M.C. Ross, et al, *Proc XVIII Int. Linear Acc. Conf.*, **I**, 308, Geneva, 26-30 August, 1996.

9. T. Kotseroglou et al, SLAC--PUP—7511, May, 1997, *Proc. PAC'97*, Vancouver, B.C. Canada, May 1997.

10. E. Esarey, P. Sprangle, and J. Krall, *Phys. Rev. E* **52**, 5443 (1995).

11. I.V. Pogorelsky, A. van Steenbergen, J.C. Gallardo, V. Yakimenko, M. Babzien, K.P. Kusche, J. Skaritka, W.D. Kimura, D.C. Quimby, K.E. Robinson, S.C. Gottschalk, L.J. Pastwick, L.C. Steinhauer, D.B. Cline, Y. Liu, P. He, F. Camino, I. Ben-Zvi, R.B. Fiorito, D. W. Rule, R.H. Pantell, and J. Sandweiss, JAERI-Conf 98-004, 25 (1998).

12. L.D. Landau, E.M. Lifshitz "Classical Theory of Fields" §47, M. 1973.

13. A.I. Nikishov, V.I. Ritus, *JETPh* **46**, 776 (1964); same, 1768 (in Russian).

14. V.I. Ritus, *Trudi FIAN* **111**, chap. 2, §4, M. "Nauka" (1979) (in Russian).

15. M. Feldman, R. Y Chiao, *Phys. Rev. A* **4**, 352, (1971).

16. S.P. Kuo, M.C. Lee, *J. Plasma Physics* **52**, 339, (1994).

17. A.Ts. Amatuni and I.V. Pogorelsky, "Nonlinear Compton Scattering and Electron Acceleration in Interfering Laser Beams", *accepted for publication in Phys. Rev. Special Topics: Accelerators and Beams*

18. B.M. Bolotovsky, *Trudi FIAN* **140**, 95, M., "Nauka", (1982) (in Russian).

19. E. Esarey, S.K. Ride, and P. Sprangle, *Phys. Rev. E* **48**, 3003 (1993).

20. S.K. Ride, E. Esarey, and M. Baine, *Phys. Rev. E* **52**, 5425 (1995).

21. P.J. Woodward, *IEE* **93**, 1554 (1947);
    J.D. Lawson, *IEEE Trans. Nucl. Sci.* **NS-26**, 4217 (1979);
    R.B. Palmer, *Part. Accel.* **11**, 81 (1980).

22. B.M. Bolotovsky, G.V. Voskresensky, *Uspehi Phys. Nauk* **88**, 209 (1966) (in Russian).

23. L. C. Steinhauer, R. D. Romea, and W. D. Kimura, Advanced Accelerator Concepts, Lake Tahoe, CA, *AIP Conference Proceedings* **398**, ed. S. Chattopadhyay, AIP, NY, 673 (1997).

# Bunch-Length and Beam-Timing Monitors in the SLC Final Focus[1]

F. Zimmermann, G. Yocky, D.H. Whittum, M. Seidel[2],
C.K. Ng, D. McCormick, K.L.F. Bane

*Stanford Linear Accelerator Center*
*Stanford University, Stanford, California 94309*

**Abstract.** During the 1997/98 luminosity run of the Stanford Linear Collider (SLC), two novel RF-based detectors were brought into operation, in order to monitor the interaction-point (IP) bunch lengths and fluctuations in the relative arrival time of the two colliding beams. Both bunch length and timing can strongly affect the SLC luminosity and had not been monitored in previous years. The two new detectors utilize a broad-band microwave signal, which is excited by the beam through a ceramic gap in the final-focus beam pipe and transported outside of the beamline vault by a 160-ft long X-Band waveguide. We describe the estimated luminosity reduction due to bunch-length drift and IP timing fluctuation, the monitor layout, the expected responses and signal levels, calibration measurements, and beam observations.

## MOTIVATION

At the SLC, the IP bunch length is sensitive to many variables, most notably the injection time into the linac, the voltage of the bunch compressor RF between damping ring and linac, the $R_{56}$ value of the ring-to-linac transport line (RTL), the longitudinal bunch distribution at extraction from the damping ring and the bunch charge. Because of hourglass effect (depth of focus) and disruption (mutual contraction of the two beams during the collision), the bunch length at the interaction point can strongly affect the luminosity. Also the beamstrahlung (synchrotron radiation in the field of the opposing beam) depends on the bunch length. We have simulated the dependence of beamstrahlung and luminosity on the bunch length using the code Guinea-Pig [2], for typical 1997 SLC parameters: bunch population $N_b \approx 3.5 \times 10^{10}$, rms beam sizes $\sigma_{x,y} \approx 1.70, 0.92$ $\mu m$, rms divergences $\theta_{x,y} \approx 460, 260$ $\mu rad$, and considering realistic (non-Gaussian) longitudinal distributions. For rms bunch lengths between 0.8 mm and 1.6 mm, the simulation results are summarized by the fitted dependences: $H_D \approx 1.0 + 0.29 \times \sigma_z$ mm and

---

[2] present address: DESY, Notkestr. 85, Hamburg, Germany
[1] Work supported by the U.S. Department of Energy under contract DE-AC03-76SF00515.

CP472, *Advanced Accelerator Concepts: Eighth Workshop,*
edited by W. Lawson, C. Bellamy, and D. Brosius
© 1999 The American Institute of Physics 1-56396-889-4/99/$15.00

$\delta_B[10^{-3}] \approx 0.8 - 0.27 \times \sigma_z$ [mm], where $H_D$ is the luminosity enhancement factor [1] and $\delta_B$ the relative energy loss due to beamstrahlung.

Hitherto, beam timing at the IP was not monitored at all. Yet it is a critical factor for IP spot-size tuning, spot-size stability and the luminosity. If the collision point shifts in time, it will no longer coincide with the beam waists, implying a non-optimum IP spot size. There are many potential sources of IP timing drifts and timing jitter, such as changes in the extraction time from the damping rings, variation of the bunch-compressor RF phase relative to the damping-ring phase, and beam energy and profile fluctuation in the linac. Diurnal timing changes, even if they are large and different for both beams, are not thought to be a problem, as the four beam waists are scanned and corrected regularly. However, changes on shorter time scales—such as pulse-to-pulse variation ('jitter') or drifts over a few minutes—could seriously degrade the average luminosity. The luminosity reduction caused by an rms timing fluctuation of $\Delta t$ is roughly given by $L/L_0 \approx (1 + (c\,\Delta t/(2\beta_x^*)^2)^{-1/2}(1 + \left(c\,\Delta t/2\beta_y^*\right)^2)^{-1/2}$, where $L$ denotes the actual luminosity, $L_0$ the luminosity without timing errors, $c$ the speed of light, and $\beta_x^*$, $\beta_y^*$ the nominal IP beta functions ($\beta_x^* \approx 2.8$ mm, and $\beta_y^* \approx 1.5$ mm). An rms timing jitter of 6° X-Band then corresponds to a luminosity loss of about 1.5%, and a beam-timing monitor (BTM) with a resolution of 15° X-Band (5° S-Band) could detect fluctuations resulting in a 10% luminosity loss.

## BUNCH-LENGTH MONITOR

During the 1996/97 SLC downtime a novel RF bunch length monitor (BLM) was installed in the South Final Focus, about 45 m away from the IP. The monitor detects the longitudinal frequency spectrum of both electron and positron bunches, which pass this point with a time separation of roughly 300 ns. Previous attempts to commission a bunch-length monitor based on an RF cavity [3] at the same location failed, presumably because the crystal rectifiers and amplifiers used for RF-signal conversion could not withstand the high radiation and electromagnetic noise level in the final-focus tunnel.

Performing the RF conversion outside of the beamline vault overcomes the suspected noise problem and, in addition, affords easy access to the signal processor during SLC operation. For this reason, we installed a 160-ft long section of WR90 (brass) waveguide, extending from the South final-focus tunnel through a 60-ft deep ventilation shaft to the 'laser shack' South of the collider hall, where the signal is processed. The reasons for choosing overmoded rectangular WR90 (X-band) waveguide were twofold: (1) X-band components are easily available at SLAC, and (2) the waveguide attenuation is small compared with that in waveguides of higher-cutoff frequency, an important consideration for a 50-m length. Figure 1 (left) compares the attenuation in standard rectangular brass waveguides of various dimensions.

Figure 2 shows how, using waveguide filters and couplers, the RF signal is finally split into 4 different channels, which span the 4 frequency ranges 8–11.6 GHz, 11.6–

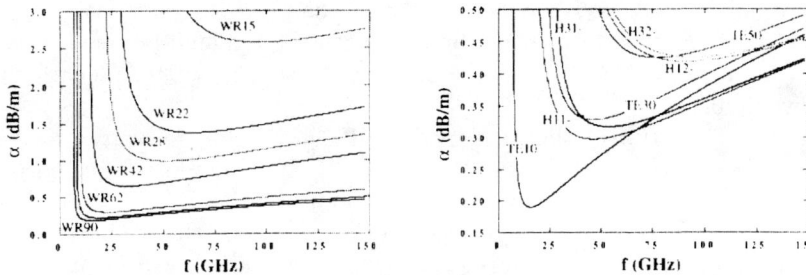

**FIGURE 1.** Left: $TE_{10}$ mode attenuation in brass waveguide of standard dimensions. At low frequencies, the microwave signal propagates more slowly in the waveguide, dissipating itself in the conductor. At high frequencies the surface resistance increases. Right: attenuation vs. frequency for the lowest-order modes of WR90 waveguide coupling to the gap. The '−' sign refers to the less attenuated of a pair of degenerate hybrid modes.

19 GHz, $\geq$21 GHz and $\geq$59 GHz. The RF power in each channel is measured with crystal rectifiers connected to gated analog-to-digital converters (GADC's), and the 8 signals so obtained (4 each for both the electron and the positron beam) are continually read out by the SLC control system. The figure also shows a photograph of the signal-processing unit assembly. Multiple channels were employed since there was no confidence in the actual value of $\sigma_z$.

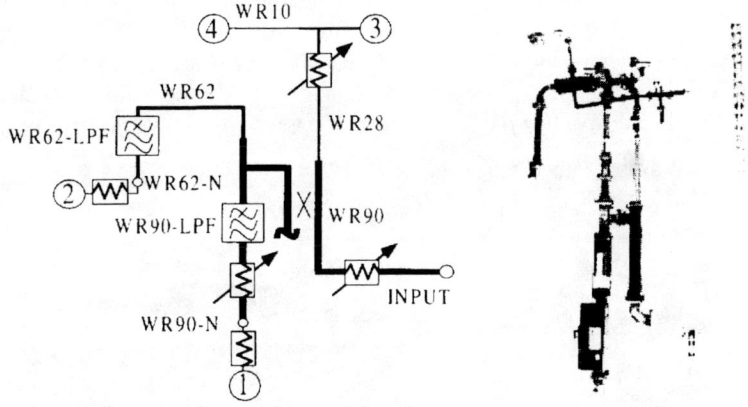

**FIGURE 2.** Schematic of RF signal processing (left) and photograph of assembly (right).

The frequency spectrum of a Gaussian charge distribution with rms length $\sigma_z$ and total charge $Q_b$, moving at the speed of light, is

$$\tilde{I}_b(\omega) = \frac{Q_b}{\sqrt{2\pi}} \exp\left(-\frac{\omega^2 \sigma_z^2}{2c^2}\right) \tag{1}$$

The power radiated by the beam into an rf pick up is proportional to the square of $\tilde{I}_b$. The principle behind an rf BLM is that the sampled beam frequency spectrum, when normalized by the square of the beam charge, correlates with the bunch length. As an example, for a 1-mm bunch length the radiated power decreases by $1/e$ at a frequency of $f = c/(2\pi\sigma_z) \approx 50$ GHz. For a 2-mm bunch length the $1/e$ fall-off would occur at 25 GHz.

In the final-focus tunnel, the open-ended WR90 waveguide is pointed at a ceramic gap in the beam pipe; see photograph in Fig. 3. This broadband pickup is more versatile than a narrow-band cavity. Inner and outer radius of the ceramic are 1.5 cm and 1.9 cm, respectively. The total gap length $g$ is 4 cm, three quarters of which are occupied by a toroid. The waveguide is situated at the last free quarter of the gap and ends about 0.5 cm above the ceramic.

The energy radiated by the beam into the waveguide can be estimated as follows. For short bunches, $\sigma_z \ll R$ (with $R$ the beam pipe radius), the energy loss $\Delta E$ of a bunch passing a gap is given by the diffraction model [4,5] as $\Delta E \approx 1/\pi\Gamma(1/4)Q_b^2 Z_0 c/(4\pi R)\sqrt{g/(\pi\sigma_z)}$, and corresponds to a loss factor of 3 V/pC. About half of the diffracted energy goes through the gap, the other half propagates down the beam pipe. To estimate the power radiated into the waveguide, we further must take into account the fractions of gap length (25%) and of solid angle (5%) occupied by the waveguide aperture. We then find that a 5 nC beam radiates about 0.75 $\mu$J onto the pick up. Since this occurs in a transit time $g/c \approx 33$ ps, the peak power incident on the waveguide is about 22 kW or 73 dBm. Attenuation over 50 m will reduce this figure to 55–65 dB. Since much of this signal is at lower frequencies, where the sensitivity to bunch length is weak, the signal is filtered in frequency. Though filtering and the effect of dispersion in the waveguide will further reduce the power level, there remains sufficient peak power above 20 GHz to produce 100 mV-level signals on a commercial crystal detector (this requires only about 100 $\mu$W power), in a passive mode without amplification.

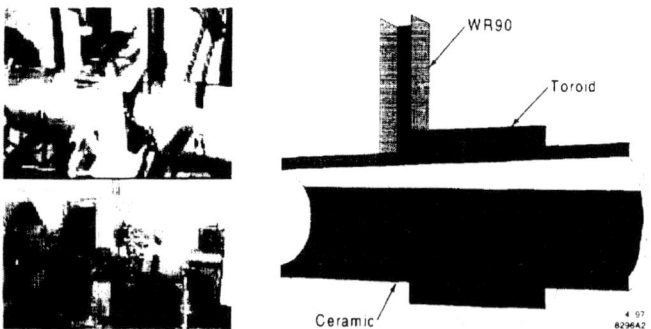

**FIGURE 3.** Left: photograph of waveguide pickup facing the ceramic gap (top picture), and final pick-up assembly with shielding box (bottom picture). Right: model for the numerical calculation of beam-waveguide coupling.

We used MAFIA [6] to more precisely calculate the waveguide excitation, for Gaussian bunches of different rms lengths. Figure 3 (right) depicts the model of the beam-pipe geometry used in these calculations. The 0.5-cm wide gap between waveguide and ceramic is neglected, and the waveguide directly borders on the ceramic. The toroid is treated as a conducting boundary. The calculations show that the energy coupled into the $TE_{10}$ mode is at least 2 times larger than that transmitted into the next important mode. Power levels at the waveguide entrance are of the order of 17–44 dBm, depending on frequency.

The interior broad dimension of the WR90 waveguide is $a = 2.286$ cm. This corresponds to a cutoff frequency $f_c = c/(2a) = 6.6$ GHz for the lowest ($TE_{10}$) mode. Using the skin depth $\delta_s \approx (2/(\omega\mu\sigma))^{1/2} \approx 4\mu m/\sqrt{f(\text{GHz})}$, where $\sigma$ denotes the conductivity and the numerical value applies to brass, the surface resistance is given by $R_s = 1/(\sigma\delta_s) \approx 15$ m$\Omega \sqrt{f(\text{GHz})}$. As the RF wave propagates through the waveguide, the power decreases exponentially from its initial value $P(0)$. After a distance $s$, the power in the $TE_{10}$ mode is $P(s) = P(0)\exp(-2\alpha s)$, with an attenuation coefficient [7] $\alpha = R_s(2b\beta_c^2 + a\beta_0^2)/(ab\beta_0\beta_z Z_0)$, where $Z_0 = 377\ \Omega$, $b \approx a/2$ the small waveguide dimension, $\beta_c = \pi/a$, $\beta_0 = \omega/c$, and $\beta_z = (\beta_0^2 - \beta_c^2)^{1/2}$. The attenuation in dB per meter is plotted in Fig. 1 (right) for the lowest modes. There are three complications. At frequencies where the skin depth becomes comparable to the (unknown) surface roughness these formulae cease to be valid [7]. Second, there are about 41 waveguide modes with a cutoff frequency below 50 GHz, and the waveguide contains several H and E bends, at which part of the energy in the $TE_{10}$ mode could be converted into other modes as the RF wave travels along the waveguide. Finally, in an overmoded waveguide, degenerate modes ($TE_{nm}$ and $TM_{nm}$ modes with the same $n$ and $m$) are coupled by wall currents in the presence of attenuation, where symmetry permits [8], giving rise to hybrid modes.

The attenuation measured at 10 GHz is about 7 dB for a wave propagating twice through the entire waveguide, or about 0.2 dB per meter. This agrees well with the theoretical prediction for the waveguide attenuation due to surface resistivity. Also accounting for the 3 couplers in the signal processing unit, we can estimate the signal level at the crystal rectifiers. For the four BLM channels, these estimates are in the range 4–27 dBm, many 10's of dB above the noise. This provides a large safety margin for additional attenuation, which might arise from mode conversion or surface roughness.

For more accurate modeling, the power flow $P$ in a single waveguide mode (index $a$) after a distance $L$ can be expressed as the product of an effective voltage and current, $P_a(L,t) = V_a(L,t)I_a(L,t)$, where the current and voltage represent integrals over frequency of the beam spectrum multiplied by various transfer functions:

$$V_a(L,t) = \int_{-\infty}^{\infty} \frac{d\omega}{\sqrt{2\pi}}\ e^{j\omega t}\tilde{I}_b(\omega)\ R_a(\omega)\ F(\omega)e^{-L\{j\beta_a(\omega)+\alpha(\omega)\}} \tag{2}$$

$$I_a(L,t) = \int_{-\infty}^{\infty} \frac{d\omega}{\sqrt{2\pi}}\ e^{j\omega t}\tilde{I}_b(\omega)\ \frac{R_a(\omega)}{Z_{ca}(\omega)}\ F(\omega)e^{-L\{j\beta_a(\omega)+\alpha(\omega)\}} \tag{3}$$

Here, $\tilde{I}_b$ is the beam frequency spectrum of Eq. (1), the last exponential factor describes the propagation and attenuation through the waveguide (in the case of the $TE_{10}$ mode, $\alpha_a = \alpha$ and $\beta_a = \beta_z$, from above), the function $F(\omega)$ is a filter which

779

approximates the bandpass charcateristics of the different channels in our signal processing unit, $Z_{ca}$ is the mode impedance ($Z_{ca} = Z_0\beta_0/\beta_a$ for $TE$ modes, $Z_{ca} = Z_0\beta_a/\beta_0$ for $TM$ modes), and $R_a(\omega)$, in units of Ohms, represents the coupling of the beam to waveguide mode $a$. At low frequencies the coupling for the $TE_{10}$ mode ($a = TE_{10}$) is given by

$$R_{TE_{10}} = -j\frac{\sqrt{8}Z_0}{\pi R(ab)^{1/2}\beta_c\beta_z}e^{-j\frac{\beta_0 b}{2}}\cos\left(\frac{\beta_0 b}{2}\right). \tag{4}$$

At high frequencies the coupling can be described by a diffraction model. Finally, the output of the crystal detector $V_d$, as viewed on a scope, is obtained from

$$\left(\frac{\partial}{\partial t} + \frac{1}{\tau}\right)V_d = \frac{k}{\tau}P_a(L, t), \tag{5}$$

where $k$ characterizes the response of the crystal detector, and the time constant $\tau$ represents, e.g., the bandwidth of the oscilloscope. The bits accumulated by a gated charge-sensitive analog-to-digital converter (GADC) are proportional to $\int_{t_1}^{t_2} dt\ V_d(t)$, with a gate width ($t_2 - t_1$) of 50 ns. The Fourier transforms and integrations can be performed numerically. If several modes are included in the calculation, the effective voltage and current are sums over these modes, $\tilde{V}(z,\omega) = F(\omega)\sum_a \eta_a \tilde{V}_a(z,\omega)$, and $\tilde{I}(z,\omega) = F(\omega)\sum_a \eta_a \tilde{I}_a(z,\omega)$, where $\eta_a$ denotes the relative coupling strength of mode $a$ to the crystal detector as compared with the coupling of the $TE_{10}$ mode ($\eta_a$ is independent of $\omega$, and $\eta_{TE_{10}} = 1$). The power measured by the detector is simply $P(L,t) = I(L,t)V(L,t)$. An exact treatment must take into account the hybridization of degenerate modes due to wall losses [8].

During BLM commissioning it was found that monitor channel 3, detecting the RF power radiated above 21 GHz, provides the most useful signal, and all measurements reported in the following refer to this channel. The BLM signal is proportional to the square of the bunch charge and it decreases with increasing bunch length. To remove the dependence on the bunch charge, we normalize the signal by subtracting the GADC pedestal and dividing by the number of particles in units of $10^{10}$. For example, at a nominal bunch population of $N_b = 4 \times 10^{10}$, a normalized signal of 10 corresponds to about 160 counts of the gated ADC, and the resolution limit due to digitization is about 0.6%.

In the SLC, one major source of IP bunch-length variation are changes in the linac injection phase, which are caused by thermal RF phase drifts. Such linac phase errors introduce a position-energy correlation along the bunch. The correlated energy spread then changes the IP bunch length via momentum compaction ($R_{56}$) in the 1.2-km long collider arcs. The BLM can be used to monitor and correct such changes. In Fig. 4 (left) we depict the measured and simulated BLM signal as a function of the linac injection phase. In the simulation, $10^4$ particles were tracked in longitudinal phase space from the damping ring to the IP, and the beam power spectrum above 21 GHz was obtained by applying an FFT to the final distribution. Over the scan range considered, the simulated IP bunch length varies by more than a factor of 3, with an associated dramatic change in the shape of the beam distribution. The source of the overall additive offset on the left axis is unclear. It could indicate high-frequency fine structure in the beam.

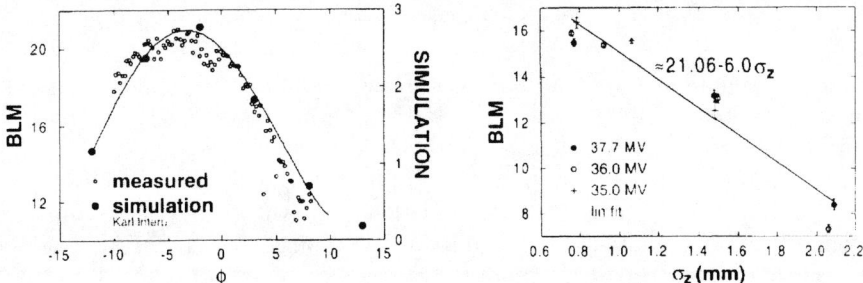

**FIGURE 4.** Left: BLM signal vs. linac injection time in degree S-Band, for $N_b \approx 1.2 \times 10^{10}$ with energy and RTL feedbacks active; each data point corresponds to one beam pulse; a simulation result (arbitrary units) is also shown. Right: calibration curve of BLM signal vs. rms IP bunch length; each data point derives from an average over 100 beam pulses, with rms variation shown by the error bars.

The BLM signal was calibrated by recording, over a few hundred pulses, both BLM signal and beam intensities for several different combinations of linac injection phase and compressor voltage. For each parameter setting, the beam energy profile at the end of the linac was also measured, using a wire scanner at a dispersive location. The corresponding IP distributions and bunch lengths were obtained from a multi-particle tracking simulation of the longitudinal beam transport. In the tracking, compressor voltage, compressor phase and the absolute offset of the linac phase were adjusted by fitting for optimum agreement between the simulated and measured energy profiles at the end of the linac [9]. The bunch compression, or anti-compression, in the arc was then calculated in a second simulation step, starting with the fitted end-of-linac distribution.

Figure 4 (right) shows the normalized BLM signal versus the simulated rms IP bunch length. The figure indicates that IP beam distributions with the same rms bunch length give rise to similar BLM signals. We attribute scatter in the data points to differences in the shape of the distributions for different compressor voltages, amounting to a 10% variation in rms bunch length.

From the measured fluctuation of the calibrated BLM signal we can estimate that the pulse-to-pulse rms bunch length variation is not larger than 5%.

## BEAM-TIMING MONITOR

The relative timing of the two colliding electron and positron bunches can be determined from the microwave signals induced by the two beams in the same X-

Band waveguide as employed for bunch-length measurements. The relative phase between the two electromagnetic pulses travelling down the waveguide is sensitive to changes in the time delay between the arrival of the first beam ($e^+$) and the second beam ($e^-$) at the pickup. Jitter in the observed time delay is a direct measure of the jitter in the collision point.

The BTM setup is illustrated in the left picture of Fig. 5. The first bandpass filter has its center at 11.39 GHz, with full width of ±50 MHz. The LO is a synthesizing signal generator, set to 11.49 GHz. Its stability, as measured with a frequency counter, is about ±0.1 MHz, after thermal insulation. The output of the first mixer is passed through a 5% bandwidth filter tuned to 100 MHz. The filtered IF signal is passed through two amplifiers (47 dB gain, 2.4 dB NF, and 27 dB gain, 11 dB NF, respectively), and then split with a tee. One arm passes through a trombone phase-shifter permitting a full phase shift of 20° with a 0.002° dial indicator. The second arm passes through 38 meters of 3 8" Heliax cable, providing about 200 ns delay. The arms of the tee are passed through variable attenuators and combined in a second mixer. The final IF output is then low-pass filtered and can be viewed on a scope, saved to disk, or passed to a gated analog to digital converter for acquisition by the SLC control system. Figure 5 (right) shows typical signals measured after the first mixer and after the final low-pass filter.

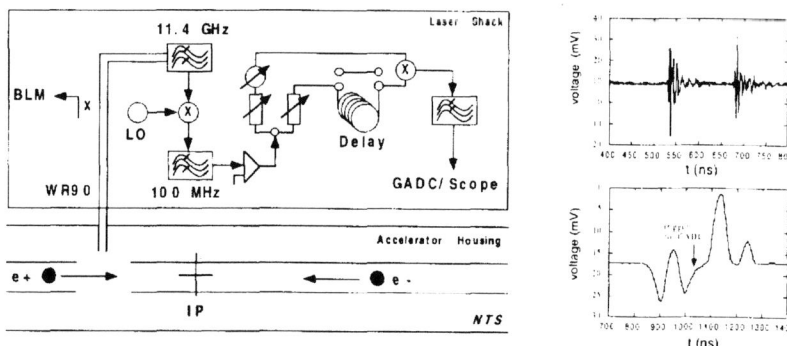

**FIGURE 5.** Left: schematic of BTM signal processing. Right: signal measured with crystal detector after the first mixer prior to second filter, where the signals induced by the two beams are clearly separated (top picture); and the low-pass filtered final two-beam signal after the 2nd mixer (bottom picture); the gate width of the GADC is 50 ns, and the trigger time (arrow) is chosen for a nominal signal close to zero.

We evaluate explicitly the limit in which the decay time of the output from the first filter is much less than the delay $(T_2 - T_1) \approx 300$ ns between the two bunches. We are considering then the time interval where the undelayed signal from the second bunch overlaps the delayed signal of the first bunch. The BTM scope waveform becomes $V_s(t) \approx k Q_1 Q_2 \sin\theta$, with $k = |V_L|^2 \cos^2\Psi$, where $|V_L|$ denotes the voltage of the local oscillator, and $\Psi = \arctan(Q(\omega_{IF}/\omega_2 - \omega_2/\omega_{IF}))$, with $Q$

$2\omega_2/\nu_2$ the quality factor of the second (narrow-band) filter, $\omega_2$ its center frequency, and $\omega_{IF} \sim (\omega_1 - \omega_L)$ the intermediate frequency which is obtained by mixing the signal after the first filter, centered around $\omega_1$, with the local oscillator signal of frequency $\omega_L$. The variables $Q_1$ and $Q_2$ are the charges of the two bunches, and the relative beam phase $\theta$ is $\theta \equiv (\omega_1 \Delta T - \omega_{IF} T_D - \phi + \pi/2)$, where $\Delta T$ is the arrival time difference, whose fluctuation we want to measure, and $T_D$ the delay time of the cable. The phase $\phi$ is a constant. Whatever the nominal beam delay $\Delta T \sim (T_2 - T_1)$ may be, using a phase shifter one may adjust $\phi$ to put the scope signal at a null. In this case, maximum sensitivity is attained and the nominal relative beam phase is 0. For small variations, one has $V_s(t) \approx k Q_1 Q_2 \theta$, and jitter in relative beam timing then correlates linearly with $V_s/(Q_1 Q_2)$. The counts registered in a gated analog-to-digital converter are given by $\int_{T_{trig}}^{T_{trig} + T_{gate}} dt\, V_s(t)$, up to an overall constant. The trigger time is $T_{trig}$ and the gate width $T_{gate}$ is 50 ns. Thus, in practice, adjusting $\phi$ also requires a choice of gate position. A refined numerical model of the signal waveforms has also been developed, taking into account the filter characteristics as measured with the network analyzer, and the numerically computed waveforms incident on the first filter (including the dispersion characteristics of the waveguide). Signals predicted by this model are in reasonable agreement with the measurement.

The raw signal is normalized by first subtracting the pedestal and then dividing by the product of the $e^+$ and $e^-$ intensities (particles per bunch in units of $10^{10}$). The two calibration curves in Fig. 6 show the BTM signal as a function of the delay (in units of degree X-Band) in one of the two arms, which was varied using the phase trombone. The left picture is an autophase measurement (one beam only, delay cable removed); the right is an actual two-beam measurement (with delay cable). Using the fitted slope we can convert the measured signal changes into degree X-Band or time. The much larger scattering of the data in the right picture represents real beam-timing jitter, as large as 14° X-Band peak to peak, or 5° rms.

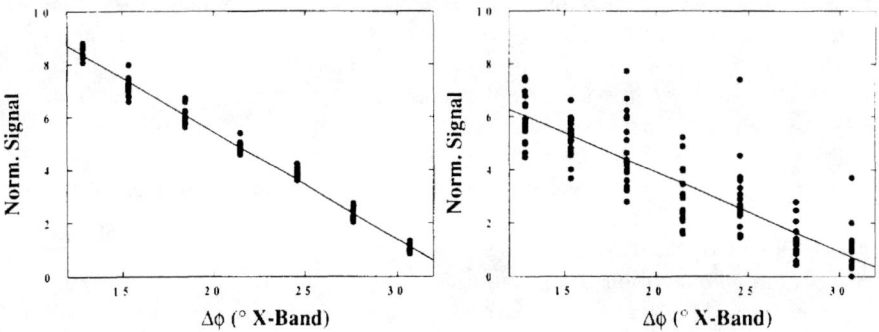

**FIGURE 6.** Calibration measurements: normalized timing-monitor signal versus the trombone phase difference in degree X-Band; (left) signal for one beam with cable delay removed; (right) signal for two beams including cable delay.

We have also measured the BTM signal as a function of the electron-beam energy at the end of the linac, holding the energy of the positron beam constant. From this measurement and from the phase calibration performed at the same time, we estimate the momentum compaction of the SLC North arc as $R_{56} \approx 120$ mm, not too far from the theoretical value (145 mm).

## CONCLUSION

The RF BLM in the SLC South Final Focus monitors pulse-to-pulse changes in the IP bunch length of both colliding beams, with a precision of a few percent. Its response to linac injection time and compressor voltage is consistent with simulations of the longitudinal beam transport. The bunch length was found to be stable to within 5% and, thus, it is not a dominant source of pulse-to-pulse fluctuation in luminosity or beamstrahlung. The RF BTM detects changes in the relative arrival time of electron and positron bunches at the SLC interaction point, with a short-term resolution better than 5° X-Band. Our measurement results indicate a luminosity degradation due to beam timing jitter of the order of 2%.

## ACKNOWLEDGEMENTS

We thank M. Ross, M. Breidenbach and N. Phinney for support, A. Menegat, S. Tantawi and S. Hanna for helpful advice, M. Woods for sharing the laser shack with us, and H. Rogers for installing the BLM shielding box.

## REFERENCES

1. K. Yokoya and P. Chen, "Beam-beam phenomena in linear colliders", *Lecture at 1990 US-CERN School on Particle Accelerators, Hilton Head Island* (1990).
2. D. Schulte, "Study of Electromagnetic and Hadronic Background in the Interaction Region of the TESLA Collider", Ph.D. thesis, University of Hamburg (1996).
3. E. Babenko, R.K. Jobe, D. McCormick, J.T. Seeman, "Length Monitor for 1 mm SLC Bunches", *Proc. of IEEE PAC93, Washington,* p. 2423 (1993).
4. J.D. Lawson, "Radiation from a Ring Charge Passing Through a Resonator", Rutherford Lab. Report *RHEL/M* **144** (1968), *Particle Accelerators* **25**, 107 (1990).
5. K. Bane, M. Sands, "Wakefields of Very Short Bunches in an Accelerating Cavity", *Particle Accelerators* **25**, 73 (1990)
6. "MAFIA RELEASE 3", *DESY* **M-90-05K** (1990).
7. R.E. Collin, "Foundations for Microwave Engineering", McGraw-Hill (1966).
8. K. Kurokawa, "Electromagnetic Waves in Waveguides with Wall Impedance", *IRE Trans. Microwave Theory and Techniques,* p. 314 (1962).
9. K.L.F. Bane, F.-J. Decker, J.T. Seeman, F. Zimmermann, "Measurement of the Longitudinal Wakefield and the Bunch Shape in the SLAC Linac", *Proc. of IEEE PAC 97, Vancouver,* and *SLAC-PUB-***7536** (1997).

# Coherent Smith-Purcell Radiation for Use in Electron Beam Diagnostics

M. C. Lampel

*G. H. Gillespie Associates, Inc.*

*P. O. Box 2961*

*Del Mar, CA 92014*

**Abstract.** Coherent Smith-Purcell Radiation (CSPR) is quite useful in the area of intense ultra-short pulse electron beams. This is because it offers the potential for non-destructive, real time measurements of high brightness pulses being developed for use in many next-generation electron beam accelerator projects, such as NLC, TESLA test facility, and various other plasma-based accelerator schemes. Both the non-destructive and real time aspects of using CSPR are significant to ultra-short pulse measurement. Preliminary calculations show that CSPR is capable of producing usable signals from realistic electron beam and electron beamline operating parameters that can be analyzed to provide pulse structure, energy, emittance, and timing information. At the same time, induced emittance growth is well below an amount that would seriously deteriorate beam quality for even ultra-high brightness beams. Timing measurements using CSPR is a critical area for development efforts, because of the lack of existing techniques for accurate single pulse sub-picosecond measurements. Injection accuracy for advanced accelerator techniques, as is the case for conventional RF structures, is dictated by getting the pulse into the accelerator at the correct phase. This implies absolute timing accuracies on the order of 10-20 femtoseconds, for a laser-accelerator system that pulses at much less than once per second. By producing CSPR at one or more locations along the beamline, interferometric techniques can then mix this radiation with a timing signal from the laser and beamline timing measurements, useful for fine tuning the transport system and laser beam delay lines, will become obtainable.

## INTRODUCTION

Coherent Smith-Purcell Radiation (CSPR) is potentially quite useful, as a beam diagnostic, in the area of intense ultra-short pulse electron beams [1,2]. This is because it offers the potential for non-destructive, real time measurements of the high brightness pulses being developed for use in many next-generation electron accelerator projects such as at the TESLA test facility [3], the BNL-ATF, [4] and various plasma-based accelerator schemes [5]. Both the non-destructive and real

CP472, *Advanced Accelerator Concepts: Eighth Workshop*,
edited by W. Lawson, C. Bellamy, and D. Brosius

time aspects of using CSPR are significant to ultra-short pulse measurements. This is due to the nature of the current development efforts, as well as to the fundamental physics of such pulses.

The ability to produce, transport, and measure both incoherent and coherent Smith-Purcell radiation has been amply demonstrated since first reported by Smith and Purcell [6], most recently at several laboratories using high brightness relativistic beams [7-9]. Current development efforts use low repetition rate equipment. In particular, plasma-laser acceleration schemes that are getting a great deal of attention from High Energy Physics in DOE use lasers pulsed at much less than once per second. This means each pulse is quite expensive, with large cost benefits for multiple diagnostics. Non-destructive measurements for as many parameters as possible during a single pulse are therefore extremely valuable to the experimentalists. Also, because of the low data acquisition rate, a single pulse measurement providing any of the key parameters such as timing, current profile, emittance, or energy spread becomes virtually indispensable. Not only is this because it allows shot to shot corrections, but also because, over periods approaching an hour, total system stability becomes an issue when trying to provide data analysis using a technique relying on multi-shot data integration.

Preliminary calculations show that CSPR is capable of producing usable signals from realistic electron beam and electron beamline operating parameters that can be analyzed to provide pulse structure, energy, emittance, and timing information. At the same time, induced emittance growth is well below the threshold for seriously deteriorating beam quality for even ultra-high brightness beams.

Timing measurements using CSPR is one promising area for development. The present lack of existing techniques for accurate single pulse sub-picosecond measurements currently holds back more rapid development of sub-picosecond pulse creation and acceleration. The proposed technique for investigation in this paper is to produce CSPR at one or more locations along the beamline, within a dielectric, and then, using total internal reflection, transport the beam to selected nonlinear optical diagnostics. Using interferometric techniques, one can then mix this radiation with a timing signal from the laser. This will produce information on several time scales. However, initially, the information of interest will be to see the relative timing of the laser pulse and electron pulse. By performing this measurement at two or more locations along the beamline (possible because of the non-destructive aspect of CSPR), beamline timing measurements, useful for fine tuning the transport system and laser beam delay lines, will become obtainable.

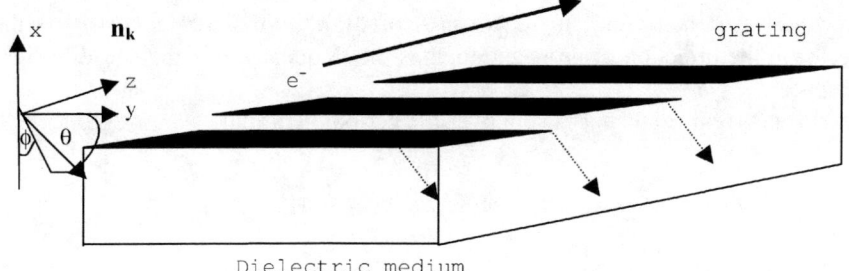

**FIGURE 1.** The electron passes over the transmission grating. The coordinate axes are displayed in the top left along with the unit vector $\mathbf{n_k}$ .

## SMITH-PURCELL RADIATION IN A DIELECTRIC MEDIUM

As do most analyses of Smith-Purcell radiation [6-10], we assume that the electrons, charge q, travel at a constant velocity v parallel to a grating surface and perpendicular to the strips, as shown in Figure 1. Let the direction of **v** be along the z-axis, so that $\mathbf{v} = \mathrm{v}\,\mathbf{n_z}$, with $\mathbf{n_z}$ the unit vector in the z direction. Let the strips define the y-axis, with unit vector $\mathbf{n_y}$. Let the strips extend to infinity, but the number of strips will be finite. If the grating is not grounded, then, by Gauss' Law, there are two induced surface charges. If a grating is a narrow compared to the width of the induced surface charge and thin relative to the width, then the induced surface charge on both sides is nearly uniform, equal and opposite with magnitude q/2A, where A is the effective area.

Following the treatment of Brownell, Walsh, and Doucas [10], we develop the equation for radiated energy off a grating. There are two modifications; 1) we treat the case of an ungrounded grating, and 2) we look at the side facing away from the electron, imbedded in a dielectric. For the case where the opposing surface is embedded in a dielectric medium, with a refractive index $n_{ref}(\omega)$, it is possible to write the equation for the radiated energy in the far field, for a surface current $J_s$, as [11]:

$$W \equiv \frac{\partial^2 I}{\partial \omega \partial \Omega} = \frac{\omega^2}{4\pi^2 c^3} \left| \int dt \int d^3 x \mathbf{n_k} \times \mathbf{n_k} \times J_s(\vec{r},t) e^{i(\omega t - \mathbf{k} \cdot \mathbf{r})} \right|^2 \qquad (1)$$

Where the equation for free space radiation has been modified by replacing $\mathbf{k} = \mathbf{n_k}\omega/c$ with $\mathbf{k} = \mathbf{n_k}\, n_{ref}(\omega)\omega/c$, $\mathbf{n_k} = [\mathbf{n_x}\sin(\theta)\cos(\phi)+\mathbf{n_y}\sin(\theta)\sin(\phi)+\mathbf{n_z}\cos(\theta)]$ is the unit vector in the direction of propagation, $\omega$ is the frequency, c is the speed of light in free space.

The surface current is taken as a sum over the periodic grating:

$$J_s(\mathbf{r},t) = \sum_{m=1}^{M} J_{groove}(\mathbf{r} - ml\mathbf{n}_z, t - ml/\mathbf{v}) \tag{2}$$

Combining the two equations and factoring out the relative phase factors generated from the sum over strips gives:

$$W = \frac{\omega^2}{4\pi^2 c^3} \left| \sum_{m=1}^{M} e^{iml\omega(\frac{1}{v} - \frac{\mathbf{n_k}\cdot\mathbf{n}_z n_{ref}}{c})} \right|^2 \left| \int dt \int d^3x\, \mathbf{n_k} \times \mathbf{n_k} \times J_{groove}(\vec{r},t)e^{i(\omega t - \mathbf{k}\cdot\mathbf{r})} \right|^2 \tag{3}$$

An important difference appears at this point between the analysis for a grating in a vacuum and in a dielectric. In the sum over phase factors, because of the appearance of $n_{ref}$ the phase term can pass through 0. That is, constructive interference occurs even at frequencies different than those given by the vacuum Smith-Purcell condition, which can be written as:

$$\omega_m = \frac{2\pi|m|c}{l(\frac{1}{\beta} - \cos\theta)} \tag{4}$$

The vanishing of frequency dependence is just the condition for emission of Cherenkov radiation. For angles greater than the Cherenkov critical angle emission into a dielectric is similar to emission into a vacuum, except that the Smith-Purcell condition is modified to be:

$$\omega_m = \frac{2\pi|m|c}{l(\frac{1}{\beta} - n_{ref}\cos\theta)} \tag{5}$$

The phase factor itself is well defined as emission passes through the Cherenkov angle, even though the frequency goes negative and the wavelength goes through 0:

$$\left| \sum_{m=1}^{M} e^{iml\omega(\frac{1}{v} - \frac{\mathbf{n_k}\cdot\mathbf{n}_z n_{ref}}{c})} \right|^2 = \frac{\sin^2[(\frac{1}{\beta} - \mathbf{n}_{ref}\cos\theta)\omega Ml/2c]}{\sin^2[(\frac{1}{\beta} - \mathbf{n}_{ref}\cos\theta)\omega l/2c]} \xrightarrow{M>>1} \sum_{m\neq 0} \frac{\omega L}{|m|l}\delta(\omega - \omega_m) \tag{6}$$

Figure 2 shows a comparison of several phase factors. TPX is a commercially available plastic which has good optical qualities in the FIR, over a range from ~60 µm<λ<300 µm, and then again in the near IR and visible. The index of refraction for TPX is about 1.46 in the FIR. Using this index, the critical angle, as measured up from the z-axis, is about 46.77° angle (which is 0.816287 to six significant figures). Phase factors for both TPX and vacuum are plotted for two angles, 0.82 radians, near the critical, and 0.89 radians. A total of 10 strips are taken for the grating, so that the maximum value of a phase factor is 100. The strip periodicity is 1 mm. The relativistic factor, γ, is 10, giving a β of ~0.995. The plots are done in two steps over four orders of magnitude in angular frequency, from $10^{10}$ to $10^{14}$. It can be seen in the plot that for TPX at 0.82 radians the coherence of the radiation remains near perfect in a continuous band from the lowest frequency out past the first order vacuum Smith-Purcell frequency. Of course this is a simplified model assuming the index of refraction is constant over the entire range. Nevertheless, given that TPX varies little over the indicated range, and that at 46.77° the free space wavelength of the Smith-Purcell radiation is 320 µm, the qualitative aspects of the figure are instructive. First, we see that radiation is emitted over a broad frequency spectrum at the critical angle, thus relatively large powers are emitted in a narrow angular range, just as for Cherenkov radiation. Second, at larger angles the spectrum begins to resemble the vacuum Smith-Purcell interference function, but shifted upward in frequency for a given emission angle. Third, this function stays real and positive for all values of emission angle, including values which, when placed into the Smith-Purcell condition give negative frequencies because $n_{ref}\cos\theta$ becomes larger than $1/\beta$. The physical interpretation of this is beyond the scope of this paper. A detailed analysis of the "Cherenkov shock front" and what happens for angles of emission beyond it requires a more formal approach to analyzing the retarded potentials with appropriate consideration of causality issues.

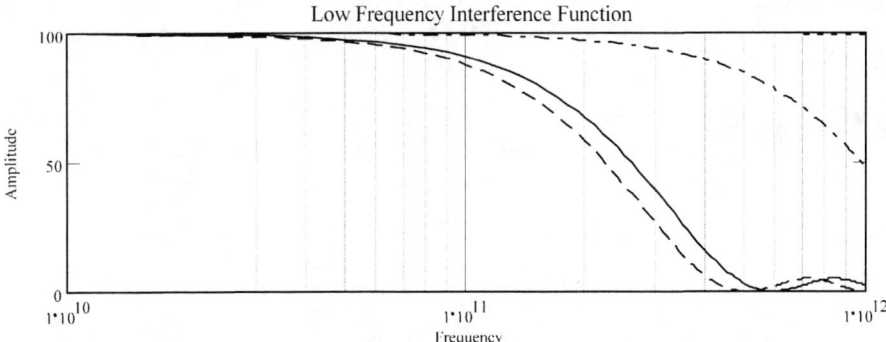

**FIGURE 2a.** Four interference functions for a 10 period strip grating emitting into a dielectric at low frequency. For both figures the different functions are calculated for: TPX at 0.82 rad - dadotted line, TPX at 0.89 rad – dashed line, VAC at 0.89 rad -dotted line, VAC at 0.82 rad – solid line.

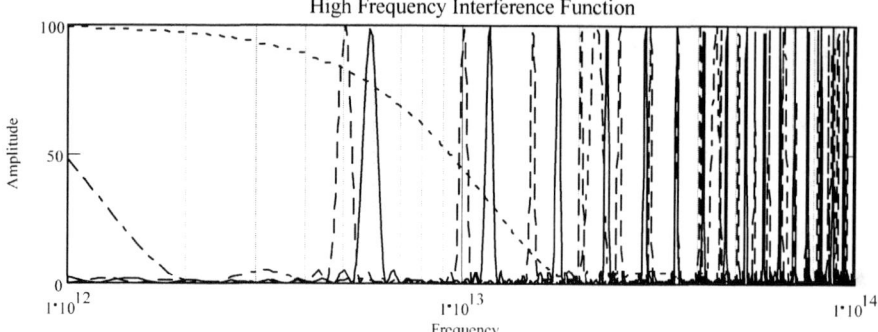

**FIGURE 2b.** Four interference functions for a 10 period strip grating emitting into a dielectric at high frequency.
For both figures the different functions are calculated for: TPX at 0.82 rad - dadotted line, TPX at 0.89 rad – dashed line, VAC at 0.89 rad -dotted line, VAC at 0.82 rad – solid line

## Surface Current and Power Radiated

The results of [10] for the expression for surface current and the single particle power radiated are easily adapted to the current case. An ungrounded conductor is slightly different than a grounded conductor, as shown in Figure 3. Instead of there being an induced surface charge only on the surface facing the particle (the inner surface), there are now surface charges on both surfaces of the conductor. Using conservation of charge it can be seen that the charge on the two surfaces must be equal and opposite. Exploiting the fact that at the surfaces, the electric field must be normal, and inside the conductor the fields must add to 0 everywhere, the charge distributions on both surfaces can be calculated. The integrated surface charge densities can be calculated from requiring that the tangential component of the total electric field is 0 on both surfaces. In the rest frame of the particle, the functions for this are given by:

$$f_1(r) \equiv \frac{1}{q}\int_0^r 2\pi\sigma(r')r'dr'$$

$$= \frac{r^3}{\left(d^2+r^2\right)^{3/2}} + \frac{r^6}{\left(t^2+r^2\right)^3 - r^6} - \frac{\left(d^2+r^2\right)^{3/2}}{\left((d+t)^2+r^2\right)^{3/2}\left(\frac{\left(t^2+r^2\right)^{3/2}}{r^3} - \frac{r^3}{\left(t^2+r^2\right)^{3/2}}\right)} \quad (7)$$

$$f_2(r) \equiv \frac{1}{q}\int_0^r 2\pi\sigma(r')r'dr' = \frac{r^3\left(t^2+r^2\right)^{3/2}\left[\left(t^2+r^2\right)^{3/2}\left(d^2+r^2\right)^{3/2} - r^3\left((d+t)^2+r^2\right)^{3/2}\right]}{\left((d+t)^2+r^2\right)^{3/2}\left(d^2+r^2\right)^{3/2}\left[r^6 - \left(t^2+r^2\right)^3\right]}$$

790

Where $f_1(r)$ and $f_2(r)$ are the normalized integrated surface densities on the inner and outer surfaces, respectively. Examination of $f_2(r)$ shows that the variation across the surface is slow, as argued in the previous section. Thus, it is not a bad approximation to take the surface charge on the back face of the strip as being uniform and equal to $q/2A$, where A is the effective strip area. Transforming to the lab frame, the equation for the charge on the outer surface of the grating is given approximately as:

$$\rho(\mathbf{r},\mathbf{r_0},t,s) = \frac{q\gamma}{4\pi} \frac{|x-x_0|}{[(x-x_0)^2 + (y-y_0)^2 + \gamma^2(z-z_0 - vt)^2]^{3/2}} \delta[x-x_1] \qquad (8)$$

The distance $x_1$ is measured from the inner surface of the grating. $x_0$, $y_0$, and $z_0$ define the position of the electron at $t = 0$. $\beta = v/c$ and $\gamma = (1-\beta^2)^{-1/2}$. Using the expression for current given in [10], with the modified expression for surface charge density, the results for the power radiated are down by ¼ from previously published results. Therefore, the result for a single electron is:

$$W = \frac{q^2\omega^2 l^2 e^{-x_0/\lambda_e}}{16\pi^2 c^3} \frac{\sin^2[(1/\beta - n_{ref}\cos\theta)\omega lM/2c]}{\sin^2[(1/\beta - n_{ref}\cos\theta)\omega l/2c]} \left| \mathbf{n_k} \times \mathbf{n_k} \times \sum_{f=1}^{M} G(\omega,\mathbf{n_k},s_f) \right|^2 \qquad (9)$$

Where, in (9) above, $s_f$ represents the coordinates of strip number f in the grating.

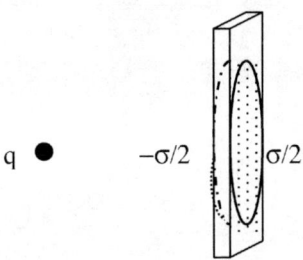

$$q \bullet \qquad -\sigma/2 \quad | \quad | \sigma/2$$

Ungrounded Conducting Strip

**FIGURE 3.** Image charge distribution on an ungrounded conductor has equal and opposite total image charges on both sides. The image charges plus the original charge produce 0 electric field inside the conductor, but there are non-zero fields on both sides of it. For a strip grating the induced current on both sides is significant.

The evanescent field length, $\lambda_e$, contains the familiar dependencies on energy and emission angle, and is defined as:

$$\lambda_e \equiv \left( \frac{2\omega}{\gamma\beta c}\sqrt{1+\gamma^2\beta^2\sin^2\theta\sin^2\phi} \right)^{-1} \tag{10}$$

The Green's function G is:

$$G(\omega,\mathbf{n_k},s_f) \equiv \left(\mathbf{n_x} \pm \mathbf{n_y}i2k_y\lambda_e\right)\frac{e^{(\pm1/2\lambda_e - i\mathbf{k_x})(x_1)}}{(+i\kappa)l}e^{+i\kappa z}\Big|_{z_1}^{z_2} \tag{11}$$

With $\kappa = \omega/v - k_z$.

It should be kept in mind that the definition of $\mathbf{k}$ is different in a dielectric, being larger than the vacuum $\mathbf{k}$ by a factor of $n_{ref}$. The other important difference is that, because the surface current is smaller by a factor of 2 than for a grounded conductor, power is down by a factor of 4. For a single strip per period, in the limit of a very long grating (M>>1) the total power per unit solid angle can be expressed as:

$$\frac{\partial I}{\partial\Omega}\Big|_{strip} = \sum_{m\neq0}\frac{1}{2\pi}\left(\frac{q}{l}\right)^2 \frac{L\sin^2\theta\sin^2(\pi md/l)}{\left(\frac{1}{\beta}-n_{ref}\cos\theta\right)^3}e^{-|x_0|/\lambda_e} \tag{12}$$

Further investigation of emission at and below the Cherenkov angle, a more precise treatment of the time dependent current on the grating, and an investigation of modifications due to finite conductivity and the quality factor of this resonant structure are some of the areas in which future work should focus. One note on coherence, because the free space wavelength of Smith-Purcell radiation is longer than in the dielectric for a given frequency, coherent radiation from a pulse will disappear more rapidly when going to higher frequencies above the Cherenkov angle.

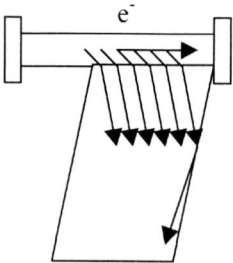

**FIGURE 4.** An ungrounded dielectric beampipe with a grating serves as a Smith-Purcell radiator. Radiation emitted into the dielectric medium can be transported using total internal reflection, creating a lightpipe.

# A PROPOSED DIAGNOSTIC CONFIGURATION

Creating Smith-Purcell radiation off a strip grating radiating away from the beam can provide a unique beam diagnostic configuration, Figure 4. The electron beam can pass through a dielectric section of vacuum pipe, perhaps shielded on the outside, which has a grating on the outer surface. The grating can be embedded into a dielectric or not, depending on the design. In either case the Smith-Purcell radiation is emitted and collected outside of the vacuum system itself. Therefore, the entire diagnostic configuration is changeable without having to break vacuum. Even the grating itself could be designed to be modular and replaceable without having to remove the beampipe.

Generation of significant coherent light [1,2,7-10] occurs when the electron pulse has a large fraction of the beam well ordered over less than the wavelength being observed. Typically this ordering is taken to be the pulse length, however for some distributions such as rectangular, the large content of higher frequencies leads to large coherent light generation for wavelengths shorter than the pulse length as well. A significant constraint for a diagnostic is how close the electron beam must be to the grating, the distance d, to generate an observable amount of light. This constraint eases as energy increases because the evanescent wavelength goes as $d/\gamma$ for angles near 90°. For implementation in a real beamline, electron beam height over the grating, grating period, angle of observation, beam energy, and the wavelength of any existing drive lasers need to be considered when configuring a coherent Smith-Purcell radiation diagnostic.

The radiation can be transported through an optical waveguide, especially easy if it is generated in a dielectric already, or by the use of mirrors and gratings. By selecting a particular line out of the spectrum, a particular scheme for laser-electron timing can be implemented. The simplest diagnostic technique conceptually is well known and straightforward, consisting of mixing a selected part of the signal with a chirped signal, $\omega_{clock}(t)$, derived from a laser/clock. Spectroscopic analysis of the mixed signal allows accurate measurement of the relative timing of the Smith-Purcell radiation to the clock pulse.

Different configurations can easily be set up by then changing gratings, dielectrics, and optical diagnostics. In addition, more than one grating could be placed on the beampipe, or the beampipe could be completely surrounded by a grating, for maximum radiation. Beampipe radius should also be minimized to achieve maximum radiated power, as the intensity goes as the square of the electric field at the surface of the conductor.

# CONCLUSION

Emission of Smith-Purcell radiation into a dielectric and off the opposing surface of an ungrounded conductor has been investigated. Both of these problems are of interest separately in terms of developing diagnostic capabilities. Taken together the possibility for an extremely flexible, ultra-short pulse diagnostic seems quite realizable. Emission off the outside of a beampipe into an optical guide made of dielectric allows quick turn-around modifications and maintenance without breaking vacuum. Thus the same diagnostic section could quickly and easily be changed from longitudinal to transverse measurements or upgraded in the future to accommodate advanced techniques such as holographic imaging of phase space.

Additional work needs to be done along several fronts. First, a more realistic model of dielectric response at the frequencies of interest is needed. Detailed analysis of radiation at the Cherenkov angle and what happens below it should be done as well. Finally, a finite resistance model of a conductor should be treated to see what changes, if any, occur to important but untreated effects such as phase lag, pulse duration, and emittance growth of the beam.

# ACKNOWLEDGMENTS

The author would like to thank Dr. George Gillespie, and Dr. Nathan Brown for support and helpful discussions, as well as the participants of Working Group 4, and especially Professor Todd Smith, the chair, for help in refining this idea. This work was supported by G. H. Gillespie Associates, Inc.

# REFERENCES

1) Lampel, M. C., *Nucl. Inst. and Meth. A,* **385,** 19-26, 1997.
2) Nguyen, D. C., *Nucl. Inst. and Meth. A,* **393,** 514-8, 1997.
3) Corsini, R., Hofmann, A., *Proceedings of the 5th European Accelerator Conference (EPAC96),* vol. 1, p.721, (1997).
4) Wang, X. J., Qiu, X., Ben-Zvi, I., *Phys. Rev. E* **54** p. 3121 (1996).
5) See for instance, Katsouleas, T., "Overview of Plasma-Based Accelerator Concepts", these proceedings.
6) Smith, S. J., and Purcell, E. M., *Phys. Rev,* **92,** 1069, 1953.
7) Shibata, Y., et. al., Phys. Rev. E **57,** pp.1061-1074.
8) Brownell, J. H., Walsh, J., Kirk, H., Fernow, R. C., Robertson, S. H., *Nucl. Inst. & Meth. A,* **393,** 323-325, (1997)
9) Woods, K. J., Walsh, J., Stoner, R. E., Kirk, H., Fernow, R. C., *Phys. Rev. Lett.,* **74,** 3808-3811 (1995).
10) Brownell, J. H., Walsh, J., Doucas, G., *Phys. Rev. E,* **57,** 1075-80, 1998.
    Jackson, J. D., *Classical Electrodynamics 2nd Ed.,* ch. 14, (John Wiley & Sons, Inc., New York, 1975).

# Synchronization of Sub-picosecond Electron and Laser Pulses[*]

J.B. Rosenzweig

*UCLA Dept. of Physics and Astronomy*
*405 Hilgard Ave., Los Angeles, CA 90095*

G.P. Le Sage

*Lawrence Livermore National Laboratory*
*P.O. Box 808, Livermore, CA 94550*

Sub-picosecond laser-electron synchronization is required to take full advantage of the experimental possibilities arising from the marriage of modern high intensity lasers and high brightness electron beams in the same laboratory. Two particular scenarios stand out in this regard, injection of ultra-short electron pulses in short wavelength laser-driven plasma accelerators (1), and Compton scattering of laser photons from short electron pulses(2, 3). Both of these applications demand synchronization, which is sub-picosecond, with tens of femtosecond synchronization implied for next-generation experiments. Typically, an RF electron accelerator is synchronized to a short pulse laser system by detecting the repetition signal of a laser oscillator, adjusted to an exact sub-harmonic of the linac RF frequency, and multiplying or phase locking this signal to produce the master RF clock. Pulse-to-pulse jitter characteristic of self-modelocked laser oscillators represents a direct contribution to the ultimate timing jitter between a high intensity laser focus and electron beam at the interaction point, or a photocathode drive laser in an RF photoinjector. This timing jitter problem has been addressed most seriously in the context of the RF photoinjector, where the electron beam properties are sensitive functions of relative timing jitter. The timing jitter achieved in synchronized photocathode drive laser systems is near, or slightly below one picosecond. The ultimate time of arrival jitter of the beam at the photoinjector exit is typically a bit smaller than the photocathode drive-laser jitter due to velocity compression effects in the first RF cell of the gun. This tendency of the timing of the electron beam arrival at a given spatial point to "lock" to the RF clock is strongly reinforced by use of magnetic compression.

In a magnetic compressor (4), electron beam pulse compression is achieved by first imparting a front-to-back energy chirp on the electron pulse by running forward of the peak accelerating phase in the RF linac, and then propagating the beam through a magnetic chicane. In the chicane, shown schematically in Figure 1, lower energy particles are forced to negotiate a longer path length than higher energy particles.

---

[*] This work was supported by U.S. Dept. of Energy grants DE-FG03-93ER40796 and DE-FG03-92ER40693, and the Alfred P. Sloan Foundation grant BR-3225.

CP472, *Advanced Accelerator Concepts: Eighth Workshop,*
edited by W. Lawson, C. Bellamy, and D. Brosius

Since ultra-relativistic particles travel at essentially the speed of light, the path length difference rearranges the relative longitudinal position of the particles in the bunch by the same differential length. For small energy spreads a purely linear chirp in the $(E,\phi)$ phase space $(\phi = k_{rf}(z - ct))$ can be mapped through the chicane so that all the particles arrive at the same position $z$ at the same time regardless of the initial time of arrival at the chicane entrance.

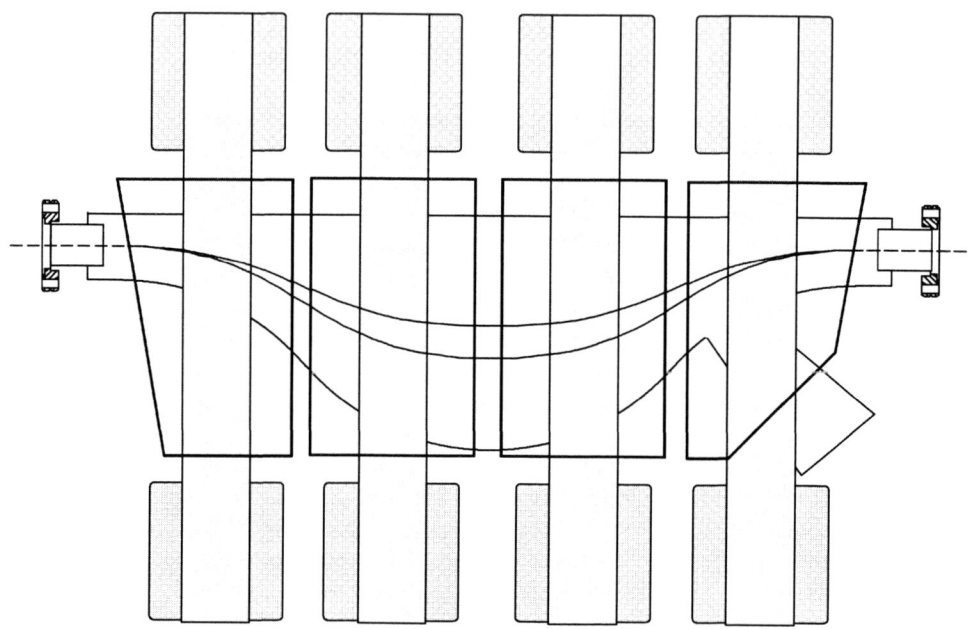

**FIGURE 1.** UCLA chicane system (5), showing path of two different energy electrons through magnetic compression system.

If the energy spectrum of the beam is entirely due to the RF accelerating forces, then the spectral dependence on phase is simply $E(\phi) \cong E_{max} \cos(\phi)$, and there is a curvature of the energy chirp, causing a non-zero longitudinal emittance to appear. For such a case, in which we ignore the space charge and wake forces on the particle distribution, the beam is optimally compressed when the slope of the $E(\phi)$ distribution is linear, so that only the quadratic and higher dependence is left. In the absence of these higher order dependencies, and ignoring the weak nonlinearity of the chicane mapping, it would be possible to compress the beam to infinitesimally short length. It was pointed out in Ref. (4) that injection timing jitters are suppressed using the linac chirp/chicane system, as in lowest order in the timing error, all nearby injection times result in the identical time of passage through the chicane exit.

For beams with non-negligible charge, this picture changes somewhat, as is illustrated in Figure 2. The effects of collective space-charge and wake-field forces can change the slope of chirp in addition to introducing new nonlinear longitudinal and radial dependencies of the longitudinal forces. These forces thus increase the longitudinal emittance of the beam in addition to moving the optimum linac phase for compression forward relative to the small charge case. Thus if one optimizes compression, by removing the linear correlation in the longitudinal phase space, one is actually overcompensating the timing error associated with the beam centroid, which is unaffected by space-charge. In this case the linear component of the timing jitter is not cancelled, but merely suppressed. This is shown in Figure 3, which presents PARMELA simulations of the equivalent spatial jitter for a 40 pC beam in the UCLA Neptune photoinjector and compressor (5). For optimum compression in this case, the initial timing jitter is suppressed by a factor of 10. If one wishes to suppress the timing jitter optimally, the chicane will undercompress beams with non-negligible collective forces. This trade-off will have to be considered when designing a given experiment.

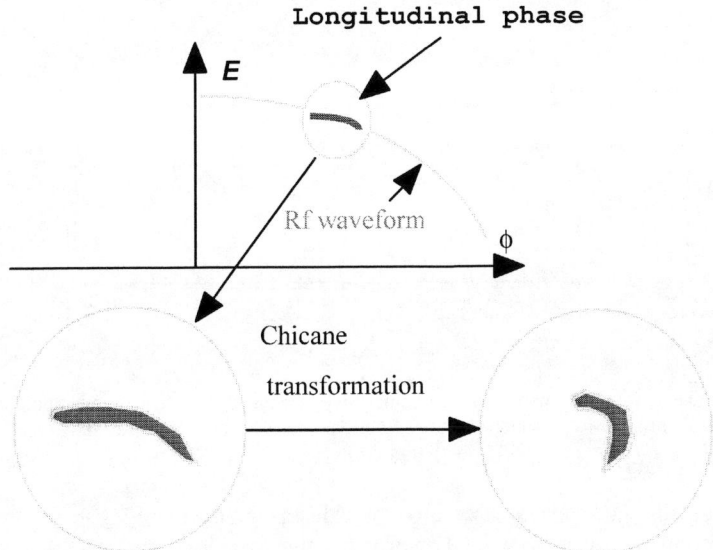

**FIGURE 2.** The transformation of a beam with non-negligible charge and phase extent from RF acceleration, collective forces and magnetic chicane. Note that collective forces cause the energy space at the rear of the to beam to sag off of the RF waveform.

To summarize, magnetic pulse-compression techniques apply a longitudinal "restoring force" which improves synchronization between electron bunches and the RF drive, partially or completely correcting errors in the exit time from the injector. While this feature is beneficial in terms of linac related phase matching issues and jitter in the photocathode laser system, laser jitter with respect to the RF clock is in fact worsened in terms of the final interaction of laser and electron pulses. That is, if one attempts to synchronize the electron bunch to an external laser, deriving external laser timing from, say, the photocathode drive laser oscillator, the relative timing of electron and the external laser will be mismatched, because the electron beam is pulled toward a given phase of the RF clock. Thus, if one wishes to take advantage of the locking of the electron beam to the (assumed stable) RF clock, then an equivalent mechanism must be found for locking the external laser to the RF clock as well.

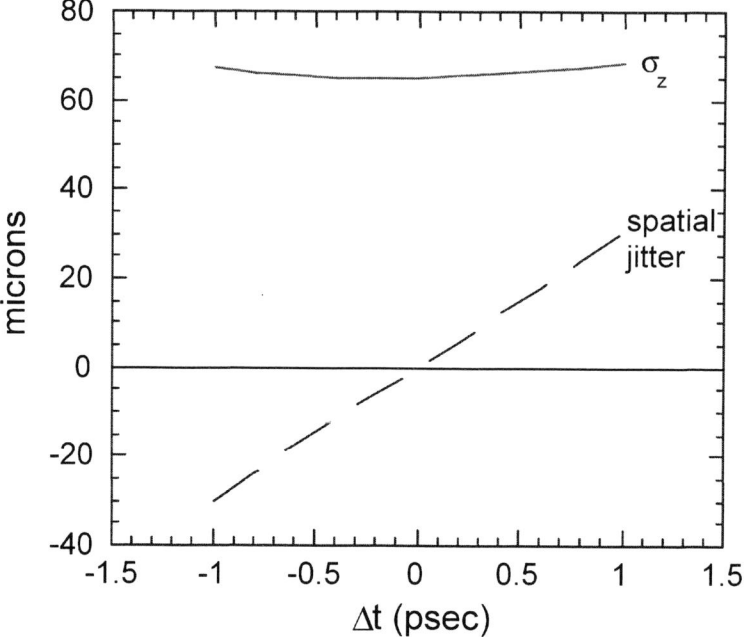

**FIGURE 3.** The equivalent spatial jitter at chicane exit in the UCLA Neptune photoinjector, with optimal compression of a 40 pC electron bunch. Note that an initial timing jitter of 1 psec (300 microns at the speed of light) is reduced to 0.1 psec (30 microns).

There are distinct causes associated with laser-electron timing in a complex high power laser and RF linac system. For propagation over long path lengths, mechanical vibration and thermal changes in the optical transport system will contribute to long time scale phase changes between the laser and electron pulses. These contributions are correctable in the kHz range using electro-mechanical feedback systems. Pulse-to-pulse jitter produced by laser oscillator pump variations as well as other sources of timing error correspond to picosecond range errors between the arrival time of a laser pulse with respect to the phase of the RF derived from the laser repetition signal. Picosecond jitter corresponds to frequencies well outside of the bandwidth delivered to a high $Q$ RF accelerator cavity, so that the RF system can not respond to deviations in the timing of a single laser pulse. An optical analog to the restoring force provided by a magnetic chicaine would therefore represent an ideal solution to the single pulse synchronization between the laser and RF systems.

We propose a new technique which produces a self correcting laser pulse delay, based on a drive signal at the fundamental or an harmonic of the linac RF (6). In short, propagation of laser light polarized along the extraordinary axis of a birefringent electro-optic (EO) crystal immersed in an electric field varying at the linac RF frequency can delay the laser pulse by an amount proportional to arrival phase or time. The fundamental physical process is analogous to a Pockels cell (EO modulator). In a standard Pockels cell, linearly polarized laser light is transmitted at 45 degrees with respect to the extraordinary and ordinary axes of an EO crystal. The linear polarization vector is divided into two vectors aligned along the e-axis and o-axis. An applied voltage changes the index of refraction by different amounts along the two axes. The vector components are delayed by different amounts, so that the net polarization vector of the laser light exiting the crystal is rotated by an amount proportional to the crystal length, applied voltage, and material EO coefficient. If the linearly polarized light is aligned directly along the e-axis, the net effect is to modulate the phase of the light proportional to the same factors that affect the Pockels cell. If the applied voltage comes from an RF signal in a resonant cavity, rather than from a pulsed DC source, the laser pulse will witness an index of refraction proportional to the arrival time of the laser pulse with respect to the RF phase. For a laser pulse with length much less than the period of the RF modulating signal, this modulation approaches a constant index. In analogy to the chicaine bunch compressor, this system provides a restoring force that can correct timing errors between the laser pulse, and master RF phase.

Using the Pockels cell analogy, a half wave Pockels cell typically has a length on the order of a centimeter, and has an applied field on the order of a few kilovolts. At 800 nm, a half wave rotation corresponds to a timing delay of one half of an optical cycle, or approximately 1.3 fs. Applying RF fields on the order of a kilovolt in a resonator cavity seems to be a reasonable requirement. Though the single pass delay time is short in this initial evaluation, compared to the desired correction of a few tenths of a picosecond, multiple passes through such a structure could allow significant timing corrections. A modulated EO crystal used in this new configuration could be integrated into an optical cavity tuned in length to have round trip time corresponding to an RF subharmonic. The basic schematic of the proposed system is shown in figure 4. In this case, the EO crystal is shown integrated into a laser amplifier cavity. The optical cavity length between each end of the crystal and end mirror, as well as the overall optical path length must correspond in this case to a sub-harmonic of the master RF frequency, so that the laser pulse witnesses the proper RF phase at each pass through the crystal. The standing wave in the RF resonator can be considered as two counter-propagating travelling waves. The laser pulse travelling in the crystal must be synchronous with the forward component on the initial pass, and the reverse component during the remainder of the round trip.

There are some immediately recognizable difficulties associated with the proposed technique, as applied to the intended application. While an UV laser pulse for photocathode applications is typically picoseconds in length, an interaction laser pulse for Thomson scattering can be much less than 50 Fs. Many passes through dispersive bulk material is not an appealing prospect for producing laser pulses with duration less than 50 fs, though short incident pulse length assures that nonlinear phase modulation of the optical pulse will be negligible. Also, the microwave loss tangent of the crystal could limit design alternatives at high frequencies. Stability of the cavity length in addition to the previously required oscillator length stability represents an additional difficulty, though in this case there are commercially available cavity length feedback systems which can overcome thermal or vibrational variations. Driving with continuous (CW) RF power could stabilize the RF resonator temperature. Since the RF relative dielectric constant of EO crystal material is high, the crystal will absorb most of the incident microwave power. This is beneficial in the sense that the required drive power is small, though the heating of the crystal will be significant. Also, if one considers the use of harmonics of the master RF to drive the EO crystal, there is an optimization between the fact that lower frequencies require higher drive power to achieve the same correction slope, while the RF loss tangent is worse at high frequencies ($\tan(\delta) \propto \omega_0$).

A stand-alone cavity, not integrated into an amplifier has the advantage of introducing a significant net delay from its many round trips in addition to the small timing adjustment. Since the photocathode laser pulse is derived from the same laser system as the high intensity final-focus interaction laser pulse, additional time delay for the high power laser arm is beneficial. Ideally, one would prefer to derive the photocathode laser and interaction pulses from the same oscillator laser pulse. In this case, if there is no delay introduced in the high power laser amplifier chain with respect to the laser pulse used for photoemission, a significant timing mismatch is present. The linac length must be traversed twice, once to deliver the laser pulse to the cathode, and once for the electrons to accelerate back to the interaction point. For a linac with length of several tens of meters or more, this is a significant delay consideration. On the other hand, losses through a multi-pass timing correction system could represent a significant disadvantage for a stand-alone, multi-pass time correction system.

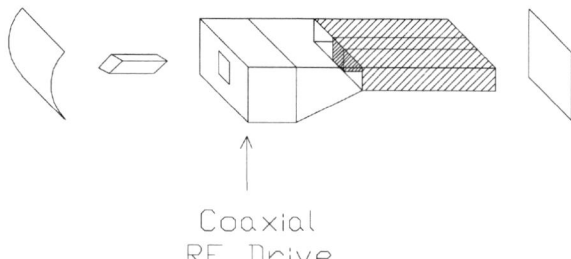

Coaxial
RF Drive

**FIGURE 4.** Microwave timing modulator integrated into laser amplifier cavity

An initial experiment that could be used to test this new concept would use the same EO crystal and RF cavity arrangement and CW laser light with incident polarization rotated to 45 degrees between the extraordinary and ordinary axes, in the more standard polarization rotation arrangement. If placed between two crossed polarizers, the modulation depth of the CW laser light would indicate the single pass delay introduced by the modulated EO crystal. This same concept can be used to diagnose synchronization between the laser pulses and RF by simply monitoring the transmission through crossed polarizers of the short laser pulse. An initial experiment will test these concepts by applying a moderate power microwave signal at 2856 MHz to a Lithium Niobate ($L_iN_bO_3$) crystal and passing a CW beam through the crystal and two crossed polarizers. Modulation depth can then be examined as a proof of the principle.

The RF electric field modulation could be accomplished using a TM cavity, driving a longitudinal field in a z-cut EO crystal such as KDP, or a TE resonator driving a transverse field in Lithium Niobate. A ridged waveguide is desirable for this application since the electric field is concentrated in the central, cutoff region of the waveguide. Because of the decreased vertical dimension, the field profile will also be more linear in the transverse direction, avoiding a transverse lensing effect. Vertical thermal lensing is still a potential problem, though at our higher operating frequency of 2856 MHz, the dimensions of the RF cavity will be smaller, allowing more uniform distribution of heat in the crystal. Forced gas or low dielectric constant liquid cooling of the crystal in the cavity could be considered. Techniques of correcting thermal lensing utilized by optical slab amplifier designers may also be applicable in some modified form (7).

A similar arrangement has been developed for the purposes of time lensing (8) and pulse length adjustment (9) by using the crystal for quadratic temporal phase modulation rather than bulk phase shift. For these applications, the optical pulse length fills a significant portion of the RF "bucket" in the synchronous travelling wave inside of the crystal, inducing a parabolic phase modulation over the pulse. For our short pulse application, the optical pulse will witness a narrow, linear region of the RF drive. We in fact want to avoid phase modulation to the largest possible extent in order to preserve the short pulse characteristics and central frequency. The referenced time lens experiment used a 40 mm Lithium Niobate crystal, in a ridged waveguide cavity, driven with approximately 30 Watts of 1.76 GHz RF power in a $TE_{101}$ mode. The achieved modulation corresponded to approximately one radian per root Watt at an optical wavelength of 632.8 nm. In terms of our application, this result would correspond to approximately 0.34 fs per pass. The time lens experiment also achieved 14 passes (7 round trips) in an optical cavity configuration. Since we require jitter correction on the order of tenths of picoseconds, we will need a longer electro-optic crystal, corresponding to a higher order longitudinal RF cavity mode ($TE_{104}$ for instance), higher microwave power level, and more passes through the crystal. There are also new materials that have demonstrated EO coefficients, which are orders of magnitude higher than Lithium Niobate (10), which warrant examination for our application. Our higher operating frequency has the advantage that the EO coefficient increases linearly proportional to frequency, assuming a constant drive power level ( $\Delta\tau \propto E_0\omega_0$ ). On the other hand, our Ti: Sapphire laser system operates at an optical wavelength of 820 nm, and the EO coefficient decreases linearly with optical wavelength. Lithium Niobate has an RF loss tangent of 0.05 and an RF relative dielectric constant of $\varepsilon_r$ of 28 at 2 GHz. For 1.06 micron light (YAG), the index of refraction along the extraordinary axis with no applied electric field is 2.15. The EO coefficient ($R_{33}$) is 30.8 picometers per Volt at 632 nm. The phase velocity of the RF guiding structure must be matched to the group velocity for our optical wavelength.

The design of this microwave timing modulator system is now being investigated in more detail. In addition to research into appropriate optical cavity geometries and EO materials, we are using three-dimensional electromagnetic simulations to aid in the design of the microwave cavity itself. Demonstration experiments at LLNL should follow shortly. In addition, we plan to study the issues associated with handling of higher power picosecond laser sources in this device, such as are found in laser-scattering and laser wake-field accelerator experiments(11).

## REFERENCES

1. Clayton, C., et al., "Second Generation Beat-wave Experiments at UCLA", Nucl. Instr. Methods A **410**, 378 (1998).
2. Schoenlein, R. W., et al., "Femtosecond X-Ray Pulses At 0.4 Angstrom Generated By 90-Degrees Thomson Scattering: a Tool For Probing the Structural Dynamics Of Materials," Science 274, 236 (1996).
3. Ride, S. K., E. Esarey, and M. Baine, "Thomson Scattering Of Intense Lasers From Electron-Beams At Arbitrary Interaction Angles," Physical Review E 52, 5425 (1995).
4. Rosenzweig, J.B., N. Barov and E. Colby, "Pulse Compression of RF Photoinjector Beams: Advanced Accelerator Applications", IEEE Trans. Plasma Sci. **24**, 409 (1996).
5. Rosenzweig, J.B., et al.. "The Neptune Photoinjector", Nucl. Instr. Methods A A **410,** 437 (1998).
6. Le Sage, Greg, private communication. Publication is forthcoming.
7. Eggleston, J. M., T. J. Kane, K. Huhn, J. Unternahrer, and R. L. Byer, "The Slab Geometry Laser — Part 1: Theory", IEEE Journal of Quantum Elect. 20, 289 (1984); T. J. Kane, J. M. Eggleston, and R. L. Byer, "The Slab Geometry Laser — Part II: Thermal Effects in a Finite Slab", IEEE Journal of Quantum Electronics 21, 1195 (1985).
8. Scott, Ryan, "Design and application of resonant electro-optic time lenses," Master's Thesis, UCLA, 1995.
9. Scott, R.P., C.V. Bennett, B.H. Kolner, "Picosecond laser source with single knob adjustable pulse width,"Lasers for RF guns, Anaheim, CA, US DOE, Brookhaven National Laboratory, UCLA, May 14-15, 1994.
10. Knopfle, G.; Schlesser, R.; Ducret, R.; Gunter, P. "Optical and nonlinear optical properties of 4'-dimethylamino-N-methyl-4-stilbazolium tosylate (DAST) crystals," ICONO'1. International Conference on Organic Nonlinear Optics, Val Thorens, France, 9-13 Jan. 1994, Nonlinear Optics, Principles, Materials, Phenomena and Devices, 1995, vol.9, (no.1-4):143-9.
11. Amiranoff, F., et al, "Observation of Laser Wakefield Acceleration of Electrons", Physical Review Letters **81**, 995 (1998).

# Electron Bunchlength Measurement
# from Analysis of Fluctuations
# in Spontaneous Emission

P. Catravas, W.P. Leemans, J.S. Wurtele and M.S. Zolotorev

*Center for Beam Physics, Lawrence Berkeley National Laboratory, Berkeley, CA 94720*

M. Babzien, I. Ben-Zvi, Z. Segalov, X. Wang and V. Yakimenko

*Accelerator Test Facility, Brookhaven National Laboratory, Upton, NY 11973*

**Abstract.** A statistical analysis of fluctuations in the spontaneous emission of a single bunch of electrons is shown to provide a new bunchlength diagnostic. This concept, originally proposed by Zolotorev and Stupakov [1], is based on the fact that shot noise from a finite bunch has a correlation length defined by the bunchlength, and therefore has a spiky spectrum. Single shot spectra of wiggler spontaneous emission have been measured at 632 nm from 44 MeV single electron bunches of 1-5 ps. The scaling of the spectral fluctuations with frequency resolution and the scaling of the spectral intensity distribution with bunchlength are studied. Bunchlength was extracted in a single shot measurement. Agreement was obtained between the experiment and a theoretical model, and with independent time integrated measurements.

## INTRODUCTION

The development of new bunchlength measurement techniques is prompted by the need to diagnose ultrashort electron beams such as those being produced by RF photocathode gun technology [2] as well as potentially by laser-based acceleration [3,4]. Recent developments in bunch monitoring are reported in references 5 through 9.

In 1996, Zolotorev and Stupakov proposed [1] that macroscopic properties of the electron bunch can be obtained by measuring fluctuations in the incoherent shot noise emissions from electrons. For a radiation pulse to be longitudinally incoherent, the spectral bandwidth $\Delta\omega$ must be much larger than the inverse of the pulse duration $\tau_b$, i.e. $\Delta\omega\tau_b \gg 1$. Using a bandpass filter, centered around $\omega_0$ and with spectral width $\delta\omega$, temporal coherence can be imposed with an associated coherence time $\tau_{coh} \propto \delta\omega^{-1}$, effectively breaking the pulse up in $N$ independent portions where $N = \tau_b / \tau_{coh}$. From shot-to-shot, the intensity will vary on the order of $1/\sqrt{N}$. Measurement of the variance of the intensity fluctuations will then give a measure for $N$ and hence $\tau_b \sim N / \delta\omega$.

CP472, *Advanced Accelerator Concepts: Eighth Workshop,*
edited by W. Lawson, C. Bellamy, and D. Brosius

This concept offers a number of desirable features. It can be implemented as an inherently single shot, non-destructive method. It is applicable to a variety of devices that are commonly found in beamlines and provide incoherent emissions from the electron beam. Furthermore, as will be shown, the technique favors shorter bunches, and hence it is a potential candidate for bunches well below 1 ps.

A proof of principle experiment is reported in which the fluctuation characteristics of wiggler spontaneous emission are studied and estimates of bunchlength are obtained from the measurements. Observed scalings, both qualitative and quantitative, are well explained by the theory. The measurements were implemented with a novel high precision microwiggler, which provides high brightness emissions in the visible wavelengths, where optical diagnostics are readily available. A variety of single shot diagnostic techniques has been developed with this microwiggler [10].

## EXPERIMENTAL SETUP

The experimental setup is presented in Figure 1. At the Accelerator Test Facility at BNL [11], an RF photocathode gun produced a variable 1-5 ps electron bunch, with corresponding charge of 100-400 pC. The bunchlength was controlled by changing the amount of compression in the RF gun. Compression depends on the relative timing of the laser producing the photoelectrons and the gun RF phase. An S-

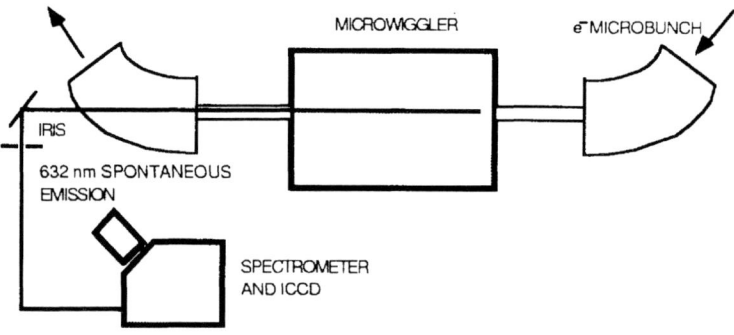

**FIGURE 1.** Experimental Setup.

band linac accelerated the electrons emitted by the gun to a mean energy of 44 MeV. Nominal beam parameters were a normalized emittance of few $\pi$ mm.mrad and an energy spread of 0.7% full width. Bunchlength was independently measured with a time integrated technique in the front end of the accelerator [12].

The electron beam was transported to the MIT microwiggler, a pulsed electromagnet with 60 periods of 8.8 mm and an on-axis peak field strength of 0.42 T [13]. Wiggler emissions at $\lambda_0 = 632$ nm were apertured and imaged onto the entrance slit of a spectrometer (model Spex 270M, with focal length 0.27 m and 1200 groove per mm visible grating). At the output, the spectrum was imaged onto a Gen IV image intensifier with 35-40% quantum efficiency in the visible. The instrumental resolution, $\delta\omega_c / \omega_c = \delta\lambda_c / \lambda_c$, was determined by measuring the spectrum of a HeNe laser. It was intensifier limited to a width (one-$\sigma$) of 0.07 nm, or 1.3 times the wavelength width of one pixel, in the images acquired by the frame grabber. Critical to this experiment was indeed to have both the required spectral resolution, roughly the inverse of bunchlength, and sufficient emission intensity in a single shot.

## RESULTS AND ANALYSIS

A representative single shot spectrum of a single electron bunch is presented in Figure 2a. As expected, there are spikes in the spectrum, with random amplitude and frequency within a bandwidth characteristic of the linewidth of the wiggler (~10 nm).

According to the theory [1], random spikes of width $\sim 1/\tau_b$ and 100% fluctuation in intensity are expected in the spectrum of a transversely coherent portion of the emissions. If the instrumental resolution, $\delta\omega_c$, is sufficient to resolve the width of the spike and there are enough spikes in the spectrum to give reliable statistics, $\tau_b$ may be estimated immediately from this width. Alternately, a bunchlength may be determined by computing the variance, $\sigma_m^2$, of the normalized intensity averaged over $m$ pixels. When the associated frequency binsize, $\delta\omega_m$, is large compared with $1/\tau_b$ and $\delta\omega_c$, (the instrumental resolution), the measured intensity in different bins becomes independent and bunchlength can be determined from a plot of:

$$1/\tau = (\sigma_m^2)(\delta\omega_m)/F \tag{1}$$

as a function of $m$. Here, $F$ is a form factor that is computed from the Fourier transform of the bunch temporal waveform. For example, the form factor for square and Gaussian bunches differ by a factor of $\sqrt{2}$, where $\tau_b$ refers to the FWHM. When the spectral bins are uncorrelated, $\sigma_m^2$ is inversely proportional to $m$. In the time domain picture, this is equivalent to the presence of $N$ independent slices with coherence time $\tau_{coh} \sim 1/\delta\omega_m$ over the pulse length, where $N$ is the ratio of $\tau_b$ to $\tau_{coh}$. It is evident that the relative fluctuations in intensity will be $\sigma_m = 1/\sqrt{N}$ and this shows that $\sigma_m^2$ scales as $1/m$. Since $\delta\omega_m$ is proportional to $m$, the quantity $1/\tau$ then becomes independent of bin size and equal to $1/\tau_b$.

The quantity $\tau$ from Eq. 1 for the data shown in Figure 2a is plotted in Figure 2b. Three distinct ranges with characteristic behavior are seen. In Region I, for small $m$, the parameter $\tau$ increases inversely with decreasing combined bin size.

805

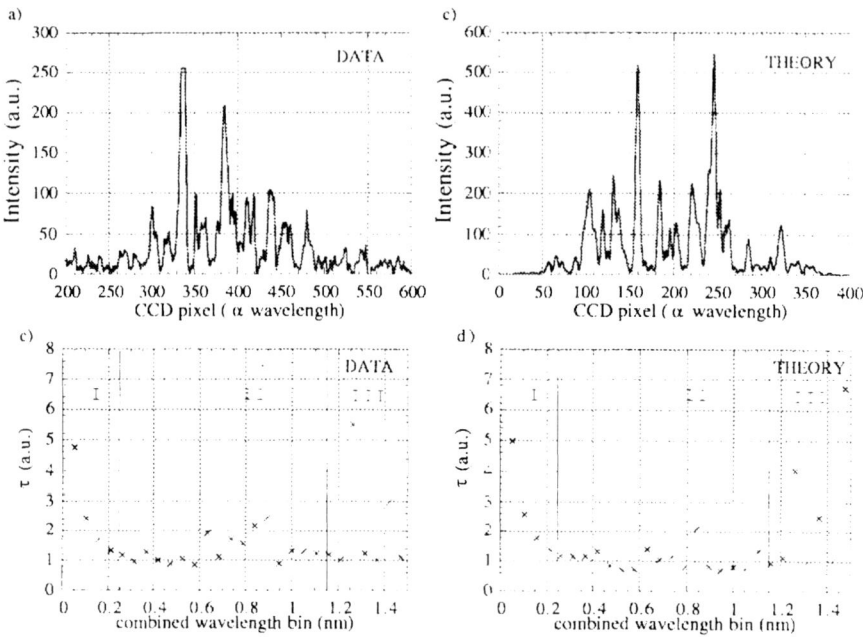

**FIGURE 2.** Typical single shot spectrum of wiggler spontaneous emission from a single electron bunch of 1-2 ps. Resolution was sufficient to record a high degree of modulation on the spectrum. A plot of the inverse of Eq. 1 vs. spectral bin size ($\alpha$ m) is shown for this spectrum in Figure 2b, from which a bunchlength of 2 ps was estimated. A simulated spectrum for a 2 ps bunch was calculated, including instrumental resolution, (Figure 2c), and is characterized by the same typical spike width and average number of spikes over the full spectrum  For comparison with Figure 2b, a similar calculation was performed on the simulated spectrum (Figure 2d). The computations show that the theoretical model well describes the data.

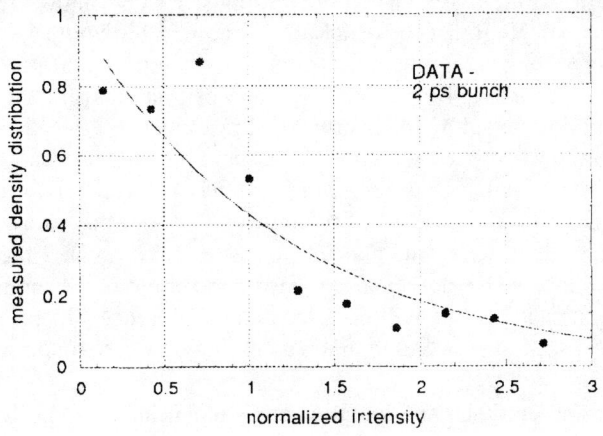

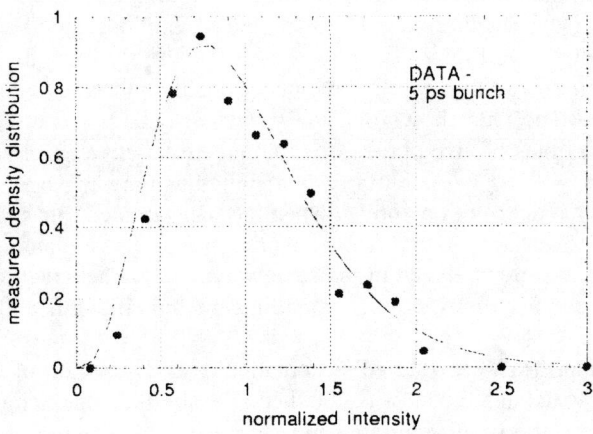

**FIGURE 3.** Comparison of measured density distributions for shorter and longer bunches. The upper data was fit with the exponential distribution, the lower with the gamma distribution. The good fits support the validity of the beam shot noise model.

In Region II, $\tau$ is independent of $m$. In Region III, $\tau$ has an increasingly large error bar. This is a vestige of the limited bandwidth of the wiggler spectral envelope, for at some point, combined bins become large compared with the width $1/N_w$ and there are few points contributing to the rms. Regions I and II extend over a useful range.

The transition from Region I to II determines the correlation length of the spectral fluctuations. In Region I, in which the combined bin width is smaller than width related to inverse bunchlength and instrumental resolution, the spectral intensity in adjacent bins is correlated. Bunchlength can be estimated from the correlation length, since it is not dominated by the instrumental resolution for this data. In Region II, combined bins are uncorrelated with each other, and bunchlength can be estimated from Equation 1 if beam emittance is negligible. We have included a correction factor for emittance, discussed in more detail in reference 14. From the information of the correlation length and the level of the plateau, we estimate $\tau_b$ to be 2 ps, assuming a Gaussian temporal shape. The statistical error for an average of 30 shots is found to be about $\pm 10\%$. Systematic errors will be discussed in reference 14. By comparison, bunchlengths of 1.5 ps $\pm$ 0.5 ps were obtained using the time integrated measurement technique [12].

For comparison with the experimental data, simulations were performed of the wiggler emission spectrum assuming a Gaussian temporal profile, and including the calibrated effect of instrumental broadening. The simulations reproduce the spike width and level of modulation seen in the experiment (Figures 2c and d).

The spectral intensity distribution was studied for both measured and simulated spectra. Results (plotted in Figure 3), show that at the shortest bunchlength, data is nearly exponential in distribution, while the longer bunchlengths are well described by the Gamma distribution. This shows that the spectrum behaves as expected for an electron bunch with times of arrival of individual electrons governed by the Poisson process: intensities will be exponentially distributed when the spectrum is fully resolved. The Gamma distribution applies when the measurement bin integrates over multiple spikes (i.e. over more than one exponential process). Simulated spectra such as shown in Figure 2b were analyzed in the same way, and are also well described by an exponential or gamma distribution, depending on bunchlength and instrument resolution.

Single shot spectra were studied systematically as a function of bunchlength for a range of 1-5 ps, and agreement was obtained between the bunchlength estimated from fluctuations and the time integrated measurement [12] for the same machine setting. These results will be discussed further in reference 14.

It is important to note that the implementation of this technique for bunchlength diagnosis as described here is sensitive to transverse coherence and requires attention to the electron beam emittance and collection optics. Resolution and detector dynamic range must be chosen to provide good signal to noise over other sources of fluctuations, such as quantum fluctuations and intensifier/CCD noise, and resolution and wavelength should be optimized according to bunchlength. Comparison with the time integrated measurements was done in assumption of isochronicity over electron beam transport.

# CONCLUSIONS

A proof of principle experiment has been performed at the ATF demonstrating electron beam bunchlength diagnosis from fluctuations in the incoherent emissions from electrons. Bunchlength information is contained in the emission spectrum through random spikes of width $1/\tau_b$, coming from shot noise on the electron beam. In principle, the technique may be implemented with a variety of sources of incoherent emission and applied to bunches created by different accelerating structures at various energies, provided that coherent enhancement and quantum effects are avoided.

The use of a long (60 period) microwiggler provided high brightness emissions over the usable spectral bandwidth and the wiggler's short period (8.8 mm) set the emissions in the visible wavelengths at 44 McV. This enabled bunches of picosecond lengths with as low as 100 pC charge to be diagnosed. Single shot spectra were recorded for bunchlengths over a range of 1-5 ps produced by single bunches of electrons from an RF photocathode gun. Nearly 100% modulation of the spectrum was observed for the shortest bunchlengths. A study of the dependence of the fluctuations on resolution differentiated correlated and uncorrelated portions of the spectrum, with scaling consistent with the theoretical model. Using this scaling, bunchlength was extracted in a single shot measurement and agreed with a time integrated measurement of bunchlength for the same machine setting. The spectral intensity distribution was investigated for two bunchlengths, and was found to be well described by exponential/gamma distributions consistent with bunchlength and resolution. The analysis indicates agreement between the experiment and theoretical model and is a practical demonstration of a new bunchlength diagnostic technique.

# ACKNOWLEDGMENTS

We acknowledge useful discussions with Gennady Stupakov, Andy Charman, Sasha Zholents, and Paul Volfbeyn. This work was supported by a Lawrence Postdoctoral Fellowship at LBNL, by the Office of Naval Research Grant No. N00014-90-J-4130, and by the US Department of Energy under contract No. AC03-76SF00098.

# REFERENCES

[1] M. S. Zolotorev and G. V. Stupakov, SLAC-PUB-7132, (March, 1996); *Ibid*, Proceedings of the 1997 Particle Accelerator Conference.
[2] P. Kung, H. Lihn and H. Wiedemann, *Phys. Rev. Lett.* **73**, 967 (1994).
[3] D. Umstadter *et al.*, *Phys. Rev. Lett.* **76**, 2073 (1996).
[4] E. Esarey et al., *Phys. Rev. Lett.* **79**, 2682 (1997).

[5] W.P. Leemans, submitted to the *LINAC98 Proceedings*, WE1001.

[6] G. A. Krafft, submitted to *DIPAC '97 Proceedings*.

[7] M. Ding, H. H. Weits, and D. Oepts, *Nucl. Instr. & Meth. A*, **393**, 504-9 (1997).

[8] E.R Crosson *et. al.*, *Nucl. Instr. & Meth. A*, **358**, 216-19 (1995).

[9] K.N. Ricci *et. al.*, these Proceedings.

[10] P. Catravas, Ph.D. thesis, MIT (1998).

[11] see http://www.nsls.bnl.gov/AccTest/Menu.html

[12] X.J. Wang, X. Qiu and I. Ben-Zvi, *Phys. Rev. E* **54** No. 4 R3121 (1996)

[13] R. Stoner and G. Bekefi, *IEEE J. of Quant. Electron*, **QE-31** (1995)

[14] P. Catravas *et al.*, *Phys. Rev. Lett.*, to be submitted.

# Working Group 5:

# Particle Beam Sources

# Advances in DC Photocathode Electron Guns

B. M. Dunham, P. Hartmann, R. Kazimi, H. Liu, B.M. Poelker, J.S. Price, P.M. Rutt, W.J. Schneider, and C.K. Sinclair

*Thomas Jefferson National Accelerator Facility*
*12000 Jefferson Ave, Newport News, VA 23606*

**Abstract**. At Jefferson Lab, a DC photoemission gun using GaAs and GaAs-like cathodes provides a source of polarized electrons for the main accelerator. The gun is required to produce high average current with long operational lifetimes and high system throughput. Recent work has shown that careful control of the parameters affecting cathode lifetime lead to dramatic improvements in source operation. These conditions include vacuum and the related effect of ion backbombardment, and precise control of all of the electrons emitted from the cathode. In this paper, we will review recent results and discuss implications for future photocathode guns.

## INTRODUCTION

At Jefferson Lab, a DC photoemission gun using GaAs and GaAs-like cathodes provides polarized electrons for the main accelerator. DC guns hold a distinct advantage for systems that require high average current as opposed to those requiring high peak current with a low duty factor. The polarized source runs at 100 kV and can provide multiple, independent beams to three experimental halls with a total average current of up to 200 $\mu$A and can deliver as much as 16 C of electrons per day.

The ability to deliver these high average currents for extended periods of time has a large impact on machine availability at Jefferson Lab. On the main accelerator, 70-80% availability is necessary to meet the full experimental schedule of the nuclear physics community so little time can be spent on cathode maintenance. Progress has been made in the last year in understanding and controlling the conditions that affect cathode lifetime: vacuum, crystal damage from ion backbombardment, and precise control of halo electrons. The polarized source now routinely reaches operational 1/e lifetimes of 100-200 hours for ~150 $\mu$A average beam currents and upgrades are being planned to increase this to over 500 hours.

In this paper, recent results from the polarized source will be presented showing details leading to dramatic improvements in the operational lifetime of the source. Properties related to the performance of future photocathode guns will be discussed.

CP472, *Advanced Accelerator Concepts: Eighth Workshop*,
edited by W. Lawson, C. Bellamy, and D. Brosius

# LIFETIME ISSUES

Photoemission electron guns utilizing GaAs (or GaAs-like) cathodes have been in operation for over 20 years as a source of polarized electrons [1]. They are notoriously difficult to operate and maintain due to the ultra-high vacuum environment necessary for activating the cathode to obtain a negative electron affinity (NEA) surface. Notably the group at SLAC [2] has made significant progress in recent years with increasing the polarization, lifetime, and operability of these sources. Their introduction of a load-lock chamber has eliminated many of the problems associated with wafer changes such as repeated high temperature bakeouts and high voltage processing. Unfortunately, many of the lessons learned there for producing high peak current, low duty cycle beams do not transfer directly to labs like Mainz, MIT-Bates and Jefferson Lab; these labs must produce high average current beams.

Lifetime is a measure of the time it takes for the QE to degrade to 1/e of its initial value. A more useful measure for labs delivering high average current is the charge delivered before the QE degrades by 1/e. Typical numbers for the Jefferson Lab polarized source are 100-200 hours for a 100 µA beam, or 36-72 C. The dark lifetime is the performance of the cathode when it is not exposed to illumination. A long dark lifetime measures the quality of the vacuum and surface chemistry is the primary mechanism of degradation. The dark lifetime with high voltage applied may be worse due to field emission from high field points. Using diamond-paste polished titanium electrodes, the maximum field emission current is less than 1 nA at 100 kV, substantially lower than reported elsewhere [2]. The overall lifetime during beam delivery is a complicated function of the vacuum conditions in the gun and the electron beam optics.

One particular physical process that limits operating lifetime at high average currents was recently recognized at Mainz [3]. They found that the QE of the cathode decayed in an unusual pattern that could only be attributed to ion damage. The process is illustrated in figure 1. An electron orbit from off-axis on the GaAs wafer is shown being accelerated towards the anode. Along its path, an electron has some probability of ionizing the residual gas in the chamber. Subsequently these particles

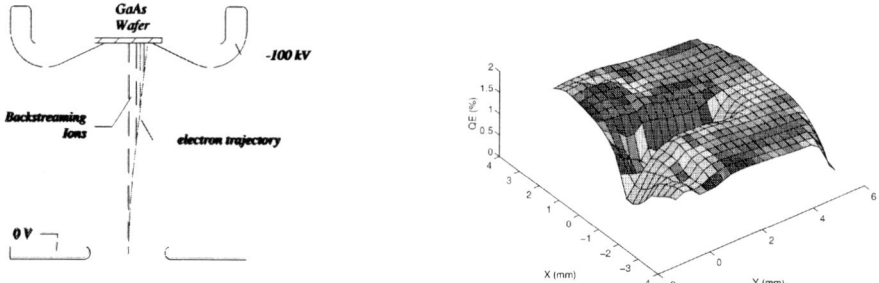

**Figure 1.** Model for ion backbombardment and a quantum efficiency scan over the wafer showing the resulting damage

are accelerated straight back (due to their much larger mass) towards the wafer where they cause crystal damage. The plot in the right side of figure 1 shows the QE profile over part of a wafer (from this lab) from which over 20 C was extracted. The laser beam was located at x=0 mm and y=2.3 mm and the QE trough from there to the center of the wafer illustrates the effect of the crystal damage.

## SIMPLE MODEL FOR ION DAMAGE

The ionization rate as a function of kinetic energy, R(E), can be calculated assuming that the residual gas in the chamber is molecular hydrogen at some pressure P as

$$R(E) = I\sigma(E)\rho\Delta z = I\sigma\rho\frac{\Delta z}{\Delta E}\Delta E$$

$$\rho(m^{-3}) = 3.54 \times 10^{22} P(Torr)$$

where $\sigma$ is the ionization cross section for $e^-$ on $H_2$ (see figure 2), I is the beam current, $\rho$ is the $H_2$ gas density, and $\Delta E/\Delta z$ is the average accelerating gradient in the anode-cathode region. From this one can calculate the total integrated ionization rate as a function of pressure or the total rate as a function of E for a given pressure (see figure 3).

For the present geometry, the anode and cathode are separated by about 60 mm with an accelerating gradient of 1 kV/mm at the wafer. From the plot of integrated rate versus energy, 50% of the ions are generated between 0 and 8 kV, or within about 8 mm of the wafer surface. These ions will be accelerated straight back and implanted into the GaAs while the higher energy ions will be distributed between the location of

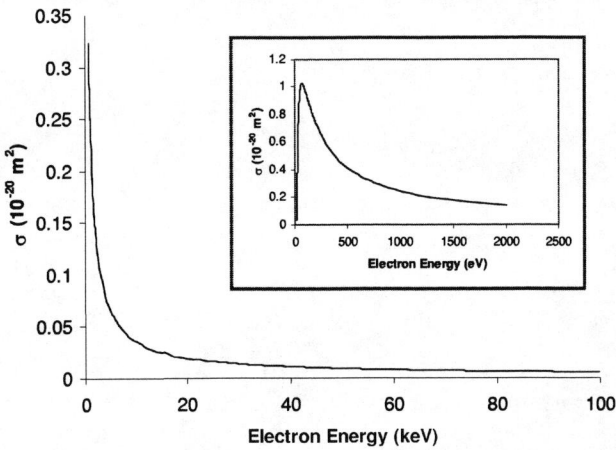

**Figure 2.** Cross-section for $H_2$ ionization by electrons [4].

the laser spot and the center of the wafer. Thus the only realistic way to reduce the number of ions (and the damage) is to reduce the pressure, since introducing any sort of fields so close to the cathode to deflect the ions away from the surface would be impractical and undesirable.

The next question is what is the damage mechanism for reducing the quantum efficiency? One possibility is that the cesium and fluorine atoms deposited on the surface to produce the NEA conditions are sputtered away by the ions. In the limited operation history of the polarized gun, adding more cesium improves the QE in the damaged region and the wafer as a whole, but it never fully recovers. This implies that a substantial portion of the damage occurs below the surface. Calculation of sputter yields for protons on GaAs using the program SRIM [5] indicate at most 1 or 2 sputtered atoms per 100 incident protons are produced, which may account for some of the photoresponse reduction.

A more likely candidate is physical damage to the bulk crystal from the implanted ions. Fortunately, since implantation of $H^+$ and $H_2^+$ in GaAs is a commonly used technique to induce damage to alter its electrical and optical properties, a rich literature exists. One important parameter is the range of the ion in the material: the range is the average distance the ion will travel before it stops. Atomic displacements near the end of the ion range give rise to point defects that absorb light and trap carriers [6]. For protons implanted in GaAs, the range varies roughly linearly from 0 (at 0 eV) to 0.8 µm (at 100 keV) with a spread of 10 to 20% in the average range (calculated using SRIM). These numbers have been experimentally verified [7] and show that most of the damage occurs when the ion stops, so a 100 keV proton will do most of its damage $0.8 \pm 0.1$ µm deep in the crystal. Using the SRIM program, other quantities of interest can be calculated and are summarized in table 1 along with other known information.

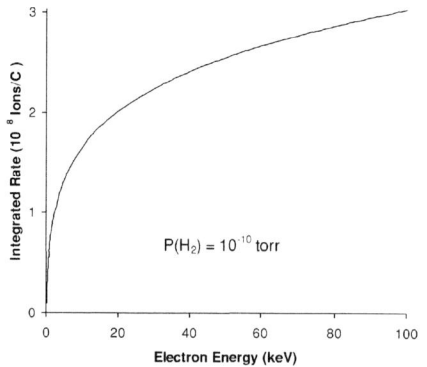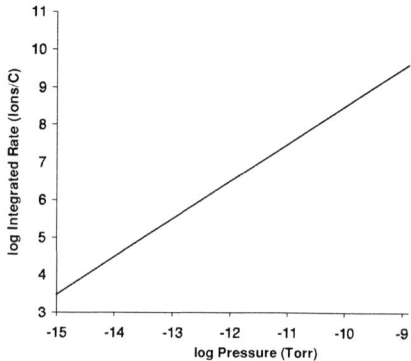

**Figure 3.** The integrated ion production rate per Coulomb of incident electrons as a function of energy for a given pressure and total rate for all energies versus pressure.

**TABLE 1. Summary of ion backbombardment properties**

| Property | Result | Ref |
|---|---|---|
| Damage produces point defects - vacancies and deep trapping levels | Reduced carrier concentration and light absorption | [6] |
| More vacancies produced as E increases | Higher energy does more damage | SRIM |
| Damage occurs at $R(E)\pm\sigma_R(E)$ | Higher energy produces deeper damage | [7] |
| Sputter yield is low | Some NEA surface modification | SRIM |
| Damage for E > 500 eV does not disappear with annealing | QE will not completely recover with annealing | [8] |
| $H_2$ dissociates on impact to yield 2 protons each with 1/2 the energy | Damage occurs closer to the surface | [7] |
| Up to 20% of the ions are backscattered out of the crystal for energies below 10 keV | Reduced number of damaging ions for low energy | SRIM |
| 50% of the ion dose occurs within 8 mm of the surface | Low energy damage occurs at the laser spot | Calc |

**TABLE 2. Experimental observations related to ion damage**

- QE degradation at the laser spot
- QE degradation along a trough towards the wafer center
- Visible wafer damage indicates a change in index of refraction
- QE away from the laser spot degrades little or not at all
- Addition of Cs improves QE over the whole wafer, but recovery is not complete
- Annealing removes most of the damage. Dimples in the QE are still visible at the previously damaged locations

These observations can be used to explain the effects seen in figure 4. A cross section through the quantum efficiency 'holes' in the cathode is shown in figure 4(d). The lines plotted on the data are inverted Gaussian distributions with a width equal to the laser spot size convoluted with the probe laser size for the QE measurements. This demonstrates that the low energy spatial ion distribution matches the initial laser size within a few mm of the wafer surface where half of the ions are produced.

The shape of the QE trough (see figure 4(c)) is more difficult to explain. Using results from PARMELA and the ionization cross section, the energy distribution of the resulting ions along the electron path can be mapped onto the surface between the location of the laser spot and the electrostatic center of the wafer. To determine the relative amount of damage along the wafer, several factors contributing to the quantum yield need to be considered. The rate calculations predict that most of the ions are produced at low energy, but the flat QE trough indicates that the total damage as a function of energy is roughly constant. By folding together the properties mentioned in Table 1 along with the fact that the absorption depth for light in GaAs is ~1 μm, a qualitative fit to the QE trough can be made (solid line in figure 4(c)). Further calculations are underway to understand the damage mechanisms in a quantitative fashion.

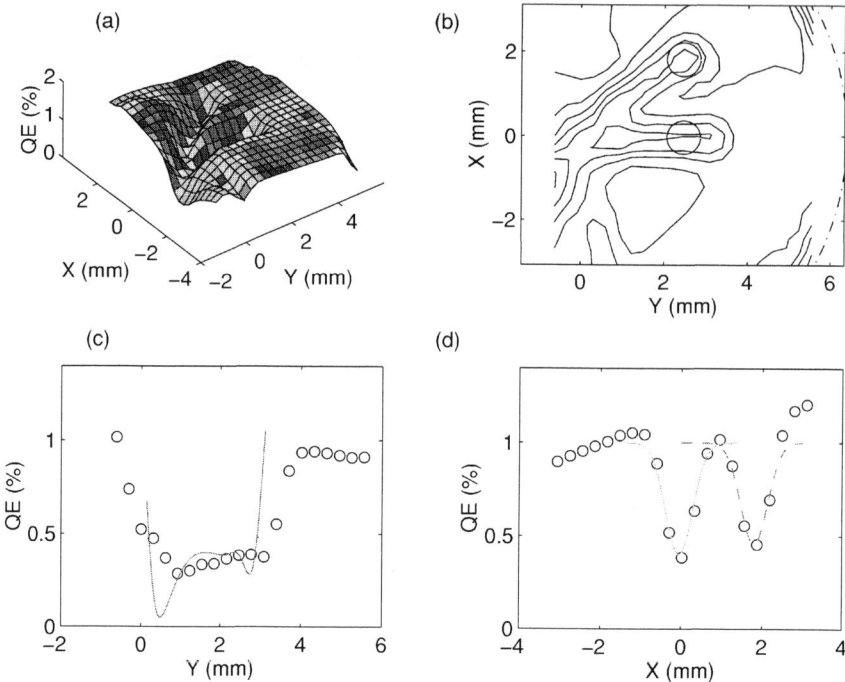

**Figure 4** Quantum efficiency scan data after extended running from two spots (a). The contour plot (b) shows the location of the two laser spot positions (small circles) and the edge of the wafer (dashed line). The bottom plots show cross sections through the trough (c) for X=0 mm and through the two spots for Y=2.3 mm (d).

From the above observations, it is clear that if ion damage is the rate limiting effect for cathode lifetime, the most straightforward way to increase the lifetime is to lower the pressure in the anode-cathode gap where the ion production occurs. There are several avenues one can take to achieve very low pressures: careful attention to ultrahigh vacuum design; careful material choices; and increased pumping speed. For our present gun, we have taken the brute force route of increasing the pumping speed as much as possible. To do this, the anode-cathode region was surrounded by ten getter pump strips (S.A.E.S Getters, model WP590 formed with ST707 alloy) with a total pumping speed of approximately 4000 L/s of hydrogen. Since the sticking probability for getters is 10-20%, the geometric arrangement of the pumps is critical. The getters are arranged symmetrically around the anode-cathode area with a mesh screen to shield the sharp edges from the high voltage (see figure 5).

There is presently no reliable way to measure the pressure in the region of interest, but an extractor gauge located outside of the getter array reads in the low $10^{-12}$ Torr range. The quality of vacuum can be ascertained by studying the QE scans in figures 1 and 4. A millimeter from the edge of the laser, the quantum efficiency does not noticeably drop relative to its initial value after drawing many tens of Coulombs from the main spot. In fact, we are in the unique position of being able to study bulk

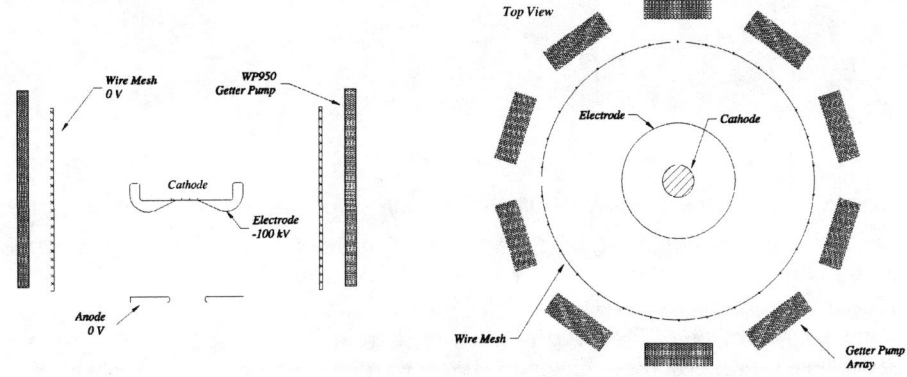

**Figure 5.** The getter pump arrangement in the anode-cathode region.

semiconductor damage mechanisms and damage rates. If the residual gas density and composition near the cathode were known, the amount of damage as a function of energy and number could be determined. Conversely, if the damage mechanisms were quantified, the pressure near the cathode could be measured.

# CONTROL OF HALO ELECTRONS

Another important issue for obtaining long lifetimes is the precise control of all electrons that leave the cathode. Any electron that can strike a surface in the gun or beamline can lead to an increased vacuum load in the anode-cathode region due to electron stimulated desorption. The problem is particularly bad here.

Many guns utilize the entire active cathode area for electron emission [2]. In the Jefferson Lab gun the laser beam size ($\sigma$ = 0.2 mm) is much smaller than the total cathode size (12.7 mm diameter) to provide the small emittance needed for the accelerator. Since the GaAs wafer has a high QE (1 to 10 %) over its entire area any stray light can cause unintended electron emission. The first clues to this problem were an increase in the gun pressure while extracting 100 µA beam and excessive radiation measured by Geiger tubes along the beamline (50 to 100 mR/hr). Simulations show that this was due to electrons emitted near the edge of the wafer.

The known sources of stray light in the gun chamber were investigated and eliminated without any effect on the stray electrons. Measurements of the remaining light (due to laser halo and recombination radiation) near the wafer but outside of the main spot were only down by a factor of 100. Thus for a main beam current of 100 µA there was as much as 1 µA of beam halo available to hit the walls.

Since the entire cathode area is not normally used, it was decided to mask the outer area of the wafer so that it could not be activated to high QE. At our lab, the wafers

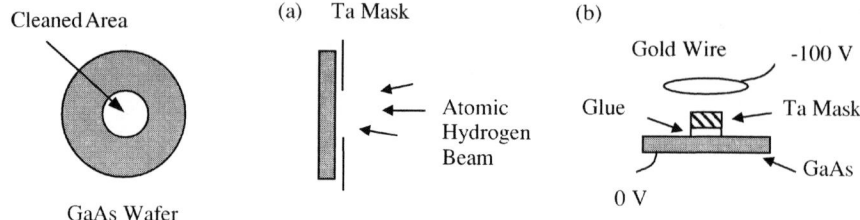

Cleaned Area

(a) Ta Mask

(b)

GaAs Wafer

Atomic
Hydrogen
Beam

Gold Wire    -100 V

Glue    Ta Mask

GaAs

0 V

**Figure 6.** Selective area cleaning using atomic hydrogen (a) and anodic oxidation (b).

are cleaned using an atomic hydrogen beam [9], so masking the outer edges of the wafer from the atomic hydrogen cleaning was tried (see figure 6).

Measurement of the cathode lifetime with the selectively cleaned wafer showed dramatic improvements. The lifetime improved from 30 minutes to 20 hours, the radiation levels dropped by a factor of 100, and the pressure in the gun chamber showed no noticeable increase. This provided direct evidence that stray electrons were one of the main causes of lifetime problems. Operationally, this method of masking is not reliable for long term accelerator operations.

To reduce the QE outside the central region to tolerable levels (down by 5 to 6 orders of magnitude), the outer region is anodized to produce a thick oxide layer. The wafer is prepared by gluing a small tantalum mask at the center and placing it in a weak phosphoric acid solution (2.8 pH). A low voltage (100 -150 V) is applied between the wafer and a loop of gold wire for several minutes; this results in a thick (0.1 - 0.2 μm) oxide layer on the exposed region (see figure 6 and [10]). The glue is easily dissolved in acetone and does not damage the masked area. The oxide layer has good stability over time and has shown no effective degradation that would cause increased photoemission.

Precise optical control of all the electrons that are emitted from the cathode is the final important aspect of improving the operational lifetime of the polarized source. For a non-space charge dominated beam, the details of the electrode geometry are very important. The focussing field from the electrode angle determines the downstream beam envelope for the main part of the beam (see figure 1). For electrons that are emitted from outside the central region of the wafer, effects of misalignments and mechanical tolerances become more important. For example, the delicate GaAs wafer is slightly set back from the electrode to reduce the chance of breakage, resulting in a discontinuity between the surface of the wafer and the electrode.

Consider the following ray tracing using PARMELA [11] in which the wafer is displaced away from the electrode by a small distance. The electrode and anode are modeled as flat disks with a small step in the electrode at the location of the wafer. The electric fields are calculated using POISSON [12] and input directly into PARMELA. Without the step, the rays exit normal to the cathode as expected while increasing the step pushes the rays towards the axis. This effect determines the mask size to use for selective area cleaning as described earlier: any electron emitted outside a radius of 3 to 4 mm will cause a beam halo that can potentially strike the vacuum walls. Thus the

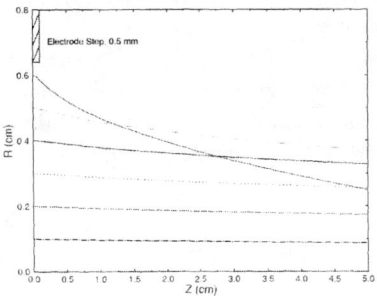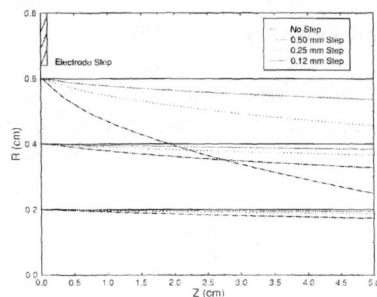

**Figure 7.** Simulations of a step near the edge of the cathode. The figure on the left shows the effect of a 0.5 mm step on the trajectories of individual electrons emitted at 1,2,3,4,5 and 6 mm from the center of the wafer. The figure on the right shows the effect on the trajectories at 2,4 and 6 mm for various step heights.

step determines the maximum useful area of the cathode and the electrode angle determines the beam envelope.

## DISCUSSION

The measurements and calculations presented here have centered on improving the operational lifetime of the polarized electron source at the Jefferson Lab. The operation of semiconductor photocathode guns with high average current present difficulties not normally encountered for low duty factor guns. For example, at the SLC the most critical factor (particularly for high polarization samples) is actually not the lifetime, but maintaining the peak quantum efficiency necessary to reach high peak currents [2]. At Jefferson Lab, QE is not as important due to the large range of available laser power and the low peak current requirements, but the lifetime is of paramount importance to maintain high accelerator availability.

Ion backbombardment (due to ionization of residual gas in the anode-cathode region) has been shown to be the rate-limiting factor in determining the lifetime. The introduction of high speed pumping close to the cathode and careful electron optical design have reduced the effects of ion backbombardment to tolerable levels and has made multi-hundred hour operational lifetimes possible. In addition, all of these results were obtained without the use of a complicated load lock chamber.

The actual crystal damage that causes the reduction in quantum efficiency is known to depend on vacuum properties, ion energy, electron optics, and crystal properties. There a several important related issues that may effect the operation of future guns planned for use in colliders or as FEL drivers [13]. For example, due to the range of ions as a function of energy, going to higher voltage guns may not cause increased damage since the range is deeper than the characteristic absorption depth for light. Also, the damage in samples with thin active layers compared to bulk samples (such as strained layer and quantum well structures) may be less for the same reasons.

The higher energy ions may effect the underlying crystal, though, especially for complicated quantum well structures [14].

In conclusion, recent improvements to the Jefferson Lab polarized electron source have led to significant increases in the operational lifetime while delivering high average currents. The understanding of the mechanisms involved has implications for future photocathode gun design and efforts are underway to model the cathode damage quantitatively.

## ACKNOWLEDGMENTS

This work was supported by the U.S. Department of Energy under contract DC-AC05-84ER40150. I would like to thank our fine technical staff consisting of P. Adderley, J. Hansknecht, J. Clark, D. Machie, A. Day and the Jefferson Lab machine shop for all of their support.

## REFERENCES

1. Sinclair, C.K. *et al.*, *AIP Conference Proceedings*, No. 35, 424 (1976).
2. Alley, R., Aoyagi, H., Clendenin, J., *et al.*, SLAC-PUB-95-6489, (1995).
3. Reichert, E., personal communication.
4. Krishnakumar, E. and Srivstava, S.K., *J. Phys. B* **27**, L251 (1994).
5. Ziegler, J.F., Biersack, J.P. and Littmark, U., *The Stopping and Range of Ions in Solids*, New York, Pergamon Press, 1985.
6. Steeples, K., Dearnaley, G., Harwell, A.E., and Saunders, I.J., *Appl. Phys Lett.*, **42** (8), 703-705 (1983).
7. Steeples, K. and Saunders, I., *IEEE Trans. Nucl. Sci.*, **NS-30**, 4087-4089 (1983).
8. Dubois, L.H. and Schwartz, G.P., *J. Vac. Sci. Technol.* **B 2** (2), 101-106 (1984).
9. C.K. Sinclair, B.M. Poelker, J.S. Price, *Proceedings of the 1997 Particle Accelerator Conference*, Vancouver, B.C., to be published.
10. Dunham, B., *Investigations of the Physical Properties of Photoemission Polarized Electron Sources for Accelerator Applications*, Ph.D. Thesis, U. of Illinois (1993).
11. Young. L.M., personal communication
12. Engwall, D.E., *High Brightness Electron Beams from a DC, High Voltage, GaAs Photoemission Gun*, Ph.D. Thesis, Univ. of Illinois, 1998.
13. Billen, J.H. and Young, L.M., *Proceedings of the 1993 Particle Accelerator Conference*, (2) 790-792 (1993).
14. Y. Kurihara, et al, *Jpn. J. Appl Phys.*, **34**, 355 (1995).

# Theoretical Study of a Diode with Dielectric-Gridded Cathode

L. Schächter,* D. Fletchner,† C. Golkowski,† J.D. Ivers†
and J. A. Nation†

*Electrical Engineering Department, Technion, Haifa 32000, Israel
†School of Electrical Engineering, Cornell University, Ithaca, NY 14853, USA

**Abstract.** We investigate the I-V characteristics of a vacuum diode with a gridded cathode. The grid is located on top of a dielectric material that on its back side is covered by a uniform electrode. Ignoring space-charge effects, the current density extracted from the grid is *proportional* to the dielectric coefficient and it is quadratic with the back electrode voltage. Considering only space-charge associated with the back electrode voltage, it is found that the anode current is proportional to the anode voltage. When all space-charge effects are considered, it is shown that the electrostatic energy *coupled* into the diode gap through the grid is responsible to the excess of current beyond the Child-Langmuir limit.

## INTRODUCTION

A series of independent experiments have indicated that in a diode with a ferro-electric cathode the current may exceed the Child-Langmuir limit by more than two orders of magnitude [1-3]. This is a result of three characteristics of the system: high dielectric coefficient of the material, non-linear constitutive $(D - E)$ relation and thin metallic grid that covers part of the ceramic. The original theoretical model developed by us [4] explains the enhanced current in terms of the *non-linear* properties of the ceramic. It shows that electrons which originally screen the polarization field in the ceramic, redistribute in the gap once the polarization is altered i.e., *negative* voltage is applied on the back of the ceramic. The predictions of this model fit the experimental data for a wide range of parameters. Geometric details at the cathode surface are considered only in a parametric way. Specifically, it was assumed that a fraction of the bound charge is redistributed in the gap; the value of this fraction was one out of two parameters of the model. It was also indicated that the electrons redistributed in the gap may originate from the metallic strips forming the grid on the ceramic. This is the case when a *positive* voltage is applied on the back of the ceramic consequently, field-emission plays an important role in the operation of such a cathode.

CP472, *Advanced Accelerator Concepts: Eighth Workshop*,
edited by W. Lawson, C. Bellamy, and D. Brosius

In the present study we disregard the non-linearity of the ferro-electric material and consider the effect of the other two characteristics on the emitted current i.e., high dielectric coefficient and thin metallic grid. The main goal is to investigate the electron emission from the metallic strips of the grid; specifically, we investigate the contribution of non-explosive field-emission to the current density generated by such a cathode. Analytic expressions for the limiting current in a diode with cathode made of a linear dielectric layer covered by a gridded electrode are developed. We show that when space-charge effects are ignored the current density emitted from the grid is proportional to the dielectric coefficient - consistent with the result in [5]. Furthermore the grid facilitates the penetration of electrostatic energy from the dielectric region into the diode gap and it is this *coupling* of energy that is "responsible", in part [4], to the excess of current beyond Child-Langmuir limit [1-3]. As the anode voltage exceeds a critical voltage, the I-V characteristic of the diode approaches the Child-Langmuir limit. These results complement the model introduced in Ref. 4 that treats primarily the non-linear characteristics of a ferro-electric ceramics.

This study is organized as follows: in the next section we examine the current density emitted from a periodic grid located on top of a dielectric layer ignoring space-charge effect. The latter is considered in the second section where we estimate the maximum current density achievable based on energy arguments.

## FIELD EMISSION FROM THE GRID

Consider a dielectric ($\epsilon_r$) layer of thickness $h$ covered by a uniform metallic electrode on the bottom surface and a metallic grid on the top. This grid has a periodicity $L$ and the width of the metallic strip is $d$; for the moment, we shall assume that the thickness of the grid is very small. The anode is located at a distance $g$ above the grid - see top frame in Figure 1. The typical dimensions of the system are: $h \sim 0.5$mm, $L \sim 0.5$mm, $d \sim L/2$, the thickness of the grid is less than $1\mu$m and $g \sim 5$mm. In the first part of this section we shall assume that both anode and bottom electrode are grounded ($V_{be} = 0$, $V_{an} = 0$) whereas the grid is held at a constant negative voltage $V_{gr} = -|V_{gr}|$; in the second part the other case ($V_{be} \neq 0$, $V_{an} \neq 0$ and $V_{gr} = 0$)will be examined.

In order to determine the extracting field on the top of the grid we have to determine first the electric potential in the entire space. Formally the latter reads

$$\Phi(x,z) = \begin{cases} \displaystyle\sum_{n=-\infty}^{\infty} A_n e^{-jk_n x} \; \frac{\sinh[k_n(g-z)]}{\sinh(k_n g)} & \text{for} \quad 0 < z < g \\ \displaystyle\sum_{n=-\infty}^{\infty} B_n e^{-jk_n x} \; \frac{\sinh[k_n(z+h)]}{\sinh(k_n h)} & \text{for} \quad -h < z < 0 \end{cases} \qquad (1)$$

where $k_n = 2\pi n/L$. Continuity of the potential at $z = 0$ implies $A_n = B_n$ and the second boundary condition is associated with the surface charge density at $z = 0$.

In order to establish the latter we denote with $\eta(x)$ the surface charge density on the strip. The Fourier transform of this function is denoted by $\eta_n$ and is given by

$$\eta_n = \frac{1}{L} \int_{-d/2}^{d/2} \mathrm{d}x\, \eta(x)\, e^{jk_n x} . \tag{2}$$

With this definition we may now establish the discontinuity in the electric field normal to the grid plane, it reads: $k_n \coth(k_n g)A_n + \epsilon_r k_n \coth(k_n h)B_n = \eta_n/\epsilon_0$ and consequently

$$A_n = B_n = \frac{1}{\epsilon_0} \frac{1}{k_n} \frac{\eta_n}{\coth(k_n g) + \epsilon_r \coth(k_n h)} . \tag{3}$$

Next we determine the surface charge density $\eta(x)$. For this purpose we write the electrostatic potential at the interface between the dielectric and the vacuum

$$\Phi(x, z = 0) = \frac{1}{\epsilon_0} \int_{-d/2}^{d/2} \mathrm{d}x'\, G(x|x')\eta(x') \tag{4}$$

where

$$G(x|x') = \sum_{n=-\infty}^{\infty} \frac{1}{k_n L} \frac{e^{-jk_n(x-x')}}{\coth(k_n g) + \epsilon_r \coth(k_n h)} ; \tag{5}$$

on the metallic strip this potential is constant $[\Phi(|x| \leq d/2, z = 0) = V_{\mathrm{gr}}]$. If we assume that $\coth(2\pi g/L) \simeq 1$, $\coth(2\pi h/L) \simeq 1$ and $\epsilon_r \gg 1$ then (5) simplifies to read

$$G(x|x') \simeq \frac{1}{\epsilon_r} \frac{h}{L} - \frac{1}{2\pi\epsilon_r} \ln \left[ 4\sin^2 \left( \pi \frac{x - x'}{L} \right) \right] . \tag{6}$$

At this stage we shall adopt a variational approach to establish $\eta(x)$. Specifically, the error associated with the actual potential across the metallic strip is given by

$$\mathrm{Error} = \int_{-d/2}^{d/2} \mathrm{d}x \left[ V_{\mathrm{gr}} - \frac{1}{\epsilon_0} \int_{-d/2}^{d/2} \mathrm{d}x'\, G(x|x')\eta(x') \right]^2 . \tag{7}$$

As a trial function we employ an "edge type solution" [5] and rather than assuming a flat surface we assume a metallic edge of a small angle $\alpha$ attached to a dielectric layer as illustrated in the middle frame of Figure 1. The potential in the upper half-space of the edge was shown [5] to have, in a cylindrical coordinate system $(r, \phi)$, the following form:

$$\Phi(r, \phi) \propto r^\nu \sin[\nu(\phi - \pi + \alpha)] , \tag{8}$$

where $\alpha$ is the angle of the edge and $\nu$ is the curvature factor of the potential; see the geometry illustrated at the bottom of Figure 1. At the limit of $\epsilon_r \gg 1$ and $\alpha \ll \pi$ the parameter $\nu$ is given by

$$\nu = \frac{1}{2} + \frac{\alpha}{2\pi} \frac{1}{\epsilon_r} . \tag{9}$$

825

Based on (8) we may assume that the charge distribution on the strip is given by

$$\eta(x) = \bar{\eta}\left(1 - \frac{2|x|}{d}\right)^{\nu-1} \tag{10}$$

and the requirement for minimum error translates into the condition that $\frac{\delta}{\delta\bar{\eta}}\text{Error} = 0$. We show elsewhere [6] that $\bar{\eta} = \epsilon_0\epsilon_r\frac{V_{gr}}{h}\frac{L}{d}\nu$.

With the electrostatic potential established we continue and calculate the field component which extracts electrons from the metallic surface i.e., *normal* to the metallic surface. Based on the solution in (1) we have

$$E_z(x, z = 0) = \sum_n A_n\, e^{-jk_n x} k_n \coth(k_n g) = \frac{1}{\epsilon_0 L}\int_{-d/2}^{d/2} dx'\, H(x|x')\, \eta(x') \tag{11}$$

where

$$H(x|x') = \sum_{n=-\infty}^{\infty} \frac{e^{-jk_n(x-x')}}{1 + \epsilon_r\coth(k_n h)\tanh(k_n g)}. \tag{12}$$

If $\coth(2\pi h/L) \sim 1$, $\coth(2\pi g/L) \sim 1$ and $\epsilon_r \gg 1$, then Eq.(12) simplifies to read $H(x|x') \simeq \frac{L}{\epsilon_r}\delta(x - x')$ hence

$$E_z(|x| \le d/2, z = 0) = \frac{V_{gr}}{h}\frac{L}{d}\nu\left(1 - \frac{2|x|}{d}\right)^{\nu-1}. \tag{13}$$

The parameter $\nu$ is determined by the edge angle ($\alpha$) and the dielectric coefficient ($\epsilon_r$) and since $\alpha \ll \pi, \epsilon_r \gg 1$ we conclude that $\nu \sim 1/2$ implying that $\eta(x)$ and $E_z(x, z = 0)$ are *singular* at the edge - as expected.

With the explicit expression for the electric field normal to the strip, we may next calculate the current density emitted from one strip, ignoring the space-charge effect. For this purpose, we now proceed and use Fowler-Nordheim formula for field-emission $J(E) = K_1 E^2 e^{-E_{cr}/E}$. The current extracted from one strip is

$$I = 2\Delta_y \int_0^{d/2} dx\, J[E(x)] = I_0 \int_0^1 d\xi(1 - \xi)^{2(\nu-1)} e^{-\chi(1-\xi)^{1-\nu}} \tag{14}$$

with $\xi = 2x/d$, $\chi \equiv E_{cr}/E_0$, $I_0 = \Delta_y dK_1 E_0^2$ and $\Delta_y$ is a unit length in the $y$ direction. Changing variables $[u = (1-\xi)^{1-\nu}]$ and after some additional algebraic steps it is possible to show [6] that the *average* current density ($J_{av}$) in one period of the grid reads

$$J_{av} \equiv \frac{I}{\Delta_y L} \simeq \frac{d}{L}K_1 E_0^2\frac{1}{2\nu - 1} \simeq K_1 E_{gr}^2 \propto \epsilon_r V_{gr}^2 \tag{15}$$

826

where $K_1 = 1.54 \times 10^{-6}/W_f$ and $W_f$ is the work function of the metal, $E_{gr} = \beta \epsilon_r |V_{gr}|/h$ is the effective electric field on the grid and

$$\beta = \sqrt{\frac{\pi}{\alpha} \frac{1}{\epsilon_r} \frac{L}{d} \nu^2} \tag{16}$$

is the field reduction factor. This expression [(15)] indicates that the current density is *proportional* to $\epsilon_r$ and it is quadratic in the applied voltage. Consider some typical values for the parameters of interest: $|V_{gr}| = 1.0\text{kV}$, $h = 1\text{mm}$, $L = 0.4\text{mm}$, $L/d = 2$, $W_f = 3eV$, $\epsilon_r = 3000$ and $\alpha = \pi/180$; for these values the average current density is $J_{av} \simeq 10^7 \text{A}/\text{cm}^2$. This is a high current density that, in practice, is expected to be reduced by the space-charge effect that will be discussed in the next section.

In order to examine the *coupling* process between the two sides of the grid it is instructive to consider the case of a grounded grid and positive voltages applied to the anode ($V_{an}$) and to the back electrode ($V_{be}$). Following a similar approach as above, the average current density emitted from one period of the grid is given by

$$J_{av} \simeq K_1 [E_{an} + E_{be}]^2 \propto \frac{1}{\epsilon_r} \left( \frac{V_{an}}{g} + \epsilon_r \frac{V_{be}}{h} \right)^2, \tag{17}$$

where $E_{an} \equiv \beta V_{an}/g$ and $E_{be} \equiv \beta \epsilon_r V_{be}/h$. This result can be interpreted as follows: two electric field components contribute to the current density at the cathode surface. The one generated by the anode ($V_{an}/g$) is reduced by a factor $\beta$ that is determined by the grid and the dielectric coefficient. The second contribution is due to the voltage applied on the back electrode ($\epsilon_r V_{be}/h$) reduced by the same factor $\beta$. The contribution of the back electrode (for $V_{an} = 0$) is identical to (15) therefore it is dominant [$\sim 10^7 \text{A}/\text{cm}^2$] since the contribution of the anode (when $V_{be} = 0$) is of the order of $15\text{mA}/\text{cm}^2$ for $V_{an} \sim 1\text{kV}$.

## SPACE-CHARGE EFFECT

All the current densities calculated in the previous section were at the *surface* of the metallic strip and the extracting field has been determined as if no charge was extracted from the metal. As electrons leave the surface they follow two major tracks: they are either attracted by the back electrode and as a result their trajectory is bended backwards or they are pulled by the anode. Electrons following the first track reach the dielectric surface, they accumulate in this region screening the metallic strip from the electric field coupled in the gap by the dielectric. If no anode voltage is applied and if the gap ($g$) is much larger than the periodicity of the grid ($L$) then although $J_{av}$ as calculated in (17) may reach very high values, the *anode* current density is zero. When the anode voltage is non-zero, a fraction of the current density reaches the anode. Based on our model we can evaluate this current density. Ignoring the space-charge effect associated with the anode we may

827

assume that the *anode* current density is the average current density at the *cathode* [(17)] reduced by the current associated with the space-charge

$$J_{\mathrm{an}} = K_1 \left( E_{\mathrm{be}} + E_{\mathrm{an}} \right)^2 - J_{\mathrm{sc}} . \tag{18}$$

Bearing in mind that when $E_{\mathrm{an}} = 0$, the anode current is zero, we may assume that $J_{\mathrm{sc}} \simeq K_1 E_{\mathrm{be}}^2$ therefore

$$J_{\mathrm{an}} \simeq 2K_1 E_{\mathrm{be}} E_{\mathrm{an}} \simeq K_1 \frac{V_{\mathrm{an}}}{g} \frac{V_{\mathrm{be}}}{h} \frac{\pi}{\alpha} . \tag{19}$$

For $V_{\mathrm{an}} \sim V_{\mathrm{be}} \sim 1\mathrm{kV}$, $g = 1\mathrm{cm}$, $h = 1\mathrm{mm}$ and $\alpha = \pi/18$ this current density is of the order of $100\mathrm{A/cm}^2$, it is proportional to the anode voltage; this linear behavior was observed experimentally.

Exact *analytic* evaluation of the space-charge effect in this complex geometry is not possible to the best of our knowledge. However we can limit the discussion to an upper bound of the extracted current. This limit may be deduced based on the following energy argument: In principle, the electrostatic field in the diode $\vec{E}(x, z)$ and the energy associated with it, may be averaged out over one period

$$E^2(z) \equiv \frac{1}{L} \int_{-L/2}^{L/2} \mathrm{d}x \vec{E}(x, z) \cdot \vec{E}(x, z) \tag{20}$$

thus the total electrostatic energy $(W_{\mathrm{es}})$ in a diode of area $S$ is $W_{\mathrm{es}} = \frac{1}{2}\epsilon_0 S \int_0^g \mathrm{d}z E^2(z)$ If we now assume that this effective electric field $[E(z)]$ may be derived from a scalar electric potential $\Phi(z)$ that satisfies the (1D) Poisson equation with the source term $\rho(z)$, then the energy stored in the space-charge $(W_{\mathrm{sc}})$ field may not exceed the electrostatic energy stored in the diode when no electrons are emitted. Thus

$$W_{\mathrm{es}} = \frac{1}{2}\epsilon_0 S \int_0^g \mathrm{d}z E^2(z) \geq W_{\mathrm{sc}} = S \int_0^g \mathrm{d}z \Phi(z)\rho(z) \tag{21}$$

Based on the (1D) continuity equation and the equation of motion (regular Child-Langmuir approach) the charge density is related to the current density by $\rho(z) = J/\sqrt{m/2e\Phi(z)}$ hence

$$J \leq \frac{1}{2}\sqrt{\frac{2e}{m}}\epsilon_0 \frac{\int_0^g \mathrm{d}z E^2(z)}{\int_0^g \mathrm{d}z\sqrt{\Phi(z)}} . \tag{22}$$

This expression may be regarded as a generalization of the Child-Langmuir law that facilitates evaluation of the scaling law when electrostatic energy is coupled into the gap. It possesses the correct scaling with the Child Langmuir law since in a planar diode we may approximate $\Phi \sim Vz/g$ and $E \sim V/g$ that imply $J \leq J_{\mathrm{M}}\frac{1}{2}\sqrt{\frac{2e}{m}}\epsilon_0\frac{V^{3/2}}{g^2}\frac{3}{2}$ which comparing to the exact Child-Langmuir limit

$$J_{\mathrm{CL}} = \frac{4}{9}\sqrt{\frac{2e}{m}}\epsilon_0\frac{V^{3/2}}{g^2} , \tag{23}$$

is a reasonable result. In the remainder of this section we shall examine the predictions of Eq. 22 for the specific configuration we are interested in this study.

Based on the formulation presented in the previous section the electrostatic energy stored in the diode gap in the case of a *grounded grid* is

$$W_{es} \simeq \frac{1}{2}\epsilon_0 S \left\{ g \left( \frac{V_{an}}{g} \right)^2 + \frac{L}{2\pi} \frac{1}{4} \left[ \sqrt{\frac{\mathcal{L}}{\epsilon_r^2}} \left( \frac{V_{an}}{g} + \epsilon_r \frac{V_{be}}{h} \right) \right]^2 \right\} \tag{24}$$

where

$$\mathcal{L} \equiv \frac{-1}{d^2} \int_{-d/2}^{d/2} dx \left( 1 - \frac{2|x|}{d} \right)^{\nu-1} \int_{-d/2}^{d/2} dx' \left( 1 - \frac{2|x'|}{d} \right)^{\nu-1} \ln \left[ 4\sin^2 \left( \pi \frac{x-x'}{L} \right) \right]$$

$$\simeq \frac{0.73}{\nu^{1.843}} \,. \tag{25}$$

$\mathcal{L}$ is the coupling coefficient between the diode gap and the region filled with the dielectric material; the last expression in (25) was evaluated assuming $d \sim L/2$. In the case of *grounded electrodes*

$$W_{es} \simeq \frac{1}{2}\epsilon_0 S \left\{ g \left( \frac{V_{gr}}{g} \frac{d}{L} \right)^2 + \frac{L}{2\pi} \frac{1}{4} \left[ \frac{V_{gr}}{g} \frac{d}{L} \sqrt{\mathcal{L} \frac{g^2}{h^2}} \right]^2 \right\} \,. \tag{26}$$

Note that in both cases the electrostatic energy has the form

$$W_{es} = \frac{1}{2}\epsilon_0 S \left[ g E_1^2 + \frac{L}{2\pi} \frac{1}{4} E_2^2 \right] \tag{27}$$

that can be derived from the following *effective* electrostatic potential

$$\Phi(z) = E_1 z + E_2 z e^{-2\pi z/L} \,. \tag{28}$$

The first term in (27) represents the energy in the diode gap as if the grid was replaced by a uniform electrode whereas the second represents the electrostatic energy attached to the grid in the diode side. The "source" of this (1D) effective potential coupled into the gap is a charge density distributed in the gap $2\epsilon_0 \frac{2\pi}{L} E_2 \left( 2 - \frac{2\pi z}{L} \right) e^{-2\pi z/L}$ as postulated by us in a previous study [4].

Comparing (27) with (24) or (26) we conclude that

$$E_1 = \begin{cases} \dfrac{V_{an}}{g} & \text{for} \quad V_{gr} = 0 \\[2ex] \dfrac{V_{gr}}{g} \dfrac{d}{L} & \text{for} \quad V_{an} = 0 \text{ and } V_{be} = 0 \end{cases} \tag{29}$$

and

$$E_2 = \begin{cases} \sqrt{\dfrac{\mathcal{L}}{\epsilon_r^2}}\left(\dfrac{V_{\mathrm{an}}}{g} + \epsilon_r \dfrac{V_{\mathrm{be}}}{h}\right) & \text{for} \quad V_{\mathrm{gr}} = 0 \\[3mm] \sqrt{\mathcal{L}}\dfrac{V_{\mathrm{gr}}}{h}\dfrac{d}{L} & \text{for} \quad V_{\mathrm{an}} = 0 \text{ and } V_{\mathrm{be}} = 0. \end{cases} \tag{30}$$

With these observations we may substitute (28-30) in (22) and therefore the upper bound on the limiting current reads

$$J \leq \frac{\sqrt{2}}{2}\frac{mc^2\,k^2}{e\eta_0}\left[\frac{1}{4}\frac{4K\mathcal{E}_1^2 + \mathcal{E}_2^2}{\int_0^K d\xi\sqrt{\xi(\mathcal{E}_1 + \mathcal{E}_2 e^{-\xi})}}\right] \tag{31}$$

where $k = 2\pi/L$, $K = kg$, $\mathcal{E}_1 = eE_1/mc^2 k$ and $\mathcal{E}_2 = eE_2/mc^2 k$. Next we shall examine the limiting current associated with the two configurations introduced previously.

In the case of grounded electrodes the limiting current scales as $|V_{\mathrm{gr}}|^{3/2}$ or explicitly

$$J < \frac{1}{2}\epsilon_0\sqrt{\frac{2e}{m}}\frac{V_{\mathrm{gr}}^{3/2}}{g^2}\frac{3}{2}\left\{\frac{2}{3}K^{3/2}\frac{1 + \mathcal{L}\dfrac{g^2}{h^2}\dfrac{1}{4K}}{\int_0^K d\xi\sqrt{\xi\left(1 + e^{-\xi}\sqrt{\mathcal{L}g^2/h^2}\right)}}\right\} \propto V_{\mathrm{gr}}^{3/2}. \tag{32}$$

The term in curl brackets is the form factor that represents the *energy coupling* through the grid; the limiting current in the diode is not dependent on the dielectric coefficient.

When the grid is grounded the dependence of the current density on the applied voltages is more complex and therefore we prefer to illustrate the behavior graphically. Figure 2 illustrates the ratio of the current density in (31) and $J_{\mathrm{M}}$ [Child-Langmuir limit in the framework of the current model]. We observe that at low anode voltages the current density may exceed the CL limit by many orders of magnitude; the larger the back electrode voltage, the current density is higher. This regime corresponds to the case when the electrostatic energy coupled from the dielectric region into the diode gap is much larger than the energy stored in the gap when only the anode voltage is non-zero. As the anode voltage increases, the current density drops and it reaches the CL limit. In this case the energy associated with the anode voltage is larger than that coupled from the dielectric region. For a 1cm diode gap, a grid period of 0.4mm and a dielectric thickness of 0.5mm, CL limit is reached at an anode voltage of 20kV when $\epsilon_r = 1000$ and $V_{\mathrm{be}} =$3kV. For a back electrode voltage of 1kV, the CL limit is reached when an anode voltage of 10kV is applied. For sufficiently high $\epsilon_r > 10$ the results are virtually independent of the dielectric coefficient.

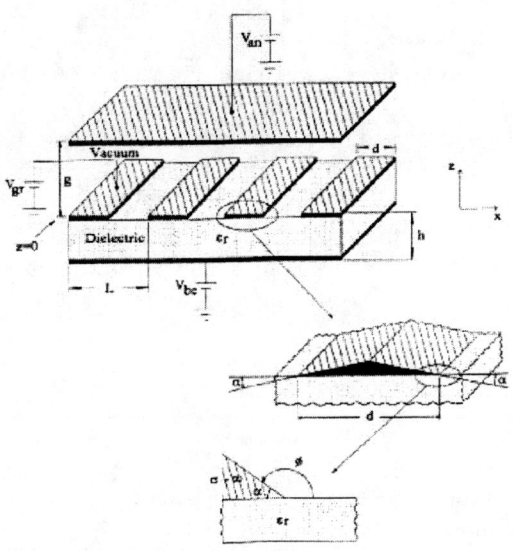

**FIGURE 1.** Diode configuration.

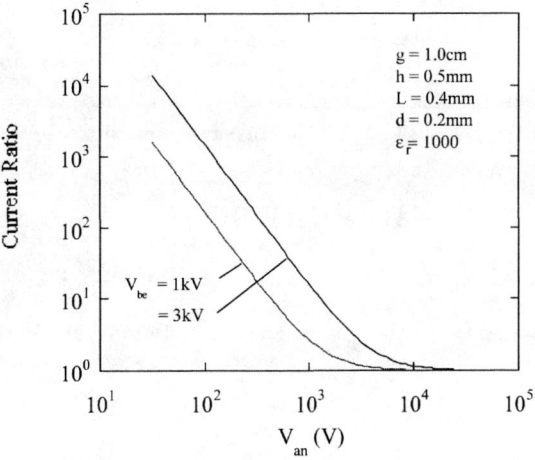

**FIGURE 2.** Normalized current vs. $V_{an}$.

# DISCUSSION

Ferro-electric cathodes have three characteristics: non-linear constitutive relation $(D - E)$, high dielectric coefficient and a thin metallic grid which faces the gap. In this study we have examined the effect of the last two characteristics on the performance of the diode. In the first part we calculated the average current density emitted from a metallic grid with special emphasis on the contribution of the edges and the effect of the dielectric. Within the framework of this part the space-charge effect was neglected. The current density was assumed to be due to field emission (Fowler-Nordheim) and we have shown that it is proportional to $(V_{an}/g + \epsilon_r V_{be}/h)^2/\epsilon_r$.

The space-charge effect is expected to reduce significantly the current density calculated in the first part. We developed an upper limit estimate for the current density achievable. This estimate relies on the assumption that the amount of energy stored in the space-charge field may not exceed the initial electrostatic energy stored in the gap when no electrons are emitted by the cathode - see Eq.(22). Based on this model we have shown that the energy coupled into the diode gap from the dielectric region is directly responsible to the excess of current beyond CL limit - at low anode voltages. As the anode voltage increases, the excess of current is reduced and at very high voltages it reaches the CL limit.

# ACKNOWLEDGMENT

This study was supported by the United States Department of Energy and by the Bi-National United States - Israel Science Foundation.

# REFERENCES

1. Ivers J.D., Nation J.A., Kerslick G.S., Advani R. and Schächter L., *J. Appl. Phys.* **73(6)**, 2667-2671 (1993).
2. Sampayan S.E., Caporaso G.J., C.L. Holmes, Lauer E.J., Prosnitz D., Trimble D.O. and Westenskow G.A., *Nuclear Instruments and Methods in Physics Research* **A340**, 90 (1994).
3. Okuyama M., Asano J. and Hamakawa Y., *Jpn. J. Appl. Phys.* **33**, 5506 (1994).
4. Schächter L., Ivers J.D., G.S. Kerslick and Nation J.A., *J. Appl. Phys.* **73(12)**, 8097–8110 (1993).
5. Schächter L., *Applied Physics Letters*, **72**, (4) 421–3 (1998).
6. Schächter L., Fletchner D., Ivers J.D. and Nation J.A. . To be published in *J. Appl. Phys.*.

# Dark Current Measurements At Field Gradients Above 1 GV/m

T. Srinivasan-Rao, J. Smedley, J. Schill

*Brookhaven National Laboratory, Upton, NY 11973*

K. Batchelor, and J. P. Farrell,

*Brookhaven Technology Group Inc., 25 E. Loop Rd, Stony Brook, NY 11970*

**Abstract.** In this paper, we report the results of dark current studies on copper cathodes and stainless steel anodes held at a field gradient > 1 GV/m. The field emission current is < 1 A for fields less than 1 GV/m. As the field is increased, the dark current increases rapidly to 150 A for applied fields of ~1.7 GV/m. Fowler-Nordheim plots in this range of applied fields indicate a field enhancement factor of 10-20 for a copper cathode with a work function of 4.6 eV.

## INTRODUCTION

In the past decade, there has been extensive research [1] in the development of low emittance, high brightness electron injectors for linear collider and free electron laser applications. RF injectors with a few nC charge in a few ps, with an emittance of ~1-5 $\pi$ mm mrad are operational in a number of facilities [2-4]. In these devices, a laser beam irradiates a photocathode embedded in an RF cavity. The photoelectrons released by the laser are immediately accelerated to relativistic velocities, thereby reducing the space charge effects. The frequency of the RF and the design of the cavity are chosen to minimize the RF and space charge effects on the electron bunch so that low emittance, high brightness electron beam could be generated. Minimization of RF effects on emittance growth require a low RF frequency while minimizing the space charge effects require high field and hence high RF frequency. The design is hence a compromise between these two conflicting requirements. Some of these limitations could be overcome by using a large pulsed electric field at the cathode rather than a RF field. The duration of the pulsed field should be chosen so that it is longer than the electron bunch length and the transit time in the accelerating region, but short enough to avoid breakdown problems.

Major issues in such a scheme is the capability of the cathode material to hold off the high field gradients without suffering electrical breakdown and the degradation

CP472, *Advanced Accelerator Concepts: Eighth Workshop,*
edited by W. Lawson, C. Bellamy, and D. Brosius

of the electron beam due to the presence of dark current. In this paper we present the preliminary results of the behavior of copper cathodes subjected to fields in excess of 1 GV/m and the characteristics of the dark current emitted from this cathode.

## EXPERIMENTAL ARRANGEMENT

A 1 MV pulse with rise and fall times of ~150 ps and duration of ~1 ns was used to bias the electrodes and measure the dark current. The high voltage pulse generator that provides this pulse is described in detail in Ref [5]. In this pulser, the voltage from a low voltage pulse generator is multiplied using a resonant transformer. The high voltage pulse is then sharpened by a pair of spark gaps and transported along a transmission line and terminated at the cathode with a matched load. The voltage can be varied by changing the gas pressure in one of the sharpening gaps. The voltage at different locations along the transmission line is measured using resistive and capacitive probes. The voltage measurements used in this paper are derived from the capacitive probe 15 cm upstream of the cathode. The field across the electrodes in the diode can be varied either by varying the applied voltage or the interelectrode gap.

The diode where the field effects are measured consists of a 6 mm diameter copper cathode biased to -1 MV. A flat stainless steel anode of 1.5 mm thickness is held at ground potential in front of the cathode. A small hole in the anode allows transport of the electron beam beyond the diode region. The size of the hole can be varied depending on the measurement requirement. For measuring the magnitude of the dark current, an anode with a central hole of 3 mm aperture was used. The interelectrode spacing can be varied from 5 mm to 0.5 mm and has been maintained at either 1 or 0.5 mm for these measurements. An ion pump attached to a stainless steel cube enclosing the electrodes maintains a vacuum level of $<10^{-7}$ Torr in the vicinity of the electrodes and surrounding diagnostic equipment. Both electrodes are removable and hence the performance of the diode for different electrode material and geometry can be investigated without changing the characteristics of the applied voltage significantly. Alternately, for a given electrode material and geometry, the performance of the diode can be studied for various shapes and amplitudes of the voltage pulse. This can be achieved by changing either the $SF_6$ gas pressure in the sharpening switch, the amplitude of the low voltage or the length of the pulse forming line, without breaking the vacuum. The pulser and the diode with its diagnostics are housed inside a RF shield to filter the EMI associated with such a system

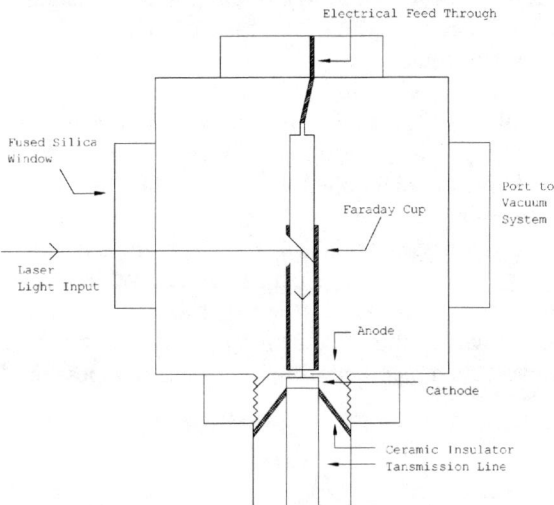

**Figure 1.** Diode with the Faraday cup inside the vacuum cell

The diagnostics for the electron beam consist of an electrically isolated Faraday cup to measure the charge, a pepper pot, phosphor screen and relay optics system to measure the spot size of the electron beam and hence calculate the emittance. The schematic of the Faraday cup along with the diode is shown in Figure 1. It is an electrically isolated copper rod of 1 cm diameter with a tubular sleeve of 1.1 cm diameter around it so that the distance between the Faraday cup and the anode can be altered. In these measurements, this distance is 3 mm so that all the primary electrons from the cathode, exiting the anode can be collected by the Faraday cup. The dimensions of the Faraday cup are such that with the grounded metallic vacuum housing surrounding it will simulate a transmission line of 90 $\Omega$ impedance. This rod is connected directly to a 7 GHz oscilloscope (Tektronix SCD 5000) for current measurements with a time resolution of 100 ps.

## MEASUREMENTS

The copper cathode was polished with diamond polishing compounds with grain sizes of 9,6 and 1 $\mu$m. The polished surface was rinsed with deionized water and ultrasonically cleaned in a hexane bath. The copper was then inserted in the vacuum system. When the background pressure was ~$10^{-7}$ Torr, the voltage on the cathode was gradually increased and the cathode was slowly conditioned. The applied voltage from the capacitive probe and the current signal from the Faraday cup were continually monitored during this conditioning process. Care was taken not to have more than one breakdown for a given voltage during this process to minimize the damage on the copper surface and deposition of copper onto the anode. When reproducible current traces were obtained from the cathode at maximum applied voltage, the cathode was assumed to be conditioned. The behavior of the cathode under high field gradients was studied by observing simultaneously the dark current and the applied field.

As mentioned before, the field gradient between the electrodes could be changed by changing either the applied voltage or the interelectrode gap. With the interelectrode gap of 1 mm, and a background pressure of $<10^{-7}$ Torr, dark current was below noise level of 1 A even at an applied voltage of 750 kV. The field between the electrodes was then doubled by reducing the interelectrode gap to 0.5 mm and the dark current was measured for various applied voltages. The applied voltage and the dark current at this voltage for 16 consecutive shots are shown in Fig. 2 a and b. The variation in the voltage curve is a measure of the shot to shot fluctuation of the applied voltage. In order to normalize this fluctuation, individual traces of voltage and corresponding current were then recorded using two oscilloscopes simultaneously.

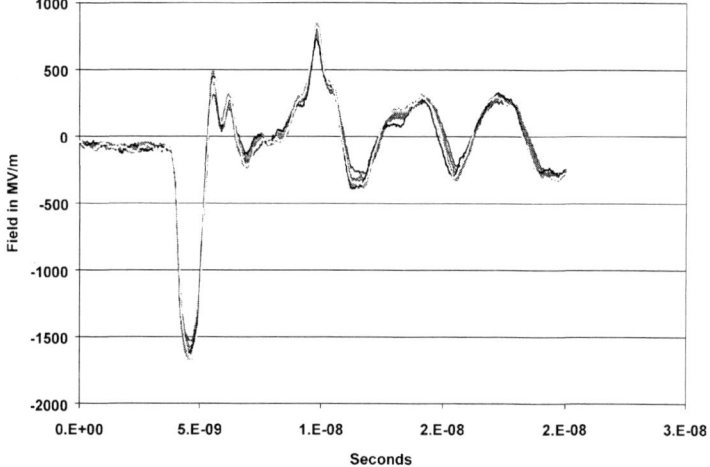

**Figure 2 a.** The applied field on the cathode, based on the voltage measured at the last probe.

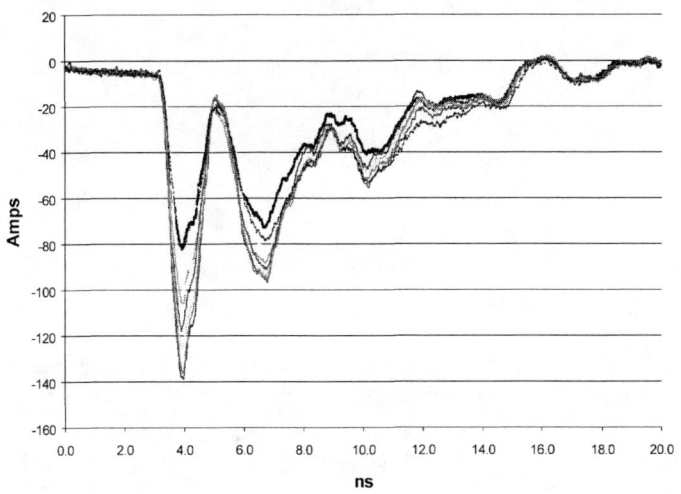

**Figure 2 b.** Corresponding current measured by the Faraday cup

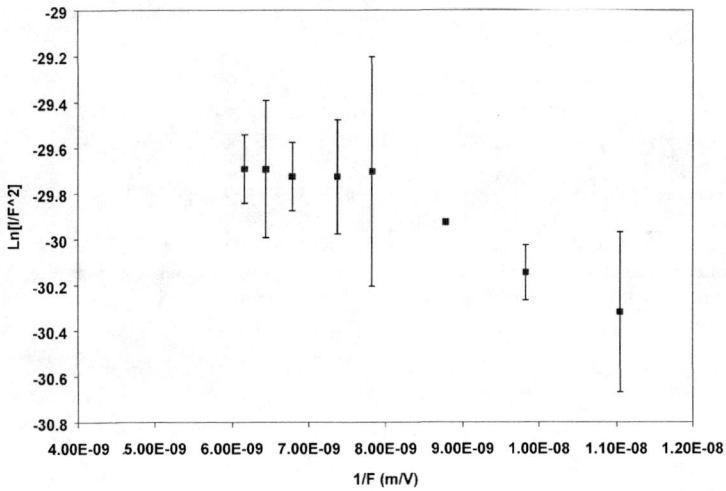

**Figure 3.** Fowler-Nordheim plot for the dark current

At fields of ~1.2 GV/m, the dark current was just above the noise level of 1 A. As the field was increased, the dark current increased rapidly to ~ 150 A for a field of ~1.7 GV/m. The Fowler-Nordheim plot for this data, in the field range of 1.2 GV/m to 1.7 GV/m is shown in Fig.3. The field induced current can be expressed as

$$J = K\ E^2 \exp(-2.89 \times 10^9\ \varphi^{3/2}/E) \tag{1}$$

The slope of ln $(j/E^2)$ vs $1/E$, the Fowler-Nordheim plot, can then be written as

$$S \quad = \quad (-2.89 \times 10^9 \times \varphi^{1.5})/\beta \tag{2}$$

If the work function $\varphi$ of copper is assumed to be 4.6 eV, the field enhancement factor $\beta$ can then be calculated to be $\sim 20$.

In conclusion, copper cathode has been tested under field gradients of 1.7 GV/m, using 1 MV pulses lasting for $\sim 1$ ns. No electrical breakdown was observed after conditioning. Up to field gradients of $\sim 1.2$ GV/m, the dark current is below 1 A. At 1.7 GV/m, dark currents of $\sim 150$ A has been measured. The Fowler-Nordheim plot of current and field yields a field enhancement factor of $\sim 20$ for a work function of 4.6 eV. This implies that properly prepared copper surfaces with minimum field enhancement could withstand applied fields >30 GV/m.

## ACKNOWLEDGMENTS

The authors would like to thank Ilan Ben-Zvi and Veljko Radeka for their support. This work was supported by DE-AC02-98CH10886 and DE-FG02-97ER82336.

## References

1. C. Travier, Nucl. Instr. and Meth in Phys. Res. A 340, (1994), 26.
2. B. Carlsten, IEEE J. Quant. Electron. QE 27, (1991), 2580.
3. K. Batchelor et.al., Nucl. Instr. and Meth. in Phys. Res. A 318, (1992), 372.
4. R. Sheffield et.al. in Proc. 1992 Linear Acceleration Conf., Ottawa, Canada, Aug. 24-28, 1992
5. T. Srinivasan-Rao, J. Smedley, Advanced Accelerator Concepts workshop, AIP Conference Proceedings 398, Ed. S. Chattopadhyay, J. McCullough and P. Dahl, AIP Press, NY 1997, P. 7307.

# Optimization of Gun Parameters for a Pulsed Power Electron Gun

T. Srinivasan-Rao, J. Smedley

*Brookhaven National Laboratory, Upton, NY 11973*

## K. Batchelor, J. P. Farrell, and G. Dudnikova

*Brookhaven Technology Group Inc., 25 E. Loop Rd, Stony Brook, NY 11970*

**ABSTRACT.** Extensive simulation work has been done to identify the optimal parameters for a pulsed power electron gun. PBGUNS, an electrostatic beam optics code, was used to optimize the electrode shape and the beam spatial distribution, including modeling the focusing effect of a curved cathode surface. MAFIA, a particle-in-a-cell code, was used to investigate those aspects that required time dependence, such as longitudinal energy spread. The range of agreement between the two codes was also investigated. The transverse phase space at a comparison plane was found to be very close (within 1 % at low currents and 4 % for higher currents), even for bunch lengths shorter than the gap transit time.

## INTRODUCTION

This paper is a result of a group of simulations used to determine the optimal parameters for a pulsed power electron gun. Two simulation codes, MAFIA and PBGUNS were utilized. PBGUNS is an electrostatic code, with a shorter average running time and less stringent computational requirements than time dependent codes such as MAFIA. It was therefore beneficial to determine in what regimes these two programs were in agreement, and what objectives of the optimization could be achieved with the DC code. It was also necessary to identify those problems that required time dependence, such as longitudinal variation in an electron bunch. PBGUNS was then used to perform the bulk of the optimization, with only those issues that required time dependence being resolved with MAFIA. The first section of this paper establishes the agreement between the two codes, in terms of the predicted transverse phase space at a comparison plane. The second section describes the optimization of the cathode curvature to provide a nearly collimated output, and the optimization of the laser spatial profile to minimize the detrimental effects of space charge. For the first two sections, the beam was assumed to have no thermal energy. The third section addresses the optimization of the bunch length to provide the highest brightness beam, a discussion of the effect of different values of thermal energy on the beam emittance, and a study of the longitudinal variation of the beam.

CP472, *Advanced Accelerator Concepts: Eighth Workshop,*
edited by W. Lawson, C. Bellamy, and D. Brosius

The pulsed power gun has been discussed elsewhere [1] and consists of a diode with a flat cathode and a flat anode mounted parallel to one another with an interelectrode spacing of 1 mm. The anode had a .5 mm radius hole allowing the bunch to escape the accelerating gap. The cathode is biased at −1 MV, yielding an accelerating gradient of 1 GV/m within the gap. The anode was 1.5 mm thick and was modeled as a perfect conductor and a perfect particle dump. For the purposes of simulation, the gun was assumed to be cylindrically symmetric. Figure 1 shows the simulated geometry of the gun in MAFIA, along with field lines and an electron bunch from a typical run. Figure 2 shows the geometry as it was used in PBGUNS, along with the equal potentials and particle trajectories from a typical run. Note that in MAFIA, the z=0 boundary is used as the cathode, and the potential of −1 MV is set on that boundary, while in PBGUNS the cathode surface is at z = 1.25 mm. The beam parameters used for comparison were extracted as close as possible to a plane 2.25 mm from the inner surface of the anode. This plane was chosen so that the particles could be taken to be in the drift region, away from any fringe fields from the accelerating gap. All of the simulations modeled emission from a .25 mm radius spot, but a variety of currents were modeled within that spot, all with uniform longitudinal current density.

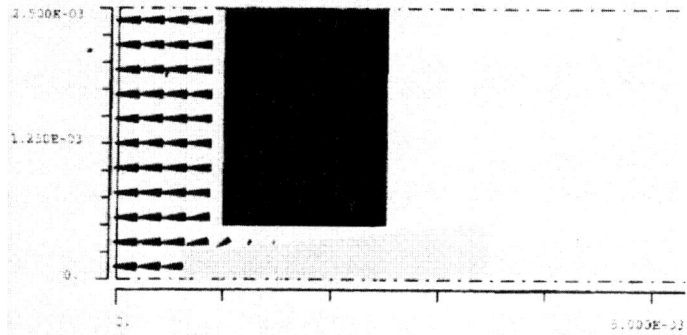

**Fig 1.** MAFIA Geometry, field lines and electron bunch

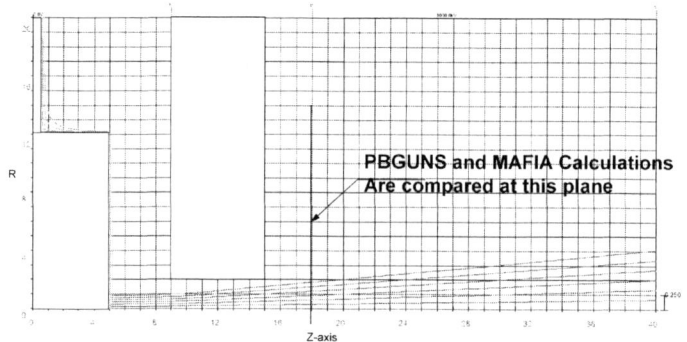

**Fig 2.** PBGUNS Geometry and particle trajectories

840

MAFIA is a software package that includes many electromagnetic simulation codes, including both time and frequency domain solvers in both 2.5D and 3D. For this problem, the mesh generator (M) was used to set up the geometry and mesh. For all of the runs discussed below, a mesh dimension of 10 μm x 10 μm was used. The static solver (S) was used to model the field present in the gun during the emission. The static solver solves Maxwell's equations for a given geometry and set of time- independent boundary conditions. This involves the implicit assumption that the applied field in the gun is constant during the electron bunch duration, which is valid for the device being modeled. The field generated in this manner was read into the 2D-time domain particle pusher (TS2). The particle pusher handles ejection and propagation of the electrons inside the gun. It is here that the bunch duration is defined. The simulation is broken into finite time steps (on the order of 10 fs for these simulations). At each time step the particles are moved under the influence of the last calculated field, while at each half time step the fields are updated due to the presence of the particles. The program combines groups of electrons into macro particles to reduce the computational resources required. Typically 75000 macro particles were used in these simulations, although fewer were used to model the shorter pulse durations. The post processor (P) was used to extract the particle positions and momenta, both longitudinal and transverse, at the plane of interest. The comparisons were made with particles taken from the center of the MAFIA bunch, except where otherwise noted.

PBGUNS [2] is a PC compatible, 2-D code that solves Poisson's equation using iterative relaxation technique on a rectangular array of squares. It computes trajectories of charged particles in electrostatic and magnetostatic focusing systems including the effects of space charge and self-magnetic fields. Either rectangular or cylindrically symmetric geometry may be used. The Poisson equation is solved by an alternate column relaxation technique known as the semi-iterative Chebyshev method. The transverse phase space plot was extracted at the measurement plane.

## PHASE SPACE COMPARISON

Fig 3 shows transverse phase space predicted by MAFIA for the center 1% slice of a 1 nC, 10 ps bunch accelerated in a field of 1GV/m across the gap of 1 mm. Fig 4 shows the DC transverse phase space predicted by PBGUNS for identical conditions. It is important to note that a DC code effectively measures only the slice phase space (since there is no long. variation in the beam). The transverse phase space does not vary significantly for bunch durations of 3 ps, 1ps, and 300 fs. The agreement between PBGUNS and MAFIA is very good, even for bunches much shorter than the gap transit time of 3 ps.

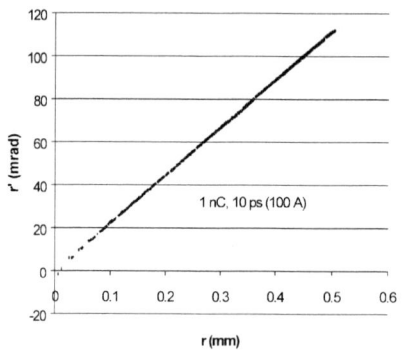

**Fig 3.** MAFIA r-r' plot 1 nC, 10 ps, center 1% of beam

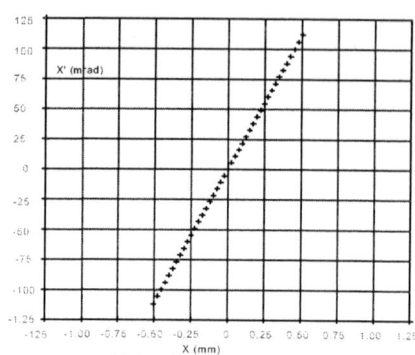

**Fig 4.** PBGUNS x-x' plot, 100 A. PBGUNS output comes as x-x', but for the cylindrically symmetric case the values in the 1$^{st}$ quadrant are equlivent to r-r'

The treatment of space charge effects in high current cases was investigated for MAFIA and PBGUNS. Table 1 gives the predicted maximum spot size and maximum divergence for various currents using these codes. For MAFIA, a bunch length of 10 ps was used. As shown on the Table, good agreement was found between MAFIA and PBGUNS for the beam radius and the max divergence. A current of 1000A is not possible in PBGUNS, as it is above the predicted DC Child's Law limit for this gun. The MAFIA results for a beam of 10 nC (corresponding to 1000 A in 10 ps) show a significant increase in bunch size, both transverse and longitudinal. While the beam is 1000A initially, by the time it leaves the gun it has increased in pulse duration to about 13 ps, thus reducing the effective current to ~750 A. The current density is further reduced due to increase in the transverse spot size.

| | **PBGUNS** | | **MAFIA** | | |
|---|---|---|---|---|---|
| CATHODE | BEAM | MAX | BEAM | MAX | NORM. |
| CURRENT | RADIUS | DIVERG. | RADIUS | DIVERG. | EMITTANCE |
| Ampere | mm | mrad | mm | mrad | π mm-mrad |
| 1 | 0.47 | 100 | 0.475 | 99 | 0.118 |
| 100 | 0.5 | 112 | 0.503 | 112 | 0.162 |
| 200 | 0.535 | 125 | 0.533 | 126 | 0.241 |
| 300 | 0.6 | 140 | 0.577 | 141 | 0.292 |
| 600 | 0.65 | 165 | (+) 0.633 | 170 | 0.617 |
| (*) 1000 | | | (+) 0.707 | 173 | 2.16 |

**Table 1.** Max Spot Size, Divergence & Emittance for 1A, 100A, 200A, 300A, 600A, 1000A for PBGUNS & MAFIA. (*) The 1000 A case is above the DC limit of Child's Law for the gun. The bunch length begins stretching immediately after emission, so that the effective current is only ~ 750A. (+) For the 600A and 1000A cases, the beam is larger than the anode aperture, and is clipped by the anode. The values shown are for the surviving particles.

The MAFIA results for the particle's r and $p_r$ are used to calculate the 1-σ normalized slice emittance (using the center 1% of the beam for the emittance calculation), using the expression:

$$\varepsilon = 2\sqrt{\left\langle r^2 \right\rangle \left\langle p_r^2 \right\rangle - \left\langle r \cdot p \right\rangle^2} \qquad (1)$$

# OPTIMIZATION USING PBGUNS

## Beam Collimating and Focusing

The electric field in the vicinity of the anode hole has a radial component, and therefore acts as a defocusing lens. To counteract this effect, studies were done into the effect of adding a curvature to the surface of the cathode. This introduces a negative radial electric field component, and acts as a focusing lens. It was found that the optimal curvature depended on the operating current of the gun. For a given current, it was possible to choose the curvature to provide a collimated beam. More severe curvature (smaller radius of curvature) could be employed to bring the beam to a focus anywhere beyond to anode, allowing the possibility to focus at the entrance to second stage linac. PBGUNS was used to shape the cathode to provide a converging, collimated or diverging beam at the exit of the anode. Fig. 5 shows the focusing effect of a 1 mm radius of curvature for a current of 100 A from a .25 mm radius spot. The beam is only convergent for cathode radii of curvature of 1.5 mm or less.

TRAJECTORIES AND EQUIPOTENTIALS

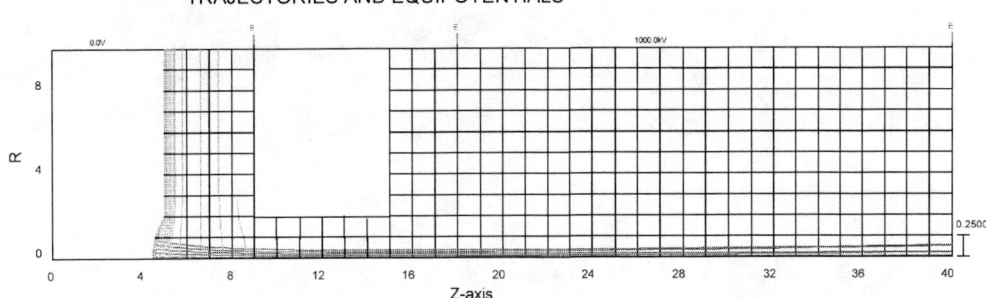

**Fig 5.** Trajectories and equipotentials for a cathode with a 1 mm radius of curvature.

## Spatial Charge Distribution

PBGUNS was also used to identify the optimal transverse spatial charge distribution to minimize the effect of space charge on the beam transverse phase space. Three cylindrically symmetric profiles were investigated: Gaussian, Uniform and Hollow. These profiles were investigated for both flat and curved cathodes. Fig. 6 shows the cathode density distributions for the curved cathode case.

843

**Fig 6.** Pictorial of the calculated current distributions.

Fig. 7 shows transverse phase space for the three initial current distributions at the simulation exit plane (E3). Also shown is the exit plane current distribution.

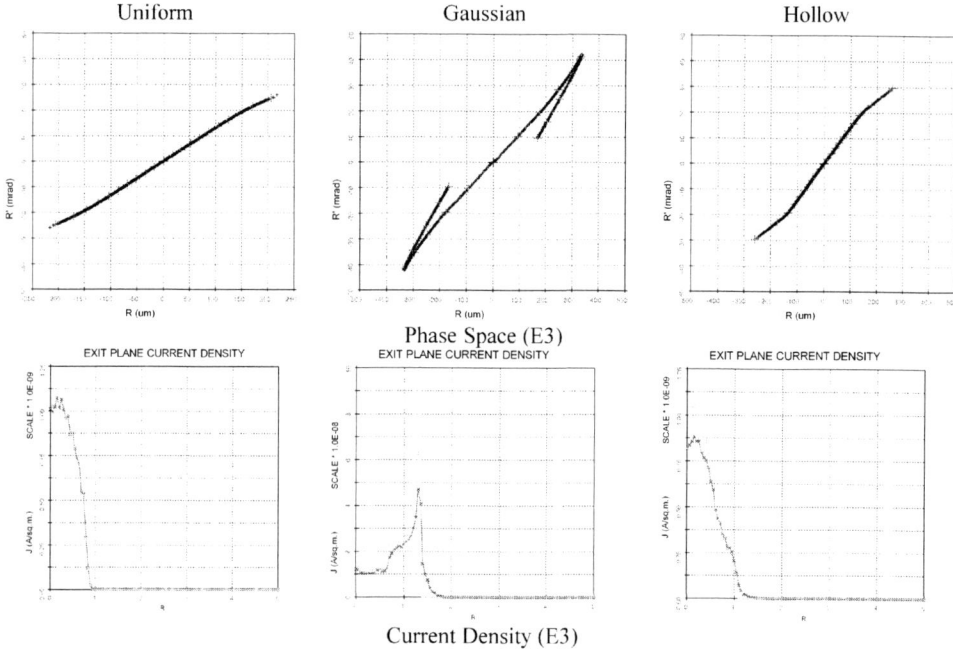

**Fig 7.** Phase space diagram and current density at exit plane for a 100 A current. The cathode has a 1 mm radius of curvature.

For a beam current of 100 A, there are clear differences in the output beam for the three density distributions. The gaussian current distribution, which gives a high current density near the axis, results in a somewhat distorted output phase space and an output (or exit plane) density distribution with a peak at the outer edge of the beam. The uniform and hollow distributions both give good exit plane density distributions and phase space plots.

Finally, in order to test the robustness of the computations, a 100 A uniform current density distribution case was executed for different anode exit apertures and different field gradients. For different anode exit apertures, it was found that the exit plane density distributions and phase space plots are quite insensitive to the anode exit aperture. This is due to the fact that the curved cathode surface tends to focus the beam at the anode plane so that the field that the beam experiences is similar in all configurations.

The results for applied field gradients of 0.5 GV/m and 1.5 GV/m show that lowering the gradient by a factor of two does not significantly alter the output beam character. However, increasing the gradient by 50% gives rise to an over focused beam and this results in some distortion of the exit plane phase space plot. The beam radius at E3 with this gradient is less than 100 μm.

# OPTIMIZATION USING MAFIA

MAFIA was used to optimize the temporal bunch length using the condition of maximum brightness. MAFIA was also used to investigate the longitudinal energy spread, the variation of the transverse phase space along the beam and the effect of the initial energy on the beam emittance.

## Bunch Length Optimization

Four bunch lengths were investigated: 10 ps, 3 ps, 1 ps, and 300 fs, all with uniform spatial and temporal profiles. The optimal pulse duration was found to be 1 ps, although the difference between 1 ps and 300 fs was within the error of the calculation. Several current values were run at 1 ps pulse length, again optimizing for the highest brightness. The highest brightness was found with a 1 ps bunch length at a current of 50 A. Table 2 shows the pulse duration, charge, current and brightness for each of these simulations. Emittance is calculated using eq. 1.

| Current | Charge | Bunch length | Emittance | Brightness |
|---|---|---|---|---|
| A | nC | ps | π mm mrad | A/(m^2rad^2) (10^14) |
| 100 | 1 | 10 | 0.16734 | 3571 |
| 100 | 0.3 | 3 | 0.0882 | 12854 |
| 100 | 0.1 | 1 | 0.062 | 26014 |
| 100 | 0.03 | 0.3 | 0.0728 | 18868 |
| 500 | 0.5 | 1 | 0.207 | 11668 |
| 250 | 0.25 | 1 | 0.132 | 14348 |
| 50 | 0.05 | 1 | 0.0386 | 33557 |
| 10 | 0.01 | 1 | 0.0446 | 5027 |

**Table 2.** Optimization of 1-σ normalized emittance and brightness with respect to pulse duration and charge. It is important to note that the calculated emittance includes no thermal component.

## Longitudinal Phase Space & Front/Back Variation

One aspect where time resolution is clearly required is the study of the longitudinal variation of the electron beam, both in terms of energy and transverse phase space. Fig. 8 shows the longitudinal momentum spread across the beam when the center of the beam is at the measurement plane for a 10 ps, 1nC bunch. It is interesting to note that this shape is independent of charge – the absolute extent of the front/back variation is charge dependent, but the shape is not.     Fig. 9 shows the longitudinal momentum spread for a 10 ps, 3nC bunch.

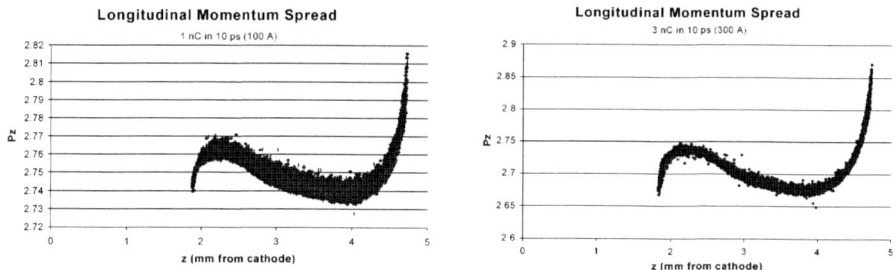

**Fig 8 & 9.** Charge induced longitudinal momentum spread. Pz is in $\beta_z\gamma$.

Fig. 10 shows the MAFIA predictions for the transverse phase of slices taken from the front, middle and back of the bunch (relative to the cathode). The variation from front to back is small. It is important, however, since most diagnostic techniques measure only the integrated phase space, which would involve drawing an ellipse, which encompassed all three slices. The emittance calculated from this ellipse would be much larger than the emittance of the individual slices.

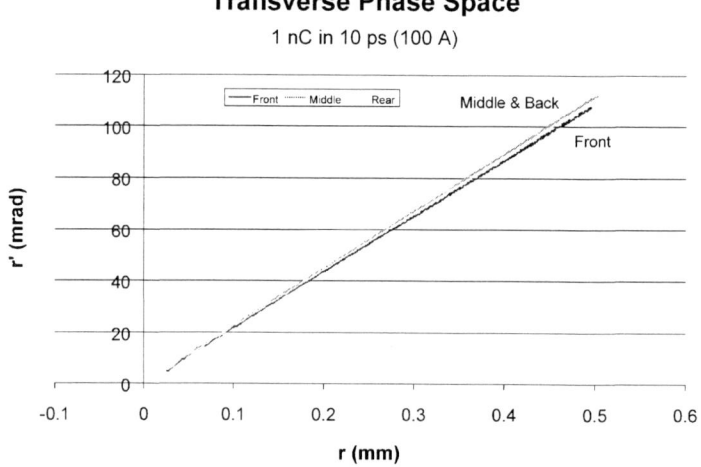

**Fig 10.** Longitudinal variation of the transverse phase space.

# Effect of Initial Energy on Beam Emittance

There exist several possible sources of random initial energy for electrons leaving a photocathode. Thermal energy would account for a random distribution with a magnitude of about .03 eV at room temperature. Photoemission results in electrons being "born" with initial energies up to (hv-φ), where hv is the photon energy and φ is the cathode work function. For metal photocathodes irradiated with UV lasers, this difference can be a 1-3 eV, especially considering the expected reduction in φ due to the Schottky effect in GV/m fields. It is also possible that surface irregularities on the cathode may give rise to field lines containing a transverse component. For a rough cathode surface, the field in the area of a protrusion will be perpendicular to the local surface of the protrusion and not perpendicular to the macroscopic surface. The skin depth of the applied voltage is considerably larger than the optical penetration depth of ~10 nm [3], thus the field experienced by an electron traveling to the surface of the cathode inside a protrusion will be the nearly the same as the field at the surface of the protrusion. With even a modest value of the field enhancement, in a 1 GV/m field an electron will gain several tens of eV before being emitted, and some fraction of this energy will be transverse. All three of these sources of initial energy can be corrected for to some extent, by cooling the cathode, choosing a photon energy well matched to the work function and carefully polishing the cathode. It is instructive, however, to investigate the expected degradation of the beam emittance due to these effects.

The geometry used for these calculations differed from the geometry shown in Fig. 1. Instead of an anode which was 1.5 mm thick with a .5 mm radius hole, a closed anode (no hole) of 50 μm thickness was used. The anode material was set to be transparent to electrons but not to electric fields, thus simulating a thick foil (no scatter occurred within the anode, however, unlike a real foil). This was done to remove the bending of the field near the anode hole and the resulting defocusing effect on the beam, allowing a more accurate calculation of the beam emittance. This new geometry is shown in Fig. 11.

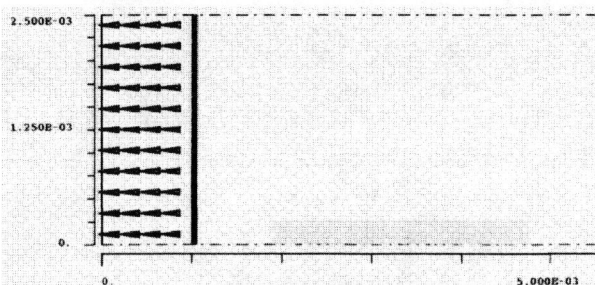

**Fig 11.** Geometry used for thermal emittance calculation, including field lines and electron bunch.

Using this geometry, simulations were performed for 0, 1, 10, and 100 eV initial energies. In each case an initial angle was randomly assigned to each macroparticle. Table 3 shows the resulting emittance values for both 1 nC and 100 pC bunch charge. A bunch length of 10 ps was used in this calculation, with uniform spatial and temporal profiles.

| Current | Charge | Initial Energy | Emittance |
|---|---|---|---|
| A | nC | eV | $\pi$ mm mrad |
| 100 | 1 | 0 | 0.251 |
| 100 | 1 | 1 | 0.287 |
| 100 | 1 | 10 | 0.597 |
| 100 | 1 | 100 | 1.704 |
| 10 | 0.1 | 0 | 0.023 |
| 10 | 0.1 | 1 | 0.171 |
| 10 | 0.1 | 10 | 0.530 |
| 10 | 0.1 | 100 | 1.682 |

**Table 3.** Effect of initial random energy distribution on emittance

The results yield the expected [4] scaling (the emittance increases as the square root of the initial energy once the initial energy is the dominant contribution). Above 1 eV initial energy, this is the dominant contribution to the beam emittance for the 100 A beam. Even at 1 eV initial energy, the 10 A beam is dominated by these effects. To preserve the emittance of the 10 A beam, the beam's initial energy needs to be kept to room temperature thermal levels (~.03 eV).

## SUMMARY

A comparison of two codes suitable for electron gun simulation was performed. Good agreement in transverse phase space values was found between the electrostatic code (PBGUNS) and the time dependent code (MAFIA) for a variety of pulse durations, even for pulse durations short compared to the electron transit time of the accelerating region. The electrostatic codes have an advantage in terms of required computational resources and run time, and are therefor a good choice for modeling jobs in which the longitudinal variation is unimportant.

PBGUNS was used to model the effects of curving the cathode surface to provide a focusing effect. The curvature of the cathode can be chosen so that it counteracts the defocusing effect of the anode aperture, allowing for a collimated beam. A deeper curvature can be chosen to provide a focus at a specific location. PBGUNS was also used to investigate the effect of different initial spatial current distributions.

MAFIA was used to optimize the bunch length and beam current to provide the highest brightness beam and to investigate the effect of varying levels of random initial energy on the beam emittance. MAFIA was also used to explore the energy spread and the longitudinal variation of the transverse phase space, both of which require time dependence.

## ACKNOWLEDGMENTS

The authors would like to thank Harold Kirk for his assistance in utilizing MAFIA and Vadim Dudnikov and Jack Boers for their assistance with PBGUNS, Ilan Ben-Zvi and Veljko Radeka for their support. This work was supported by DE-AC02-98CH10886 and DE-FG02-97ER82336.

## REFERENCES

1.  T. Srinivasan-Rao, J. Smedley, *Advanced Accelerator Concepts workshop*, AIP Conference Proceedings 398, NY 1997, P. 7307.
2.  J.E. Boers, *Proceedings of the 1995 PAC*, Dallas TX, pp. 2312-3, May 1995
3.  Sommer, A. H., *Photoemissive Materials*, New York, Krieger Publishing Co., 1980, pp. 35-41
4.  Travier, C., Nucl. Instr. and Meth in Phys. Res. **A 340**, pg. 36 (1994)

# Design, Analysis and Cold Test of a 17 GHz RF Gun

M.A.Shapiro, W.J.Brown, K.E.Kreischer, and R.J.Temkin

*Plasma Science and Fusion Center, Massachusetts Institute of Technology,*
*Cambridge, Massachusetts 02139*

**Abstract.** We analyzed and cold tested a 17 GHz 1-1/2-cell RF gun cavity excited through two coupling holes in the broad wall of a rectangular waveguide. An equivalent circuit theory and an advanced field theory were developed to describe the excitation of an 1-1/2-cell RF gun cavity. SUPERFISH was used to calculate the majority of the equivalent circuit elements as well as the field theory parameters. The matching values of magnetic polarizabilities of the coupling holes were determined by comparing the theory with the measurements. The field theory results were used to model the electric field distribution and accelerating gradient in the RF gun cavity. From this analysis we concluded that the RF gun cavity support gradients as high as 300 MV/m at the cathode without breakdown.

## 1. INTRODUCTION

The 1-1/2-cell cavity operating in a $TM_{01}$-like mode is used in the 17 GHz MIT RF gun. In this experiment [1], RF power from the klystron couples into the cavity through two coupling holes in a broad wall of a rectangular waveguide (see Fig. 1). The 0- and $\pi$-modes in a 1-1/2-cell cavity are illustrated in Fig.1 . These modes can be calculated using the 2D eigenmode solver SUPERFISH [2]. The $\pi$-mode has the proper field distribution for accelerating the electrons. Thus, the two coupling holes are designed to excite selectively the $\pi$-mode. For a number of reasons (holes placed asymmetrically in the waveguide, fabrication errors in the cells or the holes), the excited $\pi$-mode may not have equal field magnitudes in the half- and full-cells at the operating frequency (unbalanced cavity). The 0-mode can be excited as well. Such a combined excitation affects the acceleration. To analyze this effect we develop an advanced theory of RF gun cavity excitation through the coupling holes. We implement two approaches to describe the 1-1/2-cell cavity excitation: circuit theory (Sec. 2) and field theory (Sec. 3) [1].

The circuit theory is analytic, and calculates the frequency shift caused by the coupling holes. However, the circuit theory does not describe the field

---

[1] The problem of coupling hole excitation can be also solved using 3D EM codes such as HFSS, GdfidL, or MAGIC3D .

CP472, *Advanced Accelerator Concepts: Eighth Workshop,*
edited by W. Lawson, C. Bellamy, and D. Brosius
© 1999 The American Institute of Physics 1-56396-889-4/99/$15.00

distributions in the normal π- and 0-modes. This approach has been previously investigated at MIT [3]. The RF gun cavity equivalent circuit presented here has an improved iris coupling representation.

The advanced field theory of 1-1/2-cell cavity excitation gives the field distribution excited in the cavity. This theory is based on the theory of waveguide and cavity excitation through a coupling hole [4]. We generalize this theory for two mode excitation using two coupling holes.

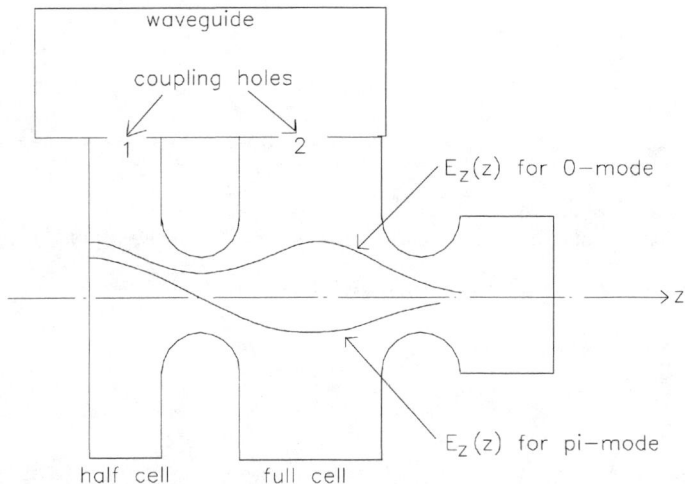

**FIGURE 1.** 1-1/2-cell cavity excited through two coupling holes and the cavity modes.

SUPERFISH is used to calculate the mode parameters of the field theory and the cell parameters of the circuit theory, such as the eigenfrequency, mode stored energy, and magnetic field at the wall. The polarizabilities of the coupling holes are used as free parameters in the theory.

We implemented both the circuit and the field theories to analyze the cavity used in the MIT experiment [1]. We calculated the $S_{11}$ scattering matrix element dependence on the frequency. By comparing the theory and the cold test (Sec. 4) we can determine the polarizabilities of the coupling holes. For these matching polarizabilities, we analyzed the field distribution and its modal content at the operating frequency.

## 2. CIRCUIT THEORY

The equivalent circuit of the 1-1/2-cell cavity (Fig. 2) includes two LC circuits: circuit $L_1$, $C_1$, $R_1$ represents the half-cell, circuit $L_2$, $C_2$, $R_2$ represents the full-cell. The LC circuit parameters can be calculated using the SUPERFISH runs

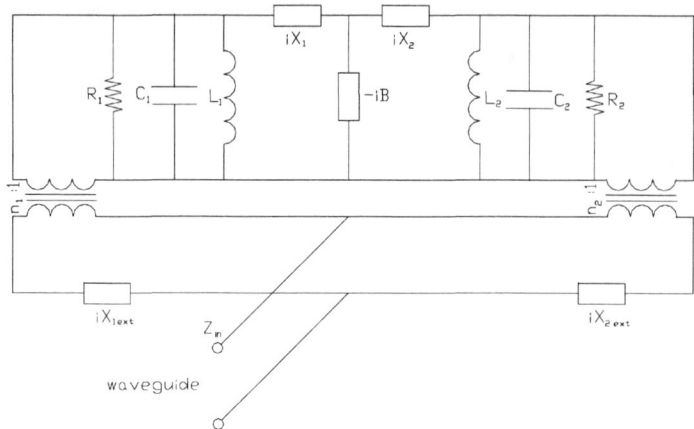

**FIGURE 2.** Equivalent circuit of the 17GHz RF gun cavity excited through a broad wall of rectangular waveguide. Circuit parameters: $C_1$=0.26pF, $L_1$=0.35nH, $X_{1ext}$=53.5k$\Omega$, $R_1$=140k$\Omega$, $n_1$=7.43, $C_2$=0.13pF, $L_2$=0.7nH, $X_{2ext}$=94.1k$\Omega$, $R_2$=440k$\Omega$, $n_2$=4.93, $X_1$= -485$\Omega$, $X_2$= -1976$\Omega$, B=0.08(1/$\Omega$) . Frequencies of LC circuits: $f_1$=16.809GHz, $f_2$=16.961GHz

**FIGURE 3a.** Half-cell plus iris with electric wall and its equivalent circuit.

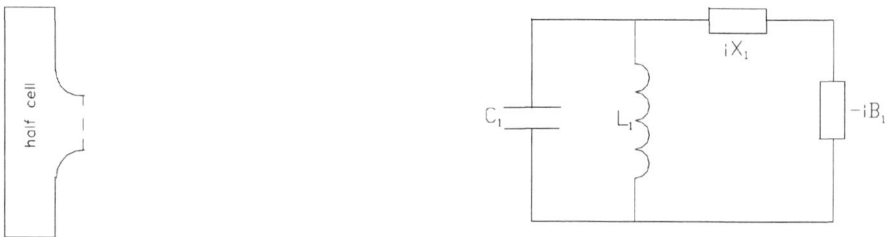

**FIGURE 3b.** Half-cell plus iris with magnetic wall and its equivalent circuit.

for the half-cell and the full-cell excluding the iris. SUPERFISH gives the following parameters of the cells: stored energies $W_{1,2}$, specific shunt impedances $r_{cy1,2}$, quality factors $Q_{1,2}$, resonance frequencies $\omega_{1,2}$. Hence the circuit parameters can be determined as follows:

$$C_{1,2} = \frac{4W_{1,2}}{(E_{av}l_{1,2})^2} \qquad L_{1,2} = \frac{1}{\omega_{1,2}^2 C_{1,2}} \qquad R_{1,2} = \frac{r_{cy1,2}l_{1,2}}{4}$$

where $l_{1,2}$ are the lengths of the cells, $E_{av}$ is an average axial electric field (always 1MV/m in the SUPERFISH simulations).

We calculate the parameters of the iris $X_1$, $X_2$ and $B$ as follows:

First, we run SUPERFISH for the half-cell with the half-iris and electric wall at the iris aperture (Fig. 3a); the impedance $X_1$ represents the half-iris. Using SUPERFISH we calculate the frequency of such a cavity. The frequency shift with respect to $\omega_1$ allows us to calculate $X_1$.

Second, we run SUPERFISH for the half-cell cavity shown in Fig. 3b where the magnetic wall is at the aperture of the iris. The frequency shift with respect to the cavity at Fig. 3a gives the admittance $B_1$. We do the same runs with the full-cell to calculate $X_2$ and $B_2$. We wind up with the T-circuit containing the elements $X_1$, $X_2$, and $B=B_1+B_2$ (Fig. 2). The T-circuit plus the $L_1C_1R_1$ and $L_2C_2R_2$ circuits represent the 1-1/2-cell cavity without coupling holes.

Third, we run SUPERFISH for the entire 1-1/2-cell structure without coupling holes to calculate the frequencies of two normal modes ($\pi$- and 0-modes). These frequencies should coincide with the normal frequencies of the circuits $L_1C_1R_1$ and $L_2C_2R_2$ coupled by the T-circuit. To match these frequencies we reduce $X_1$ and $X_2$ by 3 and 4% respectively.

To model the coupling holes we represent the transformers with transformation ratios $n_{1,2}$ and external impedances $X_{1ext}$, $X_{2ext}$ in Fig. 2 [3,4]:

$$n_{1,2}^2 = -\frac{\mu_0 H^2(1,2)Z_{wg}C_{1,2}\omega_{1,2}a^3 k_{gz}b}{2\pi^2 W_{1,2}\cos^2\left(\frac{\pi x_{1,2}}{a}\right)}, \qquad X_{1,2ext} = Z_{wg}\frac{a^3 bk_{gz}}{\pi^2 \alpha_{1,2}\cos^2\left(\frac{\pi x_{1,2}}{a}\right)} \qquad (1)$$

where $H(1,2)$ are the SUPERFISH magnetic fields calculated at the position of the holes 1 or 2 (assumed imaginary), $\mu_0$ is the permeability of free space, $k_{gz}$ is an axial wavenumber in the rectangular waveguide $TE_{10}$ mode, $Z_{wg}=Z_0 k/k_{gz}$ is the impedance of the waveguide mode, $Z_0=377\Omega$ is the free space impedance, $k$ is the vacuum wavenumber, $a$ and $b$ are the broad and narrow waveguide wall dimensions, $x_{1,2}$ are the positions of the holes 1 and 2 on the broad wall of the waveguide, $\alpha_{1,2}$ are the magnetic polarizabilities of the holes 1 and 2.

In the cold test we measure $S_{11}$ scattering matrix element of the waveguide loaded by the cavity. For the equivalent circuit (Fig. 2) we derive the following $S_{11}$ expression:

$$S_{11} = \frac{Z_{in} - Z_{wg}}{Z_{in} + Z_{wg}}$$

where the input impedance is

$$Z_{in} = \frac{\left( n_1^2 \left( 1 + \dfrac{Z_2(iX_1 + i/B)}{D_1} \right) Z_1/D_2 + iX_{1ext} \right)\left( n_2^2 \left( 1 + \dfrac{Z_2(iX_2 + i/B)}{D_1} \right) Z_2/D_2 + iX_{2ext} \right) + \dfrac{n_1^2 Z_1^2 n_2^2 Z_2^2}{B^2 D_1^2 D_2^2}}{n_2^2 \left( 1 + \dfrac{Z_2(iX_2 + i/B)}{D_1} \right) Z_2/D_2 + iX_{2ext} + n_1^2 \left( 1 + \dfrac{Z_2(iX_1 + i/B)}{D_1} \right) Z_1/D_2 + iX_{1ext} + \dfrac{2in_1 Z_1 n_2 Z_2}{B(-D_1)D_2}}$$

here

$$D_1 = (iX_1 + i/B)(iX_2 + i/B) + 1/B^2$$

$$D_2 = \left( 1 + \frac{Z_1(iX_2 + i/B)}{D_1} \right)\left( 1 + \frac{Z_2(iX_1 + i/B)}{D_1} \right) + \frac{Z_1 Z_2}{B^2 D_1^2}$$

and the 1$^{st}$ and 2$^{nd}$ cell impedances are

$$Z_{1,2} = \frac{i\omega_{1,2}/C_{1,2}}{\omega^2 - \left( \omega_{1,2} - i\dfrac{1}{2R_{1,2}C_{1,2}} \right)^2}$$

Comparison of the equivalent circuit results to the cold test is discussed in Sec. 4.

## 3. FIELD THEORY

Unlike the circuit theory, which treats the half- and full-cells individually, the field theory approach to modeling the RF gun considers the normal modes of the entire 1-1/2-cell structure.

Following coupling hole theory [4], the coupling is represented as magnetic dipoles $m(1)$ and $m(2)$ placed at the holes 1 and 2 and oriented along the magnetic field in the cells. The magnetic dipole moment $m$ is proportional to the total magnetic field at the hole (from both cavity side and waveguide side), $m = \alpha H$, where $\alpha$ is the polarizability of the hole.

We describe the actual mode excited in the RF gun by the magnetic dipoles $m(1)$ and $m(2)$ in terms of complex amplitudes ($A_0$ and $A_\pi$) of the 0- and $\pi$-modes of the unperturbed RF gun (i.e. without coupling holes). These relative amplitudes depend on the polarizabilities of both coupling holes ($\alpha_1$ and $\alpha_2$) and on the stored energy in the cavity for both the 0- and $\pi$-modes $W_{S0}$ and $W_{S\pi}$ as well as the magnetic fields at each coupling hole $H_{S0}(1)$, $H_{S0}(2)$, $H_{S\pi}(1)$, and $H_{S\pi}(2)$, which are calculated using SUPERFISH.

The complex amplitudes of these two modes are given by

$$A_0 = -2i\pi \frac{\varepsilon_0 Z_0 \omega^2 (\tilde{m}(1) H_{S0}(1) + \tilde{m}(2) H_{S0}(2))}{(\omega^2 - \omega_{S0}^2) W_{S0}} \tag{2}$$

and

$$A_\pi = -2i\pi \frac{\varepsilon_0 Z_0 \omega^2 (\tilde{m}(1) H_{S\pi}(1) + \tilde{m}(2) H_{S\pi}(2))}{(\omega^2 - \omega_{S\pi}^2) W_{S\pi}}, \tag{3}$$

where $\varepsilon_0$ is the permittivity of free space, $\tilde{m} \equiv m \dfrac{Z_0}{E_{10}}$ is a magnetic moment normalized to the waveguide electric field amplitude $E_{10}$, $\omega = 2\pi f$ is the operating circular frequency (i.e. the klystron frequency), and

$$\omega_{S0} = 2\pi f_{S0} \left(1 - \frac{i}{2Q_{S0}}\right) \quad \text{and} \quad \omega_{S\pi} = 2\pi f_{S\pi} \left(1 - \frac{i}{2Q_{S\pi}}\right)$$

are the SUPERFISH 0- and $\pi$-mode complex frequencies where $Q_{S0}$ and $Q_{S\pi}$ are the quality factors of the unperturbed 0- and $\pi$-modes calculated from SUPERFISH. For a given drive frequency, $\omega$, the actual mode excited in the RF gun can now be expressed in terms of the SUPERFISH 0- and $\pi$-modes and their corresponding complex excitation amplitudes. For example, the actual on-axis longitudinal electric field profile in the RF gun (normalized to $E_{10}$) is given by

$$E_z(z,\omega) = A_\pi(\omega) E_{S\pi z}(z) + A_0(\omega) E_{S0z}(z), \tag{4}$$

where $E_{S\pi z}(z)$ and $E_{S0z}(z)$ are the SUPERFISH on-axis longitudinal electric fields for the unperturbed $\pi$- and 0-modes respectively.

Using the theory of waveguide excitation [4], we also calculate the complex reflection coefficient, $S_{11}$, of the $TE_{10}$ mode in the waveguide from the RF gun to be

$$S_{11} = -1 + \frac{4\pi^2\omega}{ca^2bk_{gz}}\left(\tilde{m}(1)\cos\frac{\pi x_1}{a} + \tilde{m}(2)\cos\frac{\pi x_2}{a}\right), \tag{5}$$

where $x_1$ and $x_2$ are the positions of the coupling holes along the broad wall of the waveguide and $c$ is speed of light.

The expressions for the magnetic moments $\tilde{m}(1)$ and $\tilde{m}(2)$ can be determined self-consistently using the cavity and waveguide excitation formulas (2), (3) and (5) to obtain

$$\tilde{m}(1) = i\frac{c}{4\pi\omega a}\frac{-\alpha_1\Lambda_{22}\cos\dfrac{\pi x_1}{a} + \alpha_2\Lambda_{12}\cos\dfrac{\pi x_2}{a}}{\Lambda_{11}\Lambda_{22} - \Lambda_{12}\Lambda_{21}}$$

and

$$\tilde{m}(2) = i\frac{c}{4\pi\omega a}\frac{\alpha_1\Lambda_{21}\cos\dfrac{\pi x_1}{a} - \alpha_2\Lambda_{11}\cos\dfrac{\pi x_2}{a}}{\Lambda_{11}\Lambda_{22} - \Lambda_{12}\Lambda_{21}},$$

where

$$\Lambda_{11} = 1 + \frac{\mu_0\omega^2\alpha_1 H_{S0}^2(1)}{2(\omega^2 - \omega_{S0}^2)W_{S0}} + \frac{\mu_0\omega^2\alpha_1 H_{S\pi}^2(1)}{2(\omega^2 - \omega_{S\pi}^2)W_{S\pi}} - \frac{i\pi^2\alpha_1\cos^2\dfrac{\pi x_1}{a}}{a^3bk_{gz}},$$

$$\Lambda_{12} = \frac{\mu_0\omega^2\alpha_1 H_{S0}(1)H_{S0}(2)}{2(\omega^2 - \omega_{S0}^2)W_{S0}} + \frac{\mu_0\omega^2\alpha_1 H_{S\pi}(1)H_{S\pi}(2)}{2(\omega^2 - \omega_{S\pi}^2)W_{S\pi}} - \frac{i\pi^2\alpha_1\cos\dfrac{\pi x_1}{a}\cos\dfrac{\pi x_2}{a}}{a^3bk_{gz}},$$

$$\Lambda_{21} = \frac{\mu_0\omega^2\alpha_2 H_{S0}(1)H_{S0}(2)}{2(\omega^2 - \omega_{S0}^2)W_{S0}} + \frac{\mu_0\omega^2\alpha_2 H_{S\pi}(1)H_{S\pi}(2)}{2(\omega^2 - \omega_{S\pi}^2)W_{S\pi}} - \frac{i\pi^2\alpha_2\cos\dfrac{\pi x_1}{a}\cos\dfrac{\pi x_2}{a}}{a^3bk_{gz}},$$

and

856

$$\Lambda_{22} = 1 + \frac{\mu_0 \omega^2 \alpha_2 H_{S0}^2(2)}{2(\omega^2 - \omega_{S0}^2)W_{S0}} + \frac{\mu_0 \omega^2 \alpha_2 H_{S\pi}^2(2)}{2(\omega^2 - \omega_{S\pi}^2)W_{S\pi}} - \frac{i\pi^2 \alpha_2 \cos^2 \frac{\pi x_2}{a}}{a^3 bk_{gz}} .$$

Equations (2), (3) and (5) are connected by the energy conservation law:

$$|S_{11}|^2 + \frac{2\pi f_{S0} |A_0|^2 \left(\omega^2 + |\omega_{S0}|^2\right) W_{S0}}{Q_{S0} \omega \varepsilon_0 c^2 abk_{gz}} + \frac{2\pi f_{S\pi} |A_\pi|^2 \left(\omega^2 + |\omega_{S\pi}|^2\right) W_{S\pi}}{Q_{S\pi} \omega \varepsilon_0 c^2 abk_{gz}} = 1 .$$

Thus far, we have described a theory that completely describes the cavity filling and mode mixing characteristics of the RF gun provided the polarizabilities of the coupling holes are known. For holes of simple geometry, these polarizabilities can be calculated analytically. Particularly, for an elliptical hole whose dimensions are major radius $\frac{dz}{2}$ and minor radius $\frac{dx}{2}$, the polarizability is

$$\alpha = \frac{2\pi (dz/2)^3 e^2}{3[K(e) - E(e)]} f(t) \text{ in SI units. Here } e = \sqrt{1 - (dz/dx)^2}, \; K(e) \text{ and } E(e) \text{ are}$$

elliptical integrals, and the function $f(t)$ of coupling wall thickness $t$ can be determined using [5] ($f(0) = 1$ for infinitely small thickness). For the case of the RF gun dimensions, we can make an elliptical approximation with $dz_1 = dz_2 = 0.508$ cm, $dx_1 = 0.165$ cm, $dx_2 = 0.229$ cm and an average wall thickness $t=0.025$ cm. According to theory [5], the factor $f(t)$ ranged from 0.49 to 0.52 for such coupling hole dimensions. As a result, the directly calculated polarizabilities are $\alpha_1 = 10.6 \cdot 10^{-9} m^3$ and $\alpha_2 = 13.4 \cdot 10^{-9} m^3$.

However, the elliptical approximation used to describe the coupling holes is not good enough. Determination of the actual polarizabilities of the coupling holes is complicated by the non-elliptical shape of the holes as well as the non-uniform thickness across the holes. In order to obtain accurate results, the polarizabilities $\alpha_1$ and $\alpha_2$ of the holes need to be considered as free parameters of the coupling theory described above. These parameters can then be determined from comparison of an $S_{11}$ curve obtained from experimental cold tests of the cavity and the theoretically calculated $S_{11}$ curves given by Eq. (5).

## 4. COMPARISON TO COLD TEST

Figure 4 shows the $S_{11}$ frequency dependence measured using a vector network analyzer. From this measurement we determine the cavity Q-factor of 2700 and the coupling coefficient $\beta=1.56$ which is the ratio of the cavity Q-factor to the external Q-factor. Using these measured numbers and the SUPERFISH simulation we conclude that the electric field at the cathode (flat cavity wall) in the RF gun is about 300 MV/m for a klystron power of 4 MW used in our experiment.

Let us compare the measured $S_{11}$ frequency dependence with the field theory first. In the field theory, the following SUPERFISH data is used: $f_{S0}=17.281$ GHz, $W_{S0}=9.9\mu J$, $f_{S\pi}=17.451$ GHz, $W_\pi=19.4\mu J$ . The measured Q=2700 for the $\pi$-mode in the experimental cavity is 70% of the calculated SUPERFISH value. Therefore, to match the cold test and the theory, $Q_0=4200$ (70% of the SUPERFISH value) and $Q_\pi=2700$ are used in the theory.

We determine the magnetic polarizabilities of the coupling holes by matching the measured and calculated $S_{11}$ frequency curves. The theoretical $S_{11}$ curve for the matching polarizabilities $\alpha_1=13.7 \cdot 10^{-9}\,m^3$ and $\alpha_2=14.2 \cdot 10^{-9}\,m^3$ is plotted in Fig. 4.

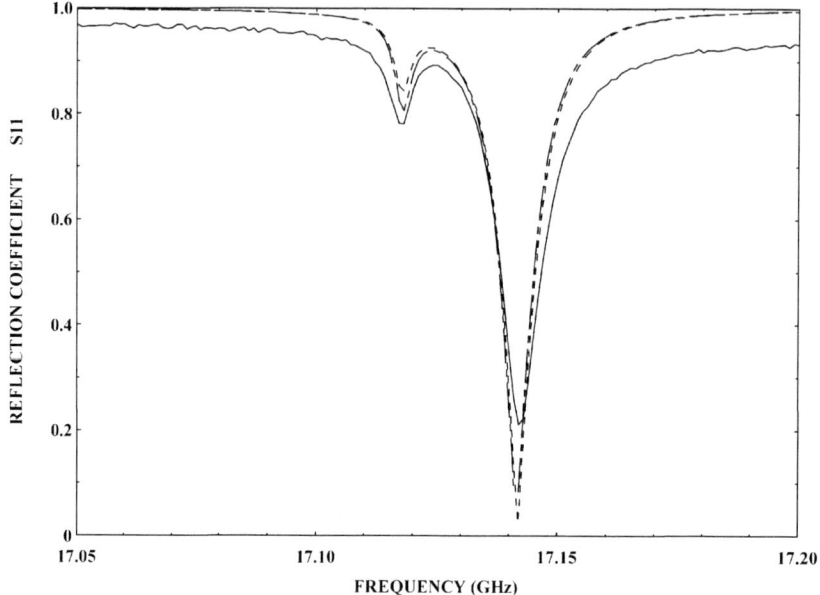

**FIGURE 4.** $S_{11}$ frequency dependence: solid line – cold test, dotted line – field theory for matching polarizabilities, broken dotted line – equivalent circuit theory.

For the matching polarizabilities of the coupling holes the coefficients of excitation of 0- and $\pi$-modes (2) and (3) are calculated. The excitation results in 93% of the power stored in the $\pi$-mode and 7% - in the 0-mode. Superposition of the SUPERFISH modes (real and imaginary parts of the $E_z$ field on axis) represented by Eq. (4) is plotted in Fig. 5.

We employ the calculated field distribution to analyze the accelerating gradient. A simple single particle simulation of electron dynamics in the RF gun indicates the average accelerating gradient of 85 MV/m for a peak electric field of 300 MV/m at the cathode.

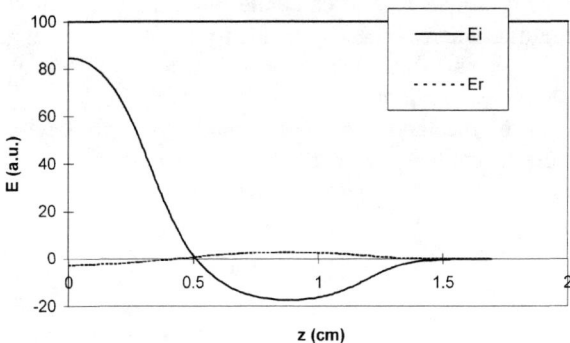

**FIGURE 5.** Actual on axis cavity field profile (real and imaginary parts of the electric field) calculated with the field theory using the SUPERFISH modes.

Before we compare the measurement and equivalent circuit theory, let us note that the circuit theory does not treat coupling through the iris and holes as accurately as the field theory. For this reason, in Sec. 2, we changed the impedances $X_1$ and $X_2$ to correct the iris coupling in the circuit theory. Coupling through the holes is described in the circuit theory by the external impedances $X_{1ext}$ and $X_{2ext}$ , and transformation ratios $n_1$ and $n_2$ given by Eqs. (1). We calculate the external impedances from Eqs. (1) for the same polarizabilities $\alpha_1=13.7 \cdot 10^{-9}$ m$^3$ and $\alpha_2=14.2 \cdot 10^{-9}$ m$^3$ used in the field theory; as a result we get $X_{1ext}=53.4$k$\Omega$ and $X_{2ext}=94.1$k$\Omega$. The parameters $n_1$ and $n_2$ calculated from Eqs. (1) should be corrected because these parameters depend on the capacitances $C_1$ and $C_2$ , and stored energies $W_1$ and $W_2$ in the cells, which do not account for the energy stored in the iris. For our geometry of the 1-1/2-cell cavity, the iris capacitance affects the transformation ratios of the coupling holes. To match the equivalent circuit theory with the field theory, we calculate $n_1$ and $n_2$ from Eqs. (1) and increase them by 3.5 and 2%, respectively. Thus, all the circuit parameters modeled for the cavity are determined. Figure 4 shows good agreement between the field and circuit theories.

## CONCLUSION

We developed a circuit theory, utilizing a simple model of the cavity, and an advanced field theory. The field theory gives more accurate analysis of excitation through the coupling holes as well as iris coupling. Both theories are based on a quasi-static approximation representing a coupling hole as a magnetic dipole. Comparing the theory and the cold test we determined the magnetic polarizabilities of the coupling holes and thus completed the determination of all the parameters of the cavity model. The equivalent circuit of the RF gun is corrected based on the field theory results.

The field theory allowed us to calculate the field distribution excited in the cavity including the iris, whereas the circuit theory gives only average electric field amplitudes in the cells. The field distribution, calculated in the existing RF gun cavity, provides a high accelerating gradient.

The next step of the theoretical study of the RF gun cavity will be a 3D analysis using an electromagnetic code. This study will allow us to build a better model not utilizing the quasi-static approximation for the coupling.

## ACKNOWLEDGEMENTS

The authors are grateful to Dr. S.Trotz and Dr. M.. Pedrozzi for useful discussions.

## REFERENCES

1. Trotz, S., Brown, W.J., Danly, B.G., Hogge, J.-P., Khusid, M., Kreischer, K.E., Shapiro, M., and Temkin, R.J., *Advanced Accelerator Concepts,* Seventh Workshop, AIP Conf. Proc. 398, AIP Press, Woodbury, NY, pp. 717-729 (1996).
2. Billen, J.H., *PARMELA/SUPERFISH User's Manual*, Los Alamos Accelerator Code Group (LAACG), LA-UR-96-1835.
3. Lin, L.C.-L., Chen, S.C., and Wurtele, J.S., *Nucl. Instr. and Meth. in Phys. Res.* **A 275**, 274-284 (1997).
4. Collin, R.E., *Field Theory of Guided Waves*, IEEE Press. 1991.
5. Radak, B., and Gluckstern, R.L., *IEEE Trans. Microwave Theory Tech.*, **43**, 194-204 (1995).

# Experimental Results of the MIT 17 GHz RF Gun

W. J. Brown, S. Trotz* , K. E. Kreischer, M. Pedrozzi[+],
M. A. Shapiro, R. J. Temkin

*Plasma Science and Fusion Center*
*Massachusetts Institute of Technology*
*Cambridge, MA  02139 USA*

*\*Present Address:  Dartmouth College, Hanover, NH*
*[+]Address after Sept. 1, 1998:  Paul Scherrer Inst., Switzerland*

**Abstract.** We report on experimental results of a 17 GHz RF photocathode electron gun. This is the first photocathode electron gun to operate at a frequency above 2.856 GHz. The 1.5 cell, $\pi$–mode, copper cavity was tested with 50 ns pulses from a 17.150 GHz klystron amplifier built by Haimson Research Corp. A Bragg filter was used at the RF gun to reduce the reflection of parasitic modes back into the klystron. Coupling hole theory in conjunction with cold test measurements was used to determine the field profile in the RF gun. The particle in cell code MAGIC was used to simulate the beam dynamics in the RF gun. With power levels of 4 MW, the on axis electric field at the cathode exceeds 300 MV/m, corresponding to an average accelerating gradient of 200 MV/m over the first half cell of the gun. Breakdown was observed at power levels above 5 MW. Electron bunches were produced by 20 µJ, 1 ps UV laser pulses impinging on the RF gun copper photocathode and were measured with a Faraday cup to have up to 0.1 nC of charge. This corresponds to a peak current of about 100 A, and a density at the cathode of 8.8 kA/cm$^2$. Multiple output electron bunches were obtained for multiple laser pulses incident on the cathode. Phase scans of laser induced electron emission reveal an overall phase stability of better than ±20°, corresponding to ±3 ps synchronization of the laser pulses to the phase of the microwave field. A Browne-Buechner magnetic spectrometer indicated that the RF gun generated 1 MeV electrons with a single shot rms energy spread of less than 2.5%, in good agreement with theoretical predictions.

## INTRODUCTION

The goal of the 17 GHz photocathode RF gun experiment is to construct an ultra high brightness source of electrons which can be used for future linear colliders or free electron lasers and to demonstrate the practicality of high gradient (several hundred MV/m) acceleration. The photocathode RF gun is a novel electron beam source intended to meet the requirements set by future high-energy linear colliders and next generation free electron lasers. A coupled pair of pillbox $TM_{010}$-like resonators is excited by sidewall coupled microwaves at 17.15 GHz (Fig. 1). Note that the axial length of the structure is determined by the microwave wavelength. A picosecond ultraviolet laser pulse illuminates the back wall of the structure at the axis of

CP472, *Advanced Accelerator Concepts: Eighth Workshop,*
edited by W. Lawson, C. Bellamy, and D. Brosius

symmetry. Electrons are released by the photoelectric effect and are accelerated by the axial electric field.

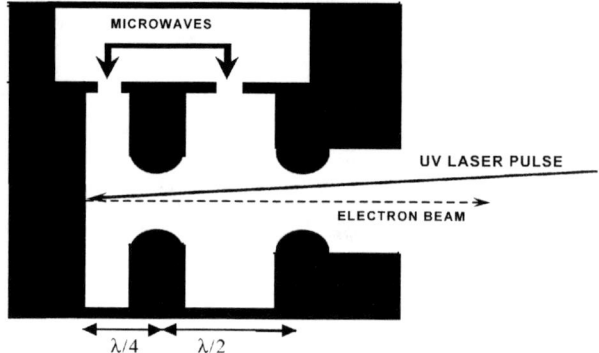

**Figure 1.** Cross Section of the side wall coupled RF Gun and waveguide.

Initial operation of the MIT 17.15 GHz RF gun was reported in 1996 [1]. Previous photocathode RF gun experiments have operated at 3 GHz or below [2-10], though an RF gun experiment at 8.5 GHz is currently operating [11]. There has also been much theoretical work in the study of RF guns in recent years [12-17]. From an analysis of the Vlasov-Maxwell equations governing the dynamics of the electron beam production in the photocathode RF gun, the scaling with RF frequency of the quality of the beam which can be produced in an RF gun has been derived [15,16]. One of the major advantages of going to higher frequency is that the emittance of the beam is inversely proportional to RF frequency, implying a quadratic increase in the beam brightness with frequency. The implication of this is that high frequency RF guns are very promising candidates for use with future high energy electron colliders and short wavelength free electron lasers, both of which will require extremely high brightness beams beyond the capabilities of more conventional electron sources. In this experiment, we have measured the bunch charge and energy of the beam produced from the 17 GHz RF gun, demonstrating successful operation of the RF gun and its robustness to repeated exposure to ultra high gradient electric fields and significant temperature rises due to power dissipation.

## THEORY AND DESIGN

The two modes of a 1 ½ cell structure like the MIT RF Gun are illustrated in Fig. 2. These were obtained from SUPERFISH [18] simulations for a cavity design slightly perturbed from the actual RF Gun dimensions. The two cells can be modeled as coupled $TM_{010}$ oscillators which have corresponding symmetric and anti-symmetric oscillations. For acceleration purposes, the mode in which the two cavities oscillate 180 degrees out of phase with one another, the $\pi$-mode, is the preferred mode of excitation. Then, provided the full cell length $L$ is equal to one half of the vacuum wavelength $\lambda$ of the accelerating microwaves, relativistic electrons will remain in phase with the RF fields in the cavity.

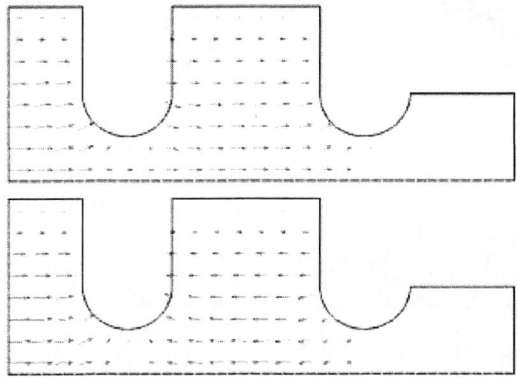

**Figure 2.** Normal modes of a 1 ½ cell RF gun (Top: 0-mode; Bottom: π-mode).

## Cold Tests and Axial Field Profile Determination

In order to completely characterize the filling characteristics of the RF gun, certain critical parameters must be experimentally determined. The most important of these parameters are the coupling coefficient $\beta_c$ of the cavity and its ohmic quality factor, $Q_{cav}$. These two quantities dictate the attainable stored energy in the RF gun for a given incident power, $P_{inc}$. For a given stored energy, the peak accelerating fields and resulting final energy of the accelerated electrons depend on the field balance between the half and full cells. Accurate determination of this field profile requires an experimental measurement of the magnetic polarizabilities, $\alpha_1$ and $\alpha_2$, of the coupling holes in the half and full cells. Once the polarizabilities are known, the theory of cavity excitation [19] along with the field profiles of the RF Gun obtained from SUPERFISH [18] simulations (i.e. without coupling holes) can be used to determine the actual field profile in the cavity.

**Table 1.** Measured RF Gun Characteristics

| Parameter | Label | Value |
|---|---|---|
| Loaded Q | $Q_L$ | 1100 |
| Unloaded Q | $Q_{cav}$ | 2700 |
| Coupling Coefficient | $\beta_c$ | 1.56 |
| Polarizabilities | $\alpha_1$ | $13.7*10^{-9} \, m^3$ |
| | $\alpha_2$ | $14.2*10^{-9} \, m^3$ |

In order to obtain values for the $Q$, $\beta_c$, and polarizabilities of the RF gun (Table 1), cold tests of the cavity were performed using an HP 8510 Vector Network

863

Analyzer. These tests consisted of the measurement of both the amplitude and phase of the reflection coefficient of the power incident on the RF gun, $|S_{11}| = \sqrt{P_{refl}/P_{inc}}$, where $P_{refl}$ is the total reflected power. The coupling coefficient is defined to be the ratio of the cavity quality factor to the external quality factor ($\beta_c = Q_{cav}/Q_{ext}$) and can be determined from the measurement of $|S_{11}|$ at resonance from the expression

$$|S_{11}|_{resonance} = \left|\frac{1 - \beta_c}{1 + \beta_c}\right|. \tag{1}$$

For our cavity, $S_{11} = 0.22$ at the resonant frequency of 17.142 GHz, which gives a coupling coefficient of $\beta_c = (1 + |S_{11}|)/(1 - |S_{11}|) = 1.56$ for an overcoupled cavity while the loaded quality factor, $Q_L$, is determined from the width of the resonance at the level of $|S_{11}| = \sqrt{1 + \beta_c^2}/(1 + \beta_c) = 0.72$ and is equal to 1060. The unloaded quality factor is then given by $Q_{cav} = Q_L(1 + \beta_c) = 2700$. The power dissipated in the cavity, $P_{cav}$, is determined by the coupling coefficient $\beta_c$:

$$P_{cav} = \frac{4\beta_c}{(1 + \beta_c)^2} P_{inc} = 0.95 P_{inc}. \tag{2}$$

The measured $Q_{cav} = 2700$ for the $\pi$ mode in the experimental cavity is 70% of the calculated SUPERFISH value.

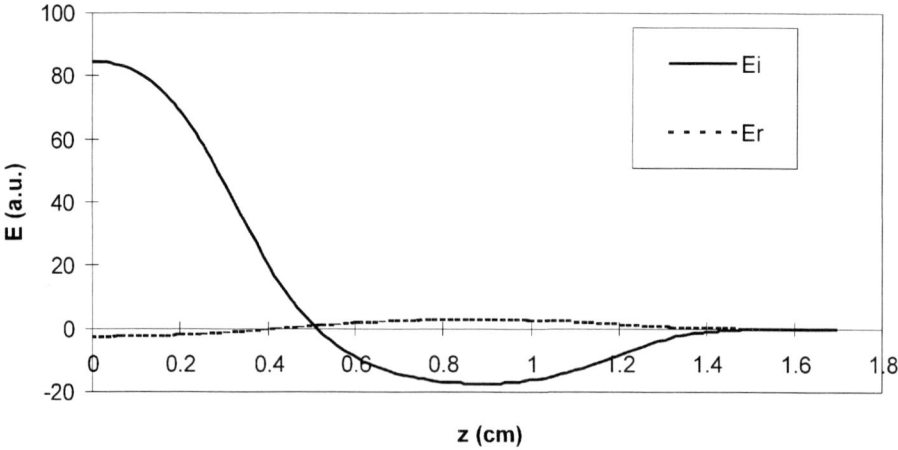

**Figure 3.** Actual on axis complex field profile (Ei = imaginary part, Er= real part) in the RF gun calculated with coupling hole theory using the SUPERFISH eigenmodes and the measured polarizabilities of the coupling holes.

The actual polarizabilities of the coupling holes are determined by matching the theoretical $S_{11}$ curve obtained from the theory of cavity excitation [19] to the measured $S_{11}$ curve. Specifically, the polarizabilities $\alpha_1$ and $\alpha_2$ are determined to match the measured resonances at 17.142 and 17.118 GHz (i.e. the $\pi$ and 0 mode respectively). This leads to the values of the polarizabilities listed in Table 1, and the resulting on axis electric field profile shown in Fig. 3. Note that the field strength in the first cell is almost 5 time larger than the field strength in the second cell. While this is not ideal for efficient acceleration, it does facilitate the production of ultra high gradients in the first cell by lowering the incident power required to reach a given electric field at the cathode.

## Effects of a Partially Reflecting Short

In the experimental determination of the cavity coupling characteristics described above, a perfectly reflecting short at the end of the waveguide was used. However, in the high power experiments, the short was replaced with a Bragg filter (see section 4) to eliminate the reflection of spurious higher frequency modes produced by the klystron. At the RF gun resonant frequency of 17.142 GHz, the reflection coefficient, $R$, of the Bragg filter was found to be about 0.67 rather than the ideal value of unity. In order to account for effects of the partially transmitted wave in the coupling theory, we modify the single-mode RF filling expressions to account for a reflection coefficient $R$ not equal to 1. For $R = 0.67$ and coupling coefficient $\beta_c = 1.56$, the expressions for $S_{11}$ at resonance, the cavity dissipated power $P_{cav}$, and the external quality factor $Q_{ext}$ are given by

$$S_{11}(\omega_{res}) = \left| \frac{R - \dfrac{1+R}{2}\beta_c}{1 + \dfrac{1+R}{2}\beta_c} \right| = 0.27 \quad , \tag{3}$$

$$P_{cav} = \frac{(1+R)^2 \beta_c}{\left(1 + \dfrac{1+R}{2}\beta_c\right)^2} P_{inc} = 0.82 P_{inc} \quad , \tag{4}$$

and

$$Q_{ext} = \frac{2}{1+R}\frac{Q_{cav}}{\beta_c} = 2070 \quad . \tag{5}$$

We can use these expressions to estimate the maximum electric field at the cathode assuming a steady state. Using the measured value for $Q_{cav}$ (Table 1), equation (4), and SUPERFISH simulation results for the electric field magnitude and stored energy we wind up with a steady state expression for the cathode electric field given by

865

$$E_{cathode}[MV\,/\,m] = 183\sqrt{P_{inc}[MW]} \quad ,\tag{6}$$

which gives $E_{cathode}$ = 366 MV/m for an incident power of 4 MW. In the experiments, however, $E_{cathode}$ is obtained as a function of time through integration.

## Simulations of Beam Dynamics

Extensive simulations of the beam dynamics in the MIT 17 GHz RF Gun have been performed. The expected beam parameters include an initial bunch length of 0.3 mm (1 ps) and a total space charge of 0.1 nC. The particle in cell code MAGIC [20] (2-D in fields, 3-D in particle motion) was employed to examine the particle dynamics with and without space charge. The results of these simulations are shown in Fig. 4.

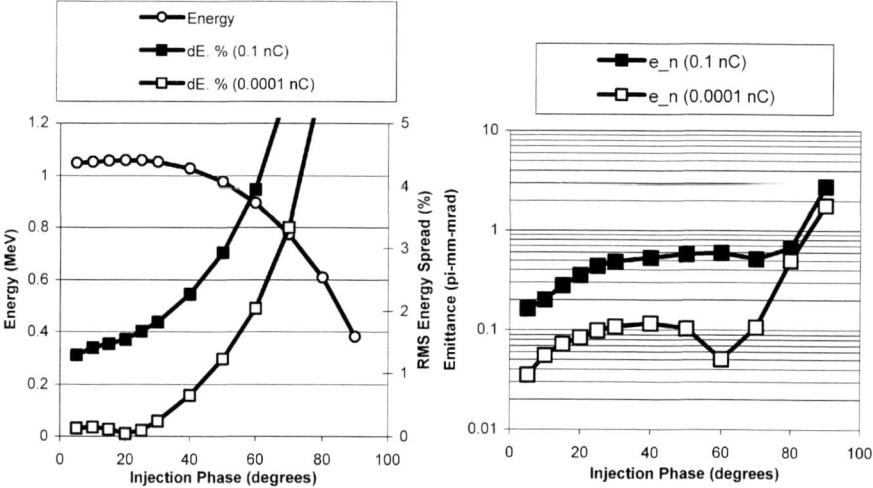

**Figure 4.** Comparison of MAGIC simulations of the beam dynamics in the RF Gun with high (0.1 nC) and low (0.0001 nC) bunch charge. The average beam energy and RMS energy spread at the exit of the RF Gun is shown on the left, while the RMS normalized emittance is plotted on the right. The maximum on axis electric field at the cathode is 300 MV/m, while the initial beam length and radius are 1ps and 0.4 mm respectively. A zero degree injection phase corresponds to the zero crossing of the electric field at the cathode.

Perhaps the most notable effect of space charge is the increase in energy spread of the beam. Also, the minimum in energy spread and emittance with respect to injection phase apparent in low space charge case disappears with increasing bunch charge, as both statistics are seen to monotonically increase with increasing injection phase. The reason for this behavior is that for space charge dominated beams, the RF effects which determine the single particle dynamics are no longer the dominant factor in the transverse dynamics of the beam propagation. For this case of high space charge (0.1 nC), the simulations suggest that for low injection phases (< *30* degrees), we should expect to produce a beam with a rms energy spread of about 1.5% and an rms geometric emittance of about 0.1 mm-mrad (0.3 mm-mrad normalized). This corresponds to a brightness, *B,* of about 1000 A/(πmm-mrad)$^2$, where brightness is expressed as

$$B=\frac{I}{\varepsilon_{xn}\varepsilon_{yn}} \tag{7}$$

where *I* is the peak current of the beam and $\varepsilon_n$ is the normalized transverse rms emittance.

## DESIGN OF THE EXPERIMENT

The major system components of the MIT experiment are shown in Fig. 5. The RF gun cavity is located in a large vacuum vessel shown at the bottom of the diagram. The RF gun cavity is a traditional Brookhaven (BNL) style sidewall coupled 1.5 cell structure [21]. The RF gun is composed of three pieces of OFHC copper which are clamped together and attached as a single piece to the WR-62 coupling waveguide.

The experiment utilizes a 17 GHz relativistic klystron amplifier constructed by Haimson Research Corporation [22] to provide a 50 ns to 1 μs pulse of up to 26 MW of microwave power. The klystron is driven by a 560 kV, 1 μs flattop modulator pulse. A 0.27 μperv Thomson CSF gun produces a space-charge limited electron beam at 560 kV with 95 A transmitted through the klystron. The amplifier chain includes a TWTA to provide up to 5 W to the klystron. The klystron gain is approximately 60 dB.

The klystron amplifier has been found to generate power in spurious modes when connected to a mismatched load. Microwaves associated with these modes are slightly higher in frequency than the operating frequency of the klystron. To eliminate these modes, which reduce the stability and gain of the amplifier, a Bragg filter was designed and installed on the RF gun coupling waveguide. This filter is reflective for frequencies between 16.7 and 17.2 GHz, and transmits for higher frequencies. The use of the Bragg filter allowed cavity coupling at 17.15 GHz, while minimizing impedance mismatches at higher frequencies that would cause the spurious oscillations to occur. The power reflectivity, $R^2$, of the Bragg filter was found to be about 0.43 at 17.15 GHz through cold tests. While this is not optimal, it is sufficient for coupling breakdown threshold power levels into the RF Gun. Up to 8 MW of klystron power was coupled into the cavity, with amplitude stability of ±5%. Breakdown in the RF gun occurring on some shots with incident power exceeding 5 MW, however, usually limited operation to a lower incident power. Phase stability was also measured using a phase discriminator. Phase variation was found to be less than 8° from shot-to-shot and less than 4° during a single shot. Dark current of 0.5 mA was observed in the RF gun when 7.5 MW of microwave power was incident on the RF gun cavity [1]. This corresponds to a peak electric field of about 450 MV/m at the cathode for a 50 ns pulse.

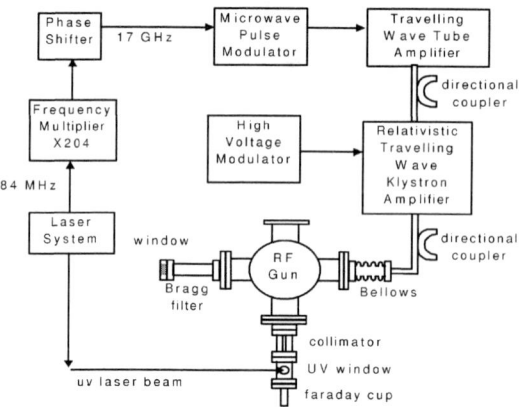

**Figure 5.** Schematic of the experiment (not to scale).

One of the advantages of going to higher frequency is that for a given field gradient, $G$, the temperature rise of the RF gun inner surface due to the power dissipated on the walls decreases (i.e. $\Delta T \propto \omega^{-1/4}G^2$) [23,24]. This is despite an increase in the power dissipated per unit area ($dP/dS \propto \omega^{1/2}G^2$) thanks to the decrease in the filling time of the structure ($t_{Fill} \propto \omega^{-3/2}$) and corresponding decrease in the RF pulse length. Using the constant surface flux solution to the heat transfer equation [25], the maximum surface temperature rise for an incident pulse of length $t_p$ is given by

$$\Delta T = \frac{2q_s\sqrt{\alpha t_p/\pi}}{k} , \qquad (8)$$

where $q_s$ is the power per unit area incident on the copper surface, $\alpha$ is the thermal diffusivity, and $k$ is the thermal conductivity. This can be shown to be consistent with the aforementioned frequency scaling law. For a typical shot (i.e. 50 ns long, 5 MW incident power or 4 MW dissipated on the cavity walls) this yields a maximum temperature rise on the RF gun walls of about 80° C. However, the RF gun has also survived incident powers up to 8 MW, corresponding to a maximum temperature rise of about 130° C, which is encouraging for future prospects of higher power operation.

The laser beam for the RF gun photocathode is generated by a Ti:Sapphire laser system, which produces 2 ps, 1.5 mJ, 2 mm diameter pulses at 780 nm after chirped pulse amplification. The pulse duration of 2 ps was verified using a single-shot autocorrelator. Pulse-to-pulse laser energy fluctuations are approximately ±10%.

These pulses are frequency tripled to 10-20 μJ of UV, and then focused on the back wall of the copper cavity. The ultraviolet spot size is approximately 1 mm in diameter. A UV pulse length of 1 ps is estimated from the measured infrared 2 ps pulse duration using nonlinear optical theory. Measurement of the frequency sidebands using a spectrum analyzer [26] indicated a laser phase jitter of less than 5 ps, or 30 degrees.

The timing scheme used in the MIT RF gun experiment is slightly different from that of other photocathode experiments. A photodiode samples the light pulses bouncing back and forth in the optical cavity of the Ti:Sapphire laser oscillator. The IR pulses of this oscillator are used as input to a regenerative amplifier. The frequency of these pulses is 84 MHz and is dictated only by the length of the cavity. No external oscillator is used. The signal from this photodiode is used as input to a frequency multiplier box, which filters and amplifies the signal. The output signal is about 17.15 GHz and is phaselocked to the laser system. This milliwatt level signal is used as input to the microwave amplifier chain which drives the RF gun experiment. Phase stability of this system has been verified by measuring the percentage of shots resulting in successful electron beam production vs. injection phase of the laser pulse into the RF Gun. Theoretically, emission should be observed over about 100 degrees of injection phase (see Fig. 4). The observed width of the experimental curve is consistent with a combined laser and microwave source phase jitter of better than $\pm 20°$.

## EXPERIMENTAL RESULTS

An example of a typical cavity filling shot is shown in Fig. 6. The forward and reflected power are measured at the directional couplers immediately before the RF Gun (see Fig. 5). The stored energy in the cavity (and hence the accelerating gradient) in the absence of beam loading can be calculated from a modified power conservation equation,

$$\frac{dW}{dt} = k_{Br}^2 P_{inc} - P_{refl} - \frac{\omega}{Q_{cav}} W , \qquad (9)$$

where $W$ is the stored energy in the cavity, $Q_{cav}$ is the unloaded quality factor, and $k_{Br}$ is a factor that accounts for the reflectivity $R$ of the Bragg filter and can be derived from equations (3) and (4) to give

$$k_{Br}^2 = \left( \frac{R - \frac{1+R}{2} \beta_c}{1 + \frac{1+R}{2} \beta_c} \right)^2 + \frac{(1+R)^2 \beta_c}{\left(1 + \frac{1+R}{2} \beta_c\right)^2} = 0.90 , \qquad (10)$$

for $R = 0.67$ and $\beta = 1.56$. These measurements demonstrate that peak accelerating fields of over 300 MV/m have been achieved at the cathode for incident power levels exceeding 4 MW. This corresponds to an average accelerating gradient of about 200 MV/m over the half cell and about 80 MV/m over both the half and full cells combined.

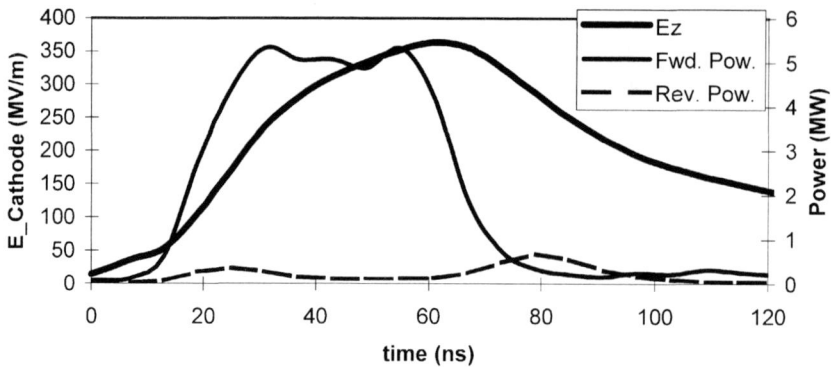

**Figure 6.** Cavity filling measurement illustrating the peak field at the cathode.

The beam diagnostics so far employed in this experiment consist of a high speed Faraday cup for beam bunch charge measurements and a Browne and Buechner style magnetic spectrometer [27] for beam energy and energy spread measurements. The Faraday cup is designed as a conical tapered 50 Ω transmission line. The central conductor is 1 cm in diameter with a 5 mm diameter center collector. The laser beam is injected into the RF gun with the use of a prism placed in the collector portion of the Faraday cup. During early experiments, it was discovered that rectified microwave signals were dominating the detected Faraday cup signals. In order to eliminate this effect, it was necessary to place a 5 mm radius cylindrical collimator between the RF gun and the Faraday cup to act as a cutoff waveguide. Integrated Faraday cup signals indicated a total bunch charge of up to 0.1 nC (Fig. 7).

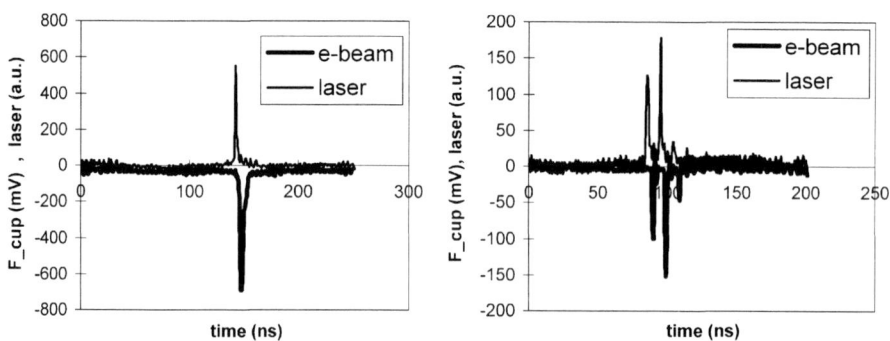

**Figure 7.** Left: Laser induced electron beam emission with a total integrated bunch charge of 0.1 nC. Right: Example of multiple electron bunch generation with multiple laser pulses.

Multiple pulse generation by sequential laser pulses has also been observed with this experimental set up. As many as three separate laser pulses can be produced through imperfect dumping of the regenerative amplifier cavity. These laser pulses are separated in time by the round trip time of the cavity (about 9 ns), which is close enough to an integer multiple of RF periods (about 155) to result in up to three corresponding electron beams (Fig. 7). The laser pulses seen in Fig. 7 are detected by picking off a small portion of the laser pulse immediately before injection into the RF Gun and detecting it with a photodiode. In the future, we plan to employ optical beam splitters to intentionally produce a train of optical pulses and a corresponding train of electron beam bunches.

The magnetic spectrometer used to perform the measurement of the beam energy is a circular magnetic field boundary, field radius $R = 10$ cm, 1 cm gap size, 1 kilo-Gauss saturation structure. The beam is bent approximately $90°$ and detected by detectors placed on the focal plane of the spectrometer. The determination of the detector placement (i.e. the location where a monochromatic divergent beam from a point source is geometrically focused back to a point) in a basic Browne-Buechner spectrometer design is determined by the ratio of the field radius $R$ to the distance $L$ between the electron source and the magnetic field boundary. The equation for the focal plane assuming the coordinate system shown in Fig. 8 is

$$\widetilde{y} = \frac{\left(2\widetilde{L} + 1\right)}{\widetilde{L}^2}\widetilde{x}^2 - 2\left(\frac{1+\widetilde{L}}{\widetilde{L}}\right)^2 \widetilde{x} + \left(\frac{1+\widetilde{L}}{\widetilde{L}}\right)^2, \qquad (11)$$

where $\widetilde{L} = L/R$, $\widetilde{x} = x/R$, and $\widetilde{y} = y/R$, and can be derived using the same procedure as in the original Browne-Buechner paper [27], but without assuming $R=L$.

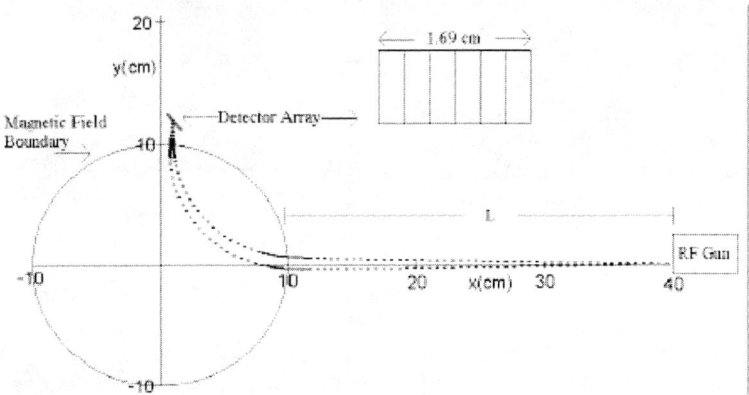

**Figure 8.** Illustration of the energy spread measurement.

The effect of increasing $L$ relative to $R$ is to move the focal plane closer to the magnetic field boundary and to decrease the slope. Ideally, one would like to avoid doing this since the dispersion on the focal plane will accordingly decrease. For our measurement, the electron source was taken to be the RF Gun exit. This simplified the beam line design by eliminating the need for a pinhole or focusing elements. The size of the RF gun housing structure and the necessity of the collimator limited the minimum drift length between the RF Gun exit and the magnetic field boundary of the spectrometer to $L = 30$ cm, or $\widetilde{L} = 3$ (Fig. 8). Simulations of the energy measurement using PARMELA [28] indicate that after traversing the distance between the RF gun and the spectrometer, the rms energy spread of a 1 MeV beam is about 2.2 %. The increase from the rms energy spread of about 1.5% at the exit of the gun is due to longitudinal space charge forces.

Two types of energy measurements were performed with this setup. The first utilized a single Faraday cup placed on a 90° arc from the beam line. The percent energy acceptance of this detector was about 10%. The low energy end of the detector was approximately located at the focal plane of the spectrometer, allowing for an accurate measurement of the maximum energy of the beam by observing the maximum magnetic field for which a beam is detected over many shots. These measurements yielded electron energies as high as 1.1 MeV. The second type of energy measurement utilized an array of up to six detectors, consisting of 2mm wide Faraday cups placed side by side along the focal plane (Fig. 8). Due to mechanical constraints, the array was placed on the low energy, low dispersion end of the focal plane, resulting in a maximum energy resolution of about 2.2%. This measurement provided similar results as the other measurement, yielding energies up to 1.1 MeV and single shot rms energy spreads as small as $\pm 2.5\%$ (Fig. 9), roughly in accordance with simulations.

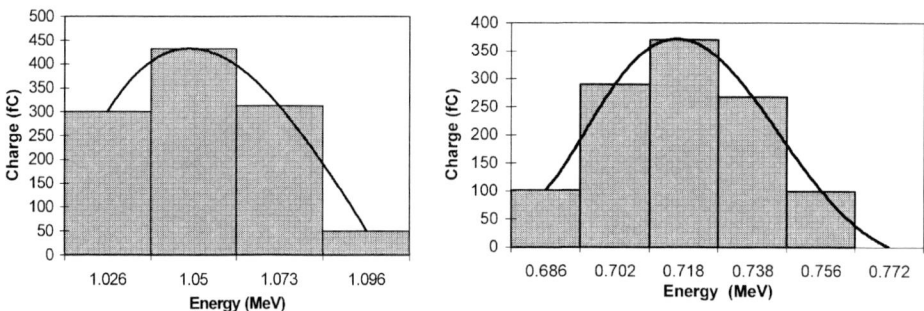

**Figure 9.** Measurements of the electron beam energy with the detector array.

Both energy measurements yielded results that are consistent with simulations in terms of the energy of the beam, indicating maximum accelerating gradients of over 300 MV/m at the cathode (200 MV/m average over the first cell) consistent with the theoretical analysis in section 2 and the measurement of the stored energy shown in Fig. 6. The 2 to 3 % measured rms energy spread at the spectrometer is also roughly in accordance with theoretical expectations. From PARMELA simulations of space charge effects during transport of the beam, these results are consistent with an energy spread of about 1.5 % at the exit of the RF gun.

## CONCLUSIONS AND FUTURE PLANS

The experimental results presented here for the MIT 17 GHz RF Gun demonstrate the potential for high gradient acceleration at high microwave frequencies. Measurements of the electron beam energy as well as high power measurements of the cavity filling of the gun suggest peak accelerating fields of about 0.3 GV/m. These measurements are consistent with theoretical studies of the beam dynamics also presented here. Up to three separate electron bunches have been produced on a single shot corresponding to the splitting of the laser pulse. Analysis of the temperature rise of the RF gun walls indicate peak temperature rises of about 80° C corresponding to peak on axis electric fields of about 300 MV/m were sustained without damage to the RF gun. In order to demonstrate the superior beam quality promised by high frequency, high gradient RF Guns, plans for measuring the emittance of the electron beam as well as plans for the development of an improved, better field balanced cavity are underway. Eventually, an emittance compensated gun will be employed with the hope that a record high brightness electron beam will be produced.

## AKNOWLEDGMENTS

This work is supported by D.O.E. H.E.P contract number DE-FG02-91ER40648. We would also like to thank Dr. Jake Haimson and Haimson Research Corporation for use of the klystron amplifier, as well as Ivan Mastovsky and Bill Mulligan, whose dedication and expertise have been instrumental in the operation of this experiment.

## REFERENCES

1. S. Trotz, W.J. Brown, B.G. Danly, J.P. Hogge, M. Khusid, K.E. Kreishcer, M. Shapiro, R.J. Temkin, "High Power Operation of a 17 GHz Photocathode RF Gun," Advanced Accelerator Concepts, Seventh Workshop, S. Chattopadhyay, J. McCullough, P. Dahl, editors, *AIP Conf. Proc.* 398, AIP Press, Woodbury, NY, pp. 717-729 (1996).
2. D.T. Palmer, X.J. Wang, I. Ben-Zvi, R.H. Miller, "Beam Dynamics Enhancement due to Accelerating Field Symmetrization in the BNL/SLAC/UCLA 1.6 cell S-Band Photocathode RF Gun," to be published in *Proceedings of the 1997 Particle Accelerator Conference*, IEEE Press (1997).
3. D.T. Palmer, X.J. Wang, R.H. Miller, M. Babzien, I. Ben-Zvi, C. Pellegrini, J. Sheehan, J. Skaritka, H. Winick, M. Woodle, V. Yakimenko, "Emittance Studies of the BNL/SLAC/UCLA 1.6 cell Photocathode rf Gun," to be published in *Proceedings of the 1997 Particle Accelerator Conference*, IEEE Press (1997).

4. K. Aulenbacher, R. Bossart, H. Braun, J. Clendenin, J.P. Delahaye, J. Madsen, G. Mulhollan, J. Sheppard, G. Suberlucq, and H. Tang, "Initial Test of an RF Gun with a GaAs Cathode Installed," Stanford Linear Accelerator Laboratory Publication, SLAC-PUB-7299 (1996).

5. S.C. Hartman, N. Barov, C. Pellegrini, S. Park, J. Rosenzweig, G. Travish, R. Zhang, C. Clayton, P. Davis, M. Everett, C. Joshi, G. Hairapetian, *Nucl. Instr. and Meth.* **A 340**, 219-230 (1994).

6. R. Bossart, H. Braun, M. Dehler, J.C. Godot, "A 3 GHz photoelectron gun for high beam intensity," *Nucl. Instr. and Meth.* **A 375**, ABS 7-8 (1996).

7. Y.C. Huang, R.H. Pantell, J.F. Schmerge, J.W. Lewellen, J. Feinstein, *IEEE Journal of Quantum Electronics* **31**, No. 9, 1637-1641 (1995).

8. M.H. Woodle, K. Batchelor, J. Sheehan, "Mechanical Design of a RF Electron Gun," *1988 Linear Accelerator Conference Proceedings*, 351-352 (1989).

9. D.W. Feldman, et. al., *IEEE Journal of Quantum Electronics* **27**, 2636-2643 (1991).

10. R.L. Sheffield, E.R. Gray, J.S. Fraser, Nucl. Instr. and Meth. **A 272**, 222-226 (1988).

11. LeSage, G.P., et. al., *Physics of Plasmas* **5**, 2048-2054 (1998).

12. Kwang-Je Kim, *Nucl. Instr. and Meth.* A **275**, 201-218 (1989).

13. L. Serafini, *Particle Accelerators* **49**, 253-271 (1995).

14. L. Serafini, J. Rosenzweig, *Physical Review E* **55**, 7565-7590 (1997).

15. J. Rosenzweig, E. Colby, "Charge and Wavelength Scaling of RF Photoinjector Designs," *Advanced Accelerator Concepts*, P. Schoessow, editor, AIP Conf. Proc. 335, AIP Press, Woodbury, NY pp. 724-737 (1995).

16. Leon C.-L. Lin, S.C. Chen, J.S. Wurtele, "On the frequency scaling of electron guns," *Advanced Accelerator Concepts*, P. Schoessow, editor, AIP Conf. Proc. 335, AIP Press, Woodbury, NY, p. 704 (1995).

17. Leon C.-L. Lin, S.C. Chen, J.S. Wurtele, *Nucl. Instr. and Meth.* A **275**, 274-284 (1997).

18. T. Weiland., *Nucl. Instr. and Meth.* **216**, 329-48, (1983).

19. R.E. Collin, *Field Theory of Guided Waves*, IEEE Press, 1991.

20. Bruce Goplen et. al. MAGIC User's Manual, 1991. Mission Research Corporation, Technical Report MRC/WDC-R246.

21. K. T. McDonald, *IEEE Transactions on Electron Devices* **35**, 2052-2059 (1988).

22. J. Haimson and B. Mecklenburg, "Initial Performance of a High Gain, High Efficiency 17 GHz Traveling Wave Relativistic Klystron for High Gradient Accelerator Research," *Pulsed RF Sources For Linear Collider*, Richard C. Fernow, editor, AIP Conf. Proc. 337, AIP Press, Woodbury, NY, pp. 146-159 (1995).

23. P. Wilson, "Scaling Linear Colliders to 5 TeV and Above," Stanford Linear Accelerator Laboratory publication, SLAC-PUB-7449, (1997).

24. D. Whittum, H. Henke, and P.J. Chou, "High-Gradient Cavity Beat-Wave Accelerator at W-Band," to be published in Proceedings of the 1997 Particle Accelerator Conference, IEEE Press, 3W011, (1997).

25. F. Incropera and D. Dewitt, *Introduction to Heat Transfer, Second Edition,* John Wiley & Sons, NY, p.260, 1990.

26. D. von de Linde, "Characterization of the noise in continuously operating mode-locked lasers," Appl. Phys. B **39**, 201 (1986).

27. C.P. Browne and W. W. Buechner, *The Review of Scientific Instruments* **27**, 899-907 (1956).

28. J.H. Billen, PARMELA User's Manual, Los Alamos Accelerator Code Group (LAACG), LA-UR-96-1835.

# Self-Bunching Electron Guns[*]

## F. Mako, L. K. Len, and W. Peter

*FM Technologies, Inc.,*
*10529-B Braddock Road, Fairfax, VA 22032*

**Abstract.** We report on three electron gun projects that are aimed at power tube and injector applications. The purpose of the work is to develop robust electron guns which produce self-bunched, high-current-density beams. We have demonstrated cold emission, long life, and tolerance to contamination. The cold emission process is based on secondary electron emission. FMT has studied this resonant bunching process which gives rise to high current densities (0.01-5 kA/cm$^2$), high charge bunches (up to 100 nC/bunch), and short pulses (1-100 ps) for frequencies from 1 to 12 GHz. The beam pulse width is nominally ~ 5% of the rf period. The first project is the L-Band Micro-Pulse Gun (MPG)[†]. Measurements show ~ 40 ps long micro-bunches at ~ 20 A/cm$^2$ without contamination due to air exposure. Lifetime testing has been carried out for about 18 months operating at 1.25 GHz for almost 24 hours per day at a repetition rate of 300 Hz and 5 µs-long macro-pulses. About $5.8 \times 10^{13}$ micro-bunches or ~ 62,000 coulombs have passed through this gun and it is still working fine. The second project, the S-Band MPG[†], is now operational. It is functioning at a frequency of 2.85 GHz, a repetition rate of 30 Hz, with a 2 µs-long macro-pulse. It produces about 150 A/cm$^2$. The third project involves the construction of a 34.2 GHz frequency-multiplied source driven by an X-Band MPG. Analytical work has been carried out on this device, and we are ready to proceed with design, fabrication, and testing.

## INTRODUCTION

High-current, short-pulse electron guns are used as injector systems in electron accelerators (industrial linacs as well as high-energy accelerators for linear colliders). Short-duration pulses are also used for microwave generation (in klystrons and related devices), for research on advanced methods of particle acceleration, and for injectors used for free-electron laser (FEL) drivers.

During the last few years, considerable effort has been applied to the development of high power linac injectors (1,2) and particularly to laser-initiated photocathode injectors (3-9). Though the best photocathodes have sufficient brightness to meet the requirements of the injector described in reference (1), the performance, reliability, and robustness of these devices depend dramatically on the choice of photocathode material. The more durable photocathodes generally suffer from low quantum efficiencies and require high power UV lasers to generate sufficient charge. On the

---

[*] This work is supported by the U. S. Department of Energy.
[†] Patents pending.

CP472, *Advanced Accelerator Concepts: Eighth Workshop,*
edited by W. Lawson, C. Bellamy, and D. Brosius

875

other hand, the higher quantum efficiency photocathodes (e.g. GaAs:Cs) have short lifetimes that significantly impact reliability. In addition, photocathode sources require expensive hardware to synchronize the laser pulse to the *rf*.

The electron gun to be described below is based on a natural bunching process which eliminates the need for bunching section(s), timing system, and laser. Also, the repetition rate can be orders of magnitude greater. In this method, short-pulse, high current pulses are formed by the resonant amplification of an electron current utilizing multipacting from the walls of a microwave cavity. One narrow bunch is transmitted every rf period. This cavity-type bunching electron gun is called the Micro-Pulse Electron Gun or MPG (10). The MPG is capable of generating high current short bunches with low emittance as required for many accelerators and microwave devices. Studies of the multipacting process (11-13) have concentrated on the theoretical aspects of the phenomenon and the rf driver interaction. Bunch stability and its effect on current saturation is treated in (14). References (10) and (15) discuss the saturation process, peak currents, and frequency scaling characteristics.

## CHARACTERISTICS OF THE MPG

Micro-pulses are produced by resonantly amplifying a current of secondary electrons in an *rf* cavity operating in, for example, a $TM_{010}$ mode (Fig. 1). Bunching occurs rapidly and is followed by saturation of the current density in ten to fifteen *rf* periods. "Bunching" occurs automatically by a natural phase selection of resonant particles. Localized secondary emission in the MPG is dictated by material selection. The reason for the name "Micro-Pulse Gun" (MPG) is the fact that the pulse is only a few percent of the *rf* period in contrast to usual *rf* guns where the pulse width is equal to half the *rf* period.

Radial space charge expansion in the MPG cavity is controlled by a combination of electric and magnetic focusing (needed for high current densities only). Radial electric focusing in the cavity is accomplished by a concave shaping of the cavity, not shown in Fig. 1. The grid not only allows transmission of bunches but provides an emitting surface for electron multiplication. A path for the *rf* current can be maintained by using a welded or electro-formed grid of wires. Emittance growth from the grid in the MPG can be very small due to a unique feature of the MPG. Briefly, the resonant particles are loaded into the wave at low phase angles. When they reach the opposite grid 180° later, they experience a reduced transverse kick from the grid wires. This reduces the emittance growth from the grid in the MPG.

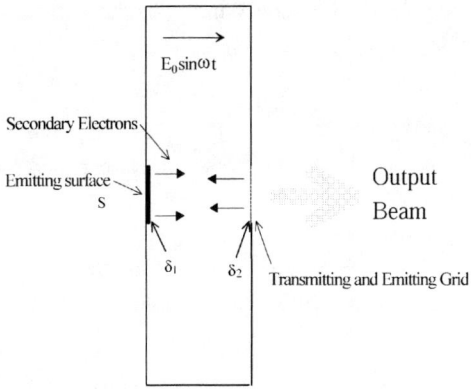

**Figure 1.** MPG concept for generation of bunched beams. Side view of MPG cavity showing emitting and transmitting grid surfaces. Figure is not to scale.

## RESONANCE AND CURRENT DENSITY GAIN

To conceptualize how rapidly the current density can build up in the MPG cavity, consider the simplified model of an *rf* cavity operating in, for example, a $TM_{010}$ mode or $TE_{101}$ mode in Fig. 1. Assume that at the gridded wall of the cavity there is a *single* electron at rest on axis, which transits the cavity in one-half the *rf* period and is in proper phase with the electric field. This electron is accelerated across the cavity and strikes the surface *S*. A number $\delta_1$ of secondary electrons are emitted off this electrode, where $\delta_1$ is the secondary electron yield of surface *S*. Assume the electrons transit the cavity in one-half the *rf* period and they are in proper phase with the field; these electrons will be accelerated back to the grid. After reaching the grid, $\delta_1 T$ electrons will be transmitted, where $T$ is the ratio of transmitted to incident electrons for the grid. The number of electrons which are absorbed by the grid is then $\delta_1(1-T)$. After one cycle the number of electrons that are produced by the grid is $\delta_2[\delta_1(1-T)]$, where $\delta_2$ is the grid secondary yield. In order to have a gain in the number of electrons, the number of secondaries produced must be greater than what we started with, that is, *one*:

$$\delta_2\, \delta_1(1-T) > 1 \tag{1}$$

It is necessary to have enough secondary emission feedback from the combination of the grid and solid wall to have gain or buildup of the current density. The electrons go back and forth between the wall and grid repeatedly, amplifying the current during the process. This is analogous to a laser cavity with the grid acting as a partially-silvered mirror. The gain of electrons after $N$ *rf* periods is $G = [\delta_2\delta_1(1-T)]^N$. If there is a "seed" current density $J_{seed}$ in the cavity initially, then after $N$ *rf* periods the current density will be given by

$$J = J_{seed}[\delta_2\delta_1(1-T)]^N \tag{2}$$

until space-charge or cavity loading limits the current. The current density limit with space charge can be found in (10) and (15). For a very low seed current density a high current density can be achieved in a very short time. For example, if $\delta_2 = \delta_1 = 8$, $T = 0.5$, and $J_{seed} = 2.6 \times 10^{-20}$ A/cm$^2$, in fifteen *rf* periods $J = 1000$ A/cm$^2$!

The resonance condition in the MPG can be calculated by using the single-particle equation of motion,

$$\frac{d^2z}{dt^2} = \left(\frac{e}{m}\right)E_0 \sin\omega t \tag{3}$$

for an electron in a cavity with an applied *rf* field of frequency $f = 2\pi/\omega$. The field has a constant amplitude $E_0 = V_0/d$, where $d$ is the gap spacing and $V_0$ is the applied voltage. It is assumed that the electron is born at one electrode surface (at $z = 0$) with a phase $\omega t = \phi$ into the *rf* cycle with an initial velocity $v_0$. To achieve resonance, this particle must reach the opposite electrode at $z = d$ at a time $\omega t = \phi + \pi$. This gives the resonant condition

$$V_0 = \frac{m\omega^2 d^2}{e}\left[\frac{1-\pi v_0/\omega d}{\pi\cos\phi + 2\sin\phi}\right] \tag{4}$$

Consider S-band at 2.85 GHz. For a gap spacing of 1.0 cm, we must use an *rf* voltage of 68 kV. Particle-in-cell simulations have been conducted to demonstrate the rapid bunching of electrons by phase selection in the cavity. The simulation code was 2-1/2 dimensional and based on Marder's algorithm (15).

Figure 2 shows how initially, at $t = 0$, electrons are emitted off the electrode at $z = 0$ and then turned off at $t = \pi/\omega$. Note that electrons are only emitted from $r < 2.7$ mm in the simulation. At a time $t = 0.161$ ns (0.46 *rf* period), the cavity is filled with a large number of particles and no bunching is seen; a short time later, at a time $t = 0.467$ ns (1.33 *rf* periods) the broad distribution of particles has diminished, and a thin bunch of particles can be seen within the cavity. Only those particles which strike the electrode at the right phase and velocity are in resonance. These particles make up the narrow electron bunch. At a time of $t = 3.589$ ns (10.2 *rf* periods), particles with the *"wrong"* phase have been filtered out of the simulation, and only particles with the *"right"* phase are present. This narrow bunch of electrons continues to amplify in density.

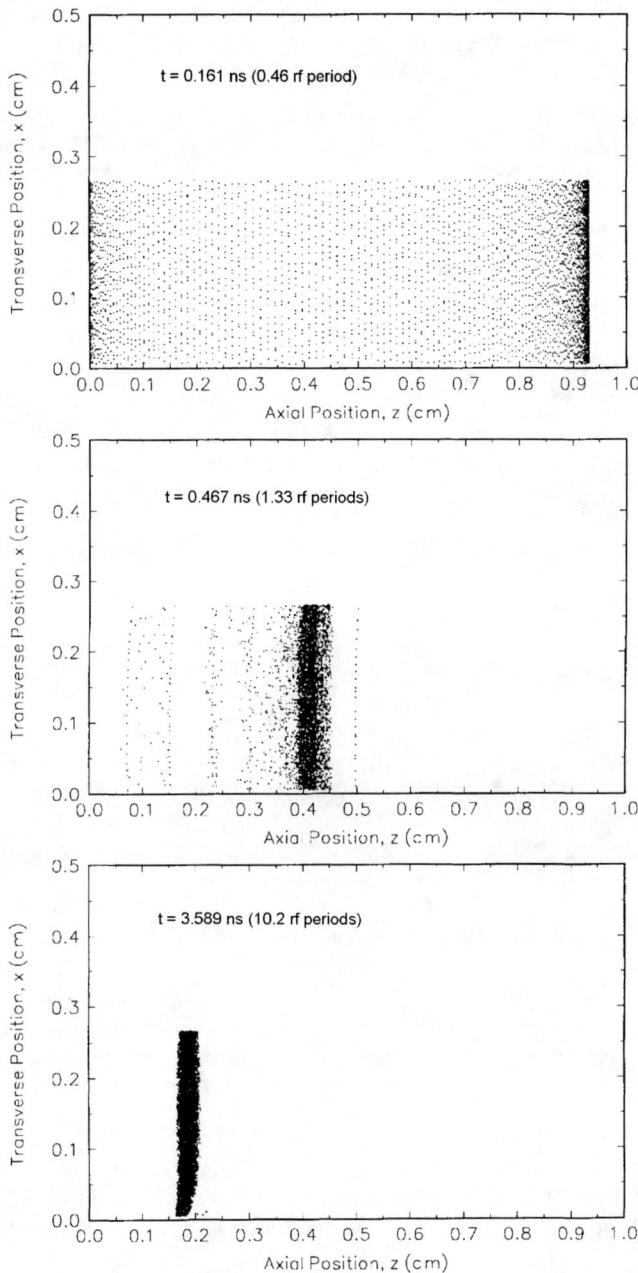

**Figure 2.** Series of time "snapshots" for the conditions $f = 2.85$ GHz, $d = 1.0$ cm and 68 kV peak cavity voltage using the 2-1/2 D PIC code with secondary emission, FMTSEC. Note the rapid particle build-up and bunching by phase selection. Electrons traverse the horizontal axis. On the vertical axis, emission is limited to the region 0 to 2.7 mm.

Figure 3 shows a plot of the current density $J_z$ as a function of time for a simulation with an *rf* frequency of 2.85 GHz and a cavity gap of 2.0 cm. The current density is measured near the second (right-hand) electrode which, in an actual experiment, would be the exit screen or grid. Hence, this is the current pulse which will exit the device. The positive current density is that current which is emitted from the second (right-hand) electrode and propagates back to the first electrode. The negative current density describes the beam that would leave the cavity. Both halves of the curve are not symmetric about $J_z = 0$ because the beam pulses have substantially different charge densities and velocities when they cross the position of the probe. In the case for which the current density is positive (i.e., electrons are propagating in the negative $z$ direction), the electrons have just been emitted from the electrode and form a highly dense bunch at a relatively low energy. In the case for which the current density is negative (i.e., electrons are propagating in the positive $z$ direction), the electrons have already crossed the gap and are at a relatively high energy and have spread somewhat due to space charge effects. At 280 kV the current density rises and saturates at about 1600 A/cm² in 10 *rf* periods.

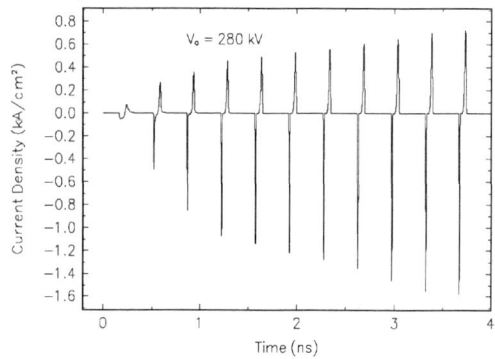

**Figure 3.** Plot of current density vs. time for simulation with *rf* frequency 2.85 GHz and peak voltage of 280 kV and $d = 2.0$ cm.

## EXPERIMENTAL DEMONSTRATION OF THE MPG

Figure 4 shows the macropulse operation of the L-band experiment. The top trace is the *rf* power in the cavity and the bottom trace is the transmitted macropulse beam current from the MPG on a 1μs/div. time scale. The *rf* power is about 50 kW and the current is about 2A or about 2 A/cm². Note the clean current trace over the full *rf* pulse length.

When a fast 50 GHz sampling oscilloscope is used and a 5 ns slice of the macropulse is examined the micropulses can be observed. Figure 5 shows a measurement of the bunches on a 500 ps/div time scale. The bunches appear with the periodicity of the *rf* field (~800 ps), in excellent agreement with simulation. More detailed measurements show that the actual bunch length is about 40 ps (FWHM) which is about 5% of the *rf* period at a current density of about 22 A/cm$^2$, in good agreement with simulation. This is about 1.1 nC or $7 \times 10^9$ electrons per *rf* period. Lifetime testing has been carried out for about 15 months at almost 24 hours per day. At this point about $5.8 \times 10^{13}$ micro-bunches or 62,000 coulombs (82,000 coulombs/cm$^2$) have passed through this gun and it is still working flawlessly.

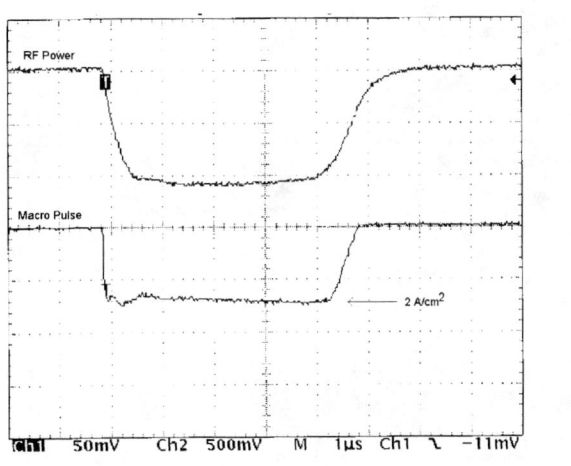

**Figure 4.** Oscilloscope traces of macropulse operation of the L-band experiment. Top trace: the *rf* power in the cavity. Bottom trace: the transmitted beam current.

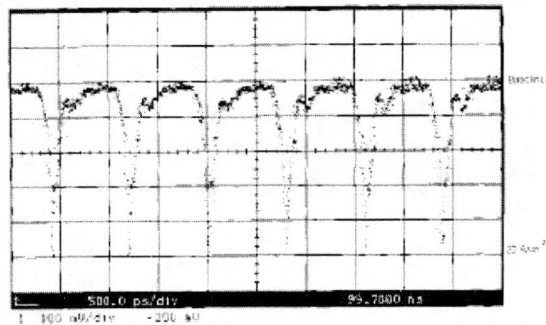

**Figure 5.** Experimental Data: Current trace of the MPG Micro Bunches. Amplitude of the micropulses is 22 A/cm$^2$.

Figure 6 shows the MPG experiment at S-band. The macro-pulse operation is shown in Fig. 7 where the top trace is the *rf* power in the cavity and the bottom trace is the transmitted macro-pulse beam current on a 0.5 μs/div. The *rf* power is about 500 kW and the current is about 45 A, or about 20 A/cm$^2$. Note also the clean current trace over the full *rf* pulse length.

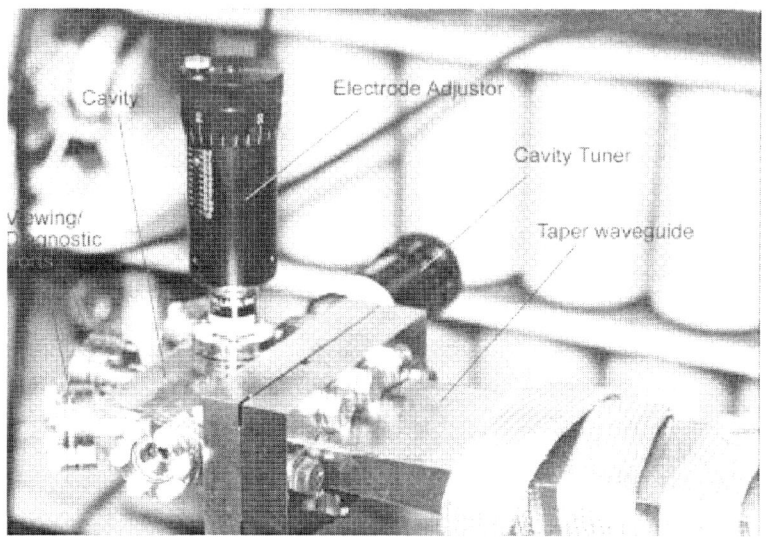

**Figure 6.** Close-up of the S-band experiment showing the taper waveguide and the MPG. The *rf* source, isolator, bi-directional coupler and vacuum pumps are not shown.

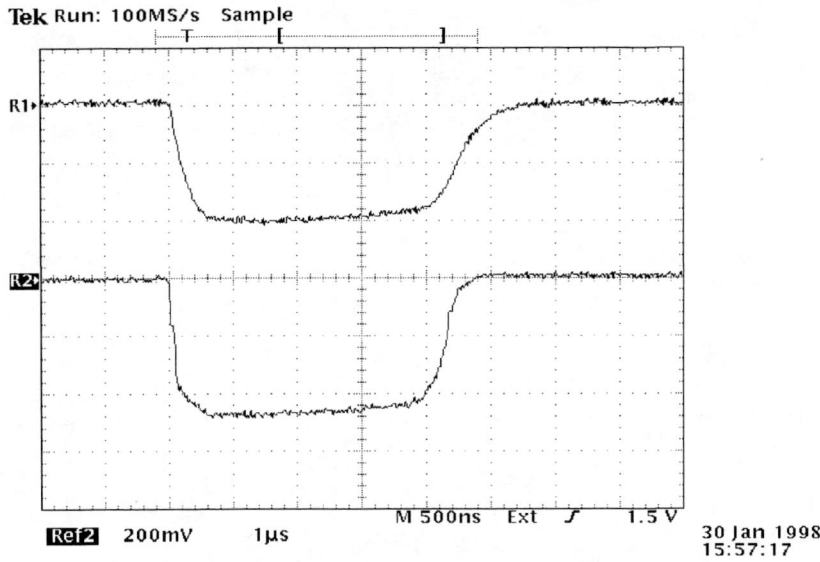

**Figure 7.** Oscilloscope traces of macro-pulse operation of the S-band experiment. *Top*: The *rf* power in the cavity. *Bottom*: The transmitted beam current macro pulse.

A new concept for a 34.2 GHz high-power microwave source would utilize a short annular bunch (<2 picoseconds) of electrons from an 11.4 GHz MPG to generate the third harmonic in an output cavity. A complete high-power amplifier consists of four components: (1) an input cavity which comprises the annular micro-pulse gun; (2) post acceleration to high energy; (3) radial compression of the annular micro bunches; and (4) an output cavity operating at an integer multiple of the input cavity frequency. This is shown schematically in Fig. 8. A point design based on numerical simulations was performed at an *rf* output power of 150 MW at 34.2 GHz. The resulting system efficiency is 53% and the gain is 60 dB. The system efficiency includes the input cavity efficiency, input driver efficiency (a 50 MW klystron at 11.4 GHz), output cavity efficiency, and the post acceleration efficiency. Breakdown in the output cavity appears to be manageable. One of the advantages of this scheme over a conventional klystron is that the beam is pre-bunched to a small fraction of the *rf* period which gives rise to higher beam to *rf* conversion efficiency. Design and fabrication is underway to establish measurements of power, efficiency and other key parameters in a prototype device.

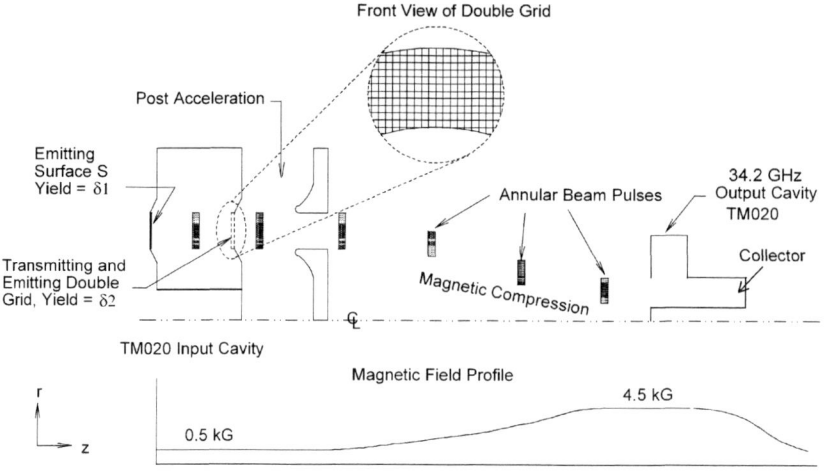

**Figure 8.** Side view of the 34.2 GHz microwave source, showing the cavity double grid and emitting and transmitting surfaces. Beam pulses, concave shaping of the MPG cavity, post-acceleration section, magnetic focusing for the $TM_{020}$ mode, and the output cavity are also shown. An inner conductor (not shown) will be used to stabilize against the diocotron instability. Figure is not to scale.

# REFERENCES

1. Adamski, J. L., Gallagher, W. J., Kennedy, R. C., Robinson, B., Shoffstall, D. R., Tyson, E. L., Vetter, A. M., and Yeremian, A. D., "The Boeing 120 MeV RF Linac for FEL Research," *IEEE Trans. Nucl. Sci.* **NS-32** (5), 3397 (1985).
2. Godlove, T. F., and Sprangle, P., "High-Power Free-Electron Lasers Driven by RF Linear Accelerators," *Particle Accelerators,* **34**, 169, (1990).
3. Jones, M. E., and Peter, W., "Particle-in-Cell Simulations of the Lasertron," *IEEE Trans. Nucl. Sci.* **NS-32** (5), 1794 (1985).
4. Schoessow, P., Chojnacki, E., Gai, W., Ho, C., Konecny, R., Mtingwa, S., Norem, J., Rosing, M., and Simpson, J., "The Argonne Wakefield Accelerator," presented at the 2nd European Particle Accel. Conf., *Proc. of the 2nd European Particle Accel.Conf.*, p. 606 (1990).
5. Batchelor, K., *et al*, "Performance of Brookhaven Photocathode RF Gun," *Nucl. Instr. and Meth. in Phy. Res.* **A318**, 372 (1992).
6. Hartman, S. C., Barov, N., Park, S., Pellegrini, C., Rosenzweig, J., Travish, G., Zhang, R., Davis, P., Joshi C., and Hairapetian, G., "Emittance Measurements of the 4.5 MeV UCLA RF Photo-Injector," presented at 1993 Part. Accel. Conf., *IEEE Cat. 93CH3279-1*, 561 (1993).

7.   Ben-Zvi, I.,   "Performance of Photocathode RF Gun Electron Accelerators," presented at 1993 Part. Accel. Conf., *IEEE Cat 93CH3279-1*, 2962 (1993).

8.   Lehrman, I. S., Birnbaum, I. A., Cole, M., Heuer, R. L., Sheedy, E.,   "Design and Construction of a High-Duty Factor Photocathode Electron Gun," Part. Accel. Conf., *IEEE Cat. 93CH3279-1*, 3012 (1993).

9.   Travier, C., Bernard, M., Dufresne, P., Mihaud, G., Omeich, M.,   Roch, M., and Rodier, J., "CANDELA Photo-Injector High-Power Test," *Nucl. Instr. and Meth. in Phy. Res.* **B89**, 27 (1994).

10.   Mako, F.,   and Peter, W., "A High-Current Micro-Pulse Electron Gun," Part. Accel. Conf., IEEE Cat. 93CH3279-1  2702 (1993).

11.   Gilmore, A. S., *Microwave Tubes* (Artech House, Norwood, MA 1986), p. 474.

12.   Vaughan, J. R. M., "Multipactor," *IEEE Trans. Electron Devices* **35 (7)**,  1172 (1988)

13.   Kishek, R. and Lau, Y. Y., "Interaction of Multipactor Discharge and rf Circuit," Phys. Rev. Lett. **75**, 6, 1218, (1995).

14.   Riyopoulos, S., Chernin, D., and Dialetis, D.,   "Theory of Electron Multipactor in Crossed Fields," Phys. Plasmas **2** (8), 3194 (1995); also IEEE, Trans. Electron Devices, **44**, 489 (1997).

15.   Mako, F., Peter, W., and Shiloh, J., "Micro-pulse gun for generation of electron bunches," submitted to Phys. Rev. Lett.

16.   Marder, B., "A Method for Incorporating Gauss' Law into Electromagnetic PIC Codes," J. Computational Physics, **68**, 48 (1987).

# Electron Injection by Dephasing Electrons with Laser Fields

E. Dodd, J. K. Kim and D. Umstadter

*Center for Ultrafast Optical Science*
*University of Michigan, Ann Arbor, MI 48109*

**Abstract.** The authors seek to review injection concepts for plasma based acceleration. It is shown that regardless of injection mechanism, resultant beams will be similar due to wave structure. Also, most schemes employ the same basic processes, namely the dephasing of electrons by laser fields, and can thus be analyzed with similar approaches.

## INTRODUCTION

Laser-plasma based acceleration coupled with CPA laser technology has become the topic of much current interest in recent years. As a part of this discussion, the question arose as to how to generate and inject femtosecond-duration electron bunches for wake-field acceleration. Electron beam quality suitable for x-ray generation or high-energy physics has yet to be demonstrated by use of laser wake-fields injected with either RF injectors or wavebreaking. A solution to this problem was recently proposed, [1] in which a second laser pulse, split from the same laser system, is used to inject an electron bunch into the wake-field. This would have several important advantages, including femtosecond-timescale synchronization and pulse durations, as well as greater simplicity. Subsequently, a number of different papers have proposed variations of this idea [2,3] or further analyzed this concept [4 6].

The concept of laser induced plasma waves to accelerate charged particles is itself nearly twenty years old [7]. Wake-field accelerators seek to take advantage of ultra-high acceleration gradients ($> 10$ GeV/m) for electrons based on laser-driven plasma waves [8–10], possible due to the invention of compact, high-peak-power lasers [11 13]. The plasma-wave electric field gradients are three-orders-of-magnitude higher than those in conventional RF linacs, because they are not limited by dielectric breakdown. In fact recently, gradients on the order of 1 GeV/cm have been demonstrated experimentally [14], and accelerate electron beams with transverse emittances that rival current electron guns. However, the plasma wave length is much shorter than that of the RF linac, hence the need for very short pulse injectors.

CP472, *Advanced Accelerator Concepts: Eighth Workshop,*
edited by W. Lawson, C. Bellamy, and D. Brosius

Electrons normally oscillate in the plasma wave and cannot be accelerated by the wake-field since they are out of phase with it, as in Fig. 1. Electrons that are not part of the plasma wave can become trapped, or continuously accelerated by

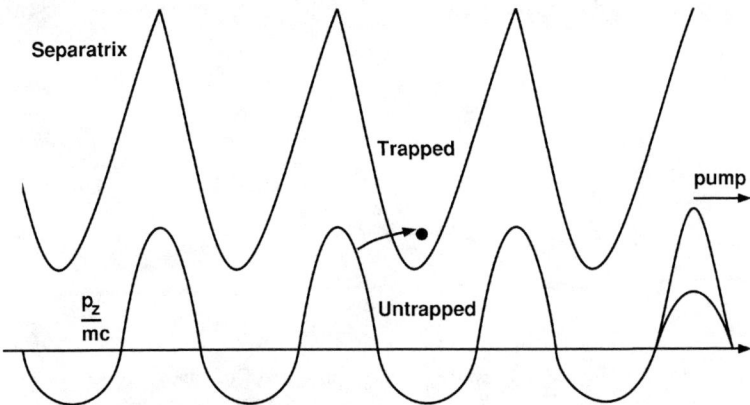

**FIGURE 1.** Basic profile for LWFA. The pump pulse creates a plasma wave to accelerate electrons. To be accelerated, electrons must cross inside the separatrix.

the wave, provided that they are moving in the correct phase at nearly the phase velocity of the wave [15]. Since this velocity is close to the speed of light, it was generally thought that such pre-acceleration can only be accomplished by external injection, such as with a conventional linac. However, the low-field gradient ($< 10$ MeV/m) [16] of a first-stage conventional linac prolongs the time during which beam emittance can grow before the beam becomes relativistic; after this point, self-generated magnetic fields can balance the effects of space charge.

## METHODS OF INJECTION

The LILAC, or Laser Injected Laser ACcelerator, was proposed in [1] to solve the injector problems. It consists mainly of three different stages. First a large amplitude wake-field is generated; second, electrons are dephased; and then thirdly, electrons are trapped in the pump's wake-field due to the dephasing and accelerated. By understanding each of the three parts involved, a description of the injection process can be derived. However, the first part is beyond the scope of this paper, and without loss of generality, no discussion of generation mechanisms will be included. The process of dephasing is quite general, using secondary laser pulses or other methods to move electrons oscillating in the wave across the separatrix for acceleration. These methods can include both the ponderomotive force or the direct field of the laser pulse acting on electrons or through collective effects of the plasma such as waves. Regardless, all methods seek to produce similar results after the injection process has ended.

The first scheme proposed, [1], used the ponderomotive force of the second, or injection pulse, with an orthogonal geometry to dephase electrons for trapping, Fig. 2 a). Besides orthogonal, other orientations of the laser pulses are also possible, collinear or counter-propagating. The next variation was the collinear LILAC shown in Fig. 2 b), where the ponderomotive force drives the wave to breaking for

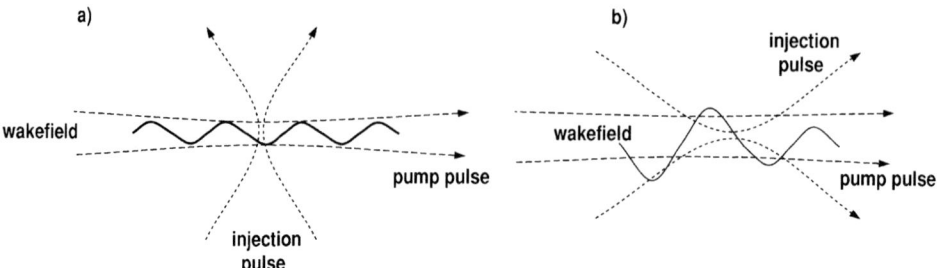

**FIGURE 2.** a) Schematic of the transverse LILAC accelerator concept. b) Schematic diagram of the collinear LILAC. Please note that only the contours of intensity are shown.

injection. Another scheme, [3], again uses the ponderomotive force, but now with two counter-propagating pulses. When overlapped, the beating produces the dephasing for injection, which only occurs when the pulses are overlapped. This way injection is gated on and off for only a short period of time.

Another way a laser pulse may inject electrons is through collective interactions with the plasma. The first example is through the use of a plasma instability, or the so-called self modulated wake field. This is based on the plasma wave generated from the Raman instability, and forward scattering accelerating the electrons. for example see [14]. It however produces electrons in a continuum and is not of interest. An injection pulse may also produce a wake. The interaction between the wakes of the various pulses will dephase electrons enough for trapping and was studied in [5]. To do this the length of the injection pulse was shorter and more highly resonant than in previous studies, [6], but still used the orthogonal orientation.

The ponderomotive force is really an average of the particle's motion over many optical cycles, as such the dephasing is **not** directly due to the electric field in the laser pulse. Since field amplitude of the pulse is larger than the ponderomotive force the use of a sub cycle pulse was proposed in [2]. This method produces a short bunch with minimal energy spread, needed for injection into plasma waves.

Besides combining laser pulse characteristics, ionization or density gradients are other possible means to produce dephased electrons. Another effect looked at briefly was trapping due to edge effects. A sharp boundary in the simulation caused the plasma frequency to change abruptly, going from zero in vacuum to full density in a few microns. Oscillating particles will see two frequencies as they move into the vacuum and return, causing them to be dephased, and possibly trapped. This was studied previously by two other groups [17]. Since this sharp boundary does not exist in the experiment, we chose to use the solution of G. Bonnaud et. al. to remove this problem from the code, and moved the particle boundary accordingly. Physical boundaries this sharp are difficult to achieve experimentally, but if some method were found it could serve as an efficient injection mechanism.

## BUNCH CHARACTERISTICS

Electrons bunches that have been accelerated have a number of important characteristics. These depend on the plasma wave as much as they depend on the injection process. For example, the first characteristic of interest is simply the number of electrons in the bunch which is limited by beam loading [18], the limit where the space charge of the bunch effectively screens out the accelerating gradient. This is independent of injection method since it depends only on the wave amplitude and plasma density. For plasma densities of $10^{19}$cm$^{-3}$ and wave amplitude of 1 GeV/cm this is $n_b = 7 \times 10^8$ electrons, any method of interest should be capable of injection this many electrons into the wave.

Also important are the energy spread of the bunch and the transverse emittance. Due to the short plasma wave length the accelerating gradient changes from zero field to more than 1 GeV/cm over only a few microns, allowing for large changes in electron energy even in short bunch lengths. To eliminate this problem electrons must be injected only over a small phase of the wave, hence the need for gating. The orthogonal laser pulse orientation also solves this problem due to short transit time of the injection pulse across the plasma wave. Wave breaking and longitudinal orientations can inject over large phase ranges giving rise to energy spreads close to 100%. This is much more dependent on injection method than bunch number.

As a bunch leaves the accelerator it begins to diverge which is characterized by the transverse emittance. Like a conventional accelerating structure, a plasma wave has limit on the largest emittance it can hold, or acceptance. This limit can be calculated from the spot size and the strength of the radial electric fields in the wave, an example of which appears in [5]. This may be used as a prediction of emittance produced in plasma wave acceleration. One finds that this value is on the same order as that of more conventional RF photo-cathode guns, however in [14] a value of $0.2\pi$mm · mrad for normalized transverse emittance was measured. These electrons were produced through wave breaking and hence have 100% energy spread in contrast to the low emittance. However this does demonstrate experimentally that with the right injection mechanism very low emittances are possible with laser-plasma based acceleration.

## CONCLUSION

In conclusion we have shown that all injection methods produce similar results, depending on the plasma wave as much as injection process. Additionally most injection schemes are variations on a few basic ideas, either directly dephasing electrons or using collective plasma effects. As yet no best method for injection to create an optimal electron bunch for acceleration in plasma waves has been found, which will only be resolved through experiment since all present work has been based on theoretical and numeric work.

We would like to acknowledge the support of the NSF and to thank Torsten Neubert and Gerard Mourou for many valuable discussions. Computing services were provided by the University of Michigan Center for Parallel Computing, which is partially funded by NSF grant CDA-92-14296.

# REFERENCES

1. D. Umstadter, J. K. Kim and E. Dodd, Phys. Rev. Lett. **76**, 2073 (1996).
2. B. Rau, T. Tajima, and H. Hojo, Phys. Rev. Lett. **78**, 3310 (1997).
3. E. Esarey, *et.al.*, Phys. Rev. Lett. **79**, 2682 (1997).
4. J. L. Bobin, *Advanced Accelerator Concepts*, AIP Conf. Proc. 398, S. Chattopadhyay, ed., (1997).
5. R. G. Hemker, *et.al.*, Phys. Rev. E.**57**, 5920 (1998).
6. J. K. Kim, E. Dodd and D. Umstadter. submitted to Phys. Rev. E.
7. T. Tajima and J. M. Dawson, Phys. Rev. Lett. **43**, 267 (1979).
8. P. Sprangle *et al.*. Appl. Phys. Lett. **53**, 2146 (1988).
9. L. M. Gorbunov and V. I. Kirsanov, Sov. Phys. JETP **66**, 290 (1987); H. Hamster *et al.*, Phys. Rev. Lett. **71**, 2725 (1993); K. Nakajima *et al.*, Phys. Rev. Lett. **74**, 4428 (1995).
10. See *e.g.*, *Advanced Accelerator Concepts, Fontana, WI, 1994*, Amer. Inst. of Conf. Proc. No. 335, P. Schoessow, ed., (AIP Press, New York, 1995) and references cited therein.
11. P. Maine *et al.*, IEEE J. Quantum. Electron. **24**, 398 (1988).
12. G. Mourou and D. Umstadter, Phys. Fluids B **4**, 2315 (1992).
13. M. D. Perry and G. Mourou, Science **264**, 917 (1994).
14. R. Wagner *et al.*, Phys. Rev. Lett. **78**, 3125 (1997).
15. E. Esarey and M. Pilloff, Phys. Plasmas **2** 1432 (1995); T. Katsouleas *et al.*, in *Advanced Accelerator Concepts*, AIP Conf. Proc. 130 (1985).
16. S. Humphries Jr., *Principles of Charged Particle Accelerators* (Wiley, New York, 1986).
17. G. Bonnaud, D. Teychenne, and J. L. Bobin, Europhys. Lett. **26**, 91 (1994); S. V. Bulanov *et al.*, Sov. J. Plasma Phys. **16**, 444 (1990).
18. T. Katsouleas *et al.*, *Particle Accelerators*, **22** 81 (1987).

# Injector for a Laser Linear Collider

## A. A. Mikhailichenko

*Wilson Laboratory, Cornell University, Ithaca, NY 14853-8001*

**Abstract**. The injector for $2 \times 1$ *km* long laser driven linac considered. Basically injector is a race track with long straight sections. These sections squeezed together for a compact size (a Kayak-Paddle like ring). In straight section the short period wigglers and RF cavities installed in series one by one for keeping the energy along the straight section practically constant. This injector is able to provide the invariant emittances of the order $5 \cdot 10^{-8} cm \, rad$ and $2 \cdot 10^{-9} cm \, rad$ for horizontal and vertical directions correspondingly. Bunch population required below $10^7$ reduces the IBS effects.

## INTRODUCTION

Injector is a most important component of a laser-based accelerator. Tiny dimensions of accelerating structure require extremely small emittances. The pass holes must have dimensions as a fraction of accelerating wavelength. For example, if this wavelength is $\lambda \cong 10 \mu m$, then the passing hole dimensions need to be made as $\delta \cong 0.2 \cdot \lambda$. Taking into account that this must have a safety margins as big as ten sigma of the beam, one can estimate $\sigma \cong \sqrt{(\gamma \varepsilon) \cdot \beta / \gamma} \cong 0.1 \cdot 0.2 \cdot \lambda$. For focusing elements displaced with period, say $3cm$, the envelope function could be $\beta \sim 12cm$, so emittance required must be $\gamma \varepsilon \cong 4 \cdot 10^{-4} \lambda^2 \cdot \gamma / \beta \cong 6 \cdot 10^{-8} cm \cdot rad$, $\gamma \cong 2 \cdot 10^3 (1 GeV)$. With *RF* focusing, effective focusing parameter could reach $k \cong 2 \cdot 10^5 \cdot Sin\varphi [m^{-2}]$ for the phase deflection $\varphi$. If this phase deflection applied along some part of the structure, the envelope function about $\beta \sim 3cm$ could be reached. Together with increased energy up to $\sim 1 GeV$ all this yields $\gamma \varepsilon \cong 5 \cdot 10^{-7} cm \cdot rad$. Emittance of $\gamma \varepsilon \cong 10^{-6} cm \cdot rad$ is probably maximal allowed emittance value.

Supposed that the beams with extreme emittances could be prepared in a damping ring [1]. Incoherent synchrotron radiation carries out momenta parallel to instant direction of motion. RF, as usual, restores the longitudinal component of momenta, what yields an effective cooling. Equilibrium emittance depends on details of damping ring optics. In particular so called dispersion invariant[1] is important parameter in minimization of emittance. Mostly efforts were spend to find a structure, having dispersion invariant as small as possible.

As the repetition rate is extensive parameter for the luminosity gain, one needs to prepare the bunches with such emittances as fast as possible. This in it's turn requires high rate of energy losses in a cooler, as the cooling is going by synchrotron radiation.

---

[1] See formula (5) below

CP472, *Advanced Accelerator Concepts: Eighth Workshop*,
edited by W. Lawson, C. Bellamy, and D. Brosius
© 1999 The American Institute of Physics 1-56396-889-4/99/$15.00

We will see, that equilibrium invariant emittances achievable *do not depend on the beam energy of the cooler*. So the working energy increase for the cooler helps much for reduction of Intra-Beam Scattering (IBS) process in a bunch. For such a tiny emittances even lowering the number of the particles in the bunch down to $N \cong 10^6$ does not help much against IBS. This could be overpassed dynamically, however.

So the process of preparing the bunch with extreme emittance, in brief, is the following. In a cooler the bunch of maximal phase density prepared. As the cooling mostly is going in straight sections filled with wigglers, the dispersion invariant is the smallest possible one. Length of the bunch is around 1- 10 *mm*. Then the bunching process arranged by the applying the inverse Free Electron Laser process in the ring, having the wiggler in it's structure. The initial bunch is subdivided into micro-bunches with the spacing of the laser wavelength. After that the bunch passed through special system where it is scrapped and all external (in phase space) particles are eliminated. After final acceleration this train of particles are directed into accelerating structure. Along the accelerating length in main linac some diaphragms required obviously.

The injector of the type described below is used in a Laser Linear Collider injection complex [2,3].

## THE COOLER

Some type of coolers able to reach extreme emittances considered in [4]. It was shown here for the first time, that there exists a strong quantum limitation for the lowest emittances in the beam like

$$(\gamma \varepsilon_x)(\gamma \varepsilon_y)(\gamma \varepsilon_s) = (\gamma \varepsilon_x)(\gamma \varepsilon_y)(\gamma \, l_b(\Delta p / p_0)) \geq (\tfrac{1}{2})(2\pi \lambda_C)^3 N ,$$

where $\lambda_C = r_0 / \alpha$ –is a Compton wavelength, $r_0 = e^2 / mc^2$ –is a classical electron radius, $\alpha = e^2 / \hbar c$ –is a fine structure constant, $\gamma = E / mc^2$, $N$–is a bunch population, $l_b$–is a bunch length, $\Delta p / p_0$–is a relative momentum spread in the bunch, $(\gamma \varepsilon_s) = \gamma \, l_b(\Delta p / p_0)$–is an invariant longitudinal emittance. There was pointed out that some among cooling systems could provide emittance not drastically bigger, than this fundamental limitation.

Linear damping system (LDS)–[5] is a sequence of wigglers and accelerating structures, installed along a straight line. The rate of losses in the wigglers is the same as the rate of energy gain given by the accelerating structures. So the particles are moving with practically constant energy. To keep so high losses high energy required. A Kayak-Paddle Cooler (KPC) [6], Fig. 1, is a racetrack, which made as a sequence of wigglers and accelerating structures, installed along straight line and having the bends at the end. The straight sections squeezed together for a compact size, so the back and forward trajectories are congruent. The longer the straight section is–the smaller influence of bends can be made. The bends itself could be made to give a small input into cooling dynamics. Here the particle needs to re-radiate its full energy a few times.

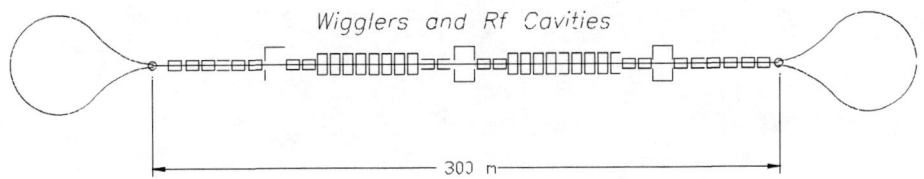

FIGURE 1. Kayak-paddle cooler. The length of straight section indicated is $300m$ .

The rate of energy losses is given by

$$dy / ds \cong -\frac{2}{3} r_0 \frac{K^2}{\lambda^2} \gamma^2 , \tag{1}$$

where $K = eH_\perp \lambda / mc^2$, $H_\perp$ –is a magnetic field value in the wiggler, $2\pi\lambda$ –is the wiggler period, $r_0 = e^2 / mc^2$, $\gamma = E / mc^2$. We would like to mention here, that $dy / ds$ *is not a function of the wiggler period.* The $\gamma$ – factor on right side of (1) needs to be considered as a constant. Substituting here for estimation $\gamma \cong 2 \cdot 10^3 (1 \ GeV)$, $K \approx 5$ ($H_\perp \cong 0.5 T$, $2\pi\lambda \cong 10 cm$), one can obtain $\dfrac{dy}{ds} \cong -\dfrac{2}{3} 2.8 \cdot 10^{-13} \cdot 10 \cdot 4 \cdot 10^6 \cong 7.5 \cdot 10^{-6}$ $[1/cm]$ or 0.00075 $[1/m]$. For so called characteristic damping length one can obtain

$$l_s = -\frac{\gamma}{dy / ds} = \frac{3}{2} \frac{\lambda^2}{K^2 \gamma} . \tag{2}$$

One can see that it has *linear* dependence on energy. For parameters under discussion $l_s = -\dfrac{\gamma}{dy / ds} \cong \dfrac{2 \cdot 10^3}{7.5 \cdot 10^{-4}} \cong 2.6 \cdot 10^6 m$. This means that after passing such a length, the energy re–radiated by the particle will be the same as it's initial one. The cooling time associated with this length is

$$\tau_{cool} \cong l_s / c = -\frac{\gamma}{cdy / ds} = \frac{3}{2} \frac{\lambda^2}{cK^2 \gamma} \tag{3}$$

One can see from here, that the cooling (damping) time *does not depend on the wiggler period.* From (3) one can obtain the estimation $\tau_{cool} \cong 2.6 \cdot 10^6 / 3 \cdot 10^8 \cong 8.6 \cdot 10^{-3} s$ , or 8.6 *ms*. This cooling time obviously does not depend on the length of straight section, as the influence of the bends was neglected –shorter the length, faster the revolution. Emittance dynamics defined by equations

$$\frac{d\varepsilon_x}{ds} = \left\langle \left( H_x + \frac{\beta_x}{\gamma^2} \right) \frac{d(\Delta E / E)_{tot}^2}{ds} \right\rangle - 2\alpha_x \varepsilon_x , \tag{4}$$

with similar equation for vertical motion, where

$$H_{x,y} = \frac{1}{\beta_{x,y}} \left( \eta_{x,y}^2 + (\beta_{x,y} \eta'_{x,y} - \frac{1}{2} \beta'_{x,y} \eta_{x,y})^2 \right), \tag{5}$$

$\eta_{x,y}$ –are dispersion functions. Derivatives are taken over longitudinal direction, which is $s$, see Fig. 2 below.

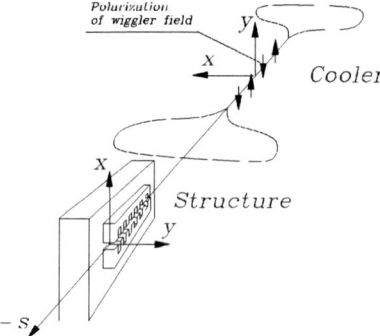

**FIGURE 2.** Relation between coordinates of the cooler and the structure. Polarization of the wiggler field is vertical, the bends of cooler are going in horizontal plane.

Partial decrements $\alpha_{x,y,s}$ are defined as $\alpha_i = \dfrac{J_i}{2 l_s}$, where $J_x \cong 1, J_y = 1, J_s \cong 2$. Partial decrement for energy spread is the same as for emittance.

The source of energy spread has two components arising from synchrotron radiation and IBS,

$$\frac{d(\Delta E / E)^2_{tot}}{ds} = \frac{d(\Delta E / E)^2_{IBS}}{ds} + \frac{d(\Delta E / E)^2_{QE}}{ds} - \alpha_s \left(\frac{\Delta E}{E}\right)^2. \tag{6}$$

Here
$$\frac{d(\Delta E / E)^2_{QE}}{ds} \cong \frac{55}{48\sqrt{3}} \frac{r_0^2 \cdot \gamma^5}{\alpha |\rho|^3}, \quad \frac{d(\Delta E / E)^2_{IBS}}{ds} \cong \frac{N r_0^2 \ln_C}{\gamma^3 \varepsilon_x \sqrt{\varepsilon_y \beta_y} l_b}, \tag{7a, b}$$

where $\ln_C = \ln(\rho_{max} / \rho_{min}) = \ln \sqrt{(v' / c)^6 / 4\pi \cdot r_0^3 n_e'}$ – is Coulomb's logarithm, $v'$ –is a transverse velocity, $n_e'$–is density of particles in it's rest frame, $l_b$– is the bunch length. For a dipole wiggler, with vertical polarization of magnetic field, the periodic solution for $\eta_x$ can be expressed as[2]

$$\eta_x = \frac{K_x \lambda}{\gamma} Sin(s / \lambda) = \frac{\lambda^2}{\rho_x} Sin(s / \lambda), \tag{8}$$

where $\rho_x = \lambda\gamma / K_x$ is the bending radius in magnetic field of the wiggler[3]. For the function $H_x$ we can estimate $H_x \cong \beta_x \eta_x'^2$. As the $\eta_x' \cong K_x / \gamma \cdot Cos(s / \lambda)$ one can obtain $H_x \cong \beta_x K_x^2 / \gamma^2 \cdot Cos^2(s / \lambda)$. So formula (4) becomes

---

[2] The wiggler must have the ending field in poles relation as ¼, -¾, 1, n(-1,1), - ¾, ¼. n=0,1,... This will give zero averaged displacement of trajectory.
[3] For simplicity we supposed that the transition region between wiggler poles is much less, than the pole width itself, so the *absolute* value of radius is a constant.

$$\frac{d\varepsilon_x}{ds} = \left\langle \left(1 + K_x^2 Cos^2(s/\lambda)\right) \frac{\beta_x}{\gamma^2} \frac{d(\Delta E/E)_{tot}^2}{ds} \right\rangle - 2\alpha_x \varepsilon_x \qquad (9)$$

For vertical emittance $K_x = 0$ (and the wiggler field has no horizontal polarization). Equilibrium emittances defined by condition $d\varepsilon_{x,y}/ds = 0$. For quantum excitation the source (7a) does not depend on emittance. Combining (2), (7), (9) for quantum excitation *only* we obtain

$$(\gamma \varepsilon_x) \cong (\tfrac{1}{2}) \cdot \lambda_C \bar{\beta}_x (1 + K_x^2/2)\gamma / \rho \cong (\tfrac{1}{2}) \cdot \lambda_C \bar{\beta}_x (1 + K_x^2/2)K_x/\lambda \qquad (10)$$

$$(\gamma \varepsilon_y) \cong (\tfrac{1}{4}) \cdot \lambda_C \bar{\beta}_y \gamma / \rho \cong (\tfrac{1}{4}) \cdot \lambda_C \bar{\beta}_y K_x/\lambda, \qquad (11)$$

where $\bar{\beta}_{x,y}$ – are averaged envelope functions in the wiggler, $\lambda_C = r_0/\alpha$ – is the Compton's wavelength. The last formulas together with (3), $\tau_{cool} \cong (\tfrac{3}{2}) \cdot (\lambda^2/cK^2\gamma)$, define the cooling dynamics under SR. One can see that both equilibrium invariant emittances *does not depend on energy*. In addition, quantum equilibrium *vertical emittance* and the cooling time do not depend on the wiggler period. Substitute here for estimation $\bar{\beta}_{x,y} \approx 1m$, $\lambda \cong 5cm$, $K \cong 5$, one can obtain for *quantum emittances* the following estimations $(\gamma \varepsilon_x) \cong 2.5 \cdot 10^{-8}$ *cm·rad*,

$$(\gamma \varepsilon_y) \cong 1 \cdot 10^{-9} \ cm \cdot rad.$$

For IBS scattering we have from (2),(7b), (9)

$$\left(1 + K_x^2/2\right) \frac{\beta_x}{\gamma^2} \frac{N \cdot r_0^2 \ln_C}{\gamma^3 \varepsilon_x l_b \sqrt{\varepsilon_y \beta_y}} = \frac{\varepsilon_x K^2 \gamma}{(\tfrac{3}{2})\lambda^2} \qquad (12)$$

$$\frac{\beta_y}{\gamma^2} \frac{N \cdot r_0^2 \ln_C}{\gamma^3 \varepsilon_x l_b \sqrt{\varepsilon_y \beta_y}} = \frac{\varepsilon_y K^2 \gamma}{(\tfrac{3}{2})\lambda^2} \qquad (13)$$

Even in absence of artificial coupling, using (12), (13), we are coming to fundamental conclusion, that there is a coupling between horizontal and vertical degrees of freedom like

$$\left(1 + K_x^2/2\right) \frac{\varepsilon_y}{\beta_y} = \frac{\varepsilon_x}{\beta_x}. \qquad (14)$$

This ratio of emittances is about the same as for quantum excitation. Substitute (14) into (12), (13) we obtain

$$\varepsilon_x^{5/2} \cong \frac{3}{2} \cdot \frac{\left(1 + K_x^2/2\right)^{3/2}}{K^2} \frac{N}{\gamma^6} \frac{\beta_x^{3/2} \cdot r_0^2 \cdot \lambda^2 \ln_C}{l_b \beta_y^2}. \qquad (15)$$

One can see that there is a strong dependence on the beam energy $\varepsilon_x \sim \gamma^{-12/5}$. For $N \cong 10^9$, $\gamma \cong 2 \cdot 10^3$, $l_b \cong 1cm$, $\bar{\beta}_{x,y} \approx 1m$, $\lambda \cong 5cm$, $K \cong 5$, we obtain for *IBS emittances* $\gamma \varepsilon_x \cong 2 \cdot 10^{-5}$ *cm·rad*,

$$\gamma \varepsilon_y \cong 1.5 \cdot 10^{-6} \ cm \cdot rad.$$

So the only way to come to emittances as required compared with ones defined by quantum process only– is increasing the energy of the beam and following *scrapping* the extra particles. For example raising energy up to, say 10 $GeV$ will drop emittance about three orders of magnitude. For successful operation of Laser Linear Collider the only $N \approx 10^6$ particles are required [2].

## OSC METHOD

From previous part we could see, that the only source of damping is the classical synchrotron radiation. Namely this radiation is responsible for damping time like (3). If some other source of damping is present in the system, then resulting damping time $\tau_{res}$ will be described

$$\frac{1}{\tau_{res}} \cong \frac{1}{\tau_{cool}} + \frac{1}{\tau_{osc}} \cong \frac{c}{l_s} + \frac{1}{\tau_{osc}} = \frac{2}{3} \frac{cK^2 \gamma}{\lambda^2} + \frac{1}{\tau_{osc}}, \tag{16}$$

where $\tau_{osc}$– is effective damping time associated with this additional cooling process. This additional cooling process might be Optical Stochastic Cooling (OSC) method [7]. This is basically a stochastic cooling with the bandwidth extended to optical one. The cooling time in this method is associated with the number of the particles in the bandwidth

$$\tau_{cool} \cong N_u \cdot T, \tag{17}$$

where $N_u$– is the number of the particles in the bandwidth, $T$–is the revolution period.

$$N_u \cong \frac{f}{\Delta f} N \frac{\lambda_u}{l_b} \cong \frac{f}{\Delta f} N \frac{L}{l_b \gamma^2}, \tag{18}$$

where $L = 2\pi\lambda$ is the wiggler period, $\lambda_u$ is the central wavelength of radiation amplified. Amplification coefficient $\kappa$ of optical amplifier can be expressed as following [7]

$$\kappa \cong \frac{\varepsilon_\parallel}{r_0} \frac{1}{N} \frac{\Delta f}{f}, \tag{19}$$

where $\gamma \varepsilon_\parallel = \gamma l_b \Delta E / E$ is an invariant longitudinal emittance $\Delta f / f$ is a relative bandwidth, $N$ is the number of the particles. One can see the only phase density in longitudinal direction $\rho_\varepsilon = \gamma \varepsilon_\parallel / N$ is important for the level of amplification. The decrease of emittance is equal to

$$\frac{\varepsilon_f}{\varepsilon_0} \cong \frac{1}{\alpha N_u}, \tag{20}$$

896

where $\quad \alpha = e^2 / \hbar c \cong 1/137$. For $\quad N \cong 10^{10}$, $\quad \lambda_u \cong 10^{-4} \ cm \ (1 \ \mu m)$, $\quad l_b \cong 1 \ cm$,

$\Delta f / f \cong 20\%$ one can expect $N_u \cong 5 \cdot 10^{10} \cdot 10^{-4} \cong 5 \cdot 10^6$ and $\varepsilon_f / \varepsilon_0 \cong 2.7 \cdot 10^{-5}$. This is of course the maximal possible value of cooling.

Simultaneous cooling of transverse and longitudinal emittances yields some threshold value in the invariant emittance level as

$$\gamma \varepsilon_x^{th} \cong \lambda \gamma \cdot (\Delta E / E) = (\lambda / l_b) \cdot l_b \gamma \cdot (\Delta E / E), \tag{21}$$

where $\gamma \varepsilon_x$ –is the invariant radial emittance, $\gamma l_B \cdot (\Delta E / E)$ – is so called *invariant longitudinal emittance*, $\lambda$ – is a central wavelength of the optical amplifier (and all system), $\Delta E / E$ – is the initial relative energy spread in the beam. The above value is of the order $\gamma \varepsilon_x^{th} \cong 3 \div 5 \cdot 10^{-4} \ cm \cdot rad$ for $\lambda \cong 1 \mu \ m$, what is not affecting our case.

One can also arrange the cooling system so that it is able to cool only the particles with lower derivatives of trajectory that satisfy the condition (21). For example one can screen the radiation from some parts of the beam to avoid heating the core by peripheral parts of emittance. One positive property of quadrupole wiggler as a pick up, is that the central parts of the beam, what have higher values of slope $x_0'$ are moving *closer* to the center of the lens, see (1), and, hence, radiate less. One can see also, that the particles with higher slope $x_0'$ in the pick up (and having maximal lengthening), will be at the center in the quadrupole wiggler used as a kicker if transformation matrix *-I* (or for cosine trajectory $C(s \to kick) = -1$). That will reduce the heating from interaction with amplified radiation. In [8] there was considered the conditions required under which the system comes to equilibrium. General output is that the system must have positive longitudinal mass, $m_s = m\gamma^3 / (1 - \eta \gamma^2)$ i.e. *negative* momentum compaction factor $\eta$. KPC satisfies this requirement, having $\eta \cong -0.0001$.

### Optical amplifier

Formulas (19), (20), (21) give an idea about amplification required and the power contained in the laser flash. Two examples considered below use these formulas.

*Example 1.*

For $N \cong 10^{10}$, $l_B \cong 15 cm$, $M = 5$, $(\Delta E / E) \cong 10^{-3}$, $\gamma \cong 10^3 \ (500 \ MeV)$, $\lambda \cong 1\mu \ m$, optical amplifier must be able to have amplification about 300, peak power about 5 $kW$, average power about 25 $W$ with repetition rate $f$ of the order of 10 $MHz$. Number of the particles in the bandwidth $N_S \cong 3 \cdot 10^5$ defines the number of the turns and the damping time $\tau_c \cong N_S / f \cong 30 ms$. Emittance reduction $\varepsilon_f / \varepsilon_0 \cong 1 / \alpha \ N_S \cong 10^{-3}$.

*Example 2.*

For $N \cong 10^8$, $l_B \cong 5 cm$, $M = 5$, $(\Delta E / E) \cong 10^{-4}$, $\gamma \cong 10^3 \ (500 \ MeV)$, $\lambda \cong 1\mu \ m$, optical amplifier must be able to have amplification about 100, peak power about 225

$W$, average power about 0.075 $W$ with repetition rate of 2 $MHz$[4]. Number of the particles in the bandwidth $N_S \cong 2 \cdot 10^3$ defines the number of the turns and the damping time $\tau_c \cong N_S / f \cong 1ms$. Emittance reduction $\varepsilon_f / \varepsilon_0 \cong 1/\alpha N_S \cong 7 \cdot 10^{-2}$. The parameters described above look as realistic from the energetics. Optical amplifier needs to be done with lowest phase distortion and minimal time delay.

TABLE 1. Parameters of amplifier

| Amplifier | Dye | Ti:Al₂O₃ |
|---|---|---|
| Wavelength | $\lambda \approx 340 \div 540$ nm | |
| Life time | $\tau_L \cong 5$ ns | $\tau_L \cong 3.5 \mu s$ |
| Absorption cross section | $\sigma_{01} \cong 2 \cdot 10^{-16} \div 4 \cdot 10^{-16}$ cm$^2$ | $\sigma_{01}(490nm) \cong 10^{-19}$ cm$^2$ |
| Emission cross section | $\sigma_{10} \cong 2 \cdot 10^{-16} \div 4 \cdot 10^{-16}$ | $\sigma_{10}(790nm) \cong 3 \cdot 10^{-19}$ cm$^2$ |
| Density | $n_0 \approx 10^{17}$ cm$^{-3}$ | $n_0 \approx 10^{20}$ cm$^{-3}$ |
| Absorption length, $l_{ab} \cong 1/n_0\sigma_{01}$ | $0.05 \div 0.1$ cm | 0.1 cm |
| Pumping area, $S_{pump} \cong \lambda \cdot l_{ab}$ | $5 \cdot 10^{-6}$ cm$^2$ | $7 \cdot 10^{-6}$ cm$^2$ |
| Saturation power density, $P_{sat} \cong \hbar\omega \, n_0 l_{ab} / \tau_L$ | $\leq 100$ $kW/cm^2$ | $\leq 1$ $MW/cm^2$ |
| Pumping power, $I = P_{sat} \cdot S_{pump}$ /stage | 0.5 W/stage | 7 W/stage |
| Pumping time, $\tau_{pump} \cong \hbar\omega S_{pump} n_0 l_{ab} / I$ | 0.15 ns | |
| Amplification[*], $\kappa \cong \exp[\sigma_{10}(\lambda) \cdot n_1 \cdot l]$ | $\approx 7$ | 20 |
| Pumping | Nitrogen laser, $\lambda = 308nm$ | Argonne laser |
| Number of stages | 3 | 2 |

[*] $l \cong l_{ab}$, $n_1 \approx n_0$.

## Installation in a cooler

An example of installation of OSC system for electron/positron cooler is represented in the Fig. 3 below.

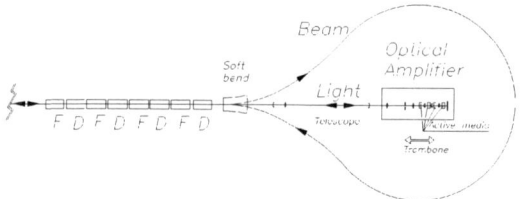

**FIGURE 3.** Example of installation of OSC system in a cooler.

---

[4] Low frequency of revolution is result of the presence of long straight sections in a Kayak-Paddle type cooler.

For the parameters we used above, the period of the lenses must be $2L \cong 2\gamma^2\lambda/(1+K^2)$ what is about 2 $m$. More hard radiation from the dipole wigglers installed in straight section has different wavelength and does not interfere with optical amplifier operating at lower wavelength around $1\,\mu m$. Optical amplifier installed at stabilized platform. For the fine phase adjustments one can use the dual prism system. Initial phase adjustments made by movement the table (trombone). Optical telescopes at both ways used for proper conjunction the radiation from the wiggler and amplifier.

## BEAM OPTICS

To confirm the parameters of a wiggler and focusing in a straight section, some numerical calculations launched. This numerical calculations took into account the IBS and quantum excitations. Few optical structures tested. Among them is the structure with separated focusing in quads and wigglers without gradients. The other one with gradient wigglers. As the trajectory is basically a straight-line radiation instability does not activated up to significant gradients. Both structures give about the same values of

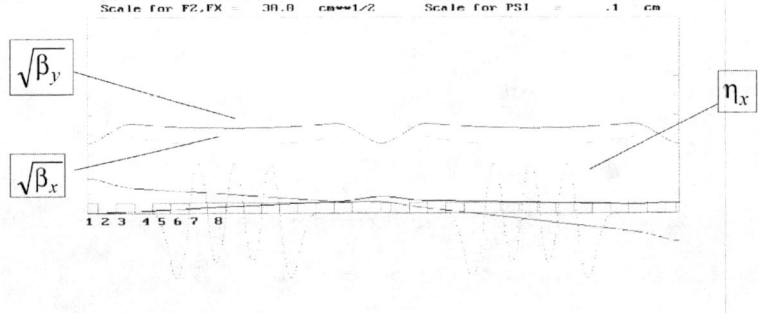

**FIGURE 4.** Example of optical functions in linear part. Vertical scale for dispersion is 0.1 $mm/div$.

emittances in agreement with calculated in section 2. For bending arcs also few optics tested. The best results give FD structure. Radiation instability compensated in straight sections. Investigation of dynamic aperture and chromaticity compensation is in progress.

## OTHER APPLICATIONS OF THE COOLER

We would like to attract attention here that the ring of the type described has a sudden and interesting application–a gamma quanta's and positronium generator. For this electron and positron beam made moved in the same direction. For proper longitudinal spacing (congruence) of $e^+$ and $e^-$ bunches, the RF cavities must be moved to the bending arcs. In the bending arcs the electrons and positrons are going face to face bypassing each other in the vertical bump, arranged with electrostatic separators. Interaction of electrons and positrons is going in a straight section. Cross-section has dependence like $\sigma \propto \pi r_0^2 c / v_{rel}$, where $v_{rel}-$ is a relative velocity of particles and formally could reach a big value. The luminosity could be estimated like

$$Lum \approx \frac{N^2 Q_x c}{L \cdot 4\pi l_b \sqrt{\varepsilon_y \beta_y}}, \tag{22}$$

where $L-$ is a length of straight section, $Q_x-$ is a horizontal betatron tune. Tiny emittances of the beams help to reach luminosity $Lum \approx 10^{31} cm^{-2} s^{-1}$. So the number of positroniums could reach $\approx 10^6 s^{-1}$. So this ring gives two rays of gammas and positroniums directed along a straight section.

## CONCLUSION

The cooler considered give extreme emittances, which could be used for preparation the beams for injection into laser accelerating structure. Quantum emittances are the smallest ones among known structures. Further investigation required for IBS rejection.

The cooler considered could be used also for addition application– generation of monochromatic gammas and positroniums.

## REFERENCES

[1] International Linear Collider Technical Review Committee Report, Editor G. Loew, SLAC-R-95-471, 1995.

[2] A.A. Mikhailichenko," A Laser Linear Collider Design", this Workshop. Also CLNS 98/1562, Cornell LNS, 1998.

[3] A.A. Mikhailichenko, "Laser Acceleration. A practical Approach", a talk on LASERS'97, Fairmont Hotel, New Orleans, LA, December 15-19; also in CLNS 97/1529, Cornell, 1997.

[4] A. Mikhailichenko, "On the physical limitation to the lowest emittance (toward colliding electron-positron crystalline beams)", 7$^{th}$ Advanced Accelerator Concepts Workshop, 12-18 October 1996, Granlibakken Conference Center, Lake Tahoe, CA, AIP Conference Proceedings 398, p.294.

[5] N.S. Dikansky, A.A. Mikhailichenko, "A linear damping system for obtaining high energy $e^{\pm}$ with extremely low emittance ", EPAC 92, Berlin, 1992, Proc., p.898; Preprint BINP 88-9, Novosibirsk, 1988.

[6] A.A. Mikhailichenko, " Damping Ring for VLEPP linear collider", III international Workshop on linear colliders LC91, Protvino, September 17-27, 1991. Proceedings, Edited by V.E. Balakin, S. Lepshokov, N.A. Solyak, Serpukhov, (IFVE). 1991. Serpukhov, USSR: BINP (1991).

[7] A. Mikhailichenko, M. Zolotorev, "Optical Stochastic Cooling", Phys. Rev. Lett., Vol. 71, No. 25, December 20, 1993.

# Design of a High Charge (10 - 100 nC) and Short Pulse (2 - 5 ps) RF Photocathode Gun for Wakefield Acceleration

W. Gai, X. Li[*], M. Conde, J. Power, and P. Schoessow

*High Energy Physics Division*
*Argonne National Laboratory, Argonne, IL 60439, USA*

**Abstract.** In this paper we present a design report on a 1-1/2 cell, L Band RF photocathode gun that is capable of generating and accelerating electron beams with peak currents > 10 kA. We have performed simulation for bunch intensities in the range of 10 - 100 nC with peak axial electrical field at the photocathode of 30 - 100 MV/m. Unlike conventional short electron pulse generation, this design does not require magnetic pulse compression. Based on numerical simulations using SUPERFISH and PARMELA, this design will produce 20 - 100 nC beam at 18 MeV with rms bunch length 0.6 -1.25 mm and normalized transverse emittance 30 - 108 mm mrad. Applications of this beam for wakefield acceleration is also discussed.

## INTRODUCTION

High current short electron beams have been a subject of intensive studies (1). One of the particular uses for this type of beam is for wakefield acceleration applications. In the case of the plasma wakefield acceleration scheme(2), the excited wakefield amplitude not only depends on the drive beam charge but is also very sensitive to the pulse length. This is particularly true in the case of nonlinear plasma wakefields in the blow out regime, because the plasma wave breaking limit is proportional to the beam density. In general, the drive pulse length (FWHM) should not exceed 1/3 of the excited wake field wavelength for optimal coupling of the beam energy to the wakefield. Dielectric structure based wakefield acceleration research has concentrated on structures in the 10 - 20 GHz range, but in order to study high gradient wakefield acceleration in high frequency 30 GHz structures, high charge (> 20 nC) and short (< 10 ps FWHM) electron beam is required.

High current (kA) short electron beam generation and acceleration did not materialize until the invention of RF photoinjector technology(3). Although most RF photocathode has been concentrated on high brightness, low charge applications such as free electron laser injectors, there have been several relatively high charge rf photocathode based electron sources built and operated(4,5,6). In general, there are

---

[*] Visiting scholar from Shanghai Institute of Nuclear Research, China.

CP472, *Advanced Accelerator Concepts: Eighth Workshop,*
edited by W. Lawson, C. Bellamy, and D. Brosius
1999 The American Institute of Physics 1-56396-889-4

two approaches to attaining high peak current. One approach is to generate an initially long electron bunch with a linear head-tail energy variation that is subsequently compressed using magnetic pulse compression. The advantage of magnetic compression is that it is a well-known technology and can produce sub-picosecond bunch lengths. Due to the strong longitudinal space charge effects, this technology is limited to somewhat lower charges (<10 nC). There have also been designs and operations of relatively high charge L-band guns at APEX(5) and TTF(6). Another approach is to directly generate short intense electron bunches at the photocathode and then accelerates them to relativistic energies rapidly using high peak axial electric fields in the gun (4). The advantage of this approach is that it can deliver very high charges, for example, 100 nC if one uses an L-band gun. This would satisfy the requirements of most electron driven wakefield experiments for both plasma and dielectric structures if the pulse length is short enough (< 10 ps FWHM). So far, the Argonne Wakefield Accelerator (AWA) has demonstrated the capability of producing 100 nC, 25 — 35 ps (FWHM) electron beams at 14 MeV. This unprecedented performance was obtained using a half cell photocathode gun cavity and two standing wave iris-loaded linac sections (7). The AWA machine has reached its design goal and has been used for dielectric wakefield (8) and plasma (9) experiments. The initial results are encouraging (10). Achieving higher gradients in wakefield experiments would require the drive electron pulse to be even shorter and have a lower emittance. In this paper, we discuss a new design for the RF photocathode gun with the capability of producing 10 - 100 nC with 2 - 5 ps (rms) pulse lengths.

## DESIGN APPROACHES

The physics of high current beam generation and transport is very complicated and some analytical work has been performed(11, 12). In general pulse lengthening is due to space charge effects, particularly in the high charge case. One can simply estimate the longitudinal space charge force of a 100 nC beam with 1 cm radius in the rest frame as 27 MV/m. Another effect is the transverse rf focusing and defocusing of the electron beam passing through the accelerating cavity, which causes both pulse lengthening and transverse emittance growth. Therefore, one needs to accelerate the electron beams generated from the cathode as fast as possible (high gradient) and to be relativistic before exiting out the RF gun. Thus in this new design, we took a brute force approach: 1) the electron beam is born in a strong axial electric field; 2) the beam is continuously accelerated with gradient as high as possible, therefore preventing bunch lengthening and emittance growth. 3) Adjusting an external focusing solenoid to minimize the emittance, eventually approaching the so called " emittance compensated" beams.

The choice for our new gun design is a Brookhaven type (13) 1 — 1/2 cell cavity scaled up to L band operation as shown in Figure 1. Although in general the beam will have lower emittance if one chooses a multi-cell gun cavity, RF power requirements will be excessive. In this study, the updated standard computer codes from Los Alamos

SUPERFISH and PARMELA (14) are used to model cavity fields and beam dynamics respectively.

Figure 1 shows a schematic diagram of the new gun and a section of the Linac. The linac section used here is an existing section from the current AWA linac(15). The drift distance between the gun and the linac is 32.3 cm designed to permit laser input and beam diagnostics. The following table summarizes the parameters used in our simulation of the new rf gun.

# The AWA new gun and linac layout

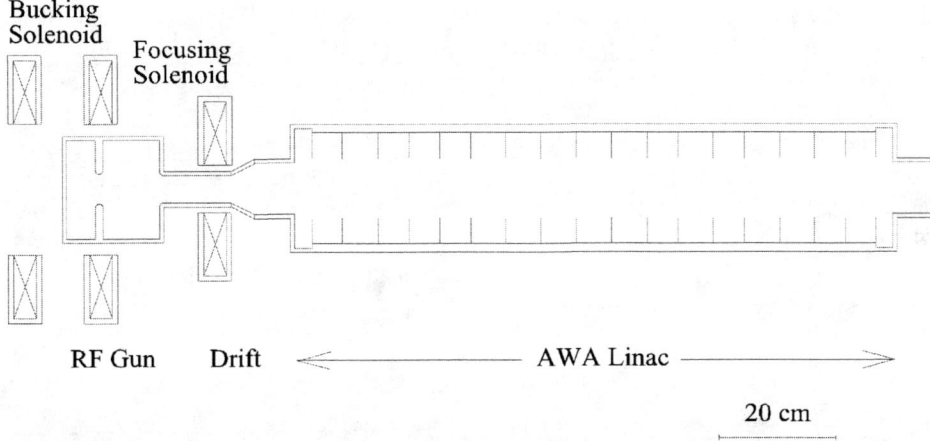

**FIGURE 1.** Schematic layout of the proposed AWA new electron gun and a section of the existing linac section.

**TABLE 1.** The gun design parameters as calculated using SUPERFISH

| | |
|---|---|
| Inner Radius of the Cell, b (cm) | 9.03 |
| Radius of the iris, a (cm) | 2.75 |
| Width of the iris, d (cm) | 1.5 |
| Aperture of the exit (cm) | 2.5 |
| Operating frequency (GHz) | 1.3 |
| Initial bunch length (ps) | 8.5 |
| Initial beam radius (cm) | 1 |
| Quality factor, Q | 26008 |
| Shunt impedance (MΩ/m) | 36.47 |

# NUMERICAL SIMULATION RESULTS

## Low Charge cases (10 - 40 nC)

As described in the above sections, for wakefield acceleration applications, one needs electron beam with charges in the range of 10 - 40 nC; also pulse length needs to be as short as 2 ps. In order to achieve such a number, we have systematically studied the optimized beam parameters for 10, 20, and 40 nC.

The simulation is done using the rf gun cavity parameters in table 1. We assume 1 cm laser radius at the photocathode with a flat top distribution. The laser pulse length used here are 2.6 ps for 10 nC, 5.2 ps for 20 nC and 10.4 ps for 40 nC beams. The choice of the laser pulse length is determined by the current AWA laser system The dependence of the electron beam rms emittance and pulse length is studied as a function of the axial electrical field on the photocathode with optimization done by varying the launch phase and gun solenoid field.

Figure 2 shows the dependence of the rms emittance and pulse length on the axial electrical field on the cathode. One can see that both quantities decrease drastically as the axial electrical field increases. At 80 MV/m, the rms emittance is 30 mm -- mrad and a pulse length of 0.33 mm is obtained. Even at 50 MV/m, σ is 0.6 mm and ε is 60 mm -- mrad. These results are much improved over the current AWA beam parameters at the same charge (7).

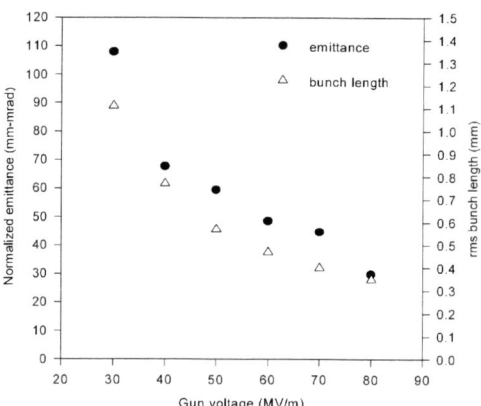

**FIGURE 2.** Dependence of the beam parameters at the linac exit on gun axial electrical field. For a 10 nC beam, one can obtain very short and low emittance bunches even at relatively low axial electric field.

The simulation results for the 20 nC and 40 nC cases are similar to the 10 nC, although both pulse length and emittance are increased due to space charge effects and also longer initial laser pulses. Figure 3 shows the dependence of pulse length and emittance as a function of the axial electrical field on the cathode for the 40 nC case. At 80 MV/m, $\sigma_z$ of 1.1 mm for 40 nC can be achieved. This is much better than the existing AWA gun which has rms pulse length of 5 mm at this intensity. We should point out that one should be able to obtain reasonable performance even at 60 MV/m cathode field. This field strength has already been achieved at the current 1/2 cell AWA drive gun.

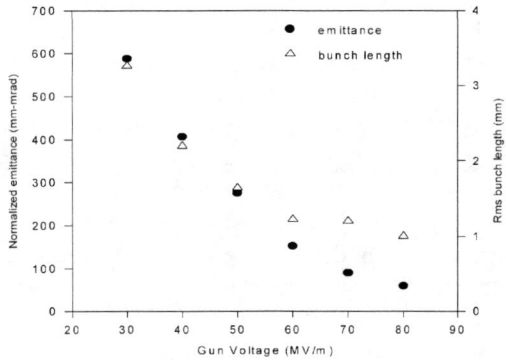

**FIGURE 3.** Dependence of the beam parameters on gun axial electrical field at the exit of linac for 40 nC beam intensity.

## High charge cases (100 nC)

In this section, we describe the simulation results for 100 nC beam charge. We assume 1 cm laser radius at the photocathode with a flat top distribution. The laser

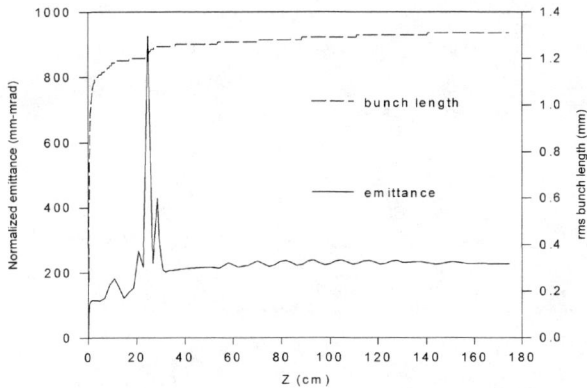

**FIGURE 4.** 100 nC beam rms bunch length and emittance evolution in the gun and linac for an optimized parameters

pulse length in all simulations is 8.5 ps also assuming a flat top distribution. The simulation parameters were varied systematically to optimize the final bunch length and emittance. We scan through solenoid field strength and rf injection phase and fields strength at cathode from 50 -110 MV/m. Our criteria for selecting the operating point is combination of parameters which gives the shortest pulse and lowest emittance.

Figure 4 shows a 100 nC bunch length and emittance evolution in the gun and linac for an optimized parameters. The effect of the peak electric field in the gun on the rms bunch length and emittance for an 100 nC beam is much stronger than at lower intensities. As expected, both emittance and pulse length decreases as the electric field increases. At 100 MV/m surface field along with axial solenoid magnetic field of 2 kG, and rf injection phase of 33°, one can achieve $\sigma_Z$ = 1.3 mm (4 ps) and normalized emittance $\varepsilon_N$ = 108 mm mrad (90% as defined in PARMELA) at the end of the linac with no beam losses. This is a very interesting result because it approaches the ~ 1 mm mrad/ nC attainable in so called "emittance compensated" beams. This is a significant improvement over the existing AWA gun design ( $\sigma_z$ = 4 mm, and $\varepsilon_N$=800 mm mrad). Figure 5 shows the beam's energy dependence on the cathode fields with energy range from 5 MeV - 11 MeV and energy spread of typically < 3 %. However, because PARMELA does not include wakefield or beam loading in its calculations, the energy spread can be much higher, particularly in the high charge cases. Typically it adds another 5% for AWA gun at 100 nC(15).

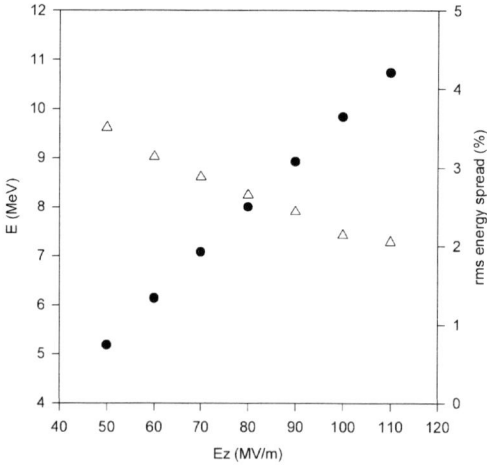

**FIGURE 5.** Output energy and energy spread dependence on the cathode field. Solid dots are energy and triangles are energy spread.

We would like to point out that for the 100 nC beam, the axial electrical field on the photocathode has to be greater than 70 MV/m. Also due to the intense radial space charge forces and higher energy of the beam, a strong solenoid field ~ 2 kG has to be maintained. This field strength is very high for this type of solenoid design.

## DISCUSSION AND SUMMARY

In the last section we showed that the 1 - 1/2 cell rf photocathode gun plus a section of the current AWA linac would yield a very low emittance and can be attained with a short electron beam. For a 20 nC electron beam, rms pulse length of 0.7mm is observed and 22 mm mrad normalized emittance with energy of 18 MeV. If this beam is used as a plasma wakefield accelerator driver, it can excite gradient in excess of 1 GeV/m with a plasma density of ~$10^{14}$ /cm$^3$ (16).

In applications to dielectric wakefield acceleration, this beam would also make a significant improvement over present attainable gradients. One can use this beam to directly demonstrate collinear wakefield acceleration gradients in excess of 50MV/m corresponding to 200 MW rf power generated in 30 GHz dielectric structures.

We have designed an RF photocathode gun for high current applications. The numerical simulation results indicate that 10 kA peak current can be obtained. This beam will enable us to study high gradient wakefield acceleration in both plasma and dielectric wakefield accelerator..

This work  is supported by the Department of Energy, High Energy Physics Division, Advanced Technology Branch under the Contract No. W-31-109-ENG-38.

## REFERENCES

1. Travier, C., in Proceedings of Advanced Acceleration Concept Workshop, Edited by P. Schoessow, AIP Proceedings, No. **335**, p57 (1994)
2. Barov, N., Ph. D. Thesis, UCLA, 1998
3. Fraser, J. *et al.*, IEEE Trans. Nucl. Sci., NS-32,p.1719 (1985)
4. Schoessow, P., *et. al.,* Proceedings of Particle Accelerator Conference, p976-978., 1995
5. Calston, B *et al.*, Proceedings of Particle Accelerator Conference, p.985., 1995
6. Colby, E. *et al.*, Proceedings of Particle Accelerator Conference, p967.,1995
7. Conde, M. *et al,.* in the Proceedings of Particle Accelerator Conference, Vancourver, May 10-14, 1997
8. Schoessow, P. *et al* in the Proceedings of Particle Accelerator Conference, Vancourver, May 10-14, 1997
9. Barov, N., M. Conde, W. Gai, J. Rosenzweig, Phys. Rev. Lett., Vol. 80, No. 1, p81, 1998
10. Schoessow, P. *et al*, to appear in Journal of Applied Physics.
11. Kim, K.J., Nuclear Instrumentation and Methods, **A275**, pp.201-218, 1989
12. Serafini, L., J. Rosenzweig, Physical Review E, Vol 55, No. 6 ,p7565 ,1997
13. Batchelor, K., et al, Proc. of European Particle Accelerator Conference, p.541-543 (1990)
14. SUPERFISH and PARMELA, Las Alamos National Lab Report LA-UR-96-1834, 1997 and LA-UR-96-1835, 1996
15. Ho, C. H., Ph.D. Thesis, 1992, University of California at Los Angeles.
16. Barov, N., private communication

# Femtosecond Electron Beam Generation by S-Band Laser Photocathode RF Gun and Linac

M. Uesaka, K.Kinoshita, T. Watanabe, T.Ueda, K.Yoshii, H.Harano,
J.Sugahara, K. Nakajima*, A. Ogata*, F. Sakai⁺, H.Dewa⁺, M.Kando⁺,
H. Kotaki⁺, S.Kondo⁺

*Nuclear Engineering Research Laboratory, University of Tokyo,
Tokai-mura, Naka-gun, Ibaraki-ken, 319-1106, Japan*

*High Energy Accelerator Research Organization,
1-1 Oho, Tsukuba-shi, Ibaraki-ken, 305, Japan*

⁺*Japan Atomic Energy Research Institute,
Tokai-mura, Naka-gun, Ibaraki-ken, 319-1106, Japan*

**Abstract.** A laser photocathode RF electron gun was installed in the second linac of the S-band twin linac system of Nuclear Engineering Research Laboratory(NERL) of University of Tokyo in August in 1997. Since then, the behavior of the new gun has been tested and the characteristic parameters have been evaluated. At the exit of the gun, the energy is 3.5 MeV, the charge per bunch 1~2 nC, the pulse width is 10 ps(FWHM), respectively, for 6 MW RF power supply from a klystron. The electron bunch is accelerated up to 17 MeV and horizontal and vertical normalized emittances of 3 $\pi$ mm.mrad are achieved. Then, the bunch is compressed to be 440 fs(FWHM) with 0.35 nC by the chicane-type magnetic pulse compressor. The linac with the gun and a new femto- and picosecond laser system is planned to be installed for femtosecond pulseradiolysis for radiation chemistry in 1999.

## INTRODUCTION

A laser photocathode RF electron gun is one of the most attractive electron sources since it can supply the relativistic electron bunch of high quality both transversely and longitudinally. Namely, low transverse and longitudinal emittances are advantageous for the brightness of synchrotron radiation and bunch compression, respectively. Especially, those features are inevitable for X-ray free electron laser(FEL) such as SASE[1]. Several works have been done for its development aiming the application to FEL and linear collider[2,3,4,5]. We installed a new S-band laser photocathode RF electron gun in the second S-band twin linac system[6] in August in 1997. The gun was constructed by KEK, Brookhaven National Laboratory(BNL) and Sumitomo Heavy Industries based on much experiences at BNL[7]. The purpose is to apply it to the joint research project on laser wakefield acceleration and femtosecond X-ray generation via Thomson scattering among Nuclear Engineering Research Laboratory of University of Tokyo, High Energy Accelerator Research Organization(KEK) , and Japan Atomic Energy Research Institute [8,9], the picosecond pulseradiolysis for radiation chemistry and the picosecond time-resolved X-ray diffraction. The main subject here is to produce a femtosecond low emittance electron bunch and to enhance the quality and stability of the gun. Technical

CP472, *Advanced Accelerator Concepts: Eighth Workshop,*
edited by W. Lawson, C. Bellamy, and D. Brosius

feasibility and reliability of the gun are totally accomplished considering the scope of the applications. Updated results are presented in this paper.

# PERFORMANCE OF LASER PHOTO-CATHODE RF GUN

Upgraded twin S-band linac linac system with the laser photocathode RF gun is depicted in Fig. 1. We also constructed the chicane-type magnetic pulse compressor. Two 6 MW S-band klystrons feed RF power to the RF gun and accelerating tube individually.

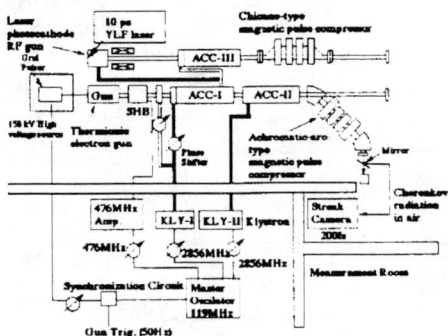

**FIGURE 1.** Upgraded subpicosecond S-band linac in the twin linac system.

The 90 kV thermionic electron gun, subharmonic buncher and two prebunchers were replaced with the RF gun so that the injector section becomes very simple. The cavity of the gun has S-band 1.6 cells. 10 ps (FWHM) light pulse is produced by the fourth harmonics(263 nm) of the YLF laser(1.05 μm) and irradiates the copper cathode at 68 degree angle at 10-50 Hz. Since the basic mechanism of electron emission is photoelectric, the lifetime of the copper cathode is intrinsically unlimited. The work function of copper is 4.6 eV(270nm). The quantum efficiency around the work function is $1 \times 10^{-4}$. 6 MW RF power is fed to the cavity to induce the field gradient close to 100 MV/m. The time-duration of the fed RF is 4-8 μs. The peak energy at the exit of the gun is 3.5 MeV. The solenoid magnet is attached to the cavity for transverse emittance compensation against space charge effect. The emittance is measured by the conventional quadrupole scanning. Horizontal and vertical emittances are uniformly 3 πmm.mrad in normalized rms. Here we controled transverse laser spot to be circular at the cathode even for oblique injection. Then, we had uniform emittance for both directions. If we do not perform the above treatment, the emittance becomes not uniform and the lowest horizontal emittance of 1 πmm.mrad is achieved while the vertical one is 7 πmm.mrad. The beam spot is $\phi 3$ mm. Maximum charge per bunch is 2 nC for 75 μJ laser energy at the cathode. Then, the low emittance electron beam is accelerated up to 17 MeV and simultaneously its energy profile is modulated for the magnetic pulse compression in the accelerating tube where the maximum field gradient is 8.5 MV/m.

We are always making efforts to reduce dark current by baking for high vacuum in the cavity and the RF aging, and observing its behavior. It is very important to reduce the dark current because it would be rather harmful for the applications such as FEL and pulseradiolysis from several aspects of noise or sample damage. The dark currents are multi-bunches existing in the traveling accelerating RF phases. Therefore, each peak current is negligible while the total charge during the whole RF pulse is more than photoelectrons. So far, its charge per 4 μs RF pulse is 2 nC at 50 Hz. When the RF pulse is elongated to 8 μs, it increase to 26 nC. We are going to continue the efforts.

## BUNCH COMPRESSION FOR FEMTOSECOND SINGLE BUNCH

The chicane-type magnetic pulse compression was designed by using PARMELA. It consists of four identical bending magnets. In order to compensate the nonlinearity of the energy modulation in the accelerating tube, we optimized the longitudinal length of the magnet and the gap between the magnets. Calculated longitudinal phase space distribution of electrons and pulse shape are shown in Fig.2. Its pulse width is 200 fs at FWHM.

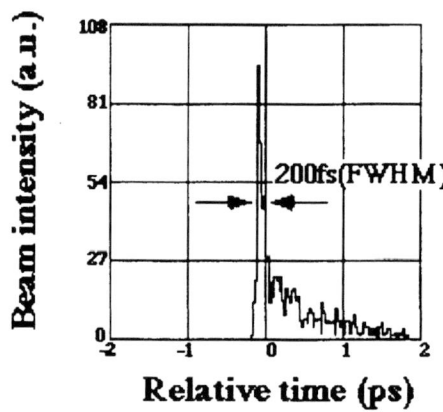

**FIGURE 2.**    Numerical result of the compressed electron bunch by PARMELA

The pulse shapes of the bunches with and without compression were measured by a single shot by the femtosecond streak camera (FESCA-200, HAMAMATSU PHOTONICS), which time-resolution is 200 fs, via Cherenkov radiation emitted in a Xe-gas chamber attached at the end of the linac. Measured pulse shapes of the bunches before and after the compression are shown in Fig.3.

It is observed that 13 ps (FWHM) bunch is compressed to 440 fs(FWHM). The average charge of the compressed bunches is 0.35 nC. This reduction of charge is mainly due unoptimized optics and alignment of the linac, which should be improved. We carried out the calibration of the time-resolution of the camera using a 100 fs Ti:Sapphire laser after the beam experiment. We found out that the error at FWHM of the camera at that time is 370 fs assuming the Gaussian error function and the law of error propagation. When we subtract the error from 440 fs, it become 238 fs at FWHM, which agrees well with the numerical result. Again here the advantage and effectiveness of the low emittance beam from the laser photocathode RF gun was confirmed.

There are several discussions about the precision of the space charge force of PARMELA as for such a ultrashort bunch. Actually the noninertial space charge force and the coherent radiation force[10,11] in a bending magnet are not considered in PARMELA. Recently, a preliminary numerical simulation of our bunch compression in the chicane was carried out and the effect of the above forces on the pulse length was calculated to be negligible[12]. We are going to measure the emittance growth due to the effect in the chicane.

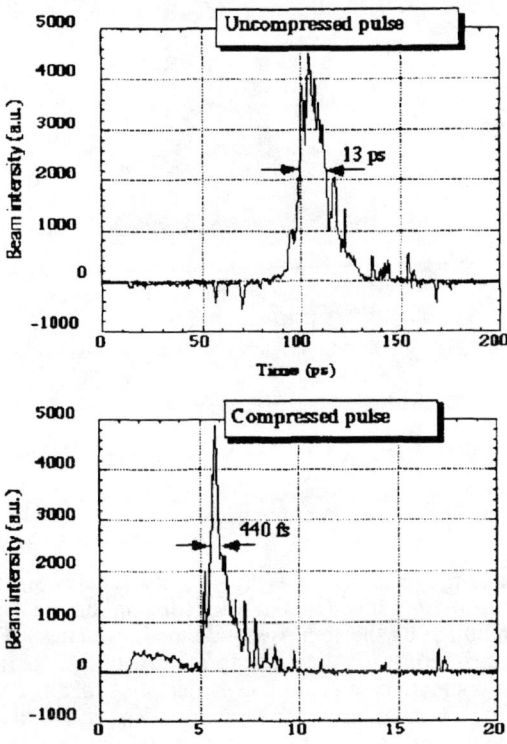

**FIGURE 3.** Measured pulse shapes with and without compression.

# SYNCHRONIZATION BETWEEN LASER AND ELECTRON

We are investigating the precision of synchronization between the laser and electron pulses. We measured it by the femtosecond streak camera. About 100 data of streak measurements were accumulated and the time-interval between the two pulses was evaluated. Its histogram is given in Fig. 4. If we assume the distribution to be Gaussian, the standard deviation is 3.5 ps. This linac can be synchronized with the femtosecond $T^3$ (Table-Top Terra-Watts) laser and has been applied to laser wakefield acceleration and femtosecond X-ray generation via the head-on Thomson scattering[13]. Further, a new femto- and picosecond laser system is going to be installed for femtosecond pulseradiolysis for radiation chemistry in 1999.

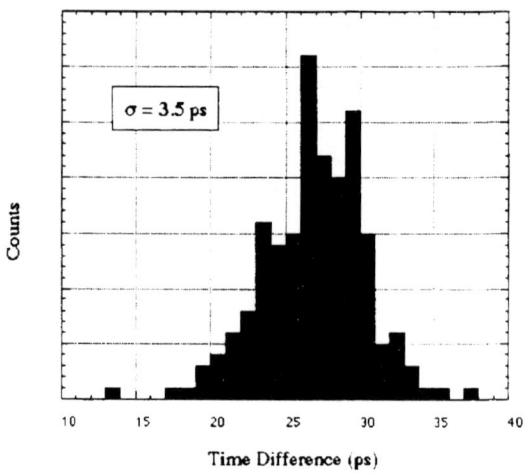

$\sigma = 3.5\ ps$

Counts

Time Difference (ps)

**FIGURE 4.** Histogram of time interval between laser and electron pulses.

# COHERENT TRANSITION INTERFEROMETER

## Theory

There are two promising methods to evaluate pulse shape of femtosecond electron bunch. The first one is to measure Cherenkov radiation or optical transition radiation emitted by the electron bunch by the femtosecond streak camera. The second one is the coherent far-infrared transition radiation interferometry [14,15,16]. It is important to compare the results by the two methods in order to confirm the precision of both methods[17,18,19]. First we performed the measurement at the first linac where the 90 kV thermionic electron gun and achromatic-arc-type compressor were installed.

Radiations from a relativistic electron bunch such as synchrotron radiation, transition radiation, Cherenkov radiation etc. have broad spectrum. In case that the wavelength of the radiation is shorter than the electron bunch length, the phase of radiation emitted by the electrons is different from one another so that the radiation is incoherent. On the other hand, in case that the wavelength is longer than the bunch length, the phase becomes almost the same so that the radiation is coherent. This is called the temporal coherence of radiation. The coherent radiation yields the interferogram when we use an interferometer such as the Michelson interferometer. The information of the electron bunch can be deduced from the interferogram. Another important feature of coherent radiation is the dependence of the power on the number of electrons in the bunch. The following theory shows that the power of the incoherent radiation is linear to the number while that of the coherent radiation is linear to the square. From the interferogram of the light intensity of two interfered coherent radiation pulses, the longitudinal bunch distribution are given in the following procedure.

When the cross section of the beam is small and the observation point is far from the source point, the intensity of the transition radiation is expressed by the analogy of the intensity of coherent synchrotron radiation as,

$$I_{total}(v) = N\left[1 + (N-1)f(v)\right]I_e(v),\tag{1}$$

where $N$ is the number of electrons in the bunch, $v$ is the wave number which is the inverse of the wavelength of the transition radiation and $I(v)$ is the transition radiation intensity emitted from a single electron. The first term of Eq.(1) expresses the incoherent transition radiation while the second term the coherent transition radiation. The quantity $f(v)$ is the bunch form factor which is given by the Fourier transform of the distribution function, $S(\bar{x})$, of the electron in the bunch,

$$f(v) = \left|\int S(\bar{x})\exp\left[i2\pi(\bar{n}\cdot\bar{x})v\right]d\bar{x}\right|^2,\tag{2}$$

where $\bar{n}$ is the unit vector directed from the center of the bunch to the observation point and $\bar{x}$ is the position vector of the electron relative to the bunch center. Since $N \gg 1$, we approximately have,

$$I_{total} \simeq N^2 f(v)I_e(v).\tag{3}$$

The form factor $f(v)$ can be divided into two parts, the longitudinal bunch form factor $f_L(v)$ and the transverse bunch form factor $f_T(v)$ as follows,

$$f_L(v) = \left|\int h(z)\exp(i2\pi z\cos\theta/v)dz\right|^2,\tag{4}$$

$$f_T(v) = \left|2\pi\int g(\rho)J_0(2\pi\rho\sin\theta v)\rho d\rho\right|^2,\tag{5}$$

where $h(z)$ and $g(\rho)$ are the longitudinal($z$) and transverse($\rho$) distribution function of the electron bunch, respectively. The transverse bunch form factor is obtained by measuring the transverse distribution of the electron bunch. When we observe the transition radiation from the on-axis or nearly on-axis direction, i.e., $\theta^2 \gg 1$, $\cos\theta$ and $\sin\theta$ can be unity and zero, respectively.

From the experiment, the interferogram of the light intensity of the two interfered coherent transition radiation pulses as a function of the moving mirror position of the interferometer is obtained. By definition the interferogram can be written,

$$S(\delta) = 4\pi\int |RT|^2\left|\bar{E}(\omega)\right|^2 e^{-i2\pi v\delta/c}dv,\tag{6}$$

where $S(\delta)$ is the intensity of the interfered radiation intensity at the detector for the optical path difference $\delta$ minus the intensity at $\delta \rightarrow \pm\infty$, $\bar{E}(\omega)$ is the Fourier transform of the electrical field of the transition radiation and $R, T$ are the coefficients of reflection and transmission at the beam splitter, respectively. Solving for $\left|\bar{E}(\omega)\right|^2$ yields,

$$\left| \tilde{E}(v) \right|^2 = \frac{1}{4\pi c |RT|^2} \left| \int_{-\infty}^{\infty} S(\delta) e^{-i2\pi v \delta / c} d\delta \right|^2. \tag{7}$$

Using Eq.( 3) and the relation $I_{total}(v) = \left| \tilde{E}(2\pi c v) \right|^2$, the bunch form factor can be obtained by,

$$f(v) = \frac{\left| \int_{-\infty}^{\infty} S(\delta) e^{-i2\pi \delta / v} d\delta \right|^2}{4\pi c |RT|^2 N^2 I_e(v)}. \tag{8}$$

Finally the Kramers-Kronig relation and inverse Fourier transform gives the longitudinal bunch distribution $h(z)$ from the longitudinal bunch form factor as follows[20],

$$h(z) = \int_{-\infty}^{\infty} g(v) \exp[i(\phi_s(v) - 2\pi v t)] dv, \tag{9}$$

$$g(v) = f_L^{1/2}(v),$$

$$\phi_s(v) = -2v \int_0^{\infty} \frac{\ln[g(v')/g(v)]}{v'^2 - v^2} dv.$$

## Experiment

We performed this comparison at the first linac where the achromatic-arc-type magnetic pulse compressor was installed. In the experiment the longitudinal bunch distribution was controlled by tuning the energy modulation of the bunch in the accelerating tube for the magnetic pulse compression. We chose femto- and picoseconds (FWHM) pulse widths

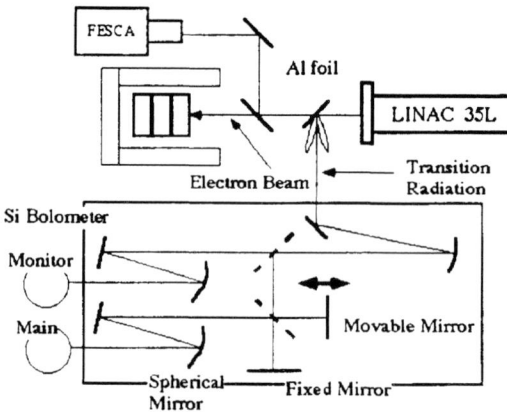

**FIGURE 5.** Experimental setup for coherent transition radiation interferometry

914

and performed the comparison between the femtosecond streak camera and the Michelson coherent transition radiation interferometry measurement as shown in Fig.5. We measured the transition radiation in the far-infrared region emitted by an electron bunch at the Al-foil put in air after the 50 μm thick Ti window at the end of the linac. We used liquid-He-cooled Si bolometers as a detector for the far-infrared radiation. The major beam parameters are as follows: the energy was 32 MeV, the pulse length is 500 fs to 1.7 ps (FWHM) and the electron charge per single pulse is 30 to 250 pC.

On the basis of the procedure of analysis as mentioned before, we have analyzed these pulses from the interferograms which we have got by the Michelson interferometer. Because of nonuniform transparency of the 100 μm-thick Mylar beam splitter and diffraction loss of long wavelength components, the bunch form factor was obtained within rather limited range. Therefore we had to use theoretical bunch form factor assuming the Gaussian or exponential distribution out of the range.

## Results and Discussion

The interferogram of the subpicosecond electron pulse is shown in Fig.6.

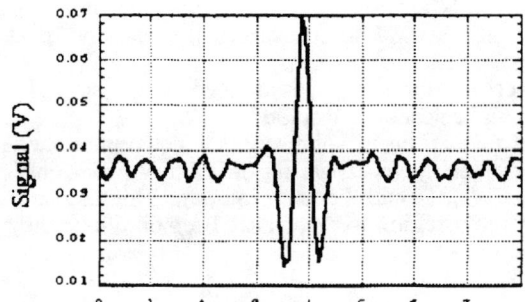

**FIGURE 6.** Interferogram of the subpicosecond electron pulse.

The experimental result of the bunch form factor is shown by the solid curve and that of theoretical by dashed curve in Fig.7. In the figure, we chose the Gaussian distribution as the theoretical curve, since the exponential function has unphysical long tails in both sizes and the observation of the bunch shapes by the streak camera done just beforehand indicates that the Gaussian is closer to the real bunch distribution.

The dashed curves in Fig.7 represent those of three bunch lengths of 400, 500 and 600 fs at FWHM. We used the measured bunch form factor from 9.5 to 18 cm$^{-1}$ range and the theoretical bunch form factor out of the range for the analysis. In this case, we adopt the bunch form factor of 500 fs bunch length and extrapolate this to the range under 9.5 cm$^{-1}$ and over 18.0 cm$^{-1}$.

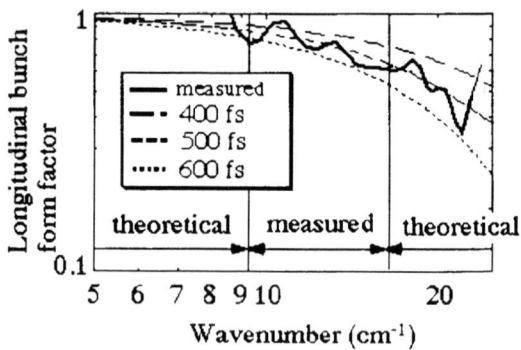

**FIGURE 7.** Bunch form factor

Finally, we got the bunch distribution from the interferogram as shown by the solid curve in Fig.8. The dashed curve in the figure is one of the pulse shape taken by the streak camera. The result by the interferometry gives 550 fs bunch length at FWHM while that by the streak camera becomes 650 fs. The calibration of the camera was also performed by using a Ti:Sapphire laser. Then the error at FWHM was found out to be 370 fs assuming the law of error propagation. After the above error is substrated, the net pulse length becomes 550 fs. These results agree with each other and it is therefore clear that the reliability of the method to measure subpicosecond electron pulse has been comfirmed.

We are going to perform the same measurement at the second linac where the laser photocathode RF gun and chicane are installed.

With the choice of thinner beam splitter which determines the appropriate spectrum window of the interfrometer, we expect the method is promising even for the shorter bunch (100 - 200 fs FWHM) with better resolution. This is also because the spectrum shifts from the far-infrared region to the infrared region where the sensitivity of the Si-bolometer becomes better.

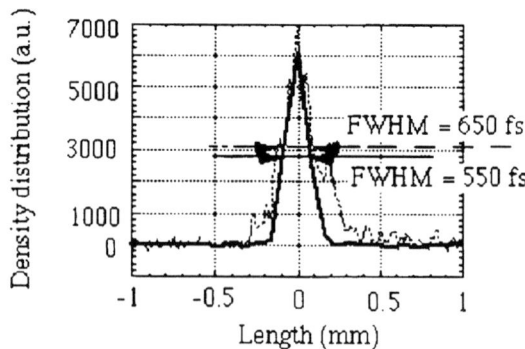

**FIGURE 8.** Reconstructed bunch distribution by the interferometry (solid curve) and the result by the streak camera (dashed curve)

# CONCLUSION

The new laser photocathode RF electron gun and the chicane-type magnetic pulse compressor were installed in the second S-band linac in the twin linac system. The details of their characteristics were measured and evaluated. Horizontally and vertically uniform emittance is 3 $\pi$ mm.mrad in normalized rms with 3.5 MeV and-1 nC. After acceleration up to 17 MeV, 13 ps bunch was compressed to 440 fs(FWHM) with 0.35 nC. The femtosecond electron single bunch of the low emittance has been used for the laser wakefield acceleration and the femtosecond X-ray generation via the head-on Thomson scattering. Both advantages and drawbacks of the gun continue to be checked including the technical feasibility and reliability for such applications.

We constructed the Michelson coherent transition interferometer for subpicosecond electron pulse shape measurement. From the comparison of the diagnostics by the interferometry with that by the streak camera, the reliability of the interferometric method to measure the subpicosecond electron pulse that are close to the time resolution of the femtosecond streak camera was confirmed. The design for 100 - 200 fs has started. Especially, a new pulseradiolysis for radiation chemistry in 1999.

# ACKNOWLEDGMENT

The authors would like to thank to Dr.M.Washio and Dr.A.Endo of Sumitomo Heavy Industries and Dr.X.J.Wang of BNL for their development of the laser photocathode RF electron gun.

# REFERENCES

[1] H.Winick et al., Nucl. Instrum. Meth. A, 347 (1994), pp.199-205.

[2] X.J.Wang, X.Qui and I.Ben-Zvi, Phys. Rev. E, 54(4) (1996), p.3121.

[3] B.E.Carlsten et al., Proc. of MICRO BUNCHES WORK SHOP(AIP Press-367, 1996), pp.21-35.

[4] H.Braun et al., ibid, pp.11-20.

[5] A.Mukoh, J.B.Rosenzweig et al., Nucl. Instrum. Meth. A(1998), in press.

[6] M.Uesaka et al., in this proceedings.

[7] I.Ben-Zvi, Proc. of ADVANCED ACCELERATOR CONCEPTS(AIP Press-398, 1997), pp.40-54.

[8] K.Nakajima et al., Proc. of the 17th Particle Accelaretor Conf., Vancouver(1997).

[9] K.Nakajima et al., ibid.

[10] B.E.Carlsten, Phys. Rev. E, 54(1)(1996), pp.838-845.

[11] L.Selafini, J.B.Rosenzweig, Phys. Rev. E, 55(6) (1997), pp.7565-7590.

[12] personal communication with R.Hajima of University of Tokyo.

[13] K. Nakajima et al, in this proceedings.

[14] A.Muroh, J.B.Rosenzweig et al., Nucl. Instrum. Meth. A(1998) in press.

[15] H.Lihn, H.Wiedemann et al., Phys. Rev. E, 53(6)(1996), p.6413.

[16] Y.Shibata et al., Phys. Rev. E, 50(2-B)(1994), p.1479.

[17] Y.Shibata, M.Uesaka et al, submitted to Nucl. Instrum. Meth.

[18] M.Uesaka et al, Nucl. Instrum. Meth. A(1998), in press.

[19] T.Watanabe, M.Uesaka et al., submitted to Nucl. Instrum. Meth.

[20] R. Lai, U. Happek and A.J. Sievers, Phys. Rev. E, 50(6)(1994), pp.R4294-R4297.

# Electron Beam Generation Using a Ferroelectric Cathode

J. D. Ivers, D. Flechtner, Cz. Golkowski , G. Liu*,
J. A. Nation, and L. Schächter

*Laboratory of Plasma Studies and School of Electrical Engineering,
Cornell University, Ithaca, NY 14853, USA.*

**Abstract.** Data is presented on the production of electron beams from a ferroelectric cathode at voltages of order 0.5 MV and current densities of order 100 A/cm$^2$. In comparison with data at lower voltages the beam current scales as the three halves power of the voltage. An interpretation of the voltage dependent scaling, based on the coupling of electrostatic energy from the ferroelectric to the gun, is presented.

## INTRODUCTION

Ferroelectric Cathodes have been extensively studied over the last several years [1]-[4] in an attempt to develop a means of emitting a high current electron beam from a robust room temperature cathode for high power microwave generation. Most of the research has focussed on two types of cathode, namely:

a. PLZT anti-ferroelectric compositions e.g. 4/95/5 in which emission occurs when an applied electric field causes the material to switch from the antiferroelectric state to the ferroelectric state. Switching occurs when an electric field of order 15-25 kV/cm is applied across the PLZT [5]. Recent work [6] has suggested that higher fields, of order 52 kV/cm are required to initiate the electron emission and,

b. PZT and PLZT ferroelectric compositions in which the emission is triggered by 'switching' around a hysteresis loop. Fields of order 10 kV/cm, typically applied across a 1 mm thick sample, result in the electron emission [7]-[10].

In this paper we present data obtained with a PZT cathode in an electron gun configuration which is used to generate an electron beam at energies in the range 200-550keV, with a beam current of up to 350 Amperes in pulses having a duration in excess of 200 ns. These results extend emission characteristics previously reported by more than one order of magnitude in voltage and by a factor of 3 in the current density. The data also presents the first reported results applicable to electron gun design. The planned application of the source is to high power microwave generation using a TWT amplifier in X

* Permanent Address: Northwest Institute of Nuclear Technology, China.
* Permanent Address: Electrical Engineering Department Technion, Haifa, Israel.

CP472, *Advanced Accelerator Concepts: Eighth Workshop,*
edited by W. Lawson, C. Bellamy, and D. Brosius

and Ka bands. In the following sections we describe the experimental arrangement used for this work, the results obtained, and their interpretation.

## EXPERIMENTAL CONFIGURATION

The electron gun used in this work employs a pulse transformer system capable of generating a 500 kV, 200A, 250 ns electron beam and uses a ferroelectric cathode as the electron source. It is designed for use in high power microwave experiments. The system operates at a repetition rate of about 0.1 Hz which is limited by the available power supplies. Vacuum levels are presently in the vicinity of $5 \cdot 10^6$ Torr.

We present a brief description of the modulator and beam generator used in this work. The modulator, which has been recently developed for this application, has been described previously in the Particle Accelerator Conference Proceedings [11] so the description given here will only summarize the system.

The primary power source consists of 12 transmission lines, each having an impedance of about 10 $\Omega$. Half of the lines are positively charged and half negatively to a voltage in the range 20-35 kV. The lines are switched at the load location as indicated in Figure 1.

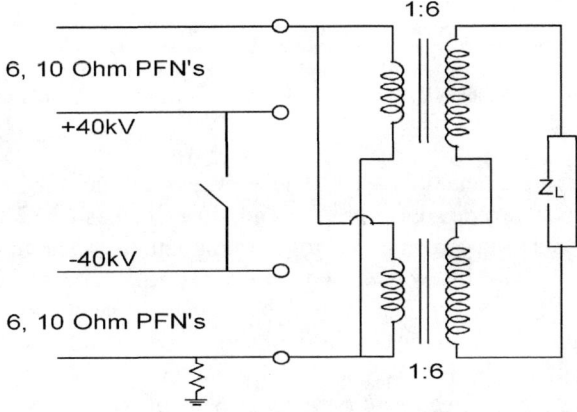

**FIGURE 1.** Primary Pulse Line Configuration and Switching.

Each line uses nine, 3.6 nF capacitors in a transmission line arrangement. On closing the switch a voltage is developed across the load with a rise time which is independent of the impedance transfer function of the line. The system is equivalent to three, 10$\Omega$ Blumlein's in parallel, so that the primary impedance is about 3.3 $\Omega$. The switch employs a rail configuration which is triggered by an 80-100 kV pulse applied to the center electrode. The transformer primary has 10 paralleltwo turn windings uniformly distributed around the circumference of each of two sets of TDK PE14 ferrite cores. The two sets of primary windings are in parallel with each other and have a total volt second product of 13.8 mVs. The ferrite cores are reset between pulses by a slow discharge of a capacitor through a single turn coil around the cores. The secondary of the transformer has 12 turns which encircle both cores, hence the overall system acts as a nominal 12:1 step-up transformer with an output impedance of about 500Ω. The final design represents a compromise between a high voltage gain and a short rise time system. Leakage inductance limits the useful gain of each section of the transformer and the use of the two parallel primaries yields a significant advantage over the use of a single stage 12:1 step up transformer. The output of the transformer is connected to a diode/electron gun configuration as shown in figure 2. This figure also indicates the amplitude of an applied external magnetic field used to confine the electron flow. The ferroelectric emitter is located in the

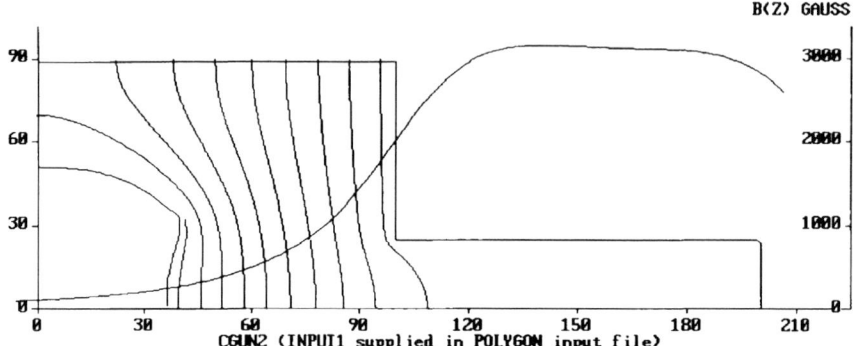

**FIGURE 2.** Diode/Electron Gun Schematic. Equipotentials for the system and the amplitude of an applied magnetic field are indicated in the figure.

cathode surface and has a diameter of 1.9 cm. The emitter is a 1 mm thick PZT sample, commercially available from Transducer products and identified as LTZ2. It is prepoled, and mounted with the Polarization vector pointing into the vacuum region. The configuration reported in this work was chosen to provide the data needed on the emission characteristics of the ferroelectric at the high anode-cathode voltages required for subsequent electron gun design. In the configuration employed in this work the axial field strength was about 3.2 kG in the center of the 15 cm long field coil system. The windings were energized by an electrostatic capacitor bank of about 2.5 mF at 150 V. The system generated the axial magnetic field in the drift tube section which, when operated at rated voltage, is capable of producing a solenoidal field, in our 5 cm bore drift section of up to about 9 kG. The present measurements were limited to injection of the beam into the 3.2 kG field where the beam was collected in the 5 cm diameter drift tube. The magnetic field penetration time into the drift section was less than the experimentally measured rise time of about 1 ms. The magnetic field at the cathode field surface is about 250 Gauss so we have significant beam compression. Since the beam dynamics are governed by the conservation of canonical angular momentum and in the present experiments no effort was made to match the beam to the cathode emission, the beam envelope scallops during propagation through the drift section.

The ferroelectric emitter has a surface polarization charge density of about $\mu C/cm^2$. Surface electric fields, derived from Gauss'Law are in excess of 1 GV/m and result in surface charge neutralization by free electrons. A thin ($<$ $\mu m$) grounded grid consisting of a number of $200\mu m$ width silver strips spaced from each other by $200\mu m$ is deposited on the front surface of the ferroelectric. Normally emission is produced by the application of a negative voltage pulse to the metalized rear surface of the ferroelectric. In these experiments however, a positive trigger pulse is applied to the rear surface of the ferroelectric. This results in electron emission from the metallic grids which drives Fowler-Nordheim field emission in the vicinity of the metallic grid ferroelectric, vacuum, triple lines. The duration of the field emission is determined by the applied pulse duration (~100ns) and by the hysteresis properties of the ferroelectric so that the total emission may exceed $1\mu s$ [7]. The use of a positive polarity pulse was found to yield a consistent electron current. The 100 ns trigger pulse used to initiate the electron emission is derived from a charged cable configuration switched by a krytron into a 50 $\Omega$ cable which is wound around the ferrite transformer core close to the secondary winding. The inductively decoupled pulse is fed via a 2:1 step down transformer to the back of the ferroelectric. The electric field applied across the ferroelectric is about 10 kV/cm.

## EXPERIMENTAL OBSERVATIONS

The pulser is typically run at 0.1 Hz for about 100 shots prior to taking data. Subsequently each data series is preceded by a sequence of about 20 or more shots. During the initial break-in of the cathode there is spiking in the emission with ~10-20ns current bursts, probably associated with out-gassing of the ferroelectric. Following use, the incidence of spiking seems to decrease and the emission pulses are similar to the data presented in figure 3. From top trace to bottom trace, the data in Figure 3a and Figure 3b show: the voltage across the secondary; the ferroelectric trigger pulse; the beam current, which is measured by a Rogowski coil after collection by a graphite collector located well into the magnetic field coil region; and finally the current through the transformer

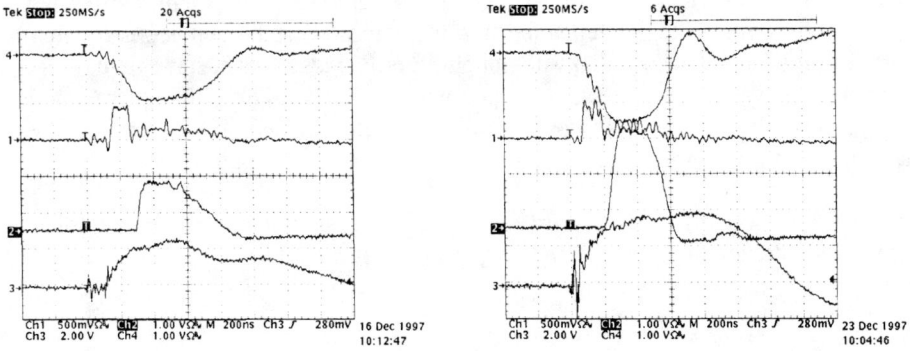

**FIGURE 3. a.** (Left) Data showing a 300 kV, 80 A beam emission from a ferroelectric disk illustrates the switching of the beam current by the ferroelectric trigger pulse. The current rise time is much less than for the secondary voltage. **b.** (Right) Similar data at an output voltage and current of 500 kV, 200 A.

primary. In both sets of data the cathode-collector separation was typically 11 cm. The two sets of data shown in the figure yielded output voltages of 300 and 500 kV and beam currents of 80 and 200 A respectively for a cathode to drift tube spacing of 6.6 cm. A copper sulfate resistor was mounted in parallel with the anode-cathode gap to improve the impedance match between the beam output and the primary. The impedance of the parallel resistor was varied with the operating conditions and was typically in the range 500-1000 $\Omega$. Although the traces are not time synchronized they are approximately time correlated with the exception that the trigger pulse shown in the figures should be delayed by 110 ns to show the correct timing relative to the other traces. Consistent beam emission as shown in the figures was obtained with the ferroelectric trigger pulse occurring at the start of the voltage maximum on the secondary. The beam emission varied considerably with the timing and amplitude of the ferroelectric trigger pulse with respect to the output voltage. While the output voltage rises in about 200 ns the current rise time is less than or equal to 20 ns and is instrument limited in the data shown. With earlier initiation of the beam current the rise time of the current is degraded and is comparable to or longer than that of the secondary voltage pulse. With both 4.5 cm and 6.5 cm cathode-drift tube spacings the electron emission from the ferroelectric did not occur until after the ferroelectric trigger pulse was applied. Computer modeling shows the vacuum electric field at the surface of the cathode reaches 100 kV/cm with an applied secondary gun voltage of 500 kV and with a 45 cm cathode-drift tube spacing.

The primary beam current and the secondary voltage waveforms show clearly the effects of the core saturation, namely an increase in the primary current and a decrease in the secondary pulse duration. The beam design output parameters required for the microwave experiments are 500 kV, 200 A with a pulse duration of 250 ns.

In figure 4 we show plots of the gun current versus the three halves power of the gun voltage for gap spacings of 6.5 and 4.5 cm respectively. The dashed line on each curve represents the results found for space charge limited emission, as measured using the EGUN code, with the actual geometry and magnetic field arrangement. The experimental data are based on representative results obtained over several thousand events with most of the data obtained with a 6.5 cm anode cathode spacing. The shorter gap data was used to illustrate the dependence of the emission on the gun geometry and to illustrate that the ferroelectric disks do not emit until triggered at surface fields of 100 kV/cm. We have as yet not operated the gun with shorter gaps or with higher cathode electric fields. The exposed cathode area of the ferroelectric disk was 2.8 cm$^2$ i.e. the outer 3 mm of the 25 mm diameter disk was lost in the mounting of the cathode. This data indicates that emission current densities of up to 125 A/cm$^2$ were obtained while still yielding reasonably shaped beam current pulses. It should be noted that a linear V-I scaling was obtained in previous diode experiments at V < 50 kV.

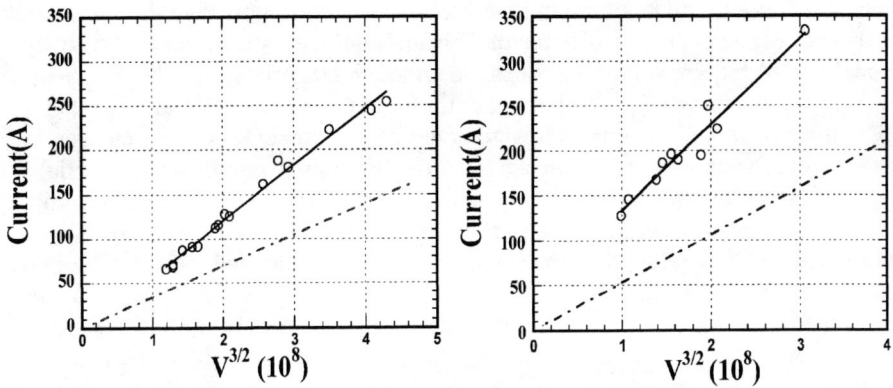

**FIGURE 4.** (Left) V-I Characteristics of the electron gun for a cathode to drift tube (anode) spacing of 6.5 cm. (Right ) V-I Characteristics for a cathode to drift tube (anode) spacing of 4.5 cm.

## DISCUSSION OF RESULTS

The emission data reported in this article considerably expands the range of the emission characteristics previously reported for PZT and provides the data needed for an electron gun design. The emission scaled with the three halves power of the gun voltage and was found to be within a factor of about 2 of that expected in space charge limited flow as predicted by the EGUN code. This result should be compared with the linear voltage scaling reported at low (< 50 kV) voltages. The large spacing between the cathode and the collector surfaces make it very unlikely that plasma closure plays any significant role in the emission process but does not preclude the possibility of explosive field emission on the cathode surface. In fact ion Faraday cup data obtained at lower voltages in a similar geometry confirm the presence of ions in the diode but show that their velocity is always of order of or less than 2-3 cm/μs. Krasik has recently obtained time resolved photographs of a ferroelectric cathode which show plasma formation at the triple points [12] followed by gap closure with a plasma velocity of 1 – 2 cm/μs [13]. In his experiments the regions of high luminosity are restricted to the immediate vicinity of the triple points and do not apparently close the space between adjacent cathode grid rulings. It should be noted however that an isolated ferroelectric cathode was used and triggered, but no anode was used. In this case the emission decayed significantly within 250ns of the end of the trigger pulse. In our experiments the emission is due to the trigger pulse applied to the rear surface of the ferroelectric, and is consistent with field emission from the metallic grid at the vacuum/grid/dielectric boundary. This process does not require explosive emission as described by Mesyats [14].

Schächter [15] has calculated the emission from a metallic wedge on the front surface of a high dielectric constant material when a negative voltage is applied to the wedge. He showed that the triple point electric field is increased by a factor of $\varepsilon_r$, the dielectric constant of the substrate, over that calculated in the absence of the dielectric. As a result of this enhancement the Fowler-Nordheim emission is increased by a factor which may be several orders of magnitude greater than that found from the same wedge in the absence of the dielectric. The result depends solely on Fowler-Nordheim emission and does not require explosive emission, although this process might be expected to develop at high enough current densities emitted for a sufficiently long time to locally volatilize the metal. It is of interest to note that our processing of the ferroelectric cathode involves sputtering a 1-2 micron layer on the ceramic and then preferentially etching the surface to form the grid using standard photo-resist techniques. To initiate the emission it is usually necessary to further etch the metallic grid. This process will produce a tapered edge rather than the standard undercut associated with etching in the presence of the photo-resist. The resulting grid surface is therefore similar to that employed in the modeling outlined above and will enhance the field emission process.

More recent work [16] has examined the emission from an array of ferroelectric/metallic strips in a geometry similar to that used in the experiment described above. This work shows that the coupling of energy stored in the ferroelectric material into the diode gap can, on application of the trigger pulse, account for the excess energy needed to drive currents in excess of the space charge vacuum limit. The energy coupled to the gap exceeds that stored in the gap (due to the anode potential) at low( <50 kV) gun voltages by a factor of up to several thousand. At high gun voltages ~500 kV the factor drops to a value of less than unity and the emission is expected to revert to that predicted by conventional space charge limited flow. The ratio of the coupled energy from the ferroelectric to that stored in the gun is plotted in fig. 5 for a range of gun voltages and spacings. The excess energy coupled into the gap may be expected to decrease rapidly as one moves from the gridded region into the anode cathode gap, decaying on a scale length of the periodicity (~400μm) of the grid.

It is worth noting that the one dimensional result for space charge limited current flow (The Child-Langmuir current) underestimates the actual current flow for a real emitter, even when the emission is thermionic. This effect was shown by Luginsland et al. [18] and may be readily interpreted for the case where the emission area is surrounded by a non-emitting conducting surface. The electric field normal to the cathode surface is reduced in the emission region by the presence of the space charge and asymptotes to the vacuum field at the metallic surface in the surrounding region. The normal electric field varies smoothly between these two limits. The reduction of the surface field in the non-emitting region implies that more space charge can be allowed close to the cathode without reversal of the field direction, i.e. the onset of space charge limited flow. We show in figure 6, a plot of the magnitude of the ratio of the space charge limited current flow from a circular emission region of radius $r_k$ divided the product of the calculated one dimensional Child-Langmuir current density and the area of the emitter. This ratio, denoted in the figure as A is plotted as a function of the ratio of the planar gap width d to the radius of the emission region of the cathode. In realistic geometries the effect can easily increase the magnitude of the space charge limited current flow over the 1 dimensional Child-Langmuir value by a factor of 10. In these experiments the average cathode current density is increased by a factor of about 3. As in the Luginsland paper this data was obtained using a PIC code to solve for the emission current in an axi-symmetric 2D (r,z) diode.

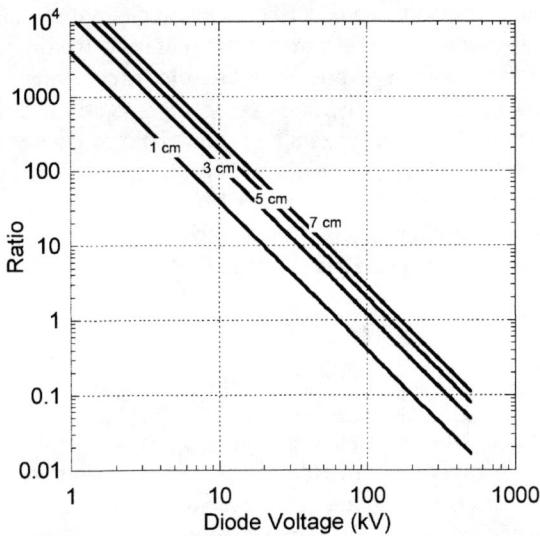

**FIGURE 5.** Ratio of the energy coupled from a 1 kV positive pulse applied to the rear surface of a 1 mm thick dielectric ($\varepsilon_r = 2,000$) to the vacuum energy stored in the gun.

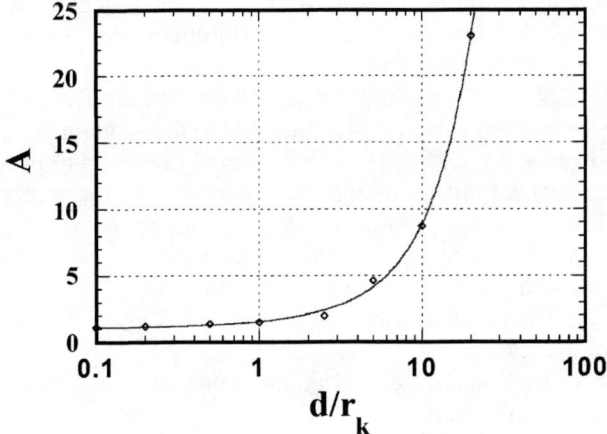

**FIGURE 6.** Enhancement of the average current density emitted from circular emitter located in the surface of a planar cathode and with a planar anode. The factor A gives the ratio of the current from the PIC code divided by the 1 dimensional Child-Langmuir current density times the emitting area. The distances d and $r_k$ represent the gap spacing and the radius of the emission region.

As expected the operation of the gun in a repetitive mode has been found to offer new opportunities for conditioning. After a few hundred shots at the 0.1 Hz repetition rate there was very little evidence of current spiking. A similar reduction in spiking of the emission current was previously seen in 50 Hz data reported earlier [19] at lower voltage and current levels.

Finally we note the analytic solution described above does not include any non-linear effects such as the hysteresis of the ferroelectric material. In spite of the fact we pulse the rear surface of the ferroelectric positively, that is from remnant polarization to saturation in the same sense of the polarization vector, there is still a significant hysteresis as was shown experimentally several years ago. The material does not return instantly to its original state but the dynamics depend on the material and are consistent with a characteristic relaxation time of the order of 1 microsecond after the removal of the trigger pulse. Since the applied pulse is over well before this time the emission is likely to continue for the duration of the gun voltage.

## CONCLUSIONS

We have, in this paper, described the electron emission process from a ferroelectric cathode in an electron gun device. Emission is at the level of a few hundred amperes at 500 keV and at a current density of about 100 A/cm$^2$. The emission was produced in a 250 ns pulse produced in a repetitive mode at a frequency of 0.1 Hz. The resulting beam will be used in high power TWT amplifier experiments.

The emission is triggered by the application of a 1 kV positive polarity pulse applied to the rear surface of the 1 mm thick PZT sample. In comparison with data obtained at lower voltages ($< 50$kV), where the current scaled linearly with the anode cathode potential difference, the emission current scaled with the threehalves power of the applied voltage. These processes are consistent with Fowler-Nordheim emission from the triple points. In a recent paper, Schächter has shown that the electric field in the vicinity of the triple point is enhanced by the dielectric constant of the substrate. The switching of the ferroelectric state by an applied pulse couples energy into the vacuum gap close to the cathode and this may account for the previously observed high current flows in low voltage diodes. At the higher voltages used in these experiments the energy coupled is less than that in the anode cathode gap due to the accelerating voltage. The triggering pulse then mainly serves to initiate the electron emission and the current flow is close to that predicted for space charge limited current flow in the gun.

The rise time of the beam current was observed to be much less than that of the gun voltage and to be less than 20 ns.

It is thought that the emission at the triplepoints results in local plasma formation caused either by explosive field emission or by electron atom collisions with desorbed gases. The latter process will decrease with operation as was observed in the experiments. In related work we, and others, have observed plasma formation on the cathode surface by monitoring visible light emission. Measurements of the ion current flow in a similar geometry, albeit at 50 kV and 10 A/cm$^2$ have indicated plasma closure velocities of less than a few cm/μs. In the current experiments this velocity is too low to have any significant effect on the beam current. The voltage scaling data reported here, combined with that at lower voltages, provides good evidence for the role of the ferroelectric in the emission process. In particular the coupling of energy originally stored in the ferroelectric into the diode gap may provide the energy required for the high current densities observed in the low voltage experiments.

# ACKNOWLEDGEMENTS

This work was supported by the Department of Energy and by the AFOSR under the High Power Microwave MURI program.

# REFERENCES

1   C. B Fleddermann and J.A. Nation, "Ferroelectric Sources and Their Application to Pulse Power: A Review," *IEEE Transactions on Plasma Science,* vol. 25, no. 2, p. 212, 1997.

2.  H. Riege, "Electron Emission from Ferroelectrics: A Review," *Nucl. Instrum. Methods Phys.Res. A,* vol. 340, p. 80-89, 1994.

3.  S.E. Sampayan, G. J. Caporaso, C. L. Holmes, D. Prosnitz, D.O. Trimble and G. A. Westenscow, "Emission from ferroelectric cathodes", *Nucl. Inst. and Methods Phys. Res. A,* vol. 340, p. 90-95, 1994.

4.  H. Gundel, "High-Intense Pulsed Electron Emission by Fast Polarization Changes inFerroelectrics," *Ferroelectrics,* vol. 184, pp. 89–98, 1996.

5.  G. Benedek, I. Boscolo, A. Moscotelli, A. Scurati, and J. Handerek, "Correlation between Emitted and Polarization Current inFerroelectric Lead LanthanumZirconate Titanate Ceramics," *J. Appl. Phys.,* vol. 83, no. 5, pp. 2766-2771, 1998.

6.  W. Zhang, private communication.

7.  D. Flechtner, C. Golkowski, J.D. Ivers, G.S. Kerslick, J.A. Nation, and L. Schächter, "Electron emission from lead-titanate-zirconate ceramics," *J. Appl. Phys.,* vol. 83, no. 2, pp. 955-961, 1998.

8.  W. Zhang, W. Hueber, S.E. Sampayan, and M.L. Krogh, "Mixed electron emission from lead zirconate-titanate ceramics," *J. Appl. Phys.*, vol. 83, no. 11, pp. 6055-6060, 1998.

9.  L. Schächter, J.D. Ivers, , J.A. Nation, and G.S. Kerslick, "The analysis of a diode with a ferrolectric cathode," *J. Appl. Phys.,* vol. 73, pp. 8097-8107, 1993.

10. M. Okuyama, J. Asano, and Y. Hamakawa, "Electron Emission from Lead-Zirconate-Titanate Ferroelectric Ceramic Induced by Pulse Electric Field," *Jpn. J. Appl. Phys.,* vol. 33, no. 9B, pp. 5506-5507, 1994.

11. J.D. Ivers, D. Flechtner, Cz. Golkowski, G.S. Kerslick, and J.A. Nation, "A Ferroelectric Cathode Electron Gun for High Power Microwave Research, "*Proceedings of the Particle Accelerator Conference*, Vancouver, in press, May 12-16, 1997.

12. Ya. E. Krasik, private communication.

13. Ya. E. Krasik, *Proceedings of the International Conference on High-Power Particle Beams,* to be published, Haifa, Israel, June, 1998.

14. G.A. Mesyats, "Electron emission from a ferroelectric ceramic," *Tech. Phys. Lett.,* vol. 20, no. 1, pp. 8-9, 1994.

15. L. Schächter, "Analytic expression for triple-point emission from an ideal edge," *Appl. Phys. Lett.,* vol. 72, no. 4, p. 421-423, 1998.

16. L. Schächter, D. Flechtner, J.D. Ivers and J. A. Nation, "Theoretical Study of a Diode with a Gridded Ferroelectric Cathode," Submitted for publication to *J. Appl. Phys.,* 1998.

17.  J.W. Luginsland, Y.Y. Lau, and R.M. Gilgenbach, "Two Dimensional Child-Langmuir Law," *Physical Review Letters,* vol. 77, no. 22, pp. 4668-4670, 1996.

18.  D. Flechtner, Cz. Golkowski, J.D. Ivers, G.S. Kerslick, J.A. Nation, and L. Schächter "New Results on Electron Emission from PZTFerroelectric Cathodes, " *Proceedings of the Particle Accelerator Conference*, Vancouver, in press, May 12-16, 1997.

# Working Group 6:

# Radiation Sources

# Recent Progress on 7 GHz Pulsed Magnicon Amplifier [1]

E.V. Kozyrev, O.A. Nezhevenko, A.A. Nikiforov, G.N. Ostreiko,
B.Z. Persov, G.V. Serdobintsev, S.V. Schelkunoff, V.V. Tarnetsky,
V.P. Yakovlev, I.A. Zapryagaev

*Budker Institute of Nuclear Physics, Novosibirsk, Russia*

**Abstract.** The report presents experimental results obtained on 7 GHz pusled magnicon amplifier with a new version of the output cavity, as well as plans of the next investigations. This magnicon was developed at INP as a prototype of a microwave power source for the next generation of linear colliders. The tube operates in frequency-doubling mode of the drive signal. At present time the following parameters were achieved: maximum output power of 55 MW, efficiency of 56%, gain of 72 dB, pulse length of 1.1 $\mu$s.

## INTRODUCTION

The work on creation of scanning-beam microwave amplifiers is carried out at INP since 1967. This new class of devices right from the start of its development was intended for high efficient power supply of accelerators. In 1970 at INP the first device of that type called gyrocon [1,2] was created. Further development of scanning-beam microwave amplifiers has been made possible with invention and development of magnicons [3]. The first magnicon – a prototype of continuous (quasi-continuous) microwave amplifier was built and tested in the 1980s at INP [3] and showed the record electron efficiency of 85% at a frequency of 915 MHz with pulse duration of 30 $\mu$s and the power of 2.6 MW.

In the beginning of 1990s at the Institute the work was started on the development of the improved magnicon version. That magnicon was proposed as a prototype of the microwave power source for linear colliders [4]. A schematic diagram of the device is shown in Fig.1. It includes: electron gun, RF-system, magnetic system, and collector. The tube operates at 7 GHz in frequency-doubling mode, so the RF system consists of the series of deflecting cavities (the first externally driven) for beam modulation at 3.5 GHz and output cavity for conversion of beam energy into the microwave power at the second harmonic (7 GHz).

---

[1] Work is supported by Russian Fund of Basic Research

CP472, *Advanced Accelerator Concepts: Eighth Workshop,*
edited by W. Lawson, C. Bellamy, and D. Brosius
© 1999 The American Institute of Physics 1-56396-889-4/99/$15.00

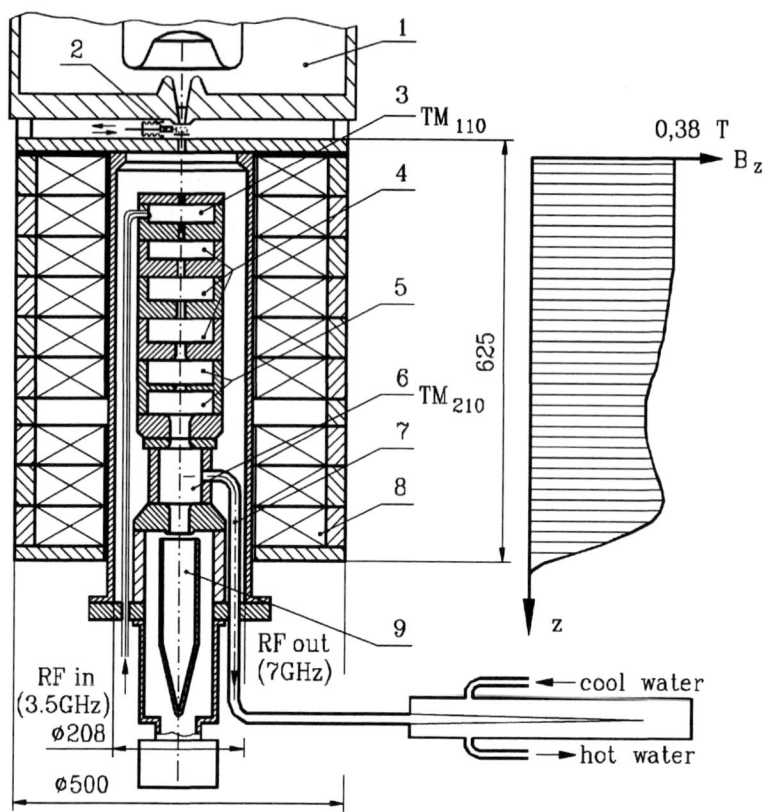

**FIGURE 1.** Schematic layout of the magnicon: 1 – electron source; 2 – vacuum valve; 3 – drive cavity; 4 – gain cavities; 5 – penultimate cavity; 6 – output cavity; 7 – waveguide ($\times 2$); 8 – solenoid; 9 – collector.

The prime object of this device investigation was to show its serviceability and ability to provide the parameters which are required from present-day microwave sources. The design parameters are listed in Table 1.

The difficulty in achieving the desired parameters was well recognized at the beginning of our work. Thus the creation of software for simulation of electron optics, electromagnetic systems, and electron beam dynamics in both steady state and time dependent regimes was carried out simultaneously with an experimental study. Computer codes SAM, SuperLANS [5,6] as well as codes for simulation of steady-state and time-dependent magnicon operation conditions [7] have been created and refined. An electron source with unique parameters [8] has been manufactured and successfully tested. Various versions of the design of the tube and its individual components [4] have also been studied.

**TABLE 1.** Design parameters

| | |
|---|---|
| Operating frequency | 7 GHz |
| Drive frequency | 3.5 GHz |
| Output power | 50–60 MW |
| Gain | 55 dB |
| Pulse duration | 1.5–2 $\mu$s |
| Efficiency | 50–60% |
| Beam voltage | 420 kV |
| Beam current | 240 A |

# EXPERIMENTAL STUDIES

From the simulation and experimental results it has been found that an electron beam size and matching its optics with the DC magnetic field are the determining factors of magnicon efficiency. In 1996, the electron source was modernized by changing the geometry of gun focusing electrode, the matching between the electron beam and tube's magnetic system was also improved. The maximum value of DC magnetic field for optimal device operating is 0.38 T. In this case scalloping of the beam transverse size in the magnetic system lies in the range from 1.9 to 2.5 mm. As a result of these improvements in 1997 an output power of 46 MW was achieved with an electron efficiency of 49% and gain of 62 dB [9].

It also has been found that the magnicon electron efficiency significantly drops due to the longitudinal inhomogeneity and azimuthal asymmetry of RF fields in the output cavity, which are caused by the coupling holes with waveguide power outputs [10]. These built-in outputs are located on the cylindrical surface and spaced 135 degrees apart by azimuth and produce both azimuthal asymmetry and longitudinal inhomogeneity of RF field. These problems were solved with the special design of the output cavity (see Ref. [10]).

With all of these improvements the present experimental results as shown in Table 2 are very close to the design goals.

**TABLE 2.** Achieved parameters

| | |
|---|---|
| Operating frequency | 7.005 GHz |
| Drive frequency | 3.5025 GHz |
| Output power | 55 MW |
| Gain | 72 dB |
| Pulse duration | 1.1 $\mu$s |
| Efficiency | 56% |
| Beam voltage | 427 kV |
| Beam current | 230 A |

FIGURE 2. 7 GHz pulsed magnicon.

Fugure 2 presents a general view of the experimental 7 GHz magnicon assembly.

Oscillograms of the pulses, experimental and calculated curves are shown in Fig.3-4. Parameters obtained at the device optimal operating regime are marked by "•" symbol. The output power calibration was carried out by calorimetric measurements of an average level of RF signal passed through the waveguide vacuum loads (there are no waveguide windows).

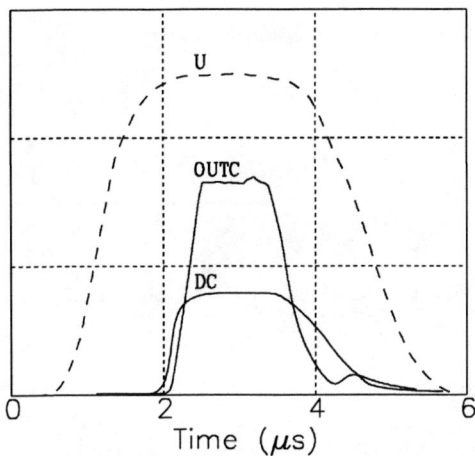

**FIGURE 3.** Oscillograms: U — beam voltage; DC — drive cavity signal; OUTC — output cavity signal.

An output cavity with a loaded Q-factor $Q_l$=230, which is somewhat higher than the optimal value $Q_l$=180, was used in experiments, (seen from the calculated curve in Fig.4a) that decreases the device efficiency by 2%. The increased Q-factor has been chosen in order to shift from the region of possible instability causing a sharp efficiency drop.

A small difference between the frequency characteristic curve and simulation results (Fig.4b, curve 2) is due to non-optimal tuning of drive cavity input circuits.

Simulation results show that increasing in efficiency can be achieved by decreasing DC magnetic field (Fig.4c). Therewith a gain drops but still remains high enough. In the given series of experiments we was not able to decrease DC magnetic field significantly, for a beam quality is impaired because of mismatching. However, since the magnetic system solenoid consists of two sections powered by individual sources, dependence between an efficiency and magnetic field value in the output cavity area has been studied (Fig.4d). It can be seen that experimental data are in good agreement with simulation results at operation regime and at minor variations of that field when change of the beam optics in the deflecting system may be ignored.

We did not achieve the designed output pulse length yet. One of the reasons is associated with a strong dependence between the efficiency and beam power. Figure 4e) shows calculated and experimental curves obtained at fixed values of

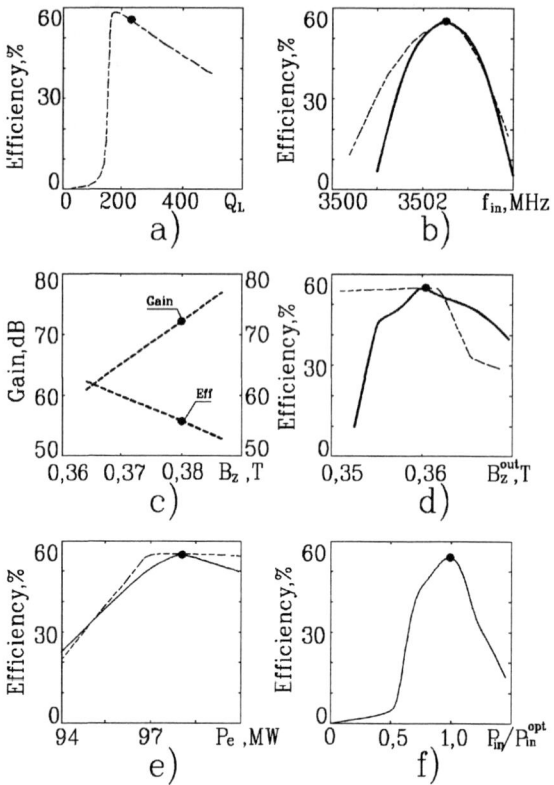

**FIGURE 4.** Experimental and calculated curves: a) loading dependence; b) input frequency dependence; c) DC magnetic field dependence of gain and efficiency; d) output cavity DC magnetic field dependence of efficiency; e) beam power dependence of efficiency; f) efficiency versus drive power. Continuous lines correspond to experimental data, dashed — to simulation results.

DC field and drive signal. Difference between them within non-optimal areas may be explained by the fact that the calculations were carried out for a fixed beam optics. That optics corresponds to the optimal operating regime.

The pulse length may be roughly estimated from the time dependence of the beam voltage (Fig.3) and beam power dependence of efficiency (Figure 4e). The real pulse length will be always shorter due to effects not taken into account.

Simulation of the signal shape at the given beam power dependence of efficiency shows, that for the voltage pulse of given shape the output signal length is limited by 1.5 $\mu$s. This problem can be overcome by improving the voltage pulse shape on the gun.

Another possible reason is related to excitation of 0-type oscillation in a penultimate cavity. The penultimate cavity consists of two cavities coupled through the central hole (coupling coefficient is 0.7%). The operating mode for this cavity is π-mode. However, simulation results show, that in spite of a high value of Q-factor, 0-mode limits a pulse length of the output signal at a level of 0.8–1.2 $\mu$s. This effect was detected in the course of 11.424 GHz magnicon development at Omega-P [11].

Figure 5 shows a typical outlook of instability growth in the penultimate cavity at 0-mode exciting, obtained by simulation. It should be noted that instability growth time is critically sensitive to the DC magnetic field value. So the change in the output pulse length may range up to 40% at the DC magnetic field change of 0.6%.

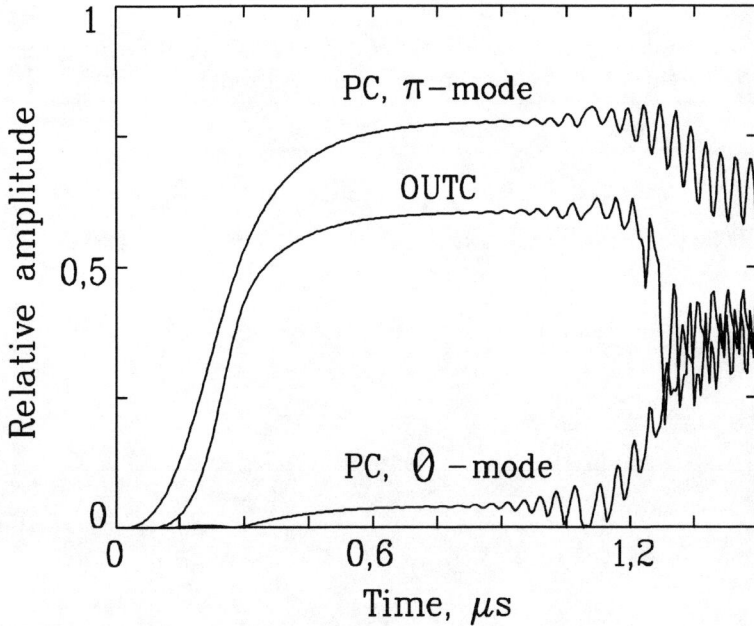

**FIGURE 5.** Transient process in the magnicon when both 0- and π- modes are excited in the penultimate cavity; PC corresponds to penultimate cavity, OUTC — to output cavity.

At present time we design a new penultimate cavity, which will allow us to remove this restriction.

## SUMMARY

In the latest series of experiments on 7 GHz magnicon almost all design parameters have been achieved. Experimental data are in an excellent agreement with simulation results. Achieved results allow to consider a magnicon as an alternative power source for linear colliders applications interchangeably with the best modern klystrons even at present time.

# REFERENCES

1. G.I.Budker et al., Particle Accelerators vol. 10, 1979, pp. 41–59.
2. O.A.Nezhevenko, in IEEE Trans. of Plasma Science vol. 22, No. 5, 1994, pp. 756–772.
3. M.M.Karliner et al., NIM, vol. A 269, No. 3, 1988, pp. 459–473.
4. E.V.Kozyrev et al.,Particle Accelerators vol. 55, 1996, pp. 55–64.
5. B.Fomel, M.Tiunov, and V.Yakovlev, in Proc. XIII Int. Conf. on High-Energy Accel., vol 1, 1987, pp. 353–355.
6. D.Myakishev and V.Yakovlev, in Proc. of Part. Acc. Conf., Dallas, 1995.
7. V.Yakovlev et al., in Proc. of Part. Acc. Conf., Dallas, 1995.
8. Y.V.Baryshev et al., NIM, vol. A 340, 1994, pp.241-258.
9. E.V.Kozyrev et al., in Proc. PAC'97 Conf. (to be published).
10. O.A.Nezhevenko et al., in AIP Conf. Proc. 398, 1997, pp. 912–919.
11. O.A.Nezhevenko, V.P.Yakovlev (private communication).

# Double Resonance in Relativistic Gyrodevices

G. S. Nusinovich and J. Zhao

*Institute for Plasma Research*
*University of Maryland, College Park, MD 20742-3511, USA*

**Abstract.** In many relativistic gyrodevices (gyroklystrons and gyrotwystrons) a beam of gyrating electrons excites standing waves in resonators. Such waves can be represented as a superposition of forward and backward waves which have opposite Doppler shifts of the operating frequency in the reference frame moving with electrons. Correspondingly, for certain axial wave numbers both forward and backward waves can be in cyclotron resonance with gyrating electrons but at different cyclotron harmonics. The theory describing the interaction between electrons and the resonator field in the case of such a double resonance is developed. It is shown that this double resonance can be beneficial when the device operates at symmetric modes while in the case of operation at non-symmetric modes it always lowers the efficiency.

## INTRODUCTION

As is known, such gyroamplifiers as gyroklystrons and gyrotwystrons driven by relativistic electron beams are promising candidates for driving future linear colliders [1]. The processes of interaction between gyrating relativistic electrons and electromagnetic fields excited in these devices are well studied ( [2]-[4]and refs. therein). However, there is one special case which was not analyzed earlier and which will be discussed below. This is the case of interaction of gyrating electrons with a resonator field at two cyclotron harmonics simultaneously, in contrast to conventional cyclotron resonance at only one harmonic.

To explain this case let us consider a resonator field with a sinusoidal axial structure $f(z) = \sin(\ell\pi z/L)$; here $\ell$ is the axial index and $L$ is the resonator length. This field can be represented as the superposition of two opposite waves with the axial wave numbers $k_z = \pm\ell\pi/L$. Substituting these axial wave numbers into the cyclotron resonance condition

$$\omega - k_z v_z \simeq s\Omega \qquad (1)$$

(where $\omega$ and $\Omega$ are, respectively, the wave and cyclotron frequencies, $v_z$ is the electron axial velocity and $s$ is the cyclotron resonance harmonic number), one can

CP472, *Advanced Accelerator Concepts: Eighth Workshop,*
edited by W. Lawson, C. Bellamy, and D. Brosius

readily find that both of the opposite waves may be in resonance with electrons: the forward wave at the $s$-th harmonic while the opposite one at the $(s+1)$th or higher harmonic. This case is illustrated in Fig. 1. Such an operation occurs when the ratio of the resonator length, $L$, to the wavelength $\lambda$. relates to the axial index. $\ell$, as

$$\frac{L}{\lambda} = \frac{2s+1}{2}\beta_{z0}\ell.$$

(2)

Here $\beta_{z0}$ is the initial axial velocity of electrons normalized to the speed of light. The corresponding Doppler up-shift of the operating frequency with respect to the cyclotron frequency is

$$\frac{\omega}{\Omega} = \frac{2s+1}{2}.$$

(3)

Note that the condition of the double resonance given by Eq. (2) can be realized in the case of fast waves ($k_z = \ell\pi/L < \omega/c$) only when the electron axial velocity is large enough:

$$\beta_{z0} > \frac{1}{2s+1}.$$

(4)

For the simplest case of the forward wave interaction at the fundamental harmonic ($s = 1$) Eqs. (2)-(4) are reduced, respectively, to $L/\lambda = (3/2)\beta_{z0}\ell$, $\omega/\Omega = 3/2$, and $\beta_{z0} > 1/3$.

It is expedient to recall that a gyrating electron can lose all its kinetic energy in the process of interacting with an electromagnetic wave when the axial phase velocity of this wave normalized to the speed of light is equal to [3]

$$\left[\frac{(1+a^2)(\gamma_0 - 1)}{\gamma_0 + 1}\right]^{1/2}.$$

Here $\gamma_0$ is the initial energy of the electron normalized to the rest energy and $a = p_{\perp 0}/p_{z0}$ is the ratio of the initial components of the electron momentum. As one can easily find. this condition of complete deceleration of one electron is consistent with the condition of the double resonance, $L/\lambda = (3/2)\beta_{z0}\ell$, when $a_{opt}^2 = 2 + 3/\gamma_0$ and $\gamma_0 > 3/2$. The latter implies an operation at voltages above 250 kV. Note that the condition of the double resonance at fast waves, $\beta_{z0} > 1/3$. for $a = a_{opt}$ means $\gamma_0 > 4/3$. i.e. voltages above 150 kV. For $\beta_{z0} > 1/3$, Eq. (2) gives $L/\lambda > \ell/2$. which in principle can be realized at any axial index $\ell$. Note that these simple estimates allow one also to evaluate the number of electron orbits in the interaction space, $N = \Omega T/2\pi$ where $T = L/v_{z0}$ is the electron transit time through the resonator. As follows from Eqs. (2) and (3), $N = \ell$. i.e. under conditions of the double resonance the number of electron orbits is equal to the axial index of the resonator mode.

940

This simple qualitative analysis just indicates some specific features of the double resonance at different harmonics which may occur in CRMs in the process of interacting with one resonator mode. The theory of this phenomenon is presented below. In Section II a general formalism is described. In Section III the results are presented and discussed. Section IV contains the summary.

## GENERAL FORMALISM

As shown by Nusinovich and Zhao [5], the electron motion in the case of the double resonance at two different cyclotron harmonics can be described by the following equations:

$$\frac{dp_\perp}{dz} = \frac{1}{3\kappa\beta_z}\mathrm{Re}\left\{iA\left[J_1'(\rho)L_1 e^{i\vartheta_1} - 2J_2'(\rho)L_2 e^{i\vartheta_2}\right]\right\}, \tag{5}$$

$$\frac{d\theta}{dz} - \frac{\mu}{p_z} = \frac{1}{3\kappa\beta_z p_\perp}\mathrm{Re}\left\{A\left[(\rho J_1'(\rho))' L_1 e^{i\vartheta_1} - (\rho J_2'(\rho))' L_2 e^{i\vartheta_2}\right]\right\}. \tag{6}$$

$$\frac{dp_z}{dz} = \frac{h\beta_\perp}{2\kappa\beta_z}\mathrm{Re}\left\{iA\left[J_1'(\rho)L_1 e^{i\vartheta_1} + J_2'(\rho)L_2 e^{i\vartheta_2}\right]\right\}, \tag{7}$$

$$\frac{d\gamma}{dz} = \frac{\beta_\perp}{2\kappa\beta_z}\mathrm{Re}\left\{iA\left[J_1'(\rho)L_1 e^{i\vartheta_1} - J_2'(\rho)L_2 e^{i\vartheta_2}\right]\right\}. \tag{8}$$

Here, the components of electron momentum are normalized to $mc$, the axial coordinate $z$ is normalized to $\omega/c$, the resonator field amplitude $A$ is normalized to $e/mc\omega$, $\kappa$ and $h$ are, respectively, transverse $k_\perp$ and axial $k_z$ wave numbers normalized to $k = \omega/c$, $\mu$ is the ratio of the nonrelativistic cyclotron frequency to the operating frequency, $\rho = k_\perp a$ is the Larmor radius of electrons normalized to $k_\perp$ ($\rho$ can also be redefined as $\kappa p_\perp/\mu$), slowly variable phases $\vartheta_1 = \omega t - k_z z - \theta$ and $\vartheta_2 = \omega t + k_z z - 2\theta$ describe the departure of the forward and opposite waves, respectively, from the exact cyclotron resonance, and $\theta$ is the electron gyrophase. Factors $L_1$ and $L_2$ describe the transverse structure of the Lorentz force with which, respectively, the forward and opposite waves act on electrons. In the case of a thin annular electron beam interacting with the TE$_{m,p}$ mode of a cylindrical cavity

$$L_s - J_{m\mp s}(k_\perp R_g) e^{i(m\mp s)\psi}. \tag{9}$$

Here $R_g$ and $\psi$ are polar coordinates of electron guiding centers. Certainly, one from Eqs. (5), (7) and (8) can be eliminated since $\gamma^2 = 1 + p_\perp^2 + p_z^2$. We presented above all three just for the completeness of our consideration. In the absence of electron velocity spread at the entrance $p_\perp$, $p_z$ and $\gamma$ obey the following boundary conditions: $p_\perp(0) = \gamma_0\beta_{\perp 0}$, $p_z(0) = \gamma_0\beta_{z0}$, $\gamma(0) = \gamma_0$. The boundary condition for the gyrophase $\theta$ in the gyromonotron is $\theta(0) = \theta_0$ where the initial gyrophase $\theta_0$ is homogeneously distributed from 0 to $2\pi$. In any gyrodevice with electron prebunching the gyrophase at the entrance to the last output cavity depends on the prebunching history. This issue will be specified below.

The interaction efficiency can be determined as

$$\eta = \frac{1}{2\pi} \int_0^{2\pi} \left\{ \frac{1}{2\pi} \int_0^{2\pi} \frac{\gamma_0 - \gamma(L)}{\gamma_0 - 1} d\theta_0 \right\} d\psi. \tag{10}$$

Here the averaging over $\theta_0$ means the averaging over initial gyrophases in one beamlet, the averaging over $\psi$ means the averaging over all beamlets having different azimuthal coordinates of guiding centers.

Using Eq. (9) one can introduce the normalized amplitude $F = iA|L_1|/2\kappa$ and phase $\vartheta' = \vartheta_1 - (m-1)\psi$. (Without a lack of generality we will consider the TE-mode rotating in the same azimuthal direction as the electron gyration. In the case of a symmetric $TE_{0,p}$ mode this does not play any role.) Correspondingly, in Eqs. (5)-(8) the ratio $|L_2|/|L_1|$ can be denoted as $R$ and the phase of the last terms in the right-hand sides of Eqs. (5)-(8) can be rewritten as

$$\theta_2' = \vartheta_2 - (m-2)\psi = 3k_z z - \omega t + 2\vartheta_1' + m\psi = 3h\beta_z - \int_0^z \frac{dz'}{\beta_z} + 2\vartheta_1' + m\psi. \tag{11}$$

Equation (11) shows that in the case of operating at symmetric ($m = 0$) modes the second averaging in Eq. (10) is redundant since all beamlets are in identical conditions while for operation at non-symmetric modes it can be very important.

In new notations, assuming that the argument $\rho$ in Bessel functions is small enough in order to expand these functions as $J_n(\rho) \simeq (1/n!)(\rho/2)^n$ and introducing $F' = F/2$, $R' = \kappa R$ (primes are omitted below), we can reduce Eqs. (5)-(8) to

$$\frac{dp_\perp}{dz} = \frac{1}{\beta_z} \text{Re} \left[ F \left( \frac{2}{3} e^{i\vartheta_1} - R\beta_- e^{i\vartheta_2} \right) \right]. \tag{12}$$

$$\frac{dp_z}{dz} = h \frac{\beta_\perp}{\beta_z} \text{Re} \left[ F \left( e^{i\vartheta_1} + \frac{3}{4} R\beta_\perp e^{i\vartheta_2} \right) \right]. \tag{13}$$

$$\frac{d\gamma}{dz} = \frac{\beta_\perp}{\beta_z} \text{Re} \left[ F \left( e^{i\vartheta_1} - \frac{3}{4} R\beta_\perp e^{i\vartheta_2} \right) \right], \tag{14}$$

$$\frac{d\vartheta_1}{dz} = \frac{1}{\beta_z} \left\{ \delta + \frac{1}{3} \left( 1 - \frac{\beta_z}{\beta_{z0}} \right) + \frac{2}{3} \left( 1 - \frac{\gamma_0}{\gamma} \right) + \frac{1}{p_\perp} \text{Re} \left[ iF \left( \frac{2}{3} e^{i\vartheta_1} - R\beta_\perp e^{i\vartheta_2} \right) \right] \right\}. \tag{15}$$

942

Here $\delta = 1 - h\beta_{z0} - \mu/\gamma_0$ is the initial mismatch of the cyclotron resonance for the forward wave, the amplitude $F$ and the ratio $R$ are renormalized by introducing $F' = F/2$, $R' = \kappa R$ and the primes for $\theta_1$, $\theta_2$, $F$, and $R$ are omitted.

These equations contain the same three parameters that describe a single-resonance interaction with one resonator mode (see, e.g. Ref. 1): the RF field amplitude $F$, the cyclotron resonance mismatch $\delta$, and the resonator length $L$. They also include the parameter $R$, which determines the ratio of coupling coefficients to the opposite and forward waves, and the initial components of the electron velocity, $\beta_{\perp 0}$ and $\beta_{z0}$ ($\gamma_0 = (1 - \beta_{\perp 0}^2 - \beta_{z0}^2)^{-1/2}$). Note that, as follows from the condition of double resonance, $h \approx 1/3\beta_{z0}$ and $L$ is determined, in accordance with Eq. (2), by $\beta_{z0}$ and the axial index of the mode, $\ell$.

# RESULTS

The double resonance interaction was studied for a 500 kV electron beam with the orbital-to-axial velocity ratio, $\alpha = 1.5$. The interaction with modes having two, three and four axial variations ($\ell = 2$, 3, and 4) was analyzed for the case of a non-prebunched electron beam and the case of ballistic prebunching. The latter implies a weak modulation in electron energies in the first resonator followed by a long drift section in which this modulation causes significant orbital phase bunching. Correspondingly, the boundary condition to Eq. (14) for the electron energy was written as $\gamma(0) = \gamma_0$ and to Eq. (2) for the phase as

$$\vartheta_1(0) = \vartheta_0 + q\sin\vartheta_0, \tag{16}$$

where the bunching parameter $q$ in the case of a non-prebunched beam is equal to zero. We analyzed both the operation at symmetric ($m = 0$) and non-symmetric ($m \neq 0$) modes. The interaction length was determined by Eq. (2) for $s = 1$ and $\beta_{z0}$ defined by $V_b$ and $\alpha$ specified above: $L/\lambda \simeq 0.72\ell$. The efficiency given by Eq. (10) was calculated as a function of the field amplitude $F$ for different values of the ratio of coupling coefficient, $R$, and for each $F$ and $R$ the cyclotron resonance mismatch $\delta$ was optimized for maximizing the efficiency.

Typical results are presented in Fig. 2 which corresponds to two axial variations ($\ell = 2$) of the resonator mode. Figures 2(a), (c) and (e) correspond to operation at the symmetric mode while cases (b), (d) and (f) correspond to non-symmetric modes (the latter set of figures implies double averaging in Eq. (10) while the former corresponds to the single averaging). Figures 2(a) and (b) correspond to the non-prebunched electron beam, figures (c) and (d) correspond to the bunching parameter $q = 1$, and figures (e) and (f) correspond to $q = 1.8$. (This value is close to the optimum for a one-cavity prebunching for operation at the fundamental resonance.)

The results presented in Fig. 2 demonstrate that in the case of operation at symmetric modes the double resonance can improve the efficiency when the coupling to the opposite wave is not too strong ($R = 0.2 - 0.4$). This effect is better pronounced at small values of the bunching parameter ($q = 0$ and $1.0$). For the optimum bunching parameter ($q = 1.8$) the maximum efficiency for the non-zero $R$'s is approximately the same or smaller than for $R = 0$. The maximum efficiency of interaction of the prebunched ($q = 1.8$) electron beam with modes having two axial variations is 38%. We also studied operation at modes with three and four axial variations and found that in these cases the maximum efficiency is 45% and 46%, respectively. Since the maximum efficiency in last two cases ($\ell = 3$ and 4) was approximately the same, we did not analyze the interaction with modes having a larger ($\ell > 4$) number of axial variations. Relatively small changes in the efficiency due to the additional resonance can be explained by the fact that, as our calculations showed, the parameters optimal for the efficient interaction with the backward wave differ significantly from the parameters optimal for the forward wave interaction.

The double resonance in the case of operation at non-symmetric modes, as follows from cases shown in Figs. 2(b), (d) and (f), always only reduces the efficiency in comparison with a single resonance interaction ($R = 0$). Moreover, at large amplitudes of the resonator field it causes the appearance of reflected particles since some electrons lose their axial momentum completely in the process of deceleration. Certainly, the absolute values of the efficiency can be enhanced when two or more prebunching cavities are used [6].

The effect of electron velocity spread on the efficiency of double resonance interaction was also studied. We considered the case when a monoenergetic electron beam has a spread in pitch-ratios $\alpha$'s, which is typical for the beams formed by magnetron injector electron guns. This spread was modelled by a tophat distribution. The results of calculations done for the 10% spread have shown that a double resonance operation at symmetric modes is as sensitive to the velocity spread as a single resonance interaction. For instance, the efficiency of the double resonance operation at a mode with two axial variations excited by a non-prebunched electron beam decreases from 24% to 19% ($R = 0.2$, c.f. Fig. 2a) while the efficiency of the single resonance operation in the same device with $R = 0$ decreases from 21.2% to 15.6%. On the contrary, the double resonance operation at non-symmetric modes is much more sensitive to the spread: for the same parameters ($\ell = 2$, $q = 0$, $R = 0.2$) the 10% spread decreases the efficiency from 19.1% to 12.4%.

Now let us discuss the relation between the optimal field amplitude $F$ (which can be found from Figs. 2 and 3) and the RF breakdown field. The normalized amplitude $F$ used in Eqs. (12)-(15) is equal to

$$J_1 \left( k_- R_g \right) \frac{1}{4\kappa} \frac{eA}{m_e c \omega}. \tag{17}$$

and its optimal value, as follows from Figs. 2-4, is on the order of 0.1. (Depending on other parameters, it varies approximately from 0.05 to 0.15). The normalized axial wavenumber $h$ in the case of the double resonance with a 500 kV, $\alpha = 1.5$ electron beam is equal to 0.694 ($h \simeq 1/3\beta_{z0}$). Correspondingly, the normalized transverse wave number $\kappa = \sqrt{1 - h^2}$ is equal to 0.72. Let us also assume that the beam radius $R_g$ corresponds to the maximum of the Bessel function $J_1(k_\perp R_g) \simeq 0.58$. This gives for the last term in Eq. (17) $eA/m_e c\omega \simeq 0.5$. From here the optimal field amplitude equals

$$A_{\text{opt}} \left(\frac{\text{MV}}{\text{m}}\right) \simeq \frac{160}{\lambda(\text{cm})}. \tag{18}$$

At high enough frequency ($f \gtrsim 10$ GHz) the RF breakdown field in the continuous wave (CW) regime can be approximated [7] by a simple expression,

$$E_{\text{br,cw}}(\text{MV/m}) \simeq 25\sqrt{f(\text{GHz})} \simeq \frac{137}{\sqrt{\lambda(\text{cm})}}. \tag{19}$$

From here it follows that the optimal amplitude does not exceed the breakdown field only at wavelengths longer than 1.36 cm. Note, however, that in many experiments it was shown that in pulse operation the breakdown field is much larger than in the CW regime. This allowed P. Wilson [8] to approximate the dependence of the breakdown field on the pulse duration $\tau(\mu s)$ by the following equation:

$$E_{\text{br}} = E_{\text{br,cw}} \left(1 + \frac{4.5}{\tau^{1/4}}\right). \tag{20}$$

So, for instance, for $\tau \lesssim 1$ $\mu$sec (which is a typical pulse duration of microwave sources intended for driving the linear colliders) the breakdown field is more than five times larger than that given by Eq. (19). This pushes the boundary of the wavelength region of stable operation to short millimeters. Note also that near the wall surface the electric field of symmetric modes is much weaker than in the interaction region that mitigates the breakdown problem.

For practical reasons, it makes sense also to estimate the resonator Q-factor required for the generation of a certain microwave power, $P_{\text{RF}}$, with the efficiency calculated above. Let us assume that the device should produce a 100 MW output power which is the goal of the present program for the development of relativistic gyroklystrons for future linear colliders [1] [9]. (For 500 kV beam voltage and 40% efficiency it implies a 500 A beam current.) In the steady state, the RF power extracted from the beam is equal to the power of microwave losses $(\omega/2Q)A^2 N$, where $N$ is the norm of the operating mode and the losses are mainly determined by the diffraction of radiation into the output waveguide. Therefore,

$$Q = \frac{\omega}{2} \frac{A^2 N}{P_{\text{RF}}}. $$

For the lowest symmetric modes in a cylindrical cavity ($\text{TE}_{01\ell}$ and $\text{TE}_{02\ell}$) it gives $N_{01} = 2.76 \times 10^{-2} L\lambda^2$ and $N_{02} = 5.2 \times 10^{-2} L\lambda^2$. (In a similar manner one can also calculate the norms of the modes in coaxial cavities [10].) Then, using Eq. (18) for

945

the optimal field amplitude one readily gets the following estimates for the required Q-factors: $Q_{01\ell} \simeq 77\ell$, $Q_{02\ell} \simeq 145\ell$.

Before closing this section, let us also briefly discuss the issue of mode competition. Certainly, in the case of operation at the lowest symmetric modes the spectrum of modes differing in transverse indices is well rarified. However, the issue can be a competition between modes having the same transverse structure but different axial indices $\ell$ when $\ell$ is large. To evaluate the frequency separation between such modes let us consider the case studied above: a mode with three axial variations is excited by a 500 kV, $\alpha = 1.5$ electron beam. As one can find using Eq. (2), the frequency of this mode is $1.39 \omega_c$ (where $\omega_c$ is the cut-off frequency) while the frequencies of modes with two and four axial variations are equal to 1.19 $\omega_c$ and $1.63 \omega_c$, respectively. So, the frequency of the closest mode is about 14% apart from the operating frequency. At the same time, the self-excitation band being estimated as $(\pi/2)T^{-1}$ (where $T$ is the electron transit time introduced in Section 1) is smaller than 6%. This allows us to conclude that the mode competition should not be a severe problem for double resonance operation.

# SUMMARY

The formalism describing the simultaneous double cyclotron resonance interaction at different harmonics between the beam of gyrating electrons and one resonator mode was developed. It was shown that in the case of operation at azimuthally symmetric modes the efficiency of the double resonance interaction can be higher than that in the standard case of a single resonance interaction. For instance, for an optimally prebunched electron beam (one-cavity prebunching) it can reach 46%. In contrast, in the case of operation at non-symmetric $TE_{m,p}$ modes (with $m \neq 0$) the additional resonance plays a destructive role only, especially when the velocity spread is significant. The estimates show that the optimal amplitude of the resonator field required for efficient operation is on the order of the breakdown field. Nevertheless, at least for a short pulse operation, the optimal field at all reasonable frequencies is smaller than the breakdown limit.

# ACKNOWLEDGMENTS

This work was supported by the U.S. Department of Energy, Division of High Energy Physics. The authors wish to thank Dr. S. Tantawi for his insightful comments.

# REFERENCES

1. Granatstein, V. L., and Lawson, W., *IEEE Trans. Plasma Sci.* **24**, 648 (1996).

2. Chu, K. R., Granatstein, V. L., Latham, P. E., Lawson, W., and Striffler, C. D., *IEEE Trans. Plasma Sci.* **13**, 424 (1985).

3. Nusinovich, G. S., Latham, P. E., and O. Dumbrajs, *Phys. Rev. E* **52**, 998 (1995).

4. Latham, P. E. and Nusinovich, G. S., *Phys. Plasmas* **2**, 3494 (1995).

5. Nusinovich, G. S. and Zhao, J., *Phys. Rev. E* **58**, 1002 (1998).

6. Nusinovich, G. S. and Dumbrajs, O., *Phys. Plasmas* **2**, 568 (1995).

7. Gold, S. and Nusinovich, G. S., *Rev. Sci. Inst.* **68**, 11 (1997).

8. Wilson, P. B., SLAC-PUB-3674 (1985); see also Granatstein, V. L. and Mondelli, A., *The Physics of Particle Accelerators*, AIP Conf. Proc. No. 153 (AIP, New York, 1987), p. 1506.

9. Saraph, G. P., et al., *IEEE Trans. Plasma Sci.* **24**, 671 (1996).

10. Vlasov, S. N., Zagryadskaya, L. I., and Orlova, I. M., *Radiotekhnika i Elektronika* **21**, 1485 (1976) [English translation, *Radio Eng. Electron. Phys.* **21**, 96 (1976)]; also Nusinovich, G. S., Read, M. E., Dumbrajs, O., and Kreischer, K. E., *IEEE Trans. Electron. Dev.* **41**, 433 (1994).

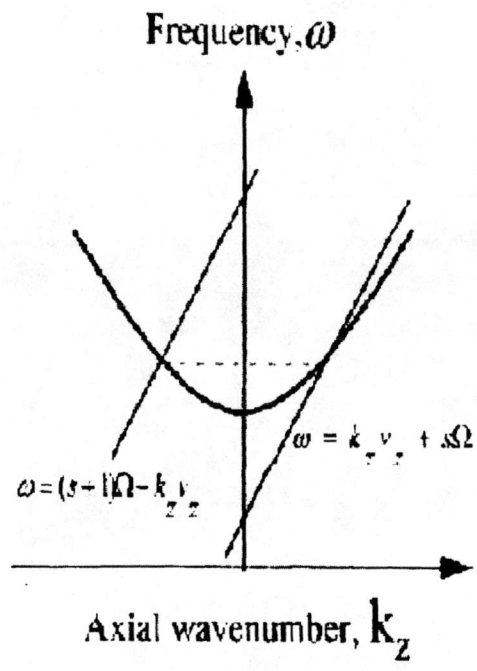

Fig. 1. Dispersion diagram illustrating the double resonance of one resonator mode formed by two traveling waves with an electron beam at two cyclotron harmonics.

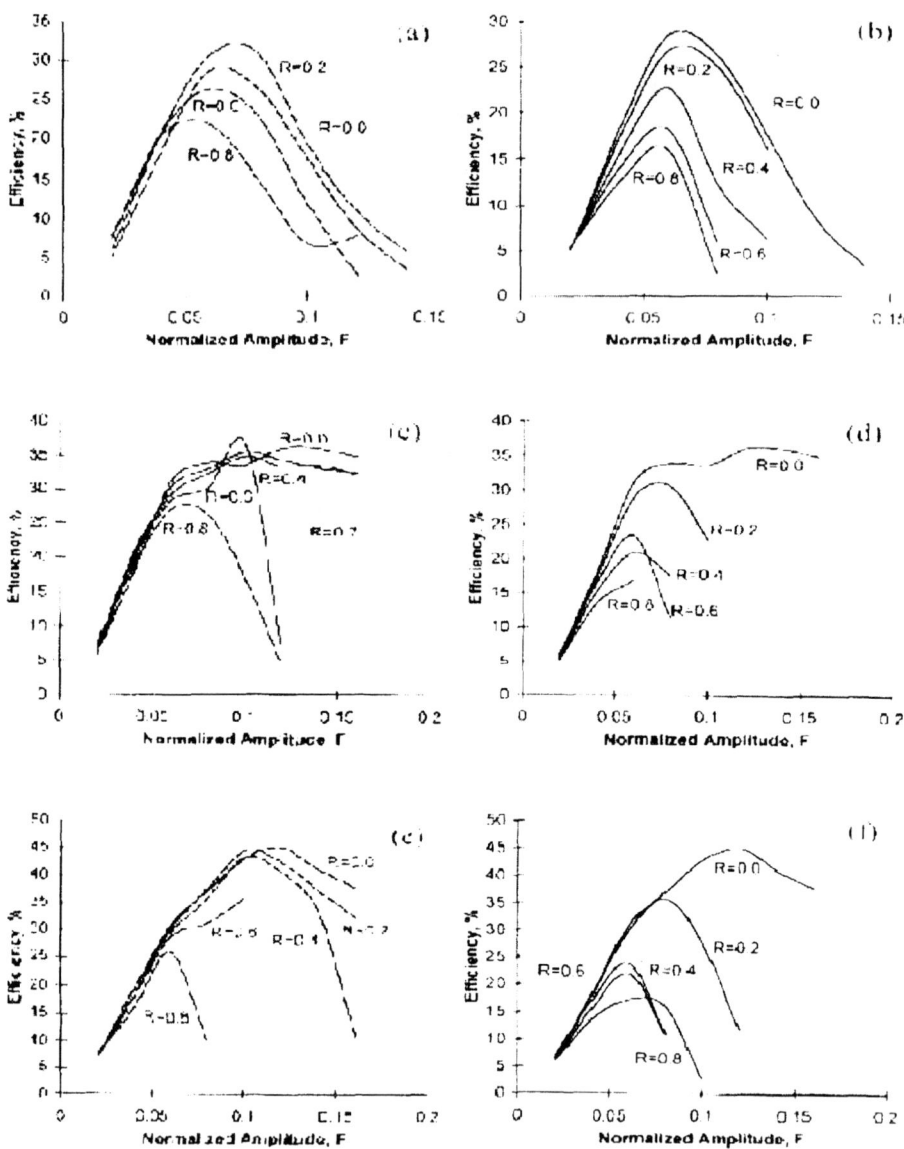

Fig. 2. Operation at a mode with two axial variations: figures (a), (c) and (e) correspond to the symmetric $TE_{m,p}$ mode ($m = 0$); figures (b), (d) and (f) correspond to non-symmetric ($m \neq 0$) modes. The bunching parameters for figures (a) and (b), (c) and (d), and (e) and (f) are equal to 0, 1.0, and 1.8, respectively.

948

# Requirements of a High-Power TWT Operating at 35GHz

L. Schächter,[*], S. Bana[*] and J. A. Nation[†]

[*] Electrical Engineering Department, Technion, Haifa 32000, Israel
[†] School of Electrical Engineering, Cornell University, Ithaca, NY 14853, USA

**Abstract.** A 2D quasi-analytic model has been developed for the investigation of the performance of a high-efficiency traveling wave amplifier operating at 35GHz. Simulations indicate that a relative energy spread of less than 5% is sufficient to reach high efficiency. It is also shown that there is an optimal guiding magnetic field for a given geometry of the slow-wave structure. Within the range of parameters of interest, high efficiency (65%) operation is feasible. We suggest a way to suppress parasitic modes which may develop in the system.

## INTRODUCTION

The next linear collider (NLC) will require acceleration gradients in excess of 100MV/m. The lower limit of the necessary gradient is achievable with the X-band klystrons as developed at Stanford Linear Accelerator Center (SLAC)[1]. For higher gradients (> 200MV/m) the source will have to operate at a higher frequency. The obvious question is what should be the operation frequency of the future system. There is no optimal value which can be established from a single or simple criterion. The complexity of the analysis and the variety of criteria involved were discussed more than ten years ago by Palmer [2] who concluded that a good trade off is operation close to 14.5 GHz. More recently Wilson [3] has analyzed the requirements of an 5-15TeV accelerator and the conclusion was in favor of operation at 34 GHz. However, higher energies would require even higher operating frequencies.

The spectrum of possibilities, with regard to what rf source should be utilized, is quite limited if we assume that conceptually NLC will operate in a similar way as the Stanford Linear Collider (SLC). One possibility is to scale the present *klystron* technology to 35GHz. This may prove to be a difficult approach since the diameter of the drift-tube between cavities has to be scaled down according to the frequency increase, in order to keep the waveguide below cutoff for all electromagnetic modes. At the same time the current in the electron-beam has to be comparable to that in present day klystrons and therefore the propagation of the electron beam imposes severe constraints on the guiding magnetic field. Furthermore, the closer the beam

CP472, *Advanced Accelerator Concepts: Eighth Workshop,*
edited by W. Lawson, C. Bellamy, and D. Brosius

is to the external wall, the larger the variation of the RF field across the beam, affecting consequently the efficiency of the system. To avoid these difficulties the beam radius is expected to be one order of magnitude smaller than the vacuum wavelength (i.e. $R_{\text{beam}} \sim \lambda/8$ [3]). In spite these constraints the klystron may still have a significant contribution to the next generation of accelerators. In fact Wilson [3], indicates that since most of the difficulties of the klystron arise from its pencil-beam, the alternative, namely a sheet-beam [4] may eliminate many of the obstacles.

A second possibility is the *gyrotron* which is under intensive investigation at the University of Maryland [5]. It has inherent advantages at high frequency and continuous operation at high power levels. An interesting concept that utilizes the advantages of the klystron at low frequency and these of the gyrotron at high frequency is being developed by Hirshfield [6]. It relies on conversion of low-frequency power from a regular klystron by bunching an electron-beam which in turn generates, after being isolated from the input RF, a higher harmonic of the input frequency. The conversion mechanism is similar to that which occurs in a gyrotron. Obviously the constraints on this conversion scheme with regard to efficiency are high since the overall efficiency is the product of that of the converter and the klystron.

The third possibility is to exploit the advantages of the interaction in a *traveling-wave* structure and this will be the topic of this theoretical study. As an acceleration structure, the traveling-wave amplifier consists of a disk-loaded waveguide but contrary to an acceleration structure where the aperture's diameter has to be reduced according to the increase in the operating frequency, in this case (35GHz) the aperture may be comparable to the dimensions required in an X-band structure. In this way we have eliminated one major constraint which is unavoidable in a pencil-beam klystron. The price we pay for using at 35GHz the same disk aperture as in an X-band structure, is operation at relatively high group velocity and low interaction impedance. The latter has two effects: for a given current and for maximum efficiency it makes the required interaction length longer which leads to a higher sensitivity to beam-quality. On the other hand, it makes the transverse motion of the particles less "violent" reducing in this way the probability of particles hitting the structure's wall at a given guiding magnetic field.

In addition to the iris aperture, that has not been reduced in the present design (35GHz), the amount of current carried by the beam remains of the same order of magnitude (as in the X-band case) implying that a typical guiding magnetic field of 0.5T will be required. This field corresponds to a cyclotron frequency $f_c \equiv eB/2\pi m\gamma$ of 14GHz (assuming $\gamma = 2$) and for sufficiently low velocities, the electrons may enable a "gyrotron-like" interaction to occur namely, $|\omega - kV| = \omega_c$ (where $k$ is the wavenumber in the structure). Consequently we may expect severe reduction of the efficiency and electron interception by the structure.

In this study we investigate several aspects associated with the operation of a high-power single-stage traveling-wave amplifier at 35GHz. For this purpose we have developed a quasi-analytic 2D model of the interaction of an ensemble of electrons with a *symmetric* transverse magnetic (TM) mode. The interaction occurs with all the components of the electromagnetic field and not only the longitudinal electric field as is the case in a 1D model [7-9]. Firstly, we compare the performance of two similar amplifiers, one operating at 35GHz and the other at X-band. We show that the sensitivity of the former to beam-quality is higher but its operation is definitely feasible. Secondly, we investigate the effect of the guiding field on the interaction of a slow wave with an ideal beam and thirdly we show that optimal disk aperture and optimal guiding magnetic field, do exist. This quasi-analytic 2D model is adequate for simulation of the interaction in a two-stage amplifier that based on our experience in X-band should enable generation of radiation with higher efficiency. The latter is a crucial requirement for an RF source designed to feed an acceleration module. The operation of a TWT driven by a modulated beam is examined and efficiencies of up to 65% were observed in simulations. Finally, since in the structure parasitic modes may develop, we suggest a scheme to suppress these modes but preserves the basic interaction features of the structure at 35GHz.

The equations that determine the dynamics of the particles and of the electromagnetic field can be simplified to read

$$\frac{\mathrm{d}}{\mathrm{d}\zeta}a = \alpha \langle \mathrm{I}_0(\bar{r}_i)\,e^{-j\chi_i}\rangle_i \,,$$

$$\frac{\mathrm{d}}{\mathrm{d}\zeta}\chi_i = \Omega \left(\beta_{z,i}^{-1} - \beta_{\mathrm{ph}}^{-1}\right) \,,$$

$$\frac{\mathrm{d}}{\mathrm{d}\zeta}\gamma_i = -\frac{1}{2}\left[a\mathrm{I}_0(\bar{r}_i)e^{j\chi_i} + \mathrm{c.c.}\right] \,,$$

$$\frac{\mathrm{d}}{\mathrm{d}\zeta}\bar{r}_i = \bar{p}_{r,i}/\gamma_i\beta_{z,i}$$

$$\frac{\mathrm{d}}{\mathrm{d}\zeta}\bar{p}_{r,i} = \Omega_c^2 \left(\bar{r}_i^4 - \bar{r}_{i,\mathrm{in}}^4\right)/4\gamma_i\beta_{z,i}\bar{r}_i^3 + \bar{r}_i\Omega_p^2/2\beta_{z,i}\gamma_i^2$$

$$\qquad\qquad - \frac{1}{2}\left(jae^{j\chi_i} + \mathrm{c.c.}\right)\mathrm{I}_1(\bar{r}_i)\Omega(1 - \beta_{z,i}\beta_{\mathrm{ph}})/\beta_{z,i}\beta_{\mathrm{ph}} \,,$$

$$\beta_{z,i} = \left\{1 - \gamma_i^{-2} - (\gamma_{\mathrm{ph}}\beta_{\mathrm{ph}}/\Omega\gamma_i)^2 \left[\bar{p}_{r,i}^2 + \left(\Omega_c(\bar{r}_i^2 - \bar{r}_{i,\mathrm{in}}^2)/2\bar{r}_i\right)^2\right]\right\}^{1/2} . \quad (1)$$

where the following definitions were used: $k$ is the wavenumber ; $\Gamma \equiv \sqrt{k^2 - (\omega/c)^2}$. The normalized phase-velocity is denoted by $\beta_{\mathrm{ph}} \equiv \omega/ck$, $\gamma_{\mathrm{ph}} \equiv 1/\sqrt{1 - \beta_{\mathrm{ph}}^2}$ and $\eta_0 \equiv \sqrt{\mu_0/\epsilon_0}$. The complex amplitude of $E_z$ is denoted by $\bar{E}_0 \equiv E_0\,e^{-j\psi}$ or in its normalized form $a \equiv e\bar{E}_0 d/mc^2$. The term $IZ_{\mathrm{int}}$ is the product of the current, $I$, representing the beam, and the interaction impedance $Z_{\mathrm{int}} \equiv \frac{1}{2}\frac{E_0^2(\pi R_{\mathrm{int}}^2)}{\langle P(z)\rangle_t}$ representing the electromagnetic system; $\langle P(z)\rangle_t$ is the average in time power which propagates along the structure. It is this product that controls the *coupling* of the beam with the wave thus $\alpha$ is referred to as the normalized coupling coefficient $\alpha \equiv \frac{I Z_{\text{int}}}{mc^2/e} \frac{d^2}{\pi R_{\text{int}}^2}$; $d$ is the interaction length. The normalized angular frequency $\Omega \equiv \omega d/c$, $K \equiv kd$, $\Omega_c \equiv ecBd/mc^2$, $\Omega_p^2 \equiv \frac{eI\eta_0}{mc^2} \frac{d^2}{\pi R_{\text{beam}}^2} \frac{1}{\langle \beta_{z,i} \rangle_i}$; $I$ is the beam average current and $R_{\text{beam}}$ is the beam radius. Based on the argument of the modified Bessel functions we define $\bar{r}_i = \omega r_i / c\gamma_{\text{ph}}\beta_{\text{ph}}$ and correspondingly, $\bar{p}_{r,i} = \gamma_i \beta_{r,i} \Omega / \gamma_{\text{ph}}\beta_{\text{ph}}$.

Several assumptions have been made in the process of developing these equations: *(a)* The interaction is with the TM mode and only harmonic the $n = 0$ is actively involved in the interaction. All the other spatial harmonics contribute indirectly via $Z_{\text{int}}$. *(b)* The radial oscillations of the beam envelope are relatively small, therefore we may neglect the coupling of the TM to TE mode throughout this analysis; simulations indicate that if this is not the case, electrons hit the structure. Explicitly we assume $|\beta_{r,i}|\gamma_{\text{ph}} \ll \beta_{z,i}$. *(c)* The guiding magnetic field, $B_0$, is uniform or at least the spatial variations are very slow on the scale of the radiation wavelength. *(d)* Single frequency ($\omega$) operation.

## SIMULATION RESULTS AND DISCUSSION

The first step when investigating the operation of a traveling-wave amplifier at 35GHz is to compare its performance with that of an X-band system. For simplicity at this stage we shall assume that an intense guiding field is applied such that we may confine our analysis to 1D motion. Adequate comparison requires that the beam is identical [500kV, 200A, 3mm radius] in both cases. By virtue of energy conservation

$$\frac{\mathrm{d}}{\mathrm{d}\zeta} \left[ \langle \gamma_i \rangle_i + \frac{|a|^2}{2\alpha} \right] = 0, \tag{2}$$

the input power in the two cases has to be the same; the latter is assumed to be $P_{\text{in}} = 20kW$ and phase velocity equal to the average velocity of the electrons. At this point we have two options: either we require that the relative change of the wavenumber is the same i.e.

$$\frac{\mathrm{Im}(k)}{\mathrm{Re}(k)} = \frac{\sqrt{3}}{2} \left[ \frac{1}{2} \frac{\alpha}{\Omega^2} \langle \frac{\beta_{\text{ph}}^3}{(\gamma_i \beta_i)^3} \rangle_i \right]^{1/3}, \tag{3}$$

or that the overall gain is the same namely,

$$\mathrm{Im}(kd) = \frac{\sqrt{3}}{2} \left[ \frac{1}{2} \alpha \Omega \langle \frac{1}{(\gamma_i \beta_i)^3} \rangle_i \right]^{1/3}. \tag{4}$$

Since both the power in the beam and RF are the same at the input in both cases, we would obviously prefer to examine the two devices such that their output power (i.e. efficiency) will be the same. Consequently we prefer the criterion in Eq.(4) rather than that in (3).

By imposing the identical output power requirement [Eq.(4)] in both systems we enforce the following relation between the length and the interaction impedance of the two systems

$$\alpha_{35}\Omega_{35} = \alpha_{8.75}\Omega_{8.75} \rightarrow d_{8.75} = d_{35} \left[ \frac{35}{8.75} \frac{Z_{\text{int},35}}{Z_{\text{int},8.75}} \right]^{1/3} . \tag{5}$$

The first topic we wish to examine is the sensitivity to *beam-quality* of an amplifier operating at 35GHz comparing to a 8.75GHz(=35/4) one. For simplicity let us assume a very strong (infinite) magnetic field guides the beam - this constraint will be released subsequently. Assuming a phase advance per cell of $2\pi/3$ ($L = 2.465$mm at 35GHz and $L = 9.86$mm at 8.75GHz), we can calculate the interaction impedance for four different internal radii: $R_{\text{int}} = 6, 6.5, 7.0 \text{ and } 7.5$mm. The corresponding result is $Z_{\text{int}}(35\text{GHz}) = 73, 52, 37 \text{ and } 26 \text{ Ohm}$. Since in this particular case an "infinite" magnetic field is assumed, the interaction impedance quoted above is calculated by averaging the electric field experienced by all the electrons i.e.

$$E_{\text{eff}}^2 = \frac{2}{R_{\text{beam}}^2} \int_0^{R_{\text{beam}}} dr \, r E_z(r)^2 . \tag{6}$$

For the dynamics of this system it is sufficient to solve the 1D equations of motion with $E_{\text{eff}}$ replacing $E_0$ and taking $\bar{r}_i = 0$. In the X-band range the corresponding interaction impedances are $Z_{\text{int}}(8.75\text{GHz}) = 2700, 2193, 1812 \text{ and } 1519 \text{ Ohm}$; these latter values are higher than generally used but they emphasize even better the difference between the two systems. For example, the significant difference in the interaction impedance emphasizes the relevance of the beam-quality question since the interaction impedance of the 35GHz system resembles that of a free-electron laser (FEL) and the sensitivity of the latter to beam quality is a well recognized fact. A comparison of FEL and TWT operation can be found in Ref. 10 [p. 305].

The two frames in Figure 1 illustrate the effect of the relative energy spread at the input on the spatial growth rate. We observe that at 35GHz the growth rate drops to virtually zero for energy spread larger than 15% whereas at 8.75GHz there is virtually no change in this range. It should be pointed out that the system is assumed to be *uniform* and the simulation is terminated when the interaction saturates. In order to understand the reason for this sensitivity we plot in Figure 2 the relative energy spread at the output as a function of the relative energy spread at the input. The result indicates that, since in the X-band system the interaction impedance is large, the energy spread at the output is also large (70%). Increasing the energy spread at the input reduces somewhat the efficiency of the interaction but the spread at the output is not even close to that at the input. This is contrary to the 35GHz case where the interaction impedance is two orders of magnitude smaller, the energy spread at the output is much more moderate (10 − 20%), and

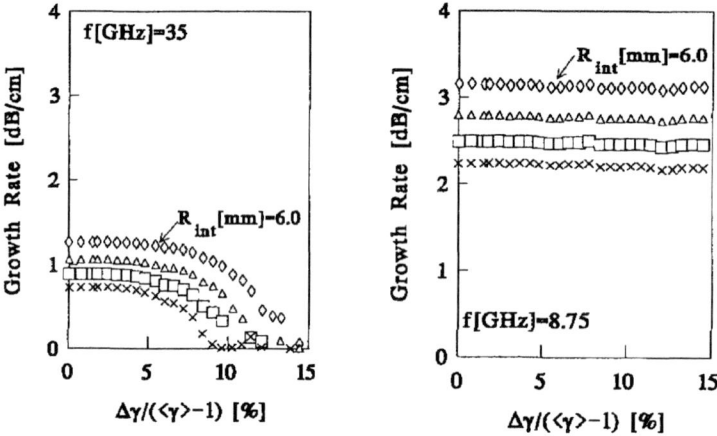

Fig. 1: Growth rate as a function of the relative energy spread at the input.

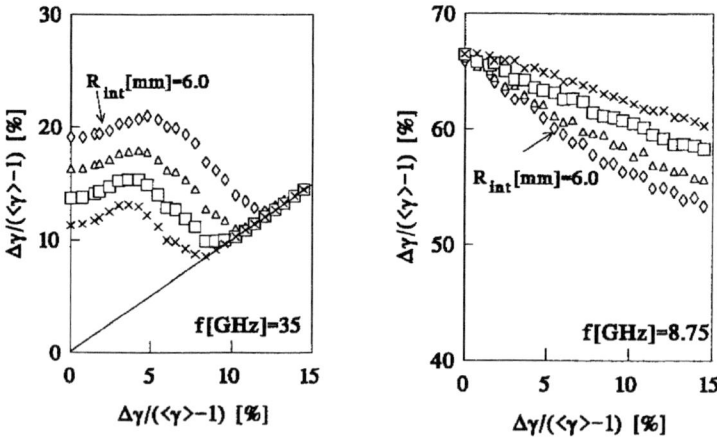

Fig. 2: Relative energy spread at the output as a function of the relative energy spread at the input.

an increase to $10 - 15\%$ in the energy spread at the input reaches rapidly the value of this quantity at the output - thus the interaction efficiency is practically zero. The line in the left frame represents the case when the energy spread at the output equals that at the input. From this comparison we conclude that the sensitivity to beam-quality is different in the two regimes. However in practice this is not an obstacle in the operation of a 35GHz traveling-wave amplifier based on a *uniform* structure and "infinite" guiding field since present technology enables to generate a beam with a relative energy spread smaller than 5%.

Our next step is to relax the constraint of "infinite" guiding magnetic field of a 35GHz amplifier but keep the relative energy spread at the input constant at 1% value. Figure 3 illustrates the efficiency in a *tapered* structure as a function of the applied magnetic field for four different disk apertures, $R_{\text{int}} = 4, 5, 6$ and 7mm. In this case the simulation is terminated either if electrons hit the external wall or they start to move backward. For this reason the interaction length is very short in the case of low intensity magnetic field and consequently the efficiency is low. According to this result, for a given internal radius (and thus interaction impedance) there is an optimal magnetic field for achieving maximum efficiency. For a 6mm internal radius this magnetic field is slightly more than 1T whereas for 7mm it is about 0.5T. It should be pointed out that in this simulation the phase velocity of the wave was tapered such that

$$\frac{1}{\beta_{\text{ph}}} = \left\langle \frac{1}{\beta_{z,i}} \right\rangle_i . \tag{7}$$

This tapering keeps the average phase of all particles relative to the wave constant, i.e. $\frac{d}{d\zeta}\langle \chi_i \rangle = 0$; this is not necessarily the optimal taper.

In order have a better understanding of the behavior as revealed in Figure 3 let us start and examine closely the behavior of the 4mm system. Its interaction impedance is the highest of the four and therefore the *radial* motion becomes significant close to the input and since our simulation is terminated if one particle hits the structure, the amount of energy transferred to the wave is small (low efficiency); each point along the curve corresponds to a different interaction length. For the next radius ($R_{\text{int}} = 5$mm) high magnetic fields $1 - 1.5$T are sufficient to confine the beam and efficiency of $15 - 25\%$ is achievable before particles hit the structure. Increasing further the internal radius ($R_{\text{int}} = 6$mm) reveals a similar behavior for magnetic fields smaller than 1.2T with efficiency up to 35%. However, increasing further the magnetic field reduces the longitudinal momentum of the beam since for the same RF field the radial swing will be larger and consequently the kinetic energy available for conversion into RF is reduced and the particles slip out of phase in spite the tapering. This behavior is even better revealed in the last case presented ($R_{\text{int}} = 7$mm) where the efficiency as a function of the applied magnetic field reveals several peaks and valleys exactly as in the case of saturation.

At saturation an additional effect may contribute to this interesting behavior of the efficiency: this is a gyrotron-like interaction. In the interaction process

955

there are sufficiently slow electrons which instead of "satisfying" the regular slow wave structure resonance condition ($\omega - k v_i = 0$), their radial motion enables a combined interaction which formally can be represented by $|\omega - k v_i| = \omega_c$. Such an interaction extracts energy from the RF field and transfers it to the electrons as can be concluded from the "valleys" in the curve corresponding to $R_{\mathrm{int}} = 7mm$. Exact description of the interaction in the vicinity of these points would require inclusion of TE modes in the analysis. In any event such a mechanism will hurt the operation of a TWT therefore it is necessary to operate as far as possible from this regime, therefore only the first peak in the curve is of interest.

So far we examined the effect of the energy spread assuming an infinite guiding field. In addition we investigated the effect of a finite guiding field on an (initially) ideal beam. Our next step is to consider the effect of energy spread at the input on the interaction in a tapered structure which has an internal radius of 6mm; the guiding magnetic field is 0.5T. Figure 4 illustrates the efficiency of such a system for four different energy spreads at the input. The results are not dramatically different to those we found when the motion is assumed to be constrained to the longitudinal direction by a strong external magnetic field. As in the other case we obtain efficiencies as high as 40% in a single-stage amplifier.

In order to examine the limits of efficiency achievable with tapered TWT we have simulated the operation of such a system when driven by a bunched beam. The parameters are similar to those in the previous simulations with two exceptions: the initial beam radius was assumed to be 2mm and the changes in the interaction impedance were taken into consideration when the phase-velocity varies in space; for $R_{\mathrm{int}} = 6mm$ and $f = 35GHz$ this relation is given by

$$Z_{\mathrm{int}} = 301 \beta_{\mathrm{ph}}^{8.6} \, [\mathrm{Ohm}] \tag{8}$$

In the simulation three different tapers were considered: *(a)* locally the phase velocity of the wave equals the average velocity of the electrons

$$\beta_{\mathrm{ph}} = \langle \beta_{z,i} \rangle_i \,, \tag{9}$$

*(b)* the condition used previously expressed in Eq. (7) that ensures that the relative average phase of the particles remains the same relative to an electromagnetic wave propagating in the structure in the absence of any interaction. *(c)* In the third case we take into consideration the *active* phase velocity of the wave. In other words the relative average phase of the particles is constant with the wave whose phase varies due to the interaction. Figure 5 illustrates the efficiency as a function of the initial phase spread (density bunching) at the input $-\Delta\chi/2 \leq \chi_i \leq \Delta\chi/2$. We observe that the maximum efficiency achievable for $B_0 = 0.5T$ and beam radius of 2mm is about 65%. For very narrow bunch, the difference between various tapering schemes is negligible but it becomes evident as the bunch width is increased when also the efficiency drops to the 40% level.

Using a $6mm$ aperture disks at 35GHz allows the propagation of other modes in the system. These may interfere with the desired mode therefore we suggest

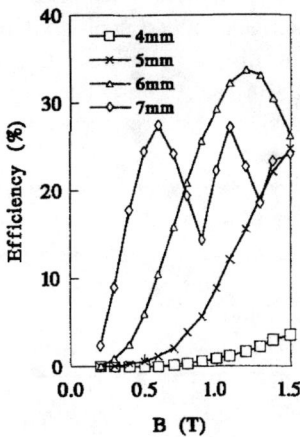

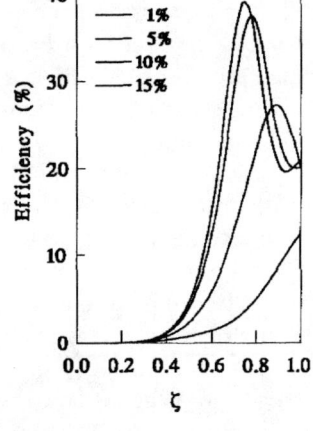

Fig.3: Efficiency as a function
of the guiding magnetic field.

Fig.4: The way efficiency
develops in the interaction
region.

Fig.5:Efficiency as a function of
the phase spread at the input of
the interaction region.

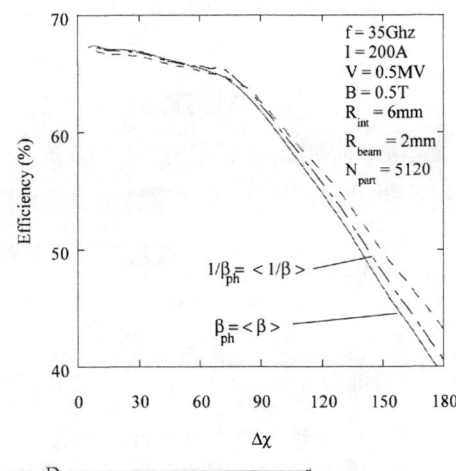

Fig.6: Superfish schematic of a
possible output traveling wave
structure design to suppress
parasitic modes.

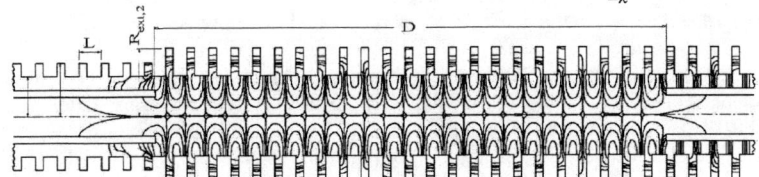

a system that its one end rather than being short circuited as is usually the case in output structures it consists of a Bragg reflector - see Figure 6. Behind the Bragg reflector there is an absorber. The Bragg reflector is designed such that at 35GHz it is a perfect reflector but at other frequencies (of interest) it transmits the waves which are absorbed. In this way we have generated a "selective cavity". At this point the design was (cold) modeled with Superfish where the axial extraction section was short circuited.

In conclusion, a 2D quasi-analytic model has been developed in order to investigate the operation of a 35GHz traveling wave amplifier. Based on this model we were able to establish constraints on beam-quality, guiding magnetic field and the radius of the aperture. Our main conclusions regarding a single stage system are: (i) as long as the relative energy spread at the input is less than 5% the 35GHz system should operate in a similar way as in X-band. (ii) A magnetic field of 0.5T should suffice to confine a beam which generates about 30% efficiency with a 6mm disk aperture and a higher efficiency ($\geq 35\%$) can be achieved if a 1.2T is applied to a 7mm aperture system; in these cases the interaction impedance is assumed to be constant and only the phase velocity varies. Considering a system driven by a density modulated beam [second of a two stage system] we found that (iii) 65% efficiency is achievable with 2mm beam radius and 0.5T guiding magnetic field. (iv) it is possible to suppress parasitic modes by using an adequate Bragg reflector that reflects at the operating frequency but transmits at the frequencies of parasitic modes.

## ACKNOWLEDGMENT

This study was supported by the United States Department of Energy and by the Bi-National United States - Israel Science Foundation.

## REFERENCES

1. R.D. Ruth; Particle Accelerator Conference 12-16 May 1997, Vancouver, Canada.
2. R. Palmer; SLAC Publication, SLAC-PUB-4295, April 1987.
3. P. Wilson; Proceedings of the Third Workshop on Pulsed RF Sources for Linear Colliders, Editor Shigeki Fukuda, Shonan Village Center, Hayama, Kanagawa, Japan, April 8-12, 1996 p.9. See also: P. Wilson, "Scaling Linear Colliders to 5TeV and Above", SLAC PUB 7449 (1997).
4. D. U. L. Yu, J.S. Kim and P. Wilson, "Design of a High-Power Sheet-Beam Klystron", Proceedings of the Advanced Accelerator Concepts Workshop, Port Jefferson, NY USA, June 1992 (AIP Conference Proceedings 279, 1993), p.85.
5. W. Lawson, B. Hogan, J. Cheng, M. Castle, G. Saraph, J. Anderson, J. Calame, M. Reiser and V.L. Granatstein; Proceedings of the Third Workshop on Pulsed RF Sources for Linear Colliders (RF'96) April 8-12, 1996 Kanagawa, Japan, p. 225.
6. C. Wang, J.L. Hirshfield and A. K. Ganguly; "Seventh Harmonic Co-Generation by Cyclotron Resonance Acceleration". Particle Accelerator Conference 12-16 May 1997, Vancouver, Canada.
7. L. Schächter, J.A. Nation and D.A. Shiffler; *J. Appl. Phys.* **70** p. 114 (1991).
8. L. Schächter and J.A. Nation; *Phys. Rev. A* **45** p. 8820 (1992).
9. L. Schächter; *Phys. Rev. E* **52** p. 2037 (1995).
10. *Beam-wave Interaction in Periodic and Quasi-periodic Structures*, L. Schächter; Springer 1996.

958

# The Design and Analysis of Multi-Megawatt Distributed Single Pole Double Throw (SPDT) Microwave Switches

Sami G. Tantawi[1]

*Stanford Linear Accelerator Center, SLAC, 2575 Sand Hill Rd. Menlo Park, CA 94025, USA*

**Abstract.** We present design methodology and analysis for an SPDT switch that is capable of handling hundreds of megawatts of power at X-band. The switch is designed for application in high power rf systems in particular future Linear Colliders (1). In these systems switching need to be fast in one direction only. We use this to our advantage to reach a design for a super high power switch. In our analysis we treat the problem from an abstract point of view. We introduce a unified analysis for the microwave circuits irrespective of the switching elements. The analysis is, then, suitable for different kinds of switching elements such as photoconductrs, PIN diodes, and plasma discharge in low-pressure gases.

## INTRODUCTION

The basic switch would have to have at least three ports. One for the incoming rf signal, port 1, and two for the out going signal, ports 2 and 3. The most straightforward idea is to put two switches that are always working in a complimentary mode, that is, when the first is off the second is on and vice versa. This way the power flow can be controlled from port 1 to either port 2 or 3. Fig.1 shows the schematic of this switch. For proper operation of the switch the splitter junction has to be matched when one of the switches is *on* and the other is *off*. This implies a specific scattering parameters for this three-port junction. At some reference plane, this scattering matrix can be written as follows (See Fig.1):

$$S = \begin{pmatrix} -1/3 & 2j/3 & 2j/3 \\ 2j/3 & 1/3 & -2/3 \\ 2j/3 & -2/3 & 1/3 \end{pmatrix} \tag{1}$$

When one of the switches is *off* it acts as a short circuit reflecting all the power back to the junction and the power flows from the input to the output that have the *on* switch.

---

[1] *Also with the Communications and Electronics Department, Cairo University, Giza, Egypt*

CP472, *Advanced Accelerator Concepts: Eighth Workshop,*
edited by W. Lawson, C. Bellamy, and D. Brosius

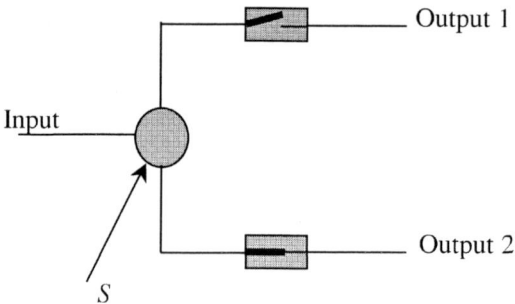

**FIGURE 1.** Simple SPDT Switch

From (1), the switch will have to handle half the power during the *off* state and all the power during the *on* state. In this case we used two switches and each switch had to control the flow of all the power.

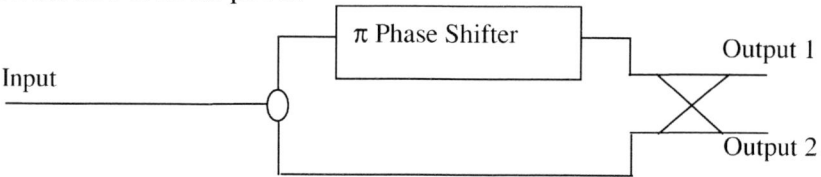

**FIGURE 2.** Dual Mode SPDT Switch

An alternative to this is the dual mode rf switch shown in Fig. 2. The power is split first between two ports and then combined again into a 3dB hybrid. A switch put at one of the arms connecting the splitter to the hybrid can switch the phase of the rf signal in that arm by a 180 degrees thus steering the rf to a particular port at the output of the hybrid. In this scheme the switch has to handle only half the power. The implementation of this switch can take different forms. For example the *Polarization Shift Switch* shown in Fig 3, which consists of a circular waveguide supporting the $TE_{11}$ mode. This mode can always be decomposed into two orthogonal $TE_{11}$ modes.

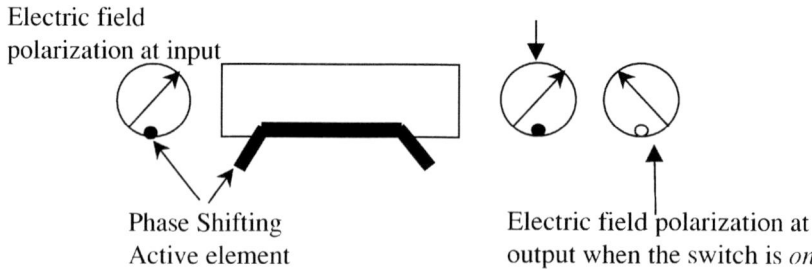

**FIGURE 3.** Polarization Shift Switch

If one has an element that shifts the phase of one of those two modes, the combined results can be made to be a $TE_{11}$ mode that has it electric field vector normal to that of the original mode. At the end of this circular waveguide one can pick up one polarization in one waveguide and the other polarization into another. This is an example of a dual mode rf switch.

It is clear from the above discussion that the basic element in a dual mode switch is a $\pi$ phase shifter. In the following sections we will describe an abstract implementation of this phase shifter. We will use several cascaded elements. We will then give an example of the design using a distributed optical switch.

## THE DISTRIBUTED PHASE SHIFTER

The phase shifting element can be considered as a series of *symmetric* three-port elements; see Fig. 4. Each element is a basic *loss-less* three-port device with two similar ports, namely, port 1 and port 2. The third port is terminated so that all the scattered power from that port is completely reflected. However, the phase of the reflected signal from the third port can be changed actively. The active element at the third port is simply a medium that changes its properties from a dielectric like material to a conductor like martial in a controlled manner. One can use bulk phenomenon in pure semi-conductors (2), a PIN diode structure or a plasma discharge in a gas (3). Thus one can shift the plane of the reflected signal from a static reflector behind the active element to a plane just in front of the active element.

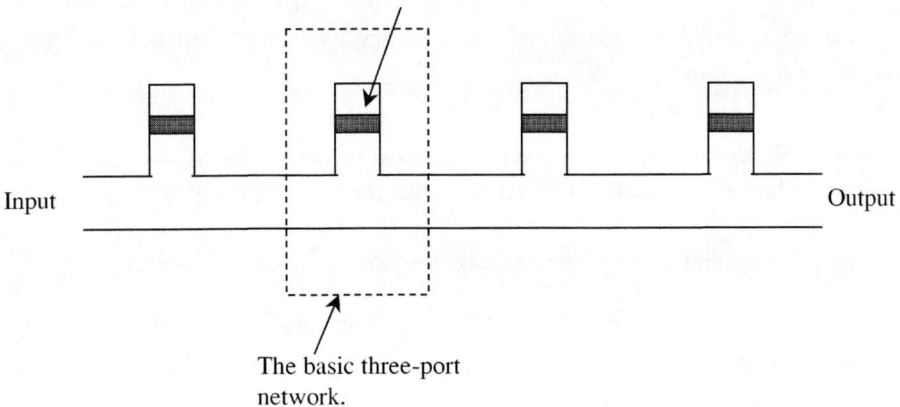

The phase of the reflected signal from the third port depends on the status of the active element

Input                                                                    Output

The basic three-port network.

**FIGURE. 4** The Distributed Phase Shifter

One can show that there is always an angle, for the reflection coefficient in the third port, which makes the device completely matched. By changing this angle and placing the elements with proper spacing one can also achieve matching conditions, however, with a total different phase shift through the elements. Under the constraints of a minimum break down fields for the elements, a reasonable switching time, and losses in the elements in both the *off* and *on* state, we derive expressions for the optimal switch design. The design is a function of the number of elements, the size and mode of operation in the third port of each element, and the parameters that specify the properties of the three port networks.

# SYNTHESIS OF THE π PHASE SHIFTER USING A SERIES OF SYMMETRIC THREE PORT NETWORKS

## The Three Port Network as a Control Element

For any loss-less and reciprocal three-port network the scattering matrix is unitary and symmetric. By imposing these two conditions on the scattering matrix $\underline{S}$ of our device and at the same time taking into account the symmetry between port 1 and port 2; at some reference planes, one can write:

$$\underline{S} = \begin{pmatrix} \dfrac{e^{j\phi} - \cos\theta}{2} & \dfrac{-e^{j\phi} - \cos\theta}{2} & \dfrac{\sin\theta}{\sqrt{2}} \\ \dfrac{-e^{j\phi} - \cos\theta}{2} & \dfrac{e^{j\phi} - \cos\theta}{2} & \dfrac{\sin\theta}{\sqrt{2}} \\ \dfrac{\sin\theta}{\sqrt{2}} & \dfrac{\sin\theta}{\sqrt{2}} & \cos\theta \end{pmatrix}. \tag{2}$$

Indeed, with the proper choice of the reference planes, this expression is quite general for any symmetric three-port network. The scattering matrix properties are determined completely with only two parameters: $\theta$ and $\phi$. The scattered rf signals $\underline{V}^-$ are related to the incident rf signals $\underline{V}^+$ by

$$\underline{V}^- = \underline{S}\,\underline{V}^+ ; \tag{3}$$

where $V_i^{\pm}$ represents the incident/reflected rf signal from the $i$th port. We terminate the third port so that all the scattered power from that port is completely reflected; i.e.,

$$V_3^+ = V_3^- e^{j\psi} . \tag{4}$$

The resultant, symmetric, two-port network, then, has the following form:

$$\underline{S}_{two-port} = \begin{pmatrix} \cos\left(\dfrac{\zeta - \phi}{2}\right) e^{j\left(\frac{\phi}{2} + \frac{\zeta}{2} + \alpha\right)} & j\sin\left(\dfrac{\zeta - \phi}{2}\right) e^{j\left(\frac{\phi}{2} + \frac{\zeta}{2} + \alpha\right)} \\ j\sin\left(\dfrac{\zeta - \phi}{2}\right) e^{j\left(\frac{\phi}{2} + \frac{\zeta}{2} + \alpha\right)} & \cos\left(\dfrac{\zeta - \phi}{2}\right) e^{j\left(\frac{\phi}{2} + \frac{\zeta}{2} + \alpha\right)} \end{pmatrix}, \tag{5}$$

where the angle $\zeta$ is given by

$$e^{j\zeta} = \frac{\cos\theta - e^{j\psi}}{\cos\theta\, e^{j\psi} - 1}, \tag{6}$$

and $\alpha$ is an arbitrary angle added to (5) so that the reference planes can be chosen at will. The signal level in the third arm is given by,

$$\left|V_3^+\right|^2 = \left|V_3^-\right|^2 = \frac{\sin^2\theta}{3 - 4\cos\theta\cos\psi + \cos 2\theta}\left|V_1^+ + V_2^+\right| \tag{6}$$

A series of these two port networks can be cascaded together to produce the desired phase shifter. When the active element in each of the three-port networks is not excited the reflection coefficient from that port is $e^{j\psi_0}$. We choose the angle $\psi_0$ such that each of the three- port networks is matched by itself, i.e., $S_{11} = S_{22} = 0$, and, in that case, $S_{12} = S_{21} = e^{j(\phi+\alpha+\pi)}$. Once the active element is excited the angle of the reflection coefficient in the third arm changes to become $\psi_1$. We choose $\psi_1$ and the angle $\alpha$ so that the train of three port networks is matched and the total phase shift across the train differs by $\pi$ when the angle of the third arm reflection changes from $\psi_0$ to $\psi_1$.

## Matching of the distributed Phase shifter

The train of three port networks can be viewed as a periodic structure. This periodic structure is inserted between two transmission lines with different characteristic impedances than that of the transmission line formed by the periodic structure. A matching condition occurs when the total phase shift across the structure is $\pi$, i.e.; the periodic structure represents a half wavelength transmission line transformer. In this case the phase shift across the periodic structure should go to zero as $\psi \rightarrow \psi_0$. This defines the proper choice of the reference planes, hence,

$$\alpha = \pi - \phi \pm \frac{2\pi}{n}; \tag{8}$$

where $n$ is the number of three port networks in the periodic structure.
Substituting from (8) into (5) we get the scattering matrix of each element of the periodic structure. We then use this scattering matrix to produce the transfer matrix $A$ (4).

$$A = \begin{bmatrix} \dfrac{1}{S_{21}} & \dfrac{-S_{22}}{S_{21}} \\ \dfrac{S_{11}}{S_{21}} & S_{12} - \dfrac{S_{11}S_{22}}{S_{21}} \end{bmatrix} \tag{9}$$

To make the total phase shift across the structure become $\pi$, $A$ should satisfy, $A^n = -I$; where $I$ is the identity matrix. Hence, the eigen values of $A$ are $e^{j(\frac{\pi+2\pi m}{n})}$; where $m$ is an integer. The sufficient and necessary condition for $A$ to have these eigen values is

963

$$\zeta_1 = \phi \pm 2\cot^{-1}\left(\frac{\cos\dfrac{\pi+2m\pi}{n} - \cos\dfrac{2\pi}{n}}{\sin\dfrac{2\pi}{n}}\right); \qquad (10)$$

where $\zeta \to \zeta_1$ as $\psi \to \psi_1$.

The eigen vectors of the matrix $A$, $v_1$ and $v_2$ are

$$v_1 = \begin{pmatrix} 1 \\ \dfrac{\sin\dfrac{(1-2m)\pi}{2n}}{\sin\dfrac{(3+2m)\pi}{2n}} \end{pmatrix}, \quad v_2 = \begin{pmatrix} 1 \\ \dfrac{\sin\dfrac{(3+2m)\pi}{2n}}{\sin\dfrac{(1-2m)\pi}{2n}} \end{pmatrix}. \qquad (11)$$

The ratio between the forward and the backward waves in the eigen vector solution has a mimuma at $m=1$ and $m=-2$.

Then the difference between the two angles $\Delta\psi = \psi_1 - \psi_0$ is given by,

$$e^{j\Delta\psi} = \frac{\left(e^{j\zeta_1} + \cos\theta\right)\left(1 + e^{j\phi}\cos\theta\right)}{\left(e^{j\phi} + \cos\theta\right)\left(1 + e^{j\zeta_1}\cos\theta\right)}. \qquad (12)$$

We then, using (7), can show that the electric field present at the active element when $\psi \to \psi_0$ is

$$E_{max} = 2\left(\frac{1 + \cos^2\theta - 2\cos\phi\cos\theta}{\sin^2\theta}\right)^{\frac{1}{2}} \times \left|\sin\frac{\Delta\psi}{2}\right|\left(\frac{P_{in}Z_3}{A_3 G_3}\right)^{1/2}; \qquad (13)$$

where $P_{in}$ is the constant level input power, $Z_3$ is the wave impedance of the mode excited in the waveguide that forms the third arm, $A_3$ is the cross sectional area of that guide, and $G_3$ is a geometrical factors that depends on the mode and the waveguide shape of the third arm.

When the switch is turned *on*, i.e., $\psi \to \psi_1$, the losses $P_l$ in the active element is given by

$$P_l = \frac{1 + \cos^2\theta + 2\cos\zeta_1\cos\theta}{2\sin^2\theta}\frac{R_s}{Z_3}\left|V_1^+ + V_2^+\right|^2; \qquad (14)$$

where $R_s$ is the surface resistance of the active element and it depends on the element type and level of excitation. The quantity $\left|V_1^+ + V_2^+\right|/(P_{in})^{1/2}$ has a week dependence on the number of elements $n$. If One chooses the total phase shift to be $3\pi$ ($m=1$) the value of $\left|V_1^+ + V_2^+\right|/(P_{in})^{1/2}$ is close to 1.0, and if the number of elements $n$ exceeds 6 this value vary slowly as a function of $n$, and approaches 1.0 as $n \to \infty$.

For a given value, for the number of elements, $n$ one can determine the proper reference planes from (8), the angle $\zeta_1$ from (10), and the angle $\Delta\psi$ from (12). To determine the best choice of the three-port network parameters we examine the normalized peak electric field $E_n = E_{max} / \left( \dfrac{P_{in} Z_3}{A_3 G_3} \right)^{1/2}$, using (13), and the normalized

losses $P_{l\,n} = P_l / \left( \dfrac{R_s}{Z_3} \right)$, using (14).

## DESIGN EXAMPLE OF AN OPTICALLY CONTROLLED X-BAND SWITCH

One of the applications of this switch is the high power pulse compression system of the Next Linear Collider (1). This system operates at 11.424GHz. We can construct the phase shifter and, hence the switch from a series of *four* three-port networks. The three-port network may be composed of a WR90 rectangular waveguide with a circular waveguide coupled to it from the broad side. A propagation of 200 MW in waveguide junctions having similar dimensions has been demonstrated (5). If the switch is to operate at a 100 MW level, the phase shifter need to handle only 50 MW.

The third arm, in this case, is composed of a circular waveguide carrying the fundamental mode $TE_{11}$. If the diameter of this waveguide is 2.54 cm, the peak field for a 50MW power level is 140 kV/cm. If the active element in these guides is a silicon wafer, which can be switched optically using a short pulse laser (2), the peak field need to be less than a 100 kV/cm at the wafer. Hence the normalized peak field need to be less than 0.714. If we assume $\phi$ to be 0; at $6 = 0.881$ the normalized peak field is 0.6, and the normalized losses is 0.914. Hence the peak electric field is 84 kV/cm. When the switch is *on* we assume a carrier density of about $10^{19}/cm^3$ which corresponds to a conductivity of 3.3 x $10^2$. Hence, the losses is 0.46% per element, i.e., a total of 230 kW is being wasted at the silicon wafer. The difficulty associated with the cooling system depends on the average power and the pulse length of the rf signal.

## CONCLUSION

We presented an abstract analysis and design methodology for a DTSP switch based on several distributed elements. We showed that such a switch, in principle, could be designed to handle a 100 MW at X-band.

## ACKNOWLEDGMENT

This work is supported by Department of Energy Contract DE-AC03-76SF00515.

The author wishes to thank Prof. Micheal Patelin and Prof. David Whittum for many useful discussions.

# REFERENCES

1 R. D. Ruth et. al., "The Next Linear Collider Test Accelerator," Proc. of the IEEE Particle Accelerator Conference, Washington DC, May 1993, pp. 543-545.

2 Sami G. Tantawi et. al. "Active High-Power RF Pulse Compression Using Optically Switched Resonant Delay Lines," IEEE Trans. Microwave Theory Tech., vol. MTT 45, No. 8, August 1997, pp 1486-1492.

3 A. L. Vikharev, et. al., "Distributed Switches of wave beams and compressors of microwave pulses," letters to JTF (Russia), Vol. 22, N0. 19, 1996.

4 R. E. Collin, "Foundations for Microwave Engineering," Second Edition, McGrow-Hill, Inc. 1992, Chapter 4, pp. 259-260.

5 Sami G. Tantawi et. al. "The Next Linear Collider Test Accelerator's RF pulse Compression and Transmission Systems," Advanced Accelerator Concepts workshop, Lake Tahoe, California, October 12-18, 1996, AIP proceedings 398, pp 805-812.

# A Multi-Moded RF Delay Line Distribution System for the Next Linear Collider

S. G. Tantawi[*], G. Bowden, Z.D. Farkas, J. Irwin, K. Ko, N. Kroll,
T. Lavine, Z. Li, R. Loewen, R. Miller, C. Nantista, R. D. Ruth, J. Rifkin,
A. E. Vlieks, P. B. Wilson, C. Adolphsen, and J. Wang

*Stanford Linear Accelerator Center, SLAC, 2575 Sand Hill Rd. Menlo Park, CA 94025, USA*

**Abstract.** The Delay Line Distribution System (DLDS) (1) is an alternative to conventional pulse compression which enhances the peak power of an rf source while matching the long pulse of that source to the shorter filling time of the accelerator structure. We present a variation on that scheme that combines the parallel delay lines of the system into one single line. The power of several sources is combined into a single waveguide delay line using a multi-mode launcher. The output mode of the launcher is determined by the phase coding of the input signals. The combined power is extracted using several mode extractors, each of which extracts only one single mode. Hence, the phase coding of the sources controls the output port of the combined power. The power is then fed to the local accelerator structures. We present a detailed design of such a system, including several implementation methods for the launchers, extractors, and ancillary high power rf components. The system is designed so that it can handle the 600 MW peak power required by the NLC design, while maintaining high efficiency.

## INTRODUCTION

During the past few years high power rf pulse compression systems have developed considerably. These systems provide a method for enhancing the peak power capability of high power rf sources. One important application is driving accelerator structures. In particular, future linear colliders, such as the proposed NLC, require peak rf powers that can not be generated by the current state-of-the-art microwave tubes. The SLED Pulse compression system (2) was implemented to enhance the performance of the two-mile linac at the Stanford Linear Accelerator Center (SLAC). One drawback of SLED is that it produces an exponentially decaying pulse. To produce a flat pulse and to improve the efficiency, the Binary Pulse Compression (BPC) system (3) was invented. The BPC system has the advantage of 100% intrinsic efficiency and a flat output pulse. Also, if one accepts some efficiency degradation, it can be driven by a single power source (4). However, The implementation of the BPC (5) requires a large assembly of over-moded waveguides, making it expensive and extremely large in size. The SLED II pulse compression system is a variation of SLED that gives a flat output pulse (6). The SLED II intrinsic

---

[*]*Also with the Electronics and Communications Department Cairo University, Giza, Egypt*

CP472, *Advanced Accelerator Concepts: Eighth Workshop,*
edited by W. Lawson, C. Bellamy, and D. Brosius
© 1999 The American Institute of Physics 1-56396-889-4/99/$15.00

efficiency is better than SLED, but not as good as BPC. However, from the compactness point of view SLED II is far superior to BPC. Several attempts have been made to improve its efficiency by turning it into an active system (7). However, the intrinsic efficiency of the active SLED-II system is still lower than that of the BPC. The DLDS is a similar system to BPC, which utilises the delay of the electron beam in the accelerator structure of the linear collider to reduce the length of the over-moded waveguide assembly. However it still uses more over-moded waveguide than that required by SLED-II. To further enhance the DLDS we introduce in this paper a variation on that system which further reduces the length of the waveguide system by multiplexing several low-loss rf modes in the same waveguide, hence the name Multi-moded DLDS (MDLDS). The system has an intrinsic efficiency of 100%, and the total over-moded waveguide length is less than that required by the compact SLED-II system.

## SYSTEM DESCRIPTION

Figure 1 shows a schematic of the proposed system. Basically, four pairs of klystrons, operating at 11.424 GHz, feed a multi-mode launcher. The launcher, then, injects one of four modes into a large (12.7cm-diamter) waveguide delay line. The choice of modes is controlled by the relative phases between the four fr power sources. The four modes are chosen to minimize the losses in the delay line. These modes are $TE_{01}$, vertically polarized $TE_{12}$, horizontally polarized $TE_{12}$, and, finally, $TE_{21}$. The $TE_{21}$ mode is quite lossy, and hence is extracted from the delay line immediately and is then converted to the $TE_{01}$ mode in the circular waveguide that feeds the closest set of accelerator structures. The power carried by any of the other three modes is extracted at the appropriate point and then converted into $TE_{01}$ mode to feed a set of accelerator structures.

The output pulse of the power sources is divided into four time bins, each with duration $\tau$. The total rf power supply pulse width is $4\tau$. During the first time bin, the phases are adjusted to inject one of the polarizations of the $TE_{12}$ mode. This signal does not get affected by any of the mode extractors. However, after the last mode extractor it gets converted into the $TE_{01}$ mode, thus, feeding the most distant accelerator structure. Then, in the second time bin the second polarisation of the $TE_{12}$ mode is injected. This signal is converted into the $TE_{01}$ mode just after the first extractor. The second extractor extracts this signal. In the third time bin the $TE_{01}$ mode is injected. The first extractor extracts that mode. Finally, in the forth time bin, the $TE_{12}$ mode is injected, and extracted immediately to feed the closest set of accelerator structures.

In this manner, each of the four accelerator structure sets will see the combined power from the four power sources during the appropriate time-bins. This is equivalent to a pulse compression system with a compression ratio of four. Since, in this scheme, the electron (or positron) beam is moving in an opposite direction to the rf power, the total delay line length required between the feed points to the different accelerator structure sets is $\sim c\,\tau/2$, where c is the speed of light in free space. For

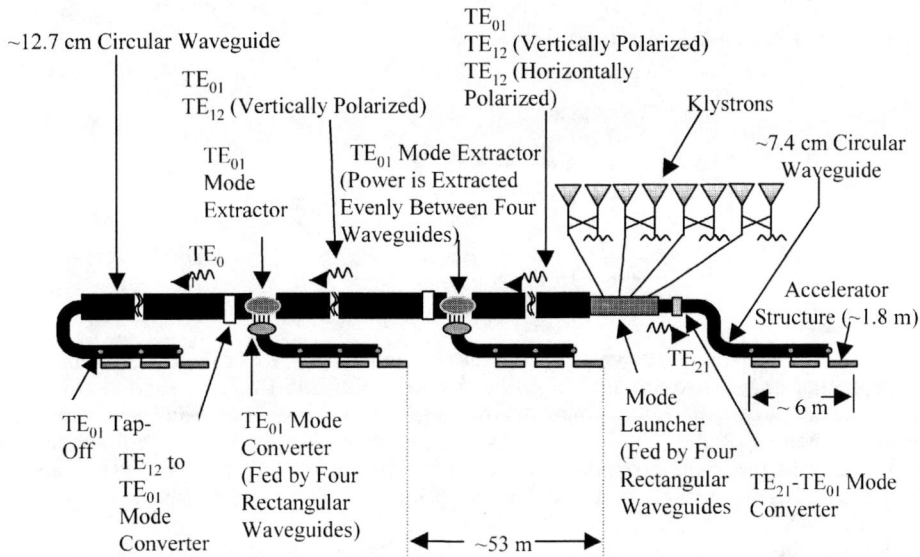

**FIGURE 1** Multi-Moded Delay Line Distribution System

simplicity, we have assumed here that the group velocity of the rf signal and the beam velocity are both approximately equal to c; a more detailed analysis is presented in the section 3.

Taking advantage of the finite time that electrons and positrons spend travelling between the accelerator structure sets reduces the total length of the waveguide required for this pulse compression system by a factor of two.

The power extracted from the rf delay line with the appropriate mode transducer is converted immediately into the $TE_{01}$ mode and fed to three different accelerator structures. The manipulation and feeding is done with circular waveguides that have a diameter of 7.4cm. Each accelerator structure is fed with a different tap-off, a mode transducer from TE01 to TE10 in rectangular waveguide. Obviously, the first tap-off is a 4.77 dB transducer, the second is a 3 dB one, and the third is a low-loss mode converter.

Each power source (a pair of klystrons) will produce 150 MW for 1.5 μs. The total amount of power in any time bin is 600 MW for a duration of 375 ns. One of the design goals is to keep the surface electric field in this rf system below 40 MV/m. The total efficiency of the system should be above 85%.

# TIMING

Because the rf power is being injected at different times into different modes that have different group velocities, one must pay a special attention to timing. The set of equation that need to be satisfied so that the each accelerator structure set get an rf pulse for a duration $\tau$ at the appropriate time are:

$$\tau = (\frac{L_1}{v_{TE01}} + \frac{L}{c}) + (\delta_2 - \delta_1),$$

$$\tau = (\frac{L_2}{v_{TE01}} + \frac{L}{c}) + (\delta_3 - \delta_2) + L_1(\frac{1}{v_{TE12}} - \frac{1}{v_{TE01}}), \tag{1}$$

$$\tau = (\frac{L_3}{v_{TE01}} + \frac{L}{c}) + (\delta_4 - \delta_3) + L_2(\frac{1}{v_{TE12}} - \frac{1}{v_{TE01}});$$

where $L$ is the distance between accelerator structure sets, $L_1$ is the distance between the launcher and first extractor, $L_2$ is the distance between first and second extractor, $L_3$ is the length of the delay line after the second extractor, $v_{TE01}$ and $v_{TE12}$ are the group velocities of the $TE_{01}$ and $TE_{12}$ modes respectively, and $\delta_1$ through $\delta_4$ are the delays due to the transmission of power from the main rf delay line system to the accelerator structure sets, i.e., the delay through and after the extractors.

There are several choices for the lengths $L$, $L_1$ through $L_3$, and $\delta_1$ through $\delta_4$ that satisfy the above set of equations. An attractive choice is to set $L_1$ through $L_3$ equal to $L$, $\delta_2 = \delta_3 = \delta_4$ and

$$\delta_2 - \delta_1 = L(\frac{1}{v_{TE12}} - \frac{1}{v_{TE01}}) \tag{2}$$

This would lead to a fairly symmetric system.

## LAUNCHER

Several ideas for the launcher have been proposed (8-9). In all of them a fundamental property of the launcher has been preserved: the launcher has only four inputs and the launcher has to launch four and only four modes. If this is preserved and the launcher is matched for all four different input conditions, because of unitarity and reciprocity the scattering matrix representing the launcher has to take the following form:

$$S = \begin{bmatrix} 0 & 0 & 0 & 0 & 1/2 & 1/2 & 1/2 & 1/2 \\ 0 & 0 & 0 & 0 & -1/2 & -1/2 & 1/2 & 1/2 \\ 0 & 0 & 0 & 0 & -1/2 & 1/2 & -1/2 & -1/2 \\ 0 & 0 & 0 & 0 & -1/2 & 1/2 & -1/2 & 1/2 \\ 1/2 & -1/2 & -1/2 & -1/2 & 0 & 0 & 0 & 0 \\ 1/2 & -1/2 & 1/2 & 1/2 & 0 & 0 & 0 & 0 \\ 1/2 & 1/2 & 1/2 & -1/2 & 0 & 0 & 0 & 0 \\ 1/2 & 1/2 & -1/2 & 1/2 & 0 & 0 & 0 & 0 \end{bmatrix} \tag{3}$$

This form forces the isolation between inputs; i.e., if one of the four power supplies drops out or fails, the rest of the power supplies will not receive any reflected power.

In all cases of launcher designs four rectangular waveguides are coupled to a circular waveguide at four different places $\pi / 2$ apart in azimuth the four waveguides supply equal amount of power with different phases, and the modes excited are $TE_{11}$, $TE_{21}$ and $TE_{01}$. The $TE_{11}$ modes are converted later to $TE_{12}$ modes using a Marie' mode converter. A circular waveguide large enough to support the $TE_{01}$ mode will support a set of TM modes and the $TE_{31}$ mode. To avoid exciting these modes, the launcher suggested in Ref. (8), perturbs the cross section of the circular guide to a cross like shape, thus allowing for only four modes to propagate. The launcher suggested in Ref. (9), uses longitudinal resonance coupling to avoid the excitation of other modes.

In all cases these launchers follow the scheme shown in Fig. 2 which consists of two parts: a $TE_{21}$ extractor and a $TE_{11}$-$TE_{01}$ launcher. The $TE_{21}$ extractor extracts the local $TE_{21}$ mode prior to the launching of the remote modes into the distribution waveguide. With the $TE_{21}$ extracted beforehand, the multi-mode launcher now only needs to launch the two polarisations of the $TE_{11}$ mode and the $TE_{01}$ mode. The $TE_{21}$ extractor has to be transparent to the modes with the $TE_{11}$ and $TE_{01}$ phase configurations, which can then bypass the $TE_{21}$ extractor, and be launched by the $TE_{11}$-$TE_{01}$ mode launcher into the cylindrical waveguide upstream. The $TE_{21}$ local mode extractor and the $TE_{11}$-$TE_{01}$ launcher in this launcher system are separate components that can be designed and tested separately. For detailed design procedures the reader is referred to (8-9).

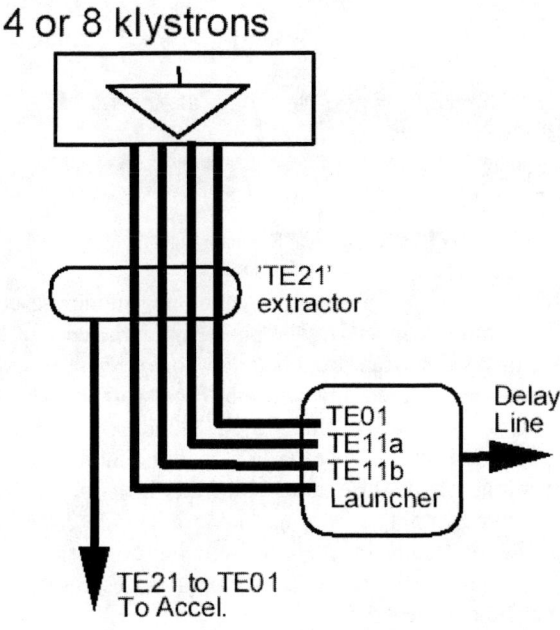

FIGURE 2 The Multi-mode Launcher System

# EXTRACTOR

The design of the $TE_{01}$ extractor is quite complicated and will be a subject of further publications. However, A design based on the wrap-around mode converter (10) is possible. In this designs a rectangular waveguide is warped around the circular guide. The power is being extracted using an azimuthal resonant coupling between the two guides. The design is shown in Fig. 3.

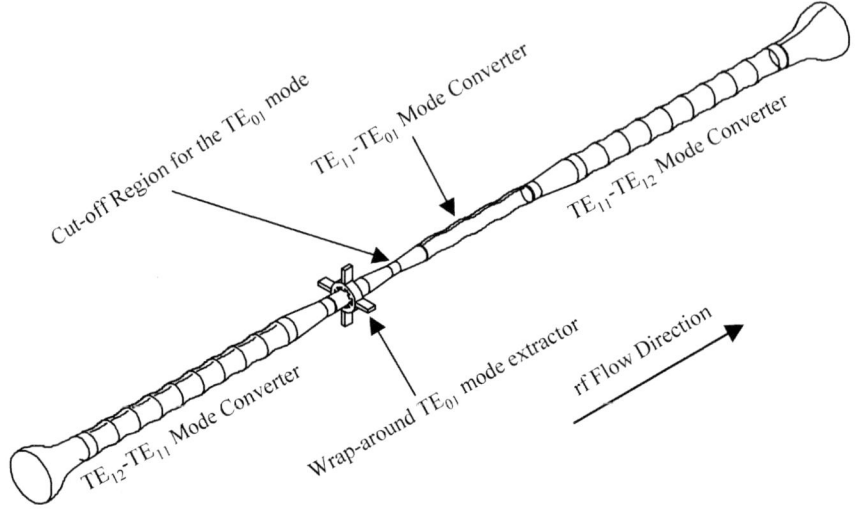

**FIGURE 3** Mode Extractor

First, we taper down so that both polarisations of the $TE_{12}$ mode are converted into $TE_{11}$ mode using a Marie' mode converter. The power is extracted into four different rectangular waveguides using the wrap-around mode converter. The exit end of the wrap-around mode converter is tapered down so that it cuts off the $TE_{01}$ mode, while allowing the $TE_{11}$ modes to go through, this way a 100% extraction of the $TE_{01}$ is possible. After the wrap-around mode converter, a serpentine mode converter converts only one of the polarisations of the $TE_{11}$ mode into the $TE_{01}$ mode. The remaining $TE_{12}$ polarisation is converted into $TE_{12}$ through a Marie' mode converter. The next extractor extracts The $TE_{01}$ and converts the only remaining polarisation of the $TE_{11}$ mode into $TE_{01}$ mode in a similar manner. Different approaches based on longitudinal resonant couplings are also possible.

After the signal is extracted into 4 different rectangular waveguides it is injected again into a circular waveguide. This is shown in Fig. 4. The 4 rectangular waveguides injects 4 signals that are equal in amplitude and phase. If the circular guide is small enough in diameter to cut-off the $TE_{14}$ mode, the only mode that can get excited, under this azimuthal symmetry condition is the $TE_{01}$ mode. Matching is accomplished by adjusting the rectangular waveguide size and by the correct positioning of the back wall, which acts as a short circuit at one side of the circular guide.

To taper up the circular waveguide diameter to a value large enough to get into the low loss regime of the $TE_{01}$ mode, we use a compact step-up structure. The diameter of the waveguide is increased in two steps. Although, the final diameter can support the $TE_{02}$ mode the structure does not excite that mode and at the same time have an excellent match at the operating frequency of the system. This type of taper is possible due to the limited frequency band requirements of the system.

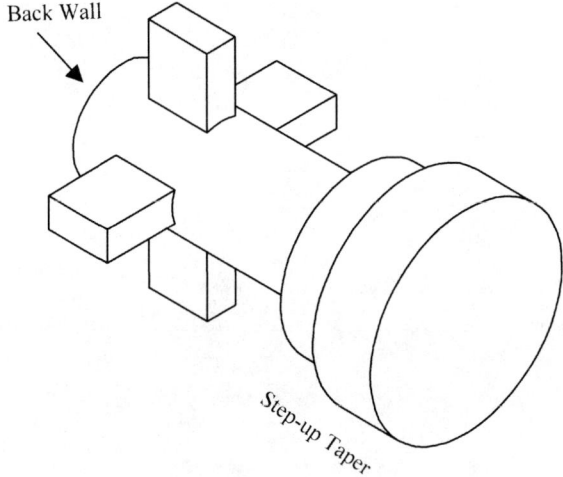

**FIGURE 4** $TE_{01}$ mode injector.

## ACKNOWLEDGMENT

This work is supported by Department of Energy Contract DE-AC03-76SF00515.

## REFERENCES

1. H. Mizuno, Y. Otake, "A New Rf Power Distribution System For X Band Linac Equivalent To An Rf Pulse Compression Scheme Of Factor 2**N," 17th International Linac Conference (LINAC94), Tsukuba, Japan, Aug 21 - 26, 1994
2. Z. D. Farkas et. al., "SLED: A Method of Doubling SLAC's Energy," Proc. of the 9th Int Conf. on High Energy Accelerators, 1976, p. 576.

3. Z. D. Farkas, "Binary Peak Power Multiplier and its Application to Linear Accelerator Design," IEEE Trans. MTT-34, 1986, pp. 1036-1043.

4. T. L. Lavine et. al., "High-Power Radio-Frequency Binary Pulse Compression Experiment at SLAC," Proceedings of the IEEE Particle Accelerator Conference, San Francisco,1991, pp. 652-654.

5. Z. D. Farkas, et . al., "Two-Klystron Binary Pulse Compression at SLAC," Proc. of the IEEE Particle Accelerator Conference, Washington DC, May 1993, p. 1208.

6. P. B. Wilson, Z. D. Farkas, and R. D. Ruth, "SLED II: A New Method of RF Pulse Compression," Linear Accl. Conf., Albuquerque, NM, September 1990; SLAC-PUB-5330

7. S. G. Tantawi, et. al, "Active RF Pulse Compression Using Switched Resonant Delay Lines," Nuc. Inst. and Meth, A, Vol. 370 (1996), pp. 297-302; SLAC-PUB 6748.

8. K. Eppley, Z. Li, R. Miller, C. Nantista, S. Tantawi, N. Kroll, "A Four Port Launcher For A Multimode DLDS Power Delivery System," these proceedings.

9. Zenghai Li, Sami Tantawi, Kwok Ko, "Mode Launcher Design for the Multi-moded DLDS," these proceedings.

10. S. G. Tantawi, "The Wrap-around mode converter: a compact TE01-TE10 Transducer," to be published.

# Active Microwave Pulse Compressors Employing Oversized Resonators and Distributed Plasma Switches

A. L. Vikharev,[1] A. M. Gorbachev,[1] O. A. Ivanov,[1] V. A. Isaev,[1]
S. V. Kusikov,[1] A. L. Kolysko,[1] A. G. Litvak,[1] M. I. Petelin[1]
and J. L. Hirshfield[2]

[1]*Institute of Applied Physics, Nizhny Novgorod, 603600 Russia*
[2]*Omega-P, Inc. 202008 Yale Station, New Haven, CT 06520-2008; and
Department of Physics, Yale University, POB 8120, New Haven, CT 06520-8120*

**Abstract.** Principles and preliminary design considerations for microwave pulse compressors using electrically-switched Bragg reflectors were presented by Petelin, Vikharev and Hirshfield at AAC'96 (1). This paper presents results of experiments to study performance of several proto-type designs of such compressors. In moderate-power tests using a plasma-switched $TE_{011}$ cavity as the active element, a 25 nsec, 1.8 MW output pulse was realized, corresponding to a peak power gain of 20:1, with an energy efficiency of 30%. A high-power prototype of a 60% energy efficient compressor is under construction that is designed to yield a 100 nsec, 100 MW output pulse with a peak power gain of 17:1. Details of this design are described.

## INTRODUCTION

During the past several years, attention has been directed towards active pulse compression as a means of achieving high peak power microwave pulses with an acceptable efficiency for driving accelerator structures (1,2). Active pulse compression, as opposed to passive pulse compression, employs rapid electrical switching of the reflectivity of a microwave circuit element, so as to allow stored microwave energy to be rapidly discharged into the accelerator structure. For rectangular pulses, the energy efficiency of a compressor is $\eta = (P_2/P_1)(\tau_2/\tau_1) = M/C$, where the power gain $M = P_2/P_1$ is the ratio of peak power output to peak power input, and the compression ratio $C = \tau_1/\tau_2$ is the ratio of input pulse width to output pulse width. Omega-P, Inc. and Institute of Applied Physics are collaborating in the design and testing of active pulse compressors at 11.424 GHz, in which microwave energy is stored in an oversized resonator operating in the $TE_{0n}$ mode, with Bragg reflectors or cavities at the ends, and with active switching achieved by rapid ignition of plasma in tubes within the Bragg reflector or

CP472, *Advanced Accelerator Concepts: Eighth Workshop,*
edited by W. Lawson, C. Bellamy, and D. Brosius

cavity. This paper provides a status report, giving results of preliminary moderate-power tests, design considerations for high power versions, and a specific design for a compressor now under construction which is expected to provide an output pulse of at least 100 MW peak power, in a 100 nsec pulse, with a peak power gain of 17:1, and an efficiency of at least 60%.

## PROOF-OF-PRINCIPLE EXPERIMENT

Preliminary evaluation of the idea of rapid electric discharge switching of microwave energy stored in an oversized $TE_{01}$-mode resonator was carried out at moderate-power levels using the experimental set-up shown in Fig. 1. The resonator was formed by an input Bragg reflector, a section of smooth copper cylindrical waveguide 100 cm long and 5.0 cm in diameter, and an output reflector (2) based on a stepped widening section of circular waveguide. This stepped widening section comprised a cylindrical $TE_{031}$-mode resonator 13.7 cm in diameter containing a quartz ring-shaped gas-discharge tube near its front wall. The quartz discharge tube is 1.1 cm

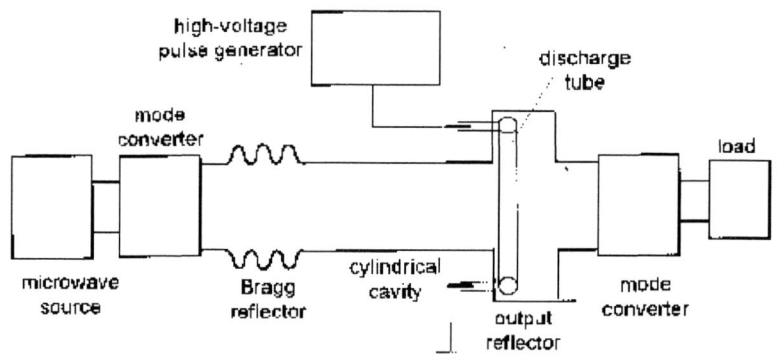

**FIGURE 1.** Schematic diagram of the experimental setup.

in minor diameter, and 12.5 cm in major diameter. The discharge tube is built with two nozzles for passage of the filling gas which, for the experiments to be described, was air in the pressure range 0.2-10 Torr. The tube was fired by applying a 40 nsec, 40 kV pulse with a rise-time of 10-15 nsec to a pair of electrodes sealed at opposite ends of the quartz ring. A detailed drawing of the cavity is shown in Fig. 2.

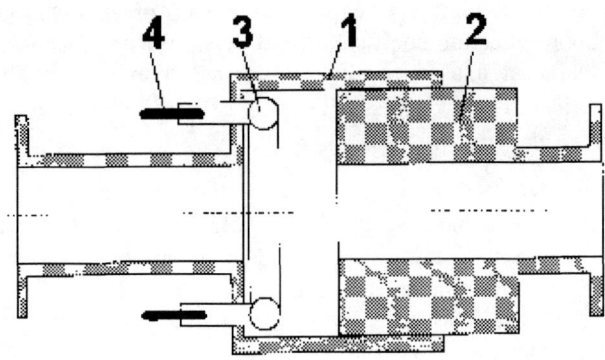

**FIGURE 2.** Output reflector based on stepwise widening of a circular waveguide: 1 - waveguide of 13.7 cm diameter, 2 - movable waveguide of 50 mm diameter, 3 - gas discharge tube, 4 - electrodes.

The moderate-power experiments were carried out at 9.4 GHz, the frequency of an available 100-150 kW magnetron. The spectral width of the magnetron is about 1-1.3 MHz, while that of the energy storage cavity is 0.6 MHz, corresponding to a $Q$ of $1.5 \times 10^4$. Thus only a portion of the magnetron ouput energy could be expected to be taken in by the cavity. The magnetron pulse width was adjustable in the range 1-10 μsec. The magnetron power was propagated in $TE_{01}$-mode rectangular waveguide, and converted to $TE_{01}$-mode circular waveguide using a Marie converter. Adjustment to resonance for the cavity was achieved by varying its length to minimize reflected energy after the initial transient. At minimum reflection, the cavity took up 70% of the incident power, a value consistent with the overly-broad spectrum of the magnetron source.

When high voltage pulses were fed to the electrodes on the discharge tube and plasma built up in the tube, the eigenfrequency of the cavity mode changed, and the transmission coefficient grew significantly. As a result, energy stored in the cavity was coupled out. The detected output shown in Fig. 3 has a peak power value of 1.8 MW, a factor-of-20 higher than the 90 kW input power level. The full-width of the output pulse at half power is 25 nsec, a factor of 80 smaller than the 2 μsec input pulse. Thus $M = P_2/P_1 = 20$, and $C = \tau_1/\tau_2 = 80$. But since the output pulse is not rectangular, integration over time is used to find the output pulse energy, giving $\eta = 30\%$. Higher efficiency can be expected when a source of better spectral purity is used, and when an energy storage cavity of higher $Q$ is used.

## ADVANCED COMPRESSOR DESIGNS

Several designs have been studied for active high-compression Bragg pulse compressors that can achieve efficiency values suitable for accelerator use ($\eta > 60\%$). The energy storage cavity in such a device must be capable of sustaining high electric fields without breakdown, the switching discharge tubes must be located in positions where the fields are not so high as to cause spontaneous breakdown, but the influence of the discharge upon reflection coefficient must be significant once the tubes are fired externally. One such design, as shown in Fig. 4, is based on use of a $TE_{02}$ mode in the energy storage cavity. The compressor consists of mode converters $TE_{01}$ (rect.) to $TE_{01}{}^{o}$ and $TE_{01}{}^{o}$ to $TE_{02}{}^{o}$ connected together via smooth tapered transitions; and the $TE_{02}{}^{o}$ resonator formed by two Bragg reflectors and a section of smooth cylindrical waveguide. The average diameter of the Bragg reflectors, equal to the diameter of the smooth cylinder, is chosen so that the $TE_{02}$ mode is the highest propagating axially-symmetric mode. This minimizes the overall length of the resonator required to achieve a given $Q$, and avoids issues with excitation of higher radial order modes. For 11.424 GHz, this diameter is 8.0 cm, for which a power reflection coefficient of 0.9990 can be obtained using a reflector about 40 periods in length. The calculated frequency characteristic of such a reflector is shown in Fig. 5.

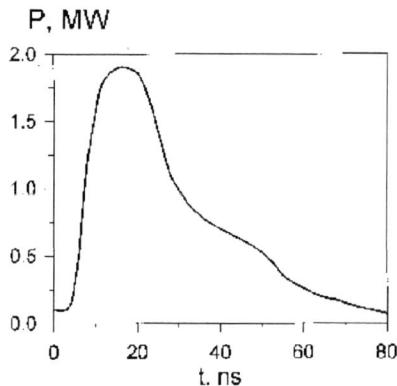

**FIGURE 3.** Envelope of compressed microwave pulse.

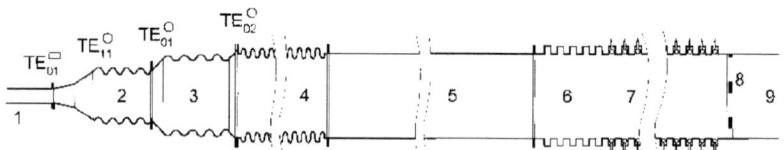

**FIGURE 4.** Schematic diagram of the Active Bragg compressor: 1 - input waveguide, 2 - $TE_{01}$ mode converter, 3 - $TE_{02}$ mode converter, 4 - input Bragg reflector, 5 - storage cavity, 6,7 - output Bragg reflector (composed of 6 - passive section, 7 - section with electrically controlled distributed gas switches), 8 - output window, 9 - output waveguide.

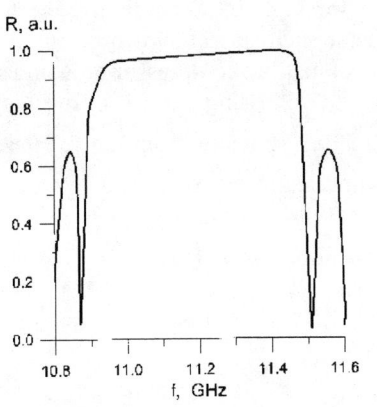

R, a.u.

f, GHz

**FIGURE 5.** Calculated frequency characteristic of the $TE_{02}$ mode reflector.

An output pulse length of 100 nsec is obtained with this resonator if its length is 150 cm. Power is switched out when the output Bragg reflector is made partially transmitting by firing the discharge tubes. For this, no tubes are installed in the first five corrugations. As a results, the coefficient of power reflection of the output reflector can be lowered to 0.65. Properties of this compressor with an input Bragg reflector 29 periods in length are summarized in Table I.

| | |
|---|---|
| duration of input pulse | 1.50 μsec |
| duration of output pulse | 0.10 μsec |
| power gain $M = P_2/P_1$ | 12 |
| cavity ohmic $Q$ | 110,000 |
| cavity diffraction $Q$ | 40,000 |
| energy efficiency $\eta$ | 60%. |

**TABLE I.** Parameters of active Bragg compressor shown in Fig. 4.

A schematic diagram of the section of Bragg reflector containing discharge tubes is shown in Fig. 6. The actual reflector consists of 20 sections containing discharge tubes and 5 sections without discharge tubes.

Reflectors such as those shown in Figs. 2 and 6 are adequate for moderate-power use, but not for output pulse power levels of 100 MW and above, as needed for accelerator applications. High power counterparts of these must be designed and built with the waveguide under high vacuum, in order to sustain the high rf fields strengths without breakdown. Vacuum-compatible designs are obviously more complex, since the discharge tube electrodes and filling-gas nozzles must be accessed from the outside. The complications involved in designing and building a vacuum-compatible version of

979

Fig. 2 are fewer than those for Fig. 6. Accordingly, the first high-power unit to be constructed will use a reflector such as that shown in Fig. 2. An engineering drawing of the high-power, vacuum-compatible compressor now under construction is shown in Fig. 7. This device contains the following elements, from left-to-right: $TE_{01}$ (rect.) to $TE_{11}^O$ transition; $TE_{11}^O$ to $TE_{01}^O$ transition; up-taper; mode filter and pumping port; $TE_{01}^O$ Bragg reflector; cylindrical waveguide; down-taper; $TE_{02}^O$ switching cavity with quartz ring discharge tube (as in Fig. 2); mode filter and pumping port; down-taper; $TE_{01}^O$ to $TE_{11}^O$ transition; and $TE_{11}^O$ to $TE_{01}$ (rect.) transition. Design parameters for this device are given in Table II. Plans are to test this unit at input power levels up to about 20 MW in early 1999 using the Omega-P/Naval Research Laboratory 11.424 GHz magnicon.

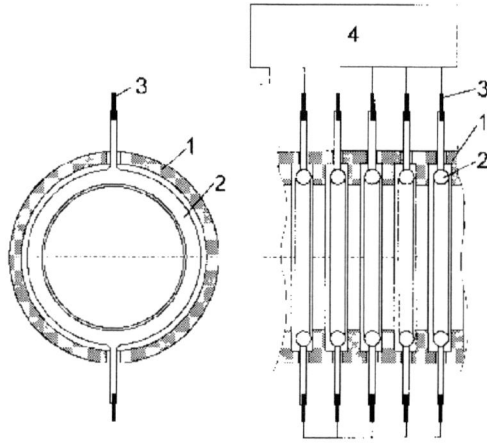

**FIGURE 6.** Schematic diagram of portion of electrically controlled Bragg reflector containing ring-shaped discharge tubes in the Bragg grooves.

| | |
|---|---|
| input pulse width | 2.10 μsec |
| output pulse width | 0.10 μsec |
| peak power gain | 17:1 |
| energy efficiency | 60% |
| ohmic cavity $Q$ | 150,000 |
| loaded cavity $Q$ | 56,000 |

**TABLE II.** Design parameters for active pulse compressor shown in Fig. 7. Loaded $Q$ value is for an input Bragg reflector with a power reflection coefficient of 98.5%.

## DISCUSSION

Practical designs of active Bragg pulse compressors have been described, including versions that have been tested at moderate power levels, and a version that is now under construction for 11.424 GHz testing at output power levels of 100 MW and

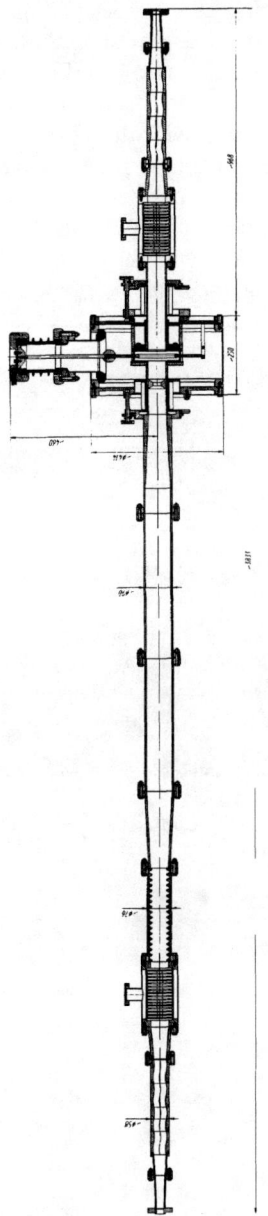

**FIGURE 7.** Active pulse compressor now under construction. See text for details.

above. Measurements at moderate input power levels up to 90 kW have shown that peak power output pulses of 1.8 MW (i.e., 20:1 peak power gain) can be produced using an active Bragg compressor comprising an energy storage cavity with a short tuned cavity as one reflector. Rapid firing of a ring-shaped discharge tube in the tuned cavity reduces its reflectivity and allows release of stored energy in a non-rectangular pulse with a full-width at half-maximum of 25 nsec (see Fig. 3). The measured energy efficiency in this test was 30%, a value consistent with the overly broad spectral width of the magnetron used as the rf source, and with the $Q$ of the energy storage cavity. Design has been completed and construction is in progress on a high-power vacuum-compatible active Bragg pulse compressor using a short tuned cavity as one reflector. This device is intended to produce an output pulse 100 nsec wide of at least 100 MW peak power with a power gain of 17:1, and with an energy efficiency of 60%. Tests using the Omega-P/NRL 11.424 GHz magnicon as the rf source are anticipated in early 1999.

## ACKNOWLEDGMENT

This work was supported by the US Department of Energy, Division of High Energy Physics.

## REFERENCES

1. Petelin, M. I., Vikharev, A. L., and Hirshfield, J. L. "Pulse Compressor Based on Electrically Switched Bragg Reflectors," in *Advanced Accelerator Concepts - Seventh Workshop, Lake Tahoe, CA 1996.* AIP Conf. Proc. **398**, pp. 822-831, 1997, Swapan Chattopadhyay, ed. (Am. Inst. Physics, Woodbury, NY, 1997).

2. Tantawi, S. G., Ruth, R. D., Vlieks, A. E., and Zolotorev, M., "Active High Power RF Pulse Compression Using Optically Switched Resonant Delay Lines," in *Advanced Accelerator Concepts - Seventh Workshop, Lake Tahoe, CA 1996.* AIP Conf. Proc. **398**, pp. 813-821, 1997, Swapan Chattopadhyay, ed. (Am. Inst. Physics, Woodbury, NY, 1997).

# Progress on the Relativistic Klystron Two-Beam Accelerator Prototype

G. A. Westenskow*, D. E. Anderson†, S. Eylon†, E. Henestroza†,
T. L. Houck*, S. M. Lidia†, D. L. Vanecek†, and S. S. Yu†

*Lawrence Livermore National Laboratory, Livermore, CA 94551
†Lawrence Berkeley National Laboratory, Berkeley, CA 94720

**Abstract.** The technical challenge for making two-beam accelerators into realizable power sources lies in the creation of the drive beam and in its propagation over long distances through multiple extraction sections. This year we have been constructing a 1.2-kA, 1-MeV, electron induction prototype injector as a collaborative effort between LBL and LLNL. The electron source will be a 3.5"-diameter, thermionic, flat-surface cathode with a maximum shroud field stress of approximately 165 kV/cm. Additional design parameters for the injector include a pulse length of over 120-ns flat top (1% energy variation), and a normalized edge emittance of less than 200 $\pi$-mm-mr. Planned diagnostics include an isolated cathode with resistive divider for direct measurement of current emission, resistive-wall and magnetic probe current monitors for measuring beam current and centroid position, capacitive probes for measuring A-K gap voltage, an energy spectrometer, and a pepper-pot emittance diagnostic. Details of the injector, beam line, and diagnostics are presented.

## INTRODUCTION

Induction accelerators are a unique source for high-current, high-brightness, electron beams. A collaboration between the Lawrence Livermore National Laboratory (LLNL) and Lawrence Berkeley National Laboratory (LBNL) has been studying rf power sources based on the Relativistic Klystron Two-Beam Accelerator (RK-TBA) concept for several years (1, 2). A major technical challenge to the successful operation of a full scale RK-TBA is the transport of the electron beam through several hundred meters of narrow aperture microwave extraction structures and induction accelerator cells. Demanding beam parameters are required of the electron source, an induction injector, to achieve the transport goals. A test facility, called the RTA, has been established at LBNL (3) to verify the analysis used in the design study. The primary effort of the facility is the construction of a prototype RK-TBA subunit that will permit the study of technical issues, system efficiencies, and costing. In this paper, we will discuss the development of the RTA electron source and it's pulsed power system, which has recently been constructed and is now undergoing testing.

CP472, *Advanced Accelerator Concepts: Eighth Workshop,*
edited by W. Lawson, C. Bellamy, and D. Brosius

# RTA GUN

A major part of our effort during the past year has been towards the design of a low emittance electron source for RTA accelerator. We expect to produce an electron source with a much lower emittance than typical induction guns. The electron source will be a 3.5"-diameter, thermionic, flat-surface W-type cathode with a maximum shroud field stress of approximately 165 kV/cm. An emission density of 20 A/cm$^2$ is required from the cathode to produce 1.2 kA beam. The RTA gun, depicted in Figure 1, has 72 induction cores, each driven at 14 kV. The voltage across the A-K gap is 1 MV. The cores are segmented radially to reduce the individual aspect ($\Delta r/\Delta z$) ratio. The lower aspect ratio reduces the variation in core impedance during the voltage pulse simplifying the pulse forming network (PFN) design. Figure 2 is a photograph of the gun undergoing initial pulsed power tests.

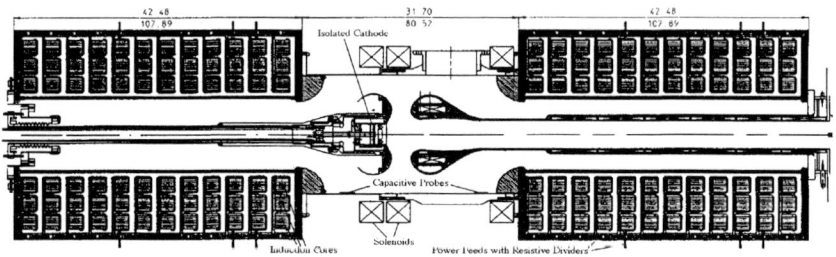

**Figure 1.** Depiction of the RTA electron gun.

**Figure 2.** RTA 1-MV, 1-kA electron gun with pulse power system.

We have done high-voltage tests on the gun. In operation a 500 kV potential is developed across the two 30-cm-ID PYREX (4) insulators producing a 5.1 kV/cm average gradient along the insulator. The maximum fields at the triple points, the intersection of insulator, vacuum, and metal, is less than 3.5 kV/cm. Maximum surface field in the cathode half of the gun electrode is about 85 kV/cm. The maximum field is about 116 kV/cm on the anode stalk.

Initial beam focusing in the gun is accomplished by three large-bore air-core solenoids installed on the central pumping spool. The first solenoid is operated to null the magnetic field from the other solenoids at the cathode front surface. There are seven smaller solenoids located within the anode stalk to provide additional focusing to transport the beam to the end of the gun.

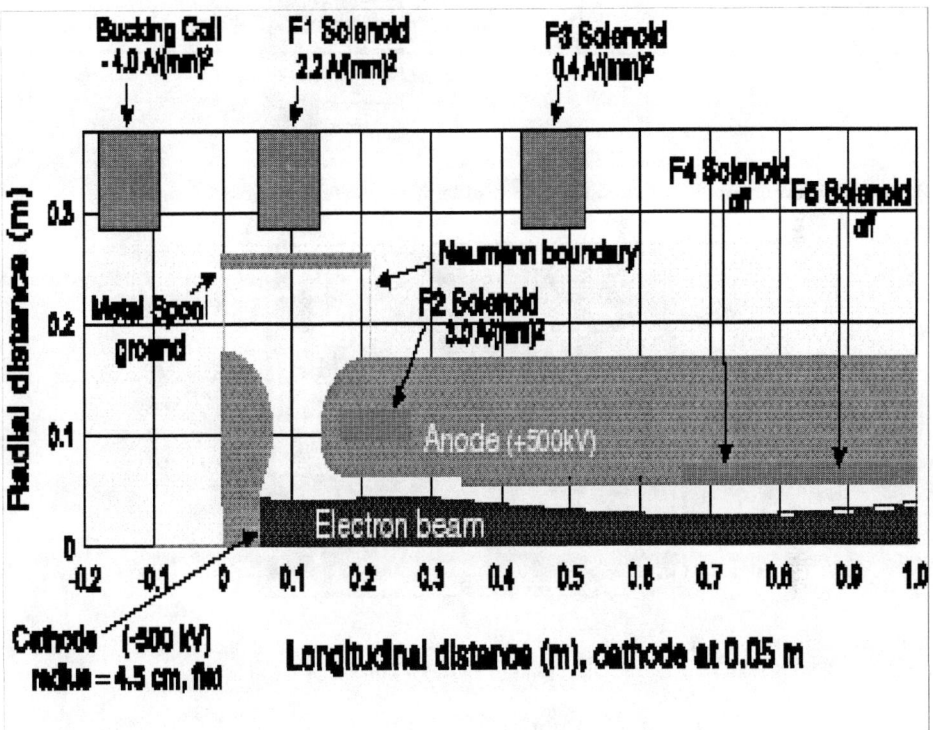

**Figure 3.** EGUN simulation of beam transport near the cathode of the RTA Gun.

## PULSED POWER SYSTEM FOR THE RTA GUN

The pulsed power system for the gun consists of a 20-kV High-Voltage Power Supply, 6-kJ Energy Storage Bank, two Command Resonant Charging (CRC) Chassis, 24 Switched Pulse Forming Networks, and four Induction Core Reset Pulsers, half of which is shown in Figure 4. Each PFN will drive a single 3-core induction cell of the gun. A sample pulse for a single cell with a 40 Ω resistor simulating beam loading is shown in Figure 5.

Segmenting the core in the induction cell and driving the individual core segments avoids a high-voltage step-up transformer. This reduces the developmental effort needed to achieve a "good" flattop pulse (minimal energy variation) with fast risetime and improves the efficiency of the overall pulsed power system. Our system of low-voltage PFNs driving multiple core induction cells is similar to the system envisioned for the extraction section in the full scale RK-TBA design. For the gun core material,

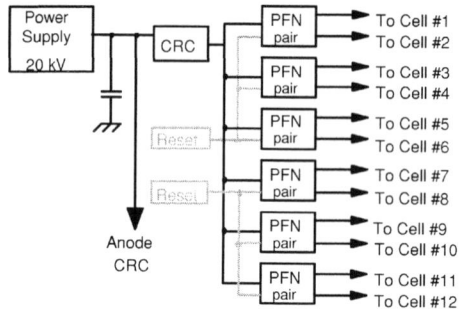

**Figure 4.** Block diagram of the pulsed power system for the cathode half of the RTA gun.

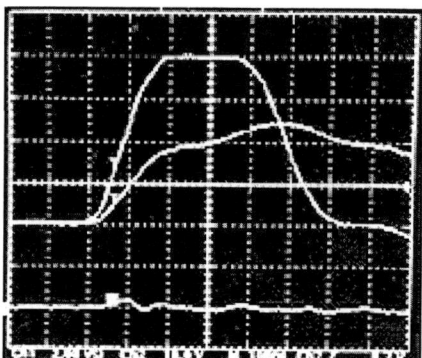

**Figure 5.** Oscilloscope trace of pulsed gap voltage applied to a single induction cell. Top trace is voltage (10 kV/div) and middle trace is current (4 kA/div). Time scale : 100 ns/div.

we choose 20-μm-thick 2605SC METGLAS. For the RTA extraction section we will use a lower loss 2714AS METGLAS for the induction cores.

Design of the switched PFNs follows easily from published METGLAS core loss data (5). For the RTA induction cells, a flux swing of 2.6 T in 400 ns (FWHM) results in a magnetization rate of 6.5 T/μs. At this rate, a loss density of 1800 J/m$^3$ translates into 30 J lost in a cell with $16.7 \times 10^3$ cm$^3$ of 2605SC METGLAS. For a cell input voltage of 14 kV applied for 400 ns, these losses require that 5900 A be supplied to the three radial cores. An additional 3600 A is required to supply beam current (1200 A x 3 cores/cell), resulting is a total current of 9 kA. The required drive impedance for a cell is then 1.5 Ω, which is provided by the PFN module shown in Figure 4.

986

An area of concern was the variation of the energy loss for different METGLAS cores. Experience (6) with the 72 cores in the gun leads us to believe that for a large relativistic klystron, matching cell cores should permit acceptable energy loss variation.

## GUN VOLTAGE WAVEFORM

Achieving the fast risetime necessary to minimize the volt-seconds required for the injector cores presented a challenge (see Figure 5). Budget constraints coupled with the large availability of EEV CX1538 thyratrons from the ATA program at LLNL made these tubes an attractive option. However, their poor time rate of current change (4 kA/µs rating) made them questionable for this application, which requires about 40 kA/µs. A variety of techniques were tried to decrease the risetime. In a 1.5 Ω system, stray circuit inductances must be maintained at or below 100 nH to achieve a 10-90% risetime of 150 ns. This was accomplished by placing the thyratron between two current sheets connecting the PFN output to the output cables. The ionization time of the thyratron was substantially reduced by applying a 1-2 A pre-pulse to the keep-alive grid 300-400 ns prior to the arrival of the main control-grid pulse. Faster risetimes were achieved with Triton F-130 ceramic thyratrons. An upgrade of the current thyratrons in the gun pulsed power system should allow us to achieve the design 100-ns risetime.

At the 1-MV, 1.2-kA operating conditions we hope to produce a ±1% gun voltage flat waveform for 120 ns. We will need to adjust the number of turns in appropriate sections of individual PFNs to achieve this goal. Insertion of ferrite material in the center of the inductors coils will allow additional small corrections to the waveform. There are 24 EEV CX1538 thyratrons used in the pulsed power system for the gun. Since the turn-on time is voltage dependent for each of the tubes, we are adjusting the trigger time of each individual tube.

## DIAGNOSTICS FOR THE RTA GUN

A variety of diagnostics will be used to determine the performance of the gun, both permanently installed monitors for general operations and temporary diagnostics specific to the injector commissioning and troubleshooting. The majority of the diagnostics will be installed after the gun. The first 1.4 m of beam line after the gun will include two beam position and current monitors to allow the offset and angle of the beam at the exit of the injector to be measured. A pop-in probe will be incorporated in a pumping port to allow the beam profile to be viewed. We plan to use a pepperpot emittance diagnostic to help characterize the beam.

### Current Measurements

An accurate measurement of the emitted current from the cathode is required both for determining the performance of the gun and benchmarking codes. We have electrically isolated the cathode from the stalk, forcing the current to flow through several parallel, 0.25 inch wide, strips of 25-µm thick nichrome foil that act as current-viewing resistors. The potential drop across the foil is measured and the current inferred. To improve the time response a parallel-strip shunt geometry is used where the foil is folded on itself to increase the resistance while lowering the series inductance.

Two different methods (7) will be used to determine the beam current and centroid position in the region between the gun and the electron dump. Magnetic pickup (B-dot) loops will determine the time derivative of the current pulse. The voltage induced on the loop can then be integrated to recover the current. The second method uses resistive wall current monitors to measure the potential drop of the return wall current across a known resistance generating a signal proportional to the current. Our monitors will consist of 25-μm thick nichrome foil that spans a short insulated break around the circumference of the beam tube wall and have a total resistance of a few milliohms. The slow decay in the output signal for this diagnostic can be corrected through compensation circuits, either passive or active. The resistive wall current monitors have the advantage of producing a signal proportional to the current. However, the frequency response is less than that of B-dot loops.

The beam dump will be electrically isolated through nichrome film (current-viewing resistors) similar to the cathode to allow for the measurement of the total current deposited in the dump..

The errors inherent in the above monitors for off-axis beams are easily calculated and not expected to be an issue. The major concern is in the calibration of the diagnostics. The nominal tolerance for commercial components used in the monitors would limit accuracy to about 5%. By calibrating complete systems, i.e. the individual diagnostic with cables, integrators, and compensation circuits, as applicable, we hope to achieve accuracies approaching 1%.

## Current Density Profile

The current density profile will be measured using Cherenkov and/or optical transition radiation from intercepting foils. A primary concern with using foils is possible damage from beam energy deposition. Average heating of the foil can be controlled by adjusting the repetition rate of the injector. The difficulty is the single shot heating where material can be melted and ejected before the heat is conducted away. If the foils are sufficiently thin that the radiation generated by bremsstrahlung escapes, and assuming the beam pulses are short compared to the thermal conduction time, the temperature rise in the foil is approximately:

$$\Delta T = \frac{J \, \Delta t}{\rho \, C_p \, e} \left[ \frac{dT_e}{dz} \right]_c$$

(1)

where $J$ is the current density, $\Delta t$ is the pulse length, $\rho$ and $C_p$ are the density and specific heat of the foil, respectively, and $[dT_e/dz]_c$ is the collisional stopping power for the electrons. For a given foil material, the minimum beam radius at intercept can be determined from equation (1). The collisional stopping power is reasonably constant with energy for relativistic electrons. To avoid damage for a thin quartz foil, the beam diameter must be larger than 2 cm for a 1-kA, 300-ns, relativistic electron beam. We anticipate using a number of foil materials, including kapton, quartz, graphite, tantalum, and tungsten.

The light generated at the beam/foil interaction will be recorded using both gated and streak cameras. The streak camera will be used principally to determine if the properties of the foil and/or beam change during the pulse. The significant levels of energy deposited in the foil could affect the dielectric constant or generate a surface plasma that could be confused as a variation in beam parameters.

## A-K Voltage and Beam Energy

Three different methods will be used to determine the A-K voltage and beam energy. The first method involves measuring the applied voltage to the induction cores at the connection of the power feeds to the induction cells. Capacitive $dV/dt$ pickup probes (7) are used for a more direct measurement of the A-K gap voltage and also to provide greater bandwidth with respect to the resistive dividers. We also hope to employ a conventional energy spectrometer comprised of an on-axis collimator, dipole magnet, scintillator, and viewing port to directly measure the beam energy.

## Emittance Measurement

Measuring the beam emittance is expected to be very difficult as the beam is highly space charged dominated. A pepperpot emittance diagnostic is being constructed. The effect of space charge can be appreciated by considering the envelope equation for a round, uniform beam in a drift region:

$$\frac{d^2 R}{dz^2} = \frac{\varepsilon^2}{R^3} + \frac{2 I}{R I_A (\beta\gamma)^3},$$

(2)

where R is beam radius (edge), $\varepsilon$ is the unnormalized edge emittance, and $I_A = 17$ kA. The cathode is in a magnetic field free region. We desire for the emittance to dominate space charge effects for the beamlets in the region following the pepperpot, i.e.

$$\frac{\varepsilon_n^2}{R_b^2} \gg \frac{2 I}{\beta\gamma I_A}, \text{ or } \quad \varepsilon_n^2 \gg \frac{I_b h^2}{2 \beta\gamma I_A}, \text{ where } \quad I = \left(\frac{h}{2 R_b}\right)^2 I_b.$$

(3)

$R_b$ and $I_b$ are the beam radius and current at the front of the pepperpot, and h is the aperture. The size of the aperture is the only variable for adjusting the relative contribution of emittance to space charge. For the designed RTA injector beam parameters, 1.2 kA, 1 MeV and 100 $\pi$-mm-mr, and using a 250-$\mu$m aperture, the emittance term is approximately an order of magnitude larger than the space charge. Our aperture plate will consist of a rectangular pattern of 121 (11x11) 250-$\mu$m apertures with 7 mm spacing on a 500-$\mu$m thick tungsten plate. The tungsten plate represents about two range thickness for 1-MeV electrons. A second 1-mm thick tungsten plate with nominal 750-$\mu$m apertures will be placed in front of the thinner plate to improve thermal conduction. At four locations, apertures will not be drilled in the thicker plate to assist in determining background and orientation. The beamlets will strike a quartz foil located 80 cm (adjustable from 40 to 160 cm) after the aperture plate. The Cherenkov light generated at the foil will be imaged with a gated camera.

Small effects such as the aperture plate thickness (vignetting), fringe fields from the upstream solenoids, and beam waist location will be accounted for in the data analysis. A more serious problem concerns the effect of the conductive aperture plate on the beam. It has been demonstrated (8) that, while the local distribution in phase space can be determined, the global x–x' curve is dominated by the non-linear focusing of the aperture plate. E-Gun simulations performed for our beam parameters indicate that the beam emittance determined from the pepperpot data will be as much as six times the actual emittance in the absence of the aperture plate. This effect can be accounted for in the analysis. However, the accuracy in the final emittance value will suffer. Note that this is only an issue for space charge dominated beams where aperturing is required.

An alternative to the pepperpot diagnostic is to vary the focusing of the beam and look at changes in the radial profile of the beam. The most straight forward method will be to insert a nonconducting foil a short distance after a solenoid. The strength of the solenoid could then be increased causing the focus to move through the foil. Matching the observed radius variation with the applied solenoidal field to computer simulations will allow the emittance to be inferred. Once again, the issue will be the minimum beam spot size the foil can tolerate. By operating in an over focus regime, i.e. the beam waist always remains in front of the foil, it may be possible to limit the minimum beam radius at the foil, but still generate significant variations in radius. We are also studying the effect of using a quadrupole magnet to focus in a single plane. This will reduce the issue of current density, but limit the emittance measurement to a single plane.

## RTA CONTROL SYSTEM

Except for the fast beam diagnostics described above, the accelerator status is monitored through several Allen-Bradley SLC-500 logic controllers. The human interface is provided by National Instrument's BridgeView software running on a PC platform. Much of the pulsed power hardware, and accelerator support electronics is located adjacent to the gun. Initial pulsed power trials (without beam) have shown that the electromagnetic noise in the accelerator area associated with pulsing the gun does not disturb the control system. In tests on ETAII (with beam) the x-ray induced noise from spilt beam did not disturbed the control electronics. Additional testing will be needed to access the viability of the layout. The equipment has been placed near the gun to reduced the cost of the facility.

## ACCELERATOR SECTION

Design of the full-scale RK-TBA system has an accelerator section after the gun that raises the beam energy from 1 MeV to 2.5 MeV before starting to bunch the beam at 11.4 GHz. In the accelerator section for a full scale RK-TBA we expect to use a pulse power system similar to that demonstrated in the gun system. The accelerator cells will be segmented to reduce the required drive. However, to reduce the cost of

RTA we are planning on using a 120 kV MARX spark-gap pulsed power system to drive 16 unsegmented accelerator cells in this section for RTA. This will still allow us to operate at a few Hz rate. The design is not acceptable for the full-scale machine because of the high gap erosion rates. The spark-gap system will provide a faster risetime than the thyratron system. The volt-second rating of the accelerator cores (operated at 120 kV) will limit the flat top of the beam pulse in the RTA. Installation of additional cells (future upgrade), and operating all the cells at a lower voltage, will increase the useful duration of the current pulse. We also plan to test a High-Gradient Insulator (9) in the first of these accelerator cells. If successful we would like to construct the 16 cells for RTA accelerator section using High-Gradient Insulators.

## ACKNOWLEDGMENTS

The work was performed under the auspices of the U.S. Department of Energy by LLNL under contract W-7405-ENG-48, and by LBNL under contract AC03-76SF00098. We thank Andy Sessler and Swapan Chattopadhyay for their support and guidance and thank Wayne Greenway, William Strelo, and Bob Candelario for their excellent technical support.

## REFERENCES

1. Sessler, A.M. and Yu, S.S., "Relativistic Klystron Two-Beam Accelerator," *Phys. Rev. Lett.* **54**, 889 (1987).
2. Westenskow, G.A., and Houck, T.L., "Relativistic Klystron Two-Beam Accelerator," *IEEE Trans. on Plasma Sci.*, **22**, 750 (1994).
3. Houck, T.L., and Westenskow, G.A., "Prototype Microwave Source for a Relativistic Klystron Two-Beam Accelerator" *IEEE Trans. on Plasma Sci.*, **24**, 938 (1996).
4. Registered name of Corning Glass Works.
5. Smith, C.H., and Barberi, L., "Dynamic Magnetization of Metallic Glasses," in Proc. of the 5th. IEEE Int'l Pulsed Power Conf., 1985.
6. Westenskow, G.A., et al., "Relativistic Klystron Two-Beam Accelerator Studies at the RTA Test Facility", Proc. of the 1997 Particle Accelerator Conference, Vancouver, Canada, 1997.
7. Houck, T.L., et al., "Diagnostics for a 1.2-kA, 1-MV Electron Induction Injector," Proc. of the 8th Beam Instrumentation Workshop, SLAC, 1998.
8. Hughes, T.P., Carlson, R.L., and Moir, D.C., *J. Appl. Phys.* **68**, 2562–2571 (1990).
9. Houck, T.L., et al., "Stacked Insulator Induction Accelerator Gaps", Proc. of the 1997 Particle Accelerator Conference, Vancouver, Canada, 1997.

991

# New Concept Input and Output Systems for High Power Gyroklystron

Xiaoxi Xu, W. Lawson, C. Liu, J. Cheng, J. Anderson,
B. Hogan and V. L Granatstein, M. Reiser

*Institute for Plasma Research and Electrical Engineering Department,
University of Maryland, College Park, MD 20742*

**Abstract.** In order to obtain the high mode purity of a $TE_{011}$ mode in an overmoded gyroklystron input cavity while maintaining high coupling efficiency, a coaxial dual-cavity input structure with an outer $TE_{411}$ mode and an inner $TE_{011}$ mode coaxial cavity has been designed to get a reasonable low Q and to avoid mode distortion due to a single coupling aperture between an input waveguide and input cavity. A quality factor of 73 and a resonant frequency of 8.570 GHz with high mode purity have been obtained for the inner $TE_{011}$ mode coaxial cavity. Furthermore, in order to inject the output power of a second harmonic gyroklystron (17.136 GHz) into our future pulse composer and accelerator system, a coaxial $TE_{021}$ output cavity with a $TE_{02}$-$TE_{01}$ mode converter is proposed and designed as the output structure of the gyroklystron. The output power can be extracted radially, and at the same time the $TE_{02}$ mode is converted to $TE_{01}$ mode into a inner coaxial waveguide.

## INTRODUCTION

In order to build a multi-TeV linear collider with a reasonable length, a high accelerating gradient and high RF frequency are required. A high power and high efficiency gyroklystron is considered to be one of the most promising RF sources to fulfill these requirements[1]. Pulsed gyrotron amplifiers have record performance at frequencies significantly above X-Band, e.g., output peak powers exceeding 30 MW were achieved with an efficiency near 30% at the frequency of 19.76 GHz at the University of Maryland[2,3]. Furthermore, the high power and high efficiency performance of a three-cavity coaxial gyroklystron has been demonstrated in a recent experiment with a peak power of 75 MW and an efficiency of 32% at the frequency of 8.6 GHz[4,5].

In the development of a gyroklystron for achieving higher power levels with higher efficiency to satisfy the requirement of the multi-TeV linear collider, one of the

CP472, *Advanced Accelerator Concepts: Eighth Workshop,*
edited by W. Lawson, C. Bellamy, and D. Brosius

most important things is to improve mode purity in the low Q input cavity of our gyroklystron. A weak or distorted operating mode in the input cavity will greatly reduce the interaction efficiency of an input RF signal and electron beam and will lower the gain of the gyroklystron amplifier since small changes or field perturbations at the input cavity can result in large amplified changes at the output cavity. However, high mode purity of a $TE_{011}$ mode is very difficult to realize in an overmoded input cavity with a single excitation aperture, especially in a low Q cavity. A possible way to excite a $TE_{011}$ mode with high mode purity is to split the input power with appropriate amplitude and phase by several azimuthally separated apertures in an intermediate coaxial cavity[6].

For injecting the output power of our second harmonic gyroklystron into our future pulse compressor and accelerator, a $TE_{021}$ output cavity with $TE_{02}$-$TE_{01}$ mode converter is proposed. The output power will be coupled radially to a inner waveguide which is inside the inner conductor of the coaxial cavity. At the same time, the $TE_{02}$ mode will be converted into $TE_{01}$ mode. This output structure will eliminate spurious oscillations in the output waveguide and should decrease the sensitivity of the tube to load mismatches. It will help to improve output cavity stability by allowing lossy ceramics to be placed after the cavity at the entrance of the beam dump. Furthermore, it will allow the implementation of a depressed collector.

In this paper, we report the numerical designs of the coaxial dual-cavity structure having an outer $TE_{411}$ mode and an inner $TE_{011}$ mode coaxial cavities and the $TE_{02}$-$TE_{01}$ output cavity structure with a 3-D electromagnetic code (HFSS)[7]. The cold test results of the input system are also presented.


## INPUT CAVITY SYSTEM

A schematic diagram of the dual-cavity coaxial input structure is shown in Fig. 1. An X-Band microwave signal with a frequency of 8.568 GHz is injected into the outer coaxial cavity of the input system through a rectangular aperture from a WR-90 waveguide. The outer coaxial cavity should be excited to resonate in the $TE_{411}$ mode at 8.568 GHz. The $TE_{411}$ mode is magnetically coupled to the inner coaxial cavity of the input structure through four rectangular slots to excite a $TE_{011}$ mode resonating at the same frequency as that of the outer coaxial cavity. The two coupling slots which are at the two sides of the input aperture are azimuthally spaced 45 degrees with respect to the input WR-90 waveguide while the whole four coupling slots are spaced 90 degrees respectively so as to align with the maxima of the $TE_{411}$ mode magnetic field. Lossy ceramics are put in the drift regions near the inner coaxial cavity in order to adjust the cavity loading for obtaining a reasonably low Q in the input system. The first requirement for this dual-cavity input structure is that the Q of the $TE_{011}$ mode in the inner coaxial cavity should be around 70 or less in order to avoid self-oscillations[4]. Another important requirement is that a major portion of the input electromagnetic energy should be stored in the inner $TE_{011}$ cavity since the energy stored in the outer cavity will not be able to be involved in the beam-wave interaction. As much of the total energy is required to be stored in the inner $TE_{011}$ cavity.

In order to complete the design of this dual-cavity input structure, there are 13 geometric parameters that have to be determined. As shown in Fig. 1, they are outer diameter $2r_{oo}$, inner diameter $2r_{oi}$ and axial length $L_o$ of the outer coaxial cavity; outer diameter $2r_{io}$, inner diameter $2r_{ii}$ and axial length $L_i$ of the inner coaxial cavity; the length $L_a$ and width $W_a$ of the coupling aperture; coupling slot length $L_s$ and width $W_s$; the inner diameter of the drift regions $2r_d$; distances between the edges of the inner coaxial cavity and ceramics in the drift regions, $d_i$ and $d_o$. The bore of our magnet system limits the outer diameter of the outer coaxial cavity $2r_{oo}$. The inner diameter of the drift regions $2r_d$ was chosen to cutoff the $TE_{011}$ mode at 8.568 GHz and

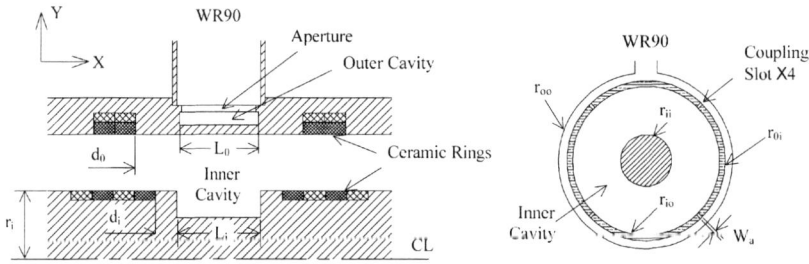

**FIGURE 1.** Schematics of the dual-cavity input structure.

allow enough clearance to the electron beam. The outer diameter of the inner coaxial cavity is equal to 3.325 cm and is determined by the size of the electron beam. The initial values of the dimensions of the outer and inner cavities were calculated separately by a scattering matrix code CASCADE[8,9]. An linear oscillation code called QPB that was developed at the University of Maryland was used to investigate the stability of the inner cavity according to the results of the CASCADE. The essential feature of this dual-cavity input structure is to transfer as much of the input power to the inner cavity resonating in the $TE_{011}$ mode as possible. The design was performed to minimize the mode impurity due to the sizing and shaping of the coupling aperture and slots, and at the same time to optimize the power transfer from the input waveguide to the dual-cavity structure and to ensure the most of the energy is stored in the inner cavities.

## NUMERICAL DESIGN OF INPUT STRUCTURE

The HFSS code is a finite-element code which evaluates in the frequency domain and is capable of calculating scattering parameters of multi port devices. The input characteristic of this dual-cavity structure was calculated from scattering matrices at the input port. Considering the symmetry of the dual-cavity structure was respect to the x-y plane, the model of the dual-cavity structure was reduced in half for HFSS simulations in order to save simulation run-time. Azimuthal symmetry of the cavity had not been taken into account to further reduce the simulation volume, because it could prevent getting better insight of possible azimuthal asymmetry modes which may be excited in this overmoded structure.

The magnitude of the reflection coefficient at the input port, $|S_{11}|$, is plotted as a function of frequency in Fig. 2. There are two minima in the frequency range from 8.0

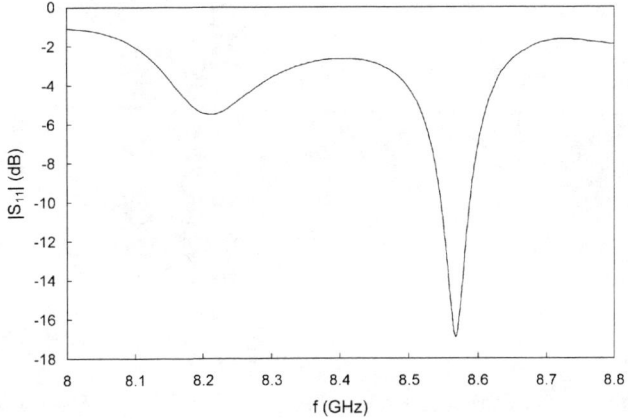

**FIGURE 2.** $|S_{11}|$ versus frequency obtained from a HFSS simulation.

to 8.8 GHz. The first is at 8.568 GHz with a magnitude of −17 dB and the other is at 8.206 GHz with a magnitude of −5.5 dB. The shaded plots of the magnitude of electric field at the frequency of 8.568 GHz through an axial cutplane (x=0) and a transverse cutplanes (z=0) of the dual-cavity structure are shown in Fig. 3 (a), Fig. 3 (b). It is obvious that both the $TE_{411}$ and $TE_{011}$ modes are established with good mode purity in the outer and inner cavities, respectively. However, the $|S_{11}|$ minimum at 8.206 GHz in Fig.2 was found mostly corresponds to the $TE_{411}$ mode in the outer cavity mode. The stored electromagnetic field energy which is proportional to $|E|^2$ was integrated over the inner cavity for the frequency of 8.568 and 8.206 GHz, respectively. The results show that the $TE_{011}$ mode purity of 98% has been realized in the inner cavity of the dual-cavity structure within the frequency range we are interested in.

Although the characteristic of the dual-cavity structure should be mostly determined by the inner cavity, the resonant frequency indicated by the $|S_{11}|$ minimum at 8.568 GHz in Fig.2 may not be the actual resonant frequency of the inner cavity. The resonant frequency and Q of the inner cavity were obtained by plotting $W_{inner}$ as a function of frequency as shown in Fig.4, where $W_{inner}$ is the time-averaged electromagnetic energy stored in the inner cavity and $W_T$ is the total energy stored in

995

the outer and inner cavities. The peak of the $W_{inner}$ versus frequency curve (circle) gives the resonant frequency of the inner cavity, which is found to be 8.574 GHz. The Q of 73 is determined from the "half-energy" bandwidth. In Fig. 4, the fraction of the energy stored in the inner cavity with respect to the total stored energy in the two cavities, $W_{inner}/W_T$, is also plotted versus the frequency (square). The curve shows that 81.3% of the total stored electromagnetic energy is in the inner cavity at the frequency of 8.568 GHz.

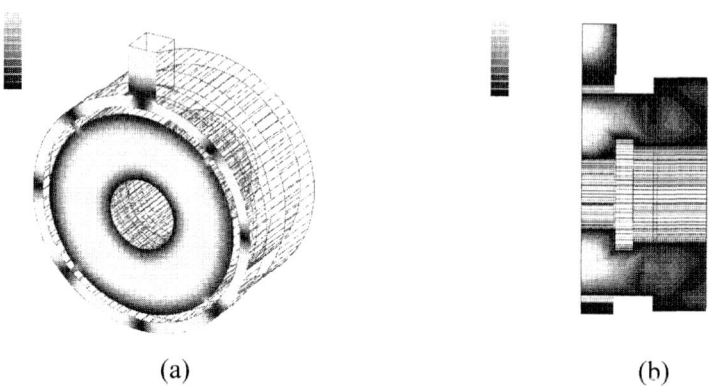

(a)                                            (b)

**FIGURE 3.** Shaded plots of the magnitude of electric field |E| at 8.568 GHZ.

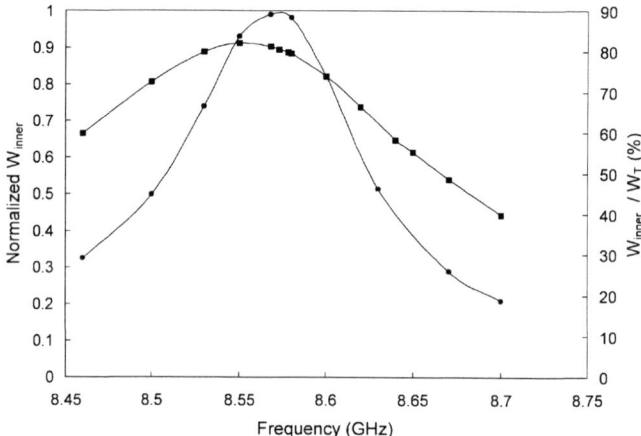

**FIGURE 4.** Normalized time-averaged electromagnetic energy stored in the inner coaxial cavity and its fraction to the total stored electromagnetic energy as a function of frequency.

## COLD TEST SET-UP AND RESULTS

A prototype of the coaxial dual-cavity input structure was built according to the above-mentioned simulation results. A schematic of the cold test setup for this coaxial dual-cavity structure is shown in Fig. 5. A microwave signal was injected into the cavity through a 10 dB dual directional coupler from an HP83050B sweep oscillator. A coaxial probe with a 0.025 cm diameter inner conductor tip was axially inserted into the inner coaxial cavity from the drift region. The thin tip was oriented to be capable of coupling to the azimuthal electrical component of the microwave signal in the inner cavity. The probe was fixed on a stand, its axial and radial position were controlled precisely by a micrometer. Input, reflection and probe signals were picked up by three HP85025B detectors respectively and measured by an HP8757C scalar network analyzer. A PC was connected to the network analyzer and sweep oscillator to record data.

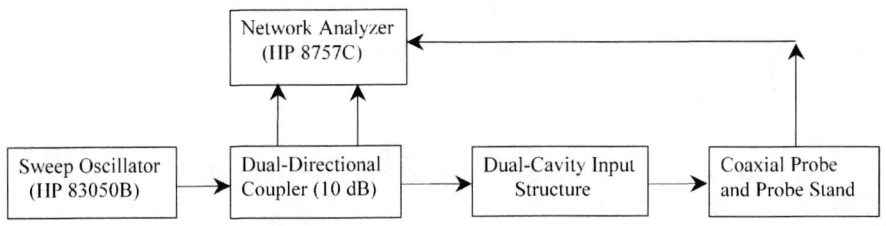

**FIGURE 5.** Cold test set-up.

In the cold test, the lossy ceramic rings of carbon-impregnated alumino-silicate (CIAS) and 80% BeO - 20% SiC were symmetrically put at the both sides of the inner coaxial cavity in the drift regions[10]. The CIAS and BeO-SiC rings were placed alternately along the inner conductor. Two layers of the CIAS and BeO-SiC rings were placed along the outer conductor with the CIAS at inner layer and BeO-SiC at outer layer. As shown in Fig. 1, there were eight ceramic rings along the inner and outer conductors at the both sides of the inner coaxial cavity, respectively. The traces of the $|S_{11}|$ and probe signals are shown in Fig. 6. It can be seen that the characteristics of $|S_{11}|$ as a function of frequency is similar to that of obtained from the HFSS simulation (refer to Fig. 2). Two minima, corresponding to the inner and outer cavity modes respectively, are observed in this frequency range. A –21 dB reflection minimum occurs at 8.582 GHz showing that more than 99% of the input microwave signal at this frequency was injected into the dual-cavity structure. The minimum due to the outer cavity mode is found at 8.322 GHz with –14.8 dB. The probe signal provides the real frequency response of the inner coaxial cavity. As shown in Fig. 6, a peak is found at 8.599 GHz, and a 3 dB bandwidth of 126 MHz gives a Q of 68. After the probe

997

loading of the coaxial probe was taken into account, the resonant frequency of 8.609 GHz and Q of 62 were found for the $TE_{011}$ mode in this input cavity.

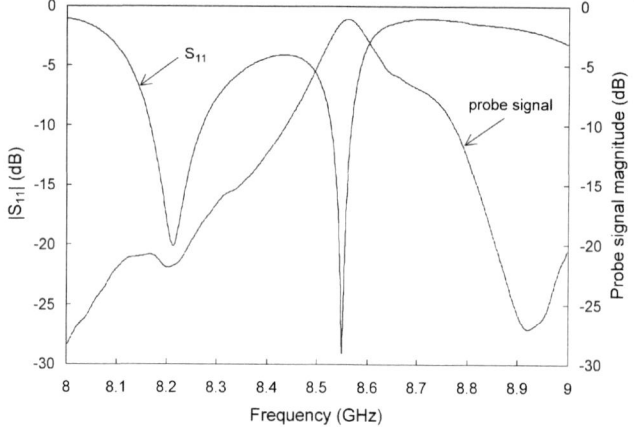

**FIGURE 6.** Traces of $|S_{11}|$ and probe signal of the dual-cavity input structure.

A perturbation method was employed to identify the field strength in the inner cavity of the asymmetry ceramic loading dual-cavity input structure[11]. A small metal nut with a diameter of 0.25 cm was put at the center plane (x-y plane) of the inner cavity and moved radially step by step by a micrometer. A frequency sweep was performed and a resonant frequency was measured every time the nut was moved by 5 mils. The frequency signal was picked up in the inner cavity at a fixed position by the coaxial probe with its center conductor tip in the center plane oriented to azimuthal direction. The resonant frequency shift of the inner cavity $\Delta f$ is shown in Fig. 7 as a

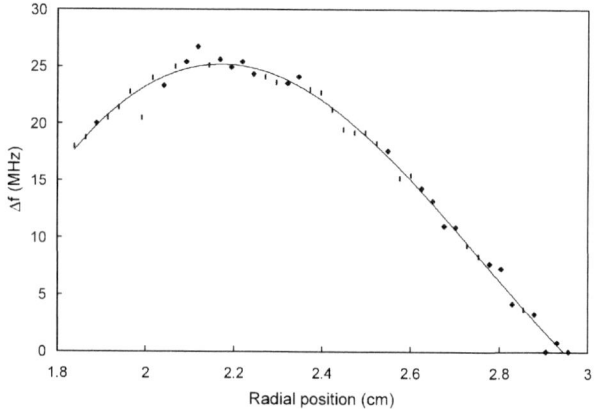

**FIGURE 7.** Frequency shift versus radial position of perturbation.

function of the radial position of the metal nut, where $\Delta f = f - f_0$, f and $f_0$ are the cavity resonant frequency with and without metal nut in the inner cavity, respectively. In our case, the profile of $\Delta f$ in Fig.7 reflects the radial distribution of the electric field of the microwave signal in the inner cavity according to the perturbation theory. It is obvious that only one peak exists between the inner conductor and wall of the inner coaxial cavity. The probe was moved around center axis azimuthally, and no frequency shift was observed. The results are consistent with the characteristic of the $TE_{011}$ mode. The peak position shown in Fig.7, r = 2.139 cm, is close to what was obtained from the HFSS simulation, r = 2.164 cm. Several orientations of the tip of the coaxial probe were used and the above-mentioned measurement was repeated, no other obvious peaks were observed.

## OUTPUT SYSTEM

The schematic of the $TE_{021}$ mode output cavity with the $TE_{02}$-$TE_{01}$ mode converter is shown in Fig. 8. An output microwave signal (17.136 GHz, $TE_{021}$ mode) in the coaxial output cavity is coupled into another inner coaxial waveguide and then converted into a circular waveguide. Eight rectangular slots which are spaced $20^0$ azimuthally on the inner wall of the output cavity (the outer wall of the inner coaxial waveguide) are used to couple the output power into output waveguide and convert the $TE_{02}$ mode into $TE_{01}$ mode. The first requirement for this output structure is to have a pure $TE_{021}$ mode at 17.136 GHz in the output cavity with a Q of 320. The second requirement is to be able to convert the $TE_{02}$ mode into $TE_{01}$ mode with high mode purity.

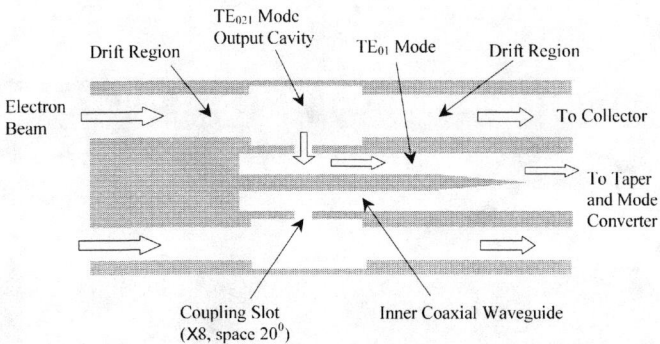

**FIGURE 8.** Schematics of the output structure.

## NUMERICAL DESIGN OF OUTPUT STRUCTURE

HFSS and almost the same design procedure which was used to the input cavity design were applied for the modeling of the output structure. A small aperture was opened on the outer wall of the output cavity. A microwave signal was injected into the output cavity through the aperture from a WR51 waveguide. The magnitude of the transmission coefficient reflecting the transmission between the input waveguide and the output circular waveguide, $|S_{21}|$, is plotted as a function of frequency in Fig. 9. Within the frequency range from 17 to 17.3 GHz, the highest peak is found at 17.136 GHz, and another peak is at 17.03 GHz. The shaded plots of the magnitude of the electric field at the frequency of 17.136 GHz at an axial cutplane (x=0) and a transverse cutplane (z=2) of the output structure are shown in  Fig. 10 (a) and Fig. 10 (b), respectively.

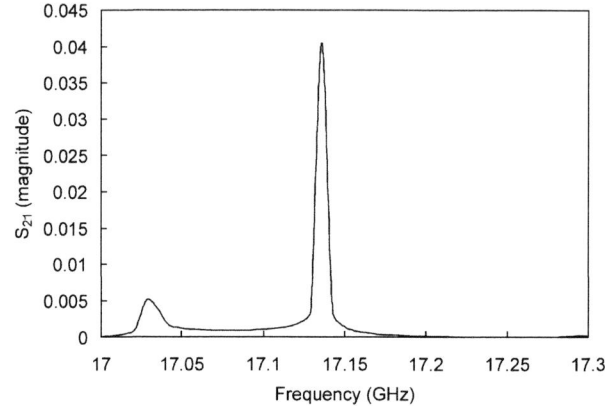

**FIGURE 9.** $|S_{21}|$ versus frequency obtained from a HFSS simulation.

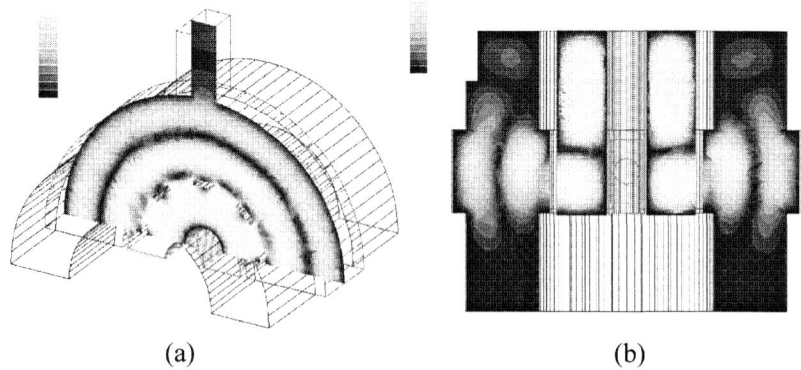

(a)                              (b)

**FIGURE 10.** Shaded plots of the magnitude of electric field |E| at 17.136 GHz.

1000

It is clear that a $TE_{021}$ mode at 17.136 GHz was excited in the output cavity with high mode purity, and the $TE_{02}$ mode was successfully converted into a very pure $TE_{01}$ mode in the output waveguide at the same time. The time-averaged electromagnetic energy stored in the $TE_{021}$ mode output cavity is plotted as a function of frequency in Fig. 11. The Q of 320 is obtained from the "half-energy" bandwidth of the electromagnetic energy stored in the output cavity.

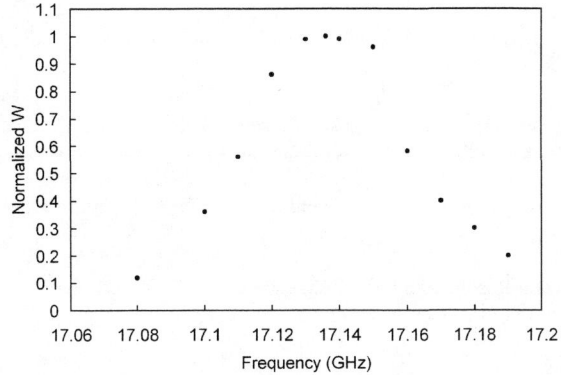

**FIGURE 11.** Normalized time-averaged electromagnetic energy stored in the output cavity.

## SUMMARY

The X-Band dual-cavity coaxial input structure was successfully designed via the HFSS simulations combined with the CASCADE codes, and the cold test input cavity was built and tested. The $TE_{011}$ mode with good mode purity and a reasonable low Q of 62 at 8.609 in the inner coaxial cavity were realized and verified experimentally through the S parameters and mode pattern measurements. The cold test and HFSS simulation results are in very close agreement. The numerical design of the $TE_{021}$ output cavity with $TE_{02}$-$TE_{01}$ mode converter has been carried out. The output cavity with high $TE_{021}$ mode purity and a Q of 320 has been obtained, and the $TE_{02}$-$TE_{01}$ mode conversion has been successfully demonstrated in the HFSS simulations. This output structure is under construction and will be tested and put in our next Ku Band gyroklystron with the dual-cavity input structure.

# ACKNOWLEDGMENTS

This work is supported by Department of Energy, contract DE-FG02-94ER40855. One of authors (X. Xu) would like to thank Dr. W. R. Fowkes at SLAC, Stanford University for his helpful discussions and advice.

# REFERENCES

1.  P. B. Wilson, et al., "RF System for a 30 GHz, 5 TeV Linear Collider Based on Conventional Technology", SLAC-PUB-7610, SLAC, Stanford, 1997.

2.  V. L. Granatstein, W. Lawson, "Gyro-Amplifier as Candidate RF Drivers for TeV Linear Colliders", IEEE Trans. Plasma Sci., vol.24, pp. 648-665, 1996.

3.  W. Lawson, et al., "High-Power Operation of a K-Band Second-Harmonic Gyroklystron", Phys. Rev. Lett., vol.71, pp. 456-459, 1993.

4.  J. Cheng, et al., "Experimental Results of a High Efficiency, High Power X-Band Gyroklystron", 25th IEEE International Conference on Plasma Science, Raleigh, NC, June 1-4, 1998.

5.  W. Lawson, et al., to be published.

6.  G. S. Park, et al., "Linearly Polarized $TE_{n,1}$ Coupler", Int. J. Electronics, vol. 78, pp. 983-990, 1995.

7.  HP 85180A High-Frequency Structure Simulator User's Reference, Release 3.0, Edition 1, 1994.

8.  J. M. Neilson, et al., "Determination of the Resonant Frequencies in a Complex Cavity using the Scattering Matrix Formulation", IEEE Trans. Microwave Theory Tech., vol. 37, pp. 1165-1169, 1989.

9.  W. Lawson and P. Latham, "The Scattering Matrix Formation for Overmode Coaxial Cavities", IEEE Trans. Microwave Theory Tech., vol. 40, pp. 1973-1977, 1992.

10. J. P. Calame and Wesley G. Lawson, "A Modified Method for Producing Carbon-Loaded Vacuum-Compatible Microwave Absorbers from a Porous Ceramic", IEEE Trans. Electron Devices, vol. 38, pp. 1538-1543, 1991.

11. L. C. Maier, Jr. and J. C. Slater, "Field Strength Measurements in Resonant Cavities", J. Applied Physics, vol. 23, pp. 68- 77, 1952.

# A Linear Accelerator Power Amplification System for High Gradient Structure Research

J. Haimson and B. Mecklenburg

*Haimson Research Corporation*
*3350 Scott Blvd., Building 60, Santa Clara, CA 95054-3104*

**Abstract.** The ongoing development of linear collider high power RF sources and pulse compression systems has resulted in substantial progress towards a goal of providing a peak RF power level of approximately 250 MW at the input of the accelerator structure. While the immediate development and the high power testing of specialized waveguide components required for power transmission at these high levels have proceeded expeditiously due to the availability of resonant ring systems, the testing of high gradient accelerator structures at very high power levels, and the investigation of coupler cavity RF breakdown problems have, typically, been curtailed due to the unavailability of suitable 200 to 300 MW RF test facilities. We describe herein a compact, high peak power amplification system based on a dual hybrid bridge configuration that avoids the need for power splitters at the accelerator dual feed couplers, and also provides a convenient interface for installing high gradient accelerator test structures. Design parameters are presented for a proposed power amplification system that makes use of a 75 MW, 1/2 μs flat-top RF source to produce 280 MW, 1/4 μs flat-top power for testing dual feed TW experimental accelerator sections.

## INTRODUCTION

The use of short (7-cavity and 2-cavity) standing wave (SW) structures and nose cone half-cavity field enhancing geometries has enabled E-field breakdown and dark current experiments to be performed with existing RF power sources at S, C and X-band frequencies [1]. Extrapolation of such SW test data to accurately predict the RF breakdown performance of long traveling wave (TW) structures is difficult, however, because of the different dark current characteristics, the statistical dependence on the plurality of cavity surfaces and the different ratios of maximum surface field to accelerating gradient.

Early operating experience [2] using a single feed, 11.4 GHz, 30-cell constant impedance TW accelerator structure, and a short RF pulse (≈40 ns) high power TW output relativistic klystron, demonstrated that accelerating gradients of approximately 85 MV/m could be sustained before encountering E-field breakdown. Subsequent borescopic inspection of these uniform impedance accelerator waveguides revealed that RF

CP472, *Advanced Accelerator Concepts: Eighth Workshop,*
edited by W. Lawson, C. Bellamy, and D. Brosius
© 1999 The American Institute of Physics 1-56396-889-4/99/$15.00

breakdown had occurred in both the input and output coupler cavities. Electric field breakdown at comparable gradient levels and longer RF pulse lengths ($\approx 250$ ns) has also been experienced in 11.4 GHz accelerating structures using dual feed coupler cavities.

Computer simulations of the field distribution in dual feed, cylindrically shaped coupler cavities [3] have confirmed that electric field enhancement occurs on the cavity disc surfaces in regions that are radially aligned with the sidewall coupling apertures (and in close proximity to the beam aperture). Recent studies of dual feed, racetrack shaped coupler cavities [4] have shown that, by correct adjustment of the aspect ratio of the racetrack cavity transverse dimensions, a highly symmetric field pattern can be achieved, and field enhancement caused by the dual feed, diametrically opposed sidewall coupling apertures can be eliminated, while still retaining the capability of readily matching and tuning the cavity. A 17 GHz high gradient accelerator structure and a 17 GHz relativistic klystron with a tapered phase velocity traveling wave output structure have recently been constructed using field symmetrized dual feed racetrack shaped couplers, and both devices are now being prepared for high power evaluation.

Separately, ongoing studies of the E-field distribution in TW accelerator structures have shown that additional factors, such as the geometry of the coupler RF cut-off nose and the tune and impedance of cavities contiguous to the coupler, can also contribute to field enhancement within the terminating cavities of an accelerator section.

The need to evaluate and avoid the above field enhancing effects presents a compelling argument for a simple means of high power testing experimental dual feed coupler and TW structure designs, without having to wait for the extensive development and fabrication of a pulse compression system and a full length, damped accelerator section. A dual hybrid junction microwave configuration that provides such a means for high power testing, and allows convenient installation and removal of high gradient accelerator test structures, is described herein.

## POWER AMPLIFICATION USING DUAL HYBRID JUNCTIONS

An outline of the proposed power amplification configuration, using a high gradient accelerator (HGA) test structure and two hybrid junctions forming a dual, paralleled RF bridge system, is shown in Figure 1. Each four port RF bridge with arms 1, 2, 3 and 4, at power levels $P_S$, $P_A$, $P_F$ and $P_L$ is shown connected to the RF source, the accelerator input, the accelerator feedback and an RF load, respectively. A variety of hybrid junction circuits [5,6] can be used for RF bridge applications. These include waveguide hybrid rings, magic tee junctions, and short branch couplers, each of which can be represented as an eight terminal network arranged so that the following specific conditions are satisfied: arms 1 and 3 should be independently matched to the bridge when arms 2 and 4 are terminated by their characteristic impedance; arms 1 and 3 should be highly isolated so that power fed into either arm is transmitted to loads in arms 2 and 4 only; conversely, arms 2 and 4 should be balanced so that power entering either arm is delivered to loads at arms 1 and 3 only; and there should be no power circulating within the bridge.

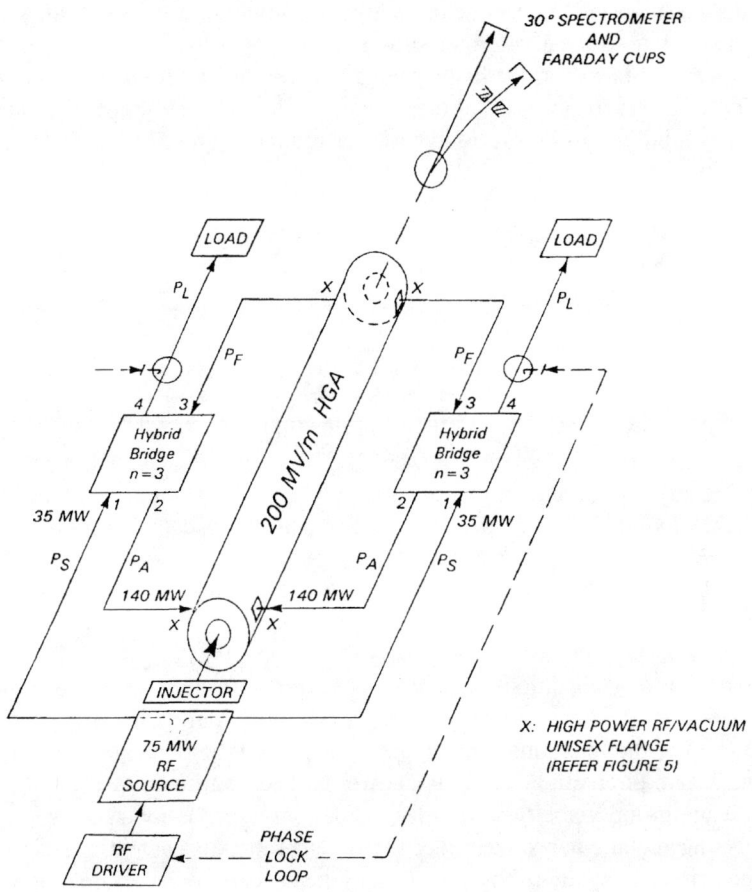

**FIGURE 1.** A dual hybrid junction configuration to provide a four-fold increase of RF source power for high gradient testing of linear accelerator coupler cavities and sections.

For a bridge ratio n, when correctly phased remnant power, $P_F$, from the accelerator is combined with the source power, $P_S$, after an initial transient build-up period, a steady state power level of $P_A = (n+1)P_S$ will appear at the accelerator input; and the load power, $P_L$, will be reduced to zero. A desirable operating feature of this power amplification technique is that the mutually conjugate properties of the bridge arms 1 and 3 ensure that, even during the RF build-up process, a constant impedance is presented to the external RF source connected to arm 1 of each hybrid junction.

Another desirable feature of the above configuration is that connection to the dual feed accelerator input (and output) coupler no longer requires the use of a rectangular waveguide power splitter (and combiner). Thus, the interconnecting waveguide components are subjected to only one half of the maximum power level in the system, and this occurs only within the accelerator structure proper.

For a correctly phased bridge having voltage coupling and transmission coefficients of C and $T_C$, respectively, the ratio of the accelerator input arm voltage, $V_a$, to the source voltage, $V_s$, will build up with successive transits around the feedback loop in accordance with the series,

$$V_a/V_s = C [ 1 + ( T_C T_{FL} ) + ( T_C T_{FL} )^2 + ( T_C T_{FL} )^3 + ... ] , \qquad (1)$$

where $T_{FL}$ is the voltage transmission coefficient of the feedback loop, giving in the limit,

$$V_a/V_s \rightarrow C / ( 1 - T_C T_{FL} ) . \qquad (2)$$

Thus, for a bridge ratio of n = 3, a voltage coefficient of C = 0.5, $[n = (1-C^2)/C^2]$, and a voltage transmission coefficient of $T_C = 0.866$, $\{T_C = [n/(n+1)]^{1/2}\}$, with a correctly phased bridge having an integral number of wavelengths ($\phi_{FL} = 2m\pi$) around the feedback loop and a feedback loop loss of 1.25 dB, $V_a$ will increase with successive transits such that,

$$V_a/V_s = \tfrac{1}{2} [ 1 + 3/4 + 9/16 + 27/64 + ... ] , \qquad (3)$$

attaining a steady state value of 2, i.e., a power buildup ( $V_a/V_s$ )$^2$ of 4.

For the n = 3 bridge conditions presented above, after 10 successive transits around the feedback loop, the linac input power level will build up to 91.7 percent ($P_a/P_s = 3.668$) of the maximum value; and, to achieve 99.5 percent of the steady state power level, 20 re-circulations of the RF power will be required.

Because the group velocity in the disc loaded structure is much lower than in the interconnecting rectangular waveguide, the transit time for each RF re-circulation is determined primarily by the filling time ($T_F$) of the accelerator structure (approximately 90 percent of the loop transit time). Thus, at full power buildup, the maximum RF pulse width achievable, namely, the flat-top pulse width of the RF source less the time required for the power buildup, will be dependent on the filling time of the accelerator structure and on the bridge ratio.

Regardless of variations of cavity impedance and group velocity along an accelerator structure, the filling time ($T_F$) for an n-cavity, $v_p = c$ waveguide can be defined explicitly by

$$T_F = L / (v_g)_{hm} , \qquad (4)$$

where L is the length of the structure and $(v_g)_{hm}$ is the harmonic mean group velocity given by

$$(v_g)_{hm} = \left[ \frac{1}{n} \sum_{i-1}^{n} \left( \frac{1}{v_g} \right)_i \right]^{-1} . \qquad (5)$$

For $2\pi/3$ mode accelerator waveguides that also comprise varying phase velocity cavities $v_p/c = \beta_w \neq 1$, the filling time can be expressed as

$$T_F = \sum_{i=1}^{n} \left( \frac{z}{v_g} \right)_i = \frac{1}{3f} \sum_{i=1}^{n} \left( \frac{\beta_w}{v_g/c} \right)_i = \frac{n}{3f} \left[ \left( \frac{v_g/c}{\beta_w} \right)_{hm} \right]^{-1}, \tag{6}$$

where

$$\left( \frac{v_g/c}{\beta_w} \right)_{hm} = \left[ \frac{1}{n} \sum_{i=1}^{n} \left( \frac{\beta_w}{v_g/c} \right)_i \right]^{-1} \tag{7}$$

is the harmonic mean $\dfrac{v_g/c}{\beta_w}$ for n cavities.

In practice, the RF filling time can be conveniently and accurately determined by performing a cold test dispersion measurement of the overall circuit and applying the simple expression,

$$\partial\theta/\partial f = (2\pi T_F), \tag{8}$$

as derived from the structure Q and power attenuation ($2\tau$) relationship,

$$\omega T_F = 2\tau Q. \tag{9}$$

Typical design parameters for 11.424 GHz and 17.136 GHz accelerator structures using dual hybrid junctions and an n=3 bridge ratio with a feedback loop loss of 1.25 dB are shown in Table 1. The 11.424 GHz system is designed to produce an average accelerating gradient of 145 MV/m with an input power level of 35 MW at each hybrid junction and with a feedback loop transit time of 23 ns. A uniform impedance structure comprising 28, $v_p = c$, $2\pi/3$ mode cavities was used for this example, and no special design features were included to provide optimum asymptotic bunch location for the injected electron beam.

The 17.136 GHz structure listed in Table 1 is designed to produce average accelerating gradients of 114 or 190 MV/m with 12.5 MW or 35 MW, respectively, applied to the input port of each hybrid junction. This 17.136 GHz linac power amplification system has a feedback loop transit time of 11 ns; and the associated growth of the accelerator input power, and the decay of the RF load power, are shown plotted in Figure 2 versus time elapsed after the commencement of the flat-top portion of the RF source pulse, for a correctly phased system and for feedback loop total loss values of 1.15, 1.25 and 1.35 dB.

**TABLE 1.** Typical Design Parameters for Linac Structure and Feedback Loop Components

| | 11.424 | | 17.136 | |
|---|---|---|---|---|
| Frequency | 11.424 | | 17.136 | GHz |
| Linac Structure Beam Aperture $\phi$ | 0.7938 | | 0.5690 | cm |
| Number of Cavities ($2\pi/3$ Mode) | 28 | | 23 | |
| Cavity Phase Velocity | c | | 6 at c, 17 at 1.015 c | |
| Linac Structure Attenuation Parameter | 0.107 | | 0.096 | Np |
| Linac Structure Harmonic Mean Group Velocity $(v_g)_{hm}$ | 0.0389 c | | 0.0475 c | |
| Total Loss in Each Feedback Loop | 1.25 | | 1.25 | dB |
| Feedback Loop Transit Time $(T_F+T_{RWG})$ | 23 | | 11 | ns |
| RF Bridge Ratio (n) | 3 | | 3 | |
| RF Power Build-up Ratio (n+1) | 4 | | 4 | |
| RF Bridge Coupling Coefficient $[c = (n+1)^{-1/2}]$ | 0.5 | | 0.5 | |
| Voltage Transmission Coefficient $\{T_C = [n/(n+1)]^{1/2}\}$ | 0.866 | | 0.866 | |
| Input Power from RF Source | 70 | 25 | 70 | MW |
| Linac Steady State Input Power | 280 | 100 | 280 | MW |
| Average Accelerating Gradient | 145 | 114 | 190 (200–182) | MV/m |
| Time to Attain 99.5% Steady State RF Power Buildup | 460 | 220 | 220 | ns |
| Loaded Beam Energy at $i_p = 200$ mA | 35.3 | 14.3 | 24.1 | MeV |
| Output Phase/Frequency Sensitivity of Linac Structure $(\partial\theta/\partial f)$ | 7.5 | 3.4 | 3.4 | deg/MHz |
| Steady State Beam Loading Derivative | 1.2 | 0.7 | 0.7 | MeV/A |

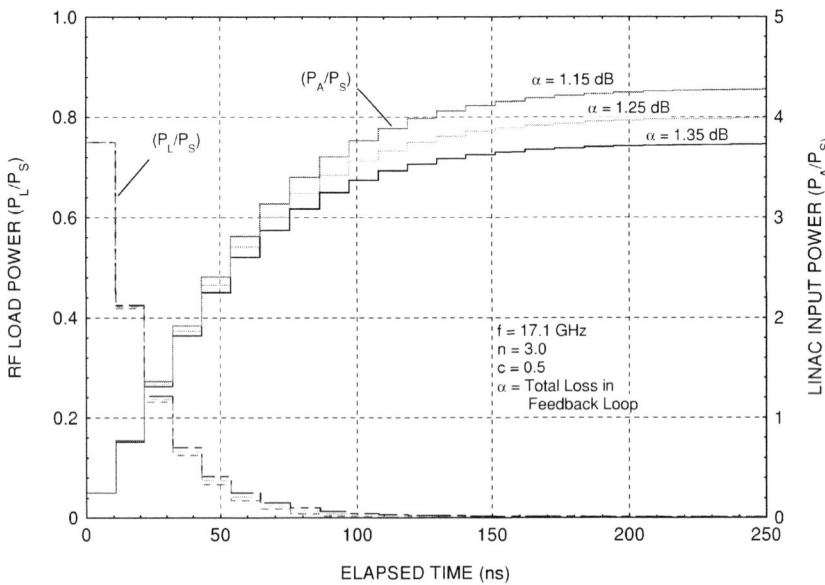

**FIGURE 2.** Linac input power buildup and RF load power decay versus elapsed time after commencement of the flat-top region of the RF source pulse ($P_S$).

The dual hybrid network shown in Figure 1 can be automatically phase locked to accurately maintain the correct feedback loop phase condition ($\phi_{FL} = 2m\pi$), as described previously for 9.5 GHz and 17.1 GHz resonant ring TW accelerators [7,8], by monitoring the power level at the bridge RF loads and maintaining a zero power flow condition in these arms by slight frequency adjustment ($<1:10^4$) of the low power oscillator used to excite the RF driver and high power RF source. For these near synchronous operating conditions, the use of cavities having a phase velocity ($v_p$) greater than the speed of light, as indicated in Table 1, enables the phase orbit performance of a high gradient capture section to be improved, as discussed in the following section.

## INJECTED ELECTRON PHASE ORBIT CHARACTERISTICS IN 17 GHz HIGH GRADIENT ACCELERATOR STRUCTURES

The presence of strong RF fringing fields and a standing wave domain at the immediate entry to a TW high gradient accelerator structure influences the initial capture and bunching of an injected electron beam and the subsequent asymptotic phase location of the accelerated bunches. It can be expected that these field interactions will have an even greater influence on the injected beam for accelerator systems that are designed to operate at very high gradients and reduced wavelengths.[9] Neglecting to carefully analyze these critical interactions can result in an incorrect choice of the electron gun operating potential, poor bunching and a substantial reduction in the energy gain of a synchronously operated accelerator waveguide.

The Figure 3(a) and 3(b) graphs show plots of emergent electron energy and phase, respectively, versus the entry phase of 500 kV electrons injected into a 17.136 GHz, $v_p = c$, uniform impedance 23 cavity accelerator structure (Design "a") operating at an average gradient of 114 MV/m. For synchronous operation at f = 17136 MHz, the solid line graphs show that an injection phase bucket of $\phi_0 = 200\pm10°$ will produce a very narrow ($1°$) emergent bunch, but this results in an emergent phase ($\delta_a$) approximately $30°$ ahead of crest and a reduction of the emergent electron energy ($V_a$) . By operating the accelerator structure asynchronously at f = 17124 MHz, as shown by the dashed line graphs in Figure 3, the circuit phase velocity can be increased, causing the electron bunch to slip back through a region of higher gradient and emerge with a phase location approximately $10°$ behind the crest. This results in a higher beam energy with a less sensitive response to injection phase jitter and, for the same injection phase bucket, the narrow emergent bunch characteristics are still maintained.

For systems that are required to operate synchronously, e.g., linacs with multiple sections or with automatic frequency phase lock controls, near optimum capture conditions can still be achieved, for a given injection energy and accelerating gradient, by fabricating and tuning the high gradient capture section to provide the desired $v_p > c$ phase slip at the normal operating frequency. The 17.136 GHz linac structure listed in Table 1, comprising a mix of $v_p = c$ and $v_p = 1.015c$, $2\pi/3$ mode cavities, is an example of this design approach. The improved performance achieved by asynchronous operation of

1009

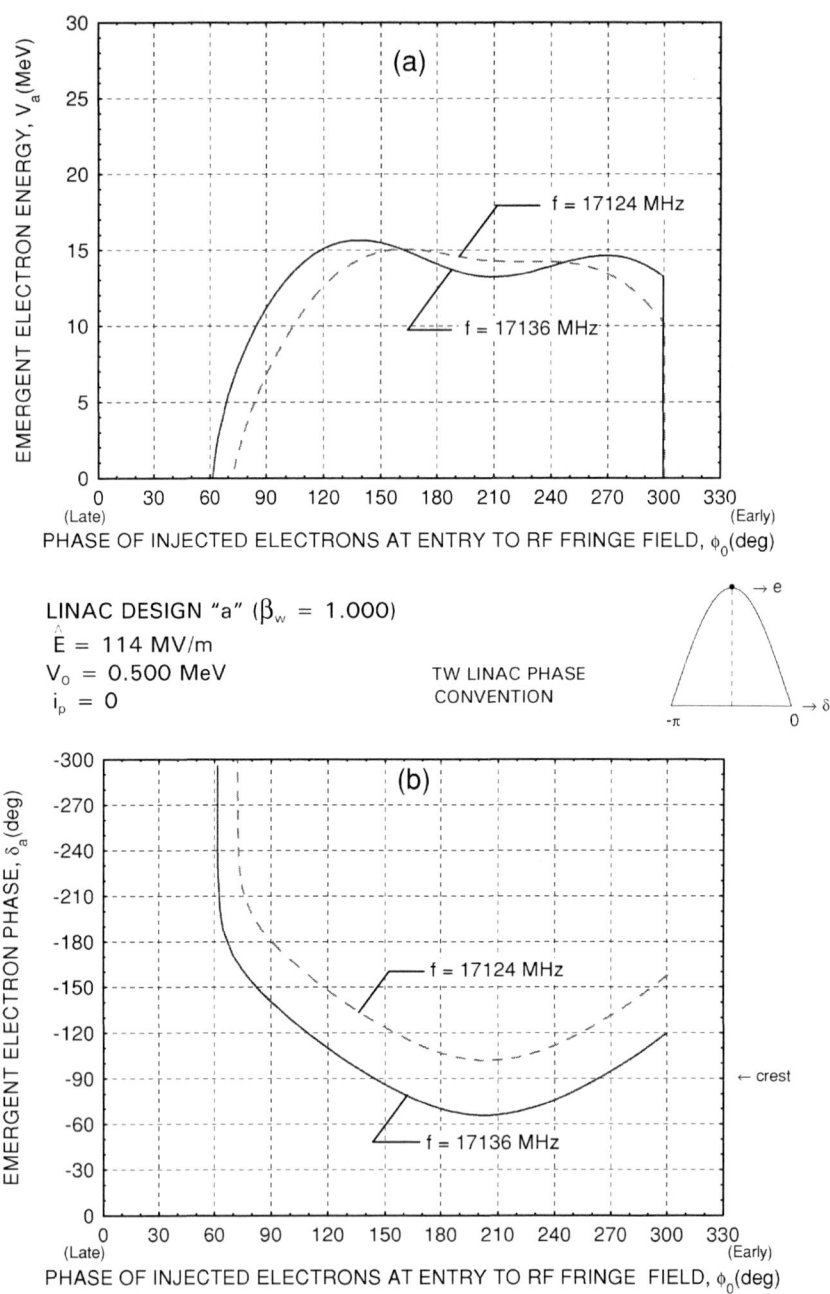

**FIGURE 3.** Energy and phase of emergent electrons versus phase of 0.500 MeV injected electrons at entry to RF fringe field for linac design "a" ($\beta_w = 1.000$). Entry phase $\phi_0$ is defined at the RF fringe field amplitude corresponding to 1% of $E_{max}$.

the $v_p = c$ structure, as illustrated by the dashed line graphs in Figure 3, is shown reproduced with synchronous operation of the $v_p > c$ structure by the solid line graphs shown plotted in Figure 4, for the same values of injection energy (0.50 MeV) and average accelerating gradient (114 MV/m).

The Figure 4 graphs compare the beam optics capture characteristics of this $v_p > c$ structure when operating at an average accelerating gradient of 114 MV/m (28 MW RF source) and 190 MV/m (75 MW RF source) with injection energies of 0.50 MeV and 2.00 MeV. An interesting feature depicted by these graphs, not seen previously with lower gradient capture sections, is a sharp rejection of the early phase injected 0.50 and 2.00 MeV electrons, rather than a gradual roll-off in energy as occurs, for example, at the late injection phase limit of the Figure 4(a) acceptance range. Early injection phases result in a full half cycle interaction with the retarding high gradient field, and the electron energy is reduced to zero in the input coupler cavity (between the midplane and the beam exit aperture). Figure 4(a) also shows that the 2.00 MeV injected particles have an adverse $V_a/\phi_o$ sensitivity, previously noted with a lower gradient 17 GHz capture section [9], and this becomes more exaggerated at higher accelerating gradients.

## REDUCTION TO PRACTICE

A practical embodiment of the dual hybrid junction linac power amplifier is illustrated in Figure 5. The dual input power feeds from the RF source are shown connected to the hybrid junctions via high conductance pump-out assemblies. High vacuum pumping ports are also located in the beam centerline contiguous to stainless steel housings that are brazed to the linac input and output coupler assemblies. These integrally machined housings provide the connections and mixing manifolds for the linac structure water cooling system, the high vacuum seals for the beam centerline connections, and the supports for the alignment mechanism.

Convenient installation of linac test structures is made possible by the use of special stainless steel RF/vacuum flanges, developed for high power applications, that are brazed to the input and output coupler bodies to provide the dual feed interface connections between the structure and the hybrid junctions. The stainless steel RF flange assemblies are of unisex design (enabling the swapping and reversing of components) and feature a narrow sealing surface that compresses and extrudes a precision machined, partially annealed copper gasket so that the radiused edges of the gasket conform to the inner surfaces of the rectangular waveguide to simultaneously produce a high vacuum seal and a high current RF joint. ($I_p$=786 A with 140 MW in WR62 waveguide at 17.136 GHz.)

The remaining arm of each hybrid junction is terminated in a well matched, ceramic window water load, ensuring that the load arm loss during the power build-up period is dissipated outside of the high vacuum envelope. By carefully matching and tuning the accelerator structure and hybrid junction, reflection coefficients of less than 0.05 can be achieved. Then, consistent with the choice of an n=3 bridge ratio, the need to include matching tuners in the feedback loop can be avoided.

1011

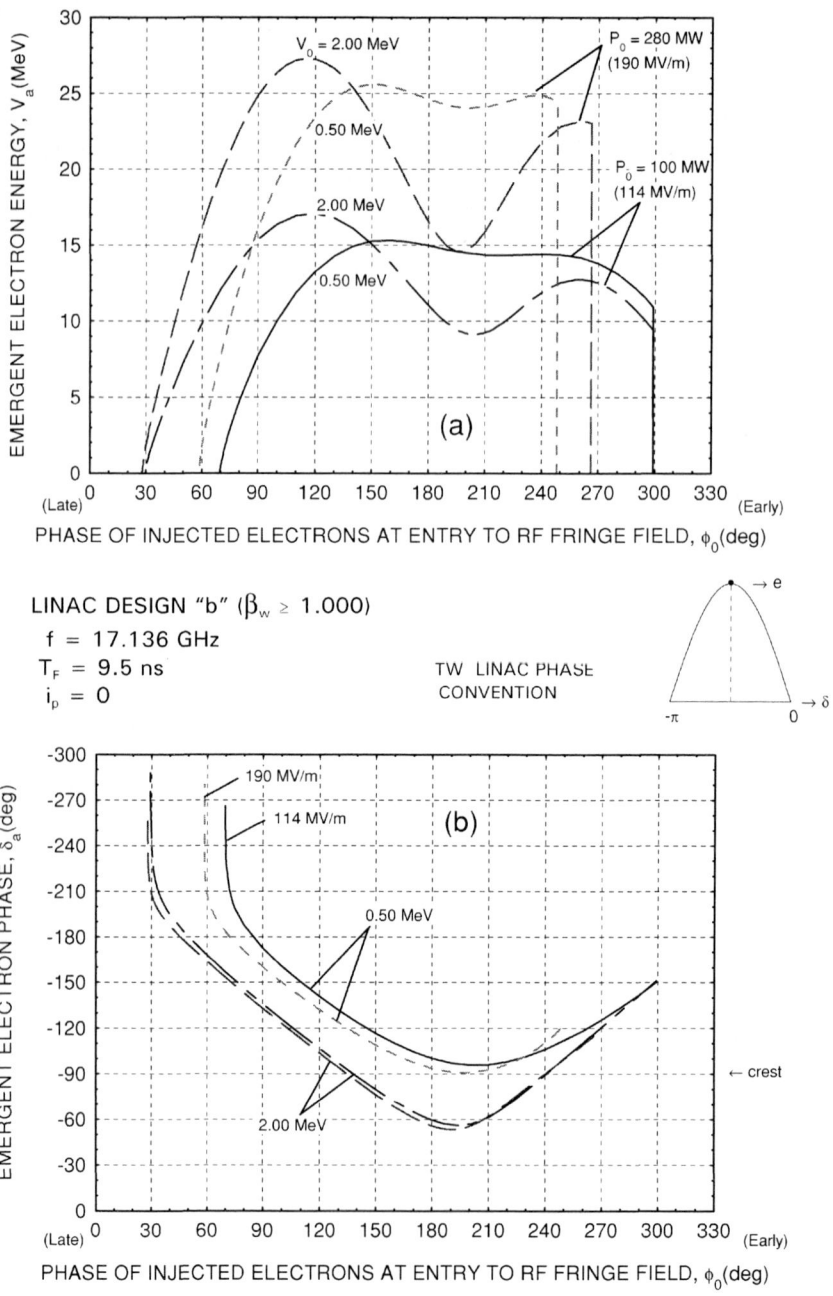

**FIGURE 4.** Energy and phase of emergent electrons versus entry phase for injection energies of 0.50 and 2.0 MeV and average accelerating gradients of 114 and 190 MV/m for linac design "b". Entry phase $\phi_0$ is defined at the RF fringe field amplitude corresponding to 1% of $E_{max}$.

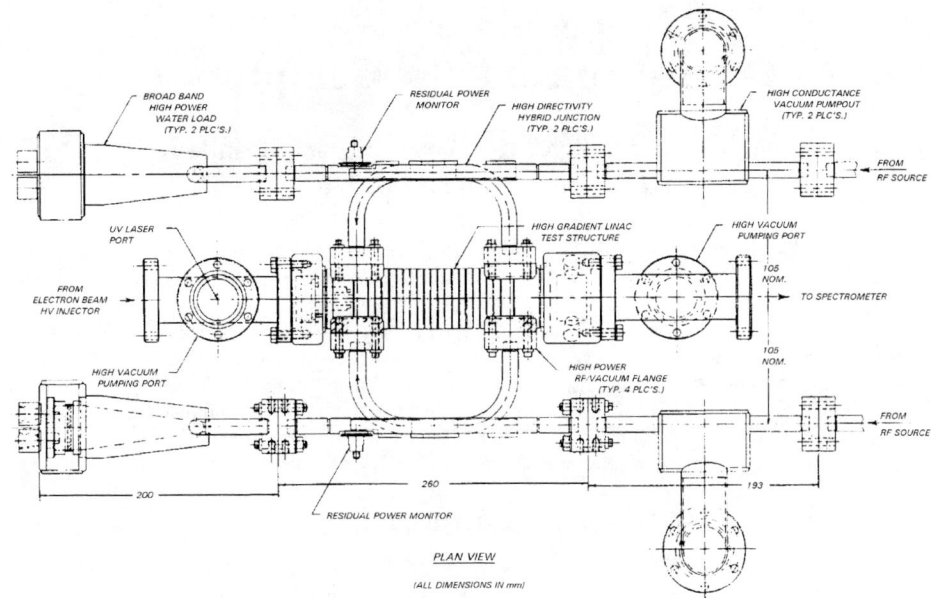

**FIGURE 5.** A 17 GHz linac power amplification system for investigating RF breakdown and beam performance of high gradient (100 to 200 MV/m) accelerator test structures.

# REFERENCES

1. Wang, J. W. and Loew, G. A., "RF Breakdown Studies in Copper Electron Linac Structures," in *Proc. 1989 IEEE Particle Accelerator Conference* **89CH2669-0**, 1989, pp. 1137–1139.
2. Allen, H. A. , *et al.*, "High Gradient Electron Accelerator Powered by a Relativistic Klystron," *Phys. Rev. Lett.* **63 No. 22**, 2472–2475 (1989).
3. Ng., C.-K. and Ko, K., "Numerical Simulation of Input and Output Couplers for Linear Accelerator Structures," presented at the *Computational Accelerator Conference (CAP 93)*, Pleasanton, California, February 1993.
4. Haimson, J., Mecklenburg, B. and Wright, E. W., "A Racetrack Geometry to Avoid Undesirable Azimuthal Variations of the Electric Field Gradient in High Power Coupling Cavities for TW Structures," in *Advanced Accelerator Concepts, AIP Conference Proceedings* **398**, New York: AIP Press, 1997, pp. 898–911.
5. Tyrell, W. A., *Proc. IRE 35*, 1294–1310 (1947).
6. Smullin, L. D. and Montgomery, C. G., *Microwave Duplexers*, New York: McGraw-Hill, 1948.
7. U.S. Patent No. 4,713,581.
8. Haimson, J. and Mecklenburg, B., "A 17 GHz 100 MW Phase Locked TW Resonant Ring High Gradient Accelerator System," in *Pulsed RF Sources for Linear Colliders, AIP Conference Proceedings* **337**, New York: AIP Press, 1995, pp. 311–321.
9. Haimson, J. and Mecklenburg, B., "HV Injection Phase Orbit Characteristics for Sub-Picosecond Bunch Operation with a High Gradient 17 GHz Linac," in *Proc. 1995 IEEE Particle Accelerator Conference* **95CH35843**, 1995, pp. 755–757.

# Design of a 17.14 GHz
# Quasi-Optical Pulse Compressor

M. I. Petelin*, S. V. Kuzikov*, Yu. Yu. Danilov*,
V. L. Granatstein**, and G. S. Nusinovich**

*Institute of Applied Physics, Russian Academy of Sciences
**Institute for Plasma Research, University of Maryland

**Abstract.** A quasi-optical version of the ring cavity pulse compressor is considered. This concept is based on the coupling of the input wave to a whispering gallery mode of a barrel-like cavity due to helical corrugations of the cavity wall. Low-power tests of the prototype were carried out at 11.4 GHz and demonstrated reasonable agree-ment between experimental data and theoretical predictions. The design of a similar pulse compressor at 17.14 GHz compatible with the 17.14 GHz, 100 MW gyroklystron currently under development at the University of Maryland is presented.

## INTRODUCTION

Progress in linear electron-positron accelerator technology is associated with shortening the wavelength of the driving RF power [1,2]. To control the short waves, it seems attractive to use quasi-optical components which are low-loss, and resistant to both breakdown and overheating [3–6].

In particular, one of the quasi-optical modifications of ring cavity pulse compressors [4] (analogous to SLED-1 [1, 7]) is a barrel-like cavity operating in a whispering gallery mode [5]. The whispering gallery mode used in such cavities has an azimuthally rotating structure, which provides the ring cavity operation, and also minimizes the volume, thereby ensuring the maximum coupling with the feeding wave. Such a cavity operates as a delay line with a time constant, $T_d$. If the phase of the input signal is reversed after $T_d$, then the addition of the pulse passed through the cavity to the directly propagating pulse with reversed phase can increase the instantaneous power up to nine times [1]. A drawback of this scheme for operation at short wavelengths is the injection and extraction of the RF power with standard waveguides. As an upgrade for such a compressor, in References 3 and 4 it was proposed to that the cavity be fed with a coaxial, oversized waveguide and that the waveguide and cavity modes be coupled by means of a helical corrugation of the wall (Fig.1).

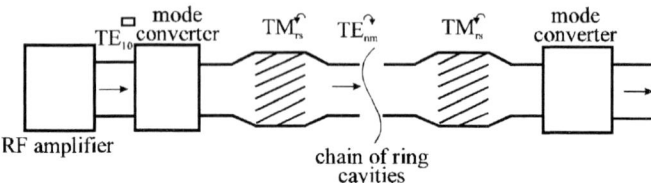

**FIGURE 1.** Pulse compressor based on a chain of barrel-like helically corrugated cavities.

CP472, *Advanced Accelerator Concepts: Eighth Workshop,*
edited by W. Lawson, C. Bellamy, and D. Brosius
© 1999 The American Institute of Physics 1-56396-889-4/99/$15.00

# THEORY OF THE COMPRESSOR

As in all ring cavities, the pulse compressor shown in Fig. 1 does not reflect the incident wave back to the source. The cavity ohmic Q-factor is assumed to be much higher than the loaded Q; i.e.,

$$Q_{ohm} \gg Q_{load} .$$ (1)

Due to the interaction with the cavity the incoming waveguide mode changes only the phase of its complex amplitude in accordance with the universal formula for the transmission coefficient [5]

$$T(\omega) = \frac{\omega - \omega_s^*}{\omega - \omega_s} ,$$ (2)

where $\omega_s$ is the cavity eigenfrequency.

The cavity is formed with a long, slightly expanded section of a cylindrical metallic tube and the operating mode is a near-cut-off trapped wave. To maximize the selectivity of coupling to the incident wave, the cavity mode should have a high azimuthal index $(m \gg 1)$ and the minimum radial index $n = 1$. To satisfy the limitation in Eq. (1), a $TM_{m,1}$- mode should be used, since its $Q_{ohm}$ is much higher than for $TE_{m,1}$- modes. The longitudinal structure of such a mode $F(z)$ is described by the nonuniform string equation

$$\frac{d^2 F(z)}{dz} + h_c^2(z) F(z) = 0 ,$$ (3)

where $h_c = \sqrt{(\omega/c)^2 - (\mu_{mn}/R)^2}$, $\mu_{mn}$ is the relevant root of the Bessel function and $R(z)$ is the metallic tube radius. The operating mode is described with the single-hump (minimum $h_c \ll \omega/c$) solution of Eq. (3), exponentially fading outside the cavity.

For the sake of coupling selectivity, the waveguide mode should be counter-rotating relative to the cavity mode. The coupling can be provided with helical corrugation of the cavity wall $R(z) = R_0 + b \cos(m_p \varphi - hz)$. The number of azimuthal variations, $m_p$, satisfies the condition

$$m_p = m_c - m_w = |m_w| + |m_c| ,$$ (4)

where $m_w$ and $m_c$ are azimuthal indices of the waveguide and cavity modes, respectively. The corrugation period, $d = 2\pi/h$, satisfies the Bragg resonance condition

$$h = h_w .$$ (5)

Here, $h_w = \sqrt{(\omega/c)^2 - (v_{mn}/R)^2}$ is the axial propagator of the waveguide mode assumed to be of the whispering gallery TE type, and $v_{mn}$ is the relevant root of the Bessel function derivative.

The RF energy radiation from the cavity into the waveguide results in the cavity mode decrement,

$$\omega_s'' = \omega_s' / 2Q_{load},$$

where $\omega_s'$ is the eigen-frequency of the non-corrugated cavity,

$$Q_{load} = [(v_{mn}/m_w)^2 - 1]/2M,$$

$$M = b_w I_1^2 / 8 R^2 I_r,$$

$$I_r = \int_{-\infty}^{\infty} F^2(z)dz,$$

$$I_1 = \int_{-\infty}^{\infty} b(z)F(z)dz,$$

and $b(z)$ is the corrugation depth.

The above formulas give the complex eigenfrequency of the cavity coupled to the waveguide, $\omega_s = \omega_s' + i\omega_s''$. Since the wave transmission coefficient given by Eq. (2) is universal for all ring cavities, the pulse compression can be provided with the input pulse modulation described in Refs. 1 and 2.

## LOW-POWER TESTS

In order to demonstrate operability of the proposed pulse compression concept, a simplified prototype (Fig. 2) was designed, fabricated and tested at IAP. The $TE_{41}$ waveguide mode excited the counter rotating $TM_{41}$ mode in a cavity with helically corrugated walls.

**FIGURE 2.** $TE_{41}$-$TM_{41}$ ring-like cavity.

In the experimental set-up (Fig. 3), a low power 11.4 GHz generator produced the $TE_{10}$ mode of a standard rectangular waveguide. To transduce this mode into the rotating $TE_{41}$ mode of the oversized circular waveguide, a cascade of mode converters was used:

- In an adiabatic taper, the $TE_{10}$ mode was converted into $TE_{11}$ mode of circular waveguide.
- A polarizer provided the mode with azimuthal rotation.
- A resonant scattering, helically corrugated waveguide section [7] with five azimuthal variations converted the rotating $TE_{11}$ mode into the counter-rotating $TE_{41}$ mode.

The input signal was modulated (Fig. 4) with a $180^0$ phase shifter based on two field-effect transistors. The output signal related to the $TM_{41}$ mode (Fig. 4) was measured with a crystal detector. The detector was not quadratic, so the relative wave power dependence on time was measured by using a calibrated attenuator, taking into account that the precursor (the first maximum in Fig. 4) is transmitted through the system without coupling to the cavity. The output power pattern is in good agreement with the theory [1]. The highest (second) maximum in Fig. 4 is related to the $180^0$ input pulse phase reversal and is approximately 9 times higher than the precursor, which is close to the theoretical limit corresponding to a very low compression efficiency [1].

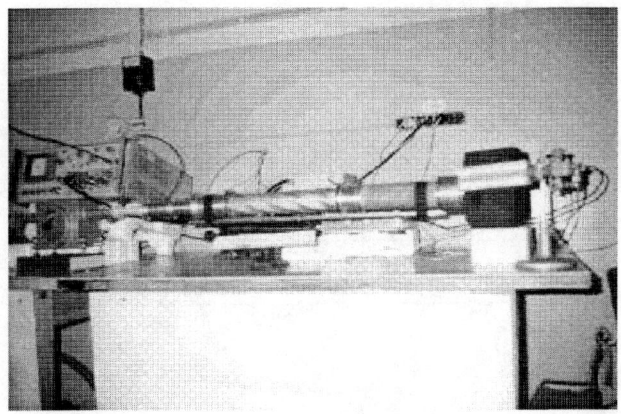

**FIGURE 3.** Experimental set-up to test the system.

# ESTIMATED PARAMETERS FOR A
# 17.14 GHz PULSE COMPRESSOR

In order to compress pulses produced by the high-power 17.14 GHz gyroklystron currently under development at the University of Maryland [8], we designed a compressor based on the helical corrugated cavity (Fig. 5) with the following parameters:

| | |
|---|---|
| Operating frequency | 17.14 GHz |
| Waveguide mode | $TE_{16.1}$ |
| Cavity mode | $TM_{15.1}$ |
| Ohmic & diffraction quality factors | $\sim10^5$ |
| Coupling quality factor | 6000 |
| Input pulse duration | 0.5 μs |
| Output pulse duration | 100 ns |
| Efficiency | 60% |
| Azimuthal corrugation number | 31 |
| Cavity length | 12 cm |
| Cavity diameter | 11.14 cm |
| Corrugation period | 4.1 cm |

The gyroklystron will feed the compressor by means of a cascade of the corrugated converters [9].

As an upgrade for the compressor, the number of cavities can be enlarged to 3 (see Fig. 1). In the latter case the input pulse modulated in accordance with Ref. 4 can be lengthened to 1 μs, other parameters of the compressor remaining the same.

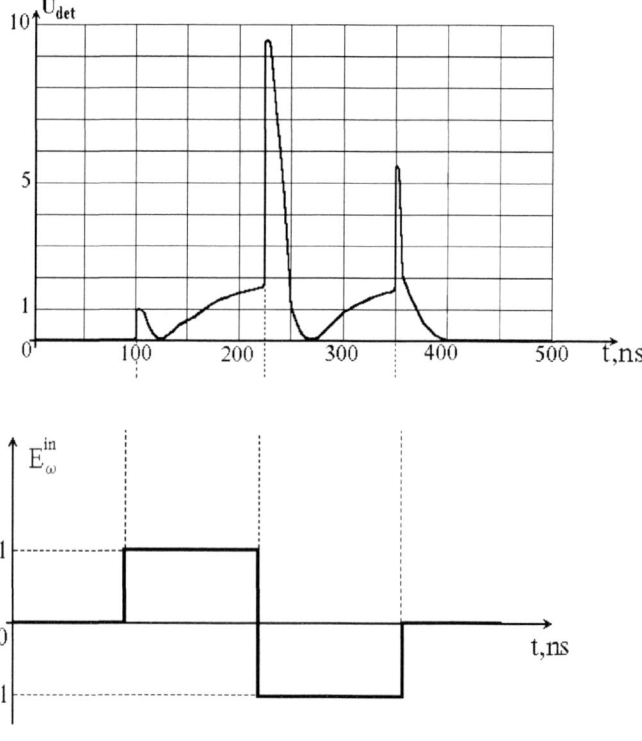

**FIGURE 4.** The input pulse complex amplitude and the detector measured output pulse

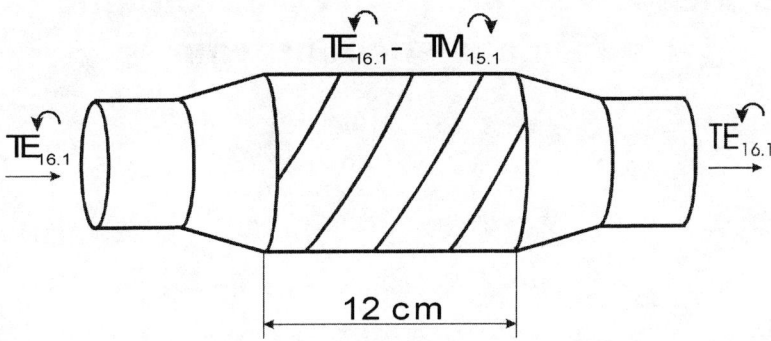

**FIGURE 5.** 17.14 GHz pulse compression configuration.

## CONCLUSION

A concept of quasi-optical pulse compressor based on coupling of a travelling wave to a whispering gallery mode of a barrel-like helically corrugated cavity is developed. Such compressors are capable of operation without RF breakdown at high power levels. Preliminary cold test measurements have been carried out at 11.4 GHz. A 17.14 GHz pulse compressor compatible with the relativistic gyroklystron currently under development at the University of Maryland has been designed.

## REFERENCES

1.  Wilson, P.B.. "Application of High–Power Microwave Sources to TeV Linear Colliders," in *Application of High–Power Microwaves*, Gaponov–Grekhov, A. and Granatstein, V.L., eds., Boston: Artech House, 1994, pp. 229–317.
2.  Wilson, P.B.. SLAC–PUB–1453, April 1997.
3.  Kovalev, N.F., Kuzikov, S.V., Petelin, M.I., Vikharev, A.L.. "Quasi-optical components for future particle accelerators," in *Annales des Journees* (Proceedings) *Maxwell'97*, Eloy, J.-F., ed., 1998, pp. 286-290.
4.  Petelin, M.I., Tai, M.L.. "Compression of phase-modulated microwave pulse by chain of ring cavities," in *AIP Conference Proceedings 337, Pulsed RF Sources for Linear Colliders*, Fernow, R., ed., 1994, pp. 303-310.
5.  Balakin, V.E., Syrachev, I.V.. "Status of VLEPP RF Power Multiplier (VPM)", in *Proceedings of the Third European Particle Acccelerator Conference*, 1992, p. 1173-1175.
6.  Kuzikov, S.V., et al. "Passive pulse compressors based on ring-like cavities," in *Proceedings of the Seventh International Workshop on Linear Colliders* (CD-ROM), 1997.
7.  Farkas, Z.D., Hogg, H.A., Loew, G.A., Wilson, P.B.. "SLED: A method of doubling SLAC's energy," in *Proceedings of the Ninth International Conference on High Energy Accelerators*, SLAC–PUB–1453, 1974, p. 576.
8.  Granatstein, V.L. and Lawson, W.. *IEEE Transactions on Plasma Science* **24**, 648-665 (1996)
9.  Kovalev, N.F., Orlova, I.M., Petelin, M.I.. *Radio Physics and Quantum Electronics* **11**, 449-450, 1968.

# Stimulated Emission of Coherent Radiation from a Relativistic Ion Beam

S. A. Bogacz

*Accelerator Physics Department,*
*Thomas Jefferson National Accelerator Facility*
*12000 Jefferson Avenue, Newport News, VA 23606, USA*

**Abstract.** We consider a relativistic beam of Hydrogen-like ions with $Z > 1$ having a single bound electron. In this two-level-system, the level population (1s and 2p) may be inverted by the application of a "π-pulse" of laser radiation tuned to the Doppler shifted 1s→2p transition. When the laser beam and ion beam move in opposite directions the required laser frequency is reduced by a factor $2\gamma$. Subsequently applied short wavelength resonant radiation moving in the same direction as the ion beam (with the inverted population) will be amplified via stimulated emission; the wavelength in the laboratory frame now being shorter than the original laser wavelength by a factor $(2\gamma)^2$.

## INTRODUCTION

Here we propose a scheme, which could result in the production of coherent electromagnetic radiation in the UV and X-ray regions. At present the most popular approach to generating short wavelength radiation involving particle beams is the free electron laser (1) The efficiency of the FEL is basically governed by the Thomson scattering cross section. Qualitatively, if one were to use a beam of particles having a resonant scattering cross section for some frequency, a greatly enhanced gain might be expected. A natural candidate is a hydrogenic positive ion, with one bound electron (a simple two level system). The fact that the ion is charged allows the beam to be accelerated and stored at relativistic energies, where one can exploit the properties of relativistic kinematics, which dictate that the "back scattered" radiation will have its wavelength shortened by a factor $(2\gamma)^2$.

The idea is illustrated schematically in Figure 1. It goes as follows. Suppose the ion beam encounters a laser beam traveling in the opposite direction. In the rest frame of the ions the frequency of the light is enhanced by a factor $2\gamma$ (extreme relativistic case $\gamma \gg 1$). Assume that the rest frame frequency matches, say, the 1s→2p transition frequency, and that the interaction path length in the lab frame corresponds to a time in the rest frame necessary to invert the level population (a so-called π pulse). If we subsequently apply a different wave, traveling parallel to the ion beam, we may achieve

CP472, *Advanced Accelerator Concepts: Eighth Workshop,*
edited by W. Lawson, C. Bellamy, and D. Brosius
© 1999 The American Institute of Physics 1-56396-889-4/99/$15.00

stimulated emission. The wavelength of this signal would be reduced by a factor $2\gamma$ from that of the resonant radiation in the rest frame, and by a factor $(2\gamma)^2$ from the original laser wavelength. Feasibility of this scheme, of course, rests on whether the necessary pump power is achievable and whether the gain is sufficient to overcome the losses associated with typical UV mirrors; both of these are governed by the spontaneous emission rate. We now proceed with a detailed discussion and evaluation of our model.

## Population inversion in H-ion beam
### (conceptual setup)

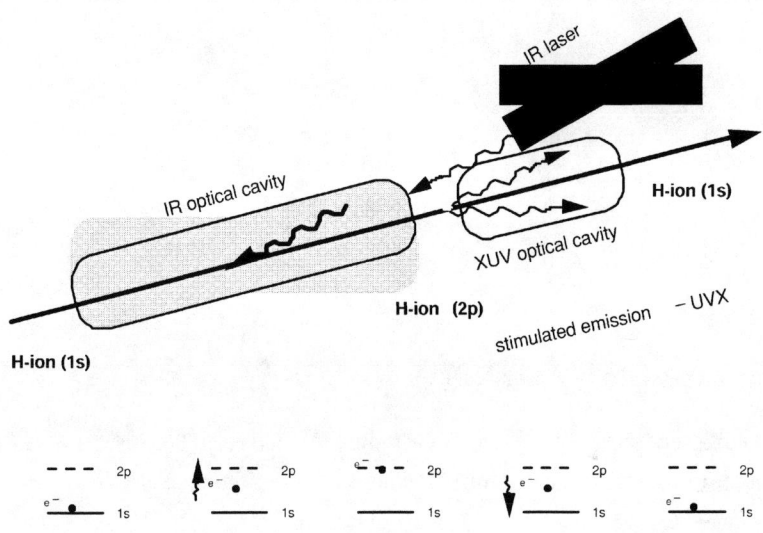

**FIGURE 1** Conceptual sketch of a setup exploring stimulated emission from a laser pumped relativistic ion beam.

## THEORETICAL APPROACH

We require expressions for the time evolution of the 1s and 2p level occupations (in the rest frame) as a function of an applied oscillatory electric field of strength E and frequency w for some initial occupation of the levels. Since we will be concerned with frequencies in the vicinity of the 1s→2p transition, we truncate our basis set to these two levels. We introduce a two-by-two density matrix of the form:

$$N = \begin{pmatrix} a_\uparrow a_\uparrow{}^* & a_\uparrow a_\downarrow{}^* \\ a_\uparrow a_\uparrow{}^* & a_\uparrow a_\downarrow{}^* \end{pmatrix},$$ (1)

where the symbols $a_\uparrow$ and $a_\downarrow$ denote the amplitudes of $|2p\rangle$ and $|1s\rangle$ states respectively. The time dependence of $N$ (in the interaction picture) is governed by the following equation:

$$\frac{\partial}{\partial t} N = ih\,[U,N] - \frac{1}{\tau}(N - N_{eq}).$$ (2)

where $U$ denotes the perturbing Hamiltonian given by:

$$U = |U| \begin{pmatrix} 0 & e^{i\varepsilon t} \\ e^{-i\varepsilon t} & 0 \end{pmatrix}.$$ (3)

Here

$$|U| = \frac{1}{2} eE\,|\langle 1s|\,z\,|2p\rangle|, \quad \langle 1s|\,z\,|2p\rangle = 4\sqrt{2}\left(\frac{2}{3}\right)^5 \frac{a_o}{Z}, \quad \varepsilon = \omega_{1s\to 2p} - \omega,$$ (4)

and

$$\omega_{1s\to 2p} = \frac{1}{h}(E_{2p} - E_{1s}) = \frac{3}{4} R Z^2,$$ (5)

where $R$ and $a_o$ are the Rydberg and Bohr radius respectively. The last term in Equation 2 accounts for spontaneous emission (2), where $\tau$ is the lifetime and

$$N_{eq} + \begin{pmatrix} 0 & 0 \\ 0 & 1 \end{pmatrix},$$ (6)

is the equilibrium density matrix (at zero temperature). By defining quantities $P_o$ and $P$ through

$$N = P_o \mathbf{1} + P_\alpha \hat{\sigma}_\alpha;$$ (7)

one can rewrite Equation 2 as follows

$$\frac{\partial}{\partial t} P_o = 0$$ (8)

1022

and

$$\frac{\partial}{\partial t} \mathbf{P} = \Omega_0 [\mathbf{P} \infty \hat{n}] - \frac{1}{\tau} (\mathbf{P} - \mathbf{P}_{eq}) , \tag{9}$$

here

$$\Omega_0 = \frac{2}{h} |U| , \quad \mathbf{P}_{eq} = (0, 0, -\frac{1}{2}) , \tag{10}$$

and the direction of the instantaneous precession axis is given by:

$$\hat{n} + ( \cos \varepsilon t , -\sin \varepsilon t , 0 ) . \tag{11}$$

From Equation 2 one can write the components of $\mathbf{P}$ and $P_o$ explicitly as follows:

$$P_o = \frac{1}{2}(N_{\uparrow\uparrow} + N_{\downarrow\downarrow}), \tag{12}$$

$$P_1 = \text{Re } N_{\uparrow\downarrow} , \tag{13}$$

$$P_2 = \text{Im } N_{\uparrow\downarrow} , \tag{14}$$

and

$$P_3 = \frac{1}{2}(N_{\uparrow\uparrow} - N_{\downarrow\downarrow}). \tag{15}$$

One can easily notice that Equation 8 guarantees probability conservation while Equation 9 is analogous to the Bloch equation and describes the precession of an electric polarization vector induced by the perturbing electric field (3), around the axis, $\hat{n}$.

A pair of equations, Equation 13 and 14, can be written in a compact matrix form as

$$\begin{pmatrix} \frac{\partial}{\partial t} + \frac{1}{t} & 0 & -W_o \sin \varepsilon t \\ 0 & \frac{\partial}{\partial t} + \frac{1}{t} & -W_o \cos \varepsilon t \\ W_o \sin \varepsilon t & W_o \cos \varepsilon t & \frac{\partial}{\partial t} + \frac{1}{t} \end{pmatrix} \mathbf{P} = \frac{1}{t} \mathbf{P}_{eq} . \tag{16}$$

It is convenient to transform the Bloch equation, Equation 16 to the rotating reference frame using the following transformation

$$P^\pm(t) = e^{\pm i\varepsilon t}(P_2 \pm iP_1) ; \tag{17}$$

The resulting inhomogenous equation can be solved in a standard way (3), assuming specific boundary conditions.

## POPULATION INVERSION

To describe the pumping stage take the solution of Equation 16 with the initial population entirely in the ground state. The required evolution of the excited state population is given by the first diagonal element of the density matrix[4]

$$N_{\uparrow\uparrow}(t) = \frac{1}{2}\frac{(\Omega\tau)^2}{1+(\Omega\tau)^2}\left\{1-e^{-t/\tau}\left[\frac{\sin\Omega t}{\Omega\tau}+\cos\Omega t\right]\right\}, \tag{18}$$

$$\Omega \equiv \sqrt{\Omega_o{}^2 + \varepsilon^2} ,$$

where $\Omega_o = 2|U|h$. Therefore population inversion, $N_{\uparrow\uparrow} = 1 - O(\Omega\tau)^{-1}$, requires $\Omega\tau \gg 1$ (strong field limit) and occurs after a time $T_{rest} = \pi/\Omega$; (this corresponds to the $\pi$-pulse in NMR). Time evolution of the population invertion, $N_{\uparrow\uparrow}$, expressed by Equation 18 is plotted for different values of $\Omega\tau$, as illustrated in Figure 2.

Evaluating $\tau$ from the Einstein relation (5),

$$\tau = \frac{3hc^3}{4e^2w_{1s\to2p^3}}|\langle 1s|z|2p\rangle|^{-2} = 1.59 \times 10^{-9}\,Z^{-4}\,\text{s}, \tag{19}$$

and using the explicit expression for $\Omega$,

$$\Omega = 4\sqrt{2}\left(\frac{2}{3}\right)^5\frac{eEa_o}{3hZ} ; \tag{20}$$

one can rewrite the strong field condition ($\Omega\tau \gg 1$) as an inequality on the required electric field (rest frame value):

$$E_{rest} \gg 0.349 \times Z^5 \text{ st. V cm}^{-1}. \tag{21}$$

In the lab-frame the E-field is reduced by a factor $\chi$ and the pulse width, $T^\pi_{rest} = \pi/\Omega$, translates to an interaction length

$$L = \beta\gamma c\, T^\pi_{rest} = 2.61 \times 10^1\, Z/E_{lab} \qquad (22)$$

and a power flux, $\langle S \rangle = (c/8\pi)\, E_{lab}^2$, of

$$\langle S \rangle \gg 3.64 \times Z^{10}/\gamma^2 \ \text{W/cm}^2, \qquad (23)$$

which is readily achievable in steady state for low Z. Were one to operate in the fundamental Gaussian mode of an optimized confocal resonator, the condition on the total power flux, P, can be written as

$$P \gg 10^{-2} \times Z^4 \ [\text{W}] \qquad (24)$$

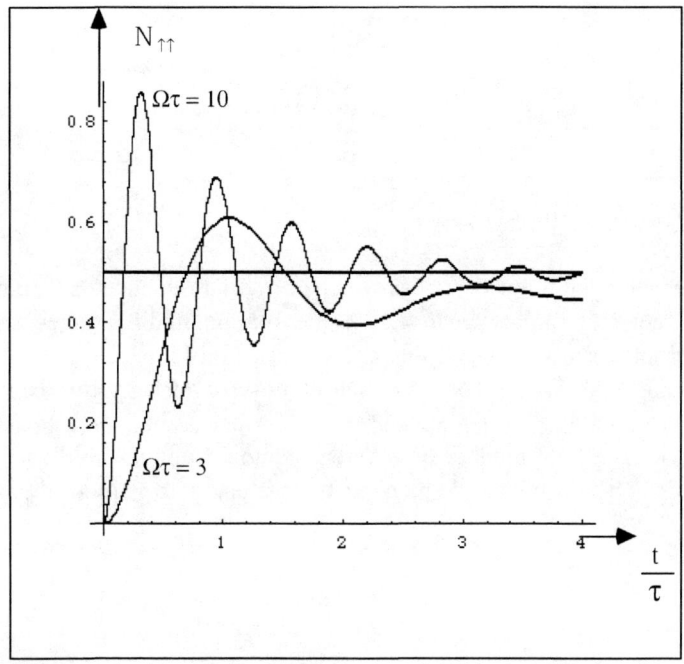

**FIGURE 2** Population inversion in a two-level-system accomplished by he application of a "π-pulse" of laser radiation tuned to the Doppler shifted 1s→2p transition.

and may be enhanced over the drive power by a factor involving the Q of the resonator. Finally, the Doppler-shifted resonance condition $2\pi c/\lambda = \omega_{1s\to 2p}/(2\gamma)$ relates $\gamma$ and $\lambda$ as follows

$$\gamma = 4.12 \times 10^4 \; Z^2 \; \lambda \; . \tag{25}$$

## STIMULATED EMISSION

We now examine the second stage of our model, where the inverted population ion beam is subjected to an incoming electromagnetic wave, which is to be amplified by stimulated emission. The field strength will be assumed small, i.e. $\Omega'\tau \ll 1$, where we use $\Omega'$ for the precession frequency associated with the amplified wave. To introduce the gain mechanism we note that the rate of production of coherent photons, $N_{coh}$, is given by a solution of Equation 16 with the assumed population inversion as the initial condition(4).

$$N_{coh}(t) = \frac{1}{2}\left(\frac{\Omega'_o}{\Omega'}\right)^2 \; \Omega'_{\text{!`}} \; e^{-t/\tau} \sin \Omega'\tau \quad + \tag{26}$$

$$+ \frac{1 + (\Omega'\tau)^2}{e^{-t/\tau}(\sin \Omega'\tau + \Omega'\tau \cos \Omega'\tau) - 1}\Big\} \; , \qquad \Omega' + \sqrt{{\Omega'}_o^2 + \varepsilon^2} \; .$$

The above solution describes a general "off resonance" situation ($\varepsilon \neq 0$) which reflects frequency spread, $\Delta\omega/\omega = \Delta\gamma/\gamma$, due to a longitudinal momentum spread in the incoming ion beam ($\Delta\gamma$). Note that this effect could be neglected in the discussion of the pumping stage where the strong field limit ($\Omega_o\tau \gg 1$) applies. However, the lasing stage is in the weak field regime and the off resonance frequency broadening has to be considered. The total number of coherent photons emitted during some time t (to be discussed shortly) results in an increment to the energy density of the initial wave, $\Delta E$, given by

$$\Delta E = hn \; \omega_{1s\to 2p} \left\langle \int_0^t N_{coh}(t')dt' \right\rangle_\varepsilon \; ; \tag{27}$$

here n is the beam density (in the rest frame) and

$$\langle \dots \rangle_\varepsilon + \frac{1}{\Delta\varepsilon} \int_{-\Delta\varepsilon/2}^{\Delta\varepsilon/2} \dots \; d\varepsilon \; , \tag{28}$$

defines the ensemble average over momenta in the incoming beam. The right hand side of Equation 27 is evaluated in two asymptotic regions: (a) cool beam limit ($\Delta\varepsilon\tau \ll 1$); (b) hot beam limit ($\Delta\varepsilon\tau \gg 1$). Evaluating appropriate $\varepsilon$–averages in the above two regimes, one can rewrite Equation 27 as follows:

$$\Delta E = \frac{\hbar}{2}\, n\, \omega_{1s\to 2p}\, (\Omega_0'\tau)^2 A ,\tag{29}$$

where

$$A = \begin{cases} 0.17 & ;\ \Delta\varepsilon\tau \ll 1 \\ 0.39\pi/\Delta\varepsilon\tau & ;\ \Delta\varepsilon\tau \gg 1. \end{cases}\tag{30}$$

The relationship between $\Omega'_0$ and the energy density E ,

$$(\Omega'_0)^2 = 2^5\pi^3\, e^2|\langle 1s|z|2p\rangle|^2\, E\,/h^2 ,\tag{31}$$

allows us to rewrite Equation 29 as

$$\Delta E = (2\pi)^2 An\, e^2\, \omega_{1s\to 2p}\, \tau^2\, E\,/h .\tag{32}$$

The total gain, G, is defined as

$$G = (E + \Delta E )/E \tag{33}$$

and lasing requires $GR^2 > 1$; here R is the mirror reflectivity of the second (lasing) cavity. Finally the gain coefficient can be written as :

$$G = 1 + (2\pi)^2 An\, e^2\, \omega_{1s\to 2p}\, \tau^2/h .\tag{34}$$

Assuming mirrors with an intensity reflection efficiency of 50% and a longitudinal momentum spread in the incoming ion beam of $\Delta\gamma/\gamma = 10^{-5}$ one can estimate the ion concentration required for lasing as

$$n_{rest} = 6.84 \times 10^6\, Z^8\ \mathrm{cm}^{-3} .\tag{35}$$

In the lab frame this requires a beam current density of

$$j_{lab} = (Z - 1)e\, n_{rest}\, c\gamma .\tag{36}$$

For the case of $Li^{2+}$ ions the threshold current density would be

$$j_{lab} = 4.31 \times 10^2 \text{ A cm}^{-2} .$$
(37)

Particle beams having the requisite current density, momentum spread and energy are within the scope of present generation high current storage rings. The discussion in this paper centers on the 1s→2p transition in a hydrogenic ion. However, other types of transitions may also be of interest; particularly those with longer lifetimes (e.g., transitions to various dipole-forbidden or metastable states). It may also be possible to exploit a nuclear transition as well as various multilevel systems.

## REFERENCES

1. J.M.J. Madey, J. Appl. Phys. **42** 1906 (1971).
2. R. Karplus and J. Schwinger, Phys. Rev. **73** 1020 (1948).
3. S.A. Bogacz and J.B. Ketterson, Appl. Phys. Lett. **49** (6) 311 (1986).
4. S.A. Bogacz, Ph.D. Thesis, Northwestern University, Evanston (1986).
5. L.I. Schiff, *Quantum Mechanics*, 3rd ed. (McGraw-Hill, 1986) 414.

| | |
|---|---|
| Mr. Ilya Alexeev | University of Maryland, College Park |
| Prof. Thomas Antonsen | University of Maryland, College Park |
| Ms. Katya Backhaus | University of California, Berkeley |
| Dr. Nikolai Barov | Argonne National Laboratory |
| Dr. Ilan Ben-Zvi | Brookhaven National Laboratory |
| Mr. Santiago Bernal | University of Maryland, College Park |
| Dr. Denis Bernard | Ecole Polytechnique |
| Mr. Zhigang Bian | University of Maryland, College Park |
| Dr. Hans Bluem | Northrop Grumman |
| Prof. J. Louis Bobin | Universite de Pierre et Marie Curie |
| Dr. Alex Bogacz | Thomas Jefferson National Accelerator Facility |
| Mr. Winthrop Brown | Massachusetts Institute of Technology |
| Dr. Robert Byer | Stanford University |
| Mr. Michael Castle | University of Maryland, College Park |
| Ms. Palma Catravas | Lawrence Berkeley National Laboratory |
| Mr. Andy Charman | Lawrence Berkeley National Laboratory |
| Dr. Swapan Chattopadhyay | Lawrence Berkeley National Laboratory |
| Dr. Chiping Chen | Massachusetts Institute of Technology |
| Dr. Pisin Chen | Stanford Linear Accelerator Center |
| Mr. Szu-Yuan Chen | University of Michigan |
| Mr. Sergey V. Cheshkov | University of Maryland, College Park |
| Dr. Christopher Clayton | University of California, Los Angeles |
| Dr. James E. Clendenin | Stanford Linear Accelerator Center |
| Dr. David B. Cline | University of California, Los Angeles |
| Dr. Patrick Colestock | Fermi National Accelerator Laboratory |
| Dr. Manoel Conde | Argonne National Laboratory |
| Dr. Thomas E. Cowan | Lawrence Livermore National Laboratory |
| Dr. Bruce Danly | Naval Research Laboratory |
| Dr. Phil Debenham | U.S. Department of Energy |
| Dr. Hideki Dewa | JAERI (Japan) |
| Mr. Evan Dodd | University of Michigan |
| Dr. Michael Downer | University of Texas, Austin |
| Dr. Bruce M. Dunham | Thomas Jefferson National Acclerator Facility |
| Dr. Eric Esarey | Naval Research Laboratory |
| Mr. Jingyun Fan | University of Maryland, College Park |
| Dr. Michael Fazio | Los Alamos National Laboratory |
| Dr. Richard Fernow | Brookhaven National Laboratory |
| Dr. Massimo Ferrario | INFN-LNF (Italy) |
| Mr. Catalin Filip | University of California, Los Angeles |
| Dr. David Finley | Fermi National Accelerator Laboratory |
| Dr. Ralph B. Fiorito | Catholic University of America |
| Dr. Nathaniel Fisch | Princeton University |
| Mr. Michael J. Fitch | Fermi National Accelerator Laboratory |
| Dr. Clifford Mark Fortgang | Los Alamos National Laboratory |

| | |
|---|---|
| Dr. Alan R. Fry | Fermi National Accelerator Laboratory |
| Dr. Wei Gai | Argonne National Laboratory |
| Mr. Li Gan | University of Maryland, College Park |
| Mr. Erhard W. Gaul | University of Texas, Austin |
| Dr. Steven H. Gold | Naval Research Laboratory |
| Ms. Richa Govil | Lawrence Berkeley National Laboratory |
| Prof. Victor Granatstein | University of Maryland, College Park |
| Dr. Bahman Hafizi | Naval Research Laboratory |
| Dr. Jacob Haimson | Haimson Research Corp. |
| Dr. Frederic Hartemann | Lawrence Livermore National Laboratory |
| Mr. Ping He | Brookhaven National Laboratory |
| Mr. Roy Hemker | University of California, Los Angeles |
| Mr. Mark Hill | Stanford Linear Accelerator Center |
| Dr. Jay Hirshfield | Yale University |
| Mr. Jerry Hoffman | University of Southern California |
| Dr. Mark Hogan | University of California, Los Angeles |
| Prof. Yen-Chieh Huang | National Tsinghua University |
| Dr. Richard F. Hubbard | Naval Research Laboratory |
| Dr. John Irwin | Stanford Linear Accelerator Center |
| Ms. Yu Jiang | University of Maryland, College Park |
| Prof. Chan Joshi | University of California, Los Angeles |
| Dr. Thomas Katsouleas | University of Southern California |
| Dr. Kwang-Je Kim | Argonne National Laboratory |
| Dr. Jin-Soo Kim | FARTECH, Inc. |
| Mr. Joon-Koo Kim | University of Michigan |
| Dr. Wayne D. Kimura | STI Optronics |
| Dr. Kenneth E. Kreischer | Massachusetts Institute of Technology |
| Mr. Karl Peter Kusche | Brookhaven National Laboratory |
| Dr. Michael C. Lampel | G.H. Gillespie Associates, Inc. |
| Dr. Michael A. Lapointe | Yale University |
| Dr. Y. Y. Lau | University of Michigan |
| Prof. Wes Lawson | University of Maryland, College Park |
| Mr. Stephen Paul LeBlanc | University of Texas, Austin |
| Ms. Seung Lee | University of Southern California |
| Dr. Wim Leemans | Lawrence Berkeley National Laboratory |
| Dr. Gregory Peter LeSage | Lawrence Livermore National Laboratory |
| Ms. Yun Li | University of Maryland, College Park |
| Dr. Xintian Eddie Lin | Stanford Linear Accelerator Center |
| Dr. Yano Liu | University of California, Los Angeles |
| Dr. Frederick M. Mako | FM Technologies, Inc. |
| Prof. Thomas C. Marshall | Columbia University |
| Dr. Shinichi Masuda | Stanford Linear Accelerator Center |
| Dr. Alexander Mikhailichenko | Cornell University |
| Dr. Howard Milchberg | University of Maryland, College Park |
| Dr. Christopher Moore | Naval Research Laboratory |
| Prof. Warren Mori | University of California, Los Angeles |

| | |
|---|---|
| Dr. Patrick Muggli | University of California, Los Angeles |
| Dr. Kazuhisa Nakajima | KEK, Japan |
| Mr. Ritesh Narang | University of California, Los Angeles |
| Dr. John Nation | Cornell University |
| Dr. David Neuffer | Fermi National Accelerator Laboratory |
| Dr. Oleg A. Nezhevenko | OMEGA-P, Inc. |
| Mr. Sergei Nikitin | University of Maryland, College Park |
| Dr. Gregory S. Nusinovich | University of Maryland, College Park |
| Dr. Patrick O'Shea | Duke University |
| Dr. Tomomi Ohgaki | Lawrence Berkeley National Laboratory |
| Dr. Renato Pakter | Massachusetts Institute of Technology |
| Dr. Dennis Palmer | Stanford Linear Accelerator Center |
| Dr. Zohreh Parsa | Brookhaven National Laboratory |
| Mr. Gerald Peters | U.S. Department of Energy |
| Mr. Tomas Plettner | Stanford University |
| Dr. Igor Pogorelsky | Brookhaven National Laboratory |
| Dr. John Gorham Power | Argonne National Laboratory |
| Prof. Martin Reiser | University of Maryland, College Park |
| Dr. James Rosenzweig | University of California, Los Angeles |
| Dr. Donald Rule | Naval Surface Warfare Center |
| Dr. Levi Schachter | Technion - Israel Institute of Technology |
| Dr. Paul V. Schoessow | Argonne National Laboratory |
| Mr. Carl Schroeder | Lawrence Berkeley National Laboratory |
| Dr. Mike G. Seidel | DESY |
| Dr. Andrew M. Sessler | Lawrence Berkeley National Laboratory |
| Dr. Bradley A. Shadwick | Lawrence Berkeley National Laboratory |
| Dr. Mikhail Shapiro | Massachusetts Insitute of Technology |
| Dr. Richard Sheffield | Los Alamos National Laboratory |
| Dr. Gennady Shvets | Princeton University |
| Prof. Todd Smith | Stanford University |
| Dr. Linda Spentzouris | Fermi National Accelerator Laboratory |
| Dr. Phillip A. Sprangle | Naval Research Laboratory |
| Dr. Triveni Srinivasan-Rao | Brookhaven National Laboratory |
| Dr. Loren C. Steinhauer | University of Washington |
| Dr. Bruce Strauss | U.S. Department of Energy |
| Dr. Hyyong Suk | University of California, Los Angeles |
| Dr. David Sutter | U.S. Department of Energy |
| Prof. Toshi Tajima | University of Texas, Austin |
| Dr. Sami G. Tantawi | Stanford Linear Accelerator Center |
| Dr. R. J. Temkin | Massachusetts Institute of Technology |
| Dr. Kathleen Thompson | Stanford Linear Accelerator Center |
| Dr. Antonio C. Ting | Naval Research Laboratory |
| Dr. Mitsuru Uesaka | University of Tokyo |
| Prof. Donald P. Umstadter | University of Michigan |
| Dr. Arie van Steenbergen | Brookhaven National Laboratory |
| Mr. Paul Volfbeyn | Lawrence Berkeley National Laboratory |

Mr. Pingshan Wang      Cornell University
Dr. J. G. Wang      University of Maryland, College Park
Dr. Changbiao Wang      Yale University
Ms. Mei Wang      Yale University
Dr. Glen Alan Westenskow      Lawrence Livermore National Laboratory
Dr. David Whittum      Stanford Linear Accelerator Center
Dr. Mark Wilson      U.S. Department of Energy
Dr. Jonathan Wurtele      University of California, Berkeley
Dr. Ming Xie      Lawrence Berkeley National Laboratory
Dr. Xiaoxi Xu      University of Maryland, College Park
Dr. Vyacheslav P. Yakovlev      OMEGA-P, Inc.
Ms. Anahid Dian Yeremian      Stanford Linear Accelerator Center
Mr. Rodney Yoder      Yale University
Dr. Tingbin Zhang      Yale University
Dr. Frank Zimmermann      Stanford Linear Accelerator Center
Dr. Max Zolotorev      Lawrence Berkeley National Laboratory
Mr. Peng Zou      Argonne National Laboratory
Mr. Yun Zou      University of Maryland, College Park

# AUTHOR INDEX

## A

Adolphsen, C., 713, 967
Alexeev, I., 434
Amatuni, A. T., 765
Amiranoff, F., 303
Anderson, D. E., 983
Anderson, J., 992
Andreev, N. E., 444
Arjona, M., 128

## B

Babzien, M., 563, 573, 755, 803
Bana, S., 949
Bane, K. L. F., 775
Batchelor, K., 833, 839
Baton, S., 303
Ben-Zvi, I., 563, 573, 803
Bernard, D., 303
Bogacz, S. A., 1020
Bowden, G., 967
Brown, W. J., 850, 861
Byer, R. L., 118, 581

## C

Campbell, L. P., 563, 573
Castle, M., 128
Catravas, P., 803
Chattopadhyay, S., 169, 315
Chen, P., 169, 260, 315, 321, 423
Chen, S.-Y., 333
Cheng, J., 992
Cheshkov, S., 153, 343
Clayton, C. E., 501, 534
Clendenin, J. E., 142
Cline, D. B., 315, 563, 573, 592, 755
Cohen, C., 394
Colestock, P., 315
Conde, M. E., 353, 626, 901
Cowan, T. E., 358
Craddock, W., 315
Cros, B., 303
Crosson, E. R., 735

## D

Danilov, Y. Y., 1014
Decker, F.-J., 315
Descamps, D., 303
Dewa, H., 368, 908
Ditmire, T., 358
Dodd, E., 886
Dong, B., 358
Dorchies, F., 303
Downer, M. C., 377, 413
Dudnikova, G., 839
Dunham, B. M., 813

## E

Ehrlich, Y., 394
Esarey, E., 57, 174, 461
Eylon, S., 983

## F

Fan, J., 434
Farkas, Z. D., 967
Farrell, J. P., 833, 839
Fazio, M. V., 220
Fernow, R. C., 233
Fiorito, R. B., 563, 573, 725, 745
Fisch, N. J., 471, 481
Flechtner, D., 823, 918
Fountain, W., 358

## G

Gai, W., 353, 626, 686, 901
Gallardo, J. C., 233, 563, 573, 755
Gaul, E. W., 377, 413
Golkowski, C., 823, 918
Gorbachev, A. M., 975
Gottschalk, S. C., 563, 573
Govil, R., 384
Granatstein, V. L., 128, 992, 1014